Parametric Modeling with SOLIDWORKS 2021

Covers material found on the CSWA exam

Randy H. Shih
Oregon Institute of Technology

Paul J. Schilling
University of New Orleans

D1200232

SDC Publications
P.O. Box 1334
Mission KS 66222
(913) 262-2664
www.SDCpublications.com
Publisher: Stephen Schroff

Copyright 2021 Randy Shih, Klamath Falls, Oregon and Paul Schilling, New Orleans, Louisiana.

All rights reserved. This document may not be copied, photocopied, reproduced, transmitted, or translated in any form or for any purpose without the express written consent of the publisher, SDC Publications.

It is a violation of United States copyright laws to make copies in any form or media of the contents of this book for commercial or educational purposes without written permission.

Examination Copies
Books received as examination copies are for review purposes only and may not be made available for student use. Resale of examination copies is prohibited.

Electronic Files
Any electronic files associated with this book are licensed to the original user only. These files may not be transferred to any other party.

Trademark
SOLIDWORKS is a registered trademark of SOLIDWORKS Corporation.
Microsoft Windows is a registered trademark of Microsoft Corporation.
All other trademarks are trademarks of their respective holders.

The authors and publisher of this book have used their best efforts in preparing this book. These efforts include the development, research and testing of the material presented. The author and publisher shall not be liable in any event for incidental or consequential damages with, or arising out of, the furnishing, performance, or use of the material.

ISBN-13: 978-1-63057-404-8
ISBN-10: 1-63057-404-X

Printed and bound in the United States of America.

Preface

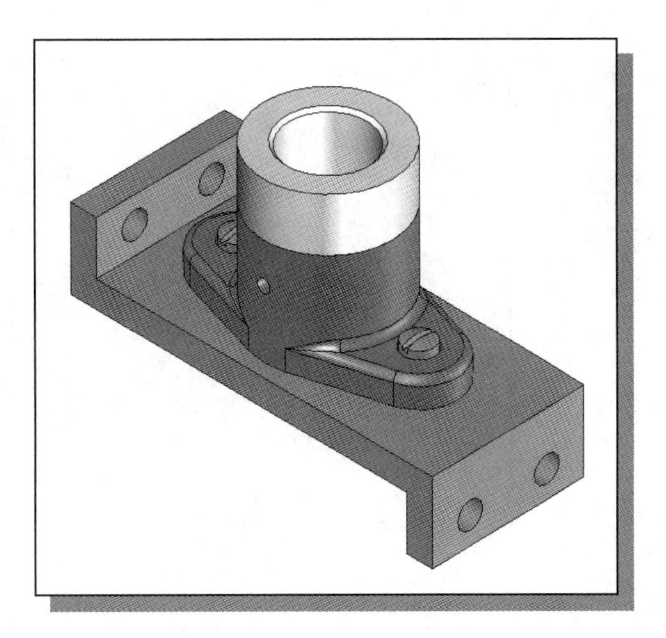

The primary goal of ***Parametric Modeling with SOLIDWORKS 2021*** is to introduce the aspects of designing with **Solid Modeling** and **Parametric Modeling**. This text is intended to be used as a practical training guide for students and professionals. This text uses SOLIDWORKS 2021 as the modeling tool and the chapters proceed in a pedagogical fashion to guide you from constructing basic solid models to building intelligent mechanical designs, creating multi-view drawings and assembly models. This text takes a hands-on, exercise-intensive approach to all the important *Parametric Modeling* techniques and concepts. This textbook contains a series of seventeen tutorial style lessons designed to introduce beginning CAD users to SOLIDWORKS. This text is also helpful to SOLIDWORKS users upgrading from a previous release of the software. The solid modeling techniques and concepts discussed in this text are also applicable to other parametric feature-based CAD packages. The basic premise of this book is that the more designs you create using SOLIDWORKS, the better you learn the software. With this in mind, each lesson introduces a new set of commands and concepts, building on previous lessons. This book does not attempt to cover all of the SOLIDWORKS features, only to provide an introduction to the software. It is intended to help you establish a good basis for exploring and growing in the exciting field of **Computer Aided Engineering**.

Acknowledgments

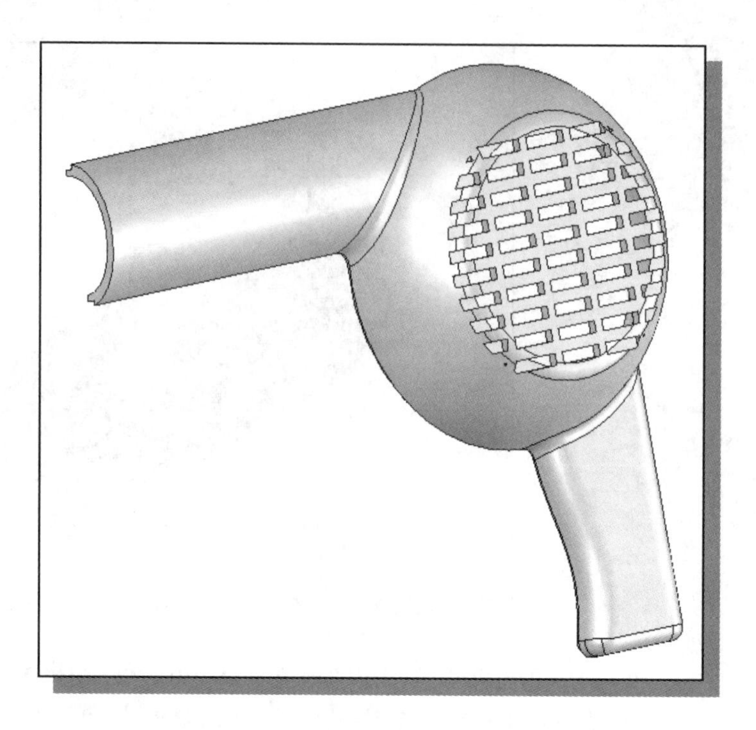

This book would not have been possible without a great deal of support. The effort and support of the editorial and production staff of SDC Publications is gratefully acknowledged. I would especially like to thank Stephen Schroff for his support and helpful suggestions during this project. Don Domes of ORTOP granting the use of their designs and resources is also appreciated. I am also very grateful that the Mechanical Engineering Technology Department of Oregon Institute of Technology has provided me with an excellent environment in which to pursue my interests in teaching and research.

Truly unbounded thanks are due to my wife Hsiu-Ling and daughter Casandra for their understanding and encouragement throughout this project.

Randy H. Shih, Spring, 2021

I would like to echo the appreciation for the support of Stephen Schroff and the staff of SDC Publications. In addition, the support of the Department of Mechanical Engineering at the University of New Orleans is gratefully acknowledged.

Paul J. Schilling, Spring, 2021

Table of Contents

Chapter 5
Geometric Relations Fundamentals

Chapter 6
Geometric Construction Tools

Chapter 7
Parent/Child Relationships and the BORN Technique

Chapter 8
Part Drawings and Associative Functionality

Chapter 9
Reference Geometry and Auxiliary Views

Chapter 10
Introduction to 3D Printing

Chapter 11
Symmetrical Features in Designs

Chapter 12
Advanced 3D Construction Tools

Chapter 15
Design Library and Basic Motion Study

Chapter 16
Design Analysis with SimulationXpress

Chapter 17
CSWA Exam Preparation

Appendix

Index

Certified SOLIDWORKS Associate (CSWA) Exam Overview

The Certified SOLIDWORKS Associate (CSWA) Exam is a performance-based exam. The examination is comprised of 10 – 20 questions to be completed in three hours. The test items will require you to use the SOLIDWORKS software to perform specific tasks and then answer questions about the tasks.

Performance-based testing is defined as ***Testing by Doing***. This means you actually perform the given task then answer the questions regarding the task. Performance-based testing is widely accepted as a better way of ensuring the user has the skills needed, rather than just recalling information.

The CSWA examination is designed to test specific performance tasks in the following areas:

Sketch Entities – lines, rectangles, circles, arcs, ellipses, centerlines

Objectives: Creating Sketch Entities.

Certification Examination Performance Task	Covered in this book on Chapter – Page
Sketch Command	2-8
Line Command	2-8
Exit Sketch	2-14
Circle Command, Center Point Circle	2-29
Rectangle Command	3-10
Edit Sketch	4-24
Sketch Fillet	4-25
Centerline	7-6
Tangent Arc	7-9
Centerpoint Arc	7-13
Construction Geometry	11-11
Construction Lines	11-13

Sketch Tools – offset, convert, trim

Objectives: Using Sketch Tools.

Certification Examination Performance Task	Covered in this book on Chapter – Page
Convert Entities	6-25
Offset Entities	6-25

Sketch Relations

Objectives: Using Geometric Relations.

Certification Examination Performance Task	Covered in this book on Chapter – Page

Boss and Cut Features – Extrudes, Revolves, Sweeps, Lofts

Objectives: Creating Basic Swept Shapes.

Certification Examination Performance Task	Covered in this book on Chapter – Page

Certified Associate Reference Guide

Fillets and Chamfers

Objectives: Creating Fillets and Chamfers.

Certification Examination Performance Task	Covered in this book on Chapter – Page

Linear, Circular, and Fill Patterns

Objectives: Creating Patterned Features.

Certification Examination Performance Task	Covered in this book on Chapter – Page

Dimensions

Objectives: Applying and Editing Smart Dimensions.

Certification Examination Performance Task	Covered in this book on Chapter – Page

Certified Associate Reference Guide

Feature Conditions – Start and End

Objectives: Controlling Feature Start and End Conditions.

Certification Examination Performance Task	Covered in this book on Chapter – Page
Extruded Boss/Base, Blind	2-15
Extruded Cut, Through All	2-30
Extruded Cut, Up to Next	3-25
Extruded Boss/Base, Mid-Plane	4-13
Extruded Boss/Base, Up to Surface	4-16

Mass Properties

Objectives: Obtaining Mass Properties for Parts and Assemblies.

Certification Examination Performance Task	Covered in this book on Chapter – Page
Mass Properties Tool	4-28
View Mass Properties of a Part	4-29
Relative to Default Coordinate System	17-17
View Mass Properties of an Assembly	17-29
Relative to Reference Coordinate System	17-29

Materials

Objectives: Applying Material Selection to Parts.

Certification Examination Performance Task	Covered in this book on Chapter – Page
Edit Material Command	4-28
Material Properties	16-8
Material in SimulationXpress	16-11

Inserting Components

Objectives: Inserting Components into an Assembly.

Certification Examination Performance Task	Covered in this book on Chapter – Page
Creating an Assembly File	14-8
Inserting a Base Component	14-9
Inserting Additional Components	14-11
Editing Parts in an Assembly	14-32
Inserting Component from Toolbox	15-16

Certified Associate Reference Guide

Standard Mates – Coincident, Parallel, Perpendicular, Tangent, Concentric, Distance, Angle

Objectives: Applying Standard Mates to Constrain Assemblies.

Reference Geometry – Planes, Axis, Mate References

Objectives: Creating Reference Planes, Axes, and Mate References.

Drawing Sheets and Views

Objectives: Creating and Setting Properties for Drawing Sheets; Inserting and Editing Standard Views.

Annotations

Objectives: Creating Annotations.

Certification Examination Performance Task	Covered in this book on Chapter – Page

Certified Associate Reference Guide

Tips about Taking the Certified SOLIDWORKS Associate (CSWA) Examination

Certified Associate Reference Guide

1. **Study:** The first step to maximize your potential on an exam is to sufficiently prepare for it. You need to be familiar with the SOLIDWORKS package, and this can only be achieved by doing designs and exploring the different commands available. The Certified SOLIDWORKS Associate (CSWA) exam is designed to measure your familiarity with the SOLIDWORKS software. You must be able to perform the given task and answer the exam questions correctly and quickly.
2. **Make Notes**: Take notes of what you learn either while attending classroom sessions or going through study material. Use these notes as a review guide before taking the actual test.
3. **Time Management**: The examination has a time limit. Manage the time you spend on each question. Always remember you do not need to score 100% to pass the exam. Also, keep in mind that some questions are weighed more heavily and may take more time to answer. You can flip back and forth to view different problems during the test time by using the arrow buttons. If you encounter a question you cannot answer in a reasonable amount of time, use the *Save As* feature in SOLIDWORKS to save a copy of the file, and move on to the next question. You can return to any question and enter or change the answer as long as you do not hit the [**End Examination**] button.
4. **Use the SOLIDWORKS *Help System***: If you get confused and can't think of the answer, remember the SOLIDWORKS *Help System* is a great tool to confirm your considerations. In preparing for the exam, familiarize yourself with the help utility organization (e.g., Contents, Index, Search options).
5. **Use Internet Search**: Use of an internet search utility is allowed during the test. If a test question requires general knowledge, for example definitions of engineering or drafting concepts (stress, yield strength, auxiliary view, etc.), remember the internet is available as a tool to assist in your considerations.
6. **Use Common Sense**: If you are unable to get the correct answer and unable to eliminate all distracters, then you need to select the best answer from the remaining selections. This may be a task of selecting the best answer from amongst several correct answers, or it may be selecting the least incorrect answer from amongst several poor answers.
7. **Be Cautious and Don't Act in Haste:** Devote some time to ponder and think of the correct answer. Ensure that you interpret all the options correctly before selecting from available choices. Don't go into panic mode while taking a test. Use the ***Arrow Buttons*** to review each question. When you are confident that you have answered all questions, end the examination using the [**End Examination**] button to submit your answers for scoring. You will receive a score report once you have submitted your answers.
8. **Relax before Exam:** In order to avoid last minute stress, make sure that you arrive 10 to 15 minutes early and relax before taking the exam.

Chapter 1
Getting Started

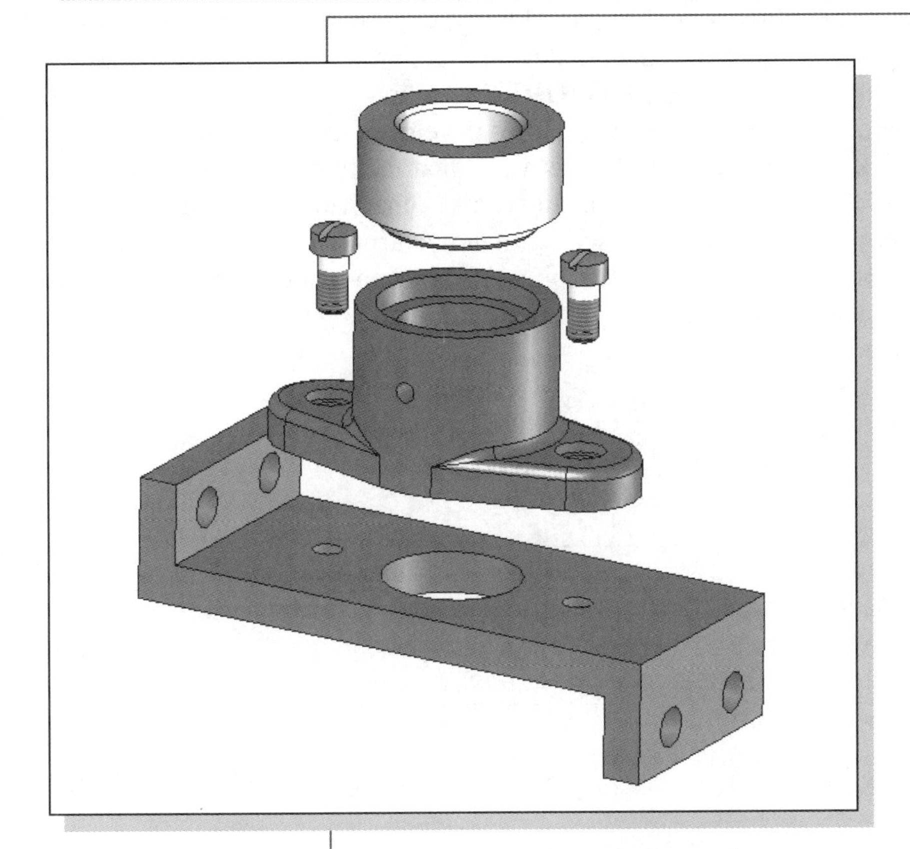

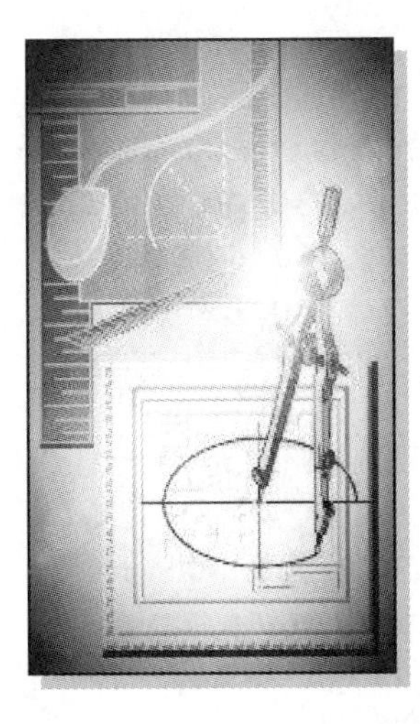

- Development of Computer Geometric Modeling
- Feature-Based Parametric Modeling
- Startup Options and Units Setup
- SOLIDWORKS Screen Layout
- User Interface & Mouse Buttons
- SOLIDWORKS Online Help

Introduction

The rapid changes in the field of **Computer Aided Engineering** (CAE) have brought exciting advances in the engineering community. Recent advances have made the long-sought goal of **concurrent engineering** closer to a reality. CAE has become the core of concurrent engineering and is aimed at reducing design time, producing prototypes faster, and achieving higher product quality. **SOLIDWORKS** is an integrated package of mechanical computer aided engineering software tools developed by *Dassault Systèmes* SOLIDWORKS *Corporation*. **SOLIDWORKS** is a tool that facilitates a concurrent engineering approach to the design and stress-analysis of mechanical engineering products. The computer models can also be used by manufacturing equipment such as machining centers, lathes, mills, or rapid prototyping machines to manufacture the product. In this text, we will be dealing only with the solid modeling modules used for part design and part drawings.

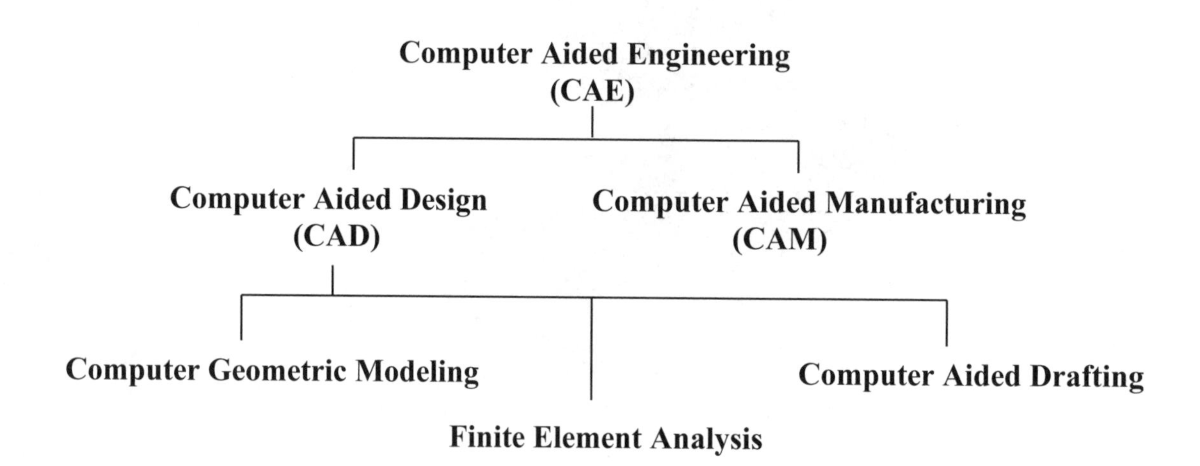

Development of Computer Geometric Modeling

Computer geometric modeling is a relatively new technology, and its rapid expansion in the last fifty years is truly amazing. Computer-modeling technology has advanced along with the development of computer hardware. The first generation CAD programs, developed in the 1950s, were mostly non-interactive; CAD users were required to create program-codes to generate the desired two-dimensional (2D) geometric shapes. Initially, the development of CAD technology occurred mostly in academic research facilities. The Massachusetts Institute of Technology, Carnegie-Mellon University, and Cambridge University were the leading pioneers at that time. The interest in CAD technology spread quickly and several major industry companies, such as General Motors, Lockheed, McDonnell, IBM, and Ford Motor Co., participated in the development of interactive CAD programs in the 1960s. Usage of CAD systems was primarily in the automotive industry, aerospace industry, and government agencies that developed their own programs for their specific needs. The 1960s also marked the beginning of the development of finite element analysis methods for computer stress analysis and computer aided manufacturing for generating machine toolpaths.

The 1970s are generally viewed as the years of the most significant progress in the development of computer hardware, namely the invention and development of **microprocessors**. With the improvement in computing power, new types of 3D CAD programs that were user-friendly and interactive became reality. CAD technology quickly expanded from very simple **computer aided drafting** to very complex **computer aided design**. The use of 2D and 3D wireframe modelers was accepted as the leading edge technology that could increase productivity in industry. The developments of surface modeling and solid modeling technologies were taking shape by the late 1970s, but the high cost of computer hardware and programming slowed the development of such technology. During this period, the available CAD systems all required room-sized mainframe computers that were extremely expensive.

In the 1980s, improvements in computer hardware brought the power of mainframes to the desktop at less cost and with more accessibility to the general public. By the mid-1980s, CAD technology had become the main focus of a variety of manufacturing industries and was very competitive with traditional design/drafting methods. It was during this period of time that 3D solid modeling technology had major advancements, which boosted the usage of CAE technology in industry.

The introduction of the *feature-based parametric solid modeling* approach, at the end of the 1980s, elevated CAD/CAM/CAE technology to a new level. In the 1990s, CAD programs evolved into powerful design/manufacturing/management tools. CAD technology has come a long way, and during these years of development, modeling schemes progressed from two-dimensional (2D) wireframe to three-dimensional (3D) wireframe, to surface modeling, to solid modeling and, finally, to feature-based parametric solid modeling.

The first generation CAD packages were simply 2D **computer aided drafting** programs, basically the electronic equivalents of the drafting board. For typical models, the use of this type of program would require that several to many views of the objects be created individually as they would be on the drafting board. The 3D designs remained in the designer's mind, not in the computer database. Mental translations of 3D objects to 2D views are required throughout the use of these packages. Although such systems have some advantages over traditional board drafting, they are still tedious and labor intensive. The need for the development of 3D modelers came quite naturally, given the limitations of the 2D drafting packages.

The development of three-dimensional modeling schemes started with three-dimensional (3D) wireframes. Wireframe models are models consisting of points and edges, which are straight lines connecting between appropriate points. The edges of wireframe models are used, similar to lines in 2D drawings, to represent transitions of surfaces and features. The use of lines and points is also a very economical way to represent 3D designs.

The development of the 3D wireframe modeler was a major leap in the area of computer geometric modeling. The computer database in the 3D wireframe modeler contains the locations of all the points in space coordinates, and it is typically sufficient to create just one model rather than multiple views of the same model. This single 3D model can then be viewed from any direction as needed. Most 3D wireframe modelers allow the user to create projected lines/edges of 3D wireframe models. In comparison to other types of 3D modelers, the 3D wireframe modelers require very little computing power and generally can be used to achieve reasonably good representations of 3D models. However, because surface definition is not part of a wireframe model, all wireframe images have the inherent problem of ambiguity. Two examples of such ambiguity are illustrated.

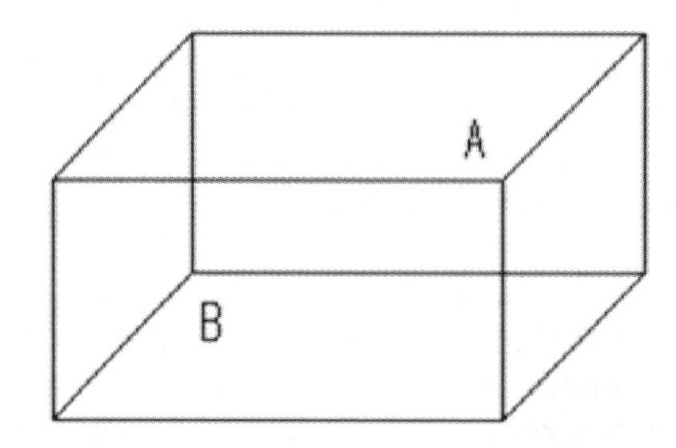

Wireframe Ambiguity: Which corner is in front, A or B?

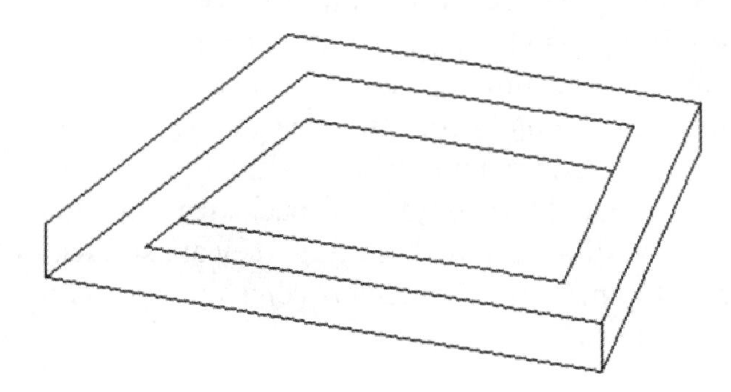

A non-realizable object: Wireframe models contain no surface definitions.

Surface modeling is the logical development in computer geometry modeling to follow the 3D wireframe modeling scheme by organizing and grouping edges that define polygonal surfaces. Surface modeling describes the part's surfaces but not its interiors. Designers are still required to interactively examine surface models to ensure that the various surfaces on a model are contiguous throughout. Many of the concepts used in 3D wireframe and surface modelers are incorporated in the solid modeling scheme, but it is solid modeling that offers the most advantages as a design tool.

In the solid modeling presentation scheme, the solid definitions include nodes, edges, and surfaces, and it is a complete and unambiguous mathematical representation of a precisely enclosed and filled volume. Unlike the surface modeling method, solid modelers start with a solid or use topology rules to guarantee that all of the surfaces are stitched together properly. Two predominant methods for representing solid models are **constructive solid geometry** (CSG) representation and **boundary representation** (B-rep).

The CSG representation method can be defined as the combination of 3D solid primitives. What constitutes a "primitive" varies somewhat with the software but typically includes a rectangular prism, a cylinder, a cone, a wedge, and a sphere. Most solid modelers also allow the user to define additional primitives, which are shapes typically formed by the basic shapes. The underlying concept of the CSG representation method is very straightforward; we simply **add** or **subtract** one primitive from another. The CSG approach is also known as the machinist's approach, as it can be used to simulate the manufacturing procedures for creating the 3D object.

In the B-rep representation method, objects are represented in terms of their spatial boundaries. This method defines the points, edges, and surfaces of a volume, and/or issues commands that sweep or rotate a defined face into a third dimension to form a solid. The object is then made up of the unions of these surfaces that completely and precisely enclose a volume.

By the 1980s, a new paradigm called *concurrent engineering* had emerged. With concurrent engineering, designers, design engineers, analysts, manufacturing engineers, and management engineers all work together closely right from the initial stages of the design. In this way, all aspects of the design can be evaluated and any potential problems can be identified right from the start and throughout the design process. Using the principles of concurrent engineering, a new type of computer modeling technique appeared. The technique is known as the *feature-based parametric modeling technique.* The key advantage of the *feature-based parametric modeling technique* is its capability to produce very flexible designs. Changes can be made easily and design alternatives can be evaluated with minimum effort. Various software packages offer different approaches to feature-based parametric modeling, yet the end result is a flexible design defined by its design variables and parametric features.

Feature-Based Parametric Modeling

One of the key elements in the SOLIDWORKS solid modeling software is its use of the **feature-based parametric modeling technique**. The feature-based parametric modeling approach has elevated solid modeling technology to the level of a very powerful design tool. Parametric modeling automates the design and revision procedures by the use of parametric features. Parametric features control the model geometry by the use of design variables. The word ***parametric*** means that the geometric definitions of the design, such as dimensions, can be varied at any time during the design process. Features are predefined parts or construction tools for which users define the key parameters. A part is described as a sequence of engineering features, which can be modified/changed at any time. The concept of parametric features makes modeling more closely match the actual design-manufacturing process than the mathematics of a solid modeling program. In parametric modeling, models and drawings are updated automatically when the design is refined.

Parametric modeling offers many benefits:

- **We begin with simple, conceptual models with minimal detail; this approach conforms to the design philosophy of "shape before size."**
- **Geometric relations, dimensional constraints, and relational parametric equations can be used to capture design intent.**
- **The ability to update an entire system, including parts, assemblies and drawings after changing one parameter of complex designs.**
- **We can quickly explore and evaluate different design variations and alternatives to determine the best design.**
- **Existing design data can be reused to create new designs.**
- **Quick design turn-around.**

Getting Started with SOLIDWORKS

SOLIDWORKS is composed of several application software modules (these modules are called *applications*), all sharing a common database. In this text, the main concentration is placed on the solid modeling modules used for part design. The general procedures required in creating solid models, engineering drawings, and assemblies are illustrated.

How to start SOLIDWORKS depends on the type of workstation and the particular software configuration you are using. With most *Windows* systems, you may select **SOLIDWORKS** on the *Start* menu or select the **SOLIDWORKS** icon on the desktop. Consult your instructor or technical support personnel if you have difficulty starting the software. The program takes a while to load, so be patient.

The tutorials in this text are based on the assumption that you are using the SOLIDWORKS default settings. If your system has been customized for other uses, contact your technical support personnel to restore the default software configuration.

Once the program is loaded into the memory, the SOLIDWORKS program window appears. In addition, the *Welcome* dialog box will open by default. The *Welcome* dialog box provides a convenient method to start new parts, drawings, or assemblies; open existing documents; or access SOLIDWORKS resources.

If the *Welcome* dialog box does not appear, it can be opened by clicking the *Welcome to SolidWorks* icon in the *Task Pane* or on the *Menu Bar*, both of which are described below.

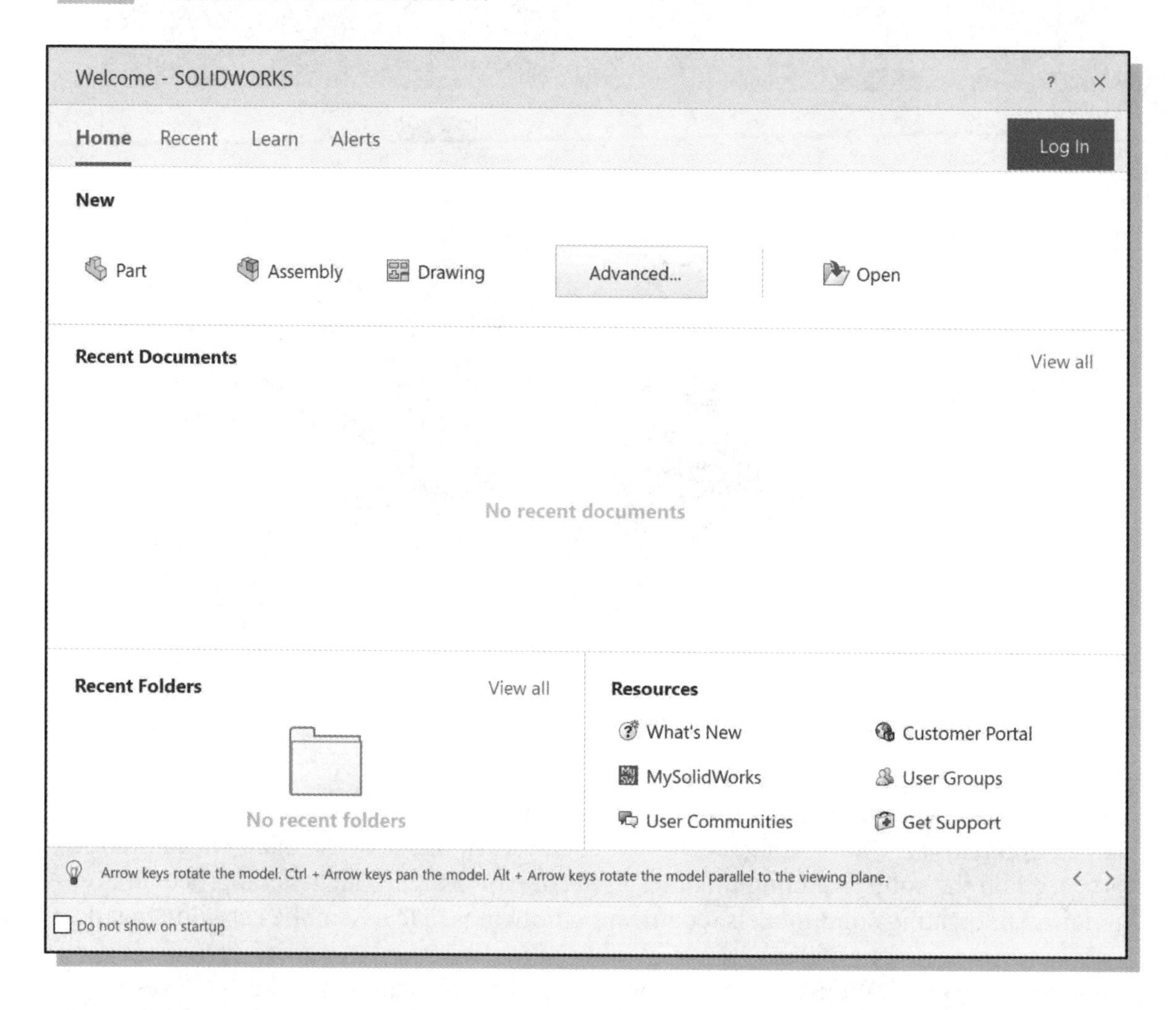

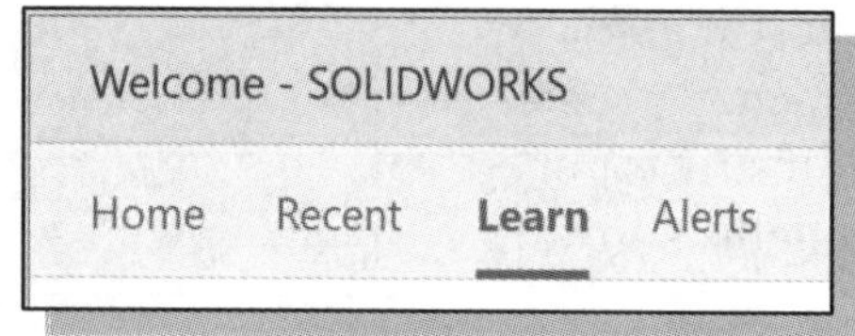

The *Welcome* dialog box has the following tabs: *Home*, *Recent*, *Learn*, and *Alerts*. Under the *Learn* tab, access is provided to SOLIDWORKS documentation, tutorials, and files provided for the tutorials. On your own, select the various tabs to reveal the options available.

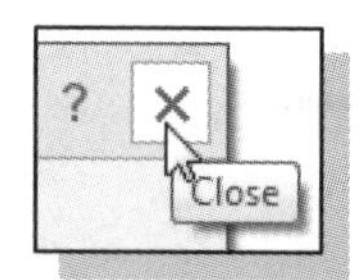

Close the *Welcome* dialog box by clicking on the **X** in the upper right corner of the box to view the SOLIDWORKS program window.

The SOLIDWORKS program window contains the *Menu Bar* and the *Task Pane*. The *Menu Bar* contains a subset of commonly used tools from the *Standard* toolbar (**New**, **Open**, **Save**, etc.), the SOLIDWORKS menus, the SOLIDWORKS *Search* oval, and a flyout menu of *Help* options. By default, the SOLIDWORKS menus are hidden. To display them, move the cursor over or click the SOLIDWORKS logo.

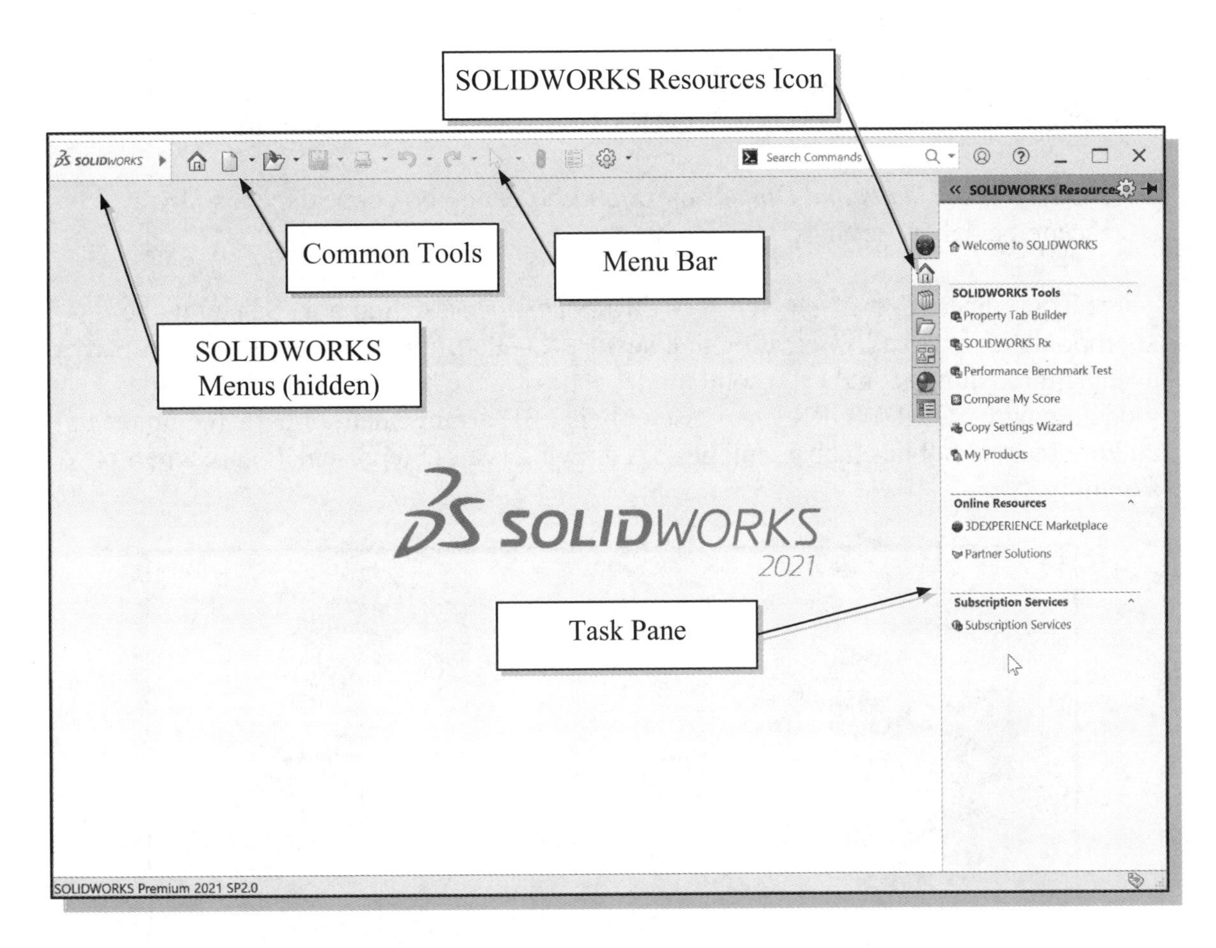

If the *Task Pane* does not appear to the right of the screen, right click on the *Menu Bar* to reveal a menu of toolbars and select the **Task Pane**, or select *View* from the SOLIDWORKS *Menus* and select **Task Pane**. Other options for the *Task Pane* include *Design Library*, *File Explorer*, *View Palette*, and *Appearances, Scenes, and Decals*. The icons for these options appear below the **SOLIDWORKS Resources** icon. The *File Explorer* duplicates *Windows Explorer* and provides access to recent documents. Other options will be used in future lessons. To collapse the *Task Pane*, click anywhere in the main area of the SOLIDWORKS window. (**NOTE:** If the *Task Pane* does not collapse upon clicking in the graphics area, it has been 'pinned' using the *Auto Show* button in the upper right corner. Simply left-click on the icon to unpin the *Task Pane* and allow it to collapse.)

The following two startup options are available: **New** and **Open**. The **New** option allows us to start a new modeling task. The **Open** option allows us to open an existing model file. These two commands can be executed in the *Welcome* dialog box or on the *Menu Bar*.

➤ Select the **New** icon on the *Menu Bar* with a single click of the left-mouse-button. The *New* SOLIDWORKS *Document* dialog box appears.

➤ **NOTE:** If the *Units and Dimensions Standard* dialog box appears, click **OK** to accept the default settings.

Three icons appear in the *New* SOLIDWORKS *Document* dialog box. Selecting the appropriate icon will allow creation of a new Part, Assembly, or Drawing file. A part is a single three-dimensional (3D) solid model. Parts are the basic building blocks in modeling with SOLIDWORKS. An assembly is a 3D arrangement of parts (components) and/or other assemblies (subassemblies). A drawing is a 2D representation of a part or assembly.

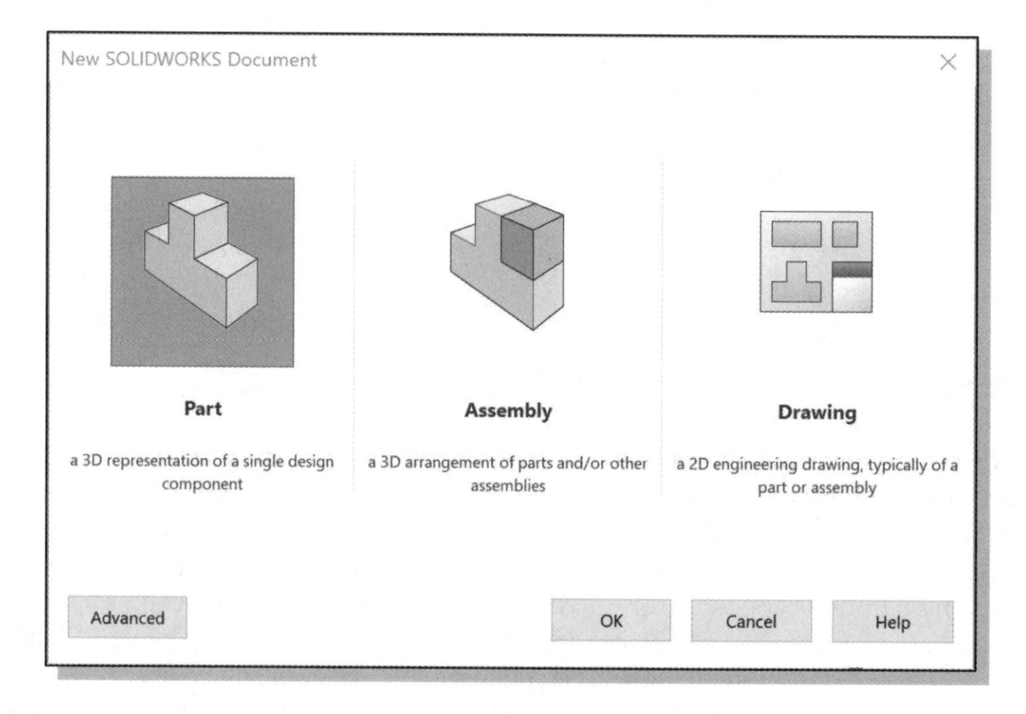

➤ Select the **Part** icon as shown. Click **OK** in the *New SOLIDWORKS Document* dialog box to open a new part file.

Units Setup

When starting a new CAD file, the first thing we should do is to choose the units we would like to use. The *Unit system* for the active document is shown on the *Status Bar* at the bottom of the SOLIDWORKS window (e.g., millimeter, gram, second - MMGS). We will use the English (inch, pound, second - IPS) setting for this example.

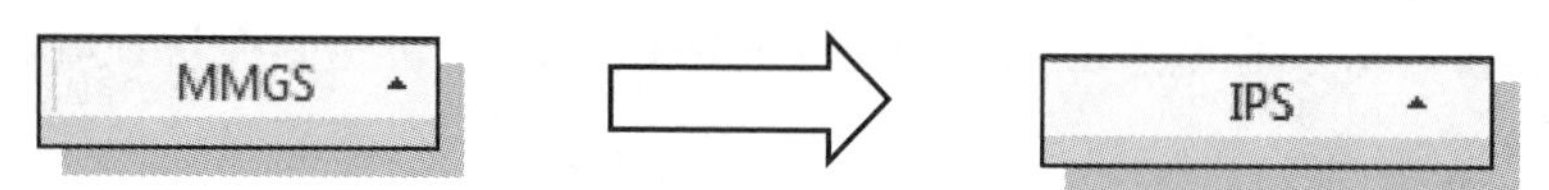

➢ Select **Edit Document Units** from the *Status Bar* OR the **Options** icon from the *Menu Bar* to open the *Options* dialog box.

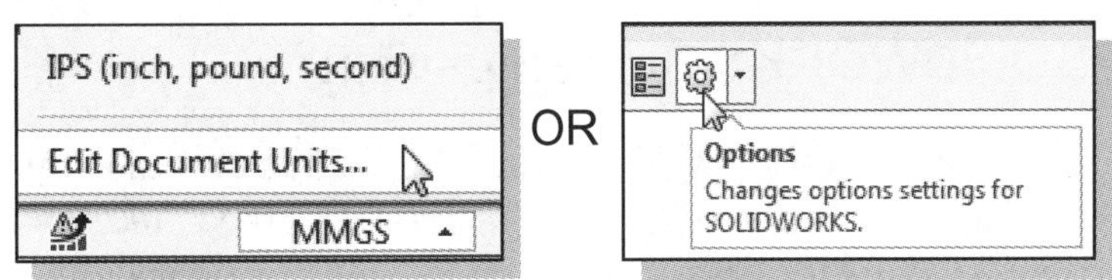

➢ When the *Options* dialog box opens, the **System Options** tab is active. The *Units* setup is located under the Document Properties tab. Select the **Document Properties** tab as shown.

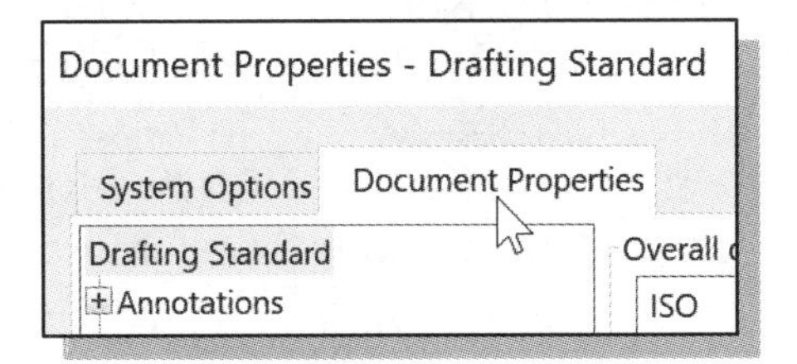

➢ Select **Units** on the left menu as highlighted below. Select **IPS (inch, pound, second)** under the *Unit system* options. Select **.123** in the *Decimals* spin box for the *Length units* as shown to define the degree of accuracy with which the units will be displayed. Click **OK** at the bottom of the *Document Properties - Units* window to set the units. (Notice IPS now appears on the *Status Bar* at the bottom of the window.)

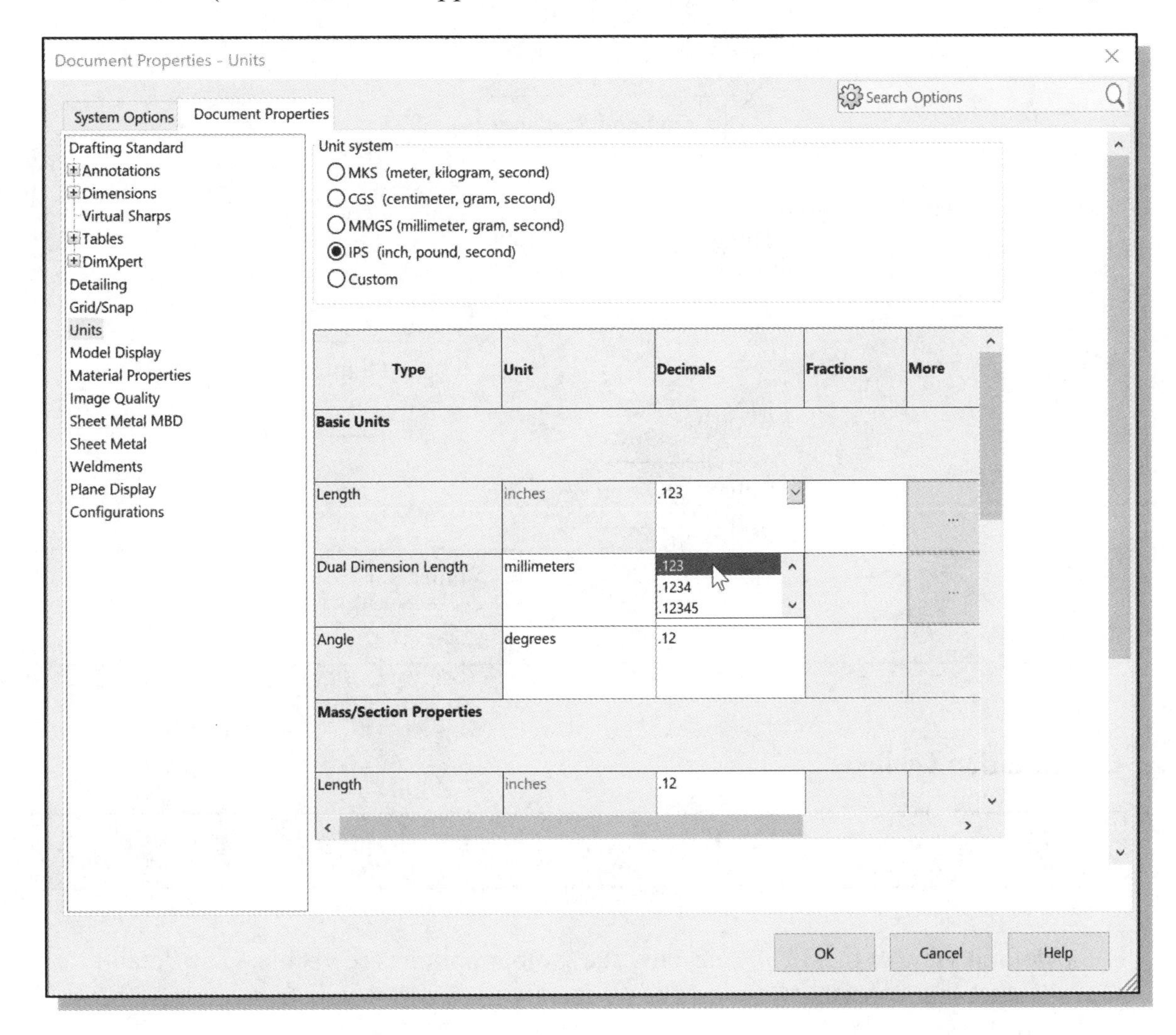

SOLIDWORKS Screen Layout

The default SOLIDWORKS drawing screen contains the *Menu Bar*, the *Heads-up View* toolbar, the *FeatureManager Design Tree*, the *CommandManager* (below the *Menu Bar*), the *task pane* (collapsed to the right of the graphics area in the figure below), and the *Status Bar*. (Note: If the *CommandManager* is inactive, the *Features* toolbar will appear vertically at the left and the *Sketch* toolbar will appear vertically at the right.) A line of quick text appears next to the icon as you move the *mouse cursor* over different icons. You may resize the SOLIDWORKS drawing window by clicking and dragging at the edges of the window, or relocate the window by clicking and dragging at the *window title* area.

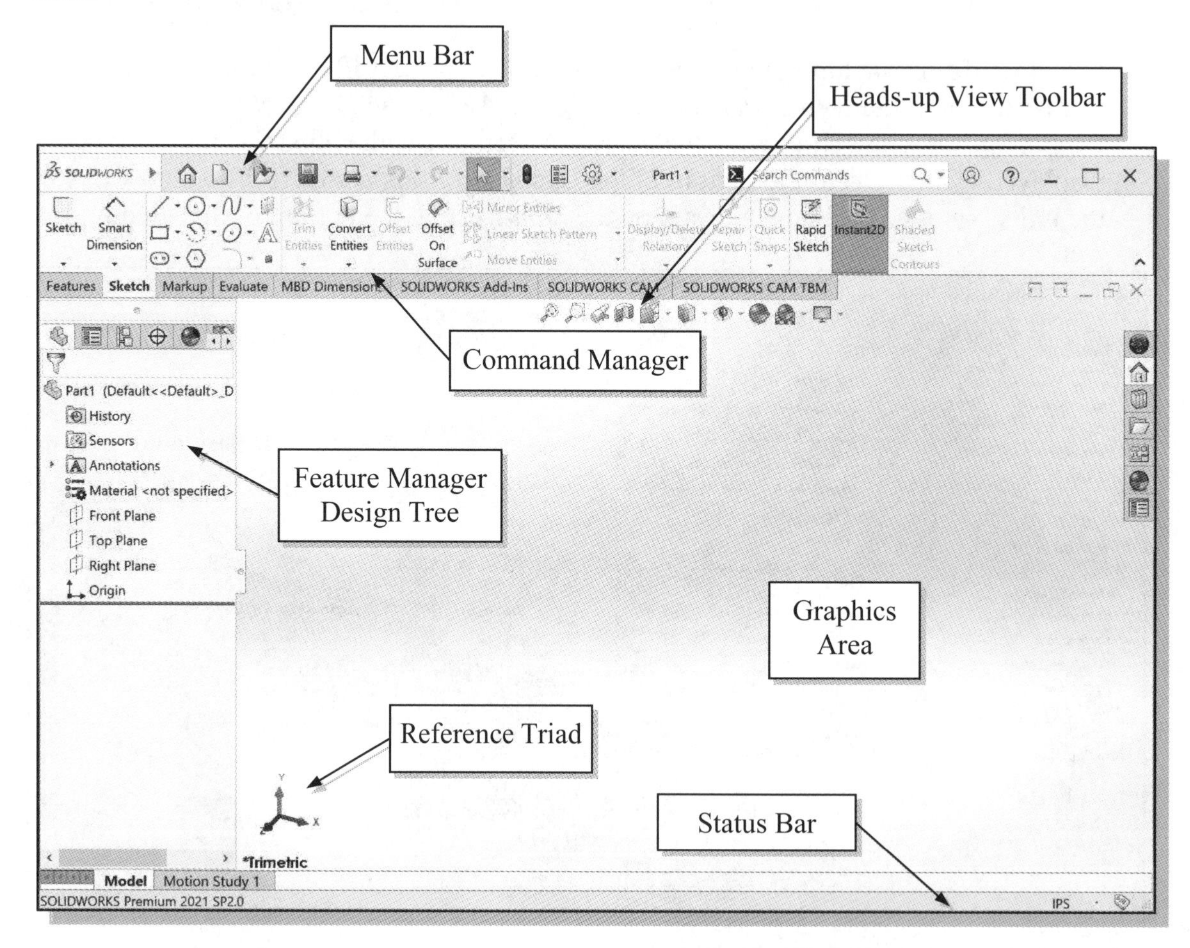

- **Menu Bar Toolbar**

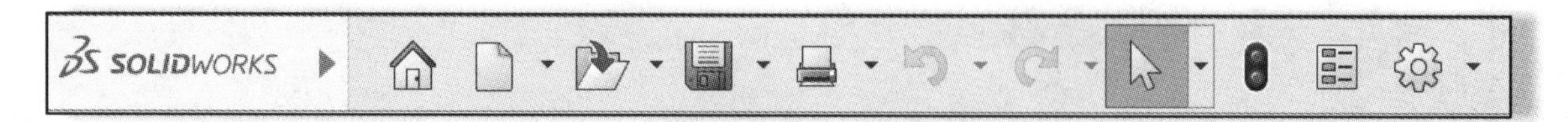

In the default view of the *Menu Bar*, only the toolbar options are visible. The default *Menu Bar* toolbar consists of a subset of frequently used commands from the *Menu Bar* as shown above.

- **SOLIDWORKS Pull-down Menus**

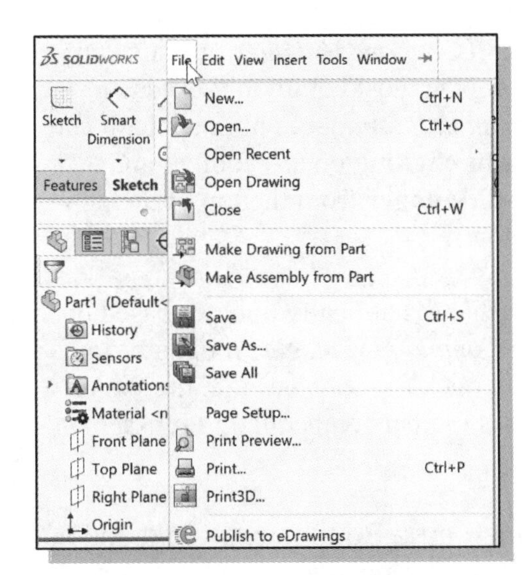

To display the *pull-down* menus, move the cursor over or click the SOLIDWORKS logo. The *pull-down* menus contain operations that you can use for all modes of the system.

- **Heads-up View Toolbar**

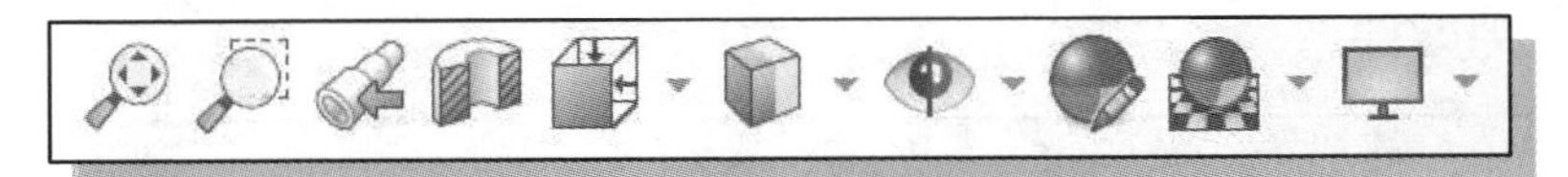

The *Heads-up View* toolbar allows us quick access to frequently used view-related commands. **NOTE:** You can hide or customize the *Heads-up View* toolbar.

- **Features Toolbar**

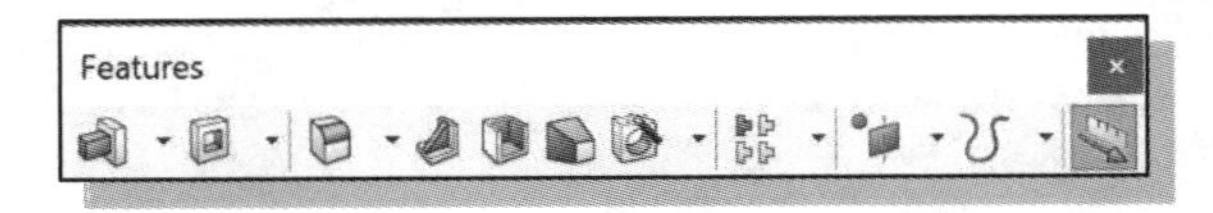

The *Features* toolbar allows us quick access to frequently used features-related commands, such as Extruded Boss/Base, Extruded Cut, and Revolved Boss/Base. When the *CommandManager* is used (with the Use Large Buttons with Text option turned *OFF*) the *Features* toolbar appears as shown above. When the *CommandManager* is turned *OFF*, the *Features* toolbar is displayed (by default) vertically at the left of the SOLIDWORKS window.

- **Sketch Toolbar**

The *Sketch* toolbar provides tools for creating the basic geometry that can be used to create features and parts. When the *CommandManager* is used (with the Use Large Buttons with Text option turned *OFF*), the *Sketch* toolbar appears as shown above. When the *CommandManager* is turned *OFF*, the *Sketch* toolbar is displayed (by default) vertically at the right of the SOLIDWORKS window.

- **CommandManager**

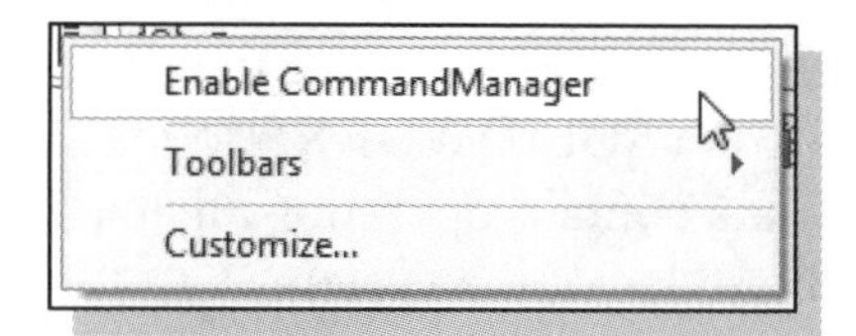

The SOLIDWORKS *CommandManager* provides one method for displaying the most commonly used toolbars. If the *CommandManager* is not visible, it can be turned *ON* by **right clicking** on any toolbar and selecting **CommandManager** from the top of the pop-up menu list of toolbars.

The *CommandManager* is a context-sensitive toolbar that dynamically updates based on the user's selection. When you click a tab below the *CommandManager*, it updates to display the corresponding toolbar. For example, if you click the *Sketches* tab, the *Sketch* toolbar appears. By default, the *CommandManager* has toolbars embedded in it based on the document type.

The display of the *CommandManager* is (with the Use Large Buttons with Text option turned on) shown below, once with the *Features* toolbar selected, once with the *Sketch* toolbar selected. You will notice that when the *CommandManager* is used, the *Sketch* and *Features* toolbars do not appear on the left and right edges of the display window.

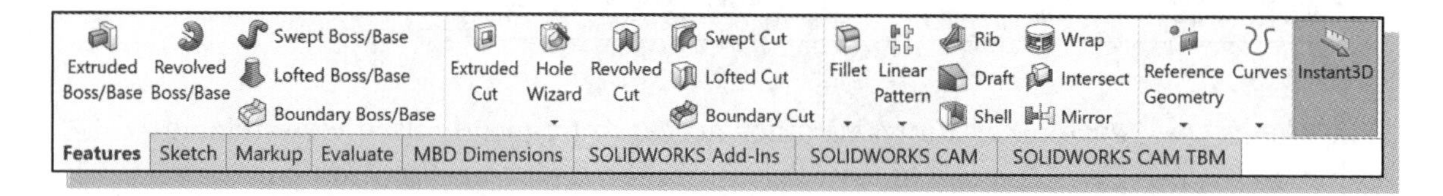

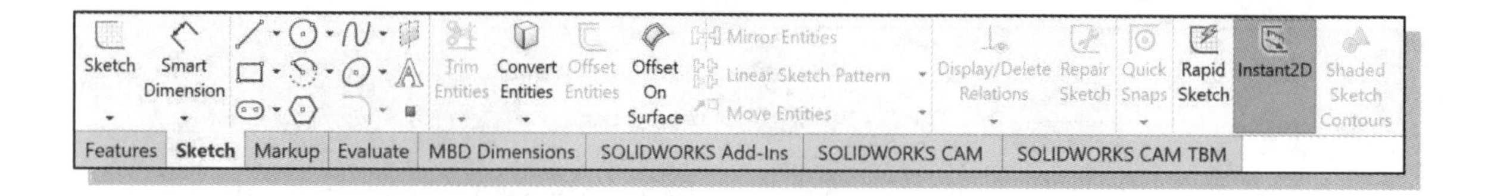

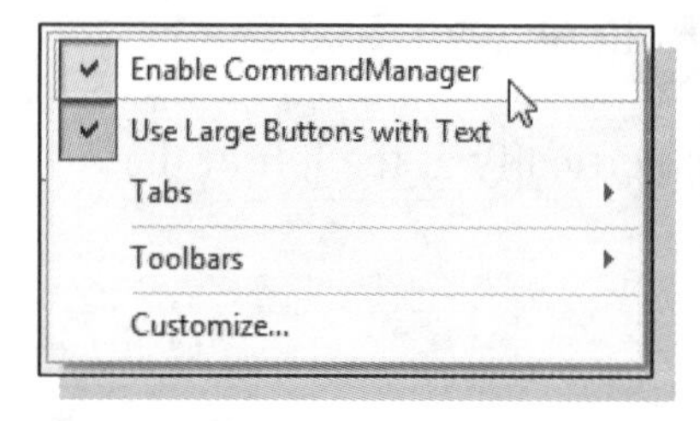

To turn *OFF* the *CommandManager* and use the standard display of toolbars, right click on the *CommandManager* (or any other toolbar) and toggle the *CommandManager* *OFF* by selecting it at the top of the pop-up menu.

IMPORTANT NOTE: Many lessons in this text use the standard display of toolbars. If a user prefers to use the *CommandManager*, the only change is that it may be necessary to select the appropriate tab prior to selecting a command. For example, if the instruction is to "select the Extruded Boss command from the *Features* toolbar," it may be necessary to first select the Features tab on the *CommandManager* to display the *Features* toolbar.

- **Standard Display of Toolbars**

The default SOLIDWORKS drawing screen using the standard display of toolbars, with the *CommandManager* turned *OFF*, is shown below. The *Features* toolbar appears at the left of the window and the *Sketch* toolbar at the right. This is the standard view used in the lessons in this text.

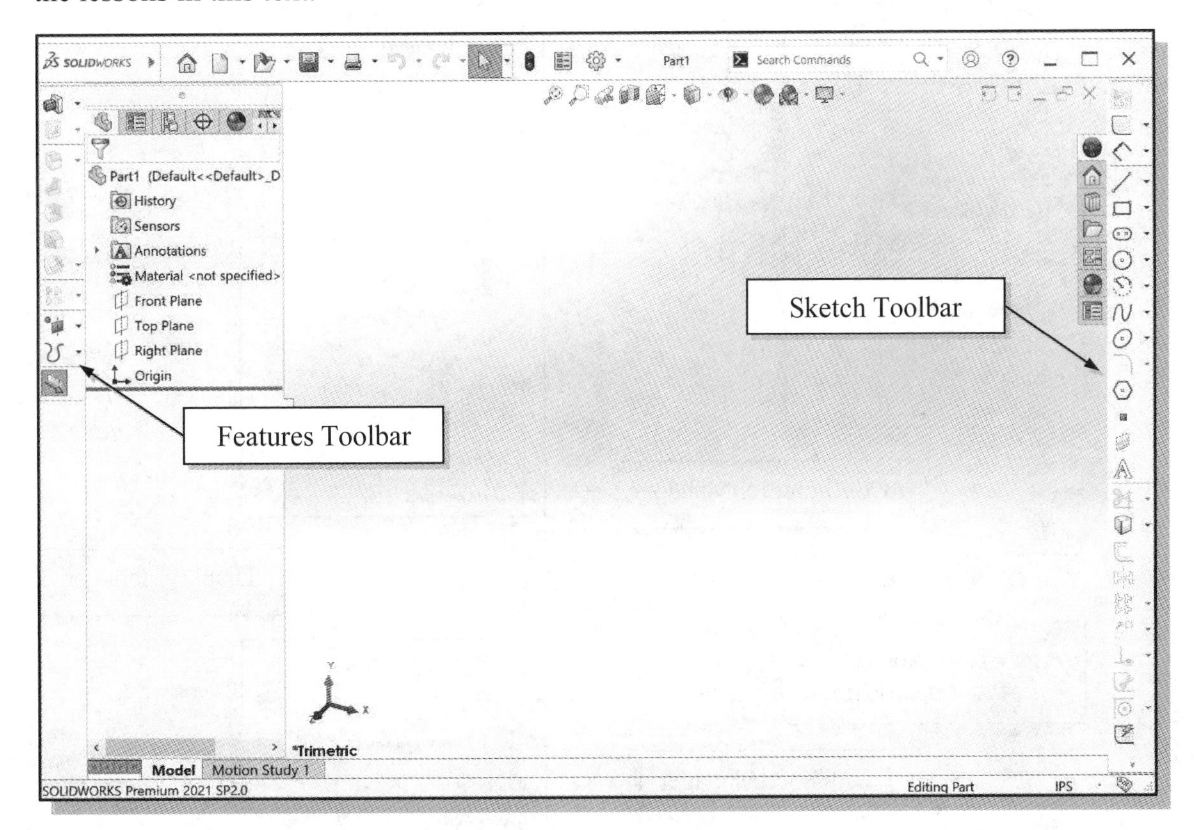

- **Mouse Gestures**

You can use a mouse gesture as a shortcut to execute some common commands. To activate a mouse gesture, move the cursor inside the graphics area, **hold the right mouse button down and drag the mouse**. A gesture guide appears showing command mappings for the gesture directions. Drag the mouse (while holding the right button down) across the appropriate button to execute the command.

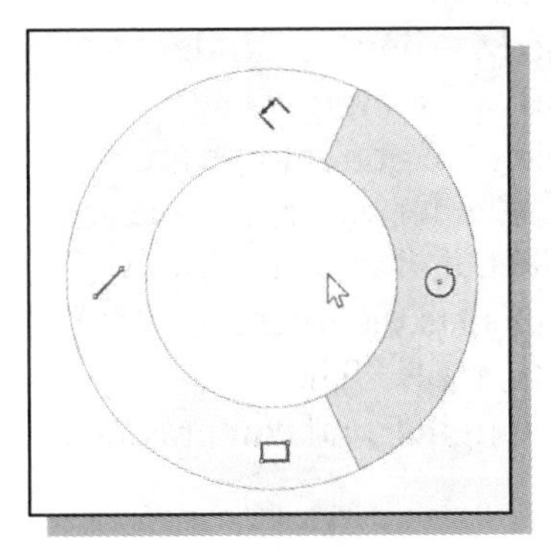

There are separate gesture guides for drawings, assemblies, parts, and sketches. The appropriate guide appears based on the current operation. For example, in sketch mode a gesture guide with sketch commands appears as shown here. You can customize up to eight gestures for each guide. To view or edit the current mouse gesture assignments, select **Customize** on the *Tools* pull-down menu, then select the **Mouse Gestures** tab.

- **FeatureManager Design Tree/PropertyManager/ConfigurationManager/DimXpertManager/DisplayManager**

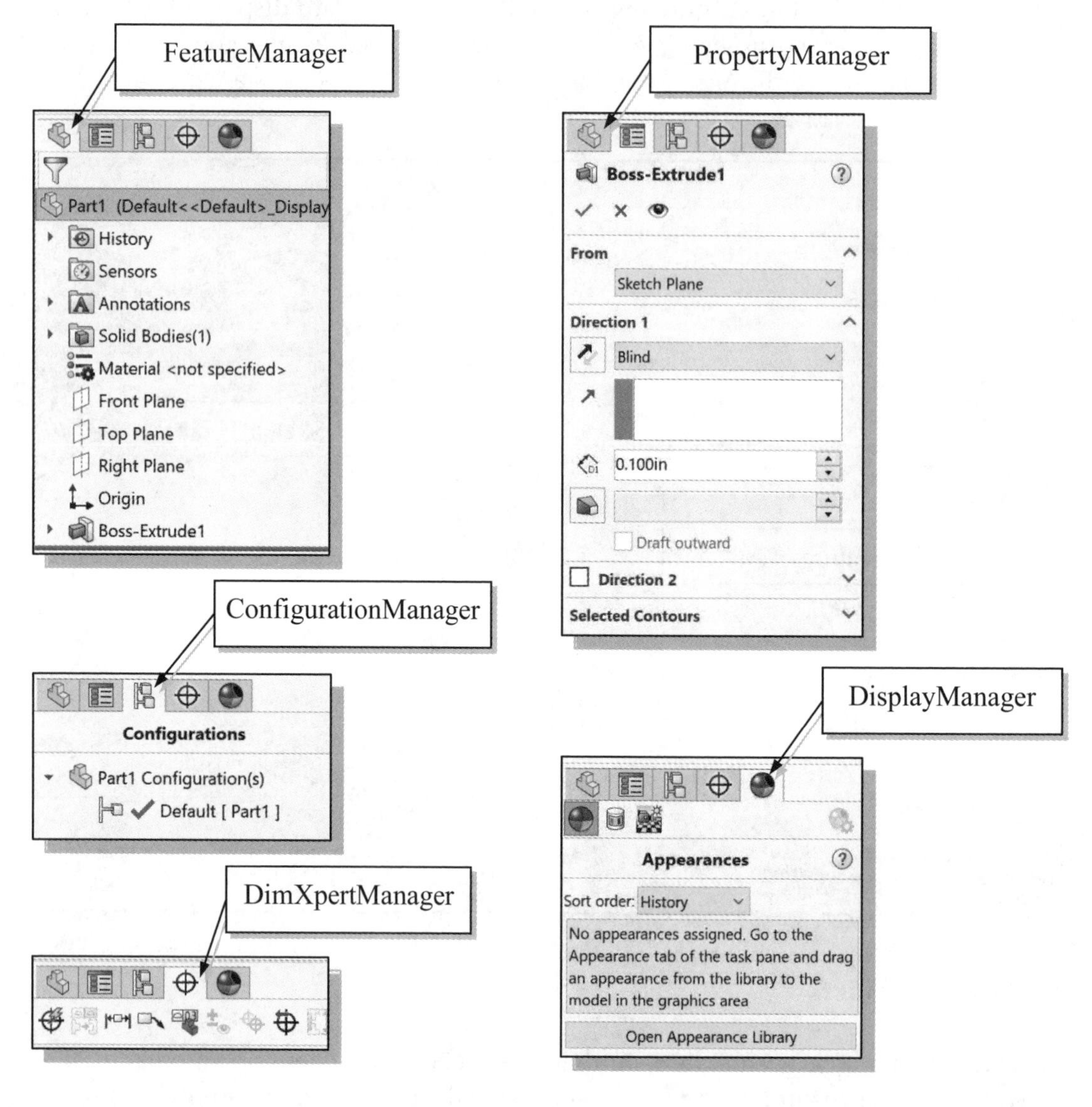

The left panel of the SOLIDWORKS window is used to display the *FeatureManager Design Tree*, the *PropertyManager*, the *ConfigurationManager*, the *DimXpertManager* and the *DisplayManager*. These options can be chosen by selecting the appropriate tab at the top of the panel. The *FeatureManager Design Tree* provides an overview of the active part, drawing, or assembly in outline form. It can be used to show and hide selected features, filter contents, and manage access to features and editing. The *PropertyManager* opens automatically when commands are executed or entities are selected in the graphics window, and is used to make selections, enter values, and accept commands. The *Configuration Manager* is used to create, select, and view multiple configurations of parts and assemblies. The *DimXpertManager* lists the tolerance features defined using the SOLIDWORKS 'DimXpert for parts' tools. The *DisplayManager* is used to control appearances, decals, scenes, lights, and cameras that are applied to the current model.

- **Graphics Area**

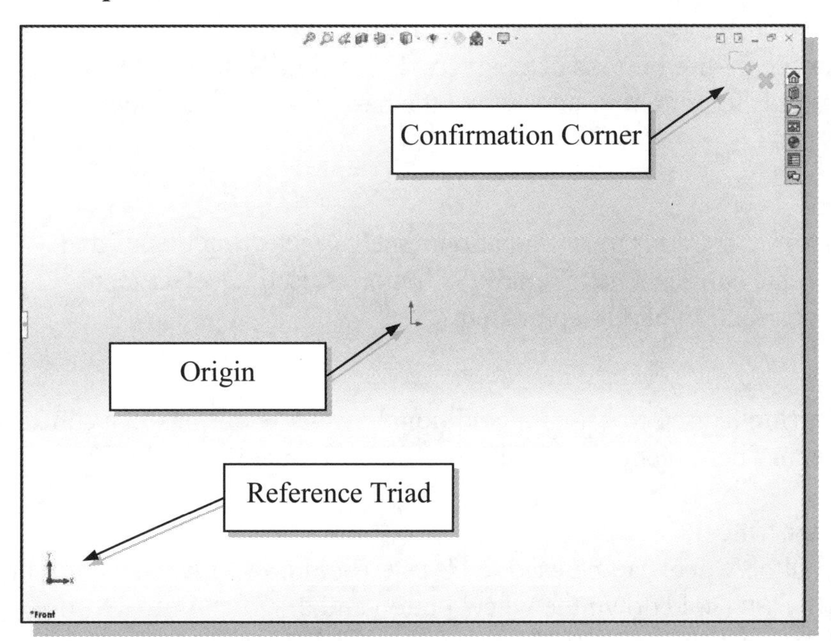

The graphics area is the area where models and drawings are displayed.

- **Reference Triad**

The *Reference Triad* appears in the graphics area of part and assembly documents. The triad is shown to help orient the user when viewing models and is for reference only.

- **Origin**

The *Origin* represents the (0,0,0) coordinate in a model or sketch. A model origin appears blue; a sketch origin appears red.

- **Confirmation Corner**

The *Confirmation Corner* offers an alternate way to accept features.

- **Graphics Cursor or Crosshairs**

The *graphics cursor* shows the location of the pointing device in the graphics window. During geometric construction, the coordinate of the cursor is displayed in the *Status Bar* area, located at the bottom of the screen. The cursor's appearance depends on the selected command or option.

- **Message and Status Bar**

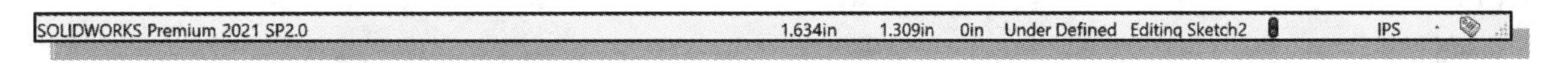

The *Message and Status Bar* area displays a single-line description of a command when the cursor is on top of a command icon. This area also displays information pertinent to the active operation. In the figure above, the cursor coordinates are displayed while in the *Sketch* mode.

Mouse Buttons

SOLIDWORKS utilizes the mouse buttons extensively. In learning SOLIDWORKS' interactive environment, it is important to understand the basic functions of the mouse buttons.

- **Left mouse button**
  The **left-mouse-button** is used for most operations, such as selecting menus and icons, or picking graphic entities. One click of the button is used to select icons, menus and form entries, and to pick graphic items.

- **Right mouse button**
  The **right-mouse-button** is used to bring up additional available options in a context-sensitive pop-up menu. These menus provide shortcuts to frequently used commands.

- **Middle mouse button/wheel**
  The middle mouse button/wheel can be used to Rotate (hold down the wheel button and drag the mouse), Pan (hold down the wheel button and drag the mouse while holding down the **Ctrl** key), or Zoom (hold down the wheel button and drag the mouse while holding down the **Shift** key) realtime. Spinning the wheel allows zooming to the position of the cursor.

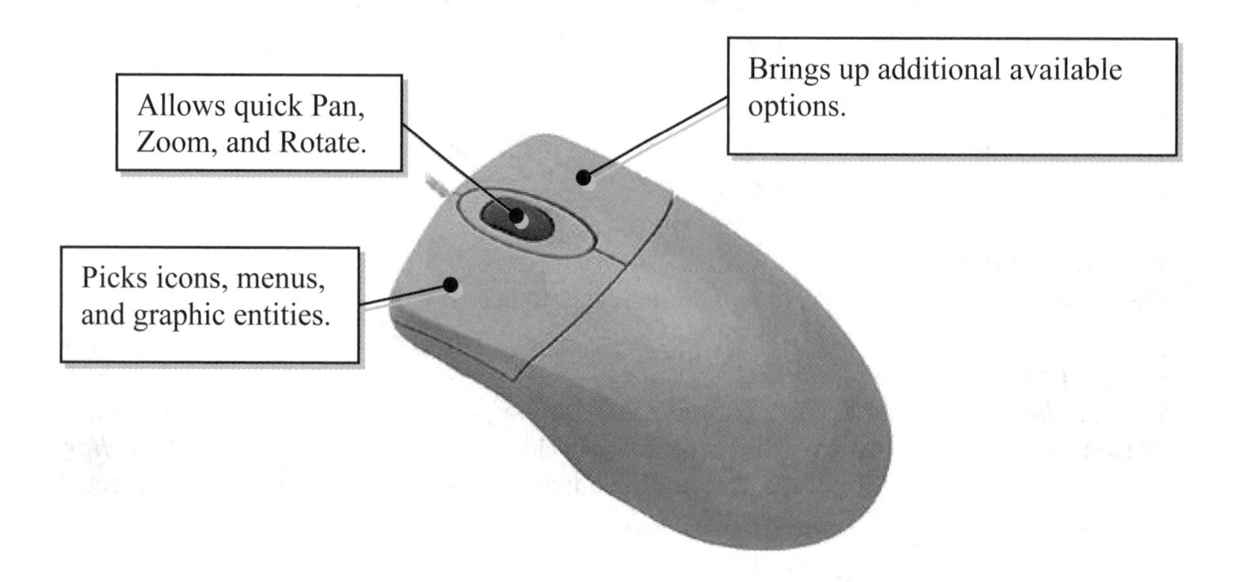

[Esc] – Canceling Commands

The [**Esc**] key is used to cancel a command in SOLIDWORKS. The [**Esc**] key is located near the top-left corner of the keyboard. Sometimes, it may be necessary to press the [**Esc**] key twice to cancel a command; it depends on where we are in the command sequence. For some commands, the [**Esc**] key is used to exit the command.

Online Help

Several types of online help are available at any time during a SOLIDWORKS session. SOLIDWORKS provides many on-line help functions, such as:

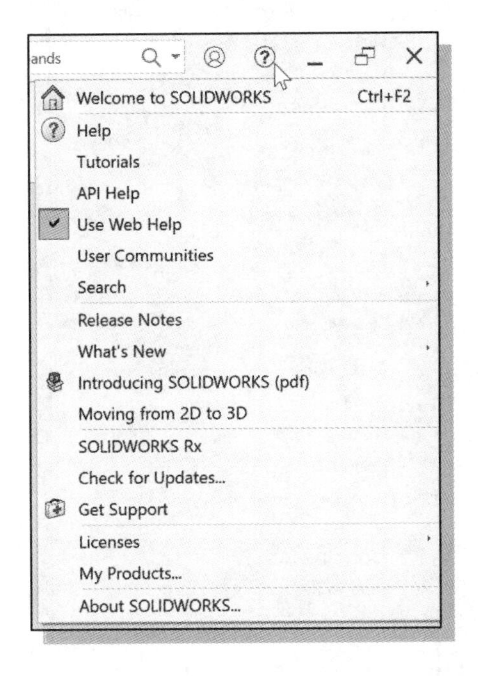

- The **SOLIDWORKS Help** option can be accessed by clicking on the **Help** icon at the right end of the *Menu Bar*. The **Help** option provides general help information, such as command options and command references. The **Tutorials** option provides a collection of tutorials illustrating different SOLIDWORKS operations.

- The **Tutorials** can also be accessed from the SOLIDWORKS *Welcome* dialog box under the *Learn* tab.

SOLIDWORKS Search

The **SOLIDWORKS Search** window, located on the *Menu Bar*, can be used to search various utilities, including searching SOLIDWORKS Help; searching for valid SOLIDWORKS commands; searching for files and models; and searching the MySolidWorks website.

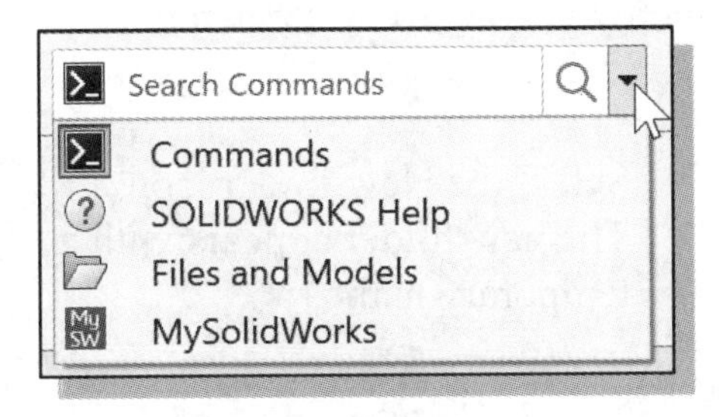

- To execute a search, expand the search menu by clicking the arrow at the right of the window; select the search utility desired, e.g. Files and Models; type the text string for the search; and press **Enter**.

Leaving SOLIDWORKS

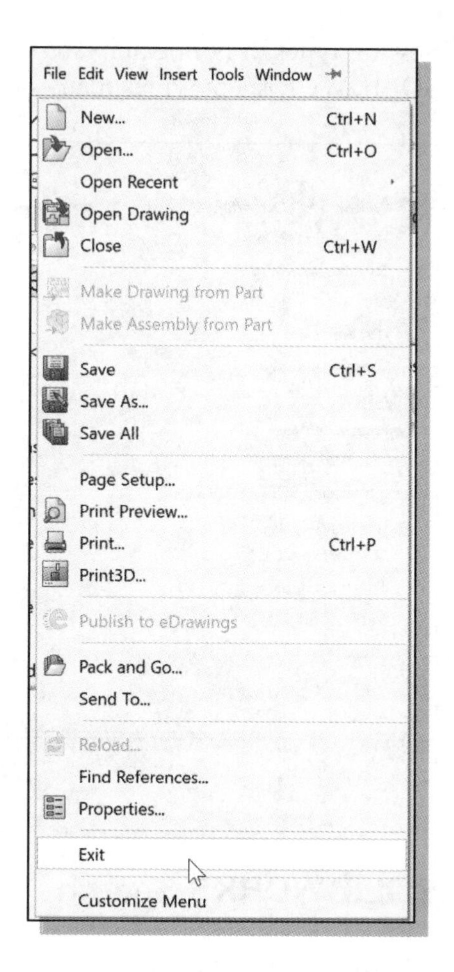

To leave SOLIDWORKS, use the left-mouse-button and click on **File** at the top of the SOLIDWORKS screen window, then choose **Exit** from the pull-down menu. (**NOTE:** Move the cursor over the SOLIDWORKS logo in the *Menu Bar* to display the pull-down menu options.)

Alternately, click the **Close** icon in the upper right corner of the window.

Creating a CAD Files Folder

It is a good practice to create a separate folder to store your CAD files. You should not save your CAD files in the same folder where the SOLIDWORKS application is located. It is much easier to organize and back up your project files if they are in a separate folder. Making folders within this folder for different types of projects will help you organize your CAD files even further. When creating CAD files in SOLIDWORKS, it is strongly recommended that you *save* your CAD files on the hard drive.

➢ To create a new folder in the *Windows* environment:

1. In *Windows File Explorer*, open the folder in which you want to create a new folder.

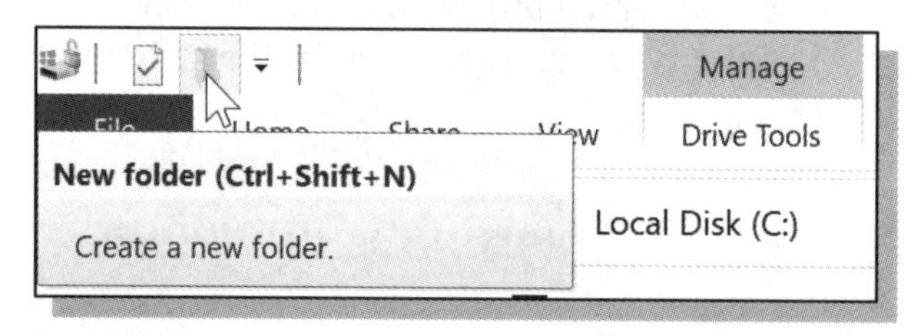

2. Select the New folder icon. The new folder appears with a temporary name.

3. Type a name for the new folder, and then press **ENTER**.

Chapter 2
Parametric Modeling Fundamentals

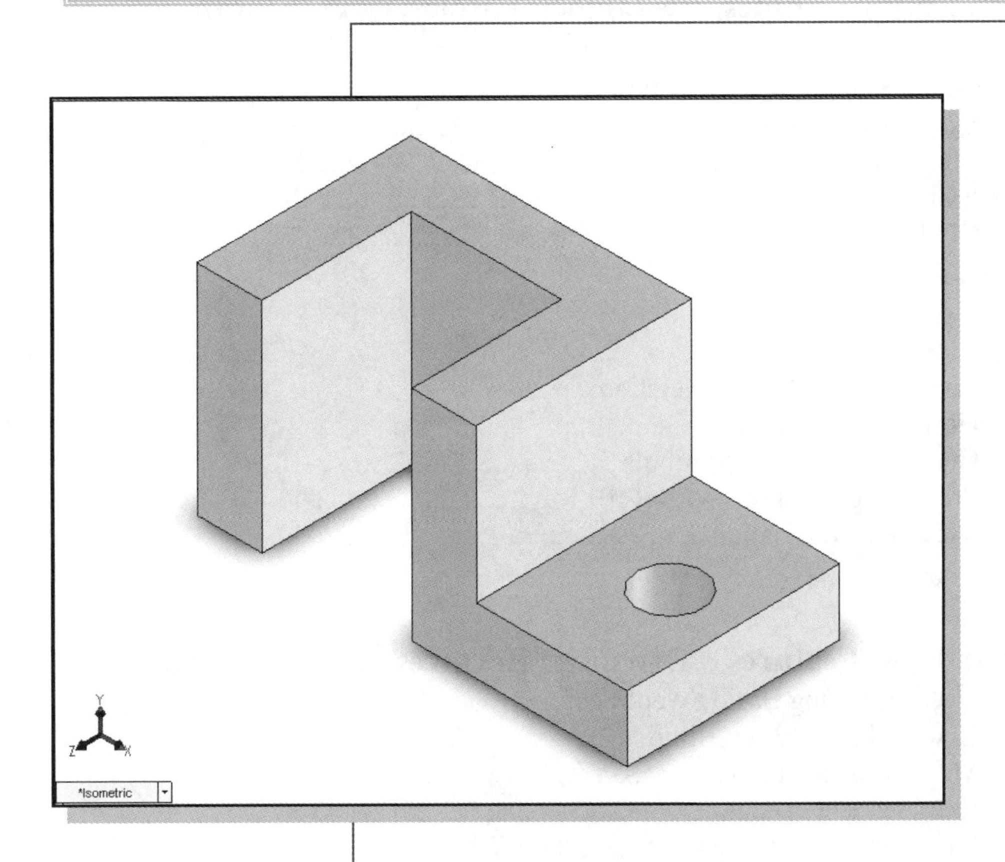

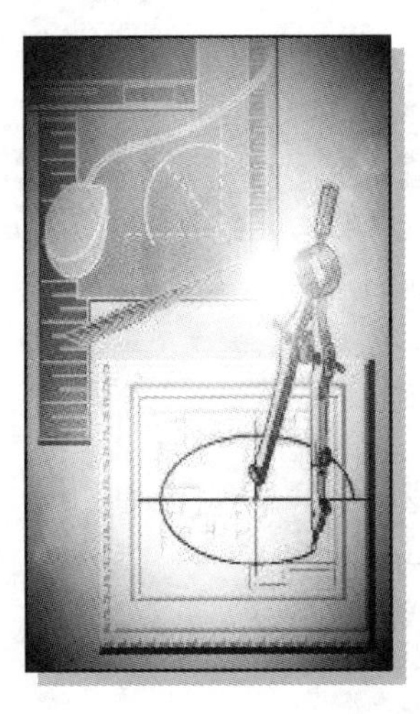

Learning Objectives

- Create Simple Extruded Solid Models
- Understand the Basic Parametric Modeling Procedure
- Create 2-D Sketches
- Understand the "Shape before Size" Approach
- Use the Dynamic Viewing Commands
- Create and Edit Parametric Dimensions

Certified SOLIDWORKS Associate Exam Objectives Coverage

Certified Associate Reference Guide

Sketch Entities – Lines, Rectangles, Circles, Arcs, Ellipses, Centerlines

Objectives: Creating Sketch Entities.

Sketch Relations

Objectives: Using Geometric Relations.

Boss and Cut Features – Extrudes, Revolves, Sweeps, Lofts

Objectives: Creating Basic Swept Features.

Dimensions

Objectives: Applying and Editing Smart Dimensions.

Feature Conditions – Start and End

Objectives: Controlling Feature Start and End Conditions.

Introduction

The **feature-based parametric modeling** technique enables the designer to incorporate the original **design intent** into the construction of the model. The word ***parametric*** means the geometric definitions of the design, such as dimensions, can be varied at any time in the design process. Parametric modeling is accomplished by identifying and creating the key features of the design with the aid of computer software. The design variables, described in the sketches as parametric relations, can then be used to quickly modify/update the design.

In SOLIDWORKS, the parametric part modeling process involves the following steps:

1. **Create a rough two-dimensional sketch of the basic shape of the base feature of the design.**
2. **Apply/modify geometric relations and dimensions to the two-dimensional sketch.**
3. **Extrude, revolve, or sweep the parametric two-dimensional sketch to create the base solid feature of the design.**
4. **Add additional parametric features by identifying feature relations and complete the design.**
5. **Perform analyses on the computer model and refine the design as needed.**
6. **Create the desired drawing views to document the design.**

The approach of creating two-dimensional sketches of the three-dimensional features is an effective way to construct solid models. Many designs are in fact the same shape in one direction. Computer input and output devices we use today are largely two-dimensional in nature, which makes this modeling technique quite practical. This method also conforms to the design process that helps the designer with conceptual design along with the capability to capture the ***design intent***. Most engineers and designers can relate to the experience of making rough sketches on restaurant napkins to convey conceptual design ideas. SOLIDWORKS provides many powerful modeling and design-tools, and there are many different approaches to accomplishing modeling tasks. The basic principle of **feature-based modeling** is to build models by adding simple features one at a time. In this chapter, the general parametric part modeling procedure is illustrated; a very simple solid model with extruded features is used to introduce the SOLIDWORKS user interface. The display viewing functions, and the basic two-dimensional sketching tools are also demonstrated.

The *Adjuster* Design

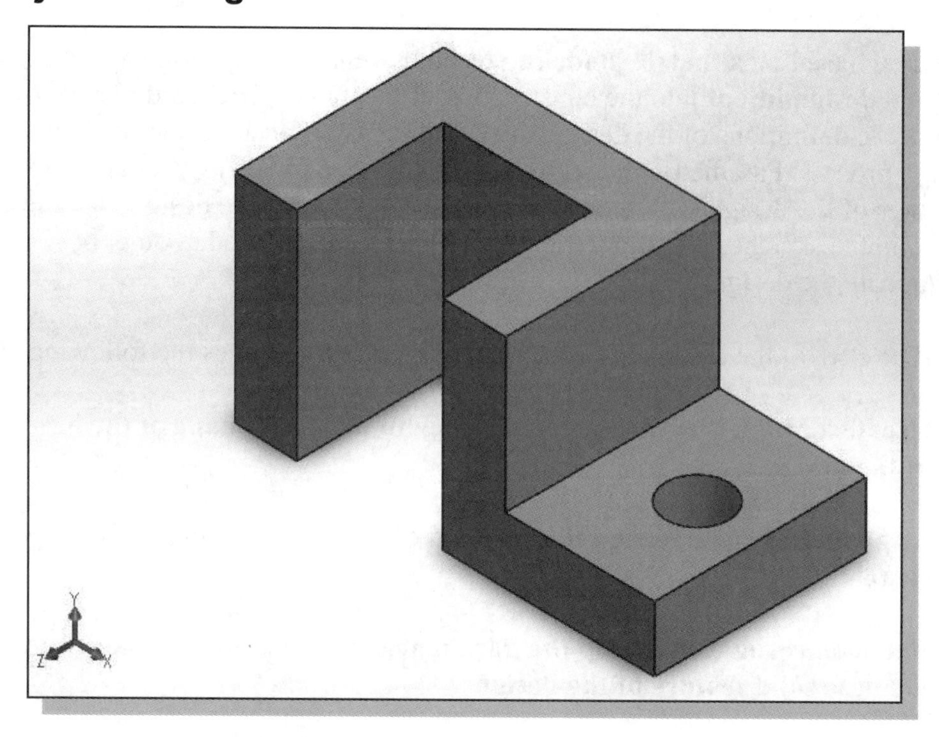

Starting SOLIDWORKS

1. Select the **SOLIDWORKS** option on the *Start* menu or select the **SOLIDWORKS** icon on the desktop to start SOLIDWORKS. The SOLIDWORKS main window will appear on the screen.

➢ We will start a new SOLODWORKS part file using the *Welcome* dialog box.

2. If the *Welcome* dialog box does not appear automatically upon opening SOLIDWORKS, it can be opened by clicking the *Welcome to SolidWorks* icon in the *Task Pane* or on the *Menu Bar*.

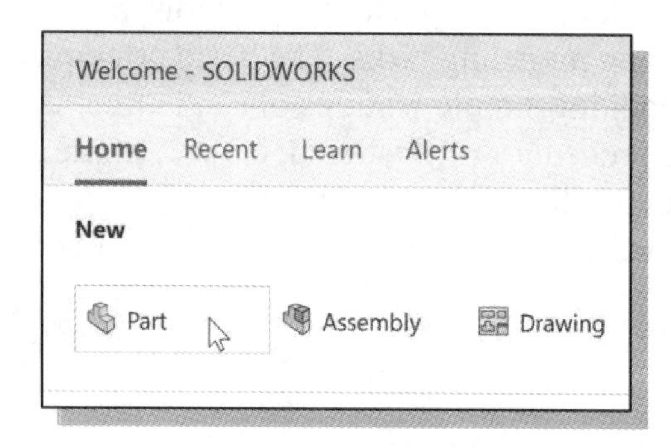

3. Select the **Part** icon with a single click of the left-mouse-button in the *Welcome* dialog box to open a new part document.

SOLIDWORKS Screen Layout

The default SOLIDWORKS drawing screen contains the *Menu Bar*, the *Heads-up View* toolbar, the *FeatureManager Design Tree*, the *Features* toolbar (at the left of the window by default), the *Sketch* toolbar (at the right of the window by default), the graphics area, the *task pane* (collapsed to the right of the graphics area in the figure below), and the *Status Bar*. A line of quick text appears next to the icon as you move the *mouse cursor* over different icons. You may resize the SOLIDWORKS drawing window by clicking and dragging at the edges of the window, or relocate the window by clicking and dragging at the *window title* area.

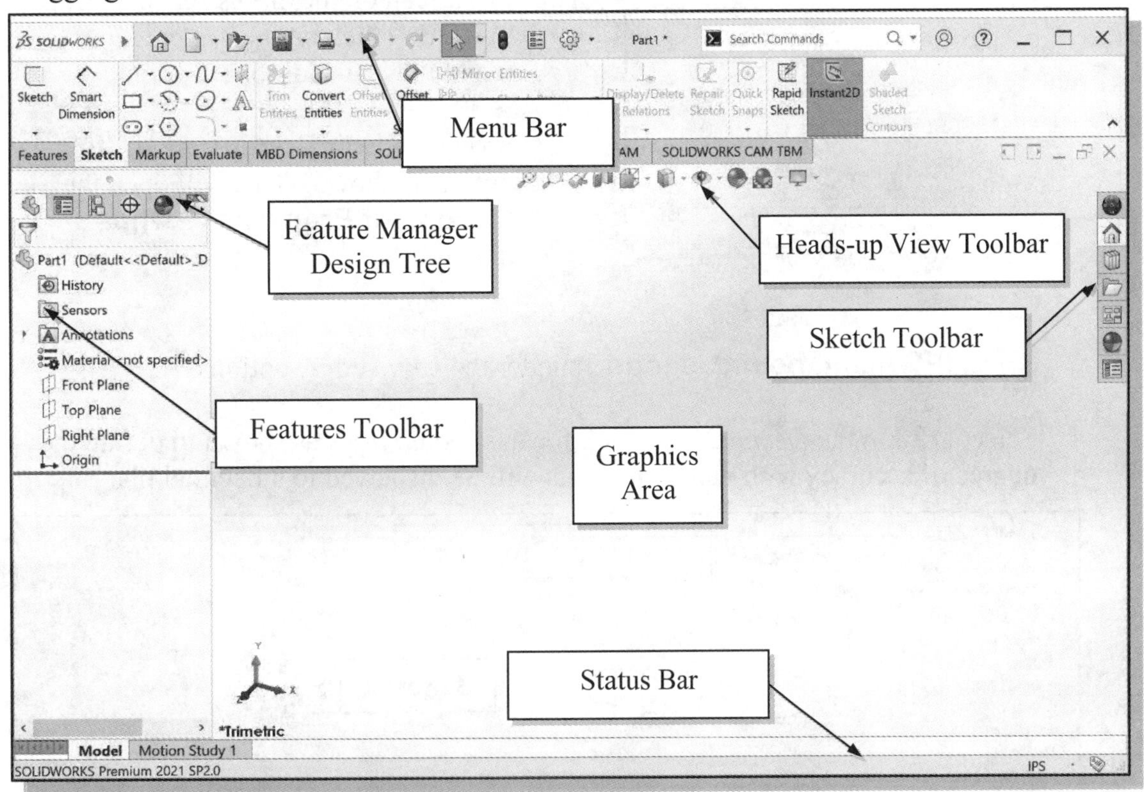

IMPORTANT NOTE: The SOLIDWORKS *CommandManager* provides an alternate method for displaying the most commonly used toolbars. If the *CommandManager* is active, the display will appear as shown on page 1-11. In this lesson, we will use the standard display of toolbars shown above. If a user prefers to use the *CommandManager*, the only change is that it may be necessary to select the appropriate tab prior to selecting a command. For example, if the instruction is to "select the **Extruded Boss** command from the *Features* toolbar," it may be necessary to first select the **Features** tab on the *CommandManager* to display the *Features* toolbar.

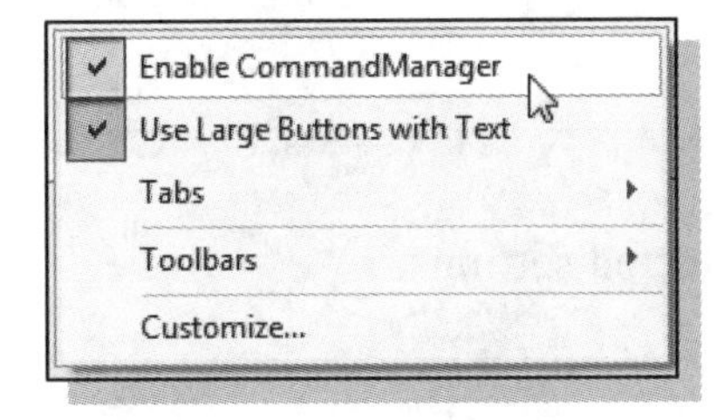

1. To turn ***OFF*** the *CommandManager* and use the standard display of toolbars, right click on the *CommandManager* (or any other toolbar) and toggle the *CommandManager OFF* by selecting it at the top of the pop-up menu.

Units Setup

When starting a new CAD file, the first thing we should do is choose the units we would like to use. The *Unit system* for the active document can be changed or customized using the **System Units** option on the *Status Bar* at the bottom of the SOLIDWORKS window. We will use English units (inches, pounds) for this example.

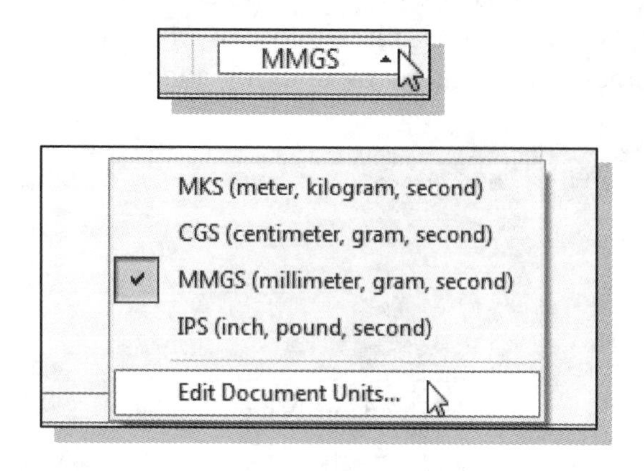

1. If the default *Unit system* is set to millimeter, gram, second, the *System Units* icon on the *Status Bar* displays **MMGS**. Click on the *System Units* icon to reveal additional options.

2. Select the **Edit Document Units** option as shown. This will open the **Document Properties - Units** dialog box.

3. Select **IPS (inch, pound, second)** under the *Unit system* options.

4. Select **.123** in the *Decimals* spin box for the *Length units* as shown to define the degree of accuracy with which the units will be displayed to 3 decimal places.

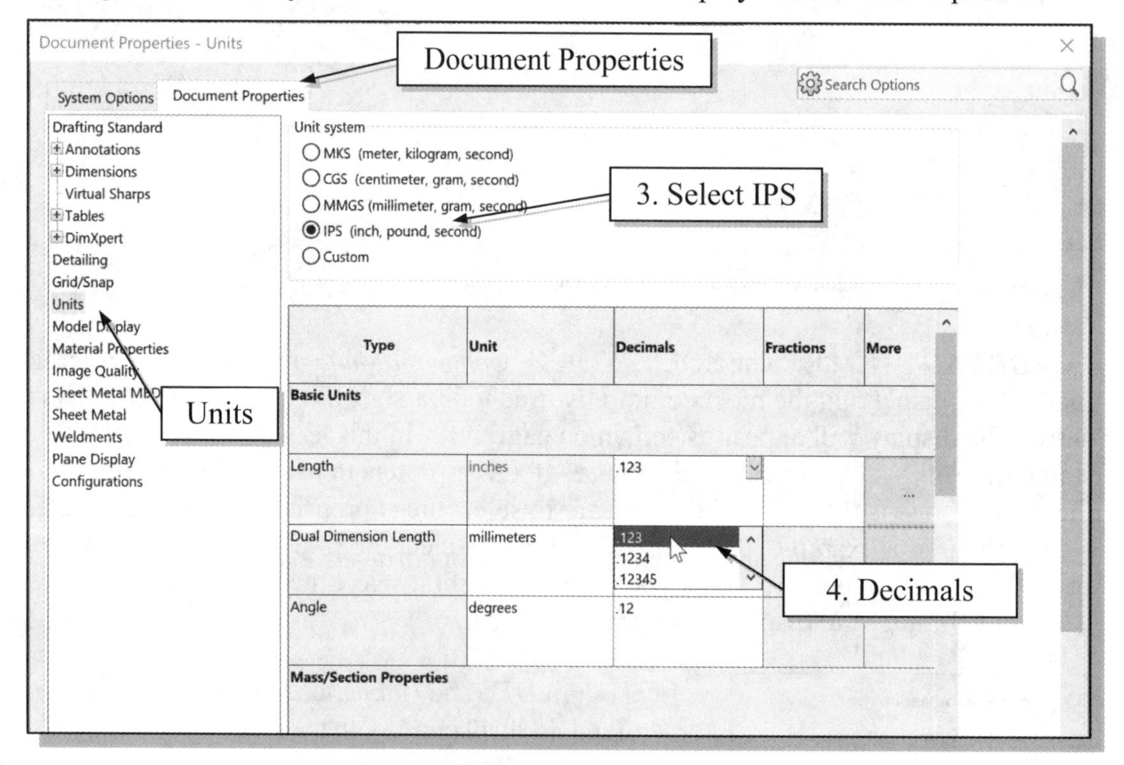

5. Click **OK** in the *Options* dialog box to accept the selected settings.

Creating Rough Sketches

Quite often during the early design stage, the shape of a design may not have any precise dimensions. Most conventional CAD systems require the user to input the precise lengths and locations of all geometric entities defining the design, which are not available during the early design stage. With *parametric modeling*, we can use the computer to elaborate and formulate the design idea further during the initial design stage. With SOLIDWORKS, we can use the computer as an electronic sketchpad to help us concentrate on the formulation of forms and shapes for the design. This approach is the main advantage of *parametric modeling* over conventional solid-modeling techniques.

As the name implies, a ***rough sketch*** is not precise at all. When sketching, we simply sketch the geometry so that it closely resembles the desired shape. Precise scale or lengths are not needed. SOLIDWORKS provides us with many tools to assist us in finalizing sketches. For example, geometric entities such as horizontal and vertical lines are set automatically. However, if the rough sketches are poor, it will require much more work to generate the desired parametric sketches. Here are some general guidelines for creating sketches in SOLIDWORKS:

- **Create a sketch that is proportional to the desired shape.** Concentrate on the shapes and forms of the design.
- **Keep the sketches simple.** Leave out small geometry features such as fillets, rounds and chamfers. They can easily be placed using the Fillet and Chamfer commands after the parametric sketches have been established.
- **Exaggerate the geometric features of the desired shape.** For example, if the desired angle is 85 degrees, create an angle that is 50 or 60 degrees. Otherwise, SOLIDWORKS might assume the intended angle to be a 90-degree angle.
- **Draw the geometry so that it does not overlap.** The geometry should eventually form a closed region. *Self-intersecting* geometry shapes are not allowed.
- **The sketched geometric entities should form a closed region.** To create a solid feature, such as an extruded solid, a closed region is required so that the extruded solid forms a 3D volume.

➢ **NOTE:** The concepts and principles involved in *parametric modeling* are very different from, and sometimes they are totally opposite to, those of conventional computer aided drafting. In order to understand and fully utilize SOLIDWORKS' functionality, it will be helpful to take a *Zen* approach to learning the topics presented in this text: **Temporarily forget your knowledge and experiences of using conventional Computer Aided Drafting systems.**

Step 1: Creating a Rough Sketch

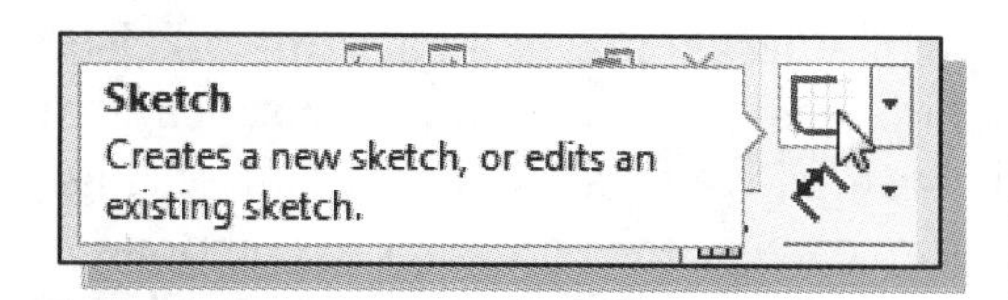

1. Select the **Sketch** button at the top of the *Sketch* toolbar to create a new sketch.

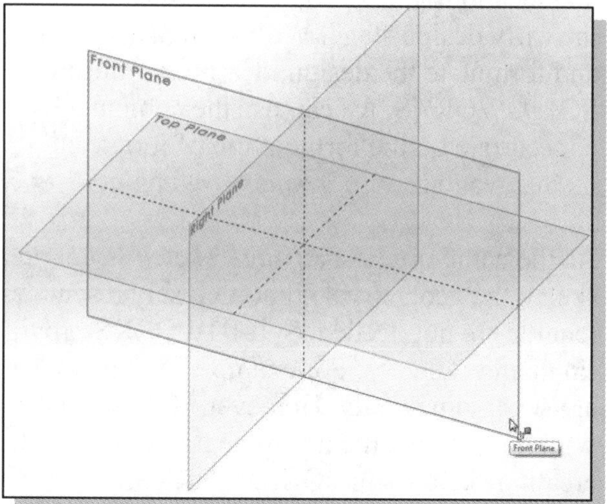

2. Notice the left panel displays the *Edit Sketch PropertyManager* with the instruction "*Select a plane on which to create a sketch for the entity.*" Move the cursor over the edge of the *Front Plane* in the graphics area. When the *Front Plane* is highlighted, click once with the **left-mouse-button** to select the *Front Plane* as the sketch plane for the new sketch.

3. Move the cursor over the **Line** icon on the *Sketch* toolbar, but do not click. Notice that a description of the Line command appears along with an animation of the execution of the command.

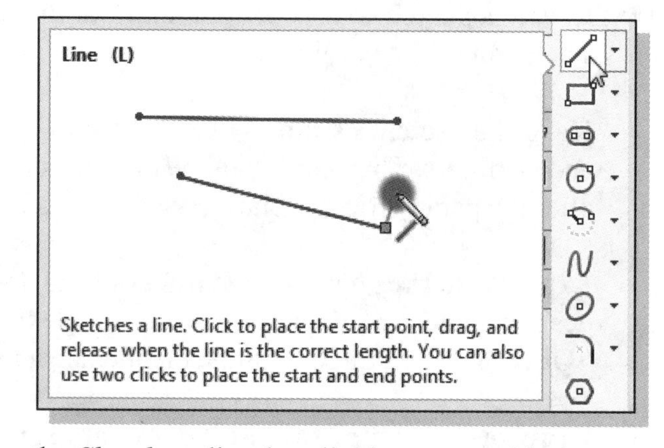

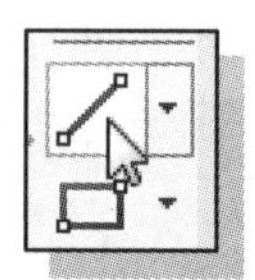

4. Select the **Line** icon on the *Sketch* toolbar by clicking once with the **left-mouse-button**; this will activate the Line command. The *Line Properties PropertyManager* is displayed in the left panel.

Graphics Cursors

Notice the cursor changes from an arrow to a pencil when a sketch entity is active.

1. Left-click a starting point for the shape, roughly near the lower center of the graphics window.

2. As you move the graphics cursor, you will see a digital readout next to the cursor. This readout gives you the line length. In the *Status Bar* area at the bottom of the window, the readout gives you the cursor location. Move the cursor around and you will notice different symbols appear at different locations.

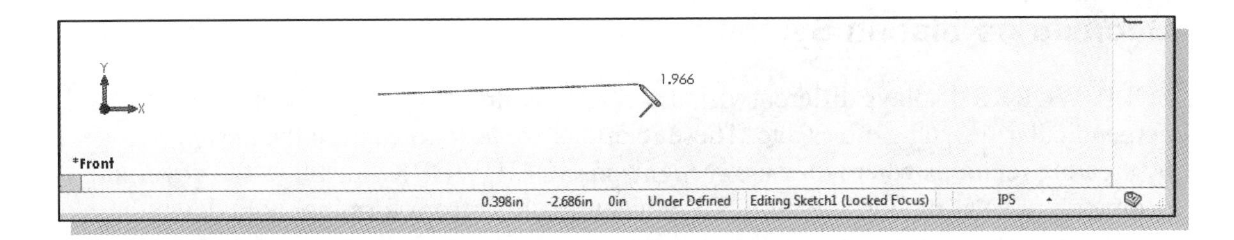

3. Move the graphics cursor toward the right side of the graphics window to create a horizontal line as shown below. Notice the geometric relation symbol displayed. When the **Horizontal** relation symbol is displayed, left-click to select **Point 2**.

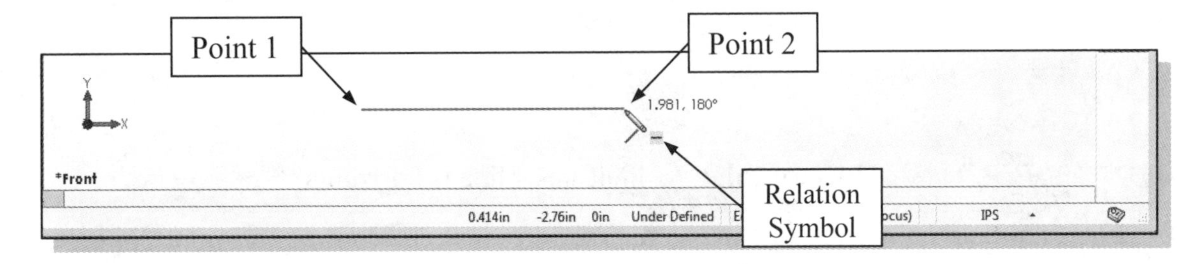

4. Complete the sketch as shown below, creating a closed region **ending at the starting point** (Point 1). Do not be overly concerned with the actual size of the sketch. Note that all line segments are sketched horizontally or vertically.

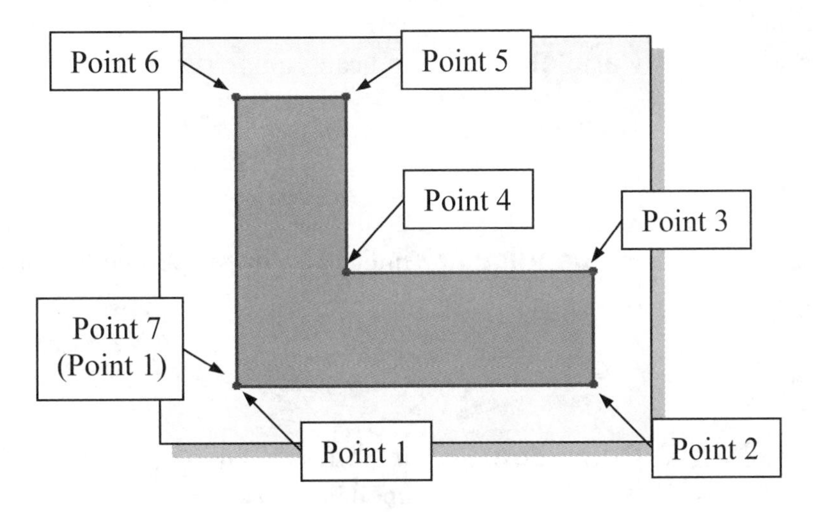

5. Click the **OK** icon (green check mark) in the *PropertyManager* to end editing of the current line, then click the **OK** icon again to end the Sketch Line command, or hit the [**Esc**] key once to end the Sketch Line command.

Geometric Relation Symbols

SOLIDWORKS displays different visual clues, or symbols, to show you alignments, perpendicularities, tangencies, etc. These relations are used to capture the *design intent* by creating relations where they are recognized. SOLIDWORKS displays the governing geometric rules as models are built. To prevent relations from forming, hold down the [**Ctrl**] key while creating an individual sketch curve. For example, while sketching line segments with the Line command, endpoints are joined with a *Coincident relation*, but when the [**Ctrl**] key is pressed and held, the inferred relation will not be created.

Vertical indicates a line is vertical

Horizontal indicates a line is horizontal

Dashed line indicates the alignment is to the center point or endpoint of an entity

Parallel indicates a line is parallel to other entities

Perpendicular indicates a line is perpendicular to other entities

Coincident indicates the endpoint will be coincident with another entity

Concentric indicates the cursor is at the center of an entity

Tangent indicates the cursor is at tangency points to curves

Step 2: Apply/Modify Relations and Dimensions

As the sketch is made, SOLIDWORKS automatically applies some of the geometric relations (such as **Horizontal**, **Parallel**, and **Perpendicular**) to the sketched geometry. We can continue to modify the geometry, apply additional relations, and/or define the size of the existing geometry. In this example, we will illustrate adding dimensions to describe the sketched entities.

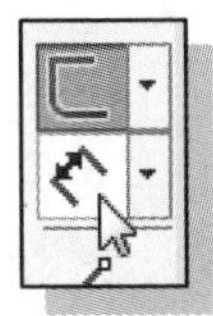

1. Move the cursor on top of the **Smart Dimension** icon on the *Sketch* toolbar. The **Smart Dimension** command allows us to quickly create and modify dimensions. Left-click once on the icon to activate the **Smart Dimension** command.

2. The message "*Select one or two edges/vertices and then a text location*" is displayed in the *Status Bar* area at the bottom of the SOLIDWORKS window. Select the bottom horizontal line by left-clicking once on the line.

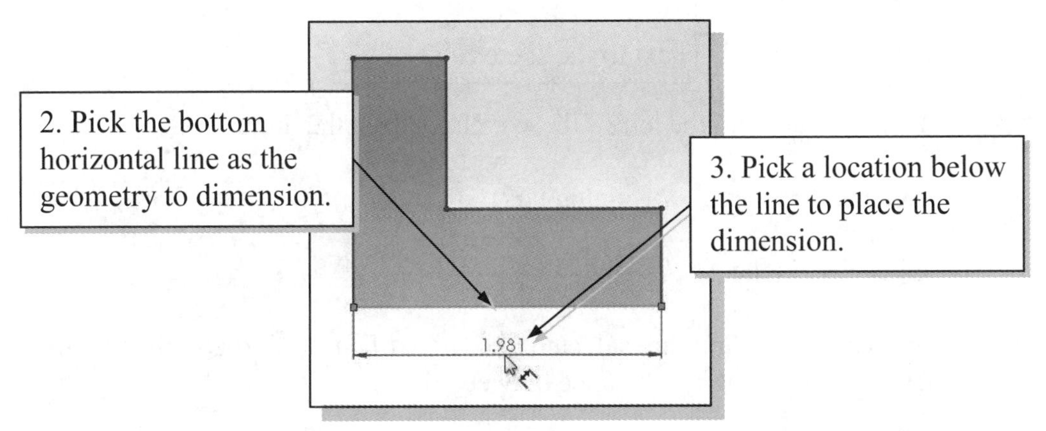

3. Move the graphics cursor below the selected line and left-click to place the dimension. (Note that the value displayed on your screen might be different than what is shown in the figure above.)

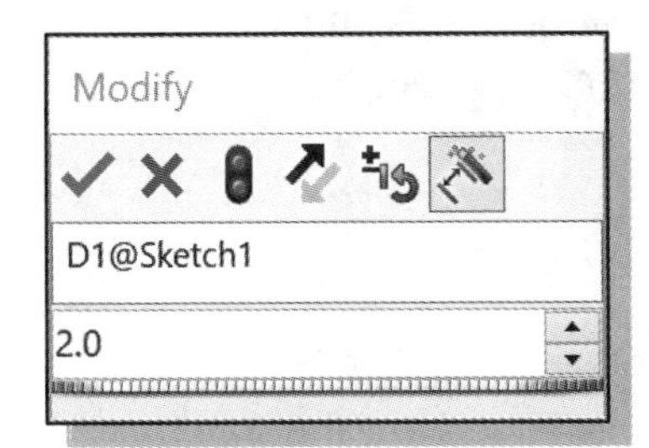

4. Enter **2.0** in the *Modify* dialog box.

5. Left click the **OK** (green check mark) in the *Modify* dialog box to save the current value and exit the dialog.

6. **On your own**, select the lower right-vertical line.

7. Pick a location toward the right of the sketch to place the dimension.

8. Enter **0.75** in the *Modify* dialog box.

9. Click **OK** in the *Modify* dialog box.

❖ The **Smart Dimension** command will create a length dimension if a single line is selected.

10. Select the top-horizontal line as shown below.

11. Select the bottom-horizontal line as shown below.

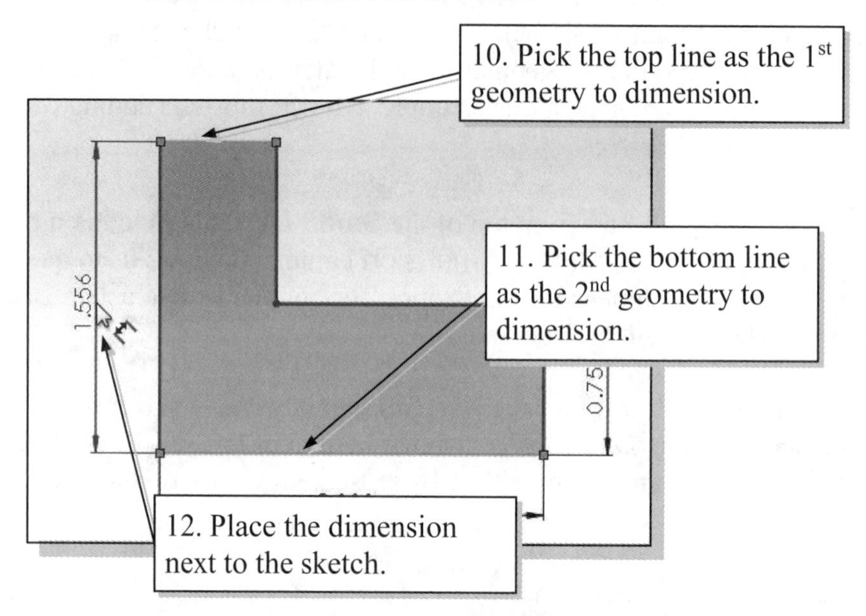

12. Pick a location to the left of the sketch to place the dimension.

13. Enter **2.0** in the *Modify* dialog box.

14. Click **OK** in the *Modify* dialog box.

❖ When two parallel lines are selected, the Smart Dimension command will create a dimension measuring the distance between them.

15. **On your own**, repeat the above steps and create an additional dimension for the top line. Make the dimension **0.75**.

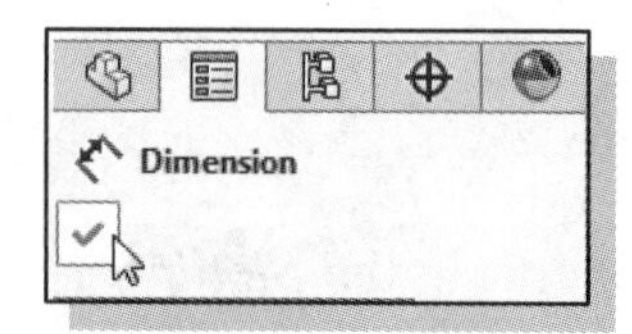

16. Click the **OK** icon in the *PropertyManager* as shown, or hit the [**Esc**] key once, to end the Smart Dimension command.

Changing the Dimension Standard

1. Select the **Options** icon from the *Menu Bar* to open the *Options* dialog box.

Document Properties - Drafting Standard
System Options   Document Properties
Drafting Standard
Annotations
Balloons
Datums
Geometric Tolerances
Notes
Overall drafting standard
ISO
ANSI
ISO
DIN
JIS

2. Select the **Document Properties** tab, then select **Drafting Standard** at the left.

3. Select **ANSI** in the pull-down selection window under the *Overall drafting standard* panel as shown.

4. Left-click **OK** in the *Options* dialog box to accept the settings.

- The sketch should now look as shown below. Notice the change in appearance of the dimensions.

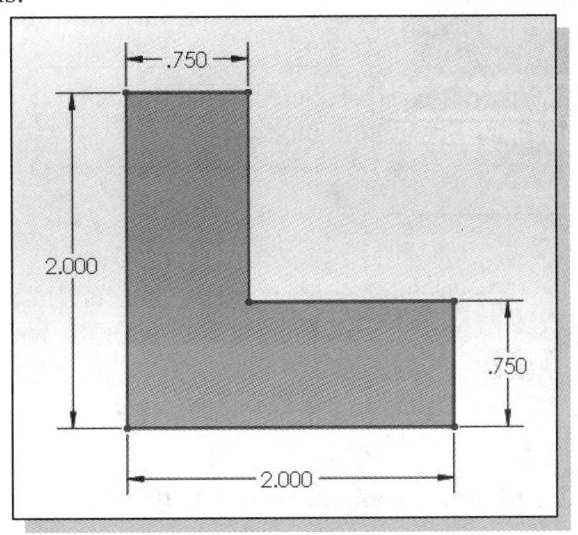

Viewing Functions – *Zoom* and *Pan*

SOLIDWORKS provides a special user interface that enables convenient viewing of the entities in the graphics window. There are many ways to perform the **Zoom** and **Pan** operations.

1. Hold the **Ctrl** function key down. While holding the **Ctrl** function key down, press the mouse wheel down and drag the mouse to **pan** the display. This allows you to reposition the display while maintaining the same scale factor of the display.

2. Hold the **Shift** function key down. While holding the **Shift** function key down, press the mouse wheel down and drag the mouse to **zoom** the display. Moving downward will reduce the scale of the display, making the entities display smaller on the screen. Moving upward will magnify the scale of the display.

3. Turning the mouse wheel can also adjust the scale of the display. Turn the mouse wheel forward. Notice the scale of the display is reduced, making the entities display smaller on the screen.

4. Turn the mouse wheel backward. Notice scale of the display is magnified. (**NOTE:** Turning the mouse wheel allows zooming to the position of the cursor.)

5. On your own, use the options above to change the scale and position of the display.

6. Press the **F** key on the keyboard to automatically fit the model to the screen.

Modifying the Dimensions of the Sketch

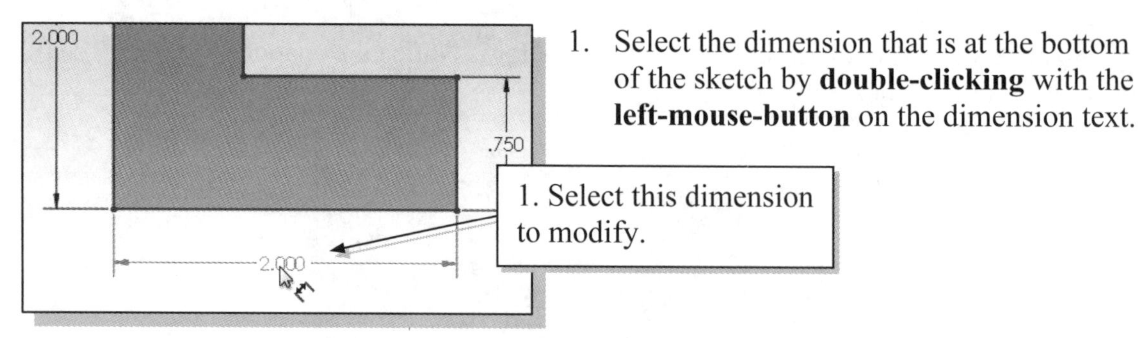

1. Select the dimension that is at the bottom of the sketch by **double-clicking** with the **left-mouse-button** on the dimension text.

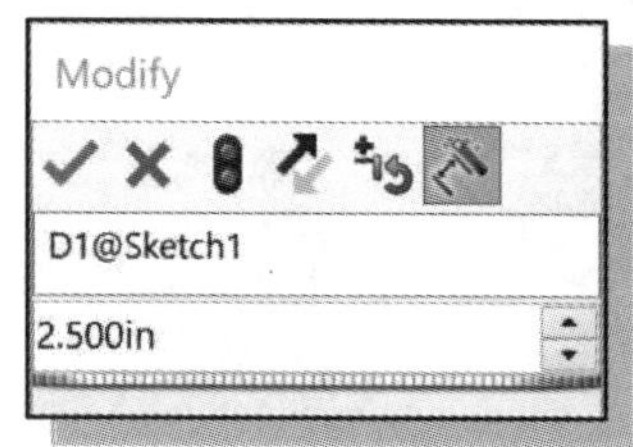

2. In the *Modify* window, the current length of the line is displayed. Enter **2.5** to reset the length of the line.

3. Click on the **OK** icon to accept the entered value.

➢ SOLIDWORKS will now update the profile with the new dimension value.

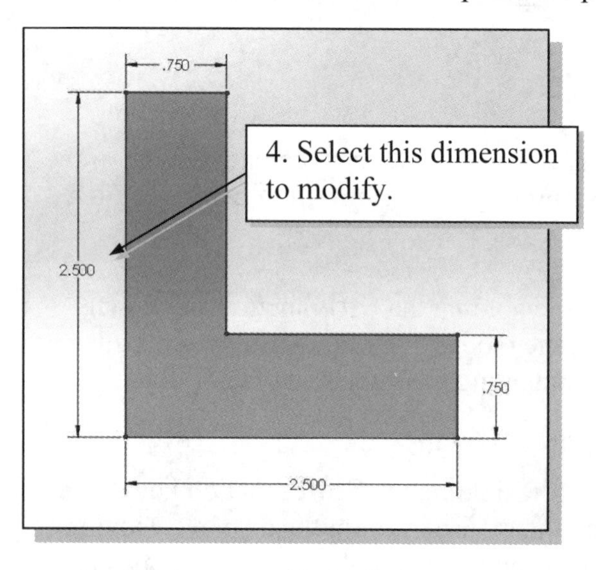

4. **On your own**, repeat the above steps and adjust the left vertical dimension to **2.5** so that the sketch appears as shown.

5. Press the [**Esc**] key once to exit the Dimension command.

6. Click once with the **left-mouse-button** on the **Sketch** icon on the *Sketch* toolbar to exit the sketch.

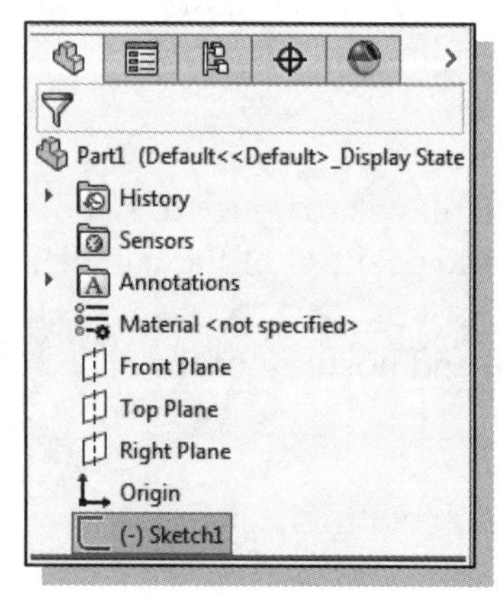

➢ Notice the newly created sketch is listed on the *Feature-Manager Design Tree* as **Sketch1**. Also notice that **Sketch1** is highlighted in the *Design Tree*, indicating that the sketch is currently 'selected'.

Step 3: Completing the Base Solid Feature

Now that the 2D sketch is completed, we will proceed to the next step: creating a 3D part from the 2D profile. Extruding a 2D profile is one of the common methods that can be used to create 3D parts. We can extrude planar faces along a path. We can also specify a height value and a tapered angle.

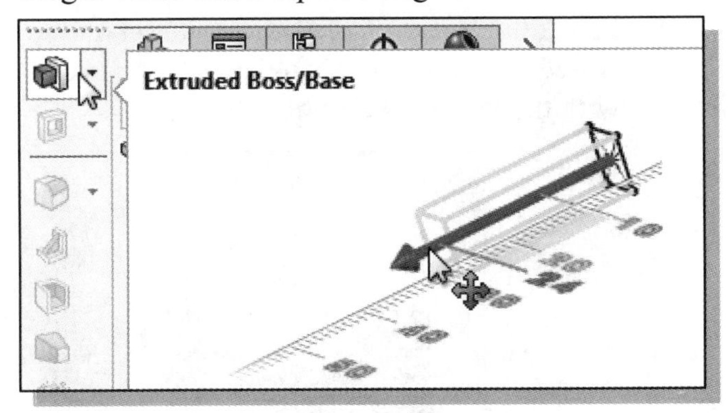

1. In the *Features* toolbar (located at the left of the window), select the **Extruded Boss/Base** command by clicking once with the left-mouse-button on the icon. The *Boss Extrude Property-Manager* is displayed in the left panel.

2. In the *Boss Extrude PropertyManager* panel, enter **2.5** as the extrusion distance. Notice that the sketch region is automatically selected as the extrusion profile.

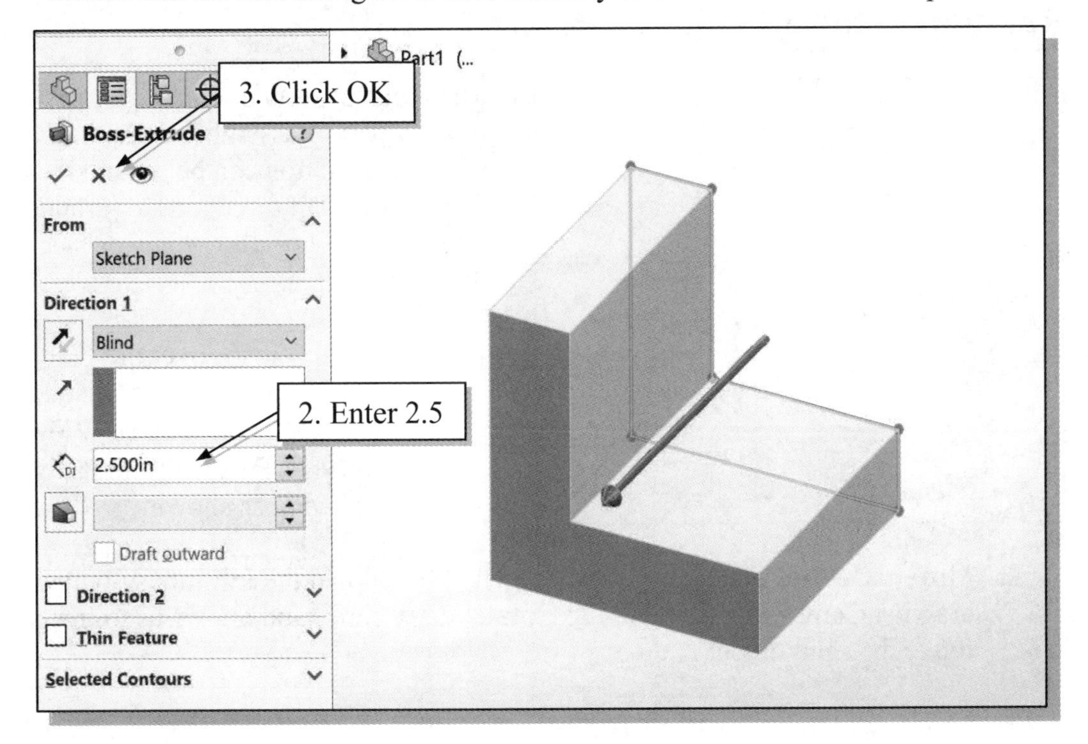

3. Click on the **OK** button to proceed with creating the 3D part.

➢ Note that all dimensions disappeared from the screen. All parametric definitions are stored in the **SOLIDWORKS database** and any of the parametric definitions can be displayed and edited at any time.

Isometric View

SOLIDWORKS provides many ways to display views of the three-dimensional design. We will first orient the model to display in the *isometric view*, by using the *View Orientation* pull-down menu on the *Heads-up View* toolbar.

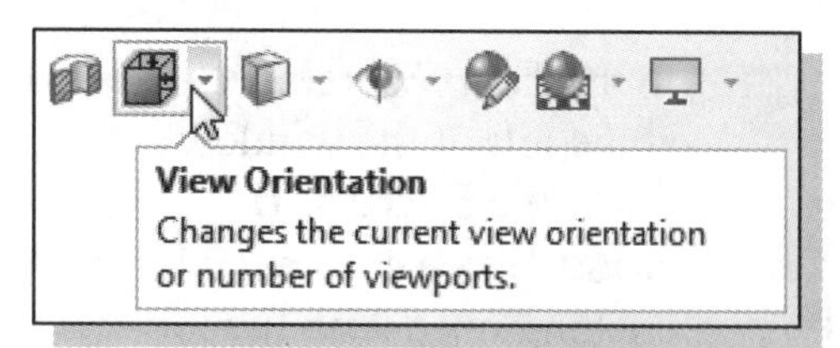

1. Select the **View Orientation** button on the *Heads-up View* toolbar by clicking once with the left-mouse-button.

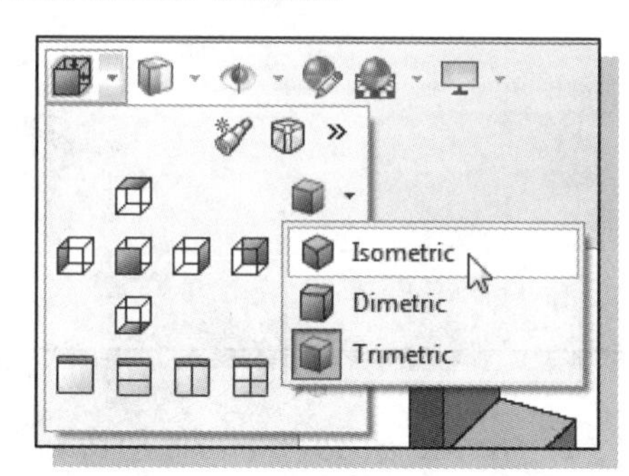

2. Select the **Isometric** icon in the *View Orientation* pull-down menu.

❖ Notice the other view-related commands that are available under the pull-down menu.

Rotation of the 3D Model – *Rotate View*

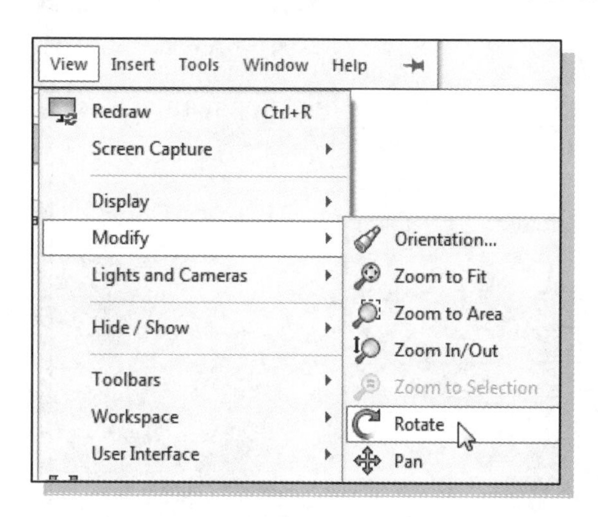

The Rotate View command allows us to rotate a part or assembly in the graphics window. Rotation can be around the center mark, free in all directions, or around a selected entity (vertex, edge, or face) on the model.

1. Move the cursor over the SOLIDWORKS logo to display the pull-down menus. Select **View → Modify → Rotate** from the pull-down menu as shown.

2. Move the cursor inside the graphics area. Press down the left-mouse-button and drag in an arbitrary direction; the Rotate View command allows us to freely rotate the solid model.

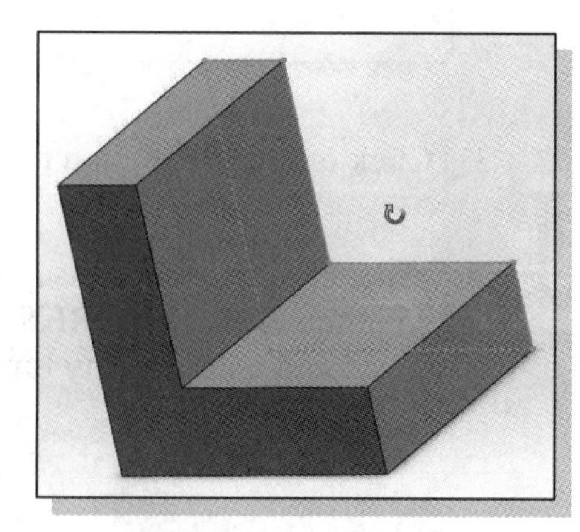

- The model will rotate about an axis normal to the direction of cursor movement. For example, drag the cursor horizontally across the screen and the model will rotate about a vertical axis.

3. Press the [**Esc**] key once to exit the Rotate View command.

4. Select the **Isometric** icon in the *View Orientation* pull-down menu (see steps 1 and 2 in the previous section) to reset the display to the isometric view.

5. Execute the **Rotate View** option from the *View* pull-down menu (see step 1).

6. Move the cursor over the left edge of the solid model as shown. When the edge is highlighted, click the **left-mouse-button** once to select the edge.

7. Press down the left-mouse-button and drag. The model will rotate about this edge.

8. Left-click in the graphics area, outside the model, to unselect the edge.

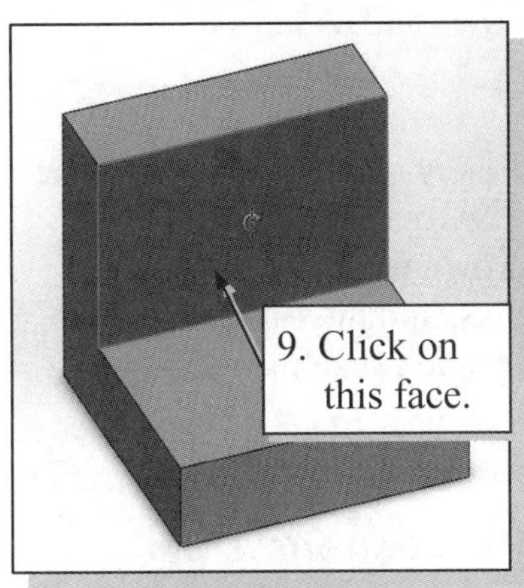

9. Move the cursor over the upper front face of the solid model as shown. When the face is highlighted, click the **left-mouse-button** once to select the face.

10. Press down the left-mouse-button and drag. The model will rotate about the direction normal to this face.

11. Left-click in the graphics area, outside the model, to unselect the face.

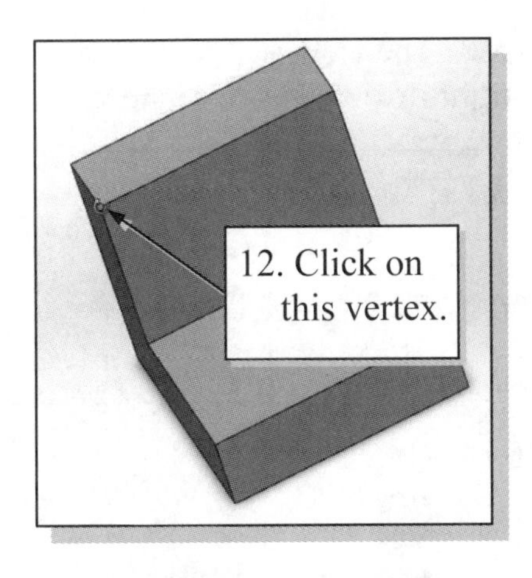

12. Move the cursor over the upper front vertex as shown. When the vertex is highlighted, click the left-mouse-button once to select the vertex.

13. Press down the left-mouse-button and drag. The model will rotate about the vertex.

14. Left-click in the graphics area, outside the model, to unselect the vertex.

15. Press the [**Esc**] key once to exit the Rotate View command.

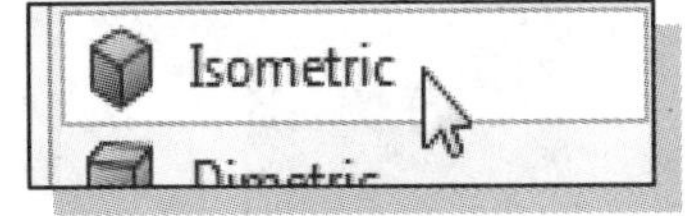

16. **On your own**, reset the display to the isometric view.

Rotation and Panning – *Arrow Keys*

SOLIDWORKS allows us to easily rotate a part or assembly in the graphics window using the arrow keys on the keyboard.

- Use the arrow keys to rotate the view horizontally or vertically. The left-right arrow keys rotate the model about a vertical axis. The up-down keys rotate the model about a horizontal axis.
- Hold down the Alt key and use the left-right arrow keys to rotate the model about an axis normal to the screen, i.e., to rotate clockwise and counterclockwise.

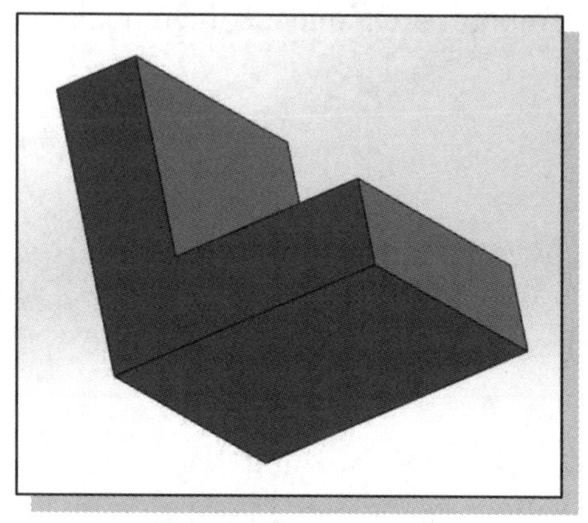

1. Hit the **left arrow** key. The model view rotates by a pre-determined increment. The default increment is 15°. (This increment can be set in the *Options* dialog box.) On your own use the left-right and up-down arrow keys to rotate the view.
2. Hold down the [**Alt**] key and hit the **left arrow** key. The model view rotates in the clockwise direction. On your own use the left-right and up-down arrow keys, and the Alt key plus the left-right arrow keys, to rotate the view.
3. Reset the display to the Isometric view.

- Hold down the [**Shift**] key and use the left-right and up-down arrow keys to rotate the model in 90° increments.

4. Hold down the [**Shift**] key and hit the right arrow key. The view will rotate by 90°. On your own use the [**Shift**] key plus the left-right arrow keys to rotate the view.
5. Select the **Front** icon in the *View Orientation* pull-down menu as shown to display the Front view of the model.

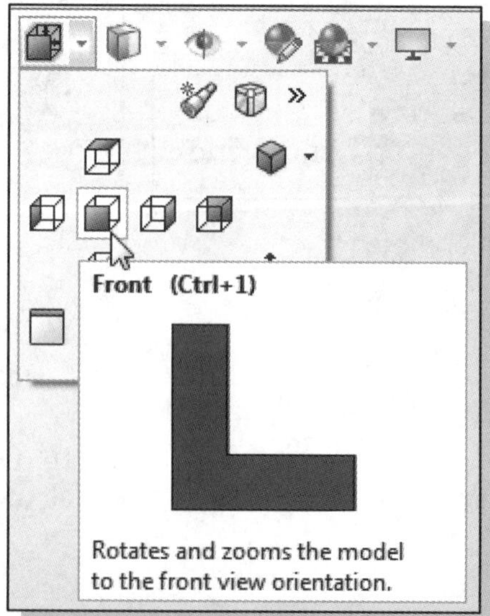

6. Hold down the [**Shift**] key and hit the **left arrow** key. The view rotates to the Right side view.

7. Hold down the [**Shift**] key and hit the **down arrow** key. The view rotates to the Top view.

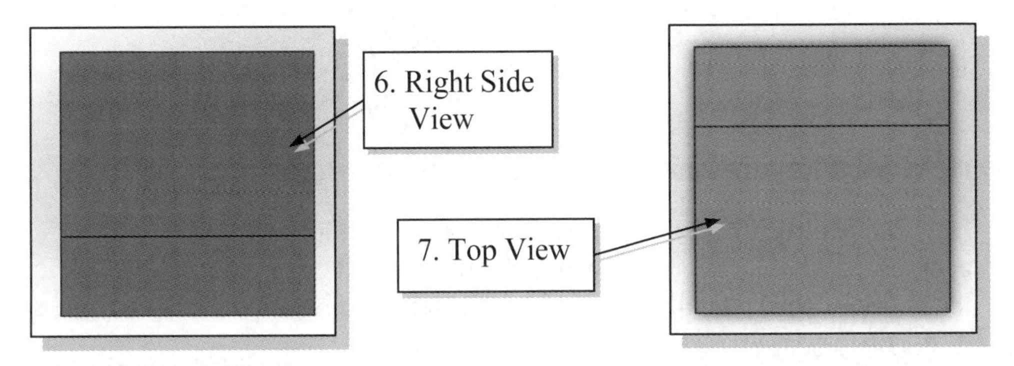

8. Reset the display to the **Isometric** view.

- Hold down the [**Ctrl**] key and use the left-right and up-down arrow keys to pan the model in increments.

9. Hold down the [**Ctrl**] key and hit the **left arrow** key. The view pans, moving the model toward the left side of the screen. On your own use [**Ctrl**] key plus the left-right and up-down arrow keys to pan the view.

Viewing – Quick Keys

We can also use the function keys on the keyboard and the mouse to access the *Viewing* functions.

❖ Panning

(1) Hold Ctrl key, press and drag the mouse wheel

Hold the [**Ctrl**] function key down, and press and drag with the mouse wheel to pan the display. This allows you to reposition the display while maintaining the same scale factor of the display.

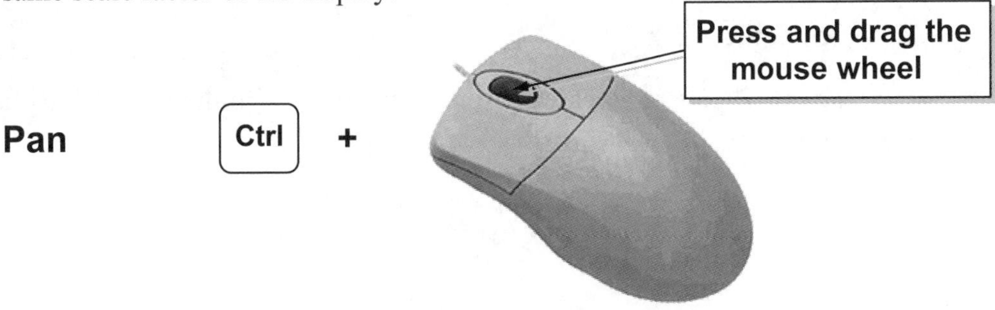

(2) Hold Ctrl key, use arrow keys

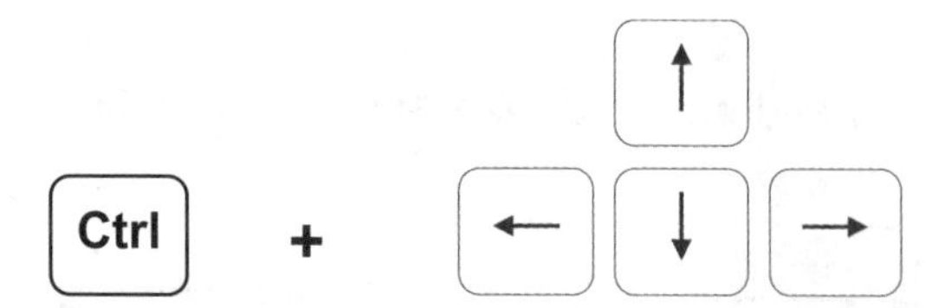

❖ Zooming

(1) Hold Shift key, press and drag the mouse wheel

Hold the [**Shift**] function key down and press and drag with the mouse wheel to zoom the display. Moving downward will reduce the scale of the display, making the entities display smaller on the screen. Moving upward will magnify the scale of the display.

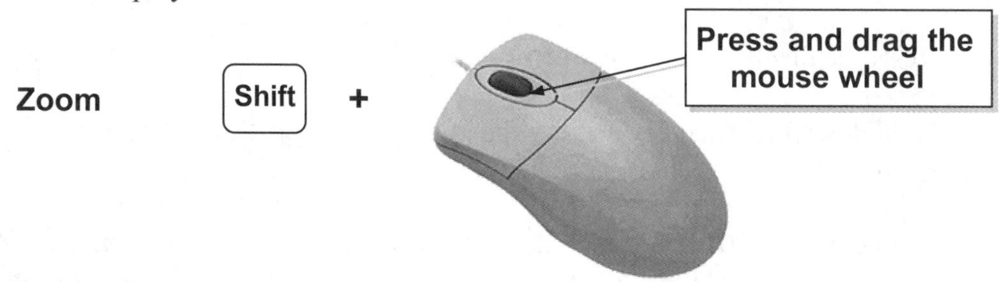

(2) Turning the mouse wheel

Turning the mouse wheel can also adjust the scale of the display. Turning forward will reduce the scale of the display, making the entities display smaller on the screen. Turning backward will magnify the scale of the display.

- Turning the mouse wheel allows zooming to the position of the cursor.

- If the cursor is outside the graphics area, the wheel will allow zooming to the center of the graphics area.

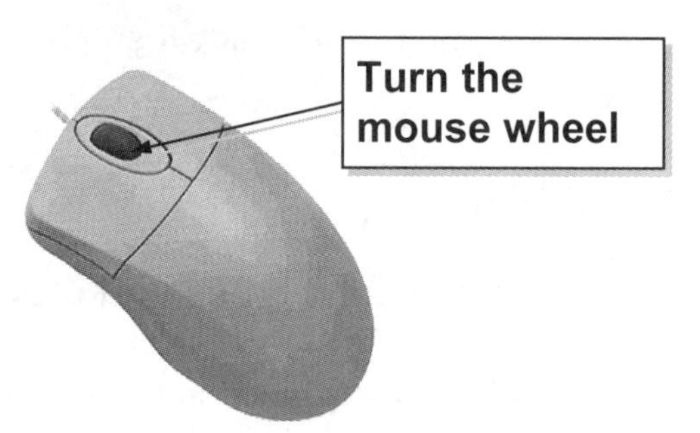

(3) Z key or Shift + Z key

Pressing the [**Z**] key on the keyboard will zoom out. Holding the [**Shift**] function key and pressing the [**Z**] key will zoom in.

Z or **Shift** + **Z**

❖ 3D Rotation

(1) Press and drag the mouse wheel

Press and drag with the mouse wheel to rotate the display.

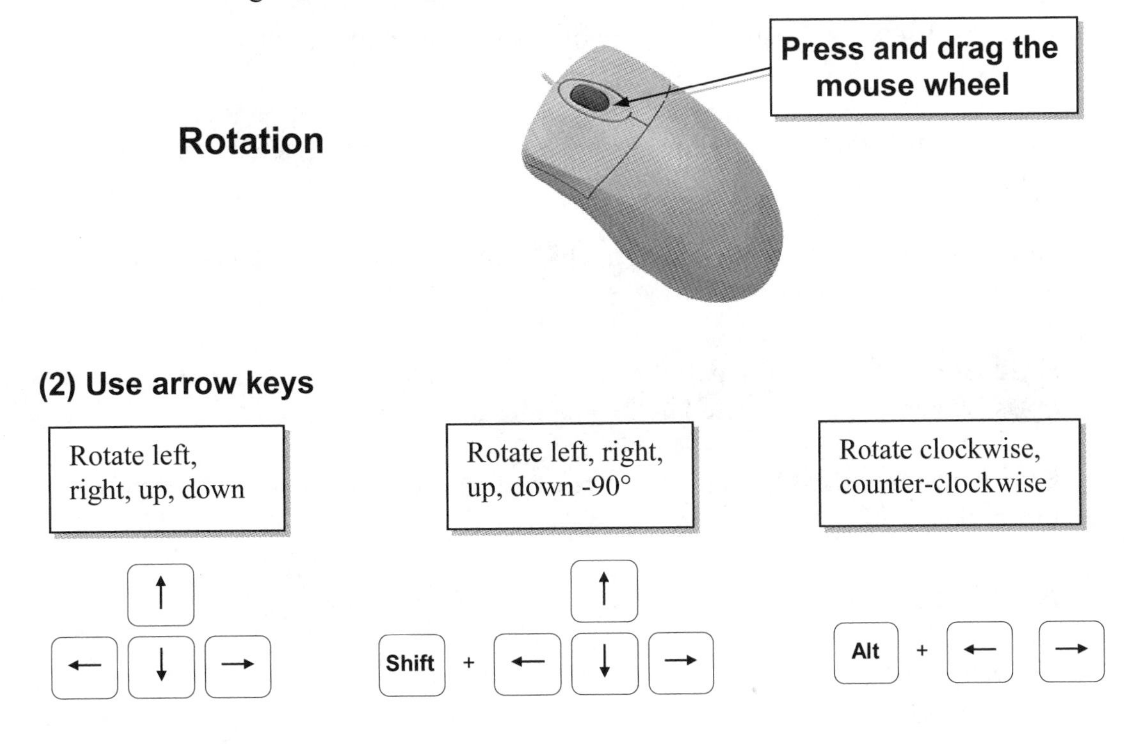

Viewing Tools – Heads-up View Toolbar

The ***Heads-up View*** toolbar is a transparent toolbar which appears in each viewport and provides easy access to commonly used tools for manipulating the view. The default toolbar is described below.

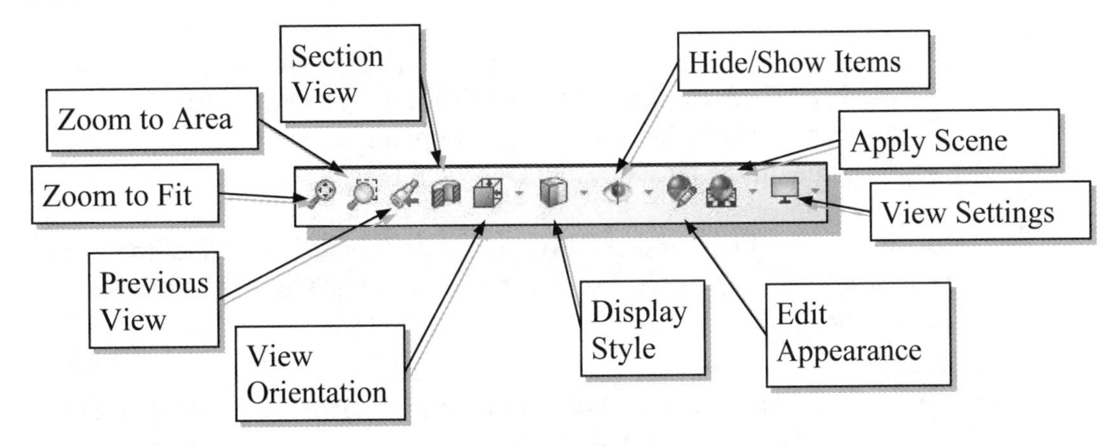

Zoom to Fit – Adjusts the view so that all items on the screen fit inside the graphics window.

Zoom to Area – Use the cursor to define a region for the view; the defined region is zoomed to fill the graphics window.

Previous View – Returns to the previous view.

Section View – Displays a cutaway of a part or assembly using one or more section planes.

View Orientation – This allows you to change the current view orientation or number of viewports.

Display Style – This can be used to change the display style (shaded, wireframe, etc.) for the active view.

Hide/Show Items – The pull-down menu is used to control the visibility of items (axes, sketches, relations, etc.) in the graphics area.

Edit Appearance – Edits the appearance of entities (e.g., parts, faces, features) in the model.

Apply Scene – Cycles through or applies a specific scene.

View Settings – Allows you to toggle various view settings (e.g., shadows, perspective).

View Orientation

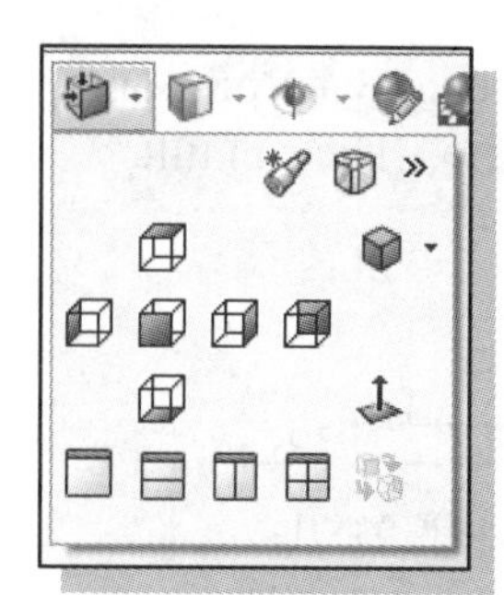

Click on the **View Orientation** icon on the *Heads-up View* toolbar to reveal the view orientation and number of viewports options.

Standard view orientation options – **Front**, **Back**, **Left**, **Right**, **Top**, **Bottom**, **Isometric**, **Trimetric**, **or Dimetric** – icons can be selected to display the corresponding standard view. In the figure to the left, the *Isometric* view is selected.

Normal to – In a part or assembly, zooms and rotates the model to display the selected plane or face. You can select the element either before or after clicking the Normal to icon.

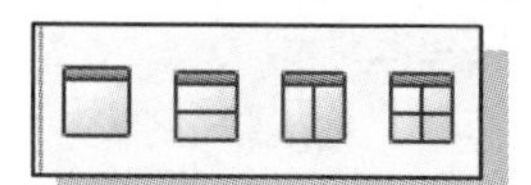

The icons across the bottom of the pull-down menu allow you to display a single viewport (the default) or multiple viewports.

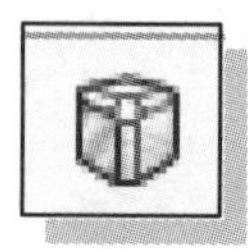

The **View Selector** provides an in-context method to select standard and non-standard views.

The **Add View** option allows you to add a custom view to the *Orientation* menu.

Display Style

Click on the **Display Style** icon on the *Heads-up View* toolbar to reveal the display style options.

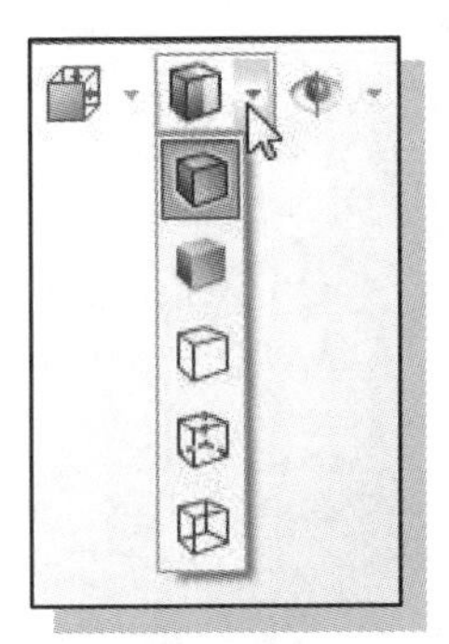

Shaded with Edges – Allows the display of a shaded view of a 3D model with its edges.

Shaded – Allows the display of a shaded view of a 3D model.

Hidden Lines Removed – Allows the display of the 3D objects using the basic wireframe representation scheme. Only those edges which are visible in the current view are displayed.

Hidden Lines Visible – Allows the display of the 3D objects using the basic wireframe representation scheme in which all the edges of the model are displayed, but edges that are hidden in the current view are displayed as dashed lines (or in a different color).

Wireframe – Allows the display of the 3D objects using the basic wireframe representation scheme in which all the edges of the model are displayed.

Orthographic vs. Perspective

Besides the basic display modes, we can also choose orthographic view or perspective view of the display. Clicking on the View Settings icon on the *Heads-up View* toolbar will reveal the Perspective icon. Clicking on the Perspective icon toggles the perspective view *ON* and *OFF*.

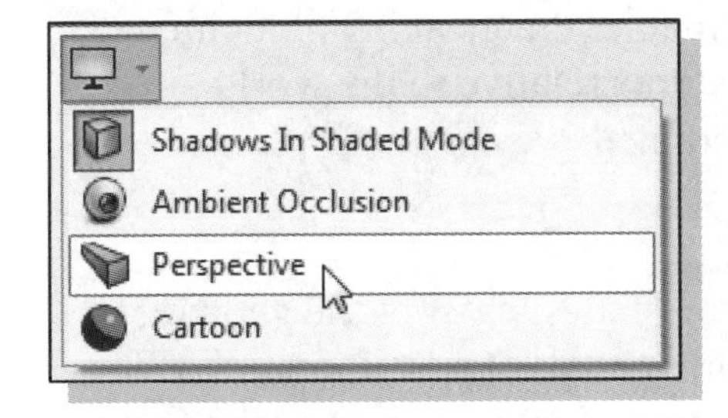

- On your own, use the different options described in the above sections to familiarize yourself with the 3D viewing/display commands. Reset the display to the standard **isometric view** before continuing to the next section.

Sketch Plane

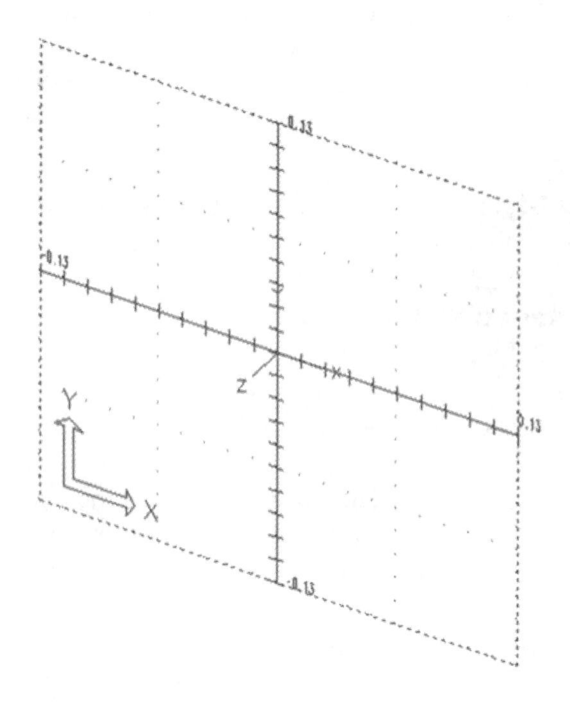

Design modeling software is becoming more powerful and user friendly, yet the system still does only what the user tells it to do. When using a geometric modeler, we therefore need to have a good understanding of what its inherent limitations are. We should also have a good understanding of what we want to do and what to expect, as the results are based on what is available.

In most 3D geometric modelers, 3D objects are located and defined in what is usually called **world space** or **global space**. Although a number of different coordinate systems can be used to create and manipulate objects in a 3D modeling system, the objects are typically defined and stored using the world space. The world space is usually a **3D Cartesian coordinate system** that the user cannot change or manipulate.

In most engineering designs, models can be very complex, and it would be tedious and confusing if only the world coordinate system were available. Practical 3D modeling systems allow the user to define **Local Coordinate Systems** relative to the world coordinate system. Once a local coordinate system is defined, we can then create geometry in terms of this more convenient system.

Although objects are created and stored in 3D space coordinates, most of the geometry entities can be referenced using 2D Cartesian coordinate systems. Typical input devices such as a mouse or digitizers are two-dimensional by nature; the movement of the input device is interpreted by the system in a planar sense. The same limitation is true of common output devices, such as screen displays and plotters. The modeling software performs a series of three-dimensional to two-dimensional transformations to correctly project 3D objects onto a 2D picture plane.

The SOLIDWORKS ***sketch plane*** is a special construction tool that enables the planar nature of 2D input devices to be directly mapped into the 3D coordinate system. The *sketch plane* is a local coordinate system that can be aligned to the world coordinate system, an existing face of a part, or a reference plane. By default, the *sketch plane* is aligned to the world coordinate system.

Think of a sketch plane as the surface on which we can sketch the 2D profiles of the parts. It is similar to a piece of paper, a white board, or a chalkboard that can be attached to any planar surface. The first profile we create is usually drawn on a sketch plane attached to a coordinate system such as the Front (XY), Top (XZ), and Right (YZ) sketch planes. Subsequent profiles can then be drawn on sketch planes that are defined on **planar faces of a part**, **work planes attached to part geometry**, or **sketch planes attached to a coordinate system**. The model we have created so far used the SOLIDWORKS **Front Plane**, which is aligned to the XY plane of the world coordinate system.

1. Select the **Sketch** button at the top of the *Sketch* toolbar to create a new sketch.

2. In the *Edit Sketch PropertyManager*, the message "*Select: 1) a plane, a planar face, or an edge on which to create a sketch for the entity*" is displayed. SOLIDWORKS expects us to identify a planar surface where the 2D sketch of the next feature is to be created. Move the graphics cursor on the 3D part and notice that SOLIDWORKS will automatically highlight feasible planes and surfaces as the cursor is on top of the different surfaces. Pick the top horizontal face of the 3D solid object.

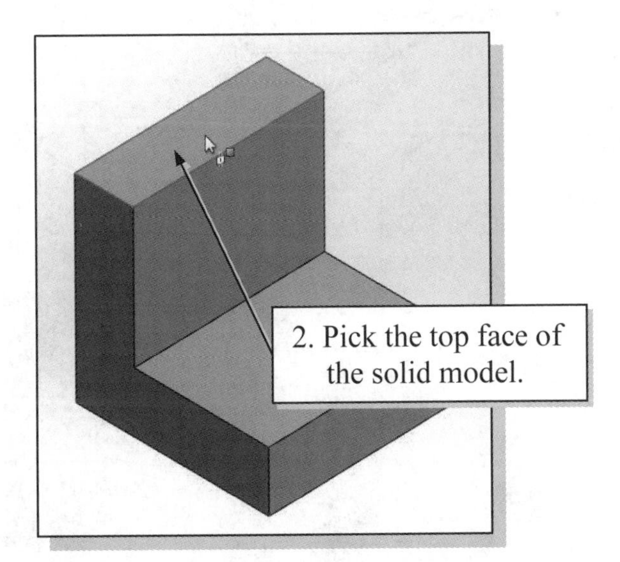

➢ Note that the sketch plane is aligned to the selected face. SOLIDWORKS automatically establishes a local coordinate system, and records its location with respect to the part on which it was created.

Step 4-1: Adding an Extruded Boss Feature

Next, we will create and profile another sketch, a rectangle, which will be used to create another extrusion feature that will be added to the existing solid object.

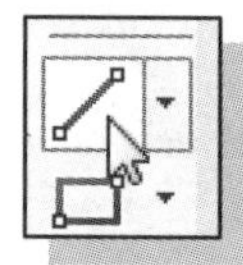

1. Select the **Line** command by clicking once with the **left-mouse-button** on the icon in the *Sketch* toolbar.

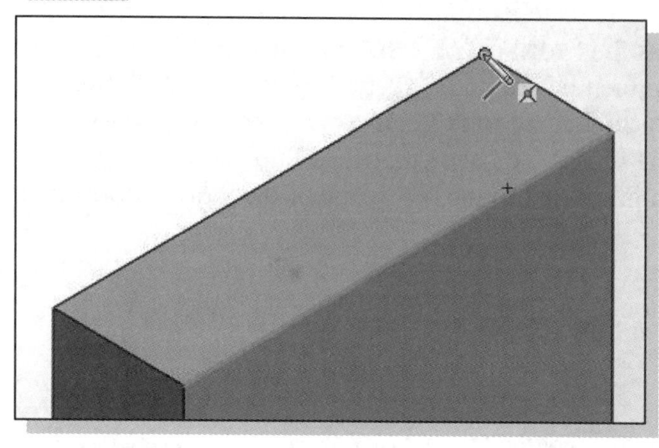

2. Move the cursor over the rear top vertex of the model. When the Coincident relation symbol appears as shown, click once with the **left-mouse-button**. This will start the first line, constraining its endpoint to be coincident with the vertex.

3. Create a sketch with segments perpendicular/parallel to the existing edges of the solid model as shown below. **Close the sketch by ending at Point 1. NOTE:** Use the Pan and Zoom options discussed earlier to control the view as needed.

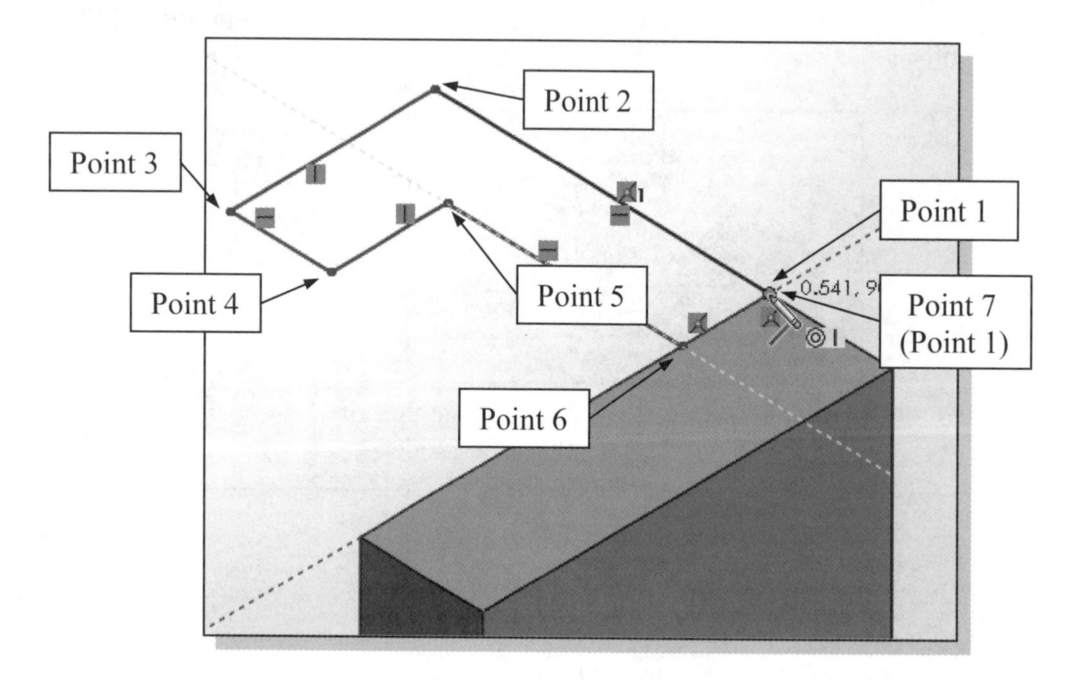

4. Click the **OK** icon (green check mark) in the *PropertyManager* twice, or hit the [**Esc**] key once, to end the Sketch Line command.

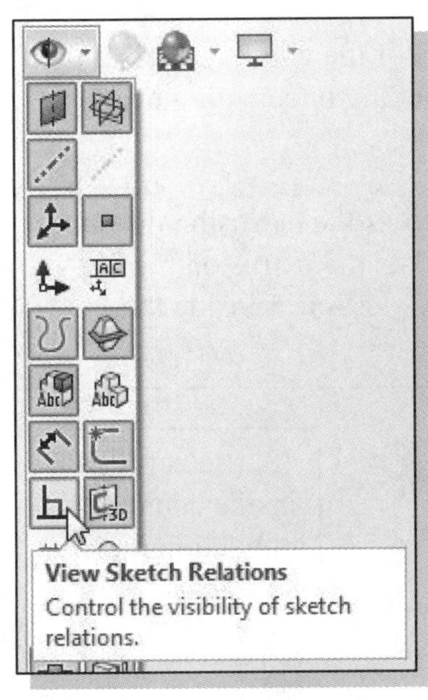

➢ We will hide the *Sketch Relation* icons in the sketch. The visibility of these and other items is controlled using the *Hide/Show Items* option on the *Heads-up View* toolbar.

5. Click on the **Hide/Show Items** icon on the *Heads-up View* toolbar to reveal the pull-down menu.

6. Click on the **View Sketch Relations** icon on the pull-down menu to toggle the sketch relation icon visibility *OFF*.

7. Click away from the pull-down menu to accept the settings.

8. Select the **Smart Dimension** command in the *Sketch* toolbar. The Smart Dimension command allows us to quickly create and modify dimensions. Left-click once on the icon to activate the Smart Dimension command.

9. The message "*Select one or two edges/vertices and then a text location*" is displayed in the *Status Bar* area, at the bottom of the SOLIDWORKS window. Create the four dimensions to describe the size of the sketch as shown in the figure, entering the values shown (**2.5**, **2.5**, **0.75** and **0.75**).

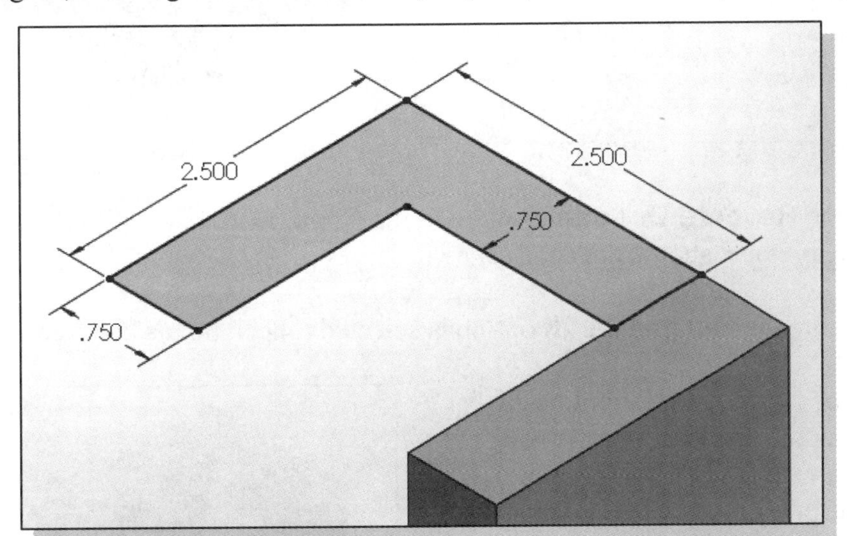

10. Click the **OK** icon in the *PropertyManager* as shown, or hit the [**Esc**] key once, to end the Smart Dimension command.

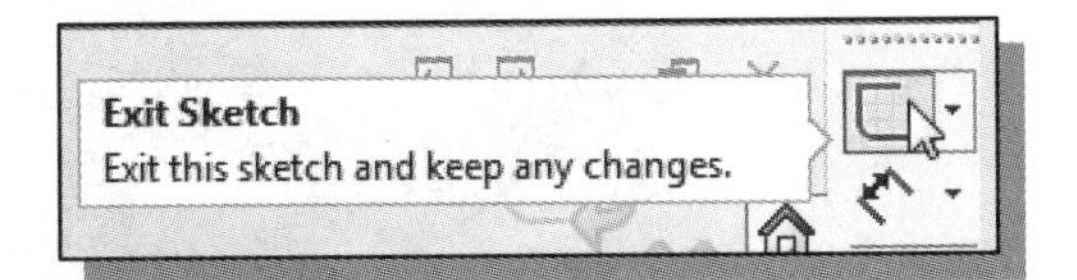

11. Click once with the **left-mouse-button** on the **Sketch** icon on the *Sketch* toolbar to exit the sketch.

12. In the *Features* toolbar (located at the left of the window), select the **Extruded Boss/Base** command by clicking once with the left-mouse-button on the icon.

13. In the *Boss Extrude PropertyManager* panel, enter **2.5** as the extrusion distance. Notice that the sketch region is automatically selected as the extrusion profile. (Note: If the *Selected Contours* window is highlighted, left-click inside the contour to select the area to be extruded.)

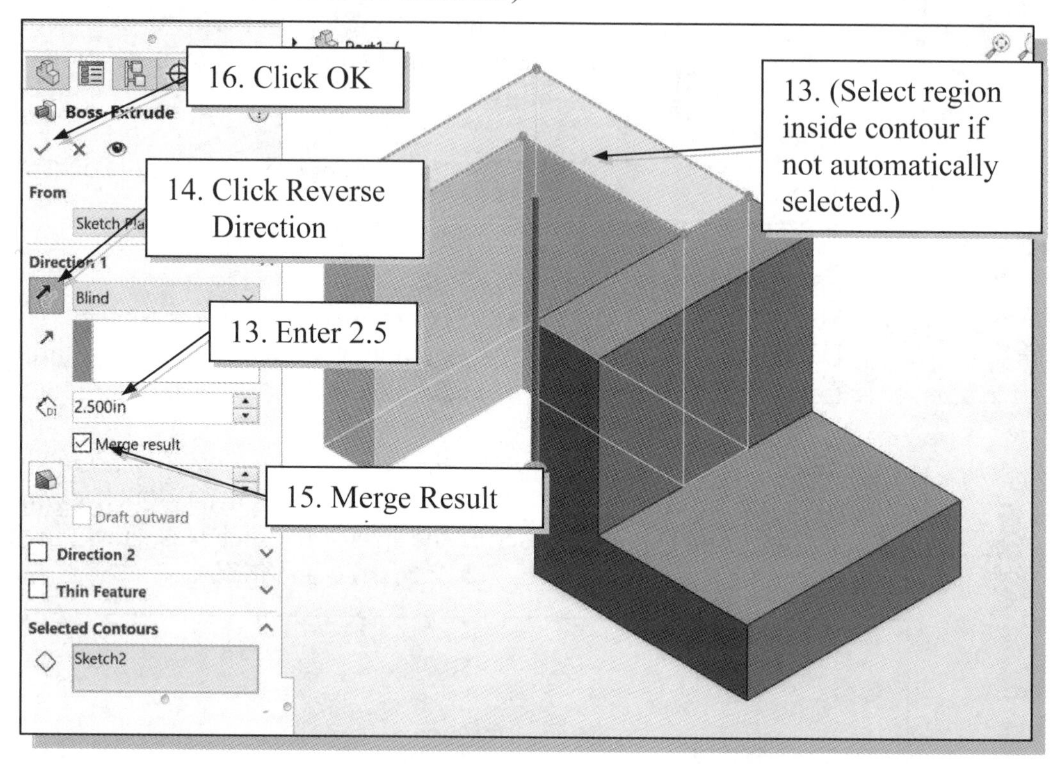

14. Click the **Reverse Direction** button in the *PropertyManager* if necessary. The extrude preview should appear as shown above.

15. Make sure the **Merge Result** option is selected (check box is checked).

❖ With the Merge Result option selected, the resultant body is merged into the existing body (if possible). If the Merge Result option is not selected, the *Extruded Boss/Base* feature creates a distinct solid body.

16. Click on the **OK** button to proceed with creating the extruded feature.

Step 4-2: Adding an Extruded Cut Feature

Next, we will create and profile a circle, which will be used to create a **cut** feature that will be added to the existing solid object.

1. Click in the graphics area, away from the model, to ensure no items are selected.

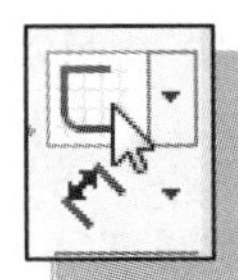

2. Select the **Sketch** button at the top of the *Sketch* toolbar to create a new sketch.

3. In the *Edit Sketch PropertyManager*, the message "*Select: 1) a plane, a planar face, or an edge on which to create a sketch for the entity*" is displayed. SOLIDWORKS expects us to identify a planar surface where the 2D sketch of the next feature is to be created. Move the graphics cursor on the 3D part and notice that SOLIDWORKS will automatically highlight feasible planes and surfaces as the cursor is on top of the different surfaces. **Pick the horizontal face of the 3D solid object.**

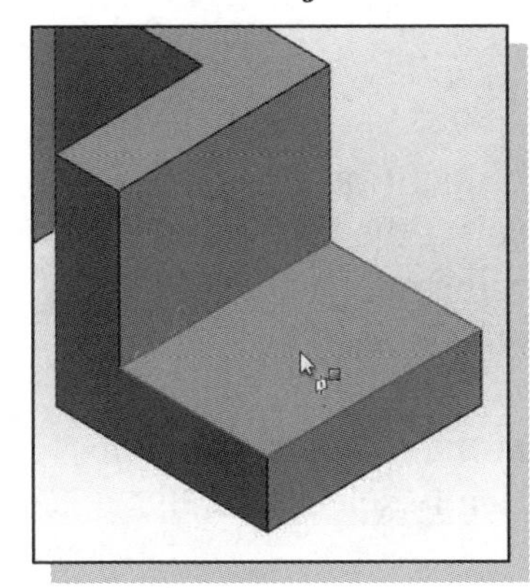

➢ Note that the sketch plane is aligned to the selected face.

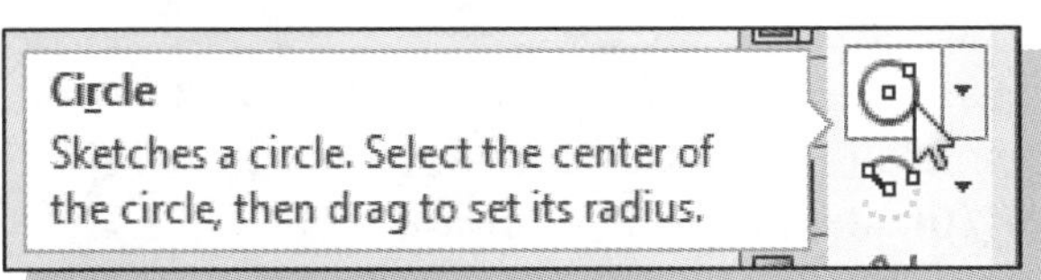

4. Select the **Circle** command by clicking once with the **left-mouse-button** on the icon in the *Sketch* toolbar.

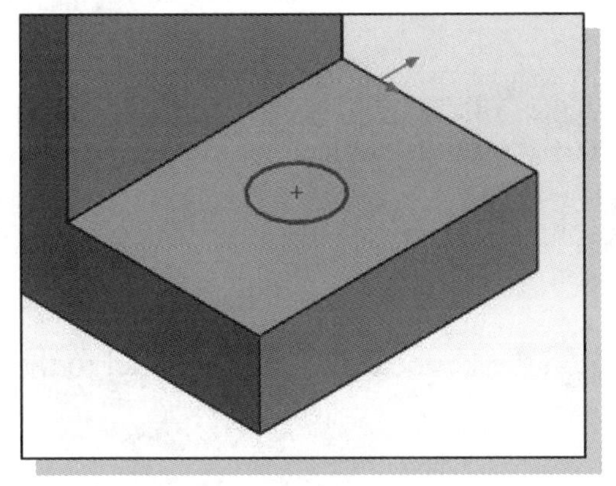

5. Create a circle of arbitrary size on the top face of the solid model as shown. Click once with the left-mouse-button to select the center of the circle, move and click again to set the radius. Press the [**Esc**] key once to end the Circle command.

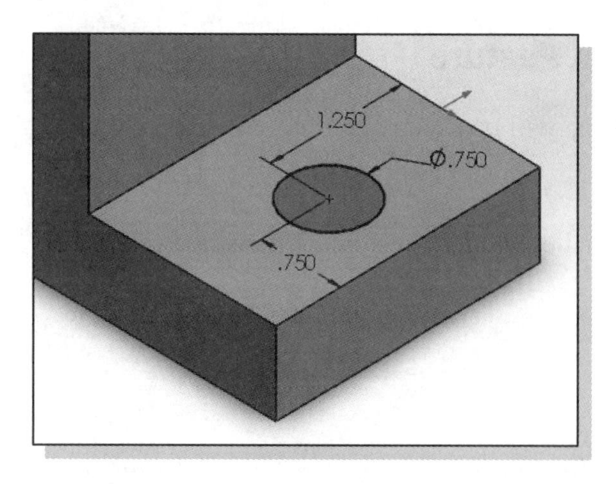

6. On your own, create and modify the dimensions of the sketch as shown in the figure. **NOTE:** Select the center of the circle, then the appropriate edge to apply the linear dimensions. Select any point on the circle to apply the diameter dimension.

7. Click the **OK** icon in the *PropertyManager*, or hit the [**Esc**] key once, to end the Smart Dimension command.

8. Click once with the **left-mouse-button** on the **Sketch** icon on the *Sketch* toolbar to exit the sketch.

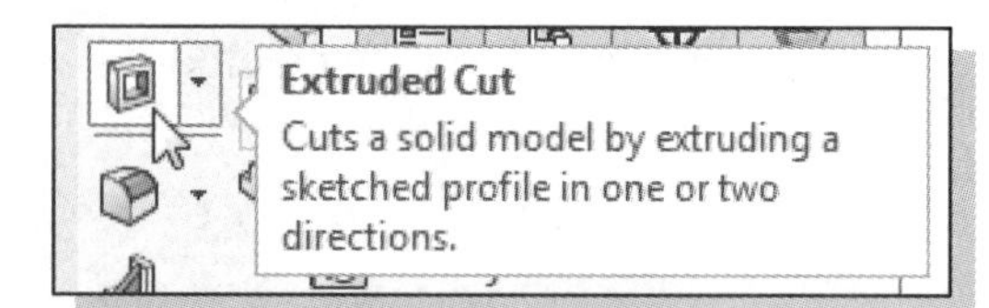

9. In the *Features* toolbar (located at the left of the window), select the **Extruded Cut** command by clicking once with the left-mouse-button on the icon. The *Extrude PropertyManager* is displayed in the left panel. Notice that the sketch region (the circle) is automatically selected as the extrusion profile.

10. In the *Cut-Extrude PropertyManager* panel, click the arrow to reveal the pull-down options for the *End Condition* (the default end condition is 'Blind"), and select **Through All** as shown.

11. Click the **OK** button (green check mark) in the *Cut-Extrude PropertyManager* panel.

12. Click in the graphics area, away from the model, to ensure no items are selected.

13. Press the **F** key on the keyboard to fit the model to the screen.

Save the Part File

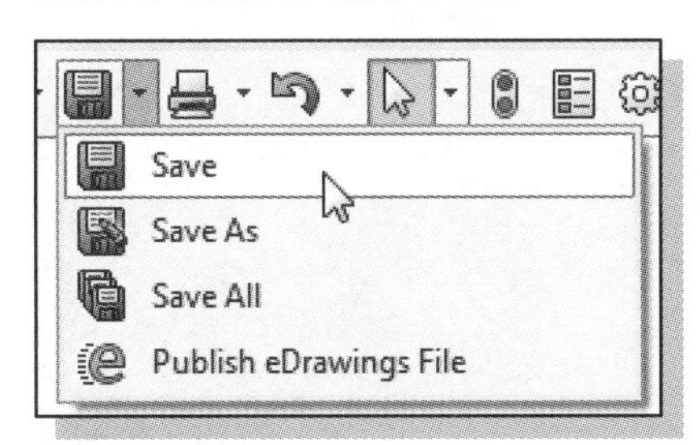

1. Select **Save** in the *Menu Bar* pull-down menu, or you can also use the "**Ctrl-S**" combination (hold down the "Ctrl" key and hit the "S" key once) to save the part.

2. In the pop-up window, select the directory to store the model in and enter ***Adjuster*** as the name of the file.

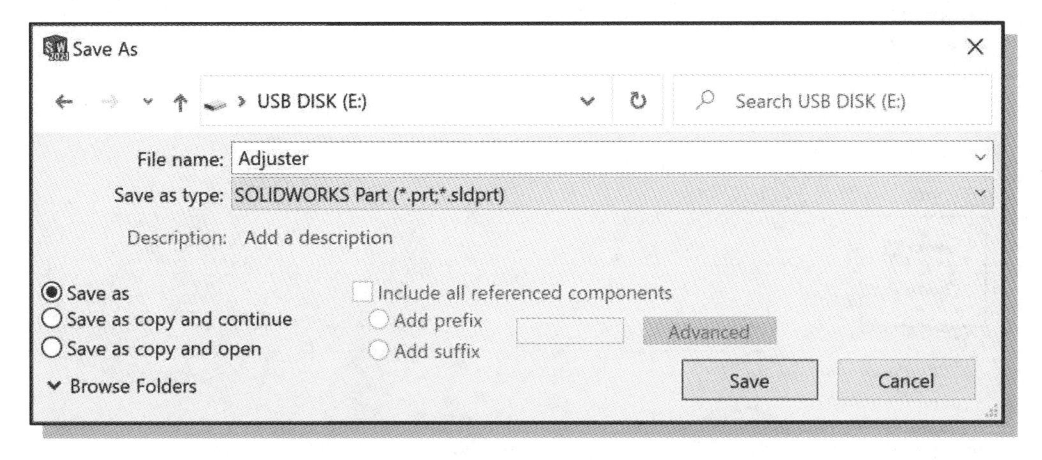

3. Click on the **Save** button to save the file.

❖ You should form a habit of saving your work periodically, just in case something might go wrong while you are working on it. In general, one should save one's work at an interval of every 15 to 20 minutes. One should also save before making any major modifications to the model.

Questions:

1. What is the first thing we should set up in SOLIDWORKS when creating a new model?

2. What is the main difference between a rough sketch and a *profile*?

3. List two of the geometric relation symbols used by SOLIDWORKS.

4. What was the first feature we created in this lesson?

5. Identify the following commands:

(a)

(b)

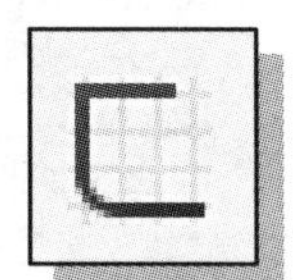

(c)

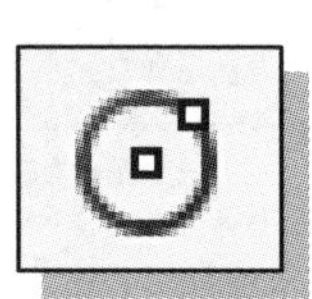

(d)

Exercises: (All dimensions are in inches.)

1. **Inclined Support** (Thickness: **.5**)

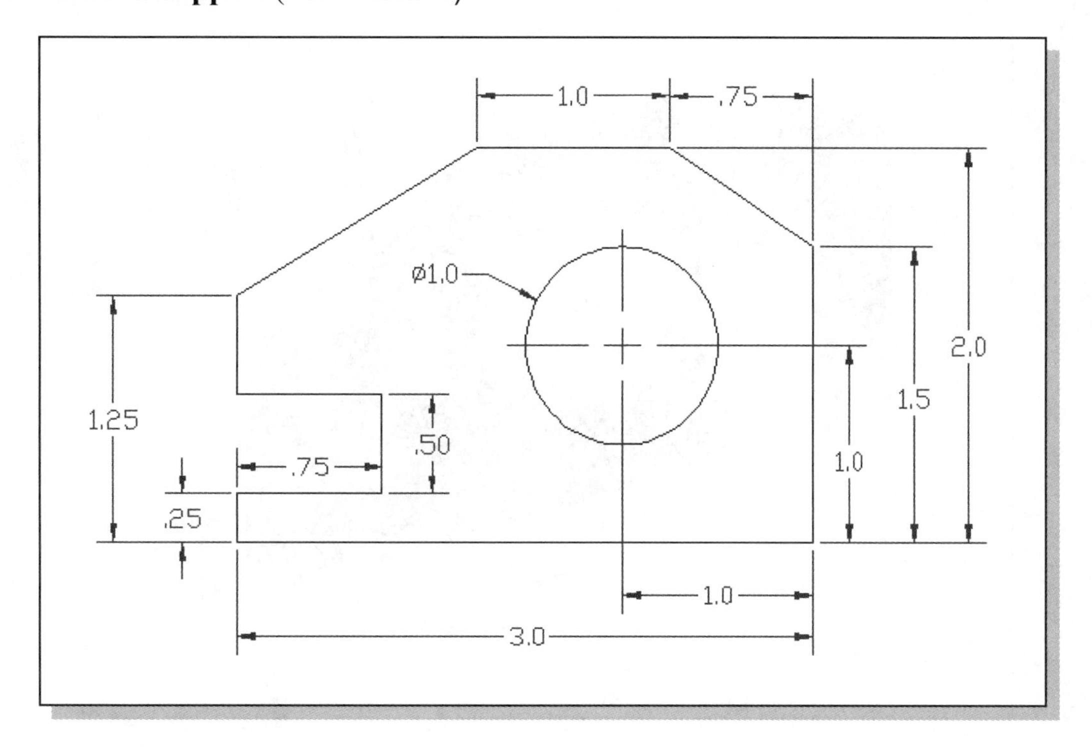

2. **Spacer Plate** (Thickness: **.125**)

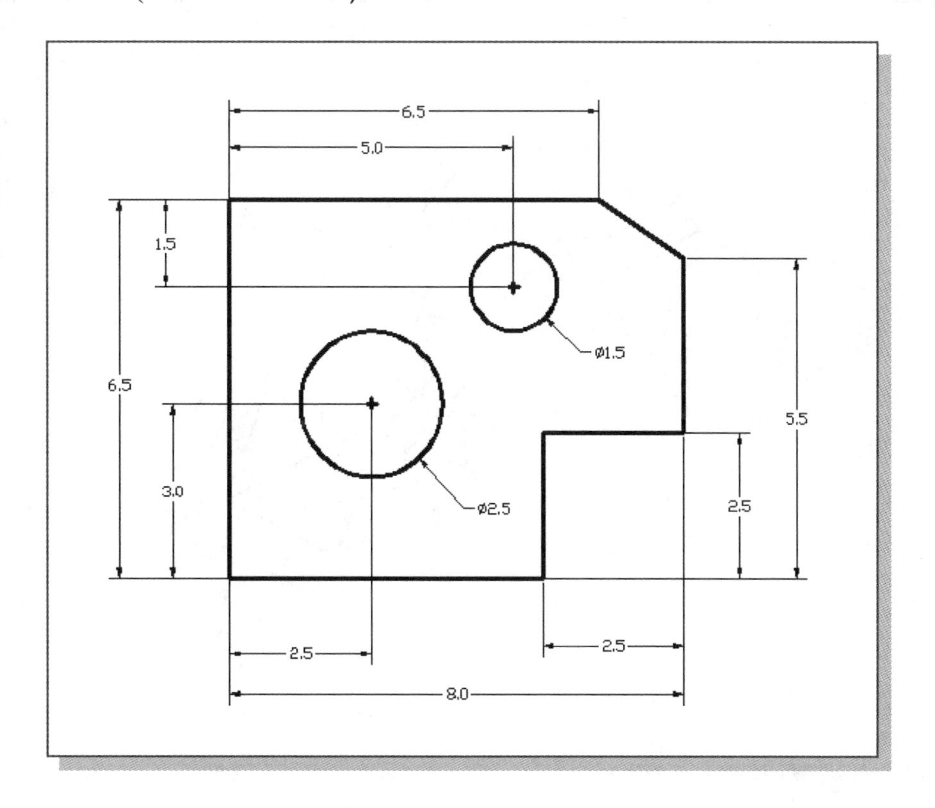

3. Positioning Stop

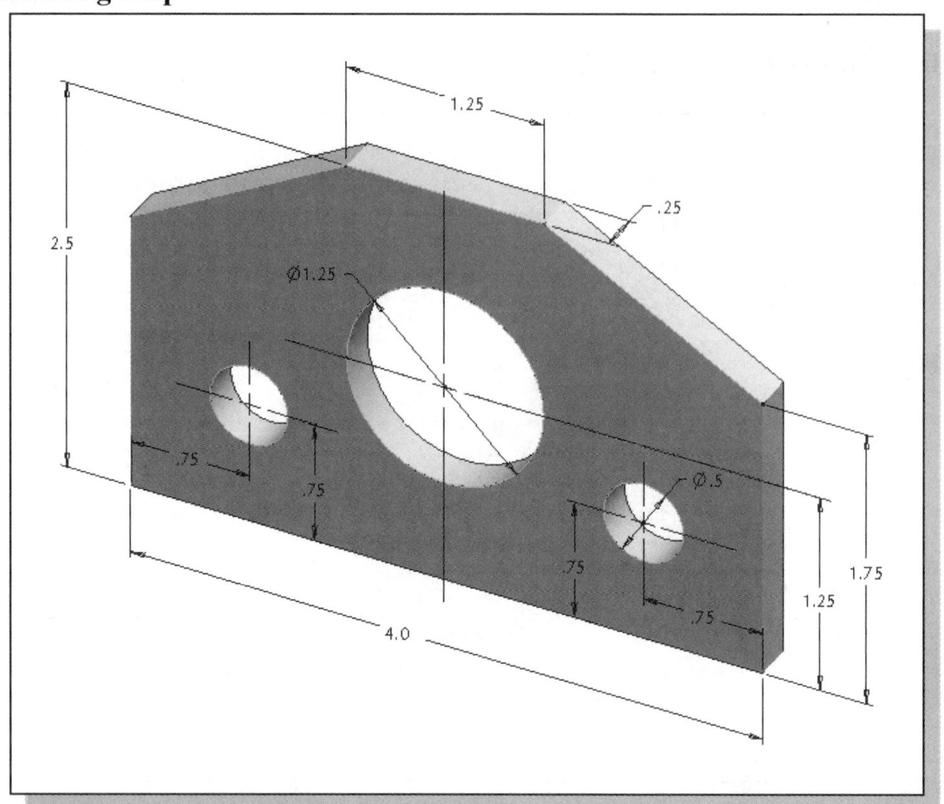

4. Guide Block

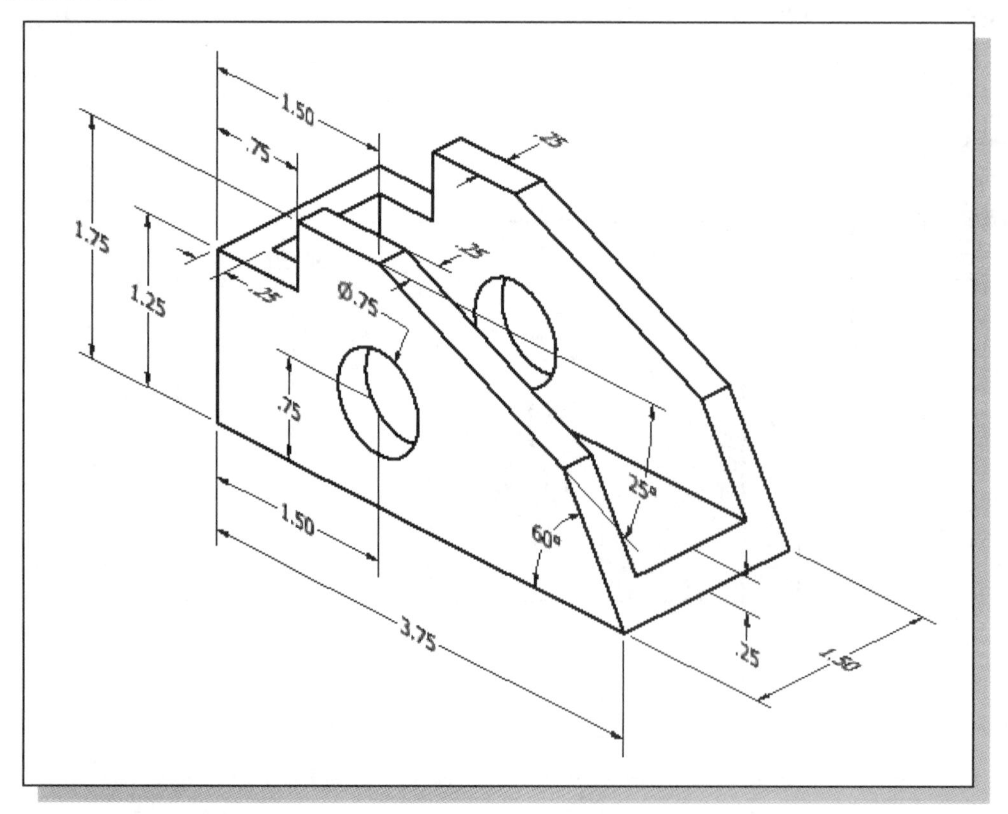

5. Slider Block

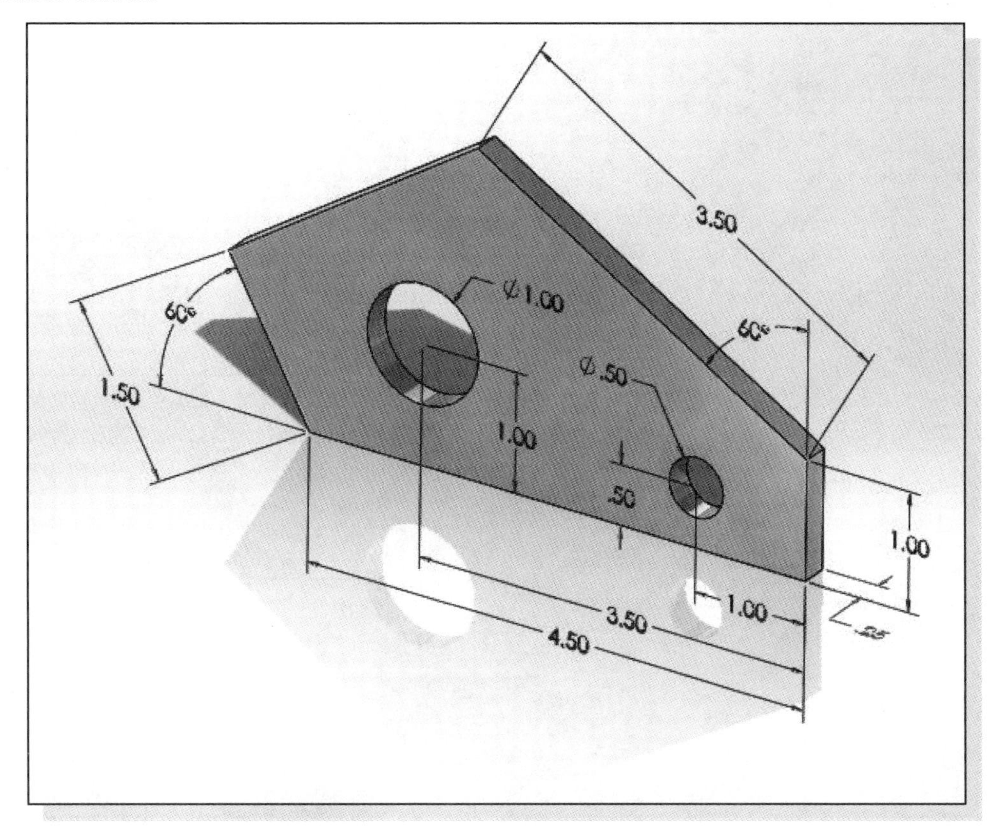

6. Angle Lock

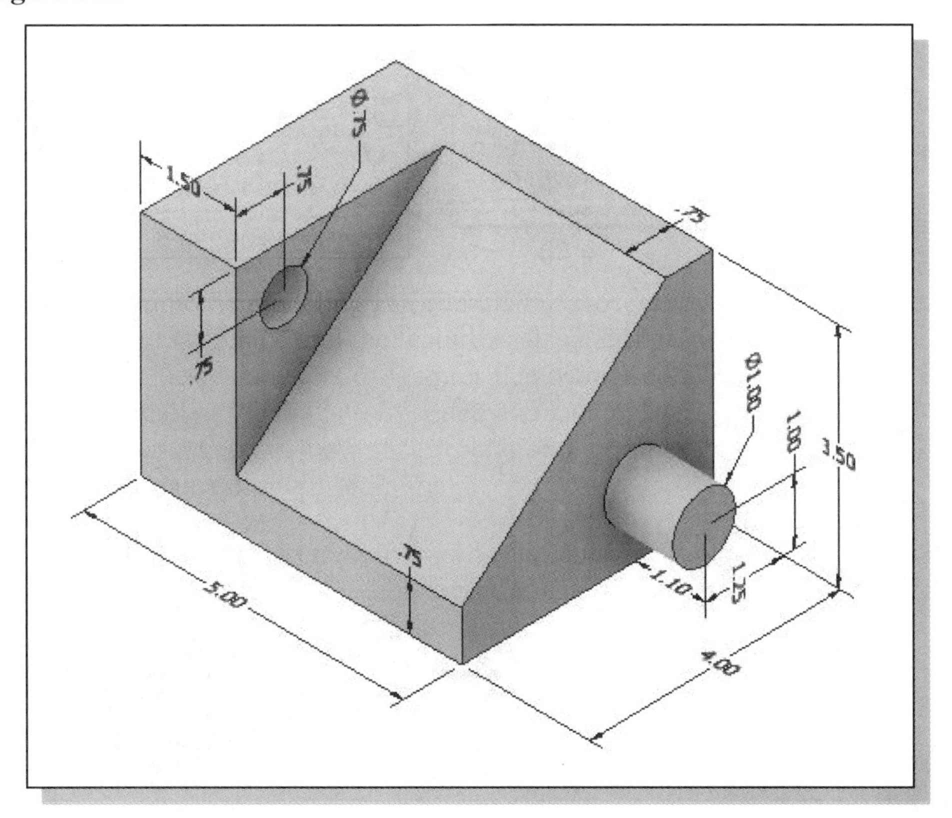

NOTES:

Chapter 3
Constructive Solid Geometry Concepts

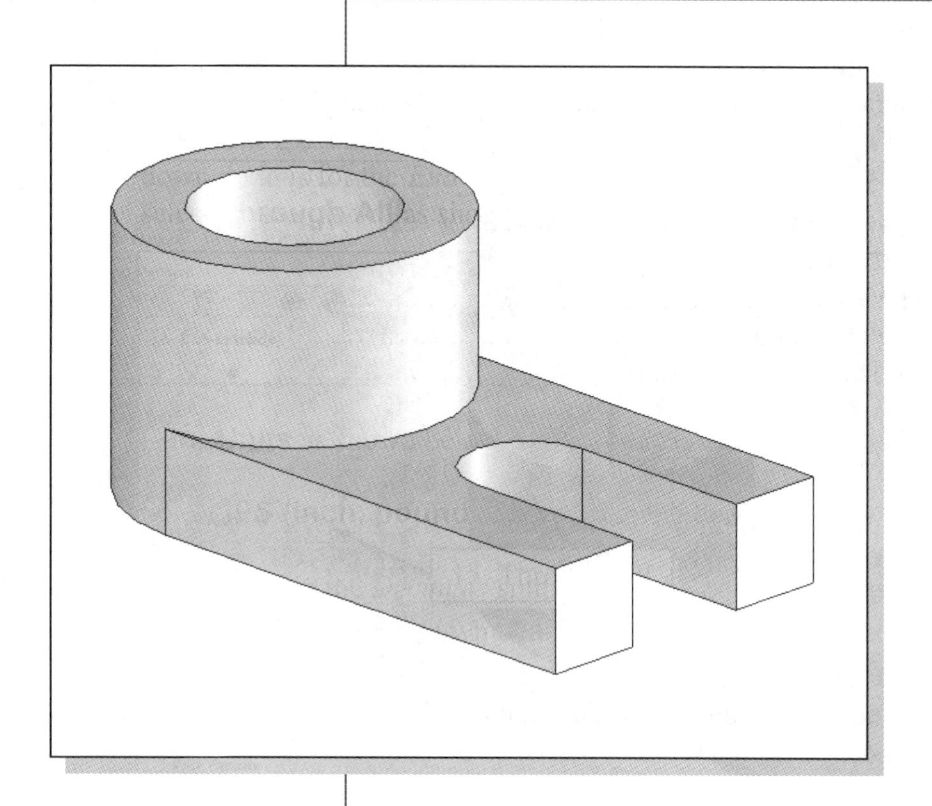

Learning Objectives

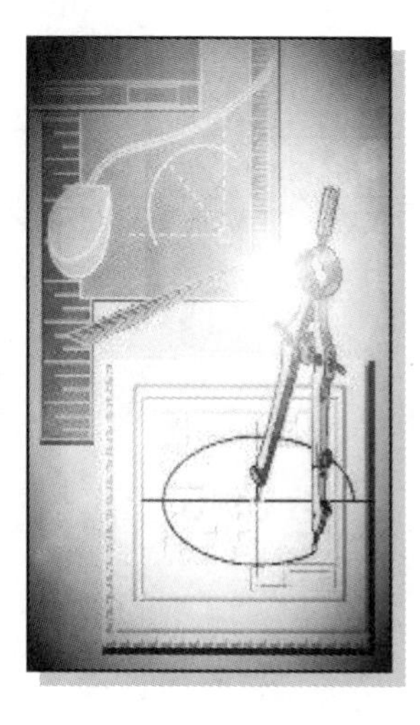

- ♦ Understand Constructive Solid Geometry Concepts
- ♦ Create a Binary Tree
- ♦ Understand the Basic Boolean Operations
- ♦ Use the SOLIDWORKS CommandManager User Interface
- ♦ Set up GRID and SNAP Intervals
- ♦ Understand the Importance of Order of Features
- ♦ Use the Different Extrusion Options

Certified SOLIDWORKS Associate Exam Objectives Coverage

Introduction

In the 1980s, one of the main advancements in **solid modeling** was the development of the **Constructive Solid Geometry** (CSG) method. CSG describes the solid model as combinations of basic three-dimensional shapes (**primitive solids**). The basic primitive solid set typically includes Rectangular-prism (Block), Cylinder, Cone, Sphere, and Torus (Tube). Two solid objects can be combined into one object in various ways using operations known as **Boolean operations**. There are three basic Boolean operations: **JOIN (Union)**, **CUT (Difference)**, and **INTERSECT**. The *JOIN* operation combines the two volumes included in the different solids into a single solid. The *CUT* operation subtracts the volume of one solid object from the other solid object. The *INTERSECT* operation keeps only the volume common to both solid objects. The CSG method is also known as the **Machinist's Approach**, as the method is parallel to machine shop practices.

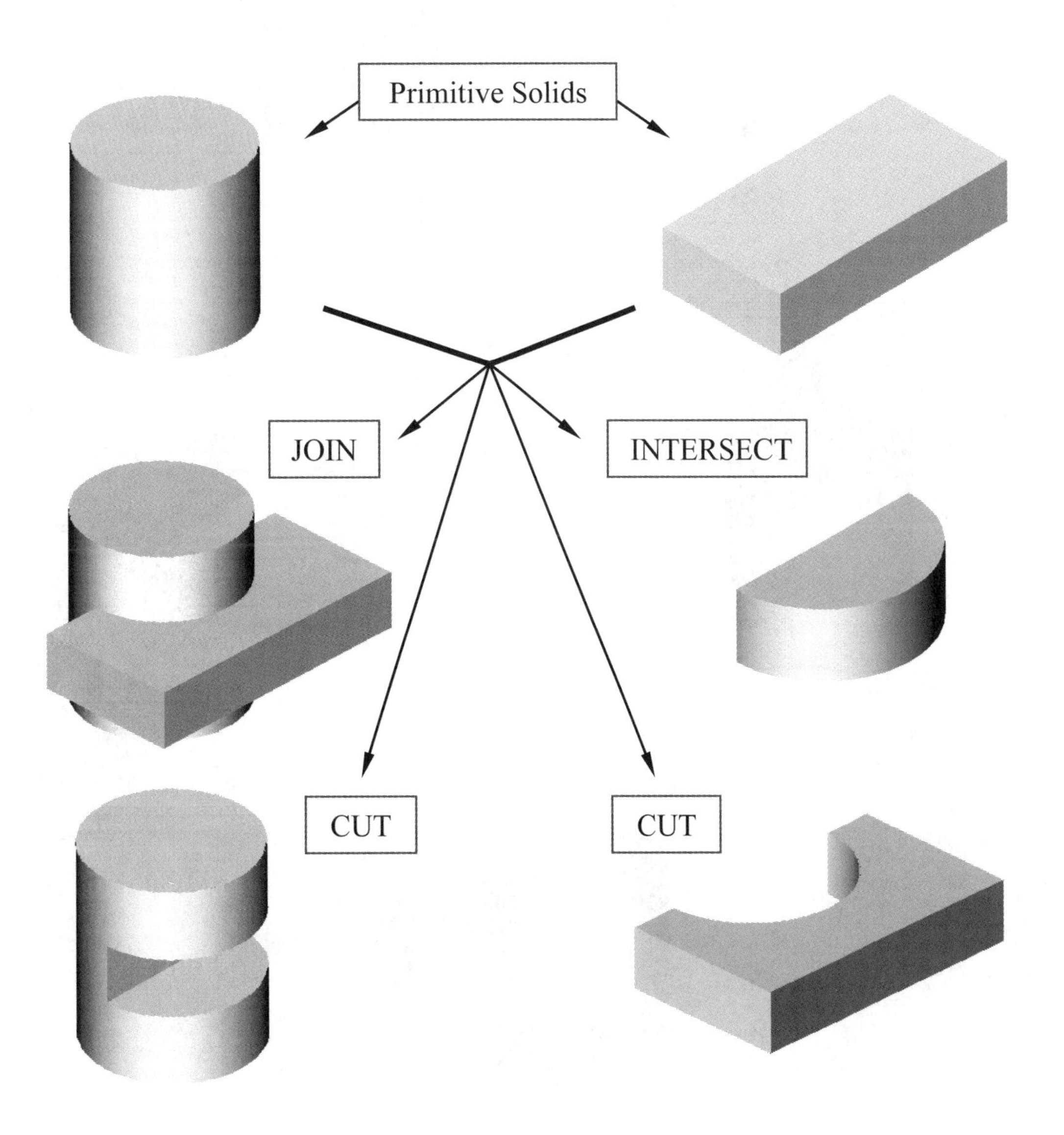

Binary Tree

The CSG is also referred to as the method used to store a solid model in the database. The resulting solid can be easily represented by what is called a **binary tree**. In a binary tree, the terminal branches (leaves) are the various primitives that are linked together to make the final solid object (the root). The binary tree is an effective way to keep track of the *history* of the resulting solid. By keeping track of the history, the solid model can be re-built by re-linking through the binary tree. This provides a convenient way to modify the model. We can make modifications at the appropriate links in the binary tree and re-link the rest of the history tree without building a new model.

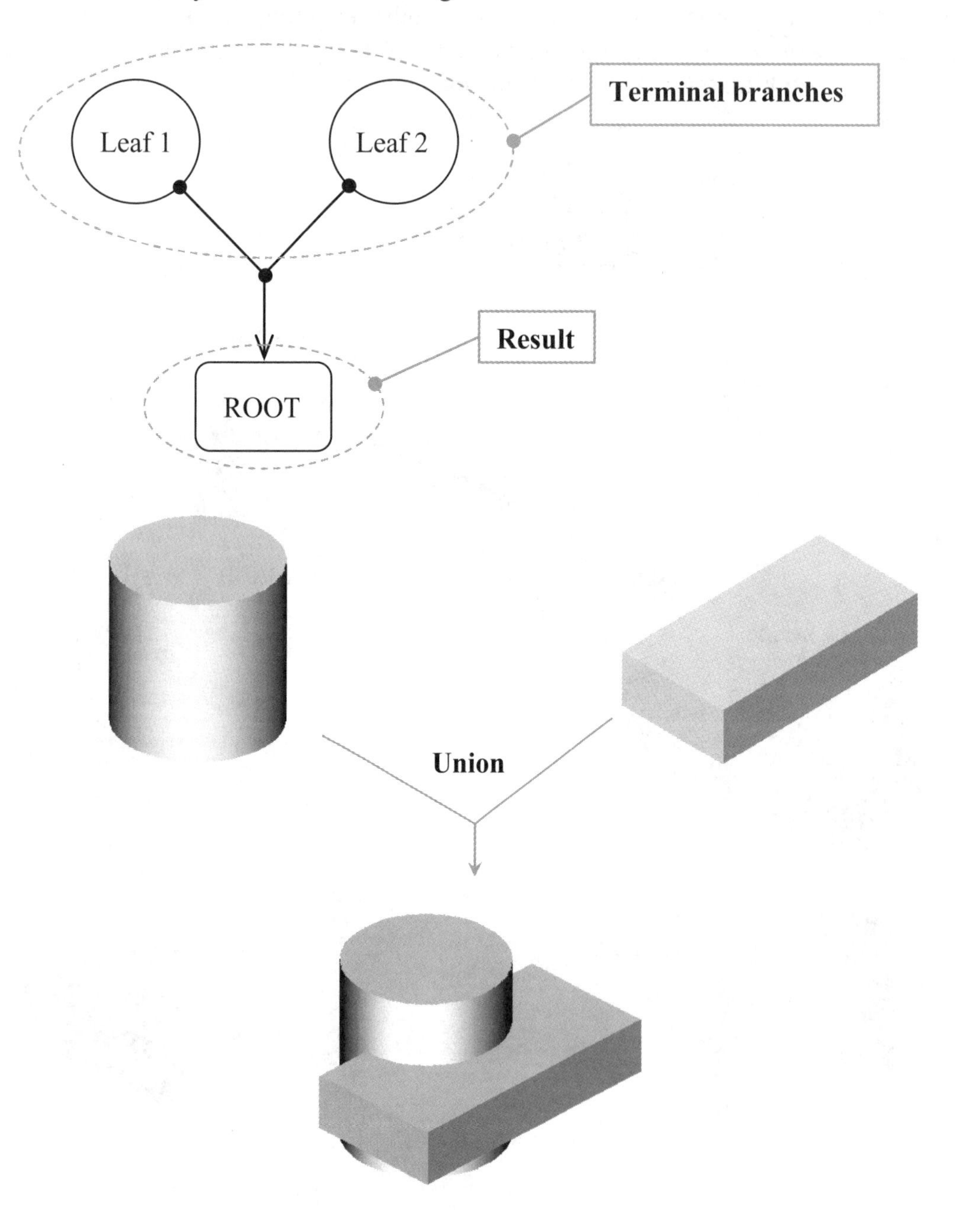

The *Locator* Design

The CSG concept is one of the important building blocks for feature-based modeling. In SOLIDWORKS, the CSG concept can be used as a planning tool to determine the number of features that are needed to construct the model. It is also a good practice to create features that are parallel to the manufacturing process required for the design. With parametric modeling, we are no longer limited to using only the predefined basic solid shapes. In fact, any solid features we create in SOLIDWORKS are used as primitive solids; parametric modeling allows us to maintain full control of the design variables that are used to describe the features. In this lesson, a more in-depth look at the parametric modeling procedure is presented. The equivalent CSG operation for each feature is also illustrated.

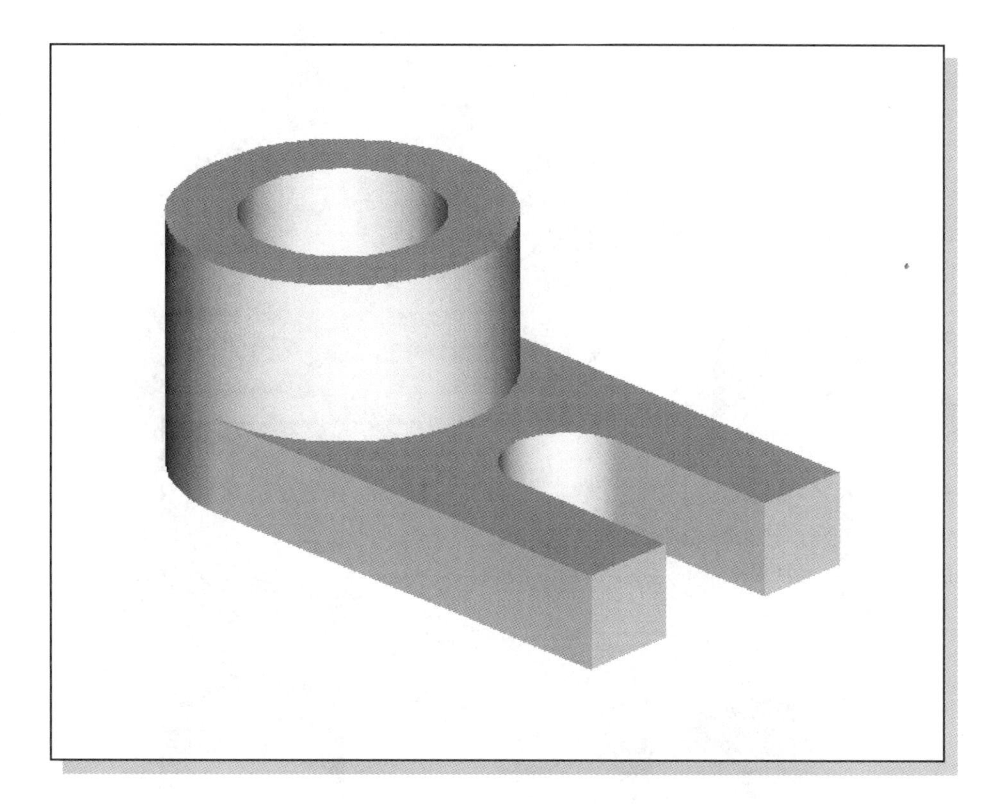

- Before going through the tutorial, on your own, make a sketch of a CSG binary tree of the ***Locator*** design using only two basic types of primitive solids: cylinder and rectangular prism. In your sketch, how many *Boolean operations* will be required to create the model? What is your choice of the first primitive solid to use, and why? Take a few minutes to consider these questions and do the preliminary planning by sketching on a piece of paper. Compare the sketch you make to the CSG binary tree steps shown on page 3-5. Note that there are many different possibilities in combining the basic primitive solids to form the solid model. Even for the simplest design, it is possible to take several different approaches to creating the same solid model.

Modeling Strategy – CSG Binary Tree

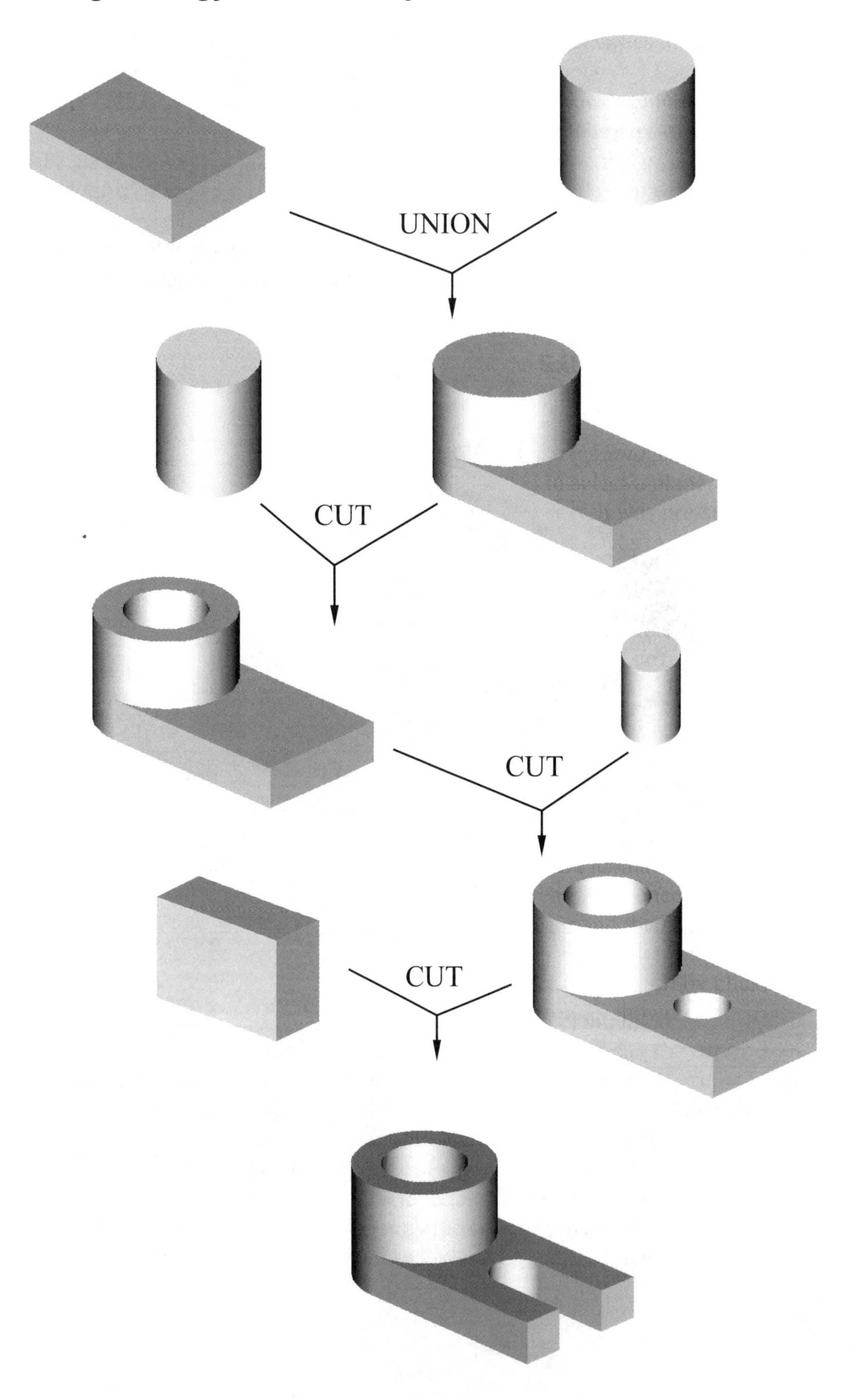

Starting SOLIDWORKS and Activating the CommandManager

1. Select the **SOLIDWORKS** option on the *Start* menu or select the **SOLIDWORKS** icon on the desktop to start SOLIDWORKS. The SOLIDWORKS main window will appear.

2. We will not use the *Welcome* dialog box in this lesson. If it appears, close the *Welcome* dialog box by clicking on the X in the upper right corner to view the SOLIDWORKS program window.

3. Select the **New** icon with a single click of the left-mouse-button on the *Menu Bar*.

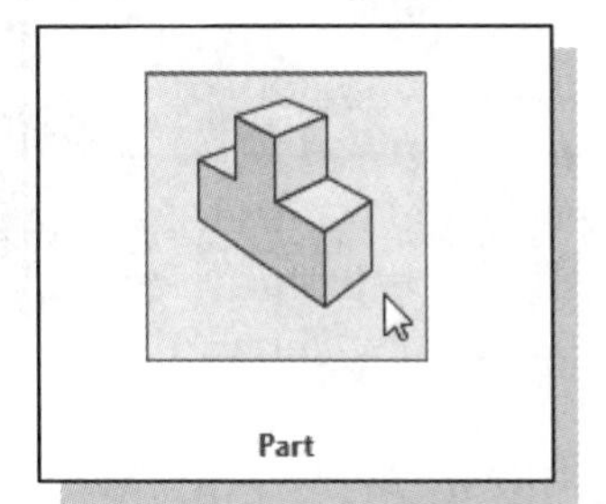

4. Select the **Part** icon with a single click of the left-mouse-button in the *New* SOLIDWORKS *Document* dialog box.

5. Select **OK** in the *New* SOLIDWORKS *Document* dialog box to open a new part document.

IMPORTANT NOTE: The SOLIDWORKS *CommandManager* provides an alternate method for displaying the most commonly used toolbars (see page 1-13 for a more complete description). We will use the *CommandManager* in this lesson.

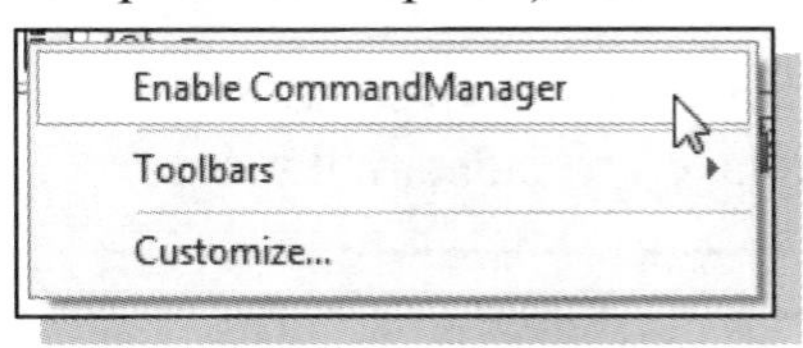

6. To turn ***ON*** the *CommandManager*, right click on any toolbar and toggle the *Enable CommandManager ON* by selecting it at the top of the pop-up menu.

❖ Notice the *Features* and *Sketch* toolbars no longer appear at the edge of the window. These toolbars appear on the *Ribbon* display of the *CommandManager*.

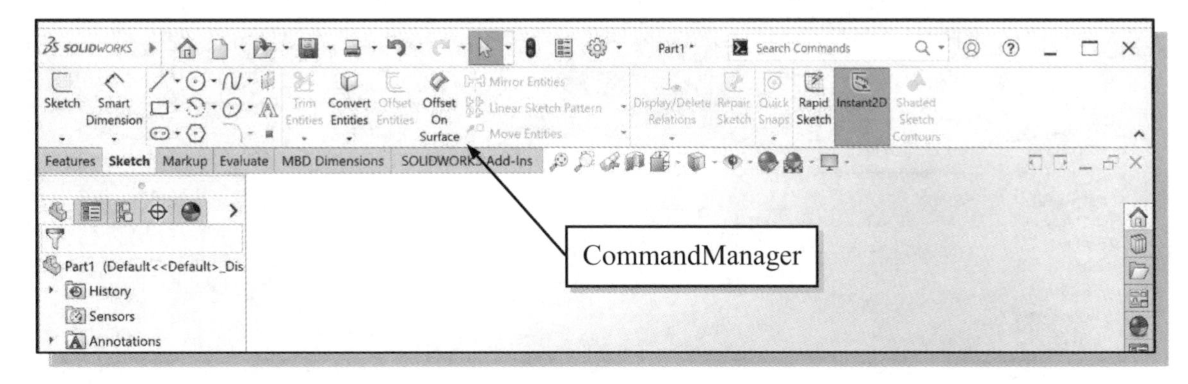

❖ Every object we construct in a CAD system is measured in units. We should determine the value of the units within the CAD system before creating the first geometric entities. For example, in one model, a unit might equal one millimeter of the real-world object; in another model, a unit might equal an inch. In most CAD systems, setting the model units does not always set units for dimensions. We generally set model units and dimension units to the same type and precision.

7. Select the **Options** icon from the *Menu Bar* toolbar to open the *Options* dialog box.

8. Select the **Document Properties** tab and the **Drafting Standard** option at the left of the *Document Properties* panel.

9. Select **ANSI** in the pull-down selection window under the *Overall drafting standard* panel as shown.

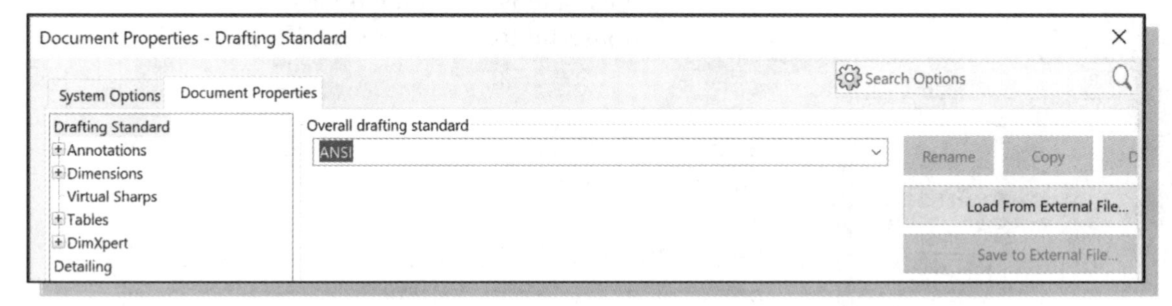

10. Click **Units** as shown below. We will use the **MMGS (millimeter, gram, second)** unit system. (NOTE: In Chapter 2, we accessed the same *Document Properties - Units* dialog box directly by using System Units access icon on the *Status Bar*.)

11. Select **.12** in the *Decimals* spin box for the *Length units* as shown to define the degree of accuracy with which the units will be displayed to 2 decimal places.

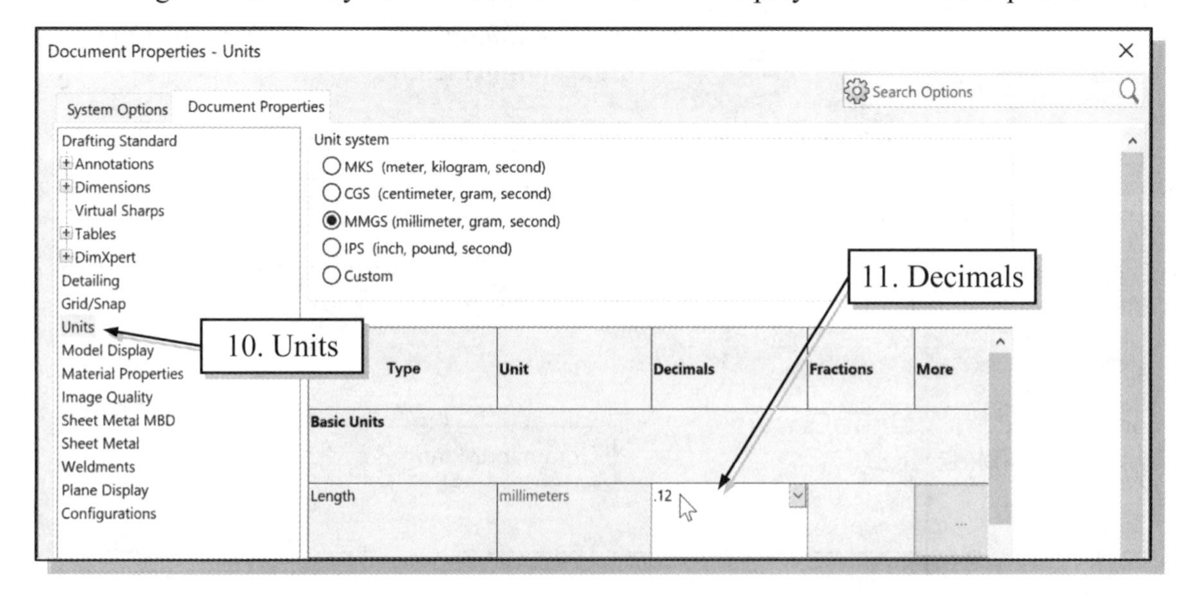

GRID and *SNAP* Intervals Setup

1. Click **Grid/Snap** as shown below.

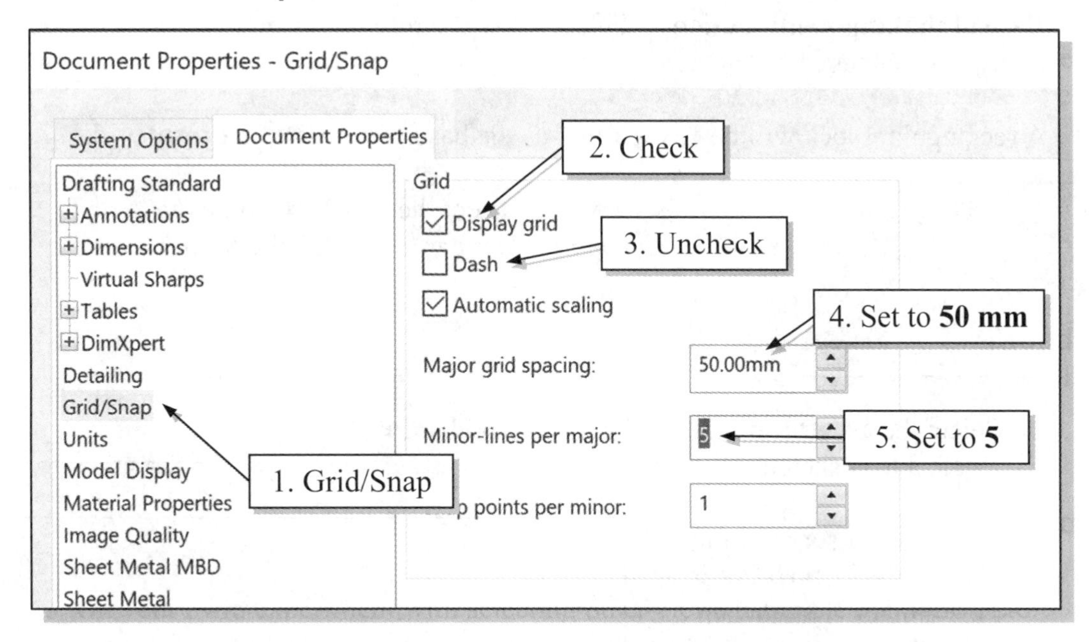

2. Check the Display grid checkbox under the *Grid* options.

3. Uncheck the Dash checkbox under the *Grid* options.

4. Set the Major grid spacing to **50 mm** under the *Grid* options.

5. Set the Minor-lines per major to **5** under the *Grid* options.

6. Click **OK** in the *Options* dialog box to accept the selected settings.

➢ Note that the above settings set the grid spacing in SOLIDWORKS. Although the **Snap to grid** option is also available in SOLIDWORKS, its usage in parametric modeling is not recommended.

Base Feature

In *parametric modeling*, the first solid feature is called the **base feature,** which usually is the primary shape of the model. Depending upon the design intent, additional features are added to the base feature.

Some of the considerations involved in selecting the base feature are:

- **Design intent** – Determine the functionality of the design; identify the feature that is central to the design.

- **Order of features** – Choose the feature that is the logical base in terms of the order of features in the design.

- **Ease of making modifications** – Select a base feature that is more stable and is less likely to be changed.

➢ A rectangular block will be created first as the base feature of the ***Locator*** design.

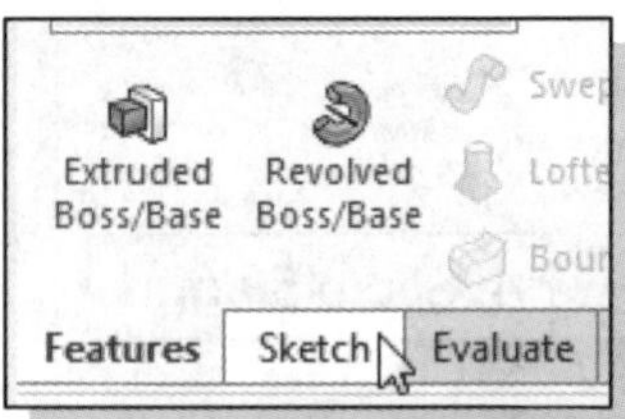

1. If necessary, select the **Sketch** tab on the *CommandManager* to display the *Sketch* toolbar.

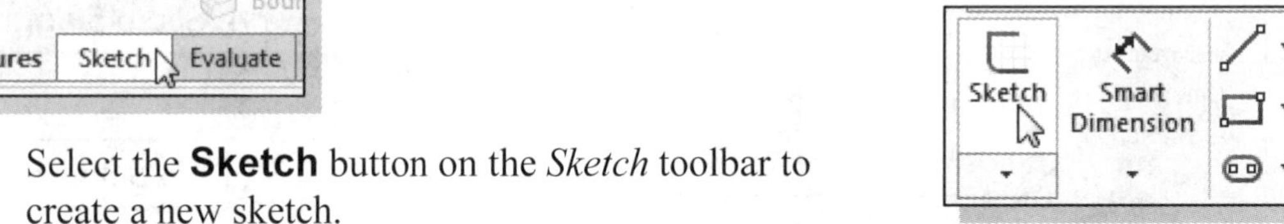

2. Select the **Sketch** button on the *Sketch* toolbar to create a new sketch.

3. Move the cursor over the edge of the Top Plane in the graphics area. When the Top Plane is highlighted, click once with the **left-mouse-button** to select the Top Plane (XZ Plane) as the sketch plane for the new sketch.

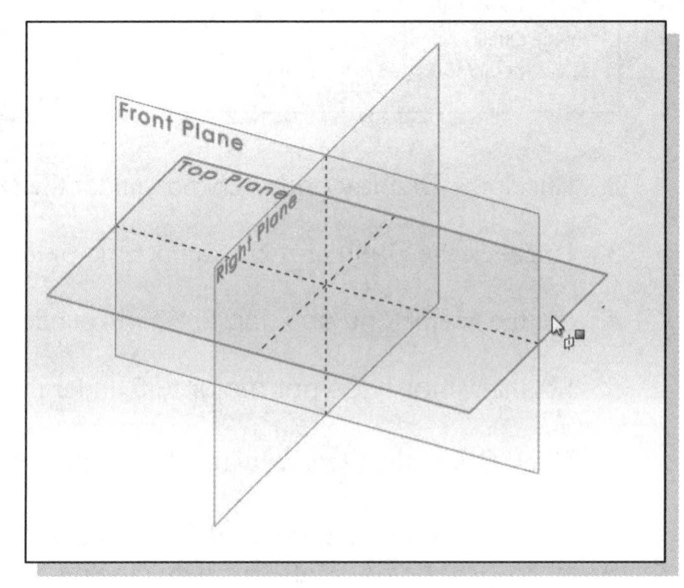

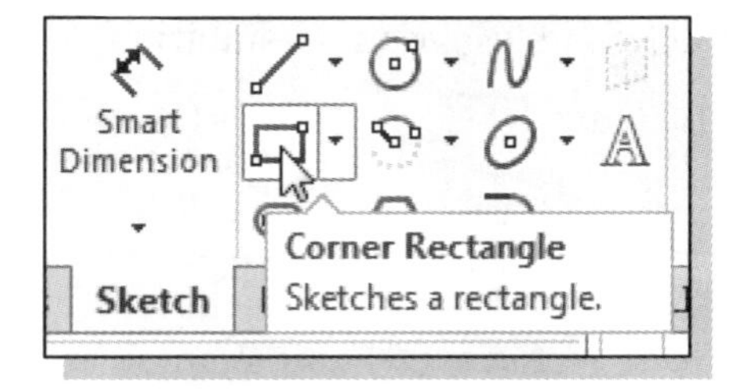

4. Select the **Corner Rectangle** command by clicking once with the **left-mouse-button** on the icon in the *Sketch* toolbar.

5. Create a rectangle of arbitrary size by selecting two locations on the screen as shown below.

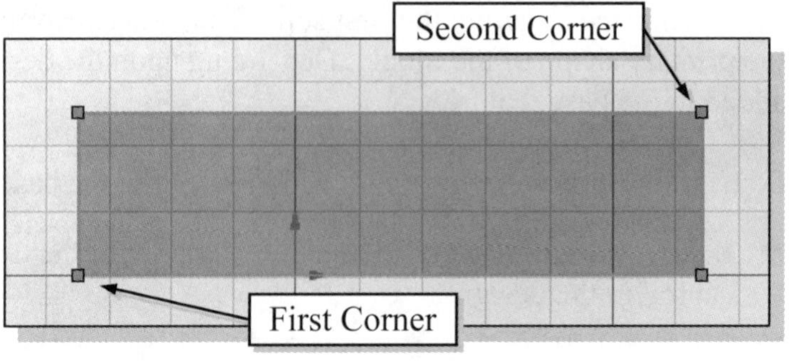

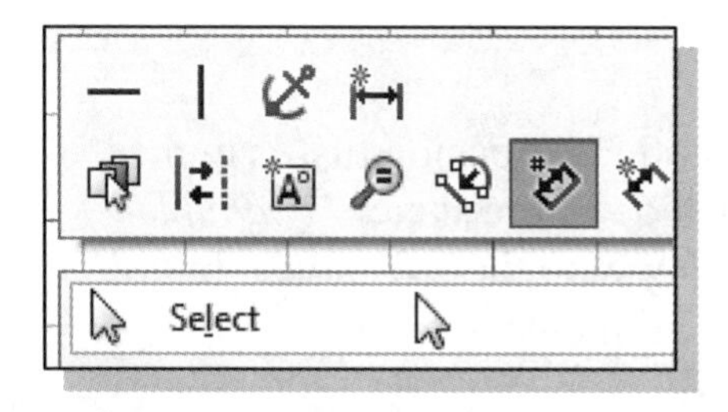

6. Inside the graphics window, click once with the right-mouse-button to bring up the option menu.

7. Click **Select** to end the Rectangle command.

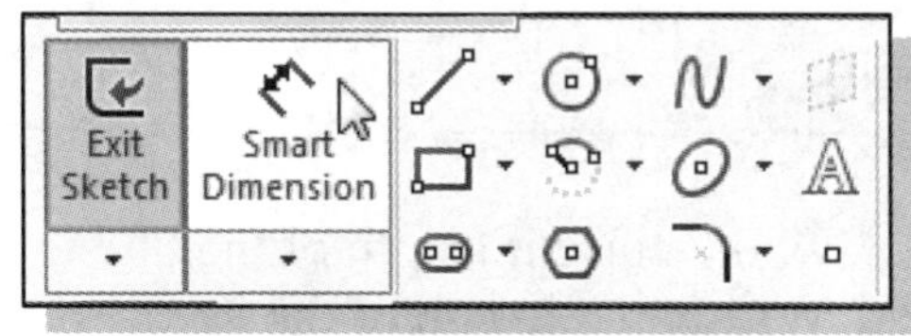

8. Select the **Smart Dimension** command by clicking once with the **left-mouse-button** on the icon in the *Sketch* toolbar.

9. The message "*Select one or two edges/vertices and then a text location*" is displayed in the *Status Bar* area at the bottom of the SOLIDWORKS window. Select the bottom horizontal line by left-clicking once on the line.

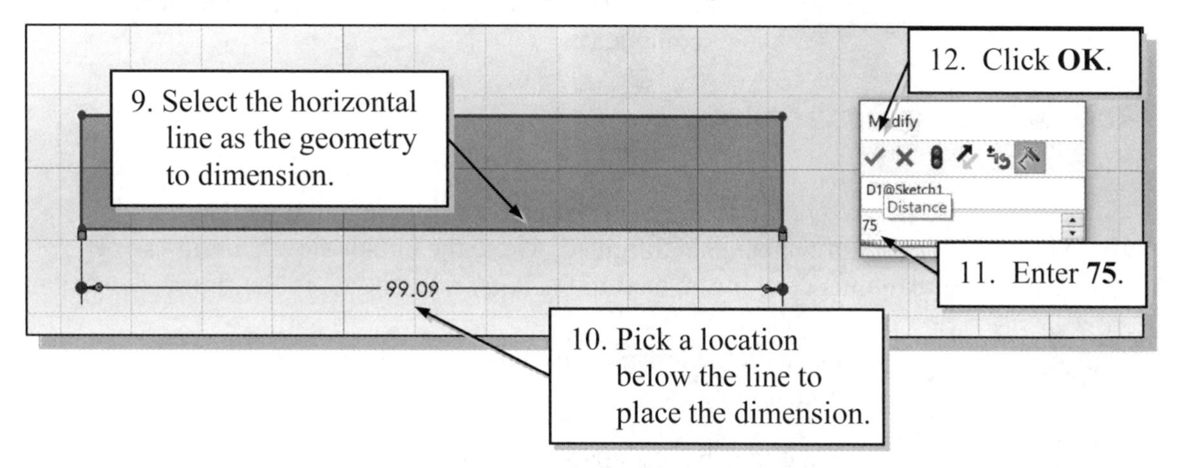

10. Move the graphics cursor below the selected line and left-click to place the dimension. (Note that the value displayed on your screen might be different than what is shown in the figure above.)

11. Enter **75** in the *Modify* dialog box.

12. Click **OK** in the *Modify* dialog box.

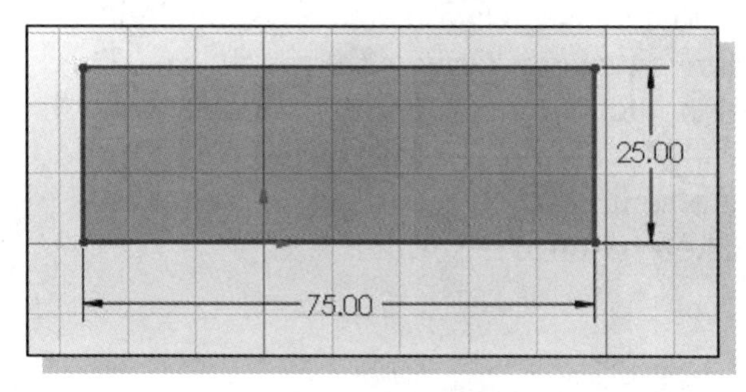

13. On your own, create the vertical size dimension of the sketched rectangle as shown.

14. Click the **OK** icon in the *PropertyManager*, or hit the [**Esc**] key once, to end the Smart Dimension command.

Modifying the Dimensions of the Sketch

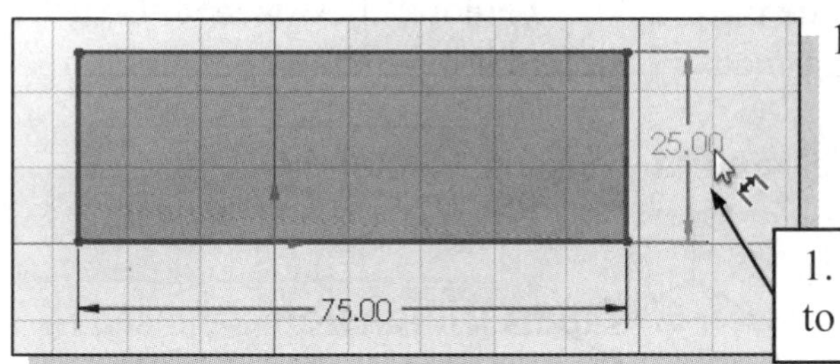

1. Select the dimension that is to the right side of the sketch by *double-clicking* with the left-mouse-button on the dimension text.

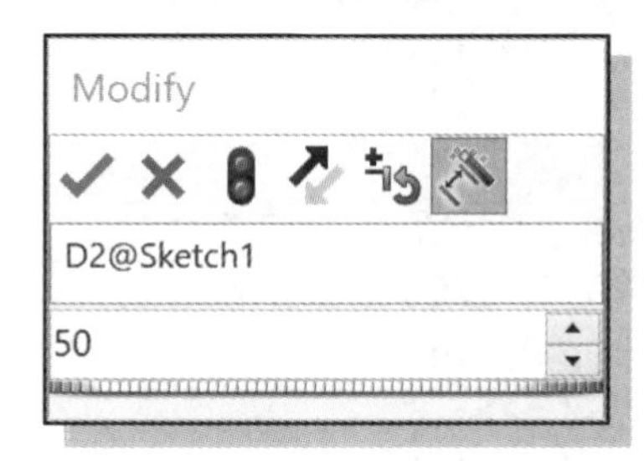

2. In the *Modify* window, the current length of the line is displayed. Enter **50** to reset the length of the line.

3. Click on the **OK** icon to accept the entered value.

4. Hit the [**Esc**] key once to end the Dimension command.

Repositioning Dimensions

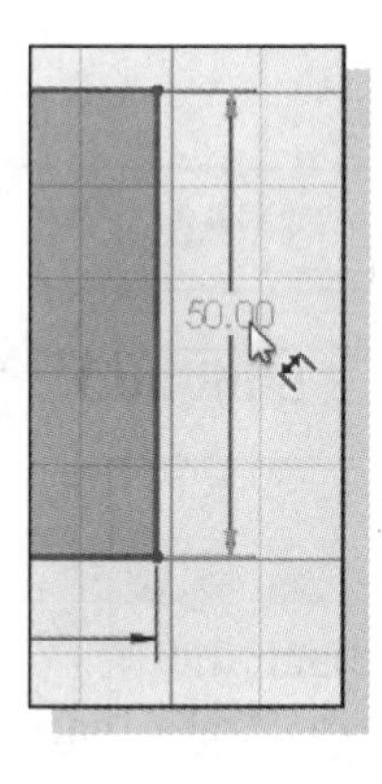

1. Move the cursor near the vertical dimension; note that the dimension is highlighted. Move the cursor slowly until a small marker appears next to the cursor, as shown in the figure.

2. Drag with the left-mouse-button to reposition the selected dimension.

3. Repeat the above steps to reposition the horizontal dimension.

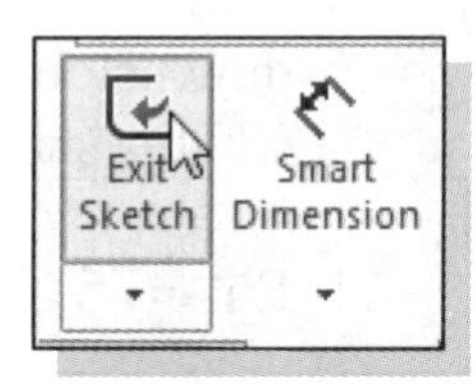

4. Click once with the **left-mouse-button** on the **Exit Sketch** icon on the *Sketch* toolbar to end the Sketch option.

Completing the Base Solid Feature

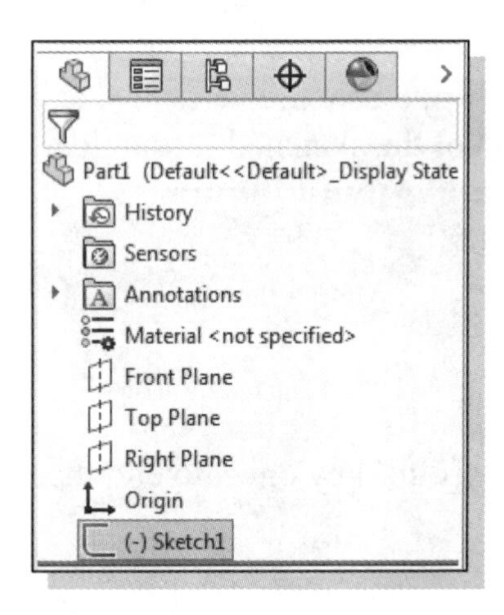

1. Make sure that the sketch you created is highlighted in the *FeatureManager Design Tree* as shown. This indicates that the sketch is currently selected and will automatically be used when we execute the Extruded Boss/Base command. If the sketch is not selected, click on Sketch1 once to select it.

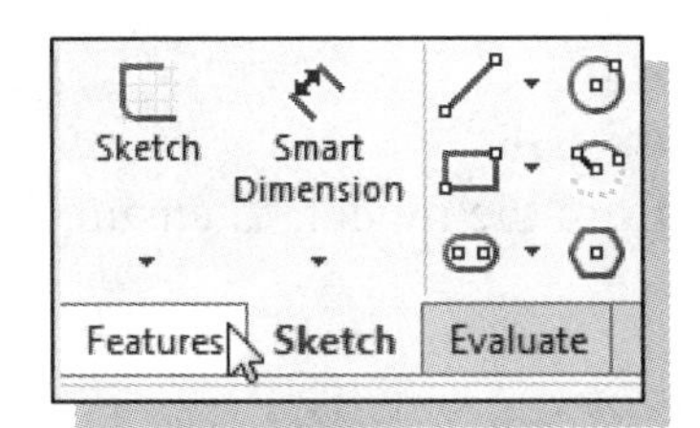

2. Select the **Features** tab on the *CommandManager* to display the *Features* toolbar.

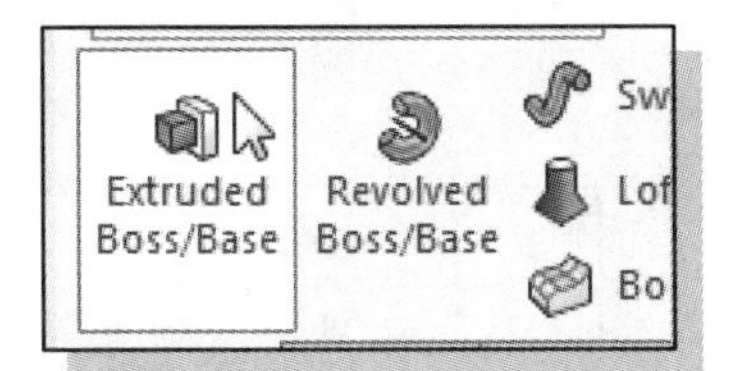

3. Select the **Extruded Boss/Base** button on the *Features* toolbar to create a new extruded feature.

➢ Notice the pre-selected sketch is automatically used for the creation of the *Extruded Boss Feature.*

4. In the *Extrude PropertyManager* panel, enter **15** as the extrusion distance. Notice that the sketch region is automatically selected as the extrusion profile.

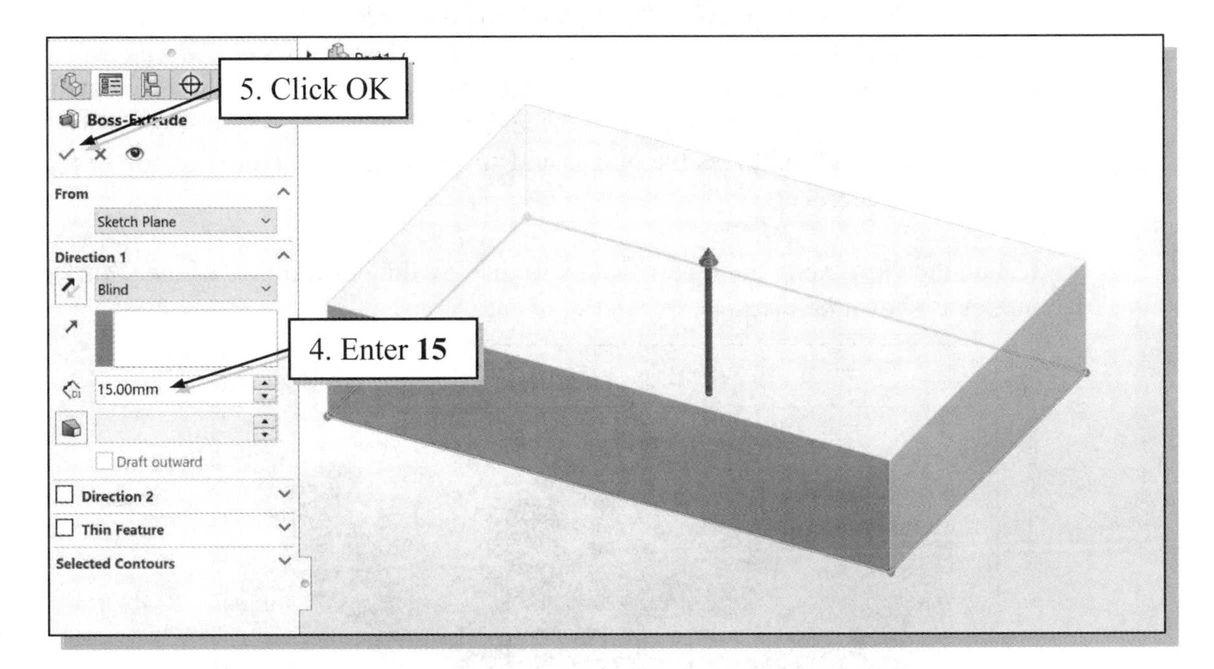

5. Click on the **OK** button to proceed with creating the 3D part.

6. Use the *Viewing* options to view the created part. On the *Heads-up View* toolbar, select **View Orientation** (to open the *View Orientation* pull-down menu) and select the **Isometric** icon to reset the display to the Isometric view before going to the next section.

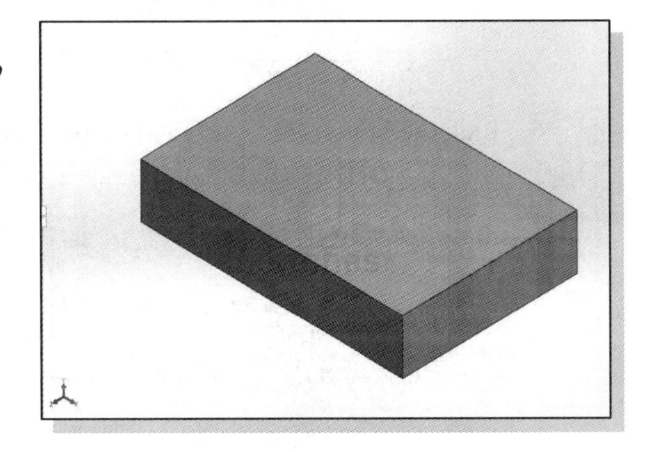

Creating the Next Solid Feature

1. Click the left-mouse-button in the graphics area, away from the model, to ensure that no features are selected.

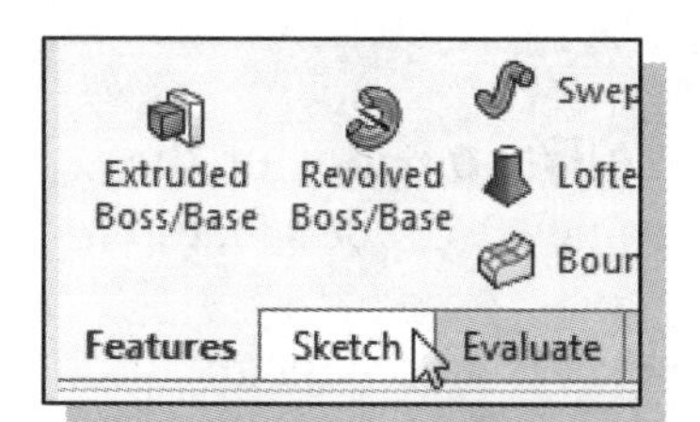

2. Select the **Sketch** tab on the *CommandManager* to display the *Sketch* toolbar.

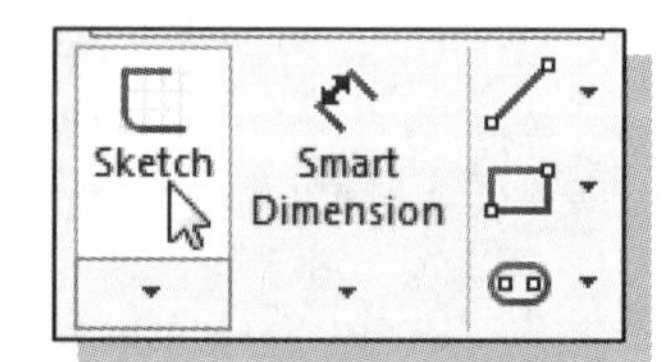

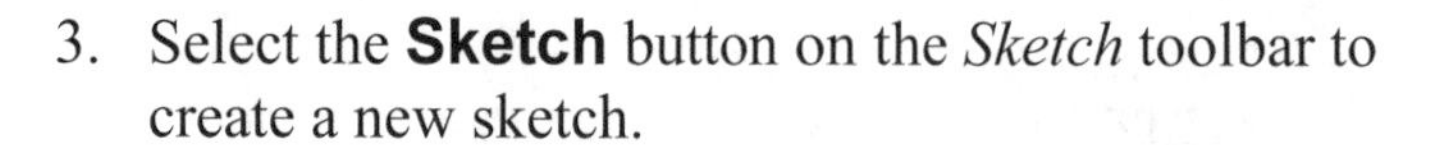

3. Select the **Sketch** button on the *Sketch* toolbar to create a new sketch.

4. In the Edit Sketch PropertyManager panel, the message "Select a plane on which to create a sketch for the entity" is displayed. SOLIDWORKS expects us to identify a planar surface where the 2D sketch of the next feature is to be created. Move the graphics cursor on the 3D part and notice that SOLIDWORKS will automatically highlight feasible planes and surfaces as the cursor is on top of the different surfaces.

5. Rotate the view using the up arrow key to display the bottom face of the solid model as shown below.

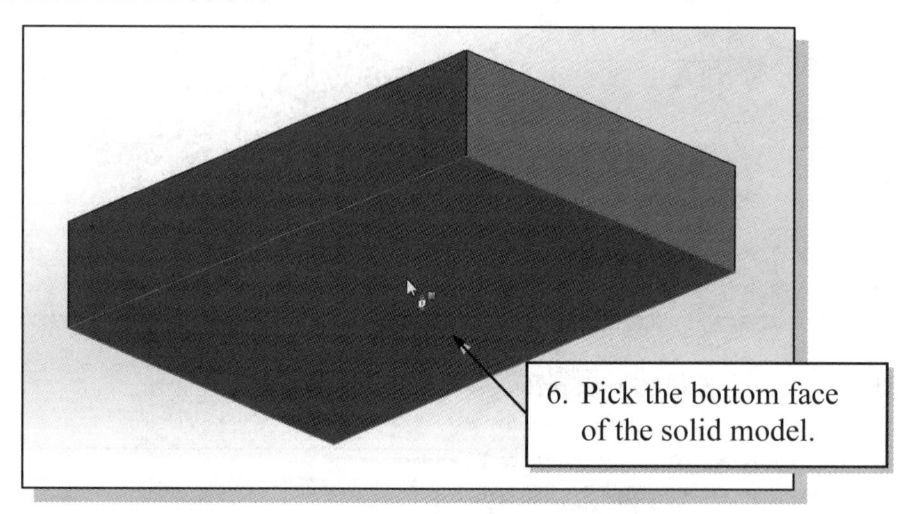

6. Pick the bottom face of the 3D model as the sketching plane.

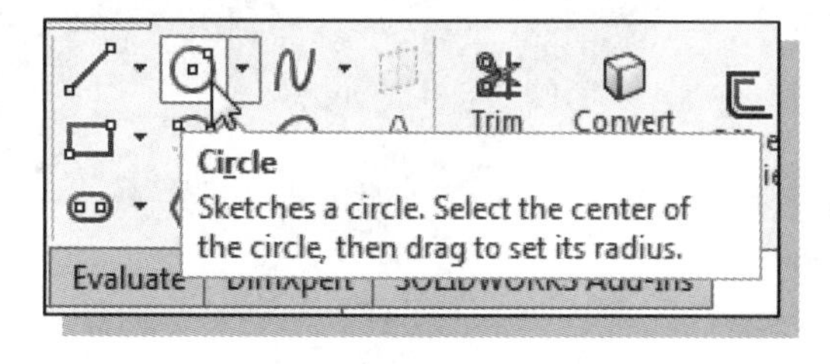

➢ Note that the sketching plane is aligned to the selected face.

7. Select the **Circle** command by clicking once with the left-mouse-button on the icon in the *Sketch* toolbar.

➢ We will align the center of the circle to the midpoint of the base feature.

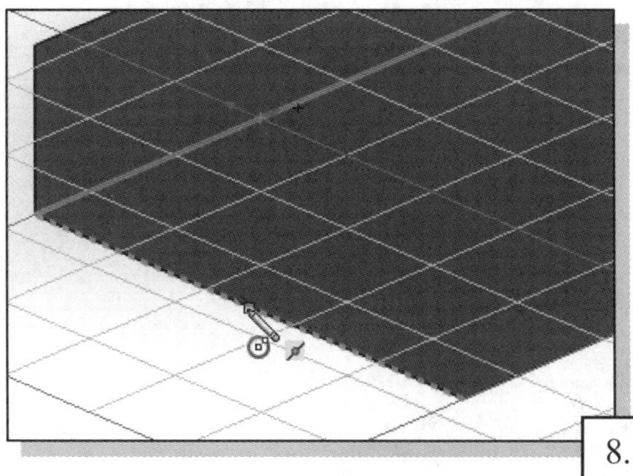

8. Move the cursor along the shorter edge of the base feature; when the **midpoint** is highlighted and the Midpoint sketch relation icon appears, click once with the left-mouse-button to select the midpoint.

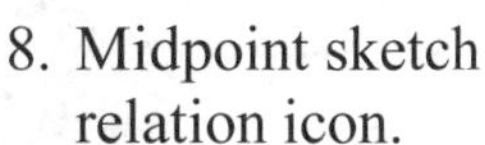
8. Midpoint sketch relation icon.

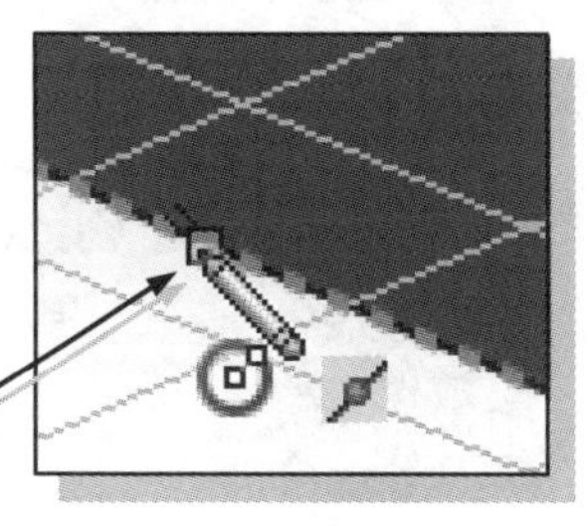

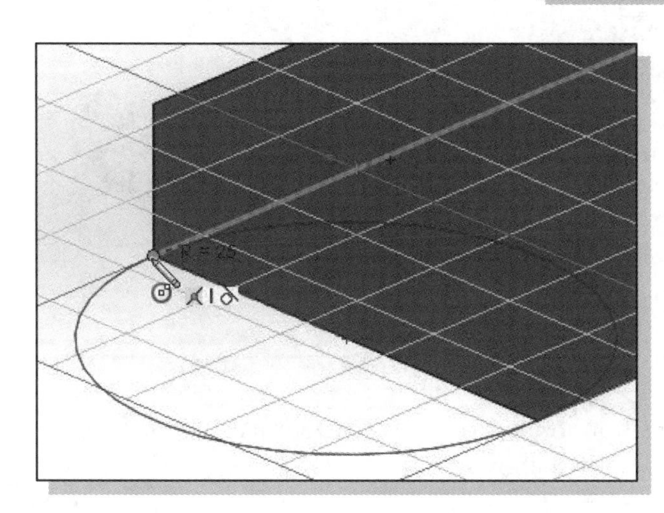

9. Move the cursor over the corner of the base feature; when the corner is highlighted, click once with the left-mouse-button to create a circle, as shown.

10. Press the [**Esc**] key once to end the Circle command.

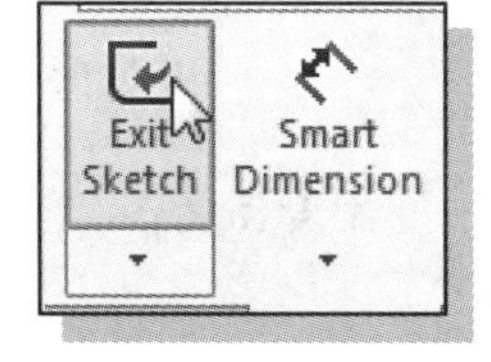

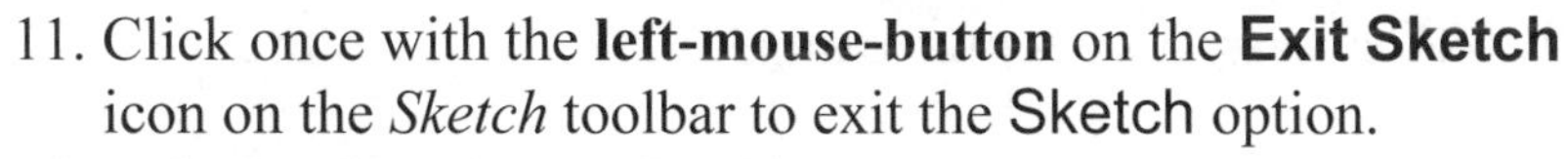

11. Click once with the **left-mouse-button** on the **Exit Sketch** icon on the *Sketch* toolbar to exit the Sketch option.

12. Select the **Isometric** icon in the *View Orientation* pull-down menu on the *Heads-up View* toolbar to reset the display to the isometric view.

13. Notice the new sketch – **Sketch2** – is highlighted in the *Design Tree*, indicating that the sketch is currently 'selected'.

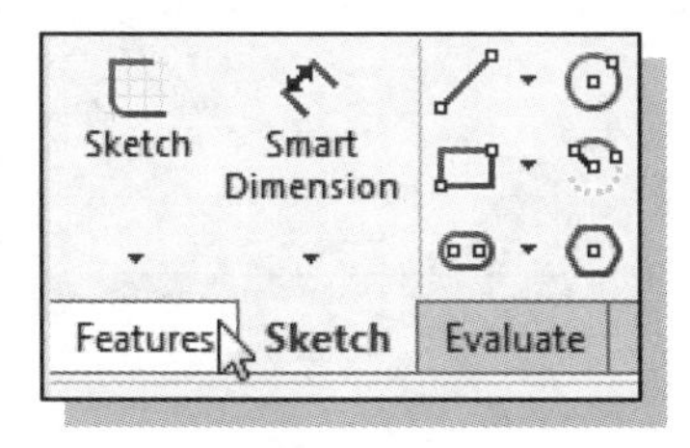

14. Select the **Features** tab on the *CommandManager* to display the *Features* toolbar.

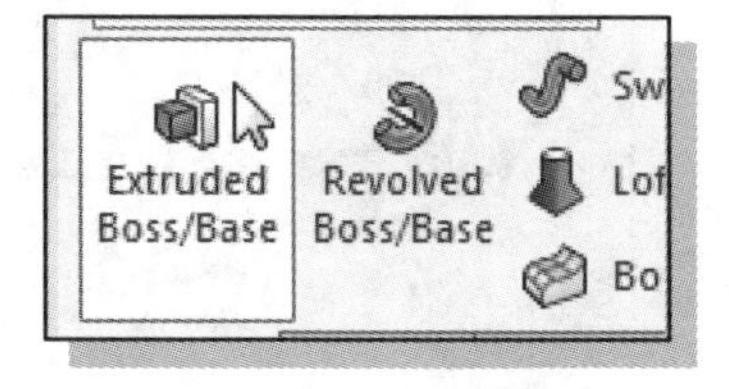

15. Select the **Extruded Boss/Base** command on the *Features* toolbar.

➢ Notice that the sketch region is automatically selected as the extrusion profile.

16. In the *Extrude PropertyManager* panel, enter **40** as the extrusion distance as shown below. Confirm the **Merge result** checkbox is checked.

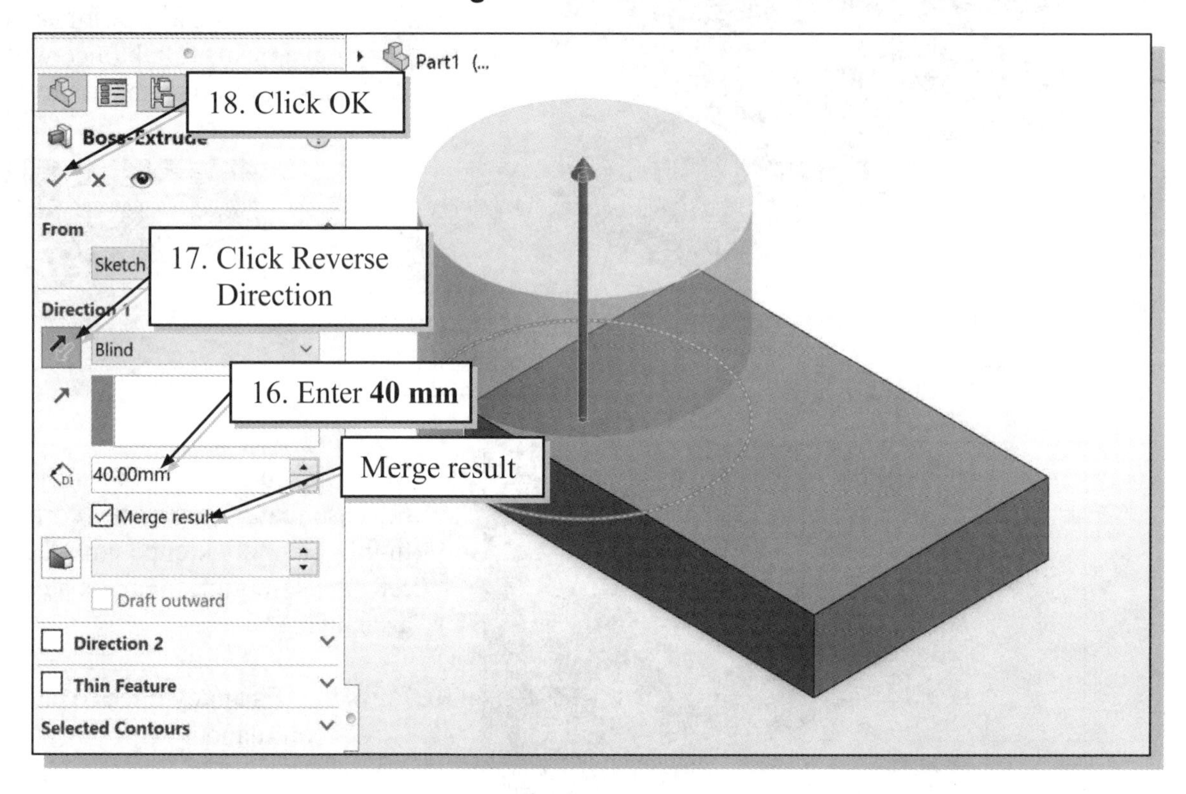

17. Click the **Reverse Direction** button in the *PropertyManager* as shown. The extrude preview should appear as shown above.

18. Click on the **OK** button to proceed with creating the extruded feature.

- The two features are joined together into one solid part; the *CSG-Union* operation was performed. This is the effect of the *Merge result* option.

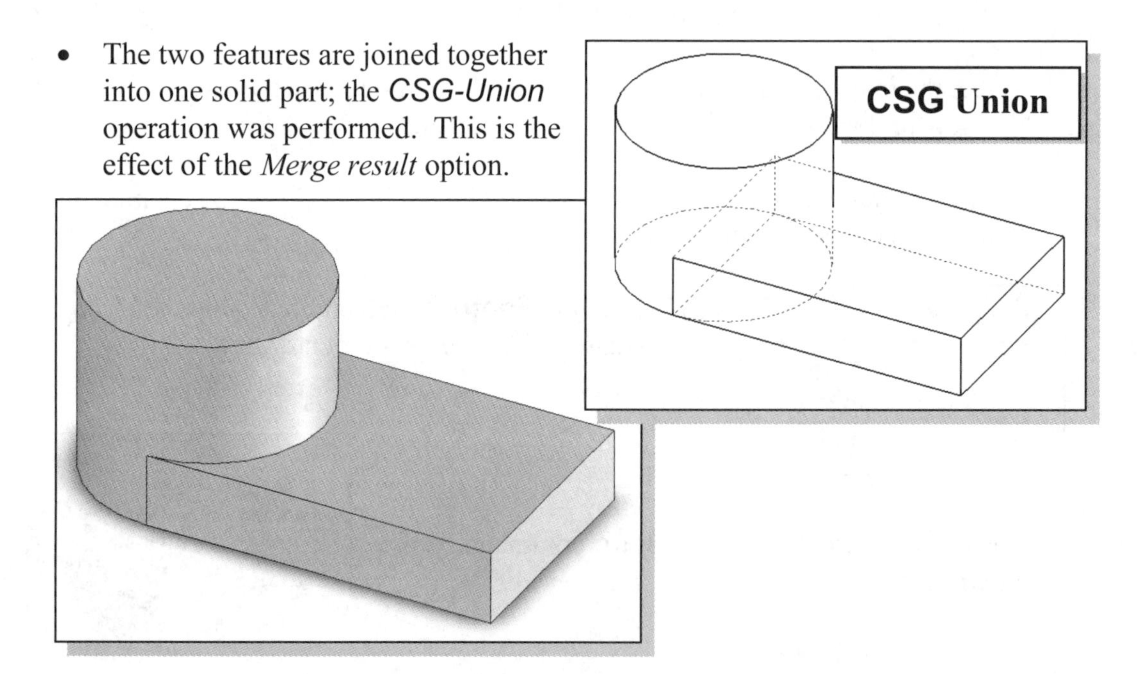

Creating an Extruded Cut Feature

We will create a circular cut as the next solid feature of the design. We will align the sketch plane to the top of the last cylinder feature.

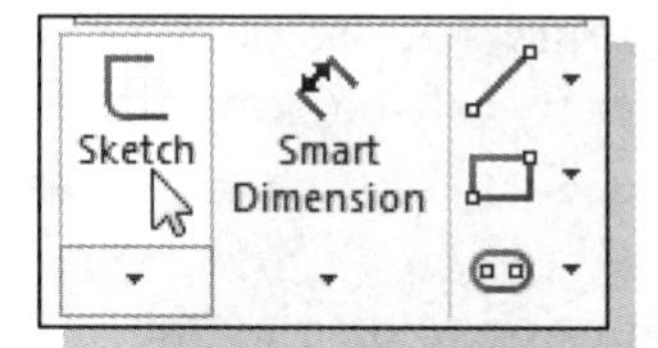

1. Select the **Sketch** tab on the *CommandManager* to display the *Sketch* toolbar.

2. Select the **Sketch** button on the *Sketch* toolbar to create a new sketch.

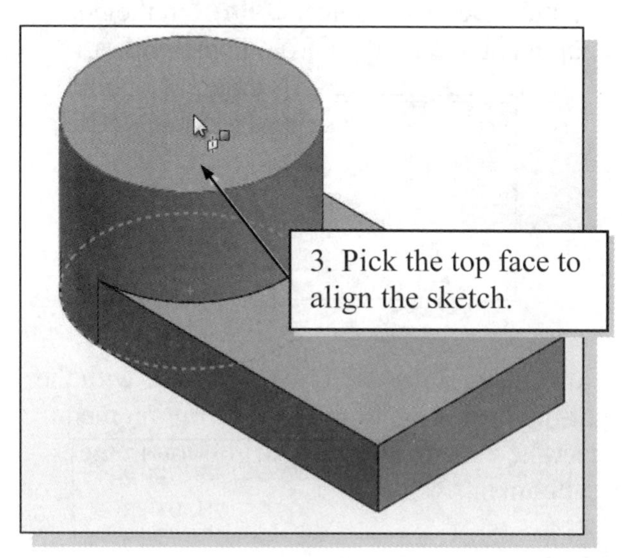

3. Pick the top face of the cylinder as shown.

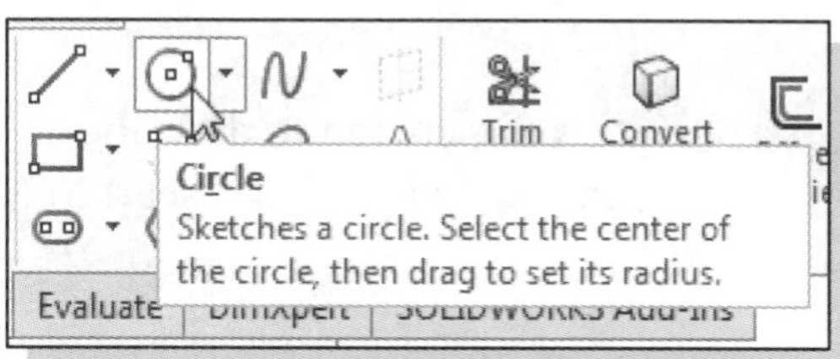

4. Select the **Circle** command by clicking once with the left-mouse-button on the icon in the *Sketch* toolbar.

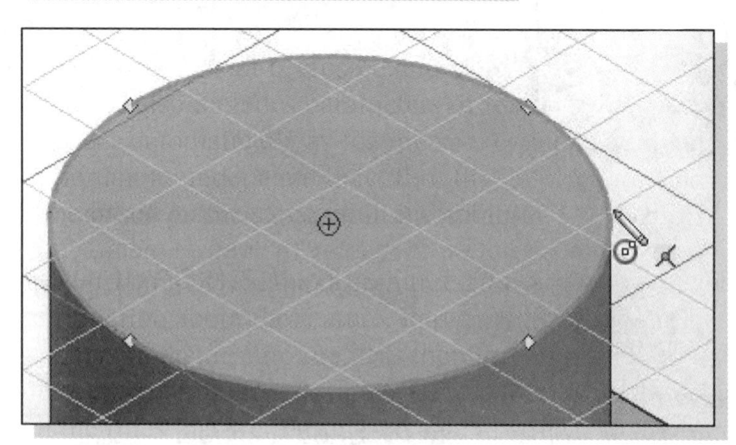

5. Move the cursor over the circular edge of the top face as shown. (Do not click.) Notice the center and quadrant marks appear on the circle.

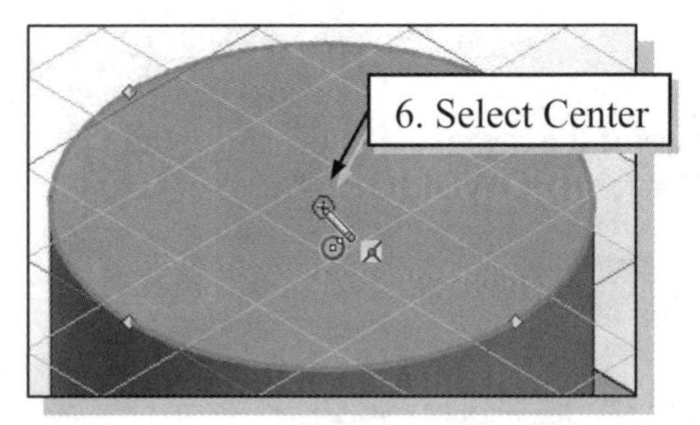

6. Select the **Center** point of the top face of the 3D model by left-clicking once on the icon as shown.

7. Sketch a circle of arbitrary size inside the top face of the cylinder by left-clicking as shown.

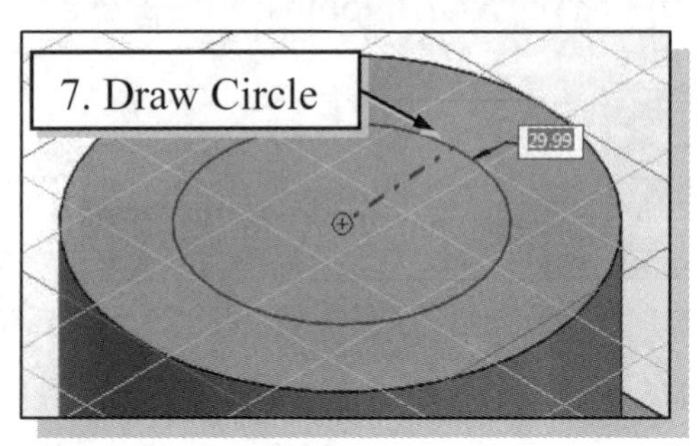

8. Use the right-mouse-button to display the option menu and select **Select** in the pop-up menu to end the Circle command.

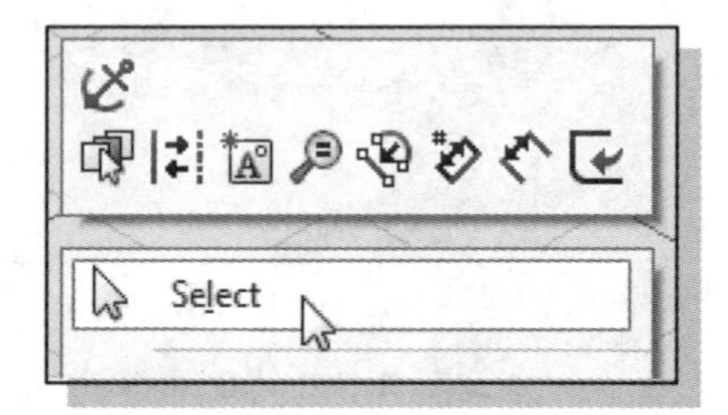

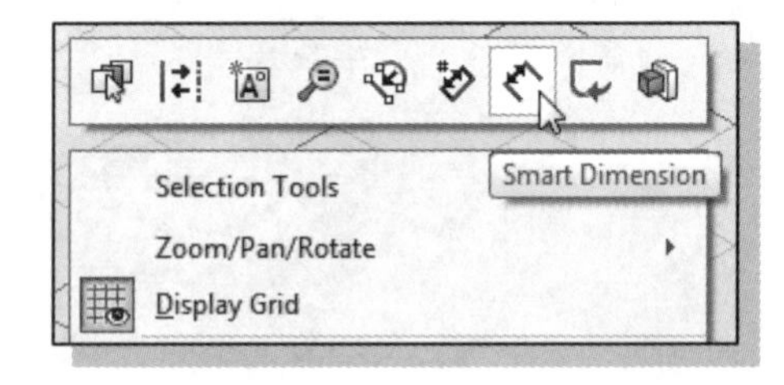

9. Inside the graphics window, click once with the right-mouse-button to display the option menu. Select the **Smart Dimension** option in the pop-up menu.

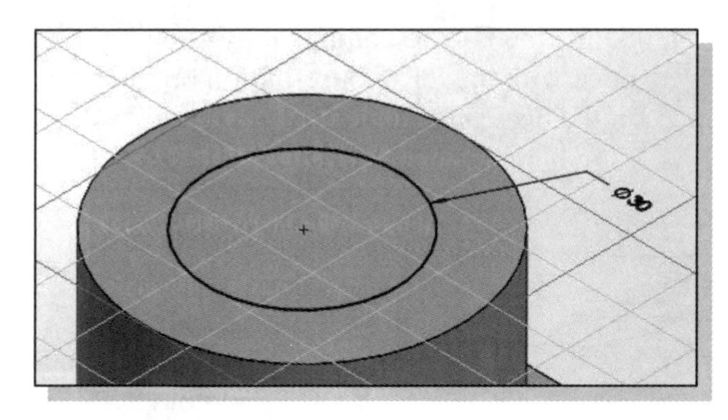

10. Create a dimension to describe the size of the circle and set it to **30mm**.

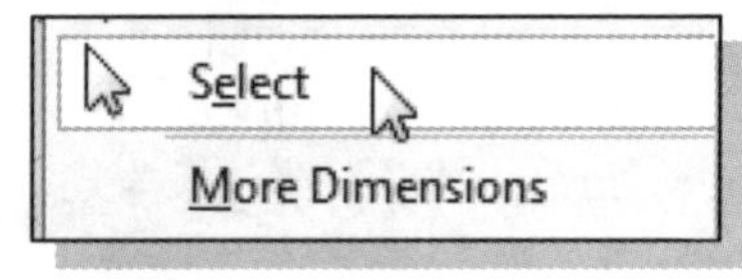

11. Inside the graphics window, click once with the right-mouse-button to display the option menu. Select **Select** in the pop-up menu to end the Smart Dimension command.

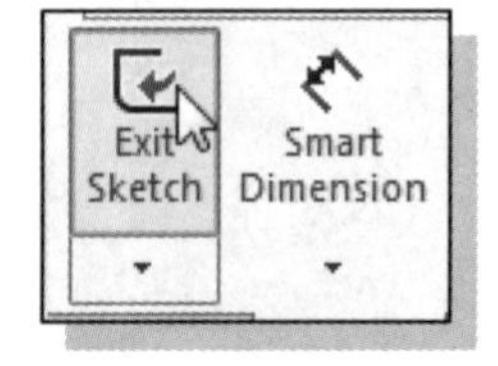

12. Click once with the **left-mouse-button** on the **Exit Sketch** icon on the *Sketch* toolbar to exit the Sketch option.

13. Select the **Features** tab on the *CommandManager*.

14. In the *Features* toolbar, select the **Extruded Cut** command by clicking once with the left-mouse-button on the icon.

➢ The *Cut-Extrude PropertyManager* is displayed in the left panel. Notice that the sketch region (the circle) is automatically selected as the extrusion profile.

15. In the *Cut-Extrude PropertyManager* panel, click the arrow to reveal the pull-down options for the *End Condition* (the default end condition is 'Blind') and select **Through All** as shown.

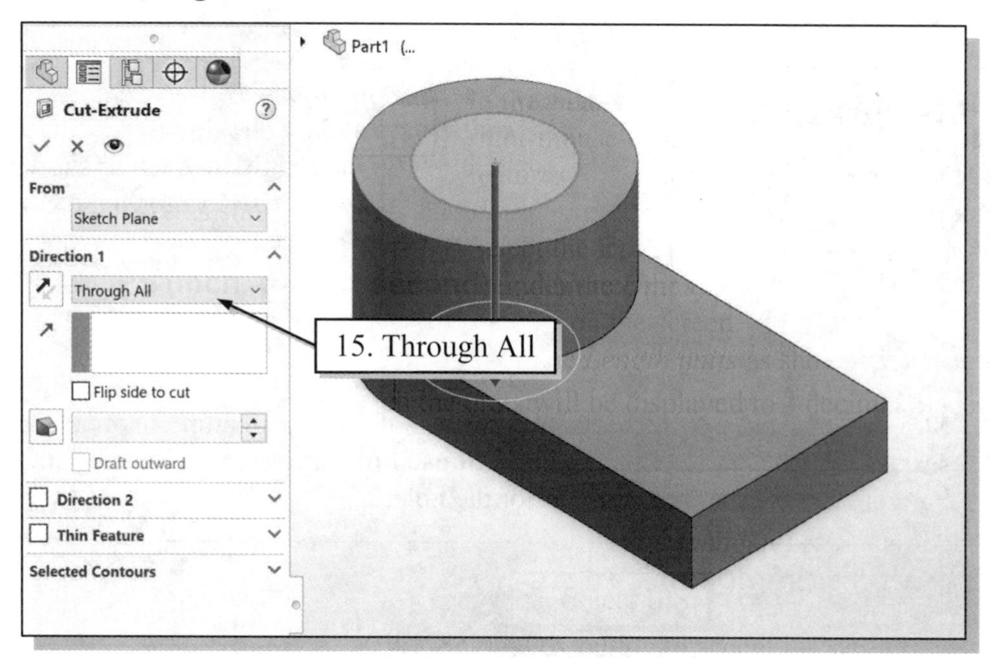

16. Click the **OK** button (green check mark) in the *Cut-Extrude PropertyManager* panel.

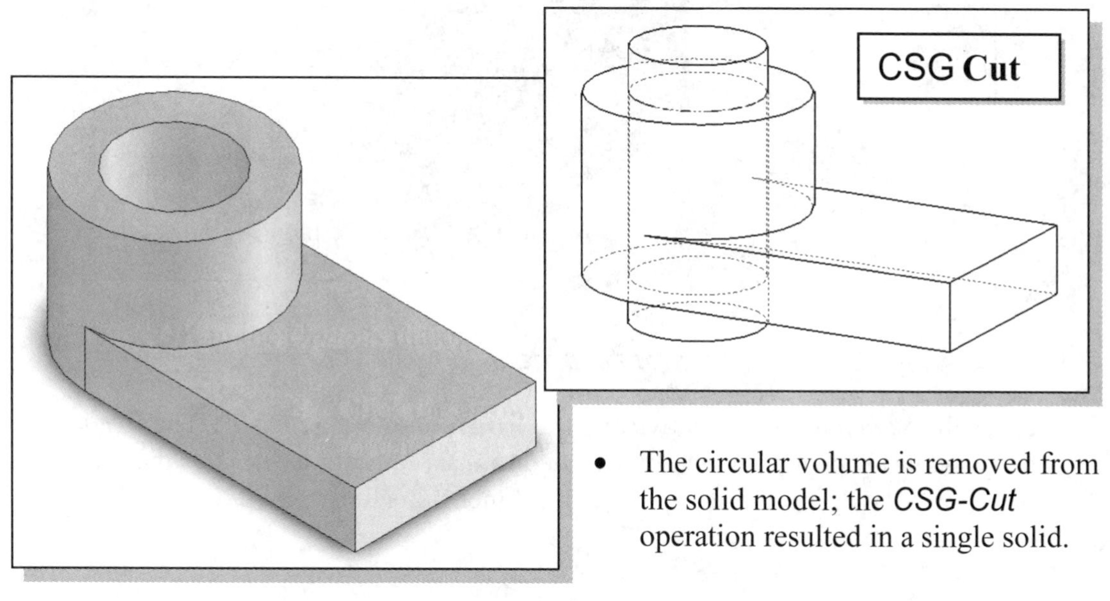

- The circular volume is removed from the solid model; the *CSG-Cut* operation resulted in a single solid.

Creating a Hole with the *Hole Wizard*

The last cut feature we created is a *sketched feature*, where we created a rough sketch and performed an extrusion operation. We can also create a hole using the SOLIDWORKS Hole Wizard. With the Hole Wizard, the hole feature does not need a sketch and can be created automatically. Holes, fillets, chamfers, and shells are all examples of features that do not require a sketch.

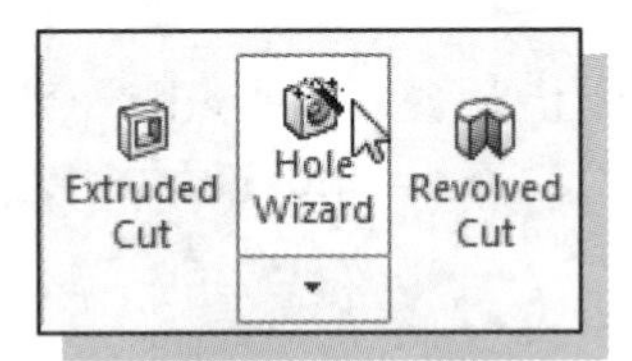

1. In the *Features* toolbar, select the **Hole Wizard** command by clicking once with the left-mouse-button on the icon as shown.

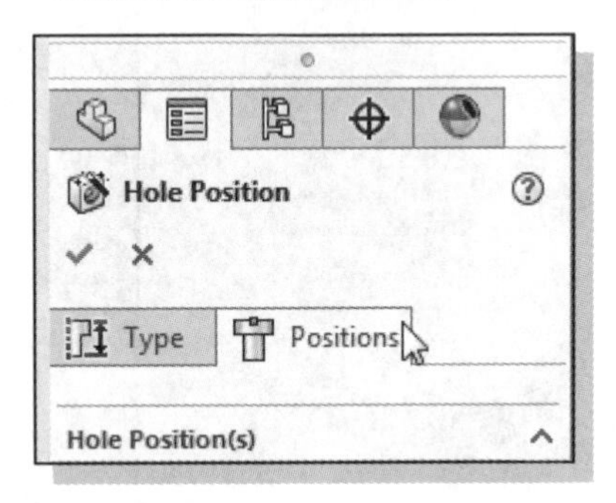

2. In the *Hole Specification PropertyManager*, select the **Positions** panel by clicking once with the left-mouse-button on the **Positions** tab as shown. The Positions tab allows you to locate the hole on a planar or non-planar face.

3. Move the cursor over the horizontal surface of the base feature. Notice that the surface is highlighted. Click the **left mouse button** to select a location inside the horizontal surface as the position for the hole.

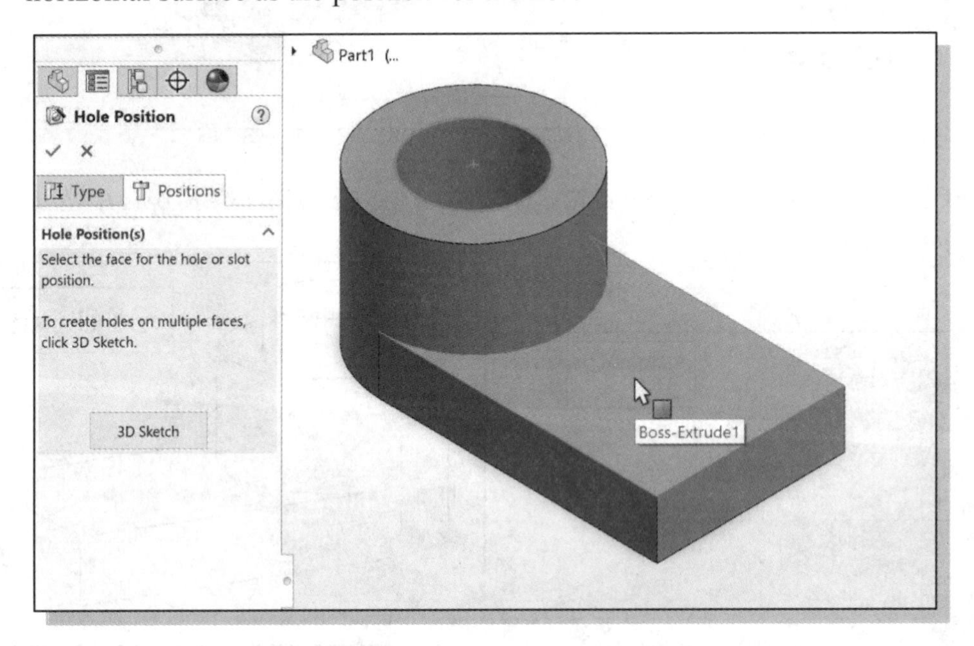

➢ Notice the *Sketch* toolbar is active and the **Point** button is selected. The Point command has been automatically executed to insert a point to serve as the center for the hole. We will insert the point and use dimensions to locate it.

4. Move the cursor to a location on the horizontal surface of the base feature as shown and click the **left mouse button** to insert the point. (Do not be concerned if your hole preview does not have the same diameter as the one shown here.)

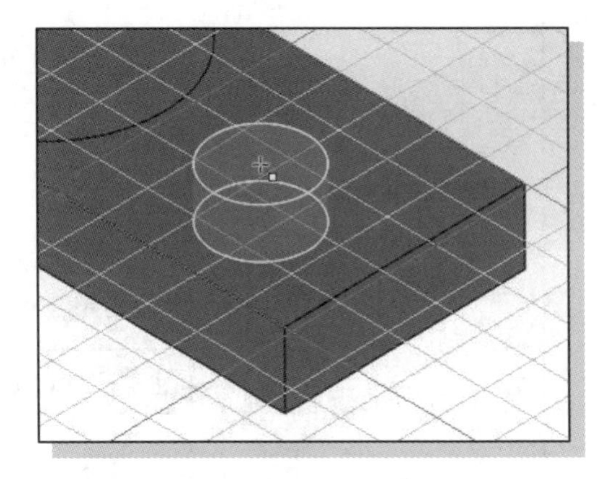

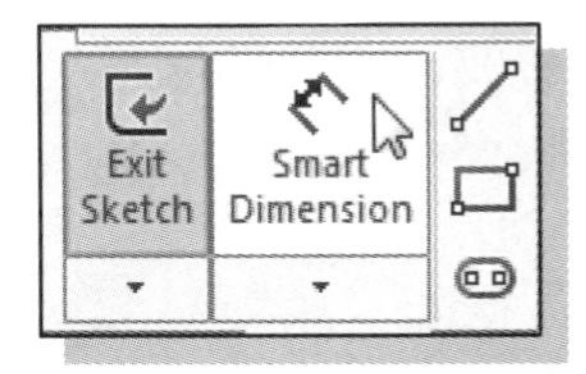

5. Select the **Smart Dimension** command by clicking once with the **left-mouse-button** on the icon in the *Sketch* toolbar.

6. Pick the center point by clicking once with the left-mouse-button as shown.

7. Pick the right-edge of the top face of the base feature by clicking once with the left-mouse-button as shown.

8. Select a location for the dimension by clicking once with the left-mouse-button as shown.

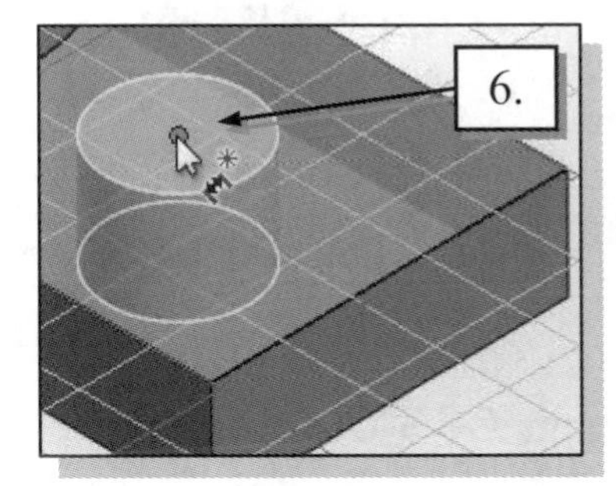

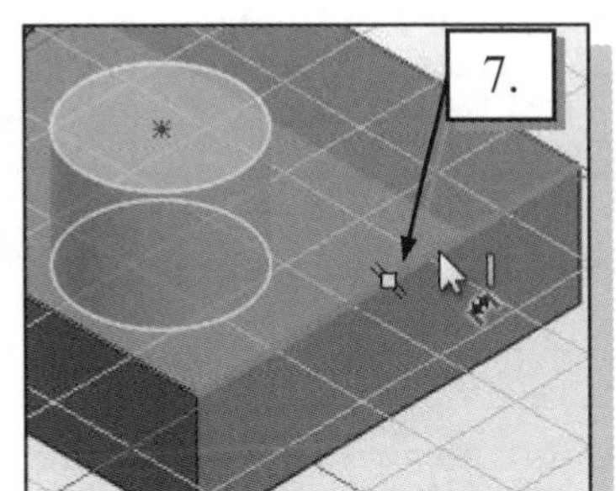

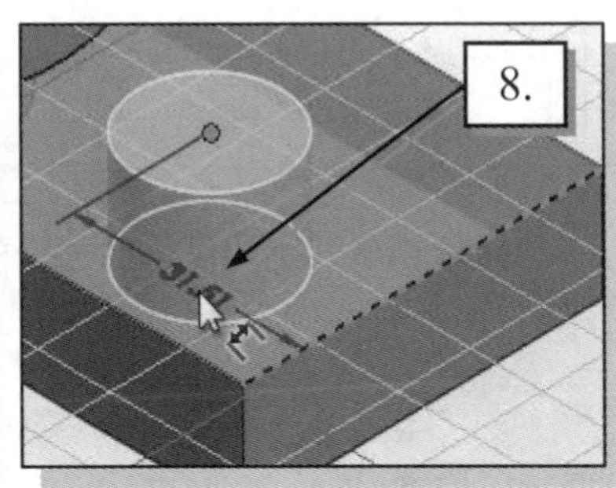

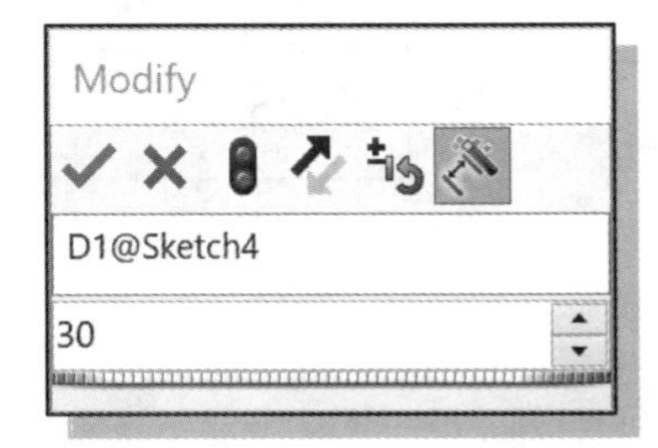

9. Enter **30** in the *Modify* dialog box, and select **OK**.

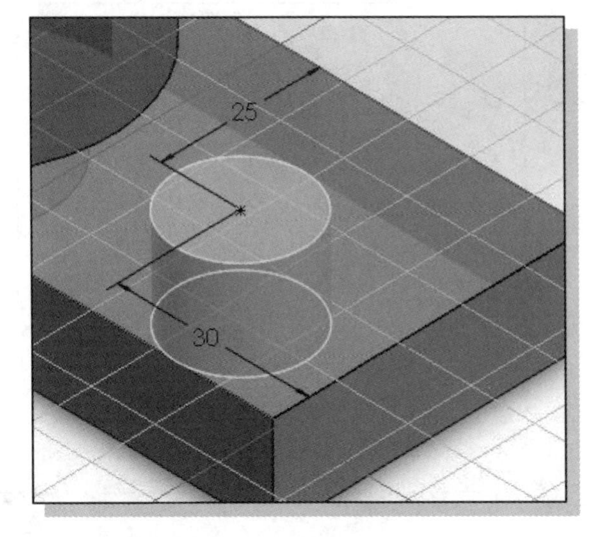

10. On your own, enter the additional dimension as shown (the dimension is **25** mm).

11. Press the [**Esc**] key once to end the Smart Dimension command.

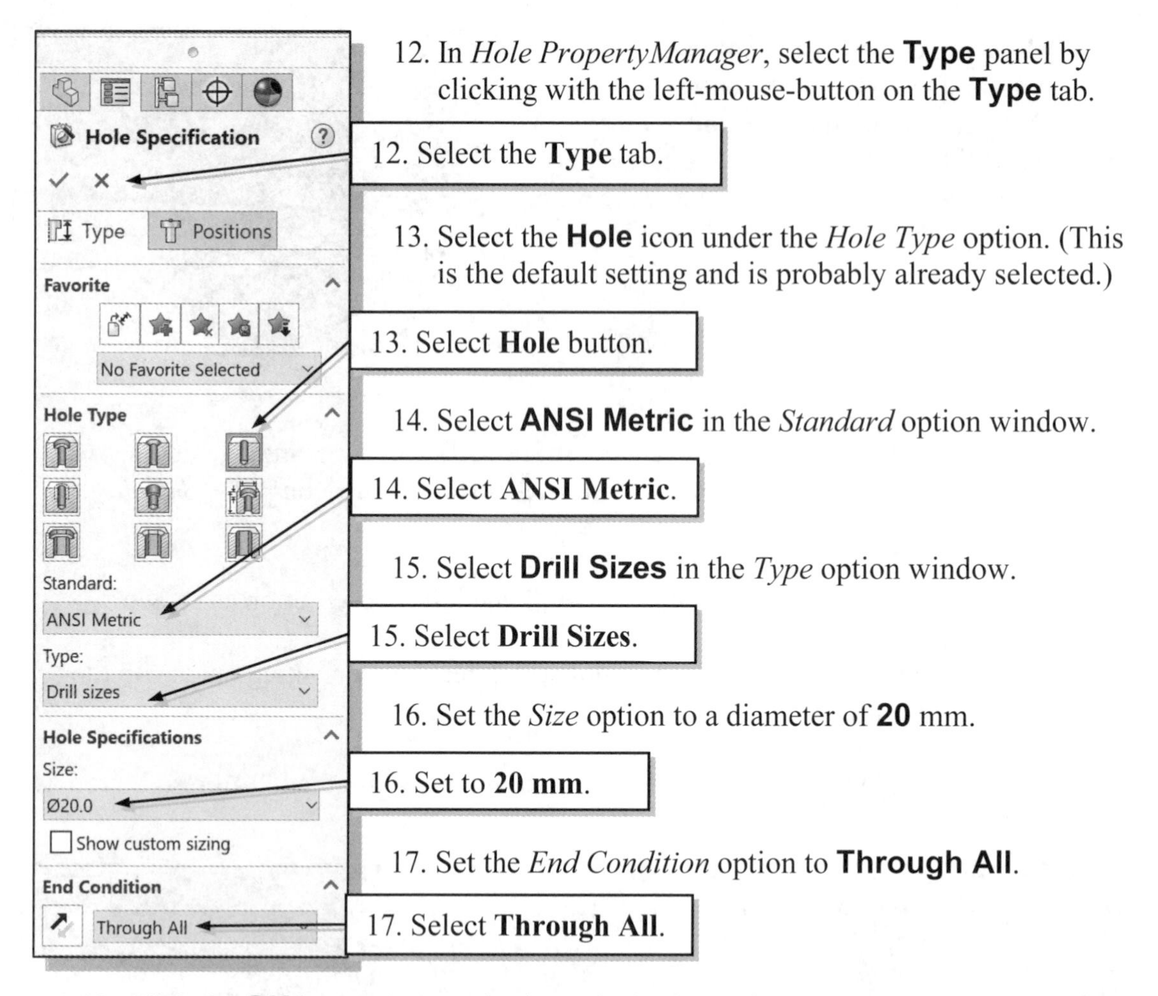

12. In *Hole PropertyManager*, select the **Type** panel by clicking with the left-mouse-button on the **Type** tab.

13. Select the **Hole** icon under the *Hole Type* option. (This is the default setting and is probably already selected.)

14. Select **ANSI Metric** in the *Standard* option window.

15. Select **Drill Sizes** in the *Type* option window.

16. Set the *Size* option to a diameter of **20** mm.

17. Set the *End Condition* option to **Through All**.

18. Click the **OK** button (green check mark) in the *Hole PropertyManager* to proceed with the *Hole* feature.

- The circular volume is removed from the solid model; the *CSG-Cut* operation resulted in a single solid.

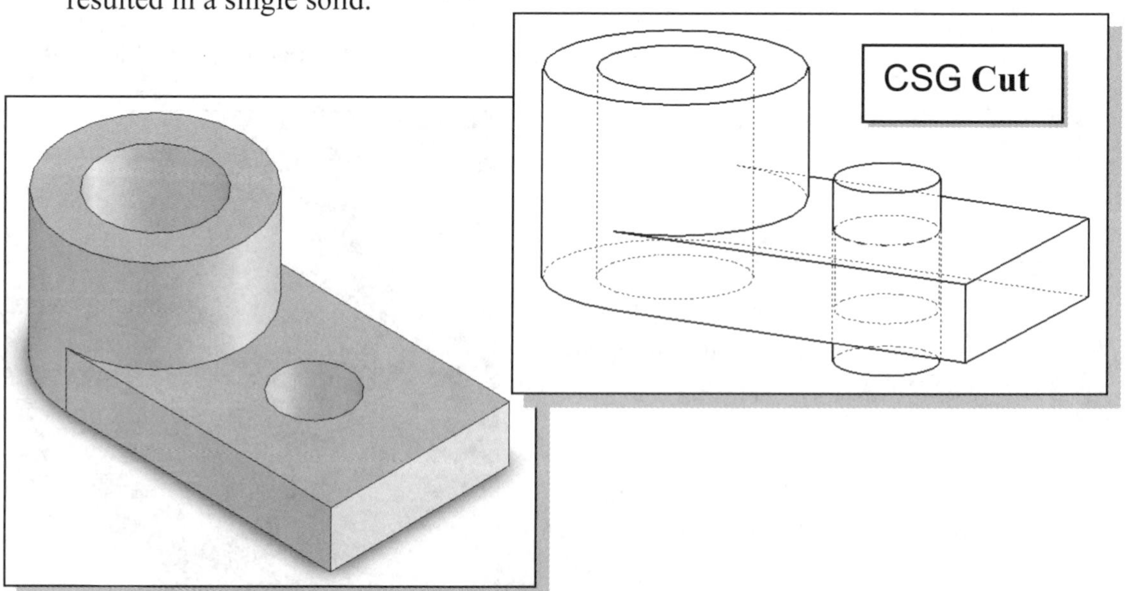

Creating a Rectangular Extruded Cut Feature

- Next create a rectangular cut as the last solid feature of the ***Locator***.

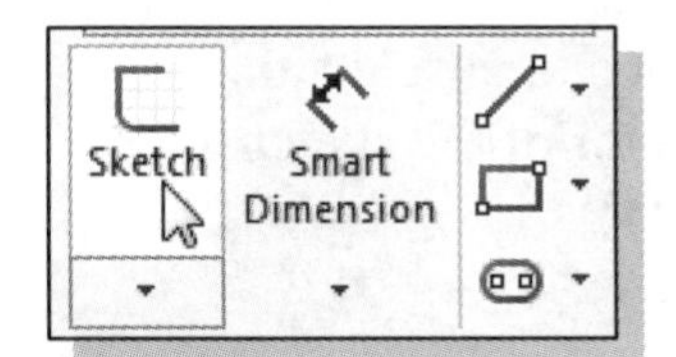

1. Select the **Sketch** tab on the *CommandManager* to display the *Sketch* toolbar.

2. Select the **Sketch** button on the *Sketch* toolbar to create a new sketch.

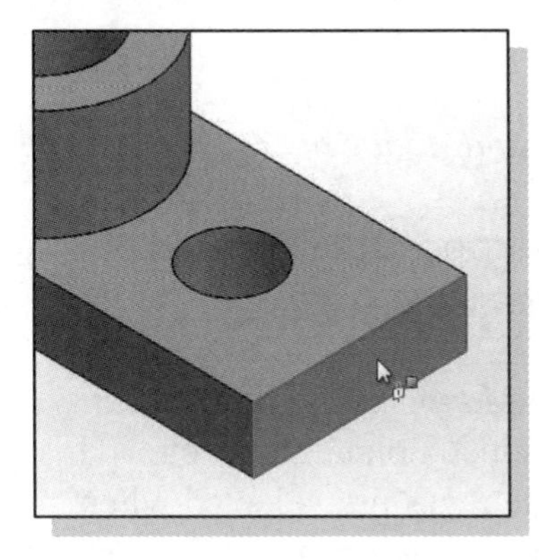

3. Pick the right face of the base feature as shown.

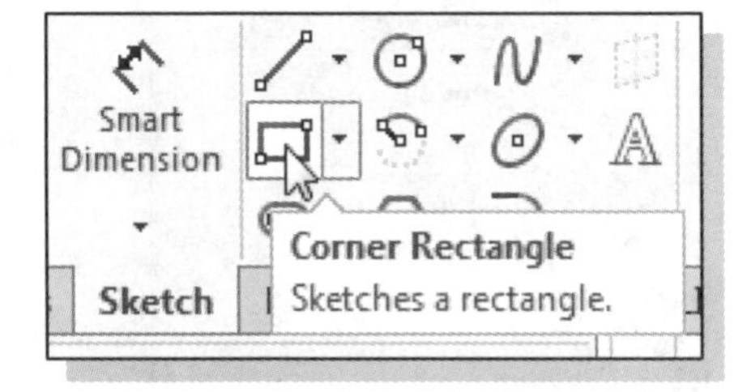

4. Select the **Corner Rectangle** command by clicking once with the left-mouse-button on the icon in the *Sketch* toolbar.

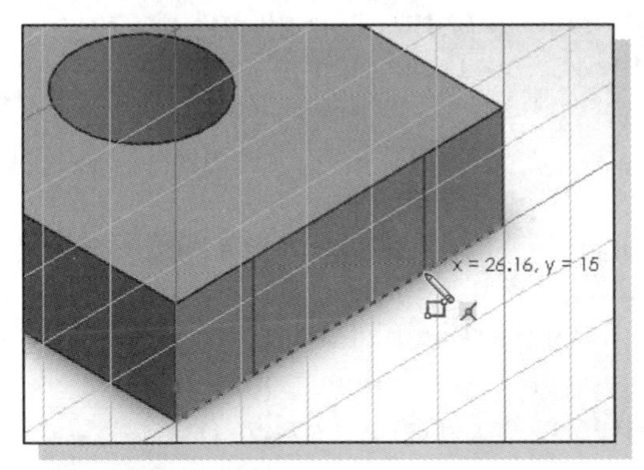

5. Create a rectangle that is aligned to the top and bottom edges of the base feature as shown.

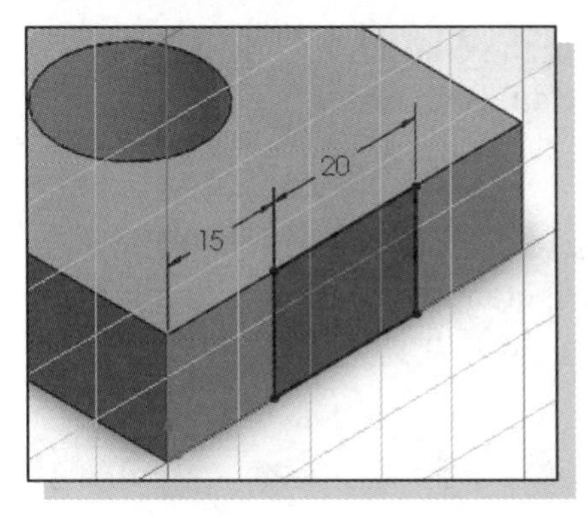

6. On your own, create and modify the two dimensions as shown. The dimensions are **15** mm and **20** mm.

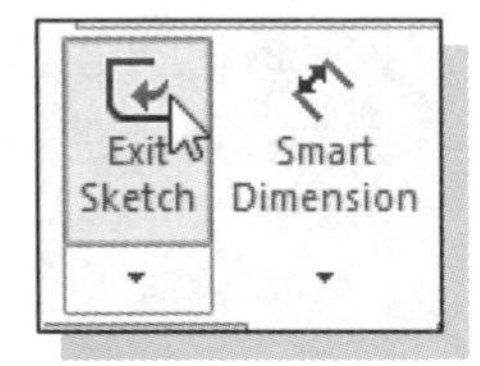

7. Click once with the **left-mouse-button** on the **Exit Sketch** icon on the *Sketch* toolbar to exit Sketch option.

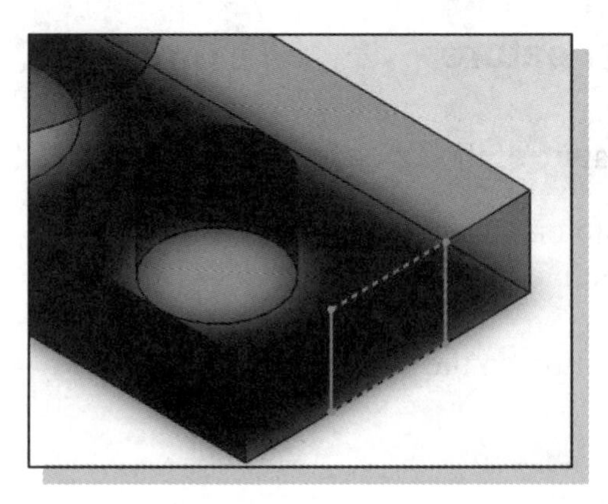

8. Notice the sketch is selected. (It is highlighted in the *FeatureManager Design Tree* and in the graphics area.)

9. Press the **Esc** key once to unselect the sketch. Notice the color of the sketch changes to grey.

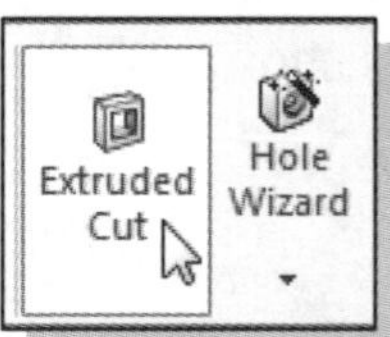

10. Select the **Features** tab on the *CommandManager*.

11. In the *Features* toolbar, select the **Extruded Cut** command by clicking once with the left-mouse-button on the icon.

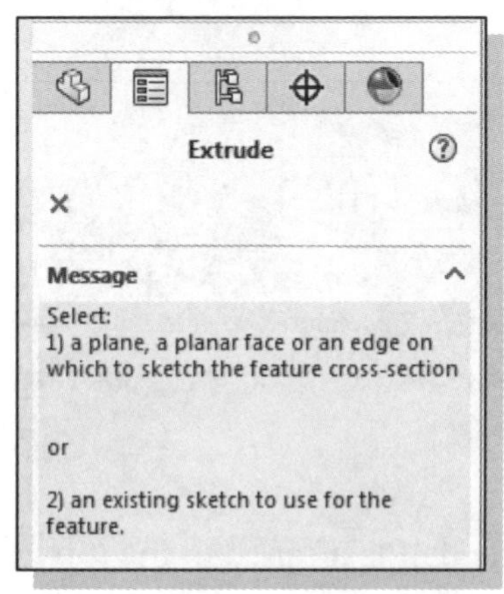

- Notice that in the *PropertyManager* and on the *Status Bar* you are prompted to select a plane, face, etc. This is because no plane or sketch was pre-selected when the Extrude-Cut command was executed.

- When creating an extruded feature, SOLIDWORKS allows the user to pre-select a sketch prior to executing the *Extrude* command or to first execute the *Extrude* command and then select the sketch.

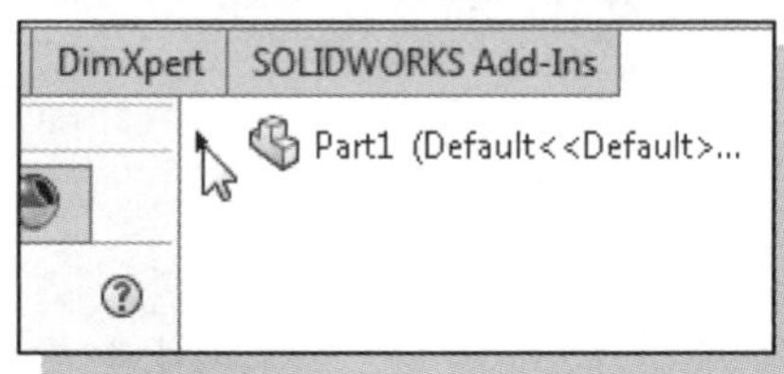

12. We will select the sketch using the *Design Tree* inside the graphics area. Move the cursor over the cross next to the **Part** icon at the upper left corner of the graphics area. Click once with the **left-mouse-button** to expand the *Design Tree*.

13. In the *Design Tree*, select the Sketch by clicking once with the **left-mouse-button** as shown.

- Notice that the sketch is selected for use in the *Cut-Extrude* feature and the sketch region (the rectangle) is automatically selected as the extrusion profile.

14. In the *Extrude PropertyManager* panel, click the arrow to reveal the pull-down options for the *End Condition* (the default end condition is Blind) and select **Up To Next** as shown.

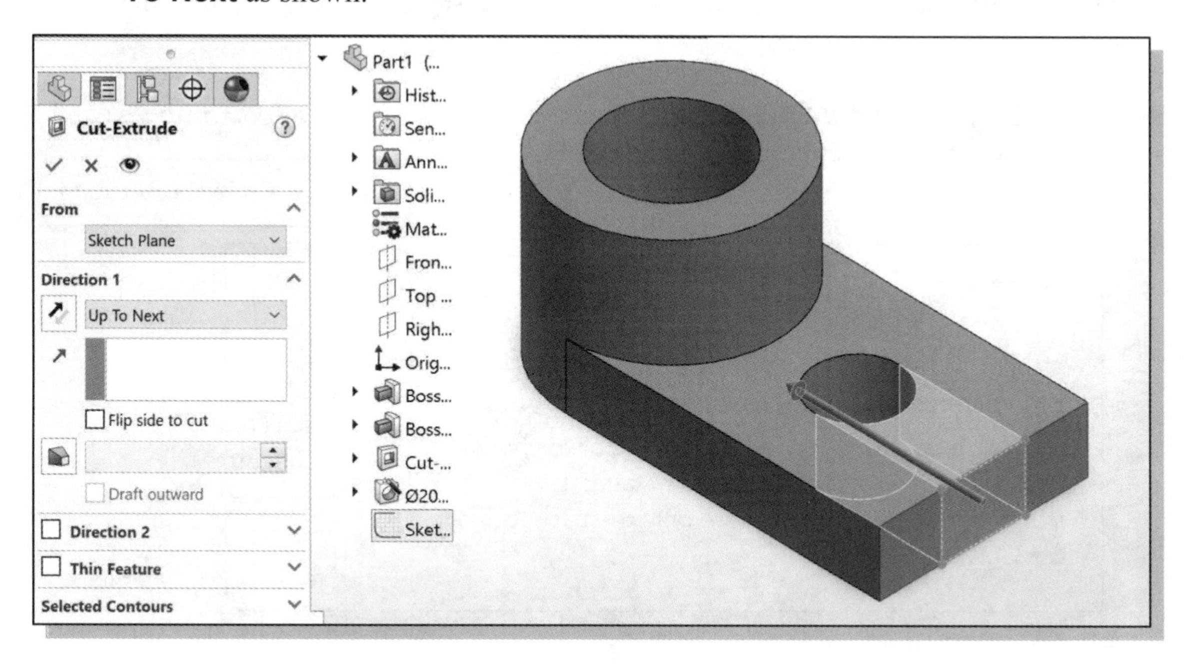

15. Click the **OK** button (green check mark) in the *Cut-Extrude PropertyManager* panel.

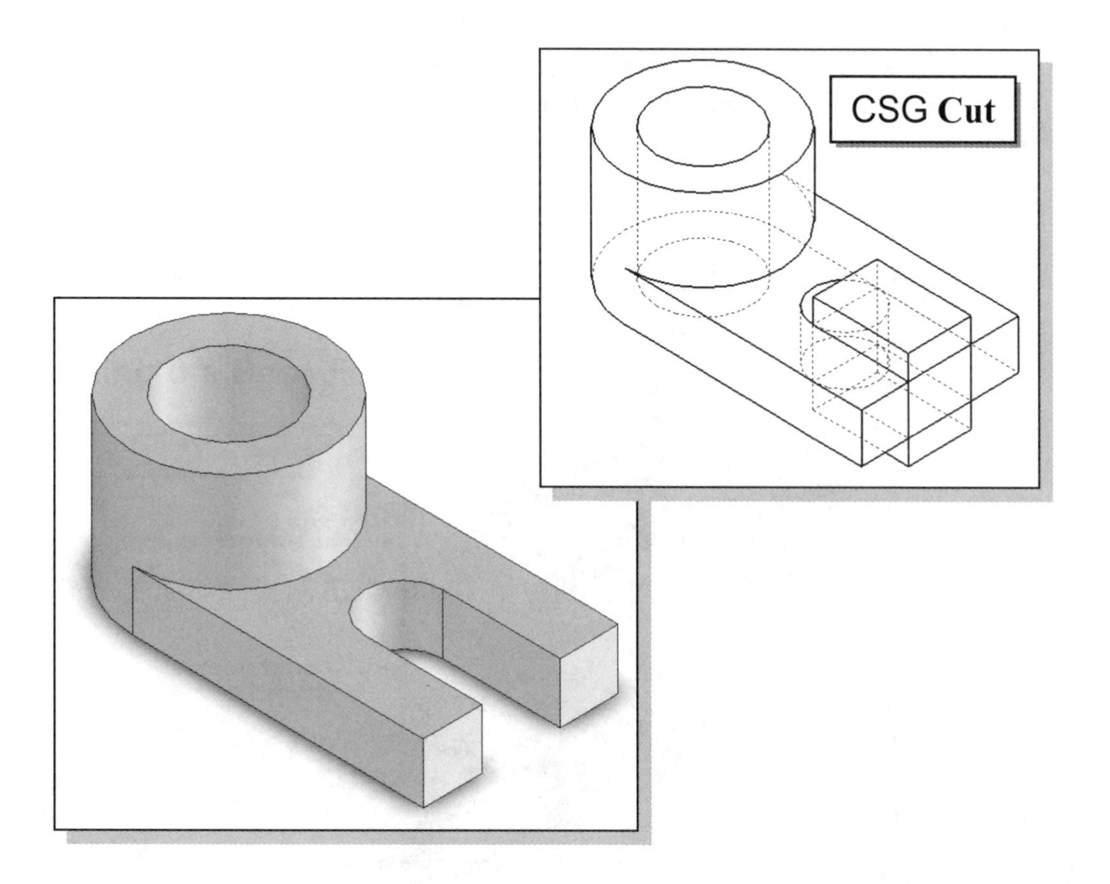

Using the View Selector

The **View Selector** provides an in-context method to select standard and non-standard views.

1. Click on the **View Orientation** icon on the *Heads-Up View Toolbar* to reveal the view orientation options.

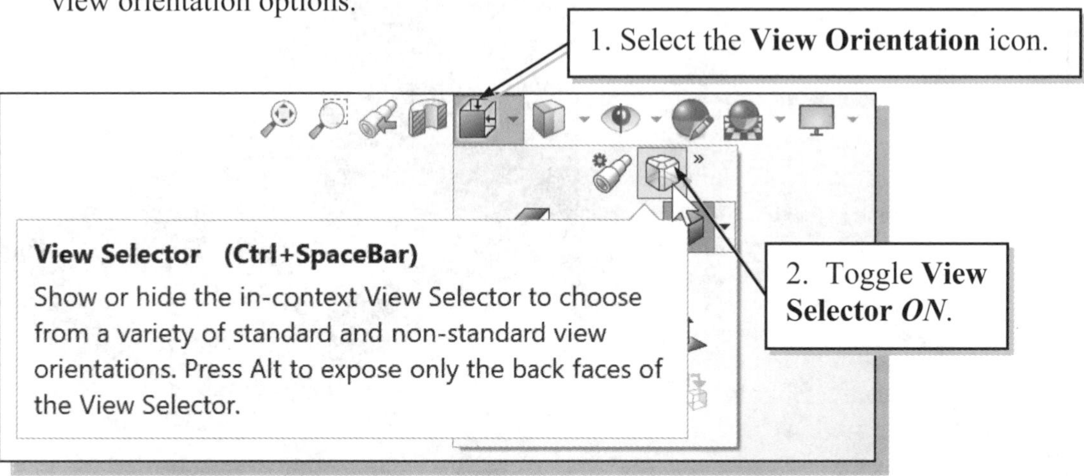

2. **Toggle** the **View Selector** ***ON*** by left-clicking on the **View Selector** icon in the *Orientation* dialog *box*.

➢ The View Selector appears in the graphics area. The View Selector provides an in-context method to select right, left, front, back, top, and isometric views of your model, as well as additional standard and isometric views.

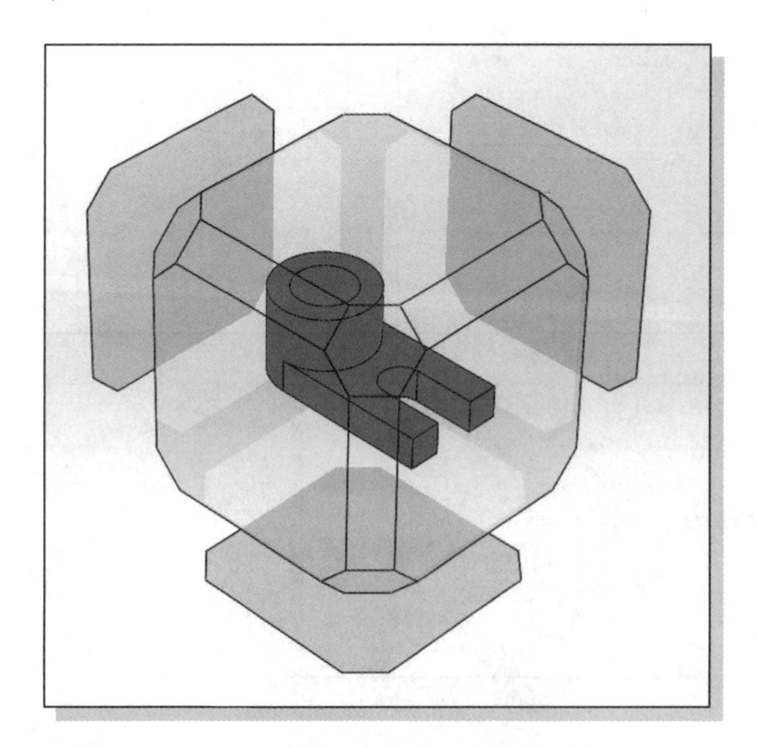

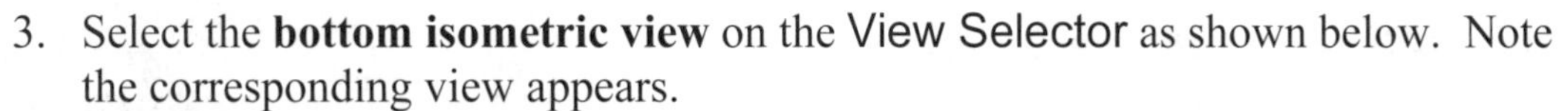

3. Select the **bottom isometric view** on the View Selector as shown below. Note the corresponding view appears.

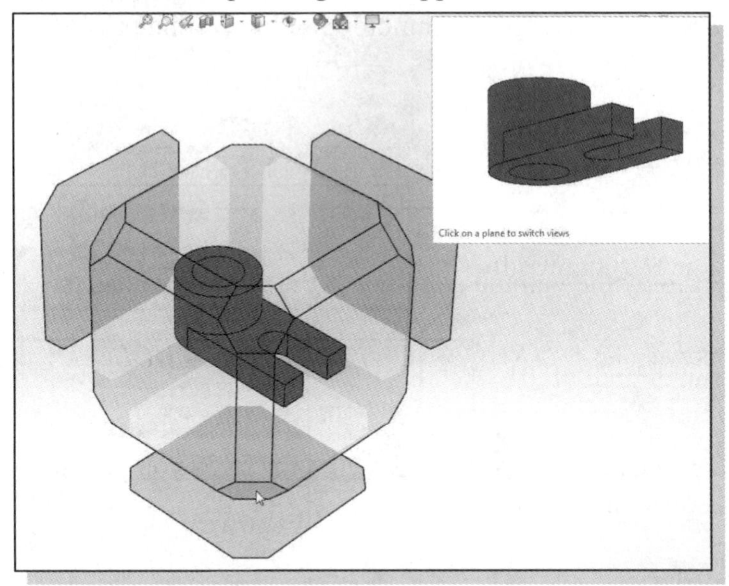

4. On your own, experiment with selecting other views using the View Selector. Notice that with the View Selector toggled ***ON***, it automatically appears when the View Orientation option is selected on the *Heads-Up View* toolbar.

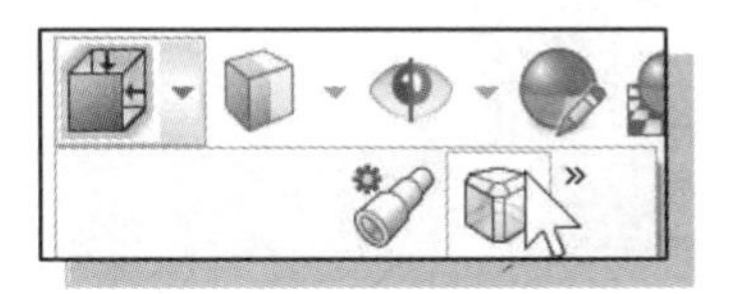

5. **Toggle** the View Selector ***OFF*** by left-clicking on the **View Selector** icon in the *Orientation dialog box*.

6. Hold down the **[Ctrl]** button and press the **[Spacebar]**. Notice the View Selector appears. This is an alternate method to activate the View Selector.

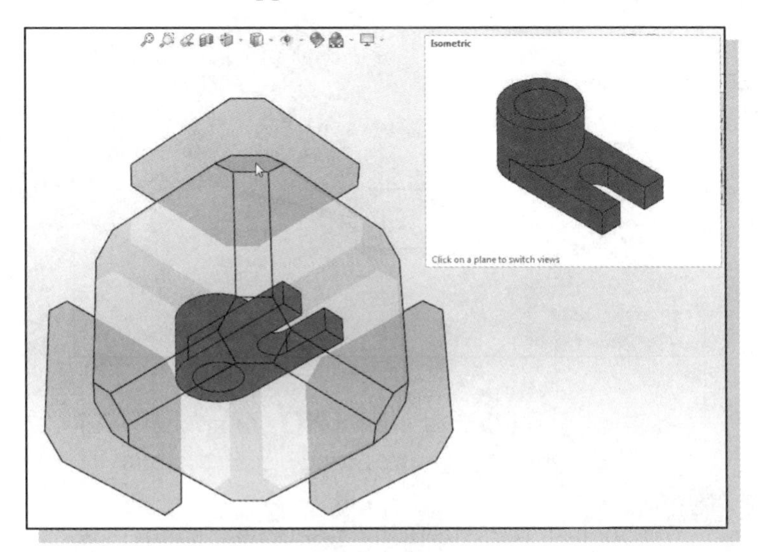

7. Select the isometric view as shown.

8. Save the model with the name *Locator*.

Questions:

1. List and describe three basic *Boolean operations* commonly used in computer geometric modeling software.

2. What is a *primitive solid*?

3. What does *CSG* stand for?

4. Which *Boolean operation* keeps only the volume common to the two solid objects?

5. What is the main difference between an *EXTRUDED CUT feature* and a *HOLE feature* in SOLIDWORKS?

6. Using the CSG concepts, create *Binary Tree* sketches showing the steps you plan to use to create the two models shown on page 3-30:

Ex.3)

Ex.4)

Exercises:

1. **Latch Clip** (Dimensions are in inches. Thickness: **0.25** inches.)

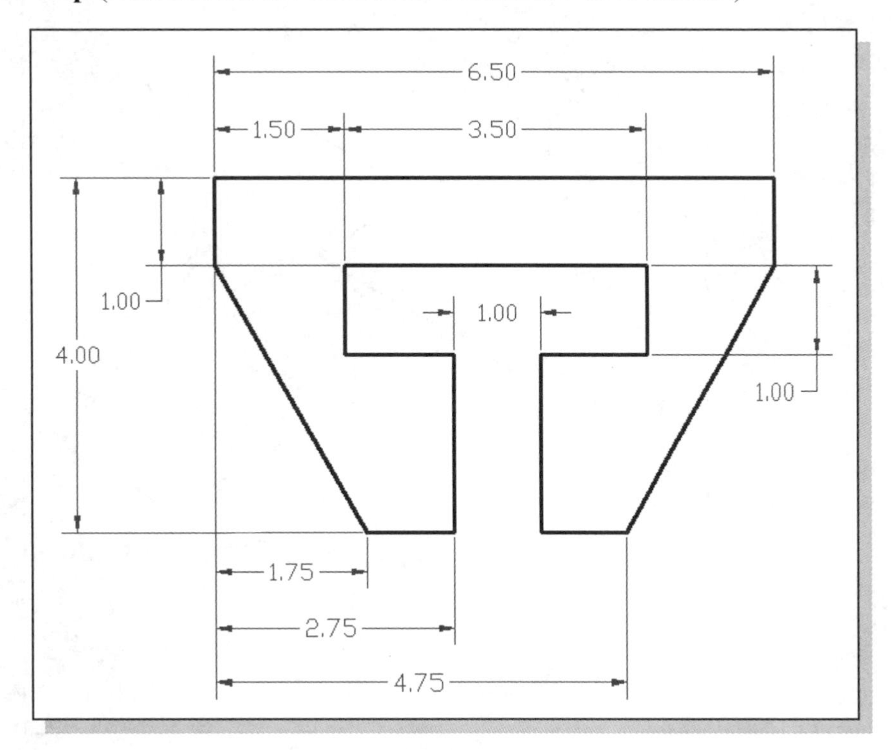

2. **Guide Plate** (Dimensions are in inches. Thickness: **0.25** inches. Boss height **0.125** inches.)

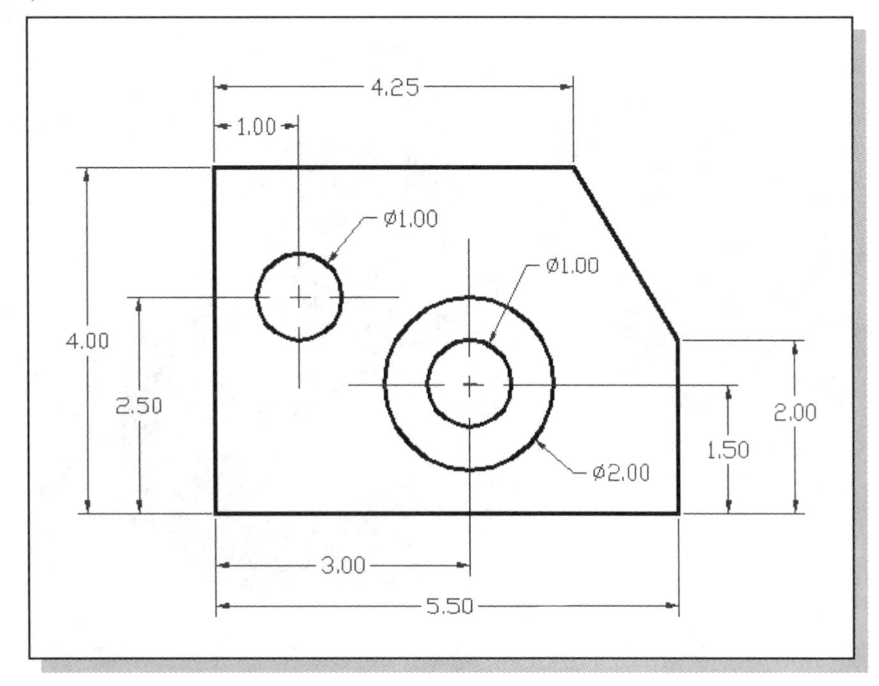

3. **Angle Slider** (Dimensions are in Millimeters.)

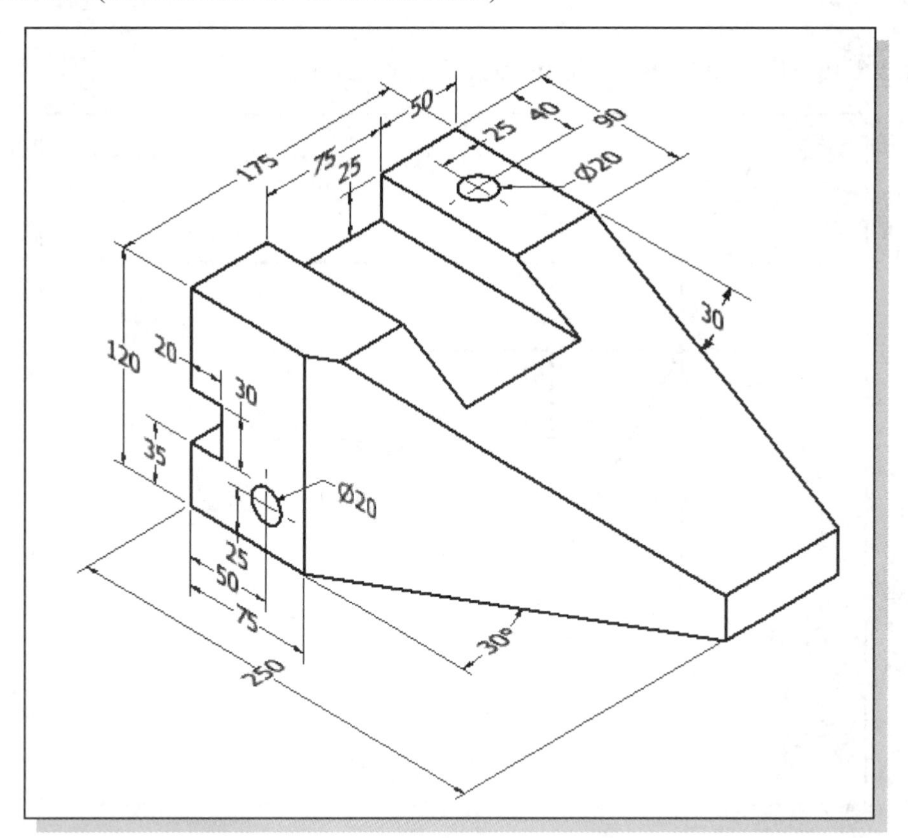

4. **Coupling Base** (Dimensions are in inches.)

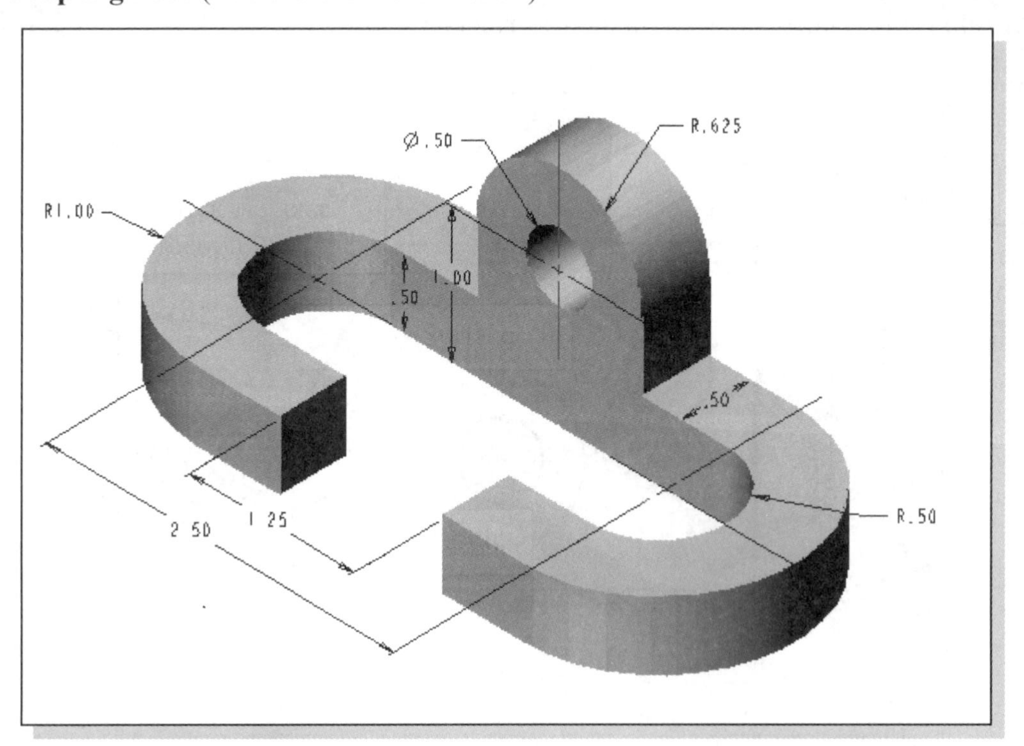

5. **Indexing Guide** (Dimensions are in inches.)

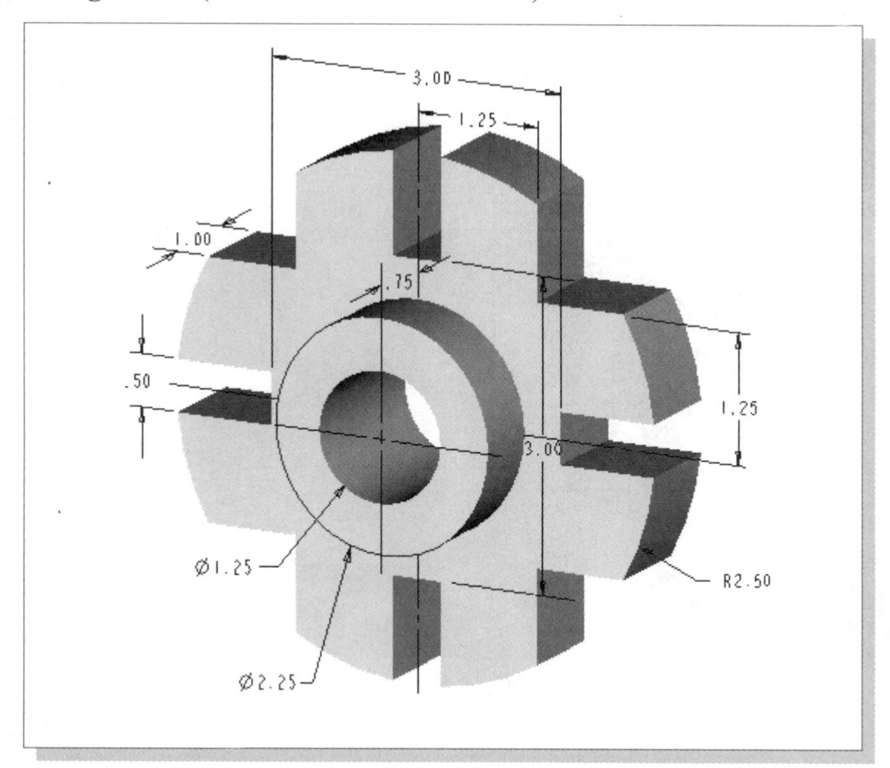

6. **L-Bracket** (Dimensions are in inches.)

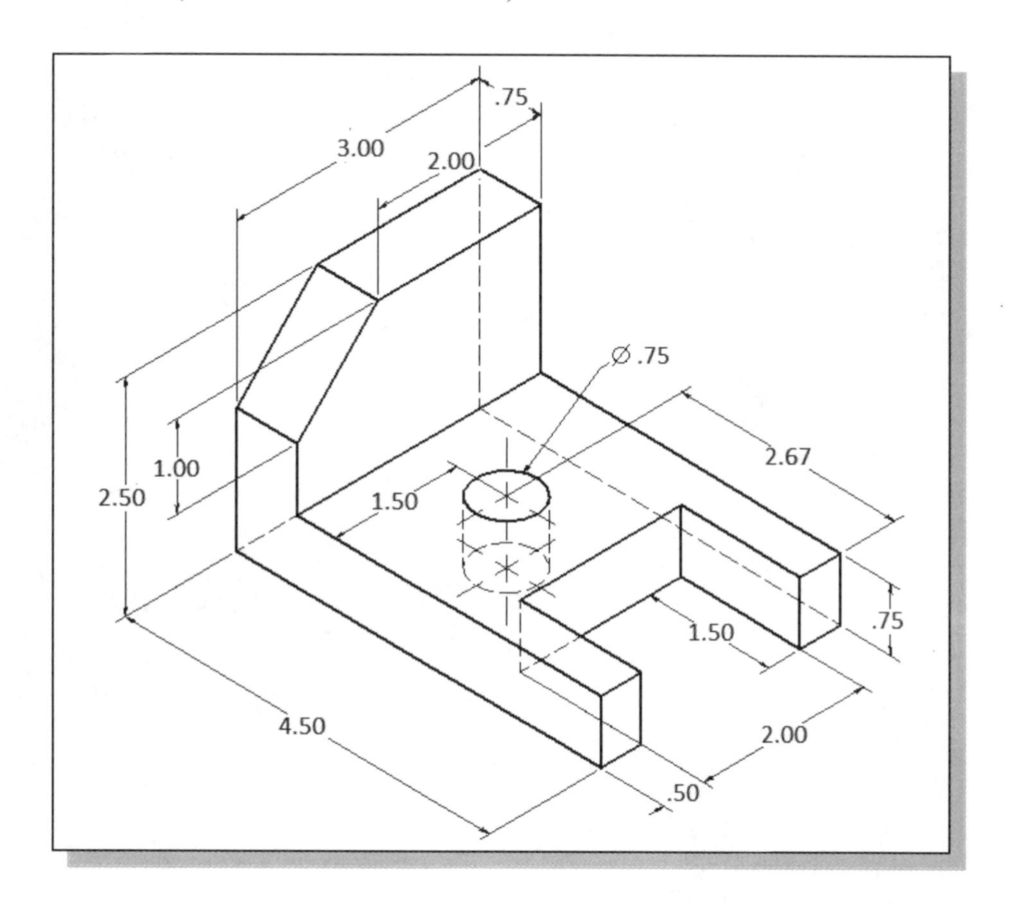

NOTES:

Chapter 4

Feature Design Tree

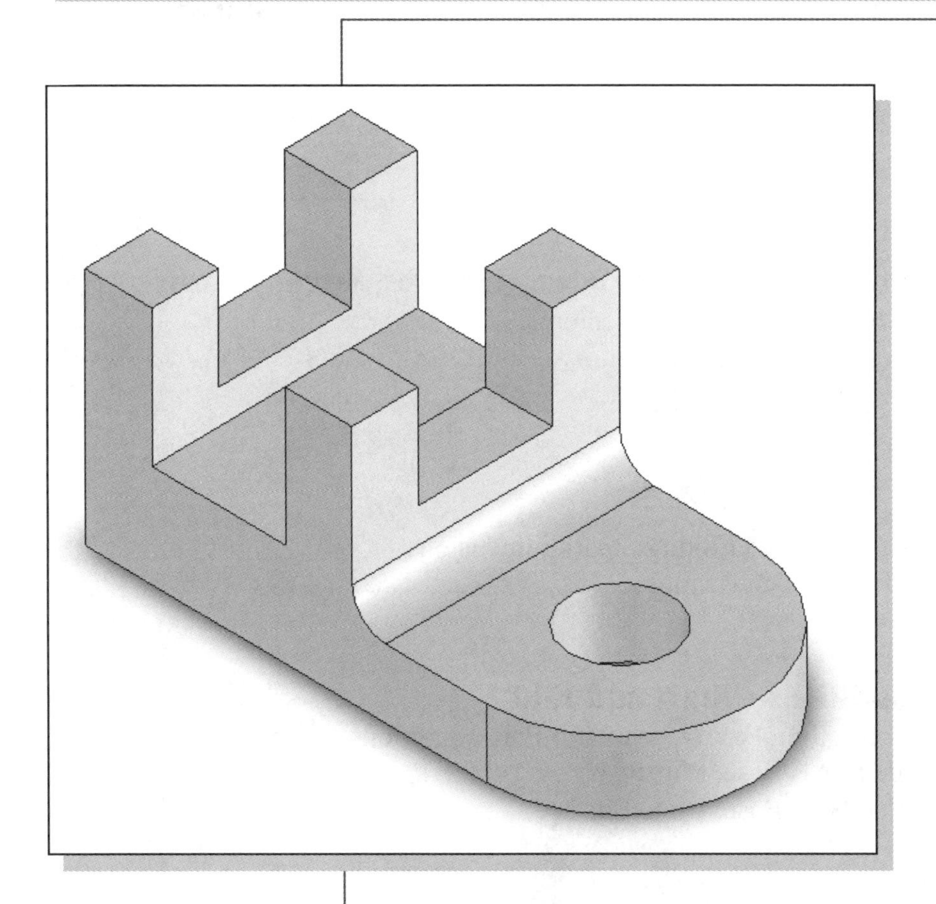

Learning Objectives

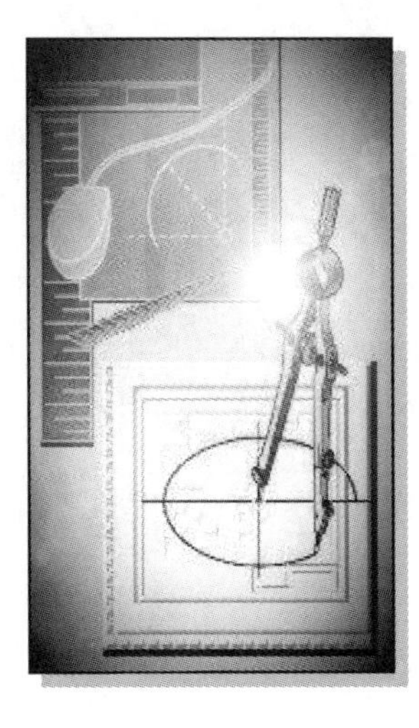

- Understand Feature Interactions
- Use the FeatureManager Design Tree
- Modify and Update Feature Dimensions
- Perform History-Based Part Modifications
- Change the Names of Created Features
- Implement Basic Design Changes
- Select a Material and View Mass Properties

Certified SOLIDWORKS Associate Exam Objectives Coverage

Certified Associate Reference Guide

Sketch Entities – Lines, Rectangles, Circles, Arcs, Ellipses, Centerlines

Objectives: Creating Sketch Entities.

Boss and Cut Features – Extrudes, Revolves, Sweeps, Lofts

Objectives: Creating Basic Swept Features.

Dimensions

Objectives: Applying and Editing Smart Dimensions.

Feature Conditions – Start and End

Objectives: Controlling Feature Start and End Conditions.

Mass Properties

Objectives: Obtaining Mass Properties for Parts and Assemblies.

Materials

Objectives: Applying Material Selection to Parts.

Introduction

In SOLIDWORKS, the **design intents** are embedded into features in the **FeatureManager Design Tree**. The structure of the design tree resembles that of a **CSG binary tree**. A CSG binary tree contains only *Boolean relations*, while the **SOLIDWORKS design tree** contains all features, including *Boolean relations*. A design tree is a sequential record of the features used to create the part. This design tree contains the construction steps, plus the rules defining the design intent of each construction operation. In a design tree, each time a new modeling event is created previously defined features can be used to define information such as size, location, and orientation. It is therefore important to think about your modeling strategy before you start creating anything. It is important, but also difficult, to plan ahead for all possible design changes that might occur. This approach in modeling is a major difference of **FEATURE-BASED CAD SOFTWARE**, such as SOLIDWORKS, from previous generation CAD systems.

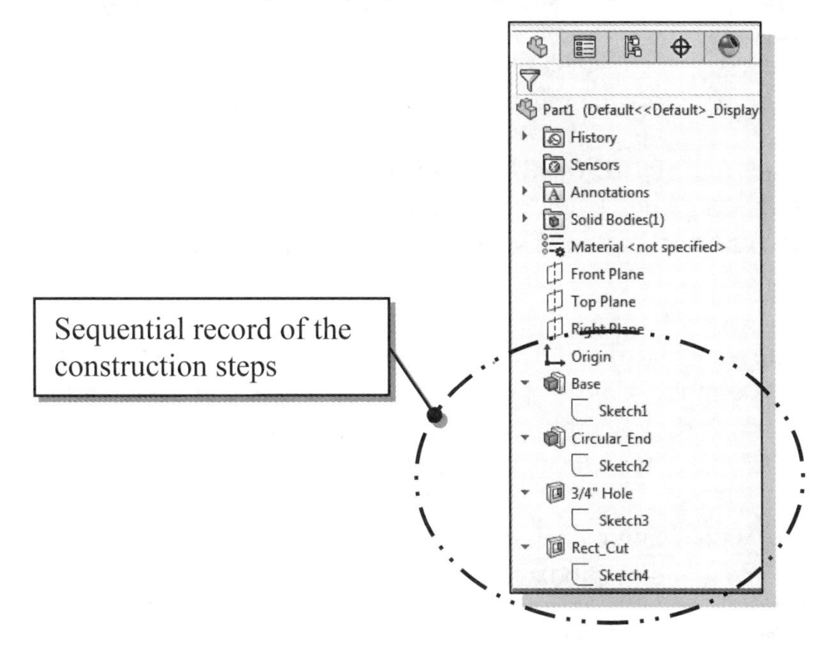

Feature-based parametric modeling is a cumulative process. Every time a new feature is added, a new result is created and the feature is also added to the design tree. The database also includes parameters of features that were used to define them. All of this happens automatically as features are created and manipulated. At this point, it is important to understand that all of this information is retained, and modifications are done based on the same input information.

In SOLIDWORKS, the design tree gives information about modeling order and other information about the feature. Part modifications can be accomplished by accessing the features in the design tree. It is therefore important to understand and utilize the feature design tree to modify designs. SOLIDWORKS remembers the history of a part, including all the rules that were used to create it, so that changes can be made to any operation that was performed to create the part. In SOLIDWORKS, to modify a feature, we access the feature by selecting the feature in the *FeatureManager Design Tree* window.

Starting SOLIDWORKS

1. Select the **SOLIDWORKS** option on the *Start* menu or select the **SOLIDWORKS** icon on the desktop to start SOLIDWORKS.

2. We will not use the *Welcome* dialog box in this lesson. If it appears, close the *Welcome* dialog box by clicking on the X in the upper right corner to view the SOLIDWORKS program window.

3. Select the **New** icon with a single click of the left-mouse-button on the *Menu Bar*.

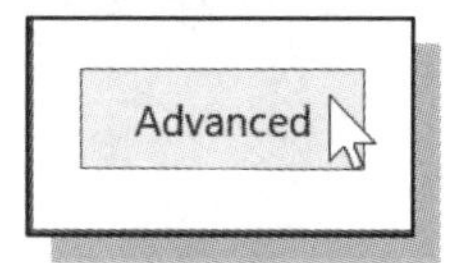

4. The *New* SOLIDWORKS *Document* dialog window appears in Novice mode. Click once with the left-mouse-button on the **Advanced** button to switch to Advanced mode.

❖ In Advanced mode, the *New* SOLIDWORKS *Document* dialog box offers the same three options as in Novice mode. These options allow starting a new document using the default templates for a Part, Assembly, or Drawing. However, the Advanced mode will allow us to start new documents with user-defined templates.

5. Select the **Part** icon with a single click of the left-mouse-button in the *New* SOLIDWORKS *Document* dialog box.

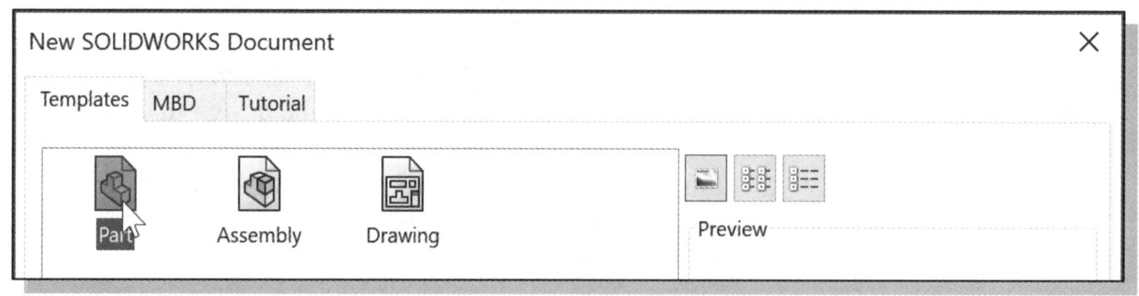

6. Select **OK** in the *New* SOLIDWORKS *Document* dialog box to open a new part document.

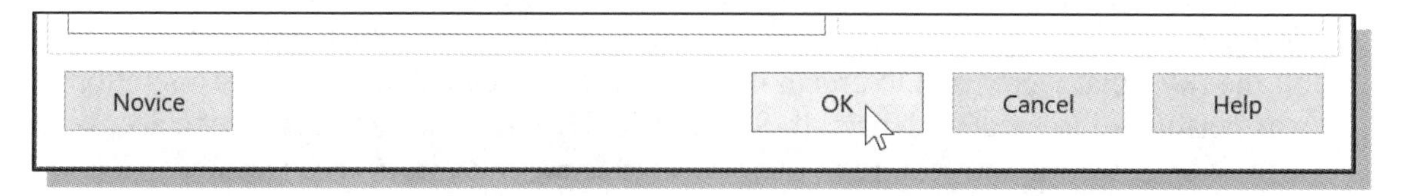

Creating a User-Defined Part Template

We will create a part template which includes a setting for the use of ANSI standards for dimensions and English (inch, pound, second) units. In the future, using this template will eliminate the need to adjust these document settings each time a new part is started.

1. Select the **Options** icon from the *Menu Bar* to open the *Options* dialog box.

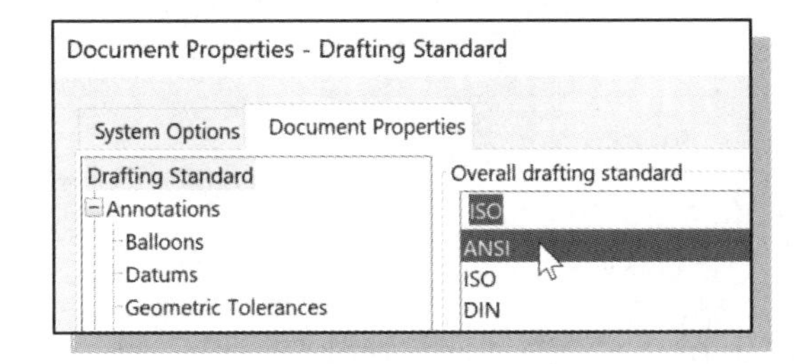

2. Select the **Document Properties** tab.
3. Select **ANSI** in the pull-down selection window under the *Overall drafting standard* panel as shown.
4. Click **Units** as shown below.
5. Select **IPS (inch, pound, second)** under the *Unit system* options.
6. Select **.123** in the *Decimals* spin box for the *Length units* as shown to define the degree of accuracy with which the units will be displayed to 3 decimal places.

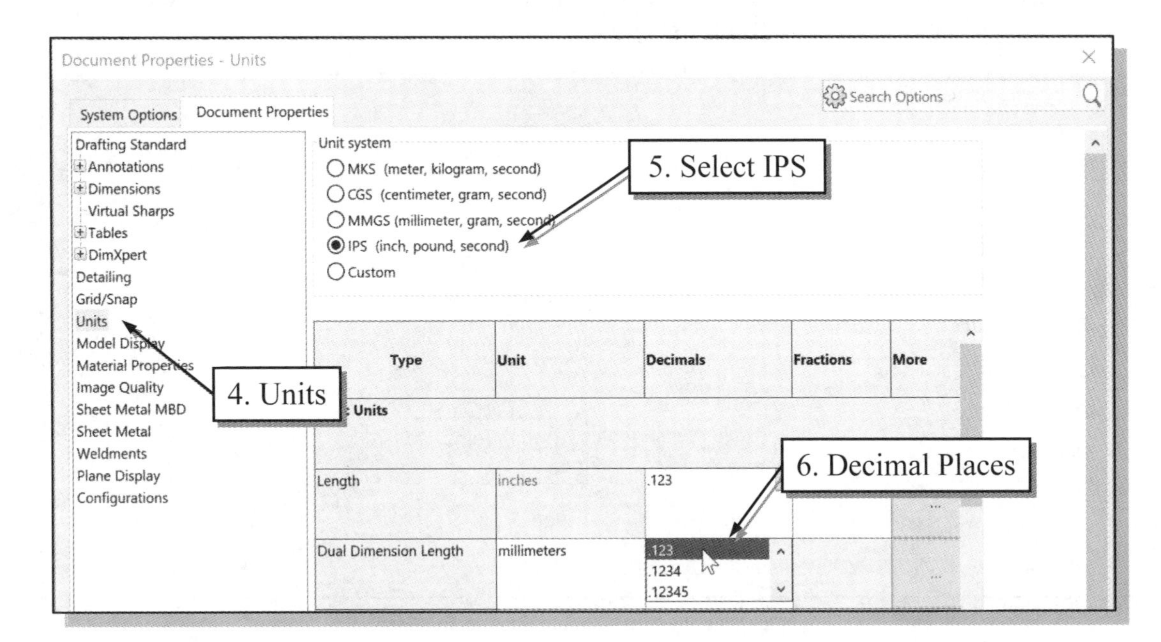

7. Click **OK** in the *Options* dialog box to accept the selected settings.

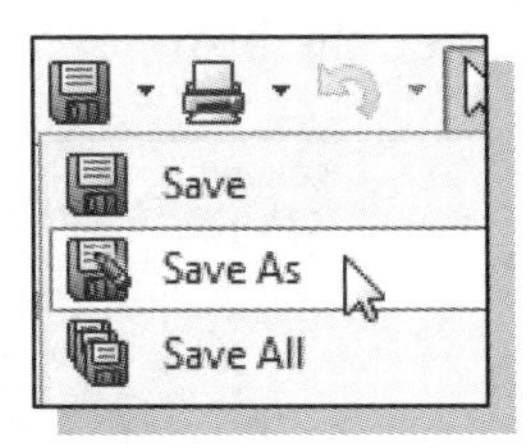

8. Click the arrow next to the **Save** icon in the *Menu Bar* to reveal the save options and select **Save As**.

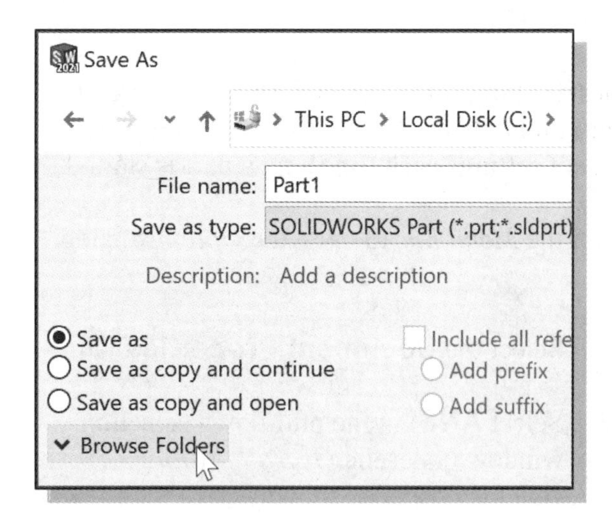

9. We will create a new folder for the user-defined templates. Decide where you want to locate this new folder and use the browser in the *Save As* dialog box to select the location. It may be necessary to expand the browser by clicking the **Browse Folders** button as shown. (**NOTE:** In the figure below, the **C:** folder is chosen.)

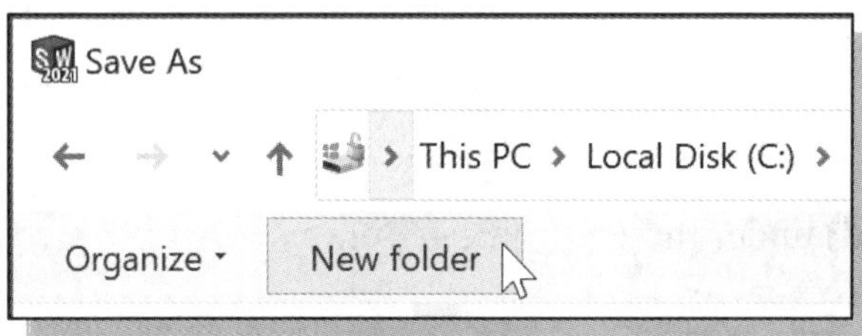

10. In the *Save As* dialog box, select the **New Folder** option by clicking once with the left-mouse-button on the icon as shown.

11. The new folder appears with the default name *New Folder*. Type the filename **Tutorial_Templates** for the new folder. (**NOTE:** This folder could also be created using Windows Explorer, etc.)

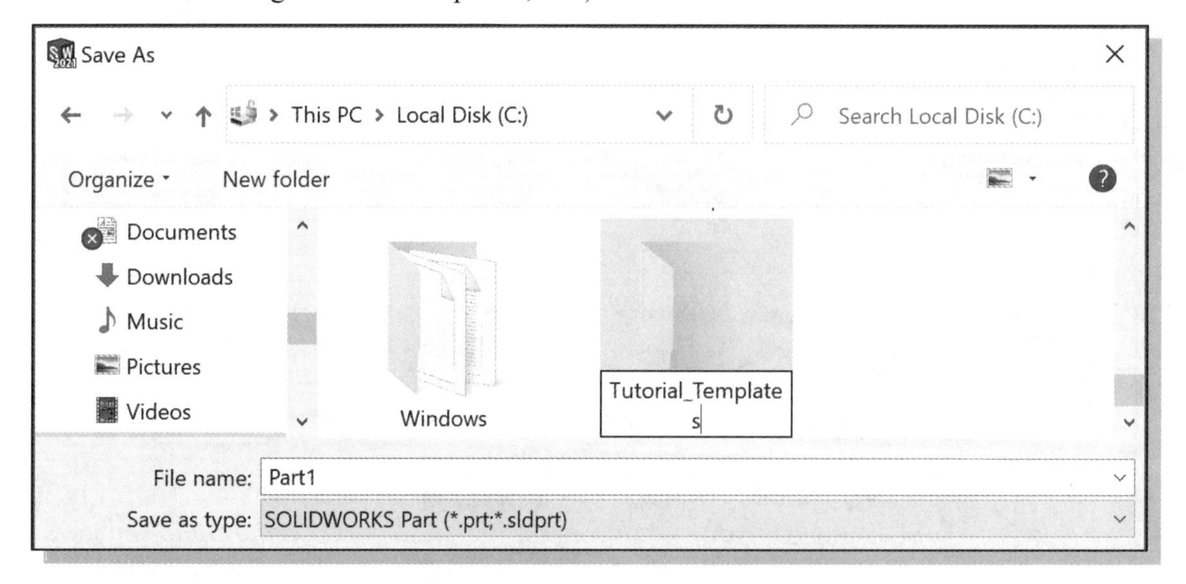

12. Left click on the **Hide Folders** button to hide the browser.

13. Under *Save as type*, select **Part Templates (*.prtdot)**. Notice the browser automatically goes to the default *templates* folder.

14. Use the *Save As* browser to select the **Tutorial_Templates** folder you created.

15. Enter the *File name* **Part_IPS_ANSI**.

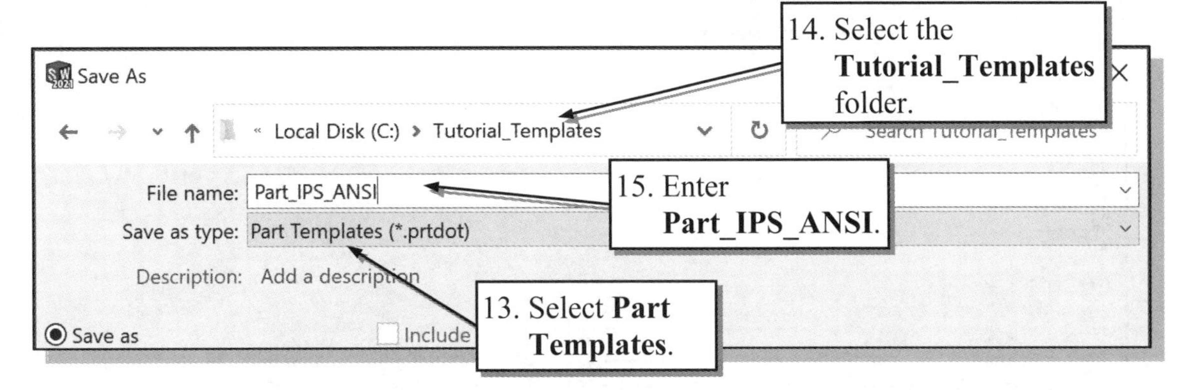

16. Click **Save** to save the new part template file.

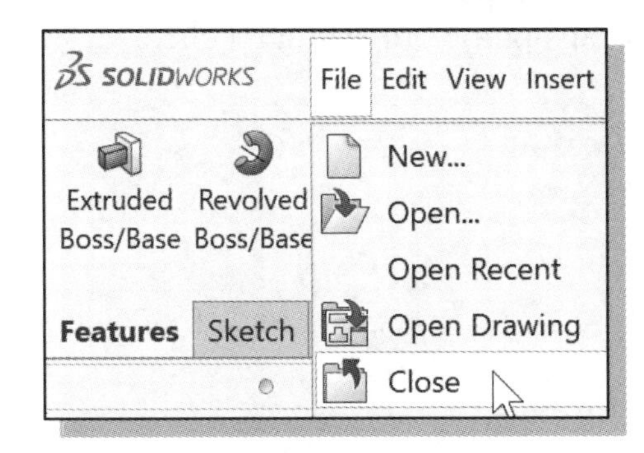

17. Select **Close** in the *File* pull-down menu to close the document. (**NOTE:** You may have to move the mouse over the SOLIDWORKS icon to reveal the pull-down menus.)

➢ We will now open a new part document using the template we just saved.

18. Select the **New** icon with a single click of the left-mouse-button on the *Menu Bar*. The *New* SOLIDWORKS *Document* dialog box appears in Advanced mode.

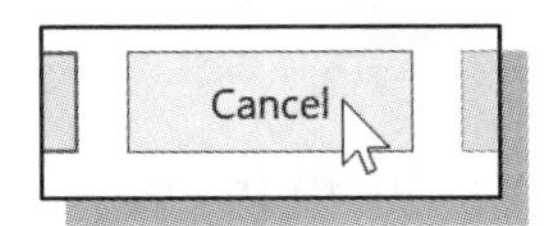

19. Notice that the new template does not appear as an option. Click **Cancel** in the *New* SOLIDWORKS *Document* dialog box.

20. Select the **Options** icon from the *Menu Bar* to open the *Options* dialog box.

21. Select **File Locations** under the *System Options* tab as shown.

22. Make sure **Document Templates** is selected as the *Show folders for:* option.

23. Click the **Add** button to add the directory with the user-defined templates to the list of folders containing document templates.

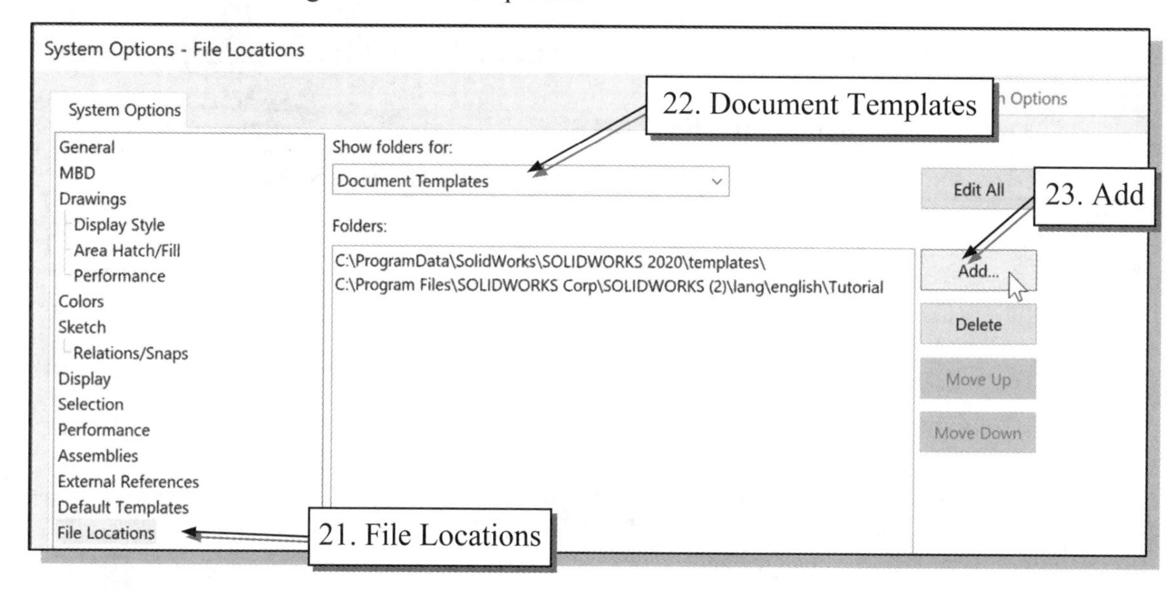

24. Locate and select the **Tutorial_Templates** folder using the browser, and click **Select Folder** in the *Select Folder* dialog box.

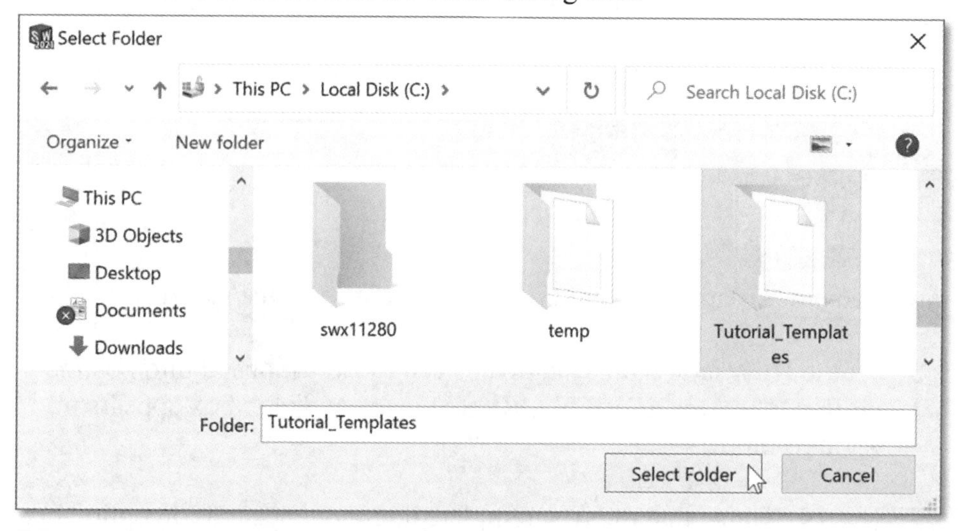

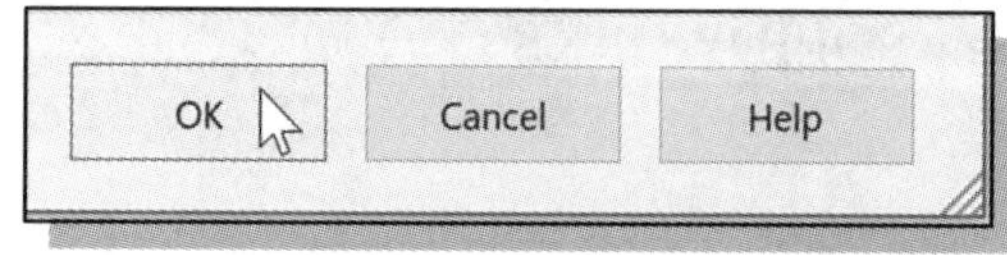

25. Select **OK** in the *System Options* dialog box.

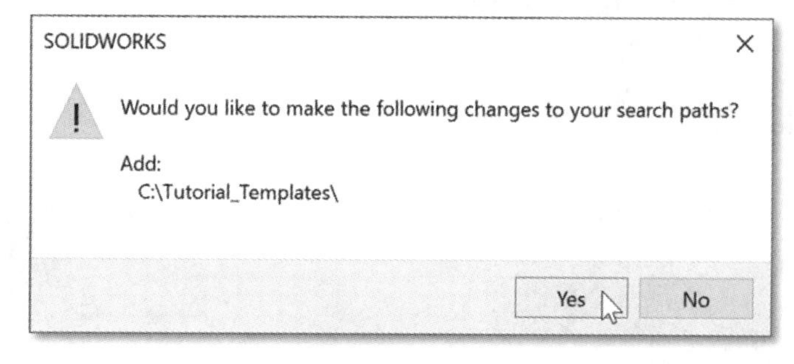

26. If a pop-up window appears with the question "*Would you like to make the following changes to your search paths?*" click **Yes**.

27. Select the **New** icon with a single click of the left-mouse-button on the *Menu Bar*. Notice the Tutorial_Templates folder now appears as a tab in the *New* SOLIDWORKS *Document* dialog box.

28. Select the **Tutorial_Templates** tab.

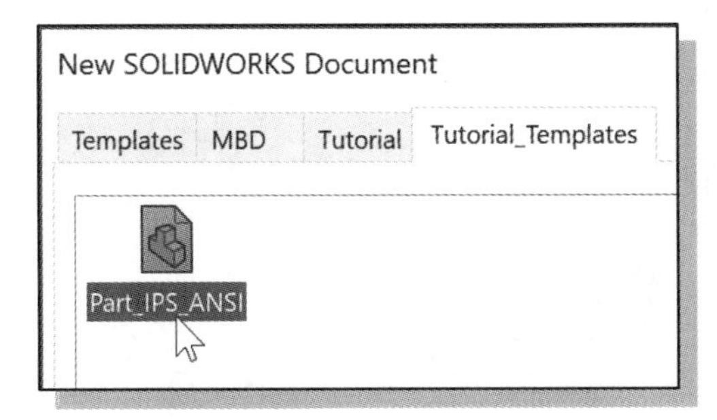

29. Notice the Part_IPS_ANSI template appears. Select the **Part_IPS_ANSI** template as shown.

30. Click on the **OK** button to open a new document.

The *Saddle Bracket* Design

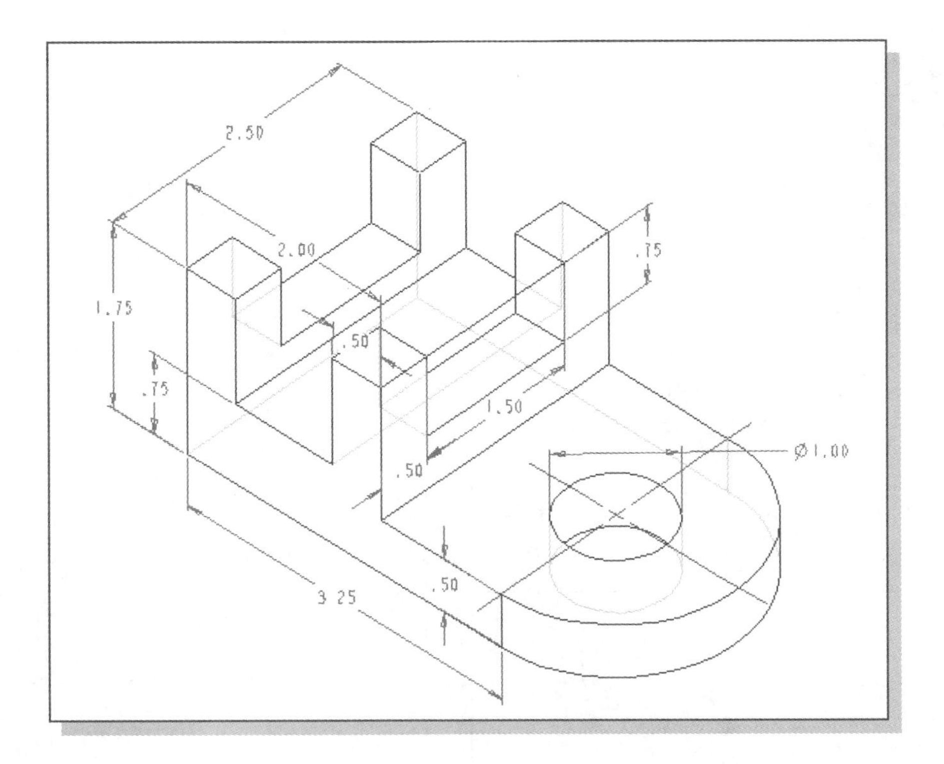

Based on your knowledge of SOLIDWORKS so far, how many features would you use to create the design? Which feature would you choose as the **BASE FEATURE**, the first solid feature, of the model? What is your choice in arranging the order of the features? Would you organize the features differently if additional fillets were to be added in the design? Take a few minutes to consider these questions and do preliminary planning by sketching on a piece of paper. You are also encouraged to create the model on your own prior to following through the tutorial.

Modeling Strategy

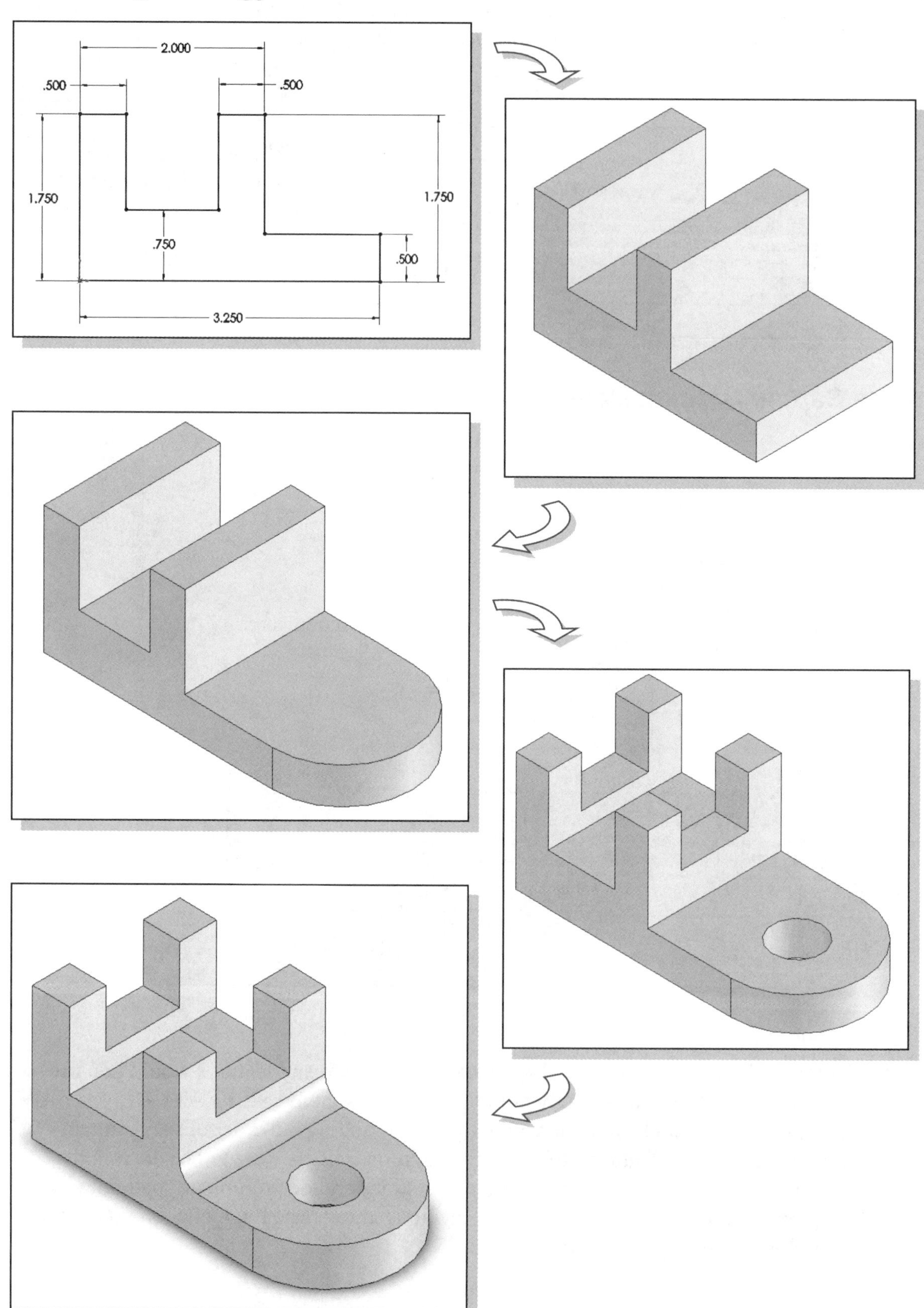

The SOLIDWORKS *FeatureManager Design Tree*

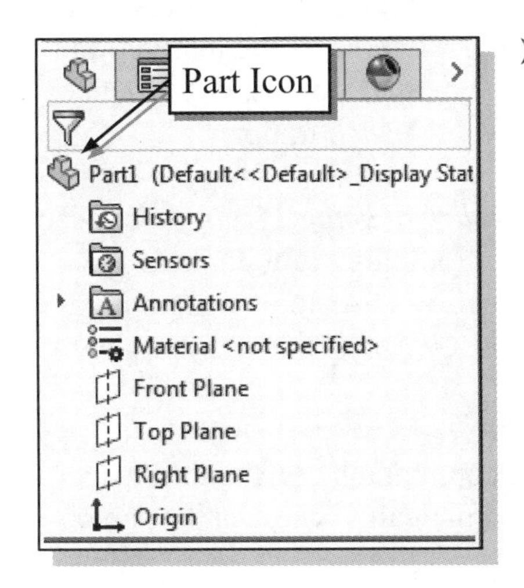

- In the SOLIDWORKS screen layout, the ***FeatureManager Design Tree*** is located to the left of the graphics window. SOLIDWORKS can be used for part modeling, assembly modeling, part drawings, and assembly presentation. The *FeatureManager Design Tree* window provides a visual structure of the features, relations, and attributes that are used to create the part, assembly, or scene. The *FeatureManager Design Tree* also provides right-click menu access for tasks associated specifically with the part or feature, and it is the primary focus for executing many of the SOLIDWORKS commands.

- The first item displayed in the *FeatureManager Design Tree* is the name of the part, which is also the filename. By default, the name "*Part1*" is used when we first start SOLIDWORKS. The parts are then numbered sequentially as new parts are created. If the part is saved with a new filename, this name will replace "*Part1.*" Do not be concerned if your part is numbered differently than the labels in the figures, e.g., "*Part3.*"

- The *FeatureManager Design Tree* can also be used to modify parts and assemblies by moving, deleting, or renaming items within the hierarchy. Any changes made in the *FeatureManager Design Tree* directly affect the part or assembly and the results of the modifications are displayed on screen instantly. The *FeatureManager Design Tree* also reports any problems and conflicts during the modification and updating procedure.

Creating the Base Feature

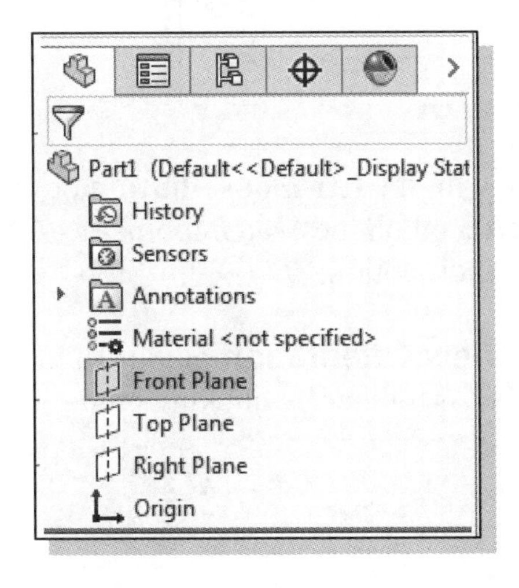

1. Move the graphics cursor to the **Front Plane** icon in the *FeatureManager Design Tree.* Notice the *Front Plane* icon is highlighted in the design tree and the *Front Plane* outline appears in the graphics area. Click once with the **left-mouse-button** to select the Front Plane.

IMPORTANT NOTE: In this lesson, we will use the standard display of toolbars. If a user prefers to use the *CommandManager*, the only change is that it may be necessary to select the appropriate tab prior to selecting a command. For example, if the instruction is to "select the Sketch command from the *Sketch* toolbar," it may be necessary to first select the Sketch tab on the *CommandManager* to display the *Sketch* toolbar.

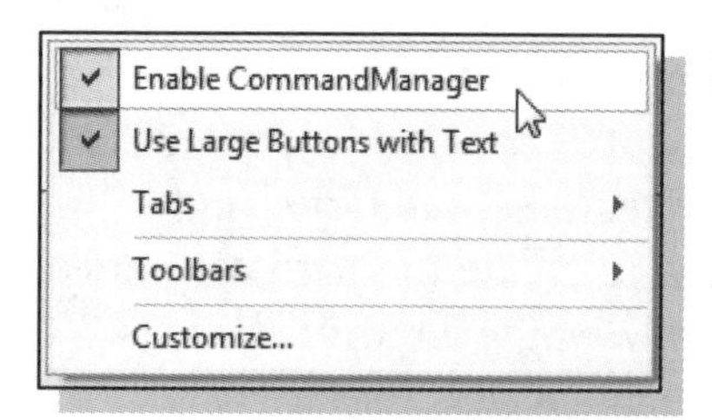

2. To turn ***OFF*** the *Enable CommandManager* and use the standard display of toolbars, right click on the *CommandManager* (or any other toolbar) and toggle the *CommandManager OFF* by selecting it at the top of the pop-up menu.

3. In the *Sketch* toolbar, select the **Sketch** command by left-clicking once on the icon. Notice the Front Plane automatically becomes the sketch plane because it was pre-selected.

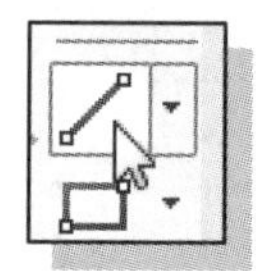

4. Select the **Line** icon on the *Sketch* toolbar by clicking once with the **left-mouse-button**; this will activate the Line command. Starting at the origin, create the geometry shown below.

5. On your own, adjust the geometry by adding and modifying dimensions as shown.

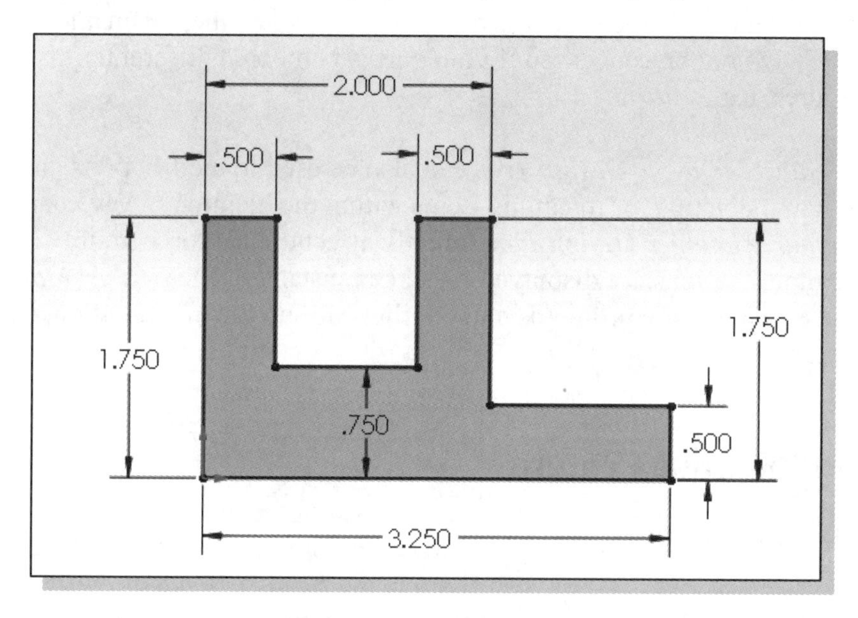

6. Click once with the **left-mouse-button** on the **Sketch** icon on the *Sketch* toolbar to exit the Sketch option.

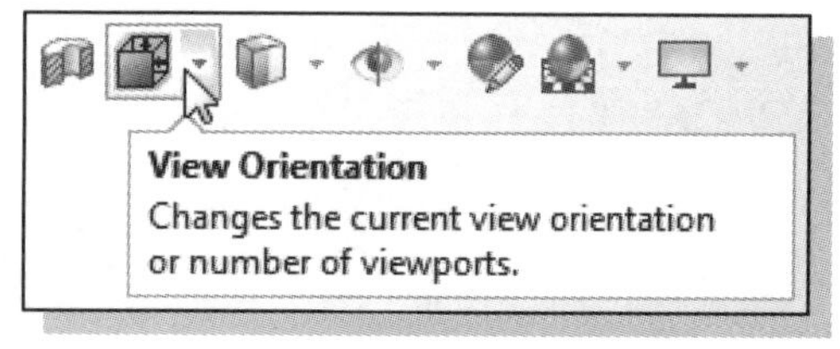

7. Select the **View Orientation** button on the *Heads-up View* toolbar by clicking on it.

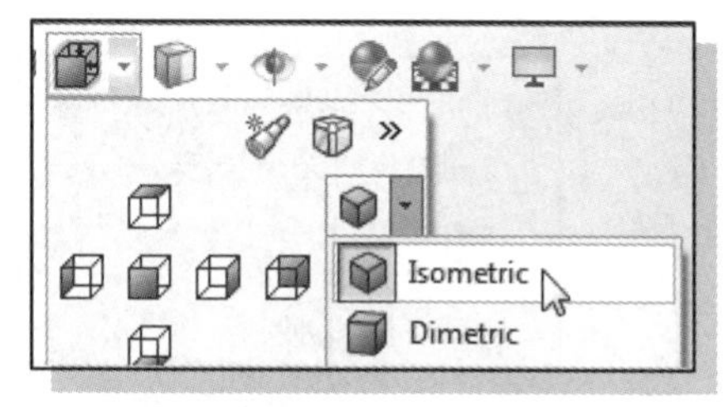

8. Select the **Isometric** option in the *View Orientation* pull-down menu.

   ➢ NOTE: If the View Selector is toggled *ON* (see page 3-26), either toggle it *OFF* or use it to select the *Isometric* view.

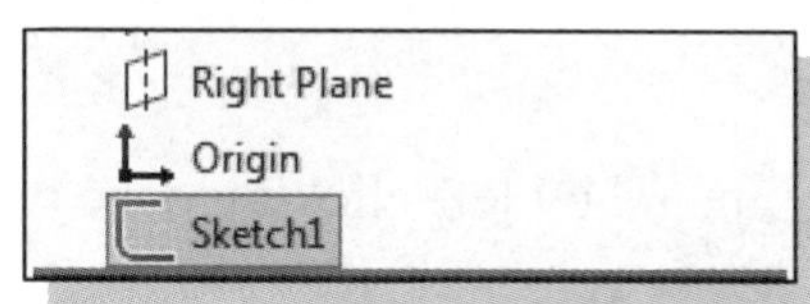

9. Make sure the sketch – Sketch1 – is selected in the *FeatureManager Design Tree.*

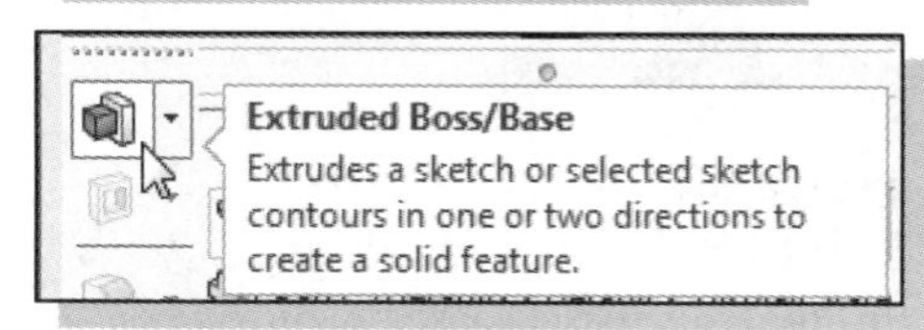

10. In the *Features* toolbar, select the **Extruded Boss/Base** command by clicking once with the left-mouse-button on the icon.

11. In the *Extrude PropertyManager* panel, click the drop down arrow to reveal the options for the *End Condition* (the default end condition is Blind) and select **Mid Plane** as shown.

12. In the *Extrude PropertyManager* panel, enter **2.5** as the extrusion distance. Notice that the sketch region is automatically selected as the extrusion profile.

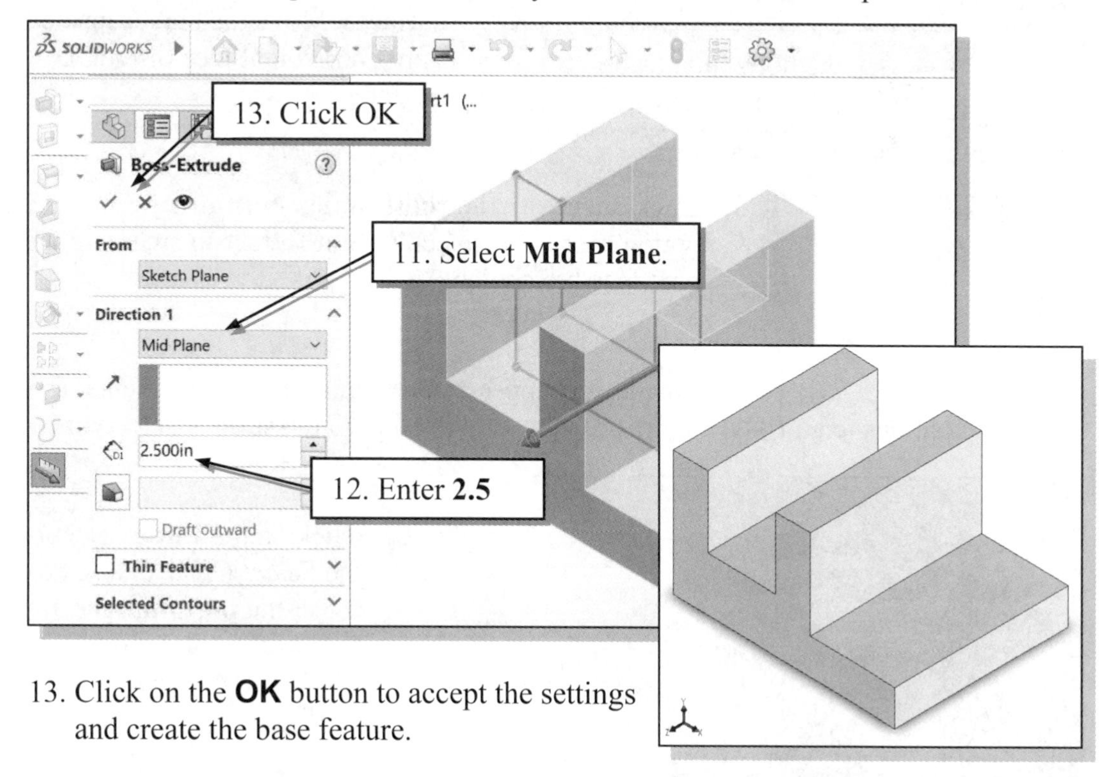

13. Click on the **OK** button to accept the settings and create the base feature.

➢ Notice the Boss-Extrude feature is added to the *Model Tree* in the *FeatureManager Design Tree* area.

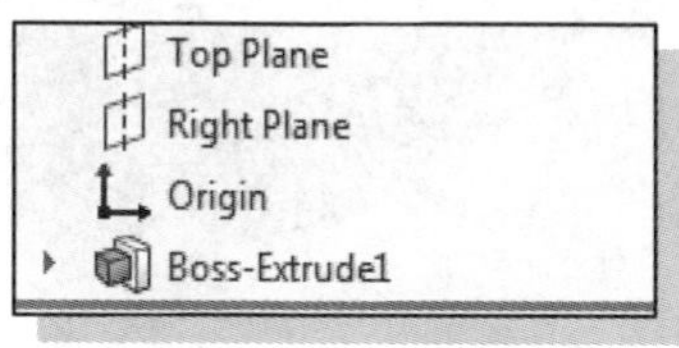

➢ Also notice, with the completion of the first extrusion, the Solid Bodies folder is added to the design tree. The folder includes one solid body as indicated by the 1 in parentheses.

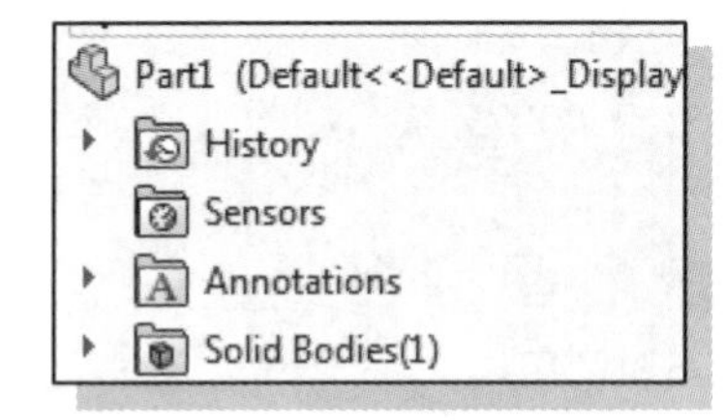

Adding the Second Solid Feature

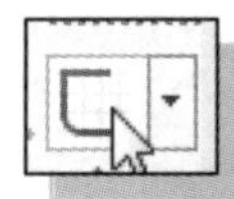

1. In the *Sketch* toolbar, select the **Sketch** command by left-clicking once on the icon.

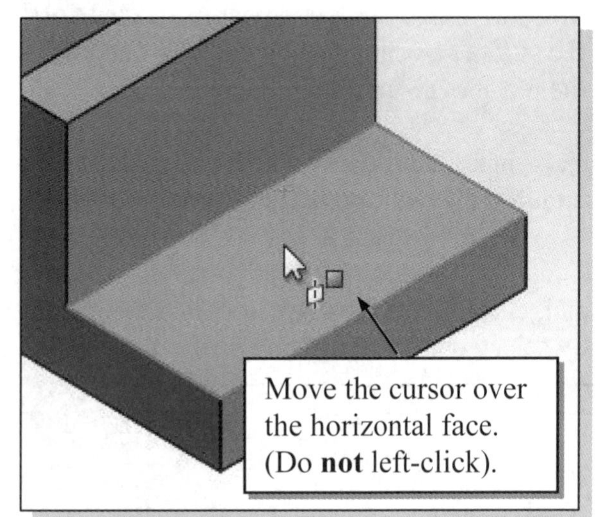

2. Notice the left panel displays the *Edit Sketch PropertyManager* with the instruction "*Select a plane on which to create a sketch for the entity.*" Move the graphics cursor on the 3D part and notice that SOLIDWORKS will automatically highlight feasible planes and surfaces as the cursor is on top of the different surfaces. Move the cursor inside the upper horizontal face of the 3D object as shown.

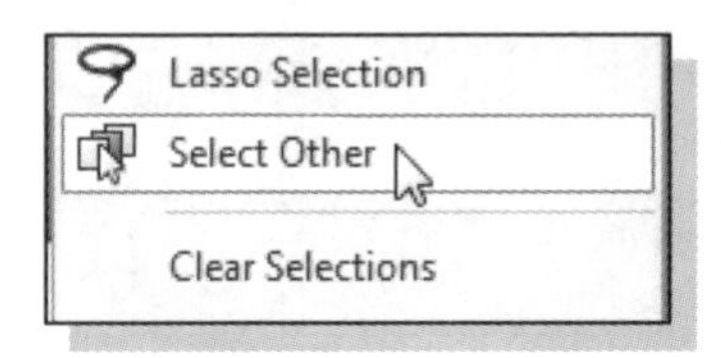

3. Click once with the **right-mouse-button** to bring up the option menu and select **Select Other** to switch to the next feasible choice.

4. The *Select Other* pop-up dialog box appears. On your own, move the cursor over the options (e.g., Face) in the box to examine all possible surface selections.

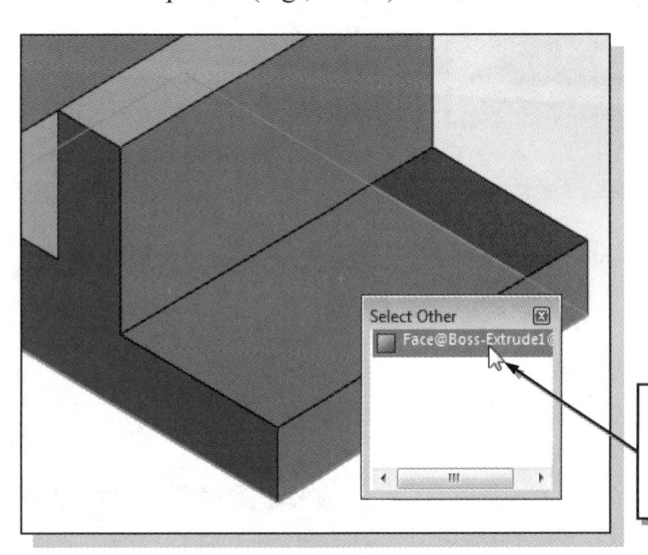

5. Click on the **Face** selection in the *Select Other* dialog box to select the **bottom horizontal face** of the solid model when it is highlighted as shown in the figure.

Creating a 2D Sketch

1. Select the **Circle** command by clicking once with the left-mouse-button on the icon in the *Sketch* toolbar.

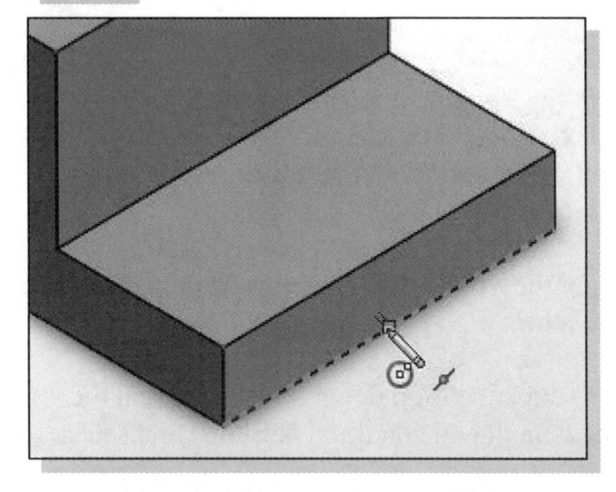

➢ We will align the center of the circle to the midpoint of the base feature.

2. Move the cursor along the shorter edge of the base feature; when the **Midpoint** is highlighted as shown in the figure, left-click to select this point.

3. Select the front corner of the base feature to create a circle as shown below.

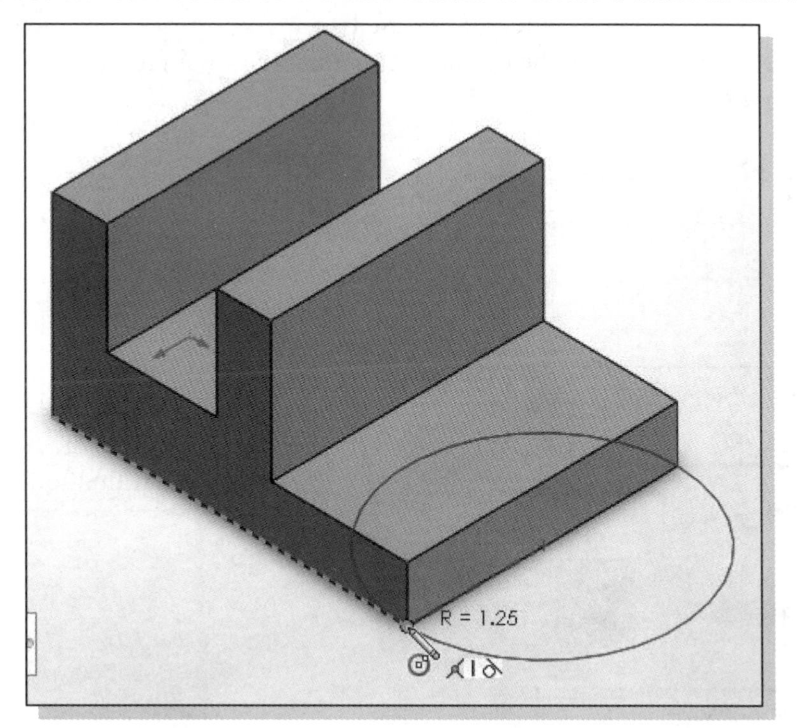

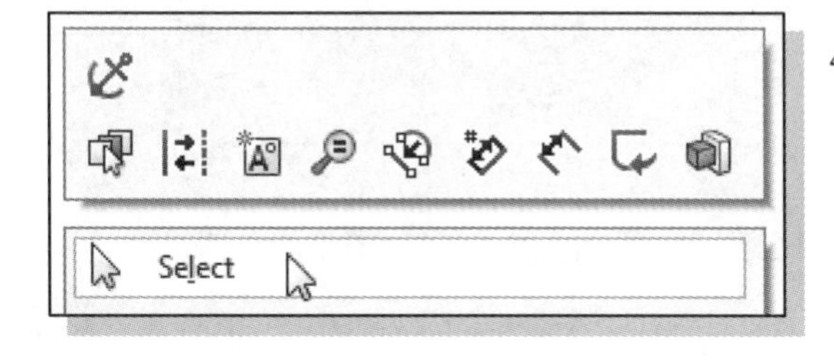

4. Inside the graphics window, click once with the right-mouse-button to display the option menu. Select **Select** in the pop-up menu to end the Circle option.

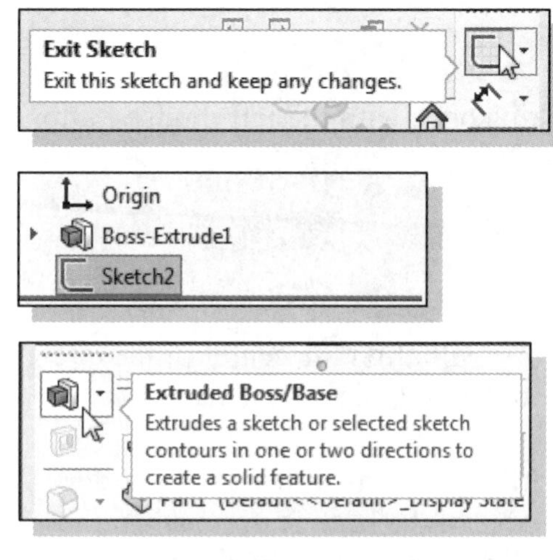

5. Click once with the **left-mouse-button** on the **Sketch** icon on the *Sketch* toolbar to exit the Sketch option.

6. Make sure the sketch – Sketch2 – is selected in the *FeatureManager Design Tree.*

7. In the *Features* toolbar, select the **Extruded Boss/Base** command by clicking once with the left-mouse-button on the icon.

8. Click the **Reverse Direction** button in the *PropertyManager* as shown. The extrude preview should appear as shown below.

9. In the *Extrude PropertyManager* panel, click the drop down arrow to reveal the pull options for the *End Condition* (the default end condition is Blind) and select **Up To Surface**.

10. Select the top face of the base feature as the termination surface for the extrusion. Notice Face<1> appears in the surface selection window in the *Extrude PropertyManager*.

11. Confirm the **Merge result** checkbox is checked.

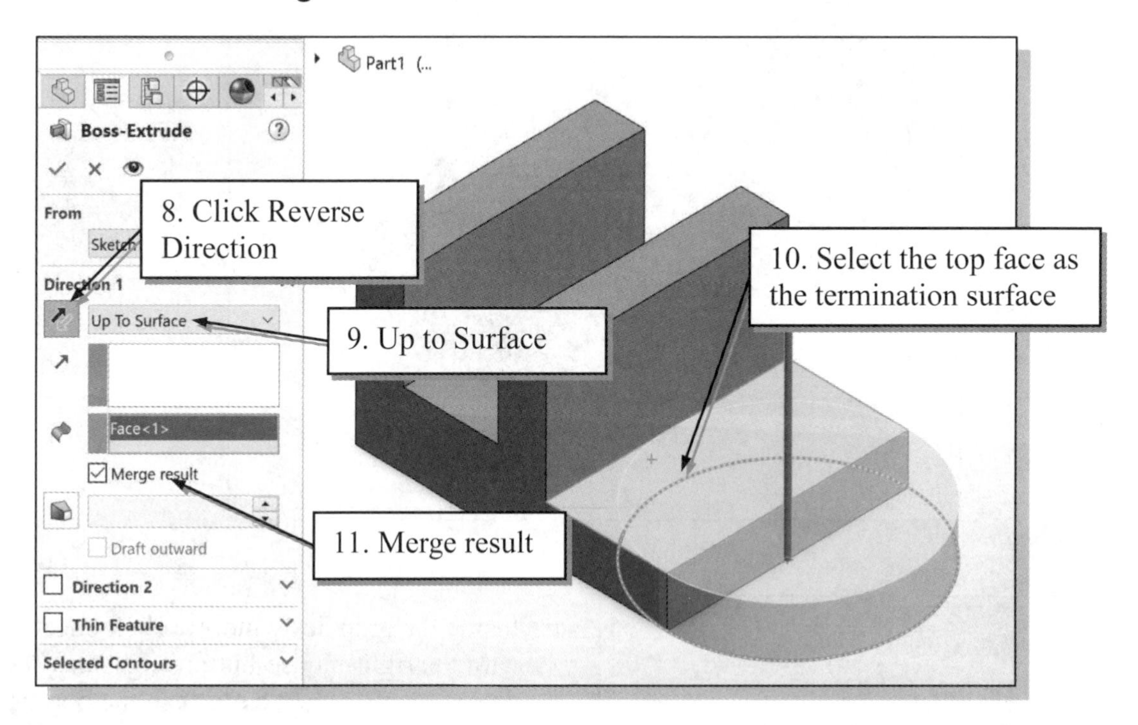

12. Click on the **OK** button to proceed with the Boss-Extrude operation.

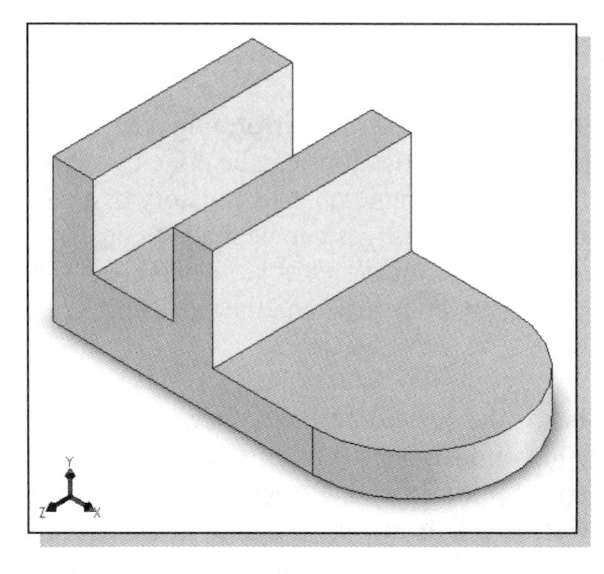

➢ Notice there remains only one solid body in the design tree. This is because the *Merge result* option was used for the second feature. If this option were not selected, a separate solid body would have been created. By default, the *Solid Body* bears the name of its last merged feature.

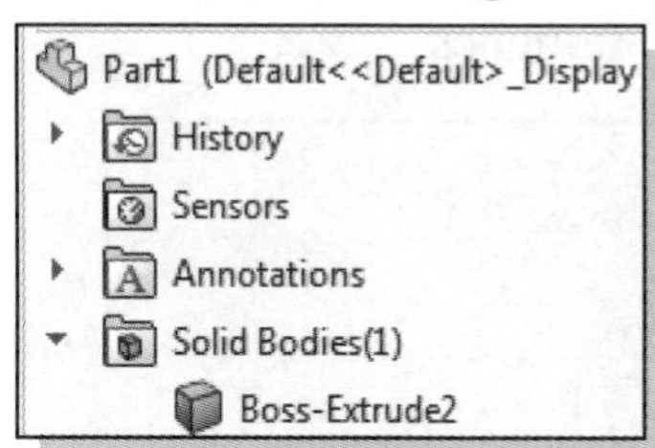

Renaming the Part Features

Currently, our model contains two extruded features. The feature is highlighted in the display area when we select the feature in the *FeatureManager Design Tree* window. Each time a new feature is created, the feature is also displayed in the *Design Tree* window. By default, SOLIDWORKS will use generic names for part features. However, when we begin to deal with parts with a large number of features, it will be much easier to identify the features using more meaningful names. Two methods can be used to rename the features: 1. **clicking** twice on the name of the feature; and 2. using the **Properties** option. In this example, the use of the first method is illustrated.

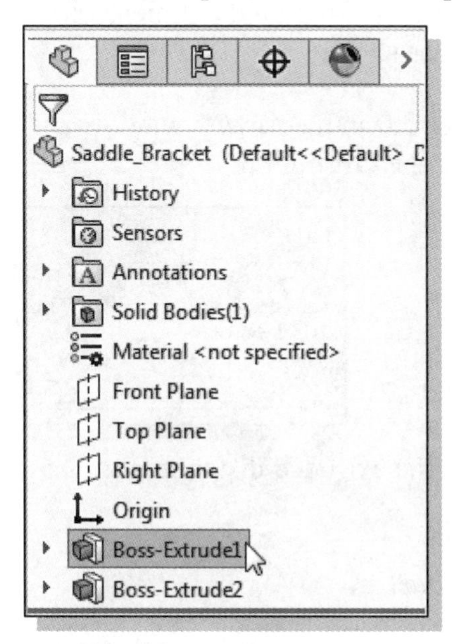

1. Select the first extruded feature in the *Model Browser* area by left-clicking once on the name of the feature, **Boss-Extrude1**. Notice the selected feature is highlighted in the graphics window.

2. Left-mouse-click on the feature name again to enter the **Edit** mode. (Alternately, right-click and select *Rename tree item* to enter **Edit** mode.)

3. Type **Base** as the new name for the first extruded feature and press the **[Enter]** key.

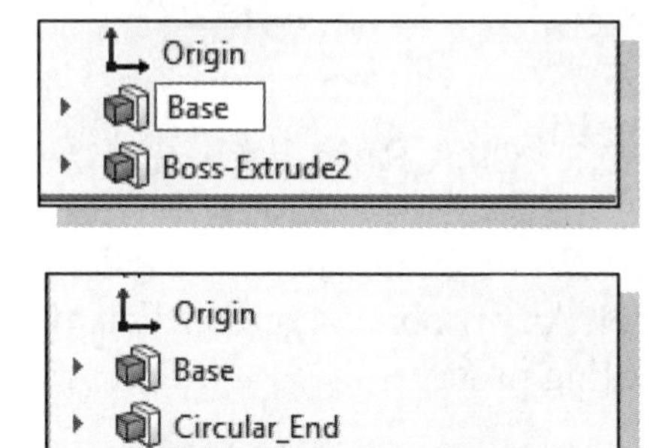

4. **On your own**, rename the second extruded feature to **Circular_End**.

Adjusting the Width of the Base Feature

One of the main advantages of parametric modeling is the ease of performing part modifications at any time in the design process. Part modifications can be done through accessing the features in the design tree. SOLIDWORKS remembers the history of a part, including all the rules that were used to create it, so that changes can be made to any operation that was performed to create the part. For our *Saddle Bracket* design, we will reduce the size of the base feature from 3.25 inches to 3.0 inches, and the extrusion distance to 2.0 inches.

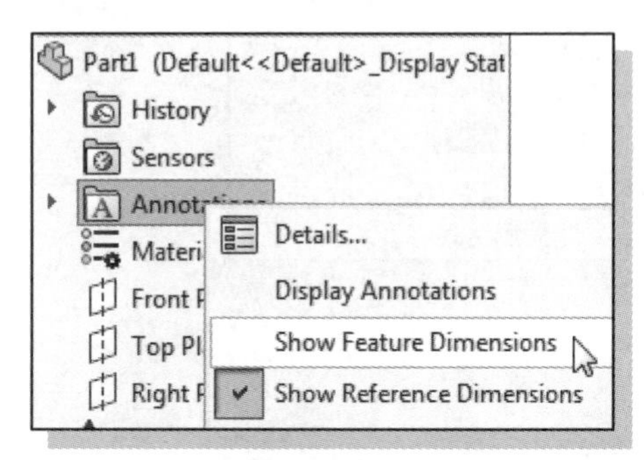

1. Inside the *FeatureManager Design Tree* area, **right-mouse-click** on **Annotations** to bring up the option menu and select the **Show Feature Dimensions** option in the pop-up menu. This option will allow us to view and edit dimensions by moving the cursor over the corresponding feature on the model.

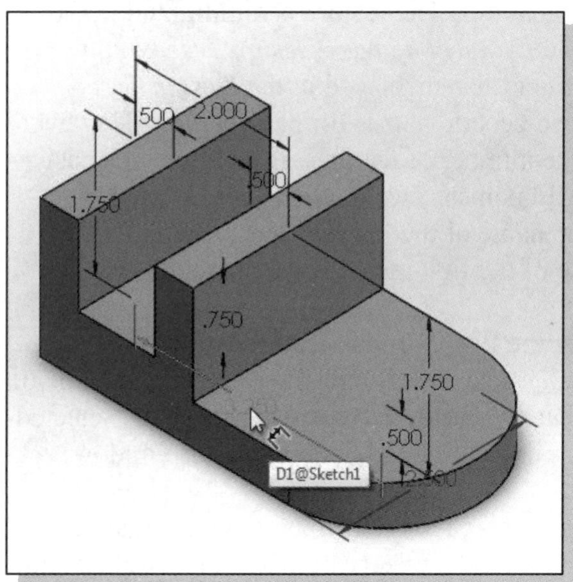

2. Move the cursor over the model to reveal the dimension indicating the **3.25** overall width of the **Base** feature.

3. Select the overall width of the **Base** feature, the **3.25** dimension value, by double-clicking on the dimension text, as shown.

4. Enter **3.0** in the *Modify* window and click the **OK** button.

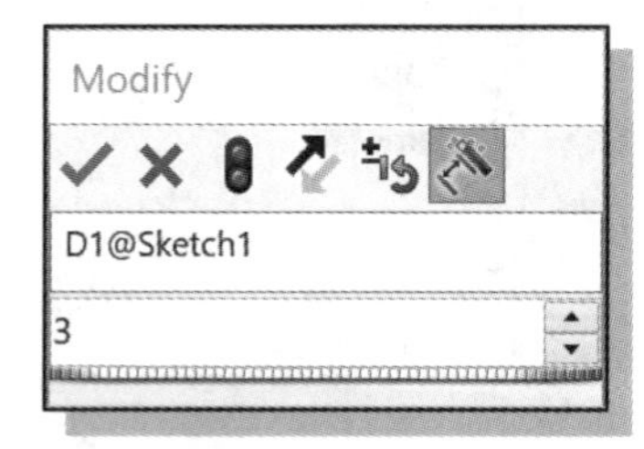

5. Hit the [**Esc**] key to exit the Dimension command.

6. **On your own**, repeat the above steps and modify the extruded distance from **2.5** to **2.0**.

7. Click **Rebuild** in the *Menu Bar*.

➢ Note that SOLIDWORKS updates the model by re-linking all elements used to create the model. Any problems or conflicts that occur will also be displayed during the updating process.

8. **Right-mouse-click** on **Annotations** to bring up the option menu and de-select the **Show Feature Dimensions** option in the pop-up menu.

Adding a Hole

1. Select the **Sketch** button on the *Sketch* toolbar to create a new sketch.

2. Pick the upper horizontal face as shown.

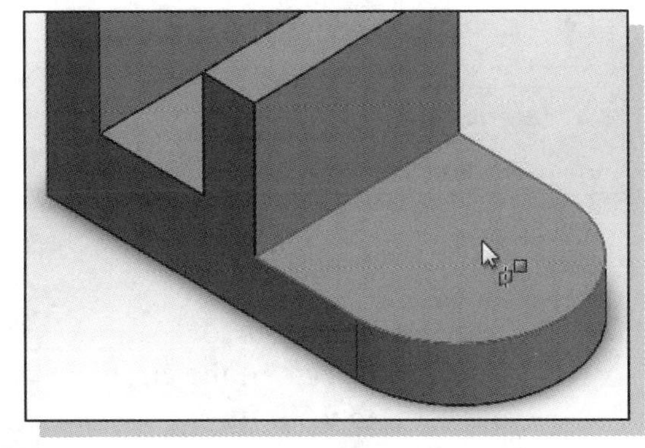

3. Select the **Circle** command by clicking once with the left-mouse-button on the icon in the *Sketch* toolbar.

4. Move the cursor over the circular edge of the face as shown. (Do not click.) Notice the center and quadrant marks appear on the circle.

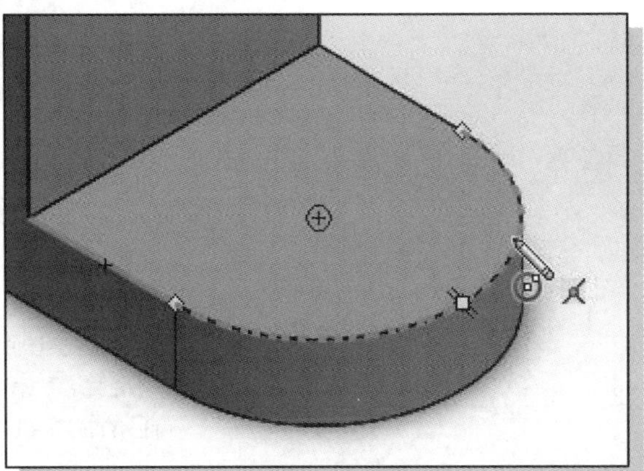

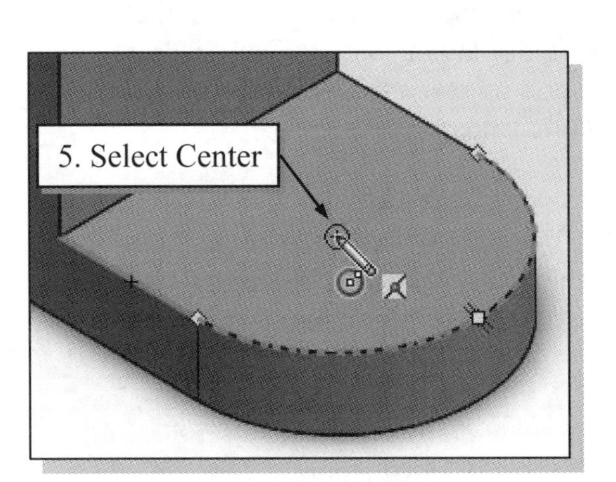

5. Select the **Center** point of the top face of the 3D model by left-clicking once on the icon as shown.

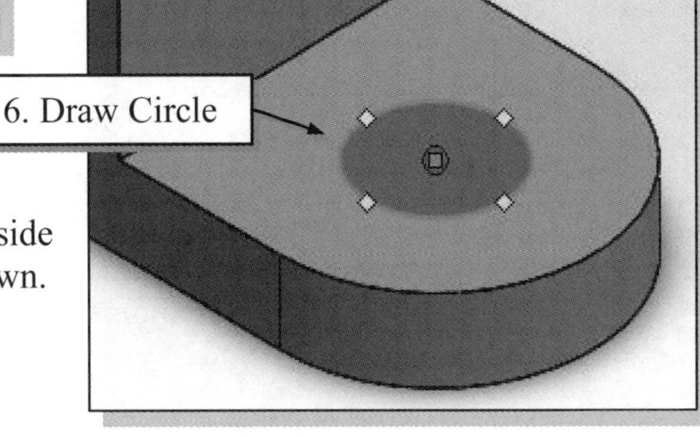

6. Sketch a circle of arbitrary size inside the top face of the cylinder as shown.

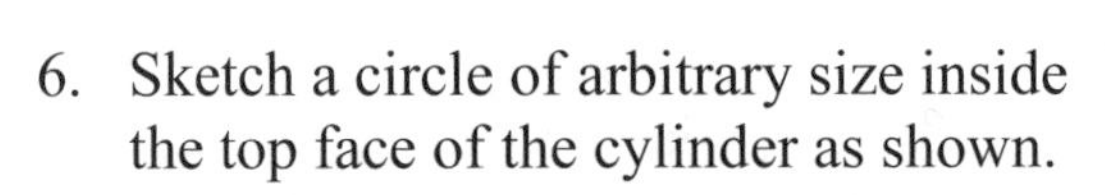

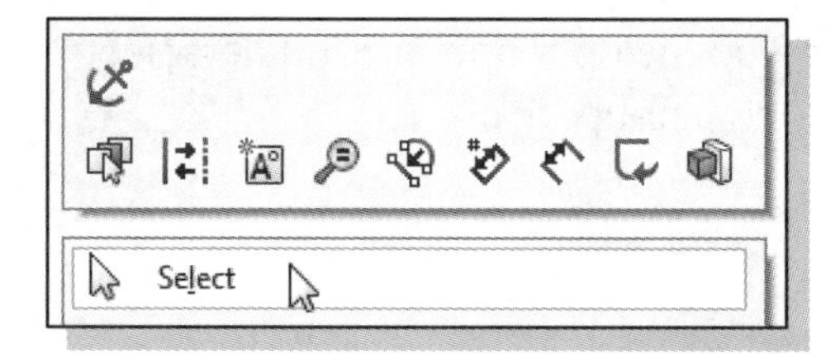

7. Use the right-mouse-button to display the option menu and select **Select** in the pop-up menu to end the Circle command.

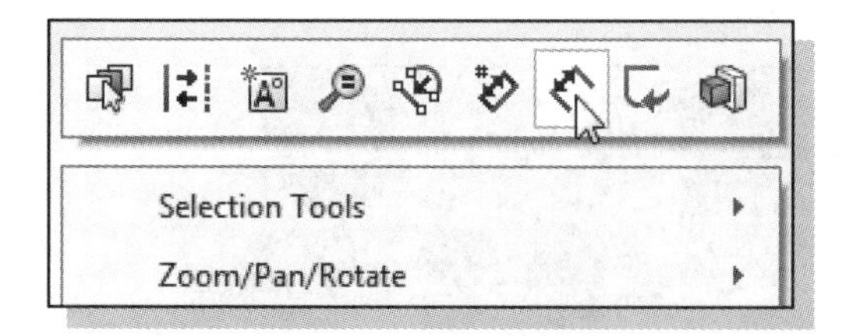

8. Inside the graphics window, click once with the right-mouse-button to display the option menu. Select the **Smart Dimension** option in the pop-up menu.

9. Create a dimension to describe the size of the circle and set it to **0.75 in**.

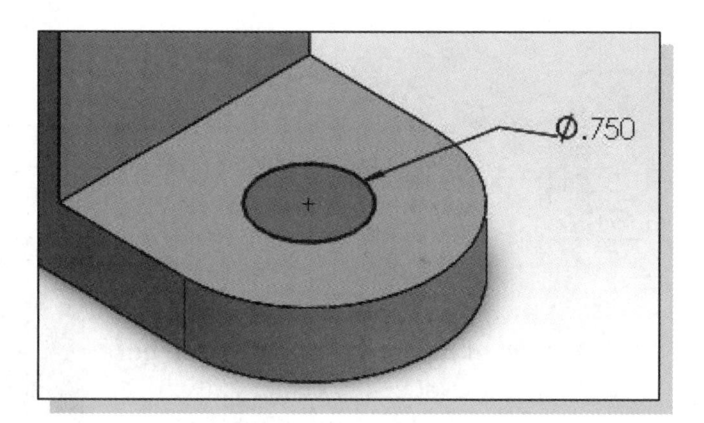

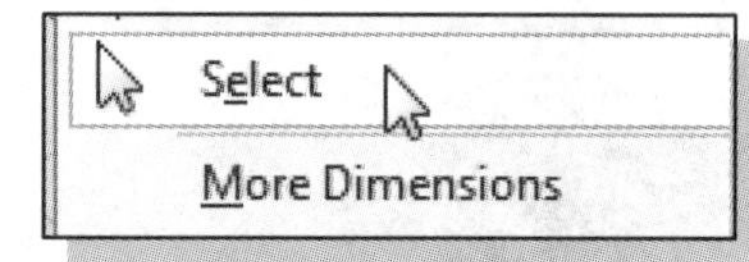

10. Inside the graphics window, click once with the right-mouse-button to display the option menu. Select **Select** in the pop-up menu to end the Smart Dimension command.

11. Click once with the **left-mouse-button** on the **Sketch** icon on the *Sketch* toolbar to exit the Sketch option.

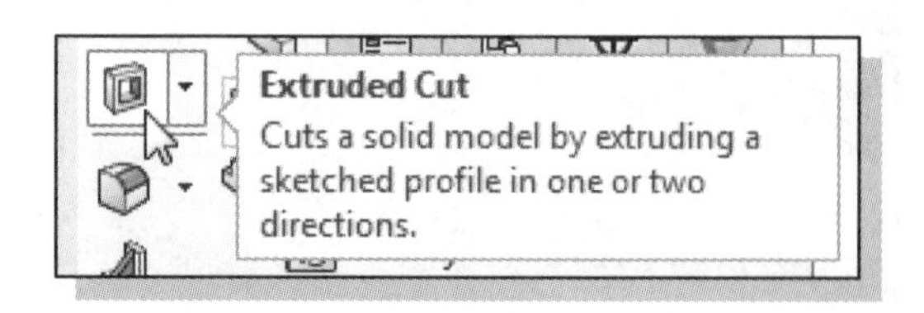

12. In the *Features* toolbar, select the **Extruded Cut** command by clicking once with the left-mouse-button on the icon.

➤ The *Cut-Extrude PropertyManager* is displayed in the left panel. Notice that the sketch region (the circle) is automatically selected as the extrusion profile.

13. In the *Cut-Extrude PropertyManager* panel, click the arrow to reveal the pull-down options for the *end condition* (the default end condition is Blind) and select **Through All** as shown.

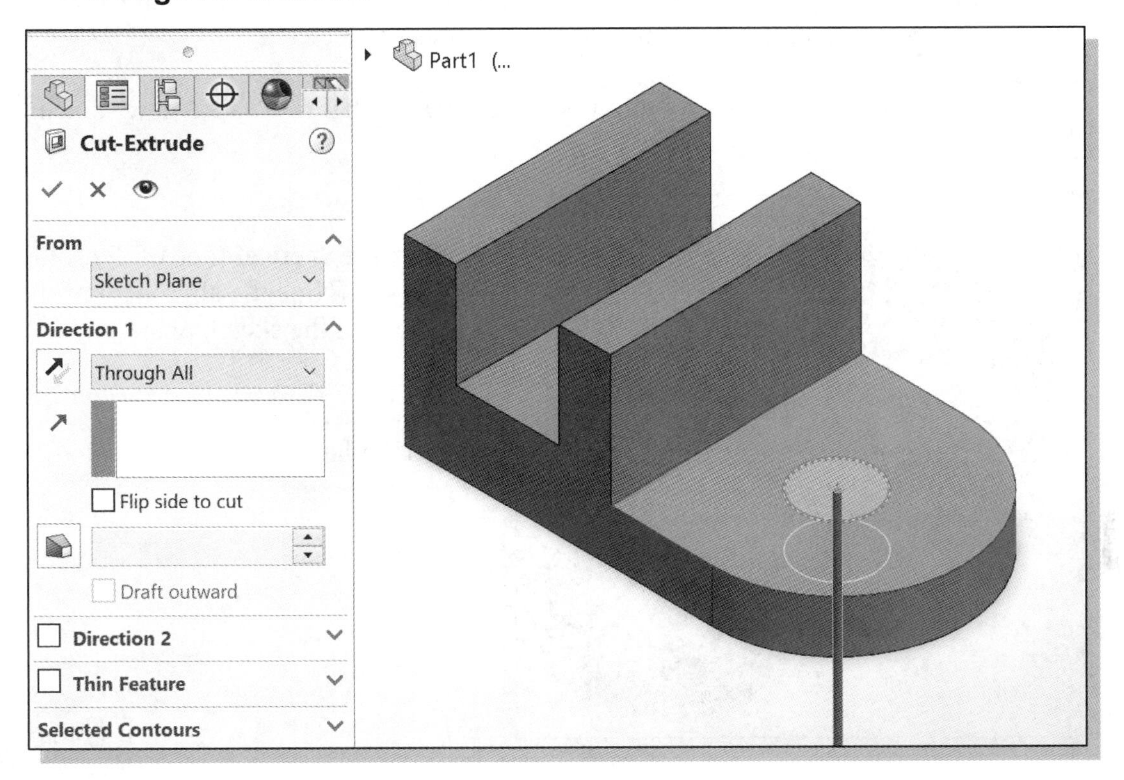

14. Click the **OK** button (green check mark) in the *Cut-Extrude PropertyManager* panel.

15. On your own, rename the feature as **3/4" Hole** in the *FeatureManager Design Tree*.

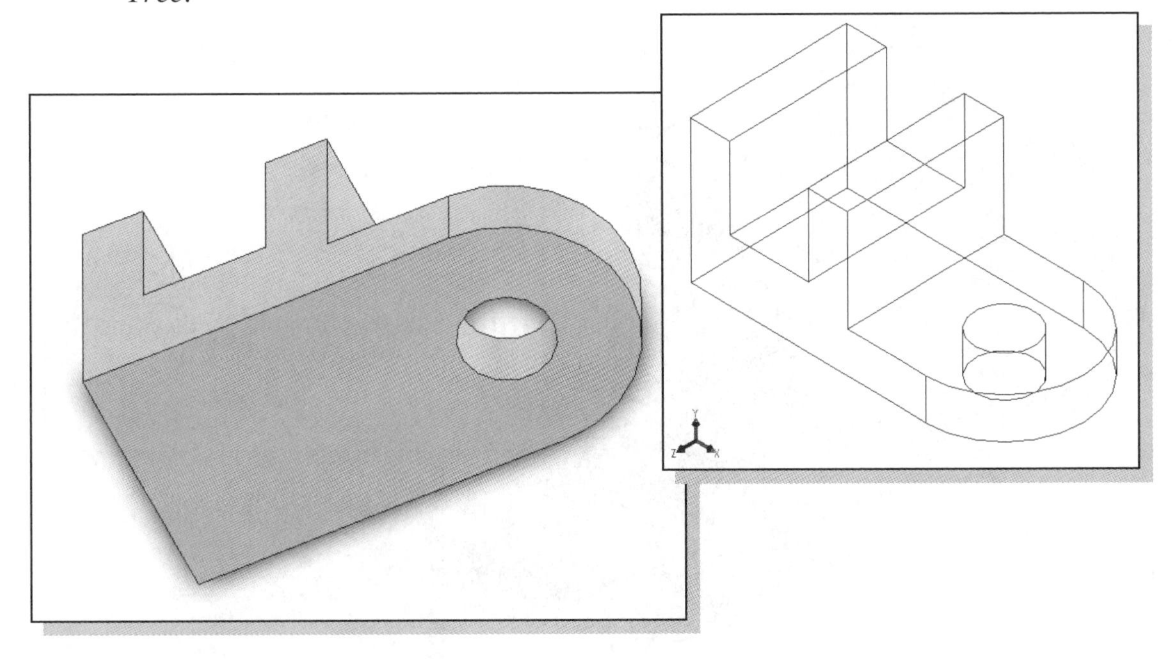

Creating a Rectangular Extruded Cut Feature

1. Move the cursor into the graphics area, away from the model, and click once with the left-mouse-button to ensure that no features are selected.

2. Select the **Sketch** button on the *Sketch* toolbar to create a new sketch.

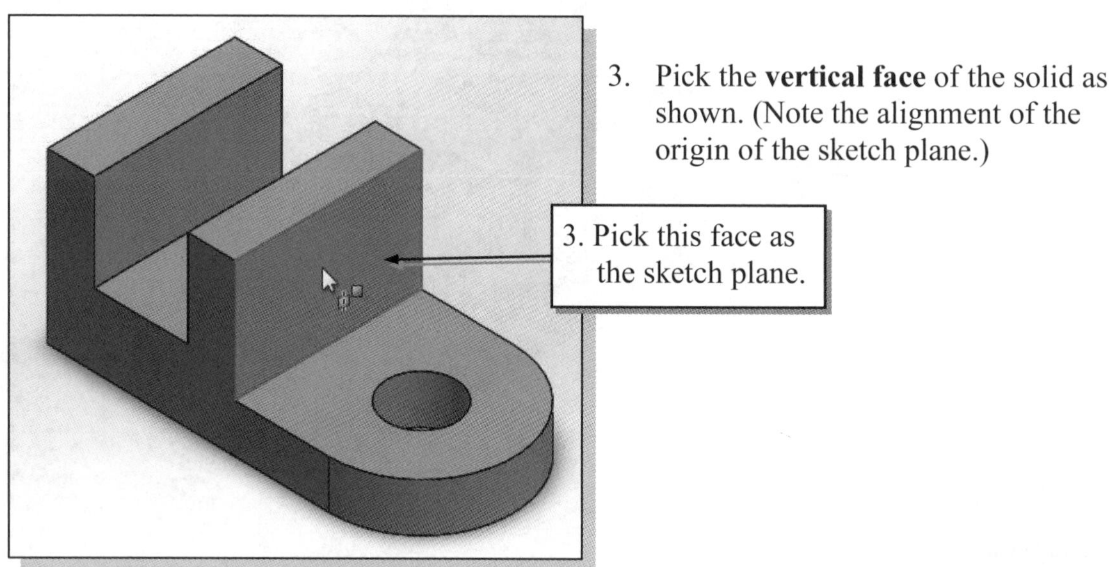

3. Pick the **vertical face** of the solid as shown. (Note the alignment of the origin of the sketch plane.)

4. On your own, sketch a rectangle (**1.0″ W x 0.75″ H** offset **0.5″** from the edge) as shown. Exit the sketch and create an **Extruded Cut** feature using the **Up To Next** option for the *End Condition* as shown. In the *FeatureManager Design Tree*, change the name of the new feature to ***Rect_Cut***.

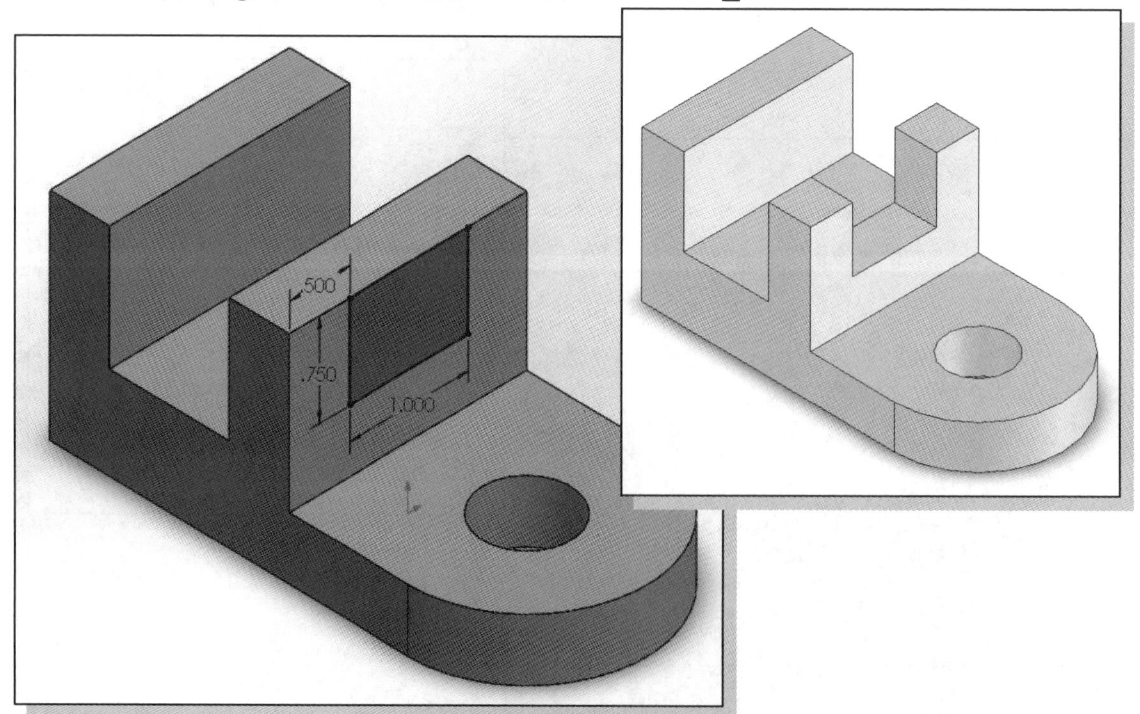

History-Based Part Modifications

SOLIDWORKS uses the *history-based part modification* approach, which enables us to make modifications to the appropriate features and re-link the rest of the history tree without having to reconstruct the model from scratch. We can think of it as going back in time and modifying some aspects of the modeling steps used to create the part. We can modify any feature that we have created. As an example, we will adjust the depth of the rectangular cutout.

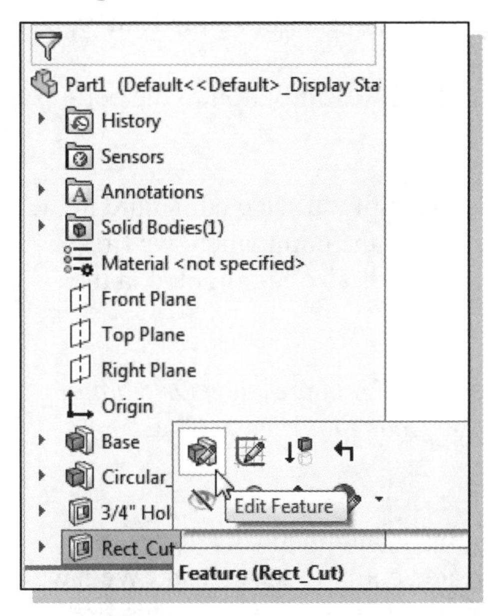

1. In the *FeatureManager Design Tree* window, select the last cut feature, **Rect_Cut**, by left-clicking once on the name of the feature.

2. In the *FeatureManager Design Tree* window, right-mouse-click once on the **Rect_Cut** feature.

3. Select the **Edit Feature** button in the pop-up menu. Notice the *Extrude PropertyManager* appears on the screen.

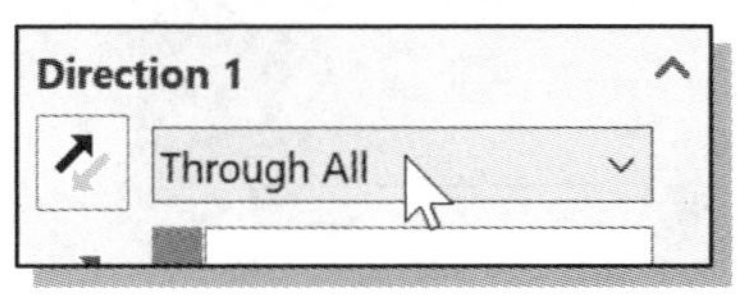

4. In the *Extrude PropertyManager*, set the termination *End Condition* to the **Through All** option.

5. Click on the **OK** button to accept the settings.

- As can be seen, the history-based modification approach is very straightforward and it only took a few seconds to adjust the cut feature to the **Through All** option.

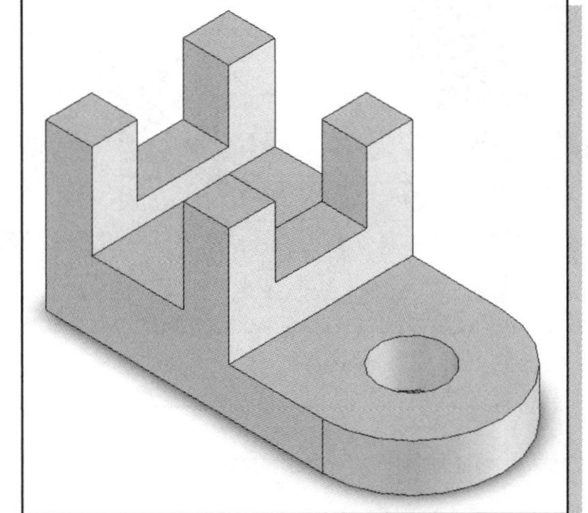

A Design Change

Engineering designs usually go through many revisions and changes. SOLIDWORKS provides an assortment of tools to handle design changes quickly and effectively. We will demonstrate some of the tools available by changing the **Base** feature of the design.

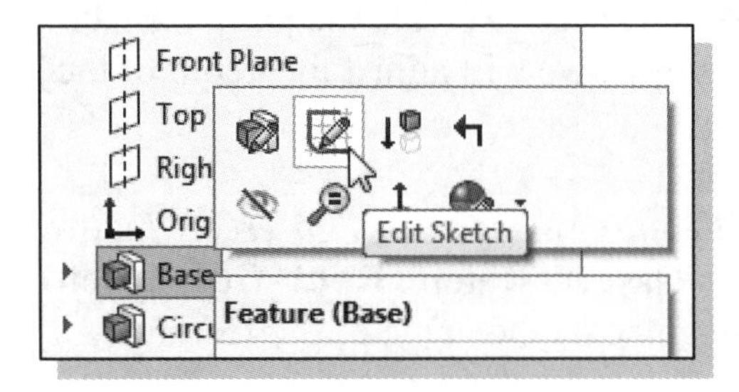

1. In the *FeatureManager Design Tree* window, select the **Base** feature by left-clicking once on the name of the feature.

2. Pick the **Edit Sketch** button in the pop-up menu.

❖ SOLIDWORKS will now display the original 2D sketch of the selected feature in the graphics window. We have literally gone back in time to the point where we first created the 2D sketch. Notice the feature being modified is also highlighted in the desktop *FeatureManager Design Tree.*

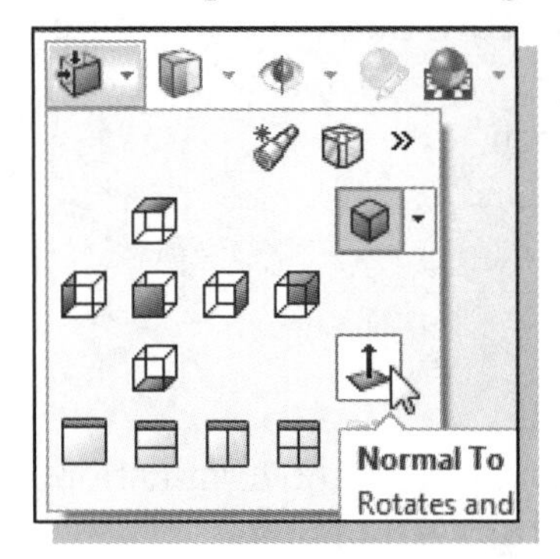

3. Click on the **Normal To** icon in the *View Orientation* pull-down menu on the *Heads-up View* toolbar.

- The **Normal To** command automatically aligns the *sketch plane* of a selected entity to the screen. We have literally gone back in time to the point where we first created the 2D sketch.

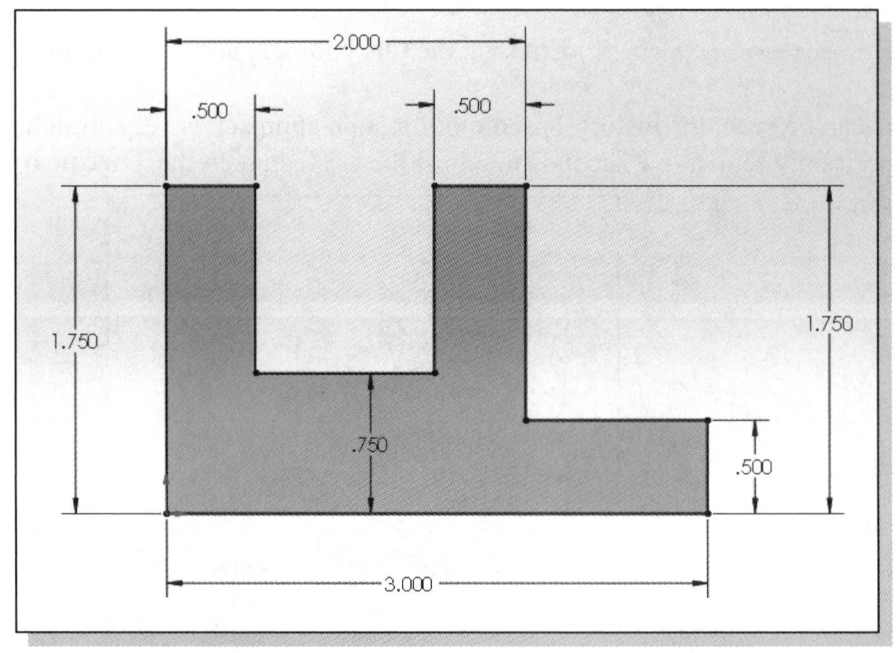

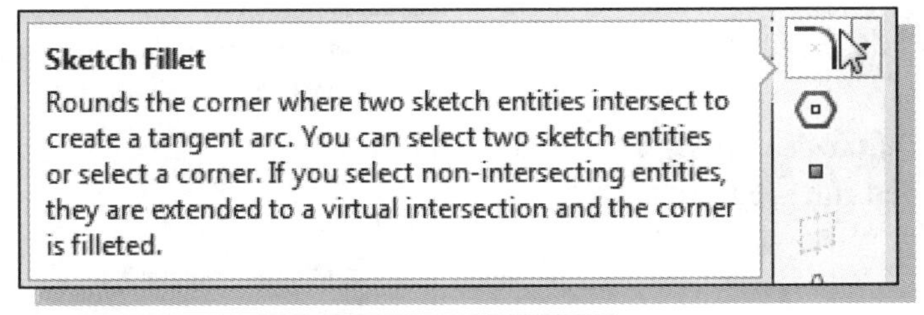

4. Select the **Sketch Fillet** command in the *Sketch* toolbar.

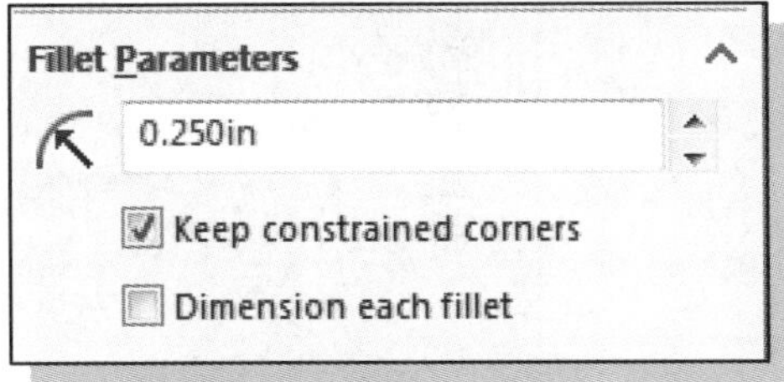

5. In the *Sketch Fillet PropertyManager*, enter **0.25** as the new radius of the fillet. (**NOTE:** If the *Fillet Parameters* panel is minimized, click on the double arrows to expand the panel.)

6. Select the two edges as shown.

7. Click the **OK** icon (green check mark) in the *PropertyManager* to create the fillet.

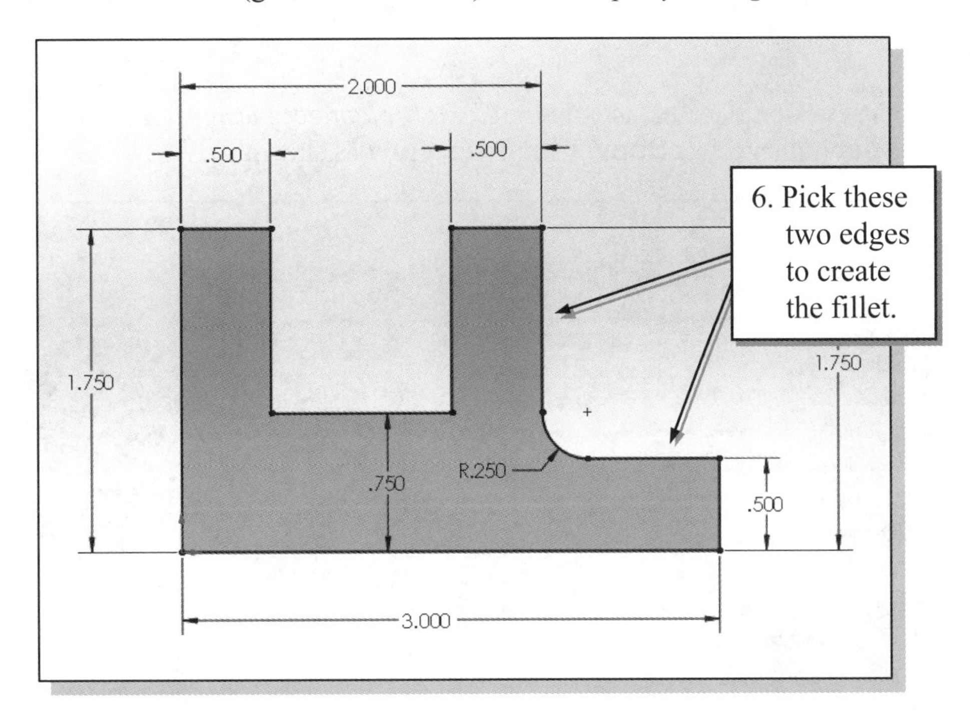

- Note that the fillet is created automatically with the dimension attached. The attached dimension can also be modified through the history tree.

8. Click the **OK** icon (green check mark) in the *PropertyManager*, or hit the [**Esc**] key once, to end the Sketch Fillet command.

9. Click on the **Rebuild** icon in the *Menu Bar*.

10. On your own, select the **Isometric** view in the *View Orientation* pull-down menu on the *Heads-up View* toolbar.

FeatureManager Design Tree Views

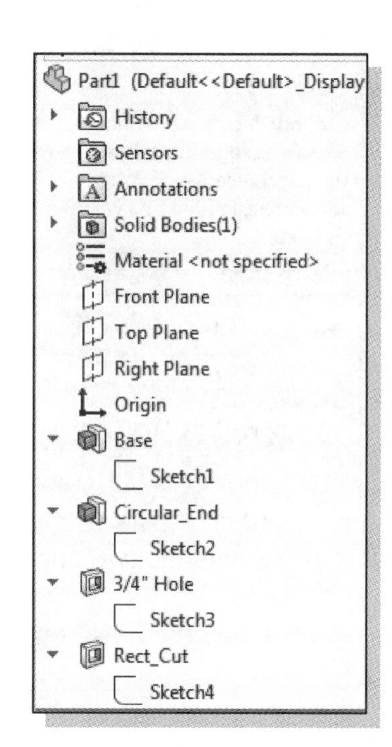

The Normal View for the *FeatureManager Design Tree* is **hierarchical**, with sketches absorbed into features.

1. Expand the *FeatureManager Design Tree* to display the absorbed sketches as shown.

➢ The *Flat Tree View* for the *FeatureManager Design Tree* displays the features in the **order they were created**, instead of hierarchically.

2. Right-click on the **Part** icon in the *FeatureManager Design Tree* and select **Tree Display**, then select **Show Flat Tree View** as shown.

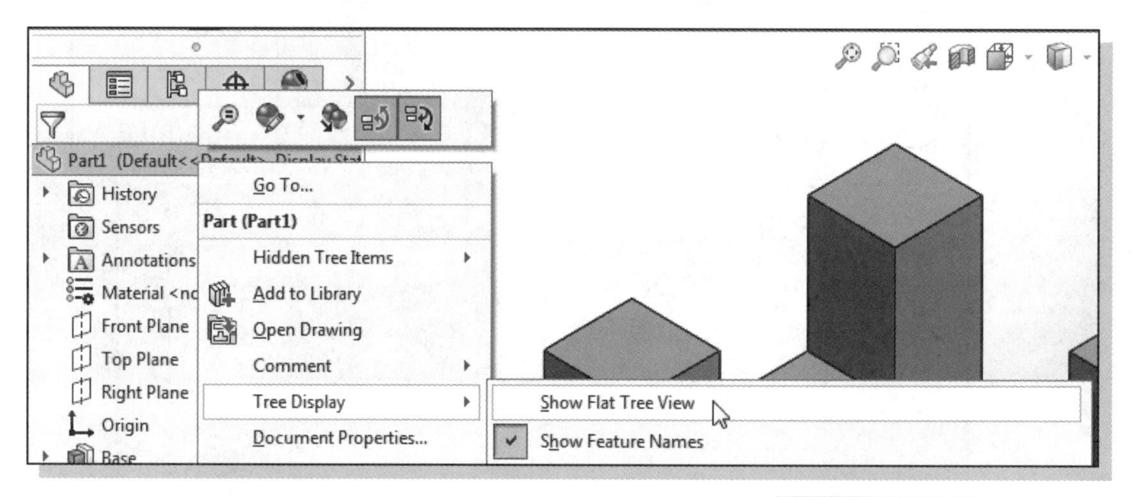

➢ Note the change in the display of the *FeatureManager Design Tree.* Items are now shown in the order of creation.

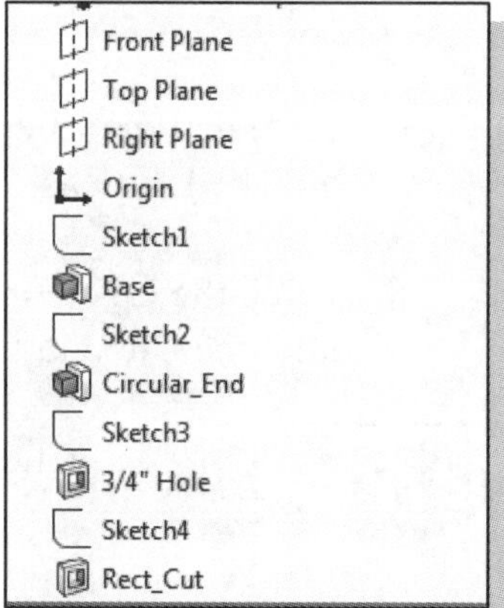

3. Right-click on the **Part** icon in the *FeatureManager Design Tree* and select **Tree Display**, then **unselect** the **Show Flat Tree View** to return to the hierarchical display.

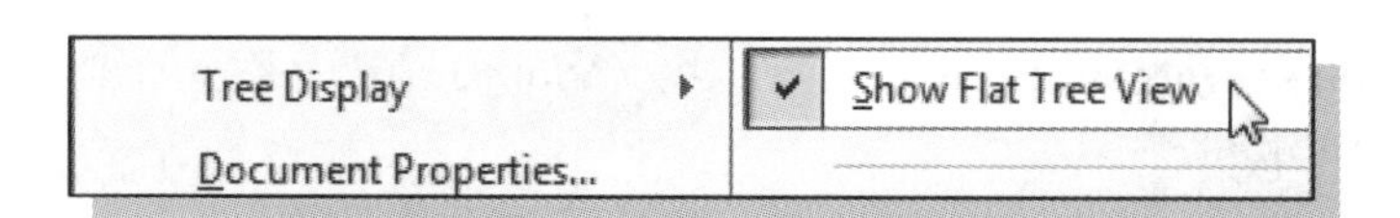

➢ The **History** folder at the top of the *FeatureManager design tree* allows you to access the features that you have most recently created or edited.

4. Expand the **History** folder by clicking on the icon to the left of the folder.

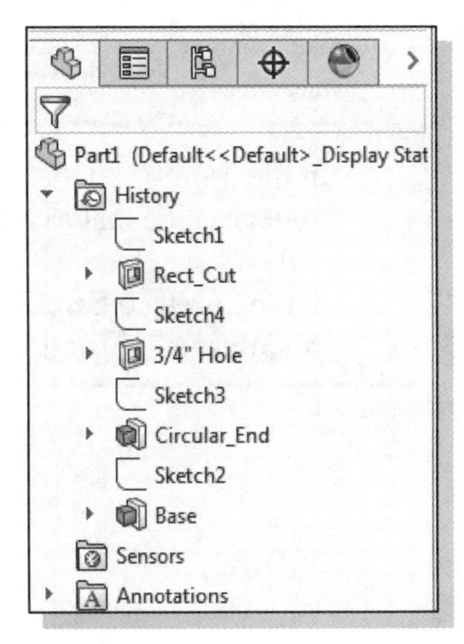

❖ In a typical design process, the initial design will undergo many analyses, testing, and reviews. The *history-based part modification* approach is an extremely powerful tool that enables us to quickly update the design. At the same time, it is quite clear that PLANNING AHEAD is also important in doing feature-based modeling.

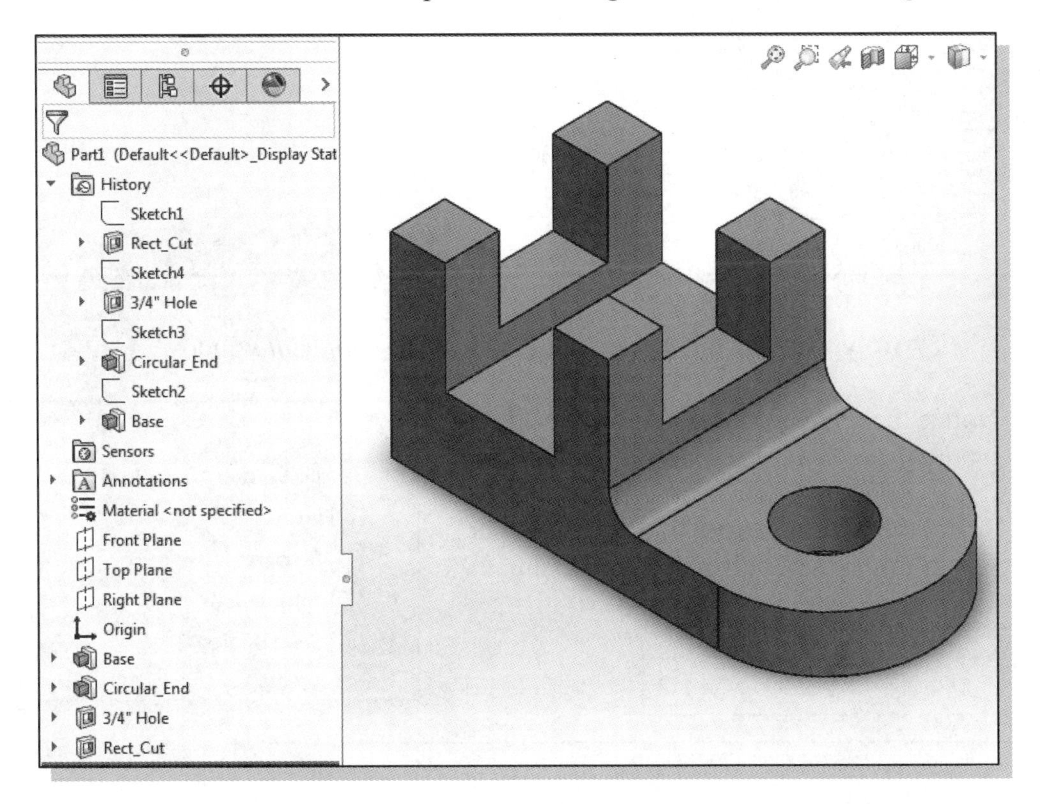

5. Save the part with the filename **Saddle_Bracket**. **NOTE:** Do **not** save the part file in the Tutorial_Templates folder.

Selecting a Material and Viewing the Mass Properties

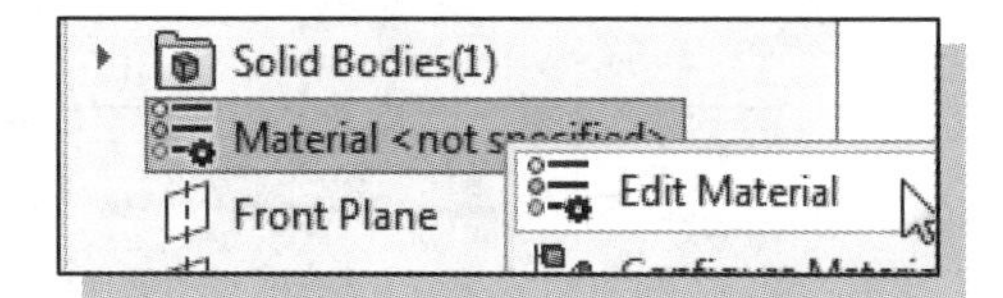

1. Right-click on the **Material** icon in the *FeatureManager Design Tree* and select **Edit Material** from the pop-up menu.

2. In the *Material* pop-up window, select **1060 Alloy** (in the *Aluminum Alloys* folder) as the material type.

3. Set the units to **English (IPS)**. Notice the material properties (Young's modulus, etc.) listed.

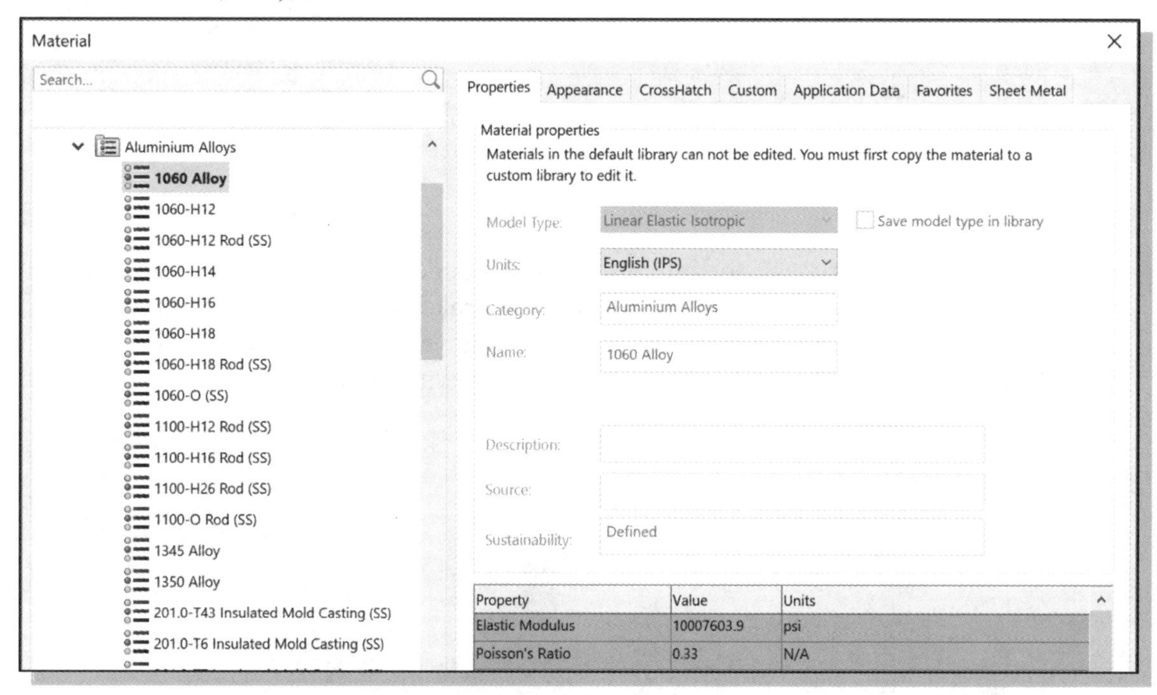

4. Click **Apply** to select the material and **Close** the *Material* window.

5. Notice the selected material now appears in the *FeatureManager Design Tree.*

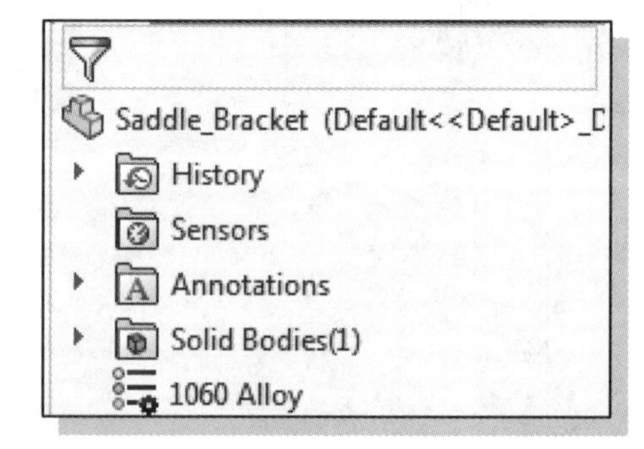

6. Select the **Mass Properties** option under Evaluate on the *Tools* pull-down menu.

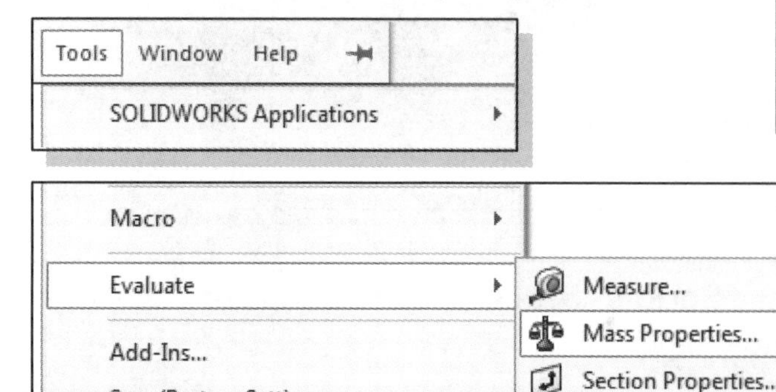

➢ The *Mass Properties* window is displayed. Notice the *Report coordinate values relative to:* window is set to the **default** system.

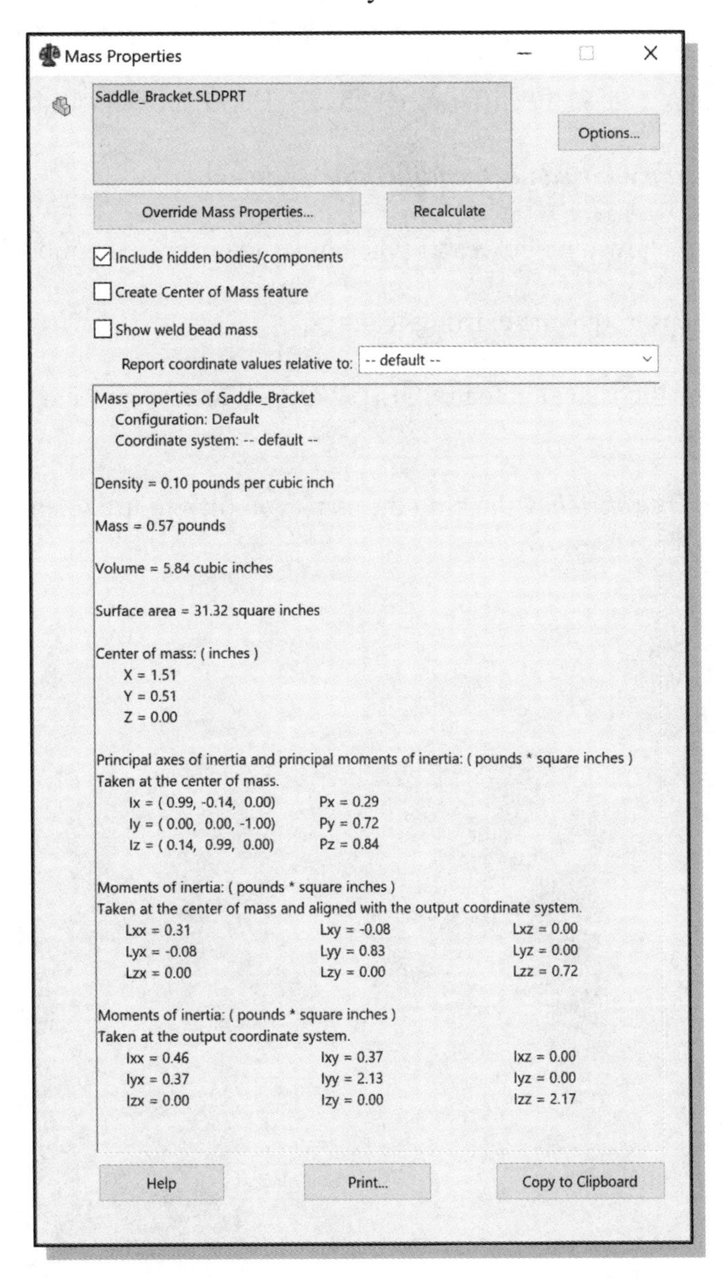

➢ Notice the *Mass* is **0.57 pounds**.

➢ Notice the *Volume* is **5.84 cubic inches**.

7. Left-Click on the **Close** icon or hit the [**Esc**] key to close the *Mass Properties* window.

8. Save the **Saddle_Bracket** part file. **NOTE:** Remember, do **not** save the part file in the Tutorial_Templates folder.

Questions:

1. What are stored in the SOLIDWORKS *FeatureManager Design Tree*?
2. When extruding, what is the difference between Blind and Through All?
3. Describe the *history-based part modification* approach.
4. What determines how a model reacts when other features in the model change?
5. Describe the steps to rename existing features.
6. Describe two methods available in SOLIDWORKS to *modify the dimension values* of parametric sketches.
7. Create *Design Tree sketches* showing the steps you plan to use to create the two models shown on page 4-32:

Ex.3)

Ex.4)

Exercises: (Dimensions are in inches.)

1. **C-Clip** (Plate thickness: **0.25 inches**.)

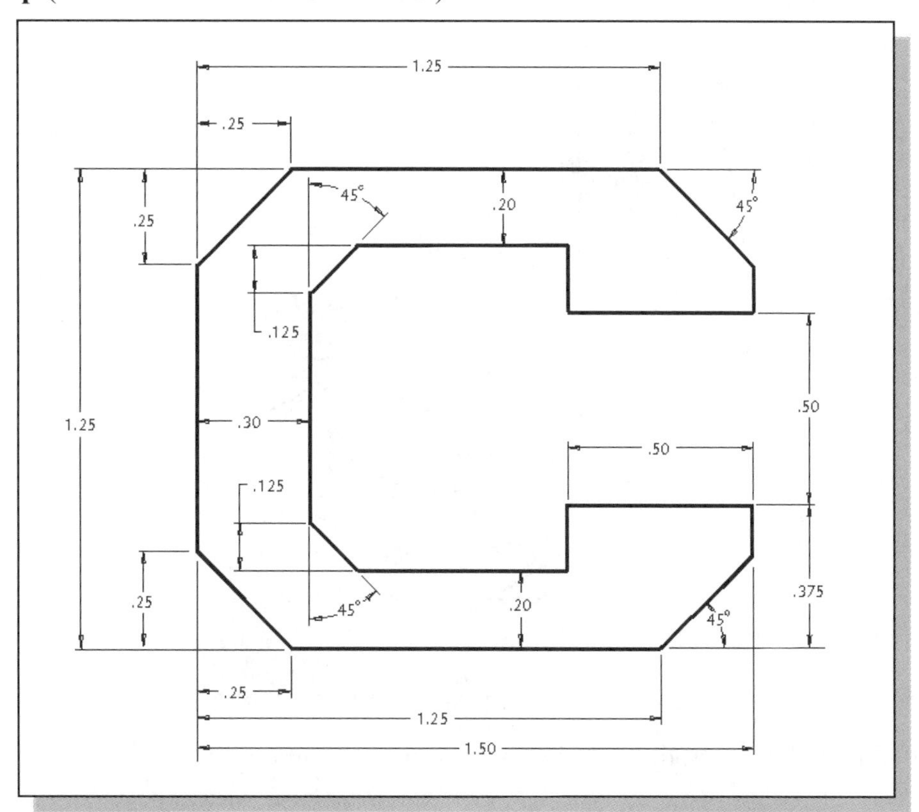

2. **Tube Mount**

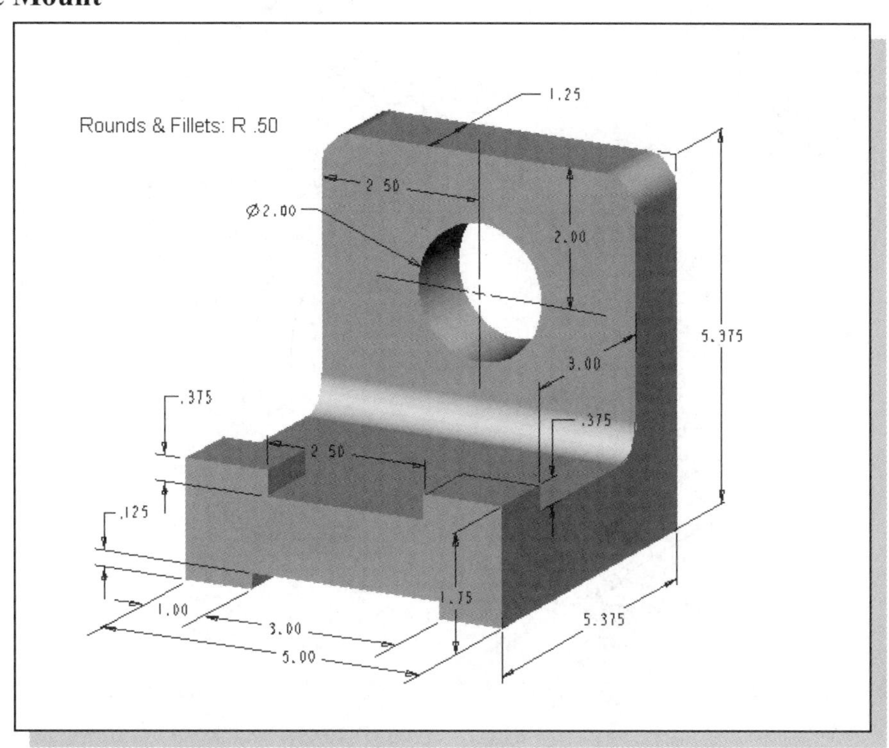

3. **Hanger Jaw** (Material: **Cast Iron**. Weight and Volume =?)

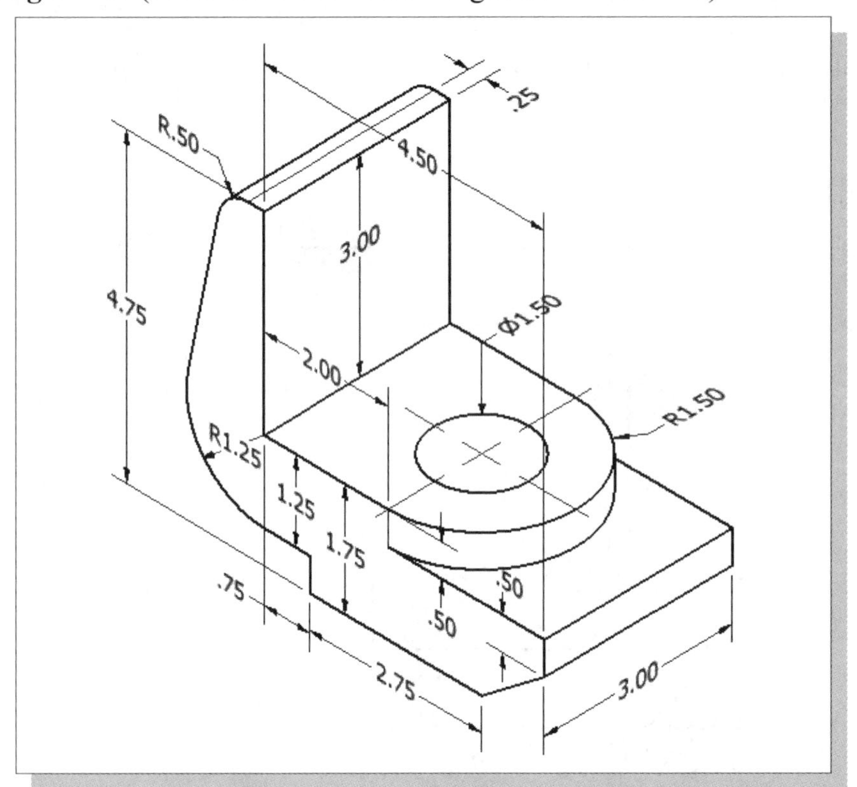

4. **Transfer Fork** (Material: **Cast Iron.** Mass and Volume =?)

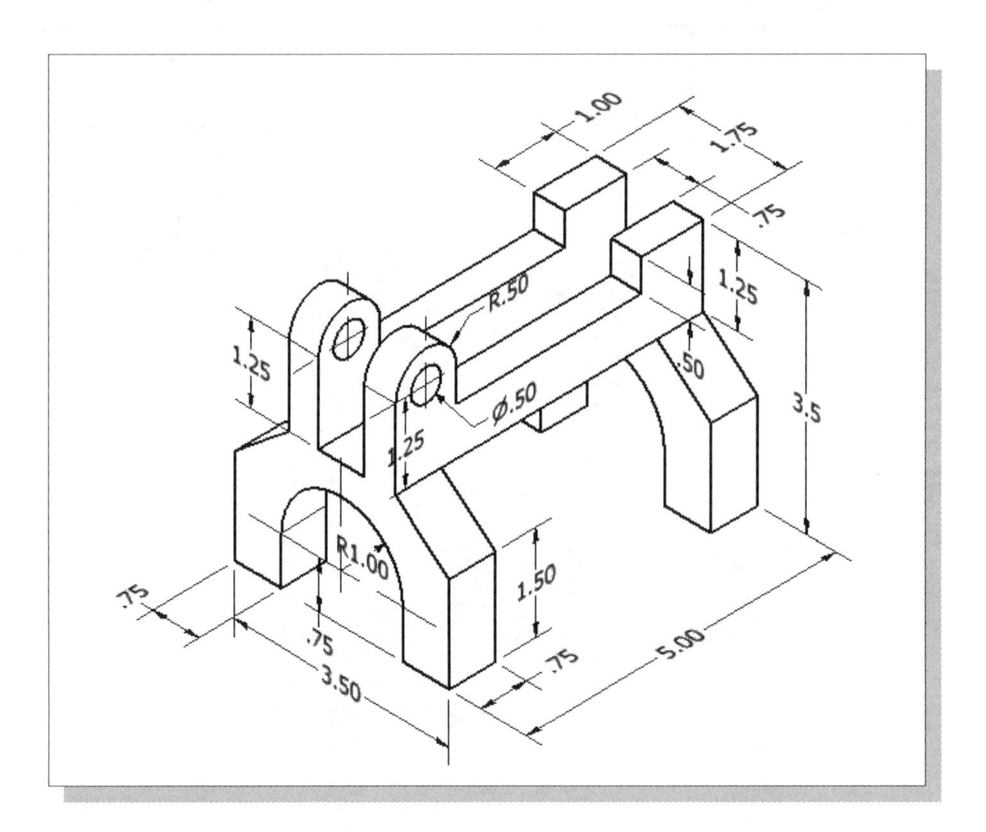

5. **Guide Slider** (Material: **Cast Iron**. Weight and Volume =?)

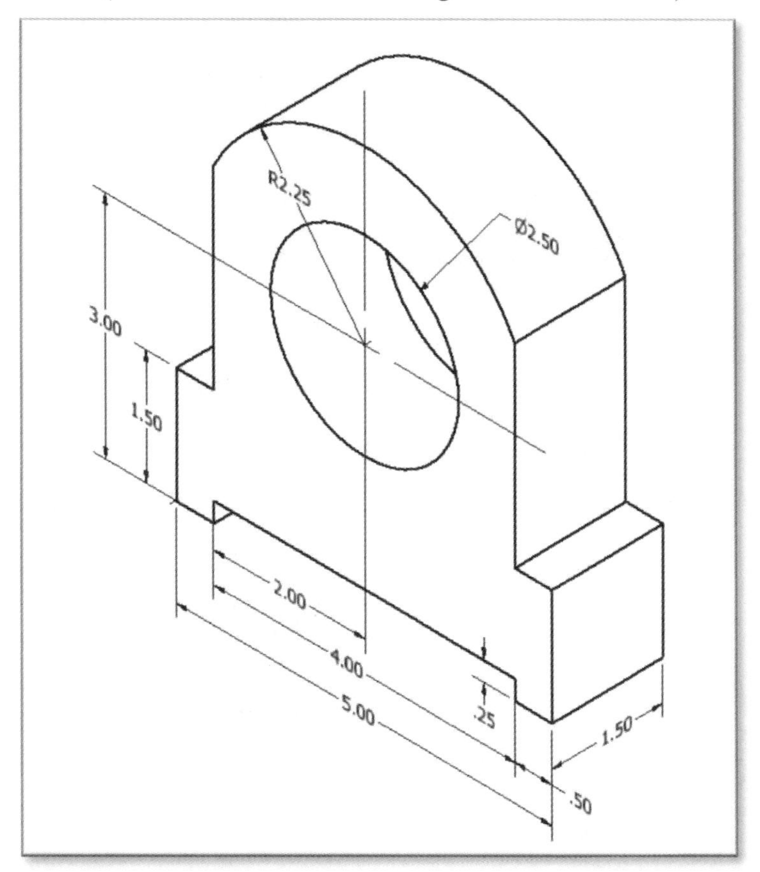

6. **Shaft Guide** (Material: **Aluminum-6061**. Mass and Volume =?)

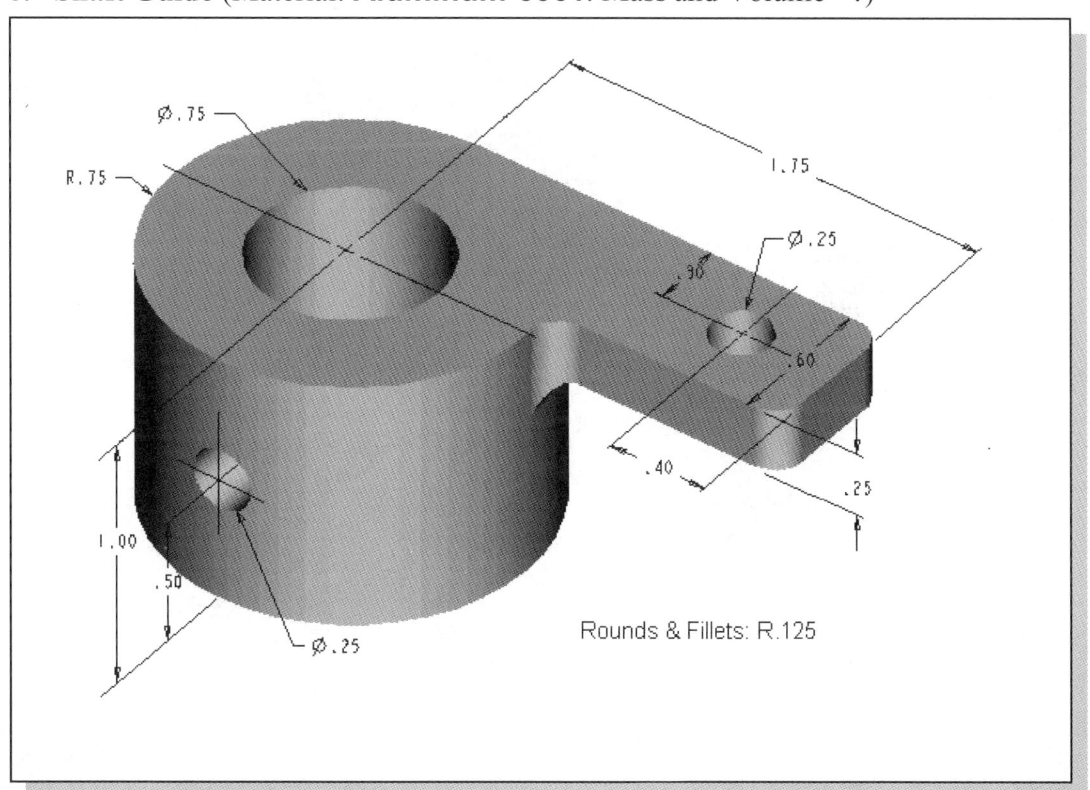

NOTES:

Chapter 5
Geometric Relations Fundamentals

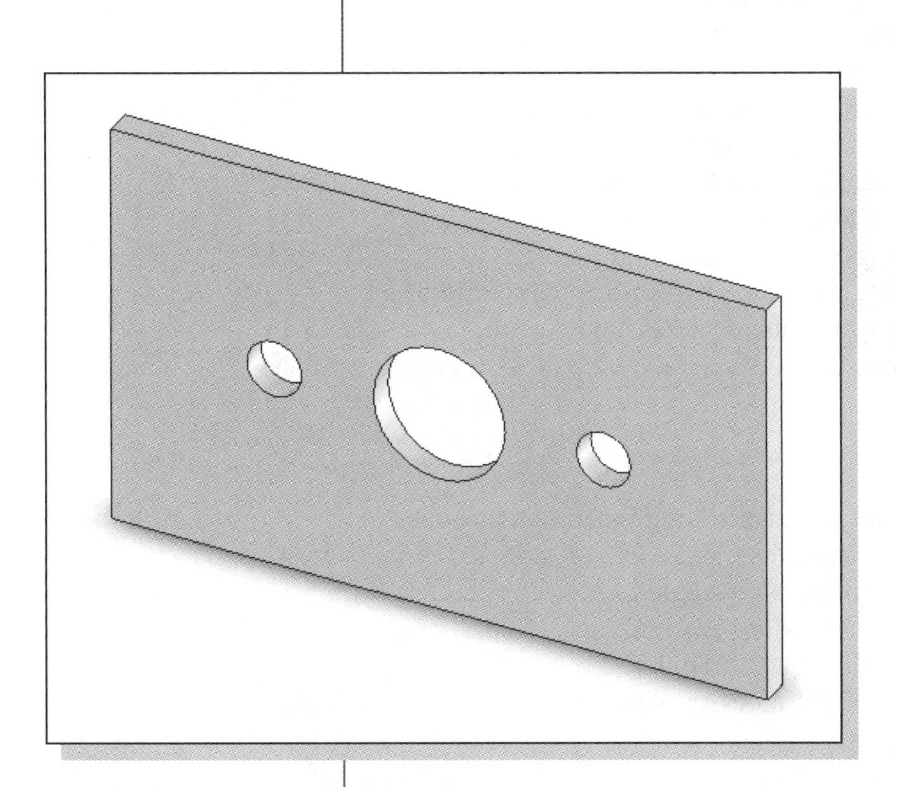

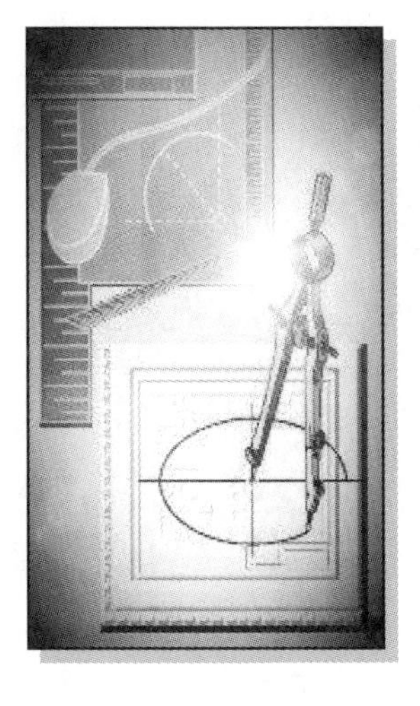

Learning Objectives

- ♦ **Create Geometric Relations**
- ♦ **Use Dimensional Variables**
- ♦ **Display, Add, and Delete Geometric Relations**
- ♦ **Understand and Apply Different Geometric Relations**
- ♦ **Display and Modify Parametric Relations**
- ♦ **Create Fully Defined Sketches**

Certified SOLIDWORKS Associate Exam Objectives Coverage

Sketch Relations

Objectives: Using Geometric Relations.

Dimensions

Objectives: Applying and Editing Smart Dimensions.

DIMENSIONS and RELATIONS

A primary and essential difference between parametric modeling and previous generation computer modeling is that parametric modeling captures the *design intent*. In the previous lessons, we have seen that the design philosophy of "*shape before size*" is implemented through the use of SOLIDWORKS' Smart Dimension commands. In performing geometric constructions, dimensional values are necessary to describe the **SIZE** and **LOCATION** of constructed geometric entities. Besides using dimensions to define the geometry, we can also apply geometric rules to control geometric entities. More importantly, SOLIDWORKS can capture design intent through the use of **geometric relations**, **dimensional constraints** and **parametric relations**. **Geometric relations** are geometric restrictions that can be applied to geometric entities; for example, *horizontal*, *parallel*, *perpendicular*, and *tangent* are commonly used *geometric relations* in parametric modeling. For part modeling in SOLIDWORKS, relations are applied to *2D sketches*. They can be added automatically as the sketch is created or by using the **Add Relation** command. **Dimensional constraints** are used to describe the SIZE and LOCATION of individual geometric shapes. They are added using the SOLIDWORKS **Smart Dimension** command. One should also realize that depending upon the way the geometric relations and dimensional constraints are applied, the same results can be accomplished by applying different constraints to the geometric entities. In SOLIDWORKS, **parametric relations** can be applied using **Global Variables** and **Equations**. Global variables are used when multiple dimensions have the same value. The dimension value is applied through the use of a named variable. SOLIDWORKS Equations are user-defined mathematical relations between model dimensions, using dimension names as variables. In parametric modeling, features are made of geometric entities with dimensional, geometric, and parametric constraints describing individual design intent. In this lesson, we will discuss the fundamentals of geometric relations and equations.

Create a *Simple Triangular Plate* Design

In parametric modeling, **geometric properties** such as *horizontal*, *parallel*, *perpendicular*, and *tangent* can be applied to geometric entities automatically or manually. By carefully applying proper **geometric relations**, very intelligent models can be created. This concept is illustrated by the following example.

Fully Defined Geometry

In SOLIDWORKS, as we create 2D sketches, geometric relations such as *horizontal* and *parallel* are automatically added to the sketched geometry. In most cases, additional relations and dimensions are needed to fully describe the sketched geometry beyond the geometric relations added by the system. Although we can use SOLIDWORKS to build partially constrained or totally unconstrained solid models, the models may behave unpredictably as changes are made. In most cases, it is important to consider the design intent, develop a modeling strategy, and add proper constraints to geometric entities. In the following sections, a simple triangle is used to illustrate the different tools that are available in SOLIDWORKS to create/modify geometric relations and dimensional constraints.

Starting SOLIDWORKS and Activating the CommandManager

1. Select the **SOLIDWORKS** option on the *Start* menu or select the **SOLIDWORKS** icon on the desktop to start SOLIDWORKS. The SOLIDWORKS main window will appear on the screen.

2. If the *Welcome* dialog box does not appear automatically upon opening SOLIDWORKS, click the *Welcome to SolidWorks* icon in the *Task Pane* or on the *Menu Bar*.

3. In order to access the custom template you created in the last chapter, click once with the left-mouse-button on the **Advanced** icon in the *Welcome* dialog box.

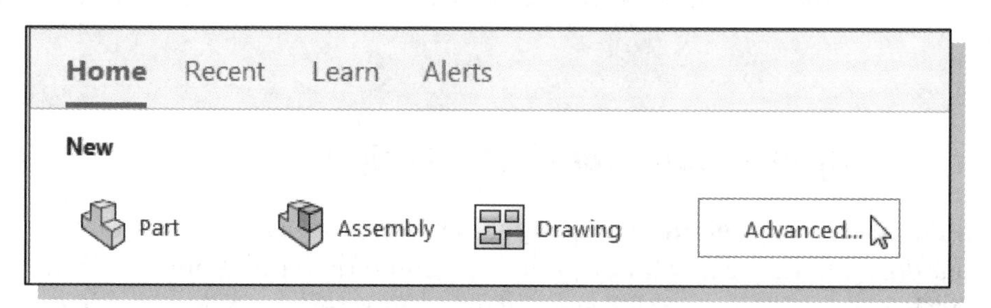

➢ **NOTE:** If the Tutorial_Templates tab and Part_IPS_ANSI template are not available, simply open a new part using the default Part template and set the *Drafting Standard* to ANSI and the *Units* to IPS (see Steps 1-7 on page 4-5). Then proceed to Step 6 below.

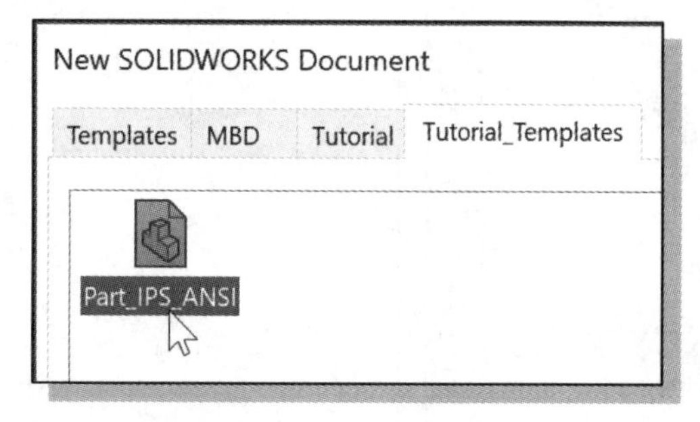

4. Select the **Tutorial_Templates** tab. (**NOTE:** You added this tab in Chapter 4.)

5. Select the **Part_IPS_ANSI** template as shown.

6. Click on the **OK** button to open a new document using the Part_IPS_ANSI template. The *Dimensioning Standard* will automatically be ANSI and the units will be set to inch, pound, second (IPS) as defined in the template.

IMPORTANT NOTE: The SOLIDWORKS *CommandManager* provides an alternate method for displaying the most commonly used toolbars (see page 1-13 for a more complete description). We will use the *CommandManager* in this lesson.

7. Right click on any toolbar and select **Enable CommandManager** at the top of the pop-up menu.

❖ Notice the *Features* and *Sketch* toolbars no longer appear at the edge of the window. These toolbars appear on the *Ribbon* display of the *CommandManager*.

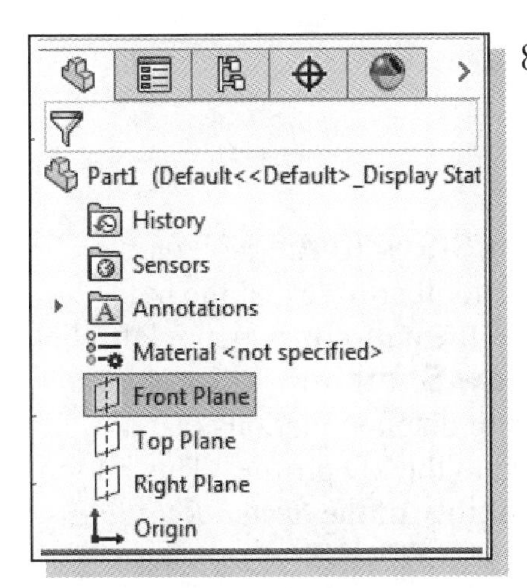

8. Move the graphics cursor to the **Front Plane** icon in the *FeatureManager Design Tree*. Click once with the **left-mouse-button** to select the **Front Plane** as the sketch plane.

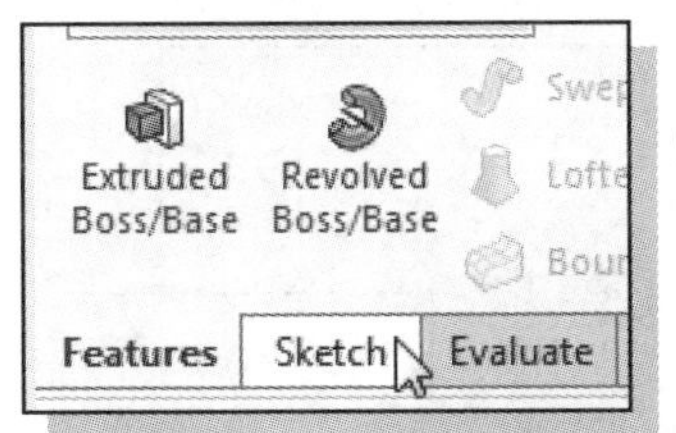

9. If necessary, select the **Sketch** tab on the *CommandManager* to display the *Sketch* toolbar.

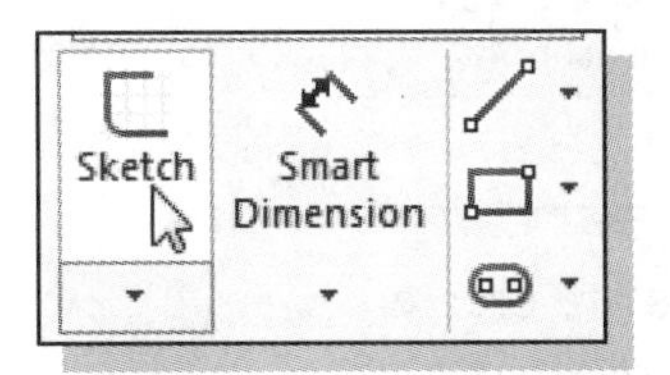

10. Select the **Sketch** button on the *Sketch* toolbar to create a new sketch.

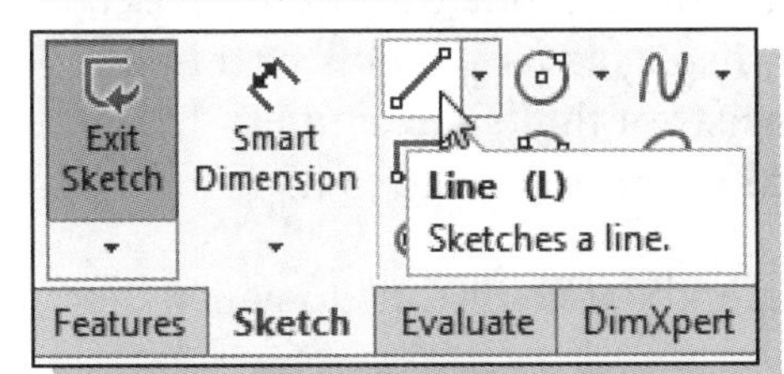

11. Select the **Line** icon on the *Sketch* toolbar by clicking once with the **left-mouse-button**.

12. Create a triangle of arbitrary size positioned near the center of the screen as shown below. (Note that the base of the triangle is horizontal.)

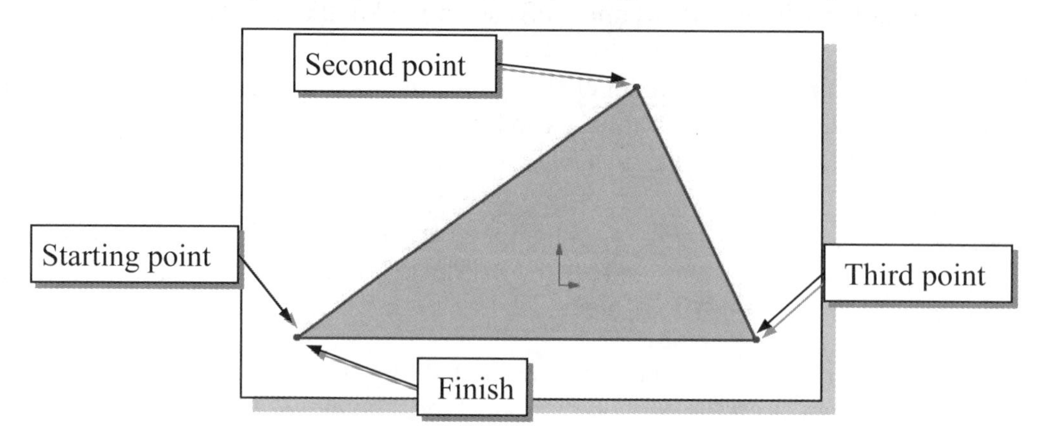

13. Press the **[Esc]** key to exit the Line command.

Displaying Existing Relations

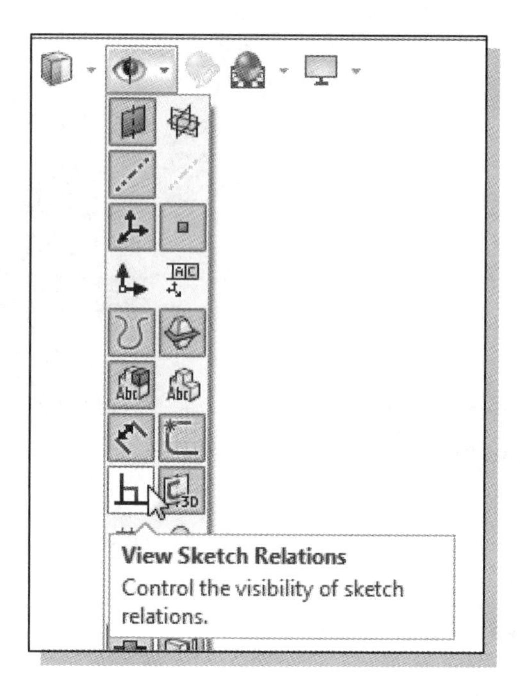

1. Select the **Hide/Show Items** icon on the *Heads-up View* toolbar to reveal the pull-down menu. In the pull-down menu, left-click once on the **View Sketch Relations** button. This allows us to display relations that are already applied to the 2D profile. This button toggles the visibility of the *Sketch Relations* symbols *ON* and *OFF*.

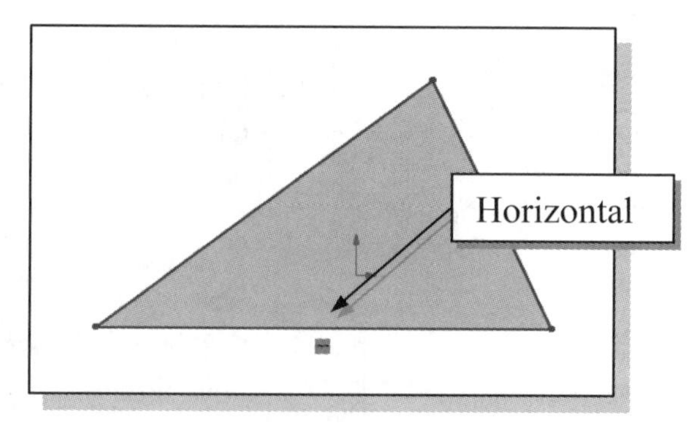

➢ In SOLIDWORKS, relations are applied as geometric entities are created. SOLIDWORKS will attempt to add proper relations to the geometric entities based on the way the entities were created. Relations are displayed as symbols next to the entities as they are created. The current profile consists of three line entities: three straight lines. The horizontal line has a **Horizontal** ***relation*** applied to it.

2. Move the cursor on top of the **Horizontal** icon and notice the line to which this relation is applied is highlighted.

3. On your own, toggle the visibility of the *Sketch Relations* symbols *OFF*.

Applying Geometric Relations/Dimensional Constraints

- SOLIDWORKS relations for 2D sketches are summarized below.

Icon	Relation	Entities Selected	Effect
	Perpendicular	Two lines	Causes selected lines to lie at right angles to one another.
	Parallel	Two or more lines	Causes selected lines to lie parallel to one another.
	Tangent	An arc, spline, or ellipse and a line or arc	Constrains two curves to be tangent to one another.
	Coincident	A point and a line, arc, or ellipse	Constrains a point to a curve.
	Midpoint	Two lines or a point and a line	Causes a point to remain at the midpoint of a line.
	Concentric	Two or more arcs, or a point and an arc	Constrains selected items to the same center point.
	Collinear	Two or more lines	Causes selected lines to lie along the same line.
	Horizontal	One or more lines or two or more points	Causes selected items to lie parallel to the X-axis of the sketch coordinate system.
	Vertical	One or more lines or two or more points	Causes selected items to lie parallel to the Y-axis of the sketch coordinate system.
	Equal	Two or more lines or two or more arcs	Constrains selected arcs/circles to the same radius or selected lines to the same length.
	Fix	Any entity	Constrains selected entities to a fixed location relative to the sketch coordinate system. However, endpoints of a fixed line, arc, or elliptical segment are free to move along the underlying fixed curve.
	Symmetric	A centerline and two points, lines, arcs or ellipses	Causes items to remain equidistant from the centerline, on a line perpendicular to the centerline.
	Coradial	Two or more arcs	Causes the selected arcs to share the same centerpoint and radius.
	Intersection	Two lines and one point	Causes the point to remain at the intersection of the two lines.
	Merge Points	Two points (sketchpoints or endpoints)	Causes the two points to be merged into a single point.

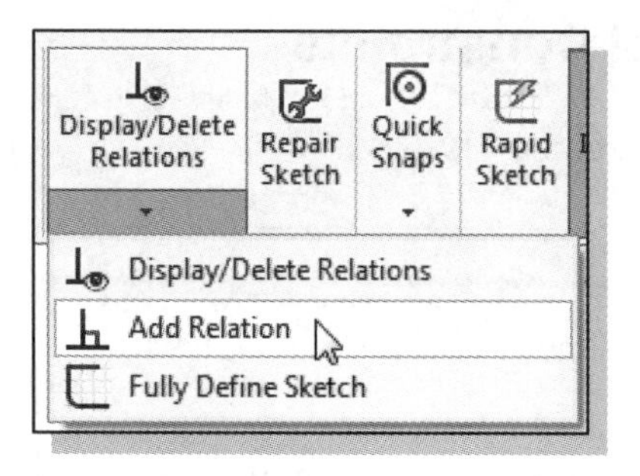

1. **Left click** on the arrow on the **Display/Delete Relations** button on the *Sketch* toolbar to reveal additional sketch relation commands.

2. Select the **Add Relation** command from the pop-up menu. Notice the *Add Relations PropertyManager* appears. The *Selected Entities* window in the *Add Relations PropertyManager* is blank because no entities are selected.

3. Select the lower right corner of the triangle by clicking once with the **left-mouse-button**.

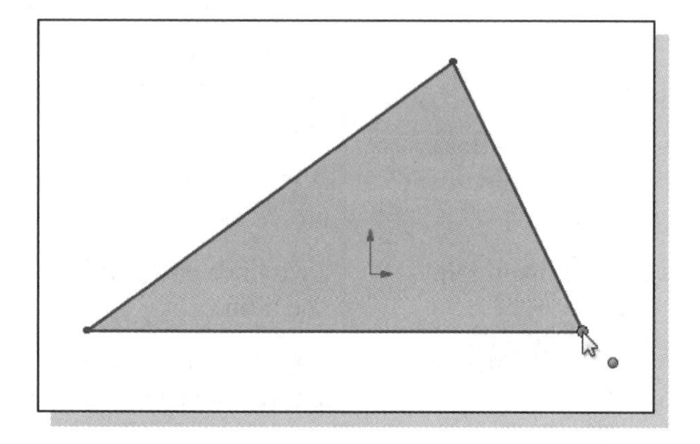

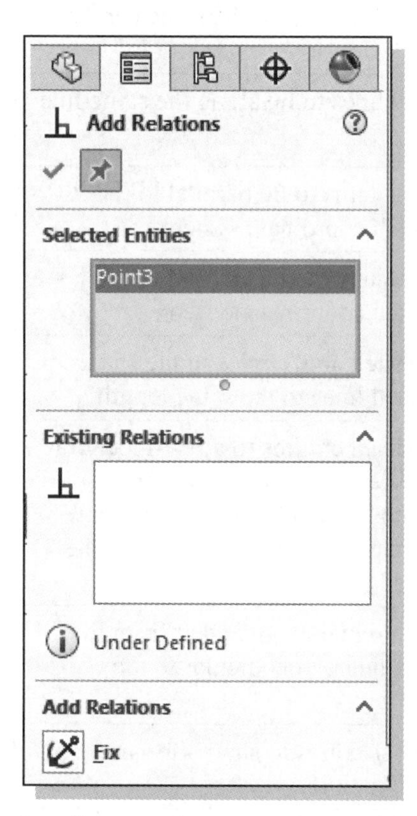

➢ Look at the *Add Relations PropertyManager*. In the *Selected Entities* window '**Point3**' is now displayed. No entries appear in the *Existing Relations* window. In the *Add Relations* menu, the **Fix** relation is displayed. This represents the only relation which can be added for the selected entity.

4. Click once with the left-mouse-button on the **Fix** icon in the *Add Relations PropertyManager* as shown. This activates the **Fix** relation.

5. Click the **OK** icon in the *PropertyManager*, or hit the [**Esc**] key once, to end the Add Relations command.

6. On your own, turn *ON* the sketch relations visibility to confirm the Fix constraint is properly applied, then turn the visibility *OFF* again.

➢ Geometric constraints can be used to control the direction in which changes can occur. For example, in the current design we will add a horizontal dimension to control the length of the horizontal line. If the length of the line is modified to a greater value, SOLIDWORKS will lengthen the line toward the left side. This is due to the fact that the Fix constraint will restrict any horizontal movement of the horizontal line toward the right side.

7. Select the horizontal line by clicking once with the **left-mouse-button**.

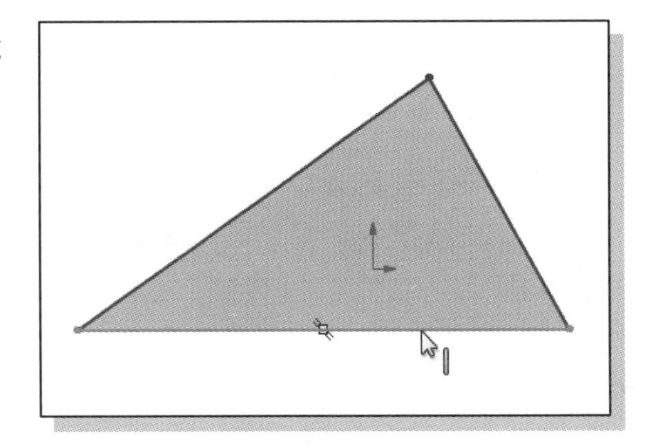

➢ Look at the *Line Properties PropertyManager*. The Horizontal relation appears in the *Existing Relations* window. In the *Parameters* window, a value is displayed for the length. This is the current (as drawn) length of the line.

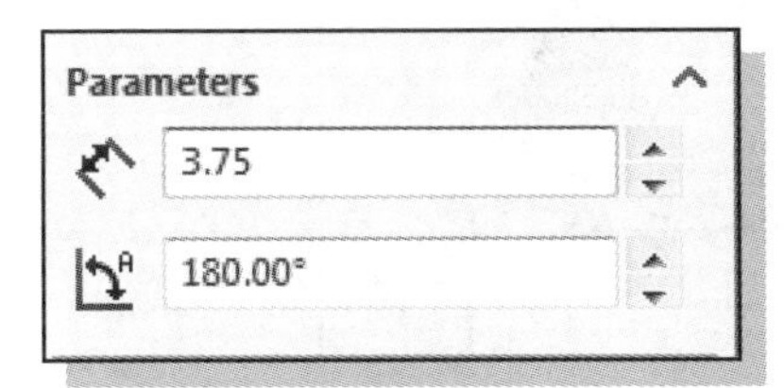

8. Enter **3.75** for the Length in the *Parameters* window as shown.

➢ Entering values in the *Parameters* panel of the *PropertyManager* will set the length of the line to the entered value. However, it **will not define or constrain the length** to remain that value. We will demonstrate this next.

9. Move the cursor on top of the **lower left corner** of the triangle.

10. Click and drag the corner of the triangle and note that the corner can be moved to a new location. Release the mouse button at a new location and notice the corner is adjusted only in the left or right direction.

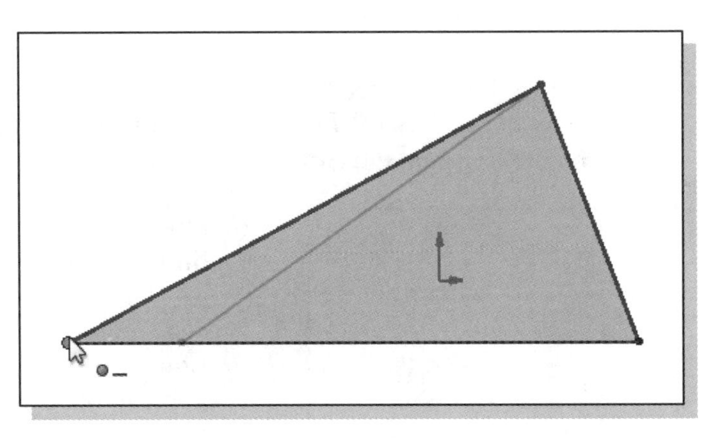

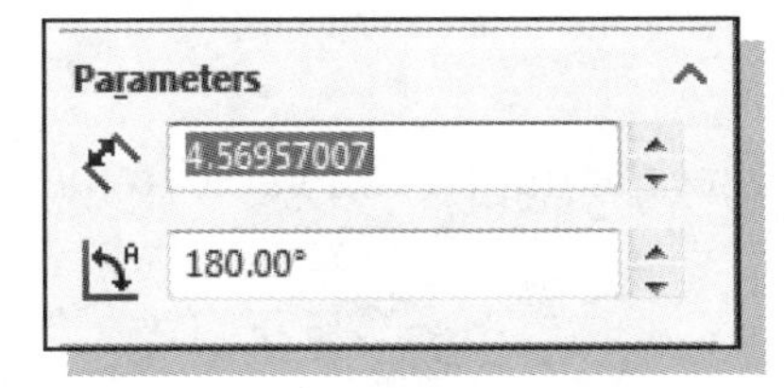

➢ Note that the two adjacent lines are automatically adjusted. Also notice that the Length entry in the *Parameter* window on the *PropertyManager* has changed to reflect the new length for the horizontal line.

➢ It is important to note the difference between entering a value in the *Parameters* window and the application of a constraining **Smart Dimension**. We will now add a Smart Dimension to define the length of the horizontal line.

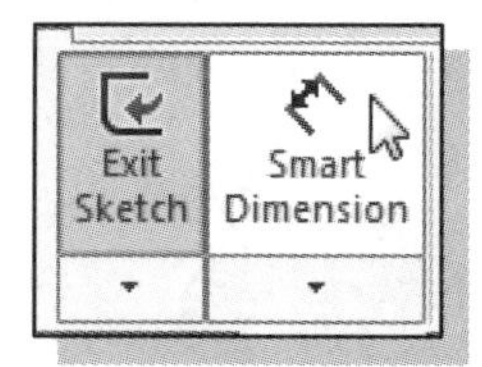

11. Select the **Smart Dimension** command by clicking once with the **left-mouse-button** on the icon in the *Sketch* toolbar.

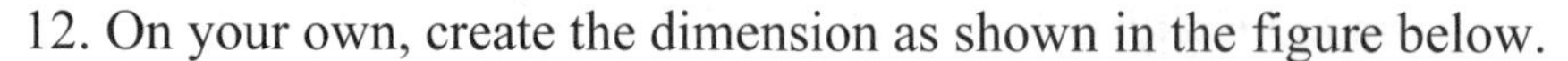

12. On your own, create the dimension as shown in the figure below.

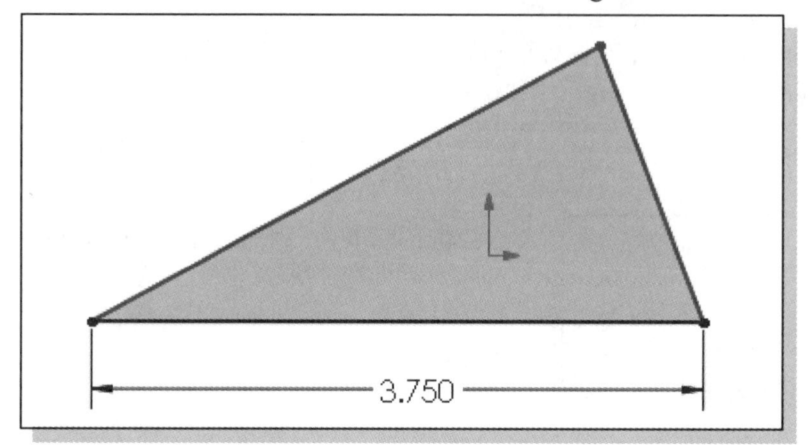

13. Press the [**Esc**] key once to exit the Smart Dimension command.

14. On your own, try to drag the lower left corner of the triangle to a new location.

➢ Notice that you cannot drag the lower left corner to a new location. Its position is fully defined by the Fix relation at the lower-right corner, the Horizontal relation on the line, and the Smart Dimension.

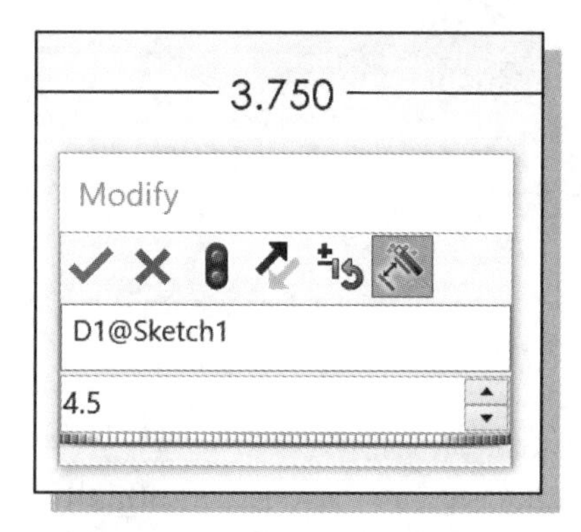

15. Double-click with the left-mouse-button on the dimension text in the graphics area to open the *Modify* dialog box.

16. Enter a value that is greater than the displayed value to observe the effects of the modification. (For example, the dimension value is 3.75, so enter **4.5** in the text box area.) Click the **OK** button in the *Modify* dialog box.

17. On your own, use the **Undo** command to reset the dimension value to the previous value.

18. **Left click** on the arrow on the **Display/Delete Relations** button on the *Sketch* toolbar to reveal additional sketch relation commands.

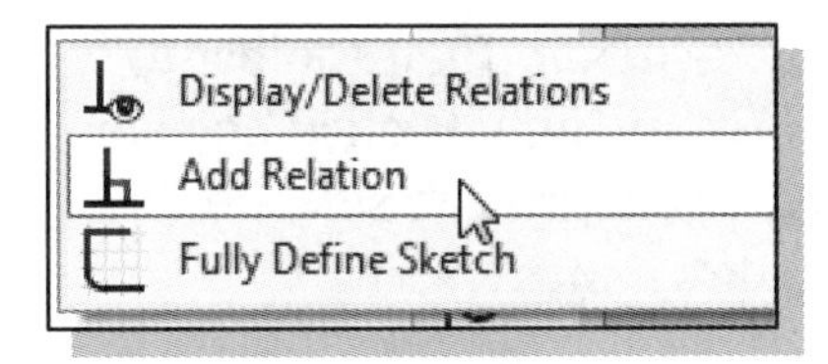

19. Select the **Add Relation** command from the pop-up menu. Notice the *Add Relations PropertyManager* appears. The *Selected Entities* window in the *Add Relations PropertyManager* is blank because no entities are selected.

20. Select the inclined line on the right by clicking once with the **left-mouse-button** as shown in the figure to the right.

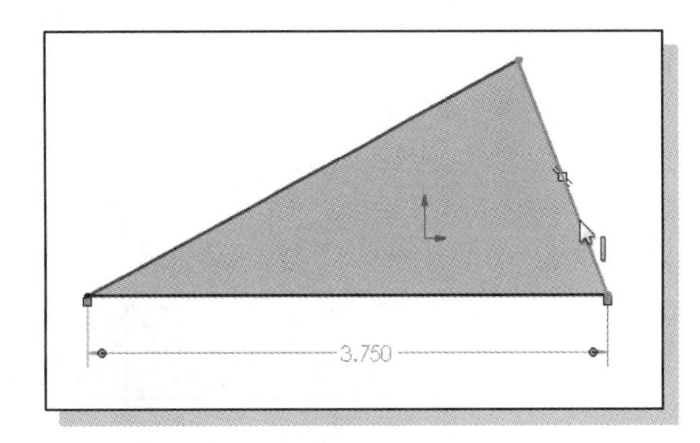

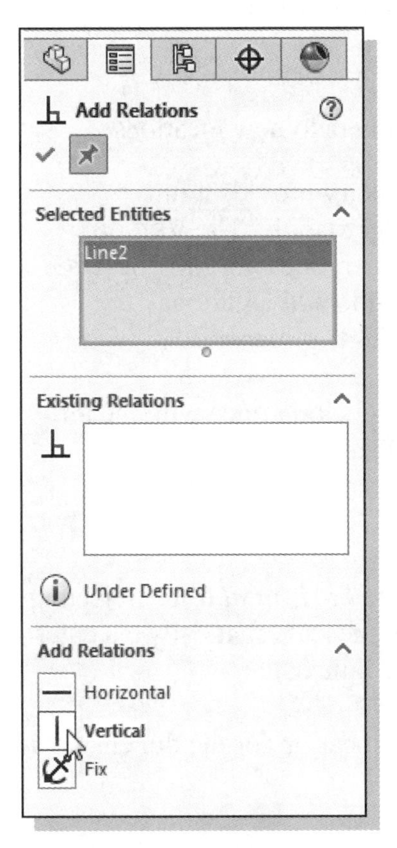

➢ Look at the *Add Relations PropertyManager*. In the *Selected Entities* text box **'Line2'** is now displayed. There are no relations in the *Existing Relations* text box. In the *Add Relations* panel, the **Horizontal**, **Vertical**, and **Fix** relations are displayed. These are the relations which can be added for the selected entity.

21. Click once with the left-mouse-button on the **Vertical** icon in the *Add Relations PropertyManager* as shown. This activates the **Vertical** relation.

22. Click the **OK** icon in the *PropertyManager*, or hit the [**Esc**] key once, to end the Add Relations command.

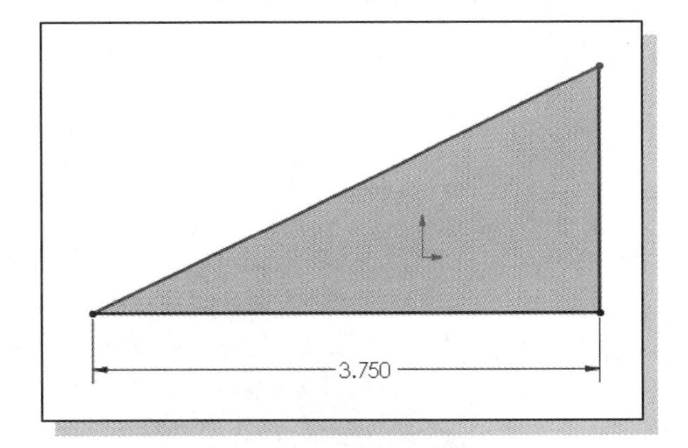

➢ One should think of the relations and dimensions as defining elements of the geometric entities. How many more relations or dimensions will be necessary to fully constrain the sketched geometry? Which relations or dimensions would you use to fully define the sketched geometry?

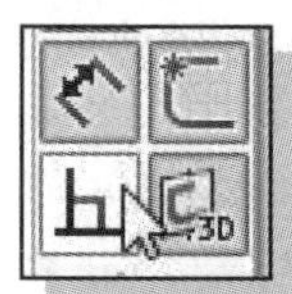

23. Select the **Hide/Show Items** icon on the *Heads-up View* toolbar to reveal the pull-down menu. In the pull-down menu, left-click once on the **View Sketch Relations** button. (After making the selection, left-click in the graphics area to close the pull-down menu.)

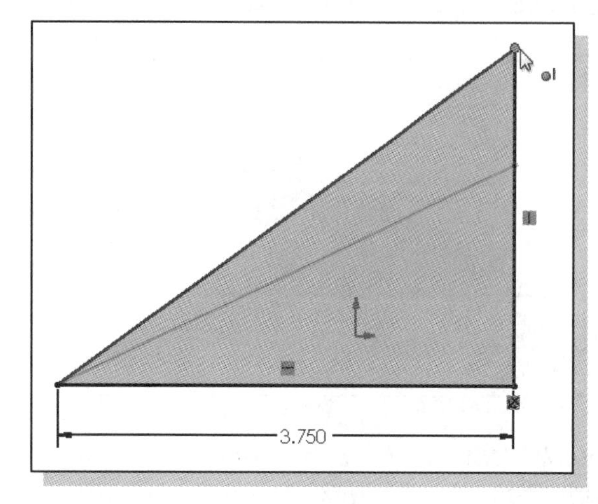

24. Move the cursor on top of the top corner of the triangle.

25. Click and drag the top corner of the triangle and note that the corner can be moved to a new location. Release the mouse button at a new location and notice the corner is adjusted only in an upward or downward direction. Note that the two adjacent lines are automatically adjusted to the new location.

26. On your own, experiment with dragging the other corners to new locations.

- The relations and dimensions that are applied to the geometry provide a full description for the location of the two lower corners of the triangle. The Vertical relation, along with the Fix relation at the lower right corner, does not fully describe the location of the top corner of the triangle. We will need to add additional information, such as the length of the vertical line or an angle dimension.

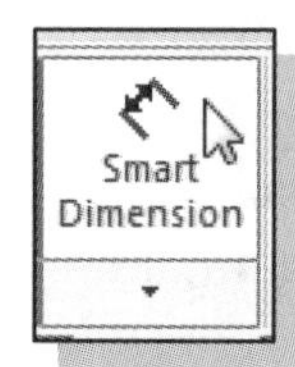

27. Select the **Smart Dimension** command by clicking once with the **left-mouse-button** on the icon in the *Sketch* toolbar.

28. Select the ***inclined line***.

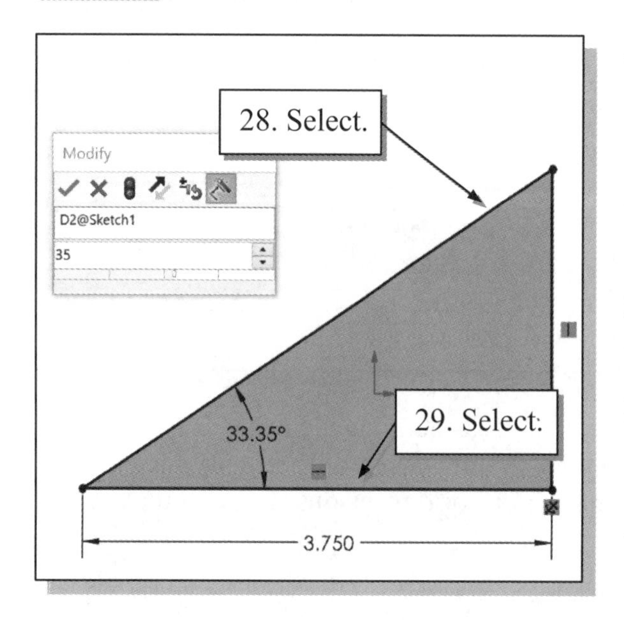

29. Select the ***horizontal line***. Selecting the two lines automatically executes an angle dimension.

30. Select a location for the dimension as shown.

31. Enter **35°** in the *Modify* dialog box and select **OK**.

32. Press the [**Esc**] key once to exit the Smart Dimension command.

- The geometry is fully defined with the added dimension.

Over-Defining and Driven Dimensions

We can use SOLIDWORKS to build partially defined solid models. In most cases, these types of models may behave unpredictably as changes are made. However, SOLIDWORKS will not let us over-define a sketch; additional dimensions can still be added to the sketch, but they are used as references only. These additional dimensions are called ***driven dimensions***. *Driven dimensions* do not constrain the sketch; they only reflect the values of the dimensioned geometry. They are shaded differently (grey by default) to distinguish them from normal (parametric) dimensions. A *driven dimension* can be converted to a normal dimension only if another dimension or geometric relation is removed.

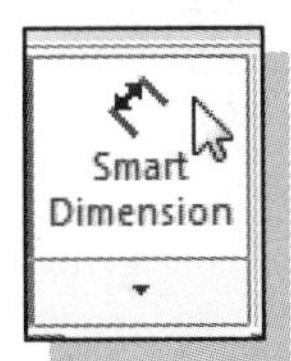

1. Select the **Smart Dimension** command in the *Sketch* toolbar.

2. Select the ***vertical line***.

3. **Pick** a location that is to the right side of the triangle to place the dimension text.

4. A *warning* dialog box appears on the screen stating that the dimension we are trying to create will over-define the sketch. Click on the **OK** button to proceed with the creation of a driven dimension.

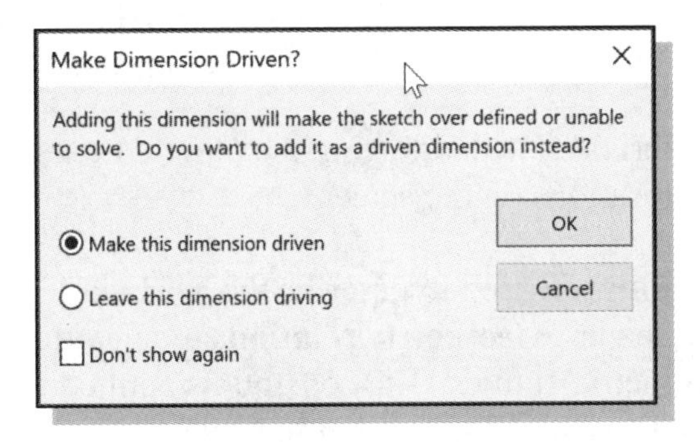

5. Press the [**Esc**] key once to exit the Smart Dimension command.

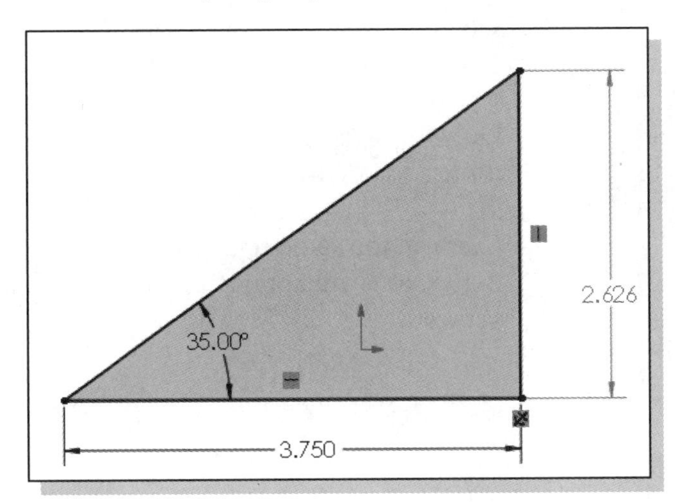

6. On your own, modify the angle dimension to **35°** and observe the changes of the 2D sketch and the driven dimension.

7. Press the [**Esc**] key to ensure that no objects are selected.

Deleting Existing Relations

1. On your own, make sure the visibility of the *Sketch Relations* is turned *ON*. (**HINT:** See Step 1 on page 5-5.)

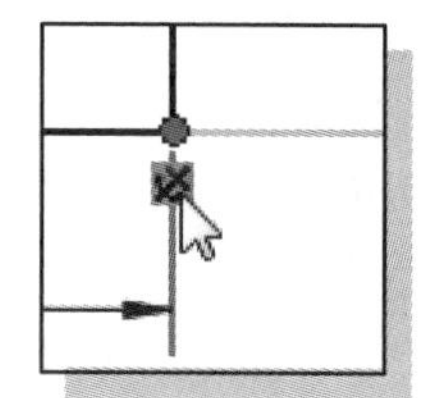

2. Move the cursor on top of the **Fixed** relation icon and right-click once to bring up the option menu.

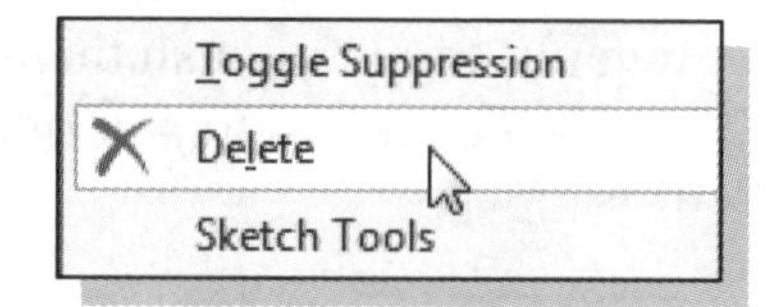

3. Select **Delete** to remove the Fixed relation that is applied to the lower right corner of the triangle.

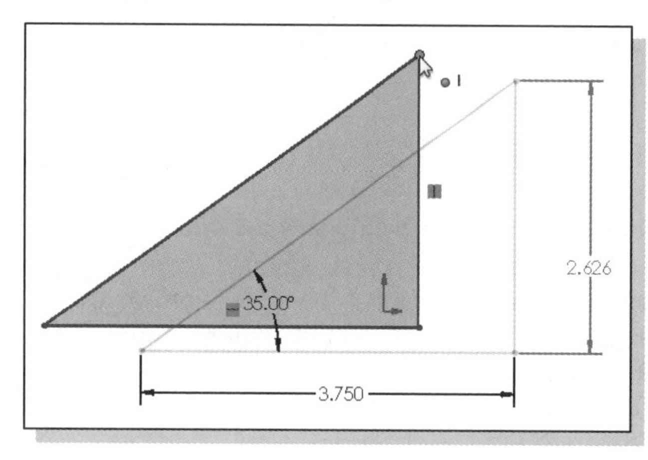

4. Move the cursor on top of the top corner of the triangle.

5. Drag the top corner of the triangle and note that the entire triangle is free to move in all directions. Drag the corner toward the top right corner of the graphics window as shown in the figure. Release the mouse button to move the triangle to the new location.

6. On your own, experiment with dragging the other corners and/or the three line segments to new locations on the screen.

❖ **Dimensional constraints** are used to describe the SIZE and LOCATION of individual geometric shapes. **Geometric relations** are **geometric restrictions** that can be applied to geometric entities. The constraints applied to the triangle are sufficient to maintain its size and shape, but the geometry can be moved around; its location definition isn't complete.

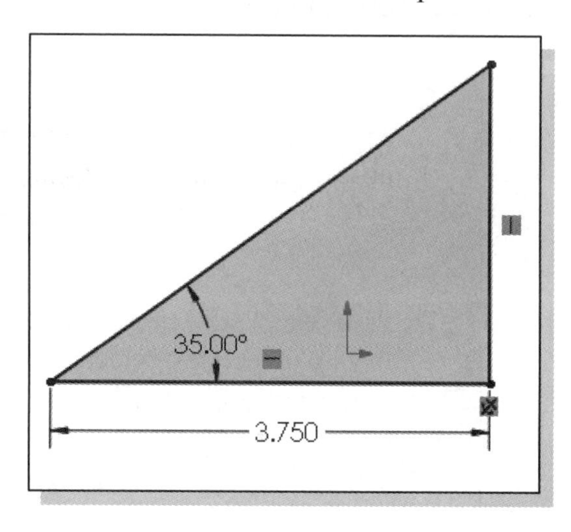

7. On your own, reapply the **Fixed** relation to the lower right corner of the triangle.

8. On your own, **delete** the reference dimension on the vertical line.

9. Confirm the same relations and dimensions are applied on your sketch as shown.

❖ Note that the sketch is fully defined and **Fully Defined** is displayed in the *Status Bar* at the bottom of the screen.

Using the Fully Define Sketch Tool

In SOLIDWORKS, the **Fully Define Sketch** tool can be used to calculate which dimensions and relations are required to fully define an under defined sketch. Fully defined sketches can be updated more predictably as design changes are implemented. One general procedure for applying dimensions to sketches is to use the **Smart Dimension** command to add the desired dimensions, and then use the **Fully Define Sketch** tool to fully constrain the sketch. It is also important to realize that different dimensions and geometric relations can be applied to the same sketch to accomplish a fully defined geometry.

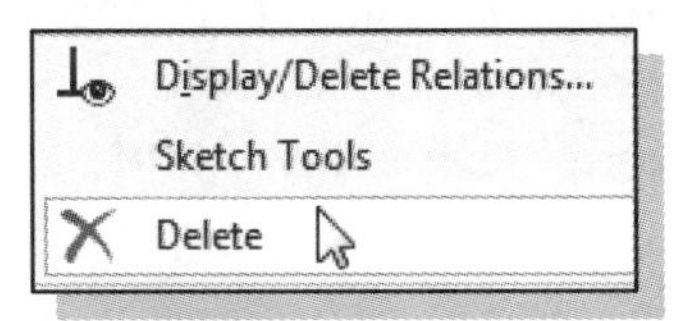

1. **Right click** once on the **3.750** linear dimension and select **Delete** in the option menu. Notice Under Defined is now displayed in the *Status Bar*.

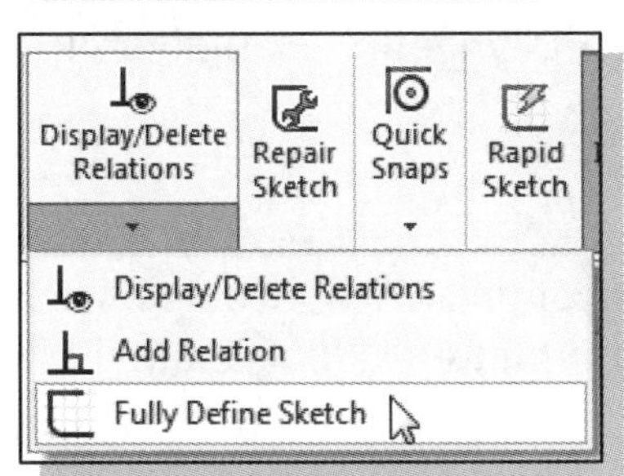

2. **Left click** on the arrow on the **Display/Delete Relations** button on the *Sketch* toolbar to reveal additional sketch relation commands.

3. Select the **Fully Define Sketch** command from the pop-up menu.

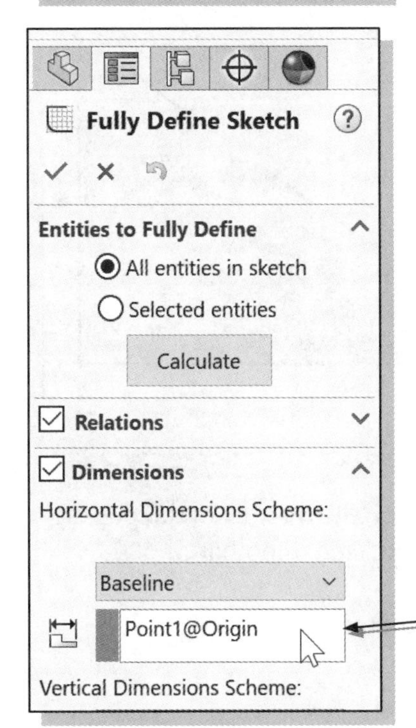

4. The *Fully Define Sketch PropertyManager* appears. **If necessary**, click once with the **left-mouse-button** on the arrows to reveal the Dimensions option panel as shown.

5. Notice that under the *Horizontal Dimensions Scheme* option, Baseline is selected. The origin is selected as the default baseline datum and appears in the datum selection window as Point1@Origin.

6. Click once with the **left-mouse-button** in the *Datum* selection text box to select a different datum to serve as the baseline.

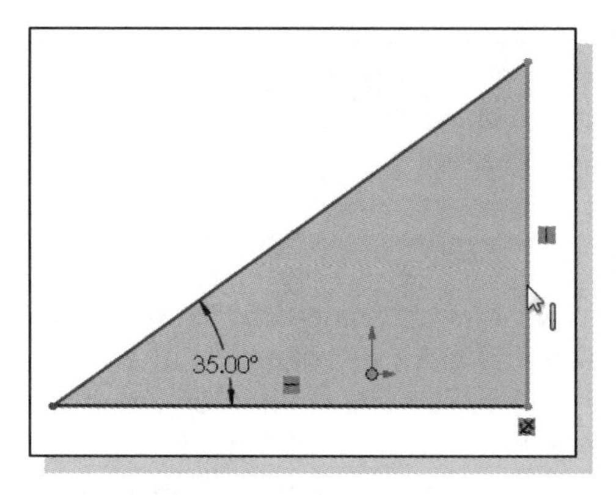

7. In the graphics window select the vertical line. Notice **Line2** is now displayed in the *Datum* selection window as the *Horizontal Dimensions* baseline.

8. **On your own**, select an appropriate baseline datum for the vertical dimensions.

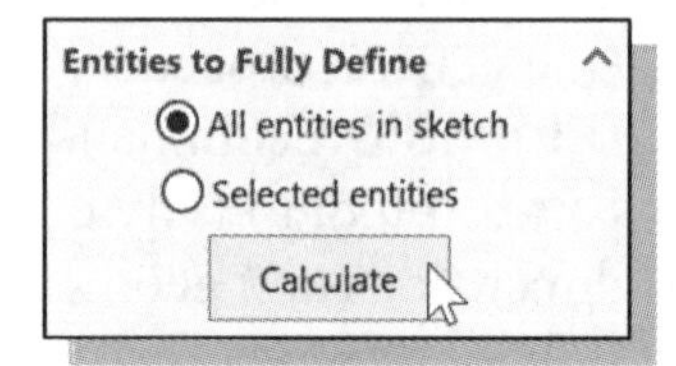

9. Select **Calculate** in the *PropertyManager* to continue with the Fully Define Sketch command.

❖ Note that SOLIDWORKS automatically applies the horizontal dimension, using the vertical line as the baseline. Since this is the only dimension necessary to fully define the sketch, it is the only dimension added. Notice that Fully Defined is now displayed in the *Status Bar*.

10. Click the **OK** button to accept the results and exit the Fully Define Sketch tool.

Adding Additional Geometry

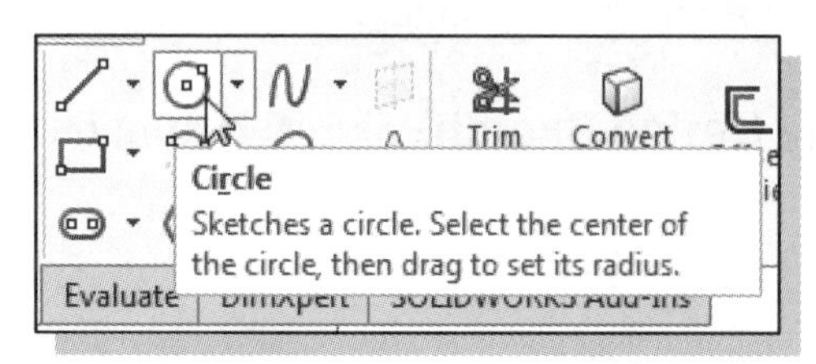

1. Select the **Circle** command by clicking once with the left-mouse-button on the icon in the *Sketch* toolbar.

2. On your own, create a circle of arbitrary size inside the triangle as shown.

3. Press the [**Esc**] key to ensure that no objects are selected.

❖ Notice that Under Defined is now displayed in the *Status Bar*.

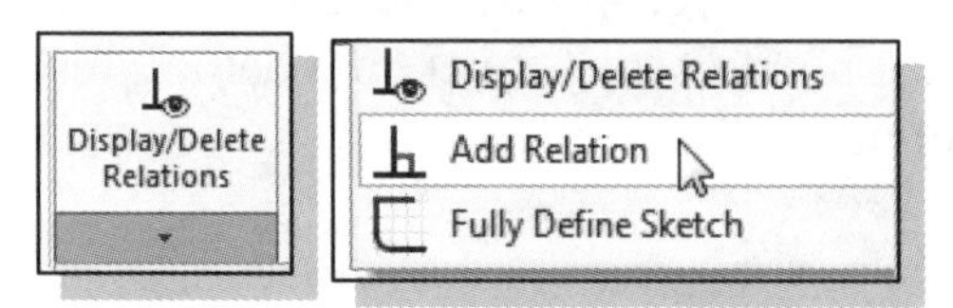

4. Select **Add Relation** from the pop-up menu on the *Sketch* toolbar.

5. Pick the circle by left-mouse-clicking once on the geometry.

6. Pick the inclined line.

➢ Look at the *Add Relations PropertyManager*. In the *Selected Entities* text box **Arc1** and **Line1** are now displayed. There are no relations in the *Existing Relations* window. In the *Add Relations* menu, the **Tangent** and **Fix** relations are displayed. These are the relations which can be added for the selected entities.

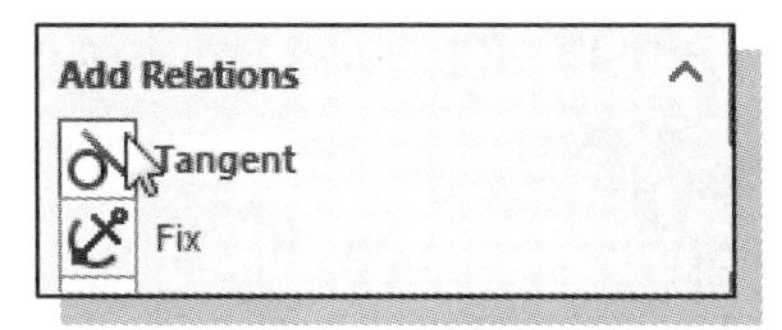

7. Click once with the left-mouse-button on the **Tangent** icon in the *Add Relations PropertyManager* as shown. This activates the Tangent relation.

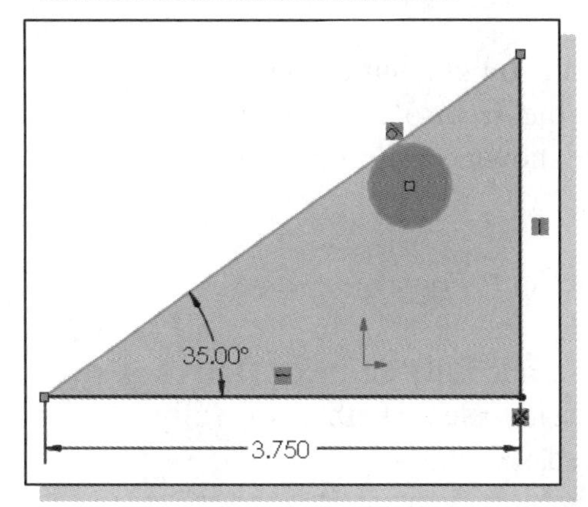

8. Click the **OK** icon in the *PropertyManager*, or hit the [**Esc**] key once, to end the Add Relations command.

➢ How many more relations or dimensions do you think will be necessary to fully define the circle? Which relations or dimensions would you use to fully define the geometry?

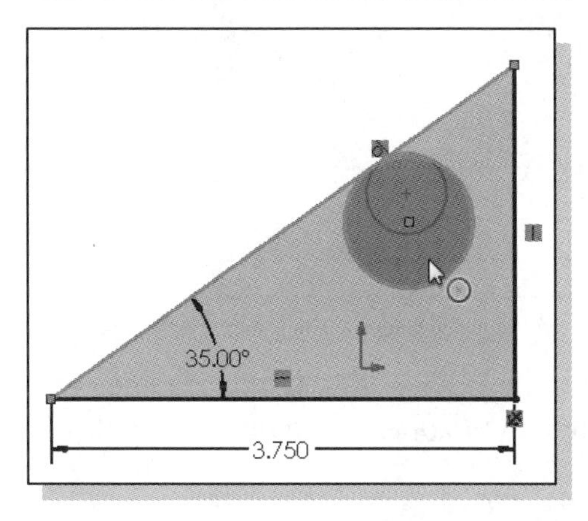

9. Move the cursor on top of the right side of the circle, and then drag the circle toward the right edge of the graphics window. Notice the size of the circle is adjusted while the system maintains the Tangent relation.

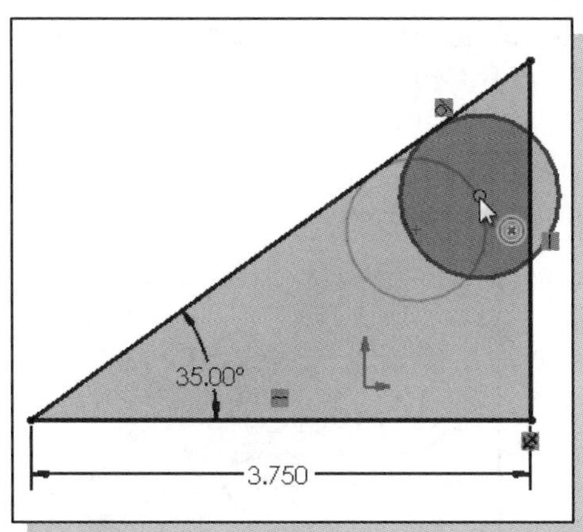

10. Drag the center of the circle toward the upper right direction. Notice the Tangent relation is always maintained by the system.

➢ On your own, experiment with adding additional relations and/or dimensions to fully define the sketched geometry. Use the **Undo** command to undo any changes before proceeding to the next section.

11. Press the [**Esc**] key to ensure that no objects are selected.

12. Inside the graphics window, click once with the **left-mouse-button** on the **center of the circle** to select the centerpoint.

13. Hold down the **[Ctrl]** key and click once with the **left-mouse-button** on the **vertical line**. Holding the [Ctrl] key allows selecting the vertical line while maintaining selection of the circle centerpoint. This is a method to select multiple entities.

- Notice the *Properties PropertyManager* for the selected entities is displayed. This provides an alternate way to control the relations for selected properties.

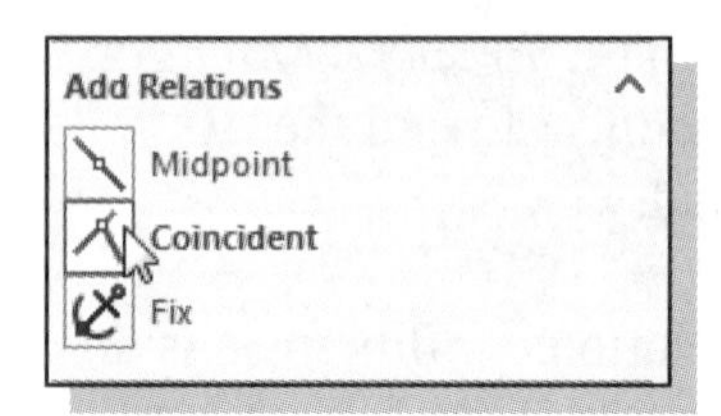

14. Click once with the left-mouse-button on the **Coincident** icon in the *Add Relations* panel in the *PropertyManager* as shown. This activates the Coincident relation.

15. Click the **OK** icon in the *PropertyManager.*

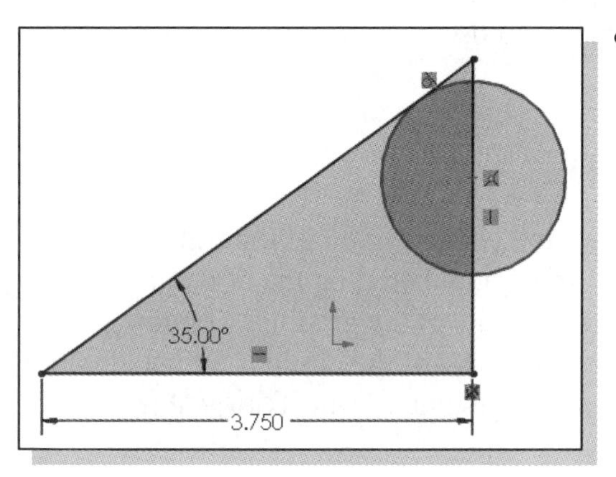

- Is the circle fully defined? (**HINT:** Drag the circle to see whether it is fully defined.)

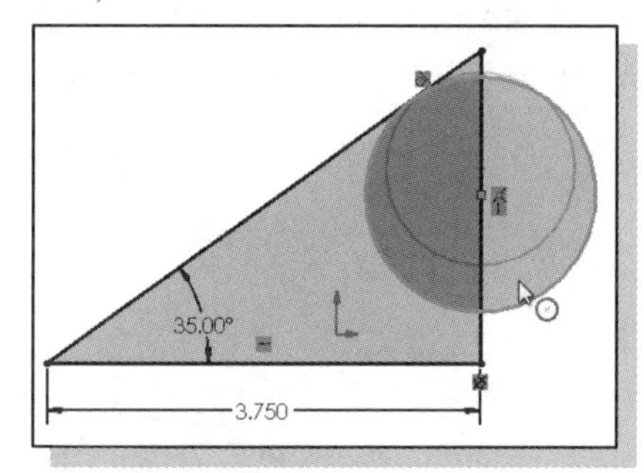

16. On your own, select the circle centerpoint and the vertical line in the graphics area. (**HINT:** Repeat steps 12 and 13.) Notice the *PropertyManager* appears.

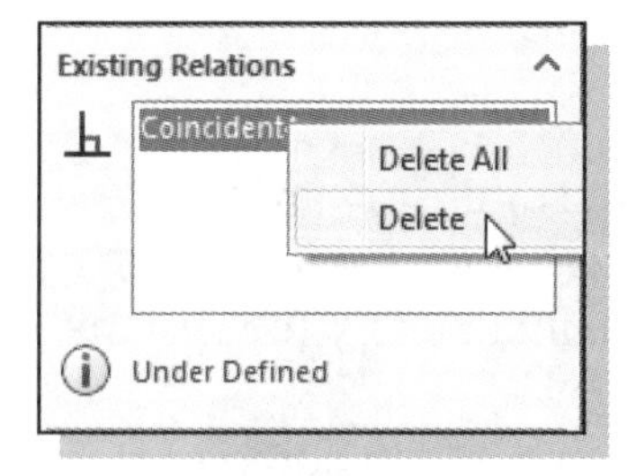

17. In the *PropertyManager*, Coincident1 is listed under *Existing Relations*. Move the cursor over **Coincident1** and click once with the **right-mouse-button**.

18. Select **Delete** in the pop-up menu by clicking once with the **left-mouse-button**.

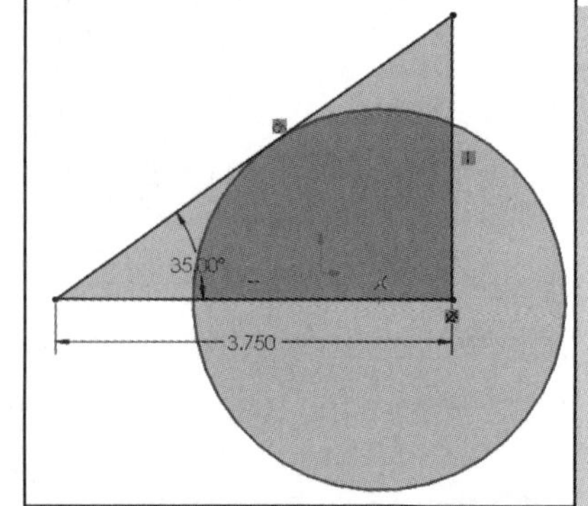

19. Click the **OK** icon in the *PropertyManager.*

20. On your own, add a Coincident relation between the center of the circle and the horizontal line.

- Which relations or dimensions would you use to fully define the geometry?

❖ The application of different relations affects the geometry differently. The design intent is maintained in the CAD model's database and thus allows us to create very intelligent CAD models that can be modified/revised fairly easily. On your own, experiment and observe the results of applying different relations to the triangle. For example: (1) adding another Fix constraint to the top corner of the triangle; (2) deleting the horizontal dimension and adding another Fix relation to the left corner of the triangle; and (3) adding another Tangent relation and adding the size dimension to the circle.

21. On your own, modify the 2D sketch as shown.

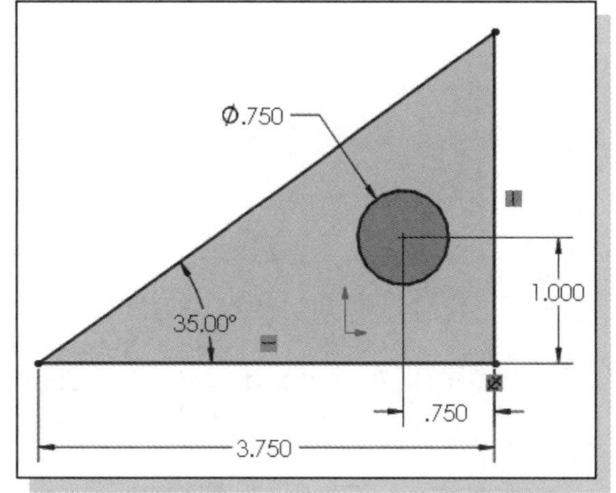

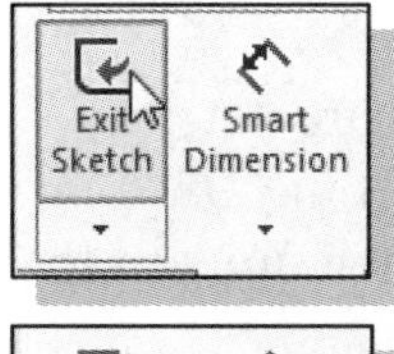

22. Click once with the left-mouse-button on the **Exit Sketch** icon on the *Sketch* toolbar to end the Sketch option.

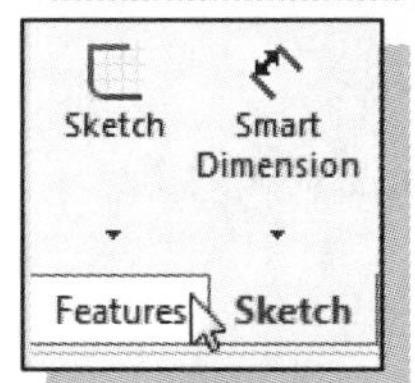

23. Select the **Features** tab on the *CommandManager* to display the *Features* toolbar.

➢ On your own, use the **Extruded Boss/Base** command and create a 3D solid model with a plate thickness of **0.25**.

Extruded Boss/Base

Relations Settings

Select **Options** in the *Menu Bar*. Click on **Relations/Snaps** under the **System Options** tab to display and/or modify the relation settings. On your own, adjust the settings and experiment with the effects of the different settings on sketching.

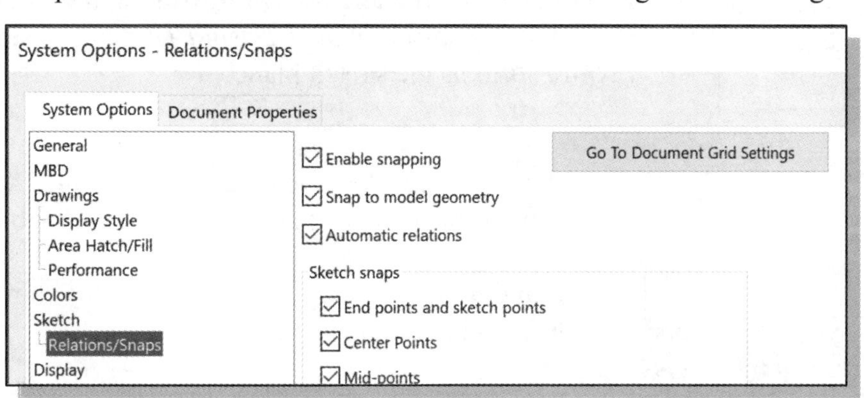

Parametric Relations

In parametric modeling, dimensions are design parameters that are used to control the sizes and locations of geometric features. Dimensions are more than just values; they can also be used as feature control variables. This concept is illustrated by the following example.

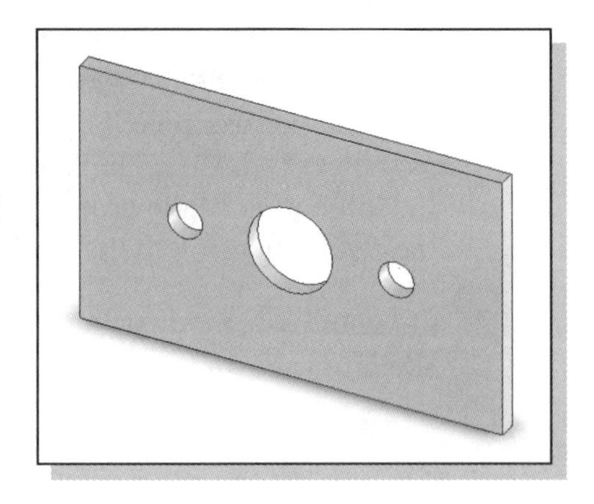

1. Select the **New** icon with a single click of the left-mouse-button on the *Menu Bar*. The *New* SOLIDWORKS *Document* dialog box should appear in Advanced mode. If it appears in Novice mode, click once with the left-mouse-button on the **Advanced** icon.

➢ **NOTE:** If the Tutorial_Templates tab and Part_IPS_ANSI template are not available, simply open a new part using the default Part template and set the *Drafting Standard* to ANSI and the *Units* to IPS (see Steps 1-7 on page 4-5). Then proceed to Step 5 below.

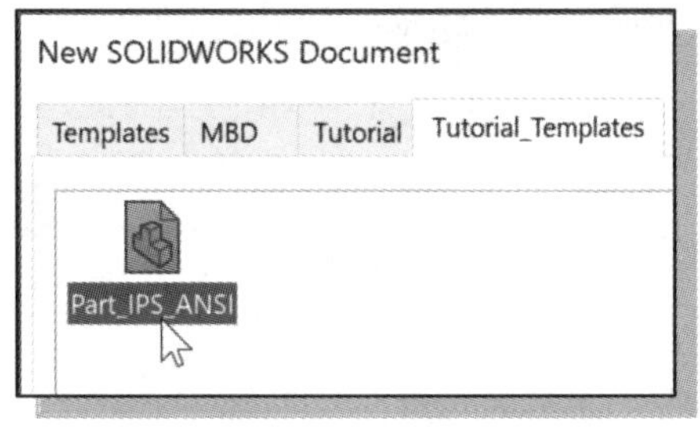

2. Select the **Tutorial_Templates** tab. (**NOTE:** You added this tab in Chapter 4.)

3. Select the **Part_IPS_ANSI** template as shown.

4. Click on the **OK** button to open a new document using the Part_IPS_ANSI template.

- Another graphics window appears on the screen. We can switch between the two models using the *Window* pull-down menu.

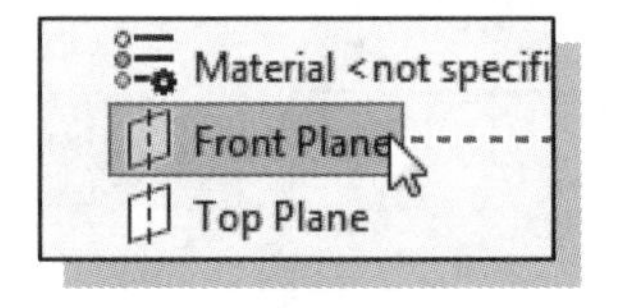

5. Click once with the **left-mouse-button** on the **Front Plane** icon in the *FeatureManager Design Tree* to pre-select the Front Plane as the sketch plane.

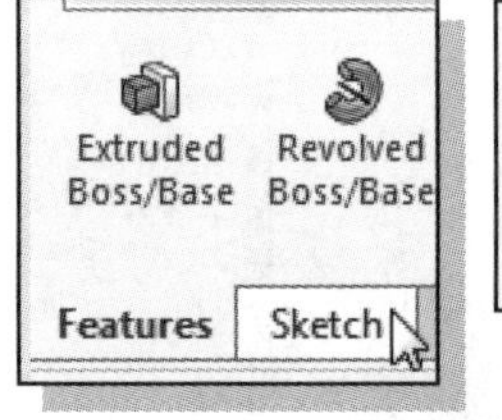

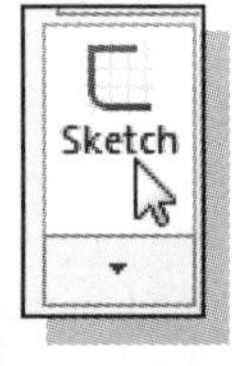

6. If necessary, select the **Sketch** tab on the *CommandManager* to display the *Sketch* toolbar.

7. In the *Sketch* toolbar, select the **Sketch** command by left-clicking once on the icon.

8. Select the **Corner Rectangle** command by clicking once with the **left-mouse-button** on the icon in the *Sketch* toolbar.

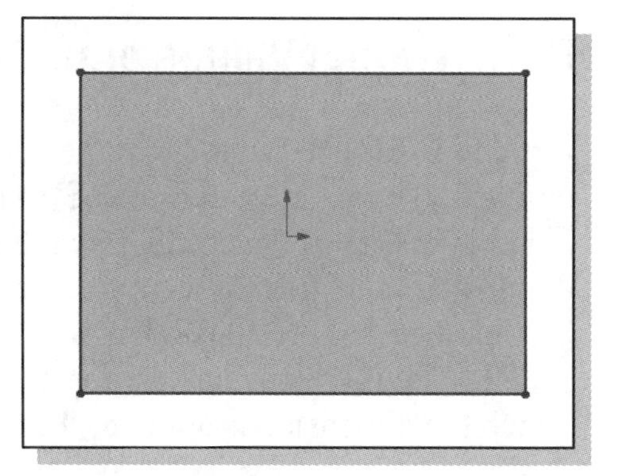

9. Create a rectangle of arbitrary size positioned near the center of the screen.

10. Select the **Circle** command by clicking once with the left-mouse-button on the icon in the *Sketch* toolbar.

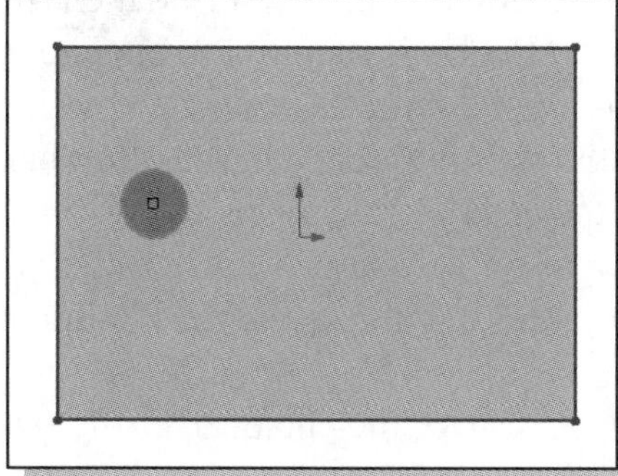

11. Create a **circle** of arbitrary size inside the rectangle as shown.

12. On your own, adjust the geometry by adding dimensions as shown below. (**NOTE:** Create and modify the overall width of the rectangle first and the overall height of the rectangle second. These two dimensions will be used as the control variables for the rest of the dimensions.)

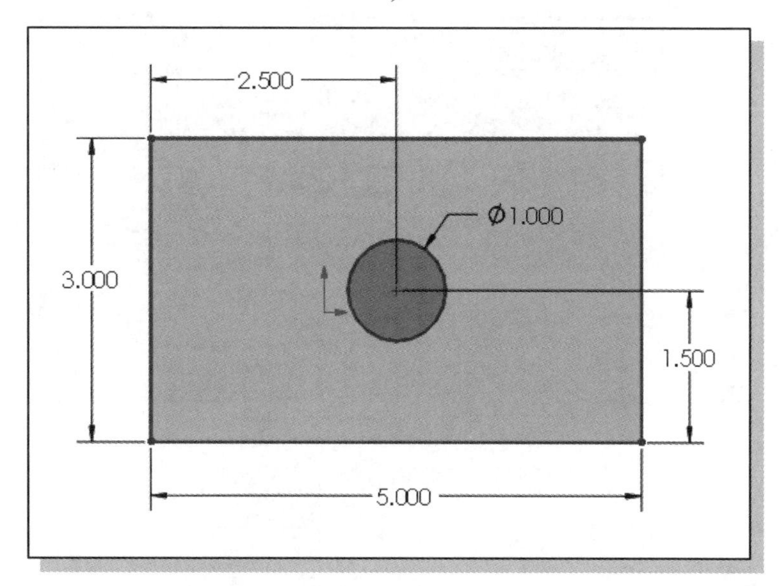

- On your own, change the overall width of the rectangle to **6.0** and the overall height of the rectangle to **3.6** and observe the location of the circle in relation to the edges of the rectangle. Adjust the dimensions back to **5.0** and **3.0** as shown in the above figure before continuing.

Dimensional Values and Dimensional Variables

Initially in SOLIDWORKS, dimension values are used to create different geometric entities. The text created by the Smart Dimension command also reflects the actual location or size of the entity. Each dimension is also assigned a name that allows the dimension to be used as a control variable. The default format is "Dxx," where the "xx" is a number that SOLIDWORKS increments automatically each time a new dimension is added. The full name has the form "Dxx@yyyyy," where the "yyyyy" is the entity in which the dimension is applied. For example, "D2@Sketch1" is the full name for the second dimension applied in Sketch 1.

Let us look at our current design, which represents a plate with a hole at the center. The dimensional values describe the size and/or location of the plate and the hole. If a modification is required to change the width of the plate, the location of the hole will remain the same as described by the two location dimensional values. This is okay if that is the design intent. On the other hand, the *design intent* may require (1) keeping the hole at the center of the plate and (2) maintaining the size of the hole to be one-third of the height of the plate. We will establish a set of parametric relations using the dimensional variables to capture the design intent described in statements (1) and (2) above.

1. Move the cursor over the **width dimension** of the rectangle to display the *Dimension Name* in the cursor pop-up box.

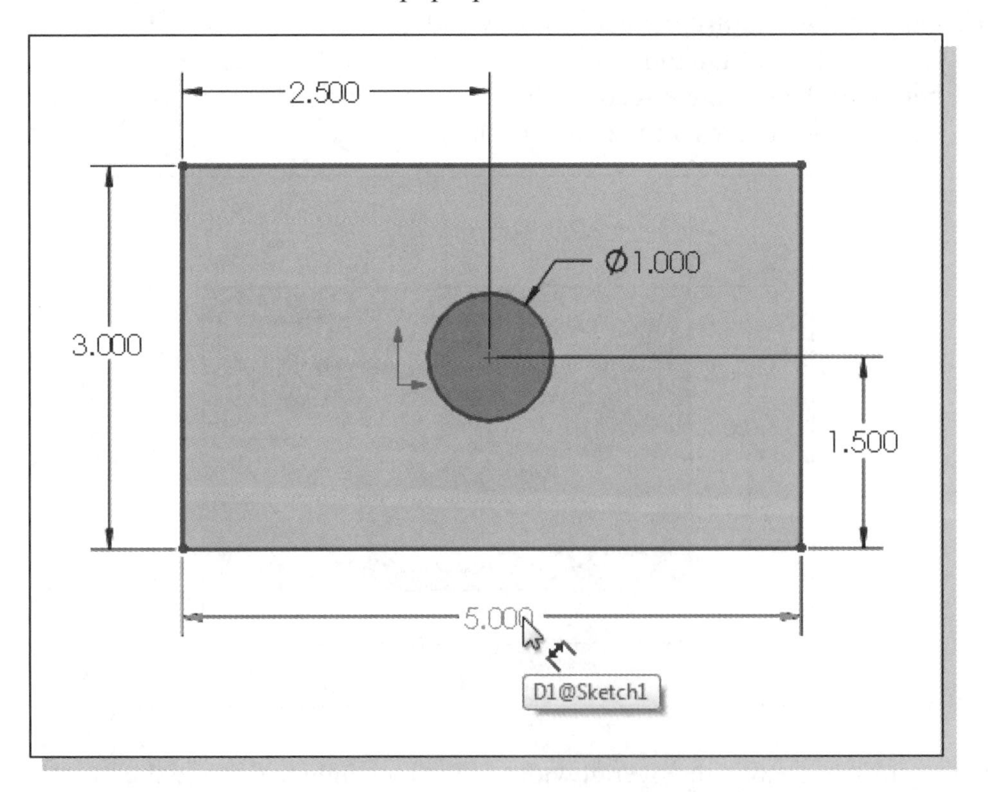

- Notice the ***variable name*** **D1@Sketch1** is displayed in the cursor box.

Parametric Equations

Each time we add a dimension to a model, that value is established as a parameter for the model. We can use parameters in equations to set the values of other parameters.

1. Double-click with the left-mouse-button on the **horizontal location dimension** of the circle to display the *Modify* window.

2. Equations can be entered directly in the *Modify* dialog box. In the *Modify* dialog box, we start with **=** to create an equation. Enter **=** in the *Modify* window in place of the 2.500in entry to initiate the creation of an equation. Finish the equation to make this dimension equal to half the width as follows: **= "D1@Sketch1"/2**.

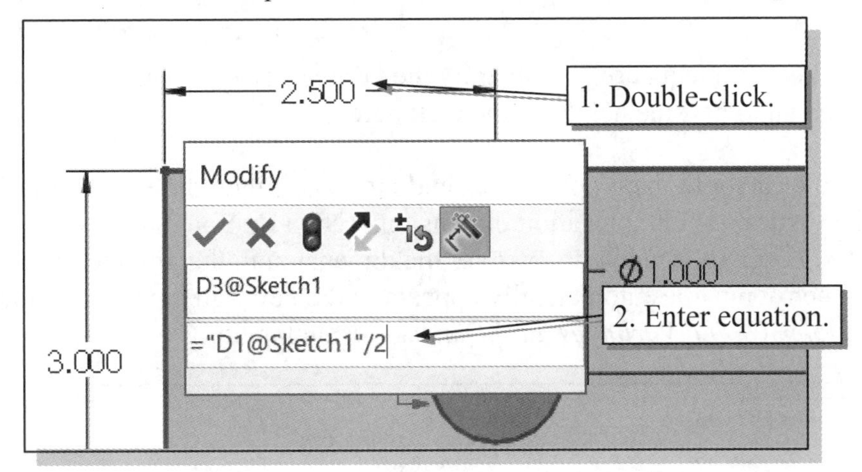

3. Click on **OK** in the *Modify* dialog box to accept the equation.

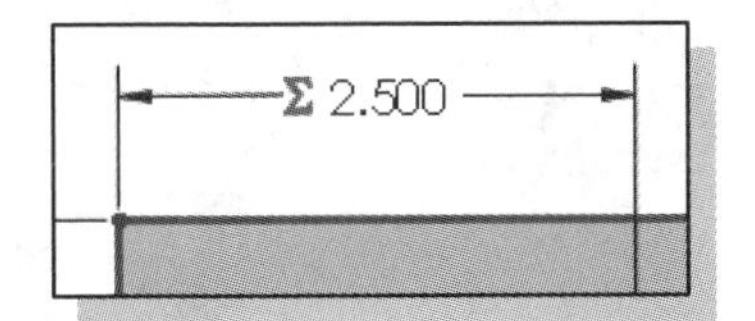

❖ Notice, the derived dimension values are displayed with **Σ** in front of the numbers. The parametric equation we entered is used to control the location of the circle; the location is based on the width of the rectangle.

Viewing the Established Equations

1. On your own, move the cursor over each dimension and note the corresponding dimension name. We will use these names to create additional equations.

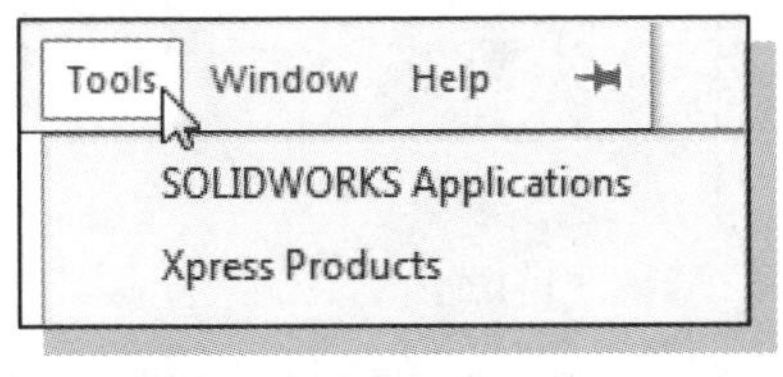

2. Move the cursor over the SOLIDWORKS icon on the *Menu Bar* to reveal the pull-down menus, click once with the left-mouse-button on the **Tools** icon and select **Equations** in the pull-down menu as shown.

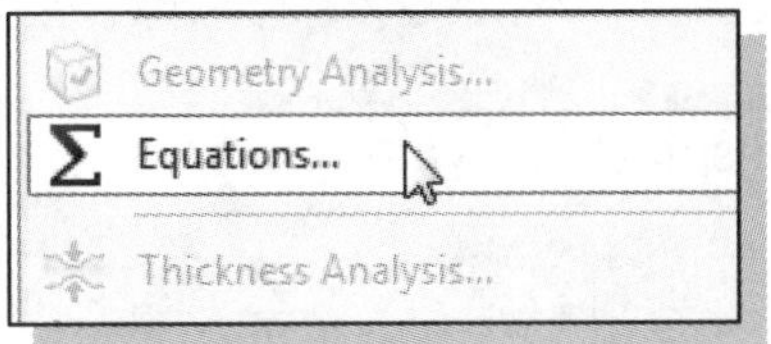

- The *Equations, Global Variables, and Dimensions* dialog box is displayed. Notice the equation we created appears in the table under **'Equations'**.

- The *Equations, Global Variables, and Dimensions* dialog box can be used to perform all tasks related to equations, including editing existing equations, and creating additional design variables and equations. We will use it here to show an alternate method to add an equation.

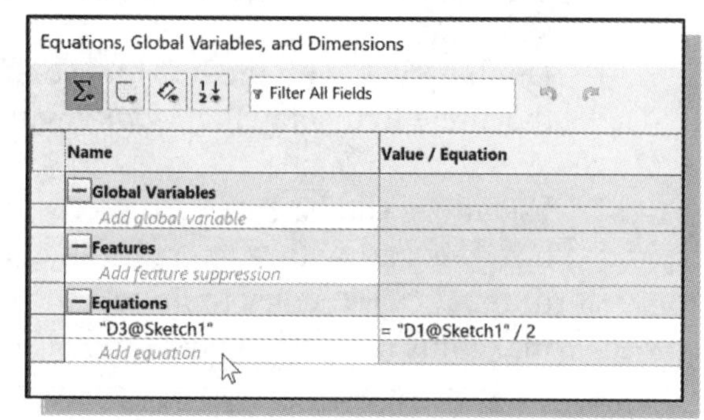

3. Click the **Add equation** cell in the *Equations, Global Variables, and Dimensions* dialog box as shown. The cell becomes active, awaiting the start of a new equation.

4. We will enter an equation to constrain the vertical location dimension for the circle equal to half the height of the rectangle.

5. Move the cursor to the graphics area and click once with the **left-mouse-button** on the vertical (1.500″) location dimension. (**NOTE:** You may have to reposition the *Add Equation* window to uncover the dimension in the graphics area.) Notice the dimension name automatically appears in the new equation line of the *Equations, Global Variables, and Dimensions* dialog box.

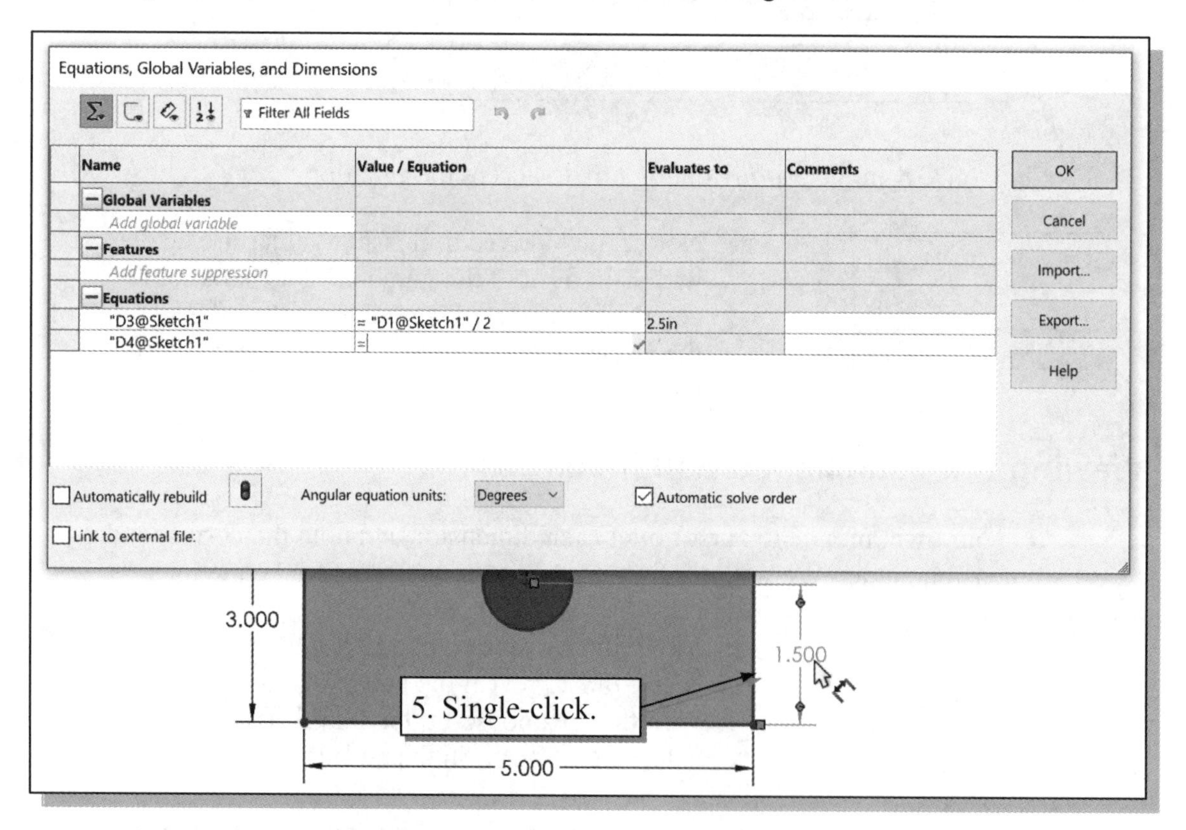

6. The *Value/Equation* column becomes active with **=** displayed, awaiting the remainder of the equation.

7. Move the cursor to the graphics area and select the vertical dimension on the rectangle.

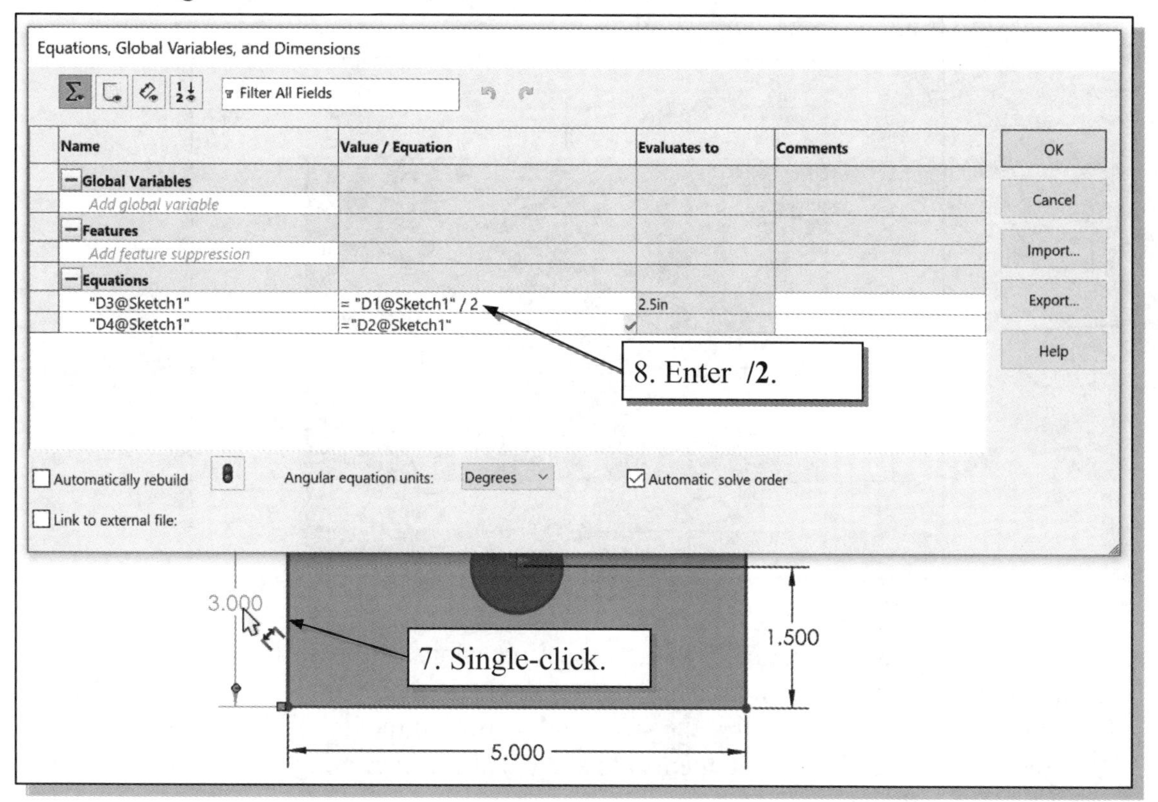

8. Notice the dimension name appears in the equation line. Enter **/2** to complete the equation as shown.

9. The equation should appear as shown below (although the dimension names may be different, depending on the order in which they were created). Click the **OK** button (green check mark) in the equation cell to accept the new equation.

Equations		
"D3@Sketch1"	= "D1@Sketch1" / 2	2.5in
"D4@Sketch1"	="D2@Sketch1"/2	
	2.5in	

10. On your own, add an equation to constrain the diameter of the circle to be ⅓ the height of the plate.

Equations		
"D3@Sketch1"	= "D1@Sketch1" / 2	2.5in
"D4@Sketch1"	= "D2@Sketch1" / 2	1.5in
"D5@Sketch1"	= "D2@Sketch1" / 3	1in

11. Click **OK** in the *Equations, Global Variables, and Dimensions* dialog box to accept the equations.

12. On your own, change the dimensions of the rectangle to **6.0** x **3.6** and observe the changes to the location and size of the circle. (**HINT:** Double-click the dimension text to bring up the *Modify* window.)

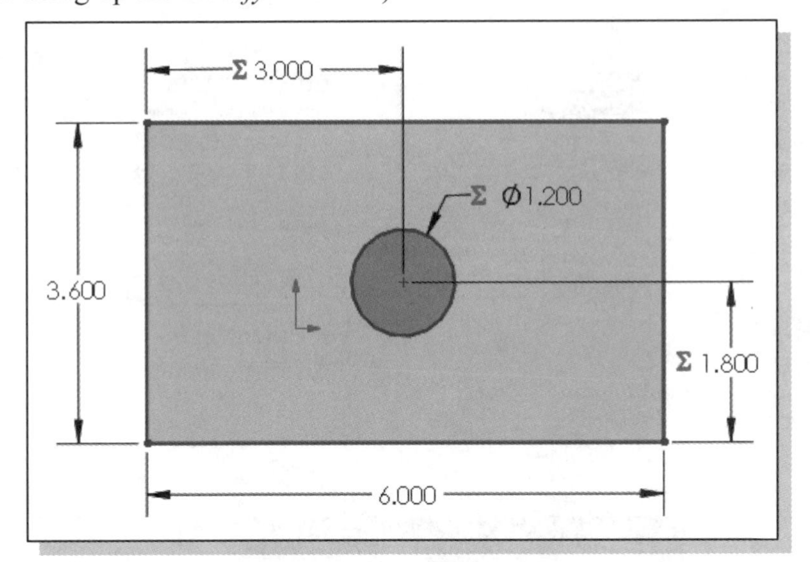

❖ SOLIDWORKS automatically adjusts the dimensions of the design, and the parametric relations we entered are also applied and maintained. The dimensional constraints are used to control the size and location of the hole. The design intent, previously expressed by statements (1) and (2) at the beginning of this section, is now embedded into the model.

Global Variables

Global Variables can be created using the *Equations, Global Variables, and Dimensions* dialog box. We will create new sketch objects, then define their location and size using Global Variables.

1. On your own, add two unconstrained circles as shown below.

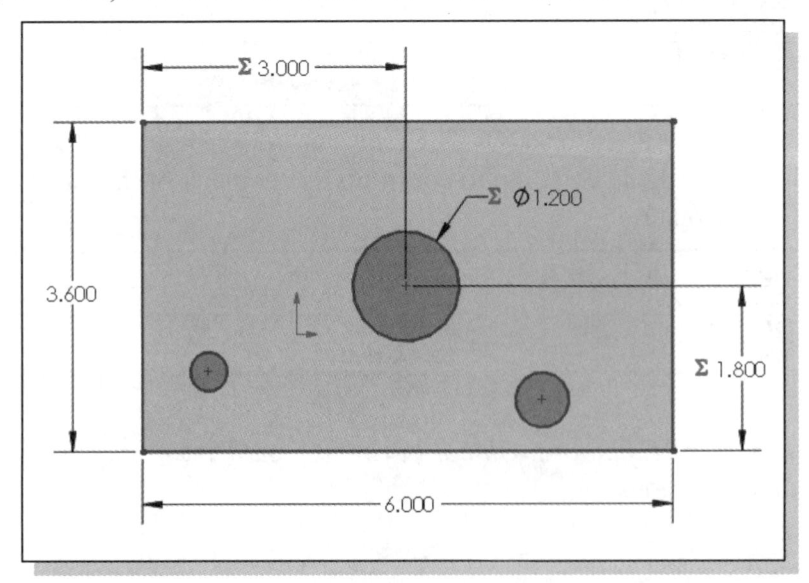

2. Press the [**Esc**] key (twice, if necessary) to ensure no items are selected.

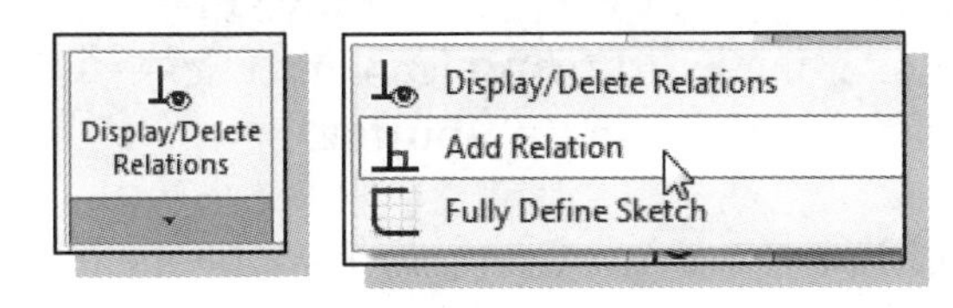

3. Select **Add Relation** from the pop-up menu on the *Sketch* toolbar. Notice the *Selected Entities* window in the *Add Relations PropertyManager* is blank because no entities are selected.

4. Select the centerpoints of the three circles.

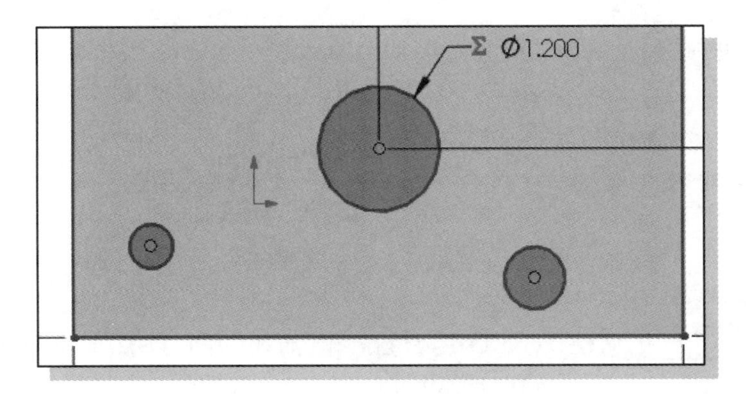

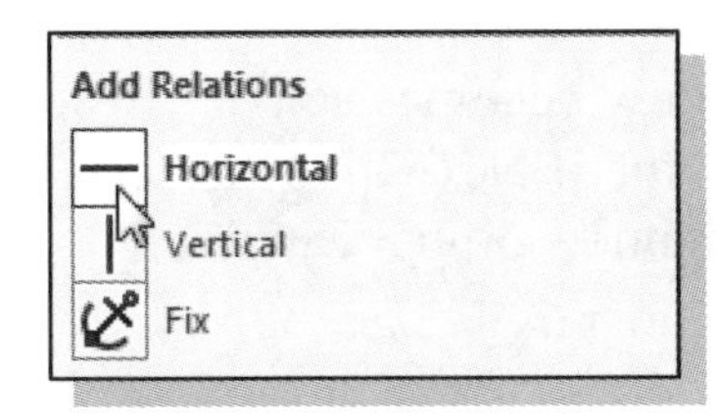

5. Click once with the left-mouse-button on the **Horizontal** icon in the *Add Relations PropertyManager* as shown. This activates the Horizontal relation. Notice the two new holes become aligned horizontally with the constrained central hole.

6. Click the **OK** icon in the *PropertyManager*, or hit the [**Esc**] key once, to end the Add Relations command.

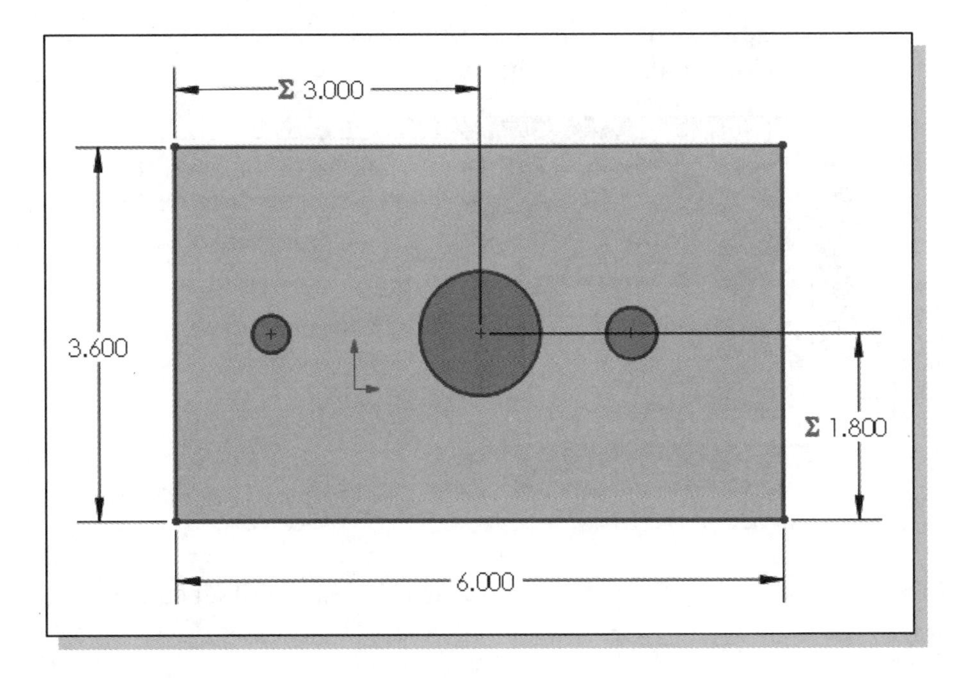

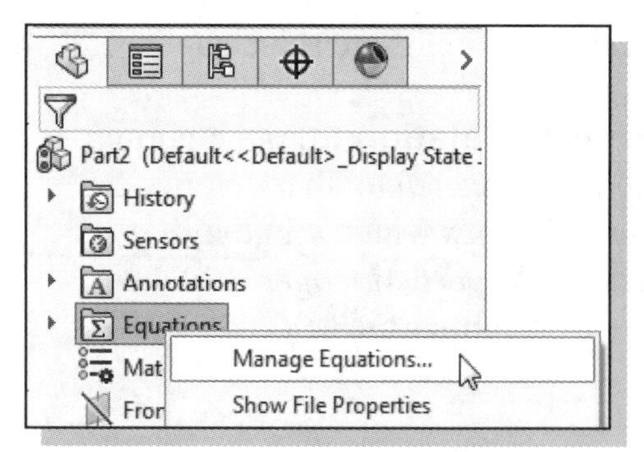

7. Notice the **Equations** icon now appears in the *FeatureManager Design Tree*. Move the cursor over the **Equations** icon and click once with the **right-mouse-button**.

8. Select **Manage Equations...** from the menu to open the *Equations, Global Variables, and Dimensions* dialog box.

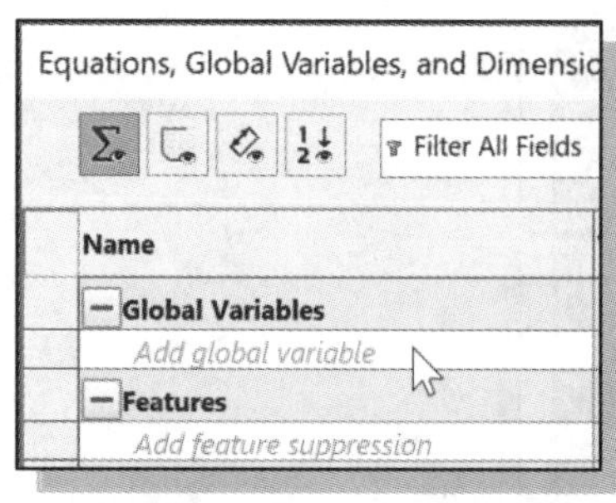

9. Click the **Add global variable** cell in the *Equations, Global Variables, and Dimensions* dialog box as shown. The cell becomes active, awaiting the start of a new global variable.

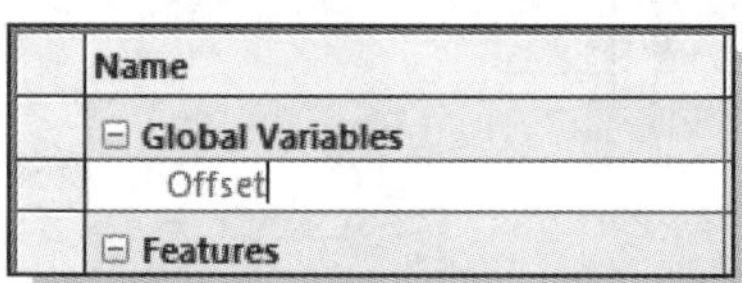

10. Type **Offset** and press the **[ENTER]** key to enter the name for the new global variable.

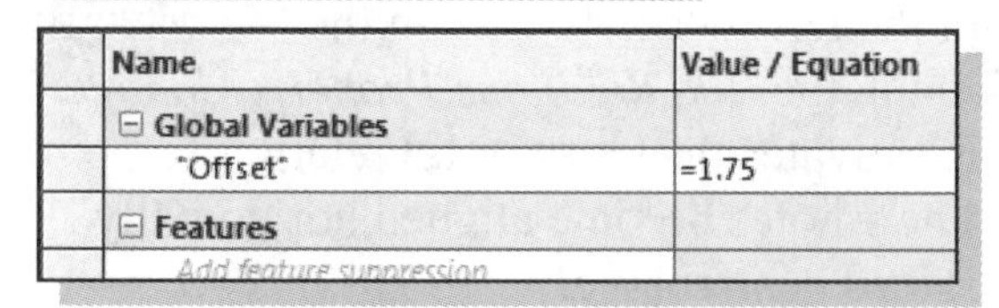

11. The value window becomes active. Type **1.75** and press the **[ENTER]** key to set the value for the new global variable.

12. Click **OK** in the *Equations, Global Variables, and Dimensions* dialog box to accept the new global variable.

13. On your own, add a **Smart Dimension** between the centerpoints of the central circle and the circle to the left as shown.

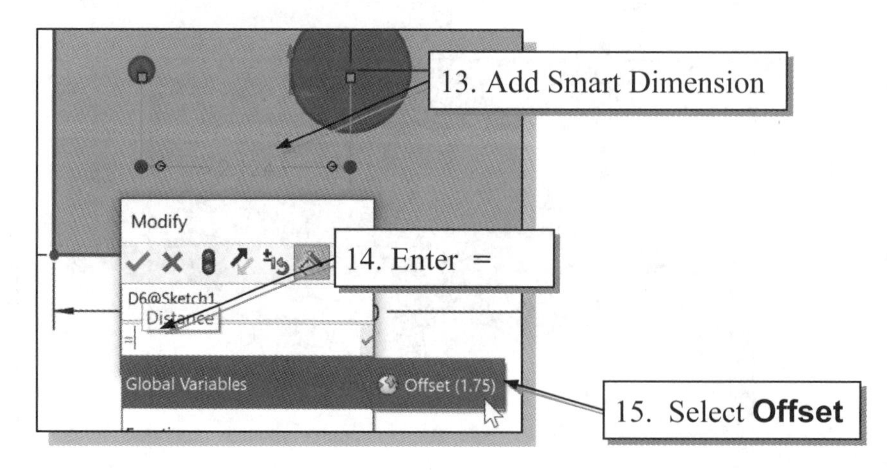

14. Enter **=** in the *Modify* dialog box to initiate the creation of an equation.

15. Notice the *Global Variables* option appears. Select **Global Variables**, then select the **Offset(1.75)** variable.

16. Click **OK** in the *Modify* dialog box to accept the new dimension.

17. On your own, add a Smart Dimension between the centerpoints for the center circle and the circle to the right and equate the value to the Offset variable.

18. On your own, add a second Global Variable. Name the variable **Dia2** and set its value to **0.5**.

Name	Value / Equation	Evaluates to
⊟ Global Variables		
"Offset"	= 1.75	1.75
"Dia2"	= .5	0.5
Add global variable		

19. On your own, add two Smart Dimensions for the diameters of the small circles and equate their values to the Dia2 variable.

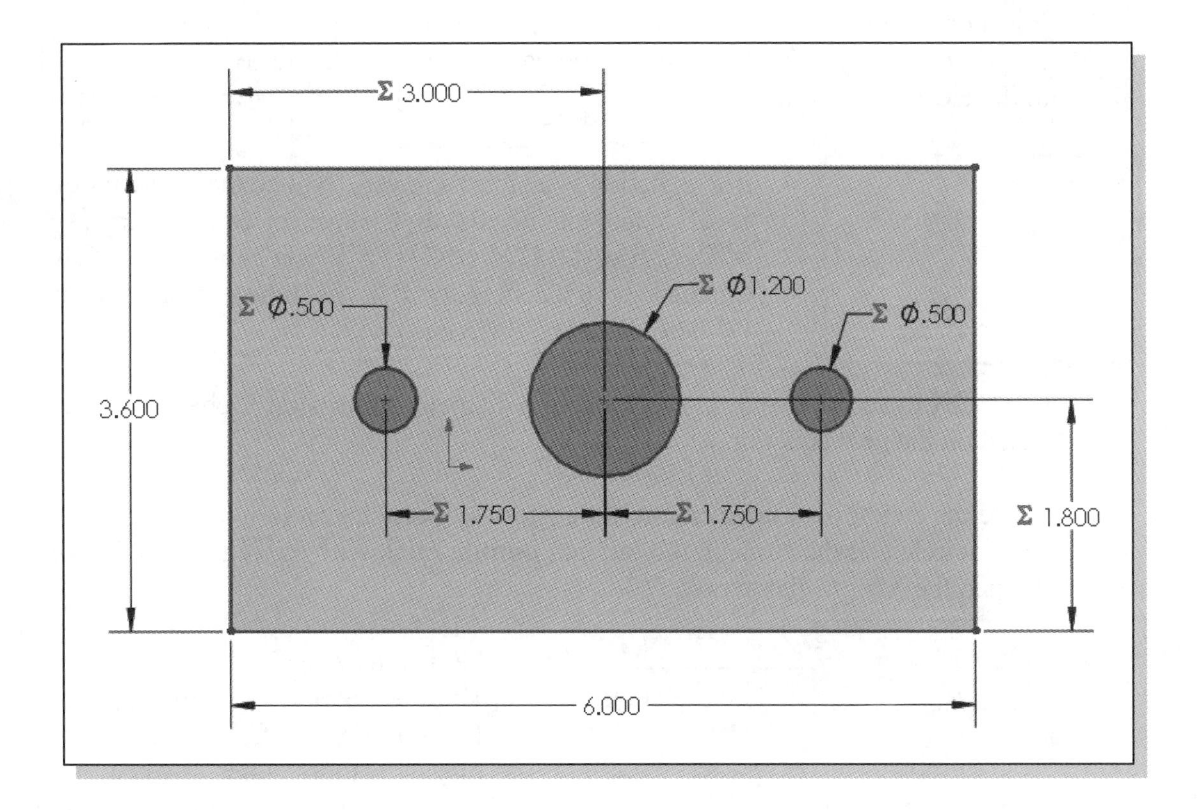

Viewing/Editing Equations and Global Variables Using the Dimension *Modify* Dialog Box

➢ The equations that have been created can be viewed and edited very easily in the regularly used *Modify* dialog box for each dimension.

1. Move the cursor over the diameter dimension for the center circle and **double-click** with the left mouse button to open the *Modify* dialog box.

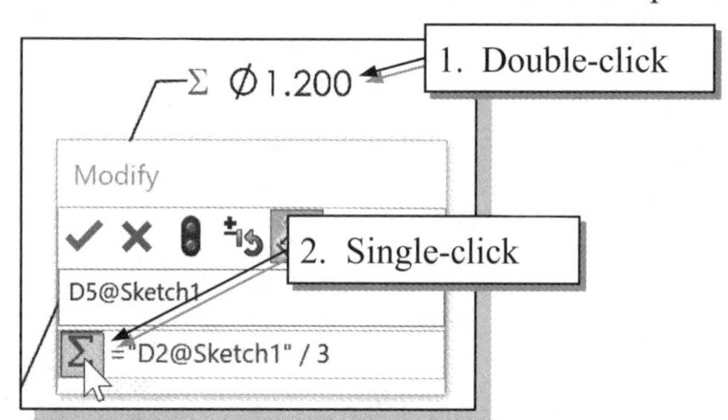

2. Notice the *Toggle Equation/Value Display* button now appears. In the figure at left, the Equation display is toggled on. Click once with the **left mouse button** to toggle to the Value display.

3. The calculated value of **1.200in** now appears. Notice the value window is not active and cannot be modified. It is a derived value based on the equation. Click the *Toggle Equation/Value Display* button again to return to the Equation display.

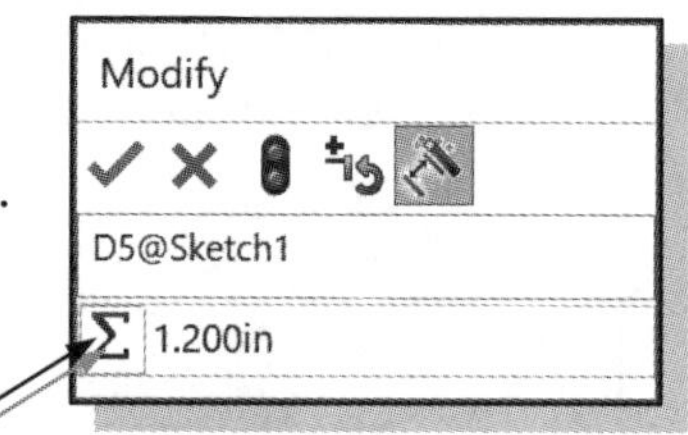

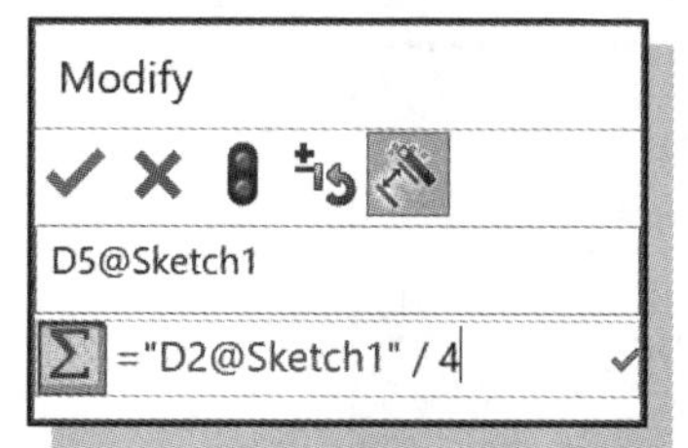

4. The equation is again displayed. Notice that the window is active and can be edited. Change the equation to **="D2@Sketch1"/4**. (NOTE: If you applied dimensions in a different order, the dimension number – D2 here – may be different.)

5. Click **OK** to accept the change. Notice the diameter dimension is changed to **9.00** based on the new equation.

6. Move the cursor over the location dimension between the centerpoints of the center circle and the circle to the left and **double-click** with the left mouse button to open the *Modify* dialog box.

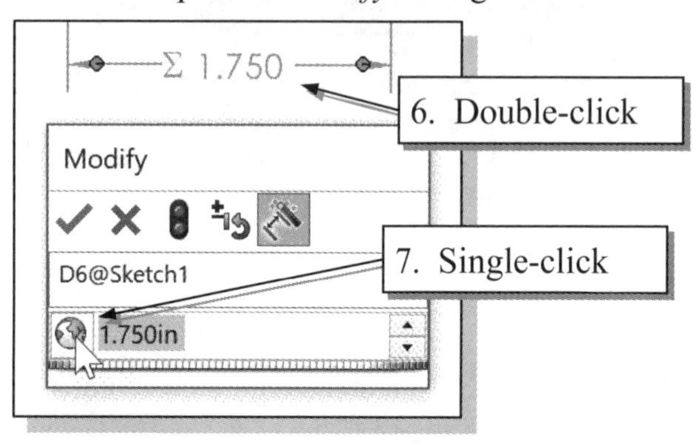

7. Notice the *Toggle Global Variable/ Value Display* button now appears. In the figure at left, the Value display is toggled on. (Notice, unlike the equation value, this value window is active.) Click once with the **left mouse button** to toggle to the Global Variable display.

8. The global variable display of **="Offset"** now appears. Click the *Toggle Global Variable/Value Display* button again to return to the Value display.

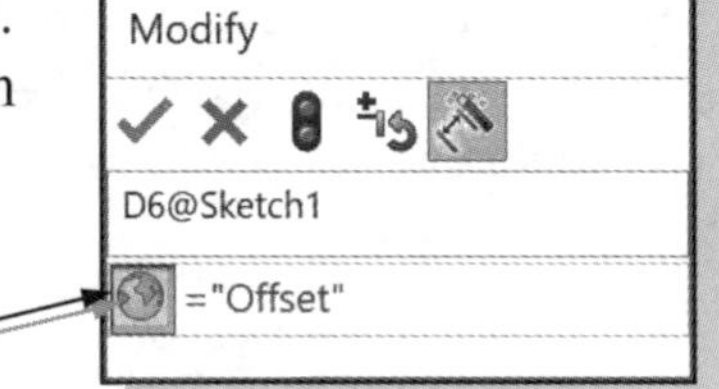

9. The value is again displayed. Change the value to **1.50**.

10. Click **OK** to accept the change. Notice the dimension is changed to **1.50** based on the change. Notice the dimension between the centerpoints of the center circle and the circle to the right is also changed. The value of the ***Global Variable*** named Offset has been changed to **1.50**.

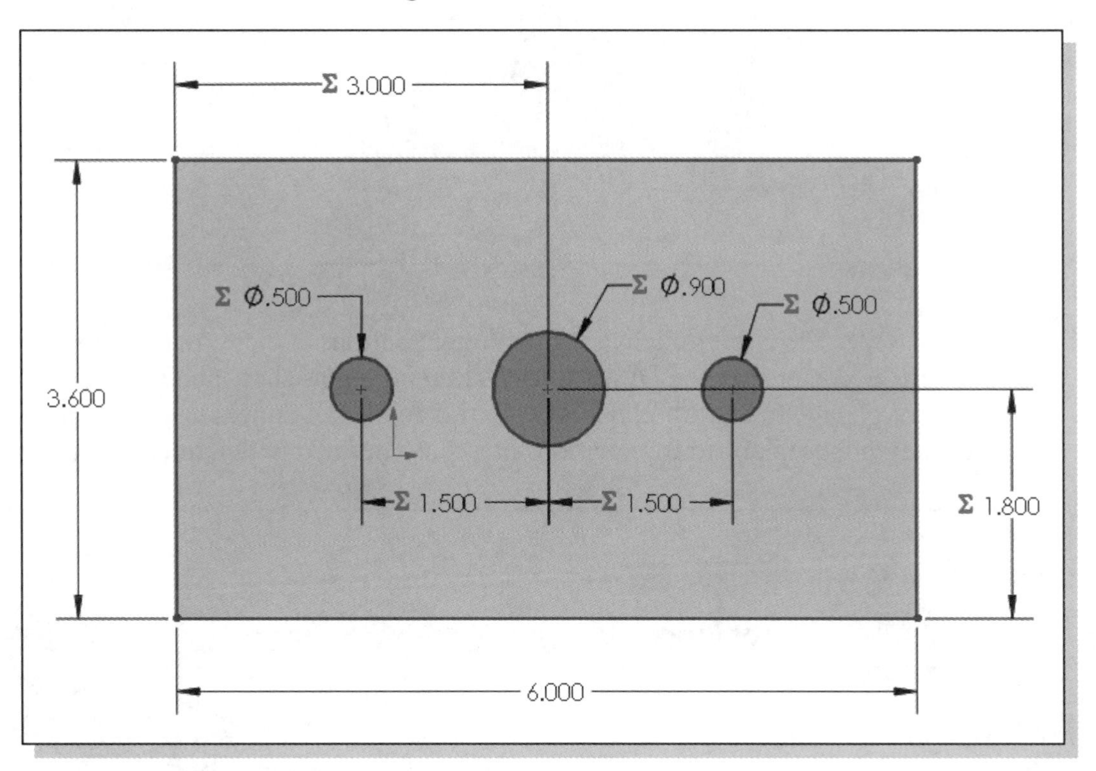

View Options in the *Equations, Global Variables, and Dimensions* Dialog Box

There are three viewing options in the *Equations, Global Variables, and Dimensions* dialog box: Equation, Dimension, and Ordered.

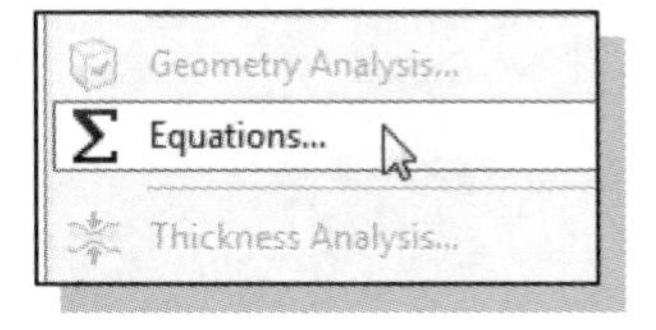

1. Move the cursor over the **SOLIDWORKS** icon on the *Menu Bar* and Select **Equations** in the *Tools* pull-down menu.

❖ The *Equations, Global Variables, and Dimensions* dialog box appears. It is shown below in Equation View mode. In this mode, all equations are shown. Notice the changes we made using the dimension *Modify* dialog boxes are now reflected.

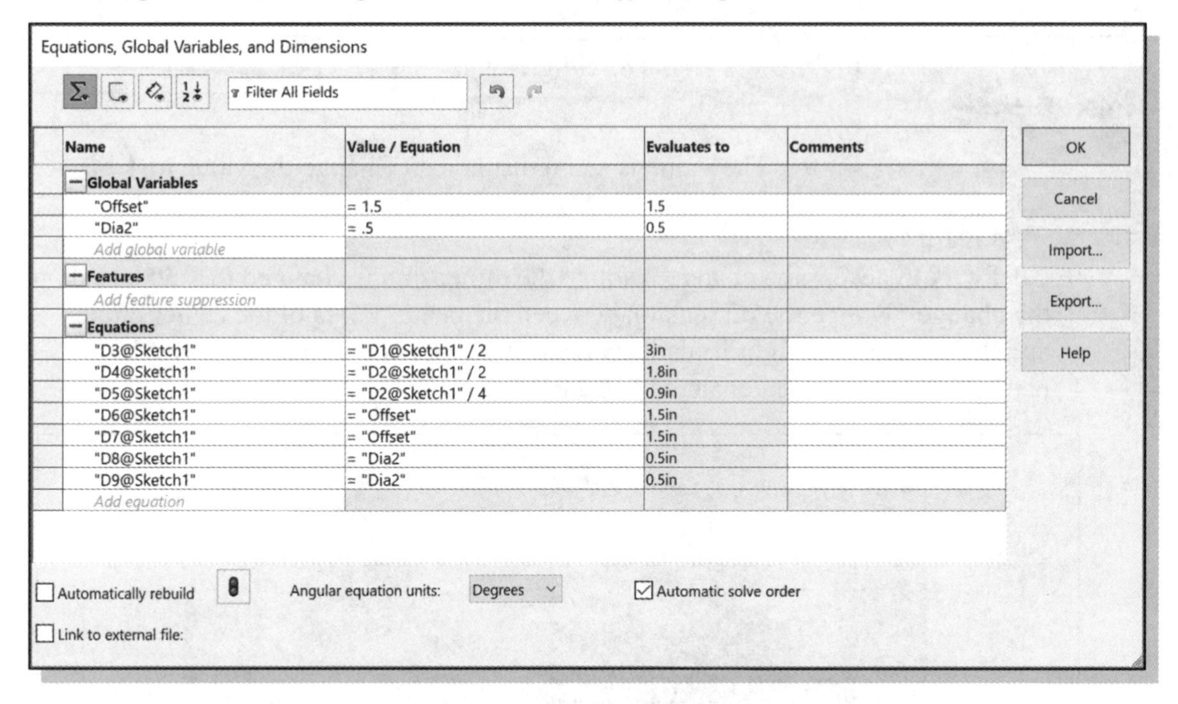

2. The view mode is controlled using the toggle buttons at the top of the dialog box. Move the cursor over the **Dimension View** button as shown below and click once to change to Dimension View. In this mode, all dimensions are shown, including those defined by equations and those defined with numerical values.

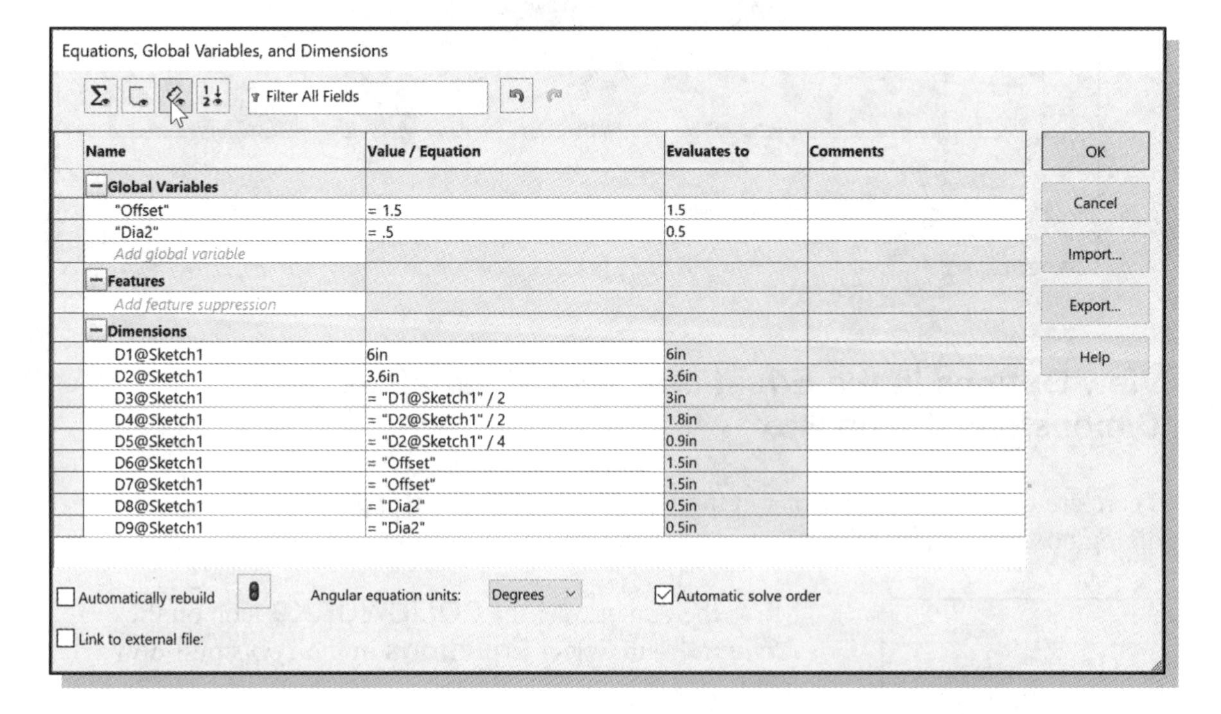

3. Move the cursor over the **Ordered View** button as shown below and click once to change to Ordered View. In this mode, all equations, including definition of global variables, are displayed in the order in which they were created.

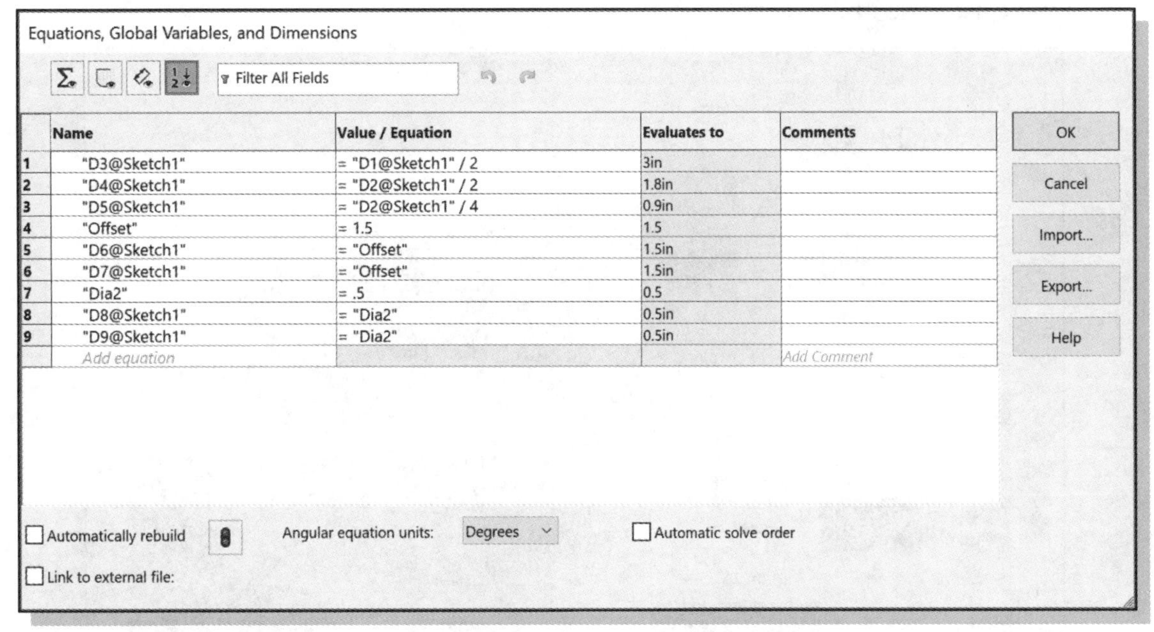

4. Move the cursor over the **Equation View** button and click once to return to Equation View, then click **OK** to close the *Equations, Global Variables, and Dimensions* dialog box.

Direct Input of Equations in PropertyManager Fields

For many features, it is possible to enter and modify equations directly in the numerical input fields of the *PropertyManager*.

1. On your own, add a third Global Variable. Name the variable **Thickness1** and set its value to **0.25**.

Name	Value / Equation
Global Variables	
"Offset"	= 1.5
"Dia2"	= .5
"Thickness1"	= .25

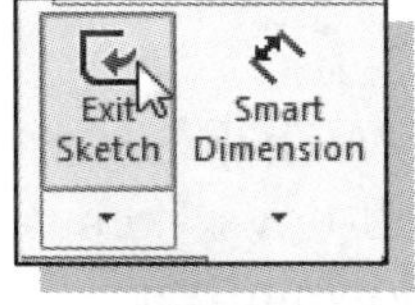

2. Click once with the left-mouse-button on the **Exit Sketch** icon on the *Sketch* toolbar to exit the Sketch option.

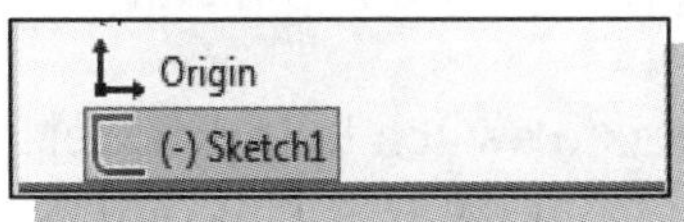

3. Make sure the sketch – Sketch1 – is selected in the *FeatureManager Design Tree.*

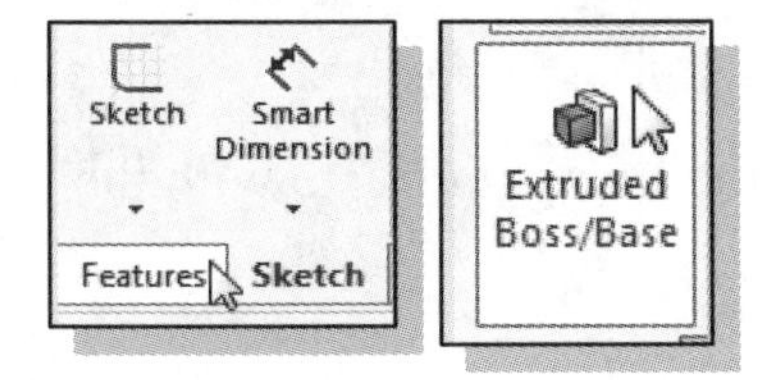

4. Select the **Features** tab on the *CommandManager*. In the *Features* toolbar, select the **Extruded Boss/Base** command by clicking once with the left-mouse-button on the icon.

5. In the *Boss-Extrude PropertyManager* panel, enter **=** in the *Depth* numeric input field to create an equation.

6. Select the **Global Variables** option and select **Thickness1** to define the extrusion depth.

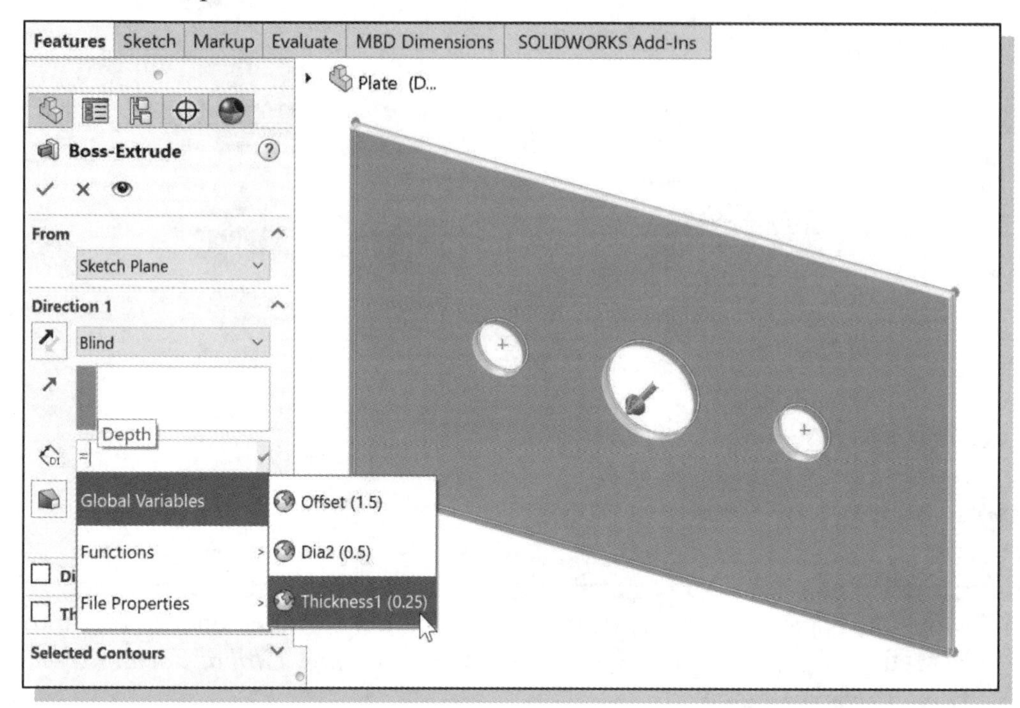

7. Click on the **OK** button to accept the settings and create the extruded boss feature.

Completing and Saving the Part File

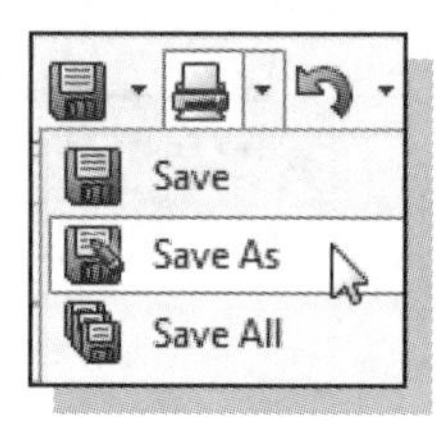

1. Click the arrow next to the Save icon in the *Menu Bar* to reveal the save options and select **Save As**.

2. In the pop-up window, enter **Plate** as the name of the file.

3. Click on the **Save** button to save the file.

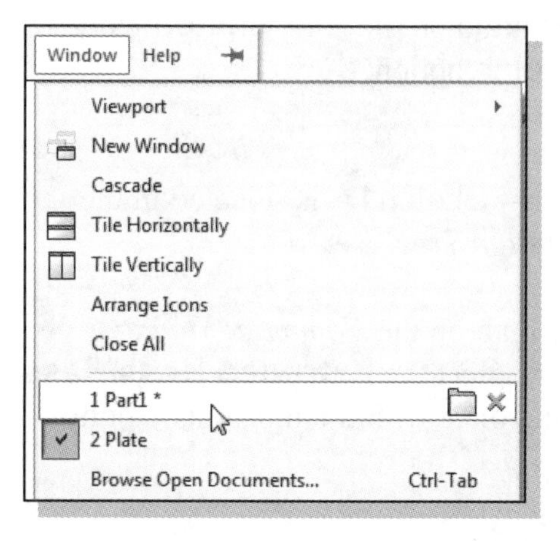

4. Move the cursor over the SOLIDWORKS icon on the *Menu Bar* to reveal the pull-down menus and click on the **Window** pull-down menu.

5. Select the **Part1** option to switch to the model of the triangular plate from the first half of this lesson.

6. On your own, save the triangular plate with the filename **Triangle**.

Questions:

1. What is the difference between *dimensional* constraints and *geometric* relations?

2. How can we confirm that a sketch is fully defined?

3. How do we distinguish between derived dimensions and parametric dimensions on the screen?

4. Describe the procedure to Display/Edit user-defined equations.

5. List and describe three different geometric relations available in SOLIDWORKS.

6. Does SOLIDWORKS allow us to build partially defined solid models? What are the advantages and disadvantages of building these types of models?

7. Identify and describe the following commands.

(a)

(b)

(c)

(d)

Exercises:

(Create and establish three parametric relations for each of the following designs.)

1. **Swivel Base** (Dimensions are in millimeters. Base thickness: **10 mm.** Boss: **5 mm.**)

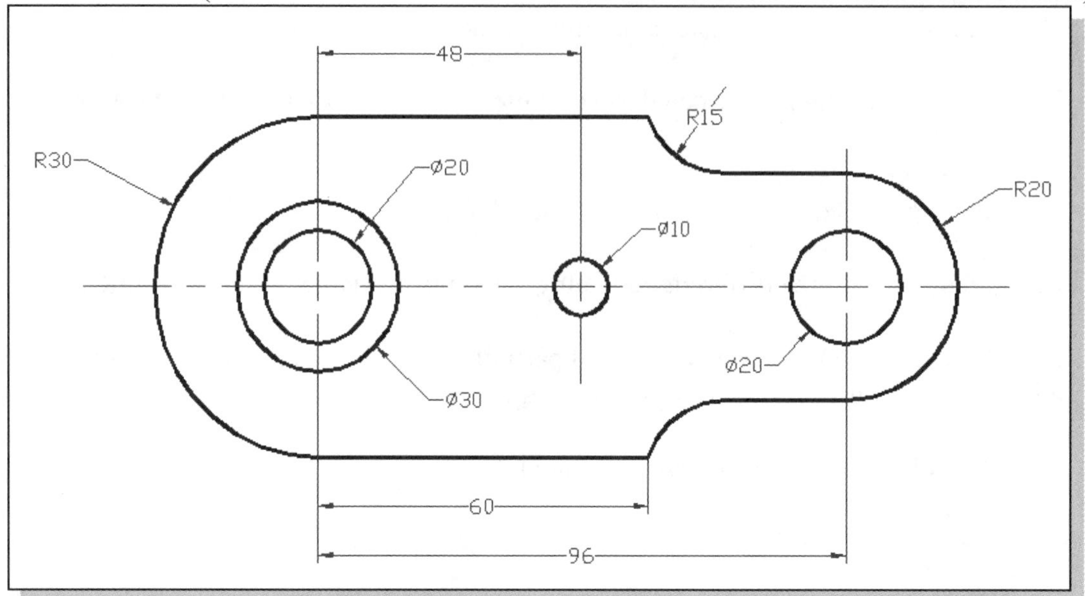

2. **Anchor Base** (Dimensions are in inches.)

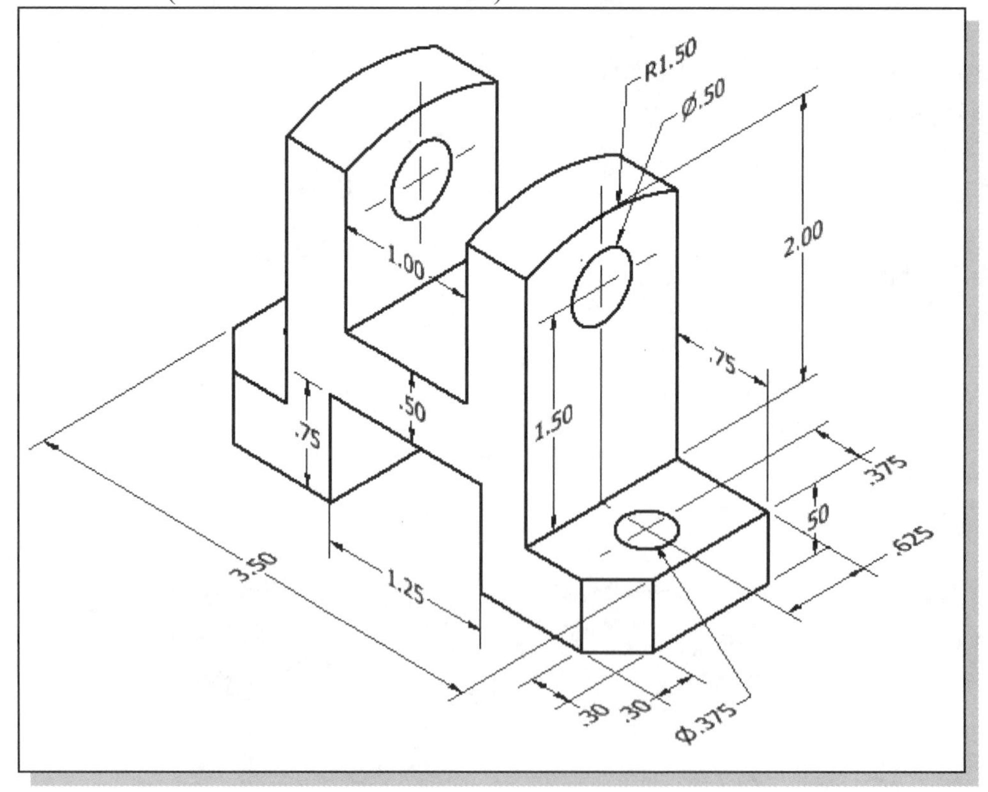

3. **Hinge Guide** (Dimensions are in inches.)

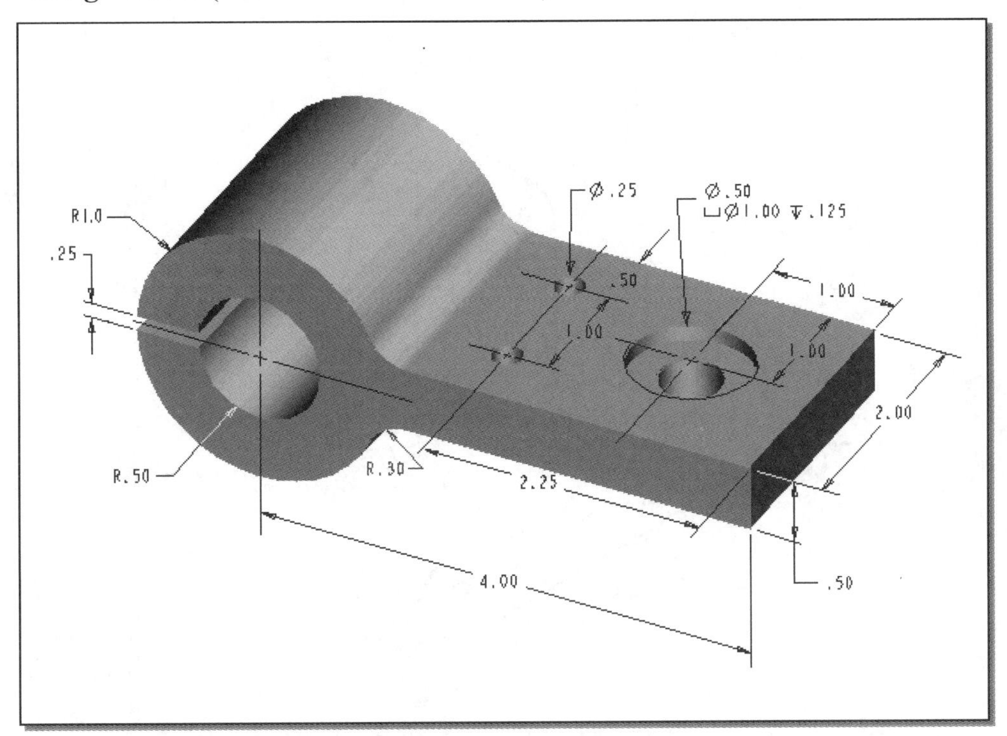

4. **Pivot Holder** (Dimensions are in inches.)

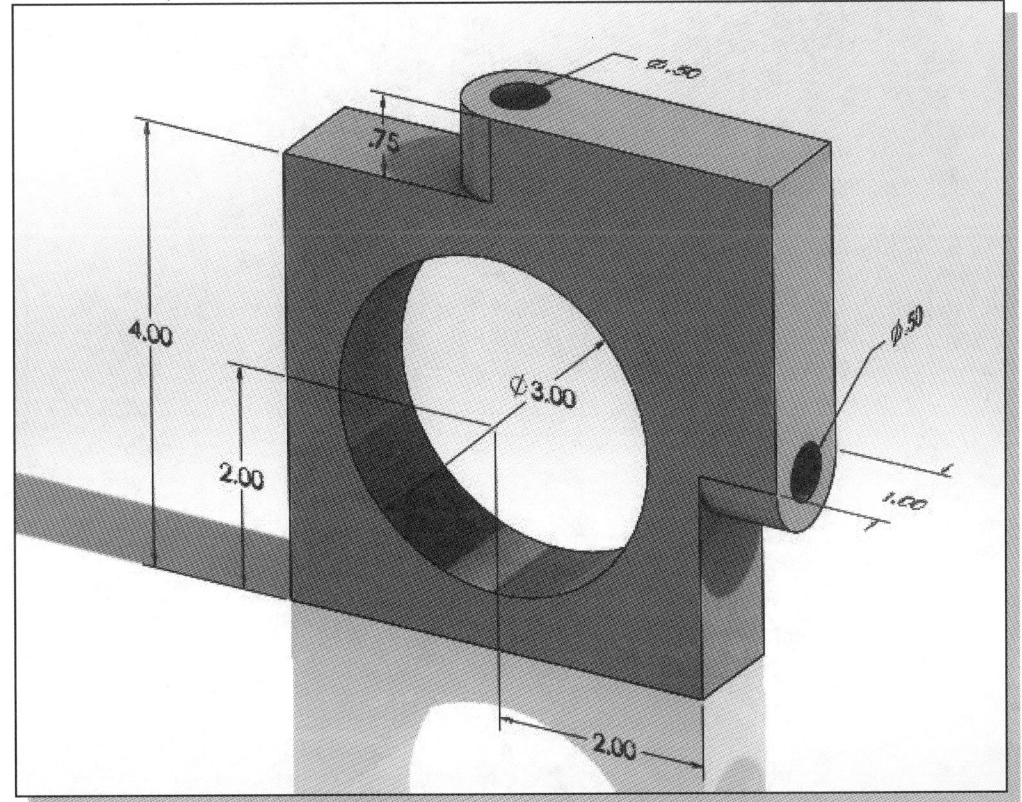

5. **Support Fixture** (Dimensions are in inches.)

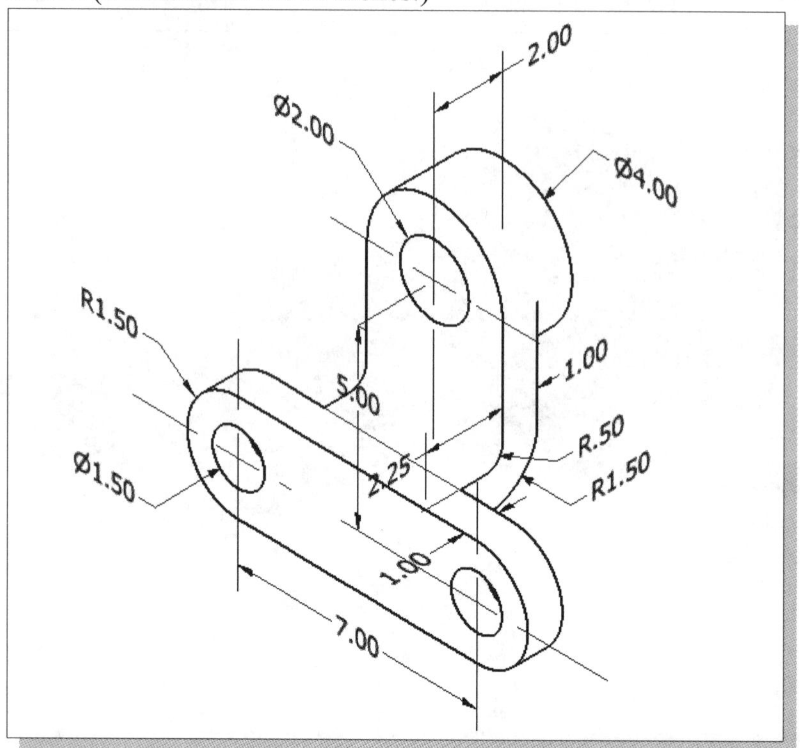

Chapter 6
Geometric Construction Tools

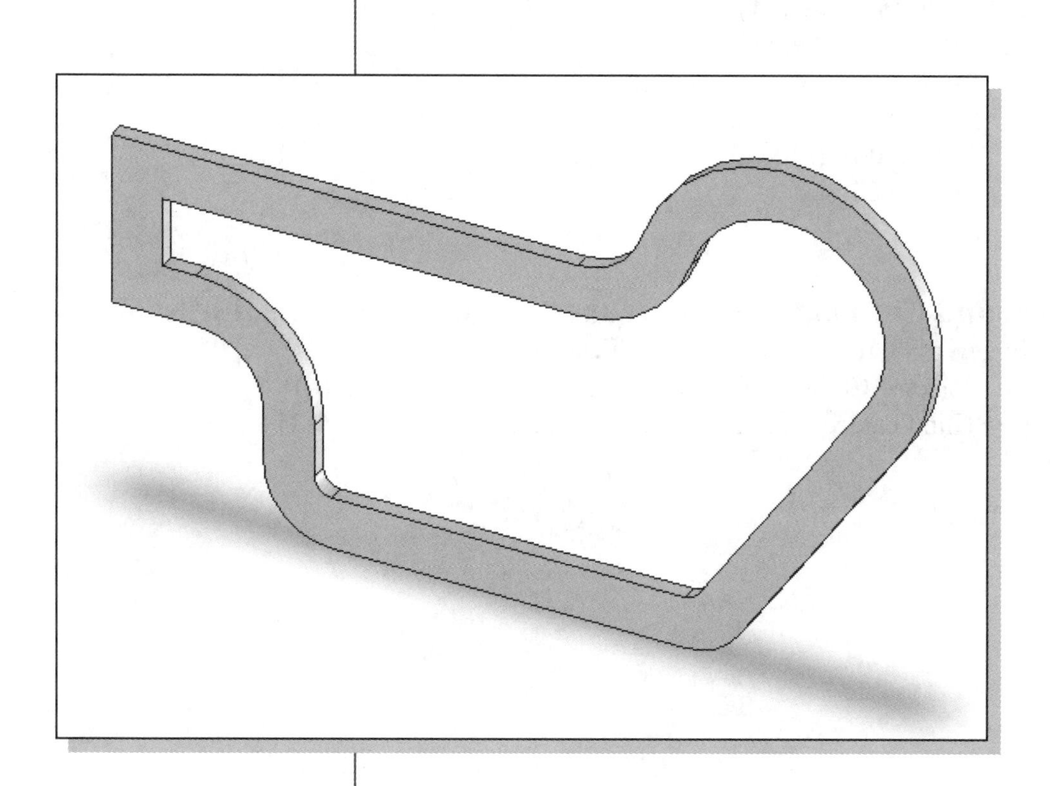

Learning Objectives

- Apply Geometry Relations
- Use the Trim/Extend Command
- Use the Offset Command
- Understand the Profile Sketch Approach
- Use the Convert Entities Command
- Understand and Use Reference Geometry
- Edit with Click and Drag
- Use the Fully Define Sketch Tool
- Use the Selected Contours Option

Certified SOLIDWORKS Associate Exam Objectives Coverage

Sketch Tools – Offset, Convert, Trim

Objectives: Using Sketch Tools.

Boss and Cut Features – Extrudes, Revolves, Sweeps, Lofts

Objectives: Creating Basic Swept Features.

Certified Associate Reference Guide

Introduction

The main characteristics of solid modeling are the accuracy and completeness of the geometric database of the three-dimensional objects. However, working in three-dimensional space using input and output devices that are largely two-dimensional in nature is potentially tedious and confusing. SOLIDWORKS provides an assortment of two-dimensional construction tools to make the creation of wireframe geometry easier and more efficient. SOLIDWORKS includes sketch tools to create basic geometric entities such as lines, arcs, etc. These entities are grouped to define a boundary or profile. A *profile* is a closed region and can contain other closed regions. Profiles are commonly used to create extruded and revolved features. An *invalid profile* consists of self-intersecting curves or open regions. In this lesson, the basic geometric construction tools, such as Trim and Extend, are used to create profiles. Mastering the geometric construction tools along with the application of proper geometric and parametric relations is the true essence of *parametric modeling*.

The *Gasket* Design

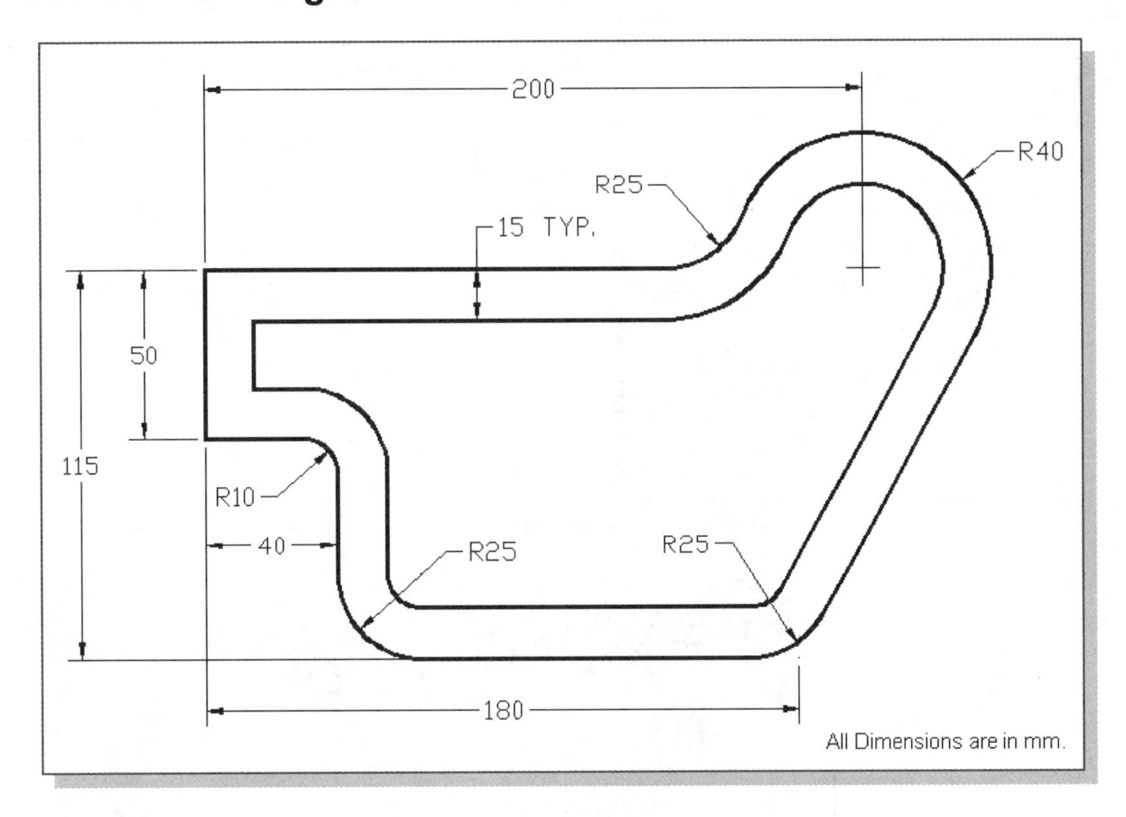

- Based on your knowledge of SOLIDWORKS so far, how would you create this design? What is the more difficult geometry involved in the design? Take a few minutes to consider a modeling strategy and do preliminary planning by sketching on a piece of paper. You are also encouraged to create the design on your own prior to following through the tutorial.

Modeling Strategy

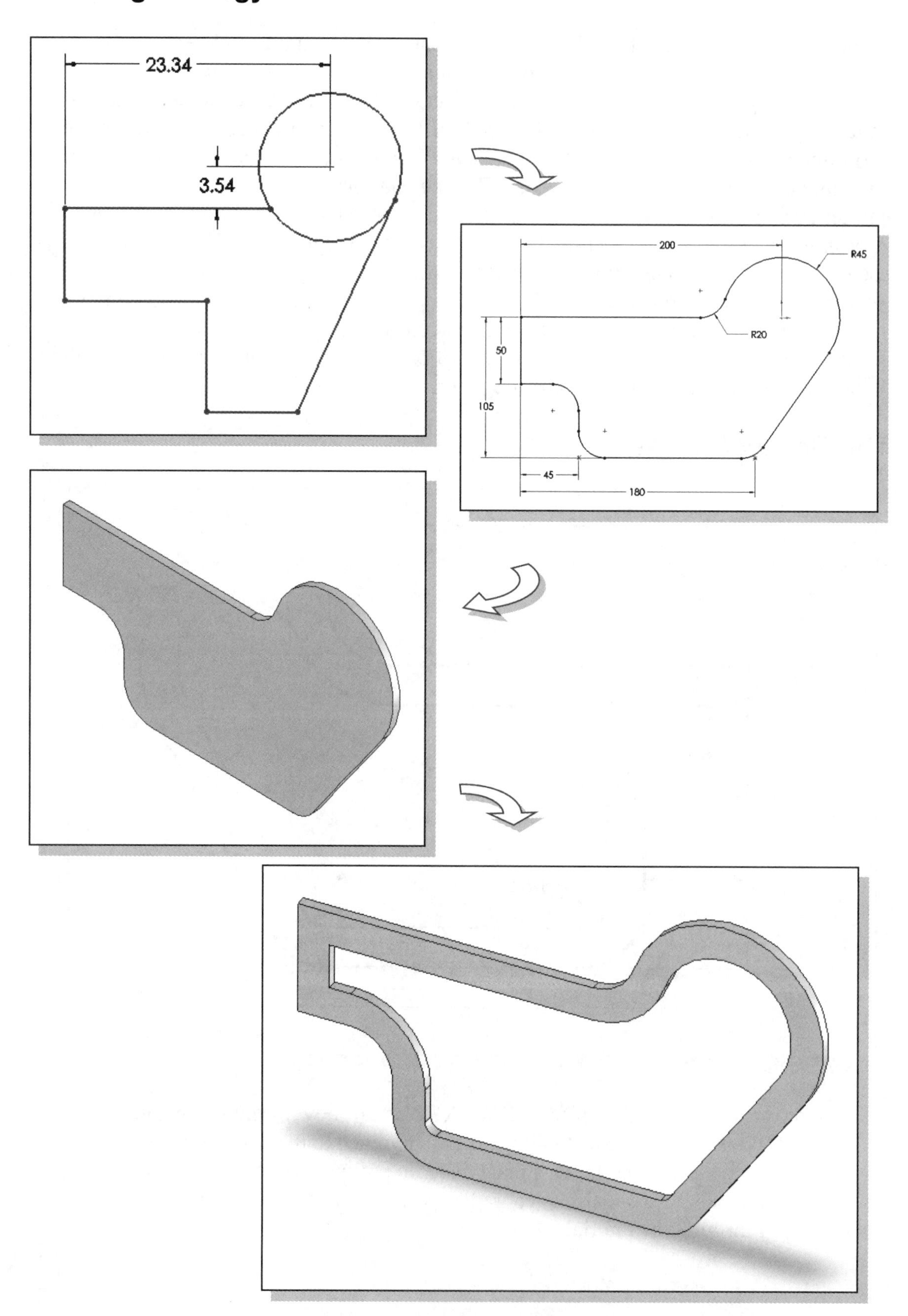

Starting SOLIDWORKS

1. Select the **SOLIDWORKS** option on the *Start* menu or select the **SOLIDWORKS** icon on the desktop to start SOLIDWORKS. The SOLIDWORKS main window will appear on the screen.

2. If the *Welcome* dialog box does not appear automatically upon opening SOLIDWORKS, click the *Welcome to SolidWorks* icon on the *Menu Bar*.

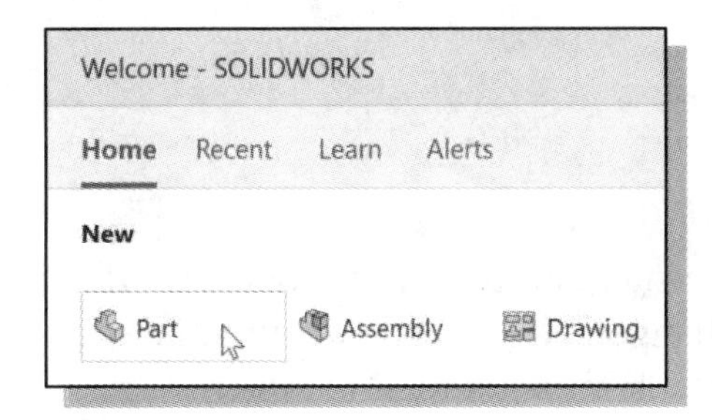

3. In the *New* panel of the *Welcome* dialog box, select the **Part** icon with a single click of the left-mouse-button to open a new part document.

4. Select the **Options** icon from the *Menu Bar* to open the *Options* dialog box.

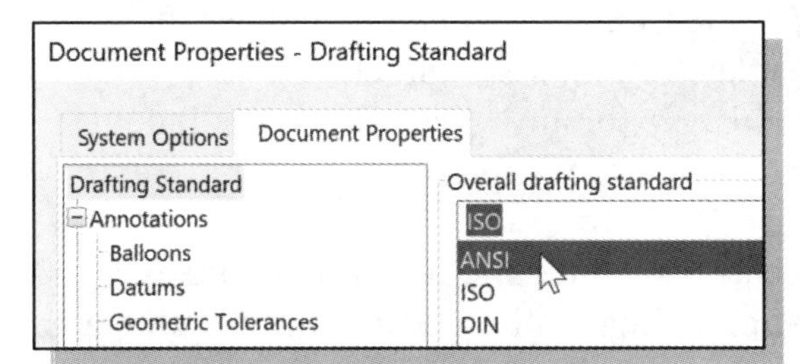

5. Select the **Document Properties** tab.

6. Select **ANSI** in the pull-down selection window under the *Overall drafting standard* panel as shown.

7. Click **Units** as shown below. We will use the default MMGS (millimeter, gram, second) unit system.

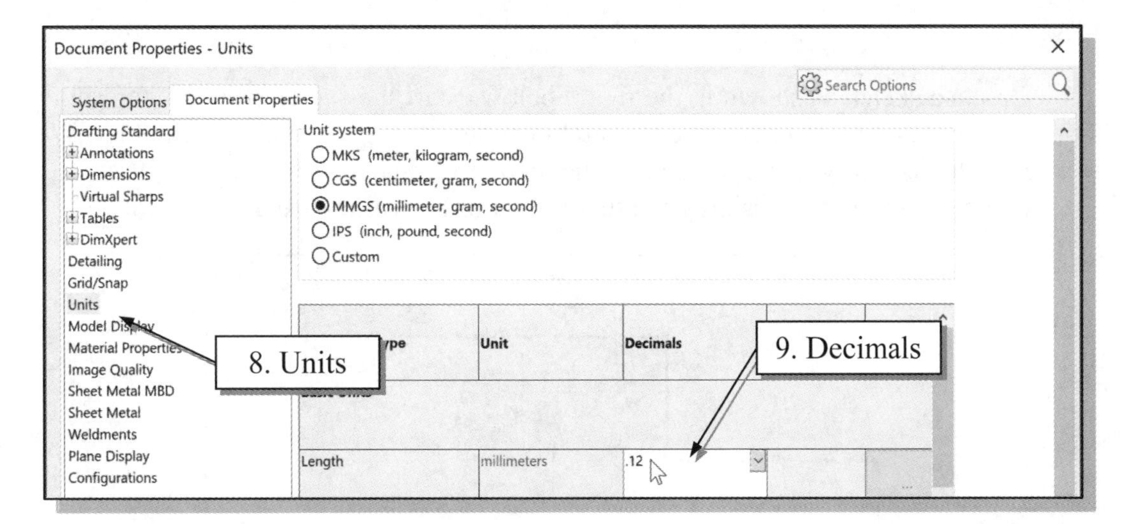

8. Select **.12** in the *Decimals* spin box for the *Length units* as shown to define the degree of accuracy with which the units will be displayed to 2 decimal places.

Creating a 2D Sketch

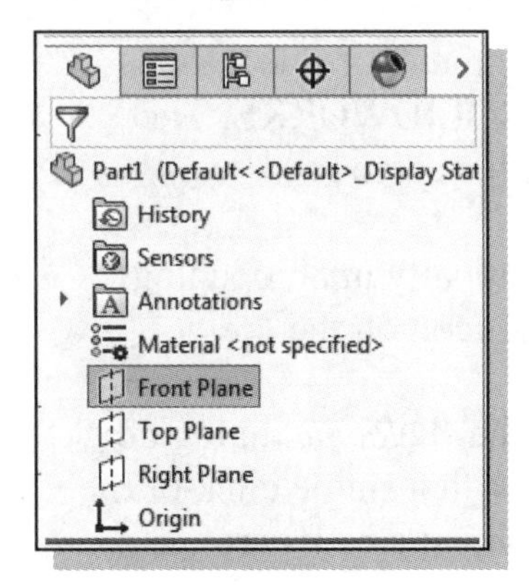

1. Click once with the **left-mouse-button** to select the **Front Plane** icon in the *FeatureManager Design Tree*. Notice the **Front Plane** icon is highlighted in the design tree and the *Front Plane* outline appears in the graphics area.

IMPORTANT NOTE: In this lesson, we will use the standard display of toolbars. If a user prefers to use the *CommandManager*, the only change is that it may be necessary to select the appropriate tab prior to selecting a command. For example, if the instruction is to "select the **Sketch** command from the *Sketch* toolbar," it may be necessary to first select the **Sketch** tab on the *CommandManager* to display the *Sketch* toolbar.

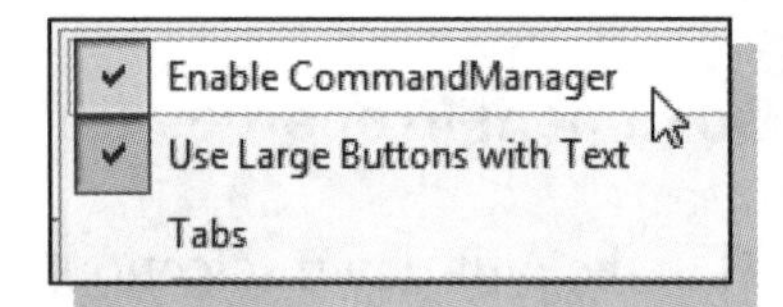

2. To turn ***OFF*** the *CommandManager* and use the standard display of toolbars, right click on the *CommandManager* (or any other toolbar) and toggle the *Enable CommandManager* option *OFF*.

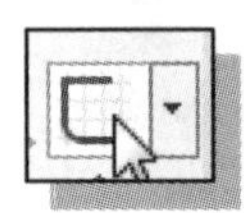

3. In the *Sketch* toolbar, select the **Sketch** command by left-clicking once on the icon. Notice the **Front Plane** automatically becomes the sketch plane because it was pre-selected.

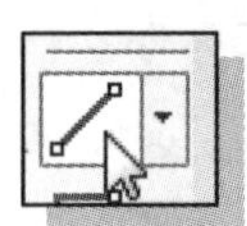

4. Click on the **Line** icon in the *Sketch* toolbar.

5. Create a sketch as shown in the figure below. Start the sketch from the top right corner. The line segments are all parallel and/or perpendicular to each other. We will intentionally make the line segments of arbitrary length, as it is quite common during the design stage that not all of the values are determined.

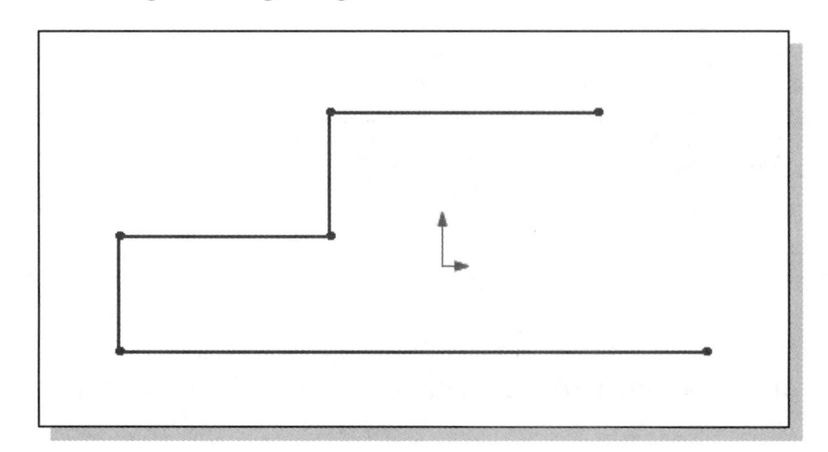

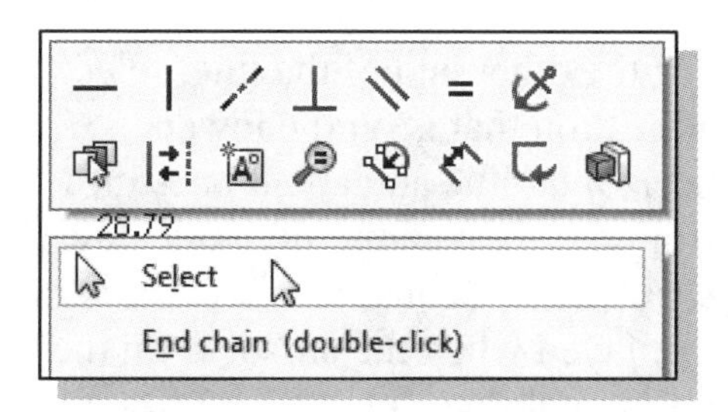

6. Inside the graphics window, right-mouse-click to bring up the option menu and select **Select** to end the Line command.

7. Select the **Circle** command by clicking once with the left-mouse-button on the icon in the *2D Sketch Panel*.

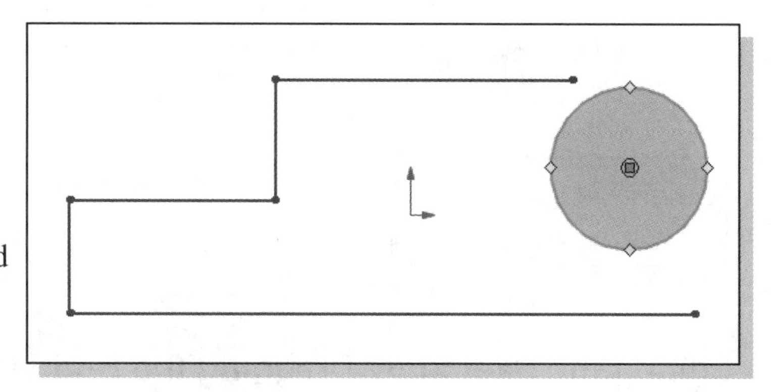

8. Pick a location that is above the bottom horizontal line as the center location of the circle.

9. Move the cursor toward the right and create a circle of arbitrary size, by clicking once with the left-mouse-button.

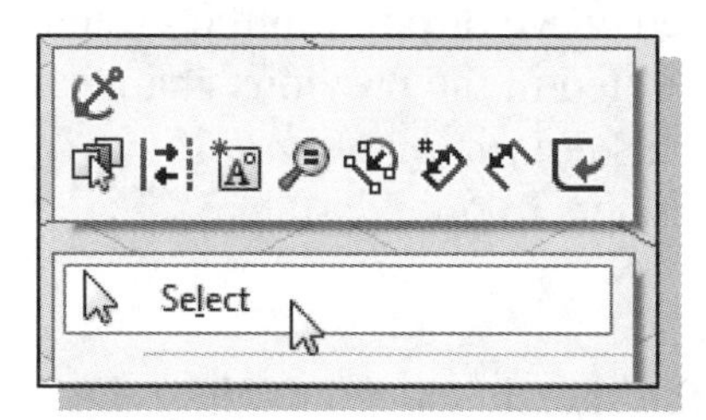

10. Inside the graphics window, click once with the right-mouse-button to display the option menu. Select **Select** in the pop-up menu to end the Circle option.

11. Click on the **Line** icon in the *Sketch* toolbar.

12. Move the cursor near the upper portion of the circle and, when the **Coincident** constraint symbol is displayed, click once with the **left-mouse-button** to select the starting point for the line.

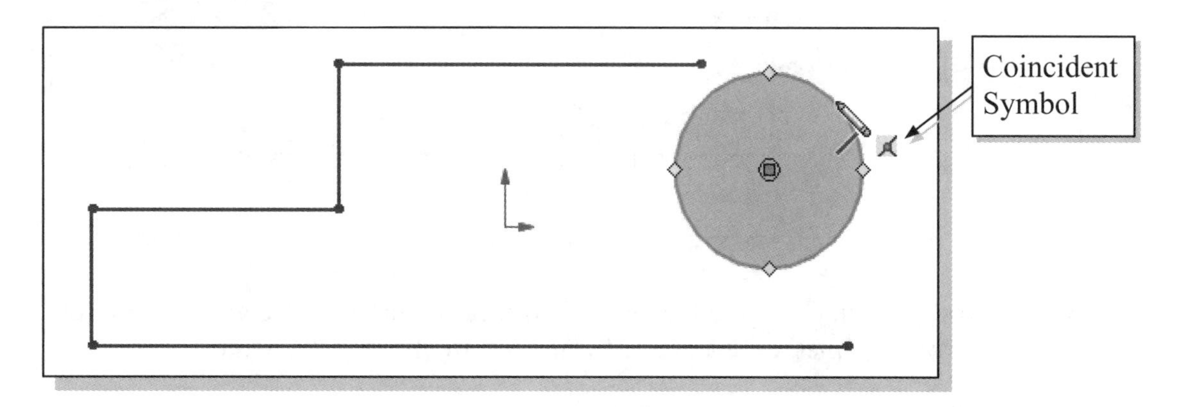

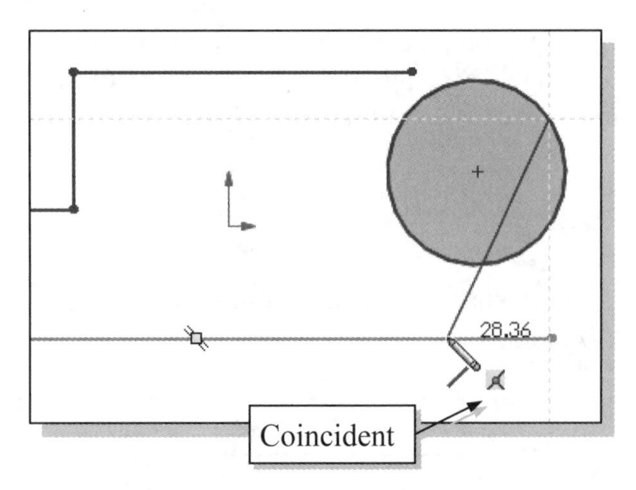

13. For the other end of the line, select a location that is on the lower horizontal line and about one-third from the right endpoint. Notice the **Coincident** constraint symbol is displayed when the cursor is on the horizontal line.

14. Inside the graphics window, right-mouse-click to bring up the option menu.

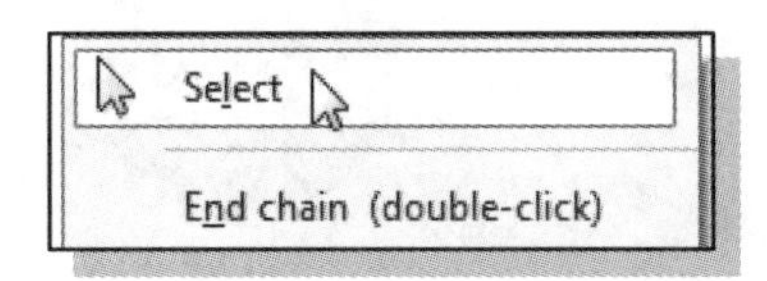

15. In the option menu, select **Select** to end the Line command.

Editing the Sketch by Dragging the Entities

In SOLIDWORKS, we can click and drag any under-defined curve or point in the sketch to change the size or shape of the sketched profile. As illustrated in the previous chapter, this option can be used to identify under-defined entities. This *editing by dragging* method is also an effective visual approach that allows designers to quickly make changes.

1. Move the cursor on the lower left vertical edge of the sketch. Click and drag the edge to a new location that is toward the right side of the sketch.

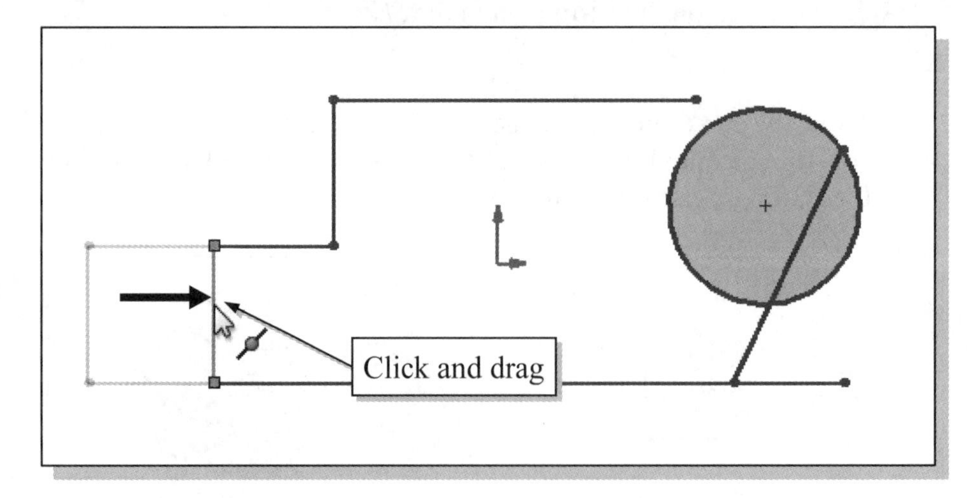

❖ Note that we can only drag the vertical edge horizontally; the connections to the two horizontal lines are maintained while we are moving the geometry.

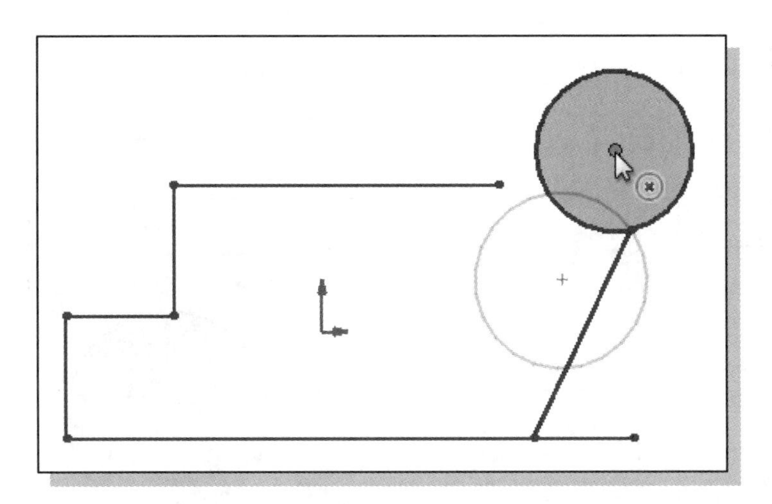

2. Click and drag the center point of the circle to a new location.

❖ Note that as we adjust the size and the location of the circle, the connection to the inclined line is maintained.

3. Click and drag the lower endpoint of the inclined line to a new location.

❖ Note that as we adjust the size and the location of the inclined line, the location of the bottom horizontal edge is also adjusted.

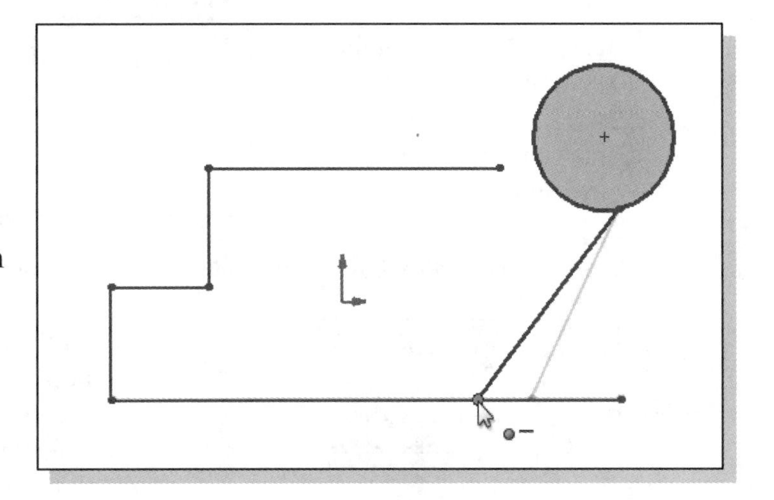

➢ Note that several changes occur as we adjust the size and the location of the inclined line. The location of the bottom horizontal line and the length of the vertical line are adjusted accordingly.

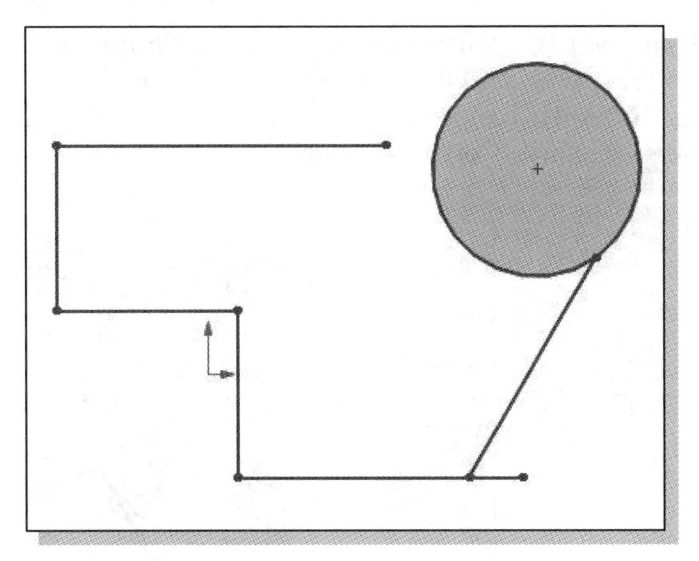

4. On your own, adjust the sketch so that the shape of the sketch appears roughly as shown.

❖ The *editing by dragging* method is an effective approach that allows designers to explore and experiment with different design concepts.

Adding Additional Relations

1. Choose **Smart Dimension** in the *Sketch* toolbar.

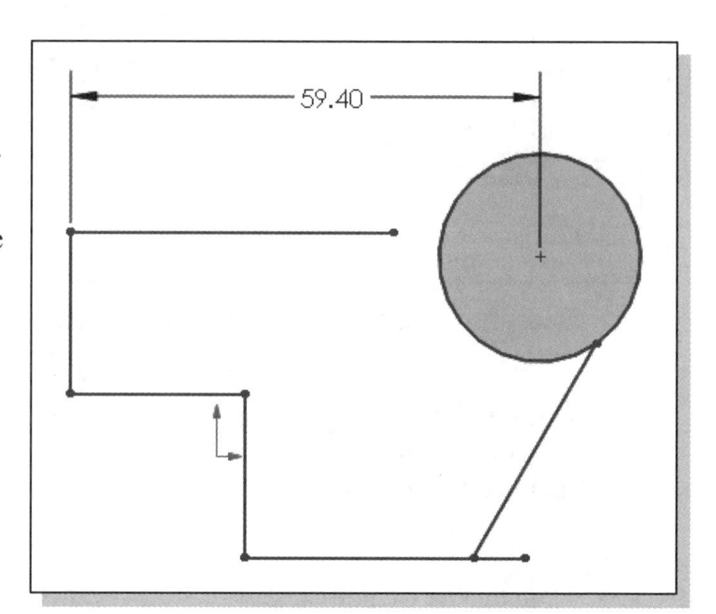

2. Add the horizontal location dimension, from the top left vertical edge to the center of the circle as shown. (Do not be overly concerned with the dimensional value; we are still working on creating a *rough sketch*.)

3. Click the **OK** icon in the *PropertyManager*, or hit the [**Esc**] key once, to end the Smart Dimension command.

4. Press the [**Esc**] key to ensure no items are selected.

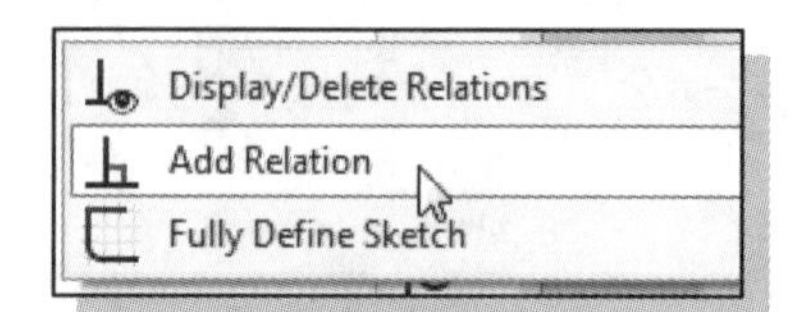

5. Select the **Add Relation** command from the pop-up menu on the *Sketch* toolbar. Notice the *Selected Entities* window in the *Add Relations PropertyManager* is blank because no entities are selected.

6. Pick the **inclined line** by left-mouse-clicking once on the geometry.

7. Pick the **circle**.

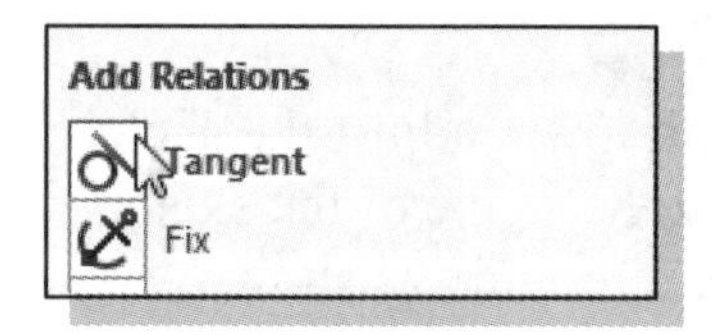

8. Click once with the left-mouse-button on the **Tangent** icon in the *Add Relations PropertyManager* as shown. This activates the Tangent relation. The sketched geometry is adjusted as shown below.

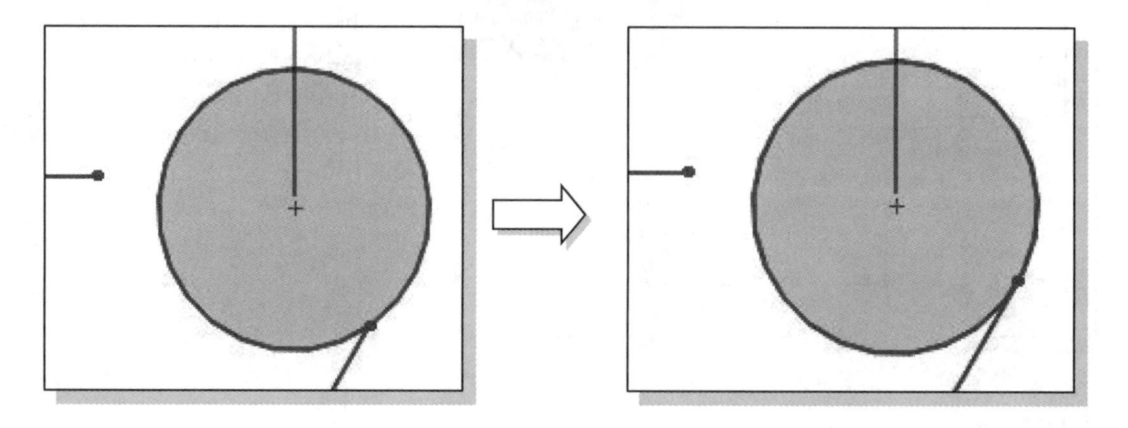

9. Click the **OK** icon in the *PropertyManager*, or hit the [**Esc**] key once, to end the Add Relations command.

10. Click and drag the circle to a new location.

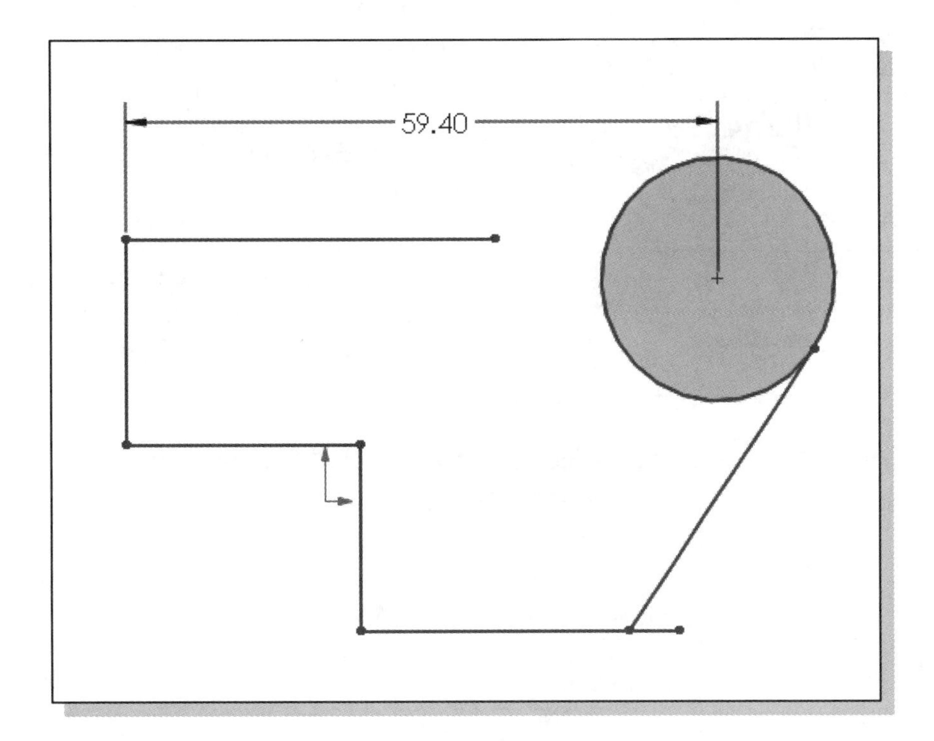

❖ Note that the dimension we added now restricts the horizontal movement of the center of the circle. The Tangent relation to the inclined line is maintained.

Using the *Trim* and *Extend* Commands

In the following sections, we will illustrate using the Trim and Extend commands to complete the desired 2D profile.

The **Trim** and **Extend** commands can be used to shorten/lengthen an object so that it ends precisely at a boundary. As a general rule, SOLIDWORKS will try to clean up sketches by forming a closed region sketch.

1. **Left-click** on the **arrow** next to the **Trim Entities** button on the *Sketch* toolbar to reveal additional commands.

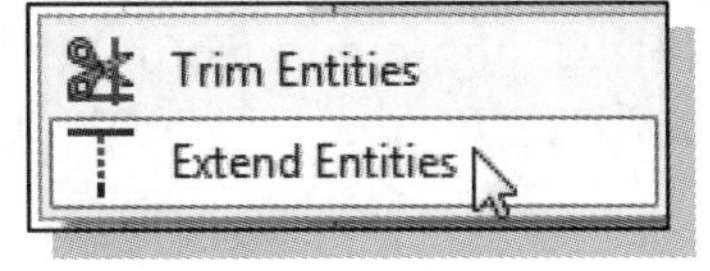

2. Select the **Extend Entities** command from the pop-up menu.

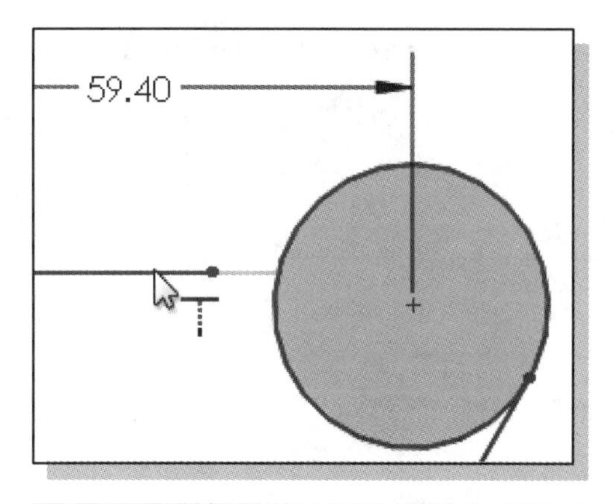

3. We will first extend the top horizontal line to the circle. Move the cursor near the right hand endpoint of the top horizontal line. SOLIDWORKS will automatically display the possible result of the selection. When the extended line appears, click once with the left-mouse-button.

4. Press the [**Esc**] key once to end the Extend command.

5. We will next trim the bottom horizontal line to the inclined line. Select the **Trim Entities** icon on the *Sketch* toolbar.

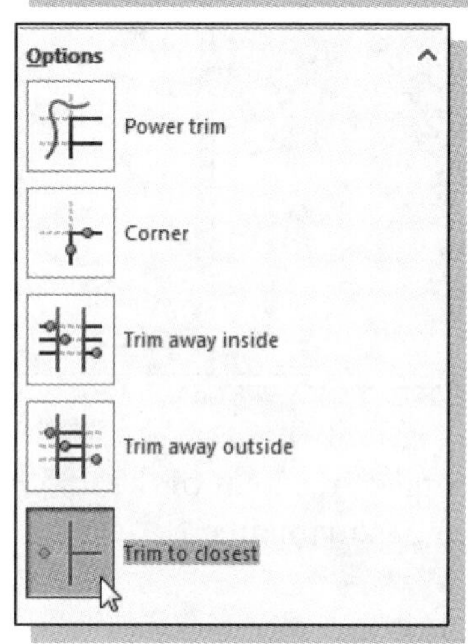

6. Select the **Trim to closest** option in the *Trim PropertyManager*.

7. Move the cursor near the right hand endpoint of the bottom horizontal line. The portion of the line that will be trimmed is highlighted. Click once with the **left-mouse-button** to trim the line.

8. Click the **OK** icon in the *PropertyManager*, or hit the [**Esc**] key once, to end the Trim command.

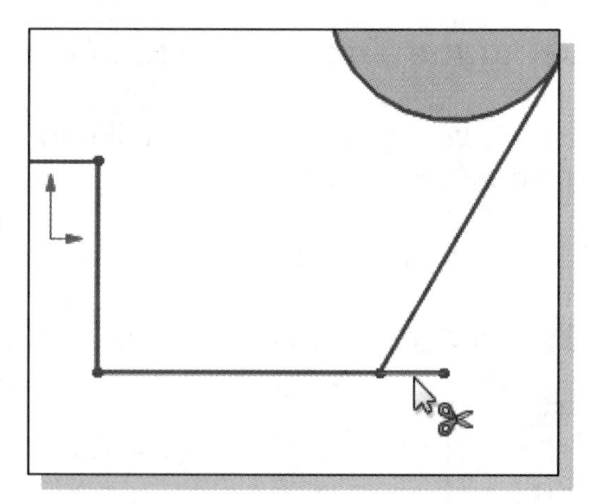

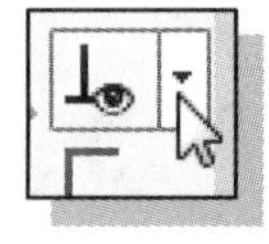

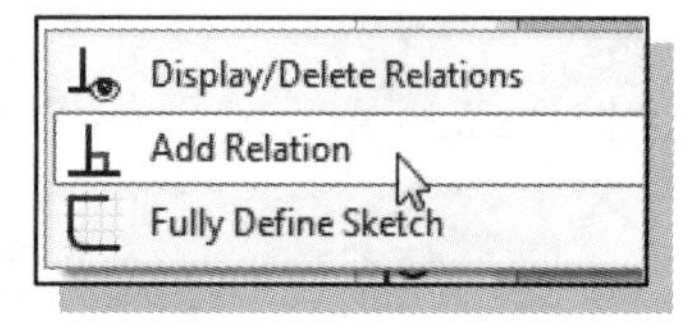

9. Select the **Add Relation** command from the pop-up menu on the *Sketch* toolbar.

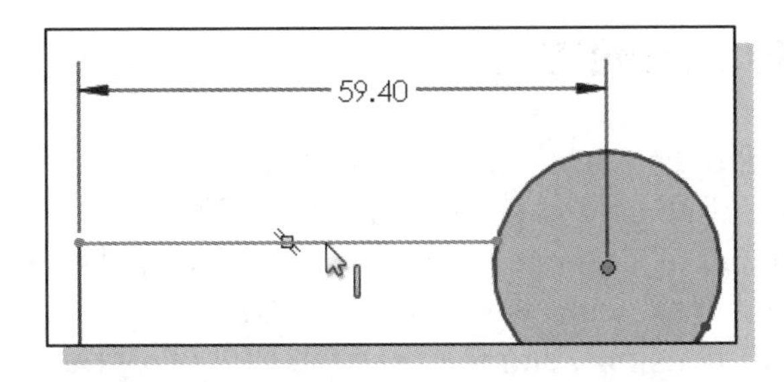

10. In the graphics area, select the **center point of the circle** and the **top horizontal line** as shown.

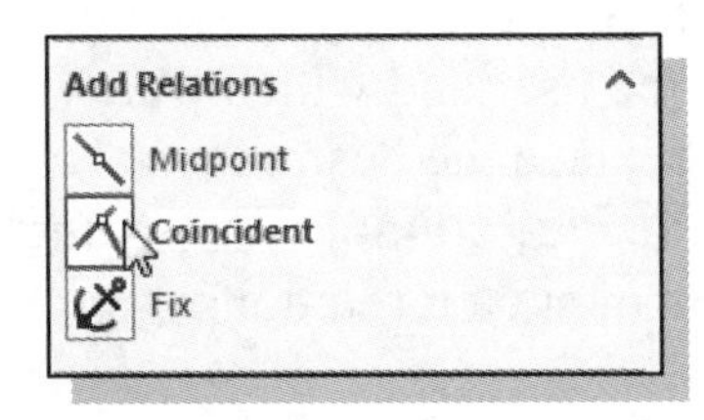

11. Click once with the left-mouse-button on the **Coincident** icon in the *Add Relations PropertyManager* as shown. This activates the Coincident relation.

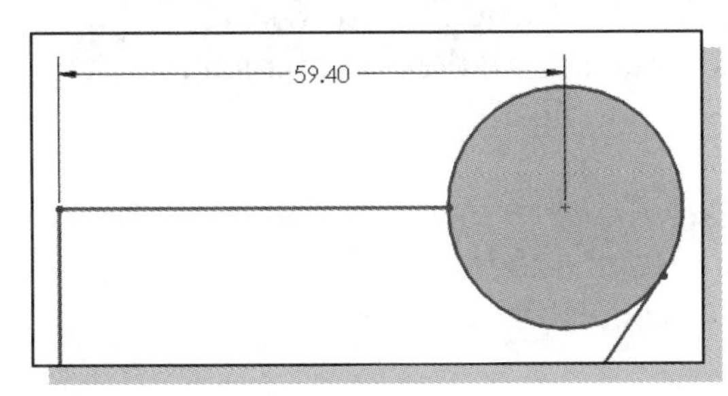

12. Click the **OK** icon in the *PropertyManager*, or hit the [**Esc**] key once, to end the Add Relations command.

13. On your own, create the **height dimension** to the left of the sketch as shown.

14. On your own, **Trim** the circle as shown.

15. Hit the [**Esc**] key to end the Trim command.

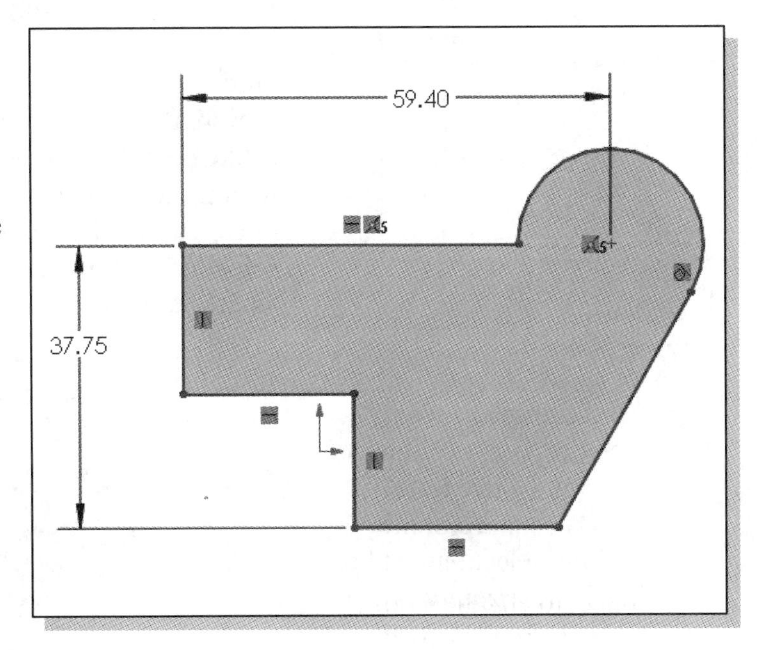

16. Turn *ON* the visibility of the *Sketch Relations* (on the *Hide/Show Items* pull-down menu in the *Heads-up* toolbar) to examine the applied constraints. Confirm that the **Tangent** and **Coincident** relations are applied as shown.

17. Turn *OFF* the visibility of the *Sketch Relations*.

Adding Dimensions with the Fully Define Sketch Tool

In SOLIDWORKS, the Fully Define Sketch tool can be used to calculate which dimensions and relations are required to fully define an under defined sketch. Fully defined sketches can be updated more predictably as design changes are implemented. The general procedure for applying dimensions to sketches is to use the Smart Dimension command to add the desired key dimensions, and then use the Fully Define Sketch tool as a quick way to calculate all other sketch dimensions necessary. SOLIDWORKS remembers which dimensions were added by the user and which were calculated by the system, so that automatic dimensions do not replace the desired dimensions. In the following steps we will use the Fully Define Sketch tool to apply the dimensions necessary but will disable the automatic application of sketch relations.

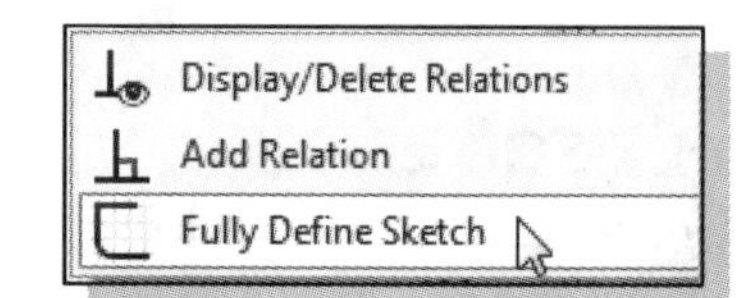

1. Select the **Fully Define Sketch** command from the pop-up menu on the *Sketch* toolbar.

2. Uncheck
3. Left-click
Relations
Dimensions
Horizontal Dimensions Scheme:
Baseline
Point1@Origin
4. Select

2. The *Fully Define Sketch PropertyManager* appears. **Uncheck** the Relations box to disable the calculation of geometric relations.

3. Click once with the **left-mouse-button** on the arrows to reveal the Dimensions option panel, **if necessary**.

4. Notice that under the *Horizontal Dimensions Scheme* option, Baseline is selected. The origin is selected as the default baseline datum and appears in the datum selection window as Point1@Origin. Click once with the **left-mouse-button** in the *Datum* selection text box to select a different datum to serve as the baseline.

5. In the graphics window, select the vertical line as shown. Notice Line2 is now displayed in the *Datum* selection text box as the *Horizontal Dimensions* baseline.

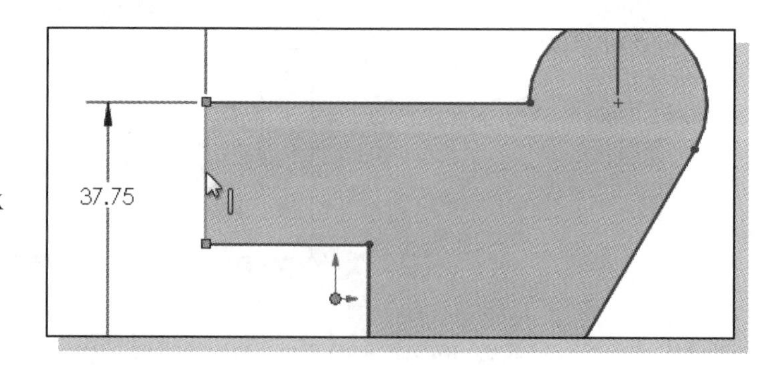

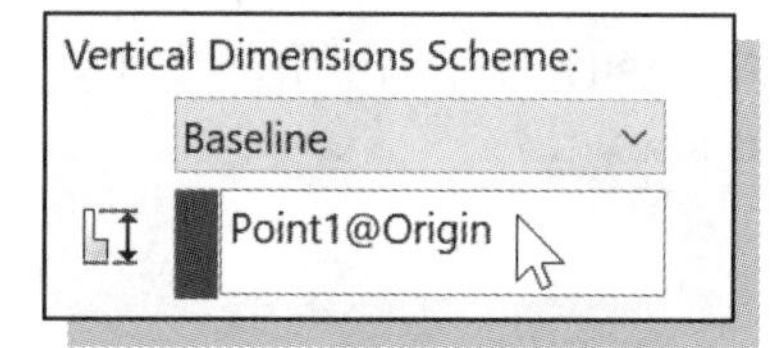

6. In the *PropertyManager*, under the *Vertical Dimensions Scheme* option panel, click once with the **left-mouse-button** in the *Datum* selection text box.

7. In the graphics window, select the horizontal line as shown. Notice Line1 is now displayed in the *Datum* selection text box as the *Vertical Dimensions* baseline.

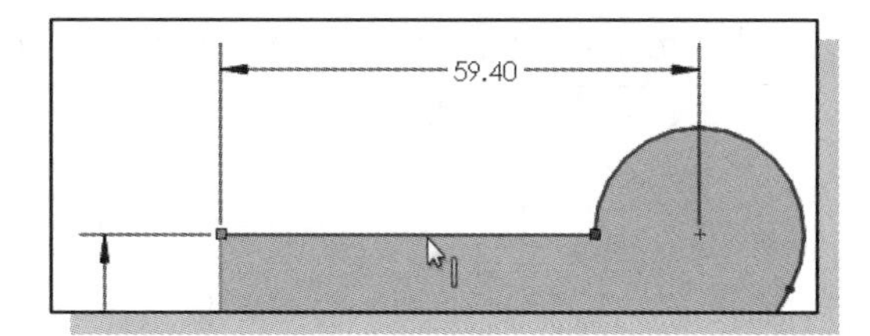

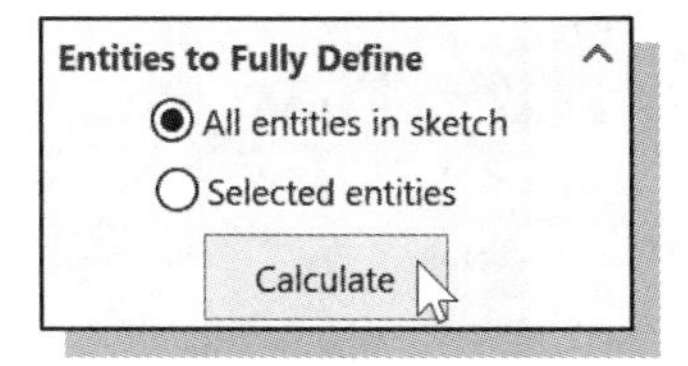

8. Select **Calculate** in the *PropertyManager* to continue with the Fully Define Sketch command.

9. A dialog box appears with the statement "*Fully Define Sketch is complete but the sketch is still under defined.*" This is because the sketch is not fixed to a location in the coordinate system. Select **OK** to close the dialog box.

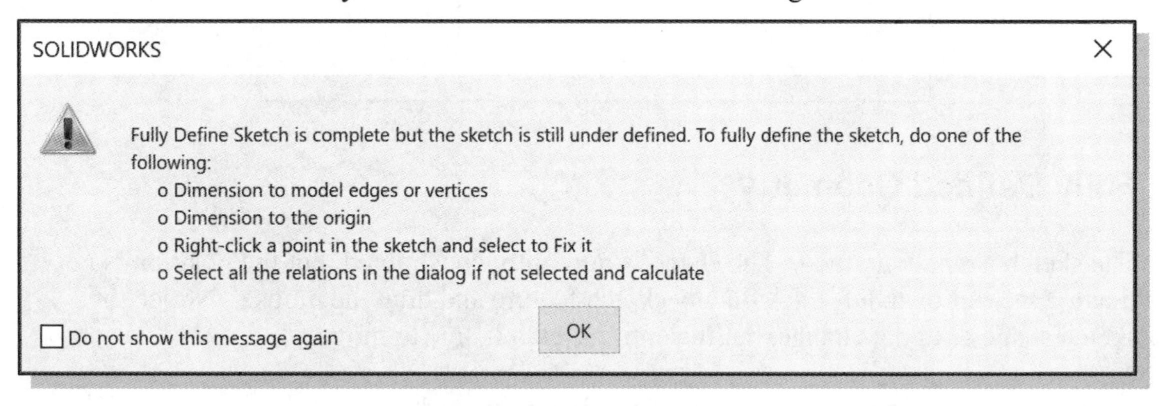

10. Click **OK** in the *PropertyManager* to accept the results and exit the Fully Define Sketch command.

➢ The *Autodimension* results should appear as shown. (**NOTE:** Dimensional values will be different; this is only a rough sketch.)

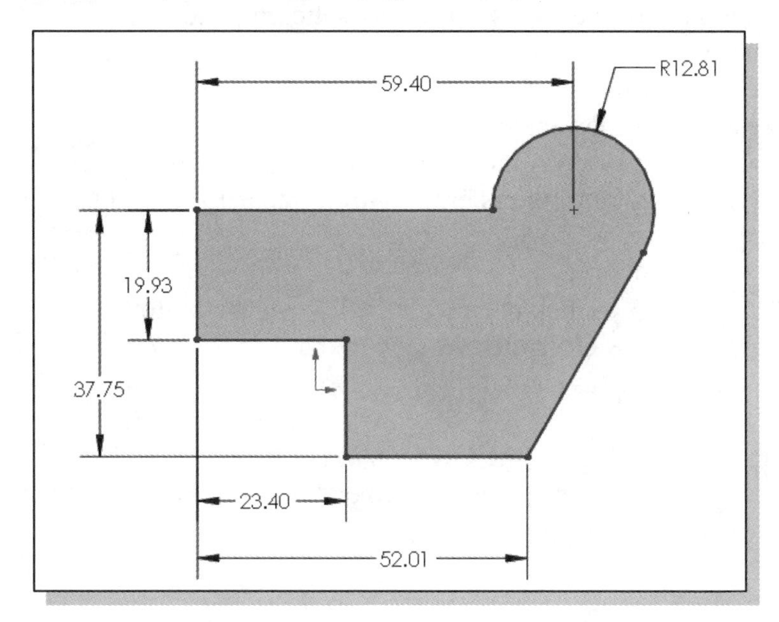

11. On your own, change the dimensions to the desired values as shown below. (**HINT:** Change the larger values first.)

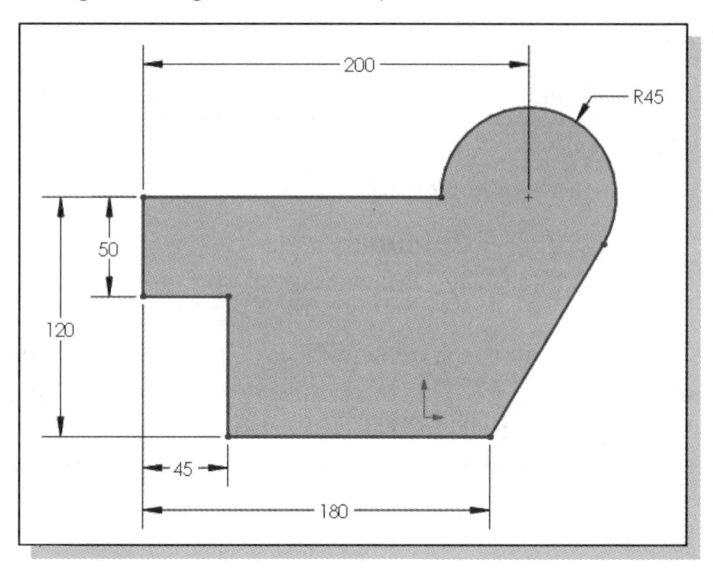

Fully Defined Geometry

The sketch is under defined. The shape is now fully constrained, but the position is not fixed. On your own, left-click on any sketch feature and drag the mouse. Notice the sketch shape does not change, but the entire sketch is free to move.

1. Press the [**Esc**] key to ensure no items are selected.

2. Select the **Add Relation** command from the pop-up menu on the *Sketch* toolbar.

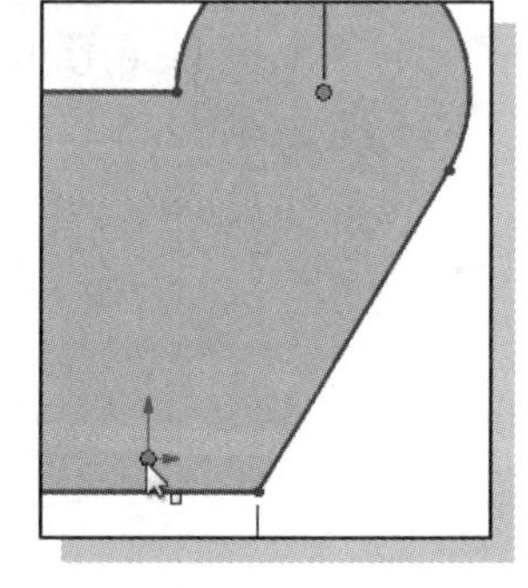

3. Pick the centerpoint of the arc by left-mouse-clicking once on the geometry.

4. Pick the sketch plane coordinate origin. (**NOTE:** If the origin symbol is not visible in the graphics area, turn *ON* the visibility by selecting Origins under the *View* toolbar.)

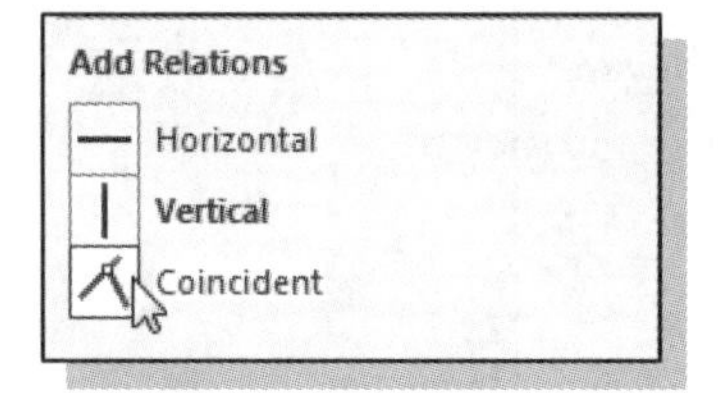

5. Click once with the left-mouse-button on the **Coincident** icon in the *Add Relations PropertyManager* as shown.

6. Click the **OK** icon in the *PropertyManager*, or hit the [**Esc**] key once, to end the Add Relations command.

7. Press the **[F]** key to zoom the sketch to fit the graphics area. Note that the sketch is fully defined with the added relation. **Fully Defined** is displayed in the *Status Bar* at the bottom of the screen, and the color of the sketch features changes to black.

Creating Fillets and Completing the Sketch

1. Click on the **Sketch Fillet** icon in the *Sketch* toolbar.

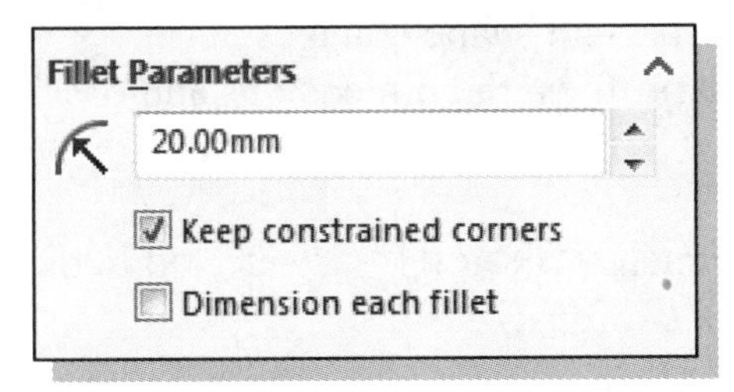

2. The *Sketch Fillet PropertyManager* appears. Enter **20 mm** for the fillet radius as shown.

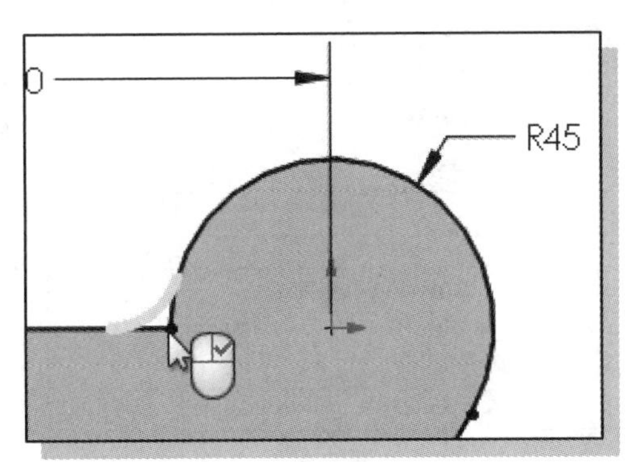

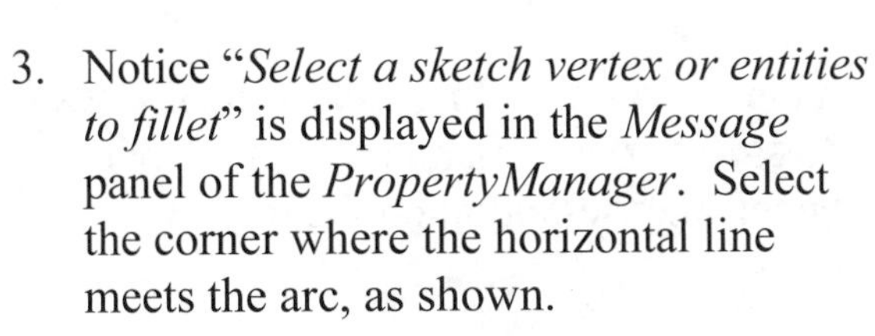

3. Notice "*Select a sketch vertex or entities to fillet*" is displayed in the *Message* panel of the *PropertyManager*. Select the corner where the horizontal line meets the arc, as shown.

4. **On your own**, create the **three additional fillets** as shown in the figure below. Note that all the rounds and fillets are created with the same radius.

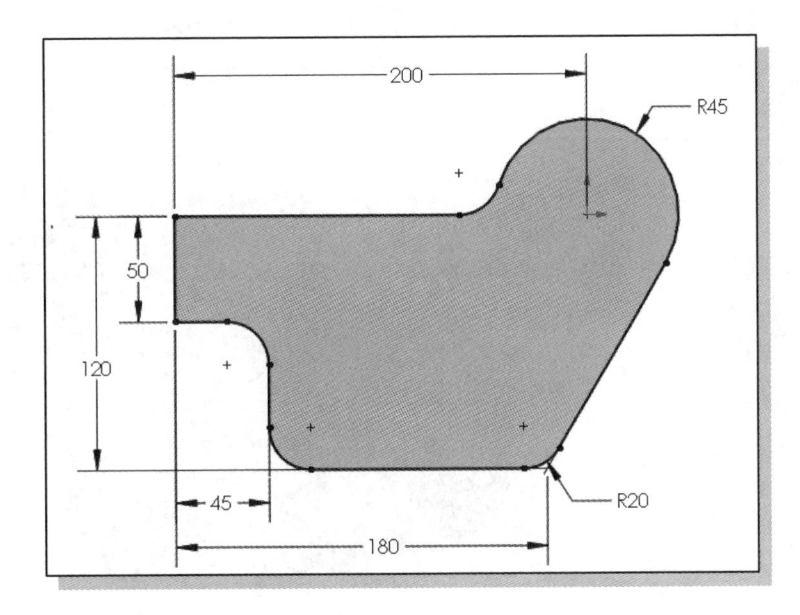

5. Click on the **OK** icon in the *PropertyManager* to end the Sketch Fillet command.

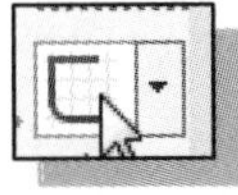

6. Click once with the **left-mouse-button** on the **Sketch** icon on the *Sketch* toolbar to exit the Sketch option.

Profile Sketch

In SOLIDWORKS, ***profiles*** are closed regions that are defined from sketches. Profiles are used as cross sections to create solid features. For example, **Extrude, Revolve, Sweep, Loft,** and **Coil** operations all require the definition of at least a single profile. The sketches used to define a profile can contain additional geometry since the additional geometry entities are consumed when the feature is created. To create a profile we can create single or multiple closed regions, or we can select existing solid edges to form closed regions. A profile cannot contain self-intersecting geometry; regions selected in a single operation form a single profile. As a general rule, we should dimension and constrain profiles to prevent them from unpredictable size and shape changes. SOLIDWORKS does allow us to create under-defined profiles; the dimensions and/or geometric relations can be added/edited later.

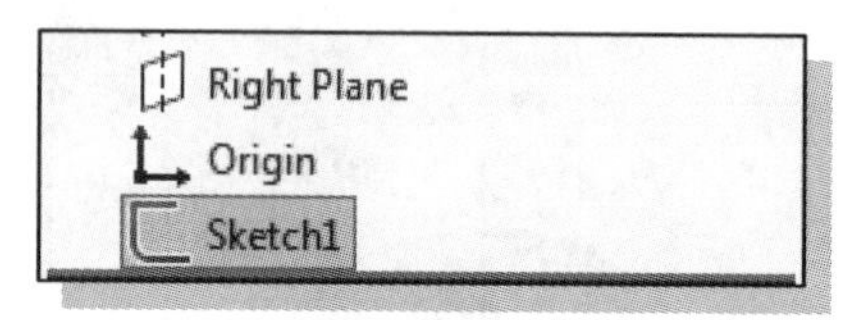

1. Make sure the sketch – **Sketch1** – is selected in the *FeatureManager Design Tree.*

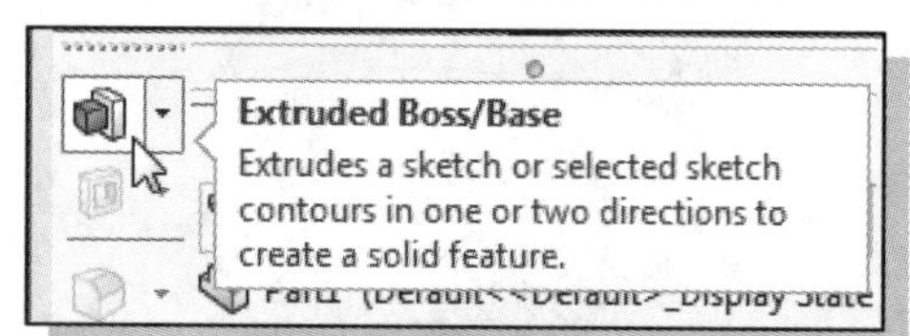

2. In the *Features* toolbar, select the **Extruded Boss/Base** command by clicking once with the left-mouse-button on the icon.

❖ SOLIDWORKS automatically highlights the selected closed region and the defining geometry, which forms the profile required for the operation.

3. In the *Extrude PropertyManager*, enter **5 mm** as the extrusion distance.

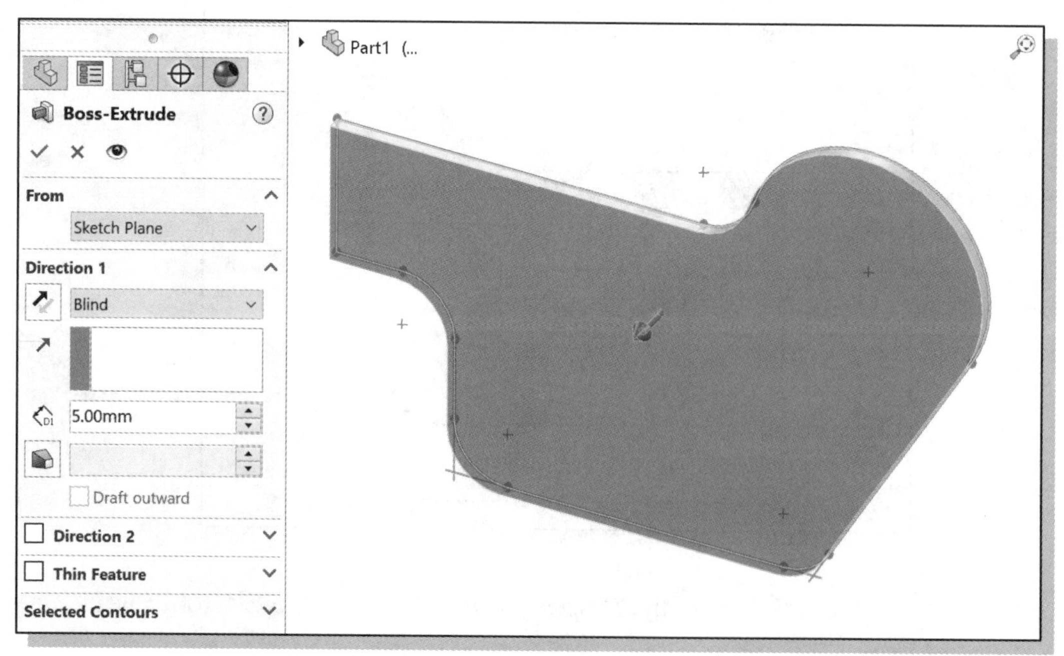

4. Click on the **OK** button to accept the settings and create the base feature.

❖ Note that all the sketched geometry entities and dimensions are consumed and have disappeared from the screen when the feature is created.

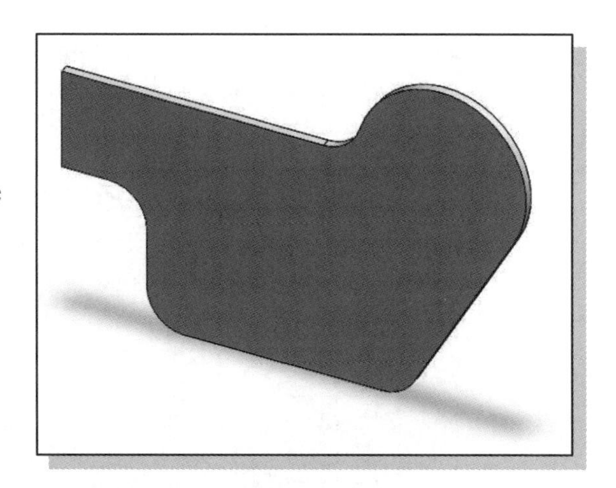

Redefining the Sketch and Profile using Contour Selection

Engineering designs usually go through many revisions and changes. SOLIDWORKS provides an assortment of tools to handle design changes quickly and effectively. We will demonstrate some of the tools available by changing the base feature of the design. The profile used to create the extrusion is selected from the sketched geometry entities. In SOLIDWORKS, any profile can be edited and/or redefined at any time. It is this type of functionality in parametric solid modeling software that provides designers with greater flexibility and the ease to experiment with different design considerations.

In the SOLIDWORKS *PropertyManager* for features requiring definition of a profile, the **Selected Contours** selection window can be used to select sketch contours and model edges, and apply features to them. Contour Selection is a grouping mechanism that allows us to use a partial sketch to create features. It is a tool that helps maintain design intent by reducing the amount of trimming necessary to build the contour. In the previous section, the Boss-Extrude feature was created using a single continuous sketch contour to define the profile. In this section we will demonstrate the utility of the Contour Selection tool to generate a similar profile from a sketch containing self-intersecting geometry and multiple closed regions.

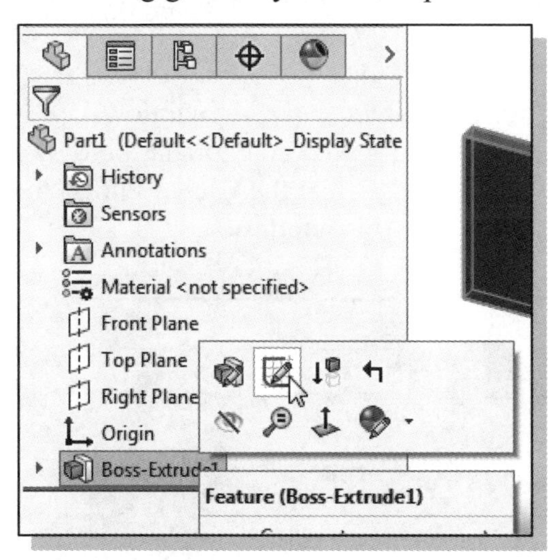

1. In the *FeatureManager Design Tree*, **right-mouse-click** once on the **Boss-Extrude1** feature to bring up the option menu; then pick the **Edit Sketch** icon in the pop-up menu as shown.

2. Select the **View Orientation** button on the *Heads-up View* toolbar, then select the **Normal to** icon to obtain a view normal to the sketch.

3. On your own, create the rough sketch as shown in the figure below. We will intentionally under-define the new sketch to illustrate the flexibility of the system. The dimensions of the new sketch elements are not important. It is important that **tangent relations** are added where the circles meet tangent to lines or other circles. (Hint: Use the **Perimeter Circle** option for the Circle command.)

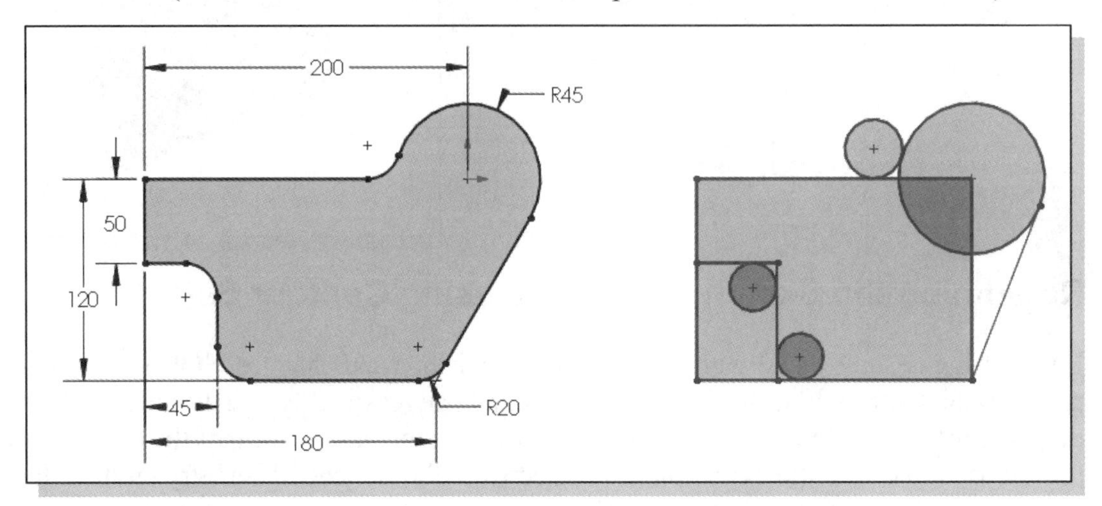

4. Select the **Sketch** icon on the *Sketch* toolbar to exit the Sketch option.

➢ The SOLIDWORKS pop-up window may appear indicating an error will occur when SOLIDWORKS attempts to rebuild the model. This is due to the multiple possible contours for extrusion now present in the sketch. We will next redefine the profile for the base feature using the **Selected Contours** tool.

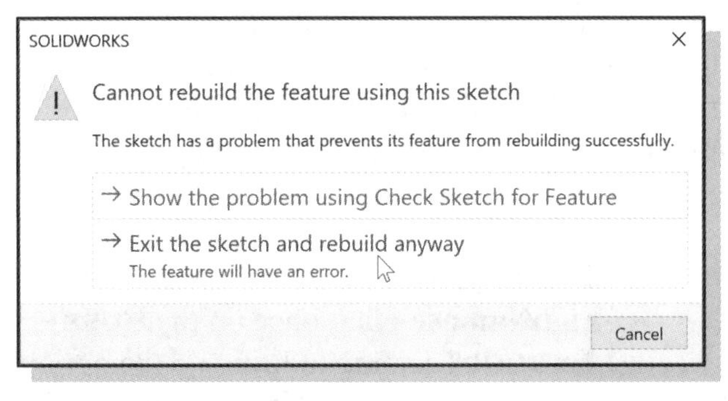

5. If the SOLIDWORKS pop-up window appears, select the option '**Exit the sketch and rebuild anyway**'.

6. The *What's Wrong* window appears with an explanation of the error. Select the **Close** button to close this window.

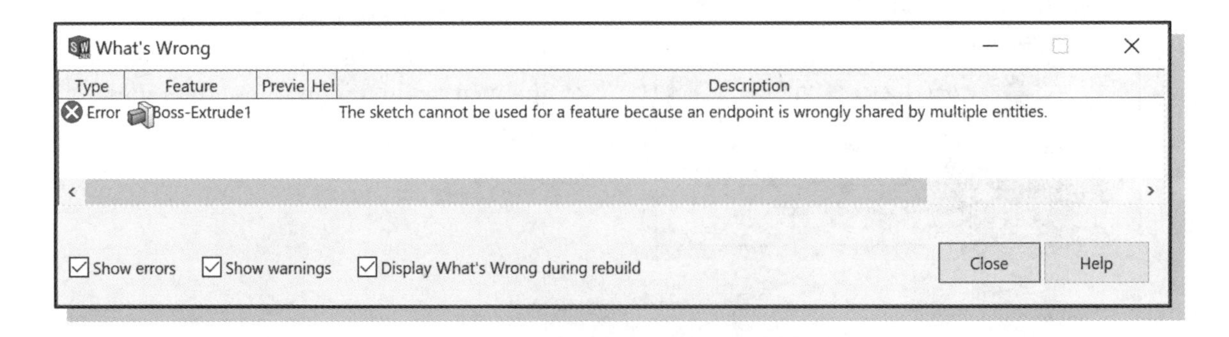

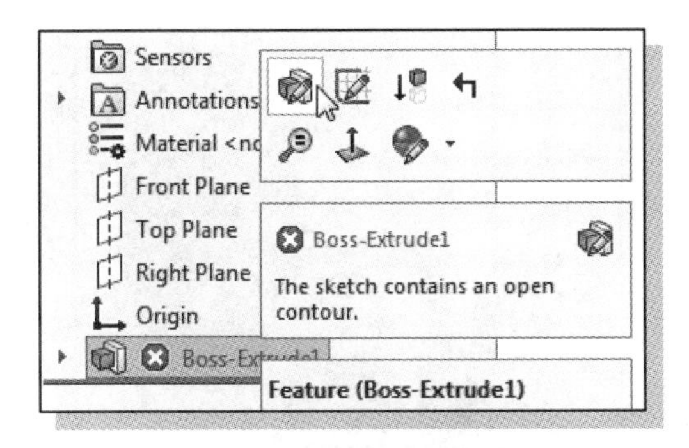

7. In the *FeatureManager Design Tree*, **right-mouse-click** once on the **Boss-Extrude1** feature to bring up the option menu; then pick the **Edit Feature** icon in the pop-up menu. (Notice the red **x** next to the icon indicating the error in the feature.)

❖ SOLIDWORKS will now display the 2D sketch of the selected feature in the graphics window. We have literally gone back in time to the point where we define the extrusion feature. The original sketch and the new sketch we just created are wireframe entities recorded by the system as belonging to the same **SKETCH**, but only the **selected profile** entities are used to create the feature. We will now change the selected profile entities.

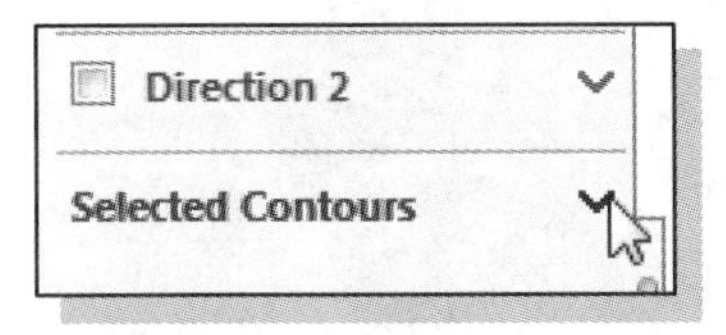

8. In the *Boss-Extrude1 PropertyManager*, click on the arrow to reveal the *Selected Contours* option panel as shown (if necessary).

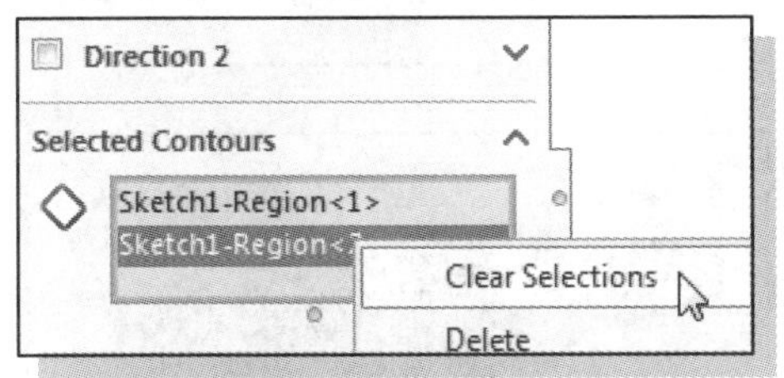

9. If the *Selected Contours* panel is not blank, move the cursor over the panel and click once with the **right mouse button**. In the pop-up option menu, select **Clear Selections**.

➢ Sketch1 is displayed in the graphics area with no entities selected to define the profile for extrusion. The instruction *"Pick a sketch entity to define an open or closed contour. To define a region, pick inside an area bounded by sketch geometry"* appears in the *Status Bar*. The profile can be defined by either selecting sketch entities (lines, circles, etc.) to define a boundary, or by selecting bound areas for inclusion. We will use the latter option here.

10. Move the cursor inside the bound area shown in the figure. When the area is highlighted, click once with the **left-mouse-button** to select the area for inclusion in the profile. Notice **Sketch1-Region<1>** now appears in the *Selected Contours* panel on the *PropertyManager*.

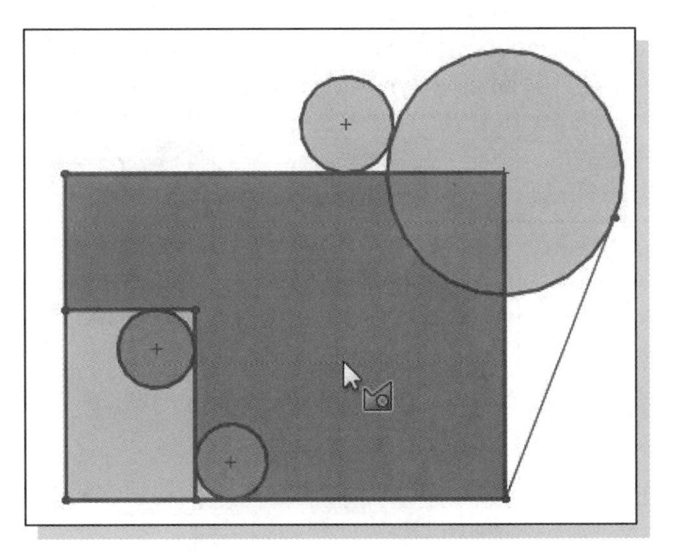

11. On your own, select additional areas for inclusion as shown. (If an incorrect area is selected, simply click on it again to unselect it. Be careful to click inside bound areas and not on the sketch entities. If difficulty is encountered, clear the selections and start over.)

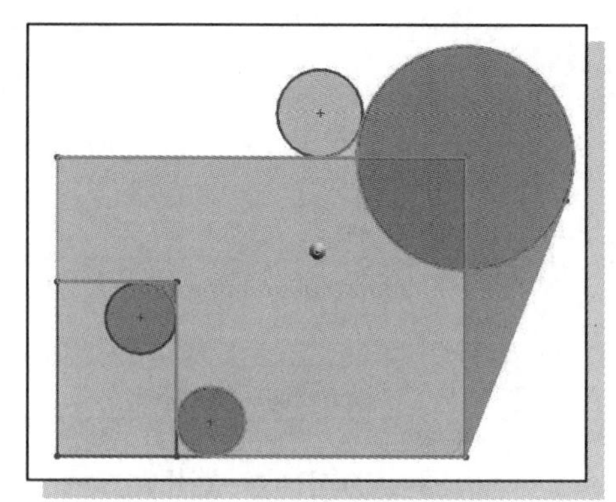

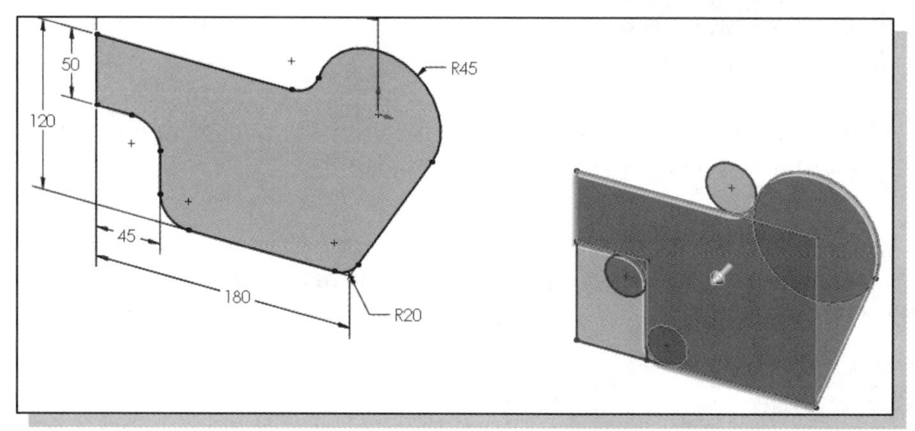

12. In the *Boss-Extrude1 PropertyManager*, click on the **OK** button to accept the settings and update the solid feature. The feature is recreated using the newly sketched geometric entities.

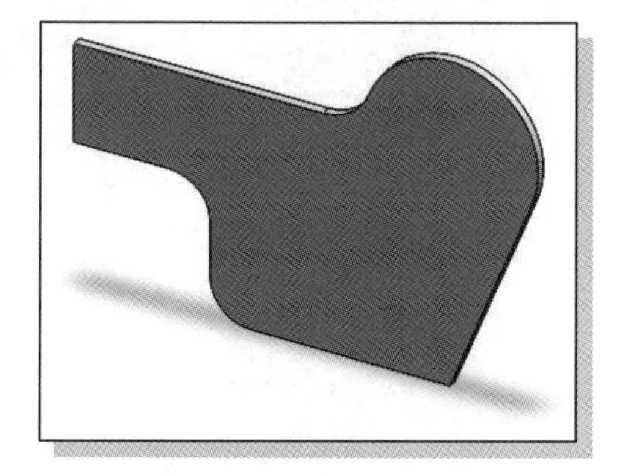

- The *Selected Contours* approach eliminates the need to trim the geometry manually. This approach encourages engineering content over drafting technique.

13. On your own, repeat steps 7 - 9 on the previous page to clear the contour selections. Then select the **original profile**. The model should again look like the one shown at the top of page 6-19.

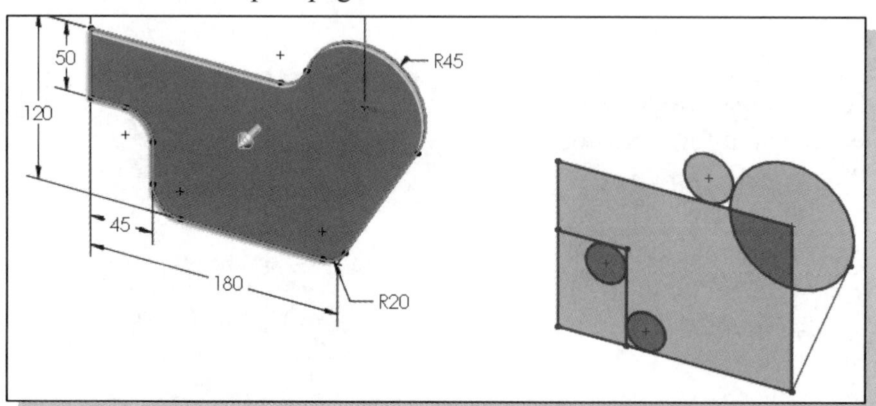

Selecting Items by Box and Lasso

SOLIDWORKS allows selection of items in the graphics area using **Box Selection** or **Lasso Selection** options. These can be used to select items in sketches, drawings, and assemblies. We will demonstrate these options and delete items in the existing sketch.

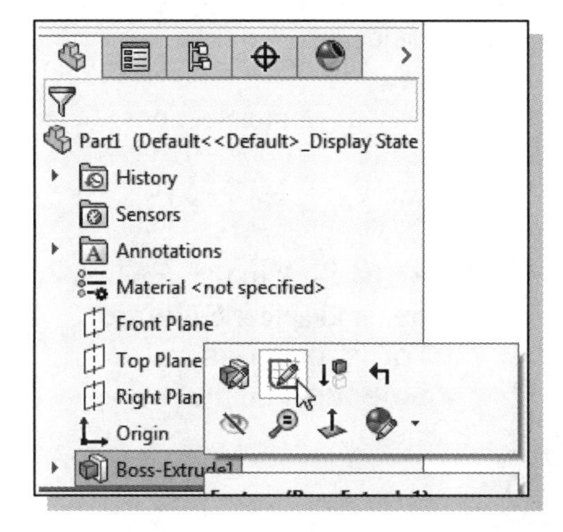

1. In the *FeatureManager Design Tree*, **right-mouse-click** once on the **Boss-Extrude1** feature to bring up the option menu, then pick the **Edit Sketch** icon in the pop-up menu as shown.

2. Select the **View Orientation** button on the *Heads-up View* toolbar, then select the **Normal to** icon to obtain a view normal to the sketch.

➢ SOLIDWORKS toggles the default selection method between lasso selection and box selection. We will demonstrate **Box Selection** first.

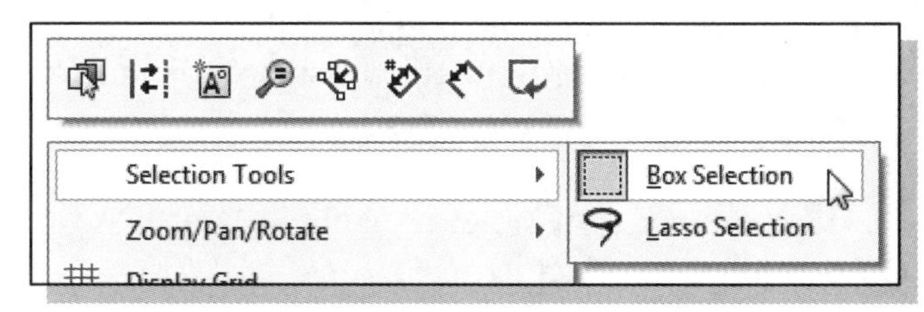

3. **Right click** in the graphics area and select **Selections Tools**, then **Box Selection** on the pop-up menu.

➢ In **Box Selection** mode, you can select entities by dragging a selection box with the pointer. When you select from left to right, all items within the box are selected. When you select from right to left, items crossing the box boundaries are also selected.

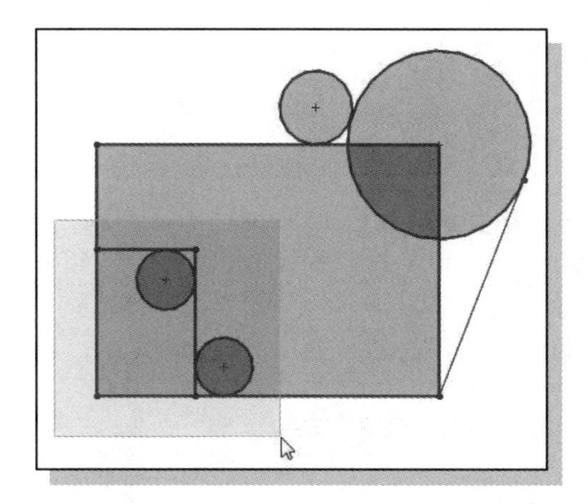

4. Create a selection box as shown by **clicking and dragging** the cursor from the upper left corner to the lower right corner.

➢ Notice that only the entities (two lines, two circles) within the box are selected.

5. Press the **[Esc]** key to clear the selections.

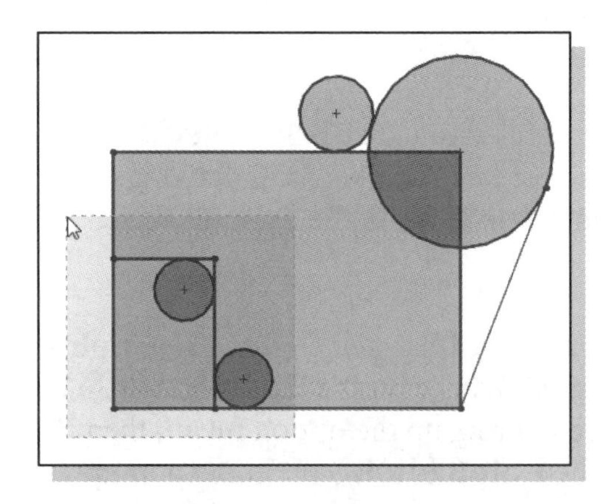

6. Create a similar selection box by **clicking and dragging** the cursor from the lower right corner to the upper left corner.

➢ Notice that the entities within the box and those crossing the box boundary (four lines, two circles) are selected.

7. Press the **[Esc]** key to clear selections.

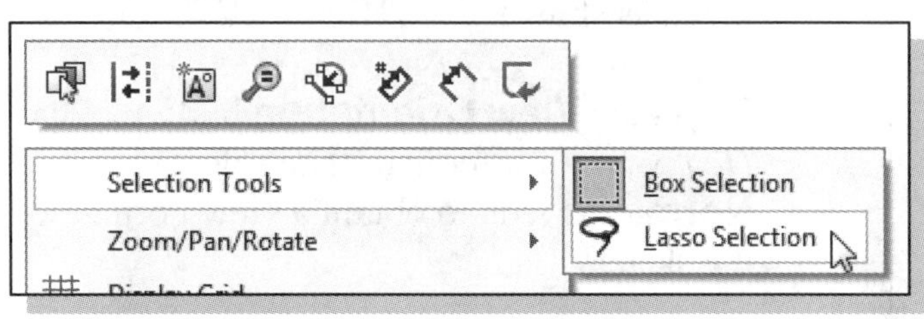

8. **Right click** in the graphics area and select **Selection Tools**, then **Lasso Selection** on the pop-up menu.

➢ In **Lasso Selection** mode, you can select entities by dragging a lasso loop with the pointer. When you drag the pointer in the clockwise direction to create the lasso, all items within the lasso loop are selected. When you drag the pointer in the counterclockwise direction to create the lasso, items crossing the lasso loop boundary are also selected. You can select items without closing the lasso.

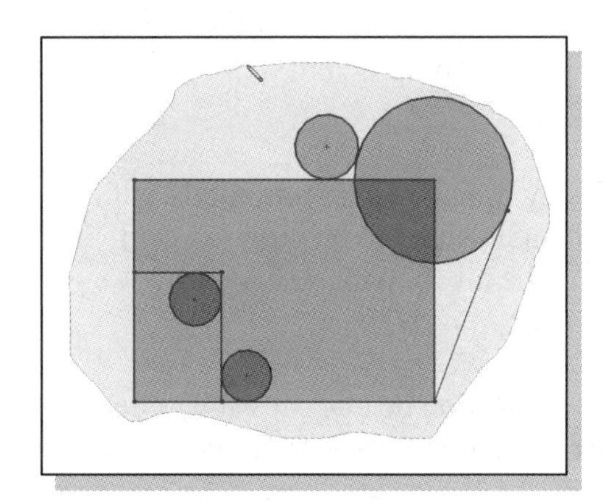

9. On your own, practice using various lasso loops to select items.

10. Create a lasso as shown by **clicking and dragging** in the clockwise direction.

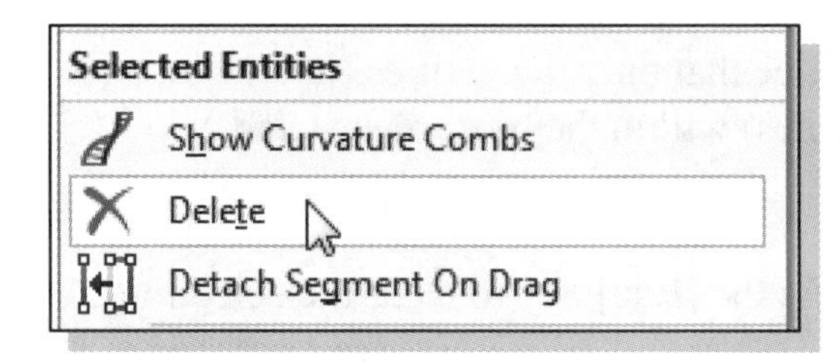

11. **Right click** and select **Delete** from the pop-up menu to delete the selected entities.

12. If the *Sketcher Confirm Delete* window appears, select **Yes to All**.

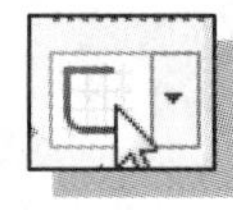

13. Click once with the **left-mouse-button** on the **Sketch** icon on the *Sketch* toolbar to exit the Sketch option.

Create an OFFSET Extruded Cut Feature

➢ To complete the design, we will create a cutout feature by using the Offset command. First we will set up the sketching plane to align with the front face of the 3D model.

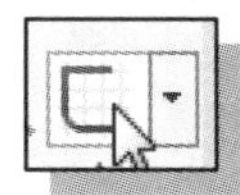

1. In the *Sketch* toolbar, select the **Sketch** command by left-clicking once on the icon.

2. Notice the left panel displays the *Edit Sketch PropertyManager* with the instruction "*Select: a plane, planar face, ...*" **Select the front face** of the 3D model in the graphics window.

3. We will now convert the edge of the front face of the model into segments in the new sketch. **Select the front face again** as shown.

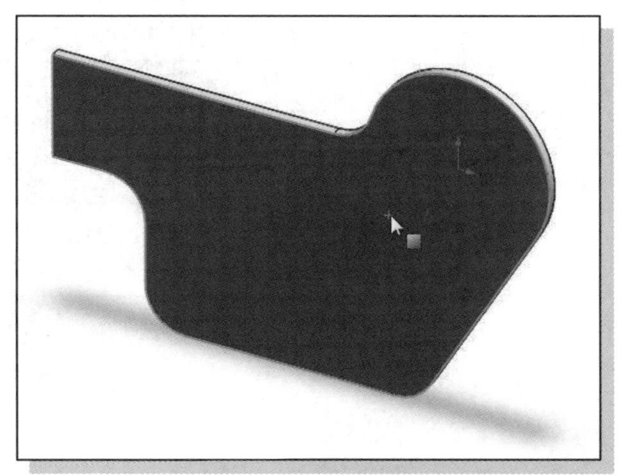

Convert Entities
Converts selected model edges or sketch entities into sketch segments.

4. Select the **Convert Entities** icon in the *Sketch* toolbar.

➢ The Convert Entities command converts selected model features into sketch segments. Notice the edges of the selected surface have been converted to line and arc segments in the current sketch.

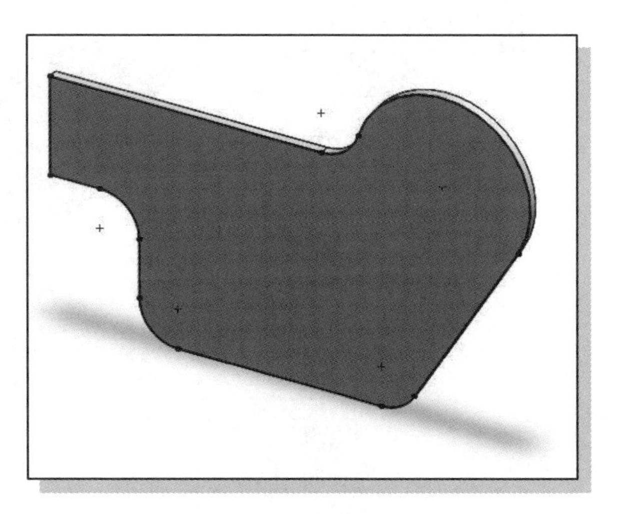

Offset Entities
Adds sketch entities by offsetting faces, edges, curves, or sketch entities a specified distance.

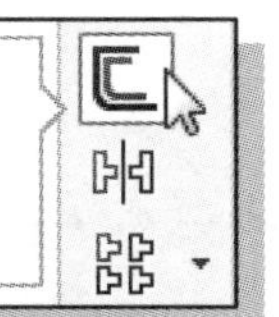

5. Select the **Offset Entities** icon in the *Sketch* toolbar.

6. In the *Offset PropertyManager*, enter **5 mm** for the offset distance.

7. In the *Offset PropertyManager*, select the options for *Add dimensions*, *Reverse*, and *Select chain* as shown.

8. In the graphics area, select any segment of the face outline which was just converted. Because the *Select chain* option is active, SOLIDWORKS will automatically select all of the connecting geometry to form a closed region.

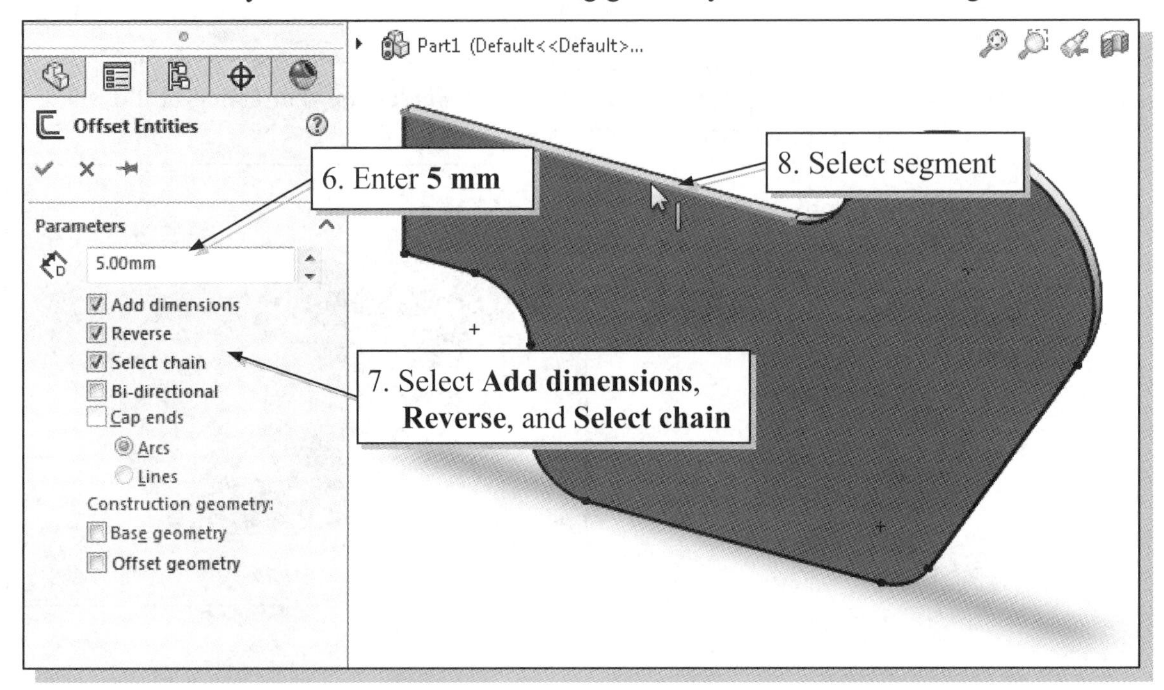

9. Click on the **OK** button to accept the settings and create the offset feature.

10. On your own, modify the offset dimension to **15 mm** as shown in the figure.

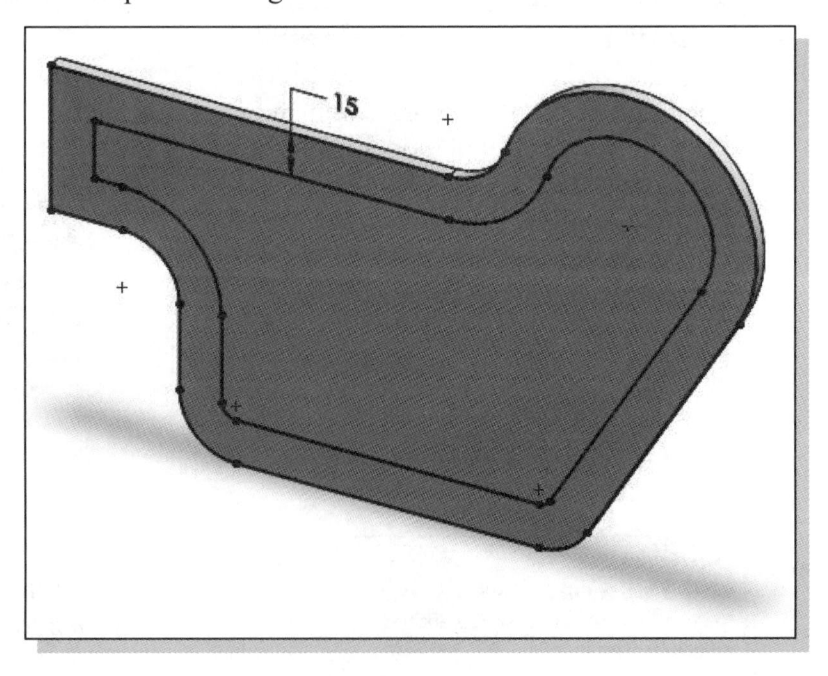

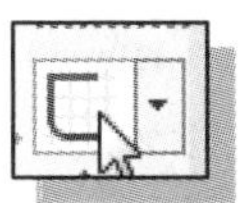

11. Click once with the **left-mouse-button** on the **Sketch** icon on the *Sketch* toolbar to exit the Sketch option.

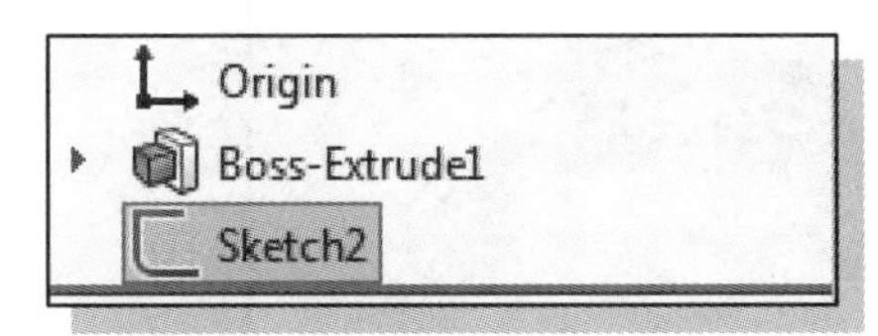

12. Make sure the sketch – Sketch2 – is selected in the *FeatureManager Design Tree.*

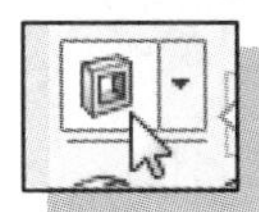

13. In the *Features* toolbar, select the **Extruded Cut** command by clicking once with the left-mouse-button on the icon. The *Cut-Extrude PropertyManager* is displayed in the left panel.

14. In the *Cut-Extrude PropertyManager* panel, click the arrow to reveal the pull-down options for the *End Condition* (the default end condition is Blind) and select **Through All** as shown.

15. In the *Cut-Extrude PropertyManager* panel, click on the arrow to reveal the *Selected Contours* option text box as shown. If the *Selected Contours* panel is not blank, move the cursor over the panel and click once with the **right mouse button**. In the pop-up option menu, select **Clear Selections**.

16. In the graphics area, move the cursor over the interior of the offset contour as shown in the figure below. Click once with the **left-mouse-button** to select the profile. Notice Sketch2-Region<1> now appears in the *Selected Contours* box on the *PropertyManager*.

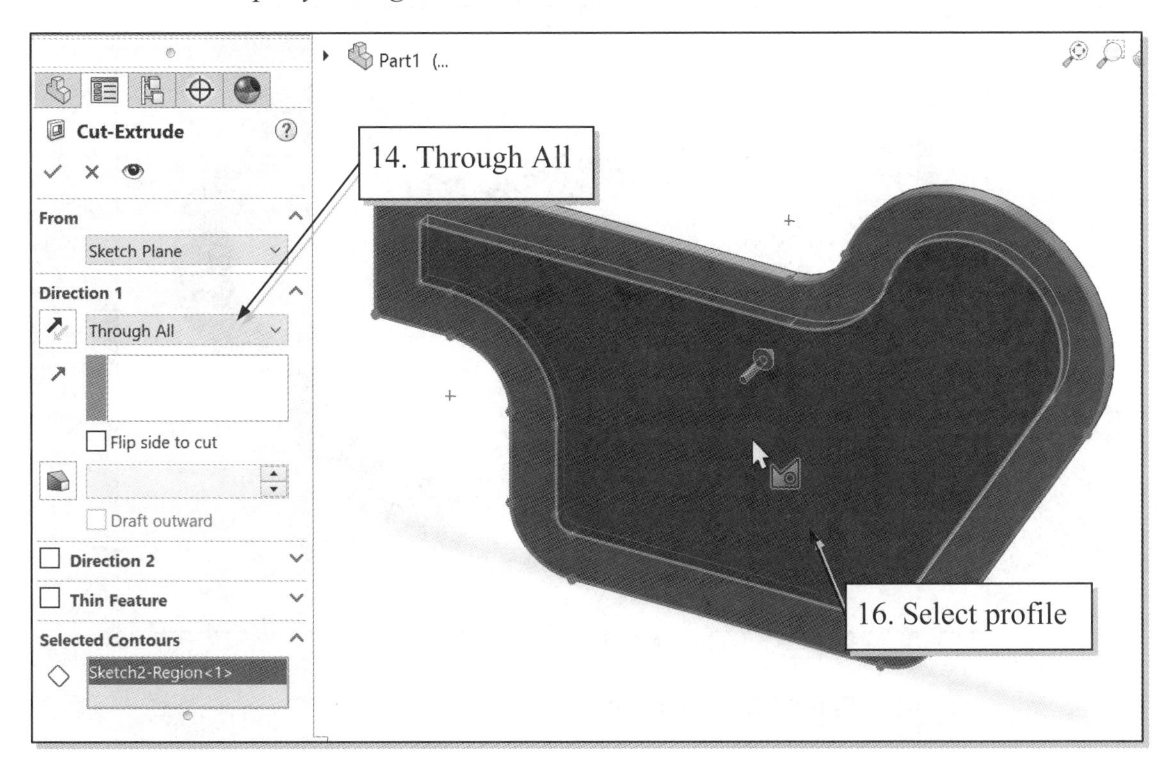

17. Click on the **OK** button to accept the settings and create the cut-extrude feature.

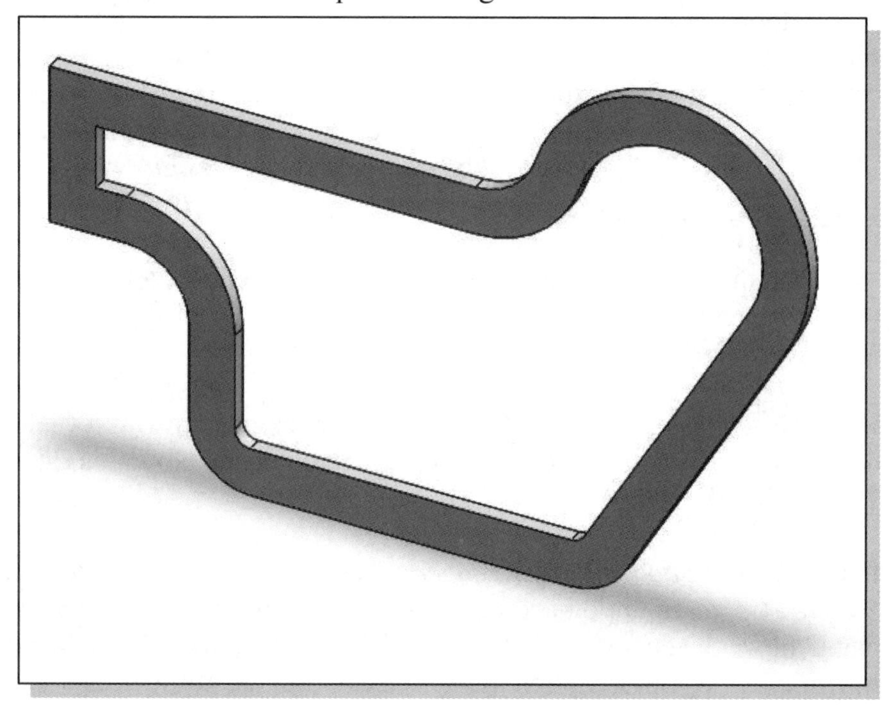

➢ The offset geometry is associated with the original geometry. On your own, adjust the overall height of the design to **150 millimeters** (i.e., change the **120 mm** vertical dimension to **105 mm**) and confirm that the offset geometry is adjusted accordingly.

18. Save the part with the filename ***Gasket***.

Alternate Construction Method – Thin Feature Option

The **Thin Feature** option is a powerful tool in SOLIDWORKS. It allows us to control the thickness (not the depth) of the extruded contour. Using this option, we can create the gasket by creating a single Extrude-Thin feature.

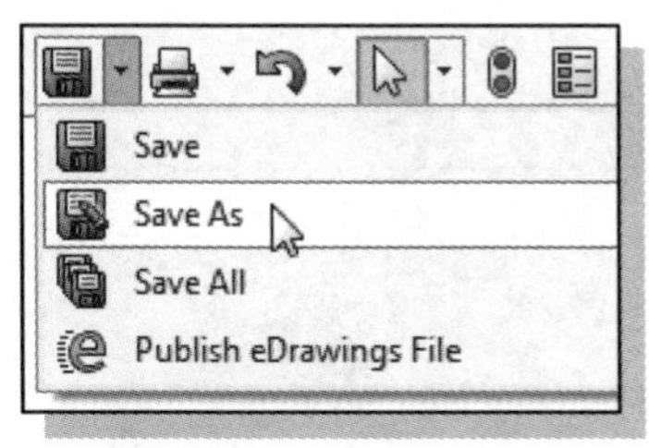

1. In the *Menu Bar*, select the **arrow** next to the Save icon and select the **Save As** option from the pull-down menu.

2. Save the part with the filename ***Gasket2***.

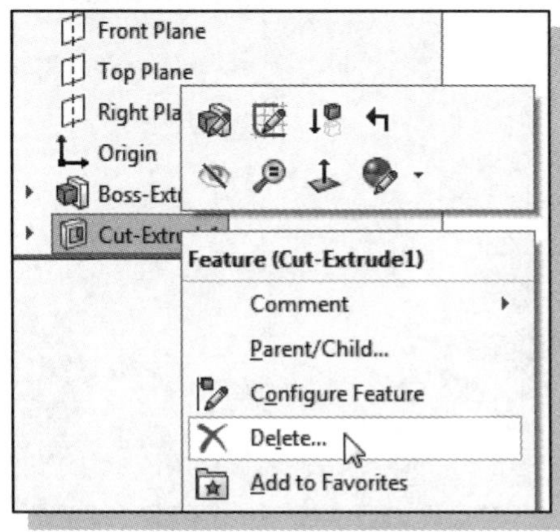

3. In the *FeatureManager Design Tree*, **right-mouse-click** once on the **Cut-Extrude1** feature to bring up the option menu, then pick the **Delete** command in the pop-up menu as shown.

4. The *Confirm Delete* pop-up window appears. Check the **Delete absorbed features** option box.

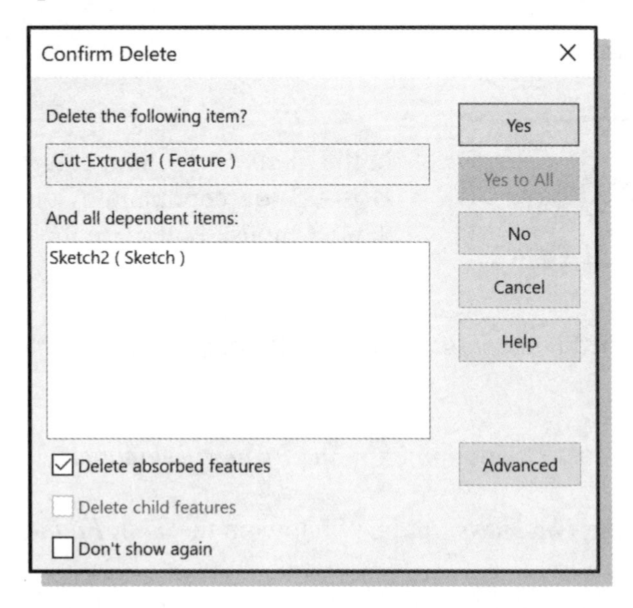

5. Notice Sketch2 appears in the *And all dependent items:* panel.

6. Select **Yes** to delete the Cut-Extrude1 feature and Sketch2.

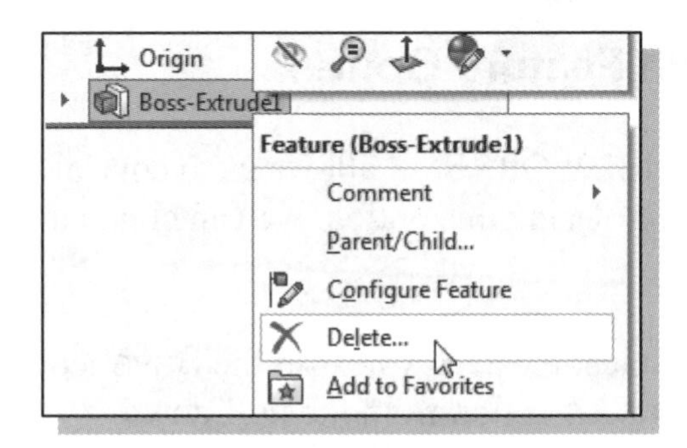

7. In the *FeatureManager Design Tree*, **right-mouse-click** once on the **Boss-Extrude1** feature to bring up the option menu, then pick the **Delete** command in the pop-up menu as shown.

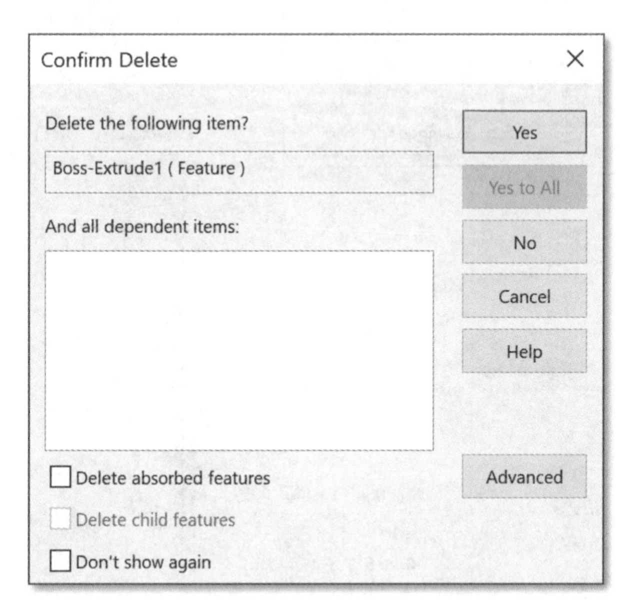

8. The *Confirm Delete* pop-up window appears. **Do not check** the Delete absorbed features option box.

9. Select **Yes** to delete the Boss-Extrude1 feature. (Do not include the dependent sketch.)

➢ Sketch1 is now the only sketch or feature in the *FeatureManager Design Tree.*

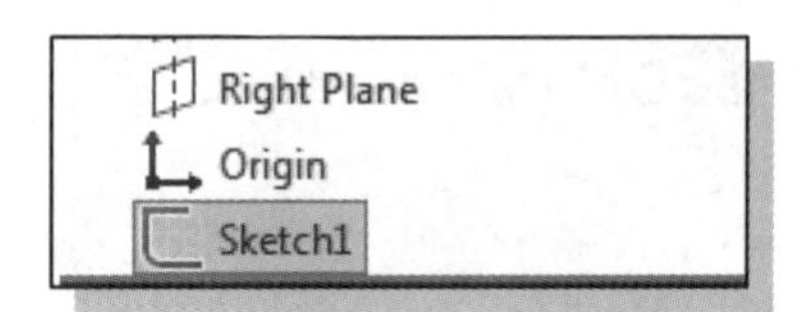

10. Select **Sketch1** in the *FeatureManager Design Tree*.

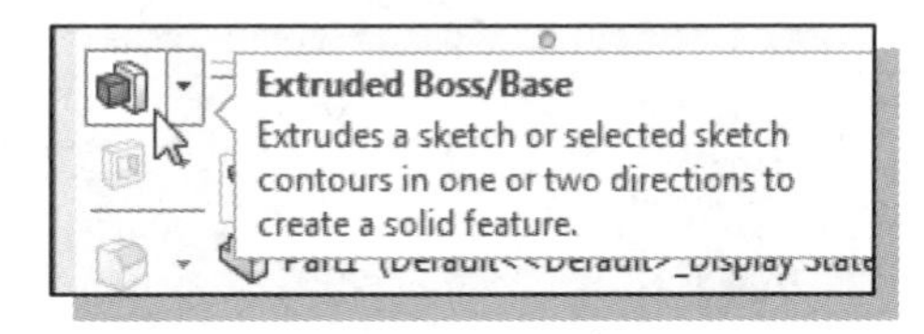

11. In the *Features* toolbar, select the **Extruded Boss/Base** command by clicking once with the left-mouse-button on the icon.

12. Verify that the **Blind** *end condition* and **5 mm** *depth* are still entered in the *Direction 1* panel.

13. Check the **Thin Feature** option box in the *PropertyManager*.

14. Enter **15 mm** in the *Thickness* option window on the *Thin Feature* panel.

15. Click the **Reverse Direction** button on the *Thin Feature* panel.

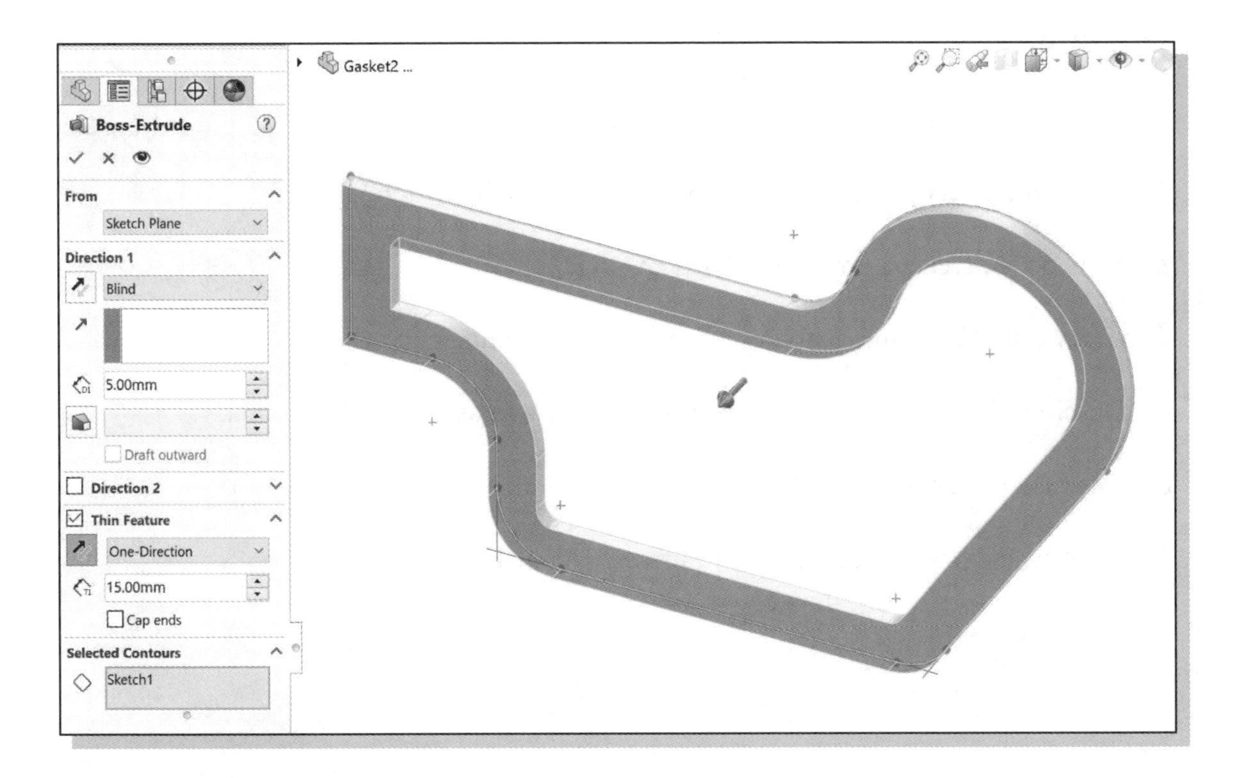

16. Click **OK** in the *PropertyManager* to accept the settings and create the *Extrude-Thin* feature.

➢ Notice the part created is identical to the part we created in the previous section.

17. Save the ***Gasket2*** part file.

➢ We now have two part files. The first was created by creating two sketches, one offset from the other by 15 mm, and two features, a *Boss Extrude* and a *Cut Extrude*. The second was created using one sketch and one *Boss Extrude* using the Thin Feature option with a thickness of 15 mm. The final parts are identical.

Questions:

1. Can we create a profile with extra 2D geometry entities?

2. How do we access the SOLIDWORKS **Edit Sketch** option?

3. How do we create a *profile* in SOLIDWORKS?

4. Can we build a profile that consists of self-intersecting curves?

5. Describe the procedure to create an offset copy of a sketched 2D geometry.

6. Describe the **Selected Contours** option in SOLIDWORKS.

7. Identify and briefly describe the following commands:

(a) 

(b) 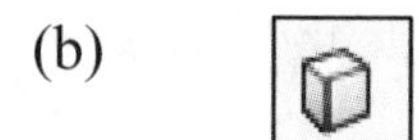

(c) 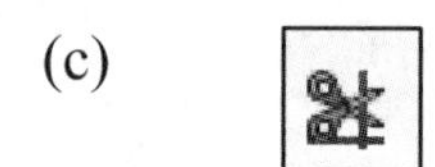

(d)

Exercises:

1. **V-slide Plate** (Dimensions are in inches. Plate Thickness: **0.25**)

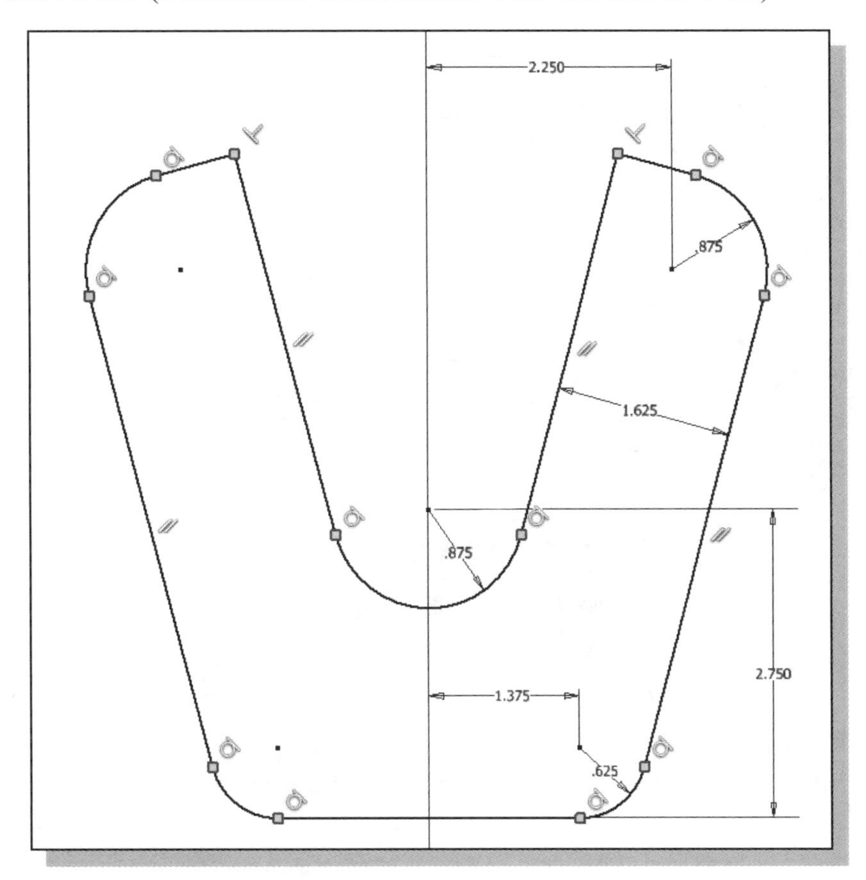

2. **Shaft Support** (Dimensions are in millimeters. Note the two R40 arcs at the base share the same center.)

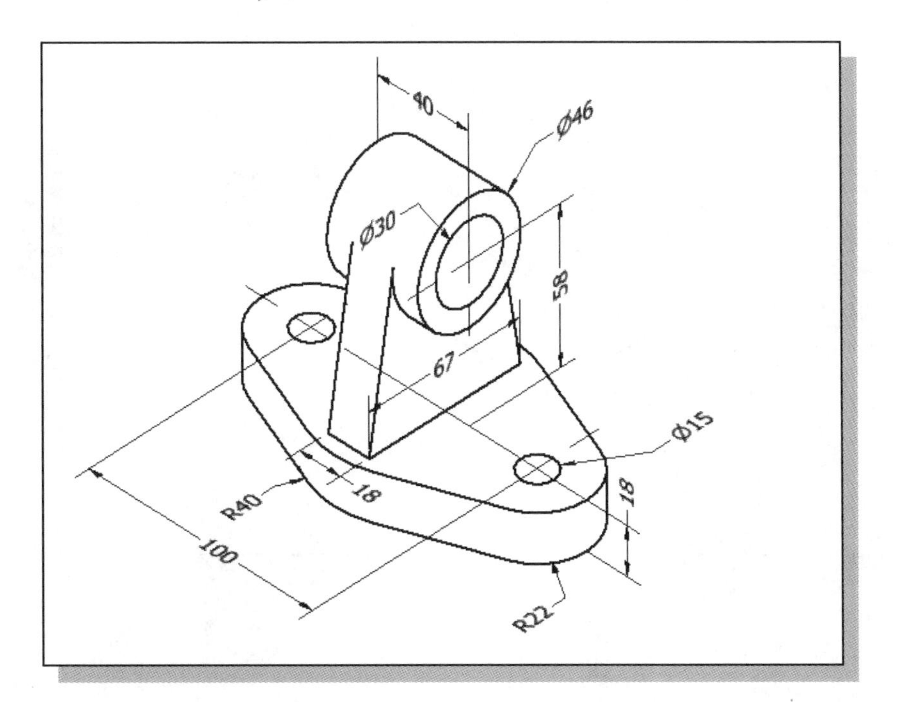

3. **Vent Cover** (Thickness: **0.125** inches. Hint: Use the Ellipse command.)

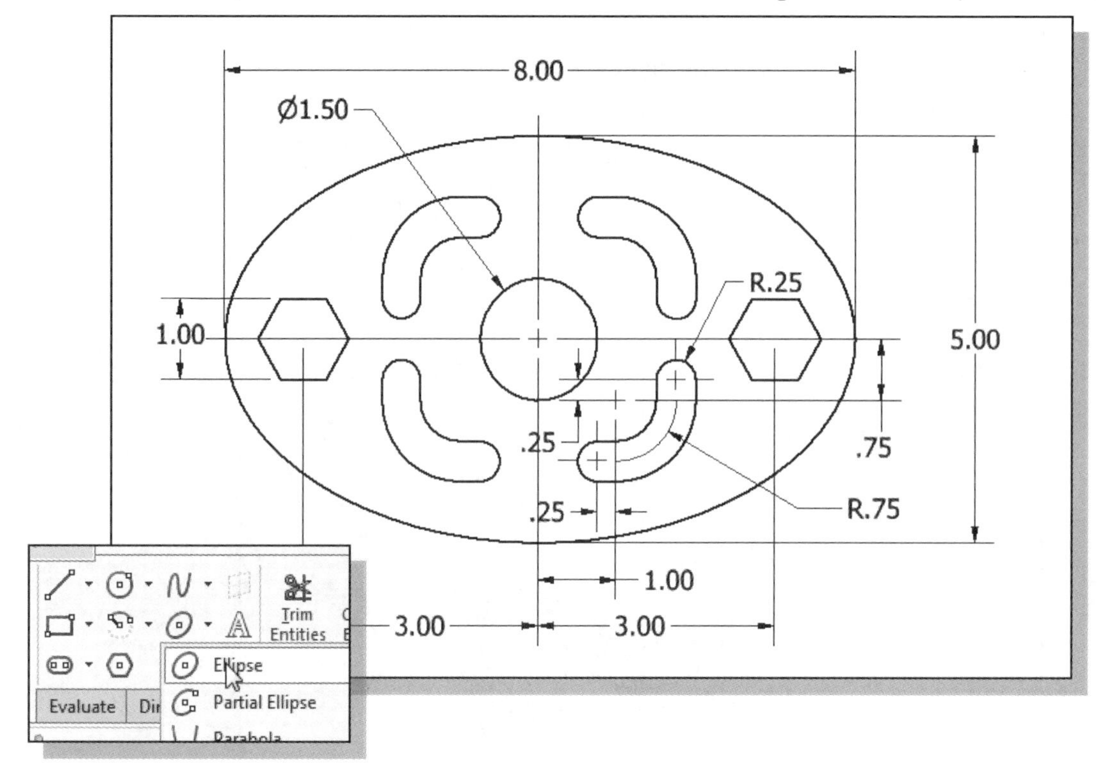

4. **Anchor Base** (Dimensions are in inches.)

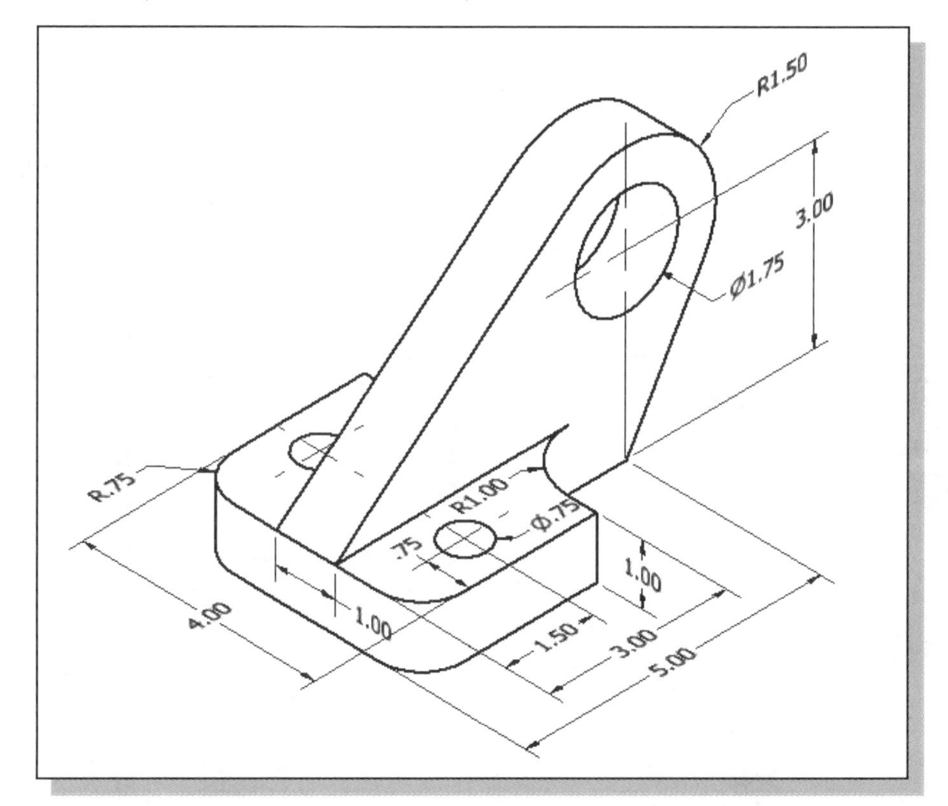

5. **Tube Spacer** (Dimensions are in inches.)

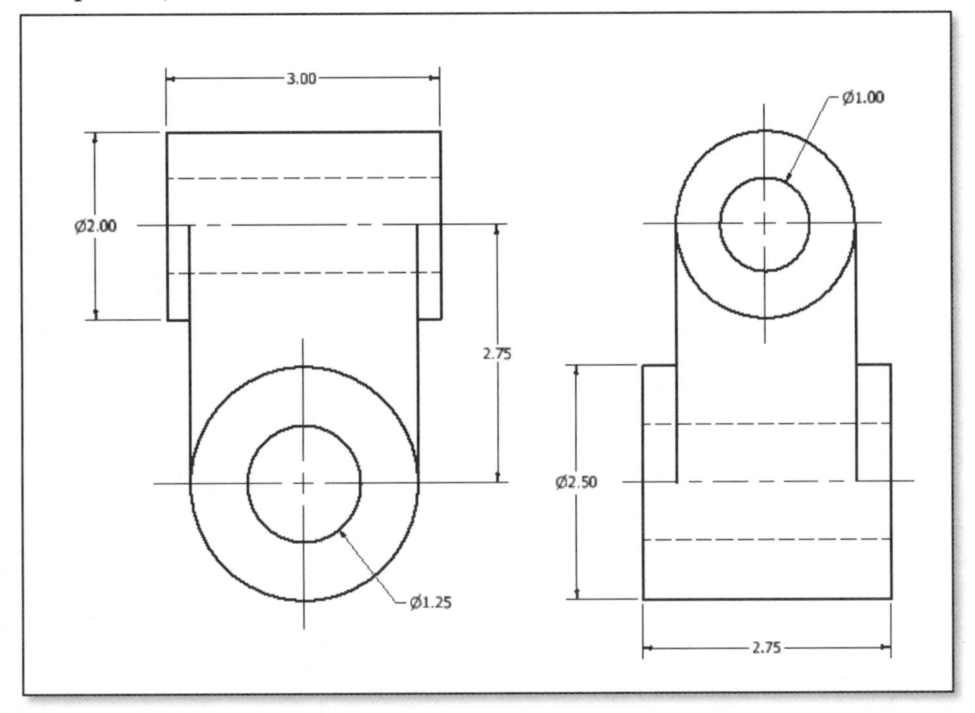

6. **Pivot Lock** (Dimensions are in inches. The circular features in the design are all aligned to the two centers at the base.)

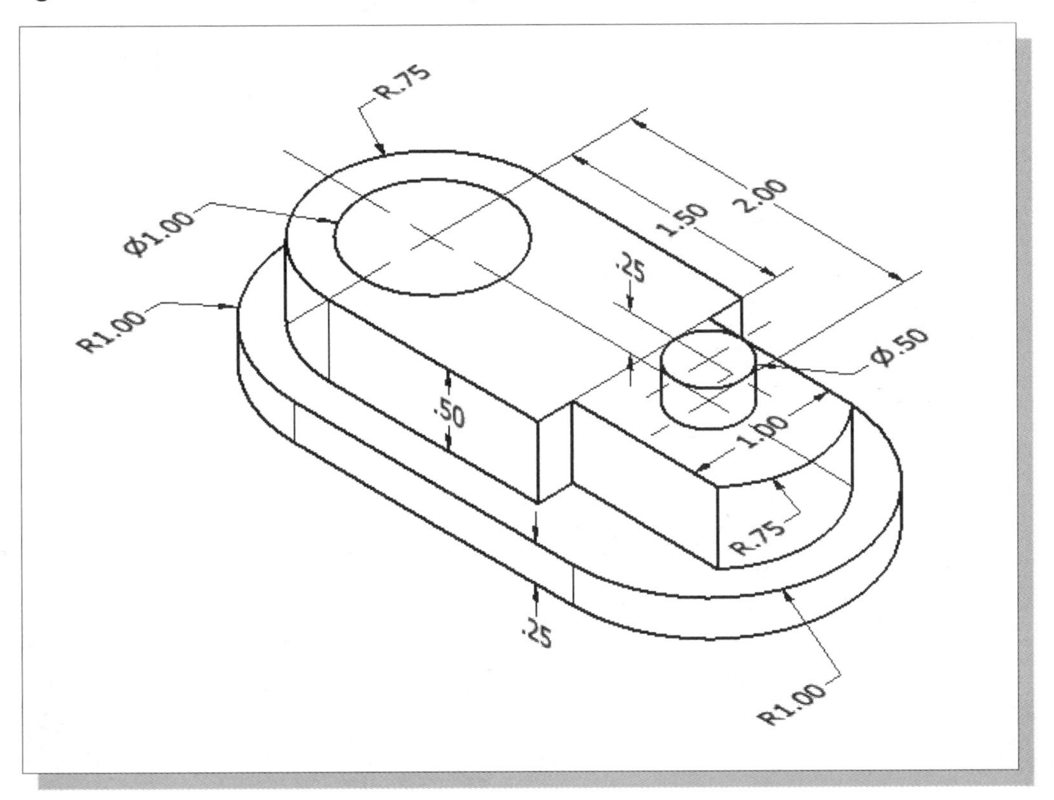

NOTES:

Chapter 7

Parent/Child Relationships and the BORN Technique

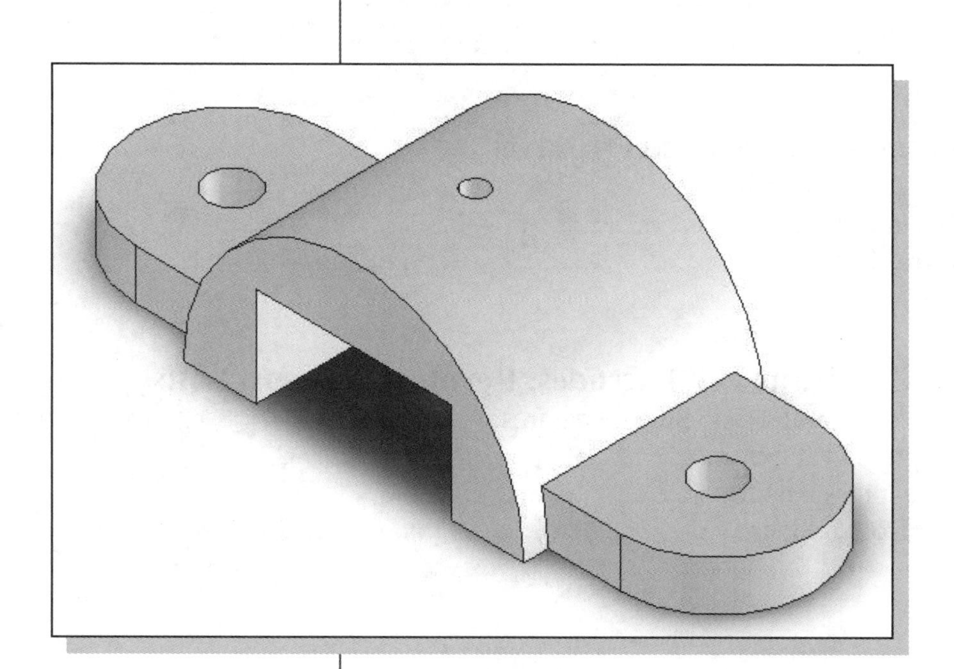

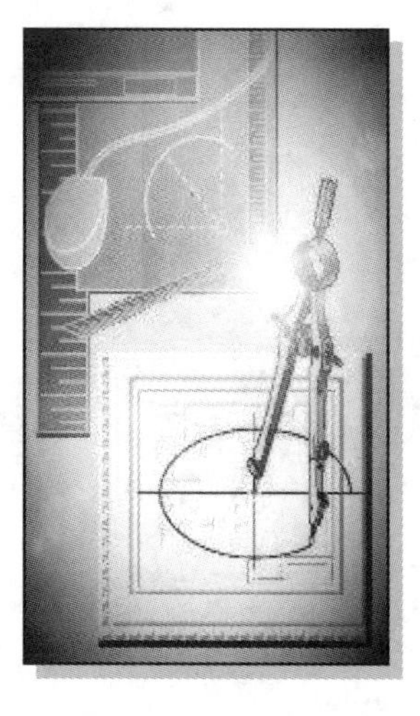

Learning Objectives

- ♦ **Understand the Concept and Usage of the BORN Technique**
- ♦ **Understand the Importance of Parent/Child Relations in Features**
- ♦ **Use the *Suppress Feature* Option**
- ♦ **Resolve Undesired Feature Interactions**

Certified SOLIDWORKS Associate Exam Objectives Coverage

Sketch Entities – Lines, Rectangles, Circles, Arcs, Ellipses, Centerlines

Objectives: Creating Sketch Entities.

Sketch Tools – Offset, Convert, Trim

Objectives: Using Sketch Tools.

Boss and Cut Features – Extrudes, Revolves, Sweeps, Lofts

Objectives: Creating Basic Swept Features.

Introduction

The parent/child relationship is one of the most powerful aspects of ***parametric modeling***. In SOLIDWORKS, each time a new modeling event is created previously defined features can be used to define information such as size, location, and orientation. The referenced features become **PARENT** features to the new feature, and the new feature is called the **CHILD** feature. The parent/child relationships determine how a model reacts when other features in the model change, thus capturing design intent. It is crucial to keep track of these parent/child relations. Any modification to a parent feature can change one or more of its children.

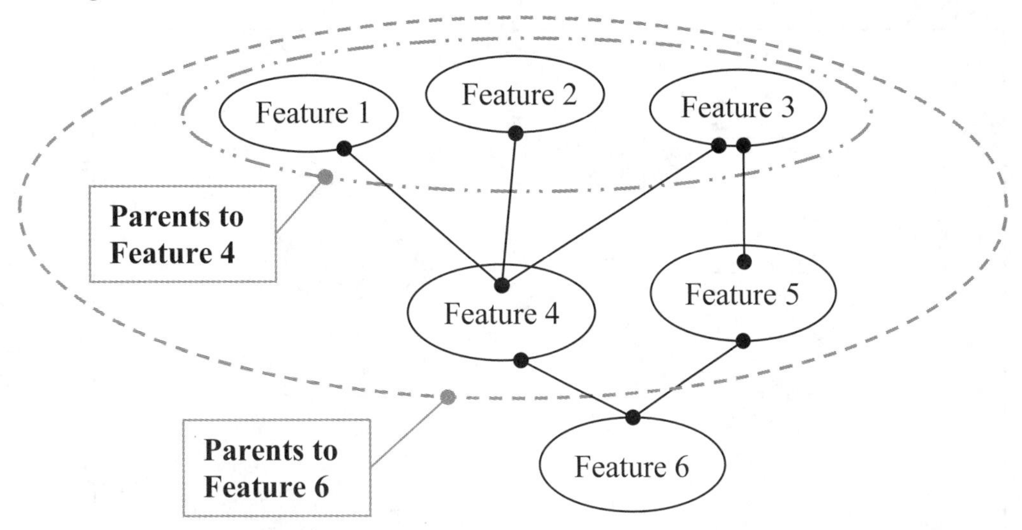

Parent/child relationships can be created *implicitly* or *explicitly*; implicit relationships are implied by the feature creation method and explicit relationships are entered manually by the user. In the previous chapters, we first select a sketching plane before creating a 2D profile. The selected surface becomes a parent of the new feature. If the sketching plane is moved, the child feature will move with it. As one might expect, parent/child relationships can become quite complicated when the features begin to accumulate. It is therefore important to think about modeling strategy before we start to create anything. The main consideration is to try to plan ahead for possible design changes that might occur which would be affected by the existing parent/child relationships. Parametric modeling software, such as SOLIDWORKS, also allows us to adjust feature properties so that any feature conflicts can be quickly resolved.

The BORN Technique

In the previous chapters, we have used a technique of creating solid models known as the "**Base Orphan Reference Node**" (**BORN**) technique. The basic concept of the BORN technique is to use a *Cartesian coordinate system* as the first feature prior to creating any solid features. With the *Cartesian coordinate system* established, we then have three mutually perpendicular datum planes (namely the *XY or 'front'*, *YZ or 'right'*, and *ZX or 'top' planes*) available to use as sketching planes. The three datum planes can also be used as references for dimensions and geometric constructions. Using this technique, the first node in the design tree is called an "orphan," meaning that it has no history to be replayed. The technique of creating the reference geometry in this "base node" is therefore called the "Base Orphan Reference Node" (BORN) technique.

SOLIDWORKS automatically establishes a set of reference geometry when we start a new part, namely a *Cartesian coordinate system* with three work planes, three work axes, and an origin. All subsequent solid features can then use the coordinate system and/or reference geometry as sketching planes. This approach is also very useful in creating assembly models, which will be illustrated in *Chapter 12*.

The *U-Bracket* Design

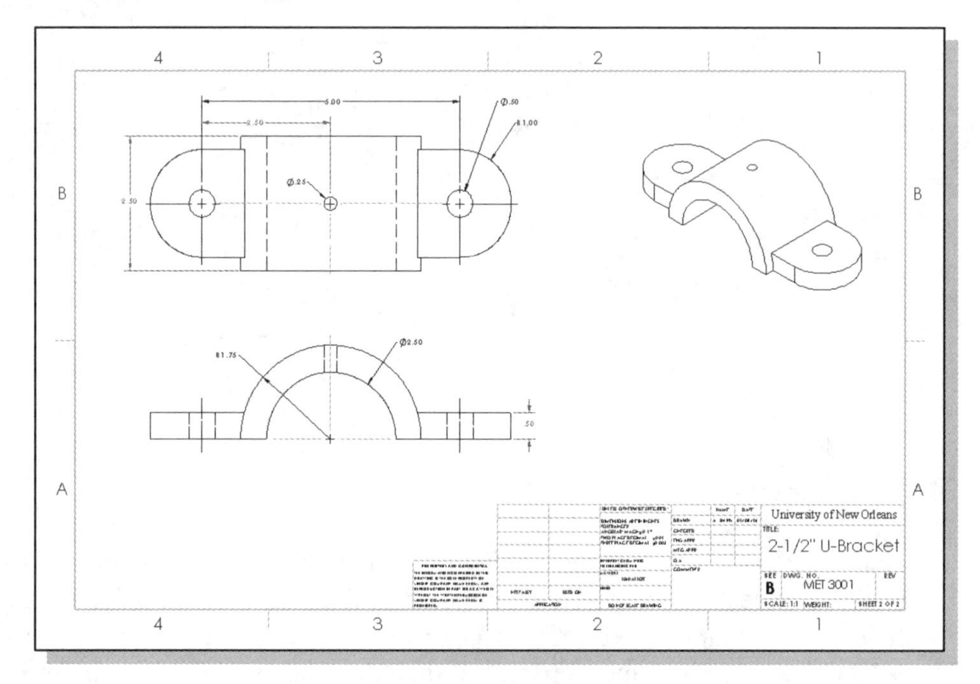

Based on your knowledge of SOLIDWORKS so far, how many features would you use to create the model? Which feature would you choose as the **base feature**? What is your choice for arranging the order of the features? Would you organize the features differently if the rectangular cut at the center is changed to a circular shape (Radius: 80mm)?

Starting SOLIDWORKS and Activating the CommandManager

1. Select the **SOLIDWORKS** option on the *Start* menu or select the **SOLIDWORKS** icon on the desktop to start SOLIDWORKS. If it appears, close the *Welcome* dialog box by clicking on the X in the upper right corner.

2. Select the **New** icon with a single click of the left-mouse-button on the *Menu Bar*. The *New* SOLIDWORKS *Document* dialog box should appear in Advanced mode. If it appears in Novice mode, click once with the left-mouse-button on the **Advanced** icon.

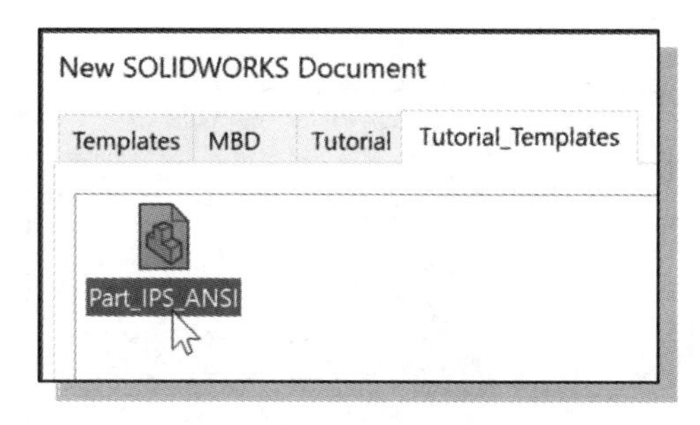

3. Select the **Tutorial_Templates** tab. (**NOTE:** You added this tab in Chapter 4. If your templates are not available, open the standard Part template, then set the *Drafting Standard* to ANSI and the *Units* to IPS.)

4. Select the **Part_IPS_ANSI** template as shown.

5. Click on the **OK** button to open a new document using the Part_IPS_ANSI template. The *Drafting Standard* will automatically be ANSI and the units will be set to inch, pound, second (IPS) as defined in the template.

IMPORTANT NOTE: The SOLIDWORKS *CommandManager* provides an alternate method for displaying the most commonly used toolbars (see page 1-13 for a more complete description). We will use the *CommandManager* in this lesson.

6. Right click on any toolbar and select **Enable CommandManager** at the top of the pop-up menu.

Notice the *Features* and *Sketch* toolbars no longer appear at the edge of the window. These toolbars appear on the *Ribbon* display of the *CommandManager*.

Applying the BORN Technique

In the *FeatureManager Design Tree* panel, notice a new part name (Part1) appears with four work features established. The four work features include three *workplanes* and the *origin*. By default, the three workplanes are aligned to the **world coordinate system** and the origin is aligned to the *origin* of the **world coordinate system**.

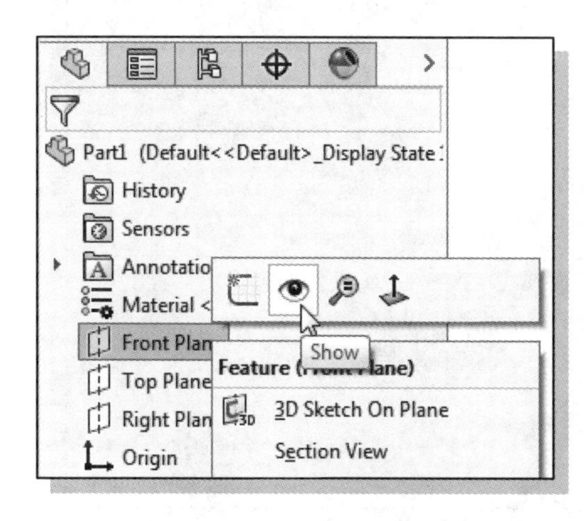

1. Inside the *FeatureManager Design Tree* panel, move the cursor on top of the first work plane, the **Front (XY) Plane**. Notice a rectangle, representing the workplane, appears in the graphics window.

2. Inside the *FeatureManager Design Tree* panel, click once with the right-mouse-button on Front (XY) Plane to display the option menu. Click on the **Show** icon to toggle *ON* the display of the plane.

3. On your own, repeat the above steps and toggle ***ON*** the display of all of the *workplanes* on the screen.

4. On your own, use the *Viewing* options (Rotate, Zoom and Pan) to view the planes.

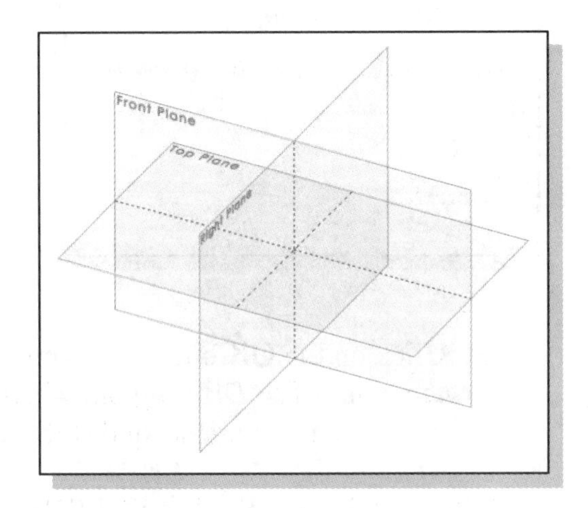

❖ By default, the basic set of planes is aligned to the world coordinate system; the workplanes are the first features of the part. We can now proceed to create solid features referencing the three mutually perpendicular datum planes. We will sketch the first feature, the base feature, in the top (XZ) plane. We will sketch the second feature in the front (XY) plane. Due to the symmetry in the shape of the part, we will use the right (YZ) plane and the front (XY) planes as planes of symmetry for the part. Thus, before starting, we have decided on the orientation and the location of the part in relation to the basic workplanes.

5. Move the cursor inside the graphics area, but away from the planes, and click once with the **left-mouse-button** to make sure no plane is selected.

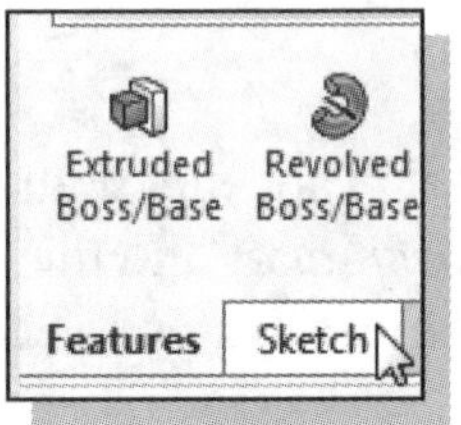

6. If necessary, select the **Sketch** tab on the *CommandManager* to display the *Sketch* toolbar.

7. In the *Sketch* toolbar, select the **Sketch** command by left-clicking once on the icon.

8. In the *Edit Sketch Property-Manager*, the message "*Select a plane on which to create a sketch for the entity*" is displayed. Move the cursor over the edge of the Top Plane in the graphics area. When the Top Plane is highlighted, click once with the **left-mouse-button** to select the **Top (XZ) Plane** as the sketch plane for the new sketch.

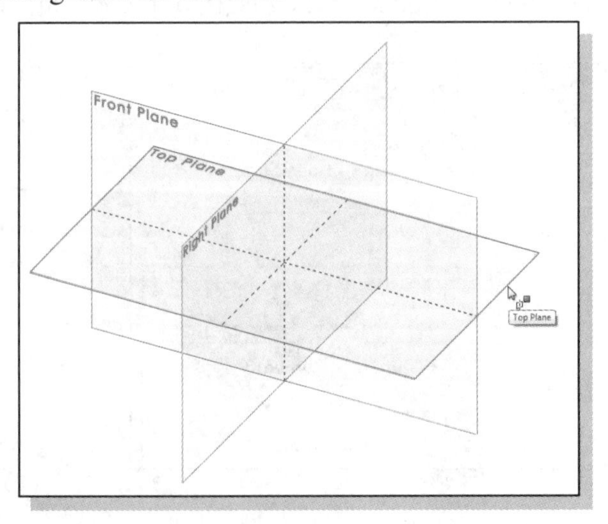

9. **On your own**, turn *OFF* the visibility of the three planes. (**HINT:** Right-click on each plane in the *FeatureManager Design Tree* and select Hide.)

Creating the 2D Sketch for the Base Feature

We will begin by sketching the base feature. The symmetry of the feature will be used in defining the geometry. Two methods, one using the Dynamic Mirror command, the other using the Mirror command, will be demonstrated. The Dynamic Mirror command allows sketch objects to be automatically mirrored as they are drawn. The Mirror command is used to mirror existing objects.

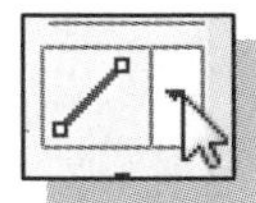

1. **Left-click** on the arrow next to the **Line** button on the *Sketch* toolbar to reveal additional commands.

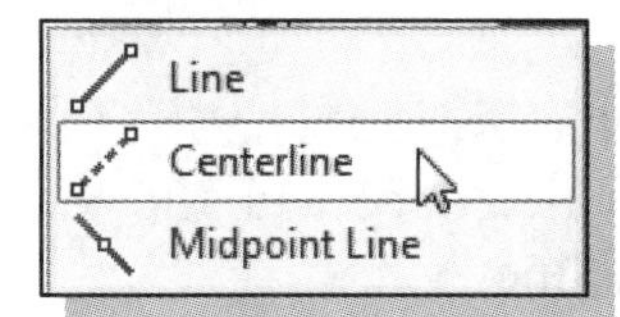

2. Select the **Centerline** command from the pop-up menu.

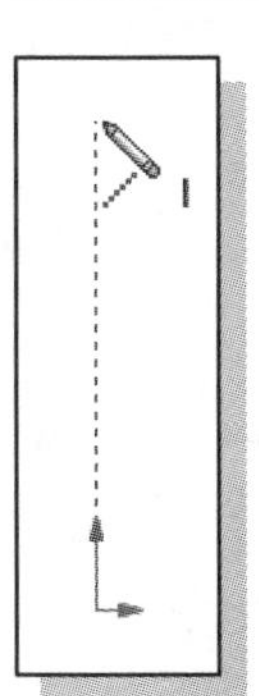

3. Move the cursor to a point directly above the *origin*. The Vertical relation will be automatically highlighted as shown. Click once with **the left-mouse-button** to select the first endpoint of the centerline.

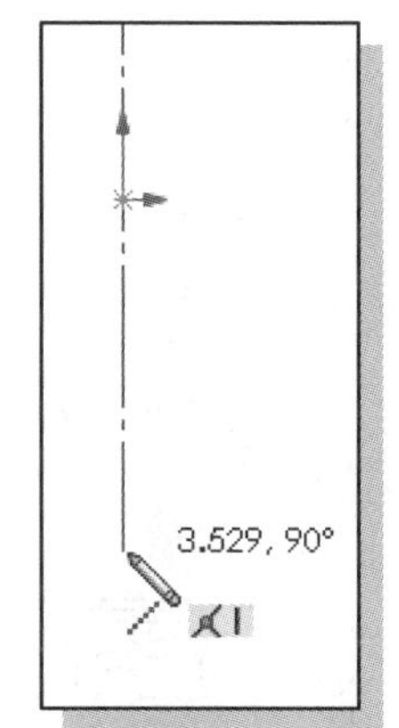

4. Move the cursor to a point directly below the *origin*. The Vertical relation will be highlighted as shown. Click once with the **left-mouse-button** to select the second endpoint of the centerline.

5. Press the [**Esc**] key once to end the Centerline command.

❖ The **Mirror** and **Dynamic Mirror** commands can be used to generate sketch objects with symmetry about a centerline.

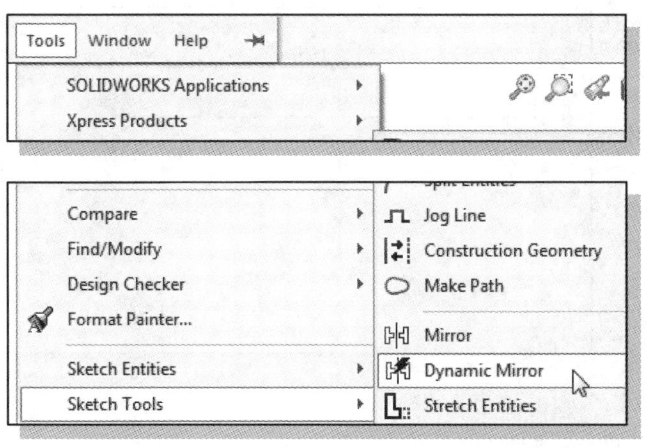

6. Select the **Tools** pull-down menu, then click on **Sketch Tools**. A complete list of sketch tools will appear. (Some of these also appear in the *Sketch* toolbar.)

7. Select **Dynamic Mirror** as shown.

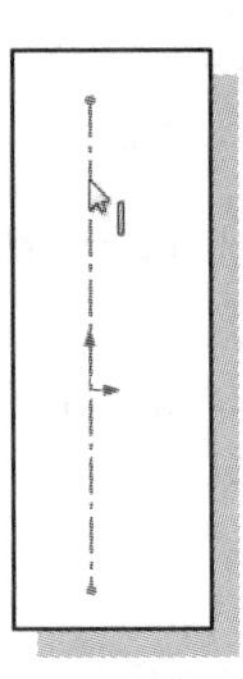

8. The message "*Please select a sketch line or linear edge to mirror about*" appears in the *Mirror PropertyManager*. Move the cursor over the centerline in the graphics area and click once with the **left-mouse-button** to select it as the line to mirror about.

➢ Notice the *hash marks* which appear on the centerline to indicate that the Dynamic Mirror command is active.

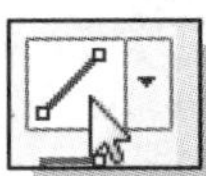

9. Click on the **Line** icon in the *Sketch* toolbar.

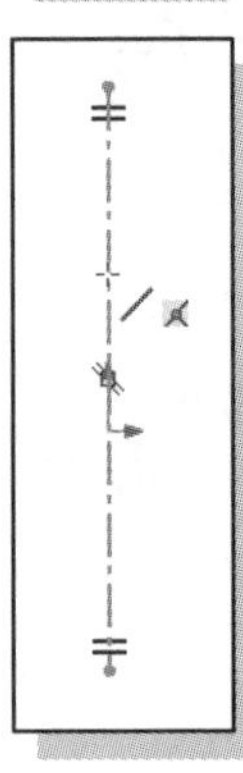

10. Move the cursor to a point on the centerline, above the *origin*. The Coincident relation will be highlighted as shown. Click once with the **left-mouse-button** to select the first endpoint of the line.

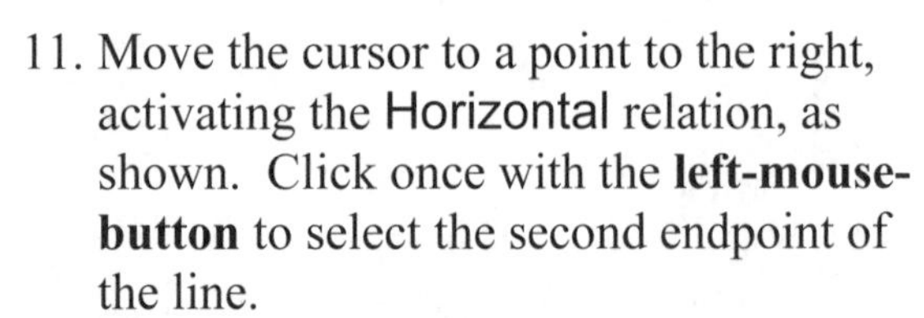

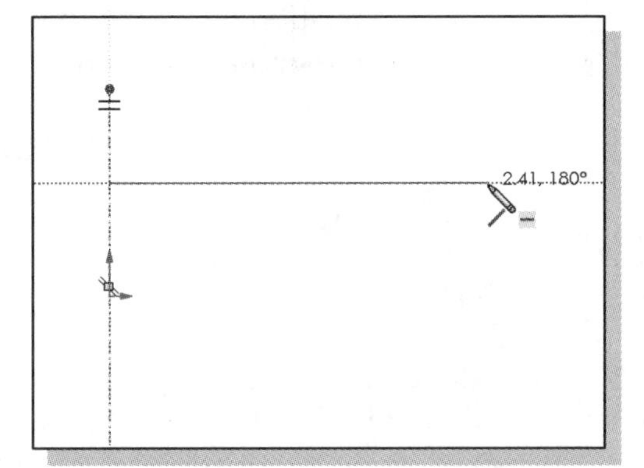

11. Move the cursor to a point to the right, activating the Horizontal relation, as shown. Click once with the **left-mouse-button** to select the second endpoint of the line.

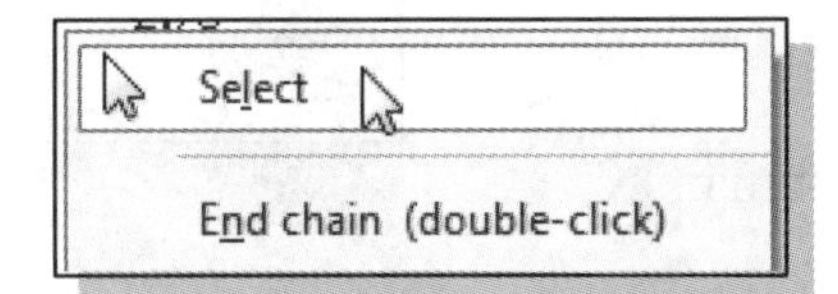

12. Inside the graphics window, right-mouse-click to bring up the option menu. In the option menu, select **Select** to end the Line command.

➢ Notice the mirror image of the line appears to the left of the centerline.

13. On your own, repeat these steps to create a similar horizontal line below the origin. Be sure that the Horizontal and Vertical relations are active for the second endpoint as shown in the image to the right.

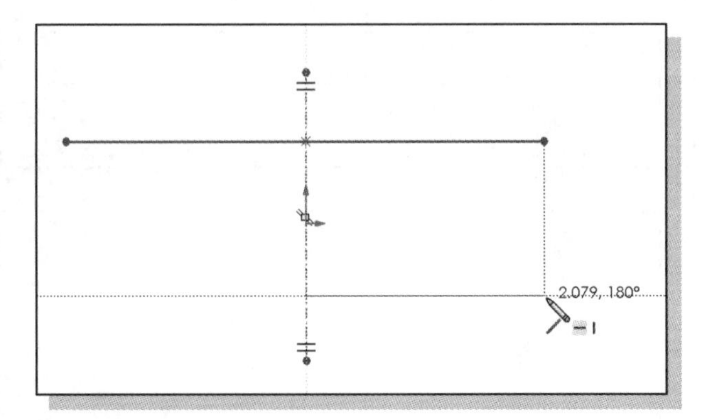

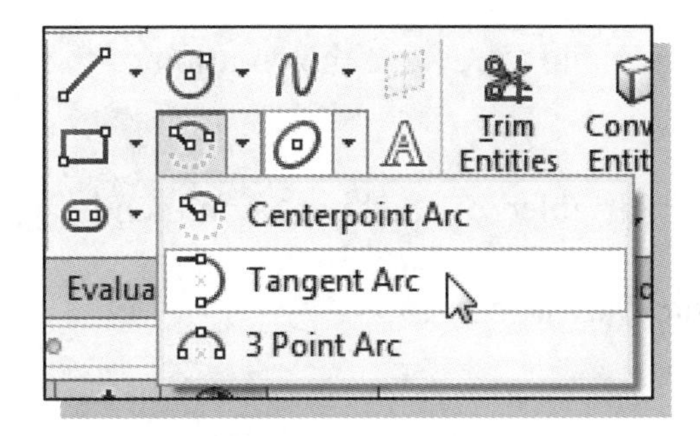

14. **Left-click** on the arrow next to the **Arc** button on the *Sketch* toolbar to reveal additional commands.

15. Select the **Tangent Arc** command from the pop-up menu.

16. Select the right endpoint of the top horizontal line, as shown below.

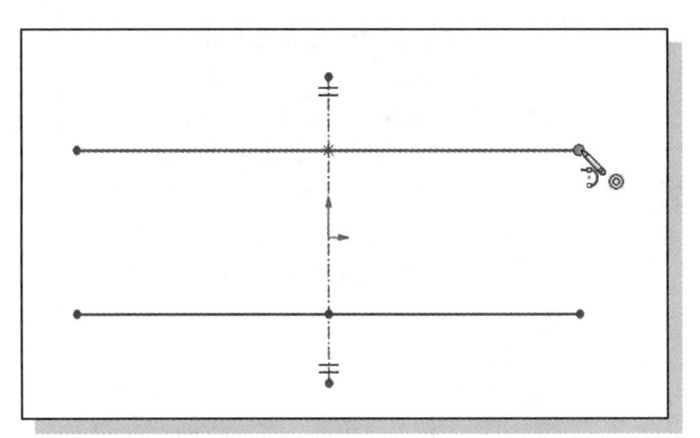

17. Select the right endpoint of the bottom horizontal line, as shown below.

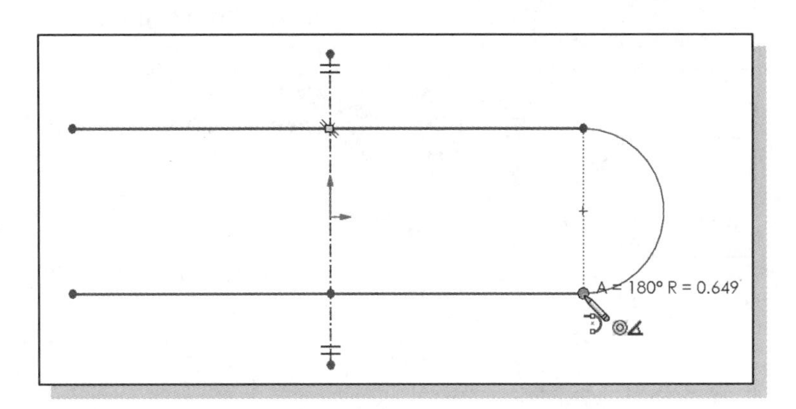

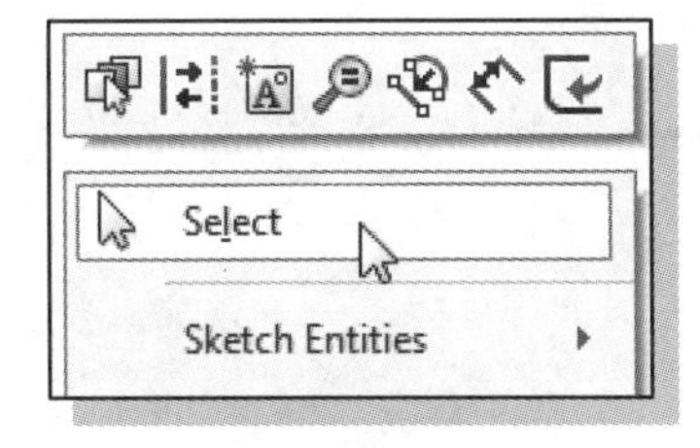

18. Inside the graphics window, right-mouse-click to bring up the option menu. In the option menu, select **Select** to end the Tangent Arc command.

❖ SOLIDWORKS offers the option to *autotransition* from the Line to the Tangent Arc command (and vice versa) without selecting the Arc command. We will redraw the second line and the arc to demonstrate this utility.

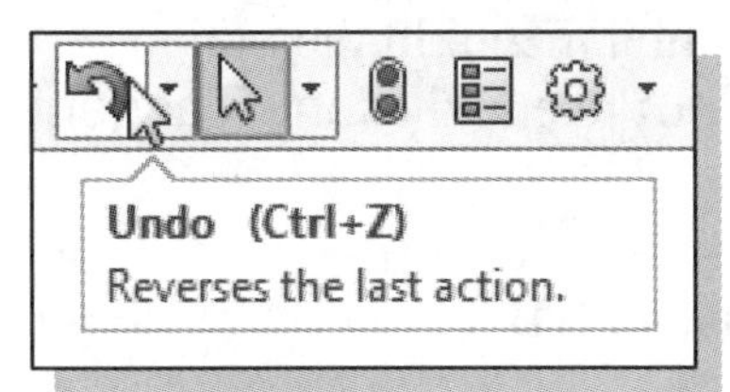

19. Click once with the left mouse button on the **Undo** command on the *Menu Bar* to reverse the last action – the creation of the arc.

20. Click a second time on the **Undo** command to reverse the creation of the line.

❖ Notice that the Dynamic Mirror tool has been disabled by the Undo command.

21. On your own, **turn *ON*** the Dynamic Mirror tool. (**HINT:** Perform steps 6 - 8.)

22. Execute the **Line** command and select a point on the centerline to restart the second horizontal line. Be sure that the Horizontal and Vertical relations are active for the endpoint as shown in the image to the right, then click once to select the endpoint.

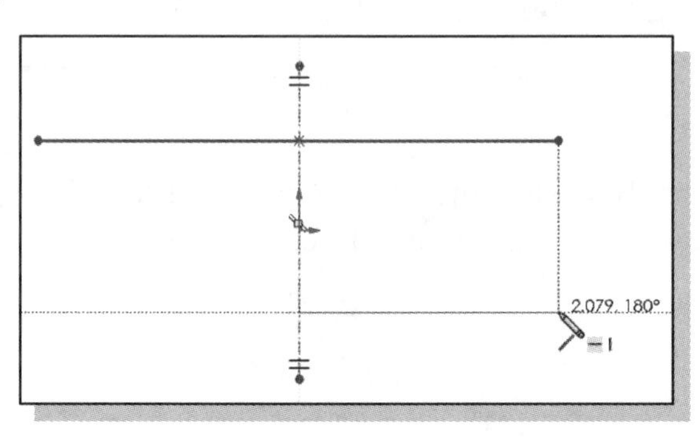

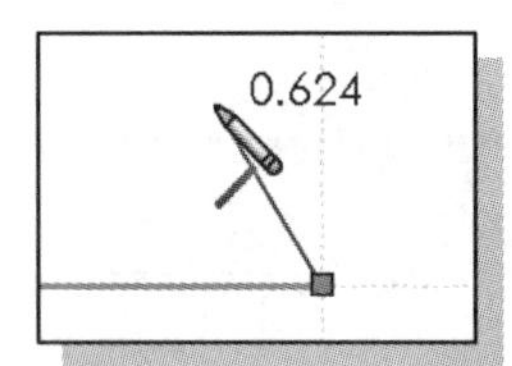

23. Move the cursor away from the endpoint of the horizontal line. **Do not click**. Notice the preview shows another line.

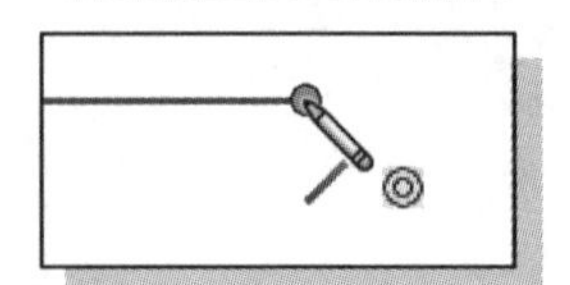

24. Move the cursor back to the endpoint. **Do not click**.

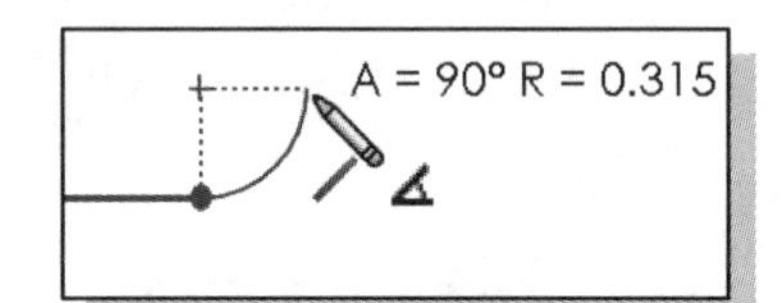

25. Move the cursor away again. Notice the preview now shows a tangent arc.

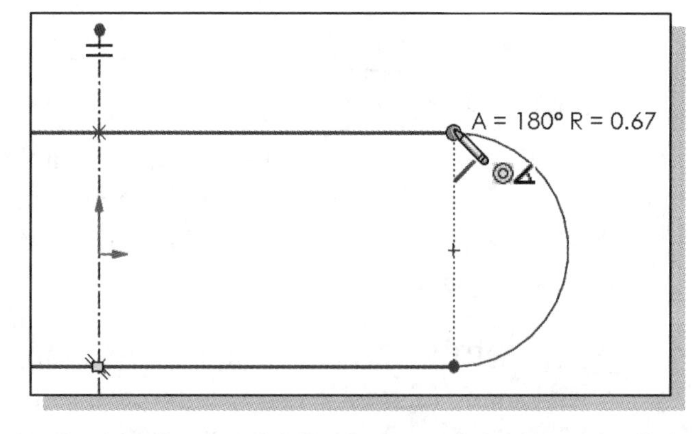

26. Select the right endpoint of the top horizontal line to end the arc.

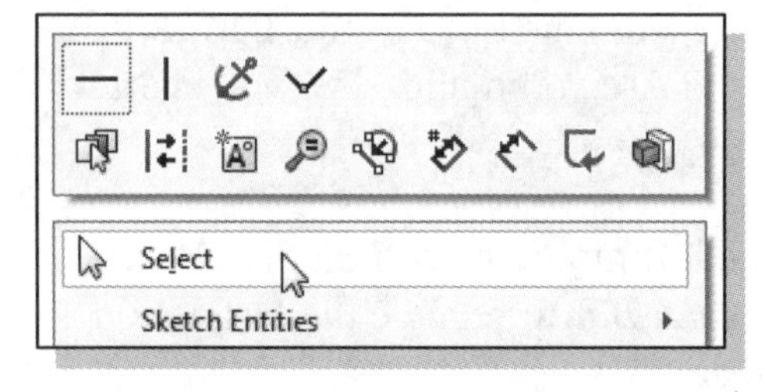

27. Inside the graphics window, right-mouse-click to bring up the option menu. In the option menu, select **Select** to end the Line/Tangent Arc command.

28. On your own, **turn *OFF*** the Dynamic Mirror command. (**HINT:** Perform steps 6 and 7.)

❖ We will now constrain the center of the arc to be on a horizontal line with the origin.

29. **Left-click** on the arrow next to the **Display/Delete Relations** button on the *Sketch* toolbar to reveal additional sketch relation commands.

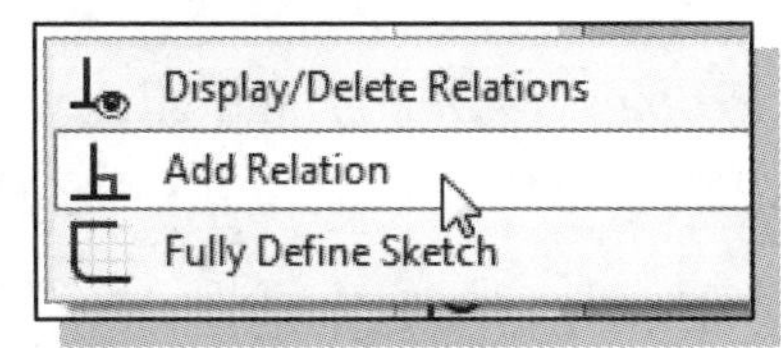

30. Select the **Add Relation** command from the pop-up menu. Notice the *Add Relations PropertyManager* appears.

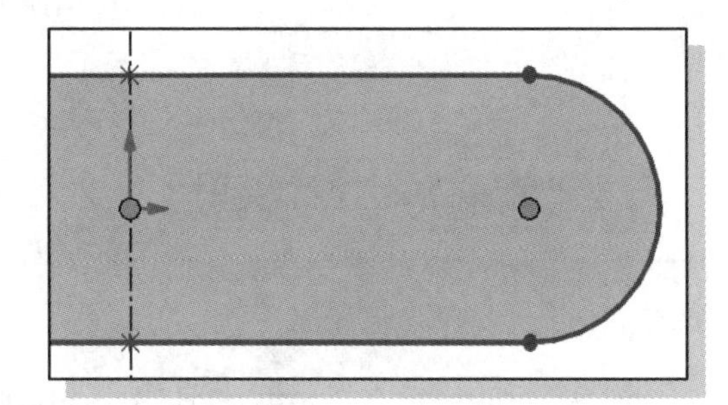

31. Pick the **centerpoint** of the tangent arc to the right side by left-mouse-clicking once on the geometry.

32. Pick the **origin**.

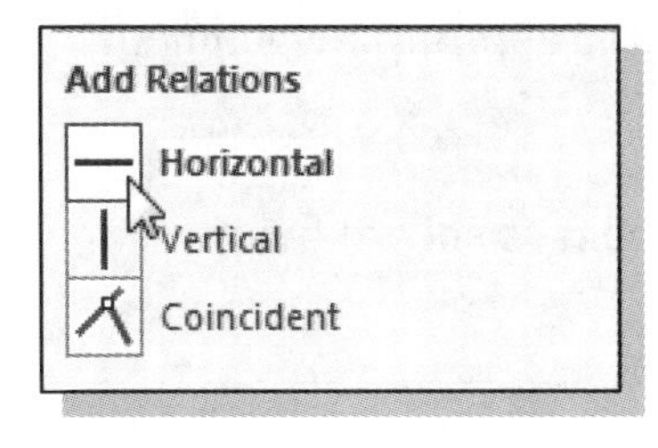

33. Click once with the left-mouse-button on the **Horizontal** icon in the *Add Relations PropertyManager* as shown. This activates the Horizontal relation.

34. Click the **OK** icon in the *PropertyManager*, or hit the [**Esc**] key once, to end the Add Relations command.

35. On your own, draw the circle as shown. Make sure the centerpoint for the circle is selected at the center of the tangent arc.

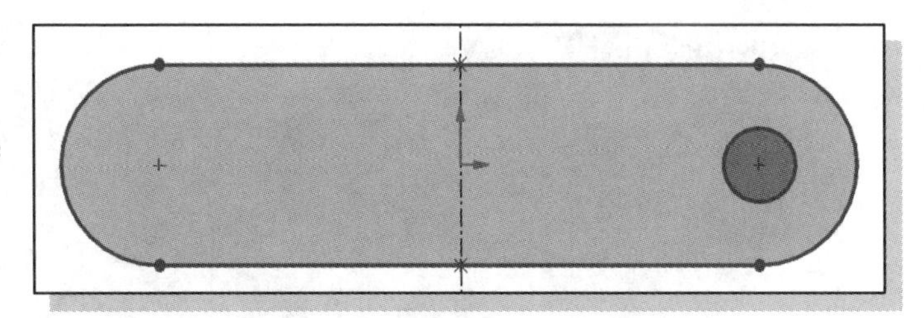

36. Press the **[Esc]** key to ensure no sketch features are selected.

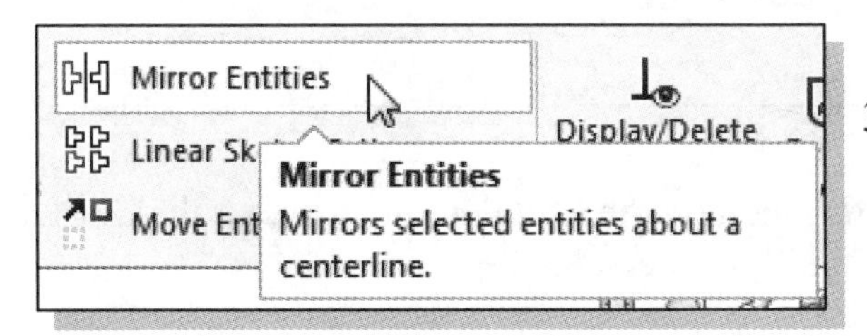

37. Select the **Mirror Entities** command on the *Sketch* toolbar.

(**NOTE:** The Mirror command could alternately be executed by selecting it with the *Sketch Tools* option on the *Tools* pull-down menu.)

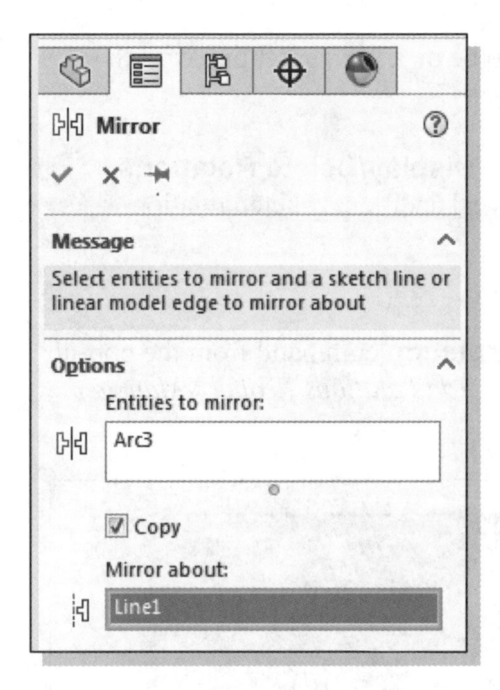

38. In the *Options* panel of the *Mirror PropertyManager*, the *Entities to mirror:* text box is highlighted. Move the cursor over the **new circle** in the graphics area and click once with the **left-mouse-button** to select it as the object to mirror.

39. Move the cursor into the *Mirror about:* text box in the *Mirror PropertyManager* and click once with the **left-mouse-button**.

40. Move the cursor over the **centerline** in the graphics area and click once with the **left-mouse-button** to select it as the object to mirror about.

41. Select **OK** in the *Mirror PropertyManager*.

❖ We have now completed the rough sketch, and applied the appropriate geometric relations. Remember the concept is 'shape before size'.

42. On your own, apply the *Smart Dimensions* and locate them as shown below. Notice the sketch is now Fully Defined.

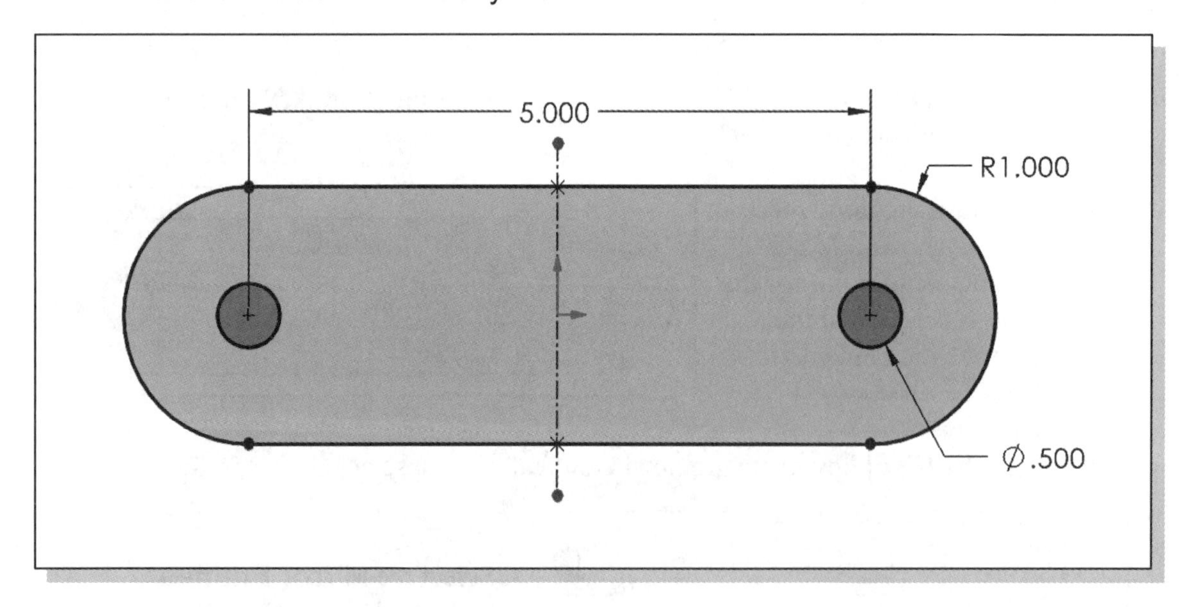

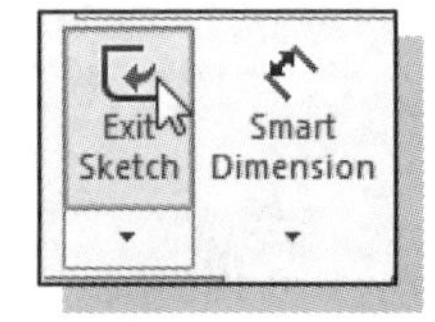

43. Select the **Exit Sketch** icon on the *Sketch* toolbar to exit the Sketch option.

Creating the First Extrude Feature

1. On your own, select the **Features** tab on the *CommandManager* to display the *Features* toolbar and use the **Extruded Boss/Base** command to create a base with a height of **0.5** inches.

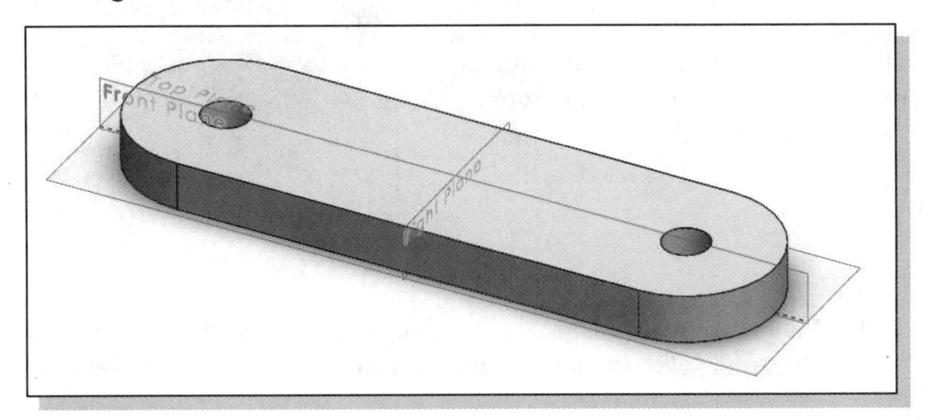

2. **On your own**, turn the visibility of the three major planes back *ON*. (**HINT:** Right-click on each plane in the *FeatureManager Design Tree* and select Show.)

➢ Notice the orientation of the base in relation to the major workplanes of the world coordinate system.

The Implied Parent/Child Relationships

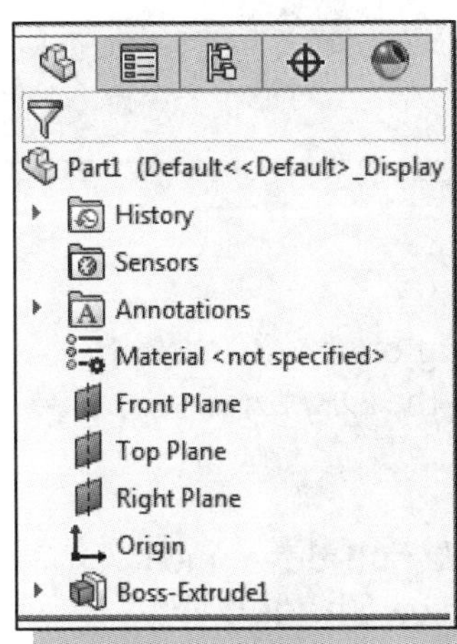

The *FeatureManager Design Tree* display includes the **Front**, **Top**, and **Right Planes**, the **Origin** (default work features) and the base feature (**Extrude1** shown in the figure) we just created. The parent/child relationships were established implicitly when we created the base feature: (1) Top *(XZ) workplane* was selected as the sketch plane; (2) Origin was used as the reference point to align the 2D sketch of the base feature.

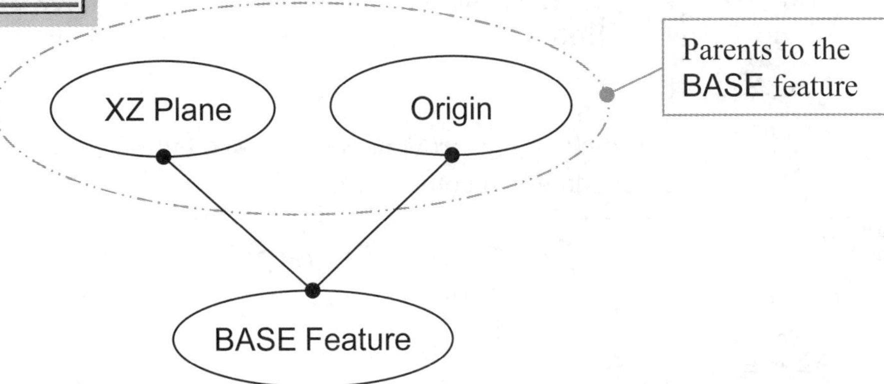

Creating the Second Solid Feature

For the next solid feature, we will create the top section of the design. Note that the center of the base feature is aligned to the **Origin** of the default work features. This was done intentionally so that additional solid features can be created referencing the default work features. For the second solid feature, the **Front** *(XY) workplane* will be used as the sketch plane.

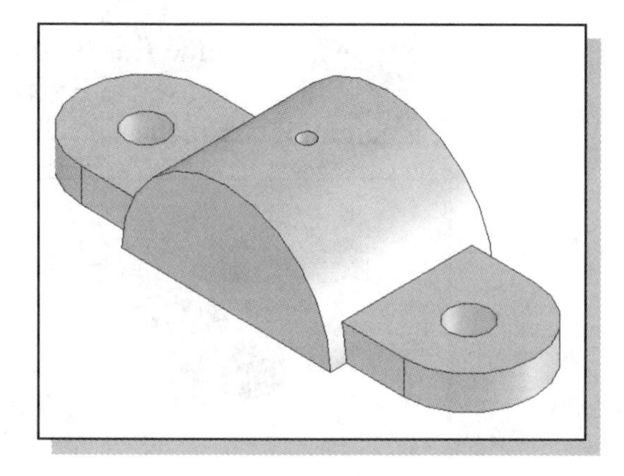

1. Move the cursor inside the graphics area, but away from the planes, and click once with the **left-mouse-button** to make sure no plane is selected.

2. Select the **Sketch** tab on the *CommandManager* to display the *Sketch* toolbar and select the **Sketch** command by left-clicking once on the icon.

3. In the *Edit Sketch PropertyManager*, the message "*Select a plane on which to create a sketch for the entity*" is displayed. Move the cursor over the edge of the Front Plane in the graphics area. When the **Front Plane** is highlighted, click once with the **left-mouse-button** to select the **Front (XY) Plane** as the sketch plane for the new sketch.

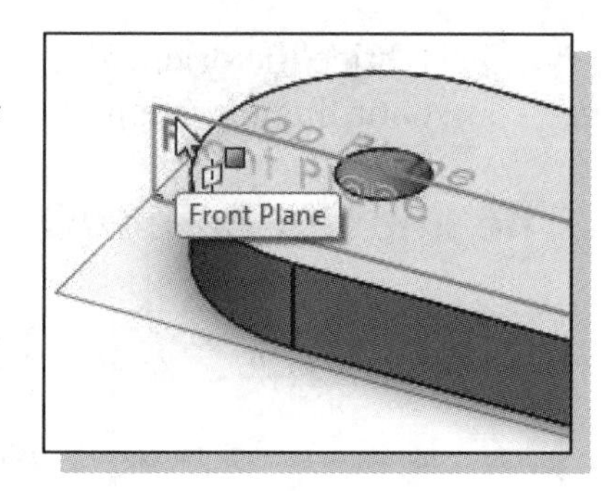

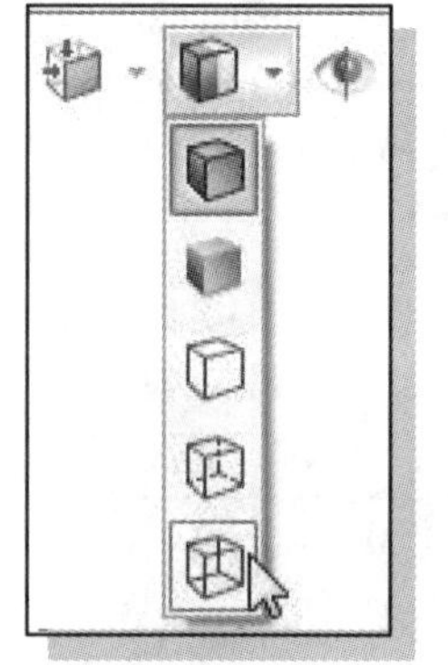

4. Change the display to wireframe by clicking once with the **left-mouse-button** on the **Wireframe** icon on the *Display Style* pull-down menu on the *Heads-up View* toolbar.

➢ NOTE: On your own, use the *Display Style* pull-down menu on the *Heads-up View* toolbar to change between **Shaded With Edges** and **Wireframe** display as needed throughout the lesson.

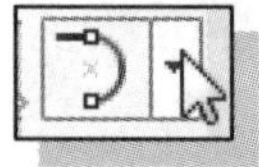

5. **Left-click** on the arrow next to the **Arc** button on the *Sketch* toolbar to reveal additional commands.

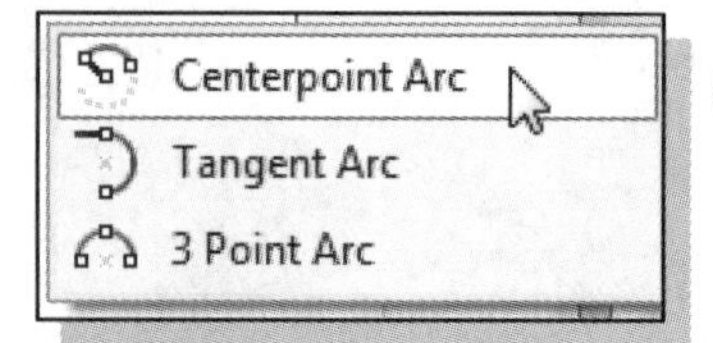

6. Select the **Centerpoint Arc** command from the pop-up menu.

7. Pick the **Origin** as the center location of the new arc.

8. On your own, create a semi-circle of arbitrary size, with both endpoints aligned to the X-axis, as shown below.

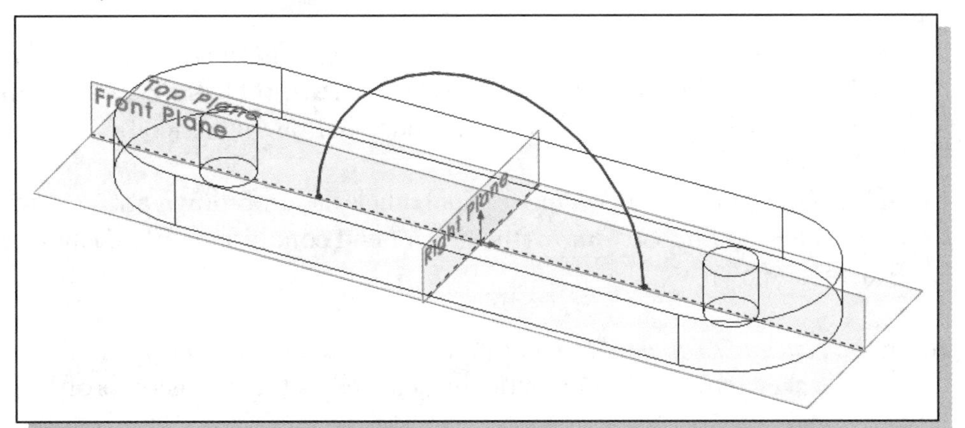

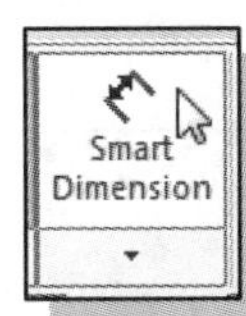

9. Move the cursor on top of the **Smart Dimension** icon. Left-click once on the icon to activate the Smart Dimension command.

10. On your own, create and adjust the radius dimension of the arc to **1.75**.

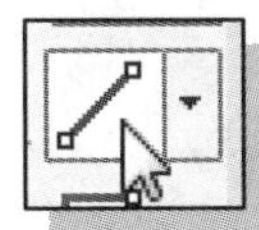

11. Select the **Line** command in the *Sketch* toolbar.

12. Create a line connecting the two endpoints of the arc as shown in the figure below.

13. On your own, add a **Coincident** relation between the origin and the horizontal line. The sketch should now be Fully Defined.

14. On your own, turn *OFF* the visibility of the workplanes.

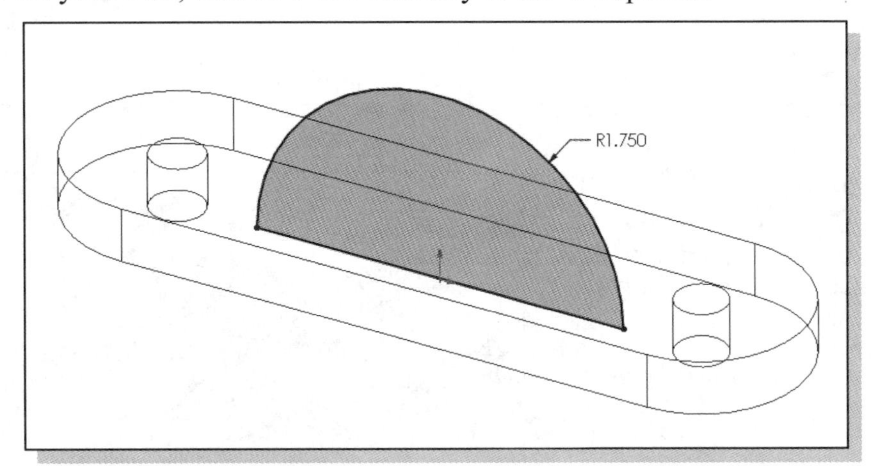

15. Select the **Sketch** icon on the *Sketch* toolbar to exit the Sketch option.

16. On your own, change the display style back to '*Shaded with edges*'.

17. Make sure the sketch – Sketch2 – is selected in the *FeatureManager Design Tree*.

18. Select the **Features** tab on the *CommandManager* to display the *Features* toolbar and select the **Extruded Boss/Base** command by clicking once with the left-mouse-button on the icon.

19. In the *Extrude PropertyManager* panel, click the drop-down arrow to reveal the options for the *End Condition* (the default end condition is Blind) and select **Mid Plane** as shown.

20. In the *Extrude PropertyManager* panel, enter **2.5** as the extrusion distance. Notice that the sketch region is automatically selected as the extrusion profile.

21. Make sure the **Merge result** box is checked.

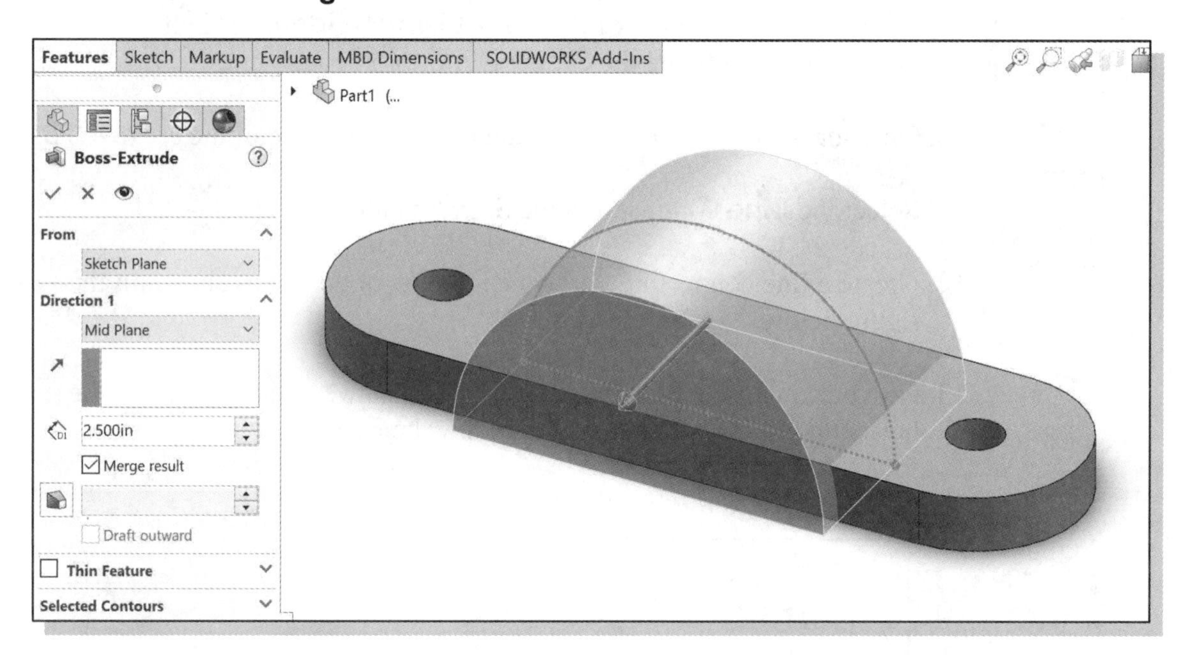

22. Click on the **OK** button to proceed with the Extrude operation.

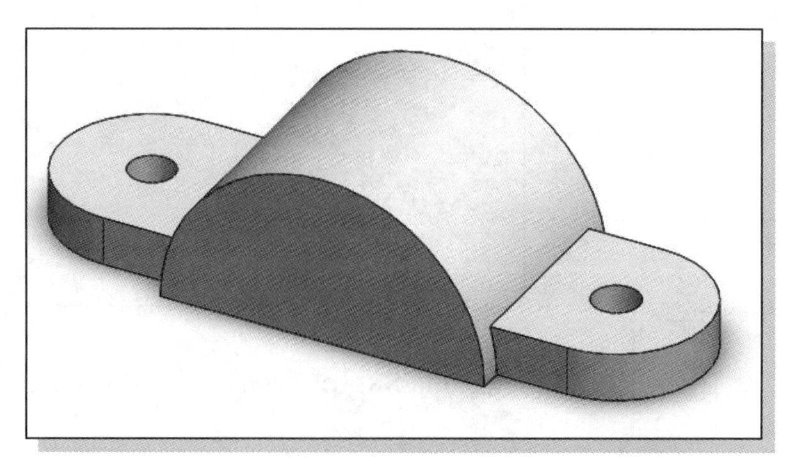

Creating the First Extruded Cut Feature

A rectangular cut will be created as the next solid feature.

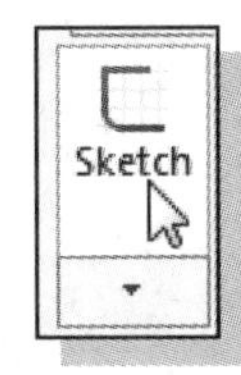

1. Select the **Sketch** tab on the *CommandManager* to display the *Sketch* toolbar and select the **Sketch** command by left-clicking once on the icon.

2. In the *Edit Sketch PropertyManager*, the message "*Select a plane on which to create a sketch for the entity*" is displayed. **Pick the front vertical face** of the solid model shown.

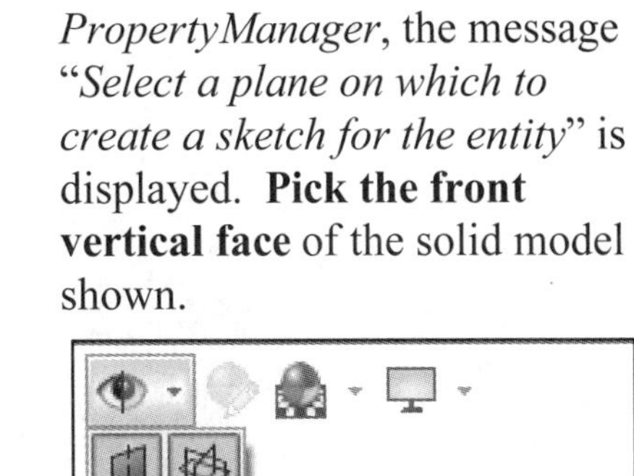

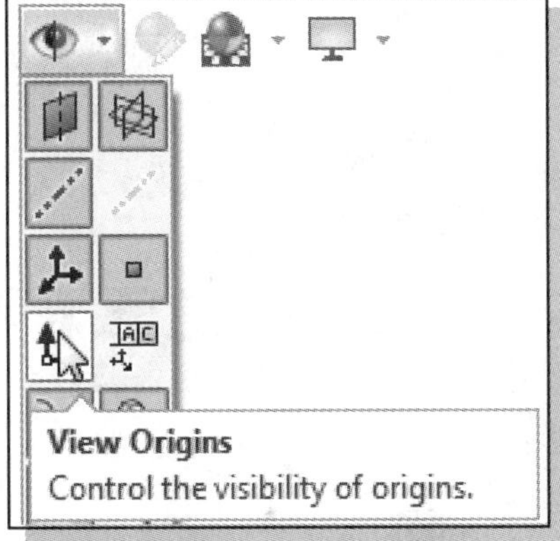

3. We will use the origin as an endpoint for one of the dimensions. Make sure the **View Origins** option is toggled *ON* using the *Hide/Show Items* pull-down menu on the *Heads-up View* toolbar.

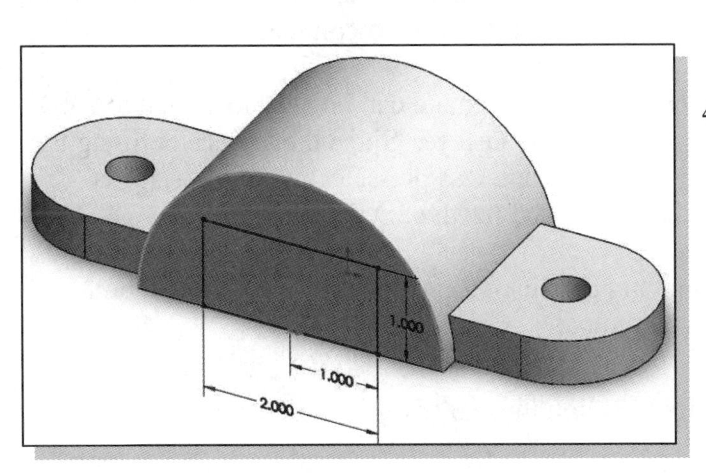

4. On your own, create a (**2.0" x 1.0" H**) rectangle and apply the dimensions as shown. (**HINT:** To apply the horizontal dimension from the midpoint of the lower edge, select the **origin** of the world coordinate system as an endpoint. SOLIDWORKS will automatically project the origin onto the workplane.)

5. On your own, exit the sketch and use the **Extruded Cut** command with the **Through All** end condition option to create a cutout that cuts through the entire 3D solid model as shown.

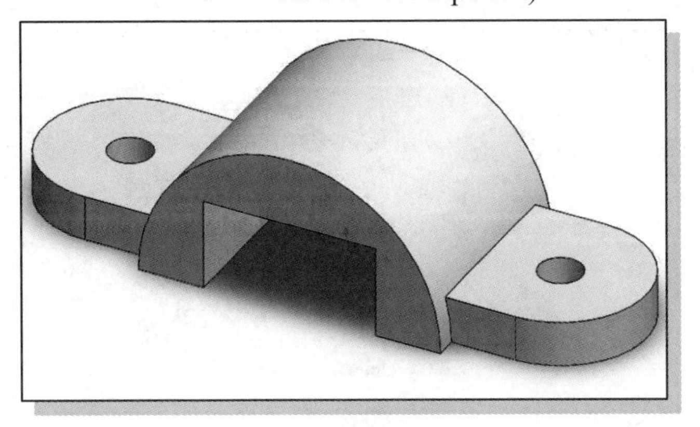

Creating the Second Extruded Cut Feature

1. Select the **Sketch** tab on the *CommandManager* to display the *Sketch* toolbar and select the **Sketch** command by left-clicking once on the icon.

2. In the *Edit Sketch PropertyManager*, the message "*Select a plane on which to create a sketch for the entity*" is displayed. Select the horizontal face of the last cut feature as the *sketching plane.*

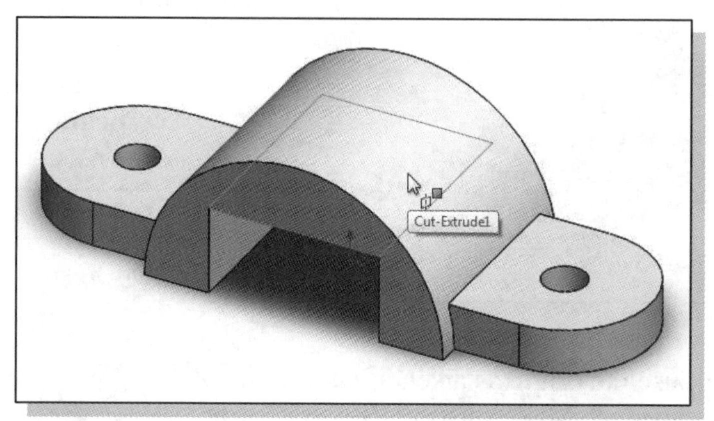

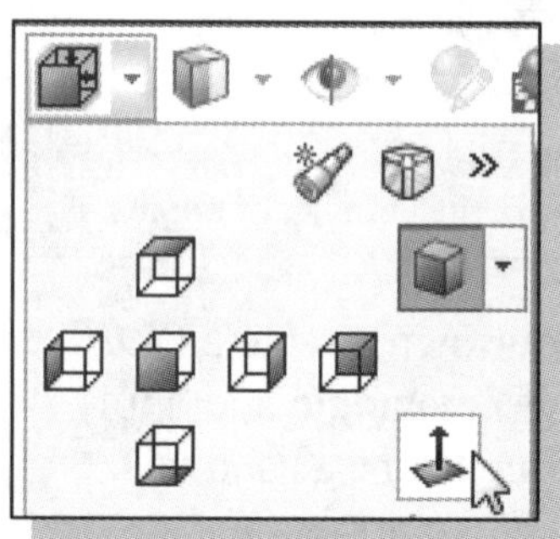

3. Left-click on the **View Orientation** icon on the *Heads-up View* toolbar to reveal the *View Orientation* pull-down menu. Select the **Normal To** command.

❖ The Normal To command can be used to switch to a 2D view of a selected surface. Since the surface defining the current sketch plane was pre-selected, this surface is automatically used for the 2D view.

4. Select the **Circle** command by clicking once with the left-mouse-button on the icon in the *Sketch* toolbar.

5. Pick the projected origin to align the center of the new circle.

6. Create a circle of arbitrary size.

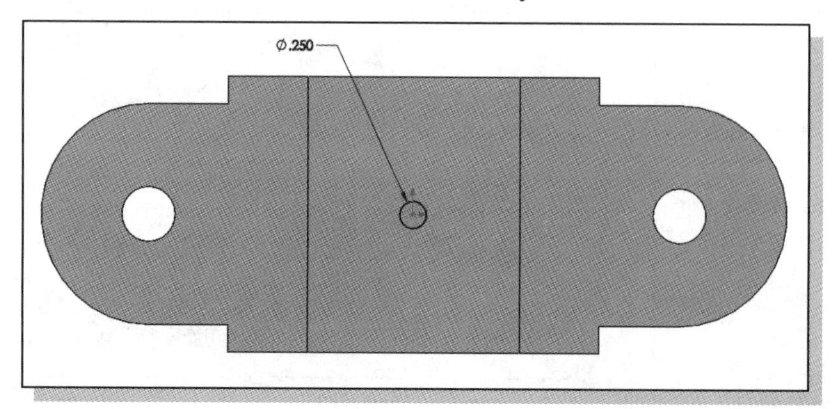

7. On your own, add the size dimension of the circle and set the dimension to **0.25**.

8. Select the **Exit Sketch** icon on the *Sketch* toolbar to exit the Sketch option.

9. Make sure the sketch – Sketch4 – is selected in the *FeatureManager Design Tree*.

10. In the *Features* toolbar, select the **Extruded Cut** command by clicking once with the left-mouse-button on the icon.

11. In the *Cut-Extrude PropertyManager* panel, click the arrow to reveal the pull-down options for the *end condition* (the default end condition is Blind) and select **Through All**.

12. Click the **OK** button (green check mark) in the *Cut-Extrude PropertyManager* panel.

13. On your own, change the view orientation to Isometric.

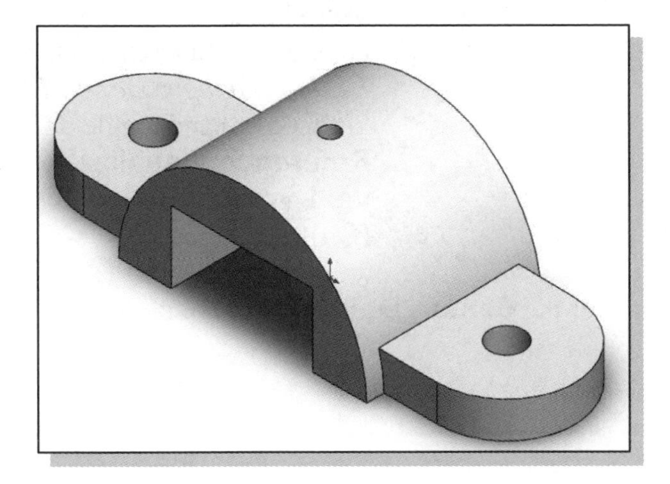

Examining the Parent/Child Relationships

1. On your own, rename the feature names to **Base**, **Main_Body**, **Rect_Cut** and **Center_Drill** as shown in the figure.

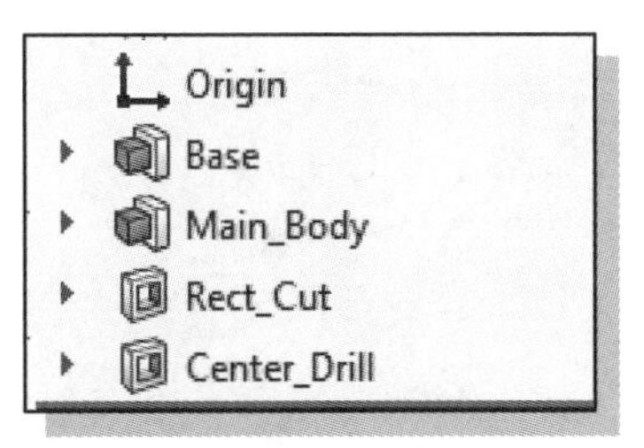

The *FeatureManager Design Tree* window now contains several items, including the major planes, the **Origin** (default work features) and four solid features. All of the parent/child relationships were established implicitly as we created the solid features. As more features are created, it becomes much more difficult to make a sketch showing all the parent/child relationships involved in the model. On the other hand, it is not really necessary to have a detailed picture showing all the relationships among the features. In using a feature-based modeler, the main emphasis is to consider the interactions that exist between the **immediate features**. Treat each feature as a unit by itself, and be clear on the parent/child relationships for each feature. Thinking in terms of *features* is what distinguishes *feature-based modeling* and the previous generation solid modeling techniques. Let us take a look at the last feature we created, the **Center_Drill** feature. What are the parent/child relationships associated with this feature? (1) Since this is the last feature we created, it is not a parent feature to any other features. (2) Since we used one of the surfaces of the rectangular cutout as the sketching plane, the **Rect_Cut** feature is a parent feature to the **Center_Drill** feature. (3) We also used the projected Origin as a reference point to align the center; therefore, the **Origin** is also a parent to the **Center_Drill** feature. (4) The Center_Drill was extruded through the Main_Body. Since we used the 'Merge Result' option when we created the Main_Body, it is merged with the Base. The merged **Base/Main_Body** is therefore a parent to the **Center_Drill**.

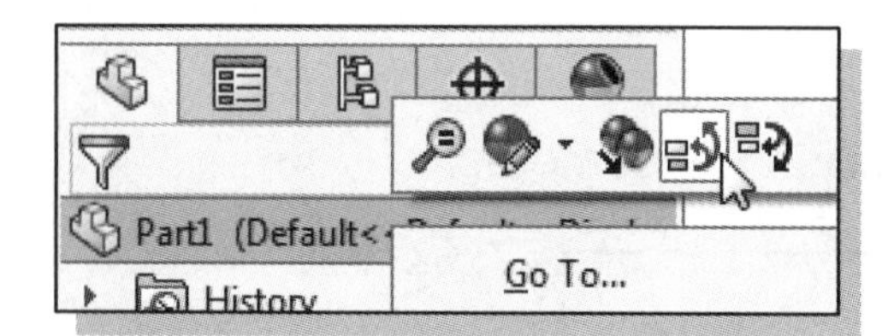

2. The *FeatureManager Design Tree* will display the parent/child relationships. This is called Dynamic Reference Visualization. To turn the tool on, right-click on the first item in the *FeatureManager Design Tree* Move (Part1 in this case) and toggle on the **Dynamic Reference Visualization (Parent)** and the **Dynamic Reference Visualization (Child)** options.

3. Move the cursor over the **Center_Drill** icon (do not click) to reveal the parent features.

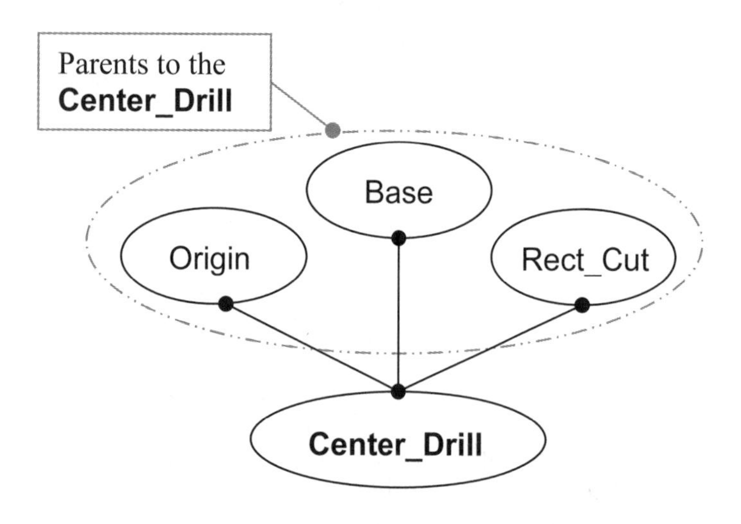

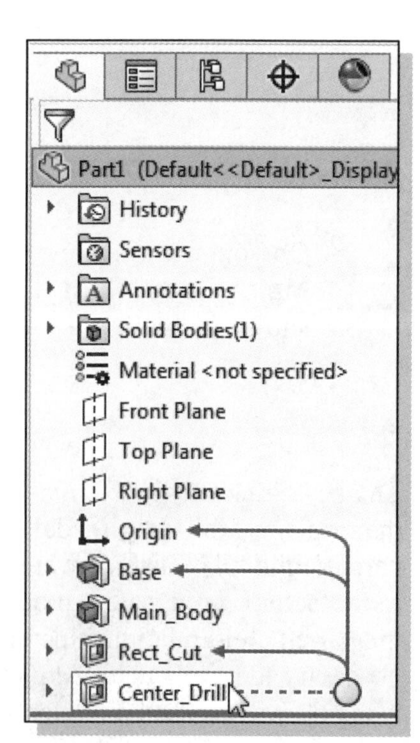

4. On your own, move the cursor over the other features in the *FeatureManager Design Tree* to explore the parent/child relationships.

Modify a Parent Dimension

Any changes to the parent features will affect the child feature. For example, if we modify the height of the **Rect_Cut** feature from 1.0 to 0.75, the depth of the child feature (**Center_Drill** feature) will be affected.

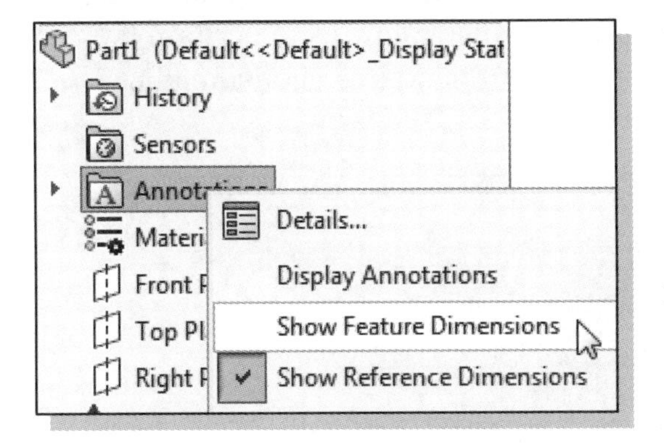

1. In the *FeatureManager Design Tree* window, move the cursor on top of **Annotations** and click once with the **right-mouse-button** to bring up the option menu.

2. In the option menu, select **Show Feature Dimensions**.

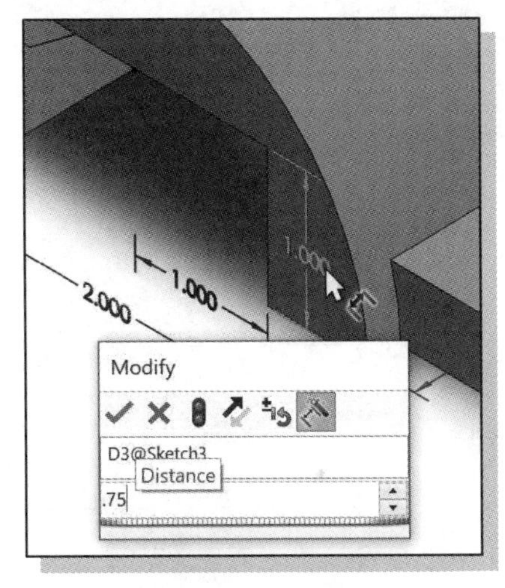

3. Select the height dimension (1.0) by double-clicking on the dimension.

4. Enter **0.75** as the new height dimension as shown.

5. Select **OK** in the *Modify* window.

6. Click on the **Rebuild** button in the *Menu Bar* to proceed with updating the solid model.

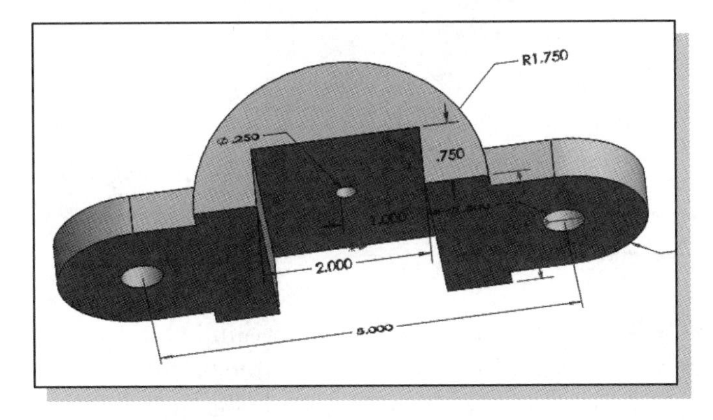

➢ Note that the position of the **Center_Drill** feature is also adjusted as the placement plane is lowered. The drill-hole still goes through the main body of the *U-Bracket* design. The parent/child relationship assures the intent of the design is maintained.

7. On your own, adjust the height of the **Rect_Cut** feature back to **1.0 inch** before proceeding to the next section.

8. Repeat steps 1 and 2 to turn off the **Show Feature Dimensions** option.

A Design Change

Engineering designs usually go through many revisions and changes. For example, a design change may call for a circular cutout instead of the current rectangular cutout feature in our model. SOLIDWORKS provides an assortment of tools to handle design changes quickly and effectively. In the following sections, we will demonstrate some of the more advanced tools available in SOLIDWORKS, which allow us to perform the modification of changing the rectangular cutout (2.0 × 1.0 inch) to a circular cutout (radius: 1.25 inch).

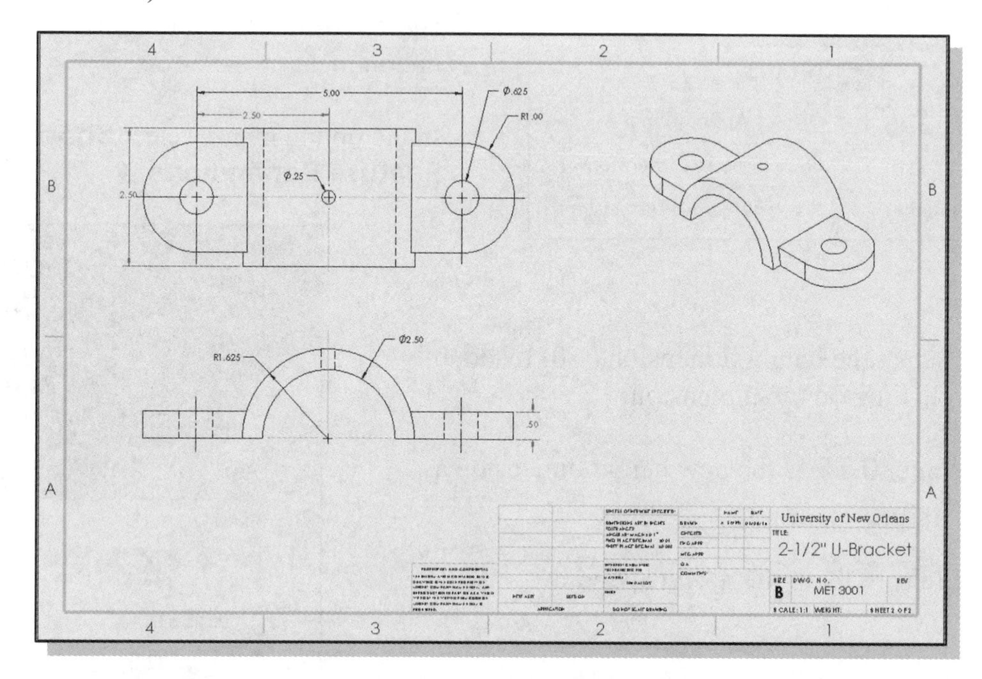

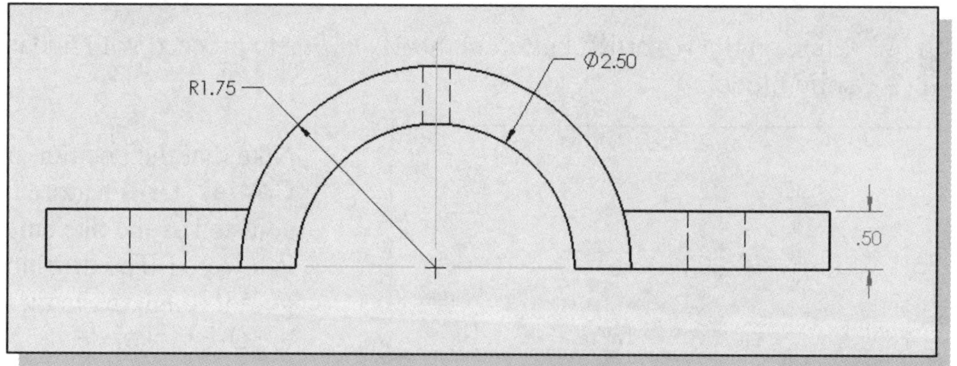

Based on your knowledge of SOLIDWORKS so far, how would you accomplish this modification? What other approaches can you think of that are also feasible? Of the approaches you came up with, which one is the easiest to do and which is the most flexible? If this design change were anticipated right at the beginning of the design process, what would be your choice in arranging the order of the features? You are encouraged to perform the modifications prior to following through the rest of the tutorial.

Feature Suppression

With SOLIDWORKS, we can take several different approaches to accomplish this modification. We could (1) create a new model, or (2) change the shape of the existing cut feature, or (3) perform **feature suppression** on the rectangular cut feature and add a circular cut feature. The third approach offers the most flexibility and requires the least amount of editing to the existing geometry. **Feature suppression** is a method that enables us to disable a feature while retaining the complete feature information; the feature can be reactivated at any time. Prior to adding the new cut feature, we will first suppress the rectangular cut feature.

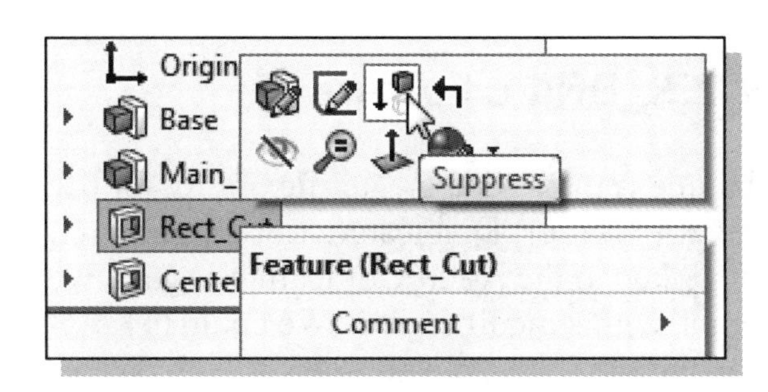

1. Move the cursor inside the *FeatureManager Design Tree* window. Click once with the **right-mouse-button** on top of **Rect_Cut** to bring up the option menu.

2. Pick **Suppress** in the pop-up menu.

❖ We have literally ***gone back in time***. The Rect_Cut and Center_Drill features have disappeared in the graphics area and both are shaded grey (meaning they are suppressed) in the *FeatureManager Design Tree*. The child feature cannot exist without its parent(s), and any modification to the parent (Rect_Cut) influences the child (Center_Drill).

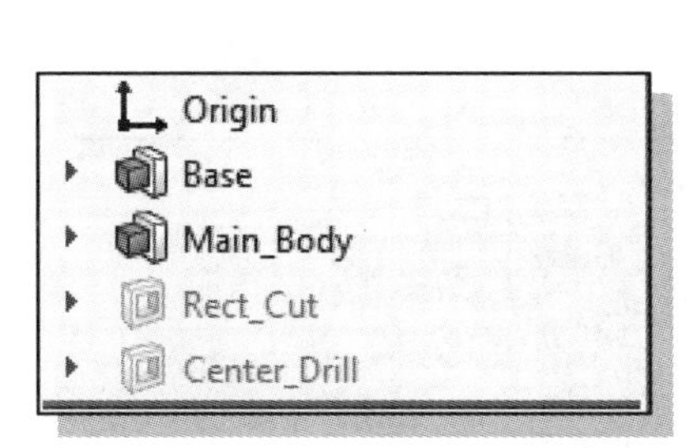

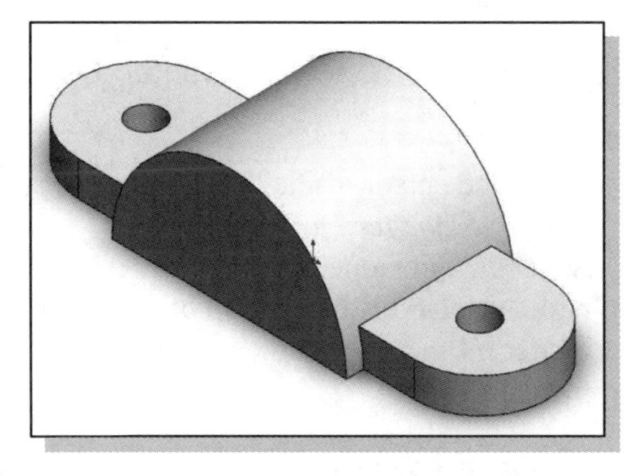

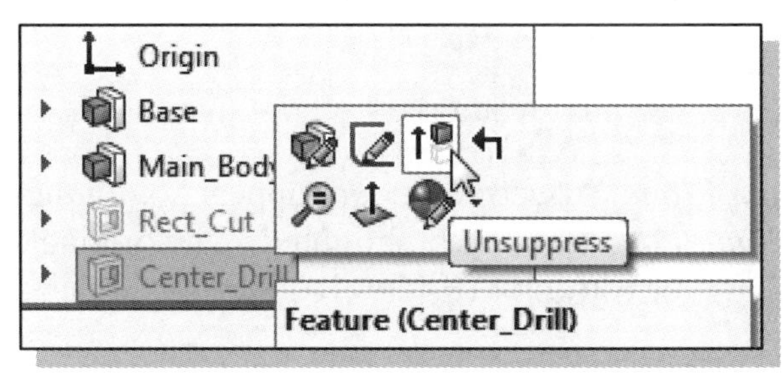

3. Move the cursor inside the *FeatureManager Design Tree* window. Click once with the **right-mouse-button** on top of **Center_Drill** to bring up the option menu.

4. Pick **Unsuppress** in the pop-up menu.

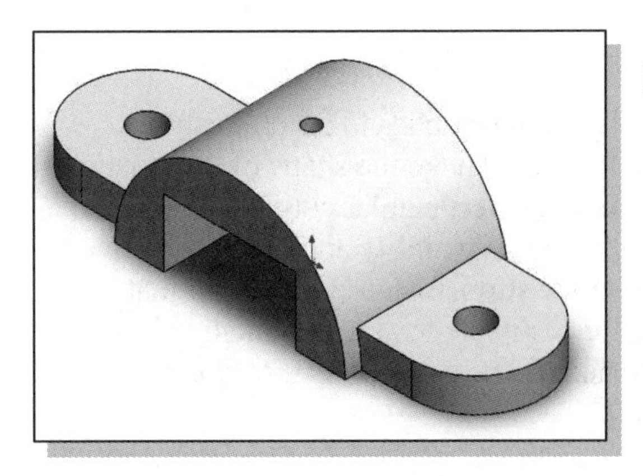

➢ In the display area and the *Model Tree* window, both the **Rect_Cut** feature and the **Center_Drill** feature are re-activated. The child feature cannot exist without its parent(s); the parent (Rect_Cut) must be activated to enable the child (Center_Drill).

A Different Approach to the CENTER_DRILL Feature

The main advantage of using the BORN technique is to provide greater flexibility for part modifications and design changes. In this case, the Center_Drill feature can be placed on the Top (XZ) Workplane and therefore not be linked to the Rect_Cut feature. Again, we can take several different approaches to accomplish this modification. We could (1) delete the Center_Drill feature and the corresponding sketch, create a new sketch on the Top (XZ) Workplane, and recreate the Center_Drill feature, or (2) redefine the sketch plane for the existing sketch associated with the Center_Drill feature. We will perform the second option.

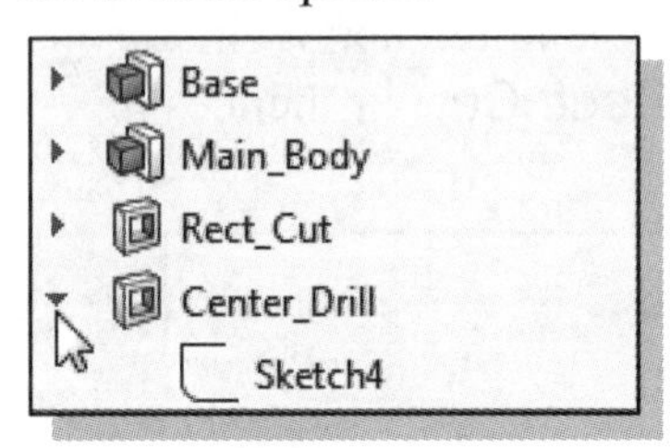

1. Move the cursor inside the *FeatureManager Design Tree* window. Click once with the **left-mouse-button** on the tree to reveal the Sketch defining the Center_Drill feature as shown.

2. Move the cursor inside the *FeatureManager Design Tree* window. Click once with the **right-mouse-button** on top of **Sketch4** to bring up the option menu.

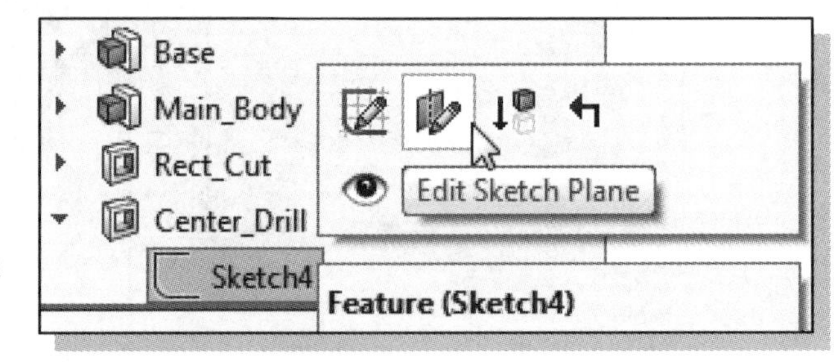

3. Pick **Edit Sketch Plane** in the pop-up menu. This will allow us to redefine the workplane for Sketch4.

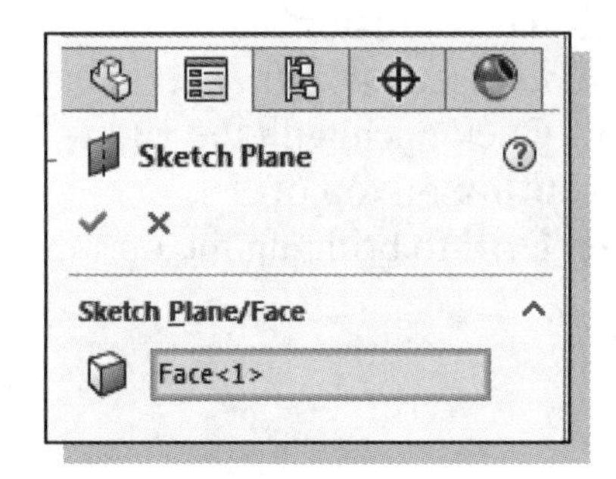

❖ The *Sketch Plane PropertyManager* appears, with the *Sketch Plane/Face* text box highlighted. The original sketch plane for Sketch4 (Face<1>) appears in the field.

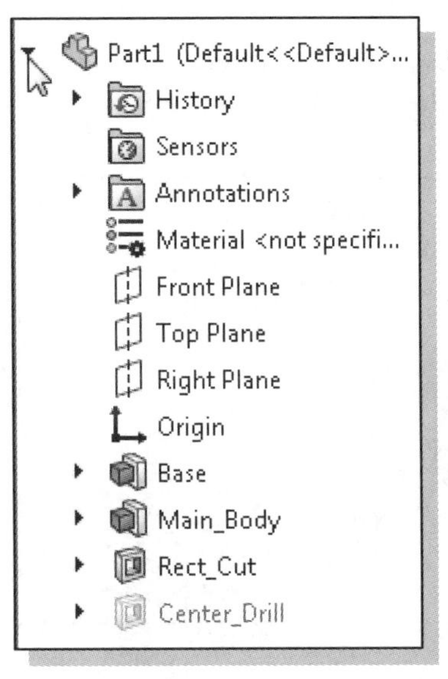

4. Notice the **Part1** icon of the design tree now appears at the upper left corner of the graphics area. **Left-click** on the box next to the **Part1** icon to reveal the design tree as shown.

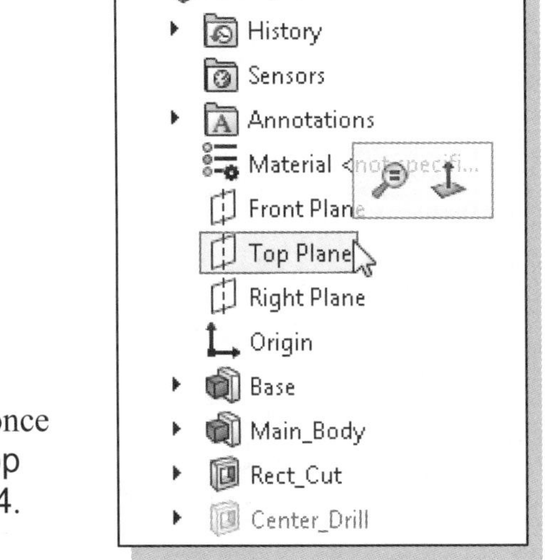

5. Move the cursor over the **Top Plane** icon. When the Top Plane is highlighted, click once with the **left-mouse-button** to select the Top (XZ) Plane as the sketch plane for Sketch4.

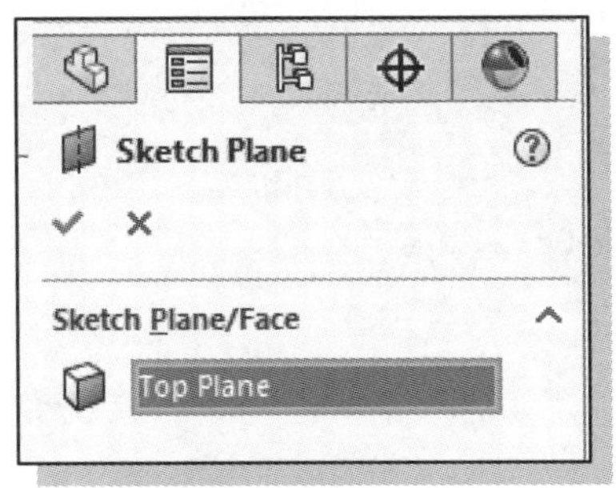

6. Top Plane now appears in the *Sketch Plane/Face* field of the *Sketch Plane PropertyManager*. Click once with the **left-mouse-button** on the **OK** icon to accept the new definition of the sketch plane.

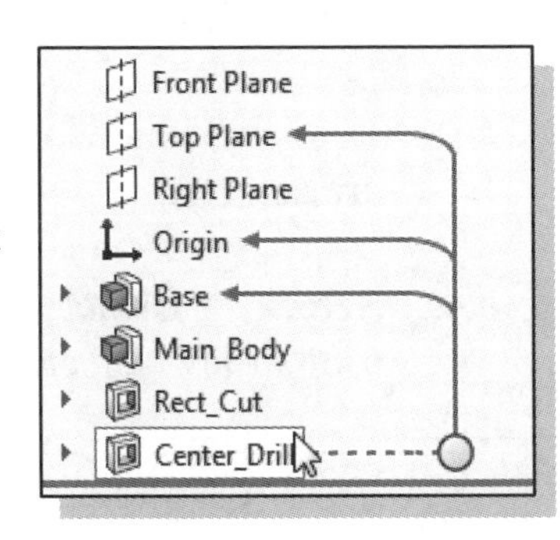

7. Move the cursor over the **Center_Drill** icon in the *FeatureManager Design Tree* to reveal the new parent relationships. Notice the Rect_Cut feature is no longer a parent to the Center_Drill feature; the Top Plane is now a parent.

Suppress the Rect_Cut Feature

Now that the **Center_Drill** feature is no longer a child of the **Rect_Cut** feature, any change to the **Rect_Cut** feature does not affect the **Center_Drill** feature.

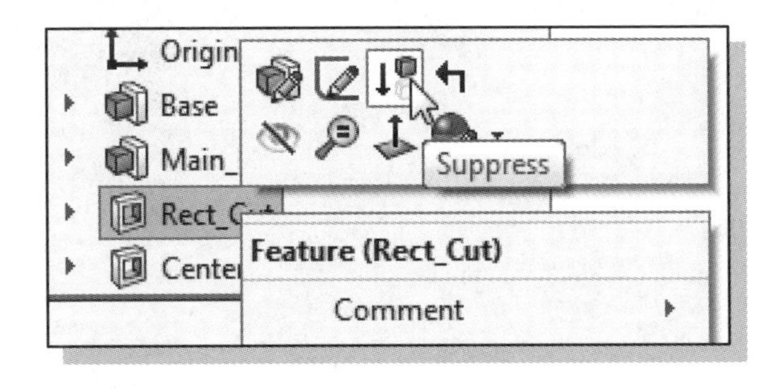

1. Move the cursor inside the *Model Tree* window. Click once with the right-mouse-button on top of **Rect_Cut** to bring up the option menu.

2. Pick **Suppress** in the pop-up menu.

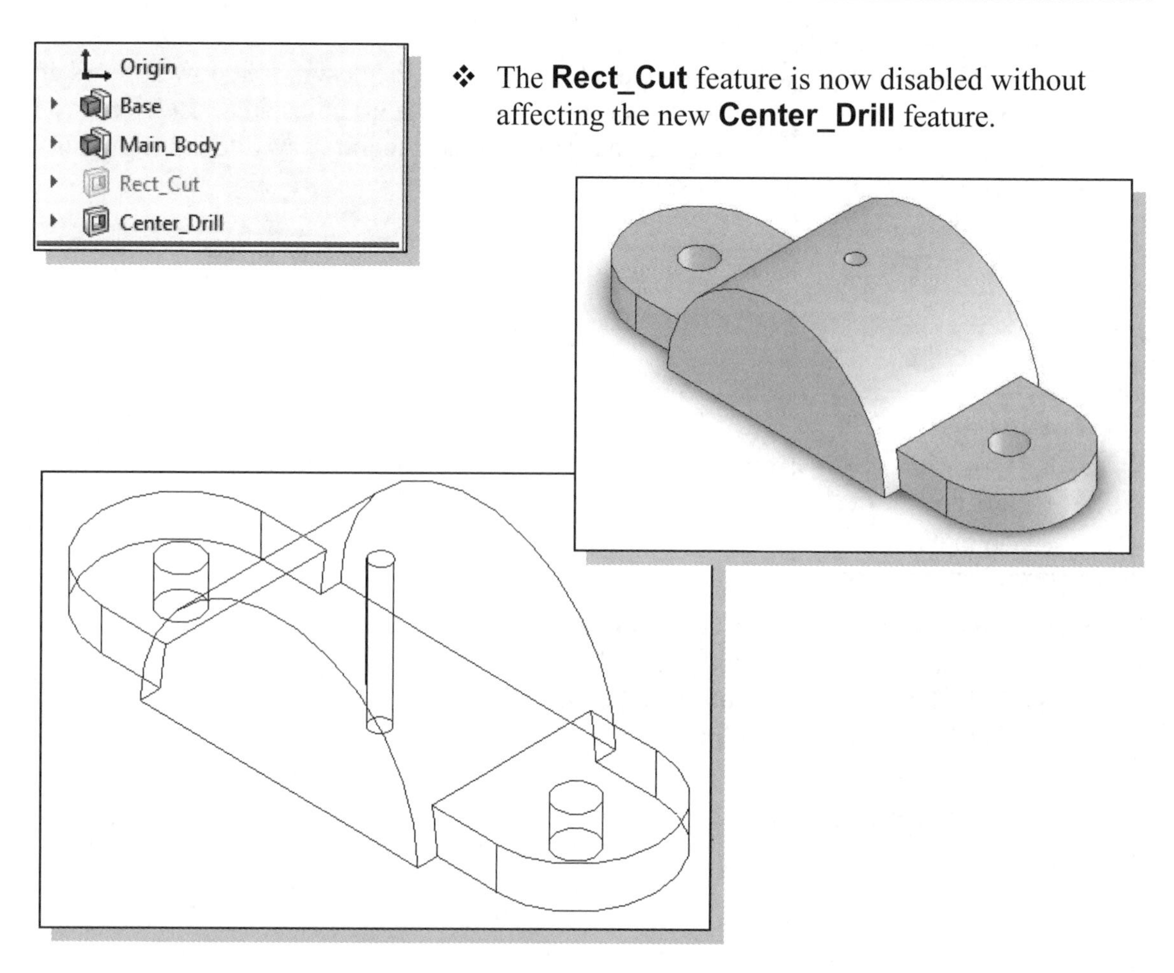

❖ The **Rect_Cut** feature is now disabled without affecting the new **Center_Drill** feature.

Creating a Circular Extruded Cut Feature

1. Left-click once in the graphics area, away from the model, to ensure no features are selected.

2. Select the **Sketch** tab on the *CommandManager* to display the *Sketch* toolbar and select the **Sketch** command.

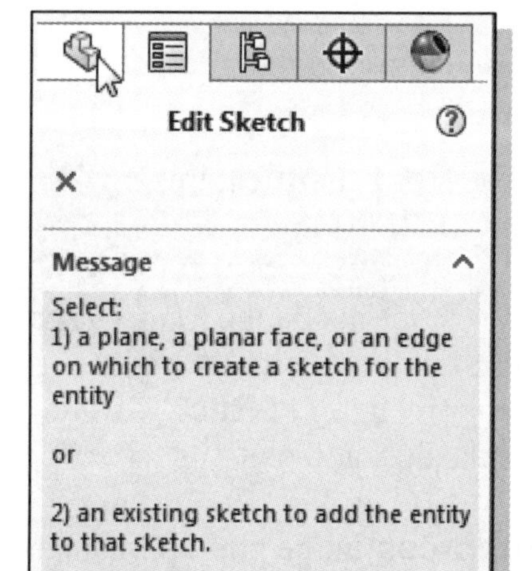

3. In the *Edit Sketch PropertyManager*, the message "*Select a plane on which to create a sketch for the entity*" is displayed. **Left-click** on the FeatureManager Design Tree tab to reveal the design tree as shown.

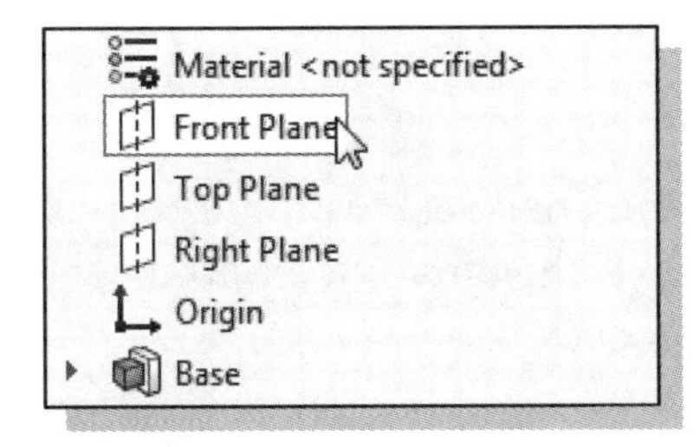

4. Move the cursor over the edge of the **Front Plane** icon in the *FeatureManager Design Tree*. When the Front Plane is highlighted, click once with the **left-mouse-button** to select the Front (XY) Plane as the sketch plane for the new sketch.

5. Select the **Circle** command by clicking once with the left-mouse-button on the icon in the *Sketch* toolbar.

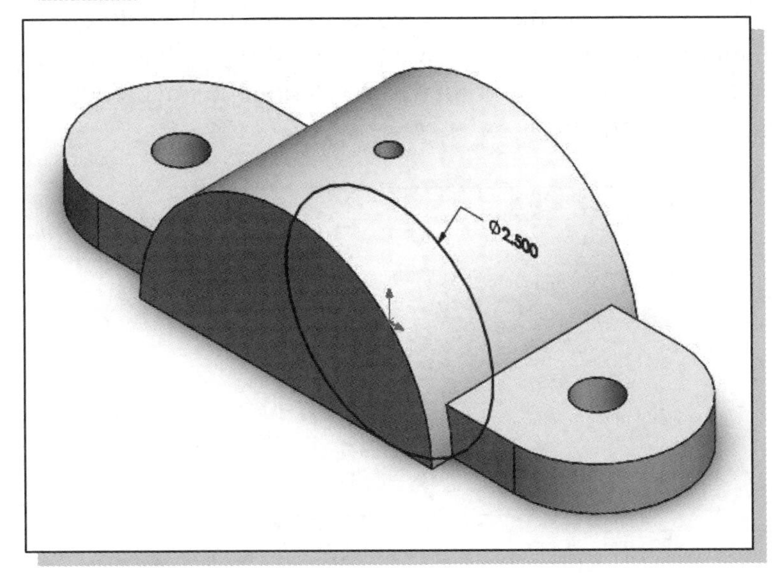

6. Pick the **Origin** to align the center of the new circle.

7. Create a circle of arbitrary size.

8. On your own, create the size dimension of the circle and set the dimension to **2.5**, as shown in the figure.

9. On your own, complete the cut feature using the **Extruded Cut** command. Use the **Mid Plane** *end condition* option and **2.5** in. (or greater) for the *depth*.

10. Rename the feature **Circular_Cut**.

❖ Note that the parents of the Circular_Cut feature are the Front (XY) Plane, the Origin, and the Base feature.

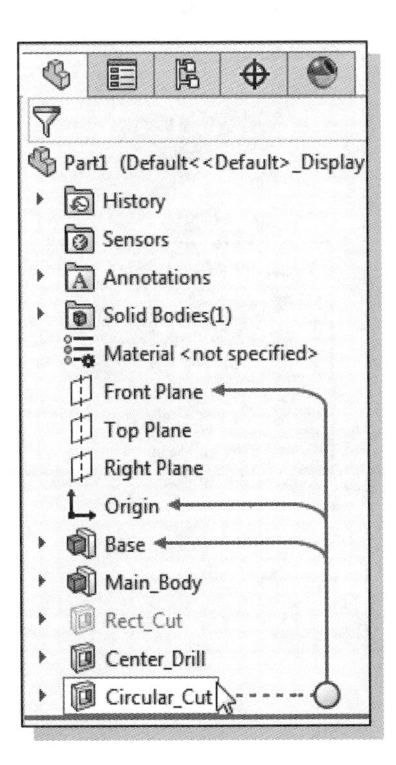

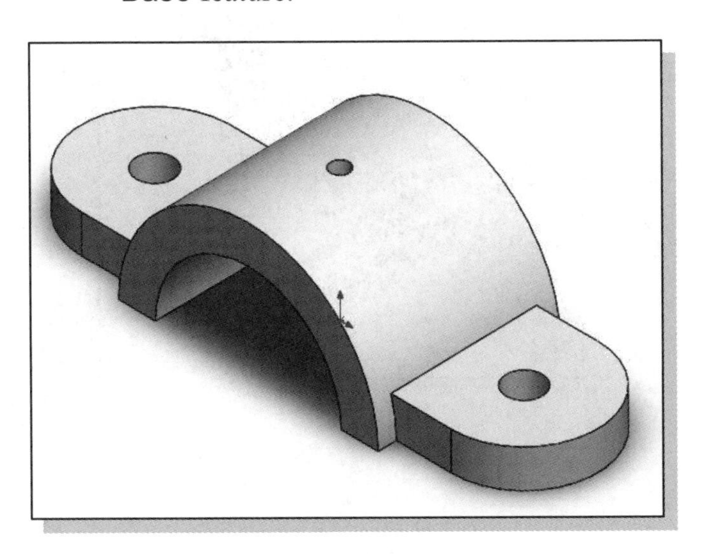

A Flexible Design Approach

In a typical design process, the initial design will undergo many analyses, testing, reviews and revisions. SOLIDWORKS allows the users to quickly make changes and explore different options of the initial design throughout the design process.

The model we constructed in this chapter contains two distinct design options. The *feature-based parametric modeling* approach enables us to quickly explore design alternatives and we can include different design ideas into the same model. With parametric modeling, designers can concentrate on improving the design and the design process to be much quicker and more effortless. The key to successfully using parametric modeling as a design tool lies in understanding and properly controlling the interactions of features, especially the parent/child relations.

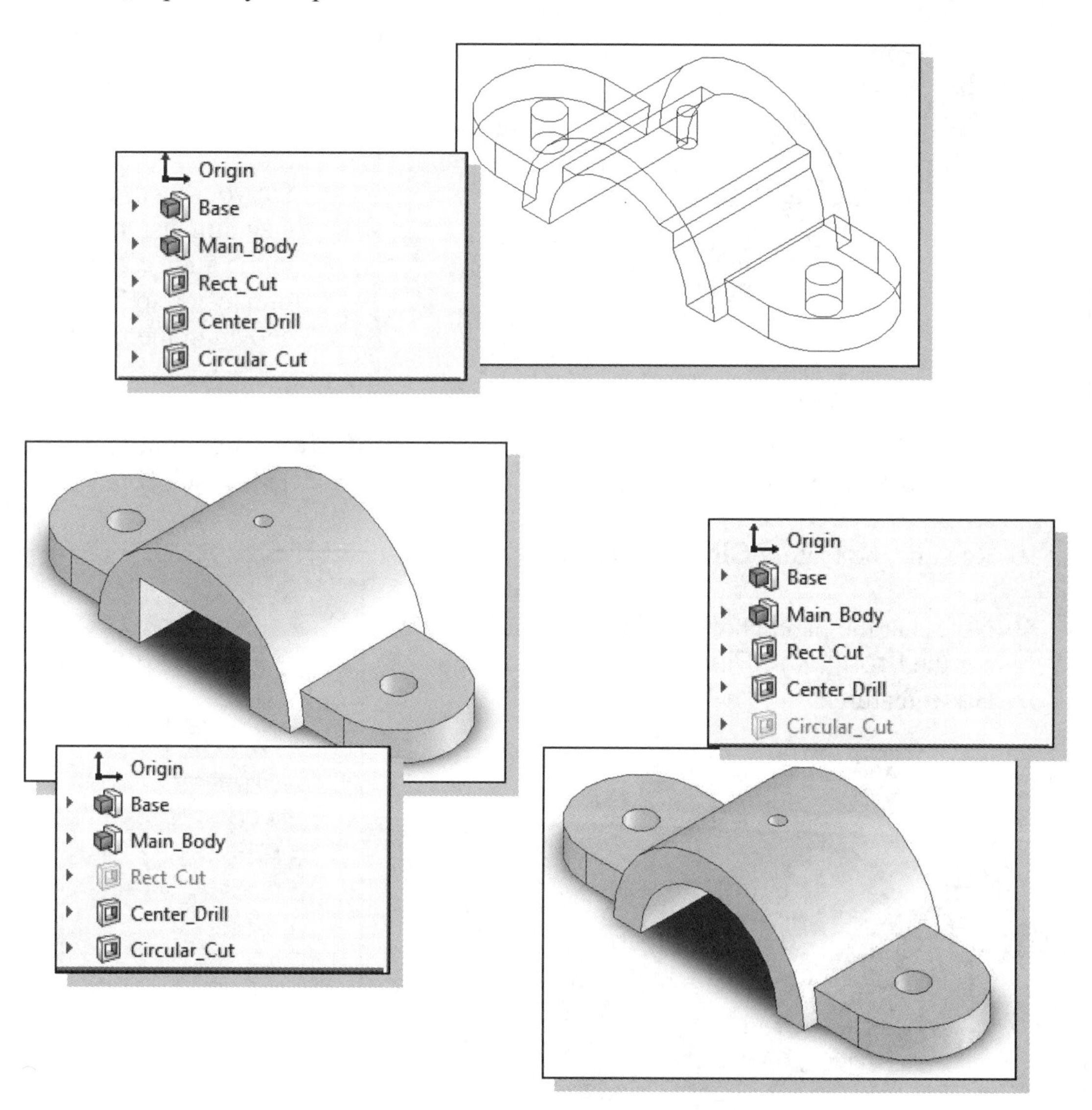

Save Part File

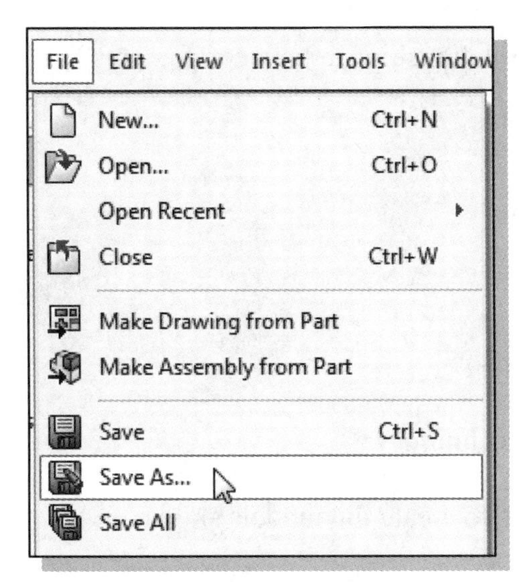

1. Select **Save as** in the *File* pull-down menu as shown.

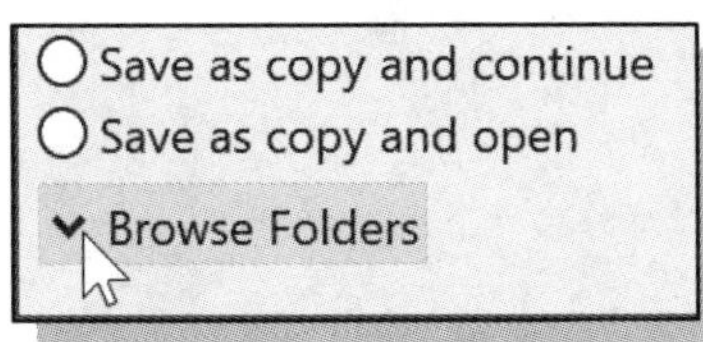

2. If necessary, click on the **Browse Folders** option.

3. Enter **U-Bracket** for the *File name.*

4. Enter **2-1/2" U-Bracket** for the *Description* as shown.

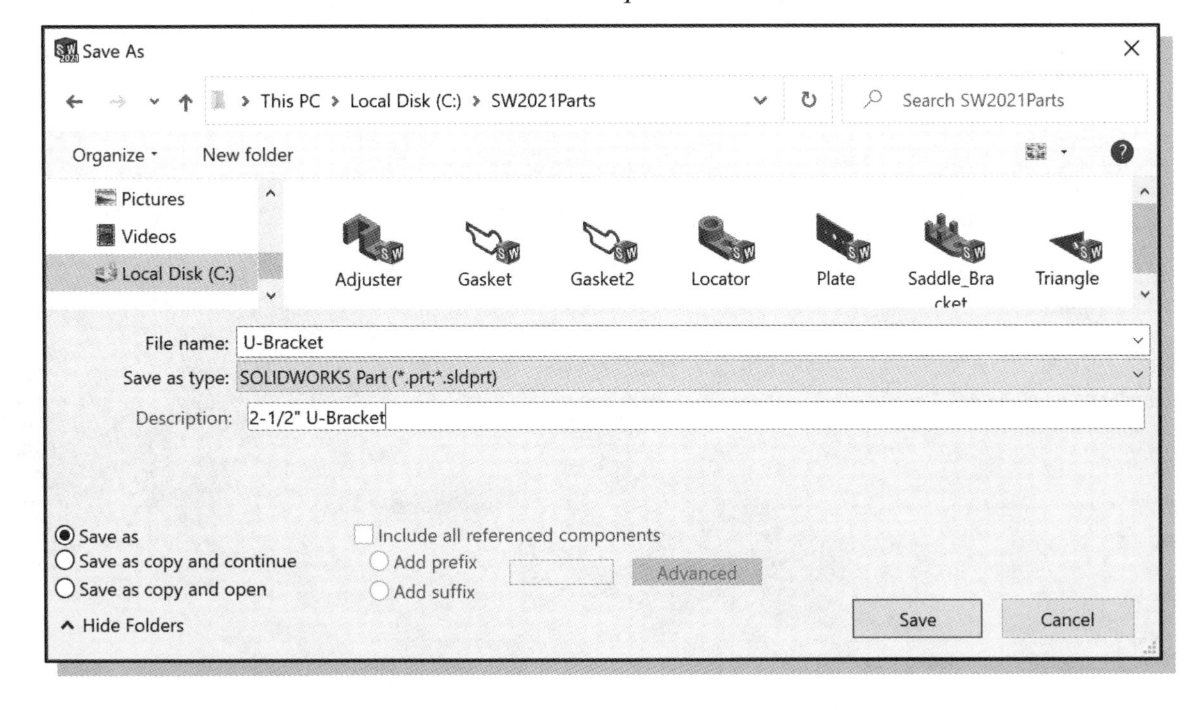

5. Click **Save**.

Questions:

1. Why is it important to consider the parent/child relationships in between features?
2. Describe the procedure to Suppress a feature.
3. What is the basic concept of the BORN technique?
4. What happens to a feature when it is suppressed?
5. How do you identify a suppressed feature in a model?
6. What is the main advantage of using the BORN technique?
7. Create sketches showing the steps you plan to use to create the models shown on the next page:

Exercises:

1. **Swivel Yoke** (Material: **Cast Iron**)

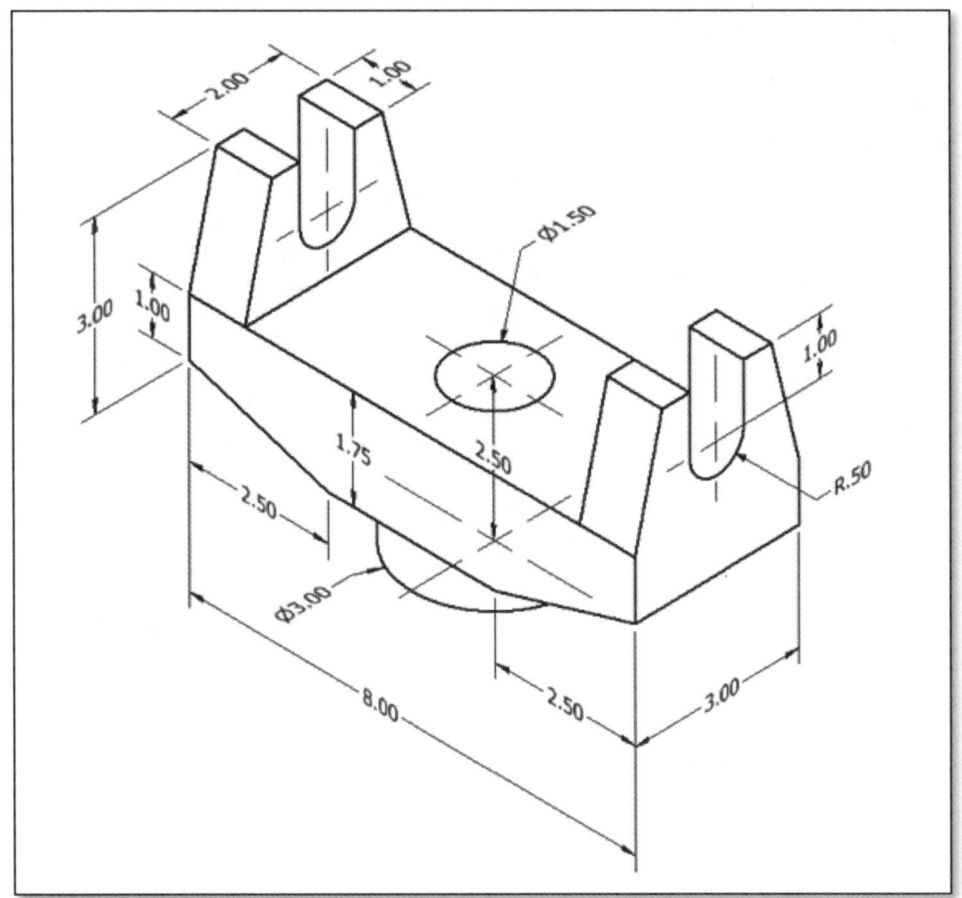

2. **Angle Bracket** (Material: **Carbon Steel**)

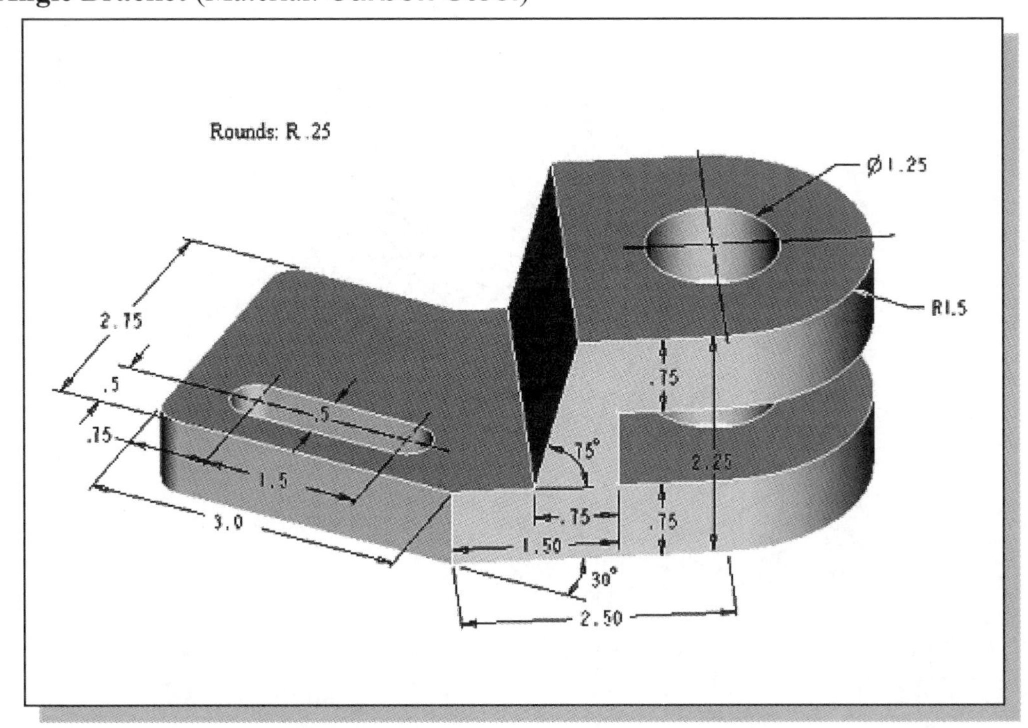

3. **Connecting Rod** (Material: **Carbon Steel**)

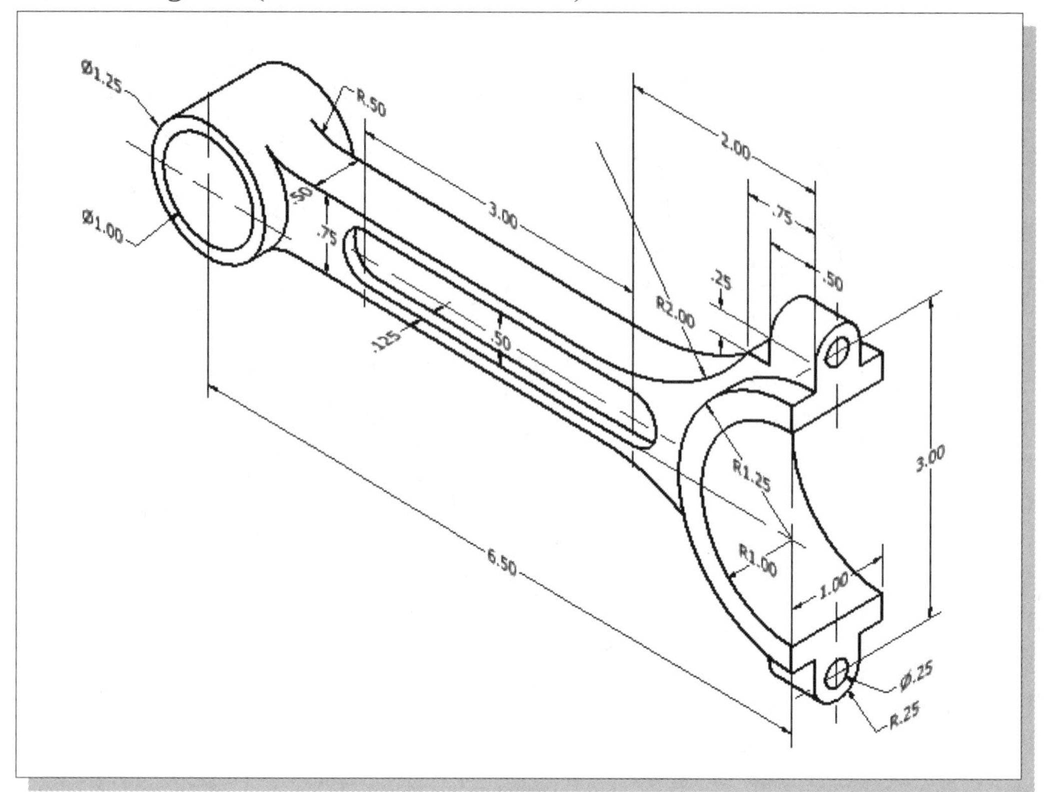

4. **Tube Hanger** (Material: **Aluminum 6061**)

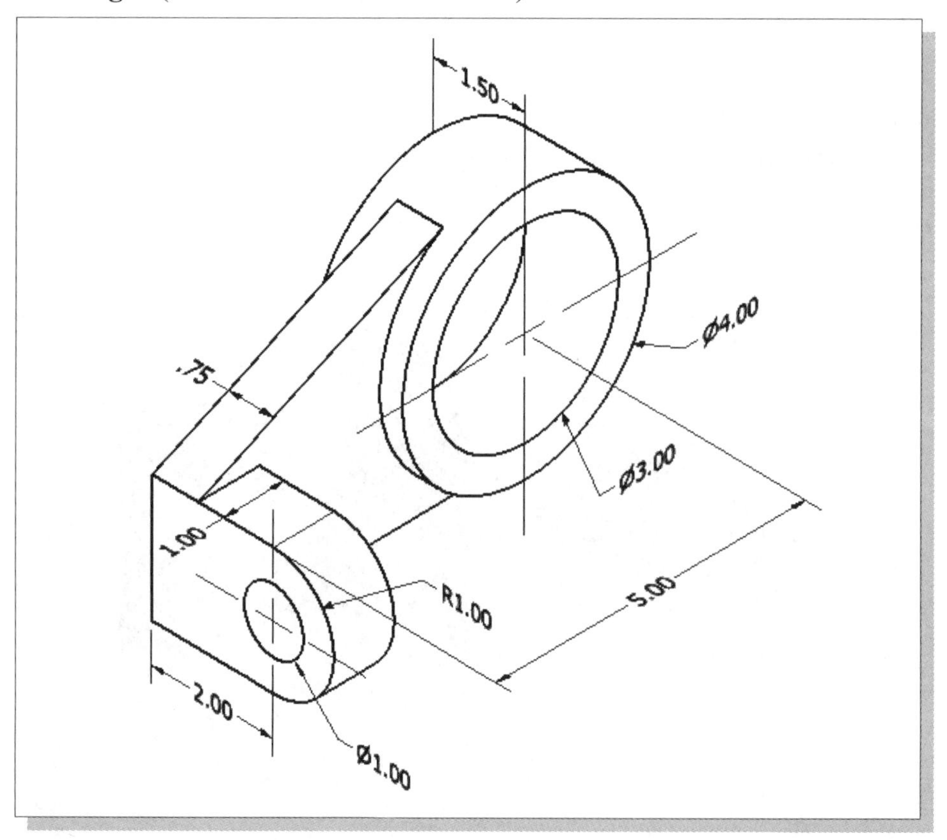

5. **Angle Latch** (Dimensions are in millimeters, Material: **Brass**)

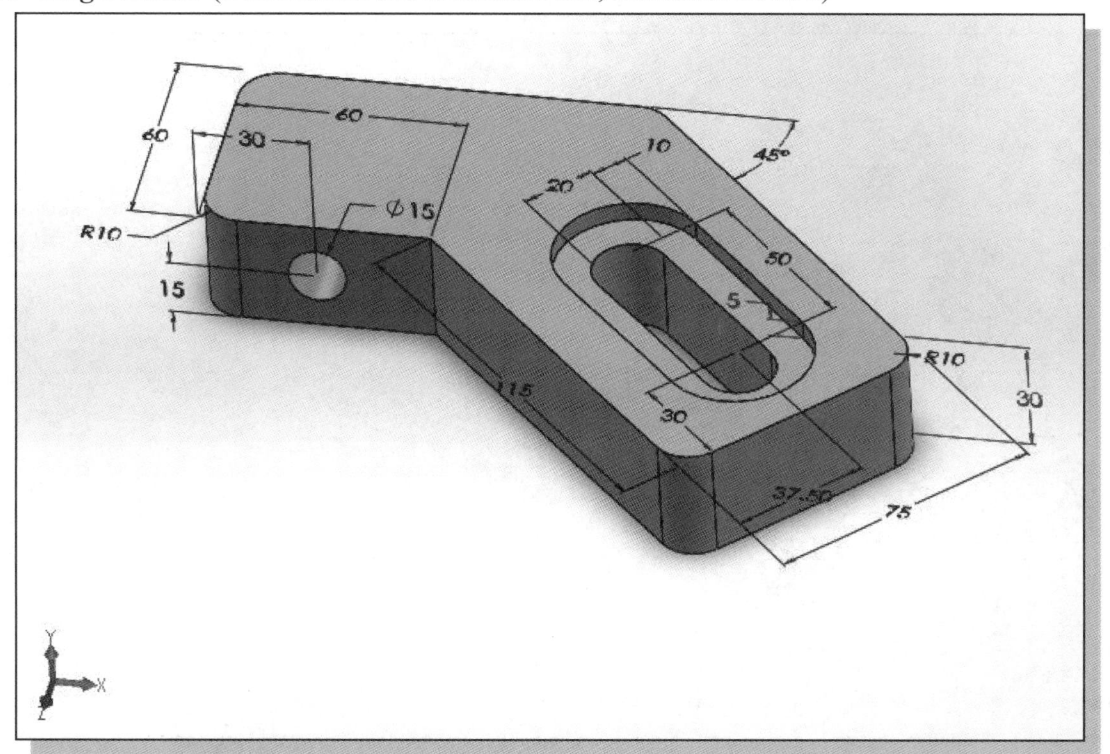

6. **Inclined Lift** (Material: **Mild Steel**)

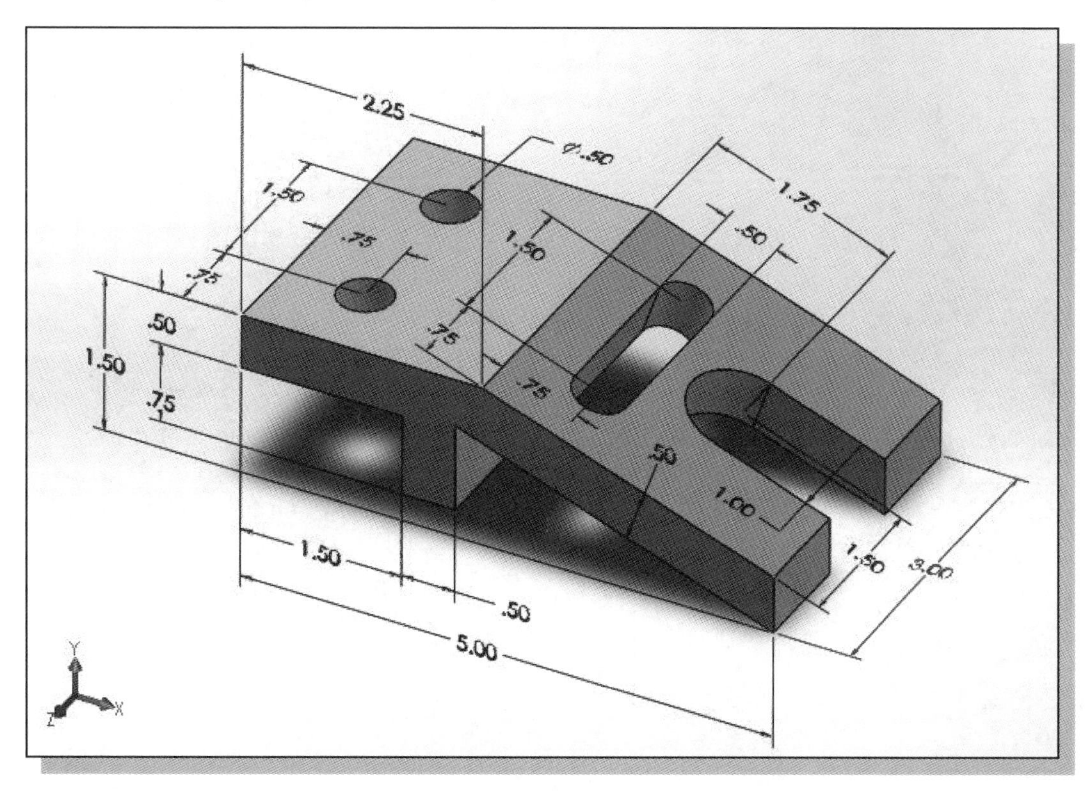

7. **Lock Ring** (Material: **Cast Iron**)

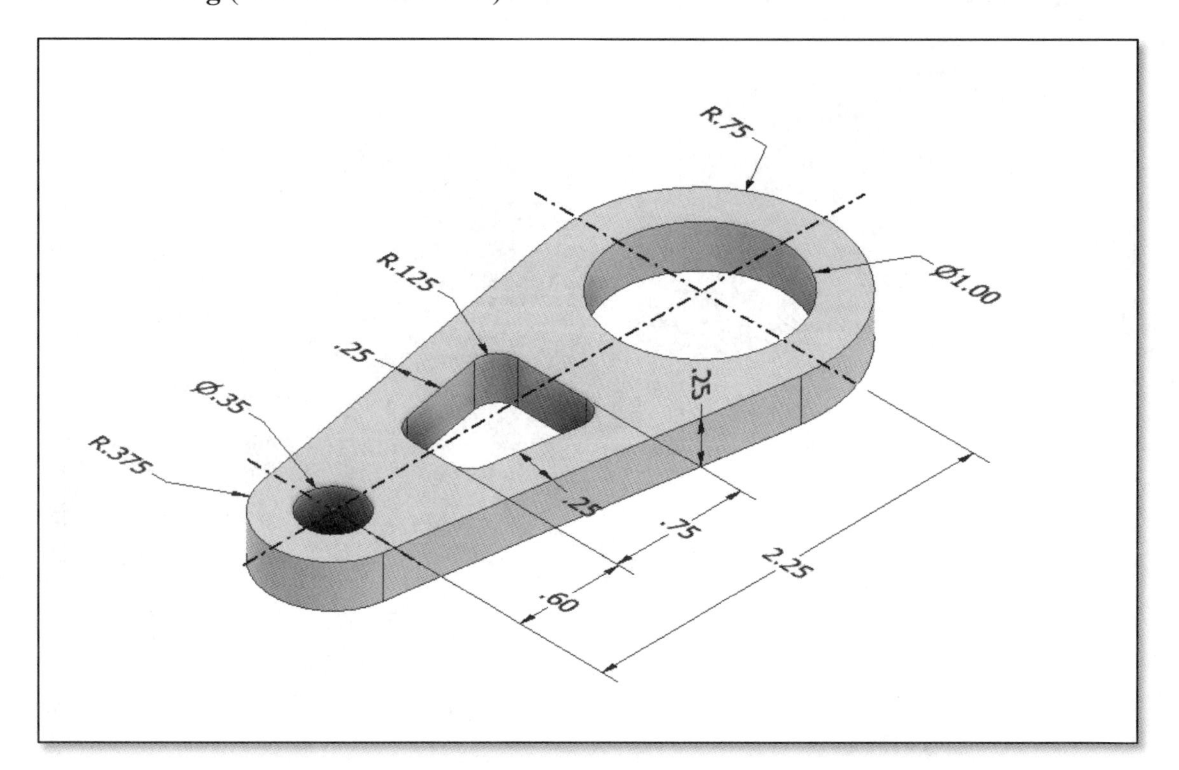

Chapter 8
Part Drawings and Associative Functionality

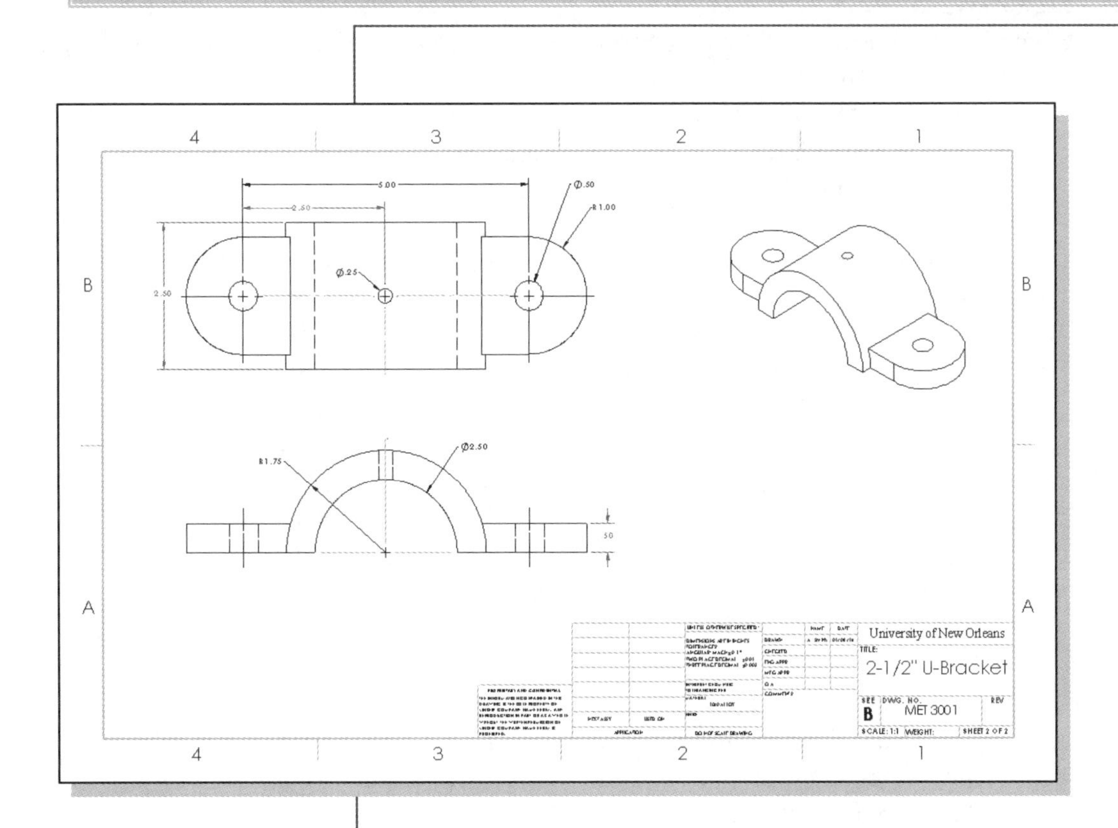

Learning Objectives

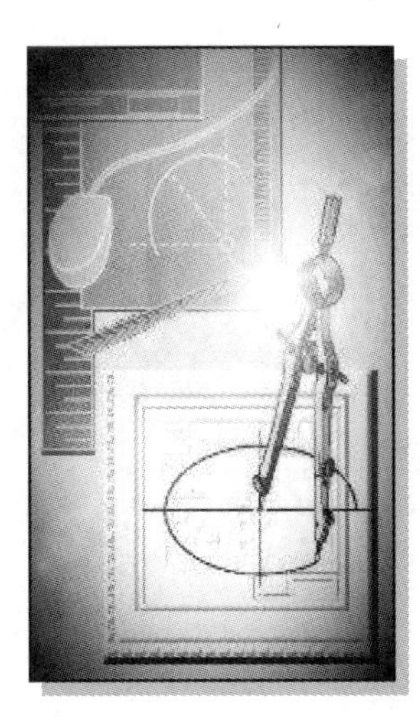

- **Create Drawing Layouts from Solid Models**
- **Understand Associative Functionality**
- **Use the default Sheet Formats**
- **Arrange and Manage 2D Views in Drawing Mode**
- **Display and Hide Feature Dimensions**
- **Create Reference Dimensions**
- **Understand *Edit Sheet* and *Edit Sheet Format* Modes**
- **Use Property Links**

Certified SOLIDWORKS Associate Exam Objectives Coverage

Drawing Sheets and Views

Objectives: Creating and Setting Properties for Drawing Sheets; Inserting and Editing Standard Views.

Annotations

Objectives: Creating Annotations.

Drawings from Parts and Associative Functionality

With the software/hardware improvements in solid modeling, the importance of two-dimensional drawings is decreasing. Drafting is considered one of the downstream applications of using solid models. In many production facilities, solid models are used to generate machine tool paths for *computer numerical control* (CNC) machines. Solid models are also used in *rapid prototyping* to create 3D physical models out of plastic resins, powdered metal, etc. Ideally, the solid model database should be used directly to generate the final product. However, the majority of applications in most production facilities still require the use of two-dimensional drawings. Using the solid model as the starting point for a design, solid modeling tools can easily create all the necessary two-dimensional views. In this sense, solid modeling tools are making the process of creating two-dimensional drawings more efficient and effective.

SOLIDWORKS provides associative functionality in the different SOLIDWORKS modes. This functionality allows us to change the design at any level, and the system reflects it at all levels automatically. For example, a solid model can be modified in the *Part Modeling* mode and the system automatically reflects that change in the *Drawing* mode. And we can also modify a feature dimension in the *Drawing* mode, and the system automatically updates the solid model in all modes.

In this lesson, the general procedure of creating multi-view drawings is illustrated. The *U_Bracket* design from the last chapter is used to demonstrate the associative functionality between the model and drawing views.

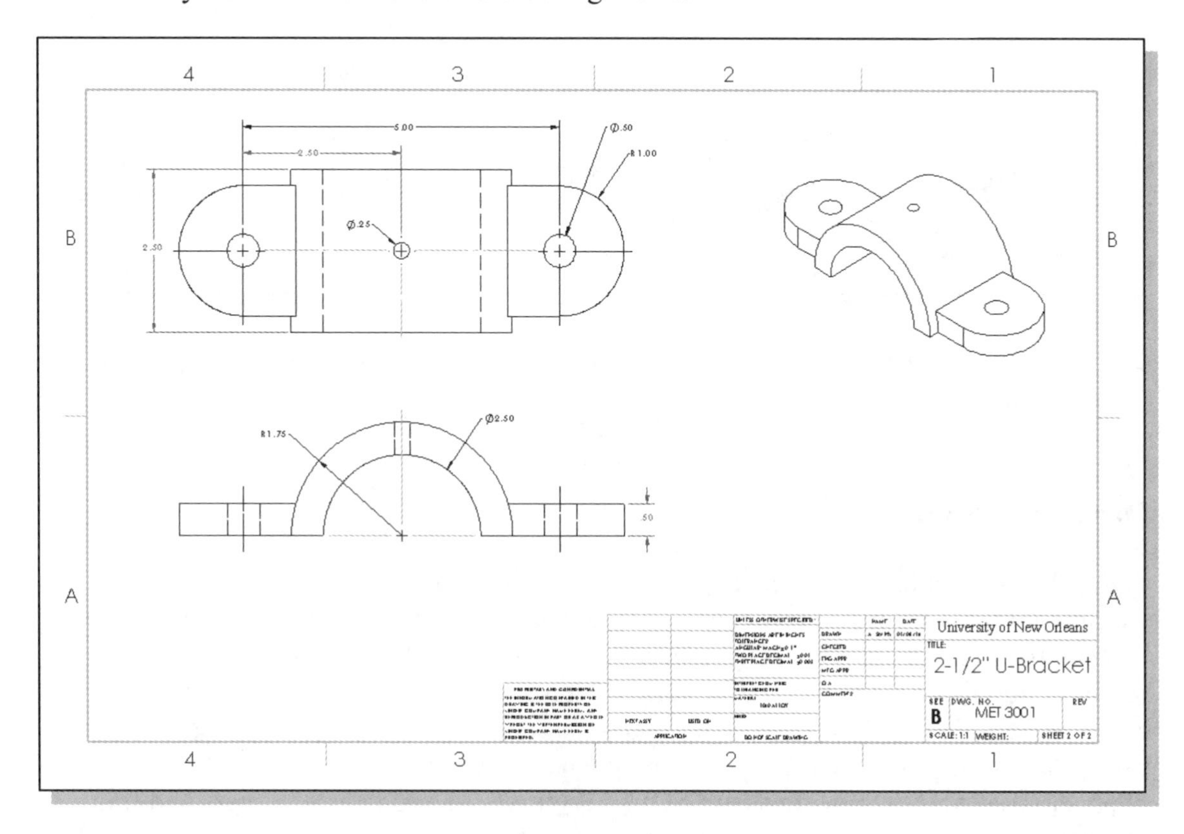

Starting SOLIDWORKS

1. Select the **SOLIDWORKS** option on the *Start* menu or select the **SOLIDWORKS** icon on the desktop to start SOLIDWORKS. The SOLIDWORKS main window will appear on the screen.

➢ We will use the Welcome dialog box to open a recently used file. An alternative is to use the Open command on the *Menu Bar*.

2. If the *Welcome* dialog box does not appear automatically upon opening SOLIDWORKS, click the *Welcome to SolidWorks* icon on the *Menu Bar*.

3. In the *Recent Documents* panel of the *Welcome* dialog box, select the **U-Bracket.SLDPRT** file with a single click of the left-mouse-button to open the part document.

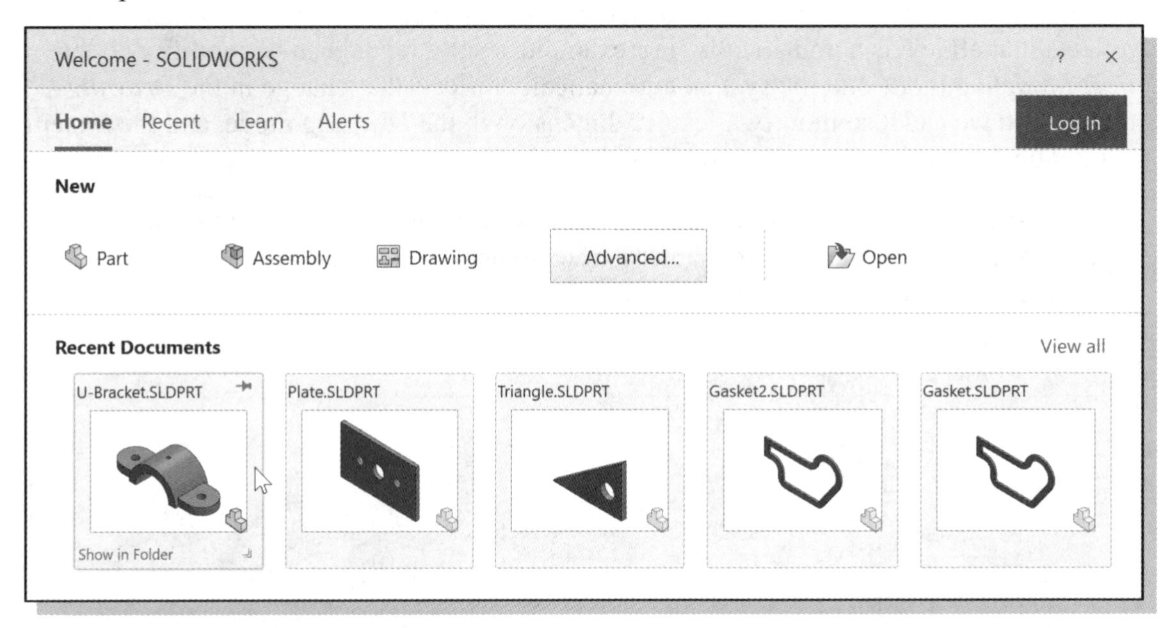

➢ Note: If the U-Bracket.SLDPRT file does not appear, click on the **Open** icon in the *Welcome* dialog box and use the browser to locate and open the file.

Drawing Mode

SOLIDWORKS allows us to generate 2D engineering drawings from solid models so that we can plot the drawings to any exact scale on paper. An engineering drawing is a tool that can be used to communicate engineering ideas/designs to manufacturing, purchasing, service, and other departments. Until now we have been working in ***Model*** mode to create our design in ***full size***. We can arrange our design on a two-dimensional sheet of paper so that the plotted hardcopy is exactly what we want. This two-dimensional sheet of paper is saved in a ***drawing*** file in SOLIDWORKS. We can place borders and title blocks, objects that are less critical to our design, in the *drawing*. In general, each company uses a set of standards for drawing content, based on the type of product and also on established internal processes. The appearance of an engineering drawing varies depending on when, where, and for what purpose it is produced. However, the general procedure for creating an engineering drawing from a solid model is fairly well defined. In SOLIDWORKS, creation of 2D engineering drawings from solid models consists of four basic steps: drawing sheet formatting, creating/positioning views, annotations, and printing/plotting.

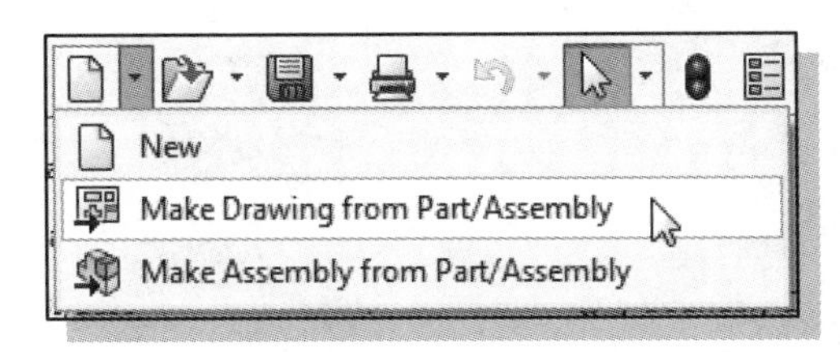

1. Click on the arrow next to the **New** icon on the *Menu Bar* and select **Make Drawing from Part/Assembly** in the pull-down menu.

➢ We must now select a drawing template. A drawing template is the foundation for drawing information in SOLIDWORKS. The template contains specifications including sheet size and orientation, and the SOLIDWORKS sheet format. The sheet format includes borders, title blocks, bill of materials table format, etc. The user can select from the SOLIDWORKS 'Standard' drawing template, the SOLIDWORKS 'Tutorial' drawing template, user-defined custom templates, or use no template. We will use the SOLIDWORKS 'Tutorial' drawing template.

2. In the *New* SOLIDWORKS *Document* window, select the **Tutorial** tab.

3. Select the **draw** template file icon.

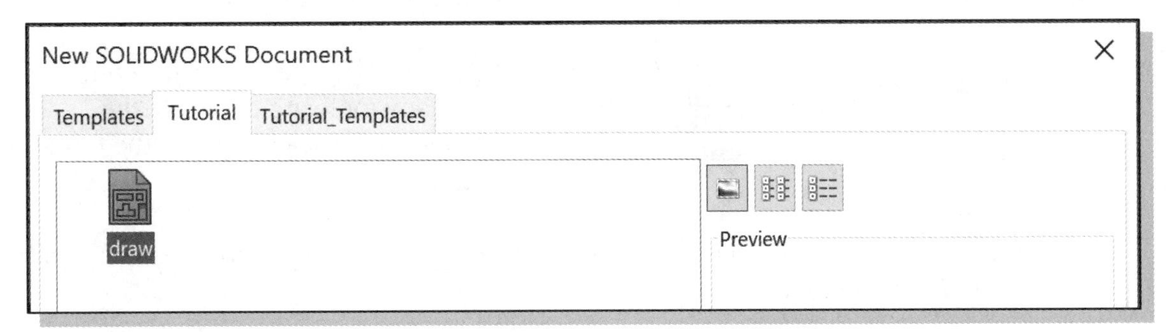

4. Select **OK** to open a drawing using the *draw* template.

5. If the *Sheet Format/Size* window appears, select any size and click **OK**. (We will reset the paper size and formatting in subsequent steps.)

➢ Note that a new graphics window appears on the screen. We can switch between the solid model and the drawing by clicking the corresponding file from the *Window* pull-down menu.

➢ The View Palette appears in the task pane. This can be used to insert drawing views. We will first set the document properties and sheet properties for the drawing.

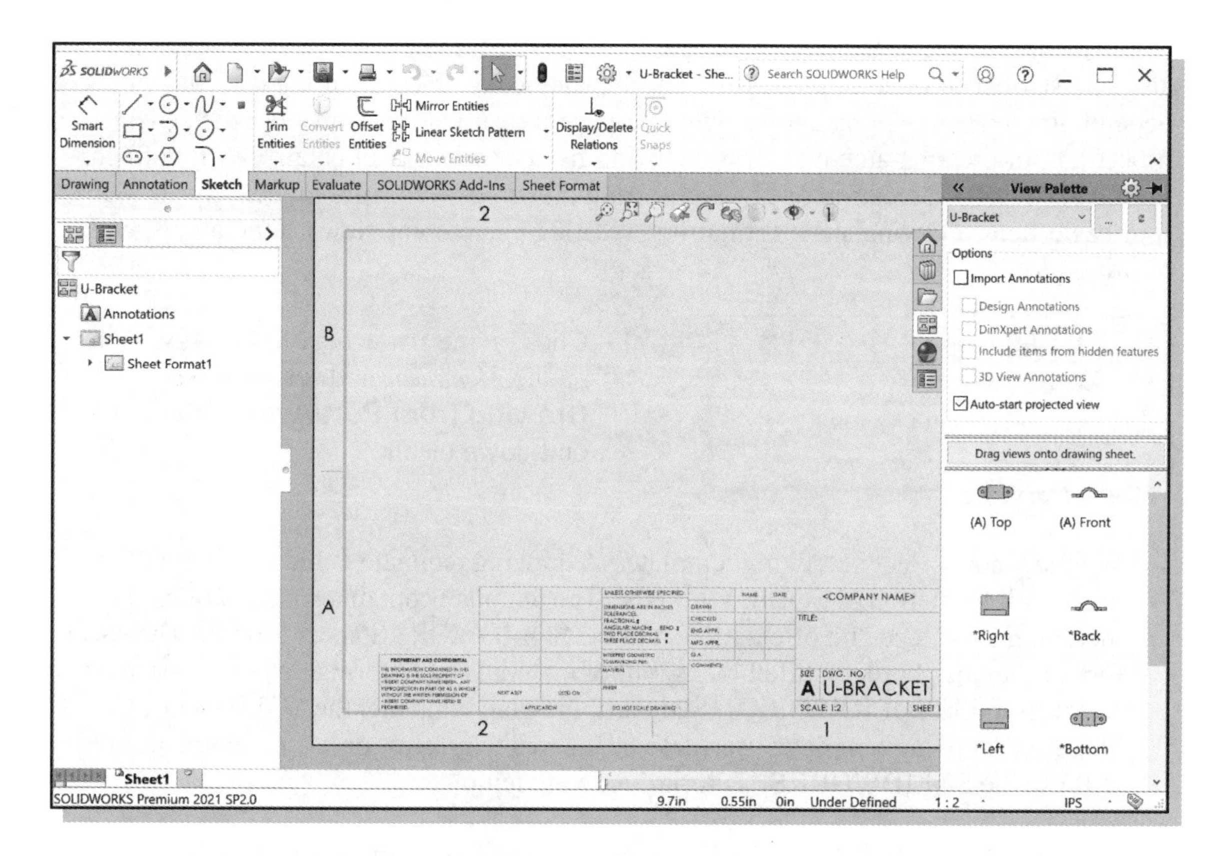

IMPORTANT NOTE: The SOLIDWORKS *CommandManager* provides an alternate method for displaying the most commonly used toolbars (see page 1-13 for a more complete description). We will use the *CommandManager* in this lesson.

6. If necessary, turn ***ON*** the *CommandManager* (right click on any toolbar and enable the *CommandManager* by selecting it at the top of the pop-up menu).

- Notice that new tabs appear on the *CommandManager* to select toolbars commonly used in ***drawing*** mode.

- In the graphics window, SOLIDWORKS displays a default drawing sheet that includes a title block. The title block also indicates the paper size being used.

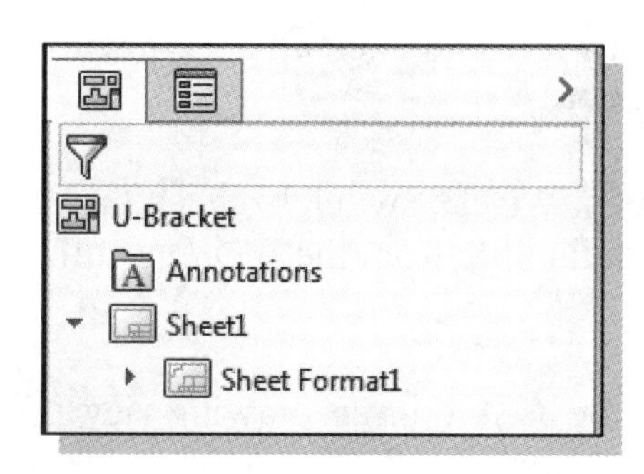

- In the *FeatureManager Design Tree* area, the Drawing icon is displayed at the top, which indicates that we have switched to ***Drawing*** mode. The default filename is the filename of the solid part file from which we created the drawing – U-Bracket. ***Sheet1*** is the current drawing sheet that is displayed in the graphics window.

Setting Document Properties

The document properties, such as dimensioning standard, units, text fonts, etc., must be set for the new drawing file. If a set of document properties will be used regularly, a new drawing file template can be saved which contains these settings.

1. Select the **Options** icon from the *Menu Bar* to open the options dialog box.

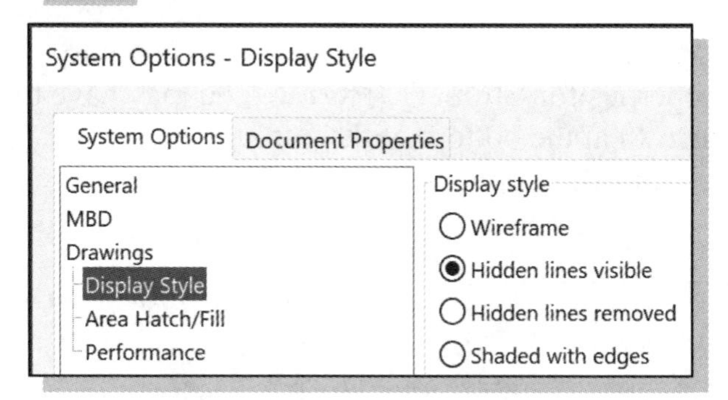

2. Under the **System Options** tab, select **Display Style** as shown.

3. Under '*Display style*', select **Hidden lines visible**. This will be the default display style for orthographic views (e.g., front view) added to the drawing.

4. Select the **Document Properties** tab.

5. Select **ANSI** in the drop-down selection window under the *Dimensioning standard* panel as shown.

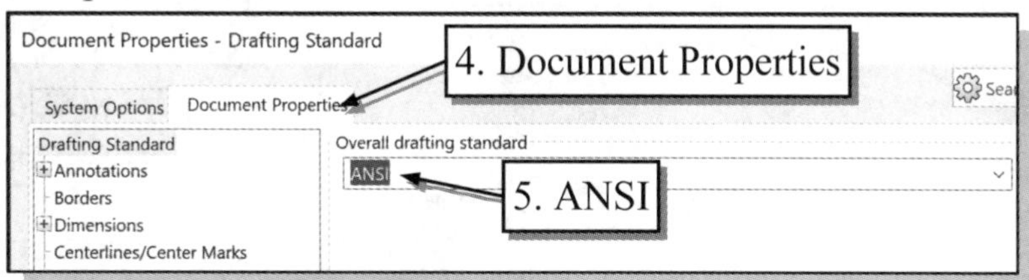

6. Click **Units** as shown below.

7. Select **IPS (inch, pound, second)** under the *Unit system* options.

8. Select **.12** in the *Decimals* spin boxes for the *Length* and *Dual Dimension Length* units to define the degree of accuracy with which the units will be displayed.

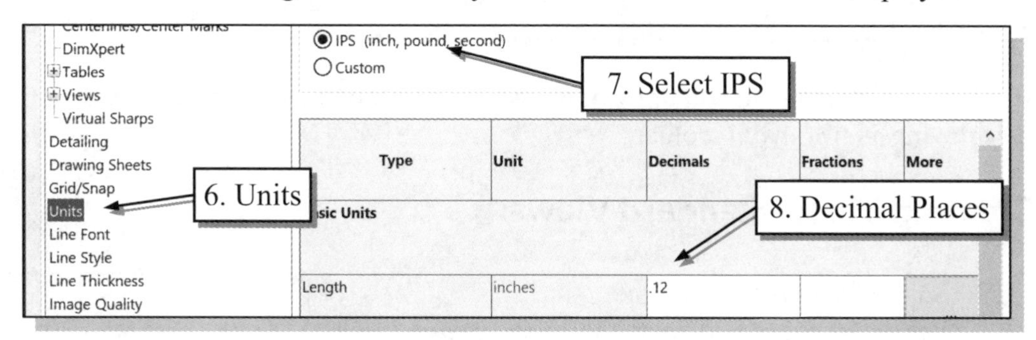

9. Click **OK** in the options dialog box to accept the selected settings.

Setting Sheet Properties Using the Pre-Defined Sheet Formats

Another set of properties are set for each sheet in the drawing file. These settings include sheet size, sheet orientation, scale, and projection type.

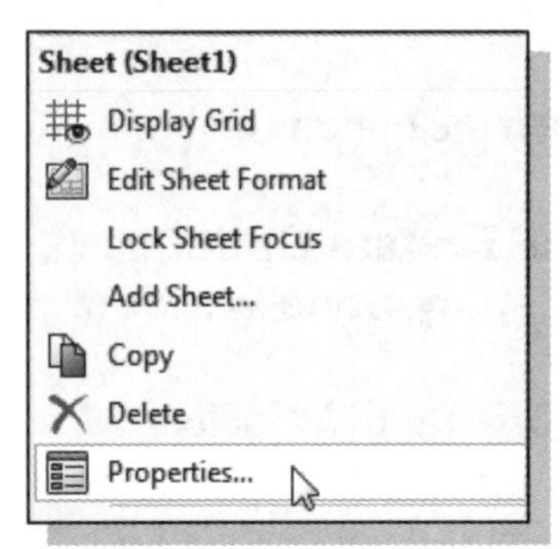

1. Move the cursor onto the graphics area and click once with the **right-mouse-button** to open the pop-up menu.

2. Select **Properties** from the pop-up menu as shown to edit the properties for Sheet1. (NOTE: You may have to click the arrows at the bottom of the pop-up menu to reveal the Properties option.)

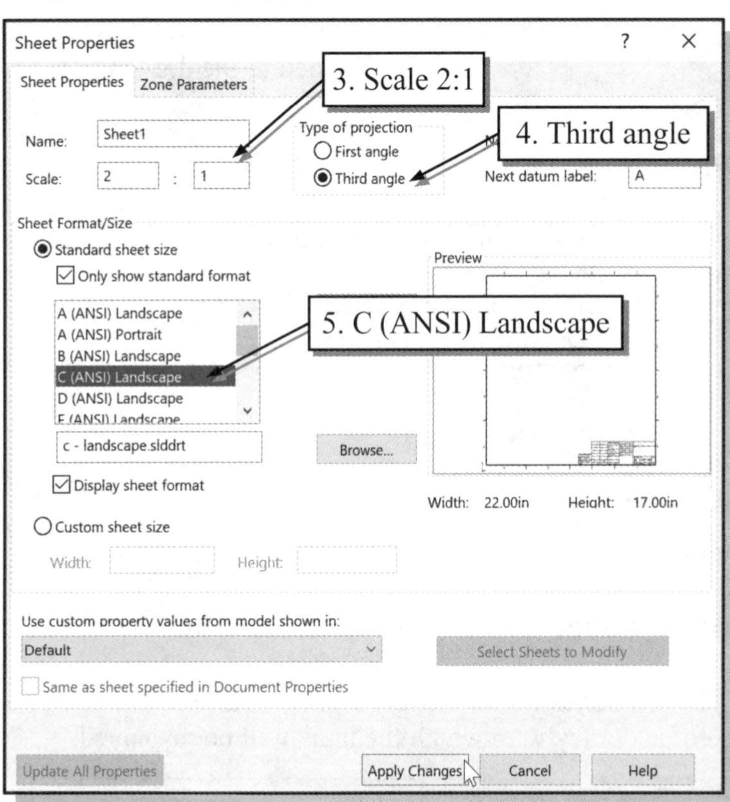

3. In the *Sheet Properties* window, set the scale to **2:1** as shown.

4. Set the *Type of projection* to **Third angle**.

5. Set the *Standard sheet size* to **C (ANSI) Landscape**.

6. Click **Apply Changes** in the *Sheet Properties* window to accept the selected settings.

7. Press the **[F]** key to scale the new sheet to fit the display area.

➤ We are now ready to add drawing views. We will do this using commands on the drawing *View Layout* toolbar.

Creating Three Standard Views

1. Select the **Drawing** tab on the *CommandManager* to display the *View Layout* toolbar.

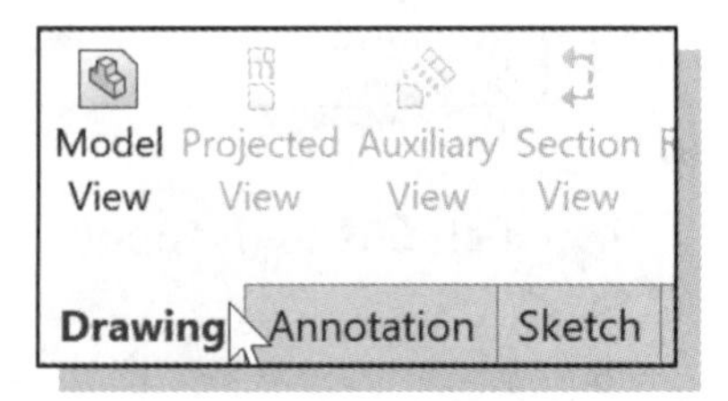

2. Select the **Standard 3 View** button on the *Drawing* toolbar.

➤ The *Standard 3 View PropertyManager* appears. In the *Open documents* field of the *Part/Assembly to Insert* option window, the **U-Bracket** part file appears and is highlighted. The *U-Bracket* model is the only model opened. By default, all of the 2D drawings will be generated from this model file.

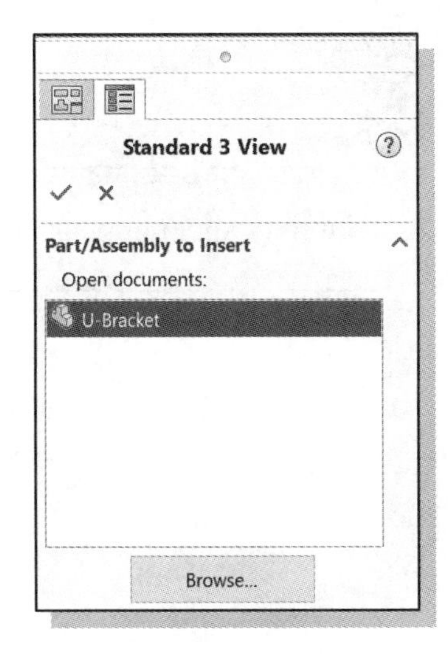

3. Left-click on the **OK** icon to accept the *U-Bracket* part file as the document used to create the three views. The three standard views for the *U-Bracket* appear on the active sheet.

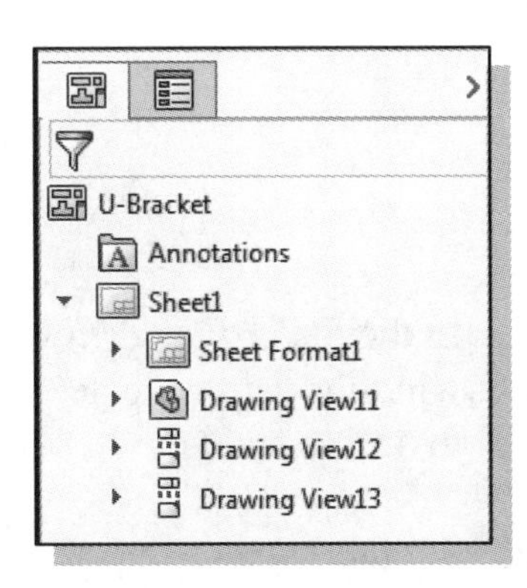

➤ The three new views now appear on *Sheet1* and in the *FeatureManager Design Tree.* Notice the icon for Drawing View11. This icon indicates Drawing View11 is a base view. The icons for Drawing View12 and Drawing View12 indicate that these views are projected from the base view.

Repositioning Views

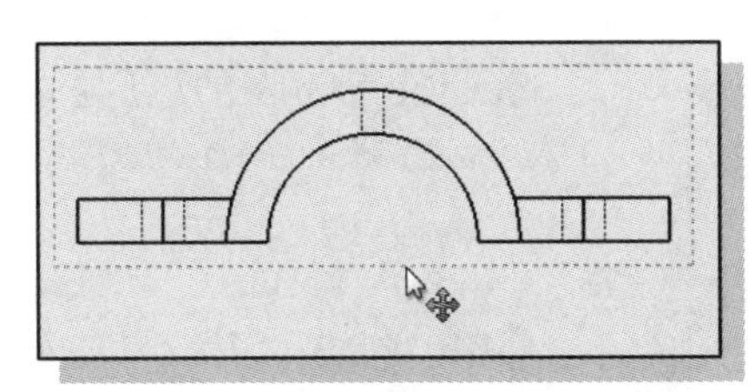

1. Move the cursor on top of the front view and watch for the **four-arrow Move symbol** as the cursor is near the border indicating the view can be dragged to a new location as shown in the figure.

2. Press and hold down the left-mouse-button and reposition the view to a new location.

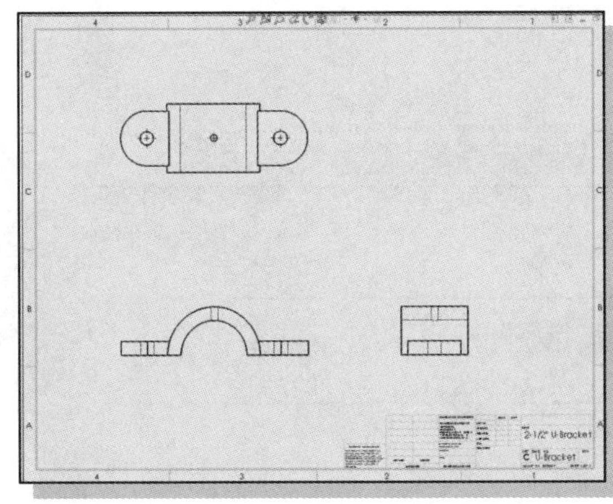

3. On your own, reposition the views we have created so far. Note that the top view can be repositioned only in the vertical direction. The top view remains aligned to the base view, the front view. Similarly, the right side view can be repositioned only in the horizontal direction.

Adding a New Sheet

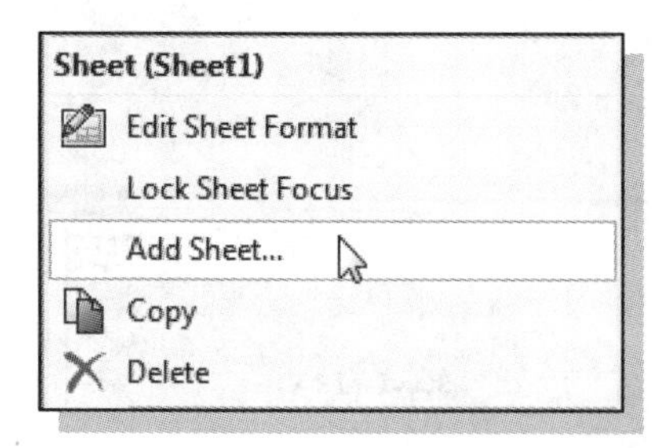

1. Move the cursor into the graphics area and click once with the **right-mouse-button** to reveal the pop-up menu.

2. Select **Add Sheet** from the pop-up menu as shown.

➢ The new sheet appears in the graphics area as the active sheet.

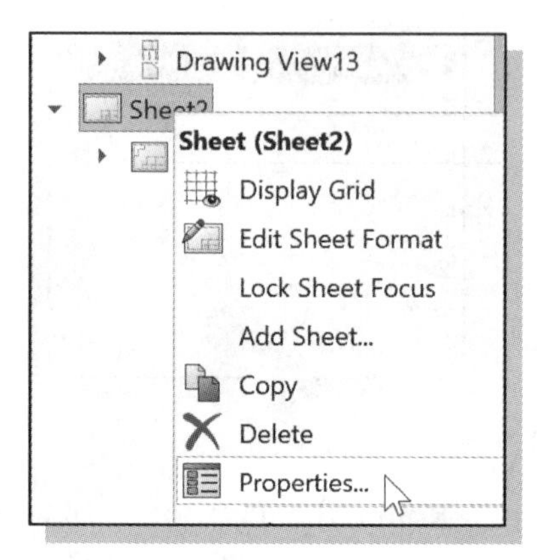

3. Move the cursor over **Sheet2** on the *FeatureManager Design Tree* click once with the **right-mouse-button** to open the pop-up menu.

4. Select **Properties** from the pop-up menu as shown to edit the properties for Sheet2.

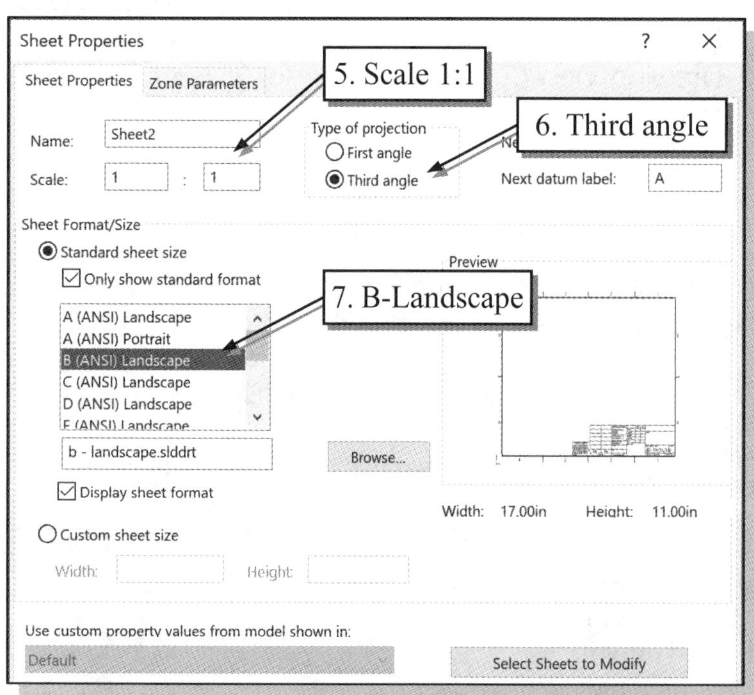

5. In the *Sheet Properties* window, set the scale to **1:1** as shown.

6. Set the *Type of projection* to **Third angle**.

7. Set the *Standard sheet size* to **B-Landscape**.

8. Click **Apply Changes** in the *Sheet Properties* window to accept the selected settings.

9. Press the **[F]** key to scale the new sheet to fit the display area.

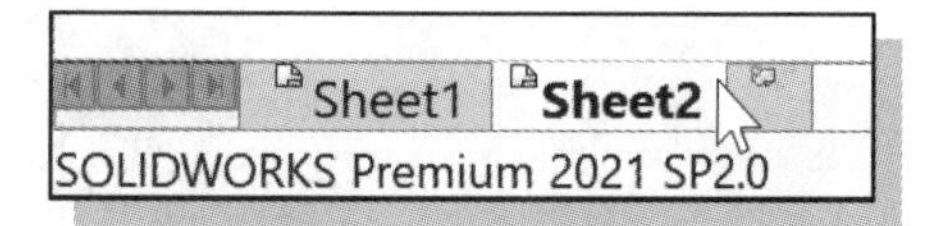

➢ The active sheet can be changed by left-clicking on the sheet tabs at the bottom of the graphics area.

Adding a Base View

❖ In SOLIDWORKS *Drawing* mode, the first drawing view we create is called a **base view**. A *base view* is the primary view in the drawing; other views can be derived from this view. When creating a *base view*, SOLIDWORKS allows us to specify the view to be shown. By default, SOLIDWORKS will treat the *world XY plane* as the front view of the solid model. Note that there can be more than one *base view* in a drawing.

1. Make sure Sheet2 is the active drawing sheet.

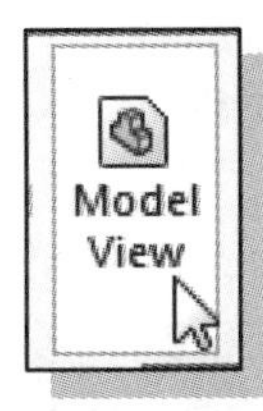

2. Click on the **Model View** icon on the *View Layout* toolbar.

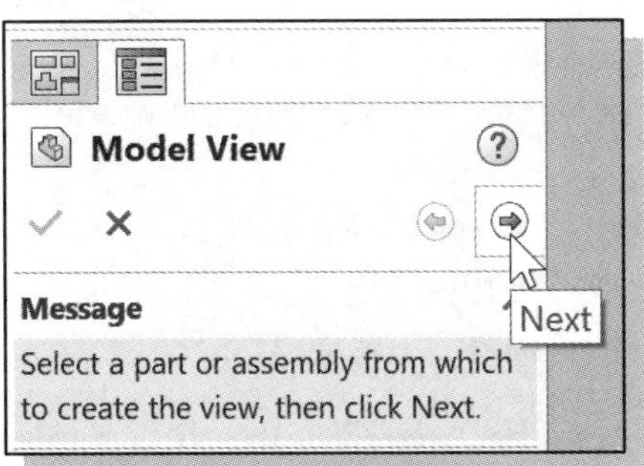

3. The *Model View PropertyManager* appears with the U-Bracket part file selected as the part from which to create the base view. Click the **Next** arrow as shown to proceed with defining the base view.

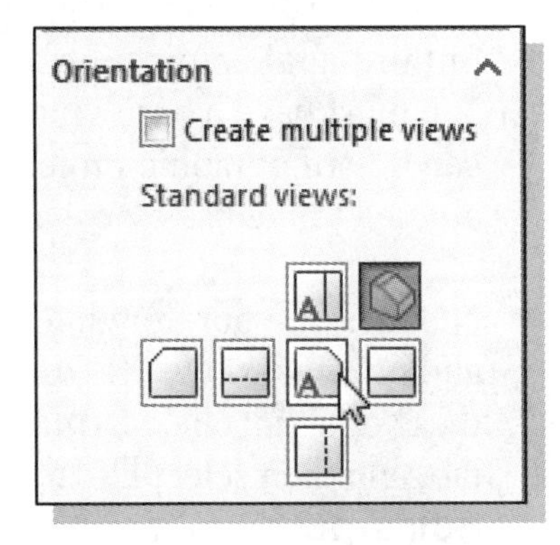

4. In the *Orientation* panel of the *Model View PropertyManager*, select the **Front View** for the *Orientation*, as shown. (Make sure the *Create multiple views* box is **unchecked**.)

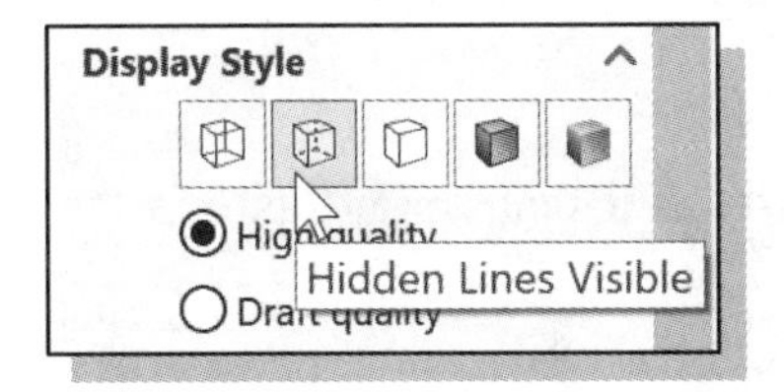

5. Scroll down in the *PropertyManager* to the *Display Style* panel. Notice the default display style is *Hidden Lines Visible.* This was defined in the *Document Settings*.

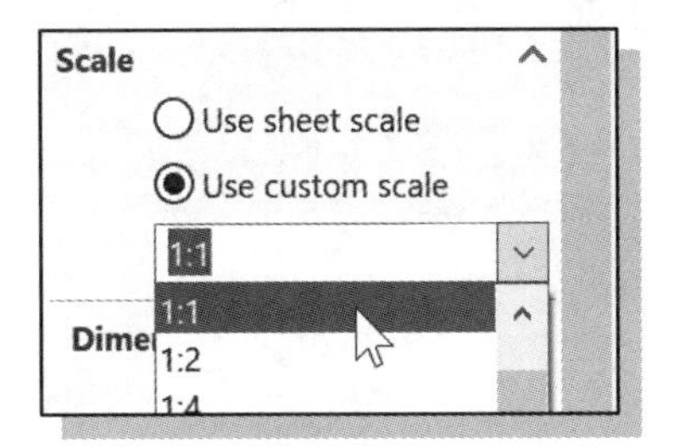

6. Under the *Scale* option, select **Use custom scale**.

7. Select **1:1** from the pull-down window, as shown.

8. Move the cursor inside the graphics area and left-click to place the ***base*** view near the lower left corner of the drawing sheet as shown below.

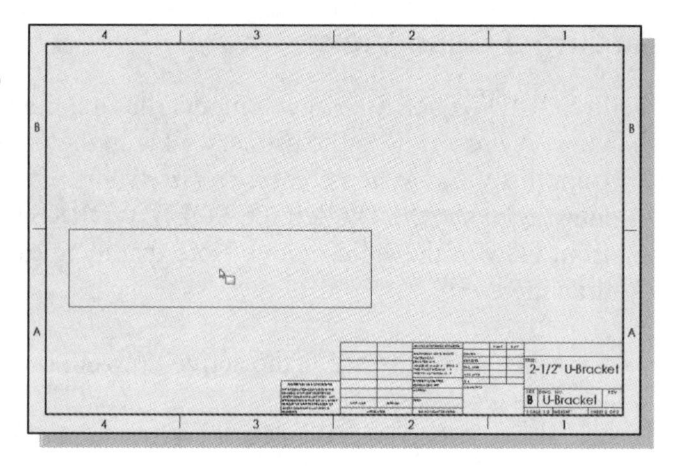

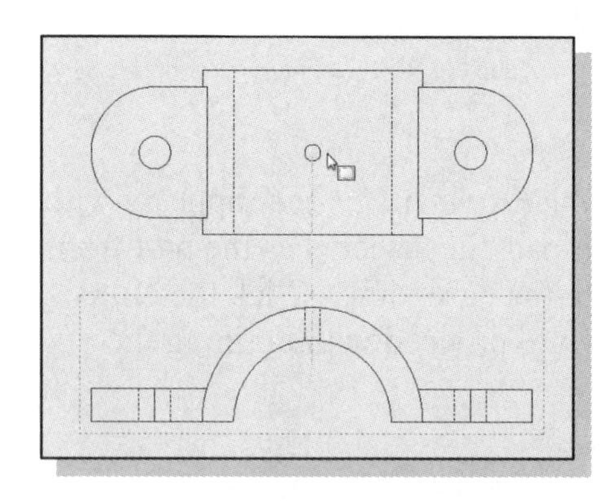

9. Upon placement of the base view, SOLIDWORKS automatically enters the **Projected View** command. Move the cursor above the base view and select a location to position the projected top view of the model.

10. Click **OK** in the *PropertyManager* to exit the Projected View command.

❖ In SOLIDWORKS *Drawing* mode, **projected views** can be created with a first-angle or third-angle projection, depending on the drafting standard used for the drawing. We must have a base view before a projected view can be created. Projected views can be orthographic projections or isometric projections. Orthographic projections are aligned to the base view and inherit the base view's scale and display settings.

➢ **NOTE:** In your drawing, the inserted views may include *center marks*, *centerlines*, and/or *dimensions*. If so, these were added with the **auto-insert** option. To access these options, click **Options** (*Standard* toolbar), select the **Document Properties** tab, and then select **Detailing**. This lesson was done with only the **Center marks-holes-part** auto-insert option selected. Other center marks, centerlines, and dimensions will be added individually.

➢ Notice the hidden lines are visible. The dash lengths and spaces can be adjusted in the document properties.

11. Select the **Options** icon from the *Menu Bar* to open the options window.

12. Select the **Document Properties** tab.

13. Select **Line Style**.

14. Select the **Dashed** line style.

15. Adjust the *Line length and spacing values* to, for example, **A,0.75,-0.25**. (This will scale the dash length by 0.75 and the space by 0.25.)

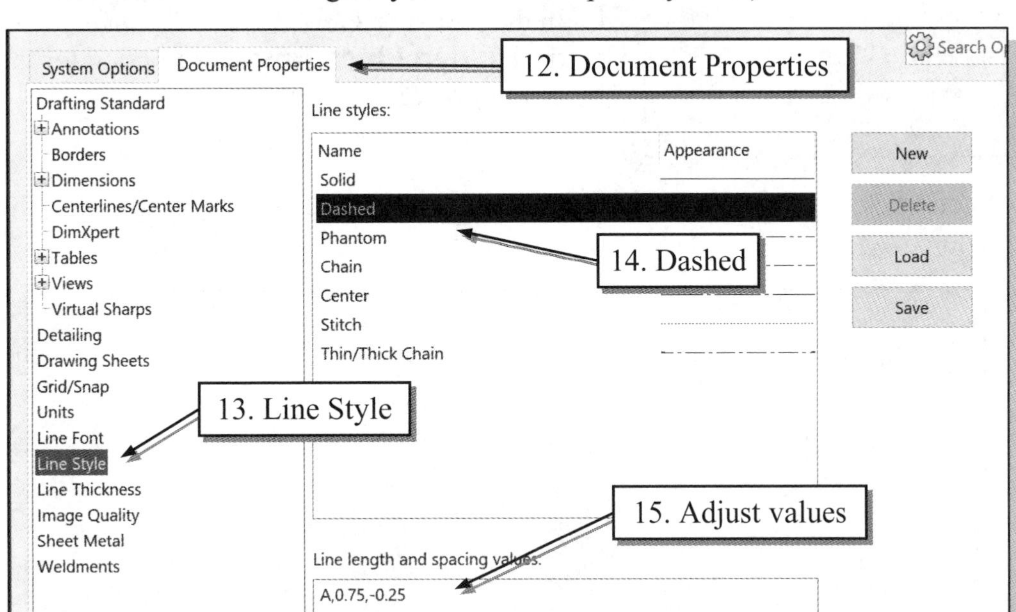

16. Left-click on the **OK** button to accept the settings.

17. Press the **[F]** key to fit the drawing to the graphics viewing area.

Adding an Isometric View using the View Palette

We will add an isometric view. The isometric view could be added as a *Projected View* based on the existing *Front View*. We will add the isometric view as a second base view. It can be added using the **Model View** command as was used for the *Front View*. An alternate method to add a base view is to use the **View Palette**. We will use this option.

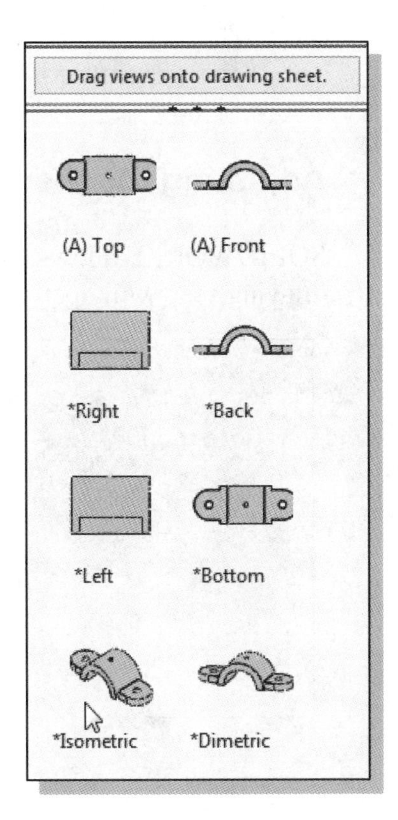

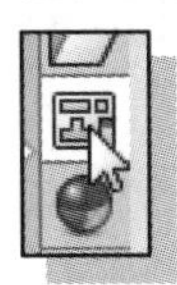

1. Click on the **View Palette** icon to the right of the graphics area.

2. The *View Palette* appears in the task pane. Select the **U-Bracket** part at the top of the View Palette.

3. Views can be dragged from the *View Palette* onto the drawing sheet to create a base drawing view. Click on the **Isometric** view in the *View Palette* as shown. Hold the **left-mouse-button** down and **drag** the view to the upper right corner of the sheet to position the isometric view. SOLIDWORKS places the view on the sheet and opens the *Drawing View PropertyManager*.

➢ In general, hidden lines are not used in isometric views.

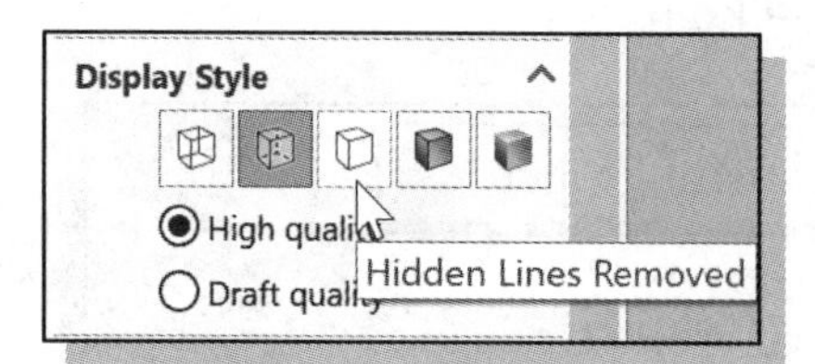

4. In the *Drawing View PropertyManager*, select the **Hidden Lines Removed** option for the *Display Style*.

5. Left-click on the **OK** icon in the *Drawing View PropertyManager* to accept the view.

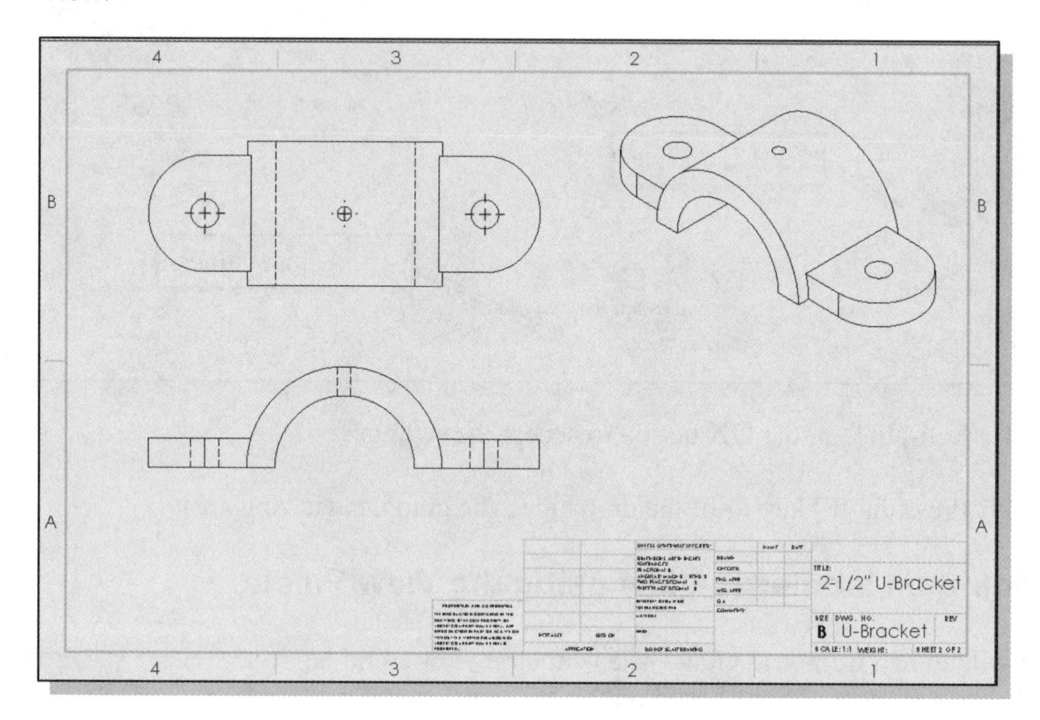

Adjusting the View Scale

SOLIDWORKS allows us to easily adjust the drawing view settings. Simply selecting a drawing view will open the corresponding *PropertyManager*.

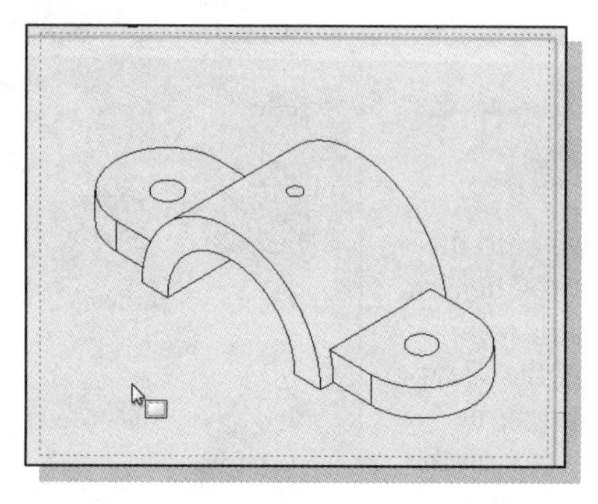

1. Move the cursor on top of the isometric view and watch for the box around the entire view indicating the view is selectable as shown in the figure. **Left-mouse-click** once to bring up the *PropertyManager*.

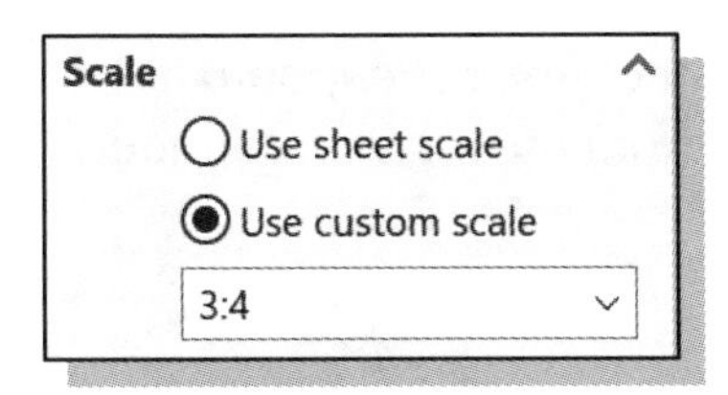

2. Under the *Scale* options in the *Drawing View PropertyManager*, select **Use custom scale**.

3. Enter **3:4** as the user defined scale.

4. Left-click on the **OK** icon in the *Drawing View PropertyManager* to accept the new settings.

5. If necessary, reposition the views to appear as shown below.

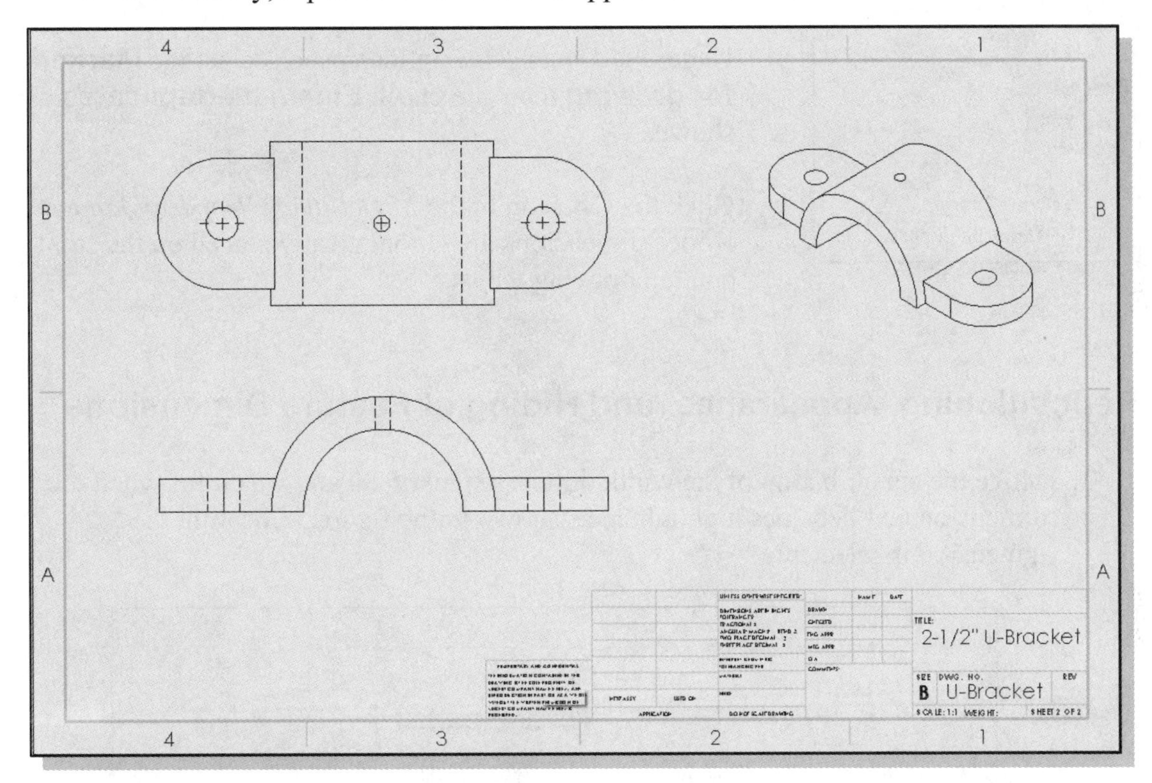

Displaying Feature Dimensions

By default, feature dimensions are not displayed in 2D views in SOLIDWORKS. The dimensions used to create the part can be imported into the drawing using the **Model Items** command. We can also add dimensions to the drawing manually using the **Smart Dimension** command. The **Model Items** command appears on the *Annotation* toolbar.

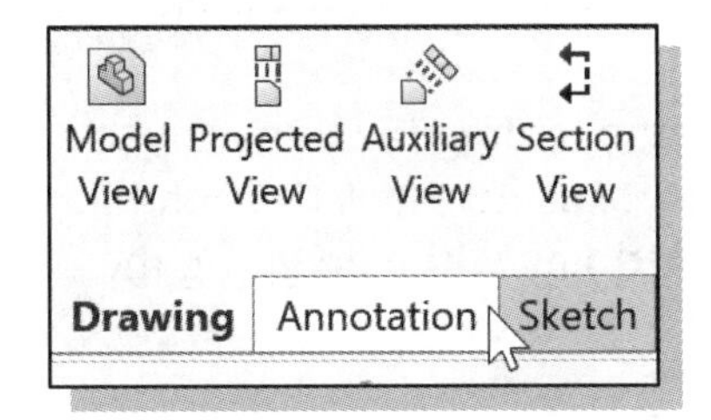

1. Select the **Annotation** tab on the *CommandManager* to display the *Annotation* toolbar.

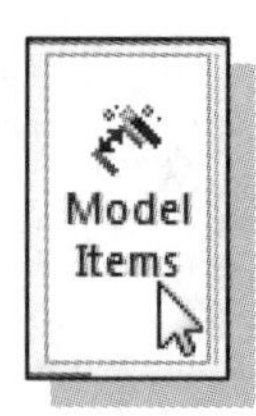

2. Left-mouse-click on the **Model Items** icon on the *Annotation* toolbar. This command allows us to import dimensions from the referenced model into selected drawing views.

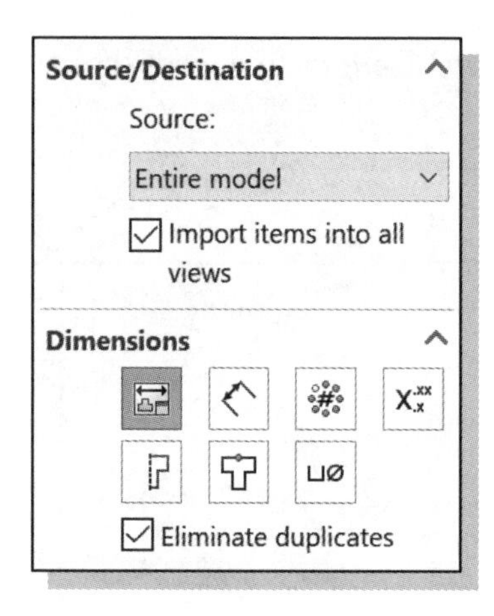

3. In the *Source/Destination* options panel of the *Model Items PropertyManager*, select the **Entire model** option from the *Import from* pull-down menu and check **Import items into all views** as shown.

4. Under the *Dimensions* options panel, select the **Marked for drawing** icon and check **Eliminate duplicates** as shown.

5. Click the **OK** icon in the *Model Items PropertyManager*. Notice dimensions are automatically placed on the front and top drawing views.

Repositioning, Appearance, and Hiding of Feature Dimensions

1. Move the cursor on top of the width dimension text **5.00** and watch for when the dimension text becomes highlighted as shown in the figure, indicating the dimension is selectable.

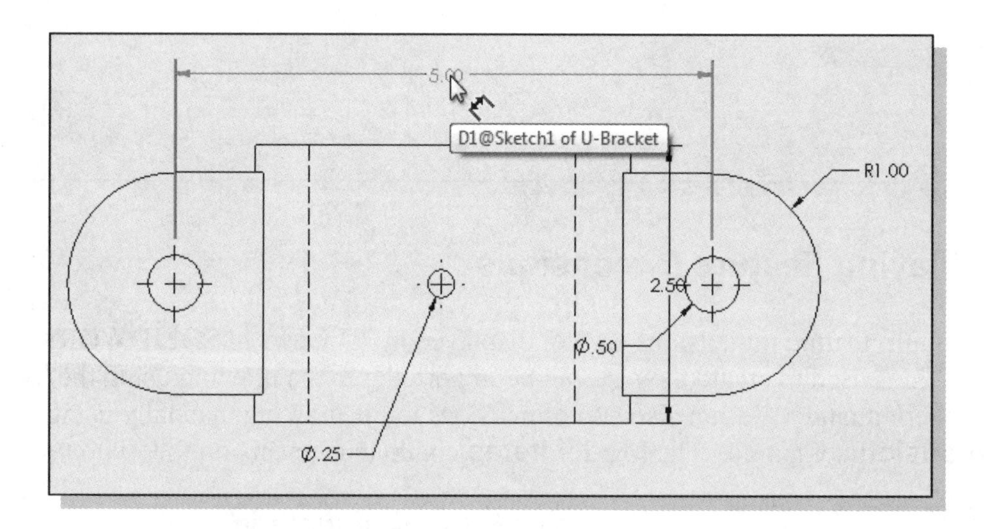

2. Reposition the dimension by using the left-mouse-button to drag the dimension text to a new location. (**NOTE:** SOLIDWORKS may execute snaps to align the text automatically. To negate these, hold down the **[Alt]** key while dragging the text and it will move freely.)

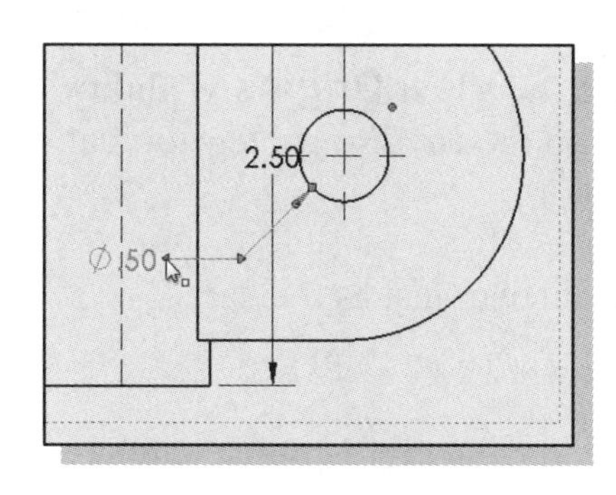

3. Move the cursor on top of the diameter dimension **0.50** and click once with the left-mouse-button to select the dimension.

4. Left-click on the arrow at the end of the leader line and drag to lengthen the bent leader line as shown.

❖ The default bent leader length can be adjusted to improve the appearance of the current view. This is done using the **Document Settings**. There are many document settings, including dimension settings, line style, etc., which can be adjusted to achieve the desired drawing display.

5. On your own, change the **Bent leader length** to **0.2 in**. (Select the **Options** icon from the *Menu Bar* to open the *Options* window. Select the **Document Properties** tab. Select **Dimensions**. In the **Leader length:** field, enter **0.2 in**. Click **OK**.)

6. Move the cursor on top of the diameter dimension **0.50** and click once with the left-mouse-button to select the dimension.

7. Select the **Leaders** tab in the *Dimension PropertyManager.*

8. Select the **Smart** display button.

9. Check the **Use document bend length** box.

10. Click **OK** in the *PropertyManager*.

➢ Notice the bent leader length (that we lengthened in step 4) now reverts to the length we set in the *Document Settings*.

11. On your own, reposition the dimensions displayed in the ***top*** view as shown.

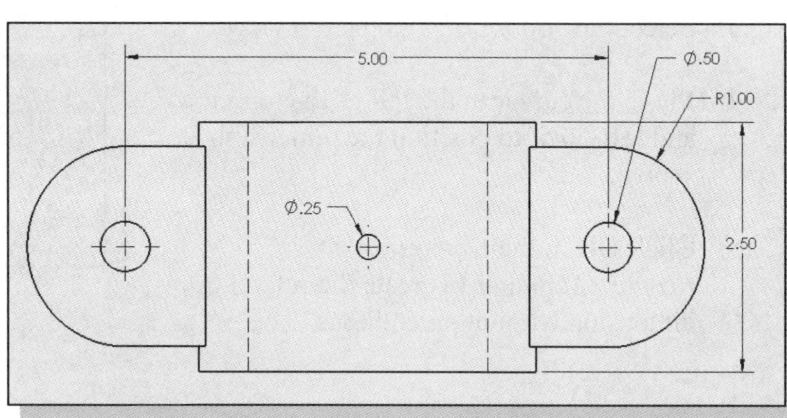

12. NOTE: If origin icons appear on each view, toggle the **View Origins** visibility *OFF* using the *Hide/Show Items* pull-down menu on the *Heads-up View* toolbar. (See Step 3 on Page 7-17.)

➢ Any feature dimensions can be removed from the display just as they are displayed.

13. Move the cursor on top of the vertical **2.50** dimension and **right-mouse-click** once to bring up the option menu.

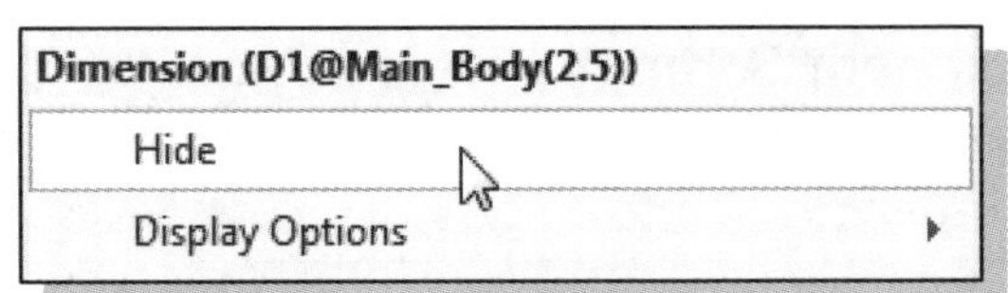

14. Select **Hide** from the pop-up option menu.

Adding Additional Dimensions – Reference Dimensions

Besides displaying the imported **(driving) feature dimensions**, dimensions used to create the features, we can also add additional **(driven) reference dimensions** in the drawing. *Feature dimensions* are used to control the geometry, whereas *reference dimensions* are controlled by the existing geometry. In the drawing layout, therefore, we can ***add*** or ***delete*** *reference dimensions*, but we can only ***hide*** the *feature dimensions.* One should try to use as many *feature dimensions* as possible and add *reference dimensions* only if necessary. It is also more effective to use *feature dimensions* in the drawing layout since they are created when the model was built. Note that additional *Drawing* mode entities, such as lines and arcs, can be added to drawing views. Before *Drawing* mode entities can be used in a reference dimension, they must be associated to a *drawing view.*

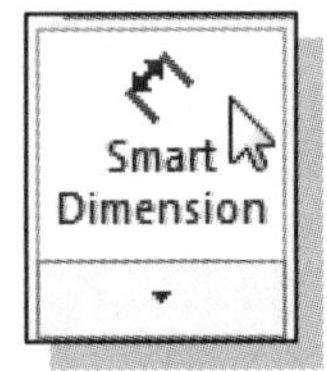

1. Left-mouse-click on the **Smart Dimension** icon on the *Annotation* toolbar.

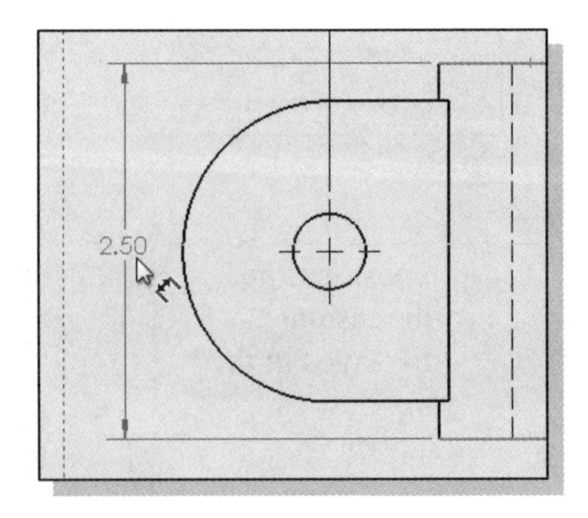

2. Select the *top edge* in the top view.

3. Select the *bottom edge* in the top view.

4. Move the cursor to the left of the top view and **left-click** to position the dimension as shown.

5. Click **OK** in the *Dimension PropertyManager* to create the reference dimension without parentheses.

6. On your own, add the **2.50** horizontal dimension and arrange the display of the top view to look like the figure below.

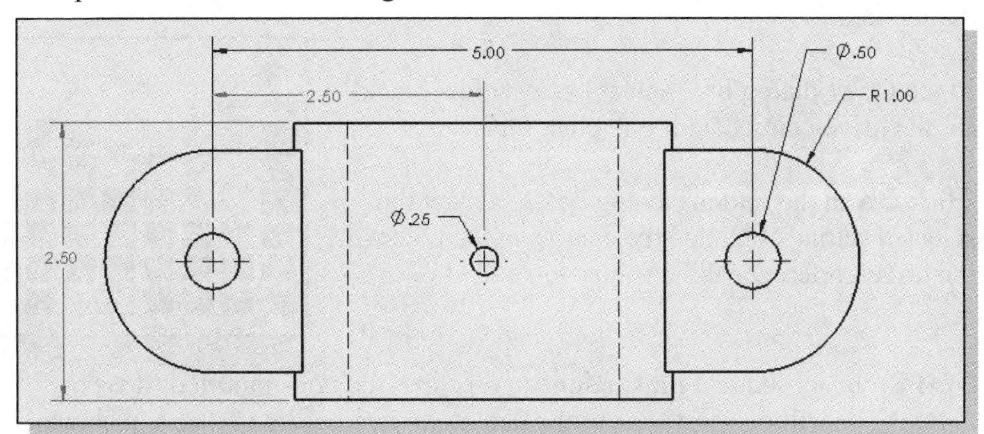

➢ NOTE: By default the non-imported (driven) reference dimensions appear grey to distinguish them from the imported (driving) feature dimensions. This can be controlled using the *Options* dialog box.

7. Select the **Options** icon from the *Menu Bar* to open the options dialog box.

8. Under the **System Options** tab, select **Colors** as shown.

9. Under '*Color scheme settings*', select **Dimensions, Imported (Driving)**. Notice the default color (black) shown to the right of the selection list.

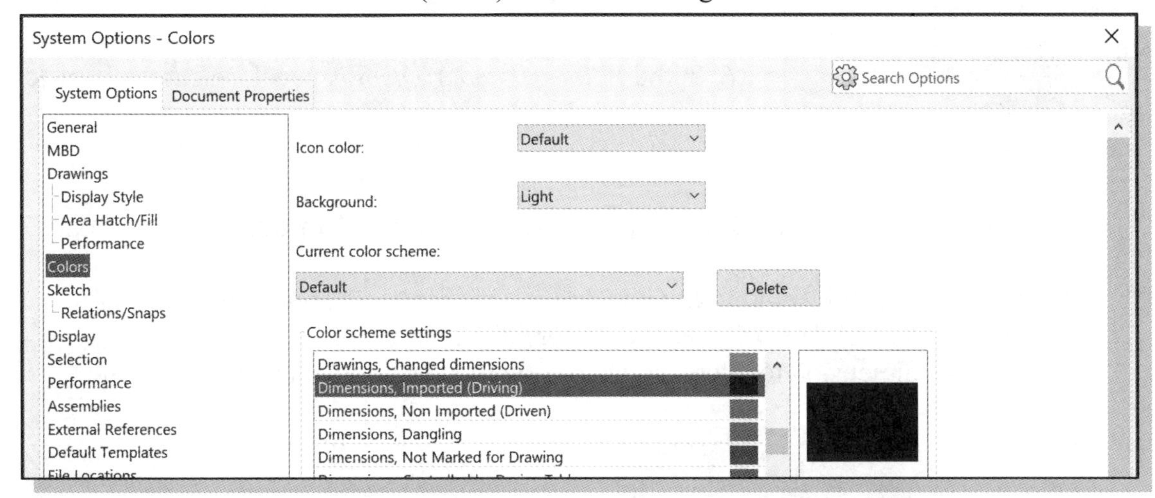

10. Under '*Color scheme settings*', select **Dimensions, Non Imported (Driven)**. Notice the default color (grey) shown to the right of the selection list.

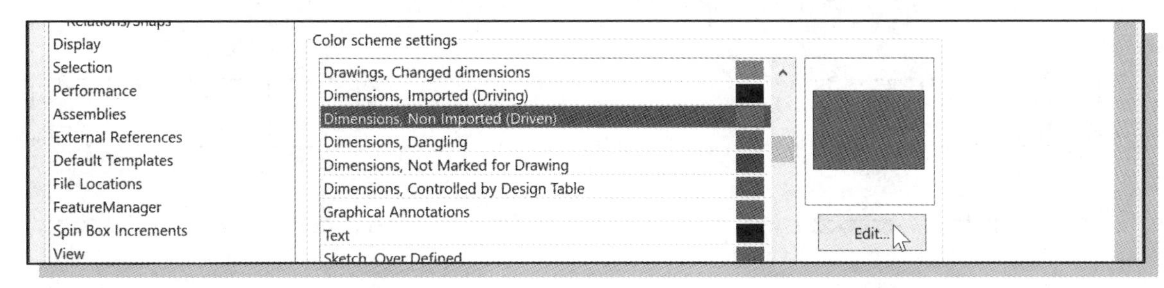

11. Click the **Edit…** button below the color illustration panel to open the *Color* dialog box.

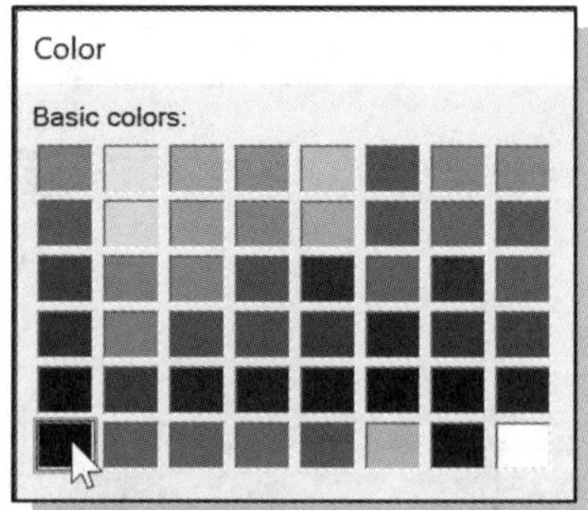

12. In the *Color* dialog box, **select** a new color (black) for the driven dimensions and click **OK**.

13. Click **OK** in the options dialog box to accept the selected settings. Notice the change in the color of the driven reference dimensions you added.

➢ **NOTE:** In subsequent illustrations in this text, the non-imported (driven) dimensions will be shown using the default color in order to distinguish which dimensions were added as reference dimensions. In your drawings, all dimensions will appear black as long as the settings above remain unchanged.

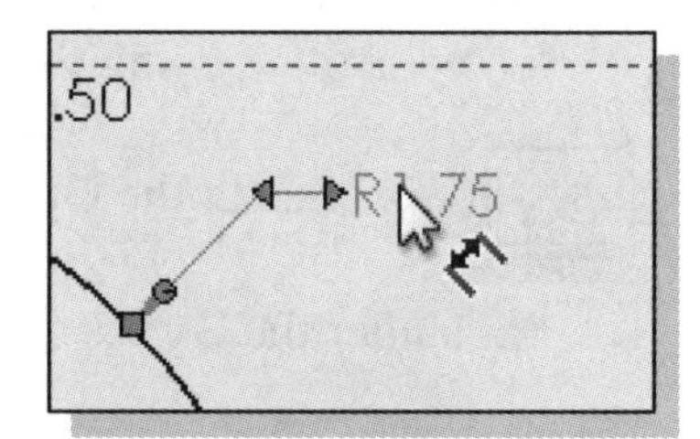

14. In the front view, left click on the **R1.75** dimension to select the radius dimension.

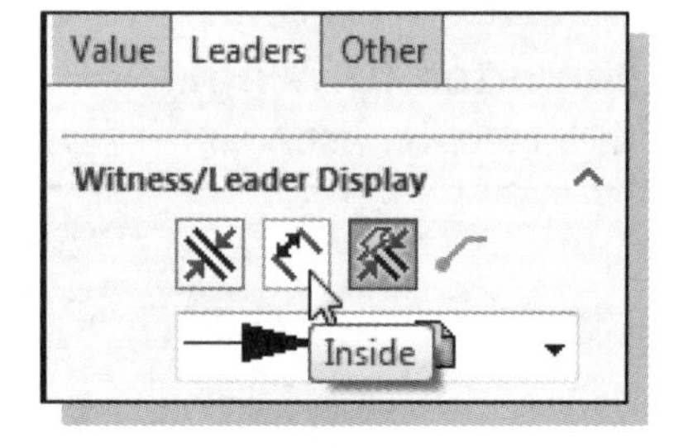

15. In the *Dimension PropertyManager*, select the **Leaders** tab.

16. In the *Witness/Leader Display* panel, select the **Inside** option as shown.

17. On your own, select the **∅2.50** dimension and select the **Smart** display option.

18. On your own, adjust the front view dimension display to appear as in the figure below. (**NOTE:** The 0.50 vertical feature dimension was replaced with a reference dimension to allow the visible gap between the object lines and the extension lines.)

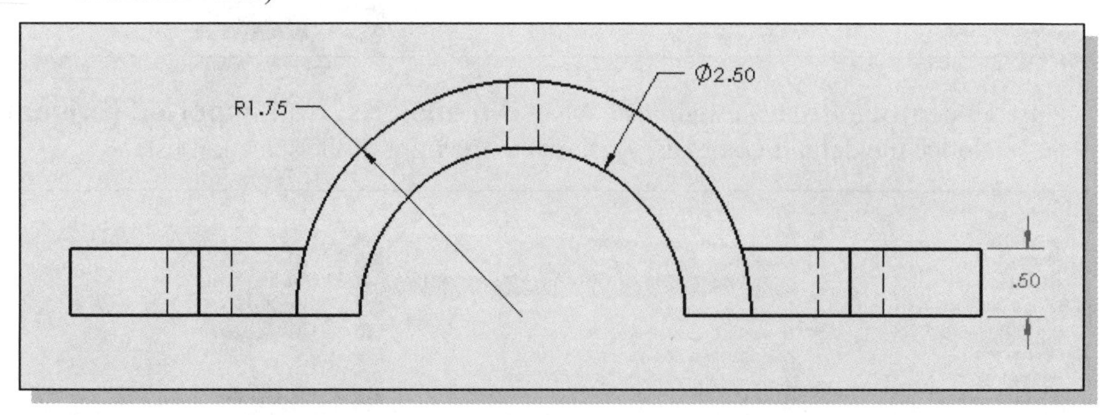

Tangent Edge Display

❖ Another important setting for drawing views is the way tangent edges – where a curved surface meets tangent to a flat surface – are represented. In the figure above, the tangent edges (the vertical lines coincident with the centerlines of the two hidden holes) are visible.

1. Press the **[Esc]** key to ensure no objects are selected.

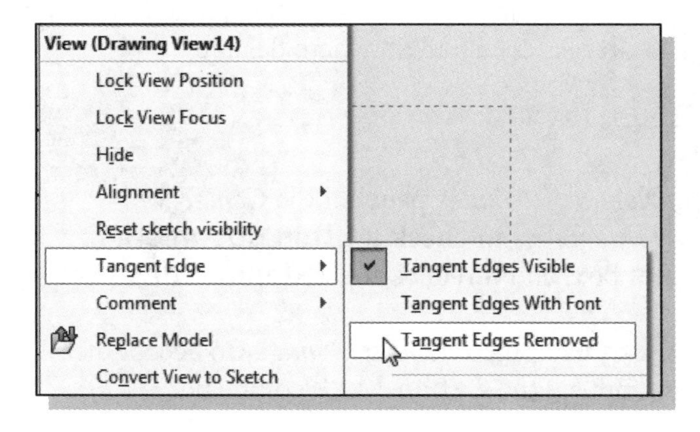

2. Move the cursor over the *front view* in the graphics area and **right-click** to open the pop-up option menu.

3. Select **Tangent Edge**, then **Tangent Edges Removed** from the pop-up menu, as shown.

4. Click **OK** in the *Property-Manager* or press the **[Esc]** key to accept the setting.

❖ Notice the tangent edges are no longer visible.

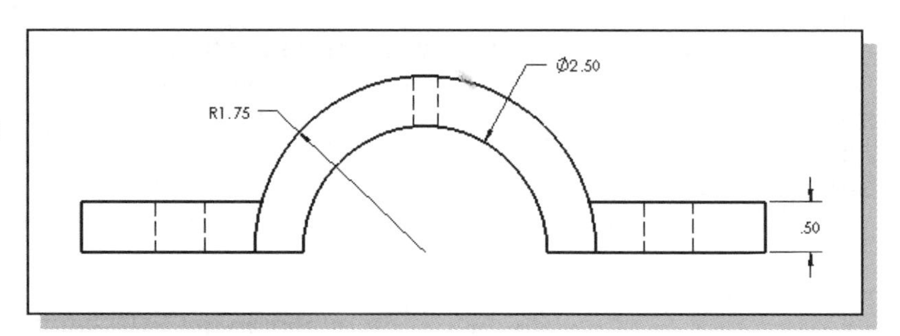

5. On your own, remove the tangent edges from the isometric view.

Adding Center Marks, Center Lines, and Sketch Objects

1. Click on the **Center Mark** button in the *Annotation* toolbar.

2. Click on the larger arc in the front view to add the center mark.

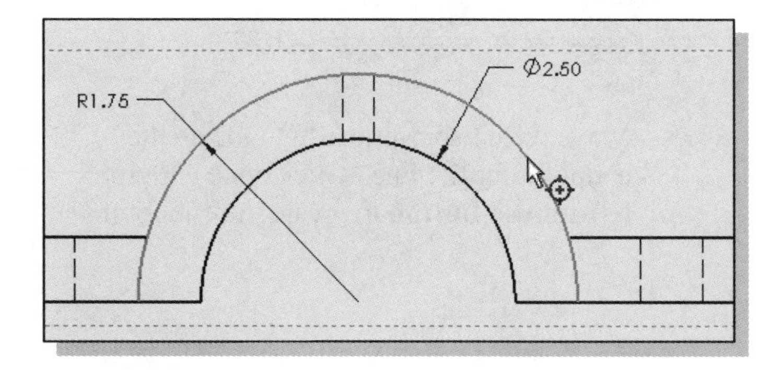

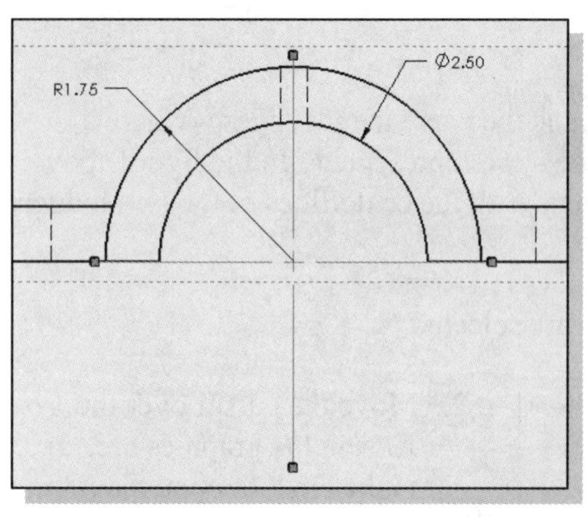

3. A preview of the center mark appears in the graphics area and the *Center Mark PropertyManager* is opened.

❖ The default center mark included extended lines in all four directions. Since the feature is semicircular, we will turn *OFF* the default extended lines and sketch them where desired.

4. In the *Display Attributes* panel of the *Center Mark PropertyManager*, **uncheck** the **Use document defaults** box, and **uncheck** the **Extended lines** box.

5. Left-Click **OK** in the *PropertyManager* to accept the settings and end the **Center Mark** command.

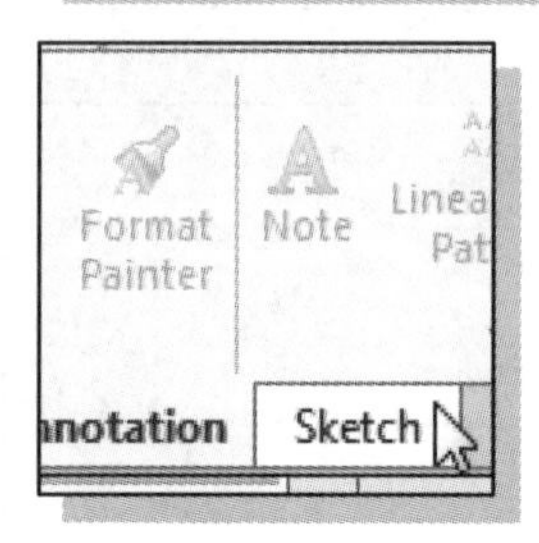

6. Select the **Sketch** tab on the *CommandManager* to display the *Sketch* toolbar.

7. Left-click on the **Line** button on the *Sketch* toolbar.

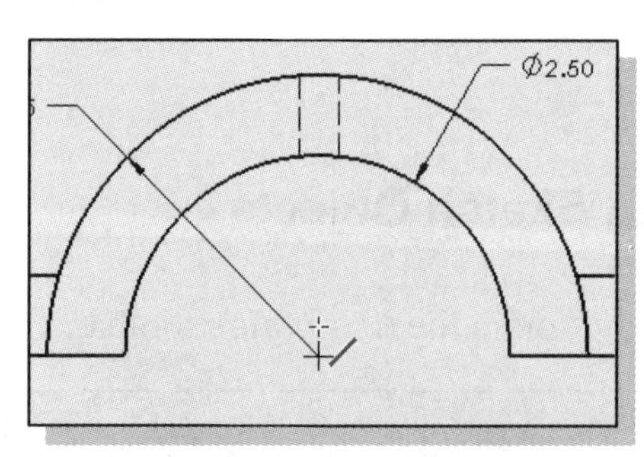

8. Move the cursor over the center mark, then move the cursor above the center to engage the **Vertical** relation. Click once with the **left-mouse-button** to locate the first point for the line.

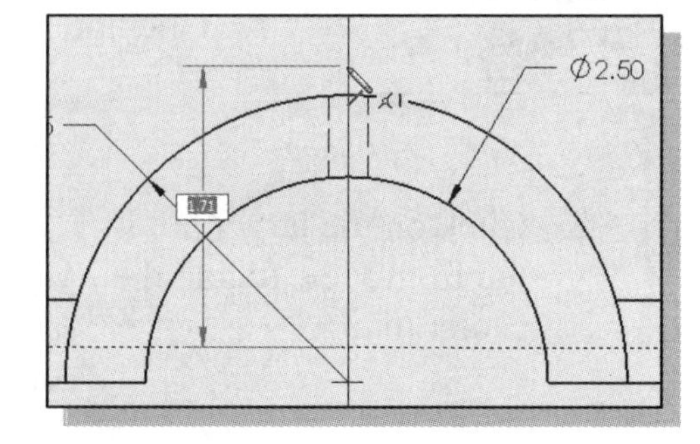

9. Move the cursor vertically to locate the endpoint for the line. Click once with the **left-mouse-button** to locate the endpoint.

10. Click once with the **right-mouse-button** to open the pop-up menu.

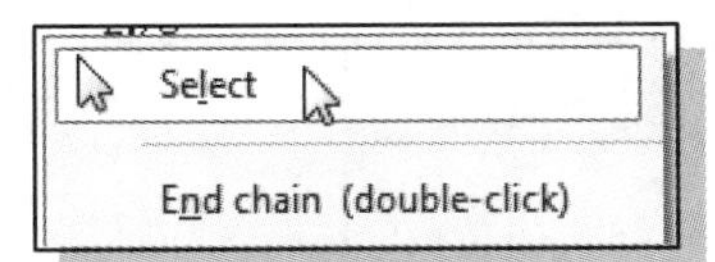

11. Left-click on the **Select** option on the pop-up menu to end the **Line** command.

12. On your own, sketch the two horizontal lines as shown in the figure.

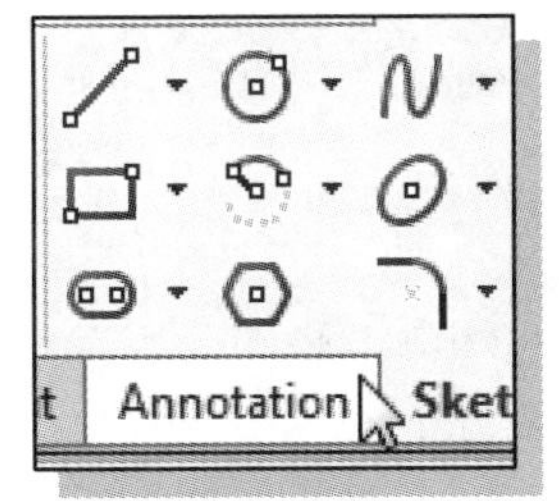

13. Select the **Annotation** tab on the *CommandManager* to display the *Annotation* toolbar.

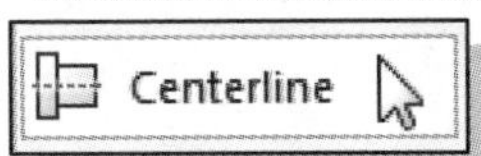

14. Click on the **Centerline** button in the *Annotation* toolbar. (**NOTE:** This is different from the Centerline command on the *Sketch* toolbar.)

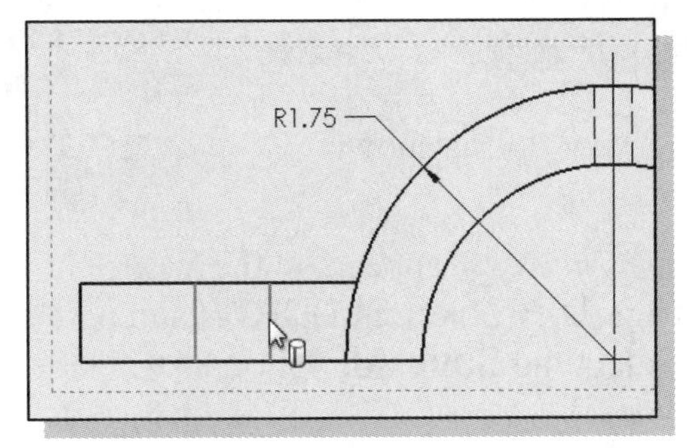

15. Inside the graphics window, click on the two hidden edges of the **Hole** on the left as shown in the figure.

16. On your own, repeat the above step and create another centerline on the right side of the front view as shown below.

17. Left-click the **OK** icon in the *Centerline PropertyManager* or press the **[Esc]** key to end the Centerline command.

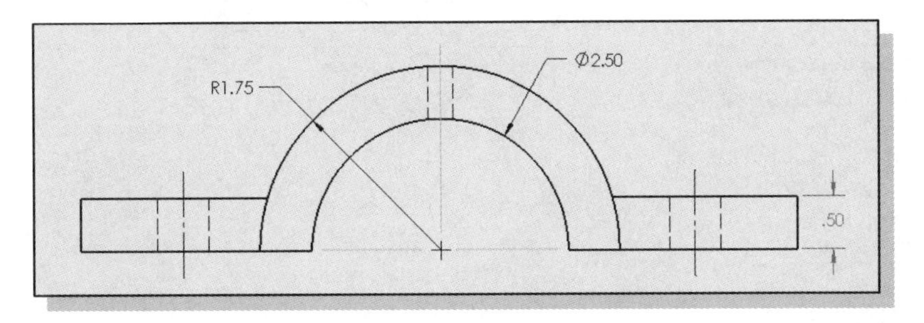

18. On your own, adjust your settings and endpoint locations until the top view appears as in the figure below. Add the three center marks if they were not added automatically. Drag the endpoint of the *Extended Lines* to desired locations. For the central hole, **uncheck** the Use document defaults option, **check** the Extended lines option, and **check** the Centerline font option in the *PropertyManager*.

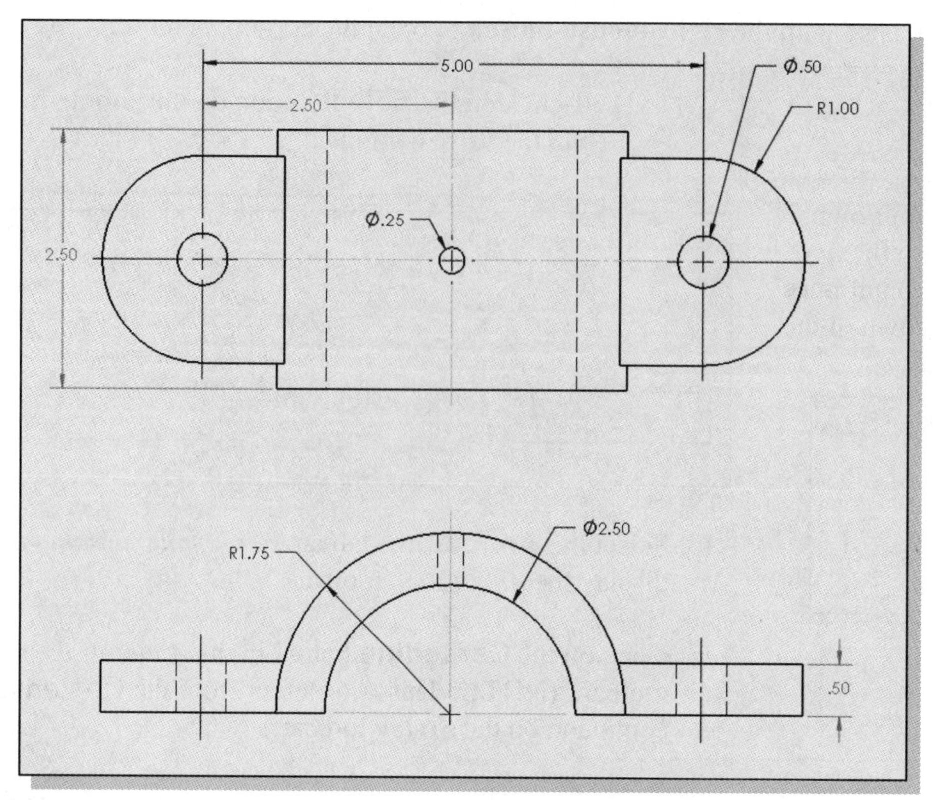

19. Click on the **Save** icon in the *Menu Bar* as shown.

20. Note the message "*The following models referenced ...*" appears in the *Save Modified Documents* window and the drawing (**U-Bracket.SLDDRW**) and part (**U-Bracket.SLDPRT**) files are selected. Select the **Save All** option.

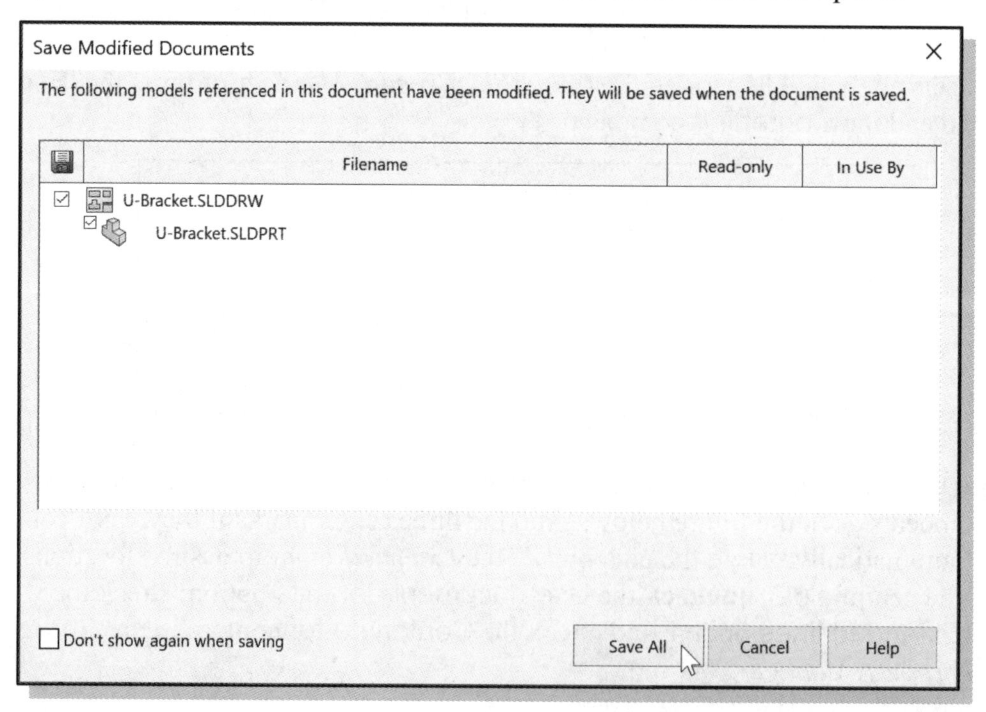

21. If the SOLIDWORKS window appears asking to *'Update the drawing?'* select **Yes**.

22. Save the drawing with the filename ***U-Bracket***.

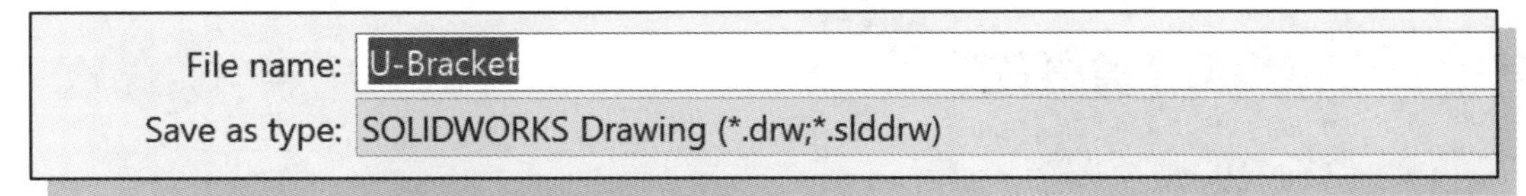

Edit Sheet vs. Edit Sheet Format

There are two modes in a SOLIDWORKS drawing: ***Edit Sheet*** and ***Edit Sheet Format***. To this point we have been operating in the *Edit Sheet* mode. The *Edit Sheet* mode is used to make detail drawings, including adding and modifying views, adding and modifying dimensions, and adding and modifying drawing notes. *The Edit Sheet Format* mode is used to add and edit title blocks, borders, and standard text that appear in every drawing. The correct mode must be selected in order to execute the corresponding commands.

1. Move the cursor into the graphics area and **right-click** to open the pop-up option menu.

2. Select **Edit Sheet Format** from the pop-up option menu to change to *Edit Sheet Format* mode.

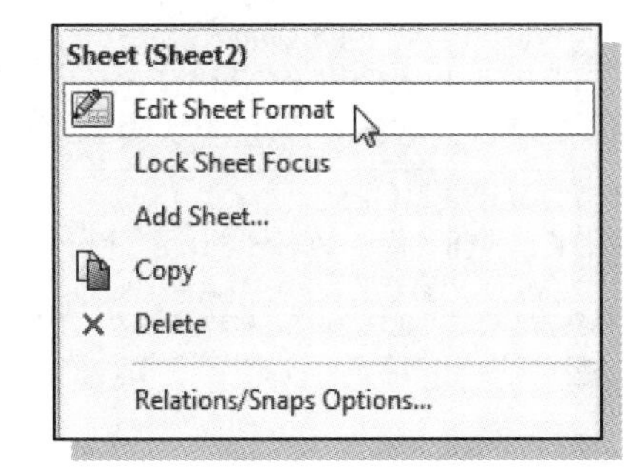

❖ Notice the title block is highlighted.

Completing the Drawing Sheet

1. On your own, use the **Zoom** and **Pan** commands to adjust the display as shown; this is so that we can complete the title block.

❖ There are many notes added as entries in the default title block. Some are simple **Text** notes; some include **Property Links** to automatically display document properties. In the following steps, we will learn to (1) edit existing notes, (2) insert new notes, and (3) control Property Links. We will first edit a simple text note.

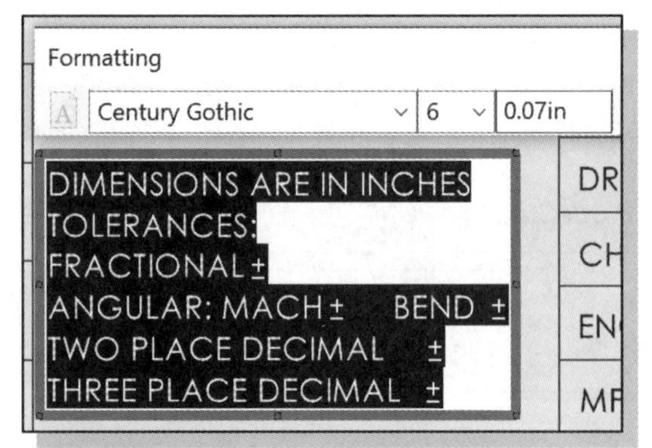

2. Zoom in on the tolerance specification box. This note contains only simple text entries. **Double-click** with the **left-mouse-button** on the text to enter the text editor.

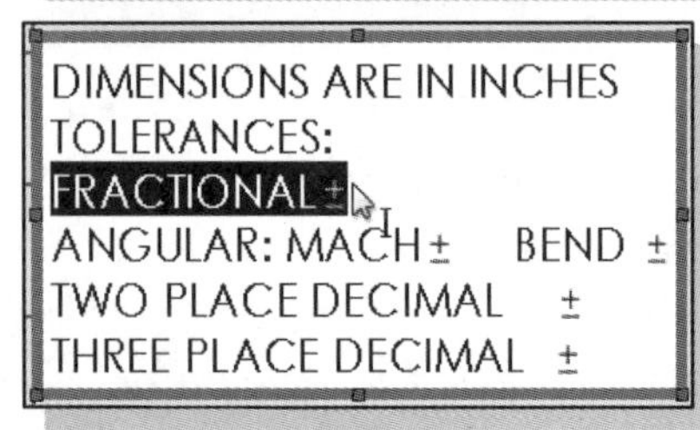

3. Highlight the FRACTIONAL ± line as shown by clicking and dragging with the left-mouse-button.

4. Press the **[Delete]** key twice to delete the line.

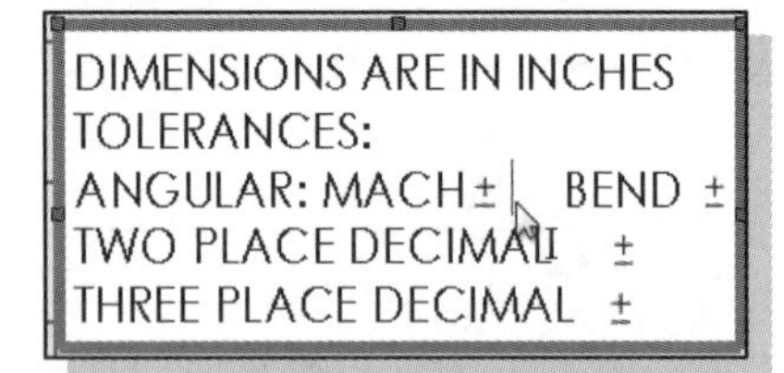

5. Move the cursor to a location to the right of the MACH ± text, and click once with the **left-mouse-button** to select a location to enter text.

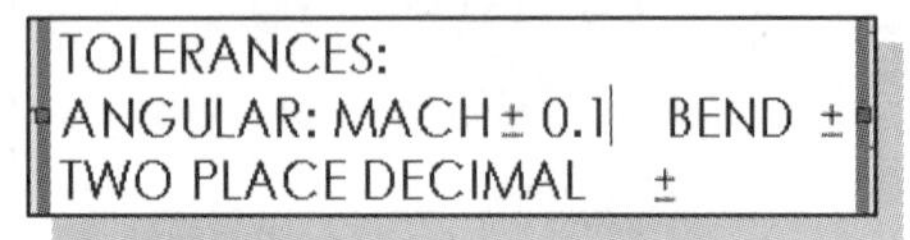

6. Enter **0.1**.

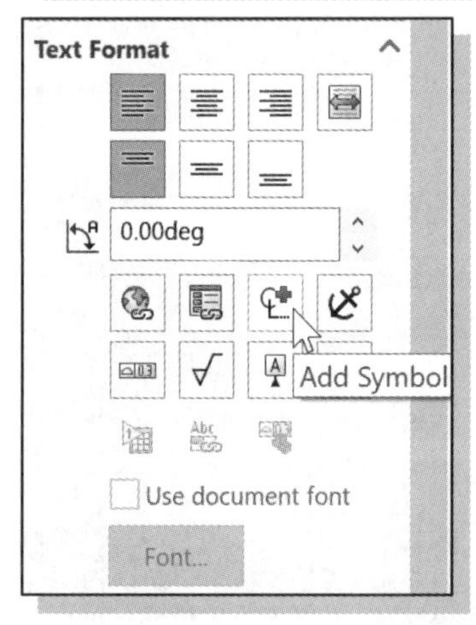

7. In the *Text Format* panel of the *Note PropertyManager*, select the **Add Symbol** icon.

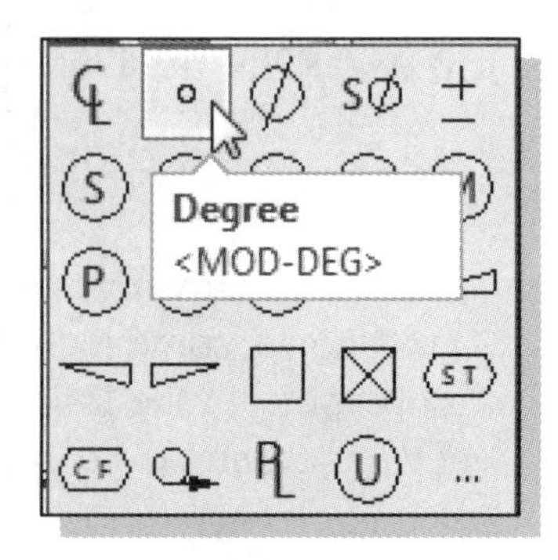

8. In the *Symbols* pop-up window, select **Degree** and click **OK** to insert a degree symbol.

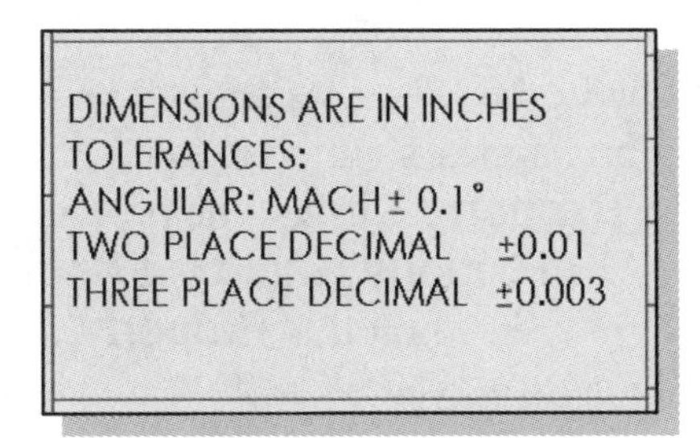

9. On your own, continue editing the text in the tolerance box until it appears as shown.

10. Select the **OK** icon in the *Note PropertyManager* or press the **[Esc]** key to accept the change to the note.

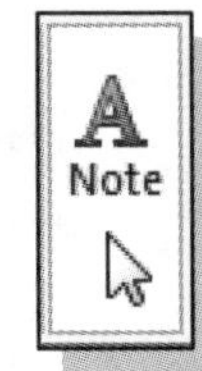

11. We will now insert a new note. Left-mouse-click on the **Note** icon on the *Annotation* toolbar.

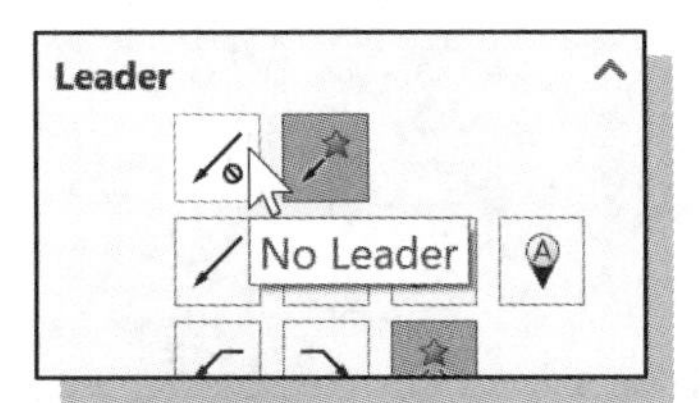

12. Select the **No Leader** option in the *Leader* panel of the *Note Property Manager*.

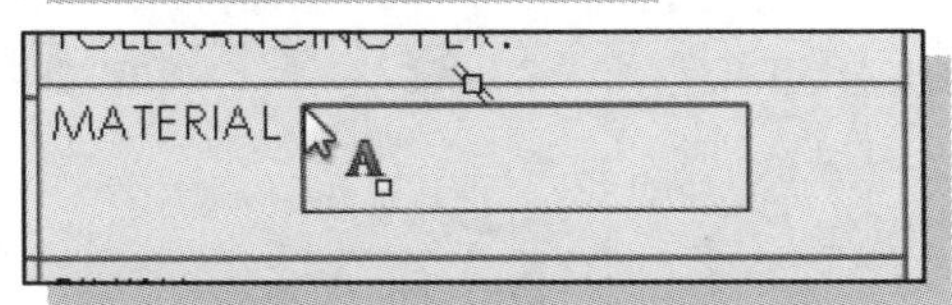

13. Move the cursor into the MATERIAL box and click once with the **left-mouse-button** to select the location for the new note.

14. Enter the text **1060 ALLOY**.

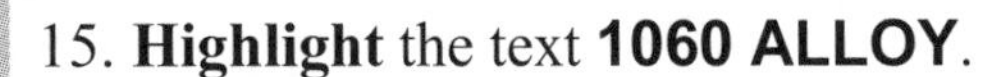

15. **Highlight** the text **1060 ALLOY**.

16. Change the font size to **6** in the *Formatting* window as shown and press the [**Enter**] key to reset the font size.

17. Select the **OK** icon in the *Note PropertyManager* or press the **[Esc]** key to accept the note.

18. Click and drag on the new note to locate it as shown. (**NOTE:** SOLIDWORKS may execute snaps to align the text automatically. To negate these, hold down the [**Alt**] key while dragging the text and it will move freely.)

Property Links

SOLIDWORKS allows notes to be linked to properties for automatic display. In SOLIDWORKS, these properties are classified as system-defined ***System Properties*** and user-defined ***Custom Properties***. Each system or custom property has a property name. There are system and custom properties associated with each SOLIDWORKS **Part**, **Assembly**, or **Drawing** file. Each **Property Link** contains (1) the definition of the source and (2) the property name. The source can be (1) the current document, (2) a model specified in a sheet, (3) a model in a specified drawing view, or (4) a component to which the annotation is attached. Several linked properties are automatically included in the default SOLIDWORKS sheet formats.

1. Move the cursor over the note in the center of the DWG. NO. box, as shown. A letter **A** appears indicating a note. A pop-up dialog box appears which reads ***$PRP: "SW-File Name."*** This is the definition of a **Property Link**. The prefix ***$PRP:*** defines the source as the current document. The Property Name is ***SW-File Name***. This is an example of a **System Property**. Based on this Property Link, the filename – **U-Bracket** – of the current file is displayed.

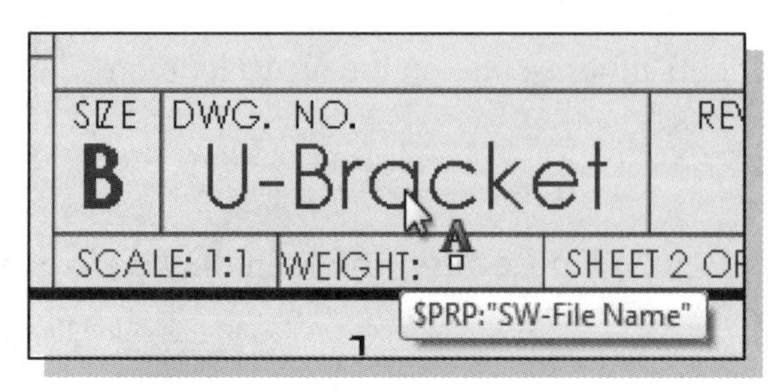

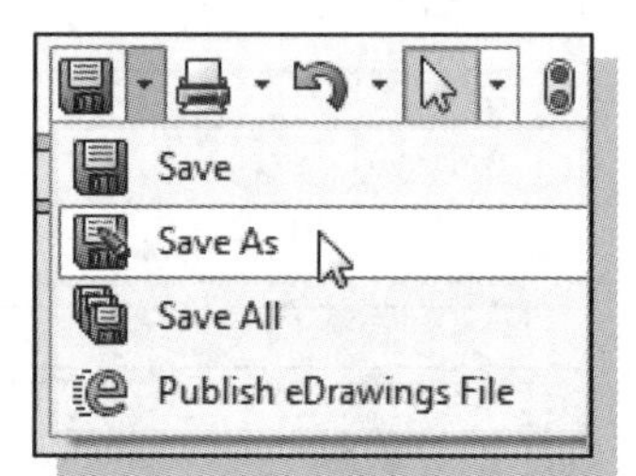

2. We will change the file name. Select **Save as** from the *Menu Bar* pull-down menu.

3. Enter **MET 3001** for the *File name.*

4. Enter **U-Bracket Detail** for the *Description.*

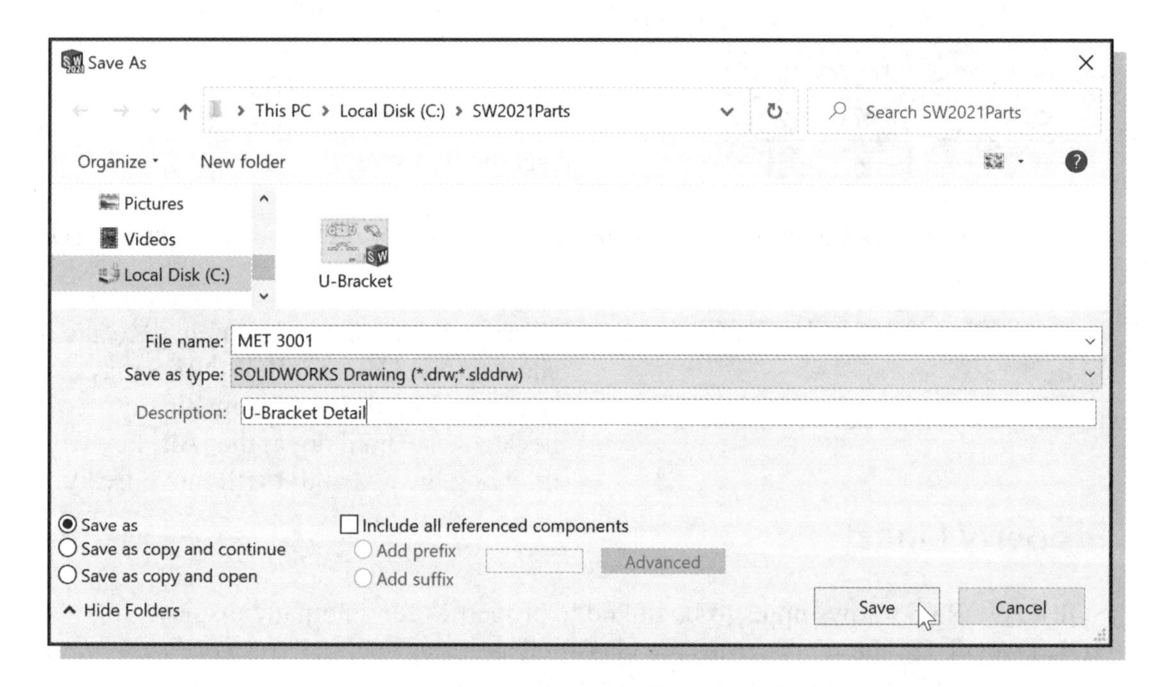

5. Click **Save**.

6. If the text in the DWG. NO. box does not change, select the **Rebuild** icon on the *Menu Bar*.

7. Notice the new file name is now displayed in the **DWG. NO.** box. Click once with the left-mouse-button on the **MET 3001** note to open the *Note PropertyManager*.

8. Click the **Font** button in the *Text Format* panel of the *Note PropertyManager*.

9. In the *Choose Font* pop-up window, select **Points** under the *Height* option.

10. Select **16** as the font size.

11. Click **OK** in the *Choose Font* pop-up window.

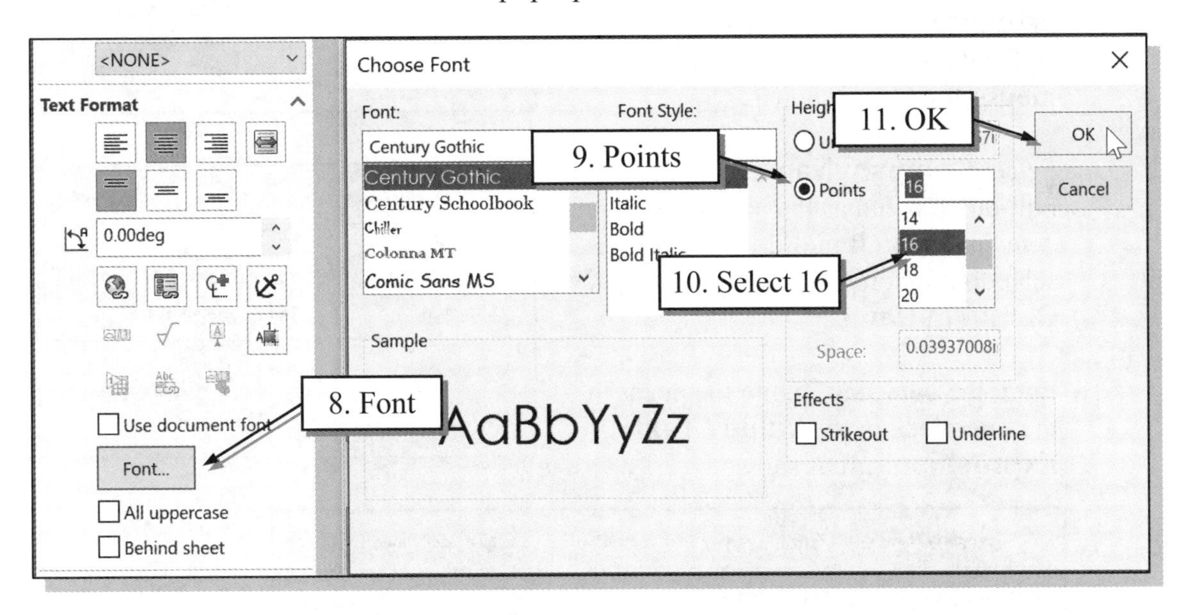

12. Press the **[Esc]** key to accept the change and unselect the note.

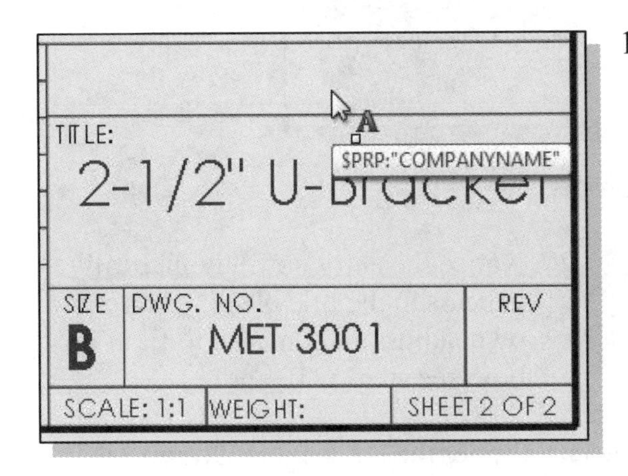

13. Move the cursor over the center of the box above the **Title:** box. A letter **A** appears indicating a note. A pop-up dialog box appears identifying a **Property Link** which reads ***$PRP: "CompanyName."*** The prefix ***$PRP:*** defines the source as the current document. The **Property Name** is ***CompanyName***. This is an example of a ***Custom Property***. The note is currently blank because the *CompanyName Custom Property* has not been defined.

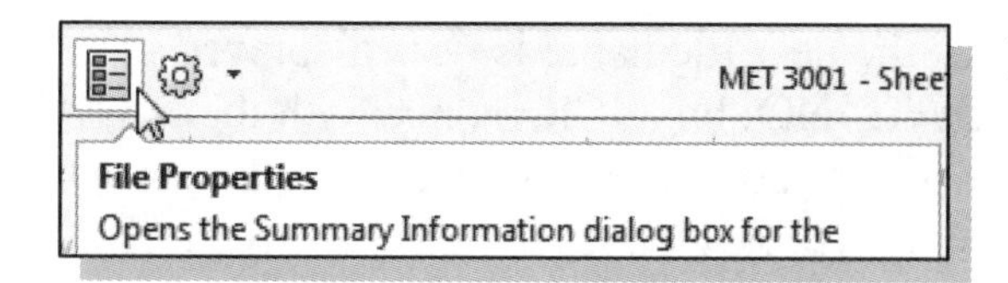

14. Select the **File Properties** option from the *Menu Bar* to open the *Summary Information* dialog box.

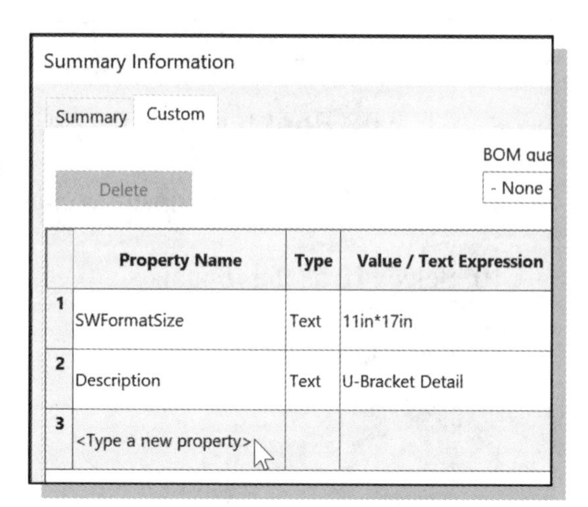

15. In the *Summary Information* window select the **Custom** tab. Notice the *Custom Property* **Description** appears in the table. We defined this property when we saved the drawing file.

16. The cell below the cell containing Description reads **<Type a new property**. Move the cursor over this cell and click once with the **left-mouse-button** as shown.

17. Select **CompanyName** from the pull-down menu as the new *Property Name*. (Note: If pull-down menu selection options do not appear, type **CompanyName**.)

	Property Name	Type	Value / Text Expression
1	SWFormatSize	Text	11in*17in
2	Description	Text	U-Bracket Detail
3	CompanyName	Text	
4	<Type a new property>		

18. Enter the **company name** to appear in the title block in the **Value / Text Expression** cell as shown.

	Property Name	Type	Value / Text Expression	Evaluated Value
1	SWFormatSize	Text	11in*17in	11in*17in
2	Description	Text	U-Bracket Detail	U-Bracket Detail
3	CompanyName	Text	University of New Orleans	University of New Orleans
4	<Type a new property>			

19. Click **OK** to accept the user-defined property.

20. Notice the entry now automatically appears in the title block. On your own, adjust the font size to fit in the box (see steps 6-10 above).

21. Move the cursor over the box next to Drawn as shown. A note containing a link to the *Custom Property* DrawnBy is contained in this box.

22. On your own, go to the *File Properties* window and enter your name or initials as the entry for the **DrawnBy** *Custom Property.*

23. On your own, determine the *Property Link* in the Drawn Date box. Enter the date for the *Custom Property.*

	NAME	DATE
DRAWN	A. Smith	01/08/2021
CHECKED		

24. Move the cursor over the **Title** (it should read **2-1/2" U-Bracket**). A letter **A** appears indicating a note. A pop-up dialog box appears identifying a *Property Link* which reads ***$PRPSHEET:*** "***Description***." The prefix ***$PRPSHEET:*** defines the source as the model (part file) appearing in the sheet. The *Property Name* is **Description**. This is a ***Custom Property***. We defined this description when we saved the U-Bracket part file. If 2-1/2" U-Bracket does not appear, you must define the description. Go to the part file (select **U-Bracket.SLDPRT** from the *Window* pull-down menu). Follow the same procedure as we did in the drawing file to change the Description property. Then return to the drawing window.

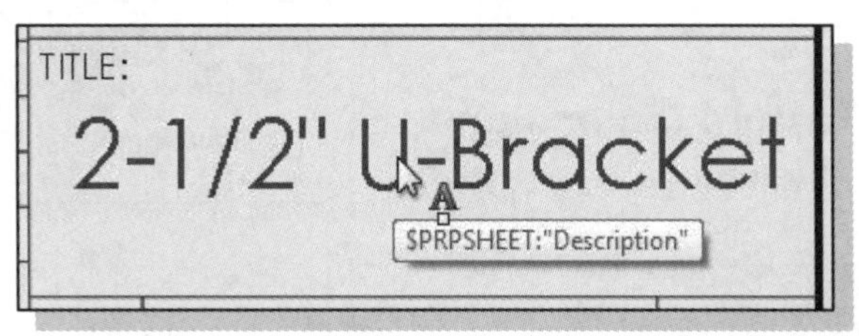

25. On your own investigate the rest of the title block to determine where other *Property Links* appear.

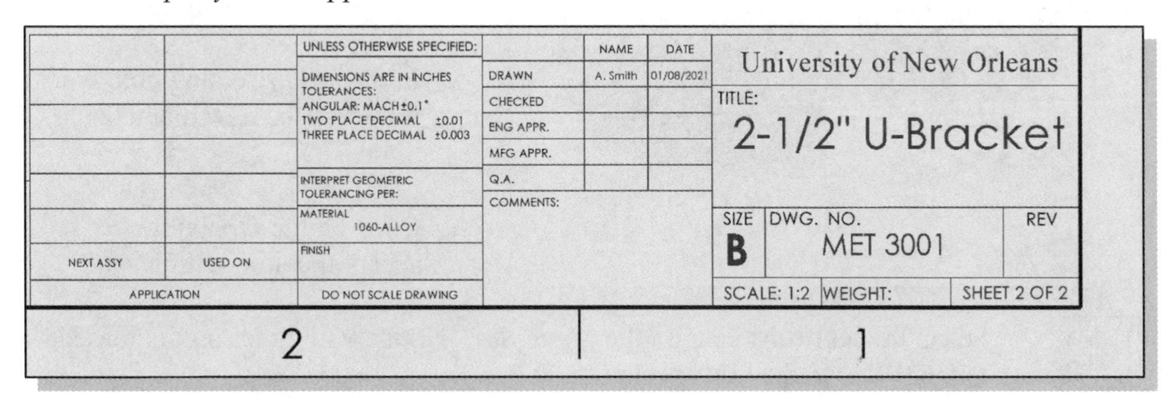

26. Press the **[F]** key to scale the view to fit the graphics area.

27. Move the cursor into the graphics area and **right-click** to open the pop-up option menu.

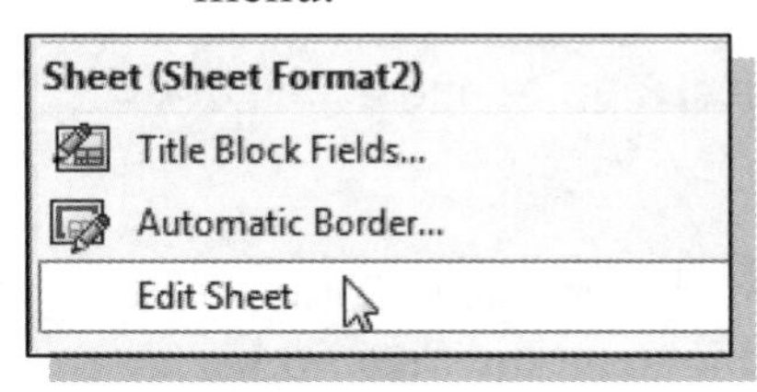

28. Select **Edit Sheet** from the pop-up option menu to change to *Edit Sheet* mode.

Associative Functionality – Modifying Feature Dimensions

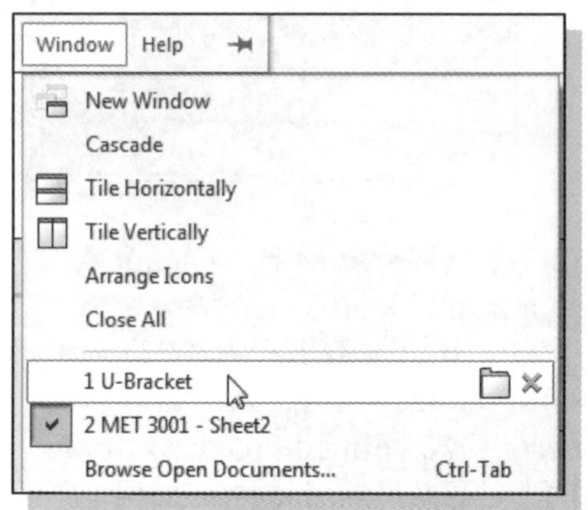

- SOLIDWORKS' *associative functionality* allows us to change the design at any level, and the system reflects the changes at all levels automatically.

1. Select the ***U-Bracket*** part window from the *Window* pull-down menu to switch to *Part Modeling* mode.

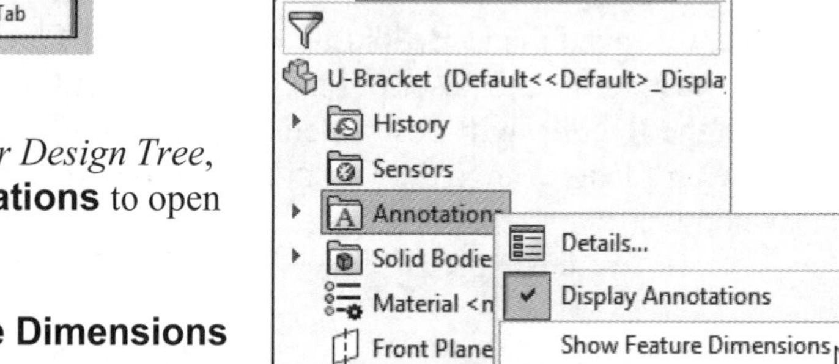

2. In the *FeatureManager Design Tree*, **right-click** on **Annotations** to open the option menu.

3. Select **Show Feature Dimensions** in the pop-up option menu.

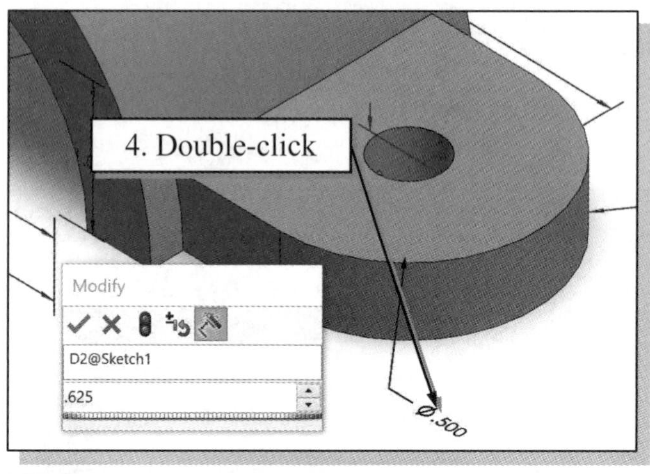

4. Double-click on the diameter dimensions (**0. 50**) of the hole on the base feature as shown in figure. (Change the display to *Wireframe* mode if necessary.)

5. In the *Modify* dialog box, enter **0.625** as the new diameter dimension.

6. Click on the **OK** button to accept the new setting.

7. Select the **Rebuild** icon on the *Menu Bar*. Notice both holes are increased to 0.625 due to the Mirror relation.

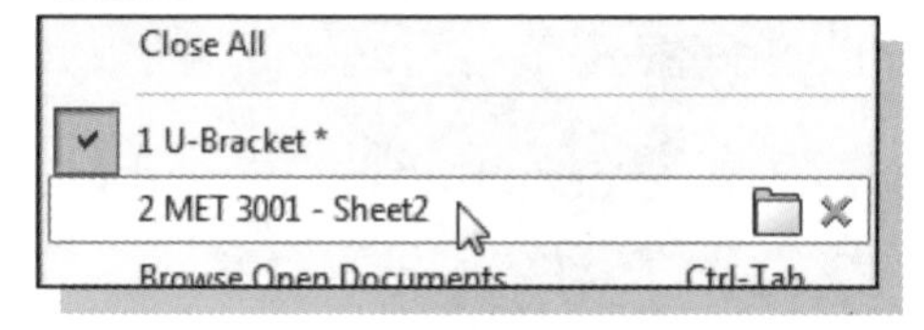

8. Select the ***MET 3001- Sheet2*** drawing window from the *Window* pull-down menu to switch to *Drawing* mode.

9. If the SOLIDWORKS pop-up window appears, select **Yes** to update the drawing file to include the changes made to the model.

10. Otherwise, the drawing views will be highlighted to indicate that the drawing file has not been updated. Select the **Rebuild** icon on the *Menu Bar*.

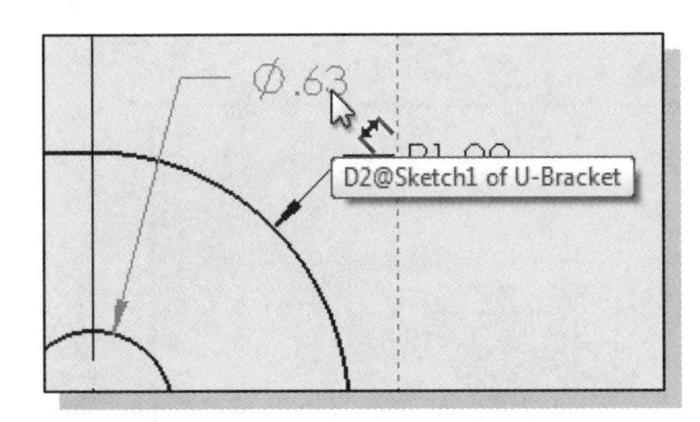

➢ Notice the change made to the model – the changed hole diameter – is reflected in the drawing.

11. Inside the graphics window, left-click on the **0.63** dimension in the *top* view to bring up the *Dimension PropertyManager*.

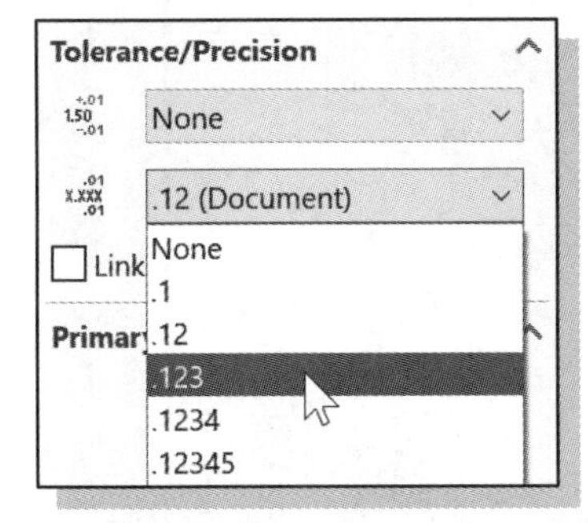

12. In the *Tolerance/Precision* panel of the *Dimension PropertyManager*, set the **Unit Precision** option to **3 digits after the decimal point** as shown.

13. Click **OK** in the *PropertyManager* to accept the new setting. Notice the accurate dimension of **.625** is now displayed.

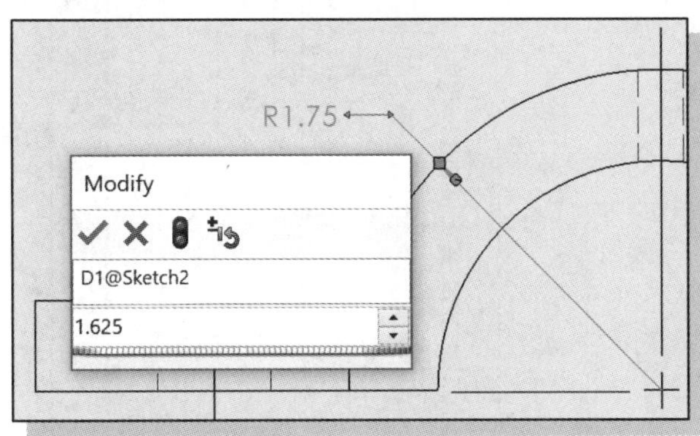

14. Inside the graphics window, **double-click** with the **left-mouse-button** on the **R 1.75** dimension in the *front* view to bring up the *Modify* pop-up option window.

15. Change the dimension to **1.625**.

16. Click on the **check mark** button in the *Modify* window to accept the setting.

17. In the *Dimension PropertyManager*, set the **Unit Precision** option to **3 digits after the decimal point**.

18. Click **OK** in the *PropertyManager* to accept the new setting.

➢ Notice the views become shaded with cross-hatching. This is an indication by SOLIDWORKS that changes have been made and a rebuild is needed to update the drawing and model.

19. Select the **Rebuild** icon on the *Menu Bar*.

❖ Note the geometry of the cut feature is updated in all views automatically.

❖ On your own, switch to the *Part Modeling* mode and confirm the design is updated as well.

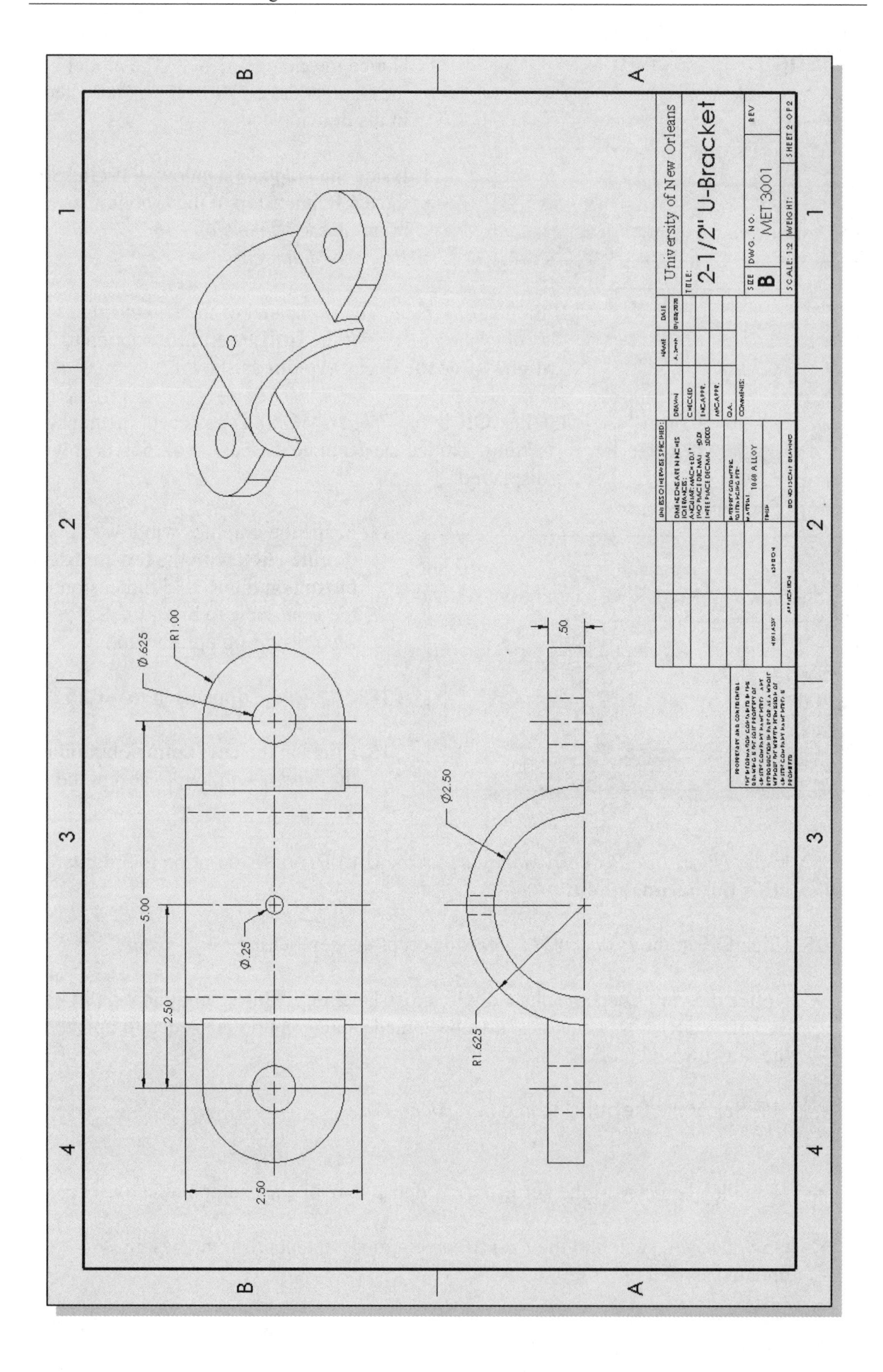
University of New Orleans
2-1/2" U-Bracket
MET 3001
SCALE: 1:2
SHEET 2 OF 2
1060 ALLOY
Ø.625
R1.00
Ø.25
5.00
2.50
2.50
Ø2.50
R1.625
.50

Saving the Drawing File

1. Go to *Drawing* mode with ***MET 3001.SLDDRW*** as the active document.

2. Select the **Save** icon on the *Menu Bar*. This command will save the **MET 3001.SLDDRW** file. This file includes the document properties, the sheet format, and the drawing views.

3. In the *Save Modified Documents* window, select **Save All**.

4. If the SOLIDWORKS window appears asking to *'Update the drawing?'*, select **Yes**.

Creating a Drawing Template

In SOLIDWORKS, each new drawing is created from a template. During the installation of SOLIDWORKS, a default drafting standard was selected which sets the default template used to create drawings. We can use this template or another predefined template, modify one of the predefined templates, or create our own templates to enforce drafting standards and other conventions. Any drawing file can be used as a template; a drawing file becomes a template when it is saved as a template (*.drwdot) file. Once the template is saved, we can create a new drawing file using the new template.

We will create a custom SOLIDWORKS drawing template file with the document properties (e.g., ANSI dimension standard, English units, etc.) and sheet format (size B paper size, etc.) we used for the U-Bracket drawing. The template can include the layout of the drawing views, but we will want to choose different views in future drawings.

1. On your own, open SOLIDWORKS. If the *Welcome* dialog box does not appear automatically, click the *Welcome to SolidWorks* icon in the *Task Pane* or on the *Menu Bar*.

2. Click once with the left-mouse-button on the **Advanced** icon in the *Welcome* dialog box.

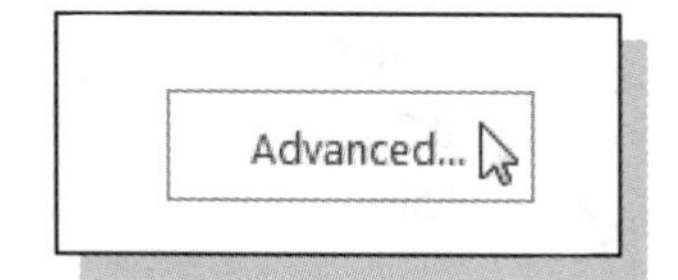

3. In the *New* SOLIDWORKS *Document* window, select the **Tutorial** tab.

4. Select the **draw** template file icon.

5. Select **OK** to open a drawing using the *draw* template.

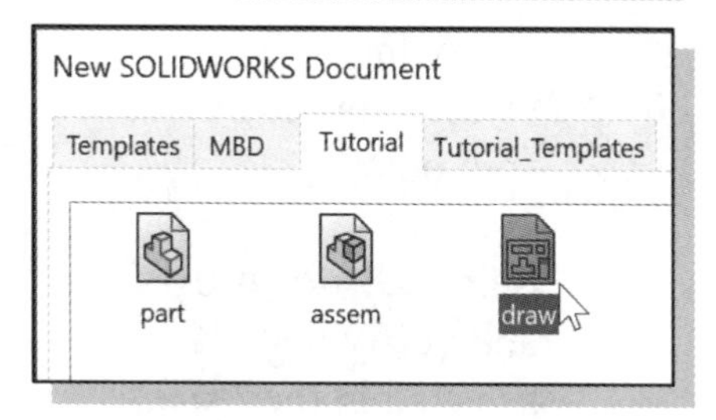

6. The *Model View PropertyManager* opens automatically. Select the **Cancel** icon or press the [**Esc**] key to close the *PropertyManager*.

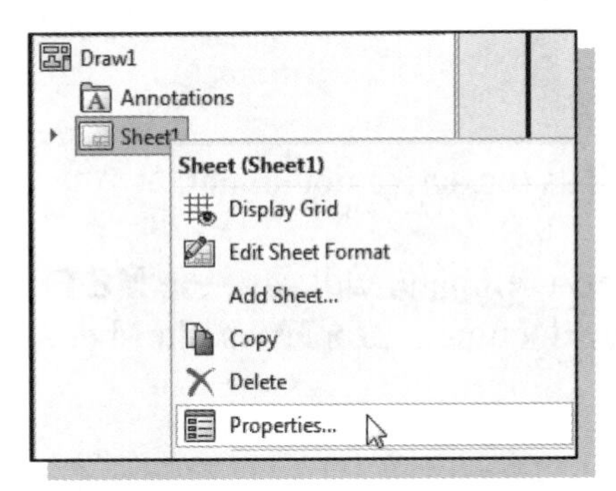

7. Move the cursor over **Sheet1** in the *FeatureManager Design Tree* and click once with the **right-mouse-button**.

8. Select **Properties** from the pop-up option menu.

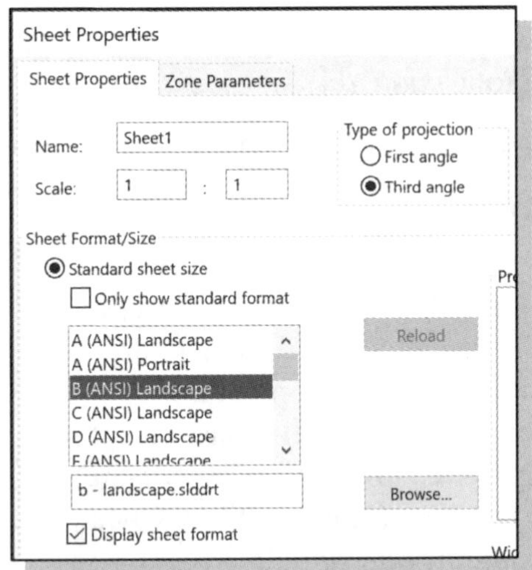

9. In the *Sheet Properties* window, set the scale to **1:1**, set the *Type of projection* to **Third angle**, and set the *Standard sheet size* to **B-Landscape.** (If necessary, uncheck the *Only show standard format* box).

10. Click **Apply Changes** in the *Sheet Properties* window to accept the selected settings.

11. Press the **[F]** key to scale the new sheet to fit the display area.

12. Select the **Options** icon, then select the **Document Properties** tab.

13. Select **Drafting Standard** and set the *Overall drafting standard* to **ANSI**.

14. Select **Units** and verify that **IPS (inch, pound, second)** is selected under the *Unit system* options. Also verify that **.12** is selected in the *Decimals* spin boxes.

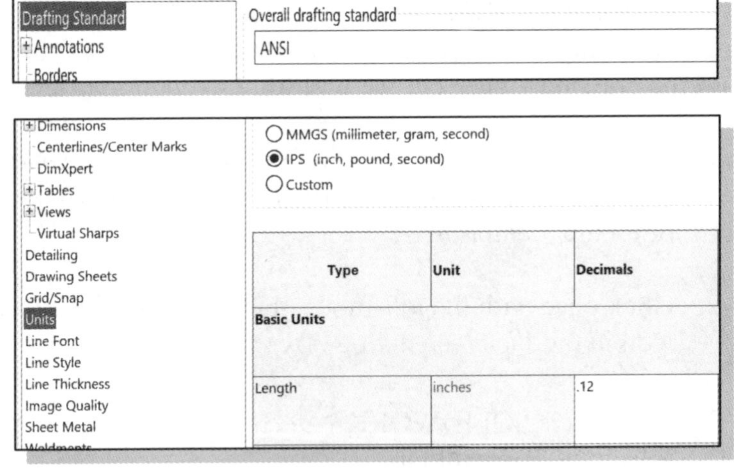

15. Select **Dimensions** and verify that **.12** is selected in the *Primary precision* spin box.

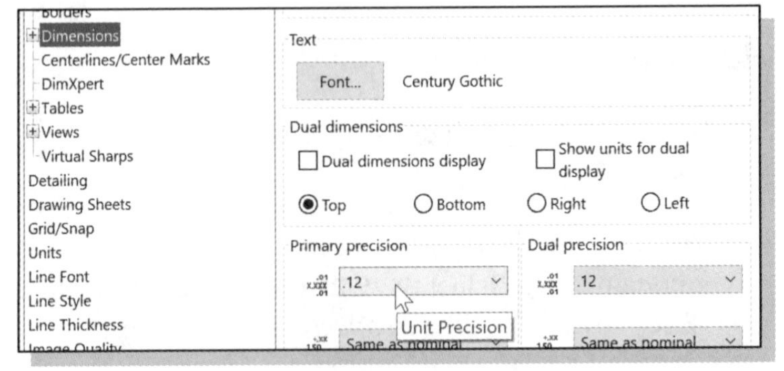

16. Select **Detailing** and **Uncheck** all of the *Auto insert on view creation* options at the right of the window as shown.

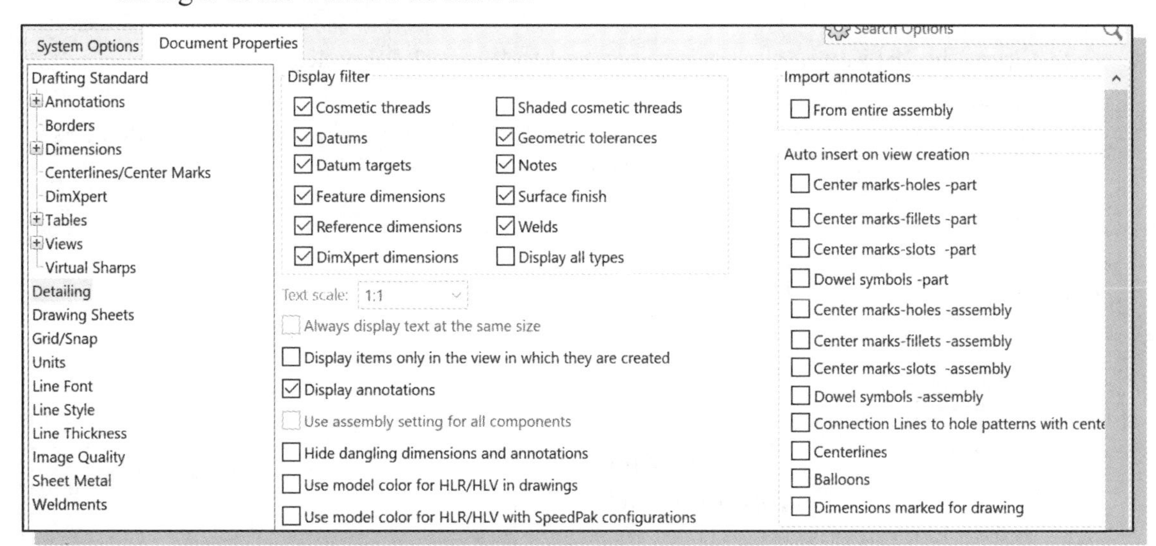

17. Select the **File Properties** option from the *Menu Bar* to open the *Summary Information* dialog box.

18. In the *Summary Information* window select the **Custom** tab. On your own, add the new properties as shown.

	Property Name	Type	Value / Text Expression	Evaluated Value	
1	SWFormatSize	Text	11in*17in	11in*17in	
2	CompanyName	Text	University of New Orleans	University of New Orleans	
3	DrawnBy	Text	A. Smith	A. Smith	
4	DrawnDate	Text	01/08/2021	01/08/2021	
5	<Type a new property				

19. Click **OK** in the *Summary Information* window to accept the change.

➢ Notice the entries now automatically appear in the title block.

20. Move the cursor into the graphics area and **right-click** to open the pop-up option menu, then select **Edit Sheet Format** to change to *Edit Sheet Format* mode.

21. On your own, adjust the font sizes of the entries to fit in the block.

22. On your own, return to **Edit Sheet** mode.

23. Press the **[F]** key to scale the new sheet to fit the display area.

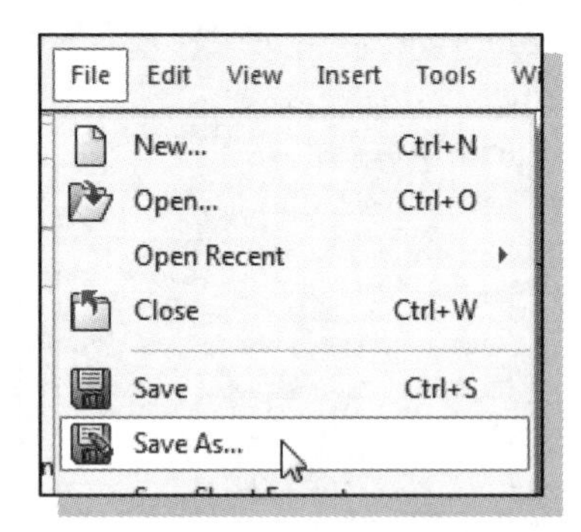

24. Select **Save as** from the *File* pull-down menu as shown.

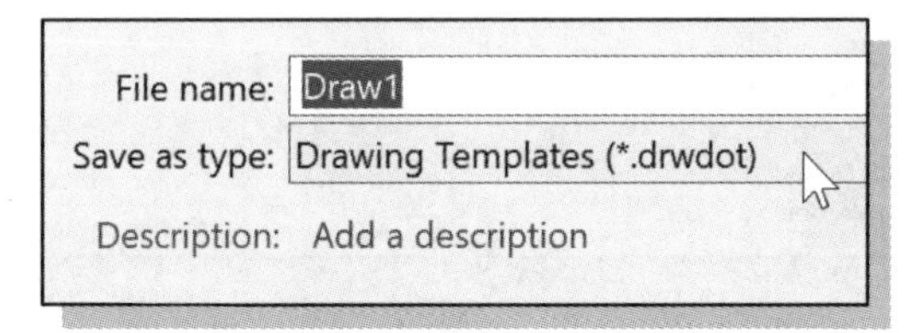

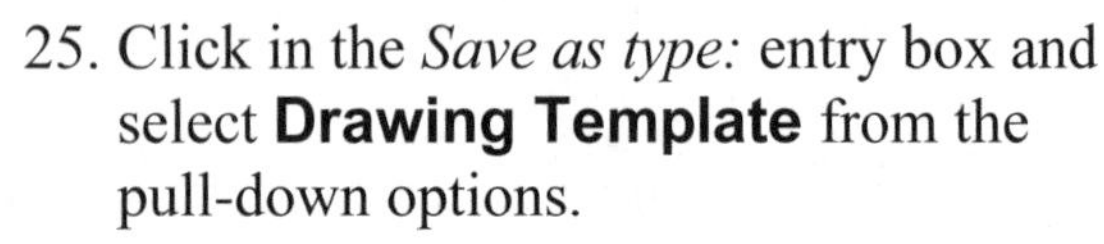

25. Click in the *Save as type:* entry box and select **Drawing Template** from the pull-down options.

26. The folder selection will automatically change to the default Templates folder. Use the browser to change the folder selection to the **Tutorial_Templates** folder you created in Chapter 4.

27. Enter ***ANSI-B-Inch*** for the *File name.*

28. Enter ***ANSI B Inch*** for the *Description.*

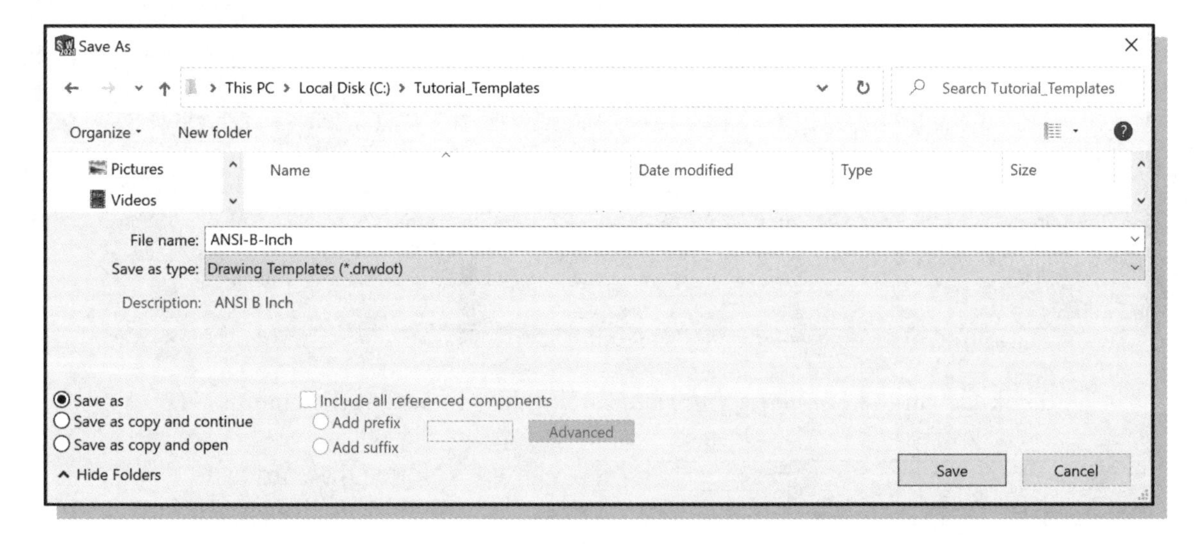

29. Select **Save**.

30. If the SOLIDWORKS pop-up window appears, click **OK**.

❖ Exit SOLIDWORKS. This drawing template will now be available for use to create future drawings.

Questions:

1. What does SOLIDWORKS' *associative functionality* allow us to do?
2. How do we move a view on the *Drawing Sheet*?
3. How do we display feature/model dimensions in the *Drawing* mode?
4. What is the difference between a *feature dimension* and a *reference dimension*?
5. How do we reposition dimensions?
6. What is a *base view*?
7. Identify and describe the following commands:

(a) 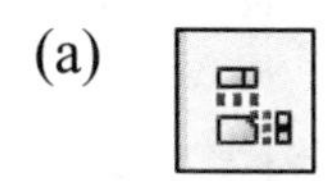

(b) 

(c) 

(d) 

Exercises: (Create the solid models and the associated 2D drawings.)

1. **Slide Mount** (Dimensions are in inches.)

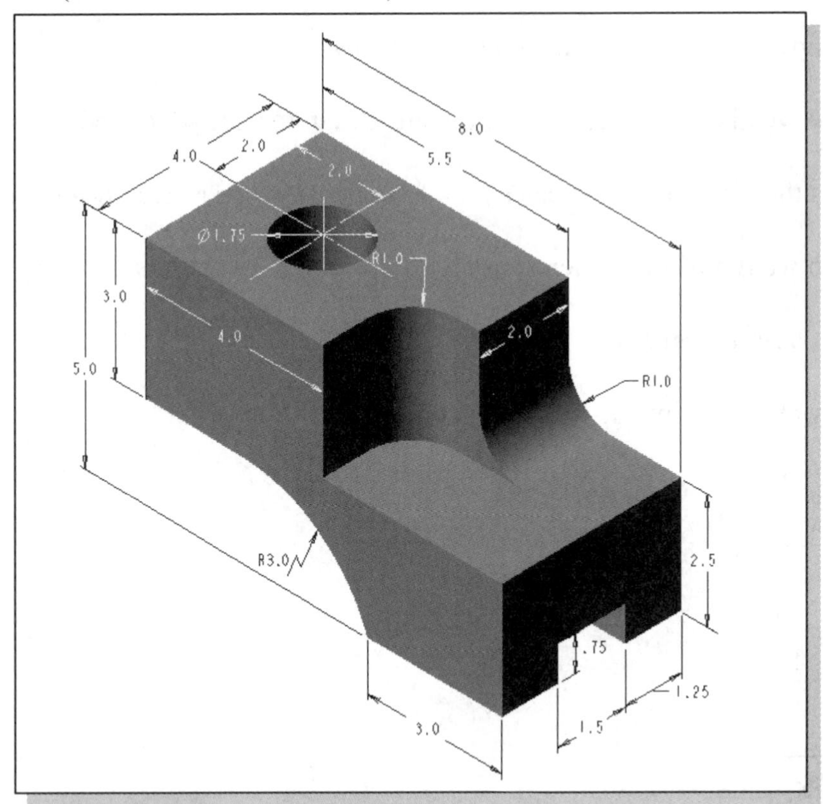

2. **Corner Stop** (Dimensions are in inches.)

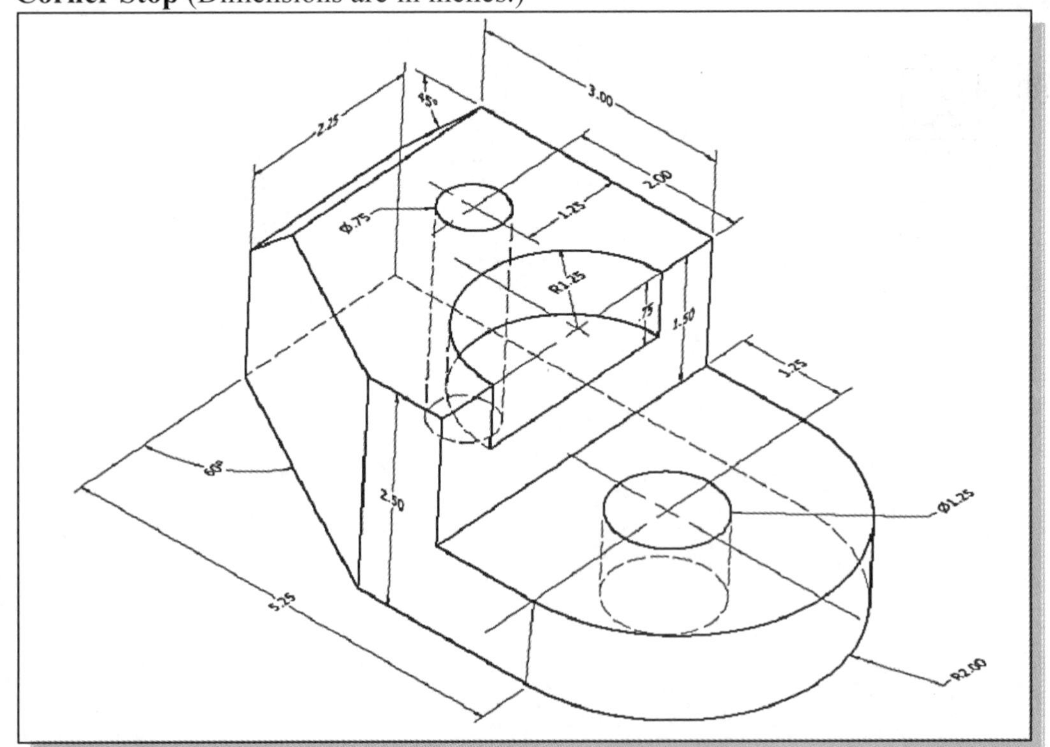

3. **Switch Base** (Dimensions are in inches.)

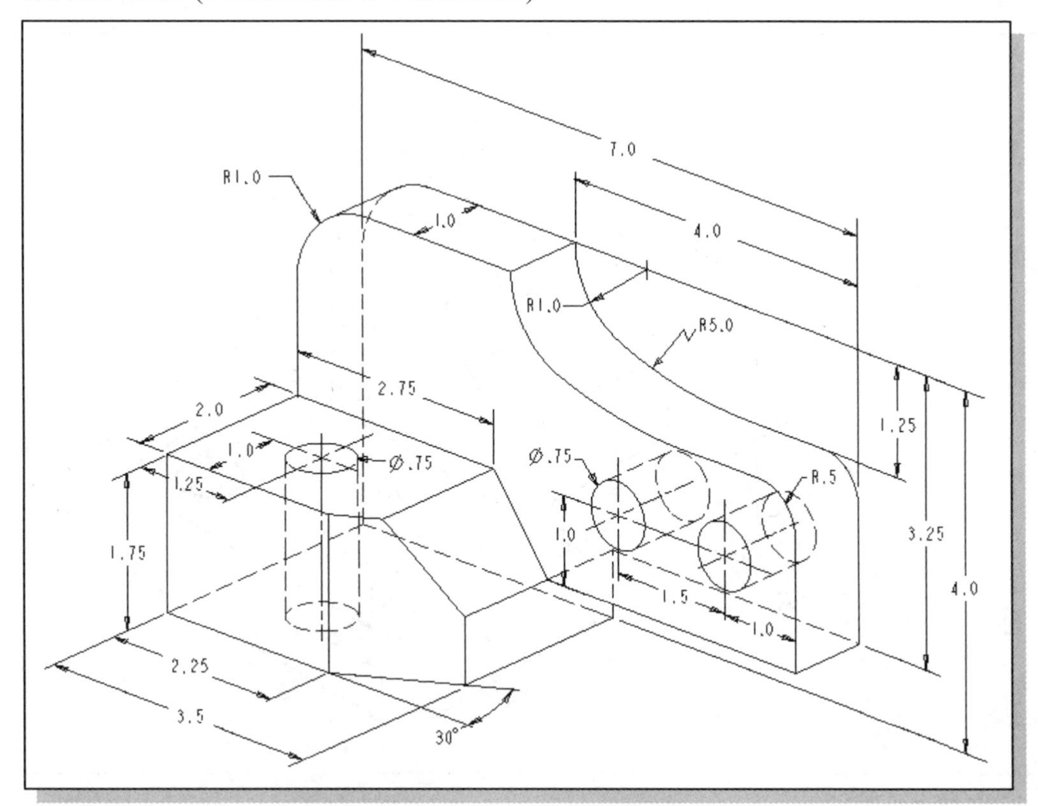

4. **Block Base** (Dimensions are in inches. Plate Thickness: 0.25)

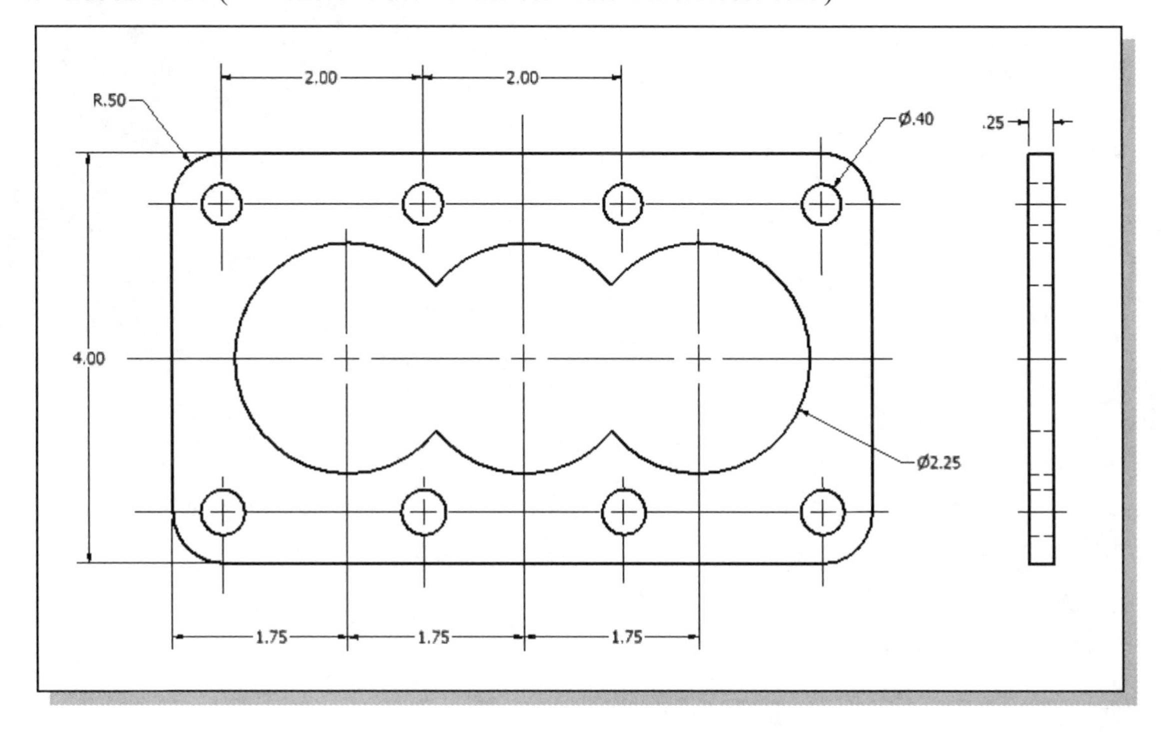

5. **Shaft Guide** (Dimensions are in inches.)

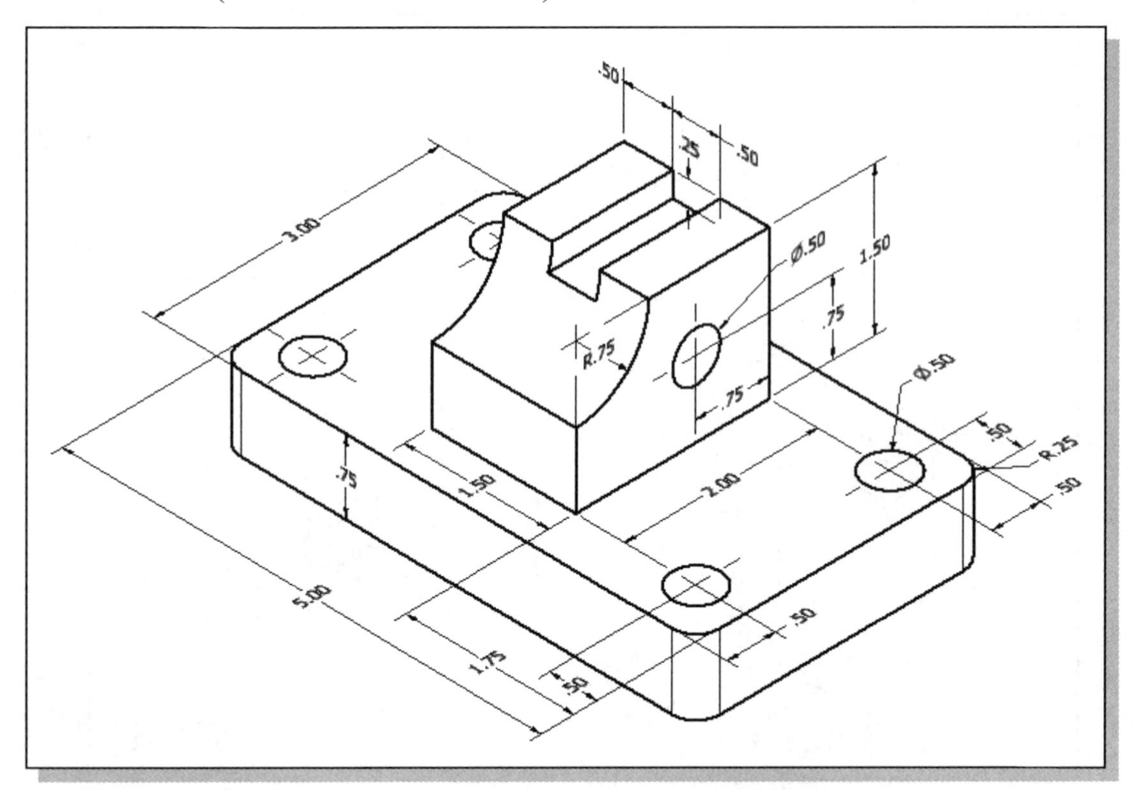

Chapter 9
Reference Geometry and Auxiliary Views

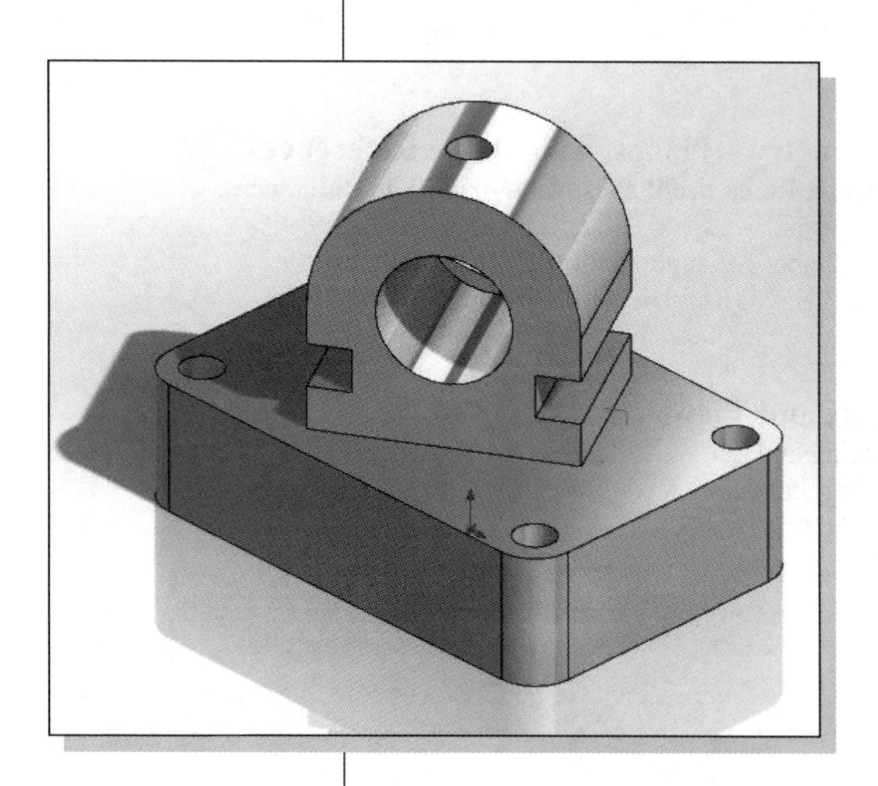

Learning Objectives

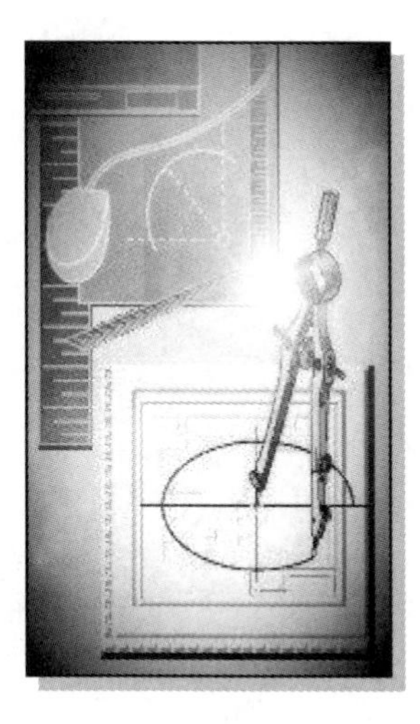

- ♦ **Understand the Concepts and the Use of Reference Geometry**
- ♦ **Use the Different Options to Create Reference Geometry**
- ♦ **Create Auxiliary Views in 2D Drawing Mode**
- ♦ **Create and Adjust Centerlines**
- ♦ **Create Shaded Images in the 2D Drawing mode**

Certified SOLIDWORKS Associate Exam Objectives Coverage

Sketch Relations

Objectives: Using Geometric Relations.

Reference Geometry – Planes, Axis, Mate References

Objectives: Creating Reference Planes, Axes, and Mate References.

Drawing Sheets and Views

Objectives: Creating and Setting Properties for Drawing Sheets; Inserting and Editing Standard Views.

Reference Geometry

Feature-based parametric modeling is a cumulative process. The relationships that we define between features determine how a feature reacts when other features are changed. Because of this interaction, certain features must, by necessity, precede others. A new feature can use previously defined features to define information such as size, shape, location and orientation. SOLIDWORKS provides several tools to automate this process. *Reference geometry* can be thought of as user-definable datum, which are updated with the part geometry. We can create planes, axes, points, or local coordinate systems that do not already exist. Reference geometry can also be used to align features or to orient parts in an assembly. In this chapter, the use of the **Offset** option and the **Angled** option to create new reference planes, surfaces that do not already exist, is illustrated. By creating parametric reference geometry, the established feature interactions in the CAD database assure the capturing of the design intent. The default planes, which are aligned to the origin of the coordinate system, can be used to assist the construction of the more complex geometric features.

Auxiliary Views in 2D Drawings

An important rule concerning multiview drawings is to draw enough views to accurately describe the design. This usually requires two or three of the regular views, such as a front view, a top view and/or a side view. However, many designs have features located on inclined surfaces that are not parallel to the regular planes of projection. To truly describe the feature, the true shape of the feature must be shown using an **auxiliary view**. An *auxiliary view* has a line of sight that is perpendicular to the inclined surface, as viewed looking directly at the inclined surface. An *auxiliary view* is a supplementary view that can be constructed from any of the regular views. Using the solid model as the starting point for a design, auxiliary views can be easily created in 2D drawings. In this chapter, the general procedure of creating auxiliary views in 2D drawings from solid models is illustrated.

The *Rod-Guide* Design

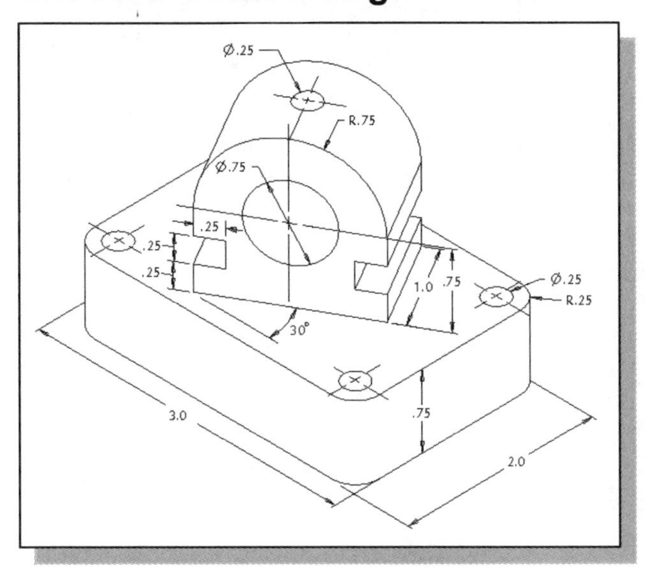

Based on your knowledge of SOLIDWORKS so far, how would you create this design? What are the more difficult features involved in the design? Take a few minutes to consider a modeling strategy and do preliminary planning by sketching on a piece of paper. You are also encouraged to create the design on your own prior to following through the tutorial.

Modeling Strategy

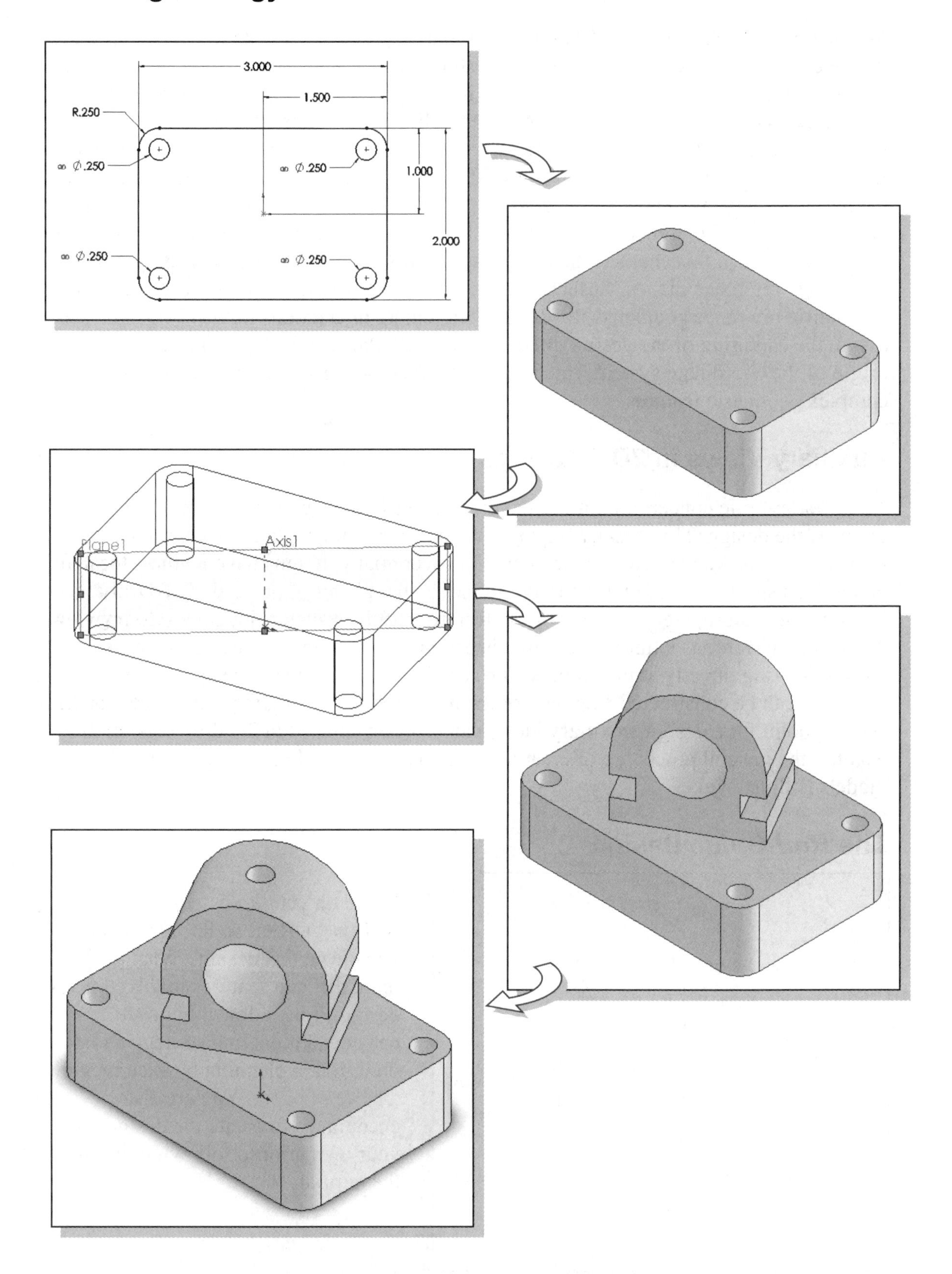

Starting SOLIDWORKS

1. On your own, open SOLIDWORKS. If the *Welcome* dialog box does not appear automatically, click the *Welcome to SolidWorks* icon in the *Task Pane* or on the *Menu Bar*.

2. In order to access the custom template you created in the last Chapter 4, click once with the left-mouse-button on the **Advanced** icon in the *Welcome* dialog box.

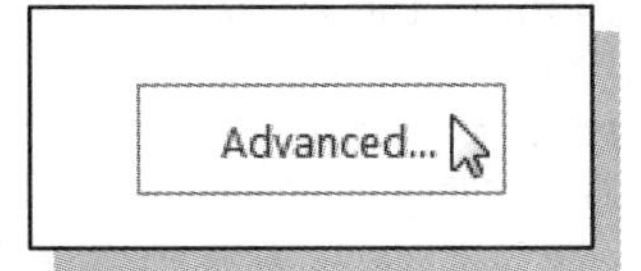

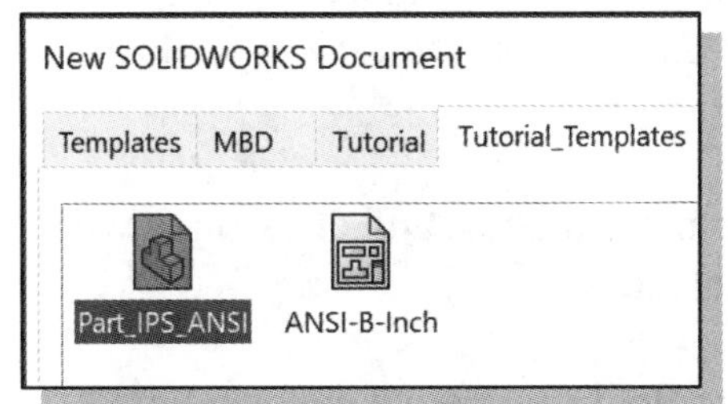

3. Select the **Tutorial_Templates** tab. (**NOTE:** If your templates are not available, open the standard Part template and set the *Drafting Standard* to ANSI and the *Units* to IPS.)

4. Select the **Part_IPS_ANSI** template as shown.

5. Click on the **OK** button to open a new document.

Applying the BORN Technique

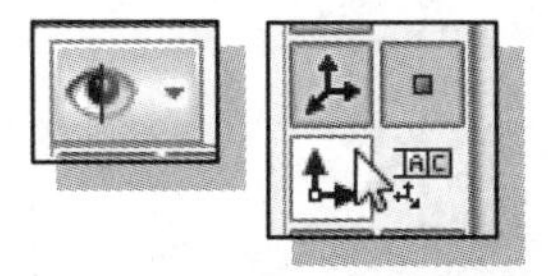

1. Select the **Hide/Show Items** icon on the *Heads-up View* toolbar to reveal the pull-down menu. In the pull-down menu, toggle on the **View Planes** and **View Origins** buttons.

2. In the *FeatureManager Design Tree* window, select the **Front, Top, and Right Planes** by holding down the **[Ctrl]** key and clicking with the left-mouse-button.

3. Click on any of the planes to display the option menu. Click on **Show** to toggle *ON* the display of the planes.

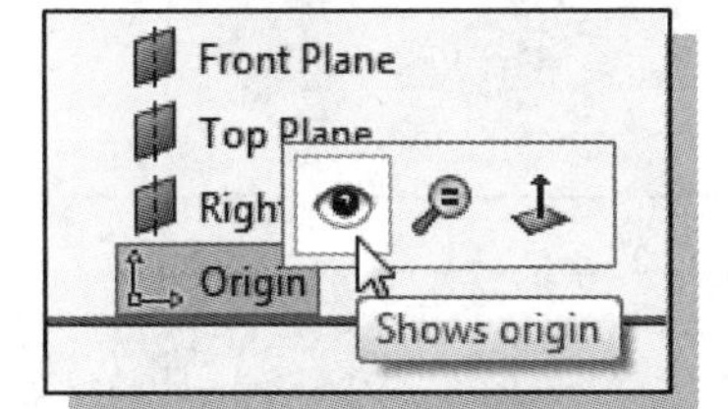

4. On your own, toggle *ON* the display of the **Origin**.

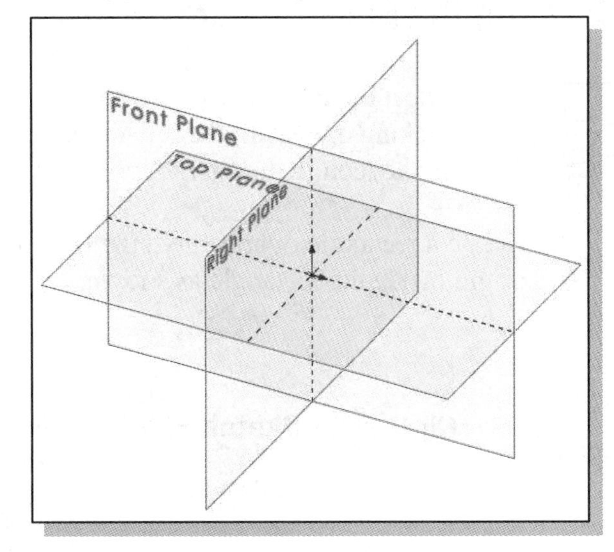

5. On your own, use the viewing options (Rotate, Zoom and Pan) to view the work features established.

IMPORTANT NOTE: In this lesson and following lessons, we will use the standard display of toolbars. If a user prefers to use the *CommandManager*, the only change that may be necessary is to select the appropriate tab prior to selecting a command. For example, if the instruction is to "select the **Extruded Boss** command from the *Features* toolbar," it may be necessary to first select the **Features** tab on the *CommandManager* to display the *Features* toolbar.

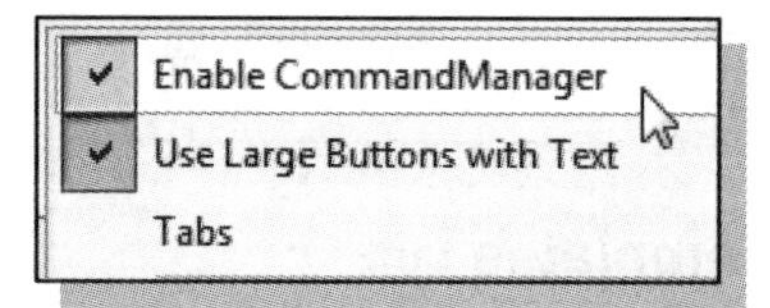

6. To turn ***OFF*** the *CommandManager* and use the standard display of toolbars, right click on the *CommandManager* (or any other toolbar) and toggle the *Enable CommandManager* option *OFF* by selecting it at the top of the pop-up menu.

7. Move the cursor into the graphics area, away from the planes, and click once with the **left-mouse-button** to ensure no planes are selected.

8. Select the **Sketch** button on the *Sketch* toolbar to create a new sketch.

9. In the *Edit Sketch PropertyManager*, the message "*Select a plane on which to create a sketch for the entity*" is displayed. Move the cursor over the edge of the **Top Plane** in the graphics area. When the Top Plane is highlighted, click once with the **left-mouse-button** to select the Top (XZ) Plane as the sketch plane for the new sketch.

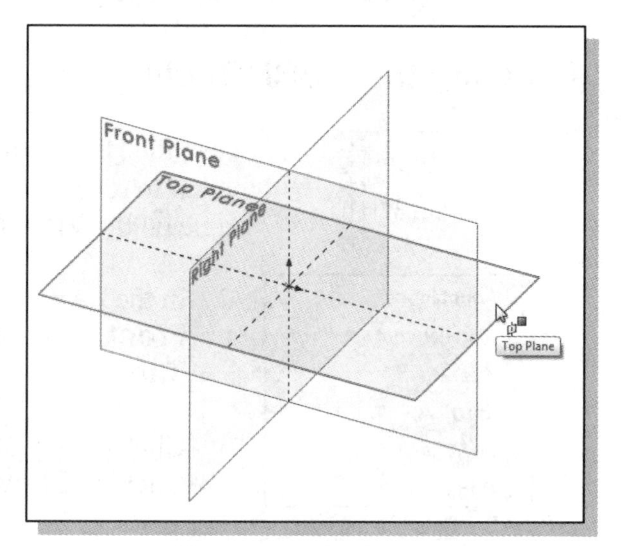

10. On your own, **hide** the three planes.

Creating the Base Feature

1. Select the **Rectangle** command by clicking once with the left-mouse-button on the icon in the *Sketch* toolbar.

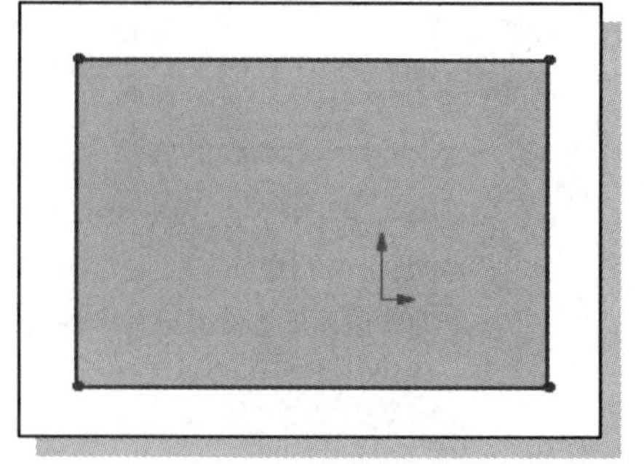

2. Create a rectangle of arbitrary size with the origin inside the rectangle as shown.

3. Click on the **Sketch Fillet** icon in the *Sketch* toolbar.

4. Enter **0.25 in** for the fillet radius in the *Fillet Parameters* panel of the *Sketch Fillet Property-Manager*.

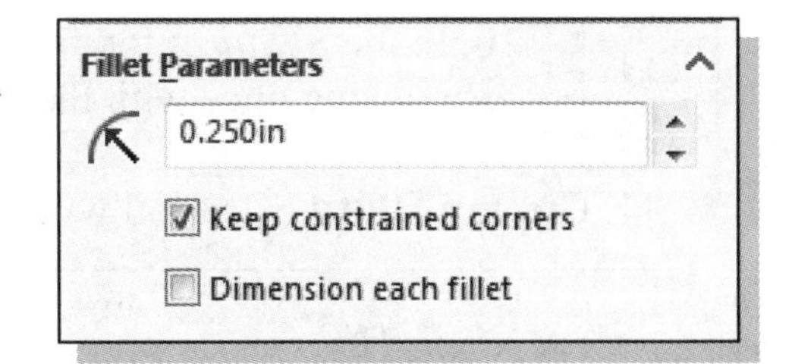

5. Create four rounded corners as shown. Each fillet can be created by moving the cursor into the graphics area and (1) clicking on the corner, or (2) clicking on the two lines forming the corner.

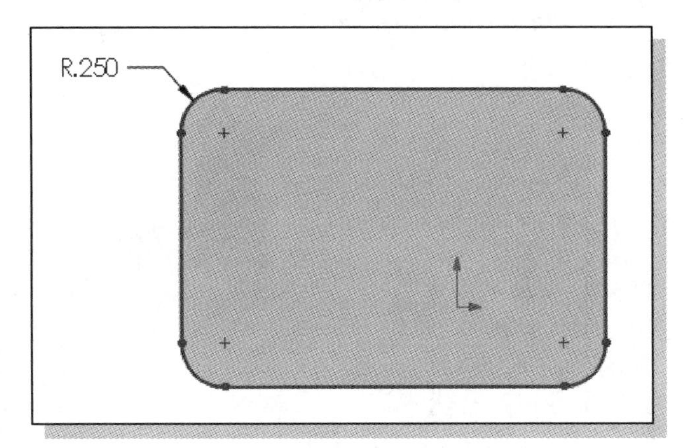

6. Select **OK** in the *PropertyManager* or press the **[Esc]** key to exit the Sketch Fillet command.

7. On your own, create four circles of the same diameter and with the centers aligned to the centers of arcs. (**NOTE:** Create a *Global Variable* for the diameter dimensions – see page 5-27.) Also create and modify the six dimensions as shown in the figure below. Note that these dimensions fix the center of the base at the origin. (**NOTE:** If you cannot select the origin, make sure the visibility is toggled *ON* in the *FeatureManager Design Tree*.)

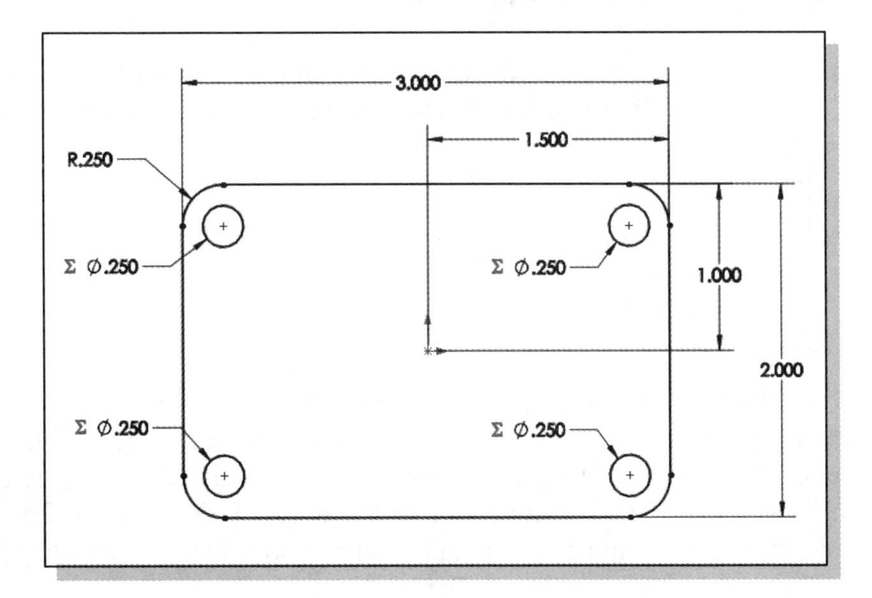

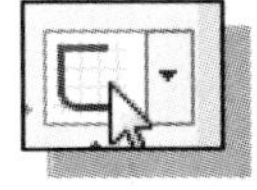

8. Select the **Sketch** icon on the *Sketch* toolbar to exit the Sketch option.

9. Make sure the sketch – Sketch1 – is selected.

10. In the *Features* toolbar, select the **Extruded Boss/Base** command by clicking once with the left-mouse-button on the icon.

11. Create a **Blind** extrusion with a distance of **0.75** in.

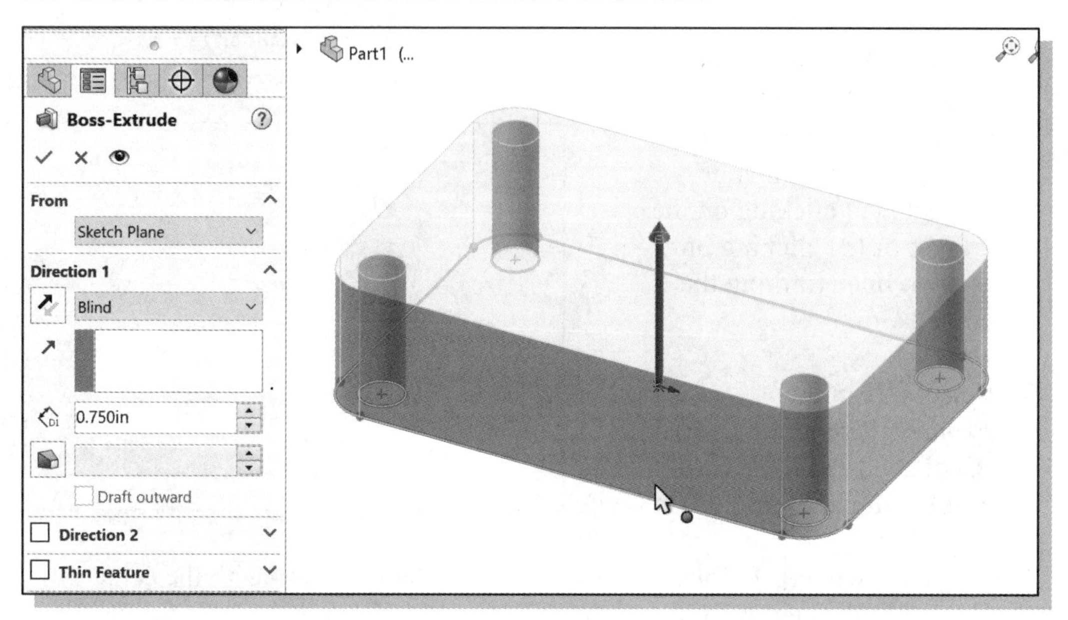

12. Click **OK** in the *Extrude PropertyManager* to accept the settings and create the extrusion.

Creating an Angled Reference Plane

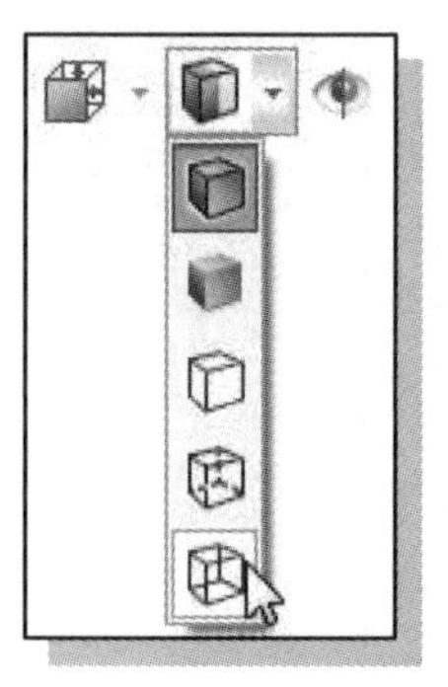

1. In the *Display Style* pull-down menu on the *Heads-up View* toolbar, select the **Wireframe** option to set the display mode to *Wireframe.*

❖ SOLIDWORKS allows the creation of a plane through an edge, axis, or sketch line at an angle to a face or plane, using the **At Angle** option of the **Plane** command. Using the **At Angle** option requires selection of an existing edge, axis, or sketch line and an existing face or plane. We will create a reference axis along the Y-direction, and create a plane that can be visualized by rotating the Front Plane about the reference (Y) axis.

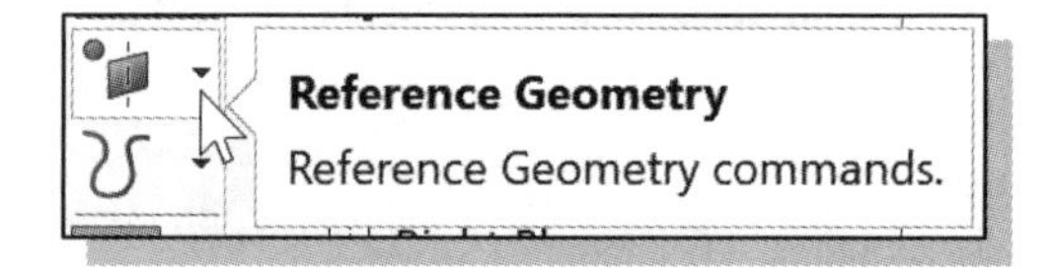

2. In the *Features* toolbar, select the **Reference Geometry** command by left-clicking the icon.

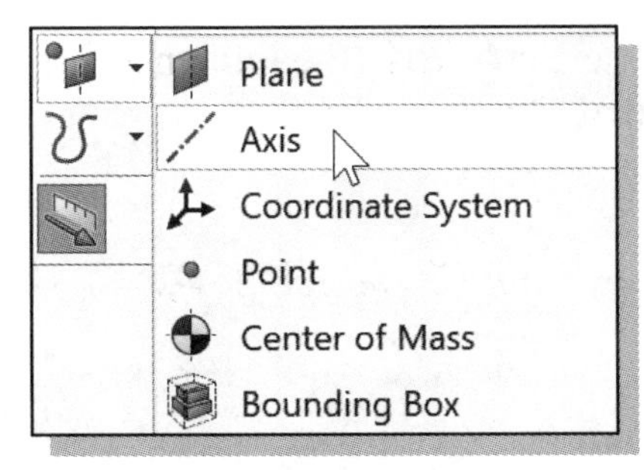

3. In the pull-down option menu of the Reference Geometry command, select the **Axis** option.

4. In the *Selections* panel of the *Axis PropertyManager*, select the **Two Planes** option. (We will define the Y-axis as the intersection of the Front and Right Planes.)

5. Select the **Front Plane** and **Right Plane** in the *Design Tree* appearing in the graphics area.

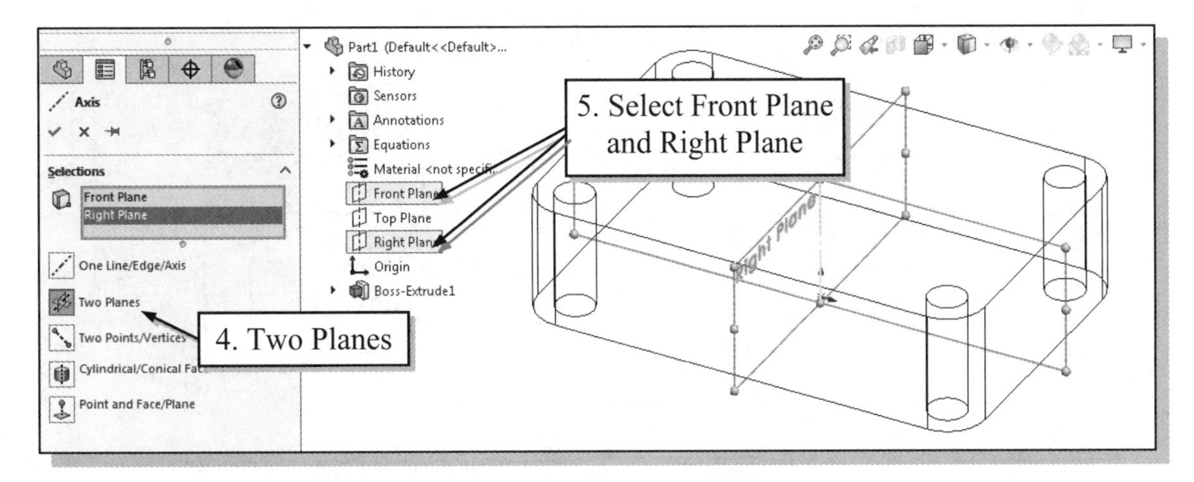

6. Click **OK** in the *PropertyManager* to create the new reference axis.

7. Press the **[Esc]** key to make sure no objects are selected.

➢ Notice the reference axis, named **Axis1**, has been created. It is aligned with the Y-axis and passes through the center of the base feature.

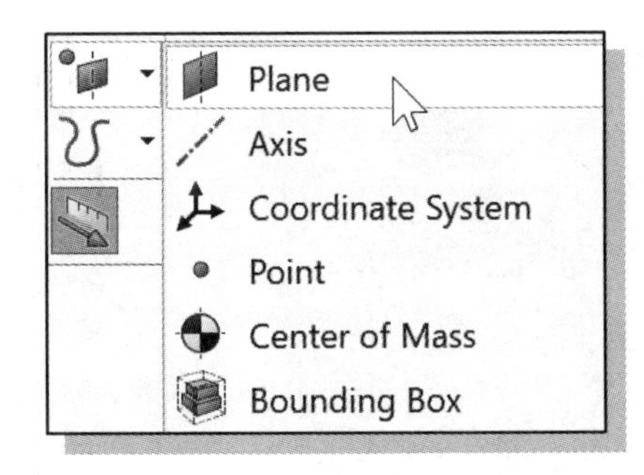

8. Select the **Reference Geometry** command from the *Features* toolbar, and select the **Plane** option from the pull-down menu.

➢ Notice the message "*Select references and constraints*" appears and the *First Reference* window is highlighted.

9. In the *Design Tree* appearing in the graphics area, select **Axis1** (the reference axis created above) as the *First Reference* and **Front Plane** as the *Second Reference*.

10. In the *Second Reference* panel of the *Plane PropertyManager*, select the **At Angle** option button.

11. Enter **30 deg** as the angle in the *PropertyManager*.

12. Check the **Flip** box.

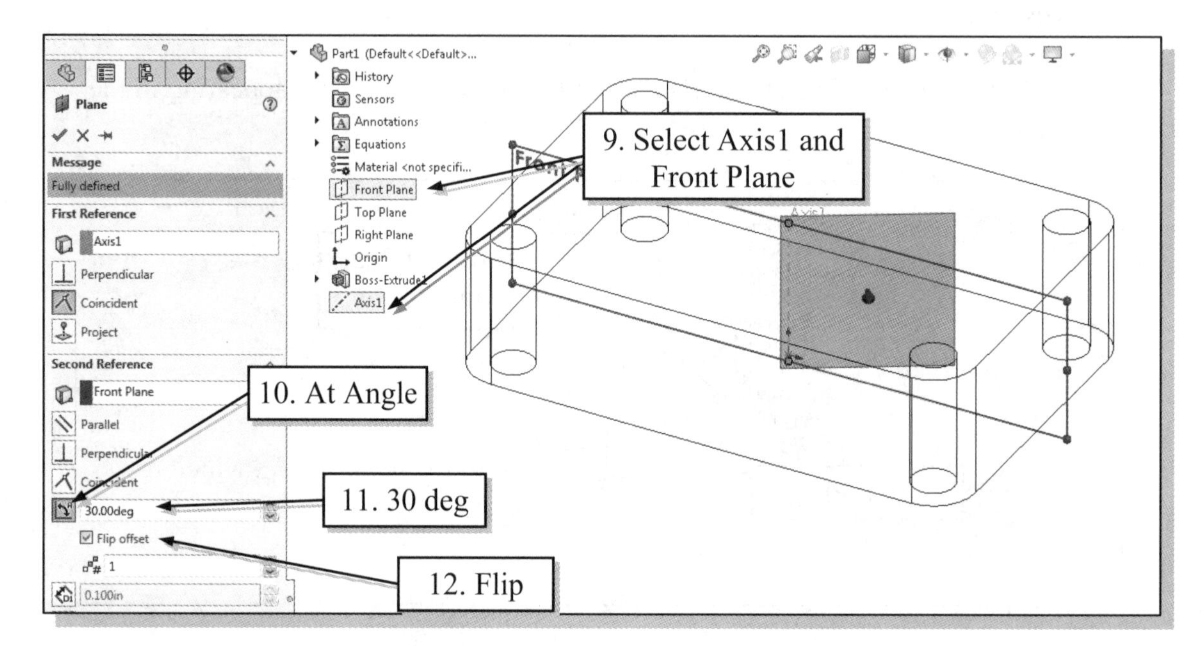

13. Check that the preview of the reference plane appears as shown above.

14. Click **OK** in the *PropertyManager* to create the new reference plane.

❖ Note that the *angle* is measured relative to the selected reference plane, the Front (XY) Plane.

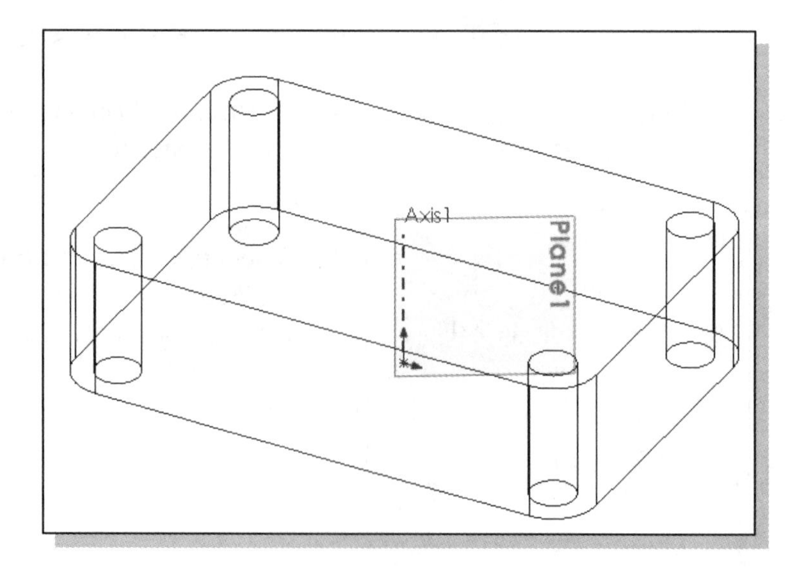

Creating a 2D Sketch on the Reference Plane

1. Press the **[Esc]** key twice to ensure no objects are selected.

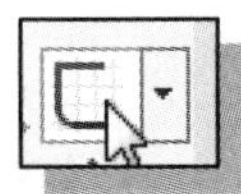

2. In the *Sketch* toolbar, select the **Sketch** command by left-clicking once on the icon.

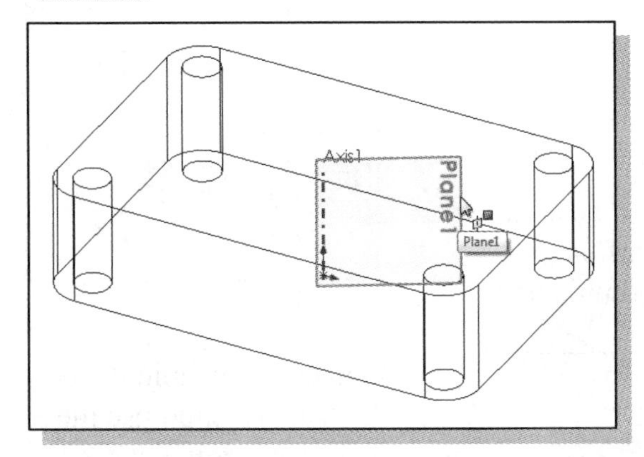

3. Select the angled reference plane **Plane1** as the sketch plane.

Using the Convert Entities Option

Projected geometry is another type of *reference geometry*. The Convert Entities tool can be used to project entities from previously defined sketches or features onto the sketch plane. The position of the projected geometry is fixed to the feature from which it was projected. We can use the Convert Entities tool to project geometry from a sketch or feature onto the active sketch plane.

Typical uses of the Convert Entities command include:

- Project a silhouette of a 3D feature onto the sketch plane for use in a 2D profile.
- Project a sketch from a feature onto the sketch plane so that the projected sketch can be used to constrain a new sketch.

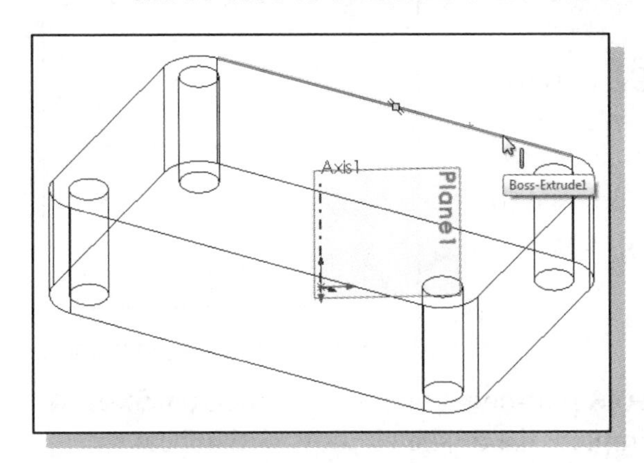

1. Move the cursor over the top rear edge of the base feature. When the edge is highlighted, click once with the **left-mouse-button** to select the edge.

2. In the *Sketch* toolbar, select the **Convert Entities** command by left-clicking the icon.

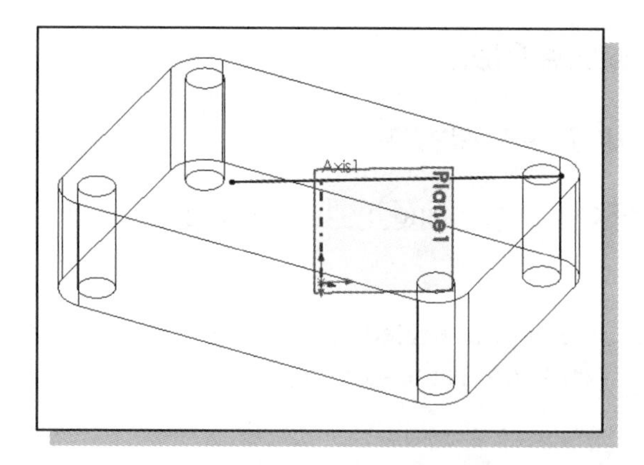

❖ Notice the selected edge is projected onto the active sketch plane as shown.

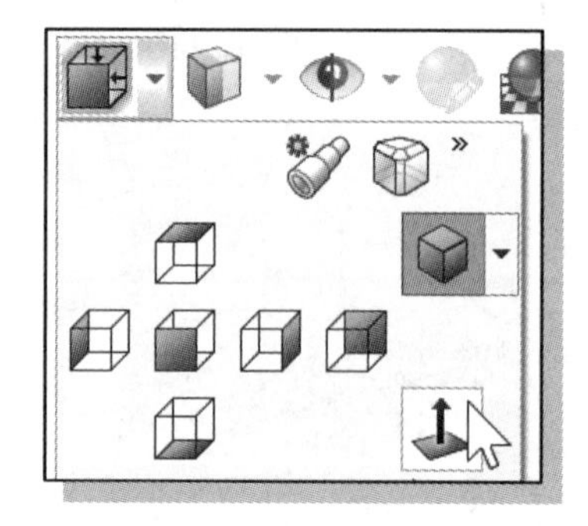

3. Select the **View Orientation** icon on the *Heads-up View* toolbar to reveal the *View Orientation* pull-down menu. Select the **Normal To** command.

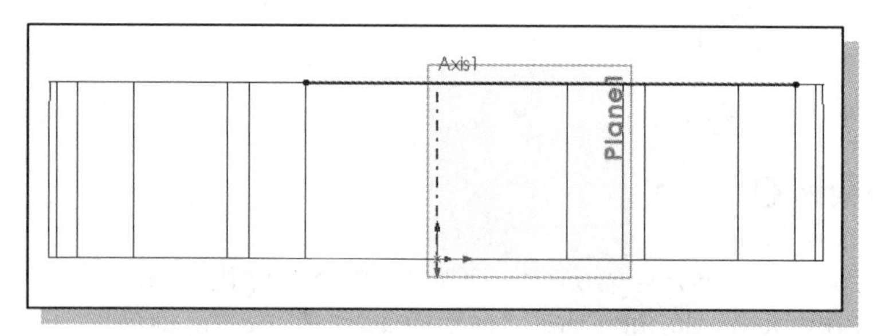

4. **If necessary**, hold down the **[Alt]** key and use the left/right arrow keys **[←]/[→]** to rotate the drawing clockwise/counterclockwise until it is oriented as shown.

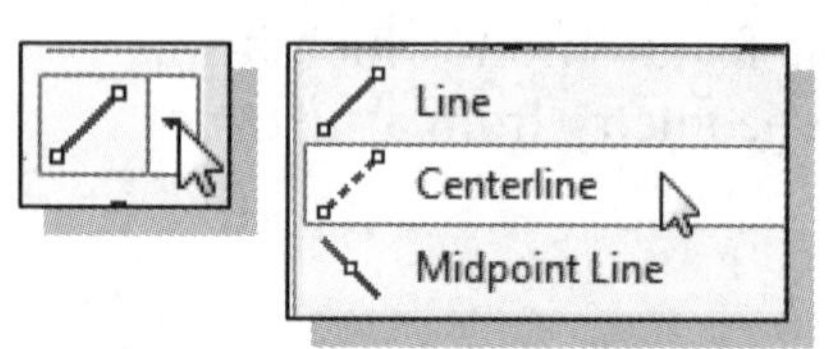

5. **Left-click** on the arrow next to the **Line** button on the *Sketch* toolbar to reveal additional commands, and select the **Centerline** command from the pop-up menu.

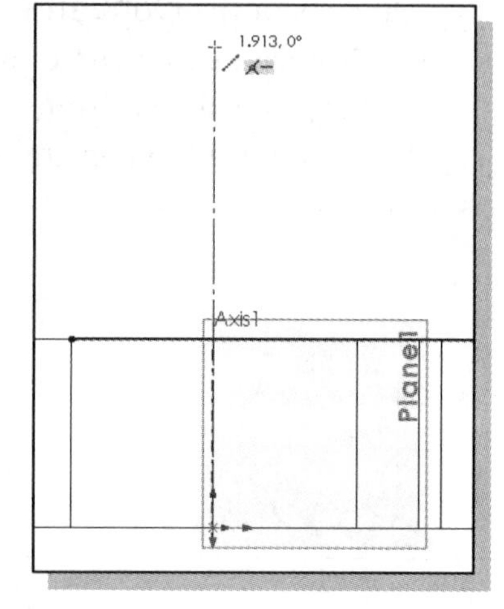

6. Draw a vertical centerline with one end coincident with the origin and the other end above the top of the base as shown.

7. Press the **[Esc]** key twice to exit the Centerline command and unselect the centerline you created.

❖ The **Dynamic Mirror** command will be used to generate sketch objects with symmetry about a centerline.

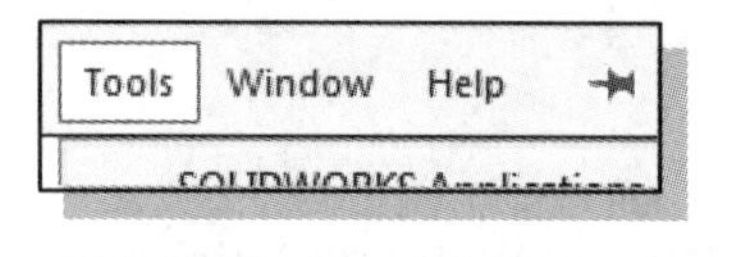

8. Select the **Tools** pull-down menu, then click on **Sketch Tools**. A complete list of sketch tools will appear. (Some of these also appear in the *Sketch* toolbar.)

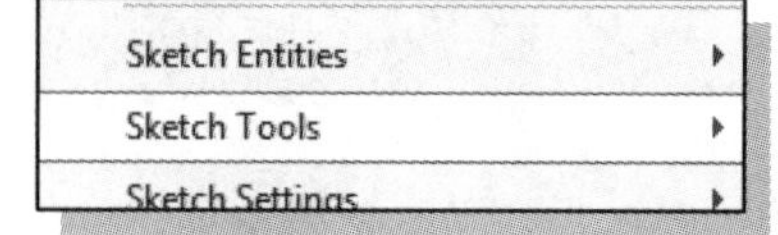

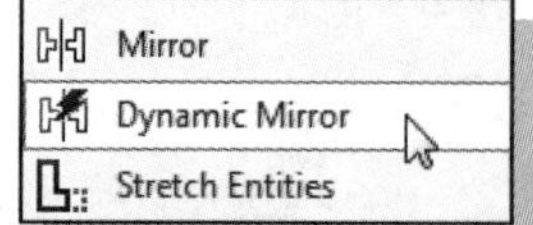

9. Select **Dynamic Mirror** as shown.

10. The message "*Please select a sketch line or linear edge to mirror about*" appears in the *Mirror PropertyManager*. Move the cursor over the **centerline** in the graphics area and click once with the **left-mouse-button** to select it as the line to mirror about.

➢ Notice the hash marks which appear on the centerline to indicate that the Dynamic Mirror command is active.

11. Click on the **Line** icon in the *Sketch* toolbar.

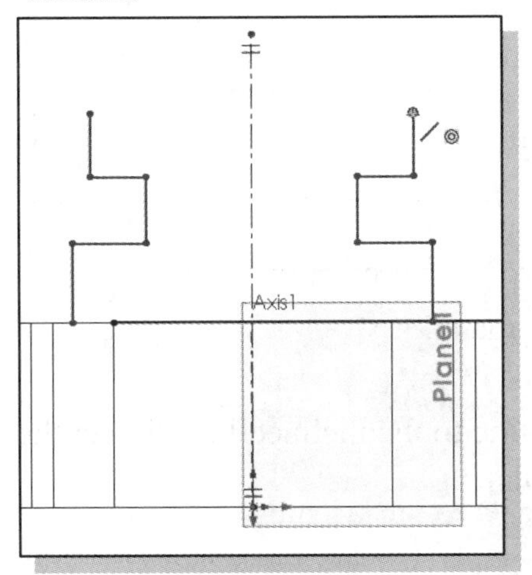

12. On your own, create a rough sketch using the projected edge as the bottom line as shown in the figure. (Note that all edges are either horizontal or vertical.)

13. Press the **[Esc]** key to exit the Line command.

14. On your own, **turn *OFF*** the Dynamic Mirror command. (**HINT:** Perform steps 9 and 10.)

15. Press the [**Esc**] key twice to ensure no entities are selected.

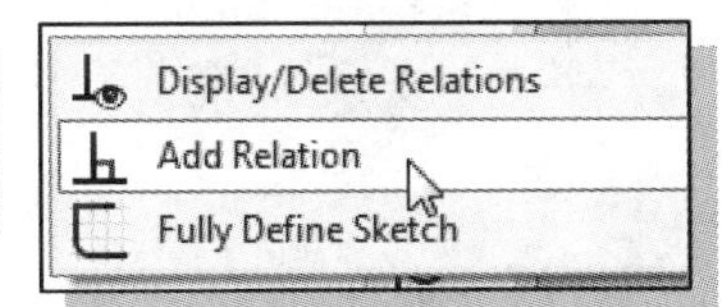

16. Select the **Add Relation** command from the *Display/Delete Relations* pop-up menu on the *Sketch* toolbar.

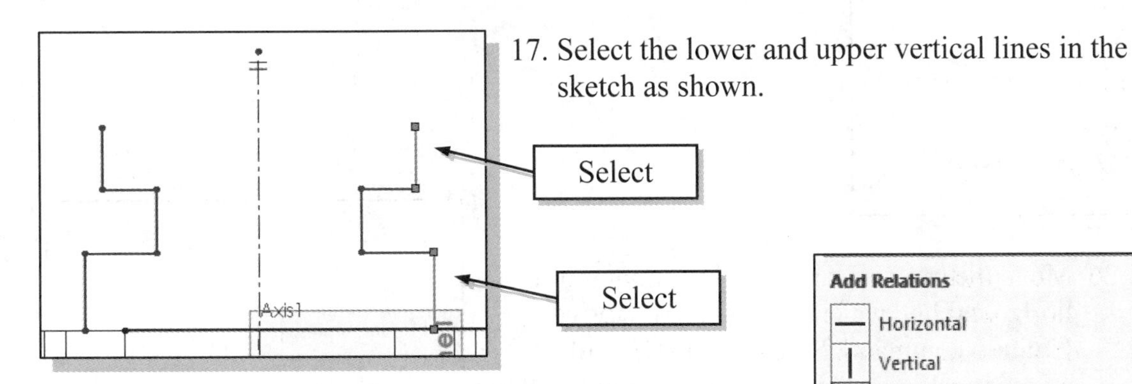

17. Select the lower and upper vertical lines in the sketch as shown.

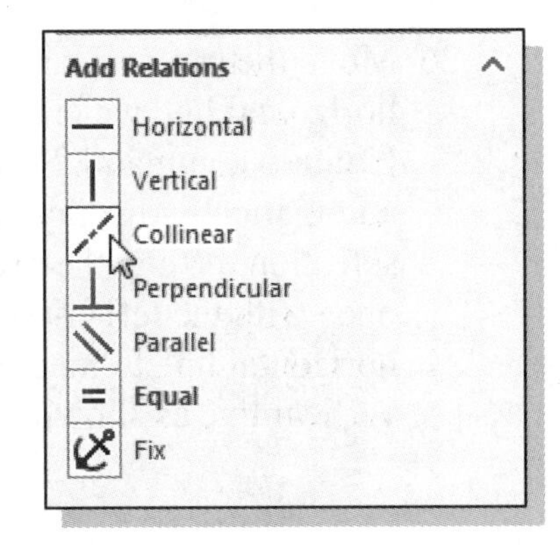

18. Select the **Collinear** option in the *Add Relations* panel of the *Add Relations PropertyManager*.

19. Press the **[Esc]** key to exit the Add Relations command.

20. **Left-click** on the **arrow** next to the **Trim Entities** button on the *Sketch* toolbar to reveal additional commands.

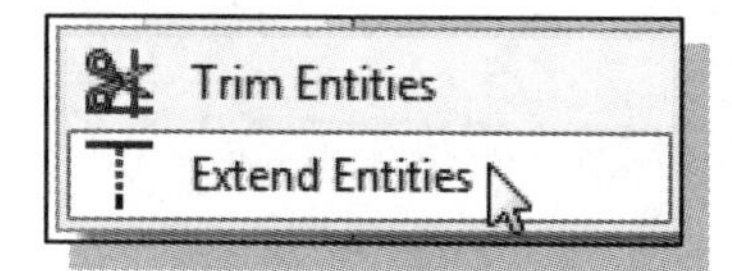

21. Select the **Extend Entities** command from the pop-up menu.

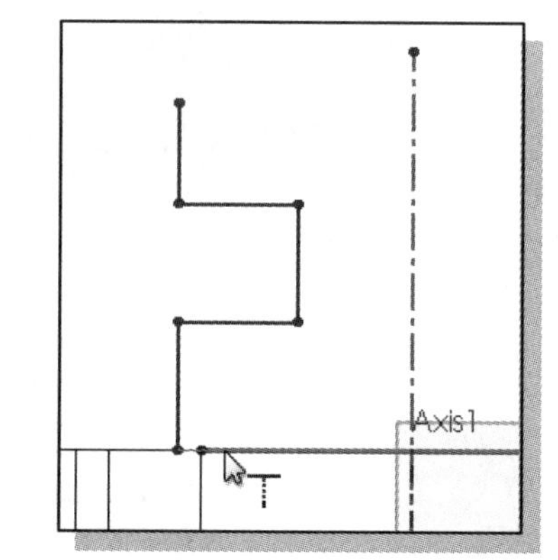

22. Move the cursor near the left endpoint of the horizontal line we created with the Convert Entities command. SOLIDWORKS will automatically display the possible result of the selection. When the extended line appears, click once with the **left-mouse-button** to extend the horizontal line to meet the bottom of the vertical line as shown.

23. Press the [**Esc**] key once to end the Extend command.

24. We will next trim the bottom horizontal line to the inclined line. Select the **Trim Entities** icon on the *Sketch* toolbar.

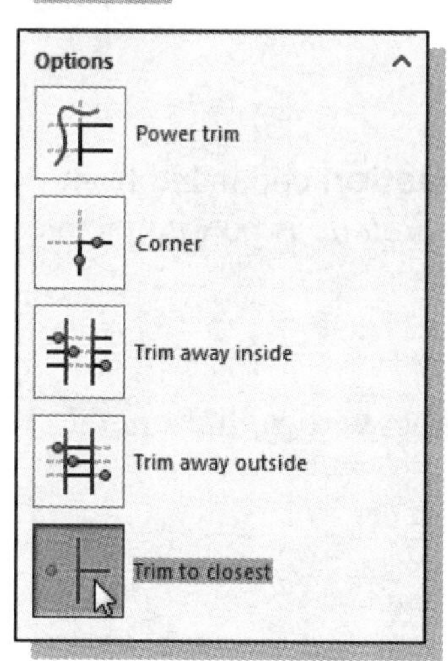

25. Select the **Trim to closest** option in the *Trim PropertyManager*.

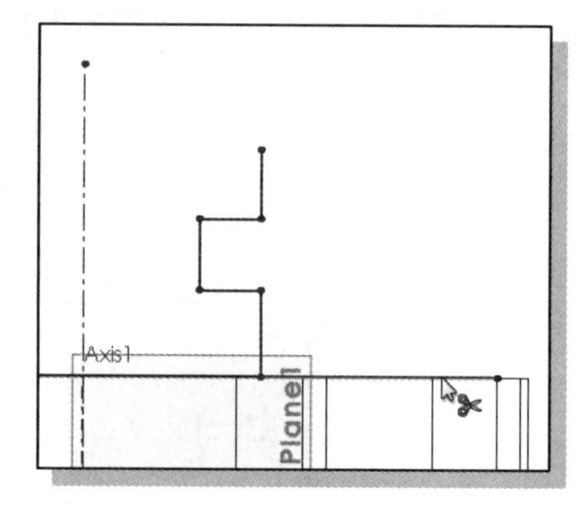

26. Move the cursor near the right endpoint of the horizontal line we created with the Convert Entities command. SOLIDWORKS will automatically display the possible result of the selection. When the preview appears, click once with the **left-mouse-button** to trim the horizontal line to meet the bottom of the vertical line as shown.

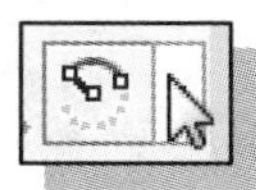

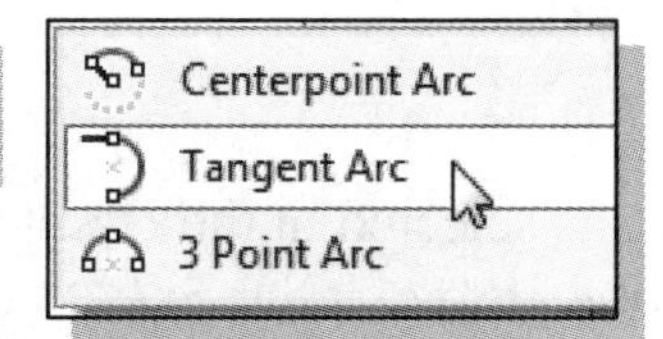

27. **Left-click** on the arrow next to the **Arc** button on the *Sketch* toolbar and select the **Centerpoint Arc** command from the pop-up menu.

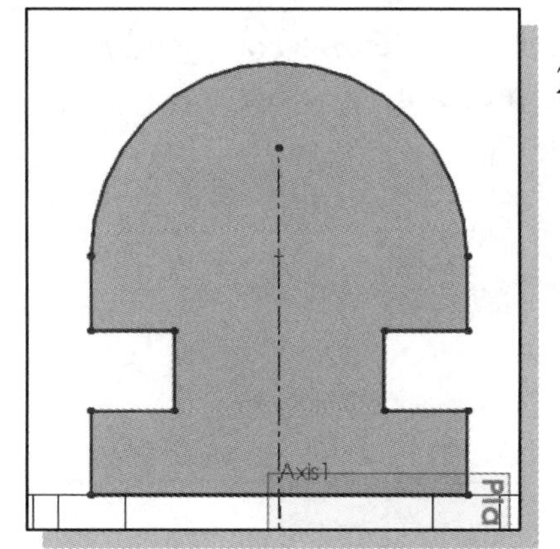

28. On your own, create the arc with the centerpoint coincident with the centerline and aligned with the top endpoints of the vertical lines as shown.

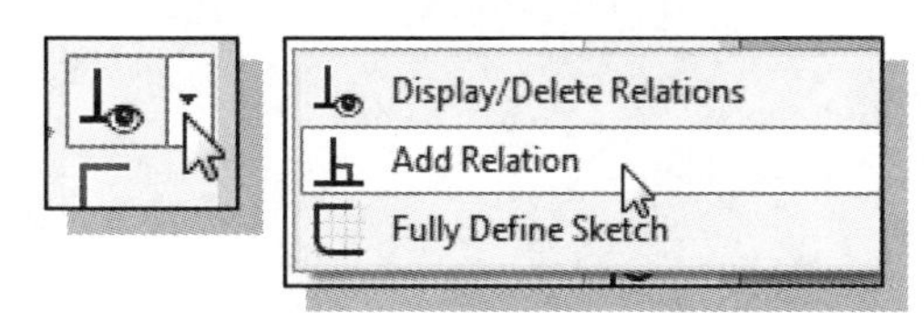

29. Use the **Add Relation** command to create a **Tangent** relation between the arc and the upper right vertical line.

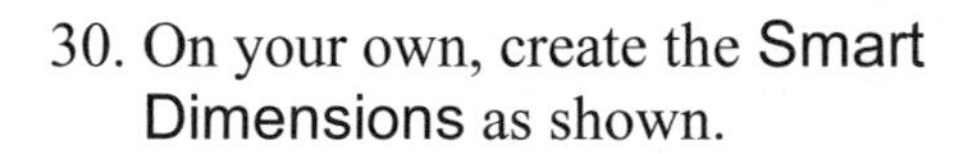

30. On your own, create the Smart Dimensions as shown.

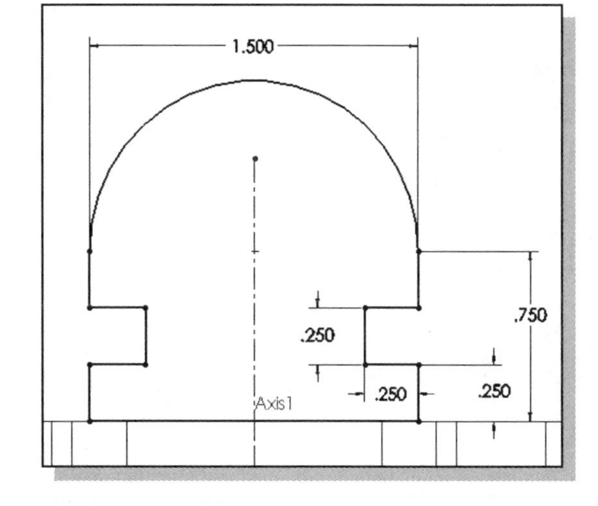

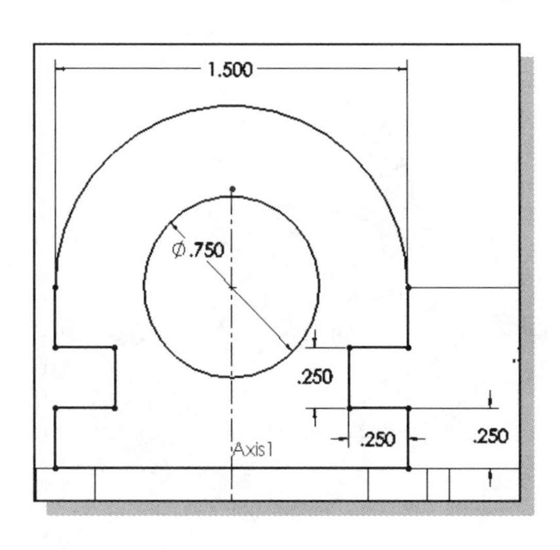

31. On your own, add a **0.75** inch circle concentric with the arc and complete the sketch as shown in the figure.

❖ The *Status Bar* should now read *Fully Defined* because the geometric and dimensional constraints fully define the sketch.

Completing the Solid Feature

1. Make sure the sketch you just completed is selected in the *Design Tree*. Select the **Sketch** icon on the *Sketch* toolbar to exit the Sketch option.

2. In the *Features* toolbar, select the **Extruded Boss/Base** command by clicking once with the left-mouse-button on the icon. The *Extrude PropertyManager* is displayed in the left panel.

3. Select the **Midplane** option.

4. Enter **1 inch** for the extrusion distance.

5. Check the **Merge result** option.

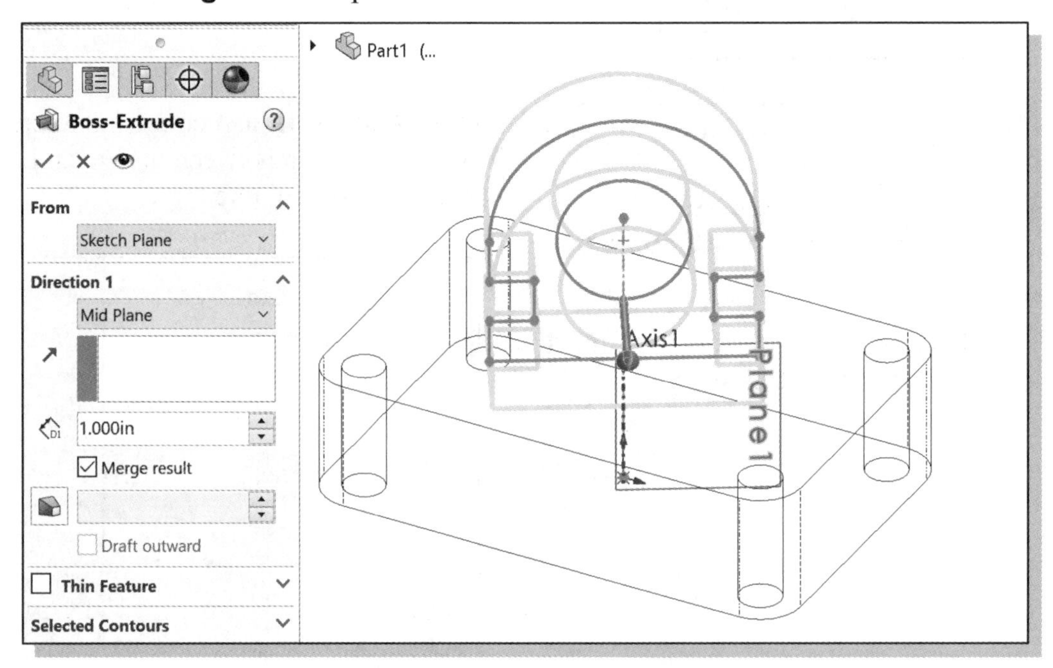

6. Click **OK** in the *PropertyManager* to accept the settings and create the extrusion.

7. Using the *Heads-up View* toolbar, set the display style to **Shaded with edges**.

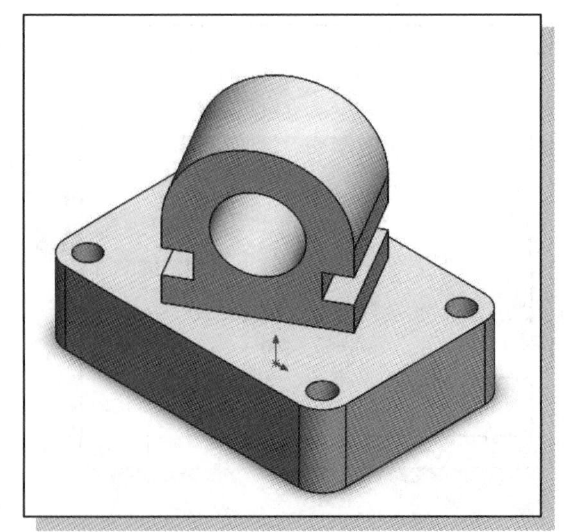

Creating an Offset Reference Plane

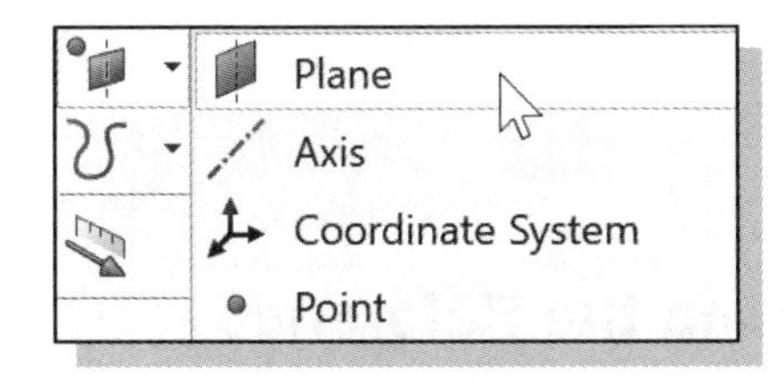

1. Select the **Reference Geometry** command from the *Features* toolbar, and select the **Plane** option from the pull-down menu.

❖ In the *Status Bar* area, the message: "*Select valid entities to define a plane (plane, face, edge, line, or point*" is displayed. SOLIDWORKS expects us to select any existing geometry, which will be used as a reference to create the new work plane.

2. Move the cursor over the top face of the base feature. Notice the top face of the base feature is highlighted. Click once with the **left-mouse-button** to select the face.

3. Notice the top face, called **Face<1>**, appears in the *First Reference* panel of the *Plane PropertyManager*, and the **Offset Distance** option is automatically selected.

4. Enter **0.75 in** as the offset distance.

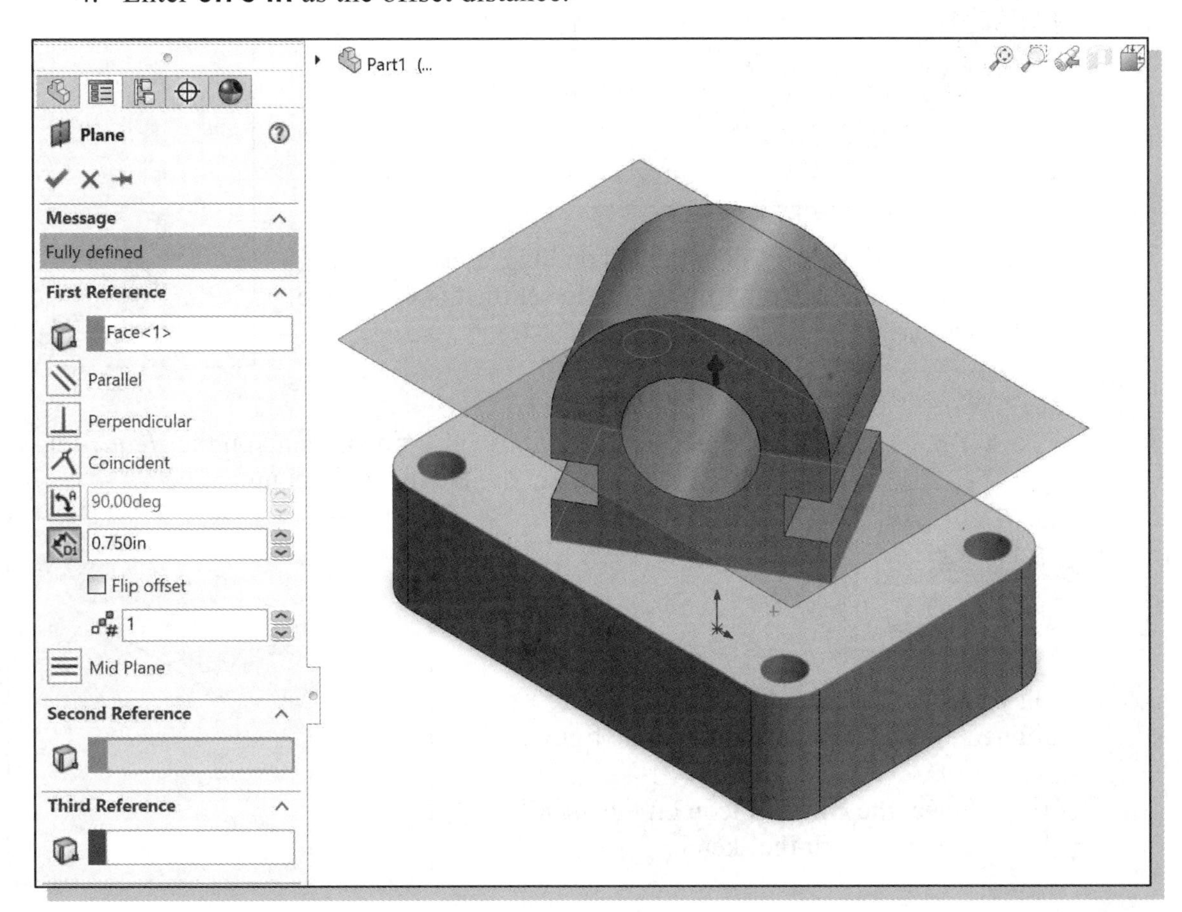

5. Click **OK** in the *PropertyManager* to accept the settings and create the new reference plane.

6. Press the **[Esc]** key to unselect the new plane.

Creating another Extruded Cut Feature Using the Reference Plane

1. In the *Sketch* toolbar, select the **Sketch** command by left-clicking once on the icon.

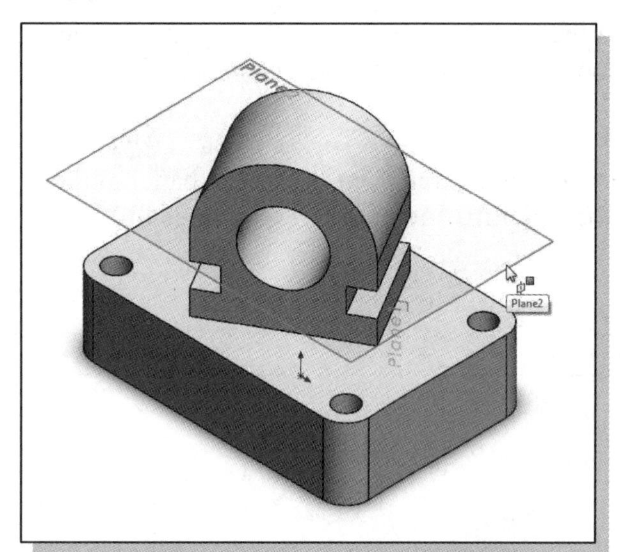

2. Select the offset reference plane **Plane2** as the sketch plane.

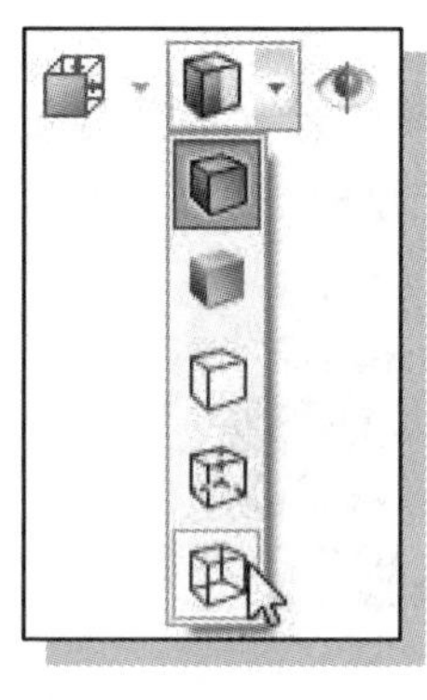

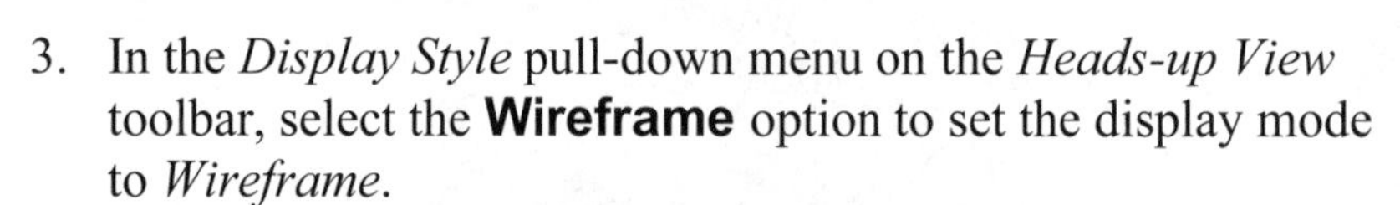

3. In the *Display Style* pull-down menu on the *Heads-up View* toolbar, select the **Wireframe** option to set the display mode to *Wireframe.*

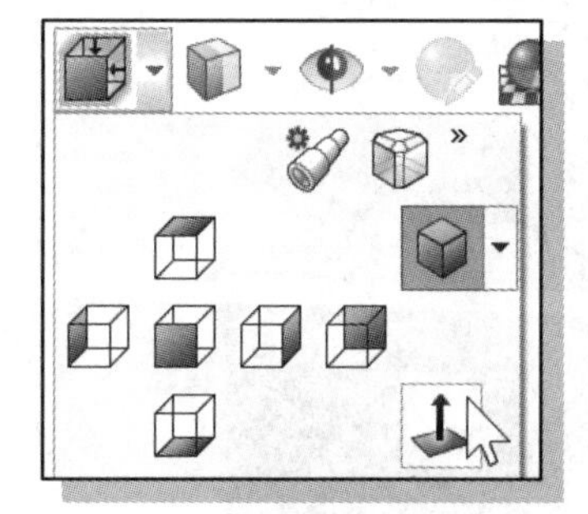

4. In the *View Orientation* pull-down menu on the *Heads-up View* toolbar, select the **Normal To** command.

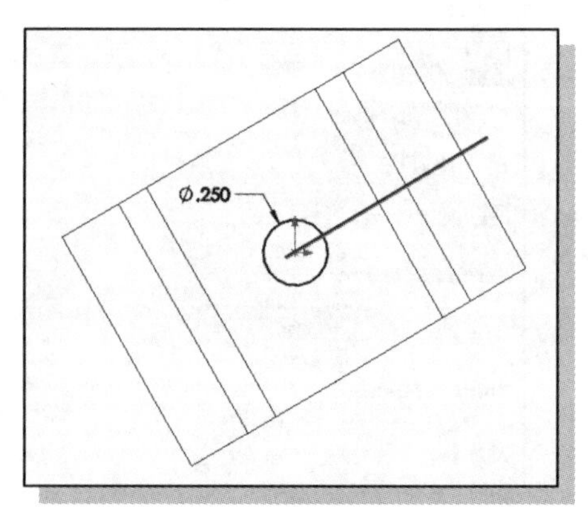

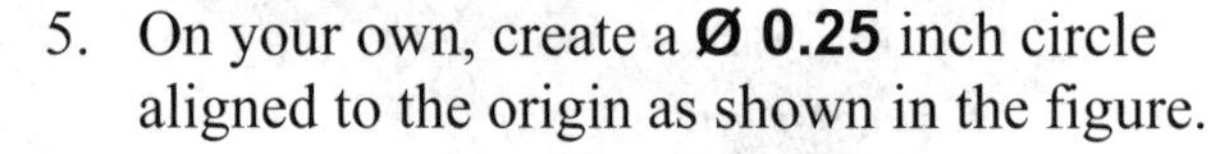

5. On your own, create a **Ø 0.25** inch circle aligned to the origin as shown in the figure.

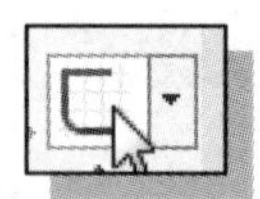

6. Select the **Sketch** icon on the *Sketch* toolbar to exit the sketch.

7. On your own, create the **Extruded Cut** feature. Use the **Through All** option and click the Reverse Direction button if necessary.

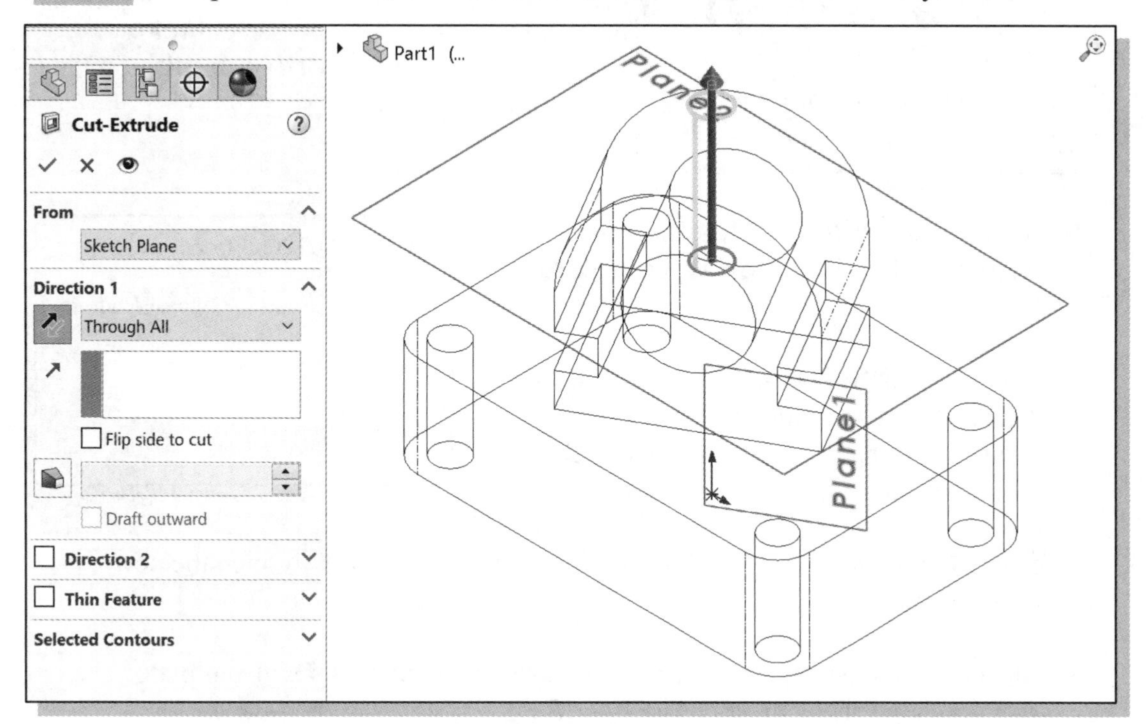

8. Click **OK** in the *PropertyManager* to create the extruded cut feature.

9. On your own, hide all planes, hide Axis1, and set the display to **Shaded With Edges**.

10. Press the **[F]** key to fit the view to the graphics area.

11. Select the **Save As** option from the *File* pull-down menu.

12. Use the *Browser* to select the folder in which you want to save the file.

13. In the *Save As* pop-up window, enter **Rod-Guide** for the *File name.*

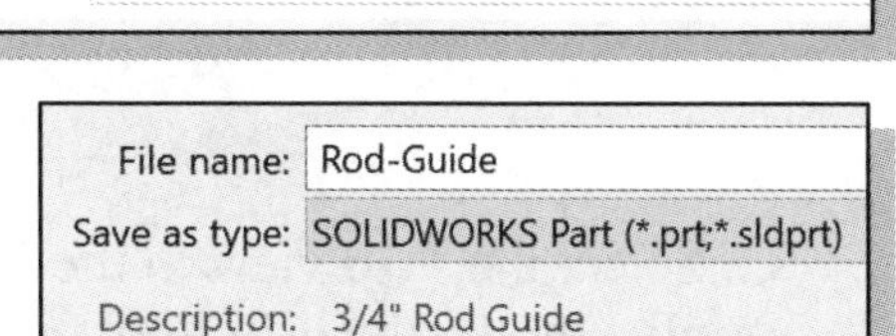

14. Enter **3/4" Rod Guide** for the *Description.*

15. Click **Save** to save the file.

Starting a New 2D Drawing and Adding a Base View

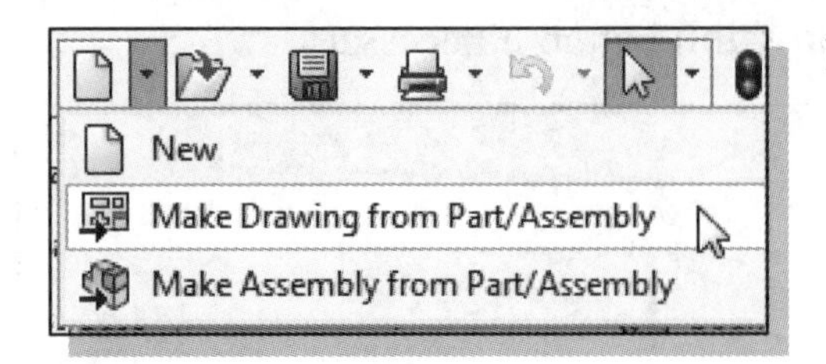

1. Click on the arrow next to the **New** icon on the *Menu Bar* and select **Make Drawing from Part/Assembly** in the pull-down menu.

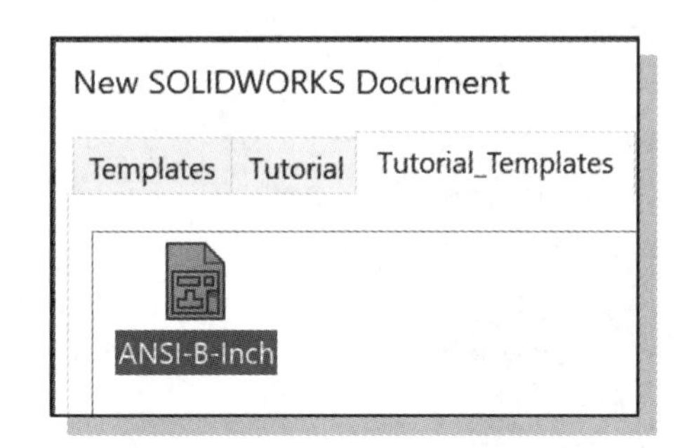

2. In the *New* SOLIDWORKS *Document* window, select the **Tutorial_Templates** tab.

3. Select the **ANSI-B-Inch** template file icon. This is the custom drawing template you created in Chapter 8. (If your ANSI-B-Inch template is not available, see note below.)

4. Select **OK** in the *New* SOLIDWORKS *Document* window to open the new drawing file.

➢ The new drawing file opens using the **ANSI-B-Inch.SLDDOT** template. The B-size drawing sheet appears. The *View Palette* appears in the *task pane.*

❖ **NOTE:** If your ANSI-B-Inch template is not available, (1) open the default Drawing template; (2) set document properties following instructions on page 8-7; and (3) set sheet properties following instructions on page 8-10.

❖ In SOLIDWORKS, the first drawing view we create is called a **base view**. A *base view* is the primary view in the drawing; other views can be derived from this view. In SOLIDWORKS, a base view can be added using the **Model View** command on the *Drawing* toolbar, or using the **View Palette**. By default, SOLIDWORKS will treat the *world XY plane* as the front view of the solid model. Note that there can be more than one *base view* in a drawing.

5. If the *View Palette* is collapsed, select the **View Palette** tab at the right of the graphics area.

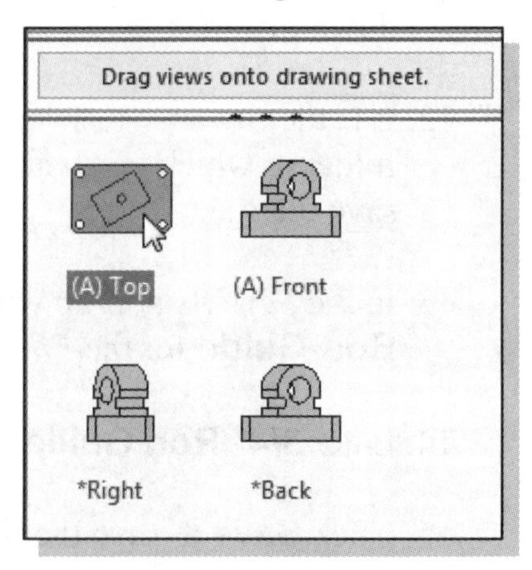

6. Select the **Top** view in the *View Palette* by clicking on the icon with the **left-mouse-button** as shown.

7. Hold down the **left-mouse-button** and **drag** the top view from the *View Palette* into the graphics window and place the **base view** near the upper left corner of the graphics window as shown below.

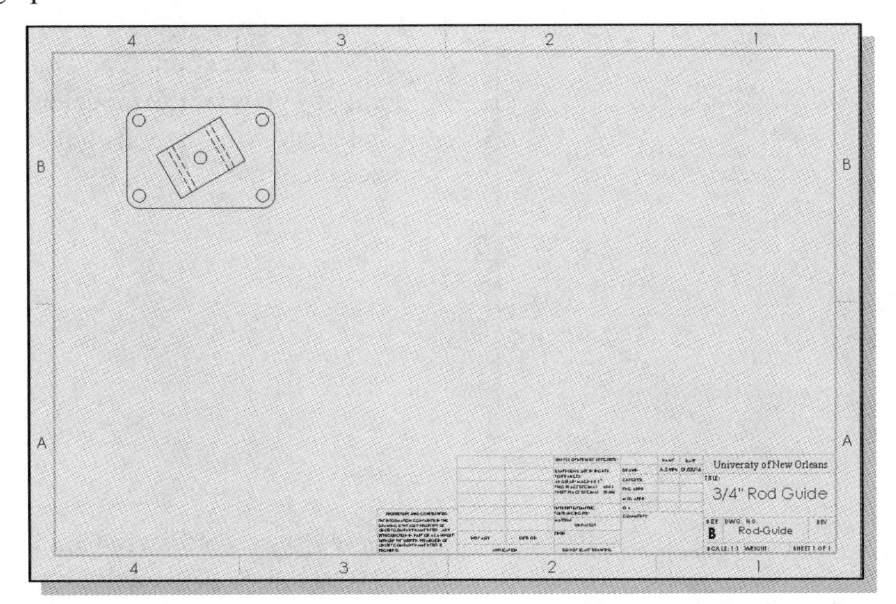

➤ When the base view is placed, SOLIDWORKS automatically executes the **Projected View** command and the *Projected View PropertyManager* appears.

8. Press the **[Esc]** key to exit the Projected View command.

Creating an Auxiliary View

In SOLIDWORKS *Drawing* mode, the **Projected View** command is used to create standard views such as the *top* view, *front* view or *isometric* view. For non-standard views, the **Auxiliary View** command is used. *Auxiliary views* are created using orthographic projections. Orthographic projections are aligned to the base view and inherit the base view's scale and display settings.

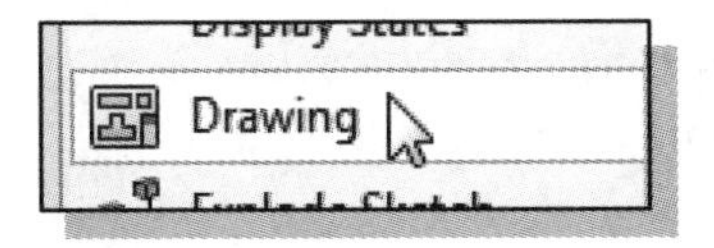

1. If the *Drawing* toolbar is not open, open it. (Move the cursor over the *Menu Bar* and click once with the **right-mouse-button** to access the menu of toolbars.)

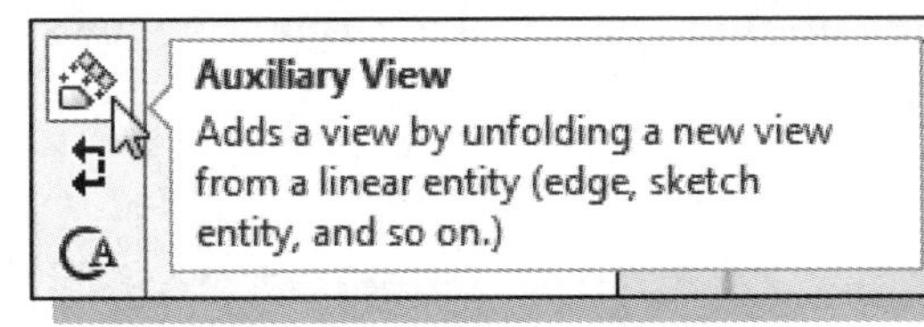

2. Click on the **Auxiliary View** button in the *Drawing* toolbar.

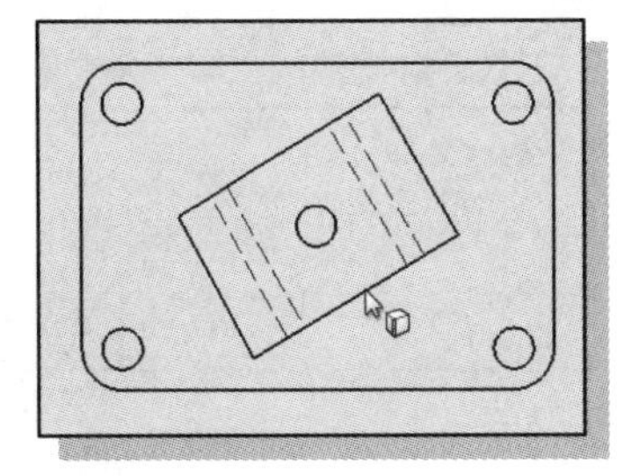

3. The message "*Please select a Reference Edge to continue*" appears in the *Auxiliary View Property-Manager*. Pick the front edge of the upper section of the model as shown.

❖ The orthographic projection direction will be perpendicular to the selected edge.

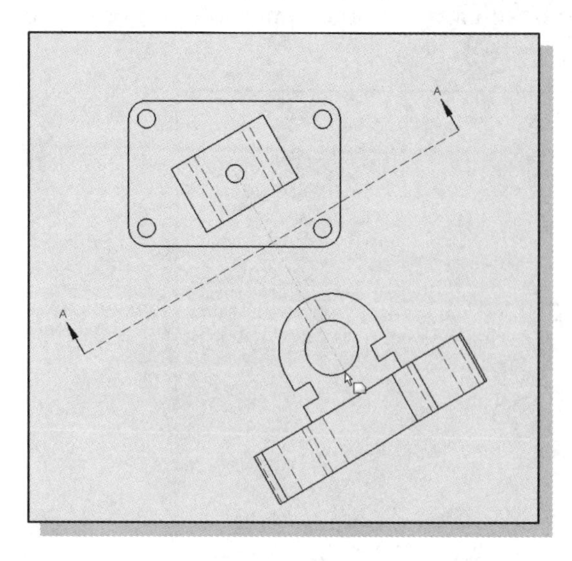

4. Move the cursor below the base view and select a location to position the auxiliary view of the model as shown. Click once with the left-mouse-button to place the view.

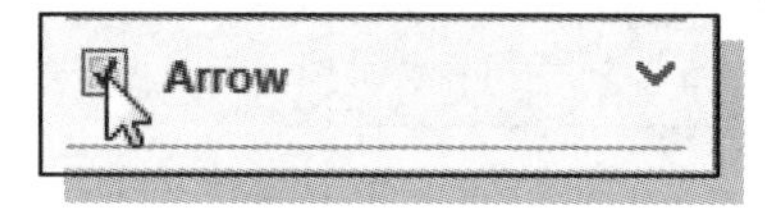

5. In the *Auxiliary View PropertyManager* **uncheck** the box next to the **Arrow** option. This will turn *OFF* the display of the direction arrow.

6. Press the **[Esc]** key to accept the settings and exit the Auxiliary View command.

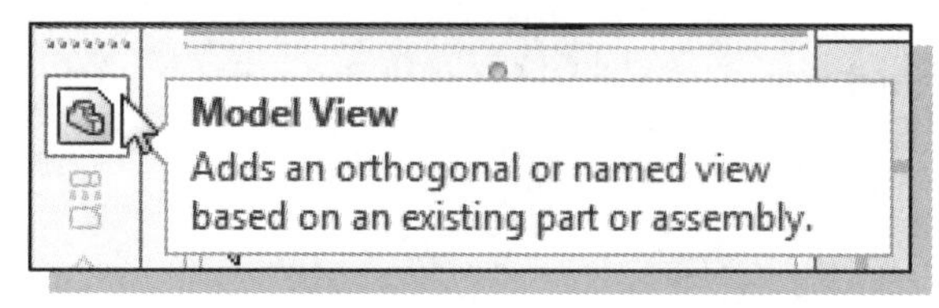

7. Click on the **Model View** icon *Drawing* toolbar.

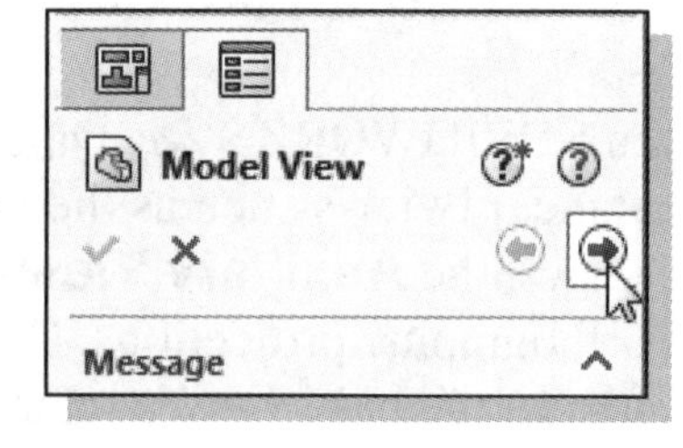

8. The *Model View PropertyManager* appears with the *Rod-Guide* part file selected as the part from which to create the base view. Click the **Next** arrow as shown to proceed with defining the base view.

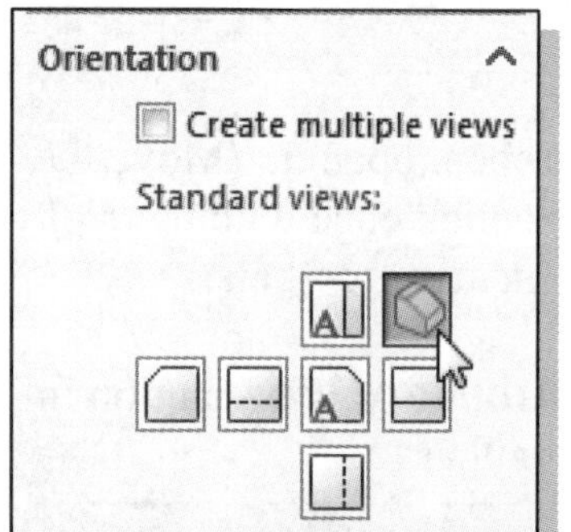

9. In the *Model View PropertyManager*, select the **Isometric** view for the *Orientation*, as shown.

➢ In general, hidden lines are not used in isometric views.

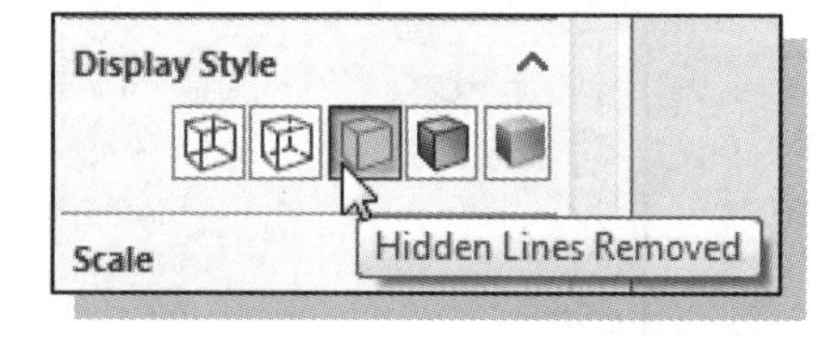

10. In the *Model View PropertyManager*, select the **Hidden Lines Removed** option for the *Display Style*.

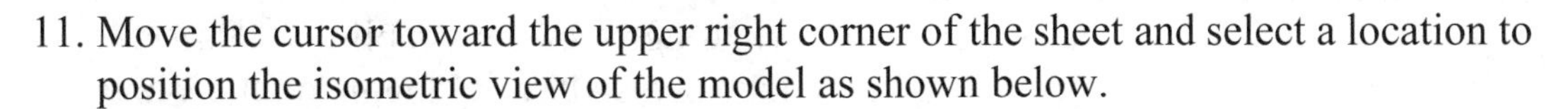

11. Move the cursor toward the upper right corner of the sheet and select a location to position the isometric view of the model as shown below.

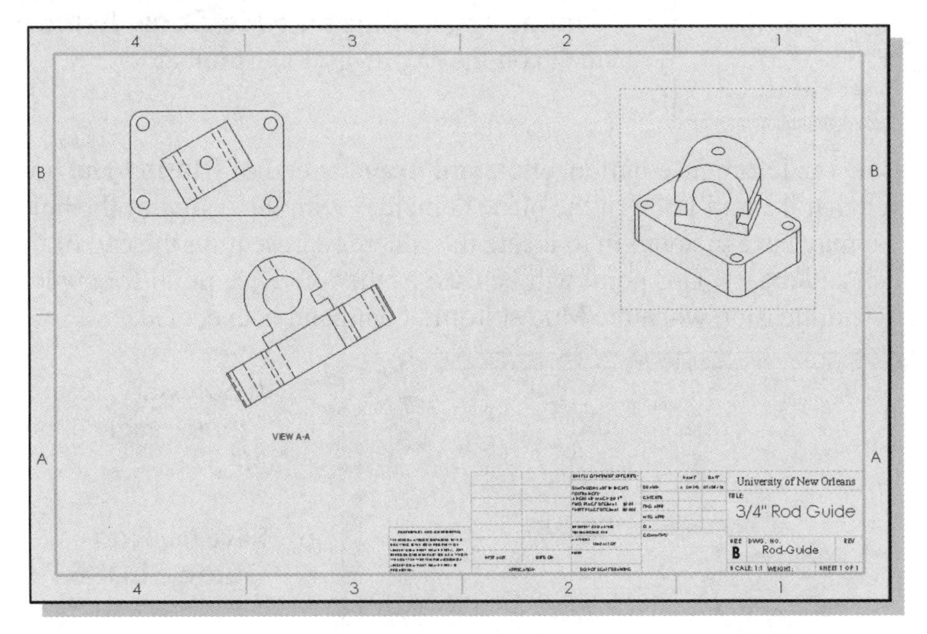

12. Left-click on the **OK** icon in the *Drawing View PropertyManager* to accept the view.

Displaying Feature Dimensions

In your drawing, the inserted views may include *center marks*, *centerlines*, and/or *dimensions*. If so, these were added with the **auto-insert** option. To access these options, click **Options** (*Standard* toolbar), select the **Document Properties** tab, and then select **Detailing**. This lesson was done with no auto-insert options selected. The center marks, centerlines, and dimensions will be added individually. The dimensions used to create the part can be imported into the drawing using the **Model Items** command. We can also add dimensions to the drawing manually using the **Smart Dimension** command.

Before we apply dimensions using the Model Items command we will adjust the size/position of the *Reference Plane 1*. The Model Items command will use the mid-line of the plane to locate the angle dimension.

1. Select the ***Rod-Guide.SLDPRT*** part window from the *Window* pull-down menu to switch to ***Part Modeling*** mode.

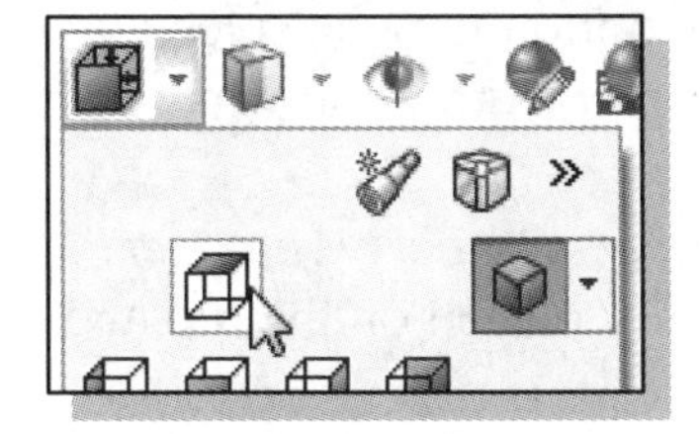

2. In the *View Orientation* pull-down menu on the *Heads-up View* toolbar, select the **Top View** command.

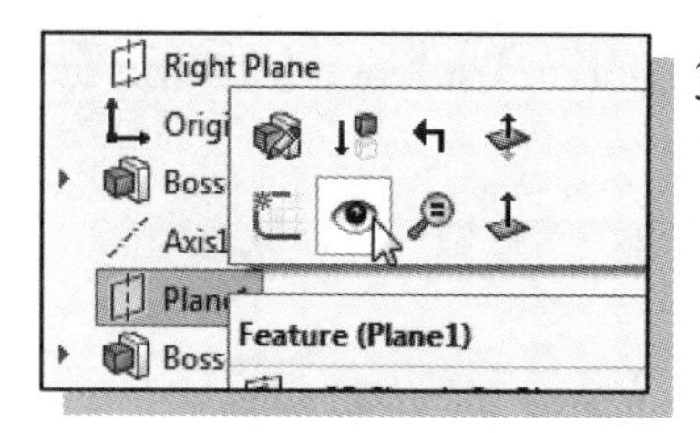

3. On your own, toggle *ON* the visibility of **Plane1**. NOTE: You may also have to toggle on the **View Planes** option using the **Hide/Show Items** pull-down menu on the *Heads-up View* toolbar.

4. Using the left-mouse-button, **click and drag** the end of **Plane1** and position it such that the mid-point of the plane coincides with the center of the hole. (NOTE: You may have to zoom in to locate the sphere representing the end of the plane to click and drag.) This point will be used as the reference point for applying the angle dimension when the Model Items command is executed.

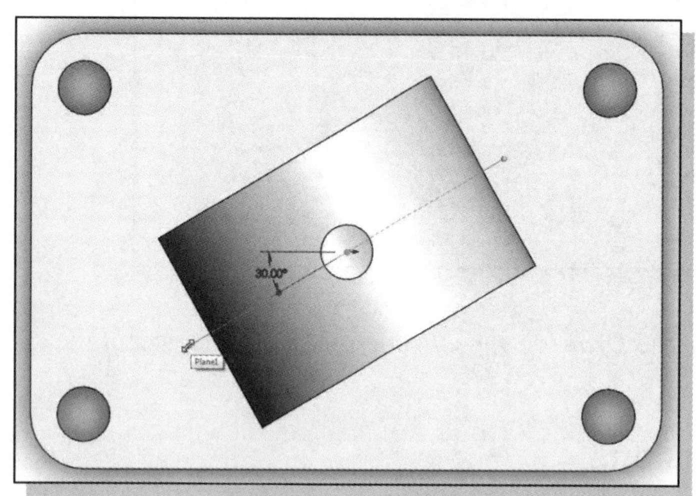

5. If necessary, reset the Plane1 angle dimension to 30°.

6. Save the **Rod-Guide.SLDPRT** file.

7. **On your own**, switch back to ***Drawing*** mode.

8. Press the **[Esc]** key to make sure no items are selected.

9. If the *Annotation* toolbar is not open, open it. (Move the cursor over the *Menu Bar* and click once with the **right-mouse-button** to access the menu of toolbars.)

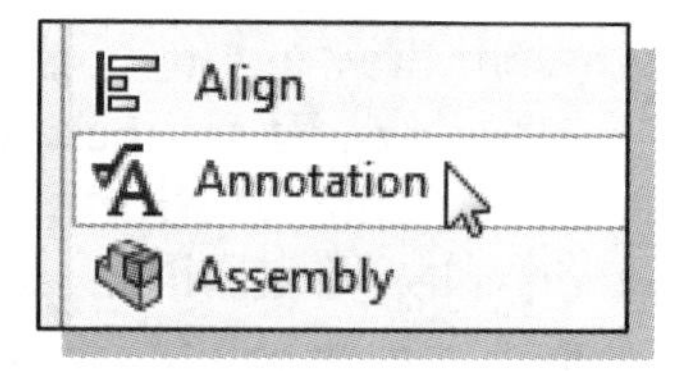

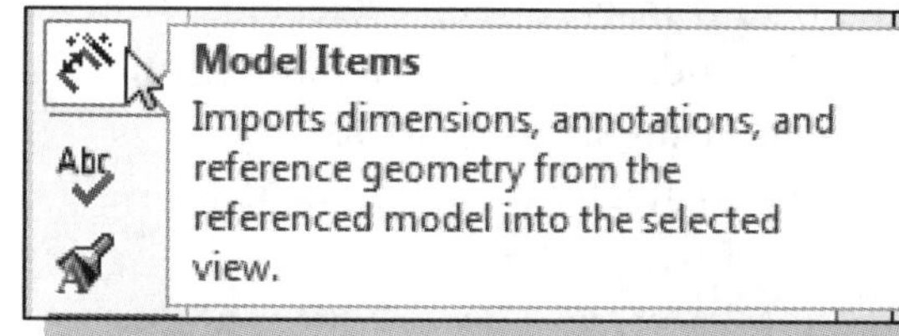

10. Left-mouse-click on the **Model Items** icon on the *Annotation* toolbar.

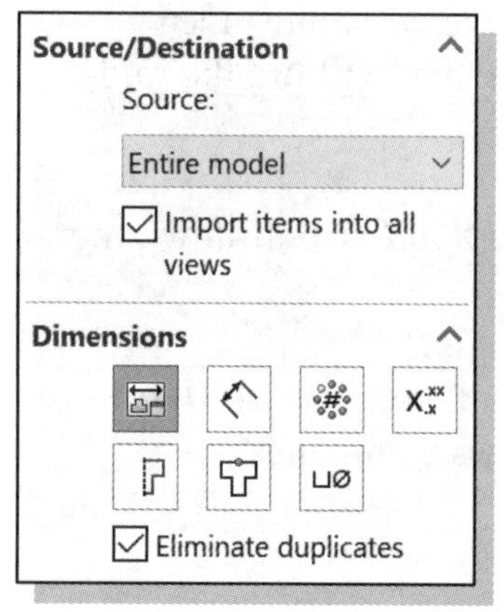

11. In the *Source/Destination* options panel of the *Model Items PropertyManager*, select the **Entire model** option from the *Import from* pull-down menu and check **Import items into all views** as shown.

12. Under the *Dimensions* options panel, select the **Marked for drawing** icon. (**NOTE:** This option may be selected by default.) Check **Eliminate duplicates** as shown.

13. Click the **OK** icon in the *Model Items PropertyManager*. Notice dimensions are automatically placed on the top and auxiliary drawing views.

Adjusting the View Scale

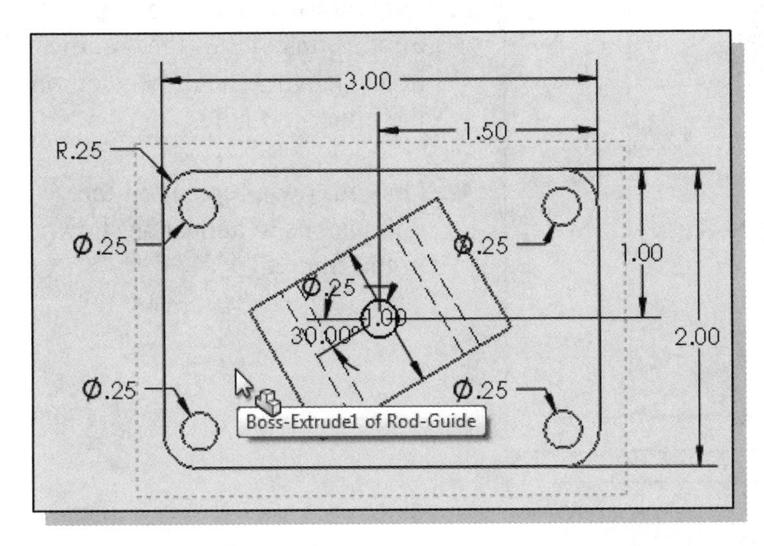

1. Move the cursor over the top view and click once with the **left-mouse-button**.
2. Left-click **OK** in the *Model Items PropertyManager* to accept the selection.
3. Move the cursor over the top view and click once with the **left-mouse-button**. The *Drawing View1 PropertyManager* appears.

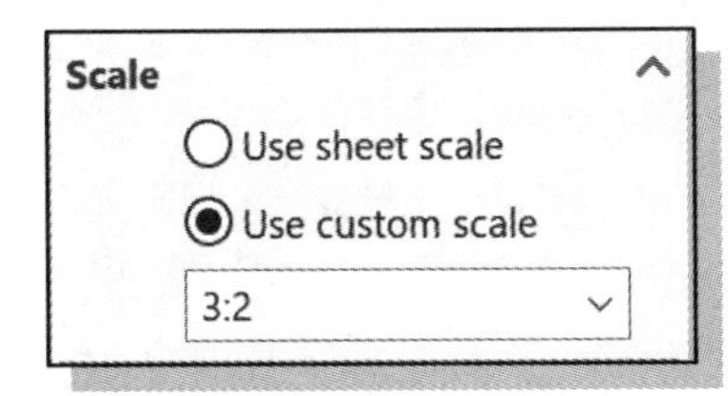

4. Under the *Scale* options in the *Drawing View PropertyManager*, select **Use custom scale**.
5. Select **User Defined** from the pull-down options, as shown.
6. Enter **3:2** as the user defined scale.
7. Left-click on the **OK** icon in the *Drawing View PropertyManager* to accept the new settings. Notice the auxiliary view is automatically scaled.
8. If necessary, reposition the views so they do not overlap.

Repositioning, Appearance, and Hiding of Feature Dimensions

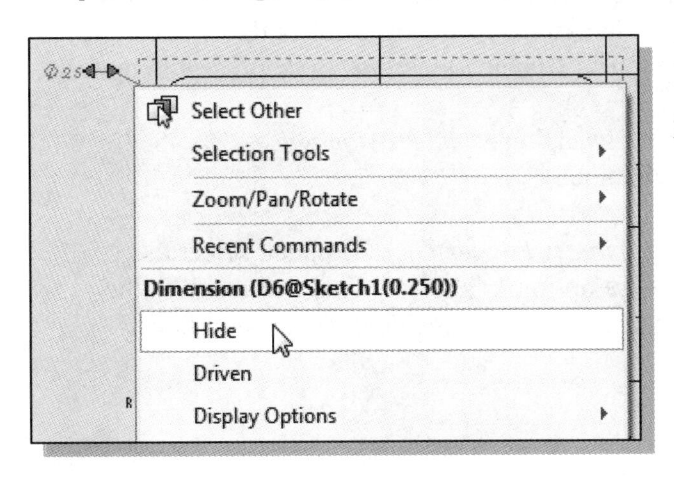

1. Zoom in on the top view.
2. **Right-click** on the **R.25** *radius* dimension of the round, and select **Hide** from the pop-up option menu.

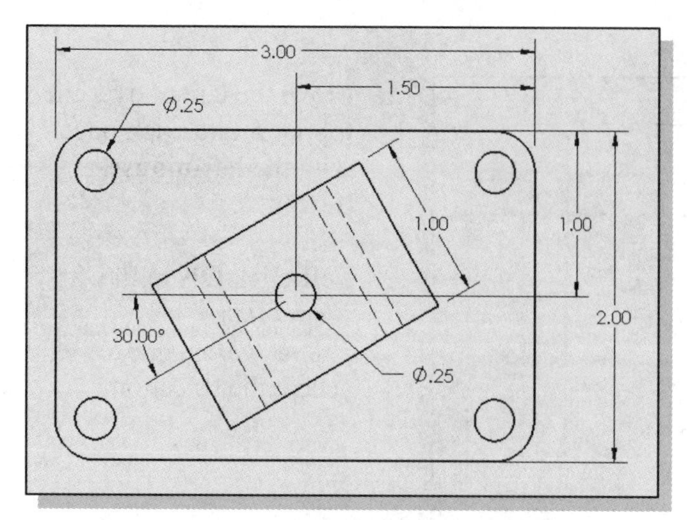

3. On your own, hide the diameter dimensions of 3 of the 4 corner holes, leaving the dimension on the upper left hole.

4. On your own, reposition the dimensions to appear as shown in the figure.

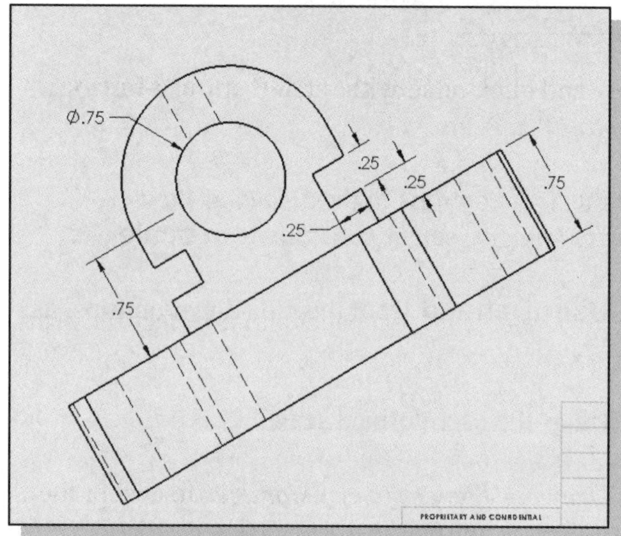

5. **Pan** to the auxiliary view. On your own, hide and reposition dimensions as shown in the figure.

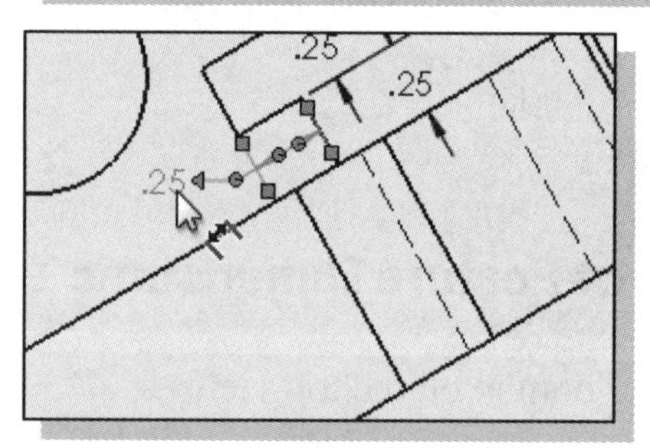

❖ The appearance of dimensions can be adjusted independently.

6. Move the cursor over the **0.25** dimension shown in the figure and click once with the **left-mouse-button** to select the dimension. The *Dimension PropertyManager* will appear.

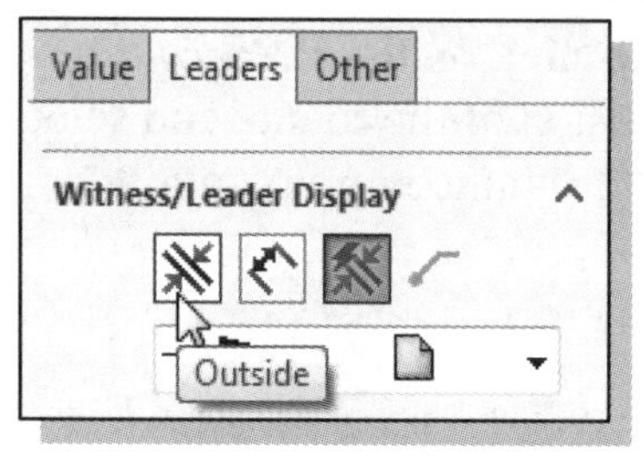

7. In the *Dimension PropertyManager*, select the **Leaders** tab.

8. In the *Witness/Leader Display* panel, select the **Outside** option as shown. This will result in the arrows being located inside the extension lines.

9. Reposition the dimension as shown.

10. On your own, return to the **Smart** option in the *Witness/Leader Display* panel. Press the **[Esc]** key to exit the Dimension command and unselect the dimension.

11. Left-mouse-click on the **Smart Dimension** icon on the *Annotation* toolbar.

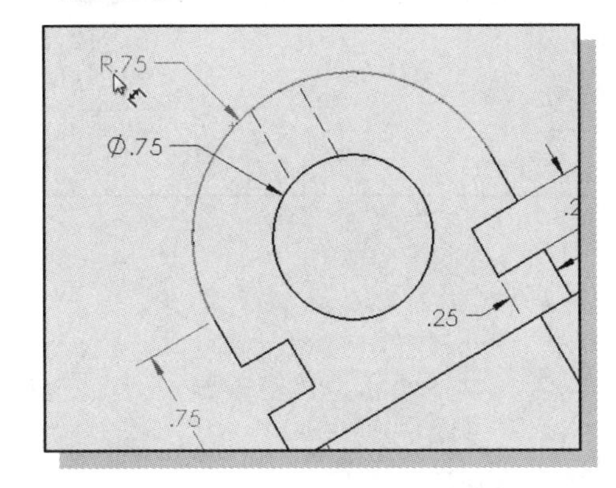

12. Select the *curved top edge* in the auxiliary view.

13. Move the cursor to and **left-click** to position the dimension as shown.

❖ By default, reference dimensions may be displayed in parentheses. However, we are using the dimension as an alternate placement for the hidden feature dimension. We will not use the parentheses.

14. If parentheses appear by default with the reference dimension, go to the *Dimension PropertyManager*, scroll down to the *Dimension Text* panel and **unselect** the **Add Parenthesis** option as shown.

15. Click **OK** in the *Dimension PropertyManager* to create the reference dimension without parentheses.

16. Adjust the dimensions to appear as shown in the figure. Also notice the note which was automatically added containing the custom scale for the auxiliary view.

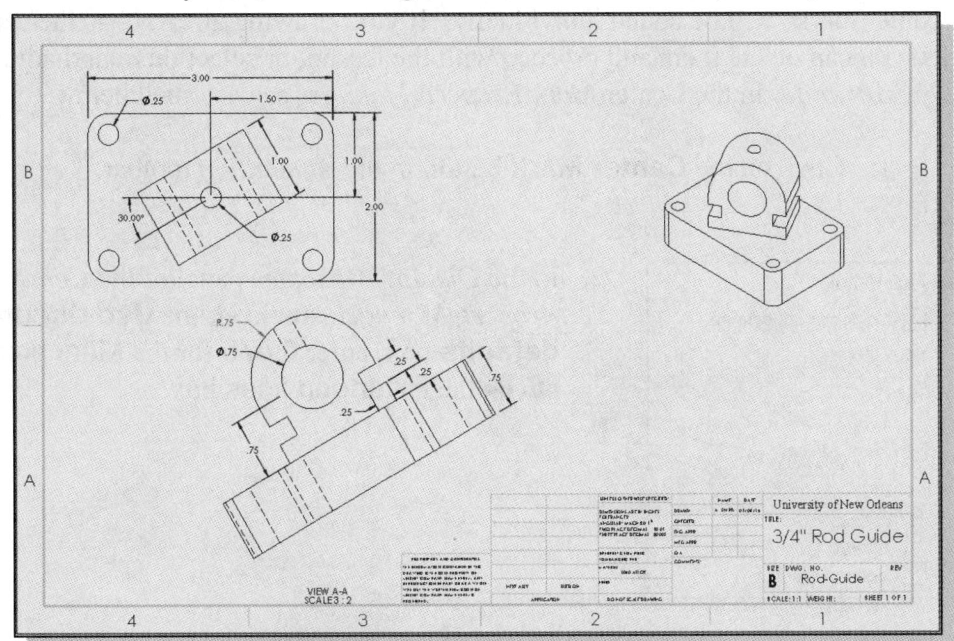

Tangent Edge Display

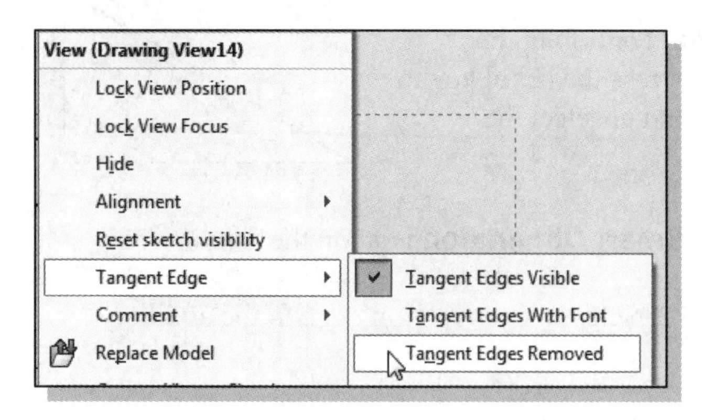

1. Move the cursor over the auxiliary view and click once with the **right-mouse-button**.

2. In the pop-up option menu, select **Tangent Edge**, then **Tangent Edges Removed**.

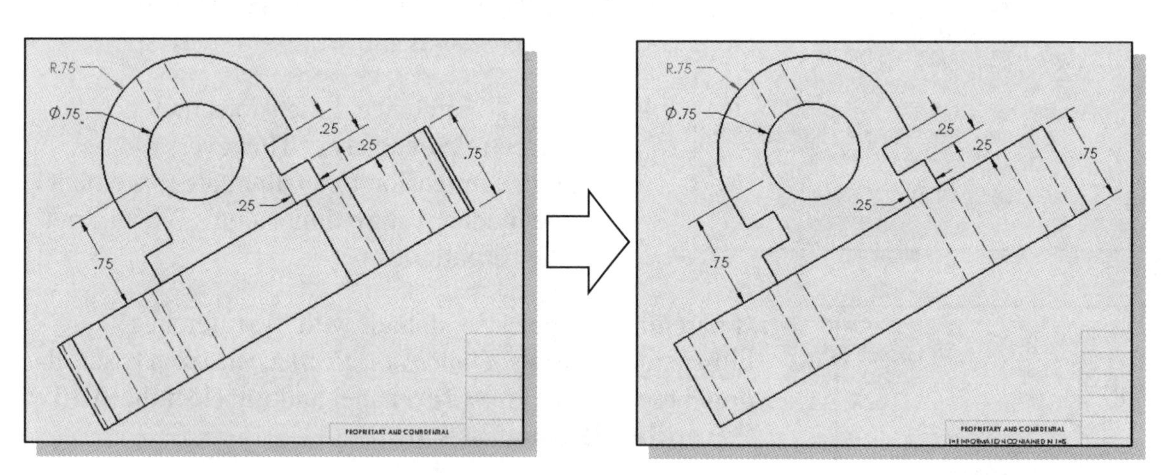

Adding Center Marks and Center Lines

In your drawing, the inserted views may include *center marks*. If so, these were added with the **auto-insert** option. This lesson was done with no auto-insert options selected. The center marks will be added individually. If your drawing already has the center marks, you can delete them and proceed with the lesson, or select them and adjust the *Display Attributes* in the Center *Mark PropertyManager* as described below.

1. Click on the **Center Mark** button in the *Annotation* toolbar.

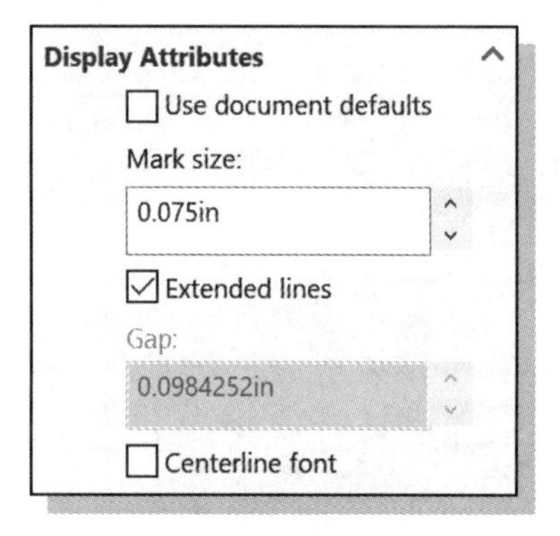

2. In the *Display Attributes* panel of the *Center Mark PropertyManager*, **uncheck** the **Use document defaults** box, enter **0.075** for the Mark size, and **check** the Extended lines box.

3. Click on the five circles in the top view to add the center marks as shown.

4. Click **OK** in the *PropertyManager* to exit the Center Mark command.

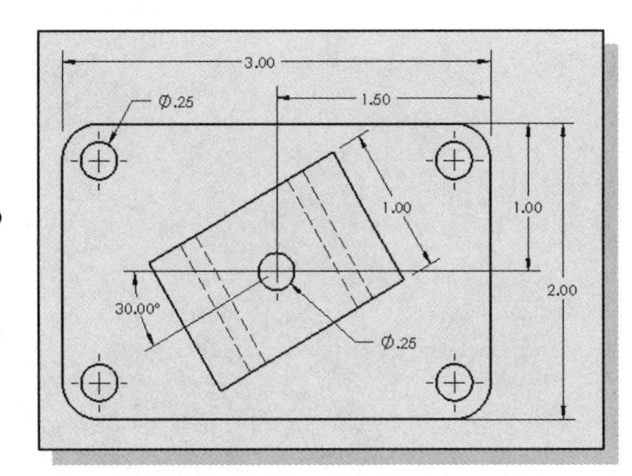

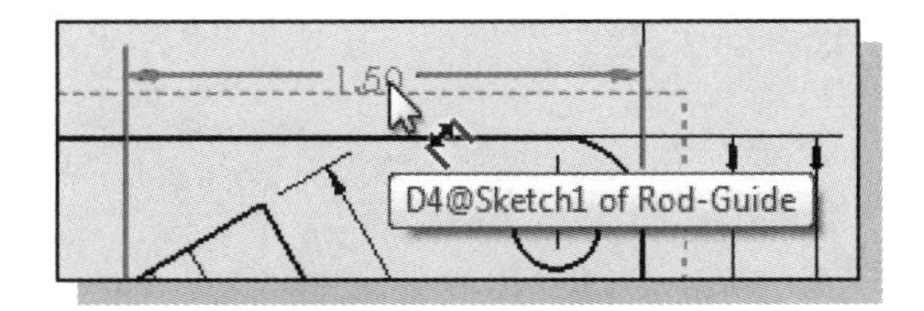

5. Notice the extension lines for some of the dimensions interfere with the center mark for the central hole. Select the **1.50** dimension as shown.

6. Move the endpoint of the extension line by **clicking and dragging** to reveal the center mark.

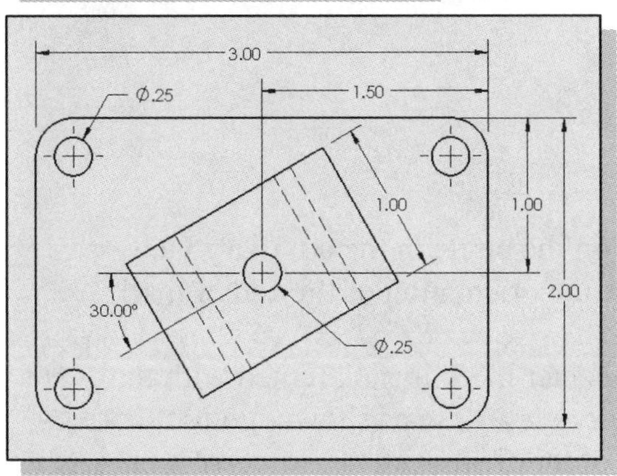

7. On your own, repeat this for the other extension lines to reveal the center mark as shown in the figure.

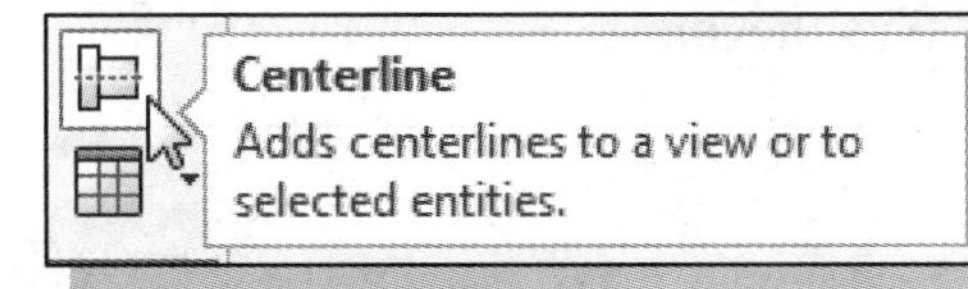

8. Click on the **Centerline** button in the *Annotation* toolbar.

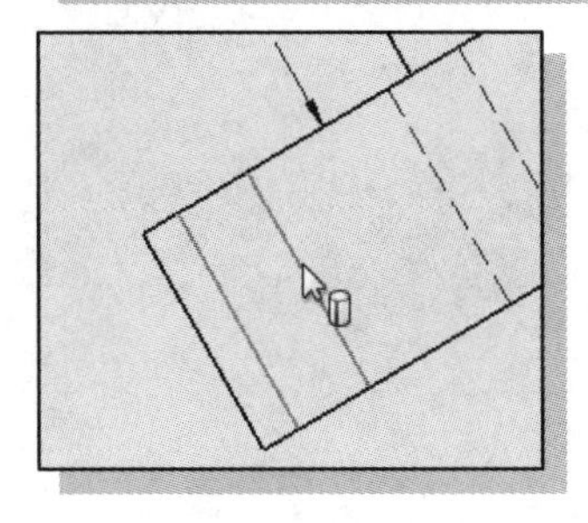

9. Inside the graphics window, click on the two hidden edges of one of the Drill features and create a center line in the auxiliary view as shown in the figure.

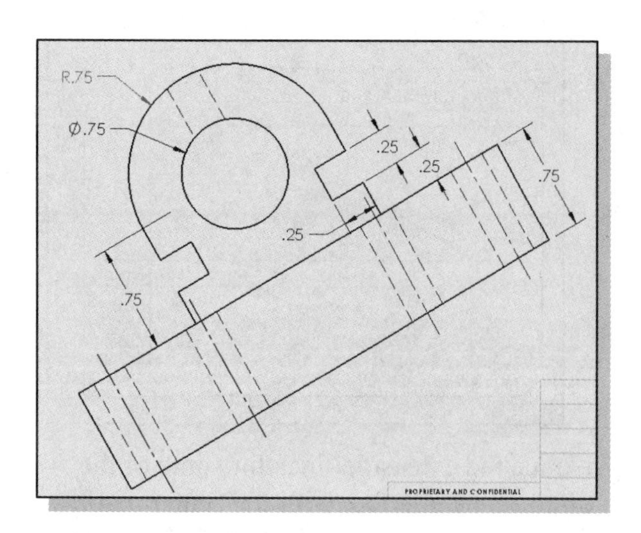

10. On your own, repeat the above step and create additional centerlines as shown.

11. Click **OK** in the *PropertyManager* to exit the **Centerline** command.

12. Click on the **Center Mark** button in the *Annotation* toolbar.

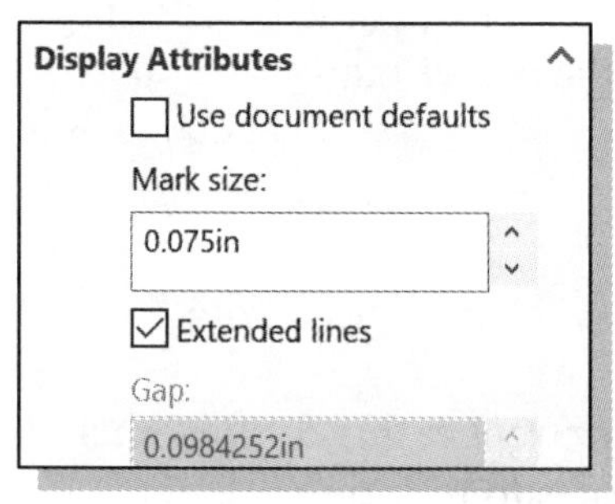

13. In the *Display Attributes* panel of the *Center Mark PropertyManager*, **uncheck** the **Use document defaults** box, enter **0.075** for the **Mark size**, and **check** the **Extended lines** box.

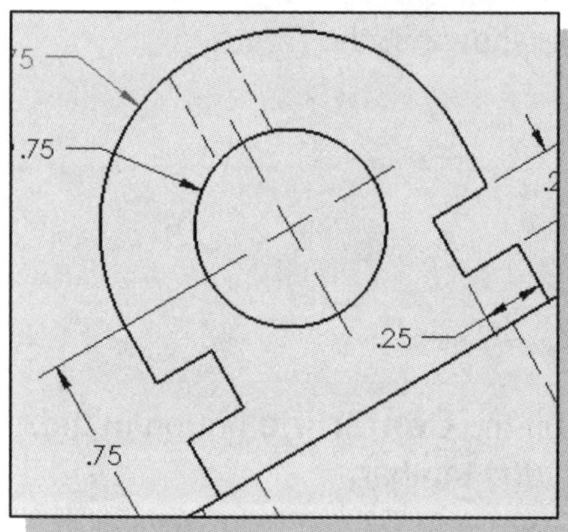

14. Click on the circle in the auxiliary view. Notice the orientation of the center mark.

15. If the center mark is not aligned with the auxiliary view base as shown, go to the *Angle* panel of the *Center Mark PropertyManager* and enter **0.0** as the angle.

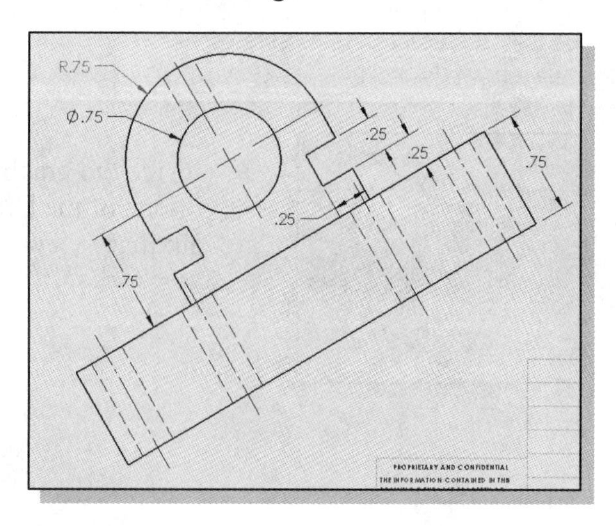

16. Click **OK** in the *PropertyManager* to exit the **Center Mark** command.

17. On your own, adjust line endpoints to appear approximately as shown in the figure.

Controlling the View and Sheet Scales

The scale for the Sheet is set at 1:1. This setting automatically appears in the title block due to the Property Link described as ***$PRP:"SW-Sheet Scale."*** The prefix ***$PRP:*** defines the source as the current document. The Property Name is ***SW-Sheet Scale***. This is a ***System Property***. Each view can be set to use the default sheet scale or to use a custom scale defined for the view. When we created the top and auxiliary views, a custom scale of 3:2 was created. For the isometric view the default sheet scale of 1:1 was used. We will change the sheet scale to 3:2.

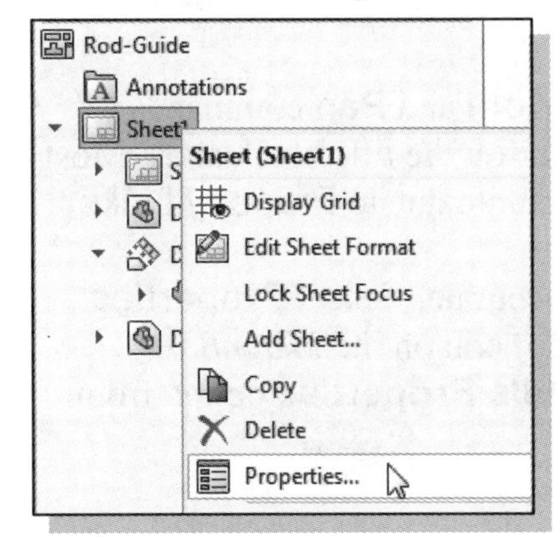

1. Move the cursor over **Sheet1** in the *FeatureManager Design Tree* and **right-click** to open the option menu.

2. Select **Properties**.

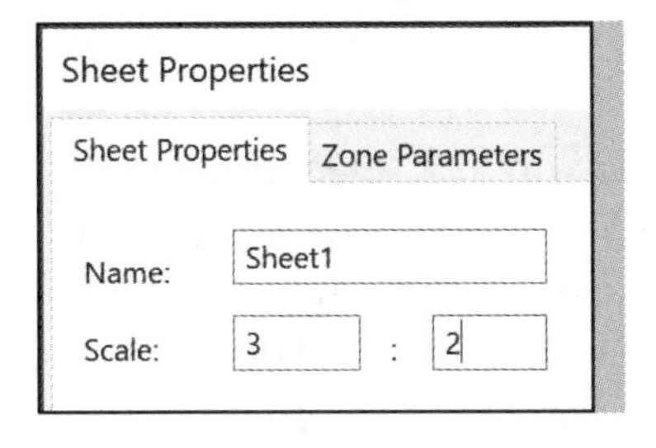

3. Enter **3:2** as the Scale in the ***Sheet Properties*** window.

4. Click **Apply Changes** in the *Sheet Properties* window.

➢ Notice the scale note in the title block now reads **3:2**. Also notice the scale of the isometric view has changed due to the Use sheet scale option.

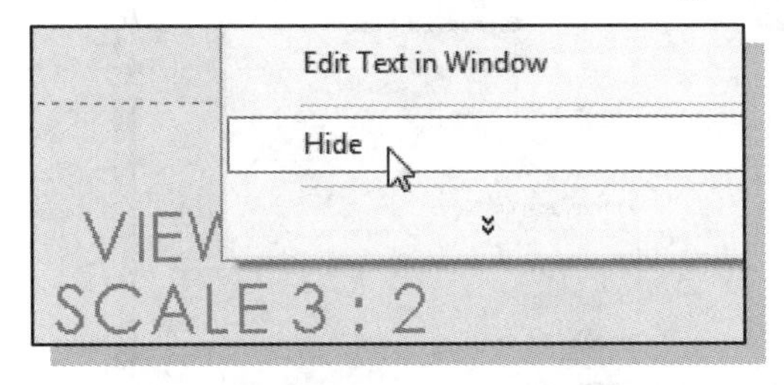

5. Since the sheet scale is labeled as **3:2** in the title block, the view scale note is not necessary. Right click on the note and select **Hide** from the pop-up option menu.

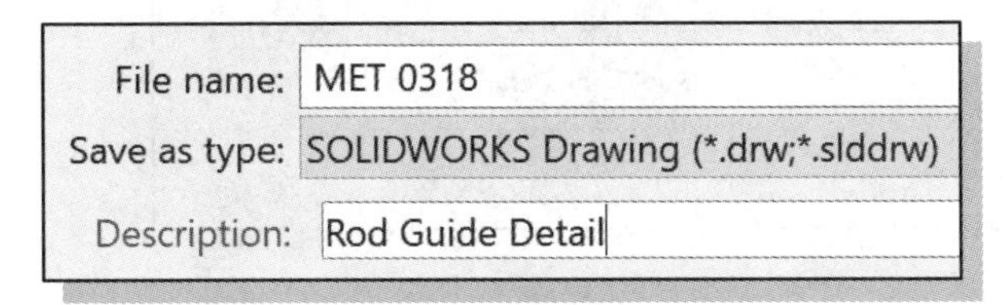

6. On your own, save the drawing file with **MET 0318** as the *File name* and **Rod Guide Detail** as the *Description*.

7. If the SOLIDWORKS pop-up window appears, click **Save All**.

Completing the Drawing Sheet

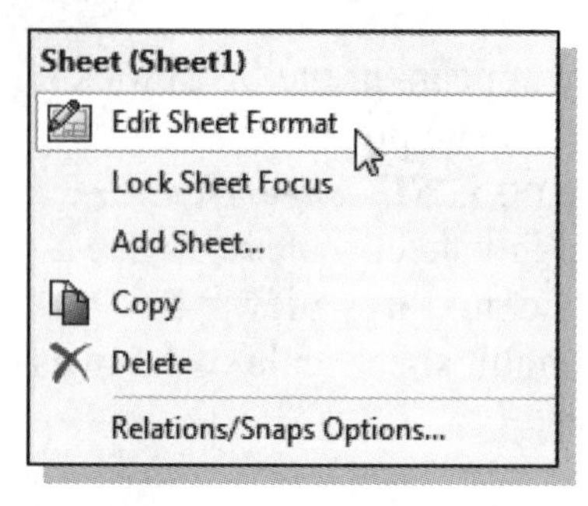

1. Move the cursor into the graphics area and **right-click** to open the pop-up option menu.

2. Select **Edit Sheet Format** from the pop-up option menu to change to *Edit Sheet Format* mode.

❖ Notice the title block is highlighted.

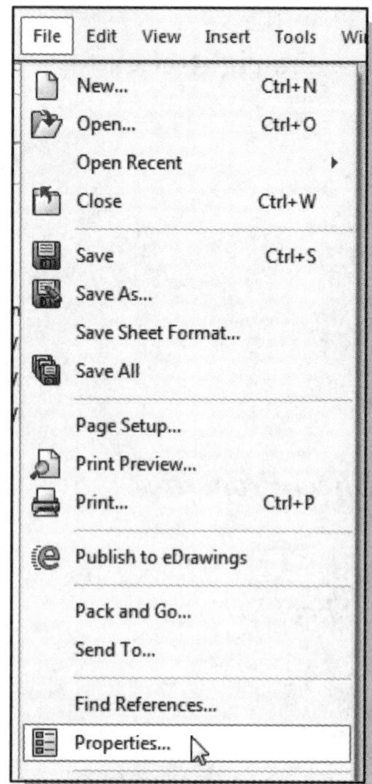

3. On your own, use the **Zoom** and **Pan** commands to adjust the display to work on the title block area. Most of the information is accurate due to Property Links.

4. The date probably is inaccurate. Select **Properties** from the *File* pull-down menu on the *Menu Bar*. (Alternately, select the **File Properties** option from the *Menu Bar*.)

5. In the *Summary Information* window select the **Custom** tab. Notice the *Custom Properties* appearing in the list. We defined these properties and saved them in the template file. (Note: If you are not working from the ANSI-B-Inch template, enter the custom properties in the table as shown.)

	Property Name	Type	Value / Text Expression	Evaluated Value	
1	SWFormatSize	Text	11in*17in	11in*17in	
2	CompanyName	Text	University of New Orleans	University of New Orleans	
3	DrawnBy	Text	A. Smith	A. Smith	
4	DrawnDate	Text	01/08/2021	01/08/2021	
5	Description	Text	Rod Guide Detail	Rod Guide Detail	
6	<Type a new property>				

6. Change the DrawnDate Value / Text Expression to the current date.

7. Click **OK** in the *Summary Information* window to accept the change.

8. On your own, correct other entries to the title block to appear as shown.

UNLESS OTHERWISE SPECIFIED:
DIMENSIONS ARE IN INCHES
TOLERANCES:
ANGULAR: MACH ± 0.1°
TWO PLACE DECIMAL ±0.01
THREE PLACE DECIMAL ±0.003
INTERPRET GEOMETRIC TOLERANCING PER:
MATERIAL 1060 ALLOY
FINISH
DO NOT SCALE DRAWING

	NAME	DATE
DRAWN	A. Smith	01/08/2021
CHECKED		
ENG APPR.		
MFG APPR.		
Q.A.		
COMMENTS:		

University of New Orleans
TITLE: 3/4" Rod Guide
SIZE B
DWG. NO. MET 0318
REV
SCALE: 3:2 WEIGHT: SHEET 1 OF 1

9. Left-mouse-click on the **Note** icon on the *Annotations CommandManager*.

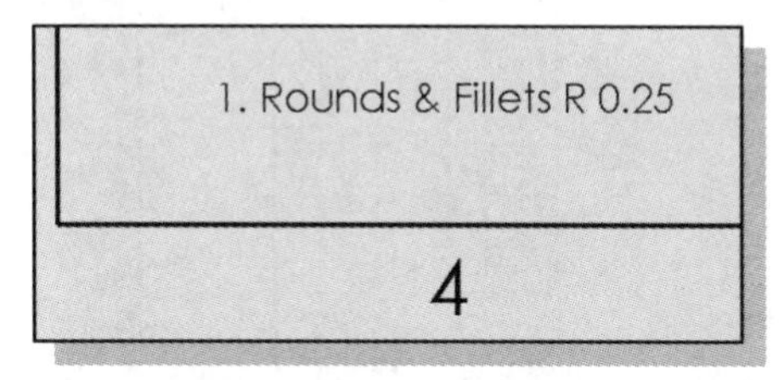

10. On your own, create a general note at the lower left corner of the border as shown.

Editing the Isometric View

1. Press the **[F]** key to scale the view to fit the graphics area.

2. Move the cursor into the graphics area and **right-click** to open the pop-up option menu.

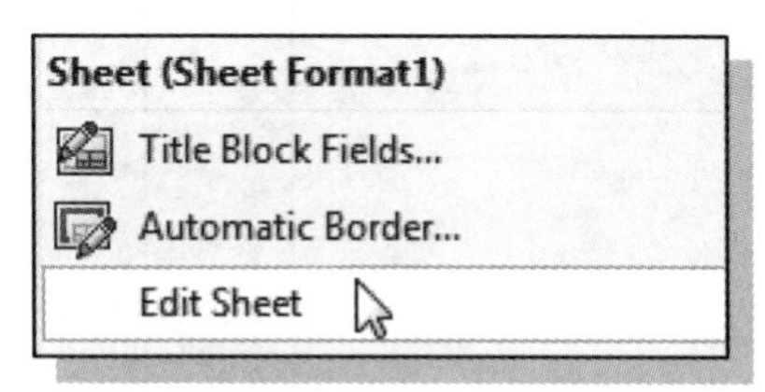

3. Select **Edit Sheet** from the pop-up option menu to change to *Edit Sheet* mode.

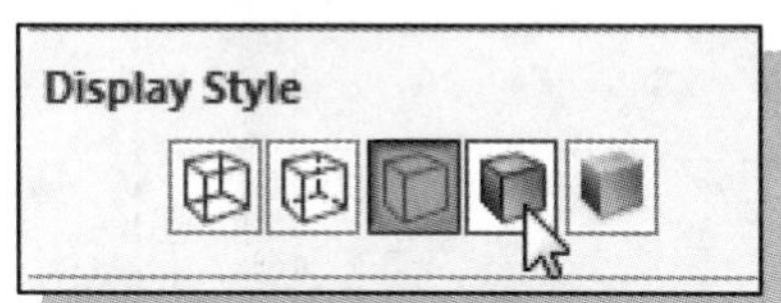

4. Left-mouse-click inside the *isometric* view to bring up the *Drawing View PropertyManager*.

5. In the *Display Style* panel of the *Drawing View PropertyManager*, select **Shaded with Edges**.

6. Click **OK** in the *PropertyManager* to accept the setting.

7. On your own, save the final version of the drawing file.

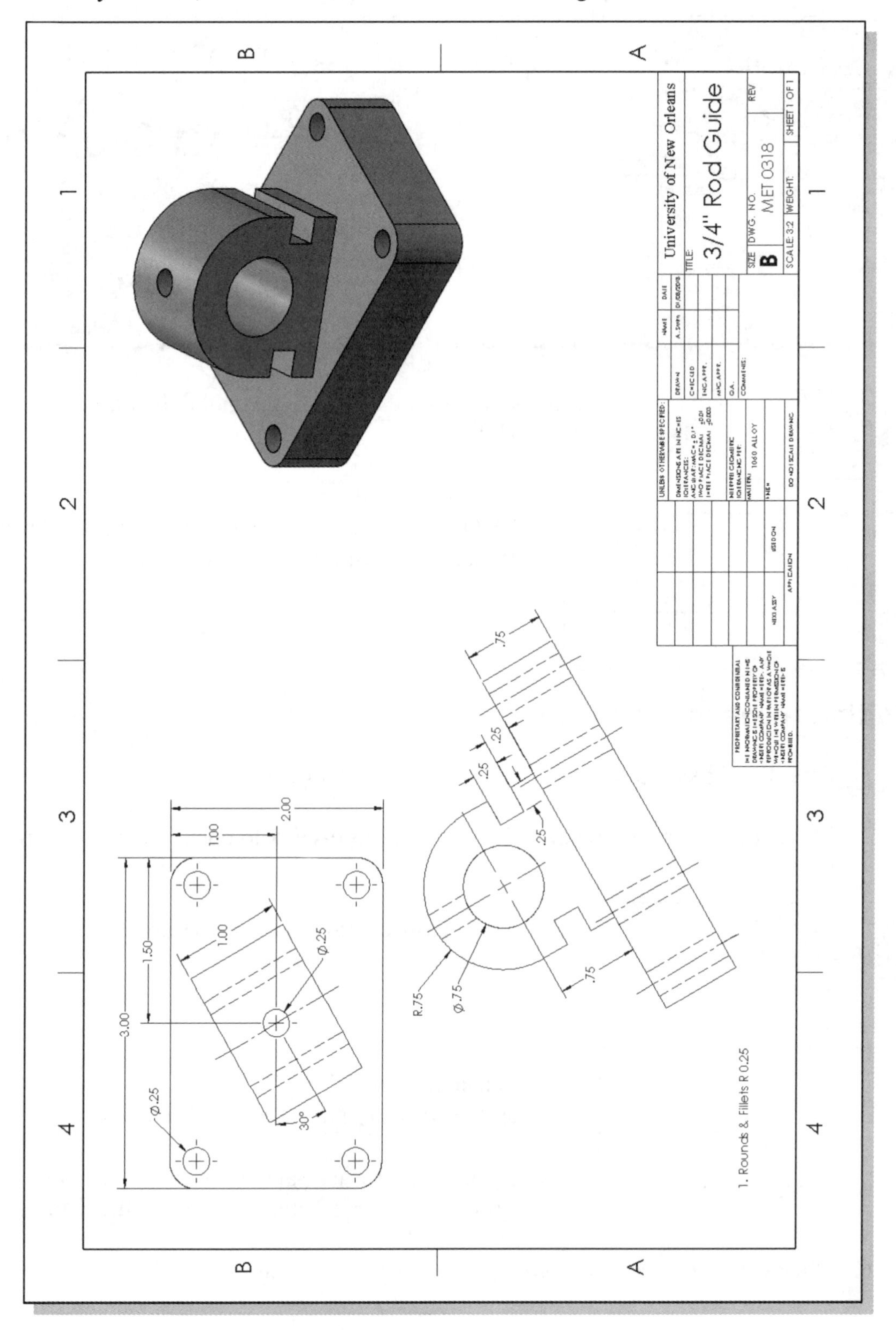

Questions:

1. What are the different types of reference geometry features available in SOLIDWORKS?

2. Why are reference geometry features important in parametric modeling?

3. Describe the purpose of auxiliary views in 2D drawings.

4. What are the required elements in order to generate an auxiliary view?

5. Describe the method used to create centerlines in the chapter.

6. Can we change the View Scale of existing views? How?

7. Identify and describe the following commands:

(a) 

(b) 

(c) 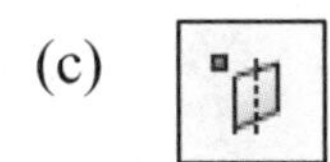

(d) 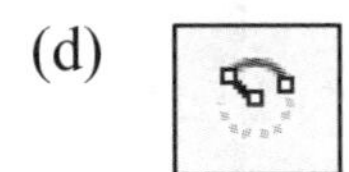

Exercises: (Create the Solid models and the associated 2D drawings.)

1. **Rod Slide** (Dimensions are in inches.)

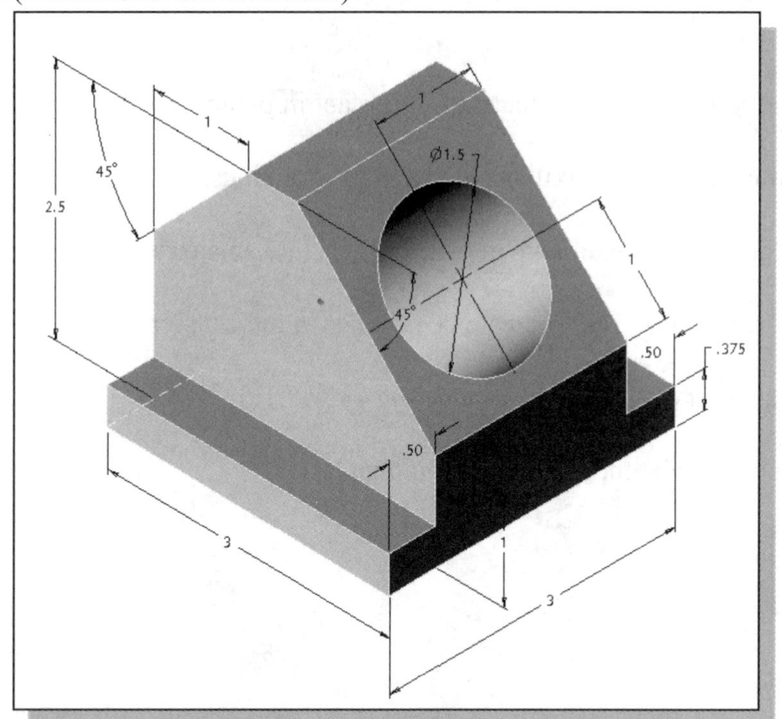

2. **Angle Support** (Dimensions are in millimeters.)

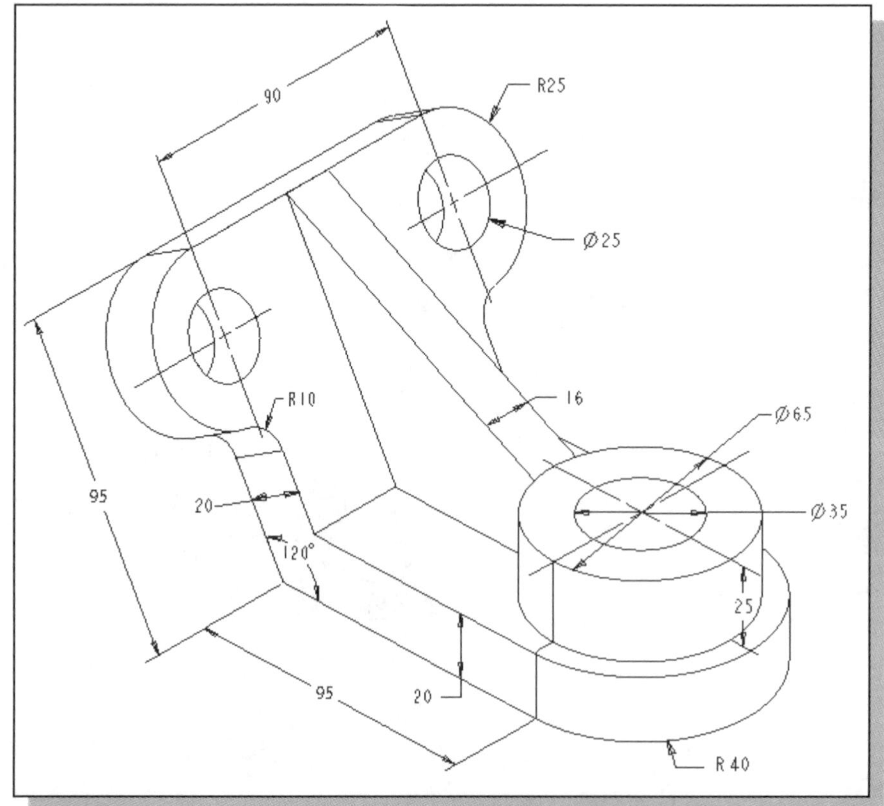

3. **Anchor Base** (Dimensions are in inches.)

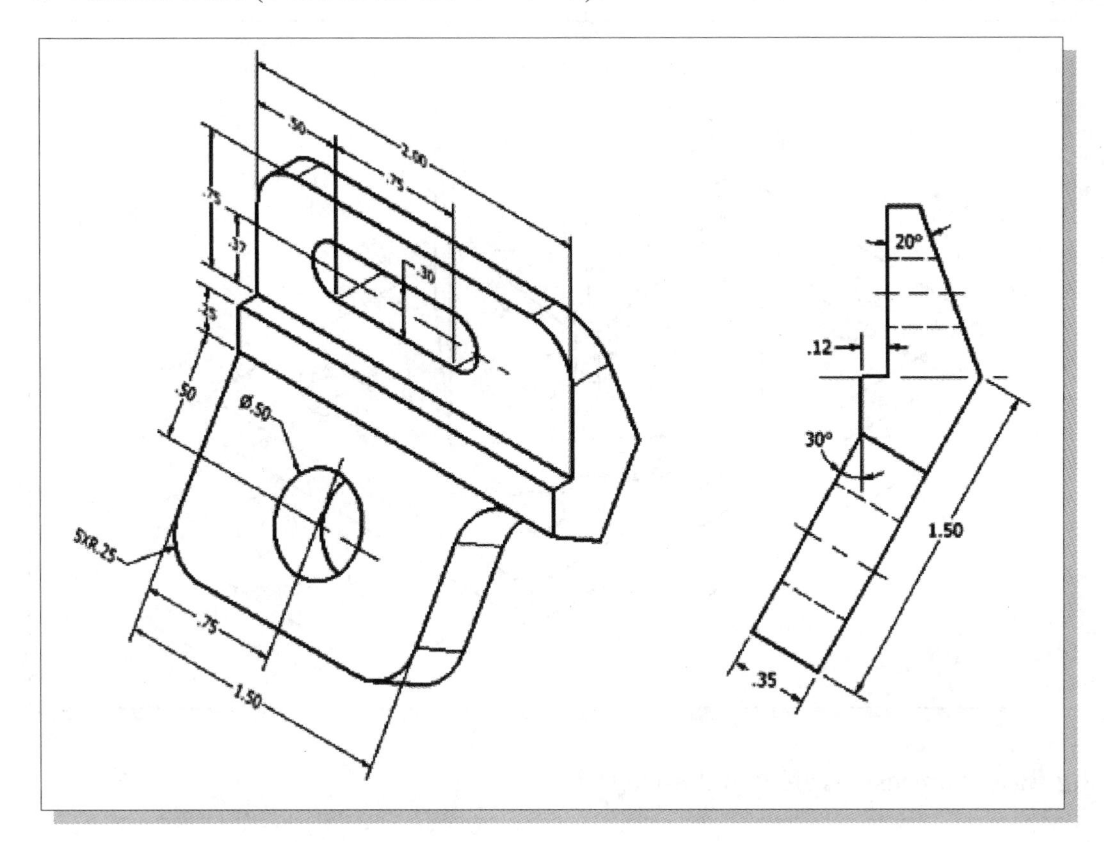

4. **Bevel Washer** (Dimensions are in inches.)

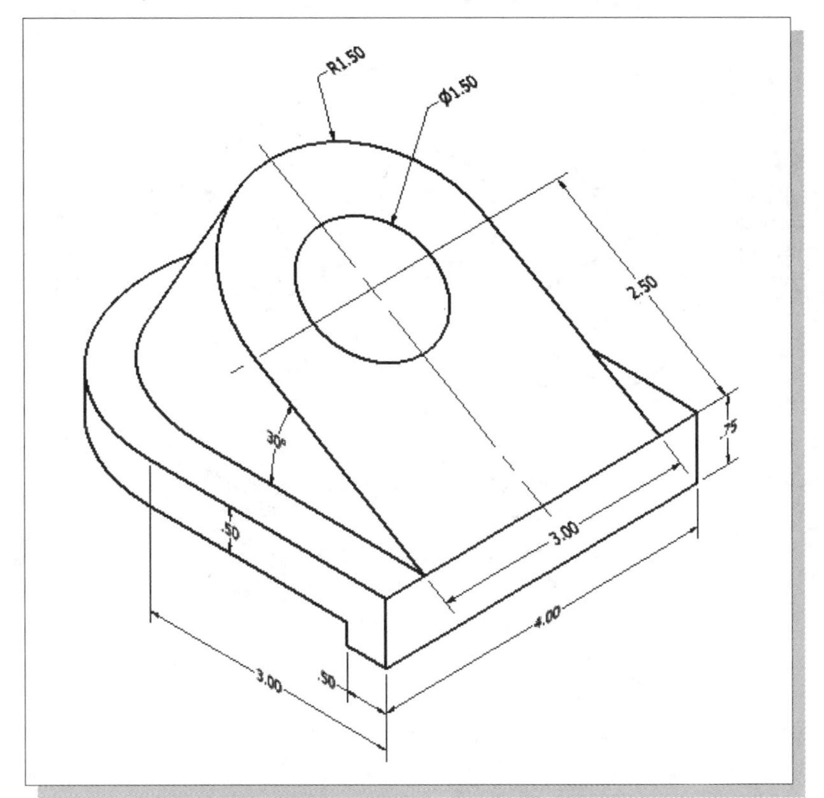

5. **Angle V-Block** (Dimensions are in inches.)

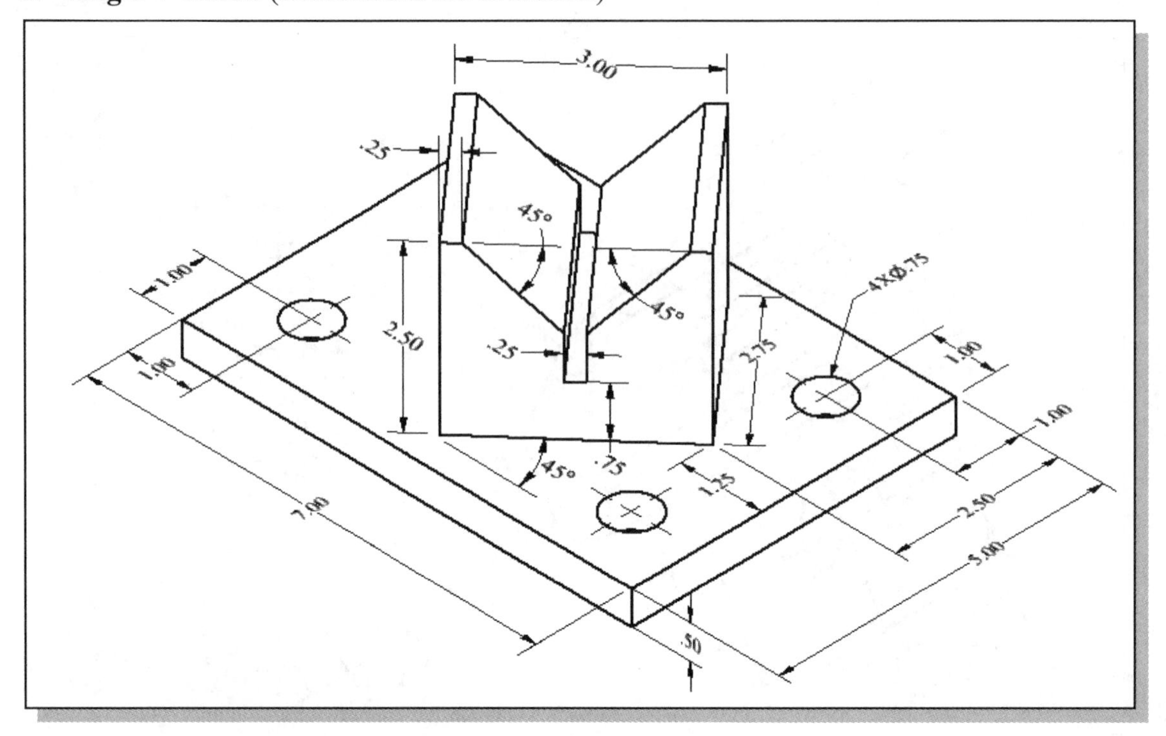

6. **Jig Base** (Dimensions are in millimeters.)

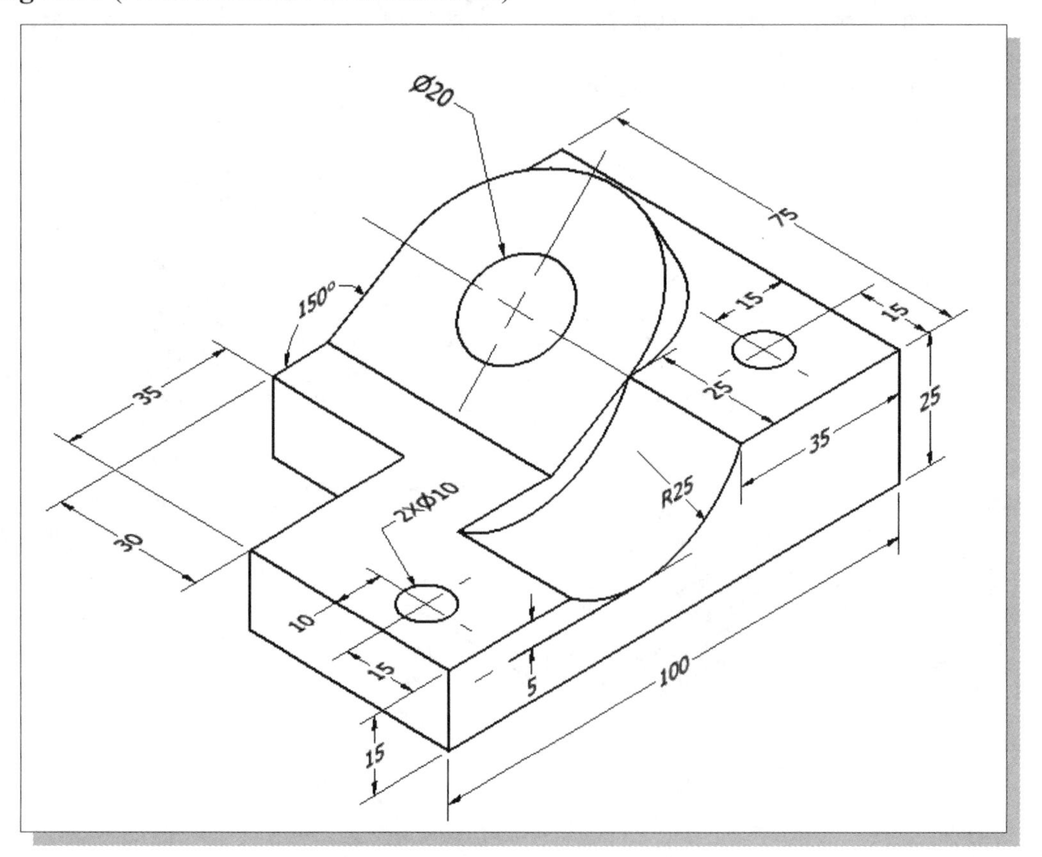

Chapter 10

Introduction to 3D Printing

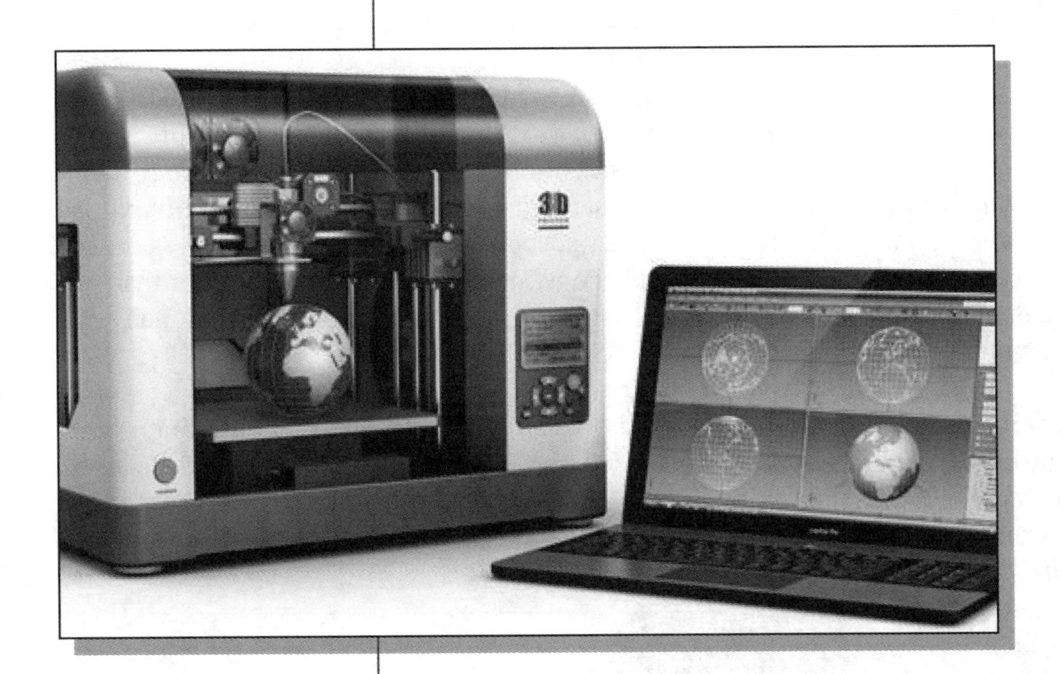

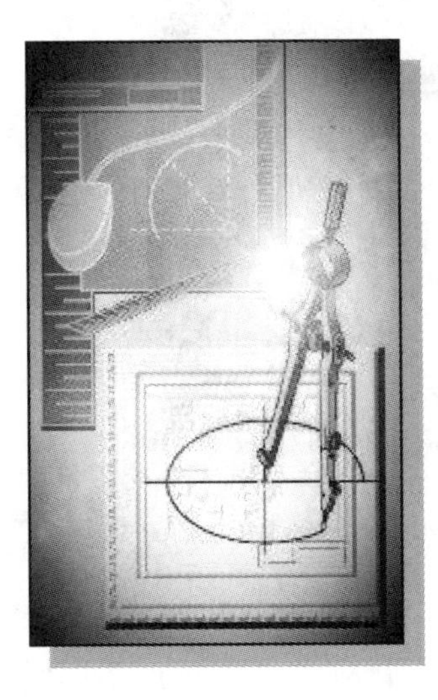

Learning Objectives

- Understand the History and Development of 3D Printing
- Be aware of the Primary types of 3D Printing Technologies
- Be able to identify the commonly used Filament types for Fused Filament Fabrication
- Understand the general procedure for 3D Printing

What is 3D Printing?

3D Printing is a type of *Rapid Prototyping* (RP) method. Rapid prototyping refers to the techniques used to quickly fabricate a design to confirm/validate/improve conceptual design ideas. 3D printing is also known as "**Additive Manufacturing**" and construction of parts or assemblies is usually done by addition of material in thin layers.

Prior to the 1980s, nearly all metalworking was produced by machining, fabrication, forming, and mold casting; the majority of these processes require the removal of material rather than adding it. In contrast to the *Additive Manufacturing* technology, the traditional manufacturing processes can be described as **Subtractive Manufacturing**. The term *Additive Manufacturing* gained wider acceptance in the 2010s. As the various additive processes continue to advance and become more mature, it is quite clear that they will compete with material removal as the main manufacturing process for many applications in the very near future.

The basic principle behind *3D printing* is that it is an additive process. 3D printing is a radically different manufacturing method based on advanced technology that create parts directly, by adding material layer by layer at the sub millimeter scale. One way to think about 3D Printing is the additive process is really performing "**2D printing over and over again**."

A number of limitations exist to the traditional manufacturing processes, which can be labor intensive, requiring expensive tooling, designing of fixtures, and the assembly of parts. The *3D printing* technology provides a way to create parts with complex geometric shapes quite easily using thin layers. The traditional *subtractive manufacturing* can also be quite wasteful as excess materials are cut and removed from large stock blocks, while the *3D printing* process basically uses only the material needed for the parts. *3D printing* is an enabling technology that encourages and drives innovation with unprecedented design freedom while being a tooling-less process that reduces costs and lead times. The relatively fast turnaround time also makes *3D printing* ideal for prototyping. Components with intricate geometry and complex features can also be designed specifically for *3D printing* to avoid complicated assembly requirements. *3D printing* is also an energy efficient technology that can provide better environmental friendliness in terms of the manufacturing process itself and the type of materials used for the product. There are quite a few different techniques to 3D print an object. *3D Printing* brings together two fundamental innovations: the manipulation of objects in the digital format and the manufacturing of objects by addition of material in thin layers.

The term **3D-printing** originally referred only to the smaller 3D printers with moveable print heads similar to an inkjet printer. Today, the term **3D-printing** is used interchangeably with **Additive Manufacturing**, as both refer to the technology of creating parts through the process of adding/forming thin layers of materials.

Development of 3D Printing Technologies

The earliest 3D printing technology was first invented in the 1980s; at that time it was generally called **Rapid Prototyping** (**RP**) technology. This is because the process was originally conceived as a fast and time-effective method for creating prototypes for product development in industry. In 1981, Dr. Hideo Kodama of *Nagoya Municipal Industrial Research Institute* invented two methods of creating three-dimensional plastic models with photo-hardening polymer through the use of a UV Laser. In 1986, the first US patent for **stereolithography** apparatus (**SLA**) was issued to Charles Hull, who first invented his SLA machine in 1983. Chuck Hull went on to co-found *3D Systems Corporation*, which is one of the largest companies in the 3D printing sector today. Chuck Hull also designed the **STL** (**ST**ereo**L**ithography) file format, which is widely used by 3D printing software performing the digital slicing and infill strategies common to the additive manufacturing processes. The first available commercial RP system, the **SLA-1** by *3D Systems* (as shown in the figure below), was made available in 1987.

The 1980s also mark the birth of many RP technologies worldwide. In 1989, Carl Deckard of *University of Texas* developed the **Selective Laser Sintering (SLS)** process. In 1989, Scott Crump, one of the founders of *Stratasys Inc*. also created the **Fused Deposition Modeling (FDM)**. In Europe, Hans Langer started *EOS GmbH* in Germany; the company focuses on the **Laser Sintering (LS)** process. The *EOS systems Corp.* also developed the **Direct Metal Laser Sintering (DMLS)** process. Today, *3D Systems*, *EOS* and *Stratasys* are still the main leaders in the *Additive Manufacturing* industry.

During the 1990s, the *3D printing* sector started to show signs of distinct diversification with two specific areas of emphasis which are much more clearly defined today. First, there was the high end of 3D printing, still very expensive systems, which were geared towards part production for relatively complex designs. For example, in 1995, *Sciaky Inc* developed an additive welding process based on its proprietary **Electron Beam Additive Manufacturing (EBAM)** technology. Many *RP* system companies, such as *Solidscape*, *ZCorporation*, *Arcam* and *Objet Geometries* were all launched in the 1990s. At the other end of the spectrum, some of the 3D printing system manufacturers started to develop smaller desktop systems in the 1990s.

The idea of creating low-cost desktop 3D printers also intrigued many technology professionals and hobby enthusiasts during the late 1990s. In 2004, a retired professor, Dr Adrian Bowyer (the person on the left in the photo below), started the **RepRap** (*Replication Rapid-Prototyper*) project of an open source, self-replicating 3D printer (**RepRap 1.0 - Darwin**). This sets the stage for what was to come in the following years. It was around 2007 that the open source *3D printing* movement started gaining visibility and momentum. In January of 2009, the first commercially available open source 3D printer, the **BFB RapMan** 3D printer, became available. *Makerbot Industries* also came out with their **Makerbot** 3D printer in April of 2009. Since then, a host of low cost desktop 3D printers have emerged each year.

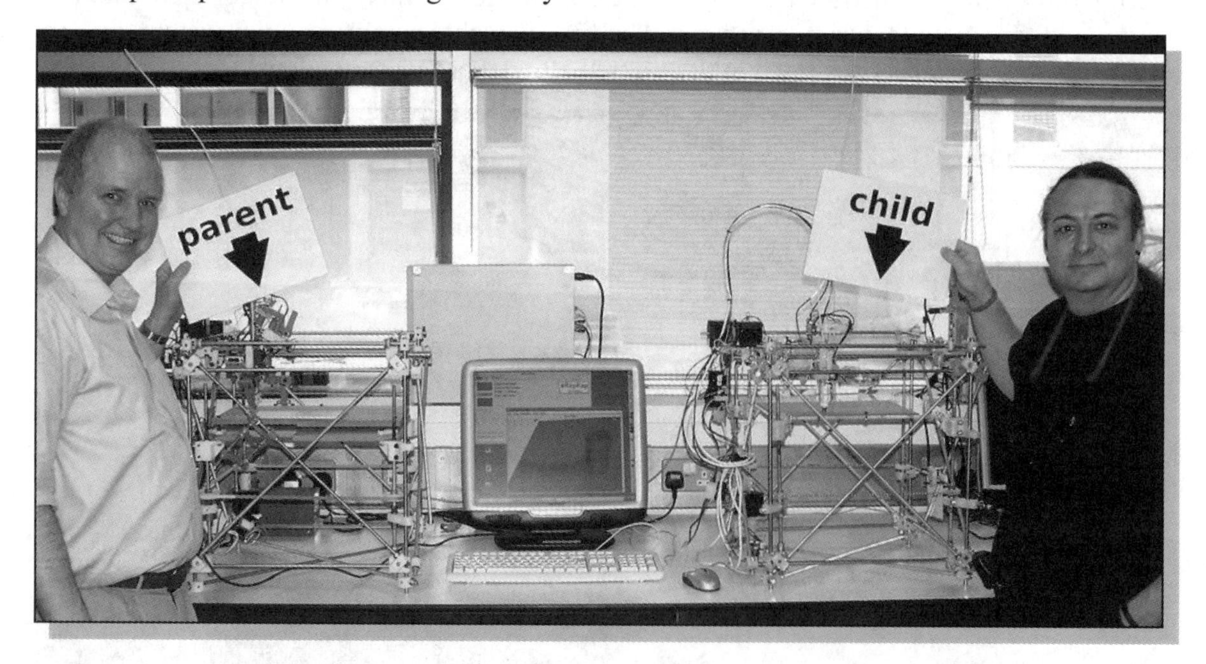

In the beginning of the 2010s, alternative 3D printing processes, such as those using **Polymer Resins**, became available at the desktop level of the market. The **B9 Creator**, using **Digital Light Processing (DLP)** technology, by *B9Creations* came first in June of 2012, followed by the **Form 1** desktop printer by *Formlabs Inc.* Both 3D printers were launched via KickStarter's crowd-funding website and both enjoyed huge success. 2012 was also the year that many different mainstream media took note of the exciting 3D printing technology, which dramatically increased its visibility and awareness to the general public. 2013 was also a year of significant growth and consolidation; one of the most notable moves was the acquisition of Makerbot by Stratasys. In 2016, the new developments in 3D printing concentrated more on multi-color, multi-material using single or multiple extruders printers and new technologies to shorten the 3D printing time.

As a result of the market divergence, the price of desktop 3D printers continues to go down each year. Today, very capable fully assembled desktop 3D printers, such as Robo3D R1+, Prusa I3 MK2, can be acquired for under $1000. Fully assembled smaller desktop 3D printers, such as XYZpring's DA Vinci mini 3D Printer and M3D's Micro 3D can be acquired for less than $350. Unassembled desktop 3D printer kits can even be acquired for under $250.

Another trend that happened in the 2010s is the availability of **3D printing Services**. 3D printing services are growing quite rapidly in the US. For example, many public libraries, especially in California, are now providing 3D printing services to the general public and UPS has started its worldwide 3D printing services in May of 2016. This trend is spreading throughout the US, with many more companies planning to provide 3D printing services in the very near future. It is now quite feasible, and perhaps more economical, to 3D print designs without owning or ever touching a 3D printer, but understanding of the technology is still needed to increase productivity.

As the exponential adoption rate continues on all fronts, more and more technologies, materials, applications, and online services will continue to emerge. It is predicted that the development of 3D printing will continue in the years to come and 3D printing will eventually become the mainstream manufacturing method in industries and at homes.

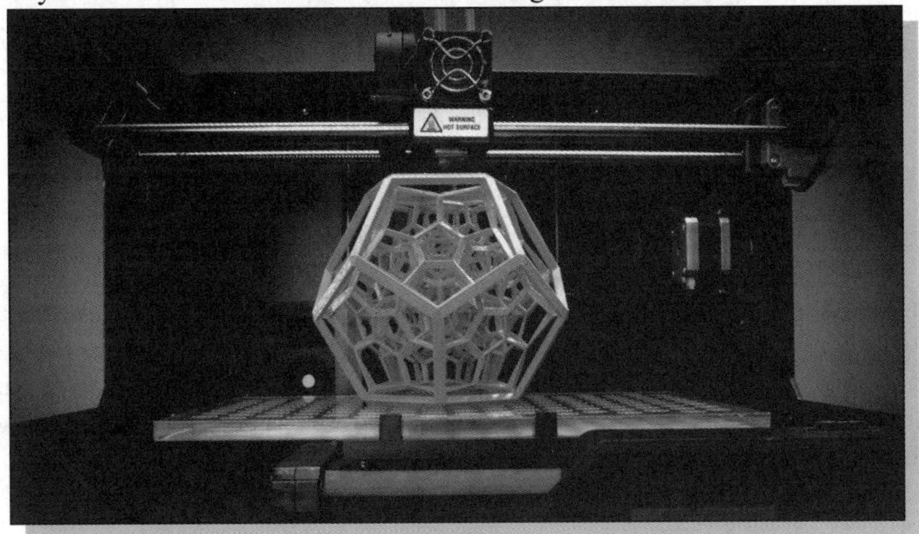

Primary Types of 3D Printing Processes

There are quite a few different techniques to 3D print an object. The different types of 3D printers each employ a different technology that processes different materials in different ways. For example, some 3D printers process powdered materials (nylon, plastic, ceramic and metal), which utilize a light/heat source to sinter/melt/fuse layers of the powder together in the defined shape. Others process polymer resin materials and again utilize a light/laser to solidify the resin in thin layers. **Stereolithography (SLA** or **SL**), **Fused Deposition Modeling (FDM** or **FFF**) and **Laser Sintering (LS** or **SLS**) represent the three primary types of 3D printing processes; the majority of the other 3D printing technologies are variations of the three main types.

Stereolithography

Stereolithography (**SLA** or **SL**) is widely recognized as the first 3D printing process; it was certainly the first to be commercialized. *SLA* is a laser-based process that works with photopolymer resins. The photopolymer resins react with the laser and cure to form a solid in a very precise way to produce very accurate parts. It is a complex process, but simply put, the photopolymer resin is held in a container with a movable platform inside. A laser beam is directed in the X-Y axes across the surface of the resin according to the 3D data supplied to the machine. The resin hardens precisely as the laser hits the designated area. Once the current layer is completed, the platform within the container drops down by a fraction (in the Z axis) and the subsequent layer is traced out by the laser. This 2D layer tracing continues until the entire object is completed.

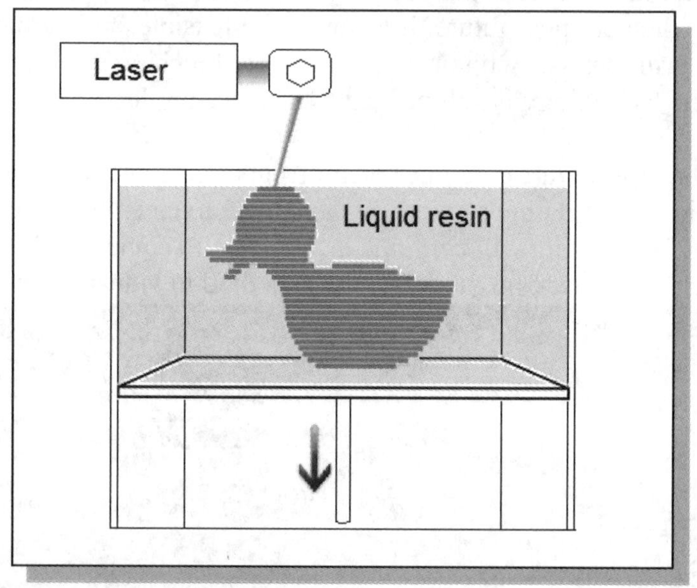

Because of the nature of the SLA process, support structures are needed for some parts, specifically those with overhangs or undercuts. These support structures need to be removed once the part is created. Many 3D printed objects using SLA need to be further cleaned and/or cured. Curing involves subjecting the part to intense light in an oven-like machine to fully harden the resin. SLA is generally accepted as being one of the most accurate 3D printing processes with excellent surface finish.

Fused Deposition Modeling (FDM) / Fused Filament Fabrication (FFF)

3D printing utilizing the extrusion of thermoplastic material is probably the most popular 3D printing process. The original name for the process is **Fused Deposition Modeling (FDM)**, which was developed in the early 1990s and is a trade name registered by *Stratasys*. However, a similar process, **Fused Filament Fabrication (FFF)**, has emerged since 2009. The majority of the desktop 3D printers, both open source and proprietary, utilize the FFF process, which is a more basic extrusion form of FDM.

The FDM and FFF processes work by melting plastic filament that is deposited, via a heated extruder, one layer at a time, onto a build platform according to the 3D data supplied to the 3D printer. Each layer hardens as it cools down and bonds to the previous layer.

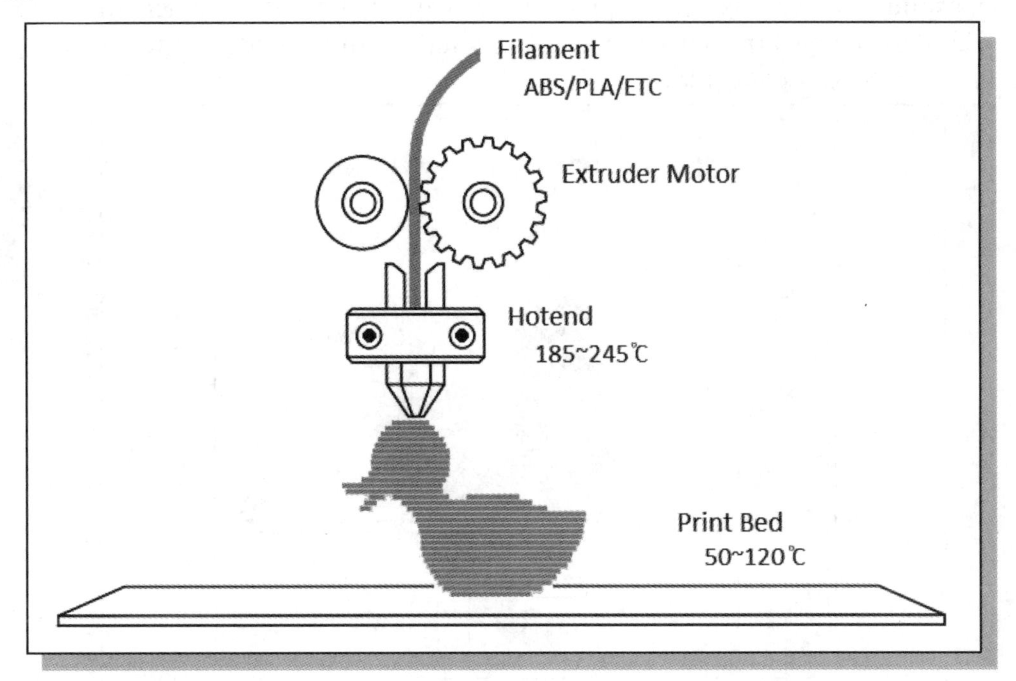

Stratasys has developed a range of proprietary industrial grade materials for its FDM process that are suitable for production applications. However, the most common materials for both FDM and FFF 3D printers are **ABS (Acrylonitrile Butadiene Styrene)** and **PLA (Polylactic Acid)**. The FDM and FFF processes require support structures for any applications with overhanging geometries. This generally entails a second, typically water-soluble or breakaway material, which allows support structures to be easily removed once the print is complete.

The FDM and FFF printing processes can be slow for large parts or parts with complex geometries. The layer to layer adhesion can also be a problem, resulting in parts that warp or separate easily. The surface finish of FDM and FFF printed parts might appear a bit rough as the thin layers are generally visible. To improve the appearance, several options are feasible, such as using acetone, sanding and/or spray paint.

Laser Sintering / Laser Melting

Laser Sintering (LS) or **Selective Laser Sintering (SLS)** creates tough and geometrically intricate parts using a high-powered CO_2 laser to fuse/sinter/melt powdered thermoplastics. The main advantage of SLS *3D printing* is that as a part is made, it remains encased in powder; this eliminates the need for support structures and allows for very complex 3D geometries to be 3D printed. SLS can be used to produce very strong parts as exceptional materials such as Nylon and metal powders are commonly used.

Laser sintering refers to a laser based 3D printing process that works with powdered materials. The laser is traced across a powder bed of tightly compacted powdered material, according to the 3D data provided to the machine, in the X-Y axes. As the laser interacts with the powdered material it sinters and fuses the particles to each other forming a solid. As each layer is completed the powder bed drops incrementally and a roller is used to compact the powder over the top surface of the bed prior to the next pass of the laser for the subsequent layer.

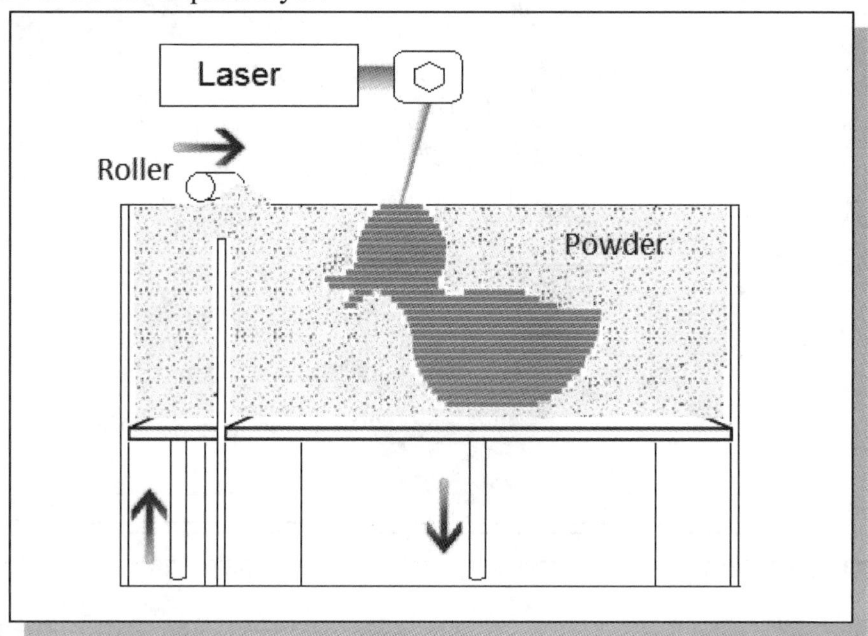

The build chamber is completely sealed as it is necessary to maintain a precise temperature during the process specific to the melting point of the powdered material of choice. One of the key advantages of this process is that the powder bed serves as an in-process support structure for overhangs and undercuts, and therefore complex shapes that could not be manufactured in any other way become possible with this process. Because of the high temperatures required for laser sintering, cooling can take a long time. Porosity is also a common issue with this process; additional metal infiltration processes may be required to improve mechanical characteristics.

Laser sintering can process plastic and metal materials, although metal sintering does require a much higher-powered laser and higher in-process temperatures. Parts produced with this process are much stronger than parts made with SLA or FDM, although generally the surface finish and accuracy is not as good.

Primary 3D Printing Materials for FDM and FFF

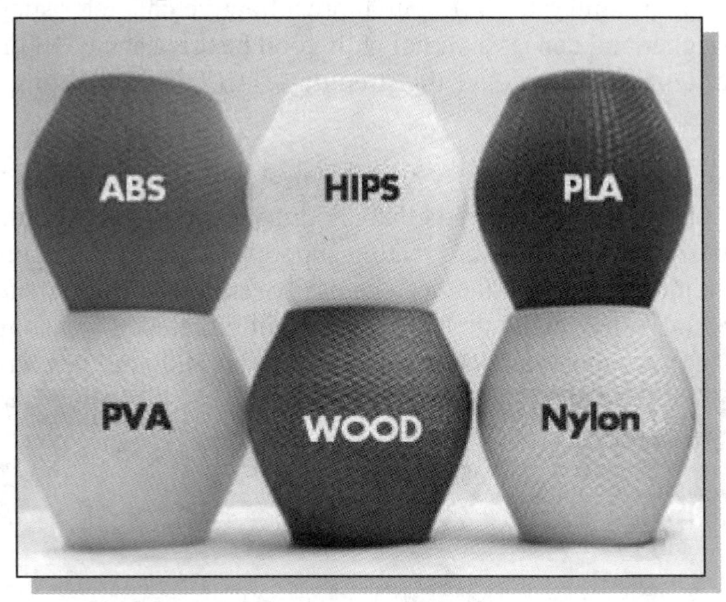

ABS (Acrylonitrile Butadiene Styrene)
ABS is a popular choice for 3D printing, it is a strong thermoplastic that is among the most widely used plastics. It is tough with mild flexibility, making it more durable to stress and has a higher heat resistance of up to 200 degrees Fahrenheit. However, this material has a tendency to shrink, which can affect the accuracy of designs. ABS has a pretty high melting point, and can experience warping if cooled while printing. Because of this, ABS objects are printed typically on a heated surface. ABS also requires ventilation when in use, as the fumes can be unpleasant. The aforementioned factors make ABS printing difficult for hobbyist printers, though it's the preferred material for professional applications.

PLA (Polylactic Acid)
PLA is a staple and it is becoming one of the most popular choices for 3D printing with good reason. In addition to the fact that it is a biodegradable thermoplastic derived from renewable resources such as corn starch, tapioca roots, chips or starch, or sugarcane, *PLA* is a very rigid material that is easy to use for 3D printing and it is able to withstand a good amount of impact and weight. It also has a glossier finish than ABS and in most scenarios PLA is the preferred material for 3D printing large objects. The main disadvantage of PLA is it's not as heat resistant as ABS, so it should not be placed in environments that exceed 140 degrees Fahrenheit.

Flexible (Thermoplastic Elastomer)
Flexible material is for applications that require incredible rubbery flex in their applications. Flexible filament goes beyond bending, it is more like rubber. When it comes to Flexible filament, it's all about finding a balance between flexibility (softness) and printability. This softness is sometimes indicated with a *Shore* value (like 85A or 60D). A higher Shore value means less flexibility. Harder filaments (less flexible) are easier to 3D print when compared to softer, more flexible filaments.

PETG (Polyethylene Terephthalate)
PETG is a material that is similar to *PLA*, with more attractive characteristics such as being generally a tougher and denser material with good heat resistance of up to 190 degrees Fahrenheit. It is reported to have the strength of *ABS*, while printing as easily as *PLA*.

HIPS (High Impact Polystyrene) and **PVA (Polyvinyl Alcohol)**
HIPS and *PVA* are relatively new materials that are growing in popularity for their dissolvable properties. They are used for creating support material. Their ability to dissolve in certain liquids means that they can be easily removed. These materials can be hard to print with because they don't stick well to the build plates. It is also important not to print *PVA* at too high a temperature, as it can turn into tar and jam the extruder.

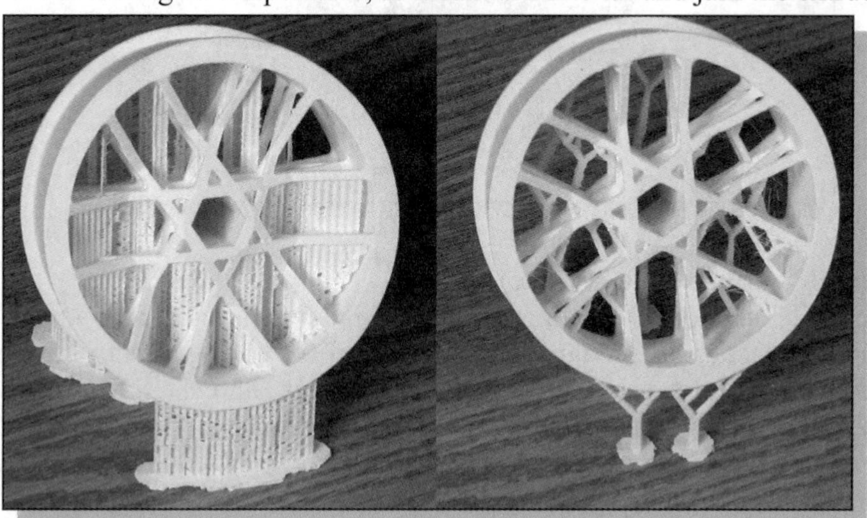

Wood Fiber
Wood Fiber filament contains a mixture of recycled wood with a binding polymer. Thus, a 3D printed object can look and smell like real wood. Due to its wooden nature, it's difficult to tell that the object is 3D printed. Using Wood filament is similar to using a thermoplastic filament like ABS or PLA. However, a 3D object having a wooden-like appearance can be created with this material.

From 3D Model to 3D Printed Part

To create a 3D printed part, it all starts with making a virtual design of the object. This virtual design may be created with a computer-aided design (CAD) package, via a 3D scanner, or by a digital camera and photogrammetry software. 3D scanning and photogrammetry software can process the collected digital data on the shape and appearance of a real object, and create a digital 3D model. The 3D virtual design can generally be modified with 3D CAD packages, allowing verification of the virtual design before it is 3D printed.

Once the virtual design is verified, the 3D data will then be transferred to the 3D printing software. There is a multitude of file formats that 3D printing software supports. However, the most popular are the STL file format and the OBJ file format. The STL file format is the most commonly used file format for 3D printing. Most CAD software has the capability of exporting models in the STL format. The STL file contains only the surface geometry of the modeled object. The OBJ file format is considered to be more complex than the STL file format as it is capable of displaying texture, color and other attributes of the three-dimensional object. However, the STL file format holds the top spot for 3D printing, as this file format is simpler to use, and most CAD packages work better with STL files than OBJ files.

Once the 3D data of the virtual design is transferred into the 3D printing software, further examination and/or repair can be performed if necessary. The 3D printing software will also process the imported 3D data by the special software known as a **Slicer**, which converts the model into a series of thin layers and produces a G-code file containing instructions tailored to a specific type of 3D printer. G-code is the common name for the most widely used numerical control (NC) programming language. It is used mainly in computer-aided manufacturing to control automated machine tools. The generated G-code file can be sent to the 3D printer and create the 3D printed part.

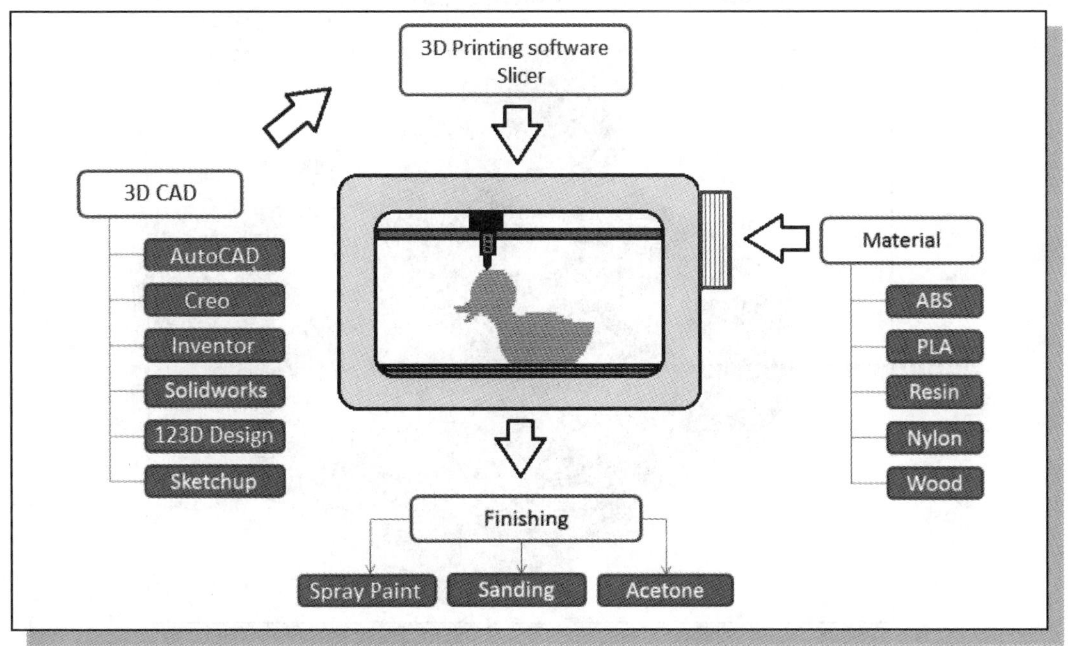

Starting SOLIDWORKS

1. Select the **SOLIDWORKS** option on the *Start* menu or select the **SOLIDWORKS** icon on the desktop to start SOLIDWORKS. The SOLIDWORKS main window will appear on the screen.

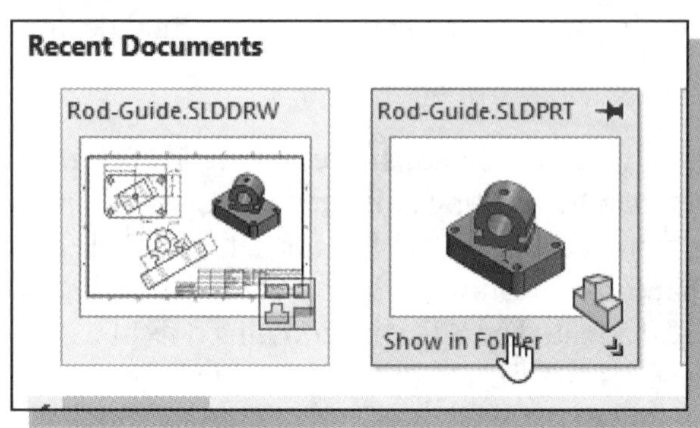

2. In the SOLIDWORKS *Startup* dialog box, select the **Rod-Guide.SLDPRT** file select **Open a Drawing** with a single click of the left-mouse-button. Use the *browser* to locate the file if it is not displayed in the *File name* list box.

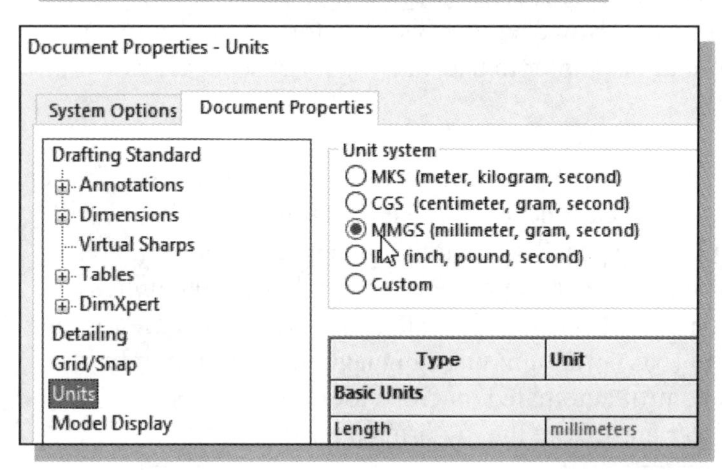

3. On your own, change the *Units* option to millimeter as shown. Note that the majority of the 3D printer settings are measured in millimeters, such as filament diameter, layer height and extruder size.

SOLIDWORKS Print3D Command

3D printers will generally accept a 3D model with the STL or OBJ file formats. SOLIDWORKS offers two options to saving the 3D model in STL format: 1. Using the SOLIDWORKS' **Print3D** command or 2. Using the **Save As** option. The Save As option can be used to very quickly export the 3D model, while the *Print3D* command provides more control review options.

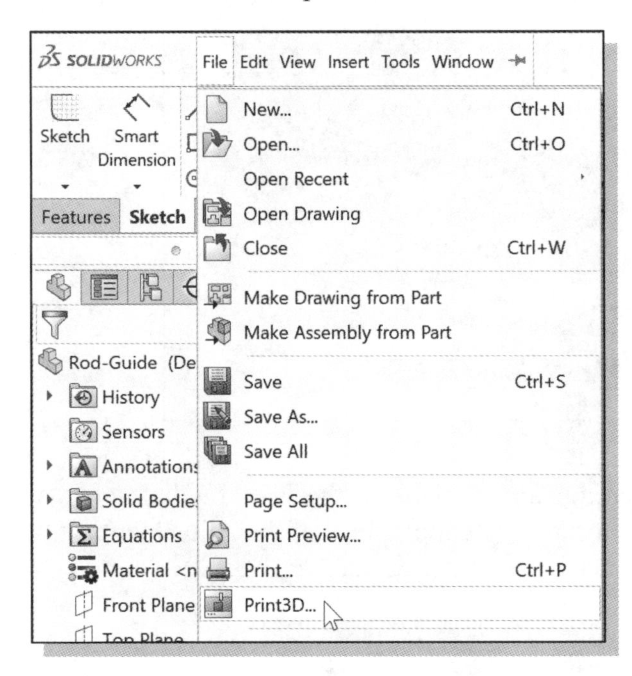

1. In the *File Toolbar,* select the **Print3D** command as shown.

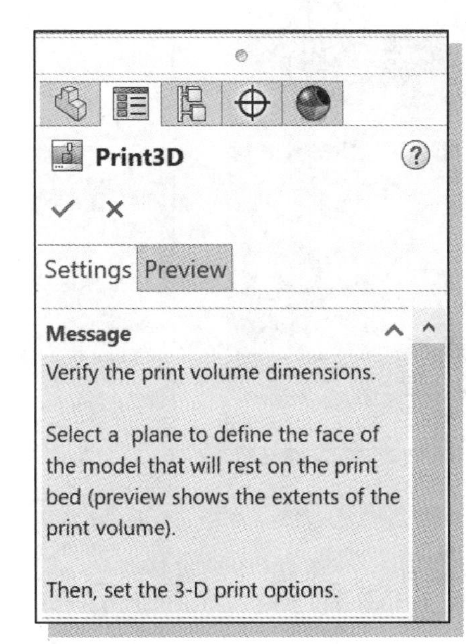

2. In the *Print3D* dialog box, SOLIDWORKS expects us to select a plane on the model that will be used to define the orientation of the 3D model to be aligned to the print bed of the 3D printer.

3. Select the bottom surface of the Rod-Guide design as shown.

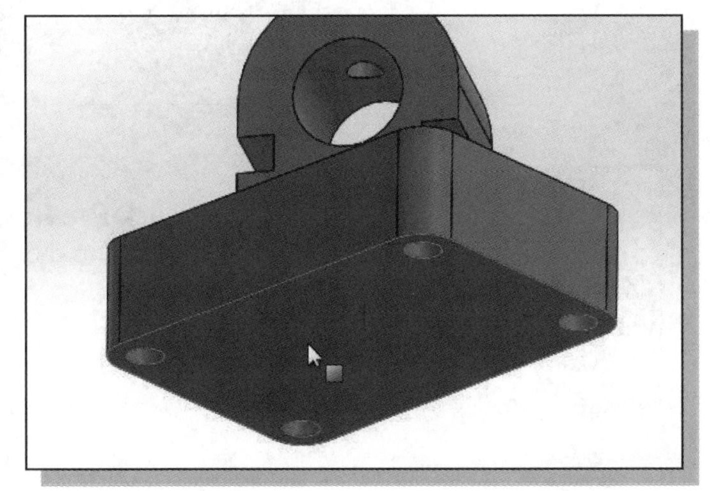

4. In the graphics window, the orientation of the 3D model is set based on the selected surface. The larger grey box indicates the max print volume of the 3D printer being used; these dimensions can be adjusted in the dialog box as shown.

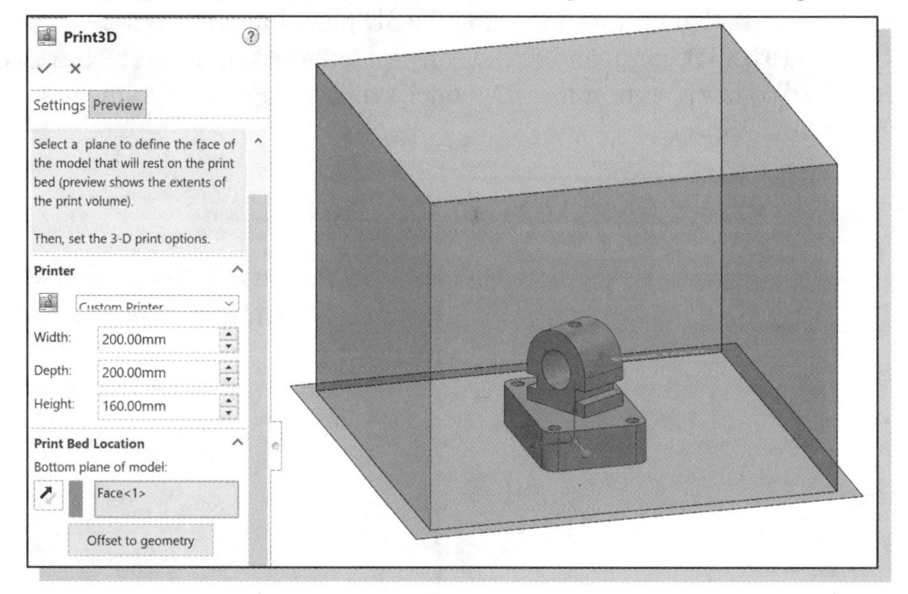

5. Additional orientation and scale settings are available in the *Print3D* dialog box. On your own, adjust some of the settings to see the effects of the adjustments.

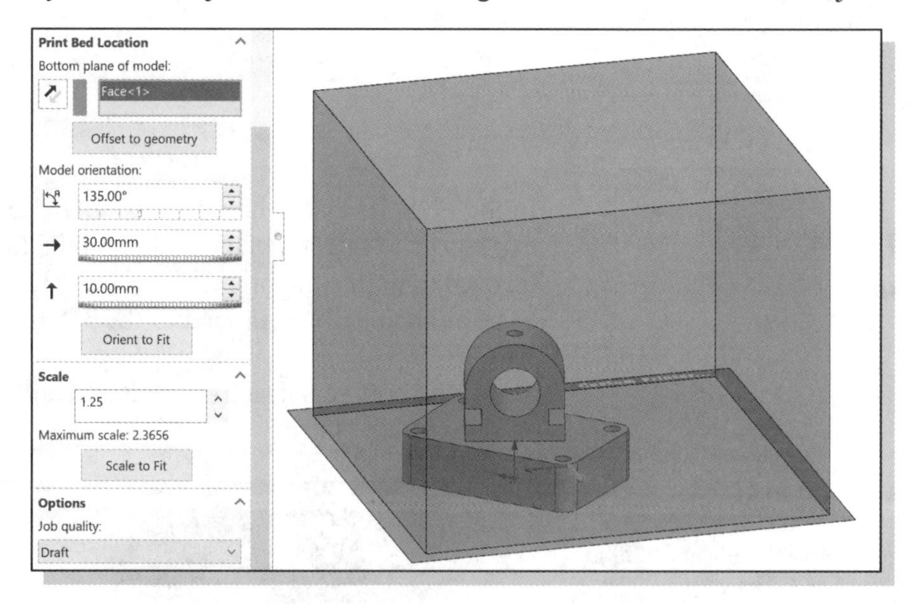

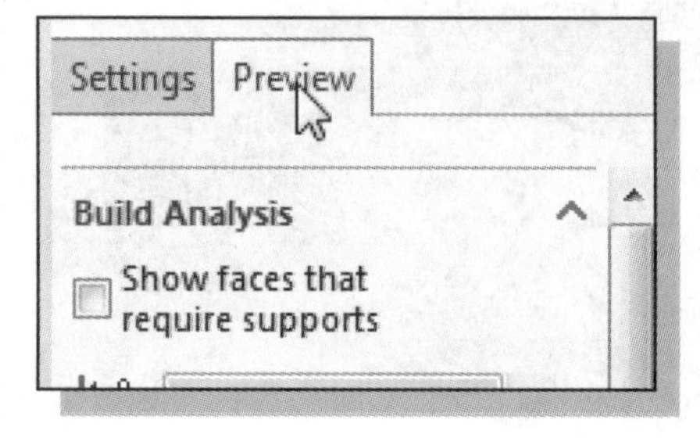

6. Click **Preview** in the *Print3D* dialog box to switch to the additional 3D print validation section.

- In SOLIDWORKS, the print validation section is now available under the Print3D **Preview** tab. The new print validation section provides additional tools to examine the 3D print model before it is actually printed, thus a **Preview**. Three options are available: 1. **Show faces that require supports**: this option will show us if any surface needs additional supports. This is typically needed for overhang features. Note that *supports* can be added in the 3D printing/Slicer software. 2. **Show striation lines**: this option can be used to determine whether the print resolution is sufficiently fine to produce the desired output. Adjusting the layer height will change the appearance of the 3D print. However, the layer height is mainly determined by the specific 3D printer in use; check your 3D printer's specs before lowering the layer height. 3. **Thickness/Gap Analysis**: one of the most common causes of a failed 3D print is because there are features in the model that are too small to print, or gaps too small to be recognized. To help prevent these failed builds, there is the *Thickness/Gap analysis check*. This check is particularly useful when scaling down a model to fit on the 3D printer. Small features and gaps can easily be overlooked when a model is scaled down. An additional benefit comes if you are not sure what value of thickness or gap to check for. For FDM/FFF 3D printers, SOLIDWORKS provides a list of materials with ideal wall thicknesses allowable based on the layer height. If the material to be used is not in the list, the Custom Thickness and Gap check box can be used to indicate specific values. This option can be used to quickly check the geometry and SOLIDWORKS will highlight any unprintable features, upfront and before sending the job to the machine. This could save hours of build time and also a lot of material.

7. In the *Print3D* dialog box, switch **on** the **Show faces that require supports** option. Note that SOLIDWORKS indicates the 3 overhang surfaces of the Rod Guide design will require additional *supports* for the 3D print.

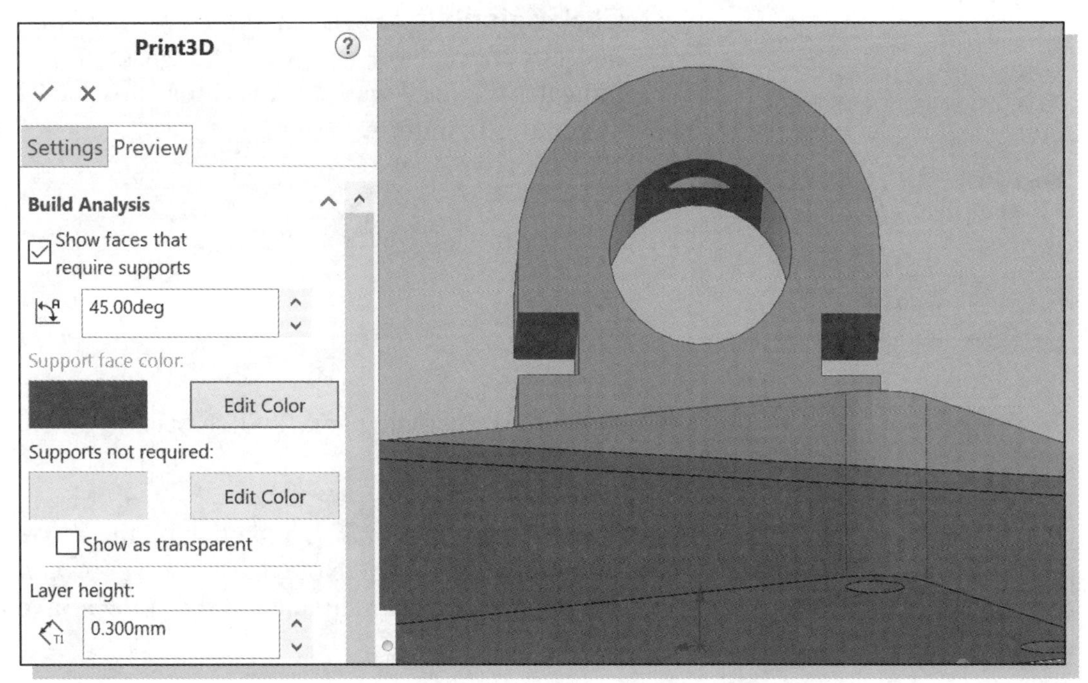

8. Turn **on** the **Show striation lines** option to view the smoothness of the model using the current layer height. (Hint: Zoom in to see the striation lines.)

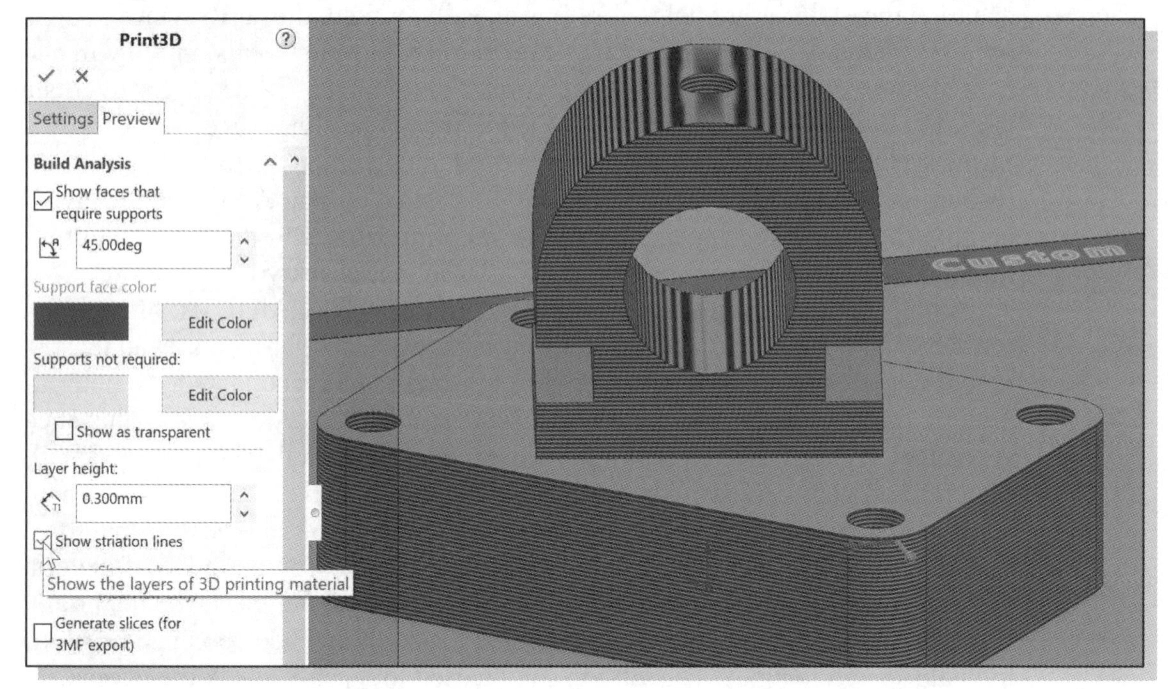

9. On your own, adjust the layer height to 0.1mm and examine the difference in smoothness of the previewed 3D print model by reducing the layer height.

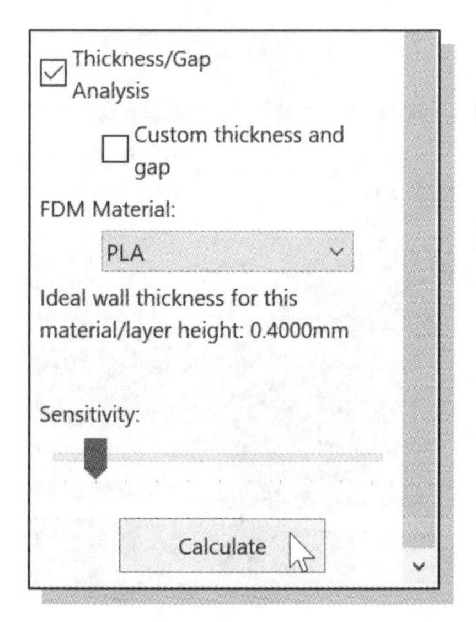

10. Turn on the **Thickness/Gap analysis** option and set the FDM material to PLA.

11. Click **Calculate** to perform the *Thickness/Gap analysis check*. Note that SOLIDWORKS also indicates the ideal wall thickness for the selected material is 0.8mm.

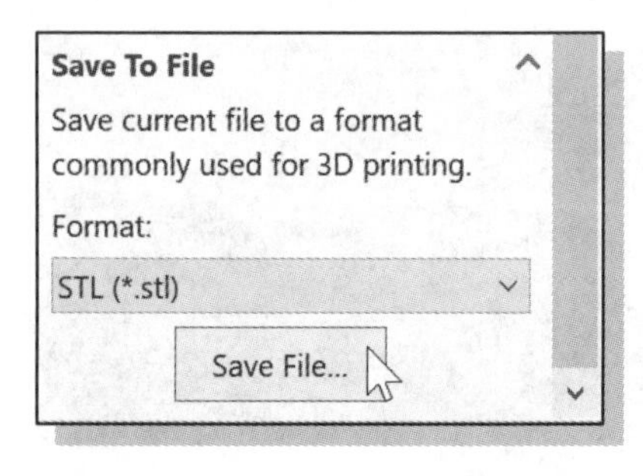

12. In the *Print3D* dialog area, switch back to the **Settings** tab.

13. Set the file format to STL and click on the **Save File** button as shown. Note that the *Save As dialog box* appears; this is redirected to the [**File→Save As**] command.

14. In the **Save As** dialog box, make sure the *Save as type:* is set to **STL**, and click on the **Options** button as shown.

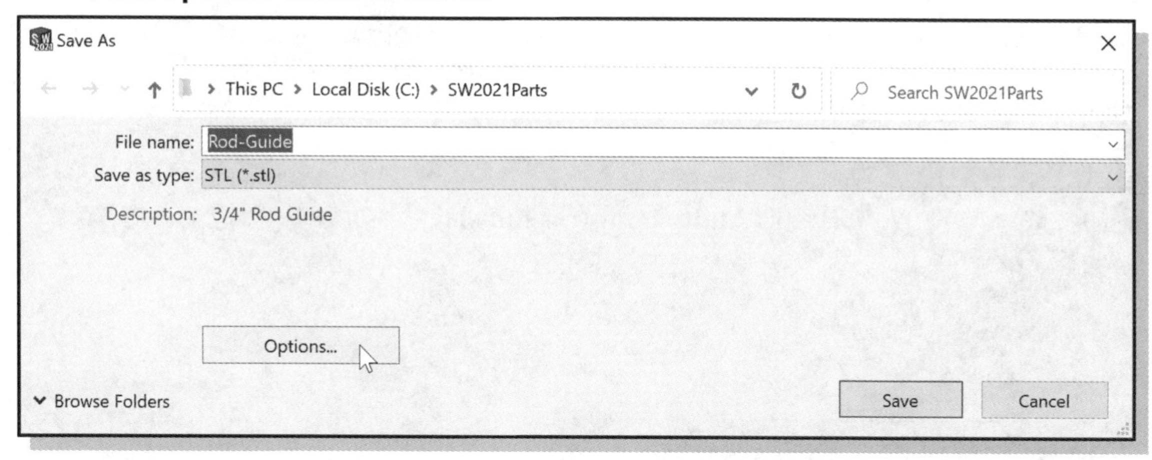

15. Note that SOLIDWORKS automatically sets the file format to STL as shown.

16. On your own, confirm the *output format* is set to **Binary** and *Resolution* to **Fine** as shown.

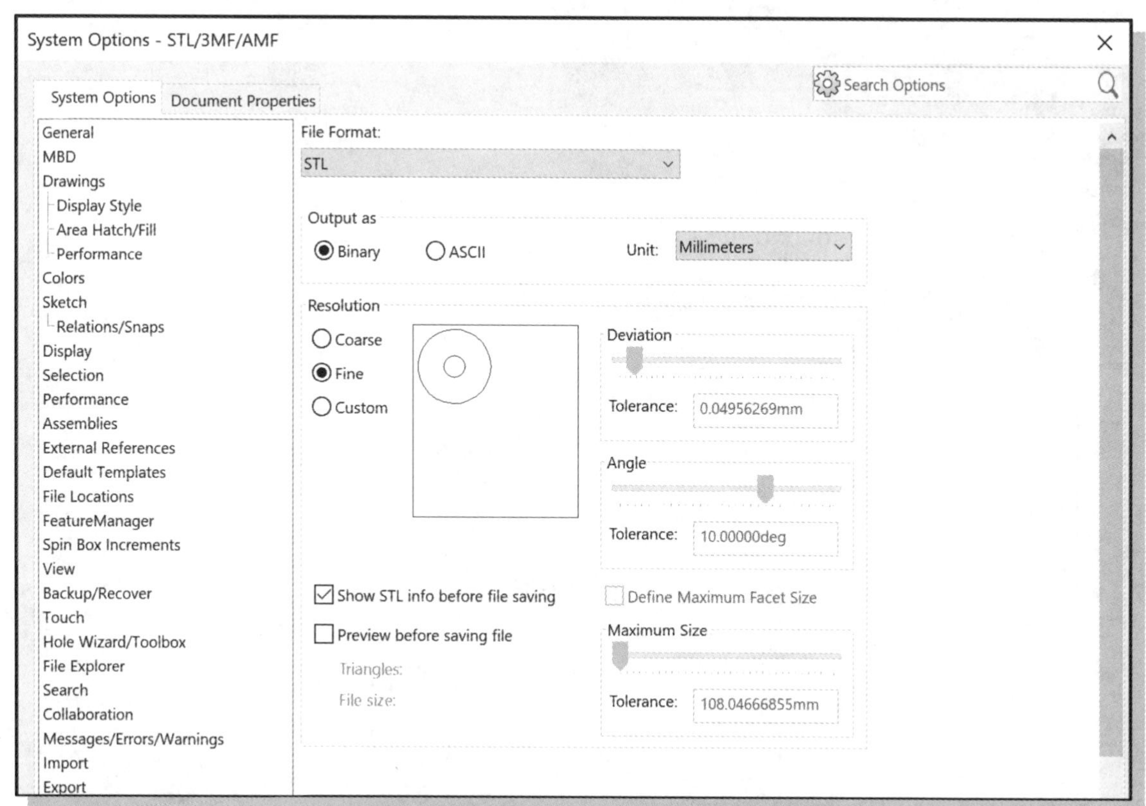

17. Click **OK** and then **Save** to proceed with saving the file.

18. Click **Yes** and **OK** to proceed with saving the Rod-Guide.STL file using the *All bodies* option.

Using the 3D Printing Software to Create the 3D Print

To 3D print the model, we will open the STL file in the 3D printing software. We will use **Matter Control** to demonstrate the procedure. Note that *Matter Control* (Freeware) supports quite a few desktop 3D printers. The procedures illustrated here are also applicable to other similar software.

1. Start the **Matter Control** software.

2. In the *toolbar area*, select **Open File** as shown.

3. On your own, switch to the saved STL file folder and select the Rod-Guide.STL file as shown.

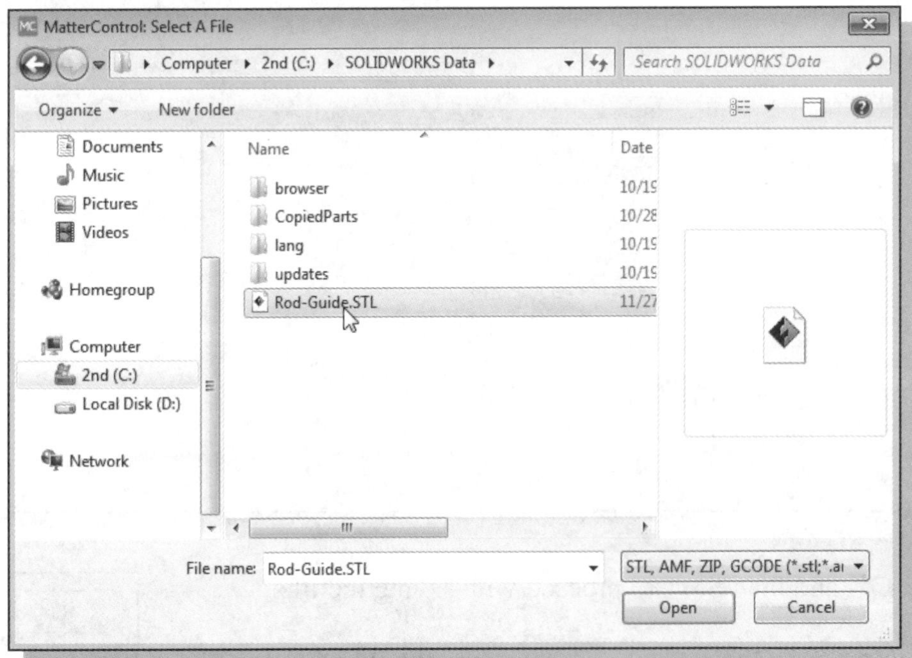

4. Click **Open** to import the STL file in to Matter Control.

5. Once the STL file is imported into the program, the STL model is displayed in the *View* window. Note that the model is imported with the incorrect orientation of the model; the bottom side of the model is not aligned to the print bed.

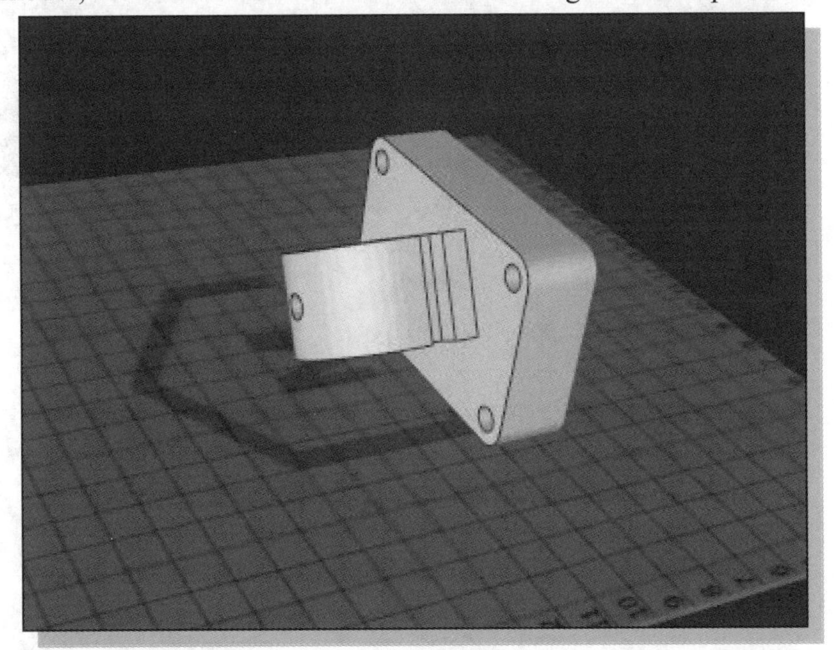

6. To adjust the display of the model, use the three mouse button to perform the Pan, Zoom and Rotate functions.

7. Note that the View Cube, located on the right side of the graphics window, is also available to control the viewing direction of the print bed.

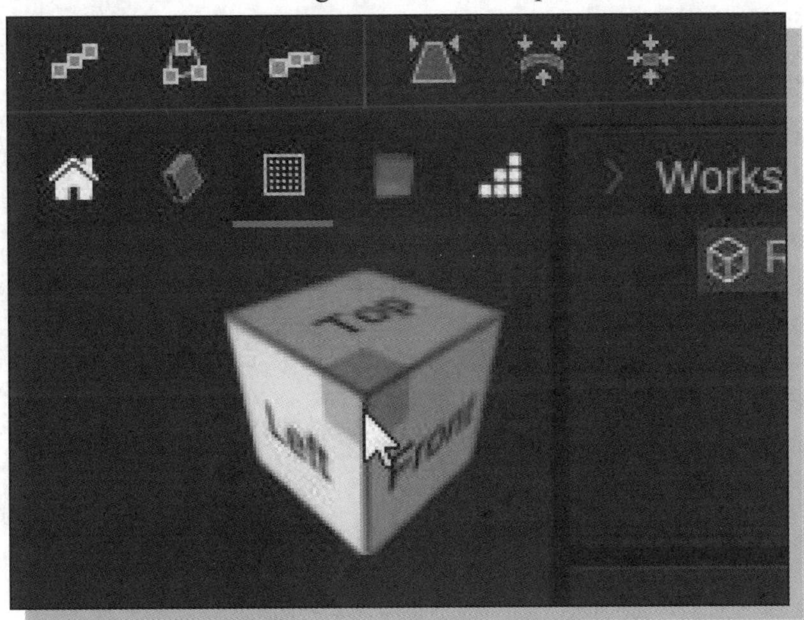

8. Click on the model, with the left mouse button, to enter the *Edit mode* and click & drag the model to reposition the model on the print-bed.

9. To rotate the model, click on the **Rotate icon** to enter the *Rotate control*; note the associated dial allows more precise rotation.

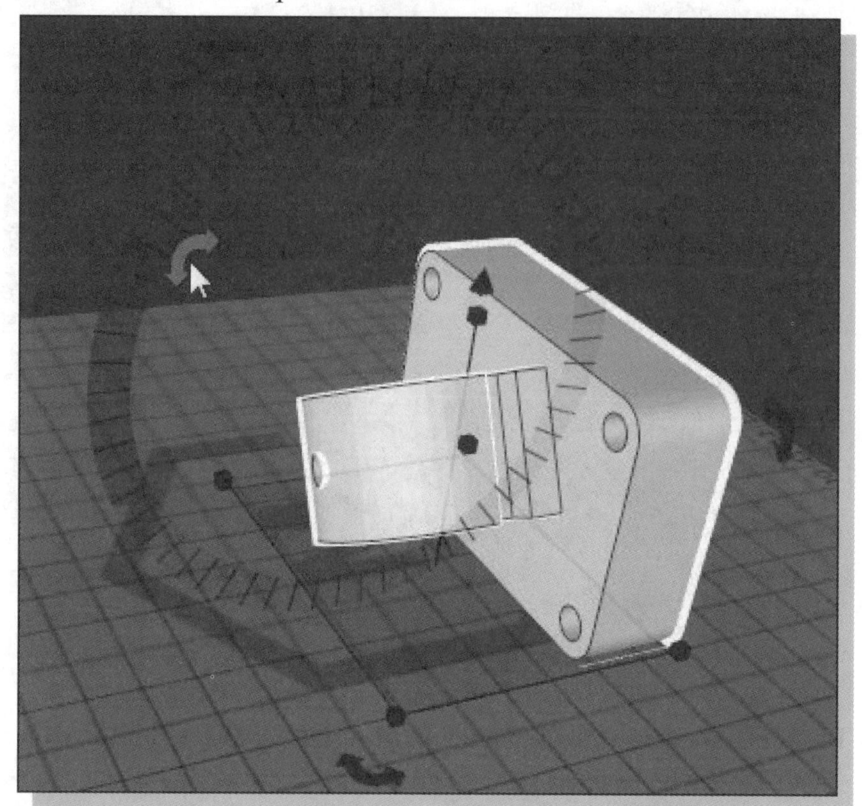

10. On your own, rotate the model so that it is setting vertically as shown.

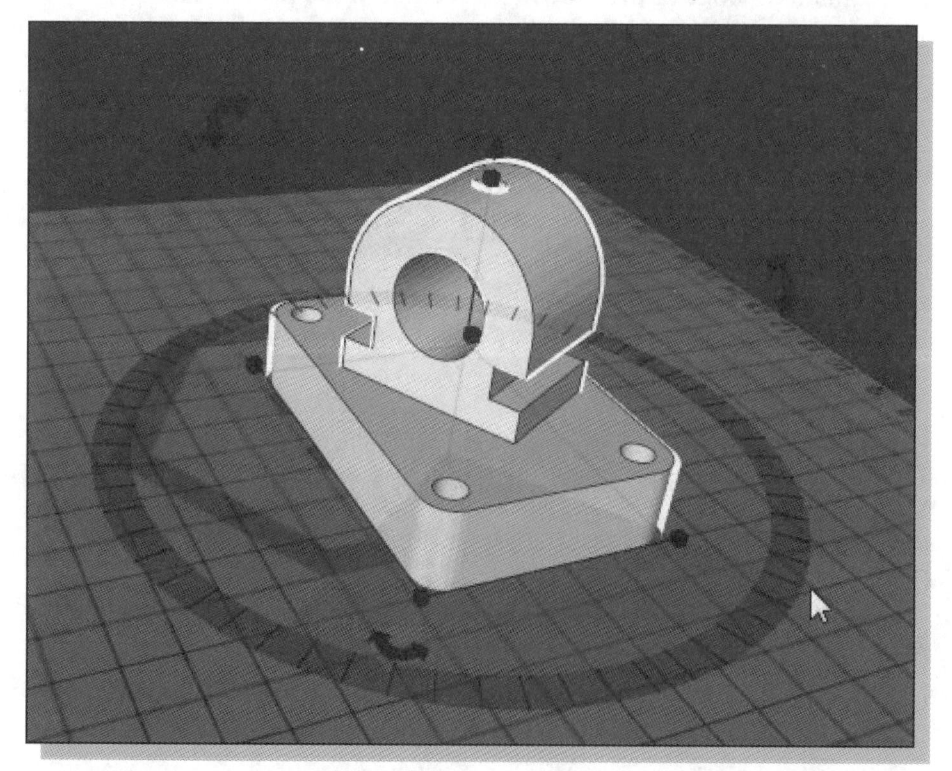

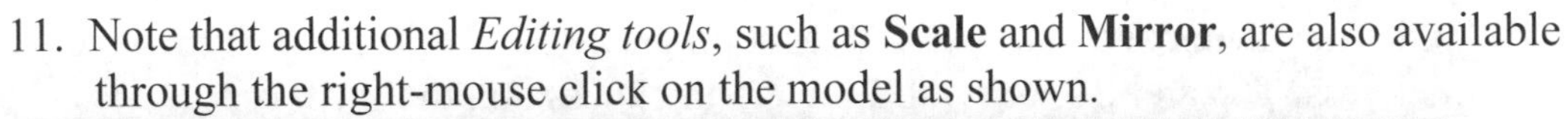

11. Note that additional *Editing tools*, such as **Scale** and **Mirror**, are also available through the right-mouse click on the model as shown.

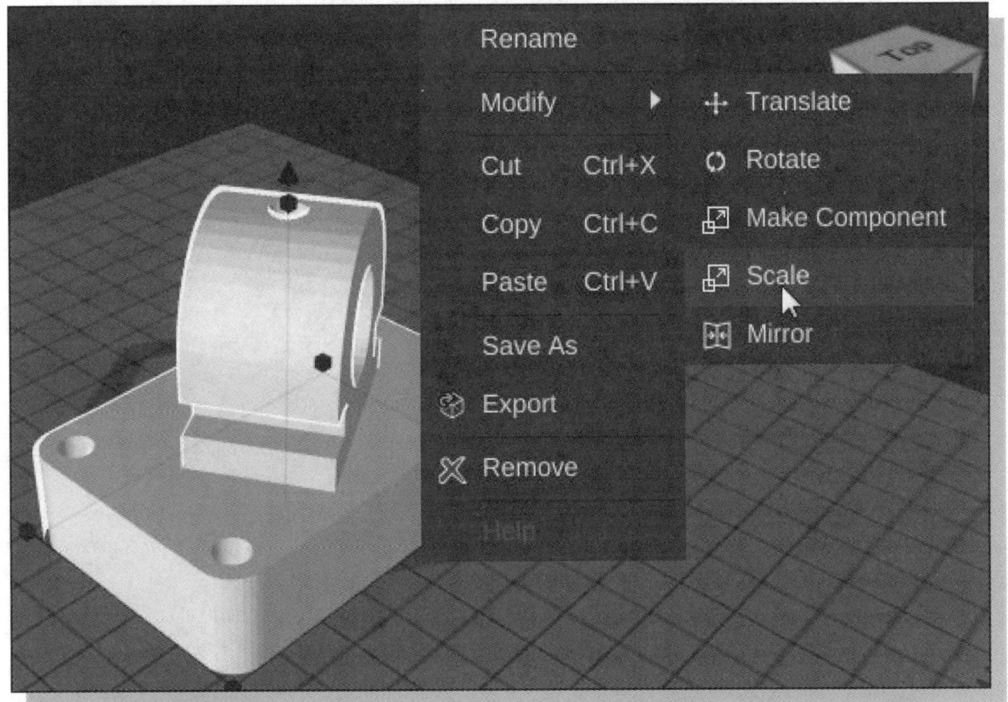

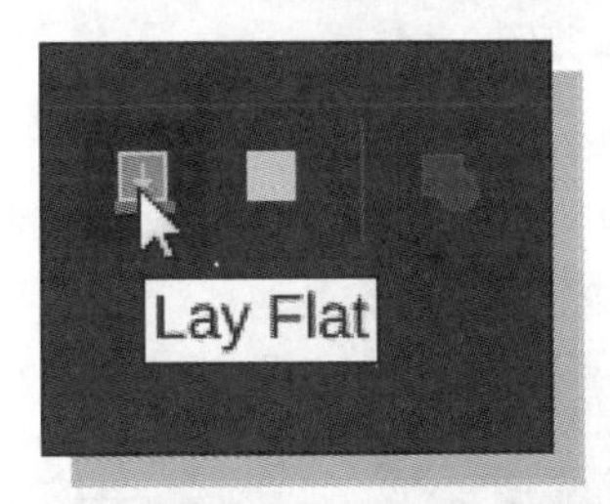

12. In the toolbar area, click **Lay Flat** to align the bottom surface of the model to the print bed.

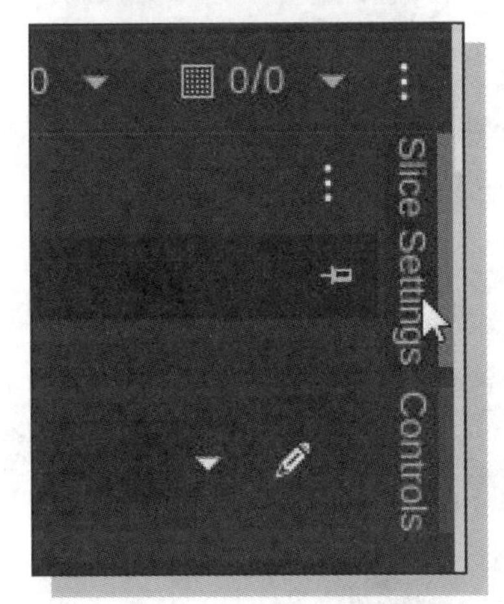

13. Click **Slice Settings** to review the 3D Printing settings. Note that the **Control tab** contain commands to directly control the movements of the 3D printer.

14. Under the **Slice Settings tab,** different settings are available to adjust the 3D printing settings.

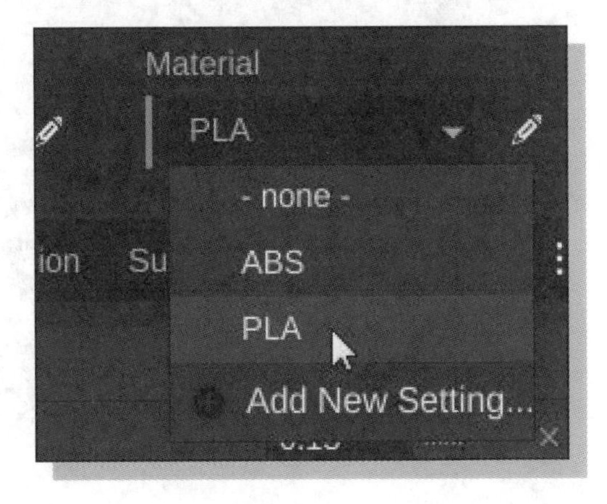

15. Under the **General** tab, a list of layer settings are available, such as the layer thickness, Top and Bottom Solid Layers thickness and the Infill type.

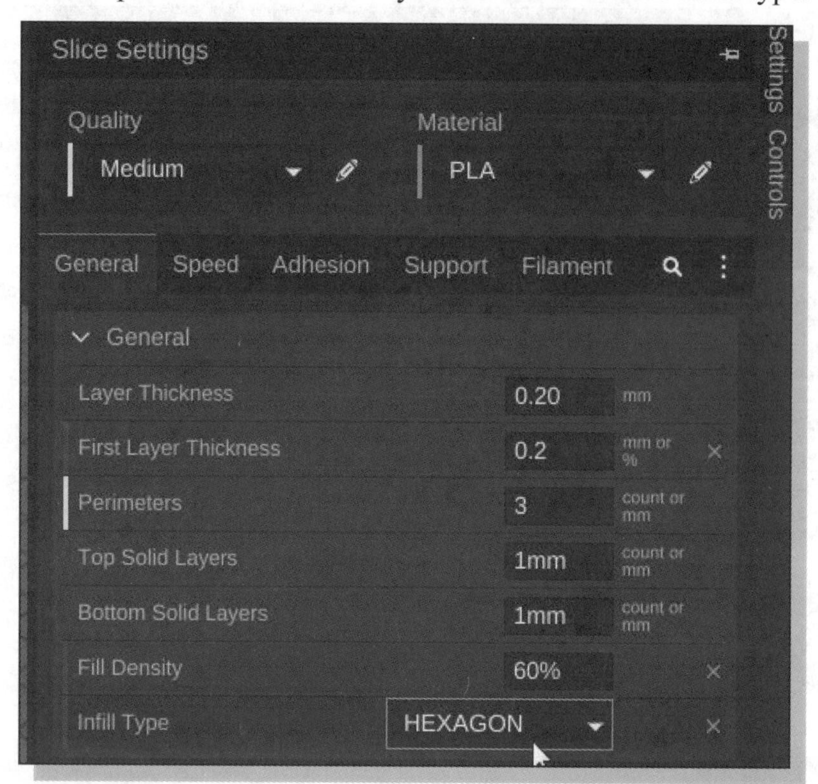

16. Under the **Speed** tab, the list of different speed settings on 3d printing are displayed and can be adjusted.

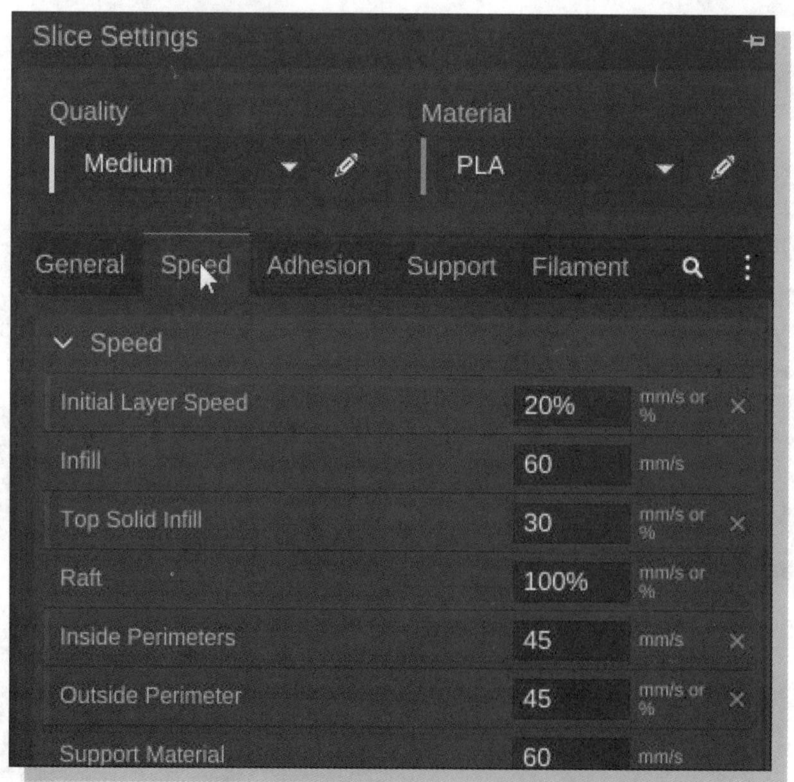

17. Under the **Adhesion** tab, the list of settings on improving adhesion on the first layer, such as Skirt, Raft and Brim, are available.

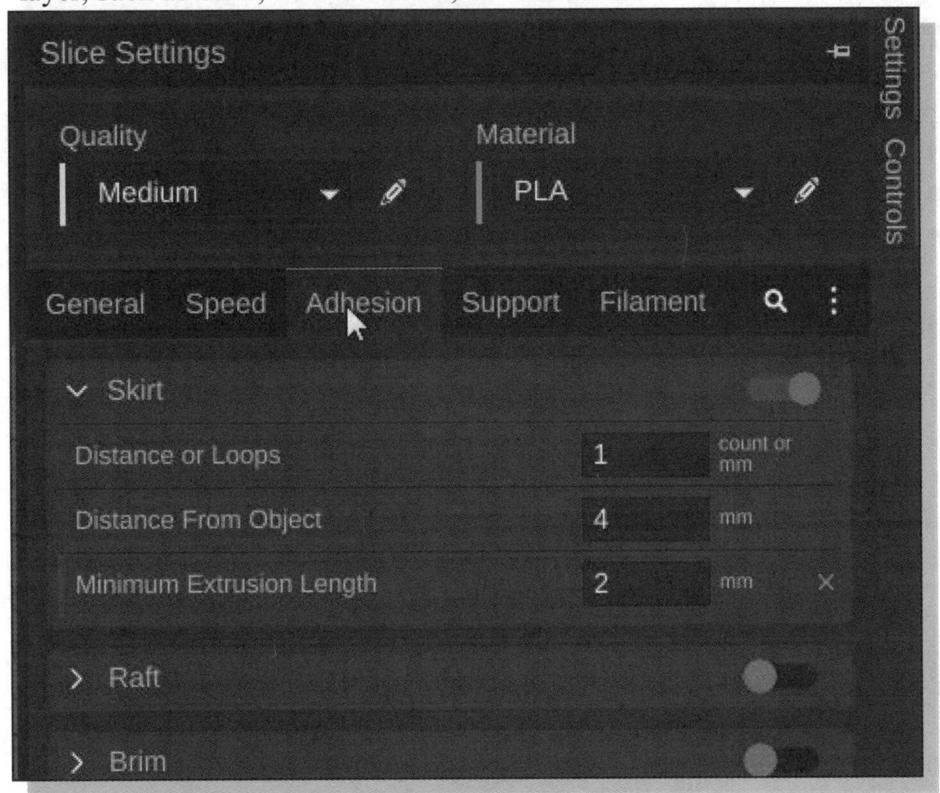

18. Under the **Support tab,** the adding support option can be turned on to *generate Support Material* as shown.

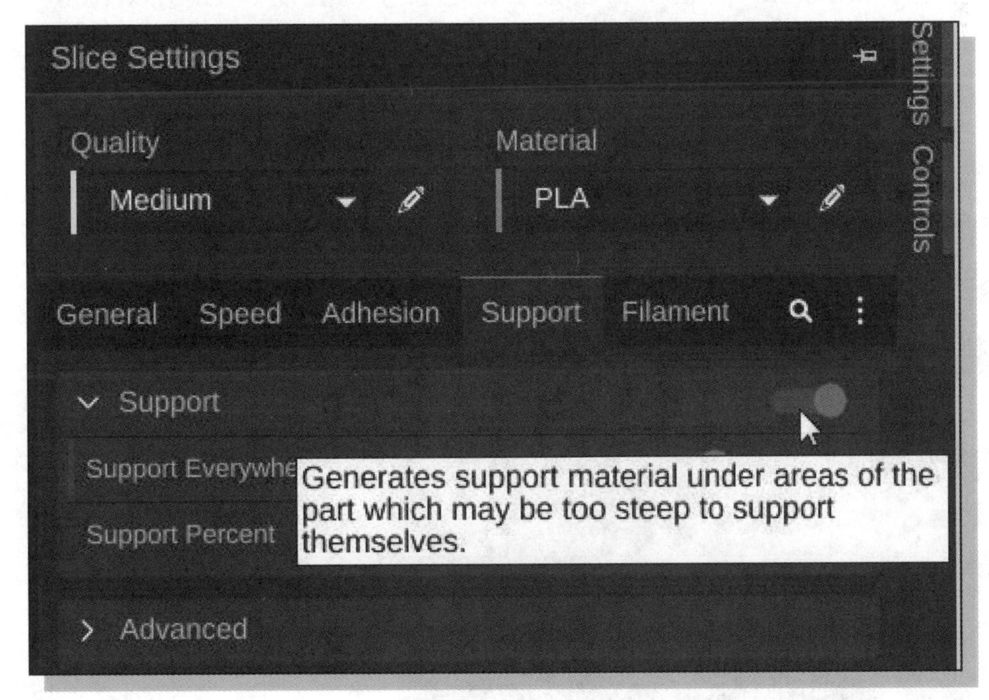

19. Switch to the **Filament tab** and in the **Material list** confirm/modify the filament properties, such as the diameter, to match the actual filament being used.

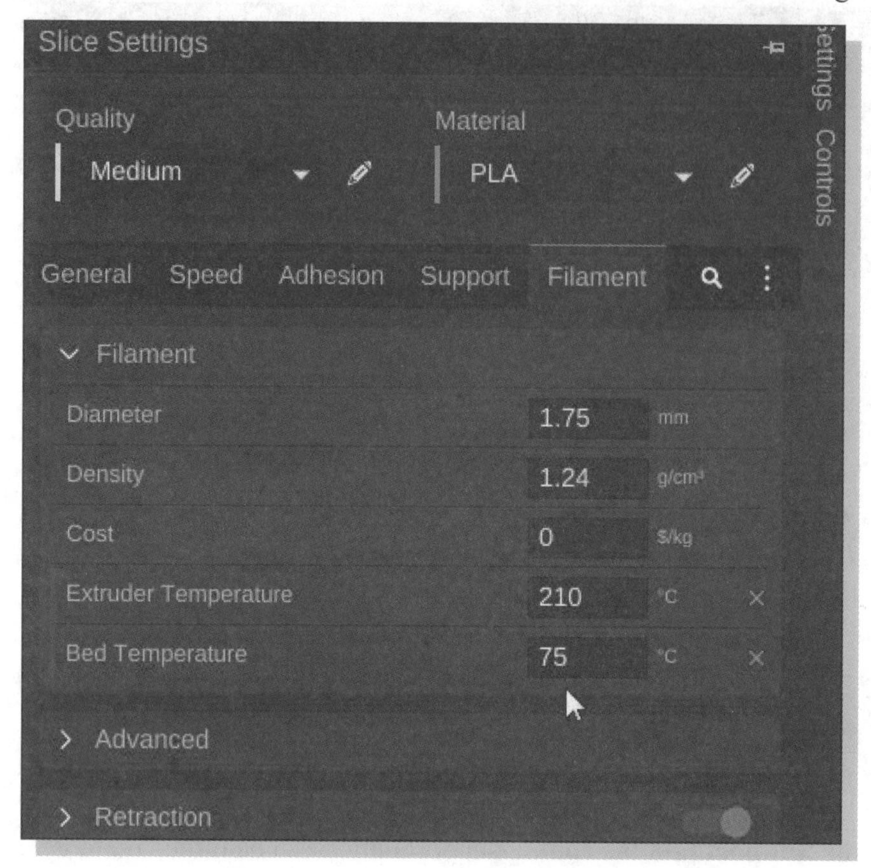

20. The temperature for the **Extruder** and the print bed can be adjusted on this page as well. Note that the temperatures are different based on the types of filaments, as well as the different brands and the type of printer bed. It is necessary to do some testing and/or experimenting when a new roll of filament is used.

21. In the toolbar area, click **Slice** to process the 3D model, which includes slicing and generating the associated G-code for the specific 3D printer.

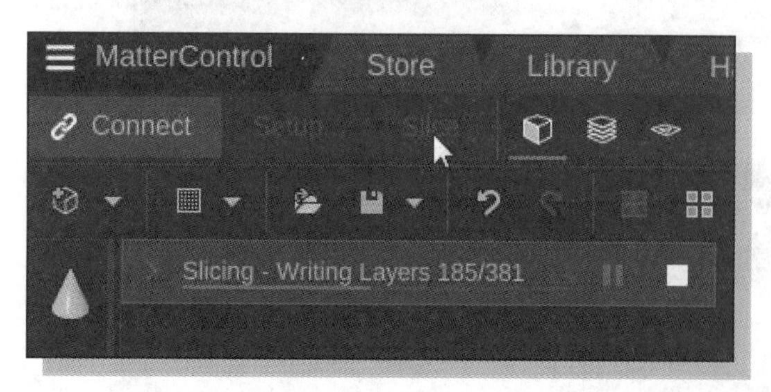

22. Depending on the size and complexity of the design, it might take several minutes to complete the process.

23. On your own, drag the vertical slider to review the thin layers generated by the slicer.

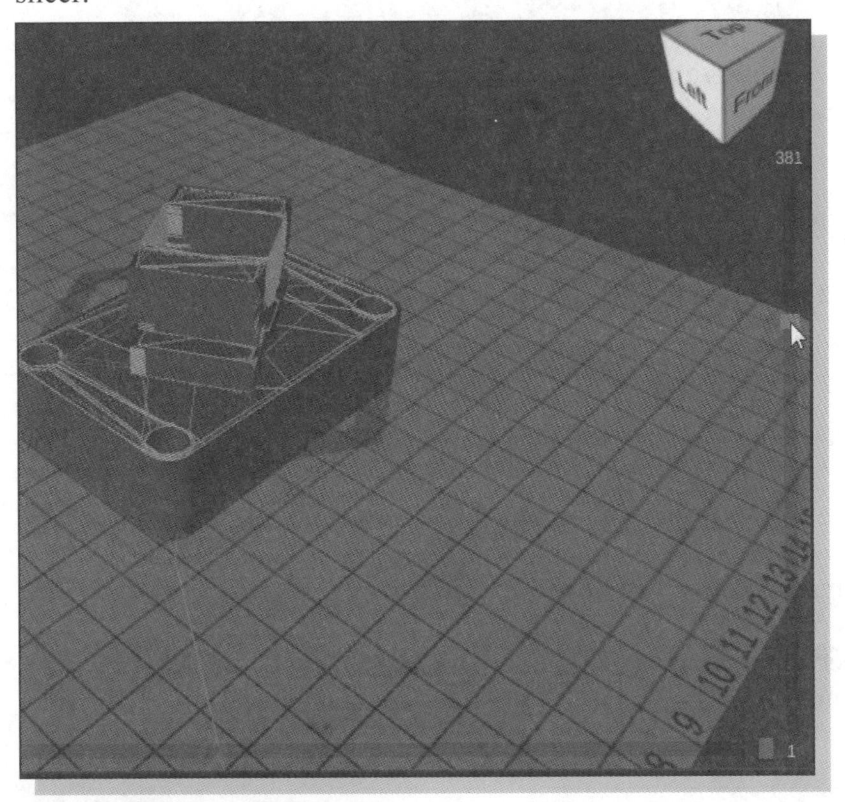

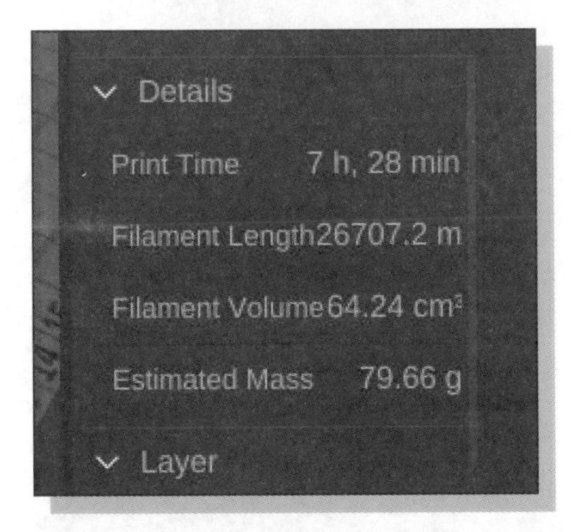

24. Note that with the current settings, it will take 7 hours and 28 minutes to complete the print using 26707.2 mm of filament, and the volume of the printed part is 79.66 gram.

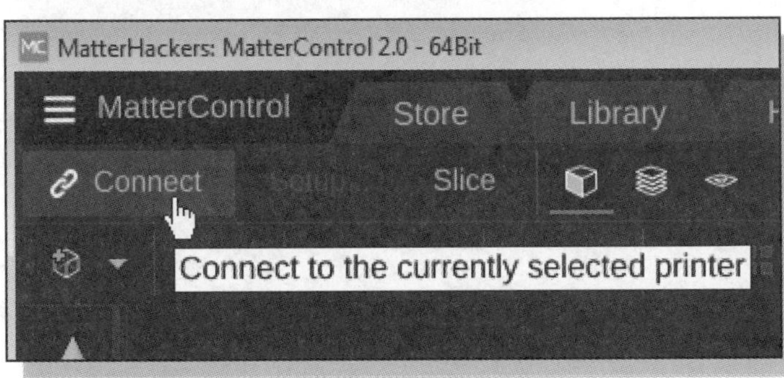

25. To start the 3D print, switch on the 3D printer and click Connect → Print to start the printing of the 3D model.

26. *Matter Control* will now begin printing.

Questions:

1. What is the main difference between the **Additive Manufacturing** and the traditional **Subtractive Manufacturing** technologies?
2. Which 3D printing process is recognized as the first 3D printing process?
3. Describe the general procedure to create a 3D printed part.
4. What are the three primary types of 3D printing processes?
5. Which 3D printing process is the most popular 3D printing process?
6. What is the main advantage of using PLA over ABS for FFF process?
7. Which are the most popular file formats for 3D printing?
8. What is the main function of a **Slicer** program?
9. List and describe the print validation options available with the SOLIDWORKS **Print3D** command.

Chapter 11
Symmetrical Features in Designs

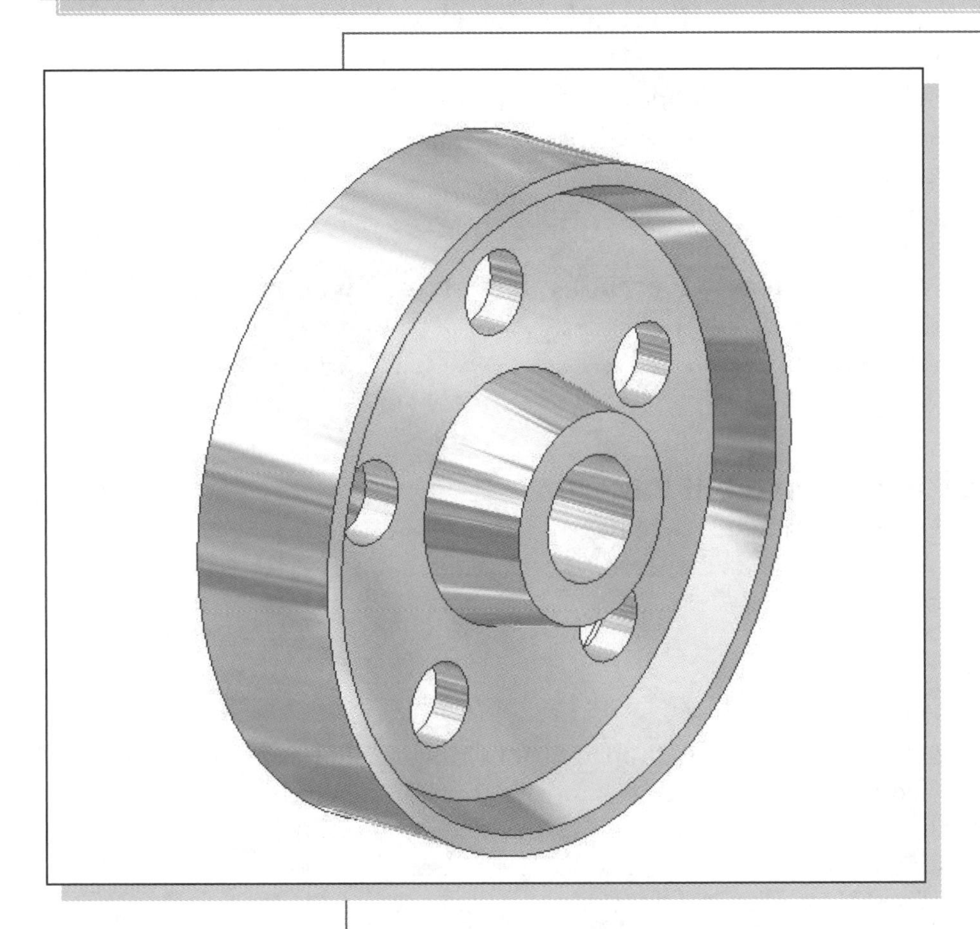

Learning Objectives

- Create Revolved Features
- Use the Mirror Feature Command
- Create New Borders and Title Blocks
- Create Circular Patterns
- Create and Modify Linear Dimensions
- Use SOLIDWORKS' Associative Functionality
- Identify Symmetrical Features in Designs
- Create a Section View in a Drawing

Certified SOLIDWORKS Associate Exam Objectives Coverage

Sketch Entities – Lines, Rectangles, Circles, Arcs, Ellipses, Centerlines

Objectives: Creating Sketch Entities.

Boss and Cut Features – Extrudes, Revolves, Sweeps, Lofts

Objectives: Creating Basic Swept Shapes.

Linear, Circular, and Fill Patterns

Objectives: Creating Patterned Features.

Drawing Sheets and Views

Objectives: Creating and Setting Properties for Drawing Sheets; Inserting and Editing Standard Views.

Annotations

Objectives: Creating Annotations.

Introduction

In parametric modeling, it is important to identify and determine the features that exist in the design. *Feature-based parametric modeling* enables us to build complex designs by working on smaller and simpler units. This approach simplifies the modeling process and allows us to concentrate on the characteristics of the design. Symmetry is an important characteristic that is often seen in designs. Symmetrical features can be easily accomplished by the assortments of tools that are available in feature-based modeling systems, such as SOLIDWORKS.

The modeling technique of extruding two-dimensional sketches along a straight line to form three-dimensional features, as illustrated in the previous chapters, is an effective way to construct solid models. For designs that involve cylindrical shapes, shapes that are symmetrical about an axis, revolving two-dimensional sketches about an axis can form the needed three-dimensional features. In solid modeling, this type of feature is called a ***revolved feature***.

In SOLIDWORKS, besides using the **Revolved Boss/Base** command to create revolved features, several options are also available to handle symmetrical features. For example, we can create multiple identical copies of symmetrical features with the **Circular Pattern** command, or create mirror images of models using the **Mirror** command. We can also use *construction geometry* to assist the construction of more complex features. In this lesson, the construction and modeling techniques of these more advanced options are illustrated.

A Revolved Design: *PULLEY*

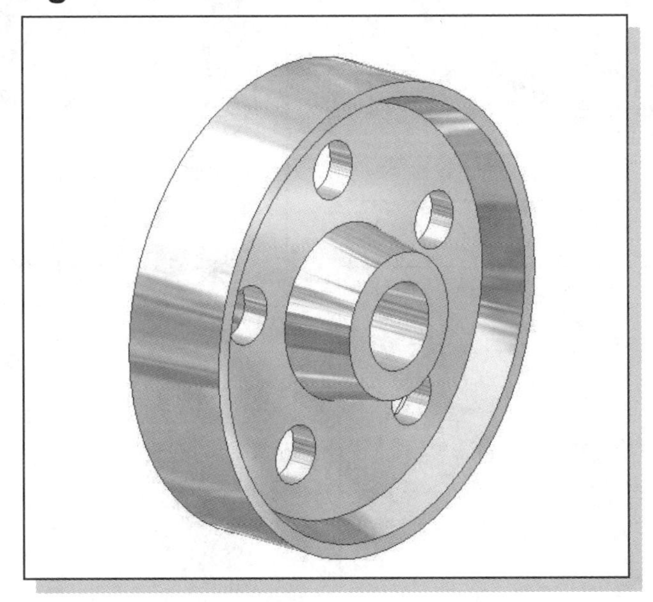

Based on your knowledge of SOLIDWORKS, how many features would you use to create the design? Which feature would you choose as the **base feature** of the model? Identify the symmetrical features in the design and consider other possibilities in creating the design. You are encouraged to create the model on your own prior to following through the tutorial.

Modeling Strategy – A Revolved Design

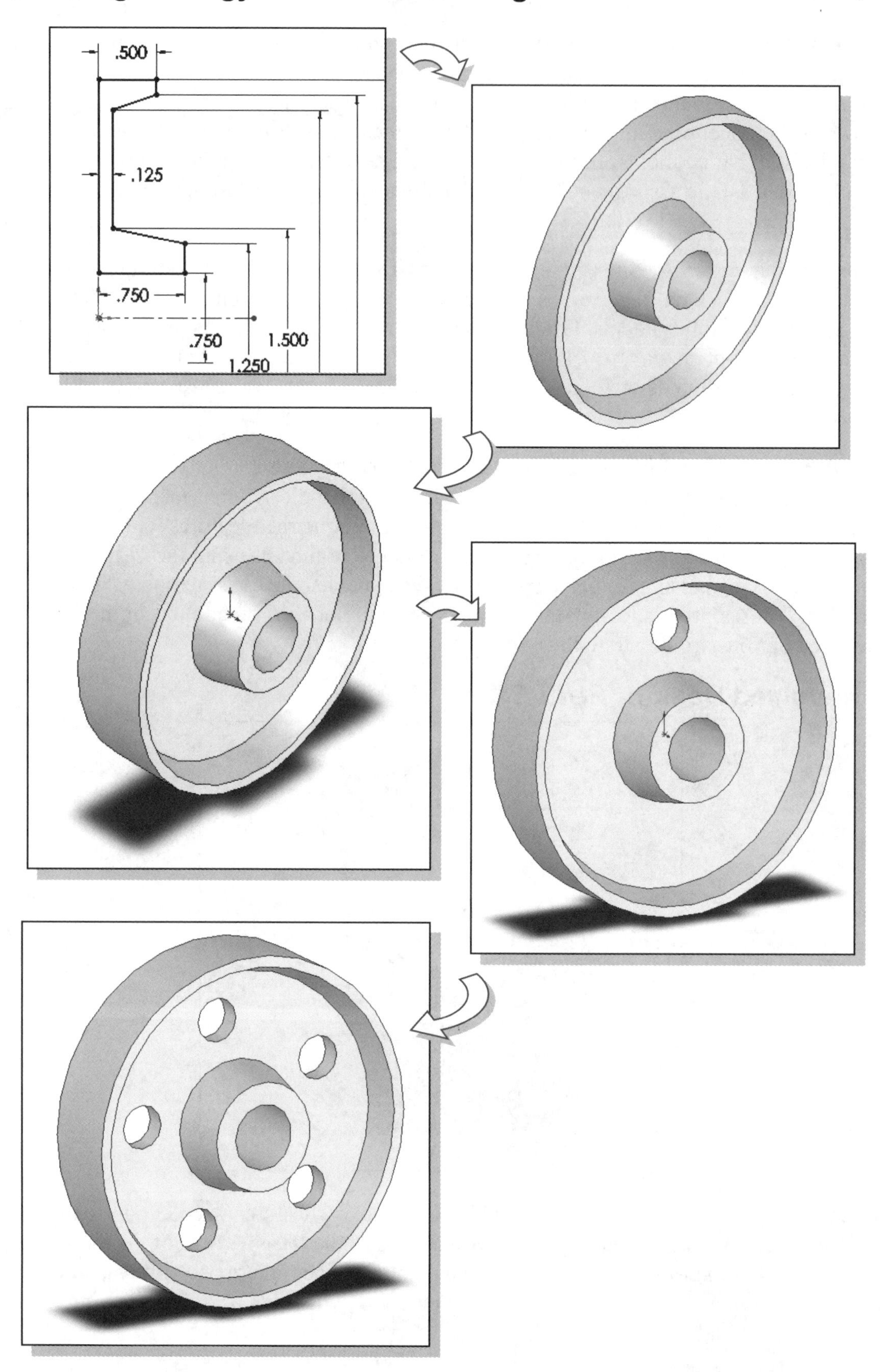

Starting SOLIDWORKS

1. Select the **SOLIDWORKS** option on the *Start* menu or select the **SOLIDWORKS** icon on the desktop to start SOLIDWORKS.

2. Select the **New** icon with a single click of the left-mouse-button on the *Menu Bar*. The *New* SOLIDWORKS *Document* dialog box should appear in Advanced mode. If it appears in Novice mode, click once with the left-mouse-button on the **Advanced** icon.

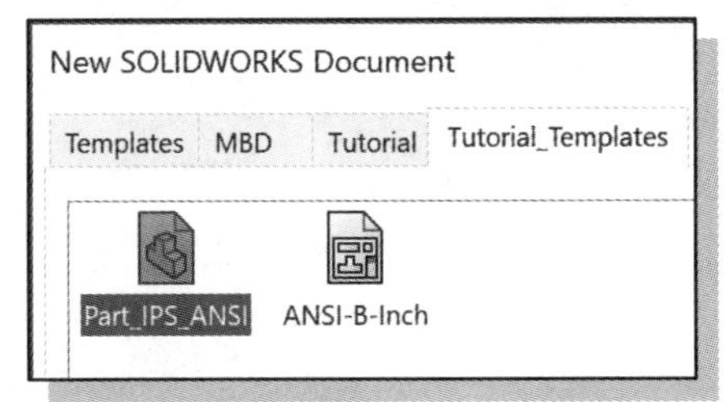

3. Select the **Tutorial_Templates** tab. (**NOTE:** If your templates are not available, open the standard Part template and set the *Drafting Standard* to ANSI and the *Units* to IPS.)

4. Select the **Part_IPS_ANSI** template as shown.

5. Click on the **OK** button to open a new document using the Part_IPS_ANSI template.

Creating the 2D Sketch for the Base Feature

IMPORTANT NOTE: In this lesson and following lessons, we will use the standard display of toolbars. If a user prefers to use the *CommandManager*, the only change that may be necessary is to select the appropriate tab prior to selecting a command.

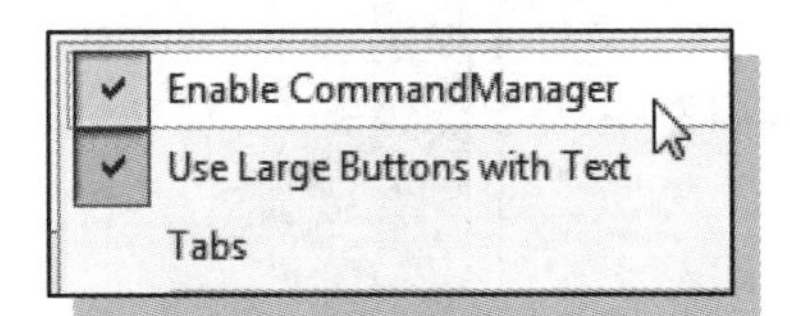

1. To turn ***OFF*** the *CommandManager* and use the standard display of toolbars, right click on the *CommandManager* (or any other toolbar) and toggle the *Enable CommandManager* option *OFF* by selecting it at the top of the pop-up menu.

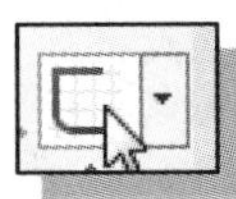

2. Select the **Sketch** button on the *Sketch* toolbar to create a new sketch.

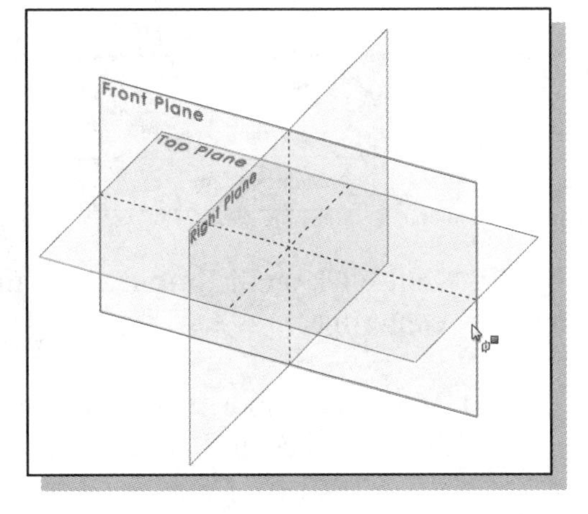

3. In the *Edit Sketch Property-Manager*, the message "*Select a plane on which to create a sketch for the entity*" is displayed. Move the cursor over the edge of the Front Plane in the graphics area. When the Front Plane is highlighted, click once with the **left-mouse-button** to select the Front (XY) Plane as the sketch plane for the new sketch.

4. Click on the **Line** icon in the *Sketch* toolbar.

5. Create a closed-region sketch as shown below. (Note that the *Pulley* design is symmetrical about a horizontal axis as well as a vertical axis, which allows us to simplify the 2D sketch as shown below.)

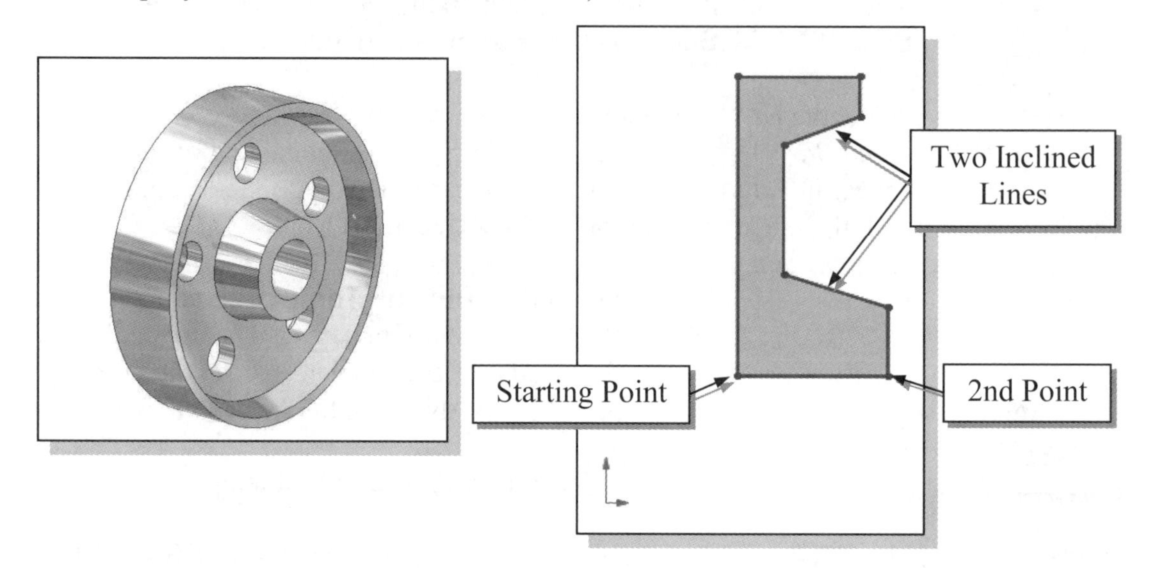

6. We want the left vertical edge to be aligned with the Y-axis. Select the **Add Relation** command from the *Display/Delete Relations* pop-up menu on the *Sketch* toolbar.

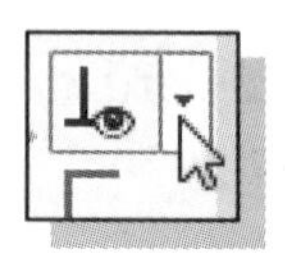

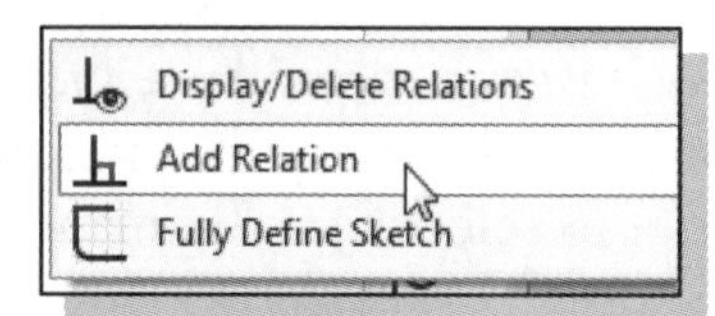

7. On your own, select the **left vertical edge** and the **origin** and add a **Coincident** relation.

➢ The X-axis is an axis of symmetry. It will serve as the rotation axis for the revolve feature, and we will use it to apply dimensions. A centerline will be created along the X-axis for these purposes.

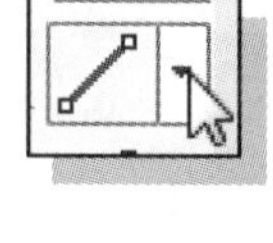

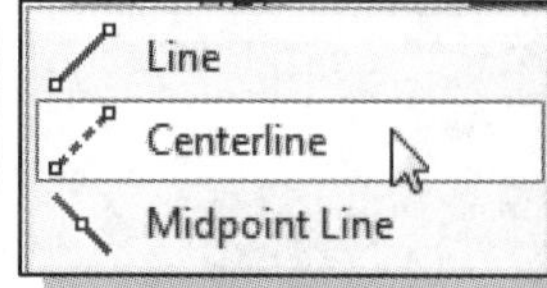

8. Select the **Centerline** command from the *Line* pop-up menu on the *Sketch* toolbar.

9. Left-click on the origin as the first endpoint of the centerline.

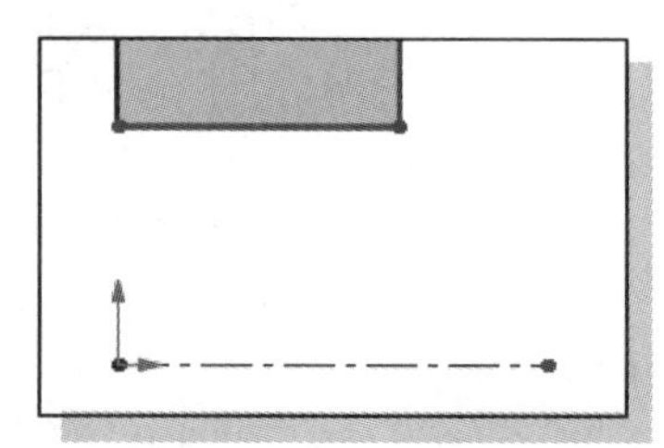

10. Move the cursor to the right and left-click to create a horizontal centerline along the X-axis.

11. Press the **[Esc]** key to exit the Centerline command.

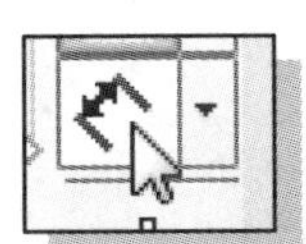

12. Select the **Smart Dimension** command in the *Sketch* toolbar.

13. Pick the **centerline** as the first entity to dimension as shown in the figure below.

14. Select the **bottom horizontal line** as the second object to dimension.

15. Locate the dimension **below the centerline**. This will automatically create a diameter dimension bisected by the centerline.

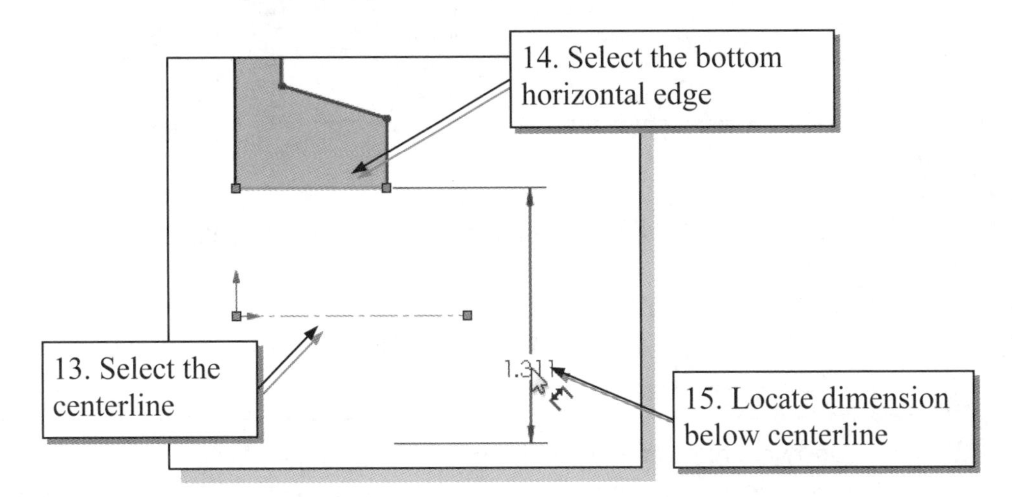

16. In the *Modify* pop-up window enter **.75** for the dimension.

17. Select **OK** in the *Modify* pop-up window.

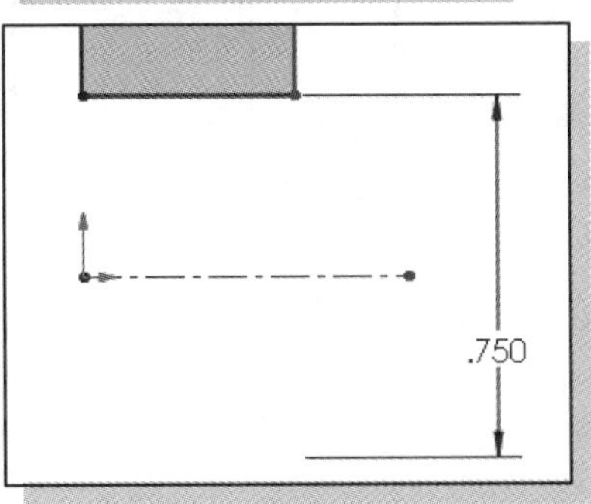

- **To create a dimension that will account for the symmetrical nature of the design pick the centerline (axis of symmetry), pick the entity, and then place the dimension on the side of the centerline opposite the entity.**

18. Pick the **centerline** as the first entity to dimension as shown in the figure below.

19. Select the **corner point** as the second object to dimension as shown in the figure below.

20. Locate the dimension **below the centerline**. This will automatically create a diameter dimension bisected by the centerline.

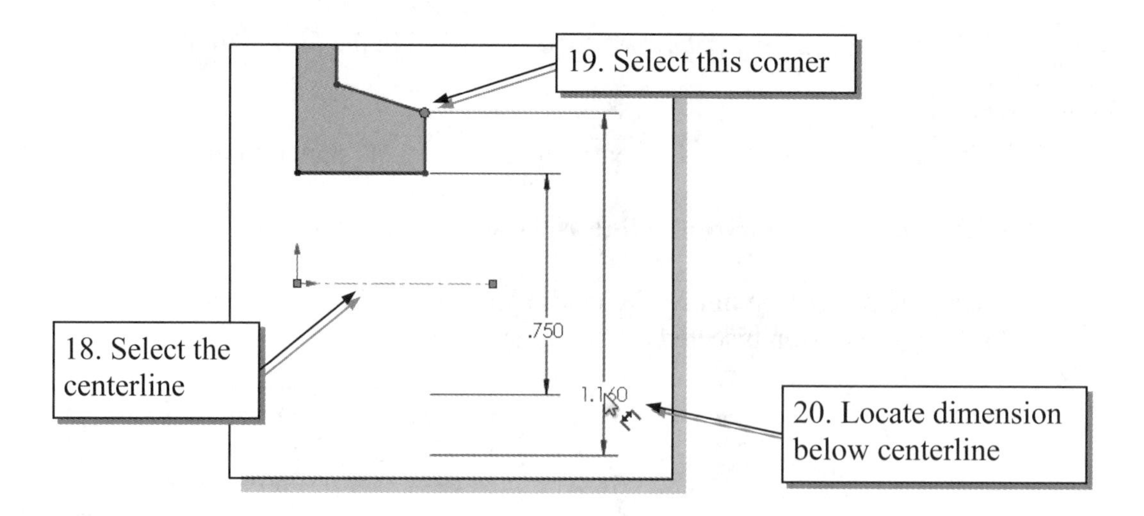

21. In the *Modify* pop-up window enter **1.25** for the dimension.

22. Select **OK** in the *Modify* pop-up window.

23. On your own, create and adjust the vertical size/location dimensions as shown.

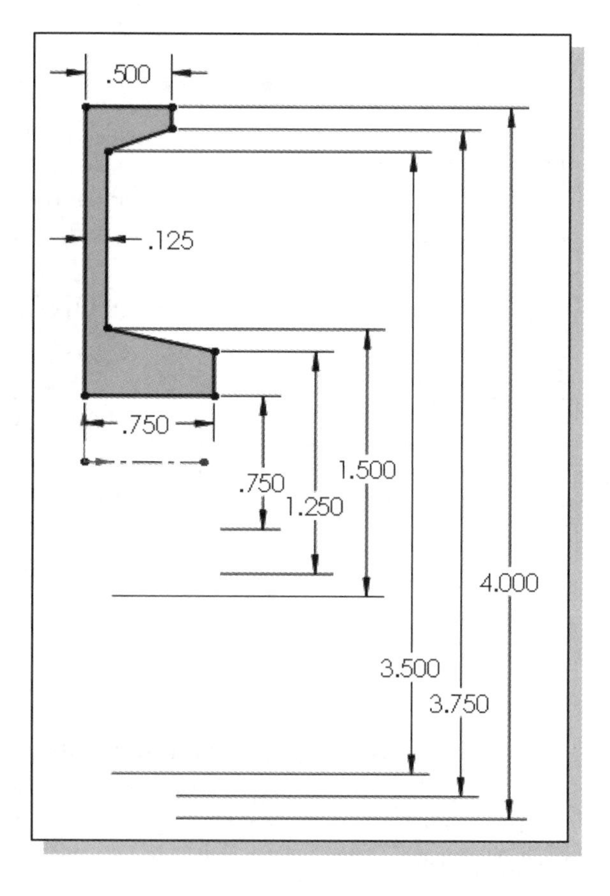

24. Select the **Sketch** icon on the *Sketch* toolbar to exit the **Sketch** option.

25. Select the **View Orientation** button on the *Heads-up View* toolbar by clicking once with the left-mouse-button.

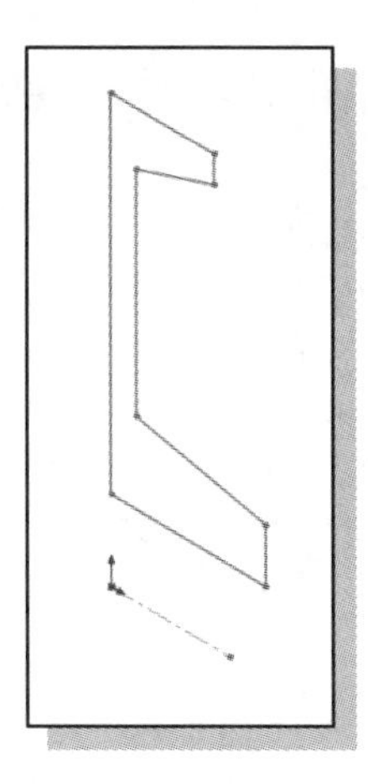

26. Select the **Isometric** icon in the *View Orientation* pull-down menu.

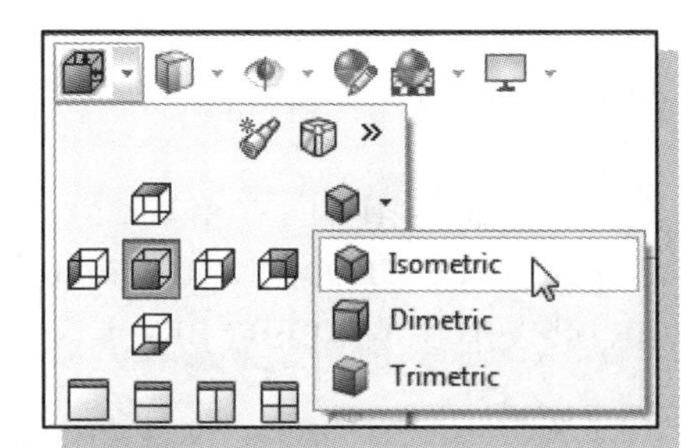

Creating the Revolved Feature

1. Make sure the sketch – Sketch1 – is selected in the *FeatureManager Design Tree*.

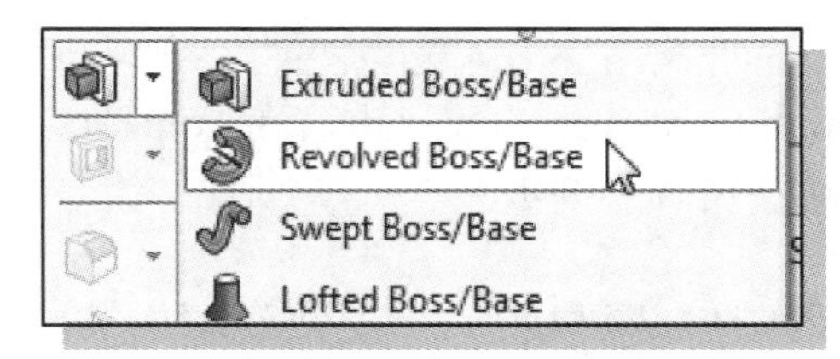

2. In the *Features* toolbar, left-click on the **arrow** next to the **Extruded Boss/Base** icon and select the **Revolved Boss/Base** command from pop-up menu.

3. The *Revolve PropertyManager* appears. In the *Revolve Parameters* panel, the centerline is automatically selected as the revolve axis. The default revolution is **360°**. Click the **OK** icon to accept these parameters and create the revolved feature.

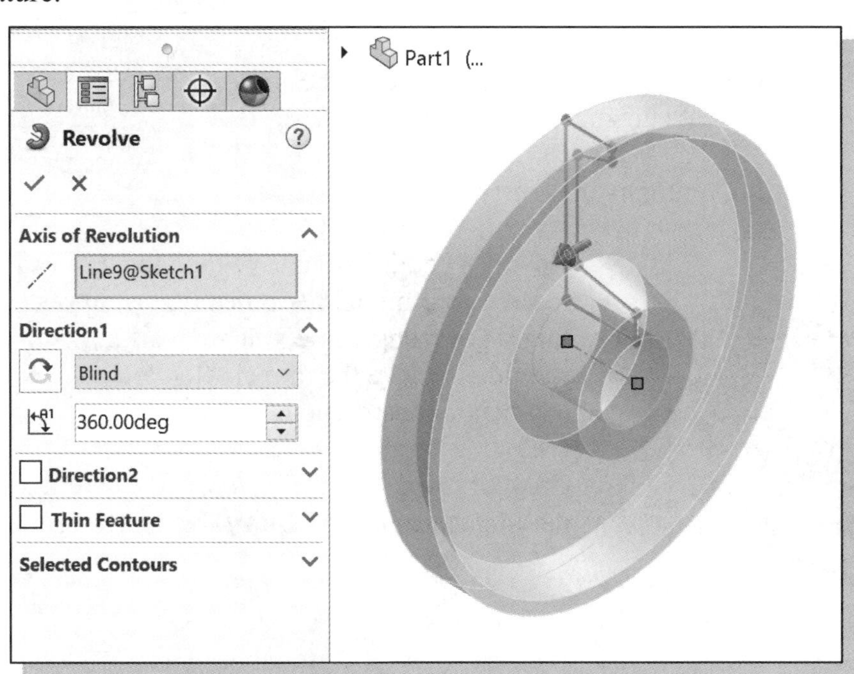

Mirroring Features

In SOLIDWORKS, features can be mirrored to create and maintain complex symmetrical features. We can mirror a feature about a work plane or a specified surface. We can create a mirrored feature while maintaining the original parametric definitions, which can be quite useful in creating symmetrical features. For example, we can create one quadrant of a feature, then mirror it twice to create a solid with four identical quadrants.

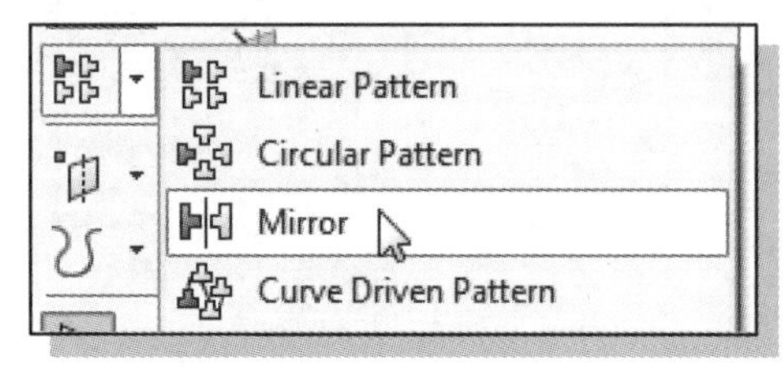

1. In the *Features* toolbar, left-click on the **arrow** next to the **Linear Pattern** icon and select the **Mirror** command from the pop-up menu.

2. The *Mirror PropertyManager* appears. The *Mirror Face/Plane* panel is highlighted and the message "*Select a plane or planar face about which to mirror* ..." appears in the *Status Bar* area. Press the right arrow button [→] several times to rotate the model so that we are viewing the back surface as shown.

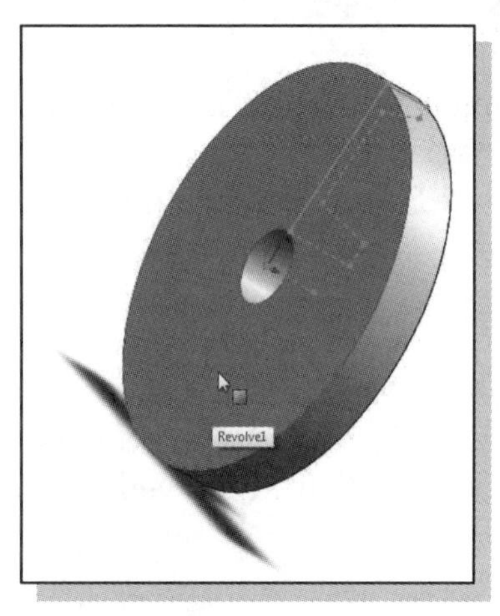

3. Select the surface as shown in the figure as the planar surface about which to mirror. The selection appears in the *Mirror Face/Plane* panel.

4. The ***Features to Mirror*** panel in the *Property-Manager* should contain **Revolve1** as the feature to mirror because this was pre-selected when we entered the Mirror command. If it does not, select the Revolve1 feature as the feature to mirror.

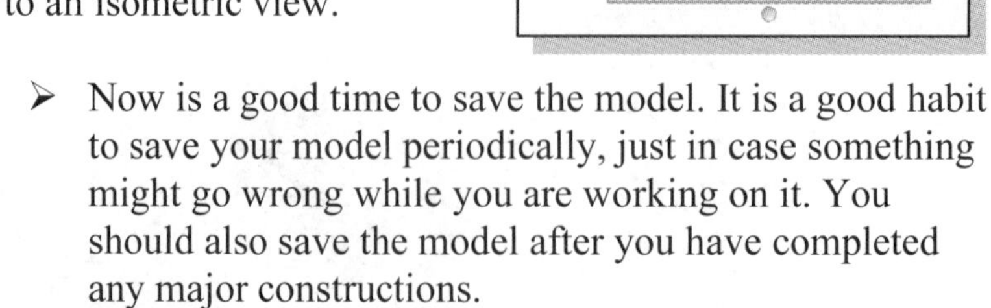

5. Click **OK** in the *PropertyManager* to accept the settings and create the mirrored feature.

6. On your own, return to an isometric view.

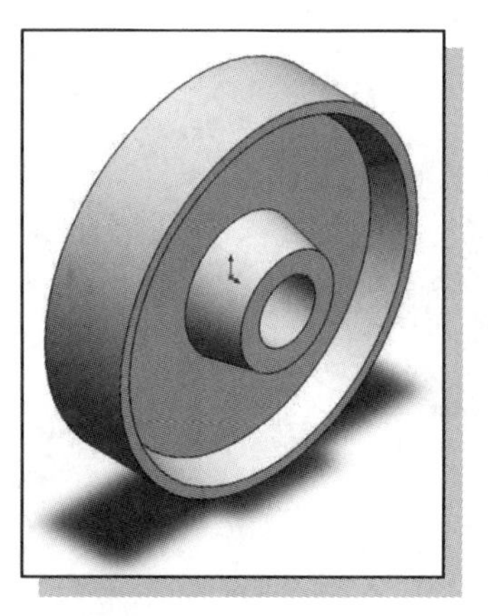

➢ Now is a good time to save the model. It is a good habit to save your model periodically, just in case something might go wrong while you are working on it. You should also save the model after you have completed any major constructions.

7. On your own, save the **.sldprt* file, entering **Pulley** for the filename and **4″ Pulley** for the Description.

Creating an Extruded Cut Feature using Construction Geometry

In SOLIDWORKS, we can also use **construction geometry** to help define, constrain, and dimension the required geometry. **Construction geometry** can be lines, arcs, and circles that are used to line up or define other geometry but are not themselves used as the shape geometry of the model. When profiling the rough sketch, SOLIDWORKS will separate the construction geometry from the other entities and treat them as construction entities. Construction geometry can be dimensioned and constrained just like any other profile geometry. When the profile is turned into a 3D feature, the construction geometry remains in the sketch definition but does not show in the 3D model. Using construction geometry in profiles may mean fewer constraints and dimensions are needed to control the size and shape of geometric sketches. In SOLIDWORKS, any sketch entity can be converted to construction geometry. Points and centerlines are always construction entities. We will illustrate the use of the construction geometry to create a cut feature.

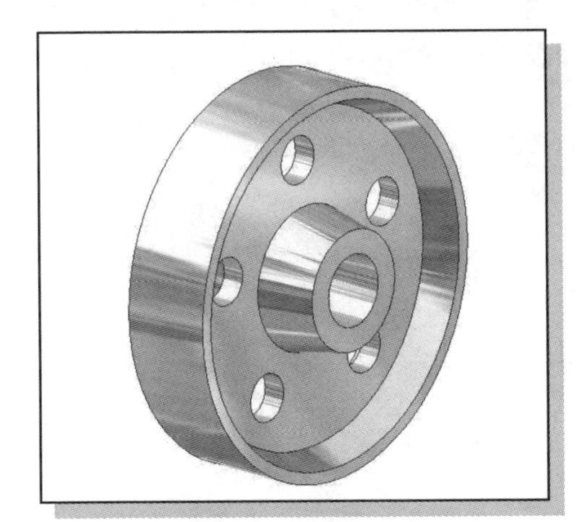

The *Pulley* design requires the placement of five identical holes on the base solid. Instead of creating the five holes one at a time, we can simplify the creation of these holes by using the **Circular Pattern** command, which allows us to create duplicate features. Prior to using the Circular Pattern command, we will first create a ***pattern leader***, which is a regular extruded feature.

1. Press the **[Esc]** key to ensure that no objects are selected.

2. In the *Sketch* toolbar select the **Sketch** command by left-clicking once on the icon.

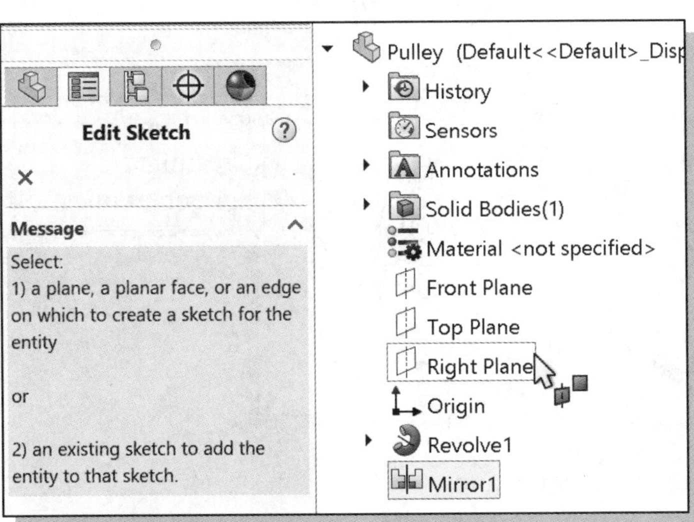

3. Use the *Design Tree* in the graphics area to select the **Right** (YZ) **Plane** as shown.

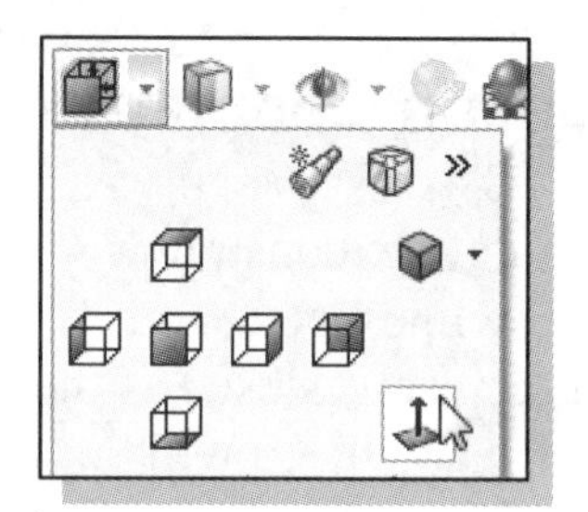

4. Select the **View Orientation** button on the *Heads-up View* toolbar and select the **Normal to** option. The view will change to be normal to the sketch plane.

5. Select the **Circle** command by clicking once with the left-mouse-button on the icon in the *Sketch* toolbar.

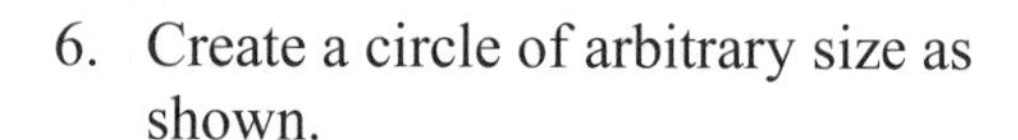

6. Create a circle of arbitrary size as shown.

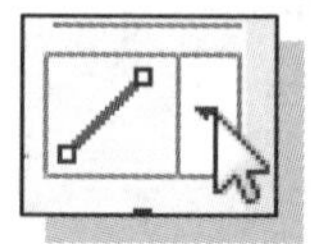

7. Select the **Line** command in the *Sketch* toolbar.

8. Create 2 *lines* by left-clicking on (1) the center of the circle we just created, (2) the ***origin***, and (3) a point to the right of the origin creating a horizontal line, as shown below.

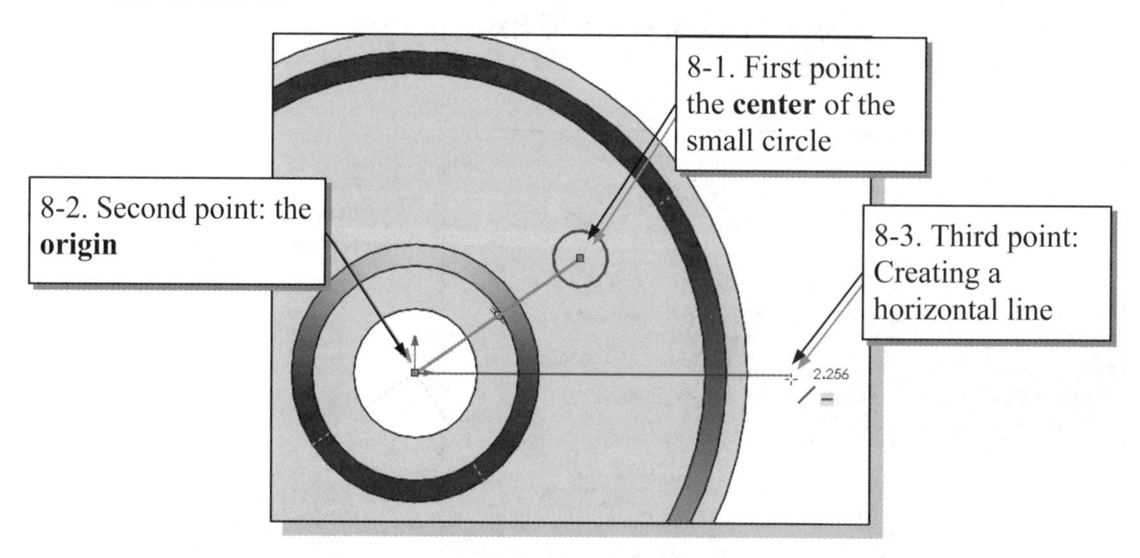

9. Press the **[Esc]** key to exit the **Line** command.

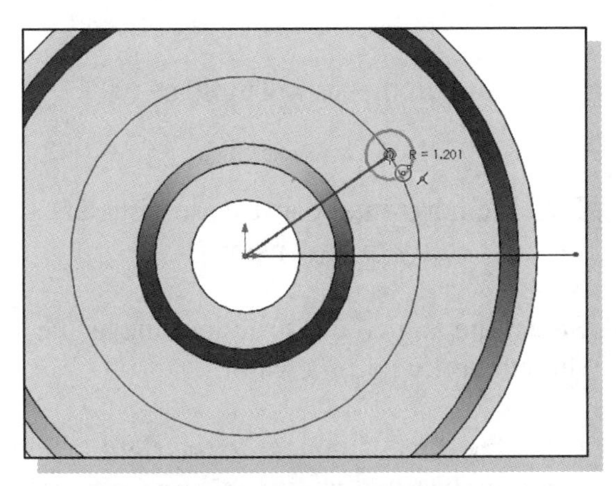

10. Select the **Circle** command on the *Sketch* toolbar.

11. Place the center at the ***origin***.

12. Pick the center of the small circle to set the size of the circle as shown.

13. Press the **[Esc]** key to exit the *Circle* command.

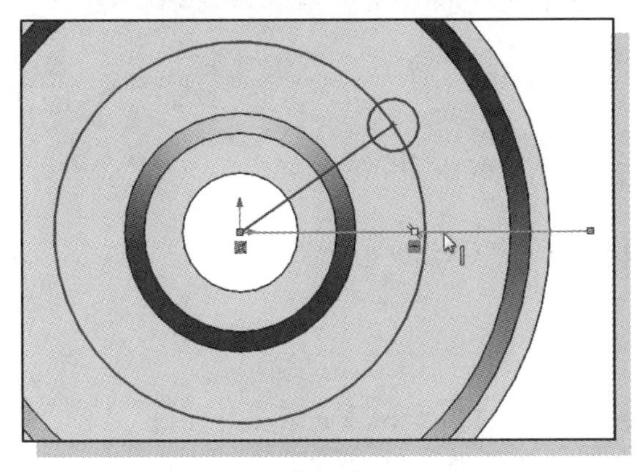

14. In the graphics area, left-click on the ***horizontal line*** to select it.

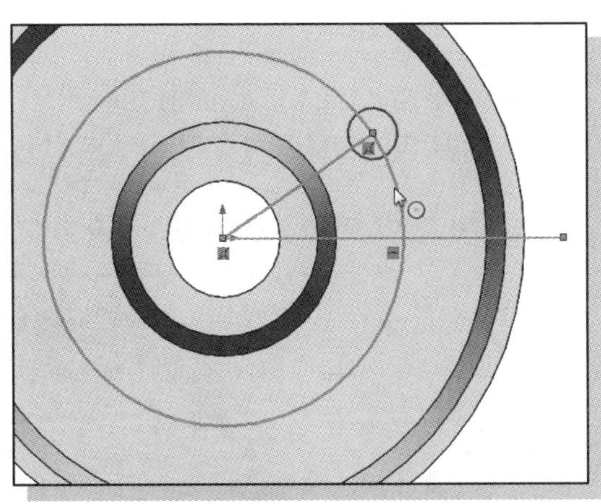

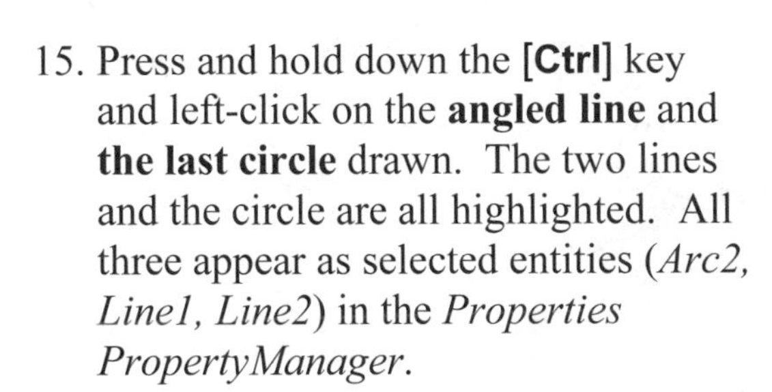

15. Press and hold down the **[Ctrl]** key and left-click on the **angled line** and **the last circle** drawn. The two lines and the circle are all highlighted. All three appear as selected entities (*Arc2, Line1, Line2*) in the *Properties PropertyManager.*

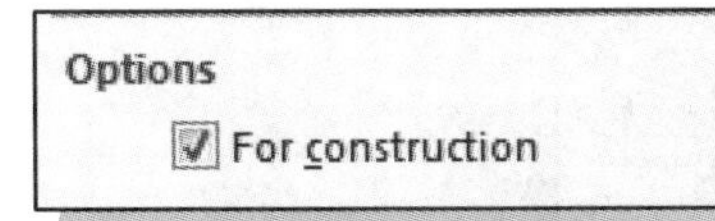

16. **Check** the box next to the **For construction** option in the options panel of the *Properties PropertyManager*.

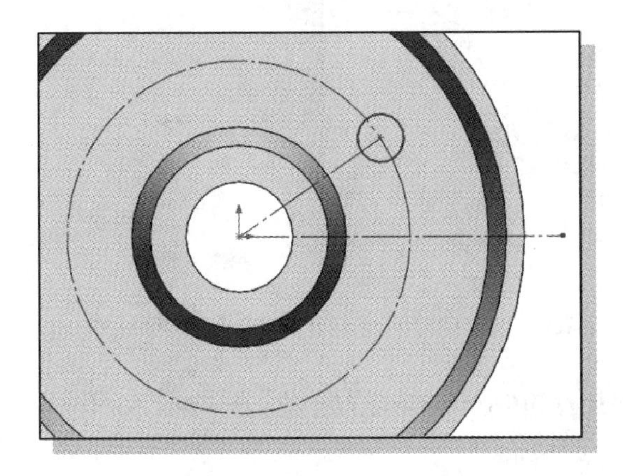

17. Click **OK** in the *PropertyManager* to accept the setting. Notice the entities have been converted to construction geometry and appear with the same line style as centerlines.

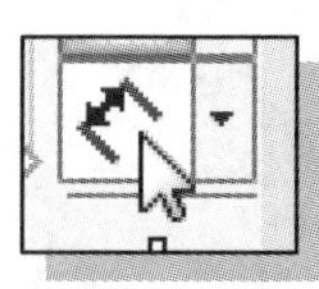

18. Select the **Smart Dimension** command in the *Sketch* toolbar.

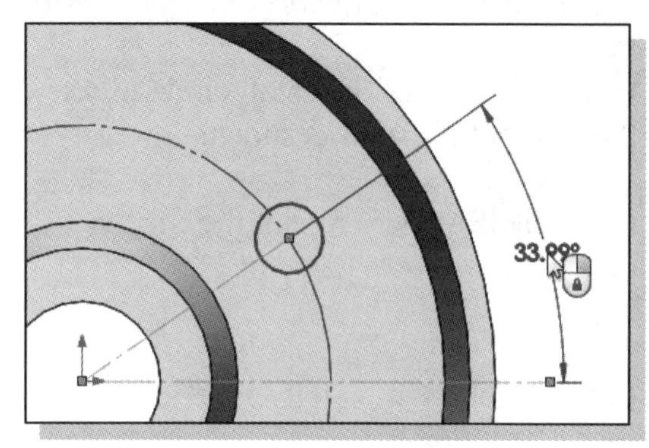

19. Pick the horizontal construction line as the first entity to dimension.

20. Select the angled construction line as the second entity to dimension.

21. Place the dimension text to the right of the model as shown.

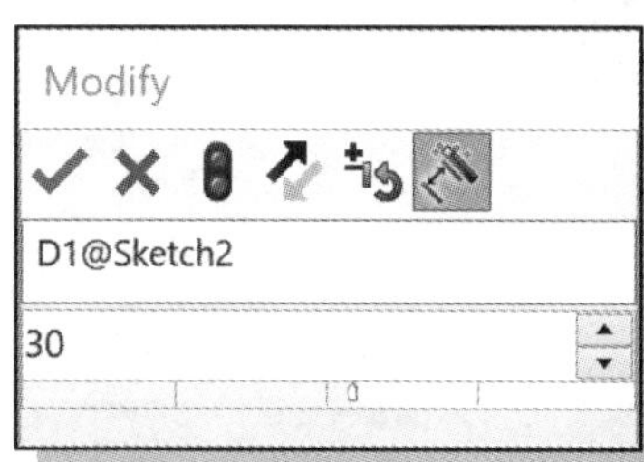

22. Enter **30 deg** for the angle dimension.

23. Click **OK** in the *Modify* window.

➢ Note that the location of the small circle is adjusted as the location of the construction line is adjusted by the *angle dimension* we created.

24. **On your own**, create the two diameter dimensions, **.5** and **2.5**, as shown below.

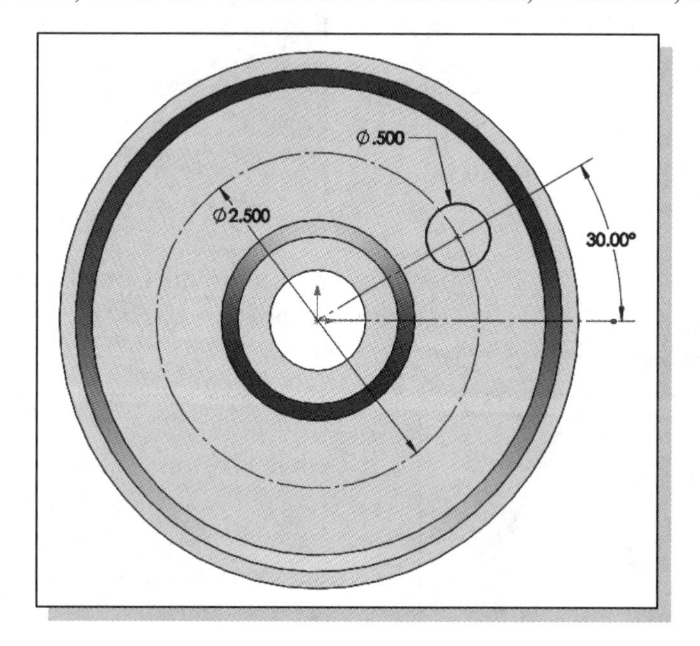

25. Select the **Sketch** icon on the *Sketch* toolbar to exit the **Sketch** option.

26. On your own, select the **View Orientation** button on the *Heads-up View* toolbar and select the **Isometric** option.

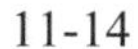

27. On your own, create the **Extruded Cut** feature. Use the **Mid Plane** option and enter **0.25** for the distance.

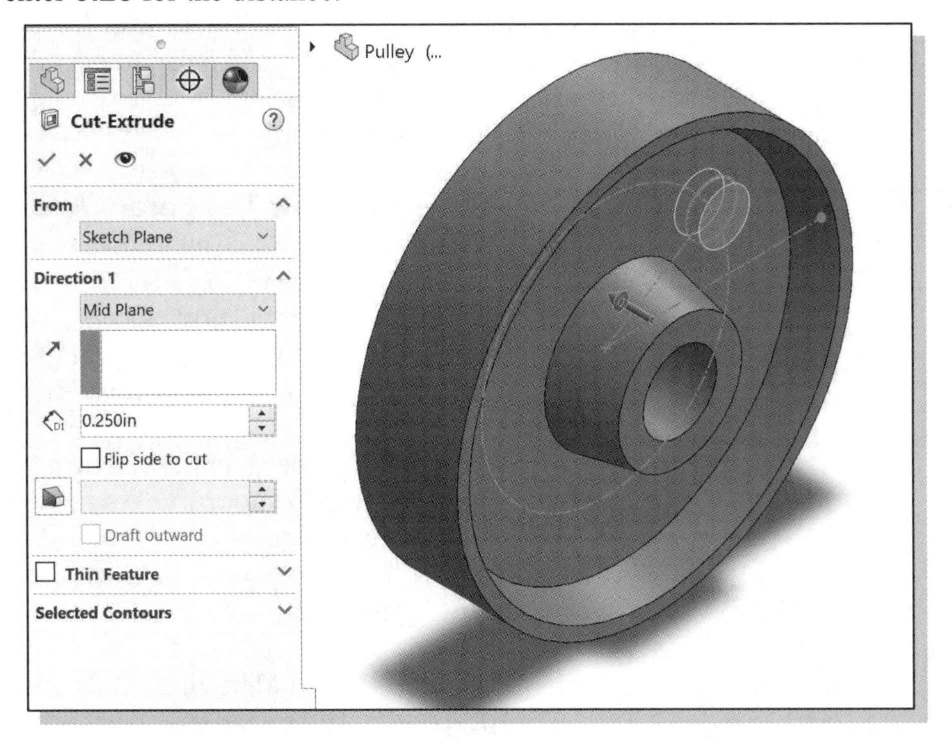

28. Click **OK** in the *PropertyManager* to create the extruded cut feature.

29. On your own, adjust the angle dimension applied to the construction line of the cut feature to **90** and observe the effect of the adjustment. (**HINT:** Access the **Show Feature Dimensions** option by right-clicking on *Annotations* in the *FeatureManager Design Tree*. Don't forget to use the **Rebuild** command.)

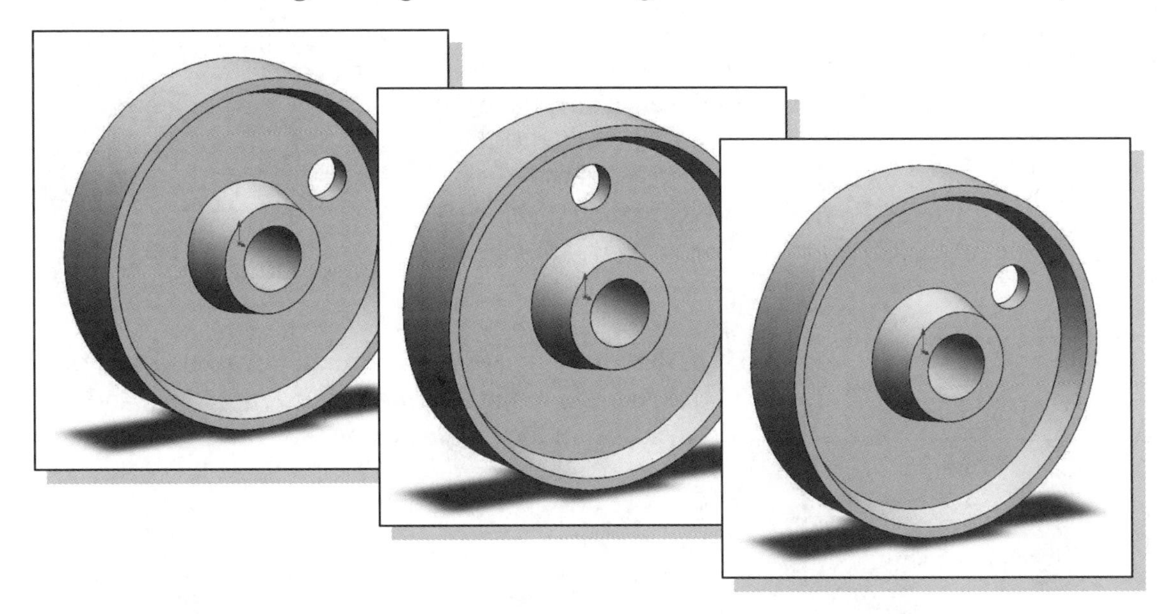

30. Reset the angle dimension to **30** degrees before continuing to the next section.

Circular Pattern

In SOLIDWORKS, existing features can be easily duplicated. The **Linear Pattern** and **Circular Pattern** commands allow us to create rectangular and polar arrays of features. The patterned features are parametrically linked to the original feature; any modifications to the original feature are also reflected in the arrayed features.

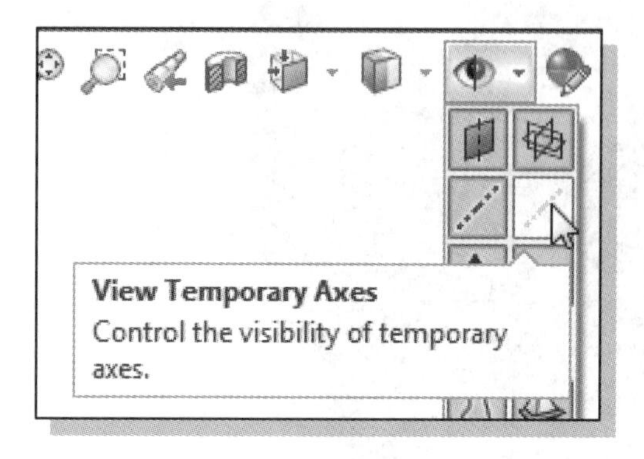

1. Select the **View Temporary Axes** option from the *Hide/Show* pull-down menu on the *Heads-up View* toolbar. Temporary axes are those created implicitly by cones and cylinders in the model. We will use the central axis for our circular pattern.

➢ Alternately, the visibility can be toggled *ON* by selecting Temporary Axes on the *View* pull-down menu.

2. Press the **[Esc]** key to ensure no feature is selected.

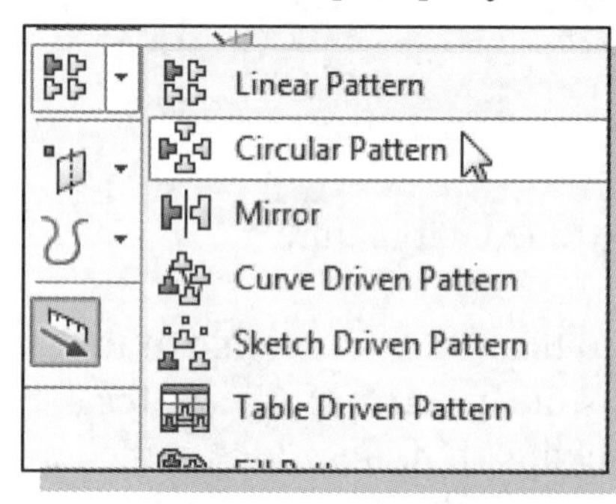

3. In the *Features* toolbar, left-click on the **arrow** next to the **Linear Pattern** icon and select the **Circular Pattern** command from the pop-up menu.

4. The message "*Select edge or axis for direction reference...*" is displayed in the *Status Bar* area. SOLIDWORKS expects us to select an axis to pattern about. Select the **axis** at the center of the *Pulley* as shown. Notice the axis (Axis<1>) appears in the *Pattern Axis* box in the *Parameters* panel of the *Circular Pattern PropertyManager*.

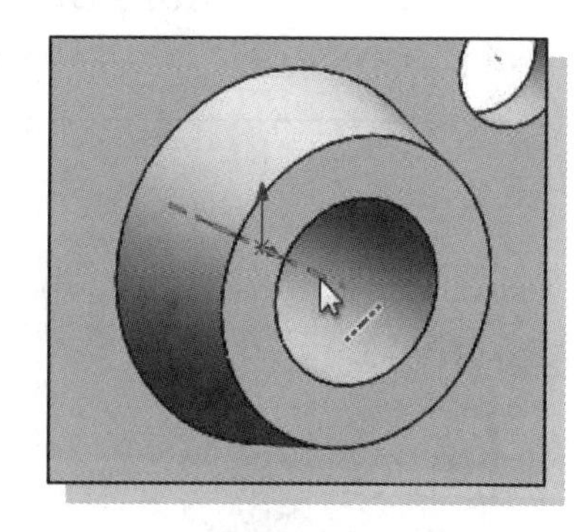

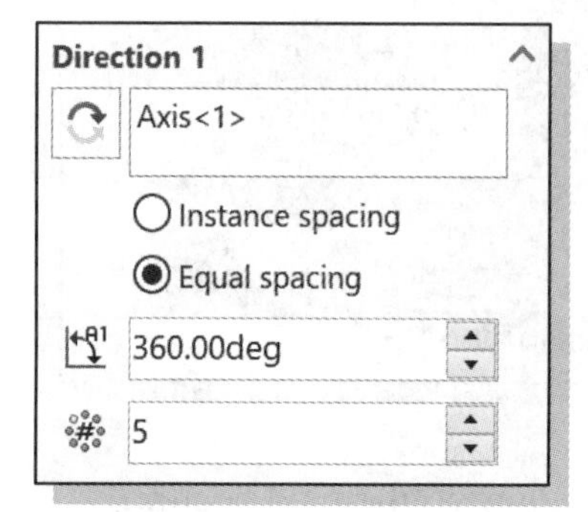

5. In the *Parameters* panel, check the **Equal Spacing** box, enter **5** in the ***Number of Instances*** box, and enter **360** in the ***Angle*** box as shown.

6. **Left-click** in the ***Features to Pattern*** text box on the *Features and Faces* panel as shown to begin selection of features to pattern.

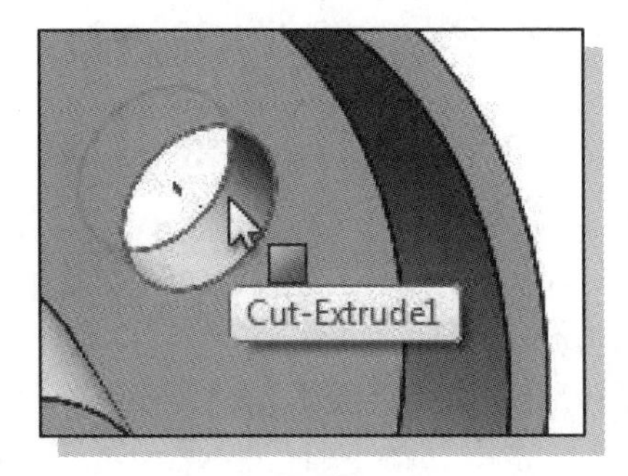

7. Select the **circular cut feature** when it is highlighted as shown.

8. Select **OK** in the *Circular Pattern PropertyManager* to accept the settings and create the circular pattern.

9. On your own, turn *OFF* the visibility of the Temporary Axes and the Origins. (See Step 1 above.)

10. Select **Save** in the *Menu Bar*; we can also use the "**Ctrl-S**" combination (press down the [Ctrl] key and hit the [S] key once) to save the part. Save the part using ***Pulley*** as the filename and **4" Pulley** as the description.

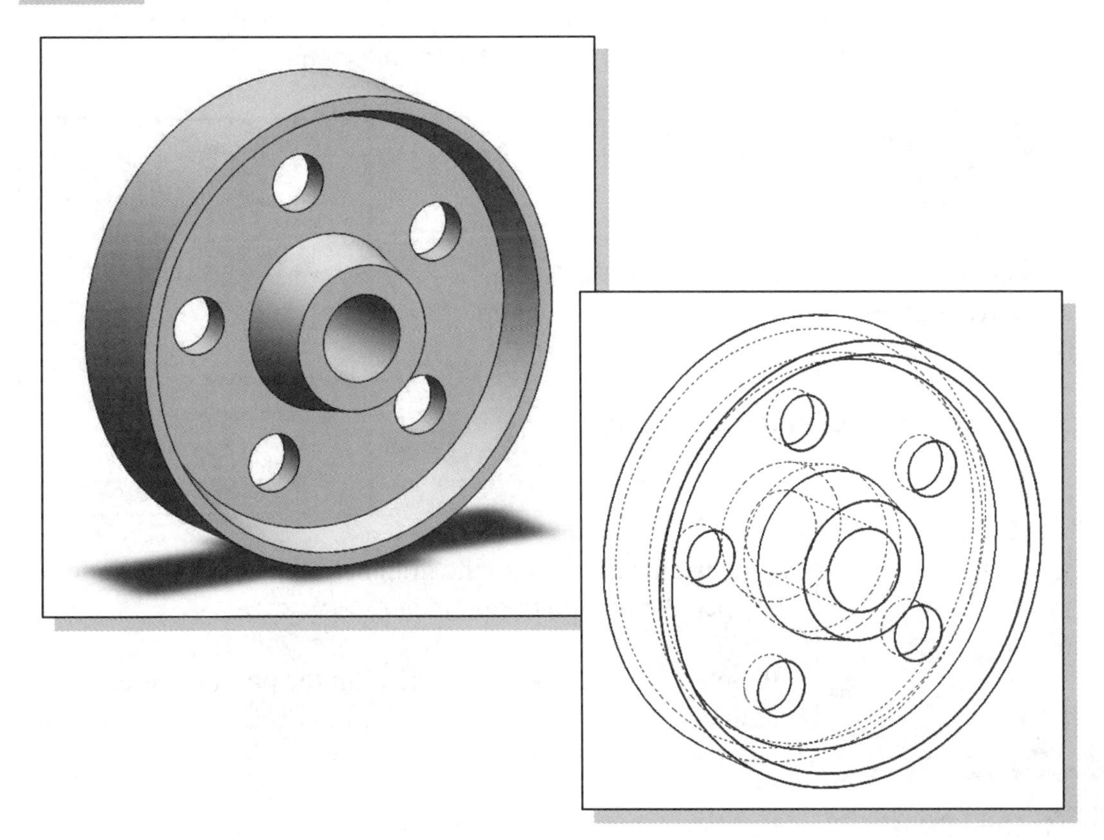

Drawing Mode – Defining a New Border and Title Block

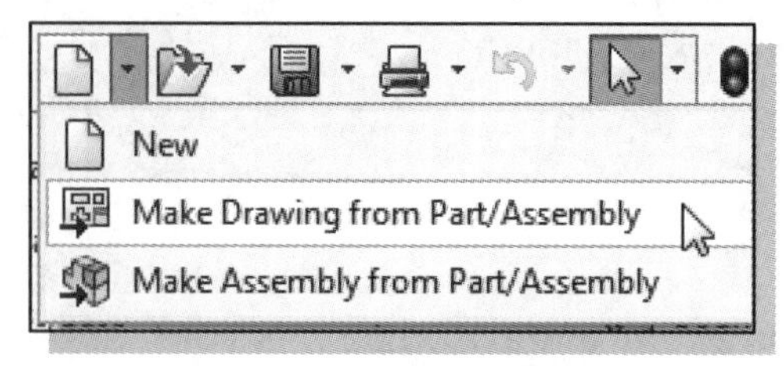

1. Click on the arrow next to the **New** icon on the *Menu Bar* and select **Make Drawing from Part/Assembly** in the pull-down menu.

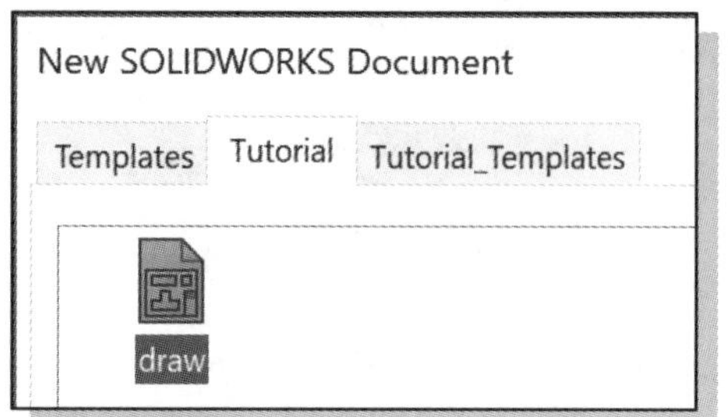

2. In the *New* SOLIDWORKS *Document* window, select the **Tutorial** tab.

3. Select the **draw** template file icon.

4. Select **OK** in the *New* SOLIDWORKS *Document* window.

5. **On your own**, set **document properties** (*Display style, Drafting Standard,* and *Units*) following instructions on page 8-7.

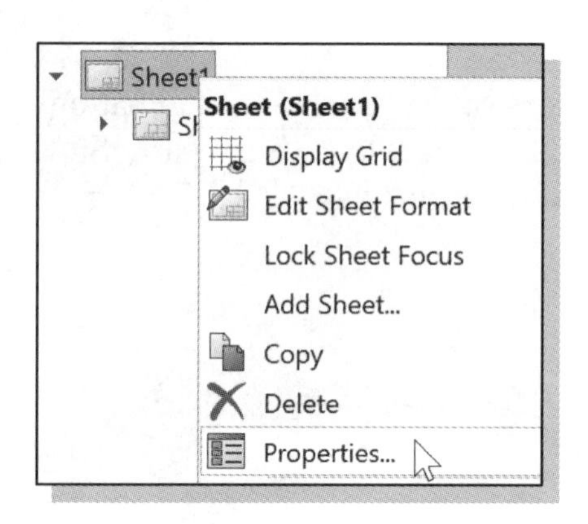

➢ We will use the *Document Properties* in this template but will create a new *Sheet Format.*

6. Right-click on the **Sheet** icon in the *Design Tree* to reveal the pop-up option menu and select **Properties** from the menu.

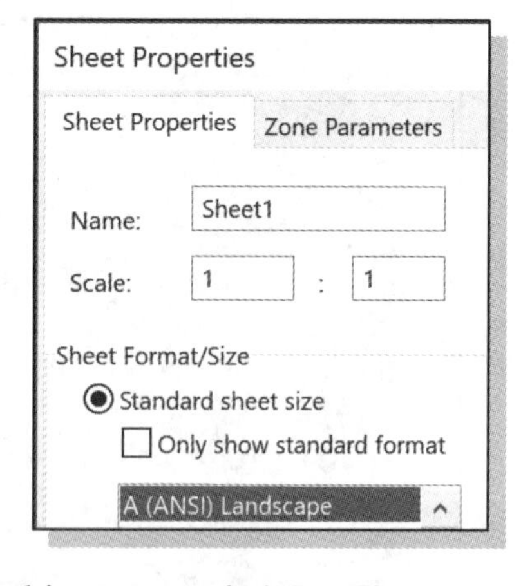

7. Set the scale to **1:1** and select **A (ANSI) Landscape** for the *Sheet Format/Size* in the *Sheet Properties* window.

8. Select **Apply Changes** in the *Sheet Properties* window.

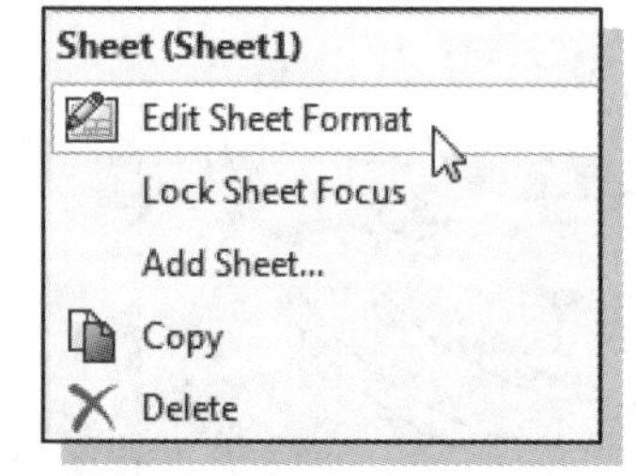

9. Move the cursor into the graphics area and **right-click** to open the pop-up option menu.

10. Select **Edit Sheet Format** from the pop-up option menu to change to *Edit Sheet Format* mode. (Notice this method could also be used to access the *Sheet Properties*.)

11. Hold the [**Ctrl**] function key down and press the [**A**] key to select all entities on the sheet, including all the entities in the title block and the border.

12. Press the **[Delete]** key to delete these entities.

13. Note: If you are working in an older version of *SOLIDWORKS*, the border may still appear. If the border remains after Step 12, expand the **Sheet Format** menu in the *Design Tree*, right-click on the **Border** icon, and select **Delete Border**.

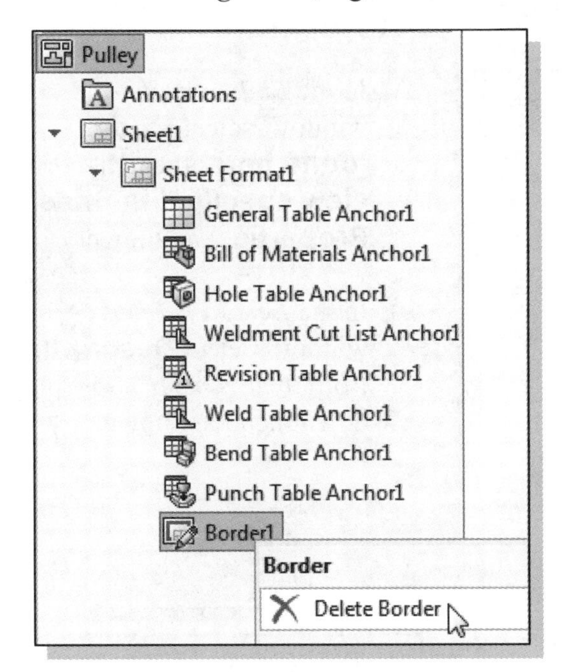

14. Select **Yes** in the *Confirm Delete* window.

15. Use the **Corner Rectangle** command on the *Sketch* toolbar to create a new border as shown below.

16. On your own, create a title block using the **Line** and **Smart Dimension** commands in the *Sketch* toolbar and the **Note** command in the *Annotation* toolbar. (If these toolbars are not open, open them by right-clicking on any open toolbar and selecting them.)

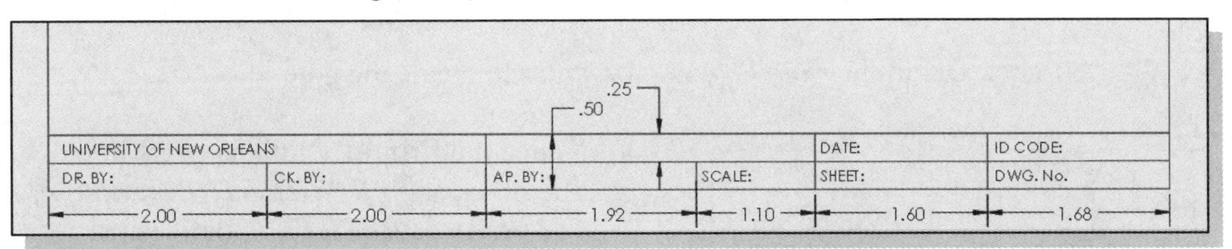

➢ We will now add a **Note** with a **Property Link**.

17. Left-mouse-click on the **Note** icon on the *Annotation* toolbar.

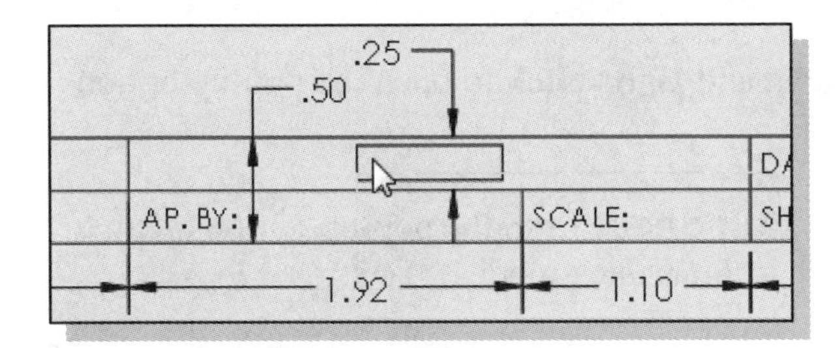

18. Select a location inside the central empty box in the new title block as shown.

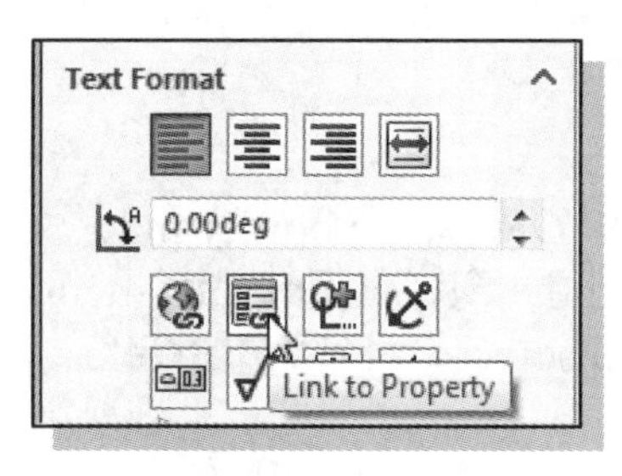

19. In the *Text Format* panel of the *Note PropertyManager*, select the **Link to Property** icon as shown.

➢ We will link to the Description entered as a user-defined property for the model part file *Pulley.SLDPRT*.

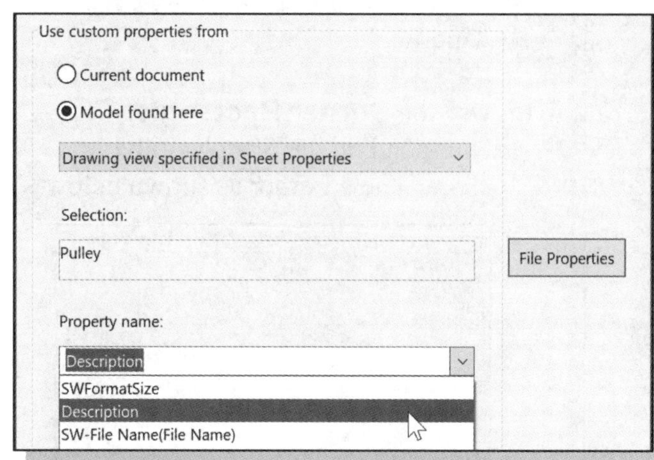

20. In the *Link to property* window, select **Model found here**, then **Drawing view specified in Sheet Properties**, as shown.

21. In the *Link to Property* window, select **Description** from the *Property name* pull-down menu as shown.

22. Note: If Description does not appear in the *Property name*: menu, click the **File Properties** button, then select the **Description** property in the *Summary Information* window, and click **OK** to add it to the list. Then perform Step 21.

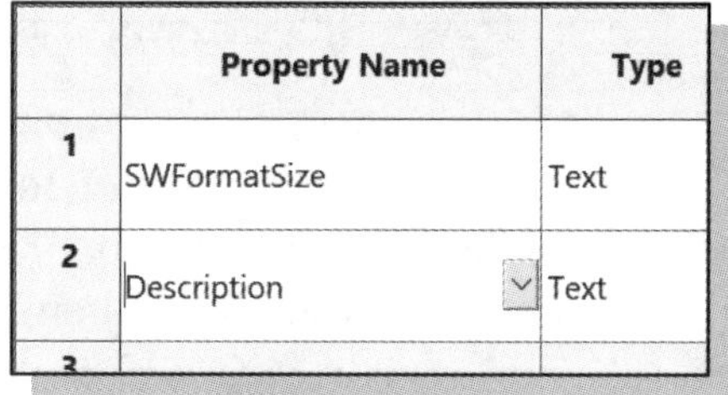

	Property Name	Type
1	SWFormatSize	Text
2	Description	Text
3		

23. Left-click **OK** in the *Link to property*.

24. Left-click **OK** in the *Note PropertyManager* to create the note.

➢ Notice the Property Link appears in the title block as ***$PRPSHEET:{Description}***. The prefix ***$PRPSHEET:*** defines the source as the model (part file) appearing in the sheet. The Property Name is **Description**. This is a ***Custom Property***.

25. On your own, **delete the dimensions** you created to set up the title block.

26. Move the cursor into the graphics area and **right-click** to open the pop-up option menu.

27. Select **Edit Sheet** from the pop-up option menu to change to *Edit Sheet* mode.

28. Press the **[F]** key to fit the view to the graphics area.

Creating a New Drawing Template

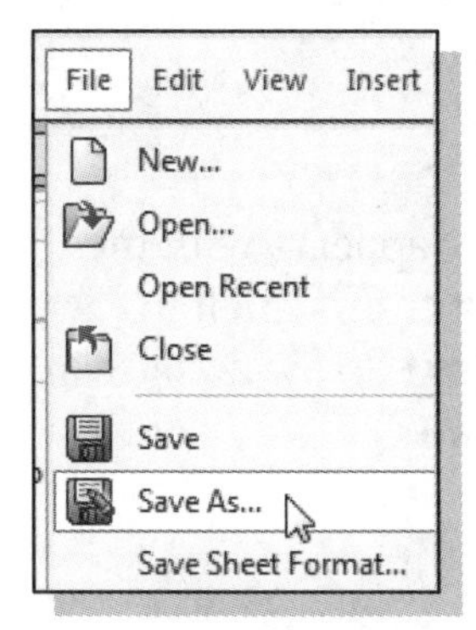

1. Select **Save As** from the *File* pull-down menu.

2. Click in the *Save as type:* entry box and select **Drawing Templates** from the pull-down options.

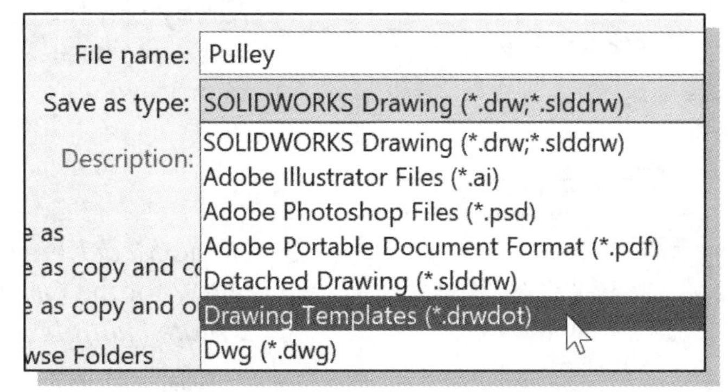

3. The folder selection will automatically change to the default Templates folder. Use the browser to change the folder selection to the **Tutorial Templates** folder you created in Chapter 4.

4. Enter ***A-Custom*** for the *File name*.

5. Enter ***ANSI A Inch Custom*** for the *Description*.

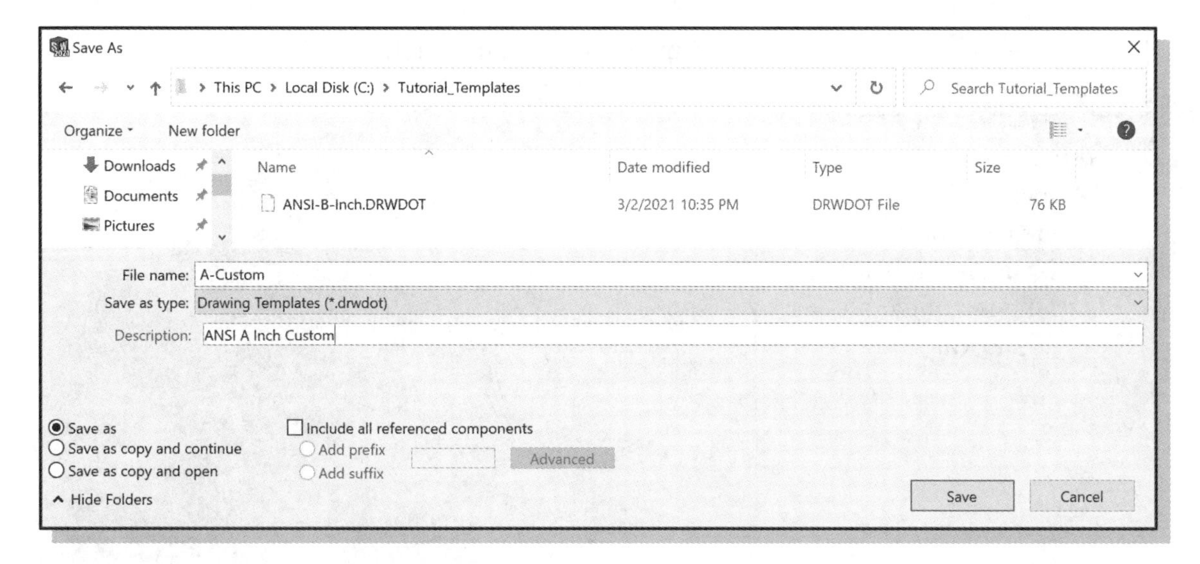

6. Select **Save**.

7. If the SOLIDWORKS pop-up window appears, click **OK**.

8. Select the **Rebuild** icon on the *Menu Bar*.

Creating Views

1. Click on the **Model View** icon in the *Drawing* toolbar.

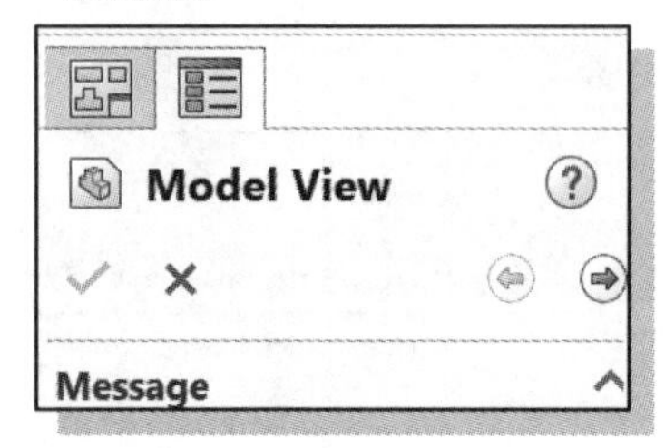

2. The *Model View PropertyManager* appears with the *Pulley* part file selected as the part from which to create the base view. Click the **Next** arrow as shown to proceed with defining the base view.

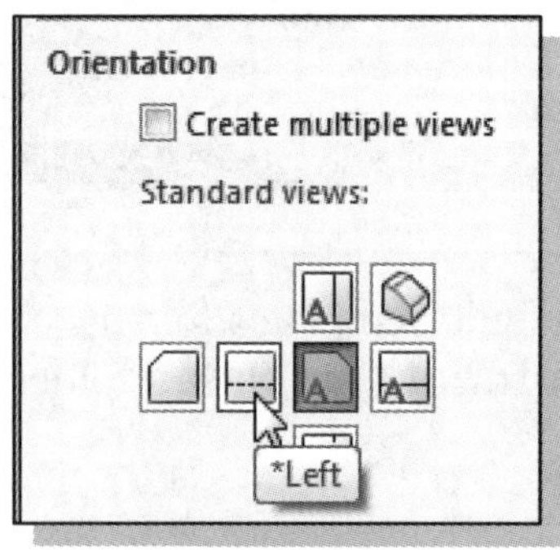

3. In the *Model View PropertyManager*, select the **Left View** for the *Orientation*, as shown. (Make sure the *Create multiple views* box is unchecked.)

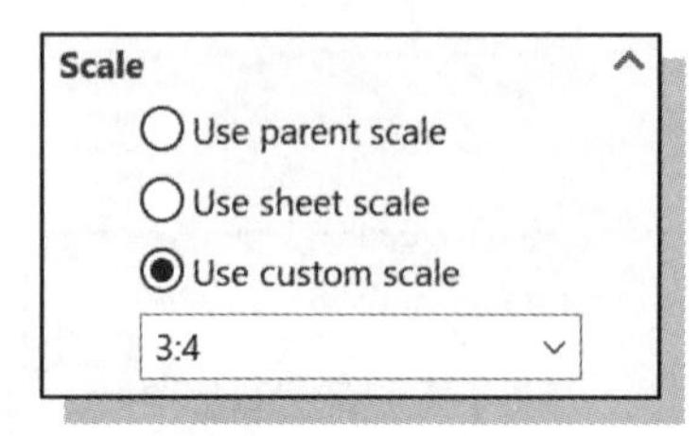

4. Under the *Scale* options in the *Drawing View PropertyManager*, select **Use custom scale**.

5. Enter **3:4** as the user defined scale.

6. Move the cursor inside the graphics area and place the **base view** toward the left side of the border as shown.

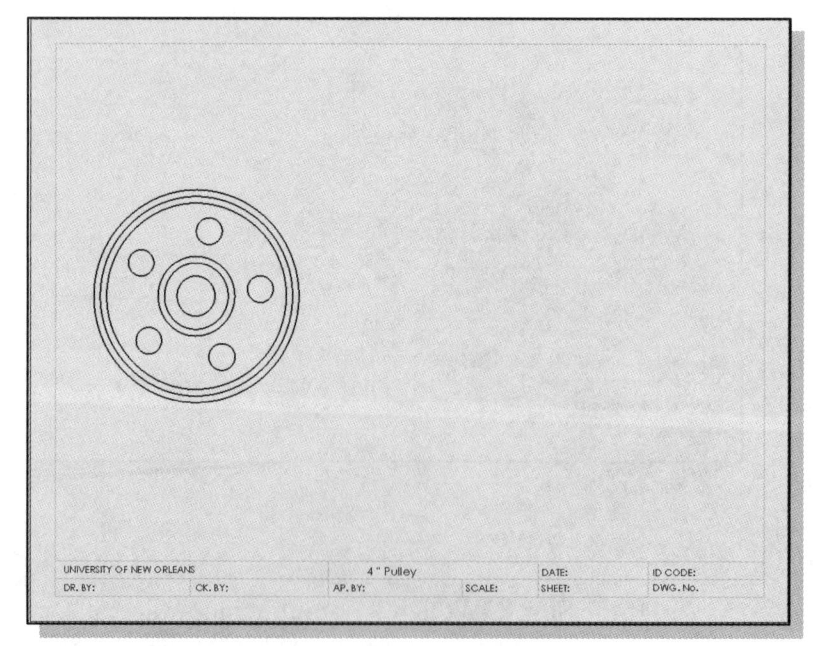

➢ When the base view is placed, SOLIDWORKS automatically executes the **Projected View** command and the *Projected View PropertyManager* appears.

7. Press the **[Esc]** key to exit the Projected View command.

8. Select the **Rebuild** icon on the *Menu Bar*.

➢ Notice the Description we entered for the *Pulley* model part file – **4″ Pulley** – appears in the title block due to the Property Link.

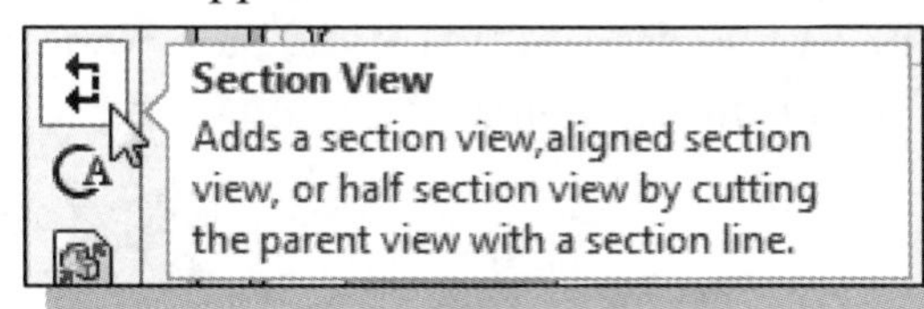

9. Select the **Section View** icon in the *Drawing* toolbar.

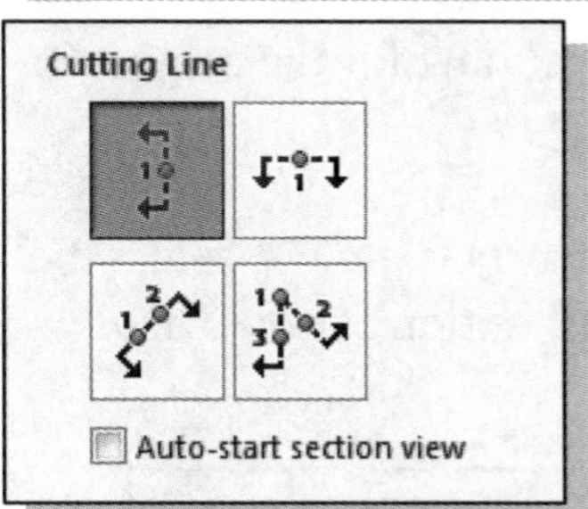

10. Verify that the **Vertical** option is selected in the *Cutting Line* panel of the *Section View PropertyManager*.

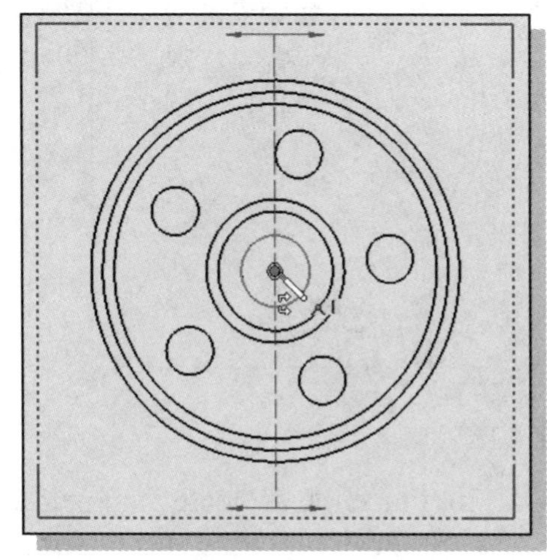

11. Inside the graphics window, align the cursor to the center of the base view and click the **left mouse button** to create the vertical cutting plane line as shown.

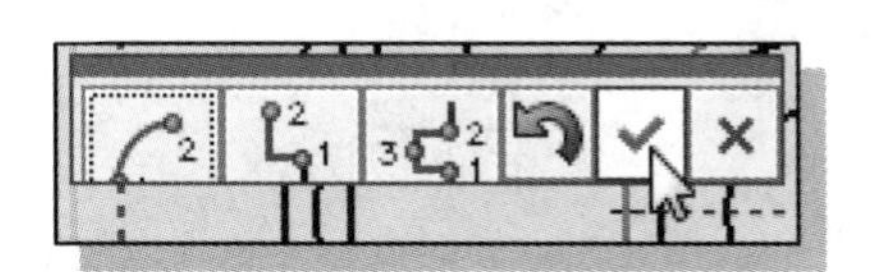

12. Select **OK** to create the vertical section line.

13. In the *Section View PropertyManager*, select the **Hidden Lines Removed** option for the *Display Style*.

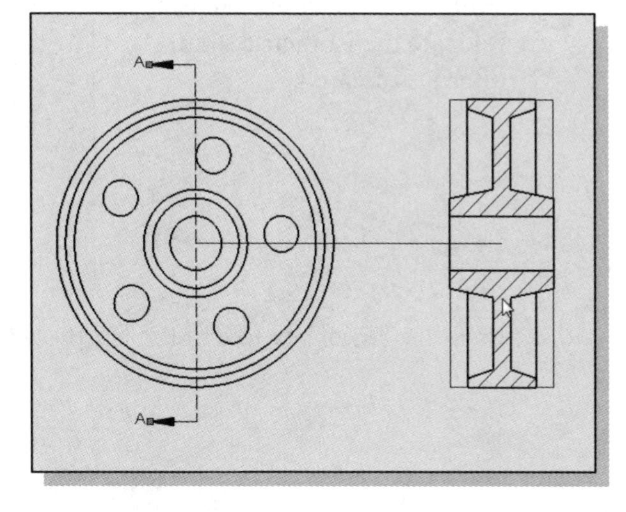

14. Next, SOLIDWORKS expects us to place the projected section. Select a location that is toward the right side of the base view as shown in the figure and click once with the **left mouse button**. (If the section view does not appear, select **Auto hatching** on the *Section View* panel of the *PropertyManager*.)

15. Select **OK** in the *PropertyManager* to create the *Section View*.

16. Press the **[Esc]** key to ensure that no view or objects are selected.

➢ We will add an isometric view using the **Projected View** command.

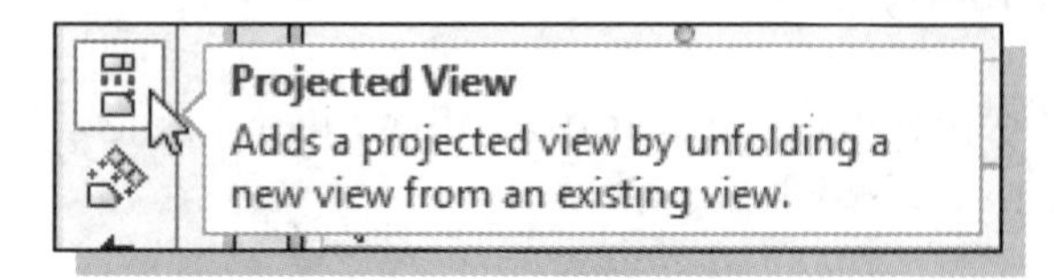

17. Select the **Projected View** icon from the *Drawing* toolbar.

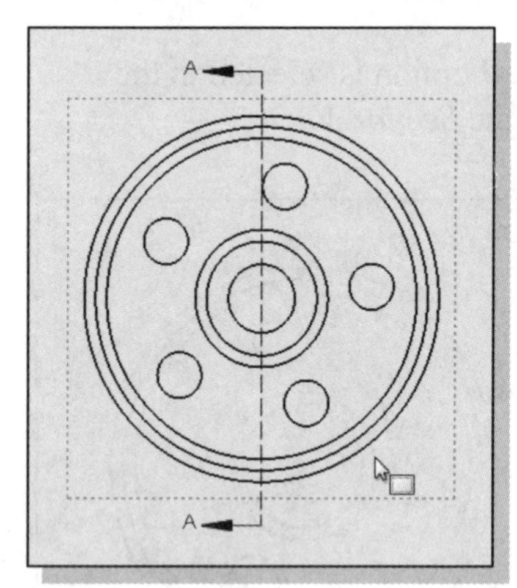

18. **Left-click** to select the base view for the projection as shown.

19. In the *Projected View PropertyManager*, select the **Hidden Lines Removed** option for the *Display Style*.

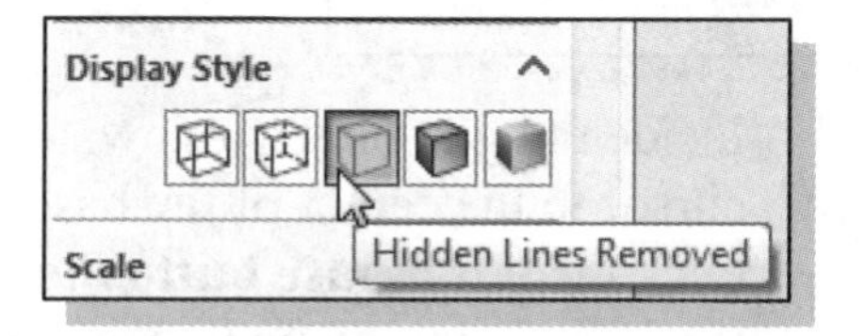

20. Place the projected isometric view by moving the cursor to the left and above the base view as shown and clicking once with the **left-mouse-button**. (**NOTE:** When it is created, the projected isometric view is automatically aligned along a line at a 45° angle from the base view. We can relocate the isometric view after it is created.)

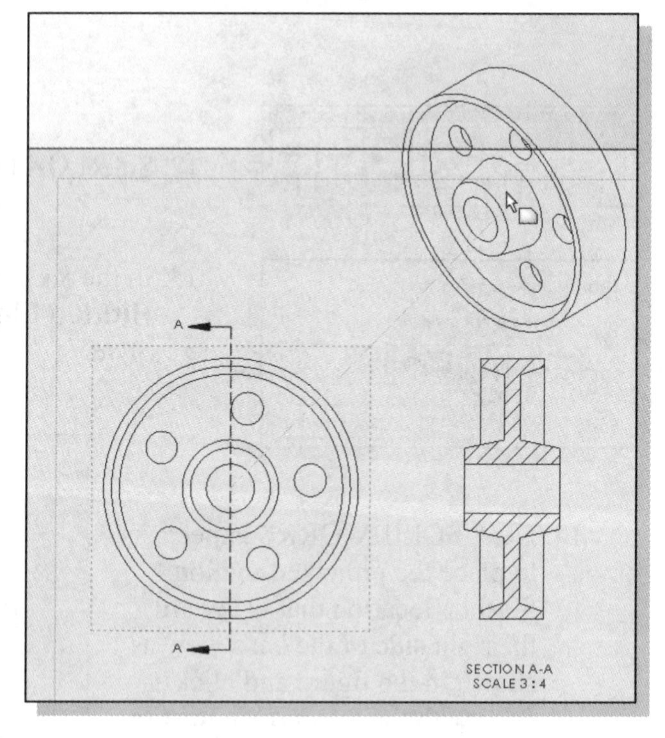

21. Press the **[Esc]** key to exit the Projected View command.

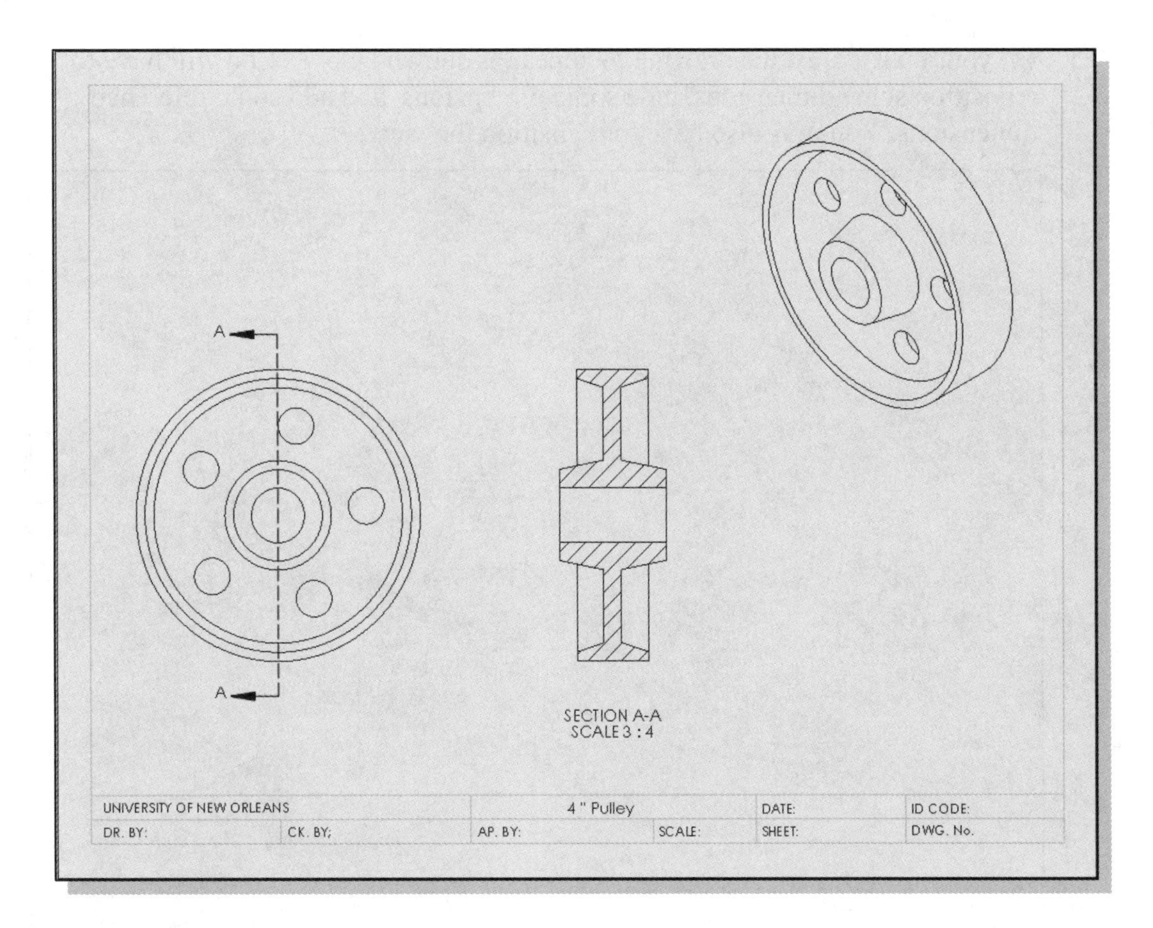

➢ On your own, reposition the views so that they appear as shown.

Retrieve Dimensions – Model Items Command

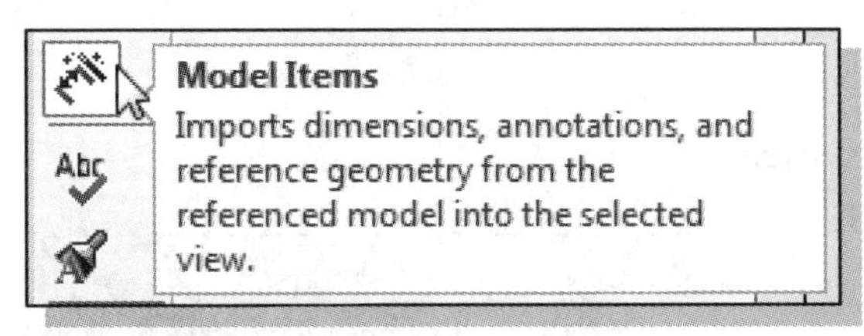

1. Left-mouse-click on the **Model Items** icon on the *Annotation* toolbar.

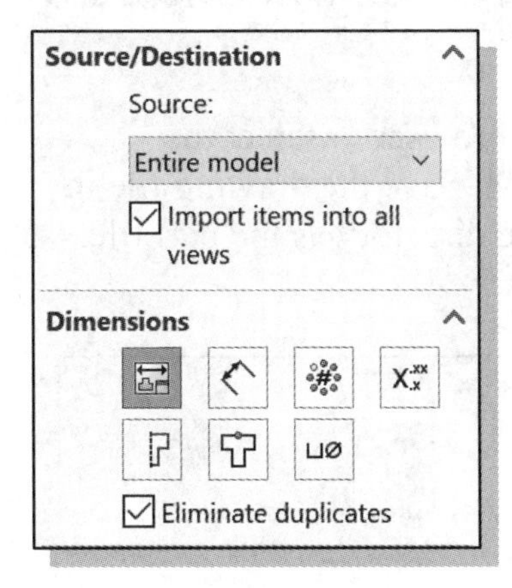

2. In the *Source/Destination* options panel of the *Model Items PropertyManager*, select the **Entire model** option from the *Import from* pull-down menu and check **Import items into all views** as shown.

3. Under the *Dimensions* options panel, select the **Marked for drawing** icon and check **Eliminate duplicates** as shown.

4. Click the **OK** icon in the *Model Items PropertyManager*. Notice dimensions are automatically placed on the side and section drawing views.

5. On your own, adjust the drawing to appear as shown below. You will have to reposition some dimensions, hide some dimensions, and add some reference dimensions. You may also have to reposition the views.

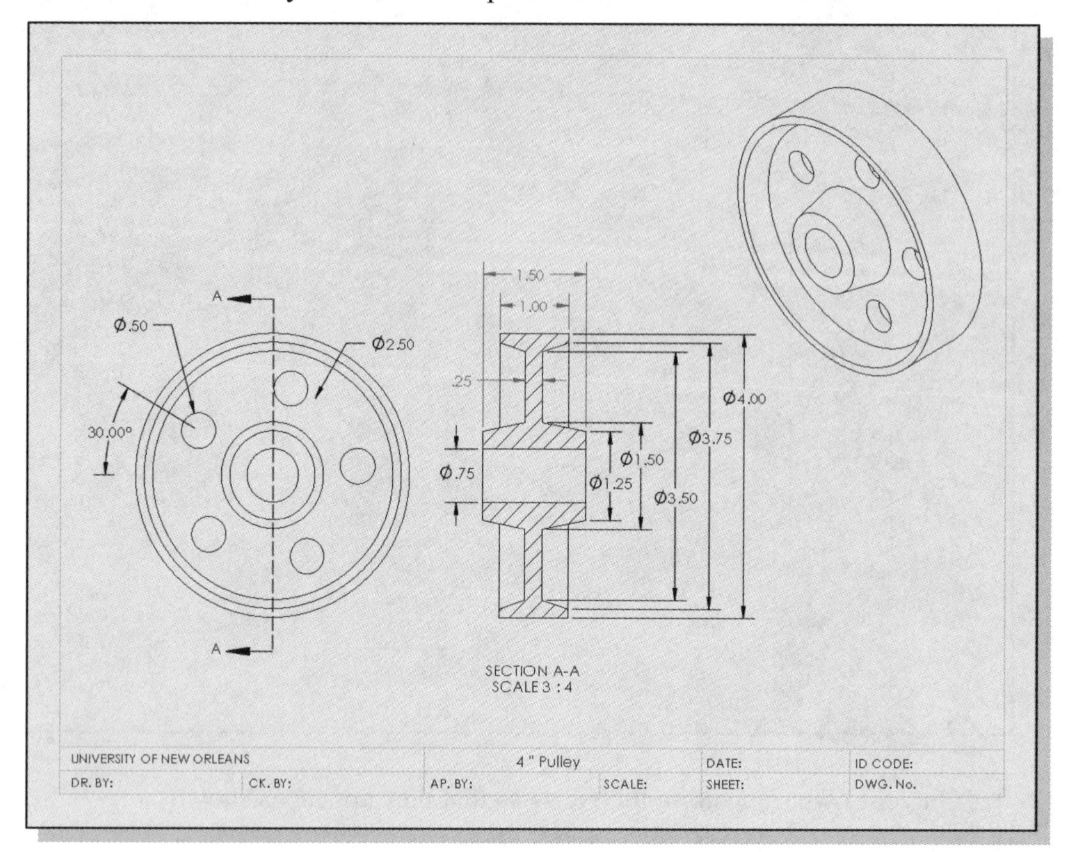

Save the Drawing File

1. Select **Save as** from the *File* pull-down menu.

2. Select **Drawing** (*.slddrw) as the file type.

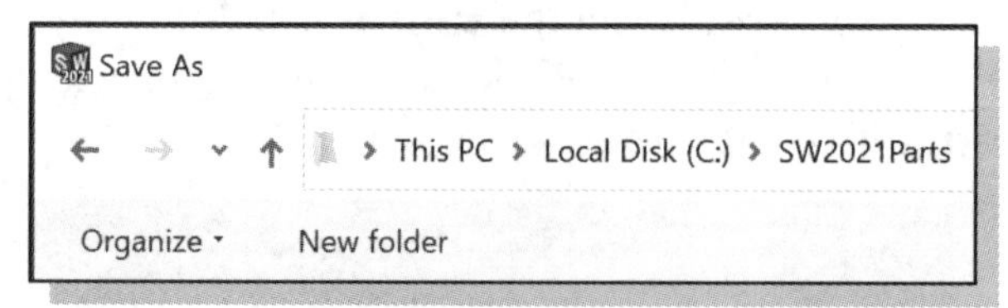

3. Use the browser to select the directory. Save the drawing file in the same directory as the part file.

4. Enter **Pulley** for the *File name* and **Pulley Detail** for the *Description.*

5. Click **Save**.

6. If the SOLIDWORKS window appears enter **Save All**.

Associative Functionality – A Design Change

SOLIDWORKS' *associative functionality* allows us to change the design at any level, and the system reflects the changes at all levels automatically. We will illustrate the associative functionality by changing the circular pattern from five holes to six holes.

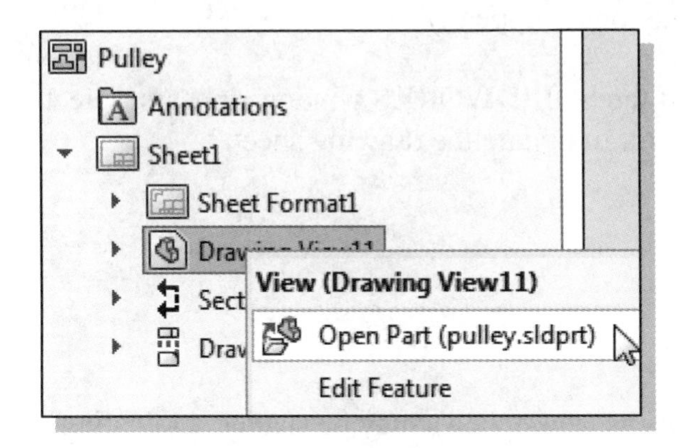

1. Inside the *FeatureManager Design Tree* window, right-click on the Drawing View icon to bring up the option menu.

2. Select **Open Part (pulley.sldprt)** in the pop-up menu to switch to the associated solid model.

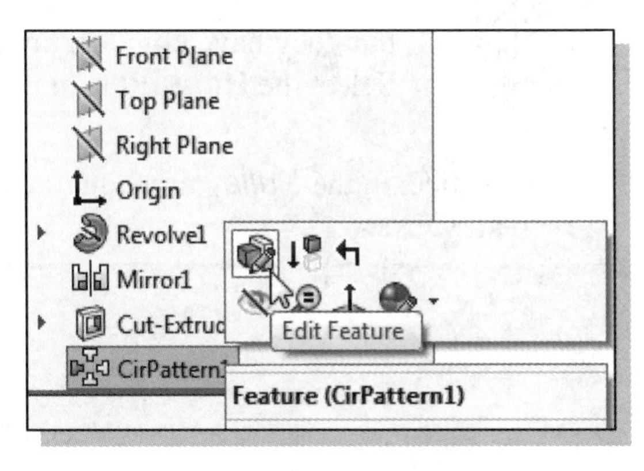

3. Inside the *FeatureManager Design Tree* window, right-click on the CircularPattern1 feature to bring up the option menu.

4. Select **Edit Feature** in the pop-up menu to bring up the associated feature option.

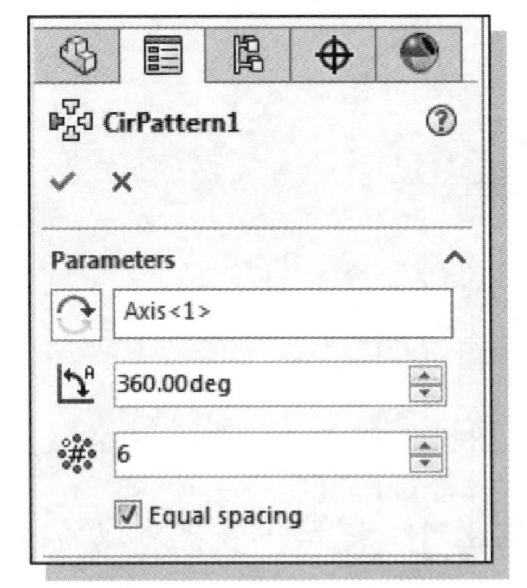

5. In the *Circular Pattern PropertyManager*, change the number to **6** as shown.

6. Click on the **OK** button to accept the setting.

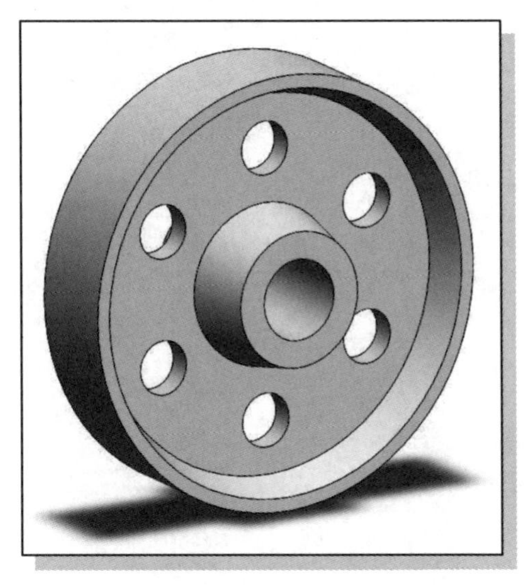

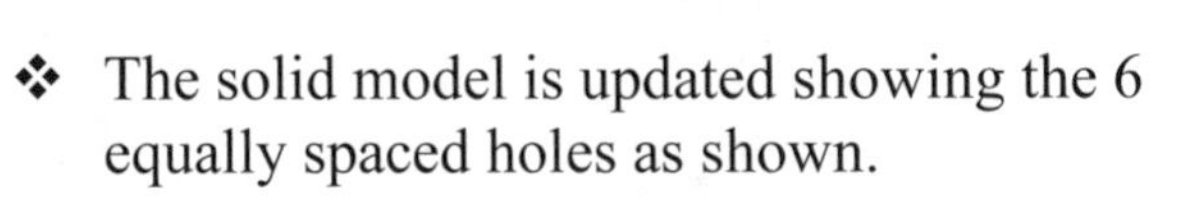

❖ The solid model is updated showing the 6 equally spaced holes as shown.

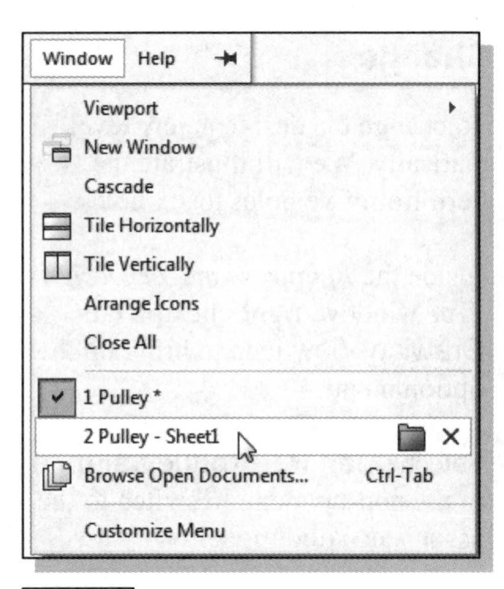

7. Switch back to the *Pulley* drawing by selecting it in the *Window* pull-down menu. (**NOTE:** The * appearing next to the filename indicates that this file has been modified but not saved. When you save the file, this * will no longer appear.)

8. If the SOLIDWORKS window appears, select **Yes** to update the drawing sheet.

9. When the drawing window opens, some views may be shaded, indicating that they have not been updated to include the changes to the model. Select the **Rebuild** icon on the *Menu Bar* to update the views.

❖ Notice, in the *Pulley* drawing, the circular pattern is also updated automatically in all views.

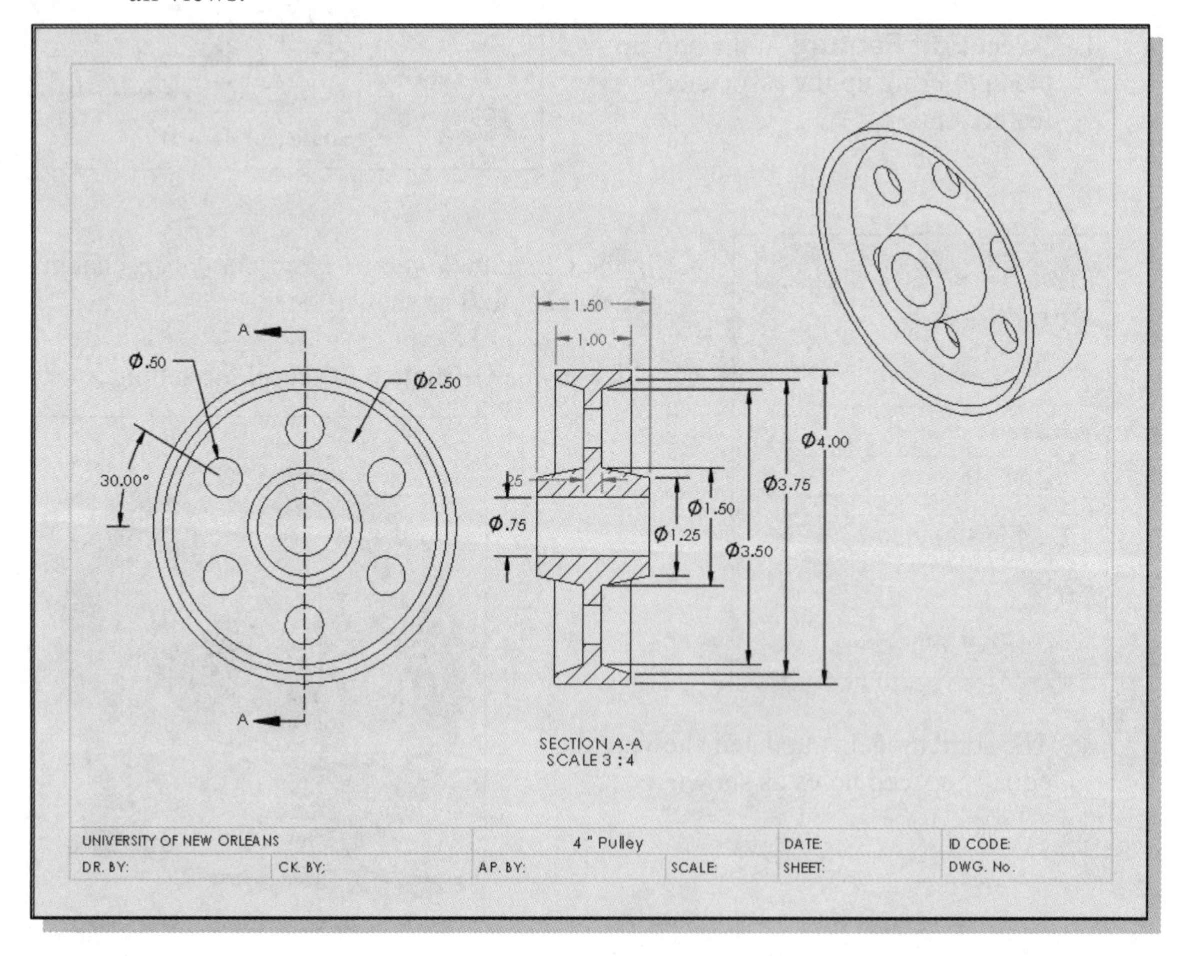

Adding Centerlines to the Pattern Feature

In your drawing, the inserted views may include *center marks*. If so, these were added with the **auto-insert** option. This lesson was done with no auto-insert options selected. If your drawing already has the center marks, you can delete them and proceed with Step 1, or select them and proceed to Step 6.

1. Click on the **Center Mark** button in the *Annotation* toolbar.

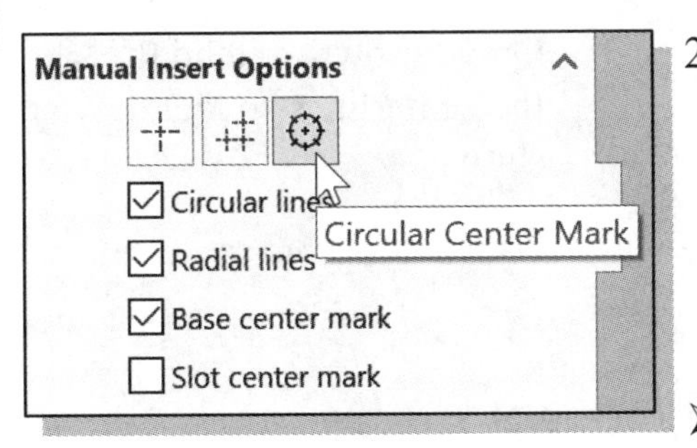

2. In the *Manual Insert Options* panel of the *Center Mark PropertyManager*, select the **Circular Center Mark** option, and **check** the **Circular lines**, **Radial lines**, and **Base center mark** option boxes.

➢ The **Circular Center Mark** option allows us to add centerlines to a patterned feature.

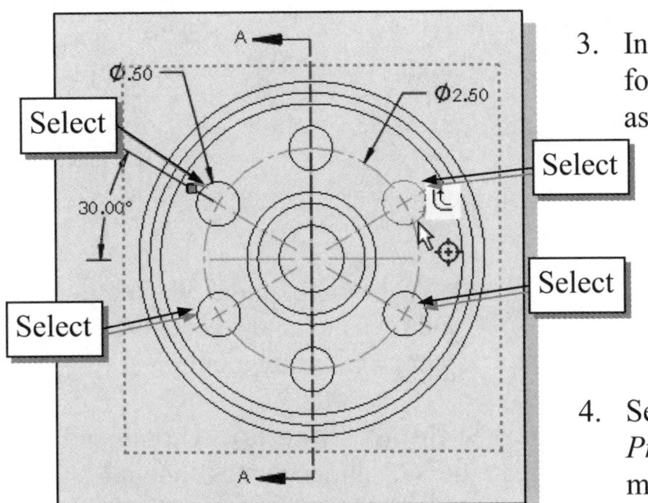

3. Inside the graphics area, select the four circular edges of the base feature as shown.

4. Select **OK** in the *Center Mark PropertyManager* to create the center marks.

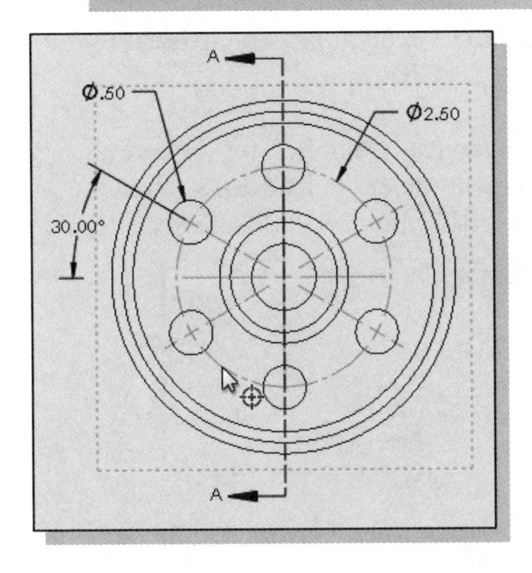

5. Notice the base circle center mark interferes with the cutting plane line. **Select** any of the newly created center marks or the associated centerlines to reopen the *PropertyManager*.

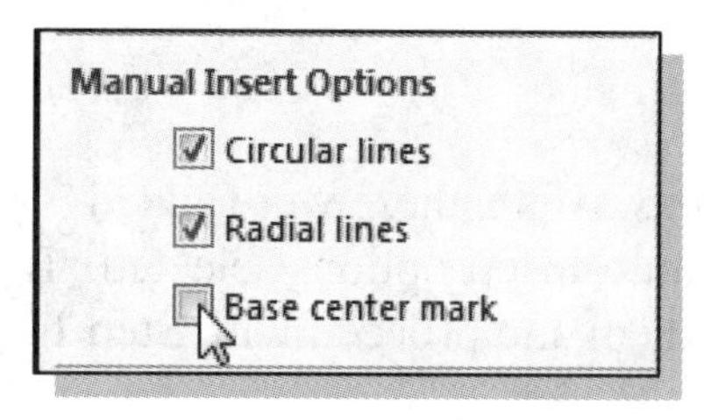

6. In the *Manual Insert Options* panel of the *Center Mark PropertyManager*, **uncheck** the **Base center mark** option box.

7. Select **OK** in the *PropertyManager* to accept the new setting.

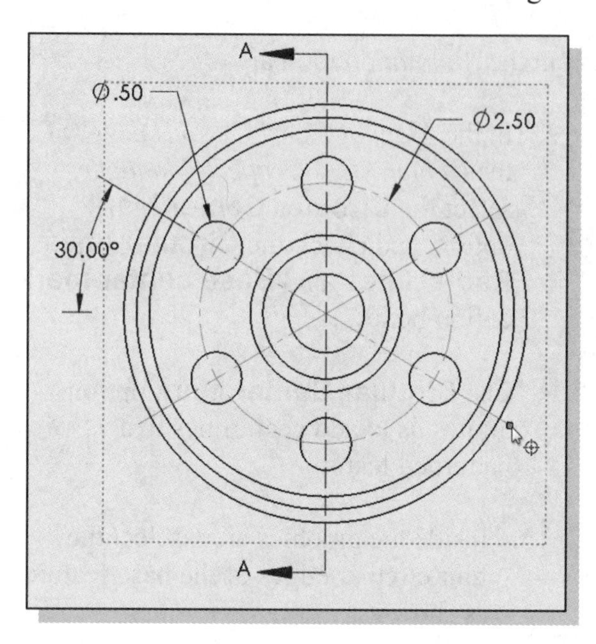

8. On your own, extend the segments of the centerlines so that they appear as shown.

Completing the Drawing

1. Right-click on the **30°** angular dimension and select **Hide** from the pop-up option menu.

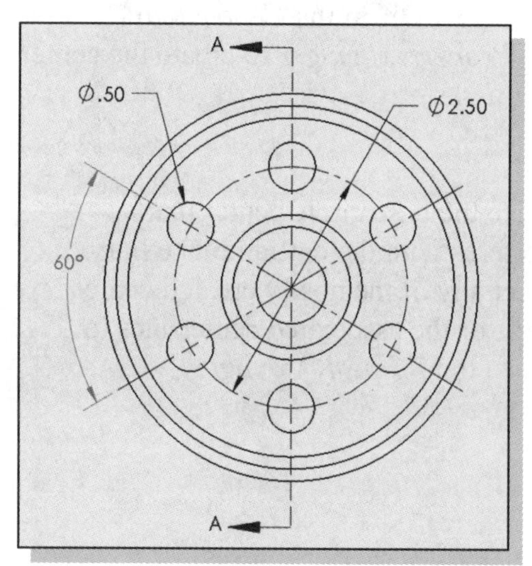

2. Use the **Smart Dimension** command to create the 60° dimension as shown.

3. Select the **Ø2.50** dimension to open the *PropertyManager*.

4. Select the **Inside** option for the arrow placement under the **Leaders** tab in the *PropertyManager* as shown below.

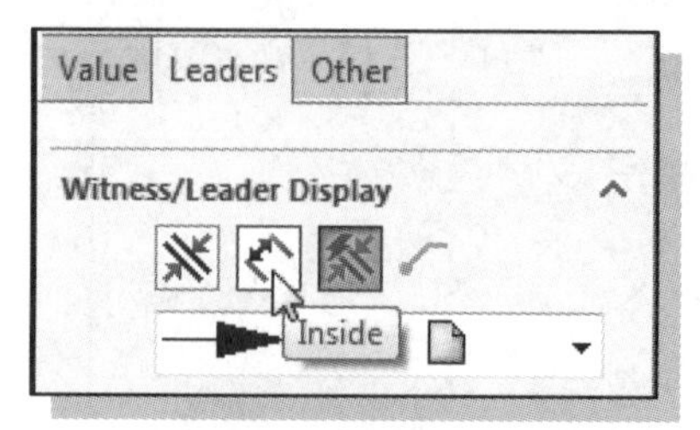

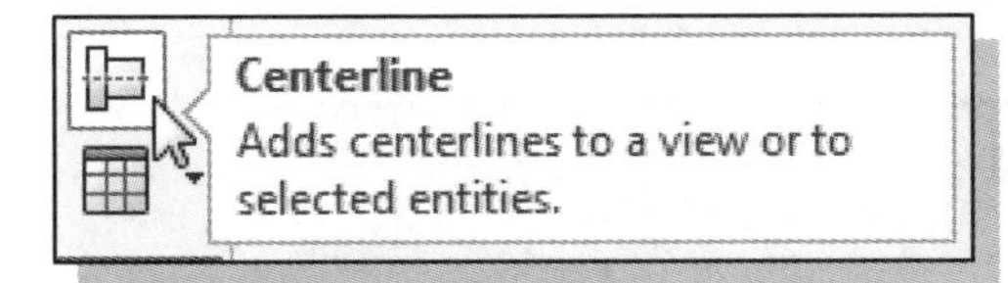

5. Click on the **Centerline** button in the *Annotation* toolbar.

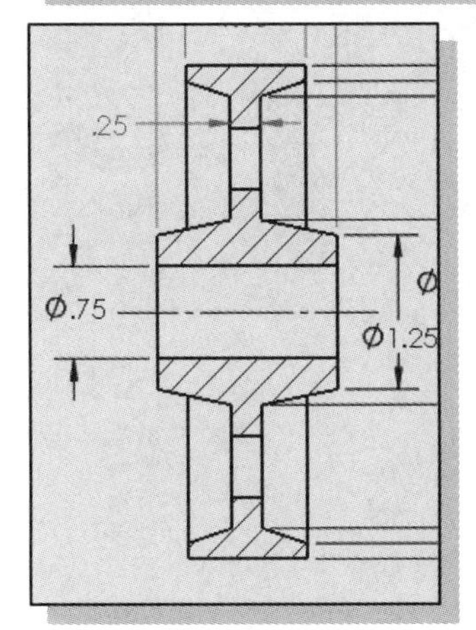

6. Inside the graphics window, click on the two edges of the *Section View* to create a centerline through the view as shown in the figure.

7. Repeat the above step and create the centerlines through the other two holes.

8. On your own, complete the title block and complete the drawing to appear as shown on the next page. Remember to switch to *Edit Sheet Format* mode to modify the title block.

9. Be sure to switch back to *Edit Sheet* mode.

10. Select the **Save** icon on the *Menu Bar*. This command will save the Pulley.SLDDRW file. This file includes the document properties, the sheet format, and the drawing views.

➢ Because the Pulley.SLDPRT model part file was not saved after the design change, the *Save Modified Documents* window will appear.

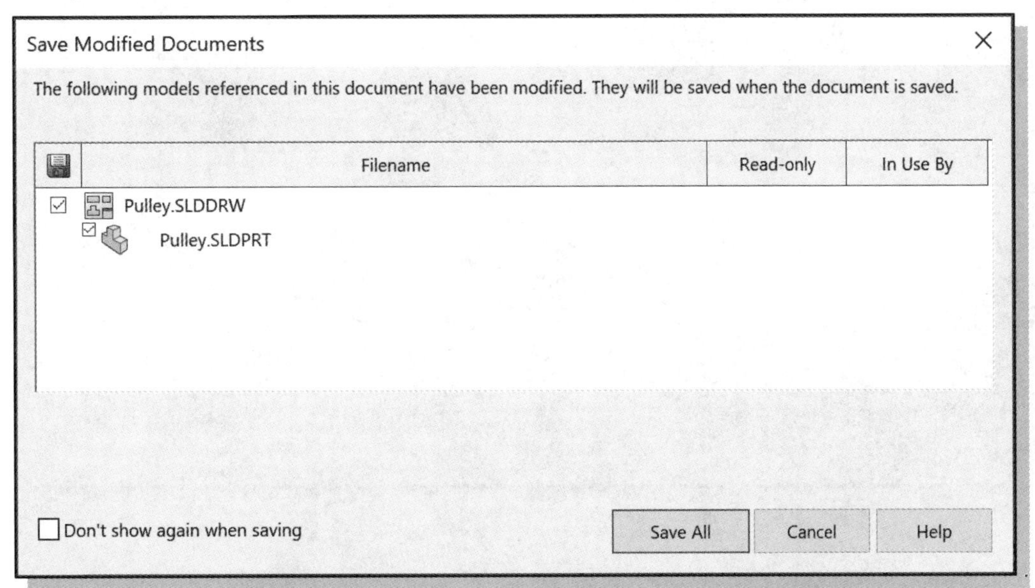

11. Select **Save All** to save the Pulley.SLDDRW drawing file and the Pulley.SLDPRT model part file.

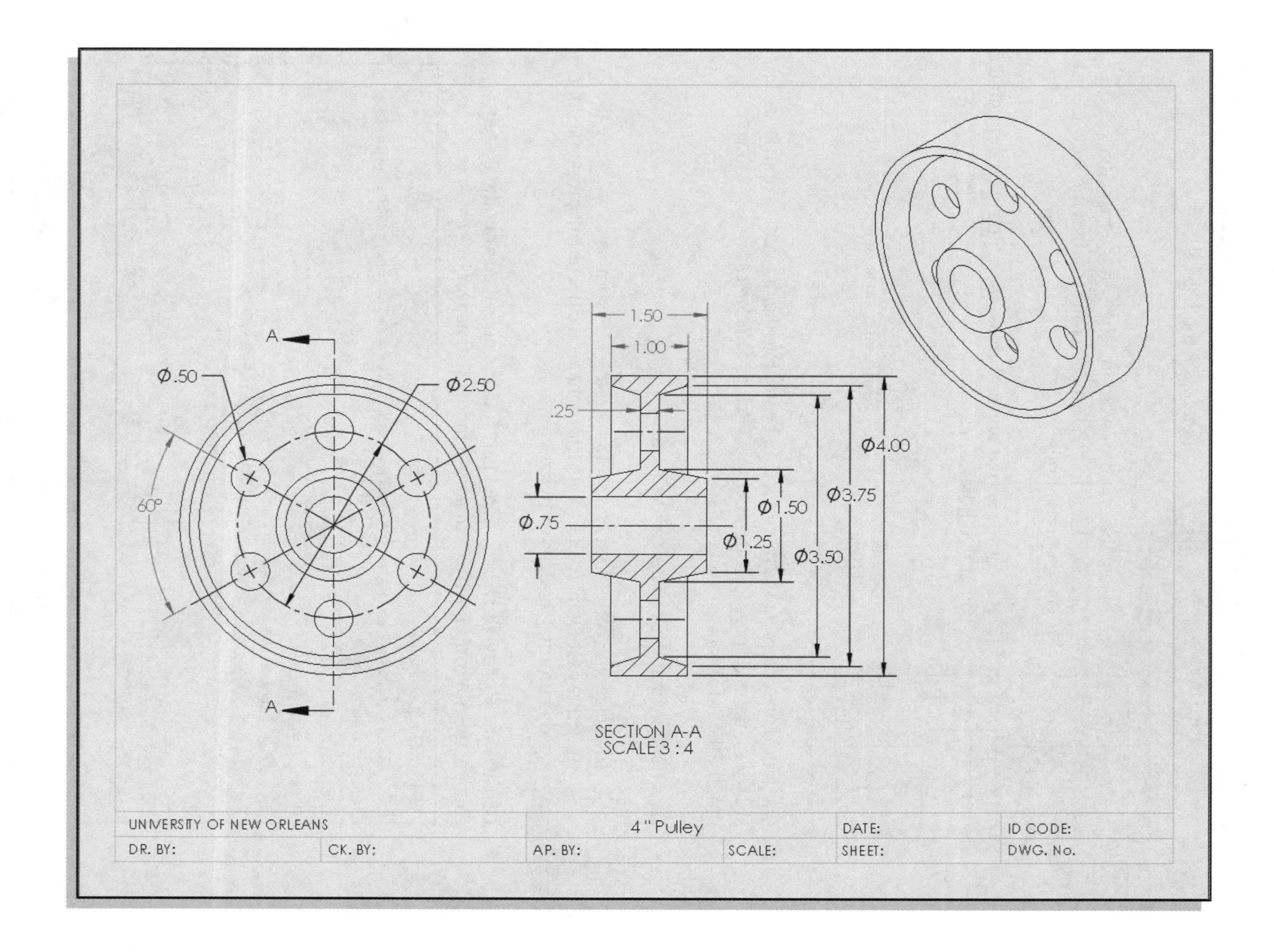
A
Ø.50
Ø2.50
60°
A
1.50
1.00
.25
Ø.75
Ø1.50
Ø1.25
Ø3.50
Ø3.75
Ø4.00
SECTION A-A
SCALE 3 : 4
UNIVERSITY OF NEW ORLEANS
4" Pulley
DATE:
ID CODE:
DR. BY:
CK. BY:
AP. BY:
SCALE:
SHEET:
DWG. No.

Additional Title Blocks

Drawing Paper and Border Sizes

The standard drawing paper sizes are as shown in the below tables. The edges of the title block border are generally 0.5 ~ 1 inches or 10~20 mm from the edges of the paper.

American National Standard	Suggested Border Size
A – 8.5″ X 11.0″	**A – 7.75″ X 10.25″**
B – 11.0″ X 17.0″	**B – 10.0″ X 16.0″**
C – 17.0″ X 22.0″	**C – 16.0″ X 21.0″**
D – 22.0″ X 34.0″	**D – 21.0″ X 33.0″**
E – 34.0″ X 44.0″	**E – 33.0″ X 43.0″**

International Standard	Suggested Border Size
A4 – 210 mm X 297 mm	**A4 – 190 mm X 276 mm**
A3 – 297 mm X 420 mm	**A3 – 275 mm X 400 mm**
A2 – 420 mm X 594 mm	**A2 – 400 mm X 574 mm**
A1 – 594 mm X 841 mm	**A1 – 574 mm X 820 mm**
A0 – 841 mm X 1189 mm	**A0 – 820 mm X 1168 mm**

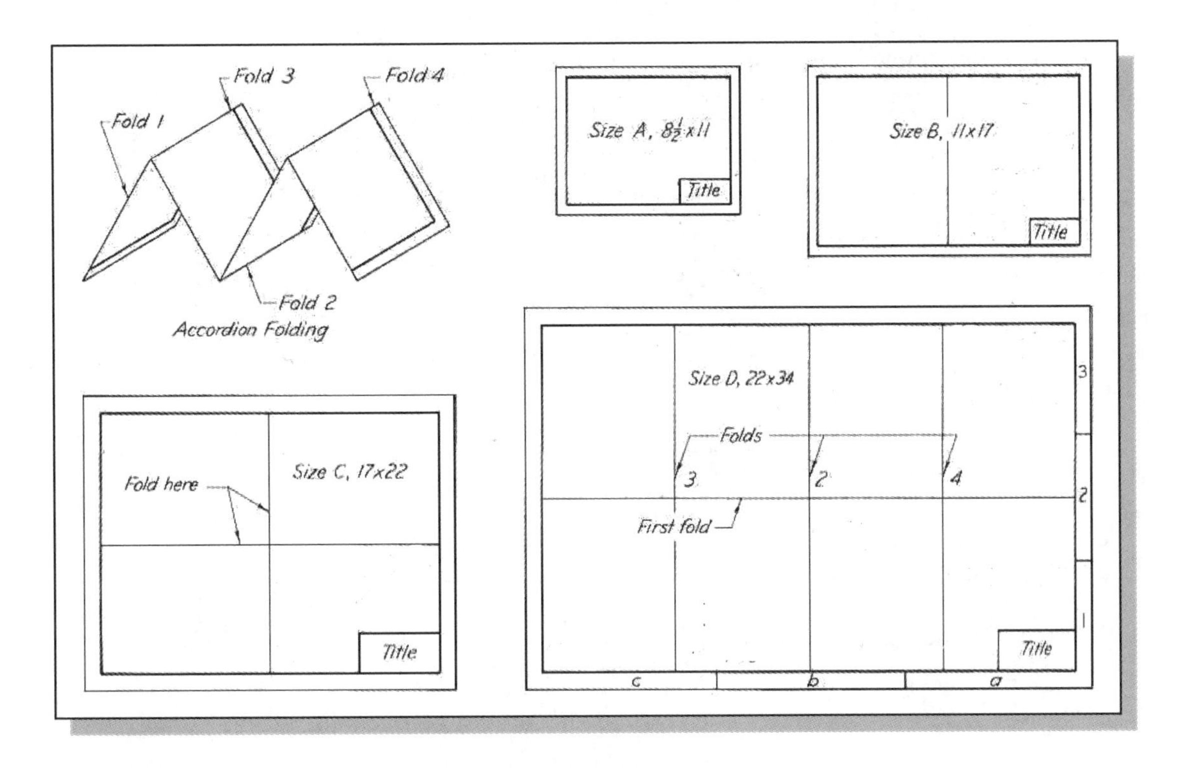

- **English Title Block** (For A size paper, dimensions are in inches.)

OREGON INSTITUE OF TECHNOLOGY				DATE:	ID CODE:
DR. BY:	CK. BY:	AP. BY:	SCALE:	SHEET:	DWG No:
1.98	1.98	1.96	1.10	1.60	1.63

.25 .50

- **Metric Title Block** (For A4 size paper, dimensions are in mm.)

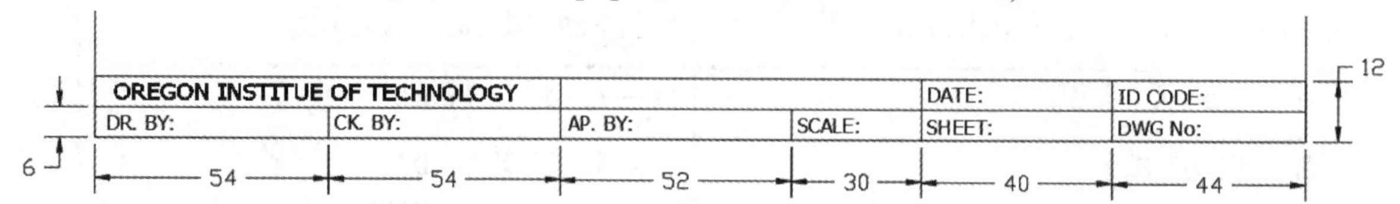

- **English Title Block** (Dimensions are in inches.)

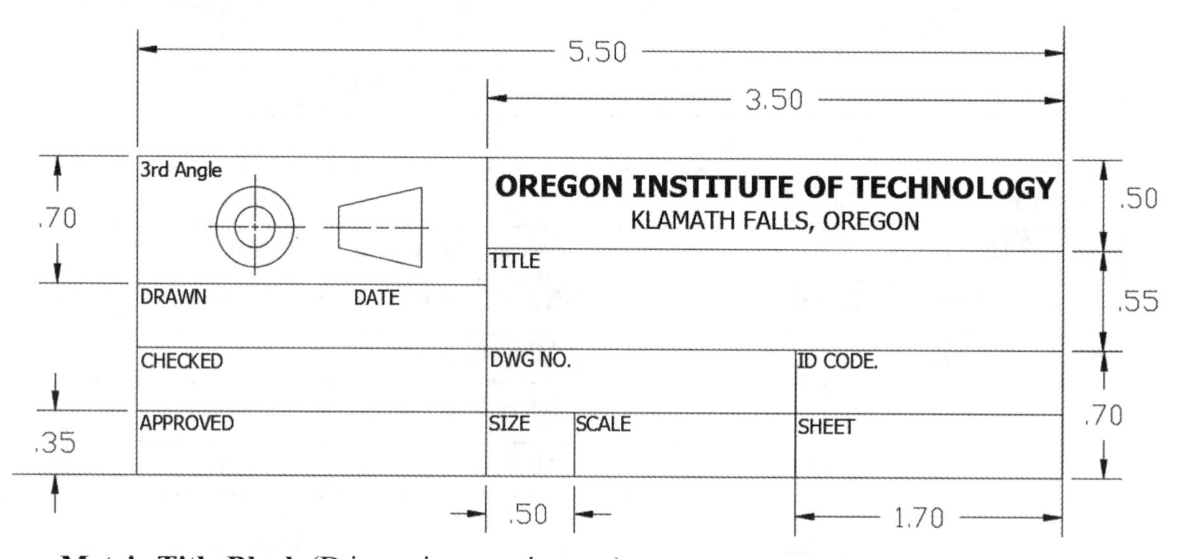

- **Metric Title Block** (D imensions are in mm.)

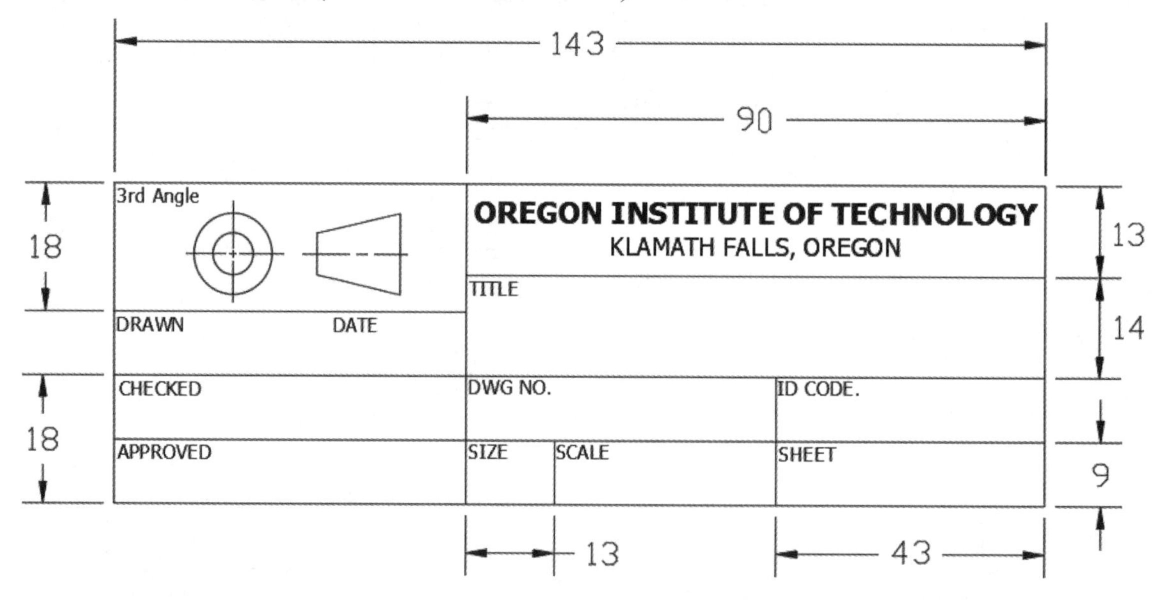

Questions:

1. List the different symmetrical features created in the *Pulley* design.
2. What are the advantages of using a *drawing template*?
3. Describe the steps required in using the Mirror command.
4. Why is it important to identify symmetrical features in designs?
5. When and why should we use the Circular Pattern command?
6. What are the required elements in order to generate a sectional view?
7. How do we create a linear *diameter dimension* for a revolved feature?
8. What is the difference between *construction geometry* and *normal geometry*?
9. Identify and describe the following commands:

(a) 

(b) 

(c) 

(d) 

Exercises: (All dimensions are in inches.)

1. **Shaft Support** (Dimensions are in inches.)

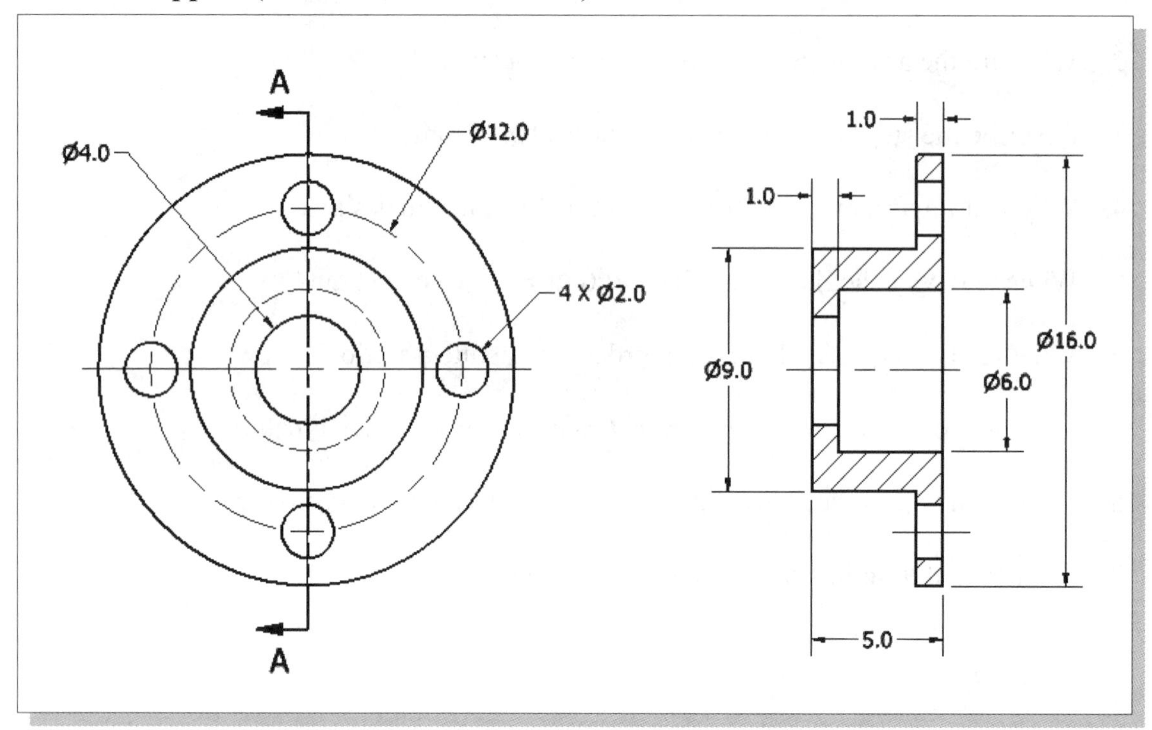

2. **Ratchet Plate** (Dimensions are in inches. Thickness: 0.125 inch.)

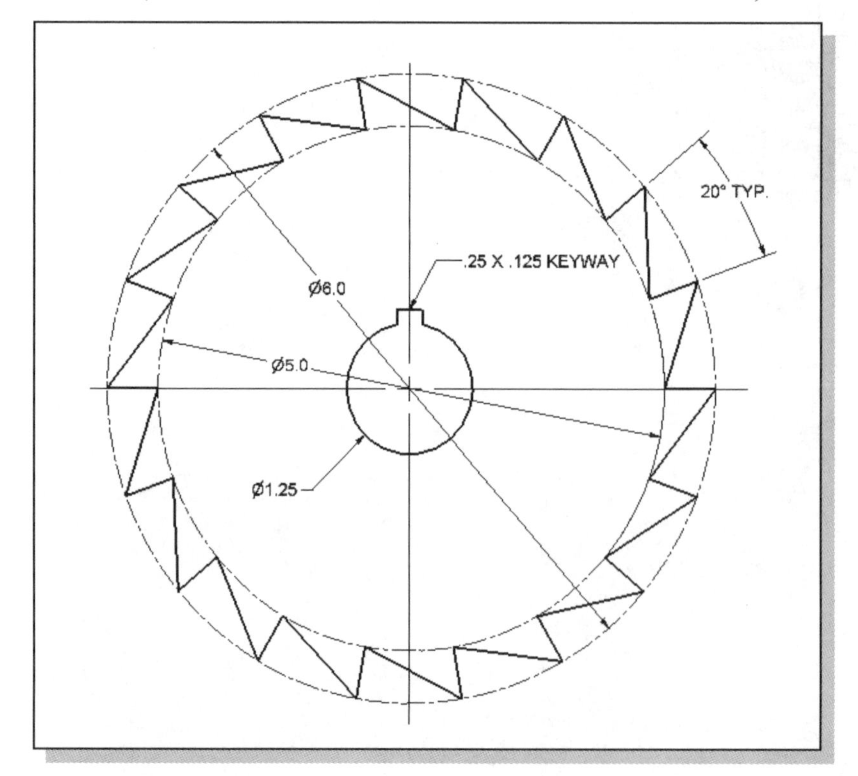

3. **Geneva Wheel** (Dimensions are in inches.)

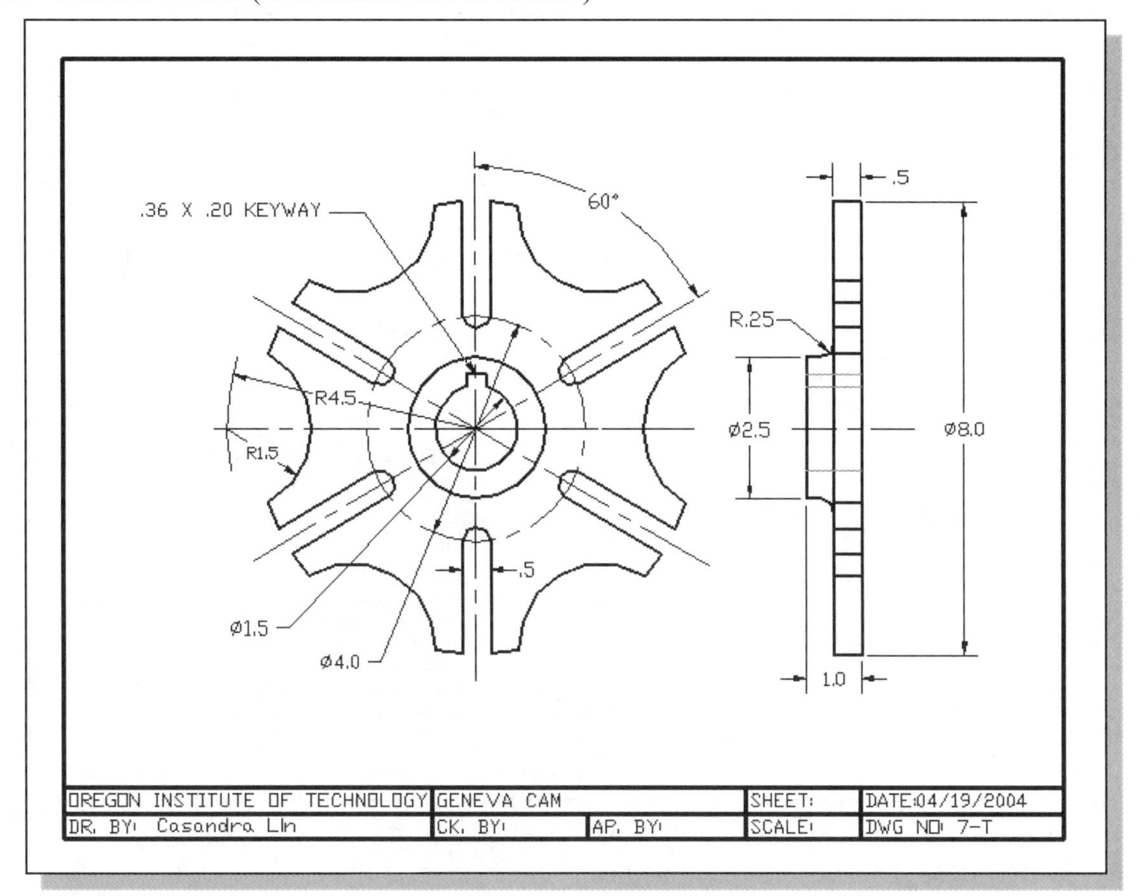

4. **Valve Knob** (Dimensions are in mm, height: 6 mm)

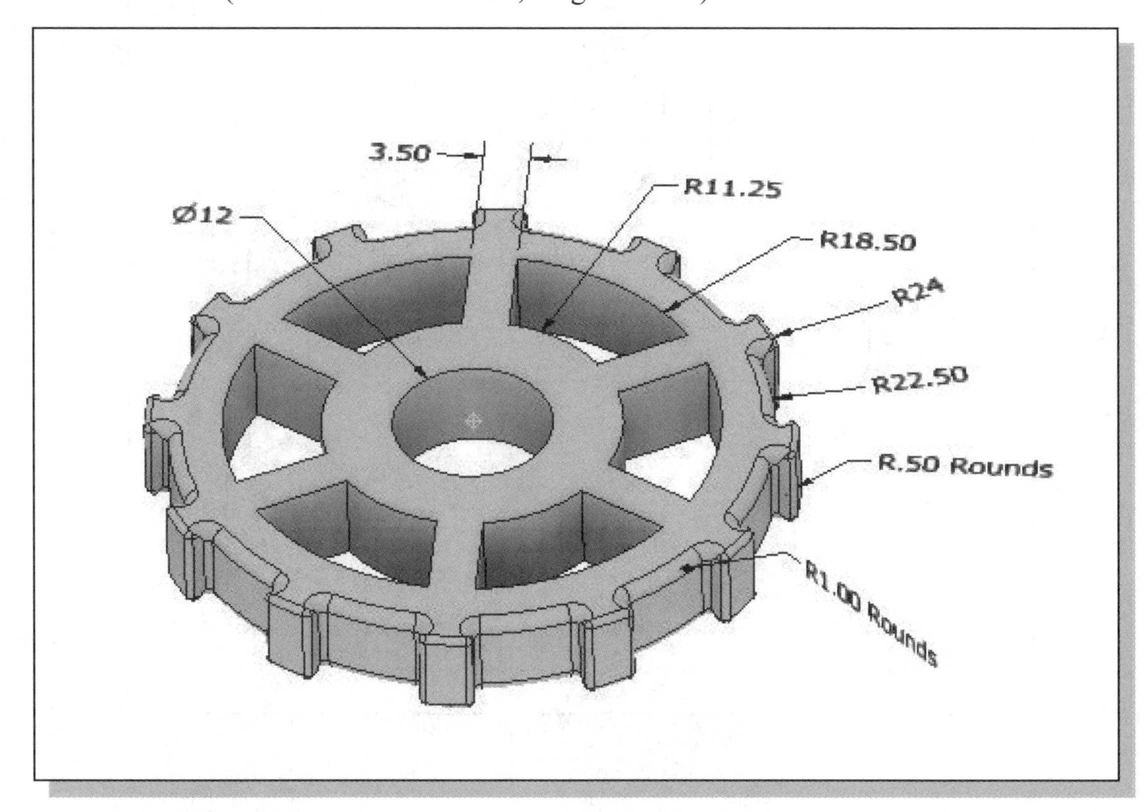

5. **Hub** (Dimensions are in inches.)

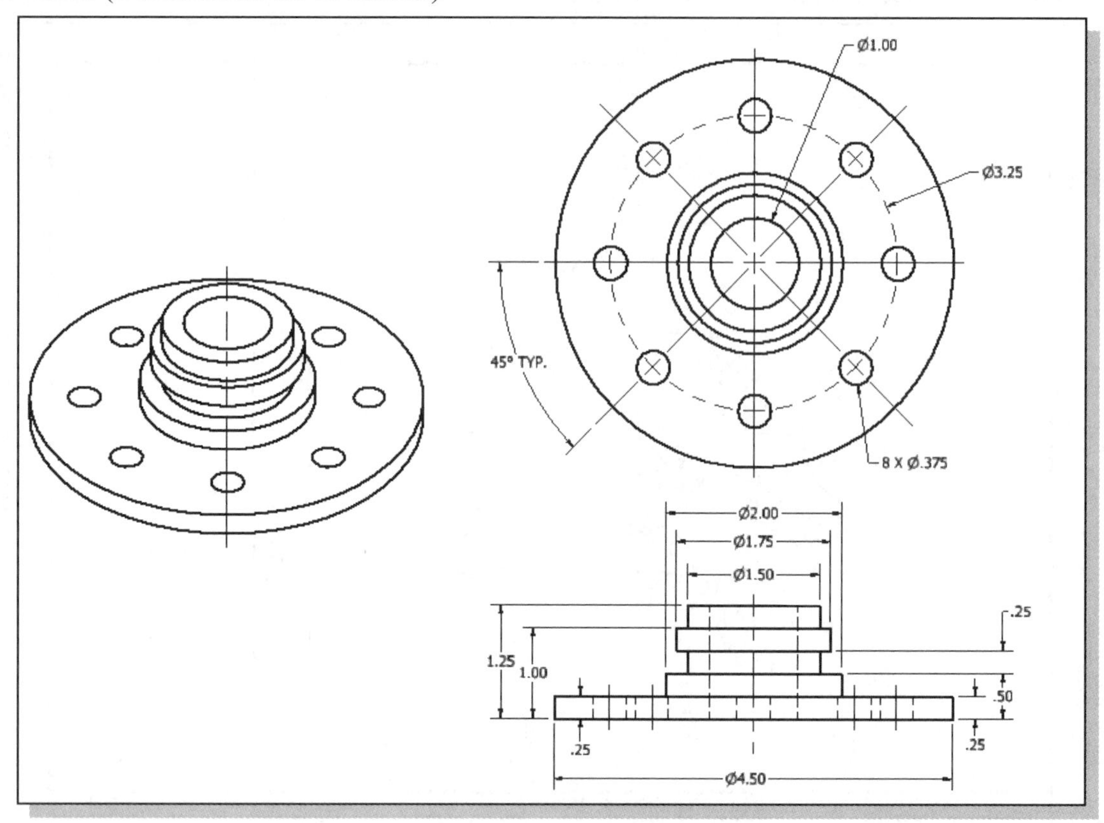

6. **Switch Base** (Dimensions are in inches.)

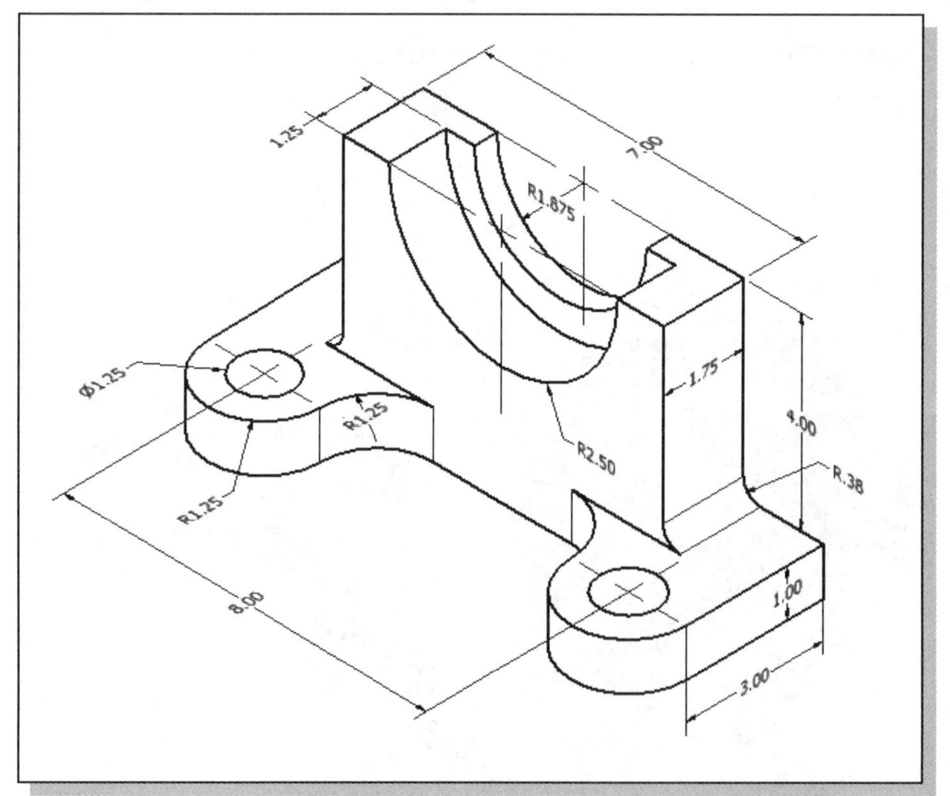

7. **Pulley Wheel** (Dimensions are in inches.)

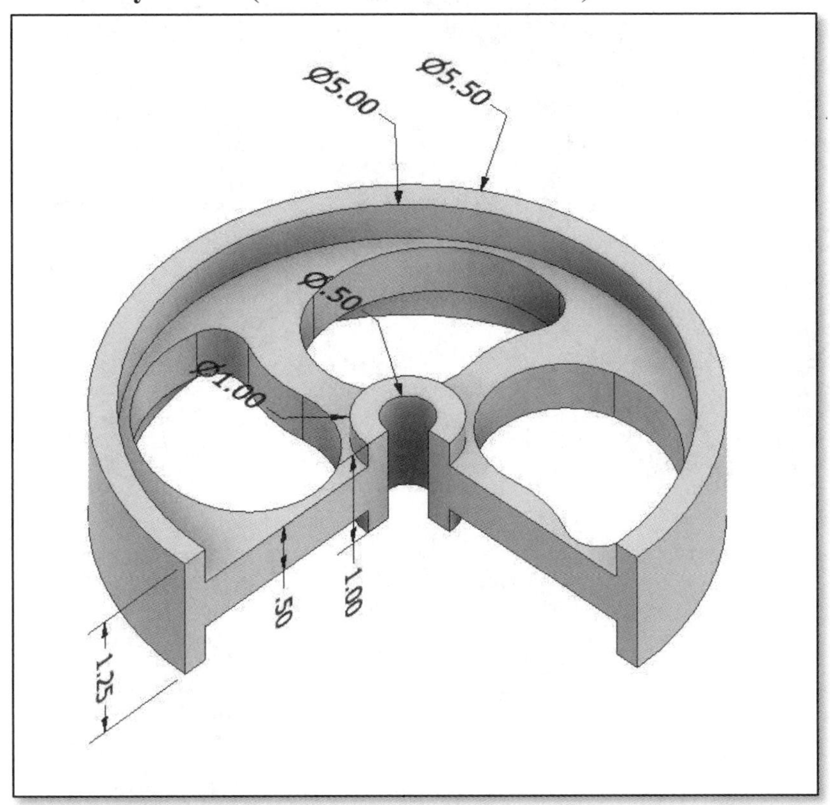

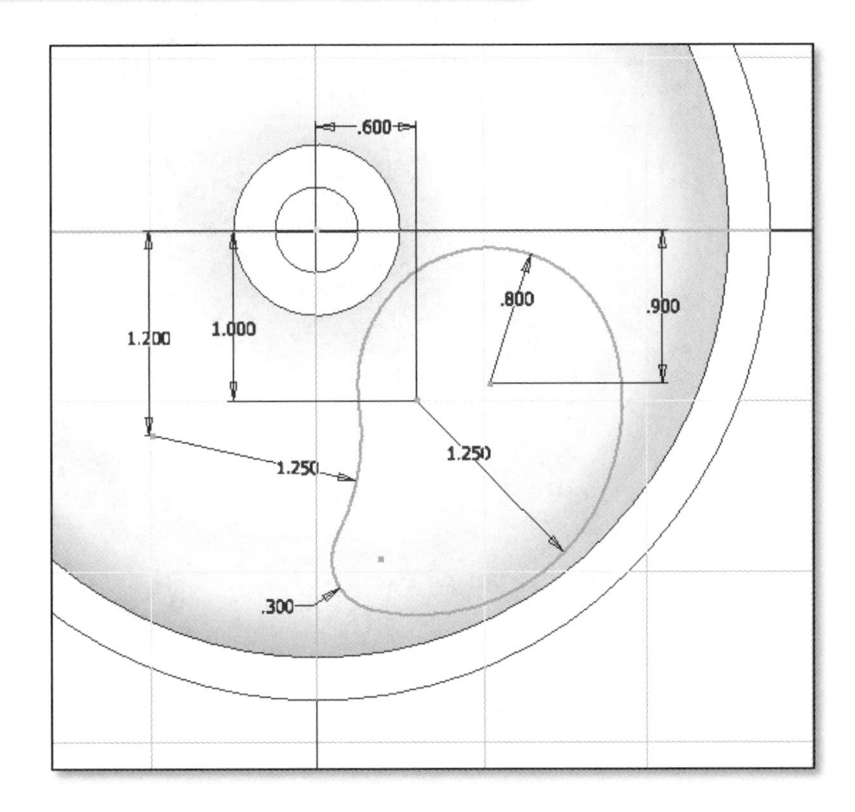

NOTES:

Chapter 12
Advanced 3D Construction Tools

Learning Objectives

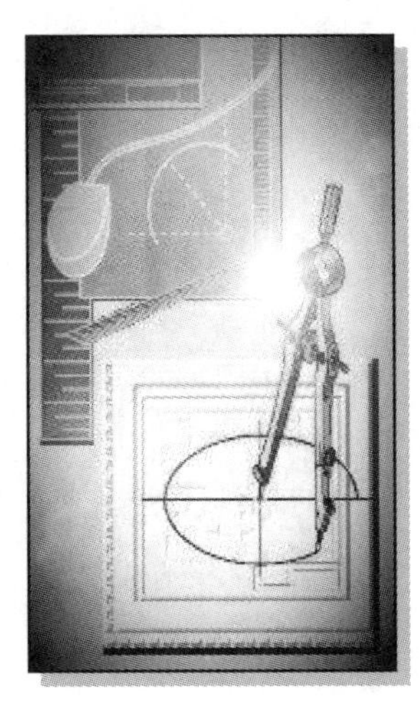

- Understand the Concepts Behind the Different 3D Construction Tools
- Set up Multiple Reference Planes
- Create Swept Features
- Create Lofted Features
- Use the Shell Command
- Create 3D Rounds & Fillets
- Create a Photorealistic Rendered View

Certified SOLIDWORKS Associate Exam Objectives Coverage

Sketch Tools – Offset, Convert, Trim

Objectives: Using Sketch Tools.

Boss and Cut Features – Extrudes, Revolves, Sweeps, Lofts

Objectives: Creating Basic Swept Features.

Fillets and Chamfers

Objectives: Creating Fillets and Chamfer Features.

Linear, Circular, and Fill Patterns

Objectives: Creating Patterned Features.

Introduction

SOLIDWORKS provides an assortment of three-dimensional construction tools to make the creation of solid models easier and more efficient. As demonstrated in the previous lessons, creating **extruded** features and **revolved** features are the two most common methods used to create 3D models. In this next example, we will examine the procedures for using the **Sweep** commands, the **Loft** commands, and the **Shell** command, and also for creating **3D rounds** and **fillets** along the edges of a solid model. These types of features are common characteristics of molded parts.

The **Sweep** options (*Swept Boss/Base* and *Cut Sweep*) are defined as moving a cross-section through a path in space to form a three-dimensional object. To define a sweep in SOLIDWORKS, we define two sections: the trajectory and the cross-section.

The **Loft** commands (*Lofted Boss/Base* and *Cut Loft*) allow us to blend multiple profiles with varying shapes on separate planes to create complex shapes. Profiles are usually on parallel planes, but non-perpendicular planes can also be used. We can use as many profiles as we wish but, to avoid twisting the loft shape, we should map points on each profile that align along a straight vector.

The **Shell** option is defined as hollowing out the inside of a solid, leaving a shell of specified wall thickness.

A Thin-Walled Design: *Dryer Housing*

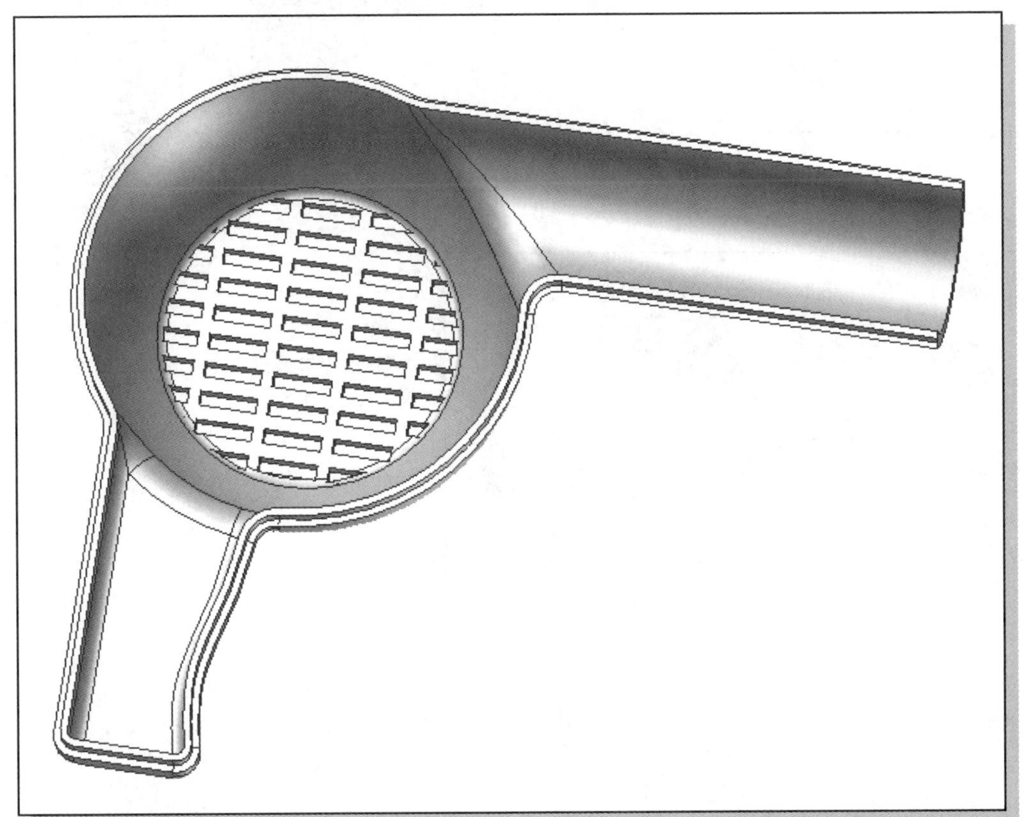

Modeling Strategy

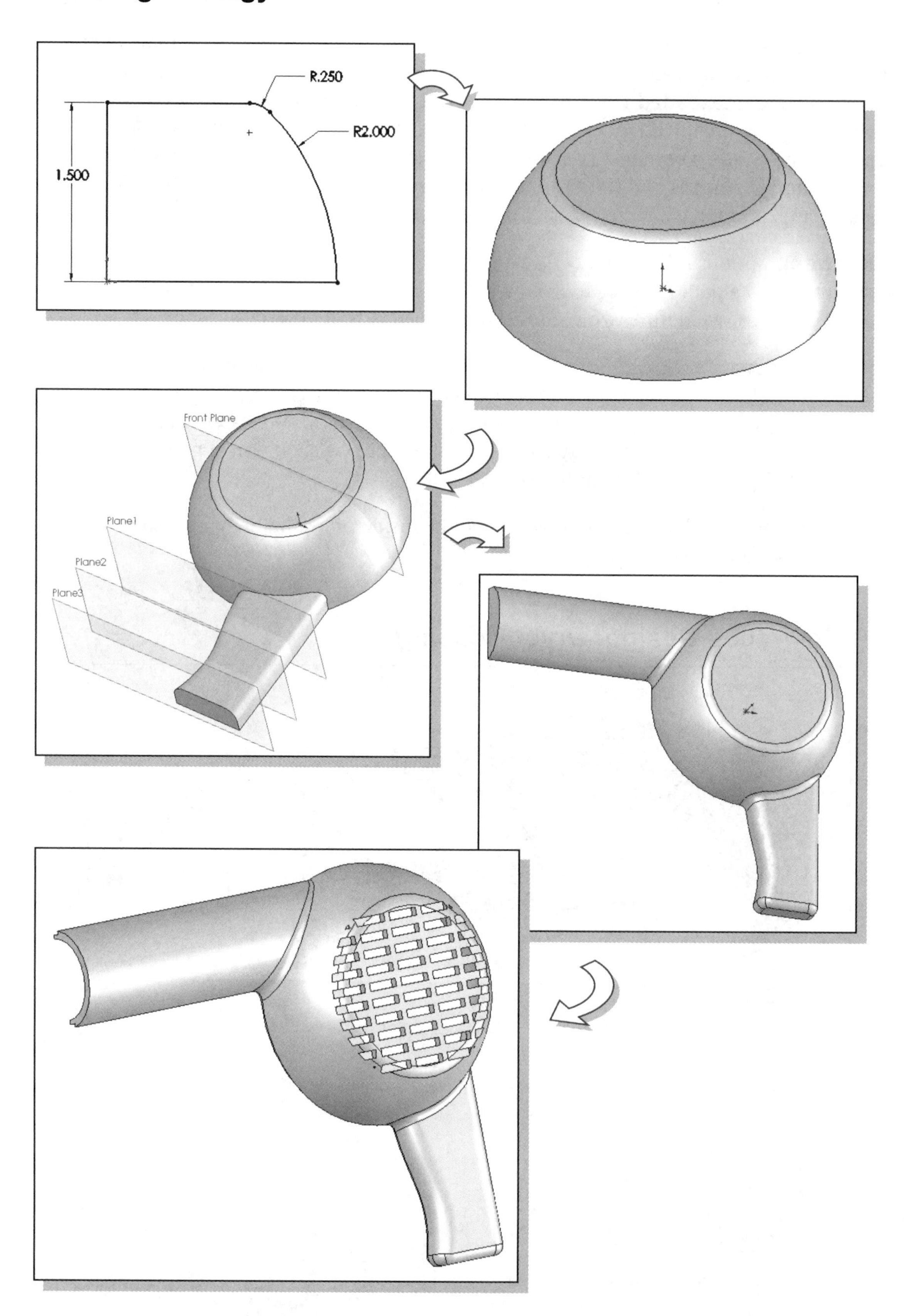

Starting SOLIDWORKS

1. Select the **SOLIDWORKS** option on the *Start* menu or select the **SOLIDWORKS** icon on the desktop to start SOLIDWORKS. The SOLIDWORKS main window will appear on the screen.

2. Select the **New** icon with a single click of the left-mouse-button on the *Menu Bar*. The *New* SOLIDWORKS *Document* dialog box should appear in Advanced mode. If it appears in Novice mode, click once with the left-mouse-button on the **Advanced** icon.

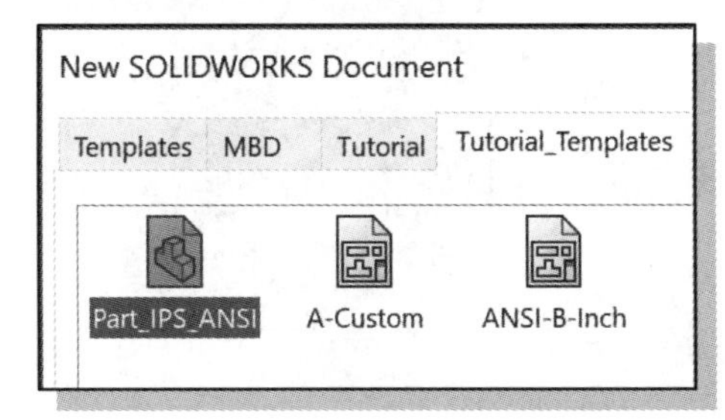

3. Select the **Tutorial_Templates** tab. (**NOTE:** You added this tab in Chapter 4.)

4. Select the **Part_IPS_ANSI** template as shown.

5. Click on the **OK** button to open a new document using the Part_IPS_ANSI template.

Creating the 2D Sketch for the Base feature

IMPORTANT NOTE: We will use the *CommandManager* in this lesson.

1. To turn ***ON*** the *Enable CommandManager* option, right click on any toolbar and select it at the top of the pop-up menu.

2. If necessary, select the **Sketch** tab on the *CommandManager* to display the *Sketch* toolbar.

3. In the *Sketch* toolbar, select the **Sketch** command by left-clicking once on the icon.

4. In the *Edit Sketch PropertyManager*, the message "*Select a plane on which to create a sketch for the entity*" is displayed. Move the cursor over the edge of the Front Plane in the graphics area. When the Front Plane is highlighted, click once with the **left-mouse-button** to select the Front (XY) Plane as the sketch plane for the new sketch.

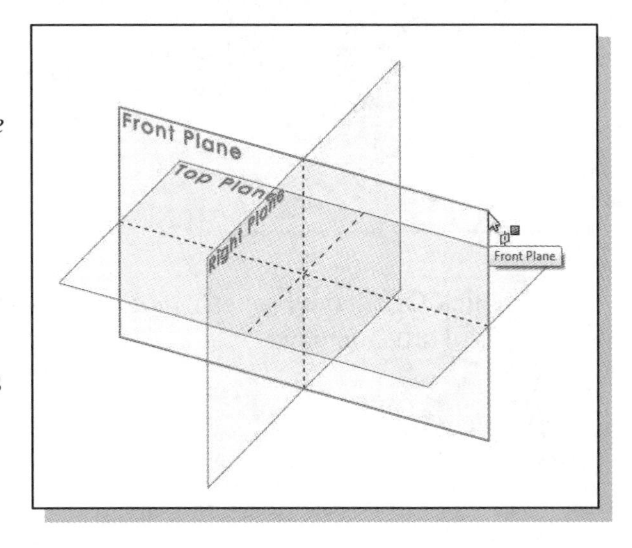

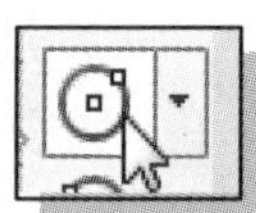

5. Select the **Circle** command by clicking once with the left-mouse-button on the icon in the *Sketch* toolbar.

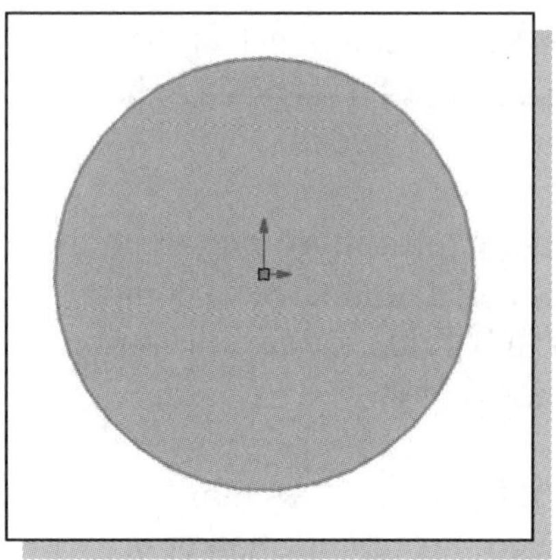

6. Create a **circle** of arbitrary size, with its center aligned to the ***Origin*** as shown.

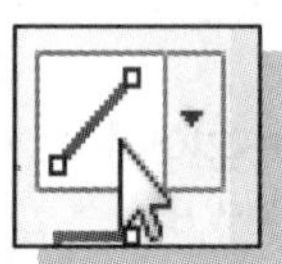

7. Click on the **Line** icon in the *Sketch* toolbar.

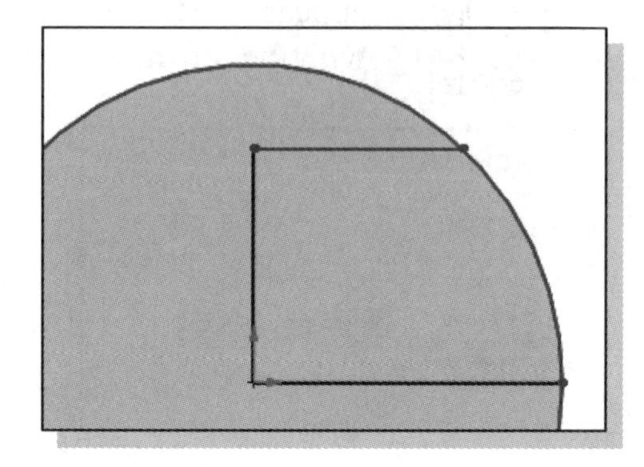

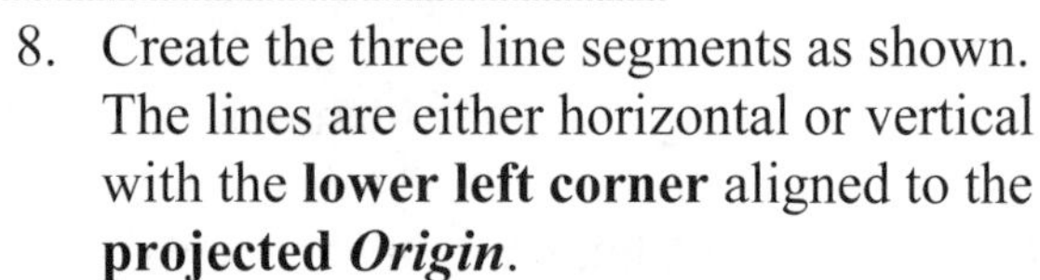

8. Create the three line segments as shown. The lines are either horizontal or vertical with the **lower left corner** aligned to the **projected *Origin***.

9. Select the **Trim** icon in the *Sketch* toolbar.

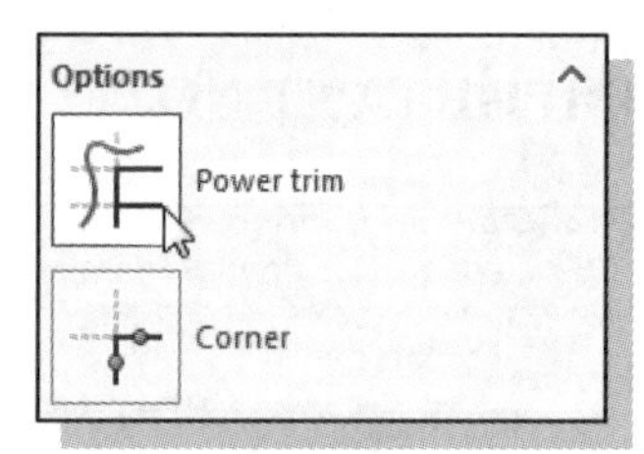

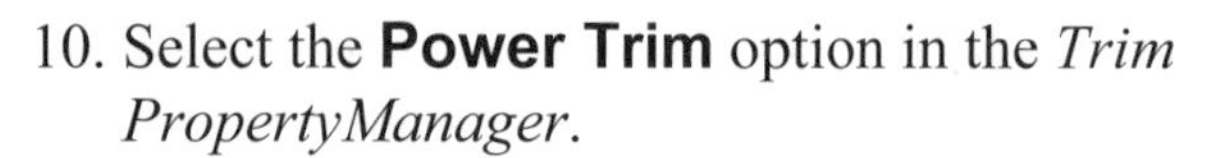

10. Select the **Power Trim** option in the *Trim PropertyManager*.

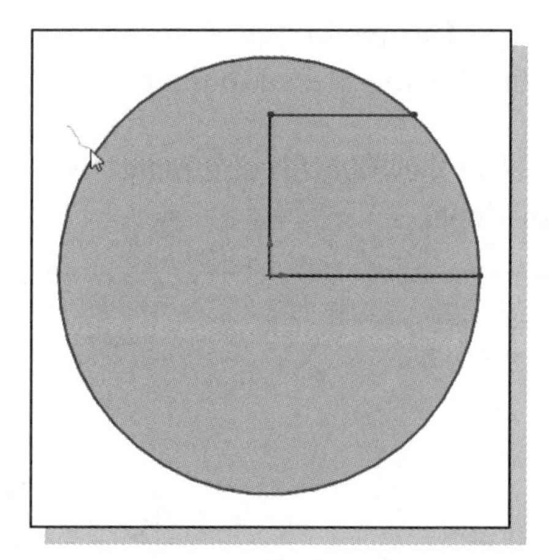

➢ To trim entities, with the Power Trim option, hold down and drag the cursor across the entity.

11. Left-click in the graphics area outside the portion of the circle to be trimmed.

12. Hold down and drag the mouse across the circle to execute the power trim.

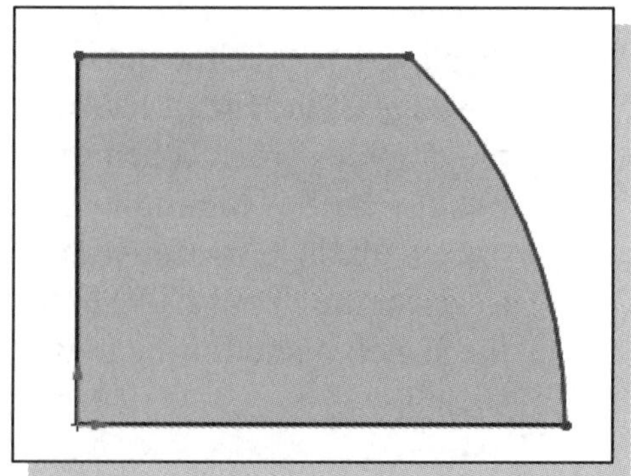

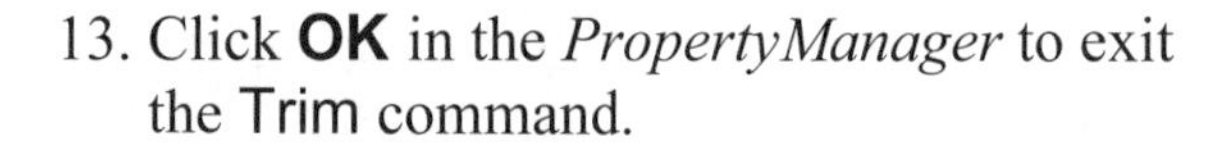

13. Click **OK** in the *PropertyManager* to exit the Trim command.

14. On your own, use the **Smart Dimension** command to create the dimensions as shown in the figure below. (**HINT:** Modify the radius dimension first.)

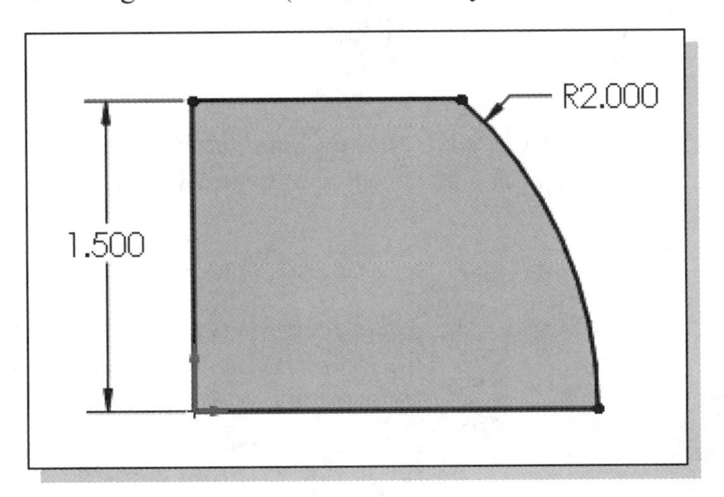

15. Click on the **Sketch Fillet** icon in the *Sketch* toolbar.

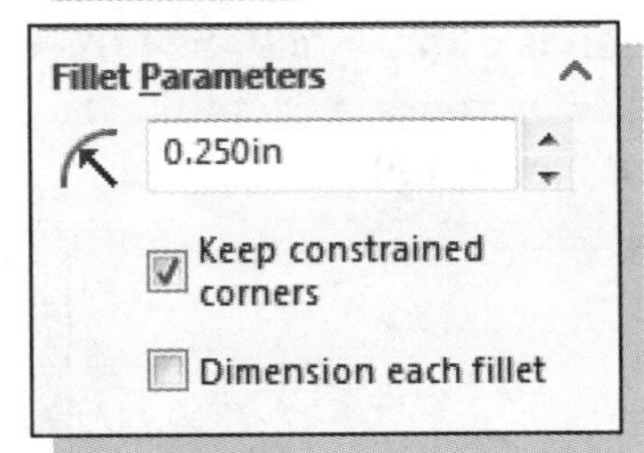

16. In the *Fillet Parameters* panel of the *Sketch Fillet PropertyManager,* set the ***radius*** to **0.25**.

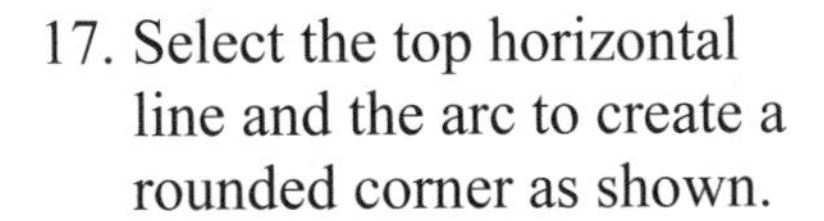

17. Select the top horizontal line and the arc to create a rounded corner as shown.

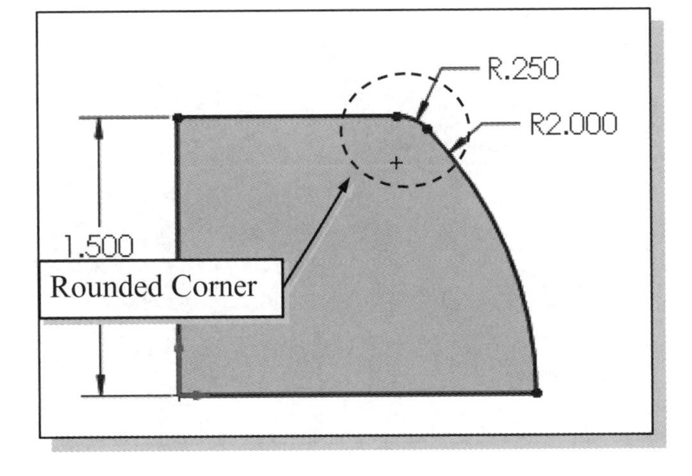

18. Select **OK** in the *PropertyManager* to exit the Sketch Fillet command.

19. Select the **Sketch** icon on the *Sketch* toolbar to exit the Sketch option.

Create a Revolved Boss Feature

1. Make sure the sketch – Sketch1 – is selected in the *FeatureManager Design Tree.*

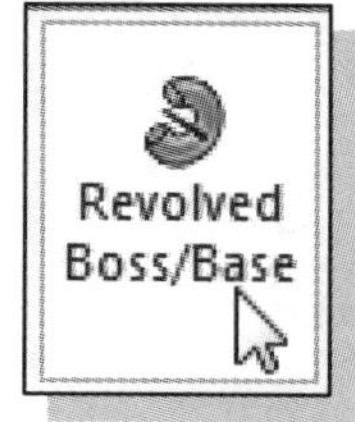

2. Select the **Features** tab on the *CommandManager* to display the *Features* toolbar and select the **Revolved Boss/Base** command.

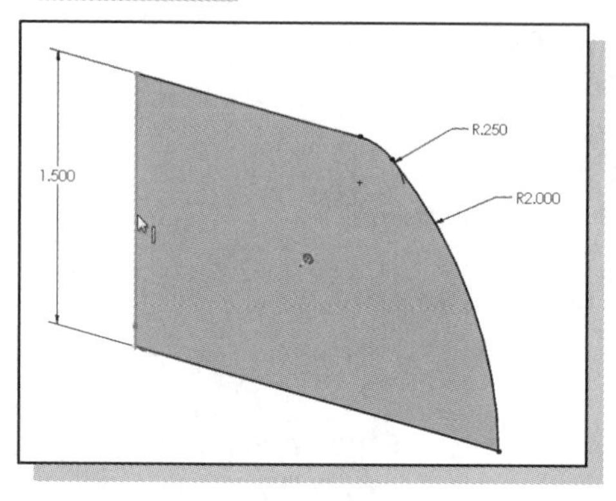

3. The *Revolve PropertyManager* appears. In the *Revolve Parameters* panel, the *Axis of Revolution* box is highlighted. SOLIDWORKS expects us to select the revolution axis for the revolved feature. Select the **vertical edge** of the sketch as the axis of rotation as shown.

4. The default revolution is **360°**. Click the **OK** icon in the *PropertyManager* to accept these parameters and create the revolved feature.

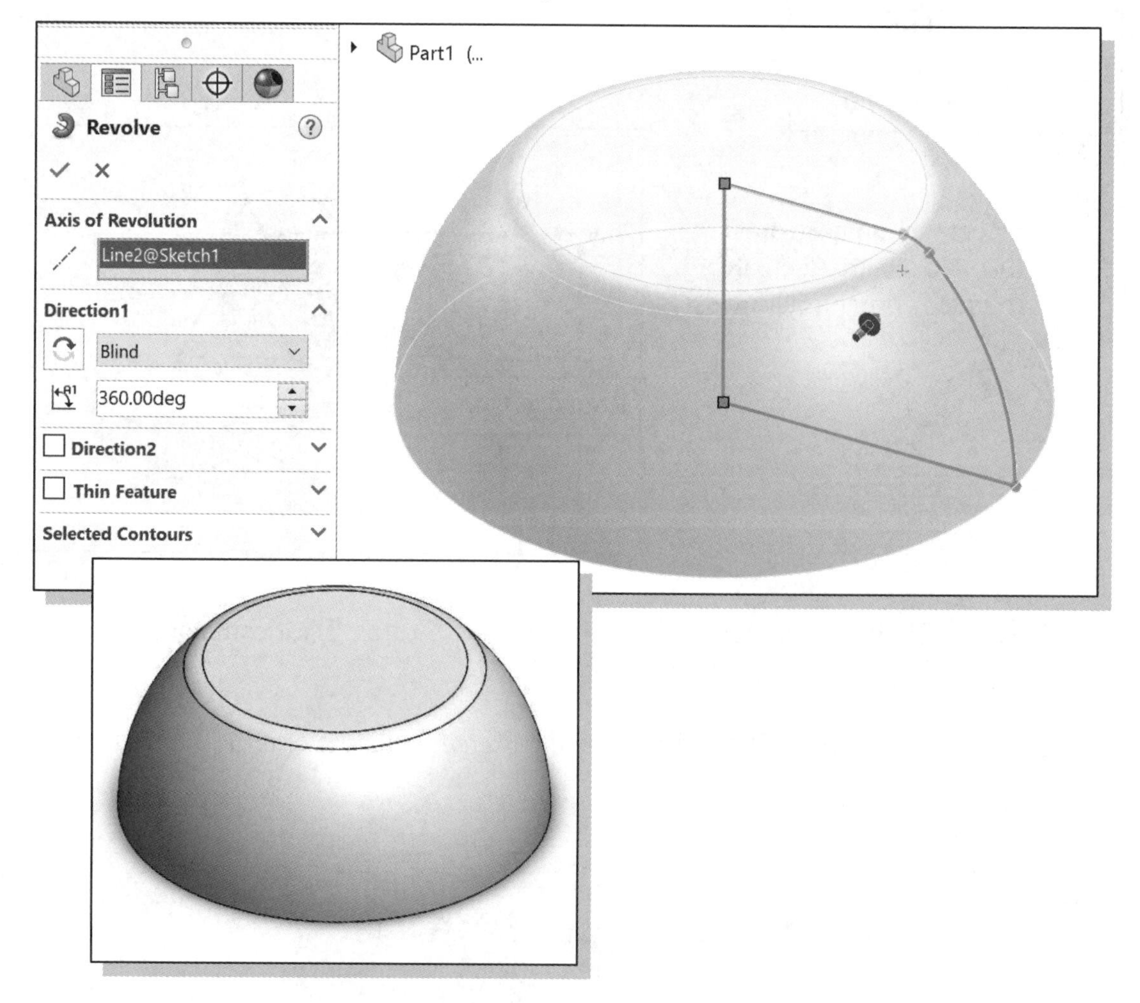

Creating Offset Reference Planes

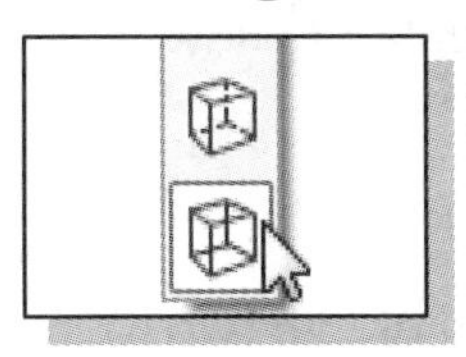

1. In the *Display Style* pull-down menu on the *Heads-up View* toolbar, select the **Wireframe** option to set the display mode to *Wireframe*.

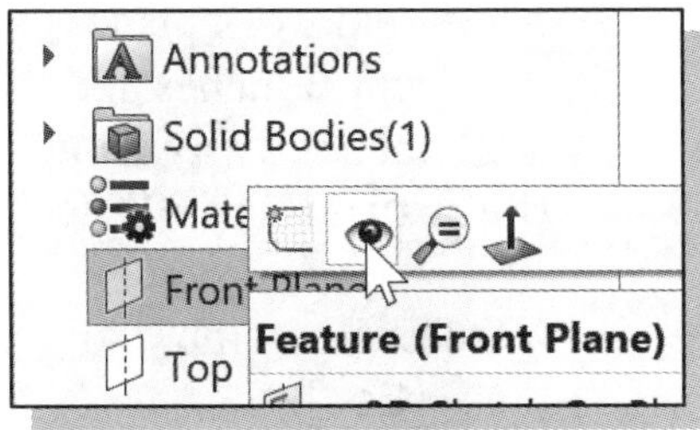

2. In the *FeatureManager Design Tree*, click on the **Front Plane** icon to reveal the pop-up option menu.

3. Select **Show** from the pop-up option menu to turn *ON* the visibility of the Front Plane.

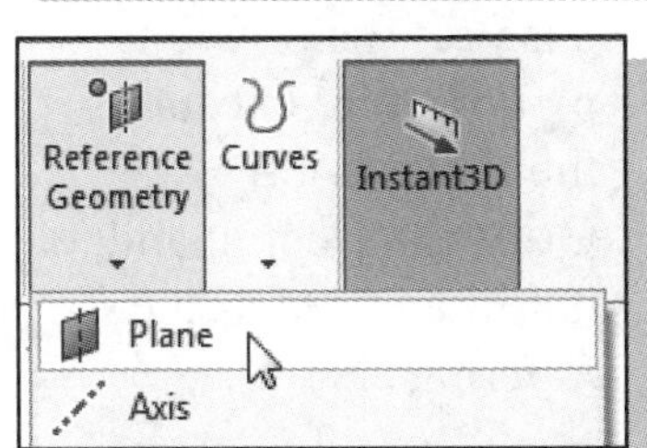

4. Press the **[Esc]** key to ensure that no objects are selected.

5. Select the **Reference Geometry** command from the *Features* toolbar, and select the **Plane** option from the pull-down menu.

6. Select the Front Plane as the *First Reference* entity by clicking on the **Front Plane** in the graphics area.

➢ Notice the **Front Plane** appears in the *First Reference* panel of the *Plane PropertyManager*, and the **Offset Distance** option is automatically selected.

7. Enter **2.5 in** as the offset distance.

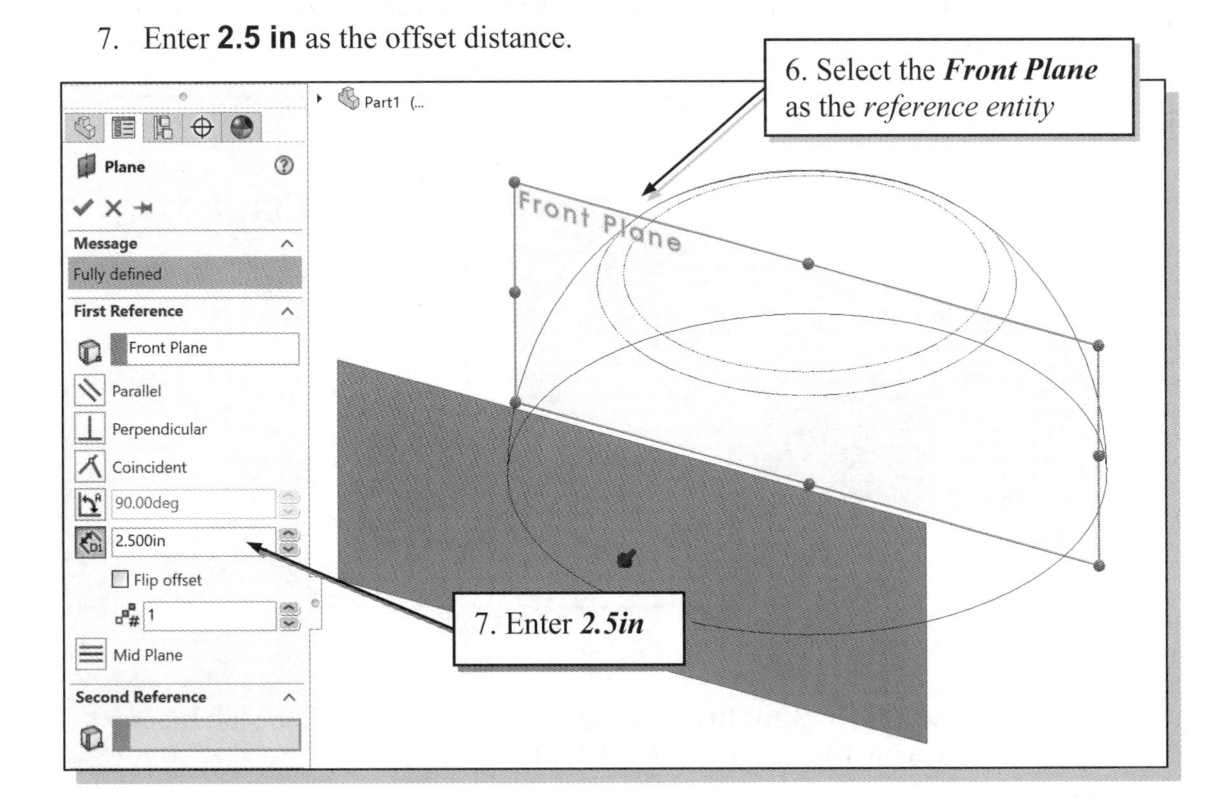

8. Select **OK** in the *PropertyManager* to accept the settings and create the new reference plane.

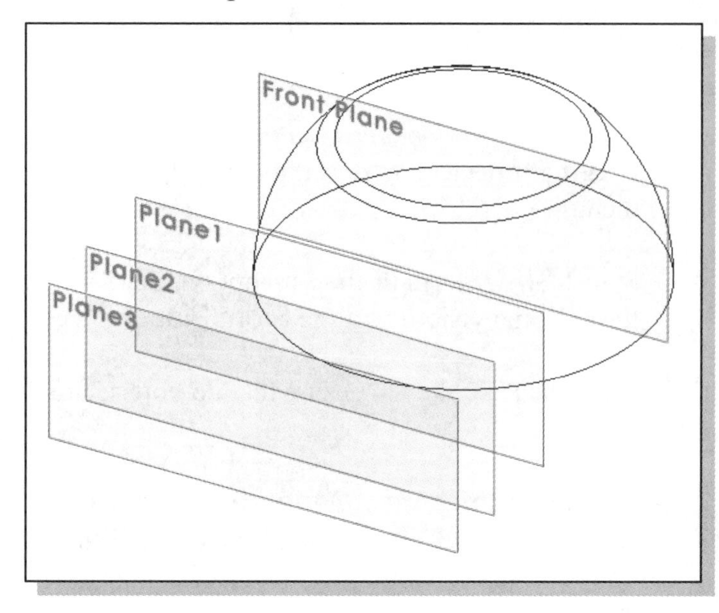

9. On your own, repeat steps 4 - 8 and create two additional work planes that are **3.5** inches and **4.25** inches **away from the Front (XY) Plane**.

➢ **NOTE:** Remember to press [**Esc**] (Step 4) to unselect the previous plane before executing the **Reference Geometry** command, or it will be used as the *First Reference*.

Creating 2D Sketches on the Reference Planes

1. Click in the graphics area, away from the model, to ensure no planes are selected.

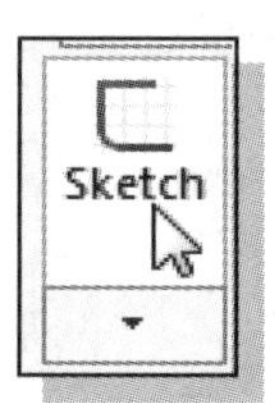

2. In the *Sketch* toolbar select the **Sketch** command by left-clicking once on the icon.

3. Select the **Front Plane** as the sketch plane.

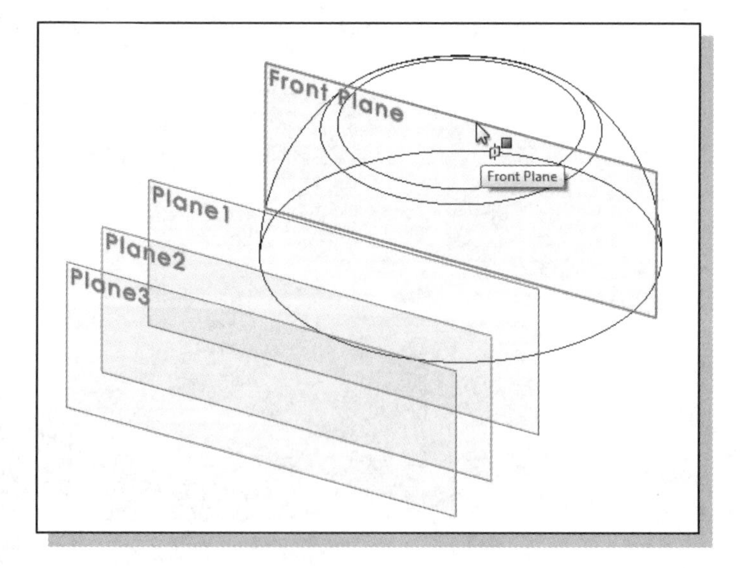

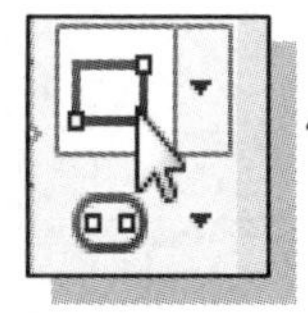

4. Select the **Rectangle** command by clicking once with the left-mouse-button on the icon in the *Sketch* toolbar.

5. Click on the projected *Origin* to align the first corner of the rectangle. (**NOTE:** If the origin is not visible, use the *Hide/Show* menu on the *Heads-up View* toolbar to turn the visibility *ON*.)

6. Create a rectangle of arbitrary size by selecting a location that is toward the right side of the graphics window as shown.

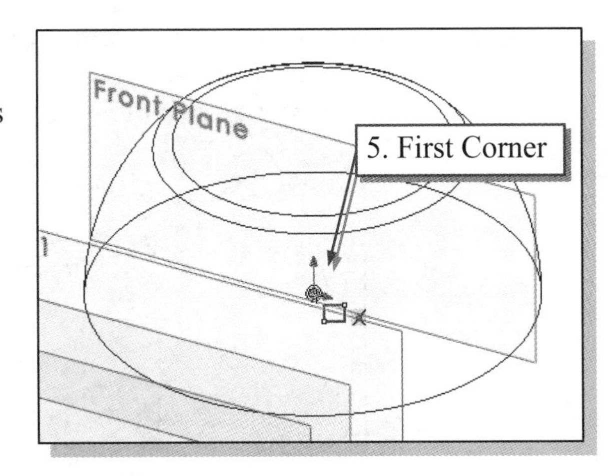

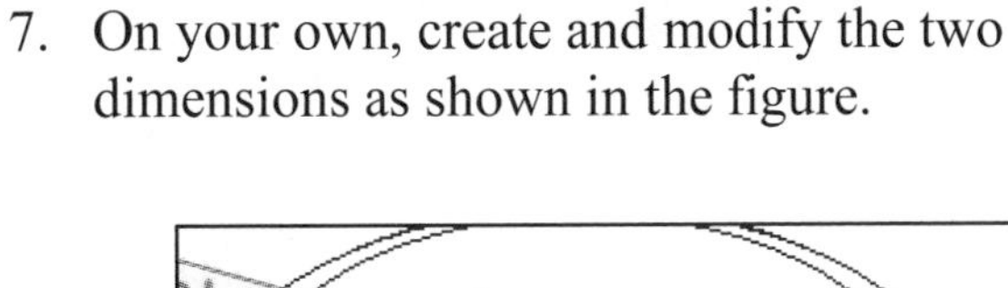

7. On your own, create and modify the two dimensions as shown in the figure.

8. On your own, use the **Sketch Fillet** command to create two rounded corners (**radius 0.25**) as shown.

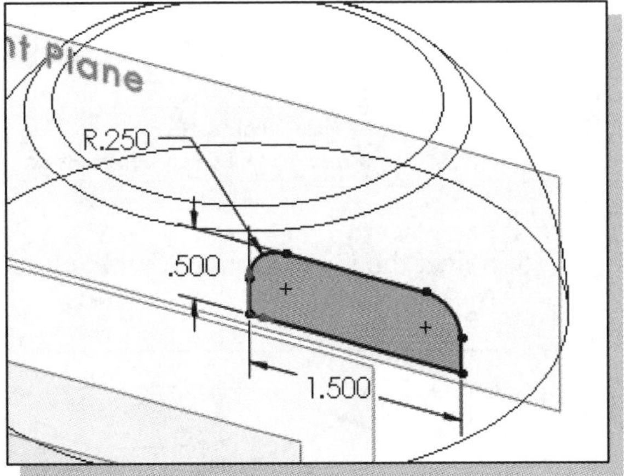

9. Select the **Exit Sketch** icon on the *Sketch* toolbar to exit the **Sketch** option.

10. Click in the graphics area, away from the model, to ensure no planes are selected.

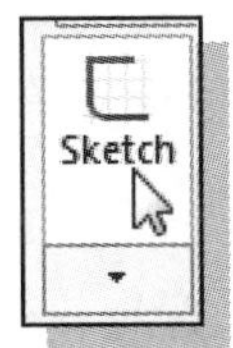

11. In the *Menu Bar* select the **Sketch** command by left-clicking once on the icon.

12. Select **Plane1**, by left-clicking once on any edge of the first offset work plane in the graphics window as shown.

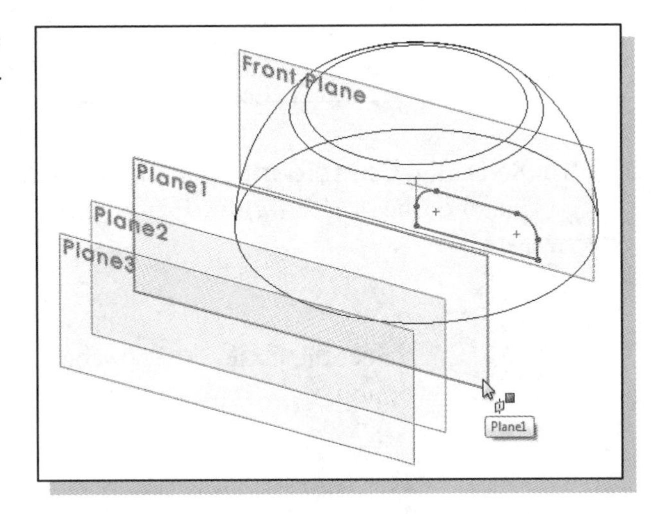

13. On your own, use the **Rectangle**, **Smart Dimension**, and **Sketch Fillet** commands to create the sketch as shown. Do **not** align the sketch to the origin.

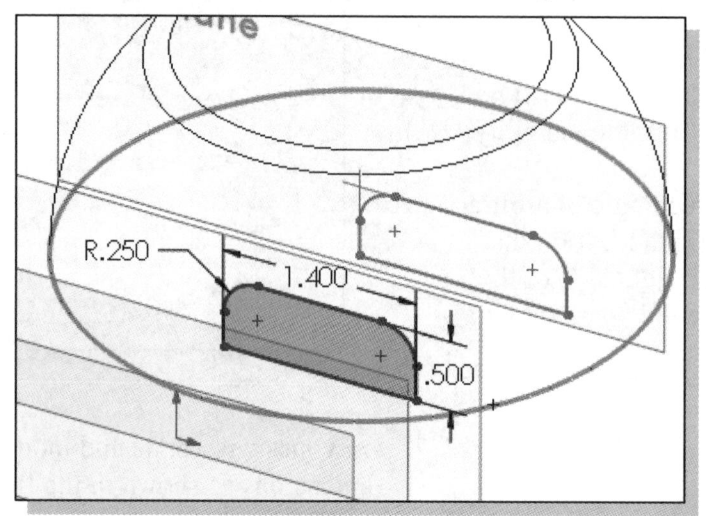

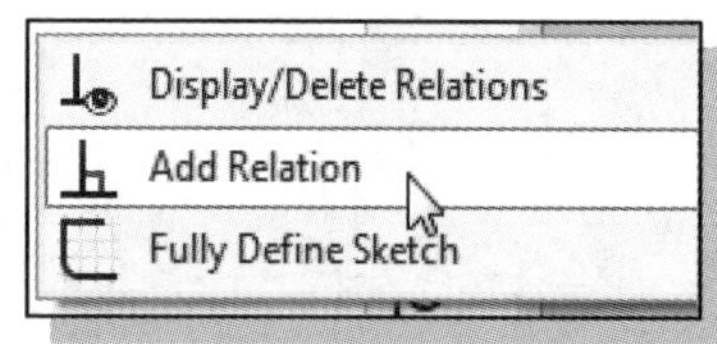

14. We will now align the lower right corners of the two sketches. Select the **Add Relation** command from the *Display/ Delete Relations* pop-up menu on the *Sketch* toolbar.

15. Select the lower right corner each of the sketches we just created, as shown in the figure.

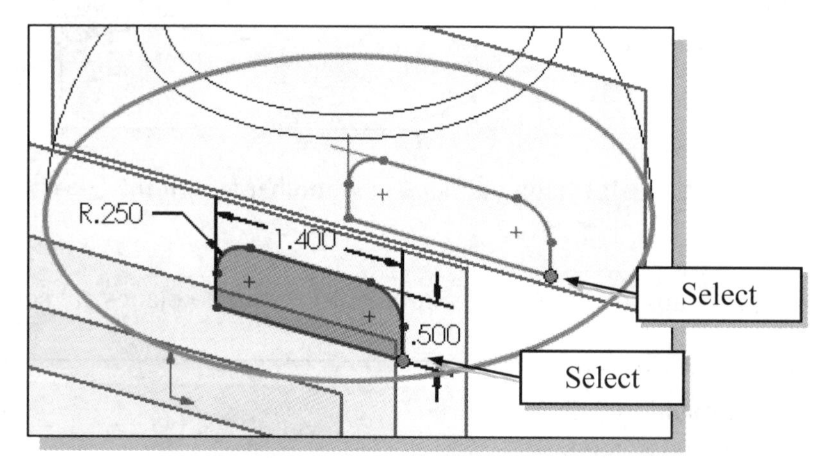

16. Select the **Coincident** option in the *Add Relations* panel of the *Add Relations PropertyManager*.

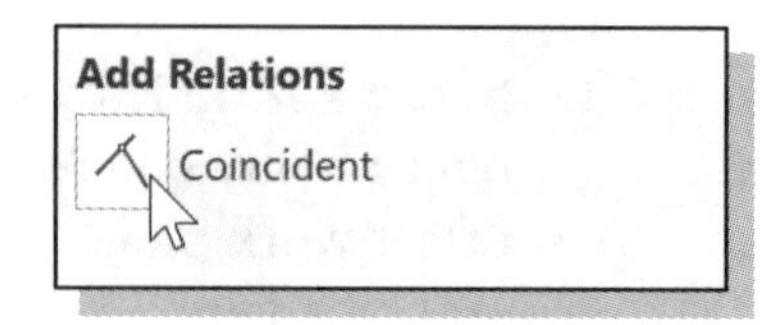

17. Select the **Exit Sketch** icon on the *Sketch* toolbar to exit the Sketch option.

18. On your own, repeat the above steps and create two additional sketches on **Plane2** and **Plane3** as shown in the figure below. Align the **lower right corners** of the four sketches in the Z-direction through the use of coincident relations. (**NOTE:** Remember to ensure the previous plane is not selected prior to beginning a new sketch.)

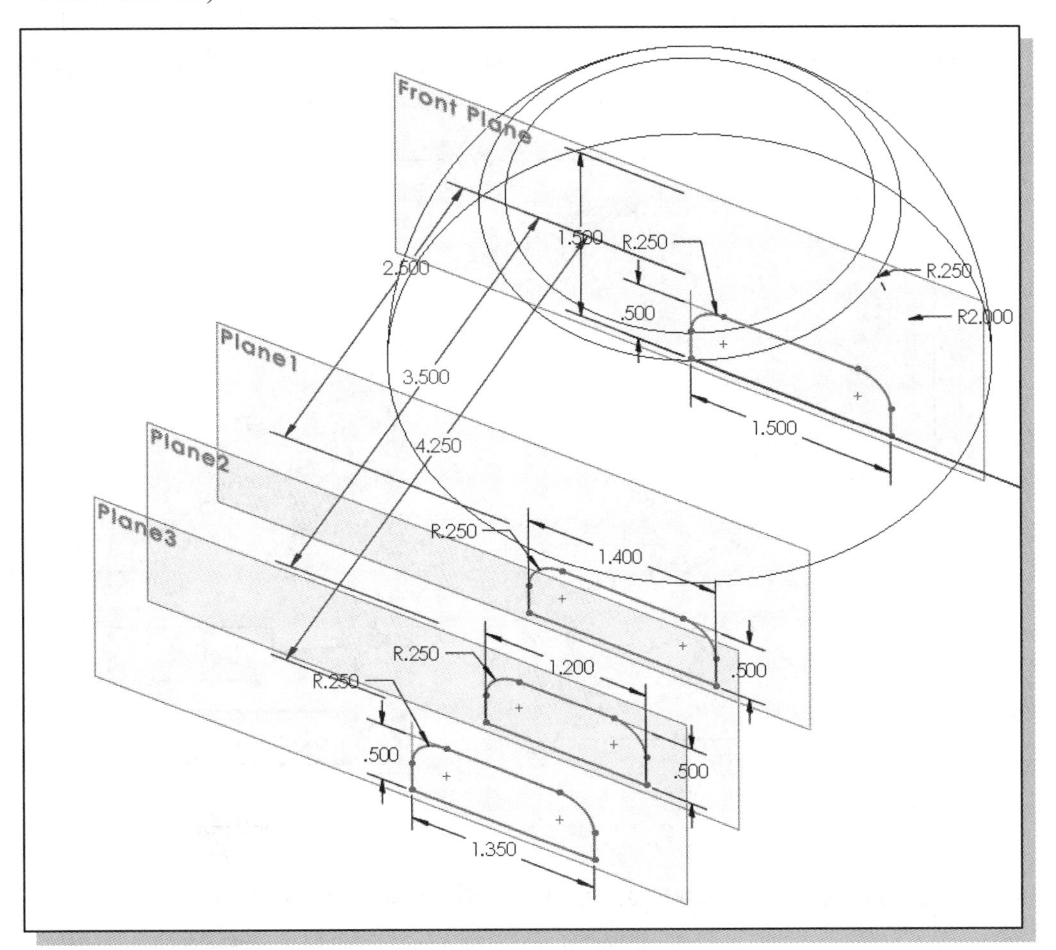

Creating a Lofted Feature

The **Loft** option allows us to blend multiple profiles with varying shapes on separate planes to create complex shapes. Profiles are usually on parallel planes, but any non-perpendicular planes can also be used. We can use as many profiles as we wish, but to avoid twisting the loft shape, we should map points on each profile that align along a straight vector.

1. Click in the graphics area, away from the model, to ensure no planes are selected.

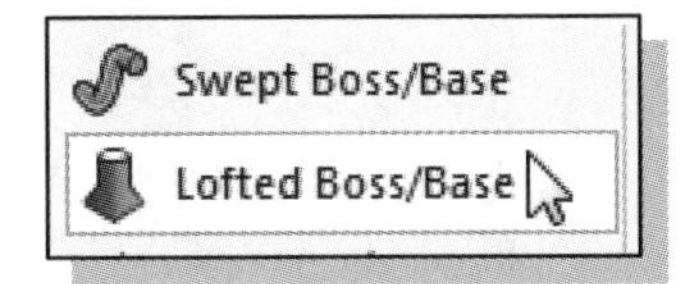

2. In the *Features* toolbar, left-click on the **Lofted Boss/Base** command. The *Loft PropertyManager* is displayed in the left panel.

3. In the *Loft PropertyManager*, the **Profiles** option is activated. SOLIDWORKS expects us to select a number of existing profiles, which will be used to create the lofted feature.

4. Pick the four sketched sections, in the order that they were created, by clicking on the lower right corners of the sketches. Three arrows are displayed showing the blending direction of the sketches. Note that, to avoid twisting the loft shape, we are mapping the same corner points on the sketches to align along a straight vector.

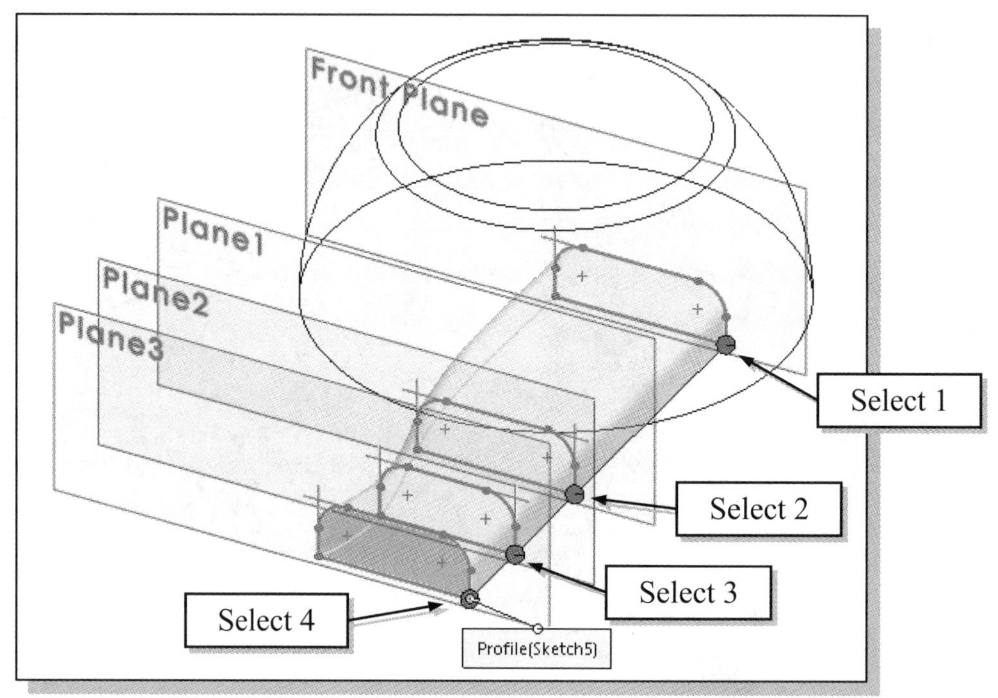

5. Click on the **OK** button to accept the settings and create the lofted feature.

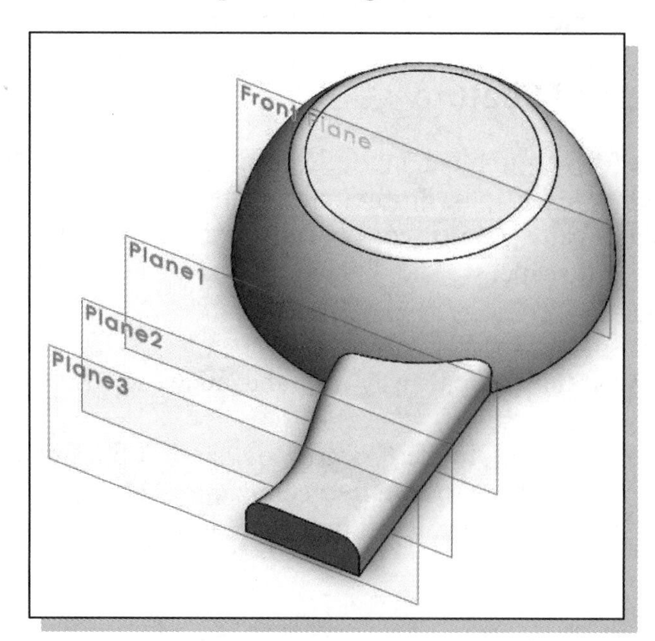

Creating an Extruded Boss Feature

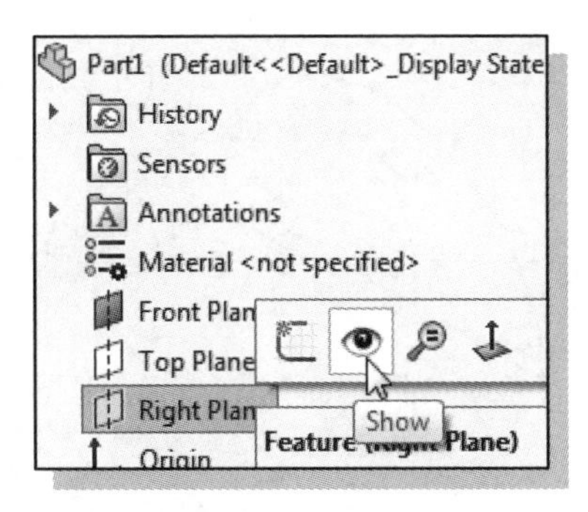

1. In the *FeatureManager Design Tree*, click on the **Right Plane** icon to reveal the pop-up option menu.

2. Select **Show** from the pop-up option menu to turn *ON* the visibility of the Right Plane.

3. Click in the graphics area, away from the model, to ensure no planes are selected.

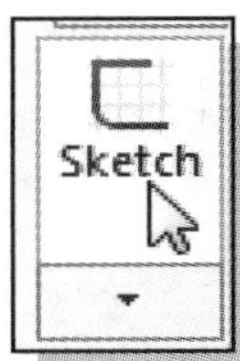

4. In the *Sketch* toolbar select the **Sketch** command by left-clicking once on the icon.

5. Select the **Right (YZ) Plane** as the sketch plane.

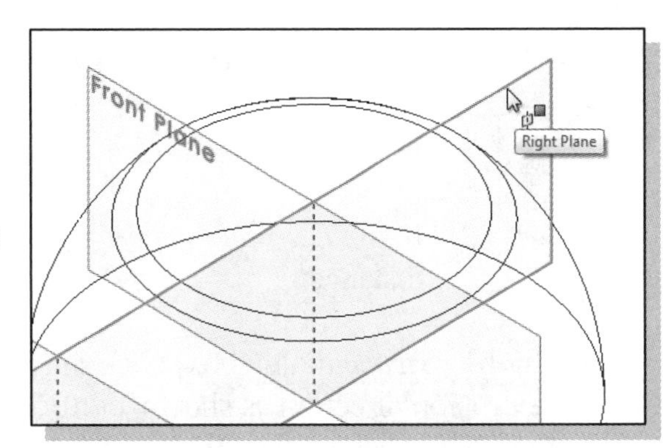

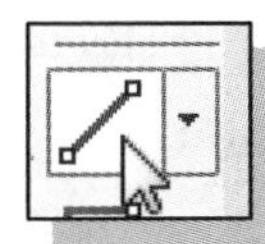

6. Click on the **Line** icon in the *Sketch* toolbar.

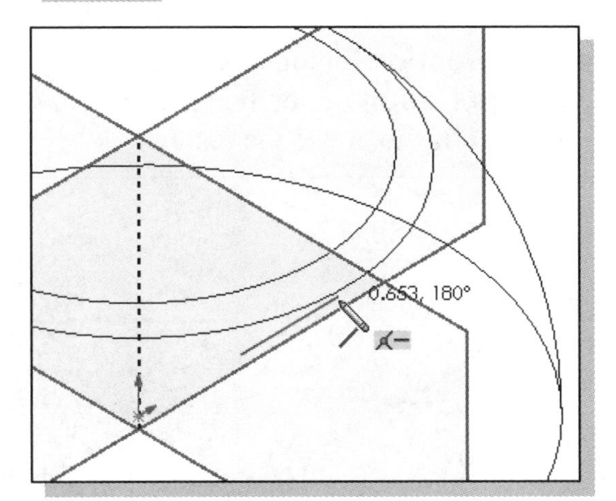

7. Create a line along the horizontal axis as shown.

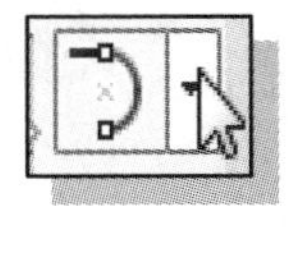

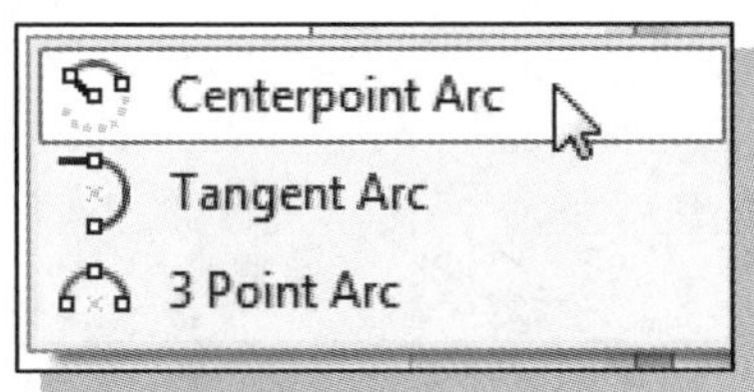

8. Select the **Centerpoint Arc** command in the *Sketch* toolbar as shown.

9. On your own, create an arc aligned to the **mid-point** and **endpoints** of the previously created line as shown in the figure.

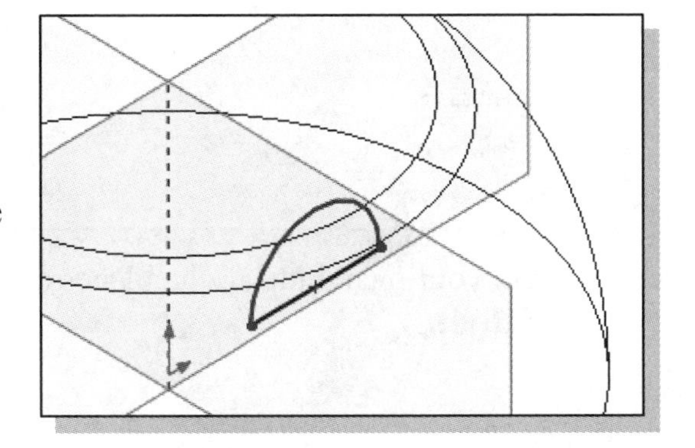

10. On your own, create and modify the dimensions as shown in the figure. The radius of the arc is **0.75 in** and the distance from the origin to the midpoint of the line is **1.125 in**.

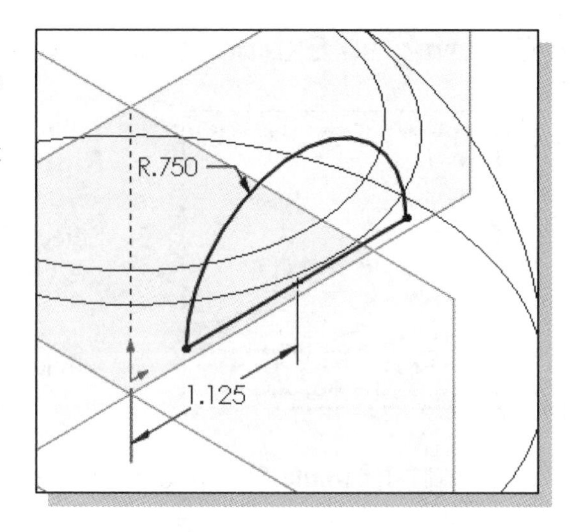

11. Select the **Exit Sketch** icon on the *Sketch* toolbar to exit the **Sketch** option.

Completing the Extruded Boss Feature

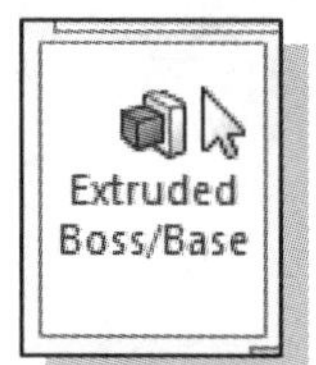

1. In the *Features* toolbar, select the **Extruded Boss/Base** command.
2. In the *Extrude PropertyManager* window, enter **5.5** as the extrusion distance.
3. In the *Extrude* dialog box, click the **Reverse Direction** button to set the extrusion direction as shown, confirm the **Merge Result** option box is checked, and then click on the **OK** button to accept the settings to create the feature.

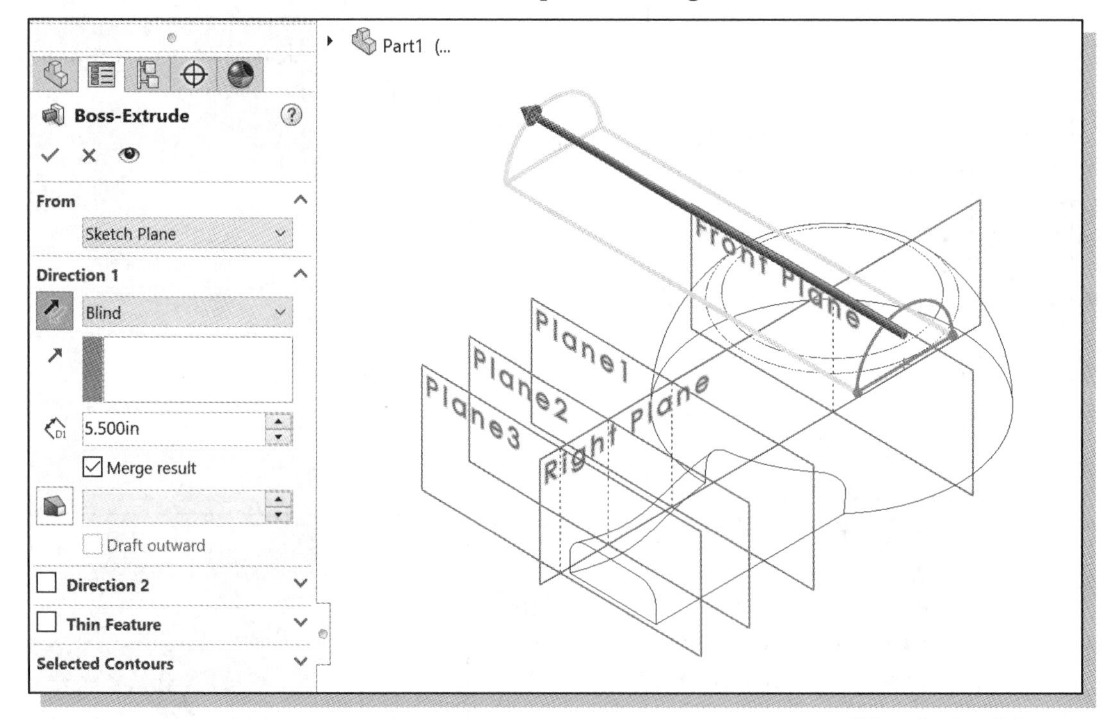

4. On your own, hide all the planes and change the display style to **Shaded with Edges**.

Creating 3D Rounds and Fillets

1. In the *Features* toolbar, select the **Fillet** command by left-clicking once on the icon.

2. In the *Fillet PropertyManager*, set the ***Radius*** option to a radius of **0.15 in** as shown below.

3. Click on the **three edges** as shown.

4. Click on the **OK** button to accept the settings and create the 3D rounds and fillets.

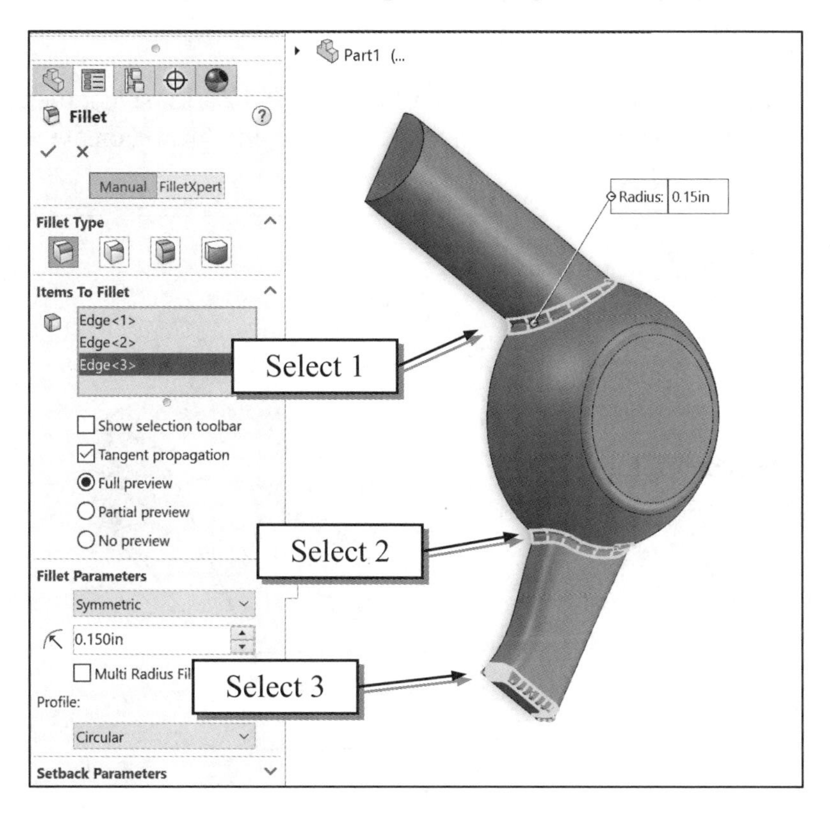

Creating a Shell Feature

The **Shell** command can be used to hollow out the inside of a solid, leaving a shell of specified wall thickness.

1. In the *Features* toolbar, select the **Shell** command by left-clicking once on the icon.

2. On your own, use the **arrow** keys to display the back faces of the model as shown below. (Use the **arrow** keys to rotate the view vertically or horizontally. Hold down [Alt] key and use the **left-right arrow** keys to rotate clockwise or counterclockwise.)

3. In the *Shell PropertyManager*, set the *Thickness* to **0.125 in**, select the **two faces** as shown, and click on the **OK** button to accept the settings and create the shell feature.

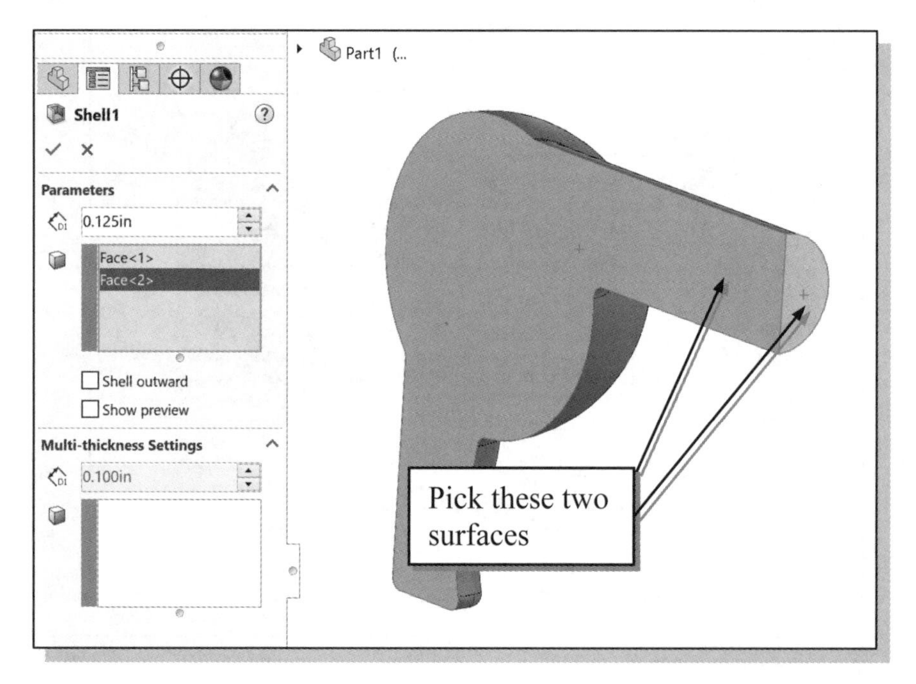

Creating a Rectangular Extruded Cut Feature

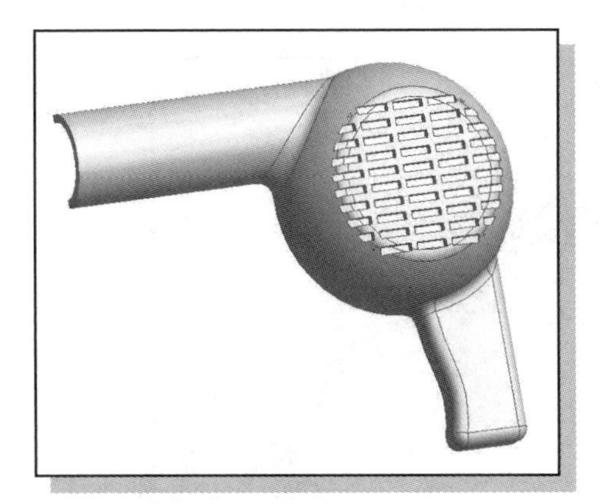

The *Dryer Housing* design requires the placement of identical holes on the top face of the solid. Instead of creating the holes one at a time, we can simplify the creation of these holes by using the **Linear Pattern** command to create duplicate features. Prior to using the Linear Pattern command, we will first create a ***pattern leader***, which is a regular cut feature.

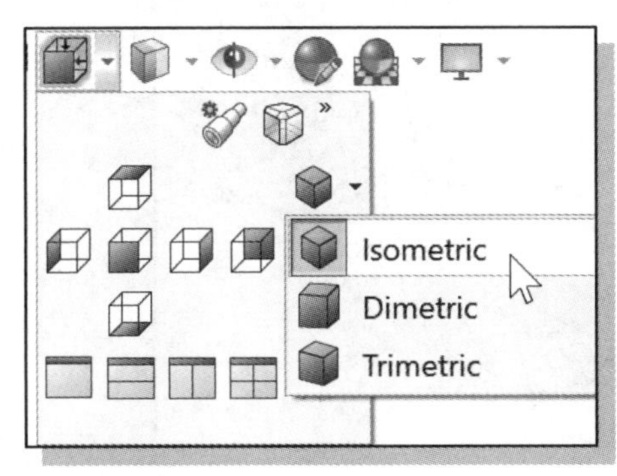

1. In the *View Orientation* pull-down menu on the *Heads-up View* toolbar, select the **Isometric** option.

2. In the *Sketch* toolbar select the **Sketch** command by left-clicking once on the icon.

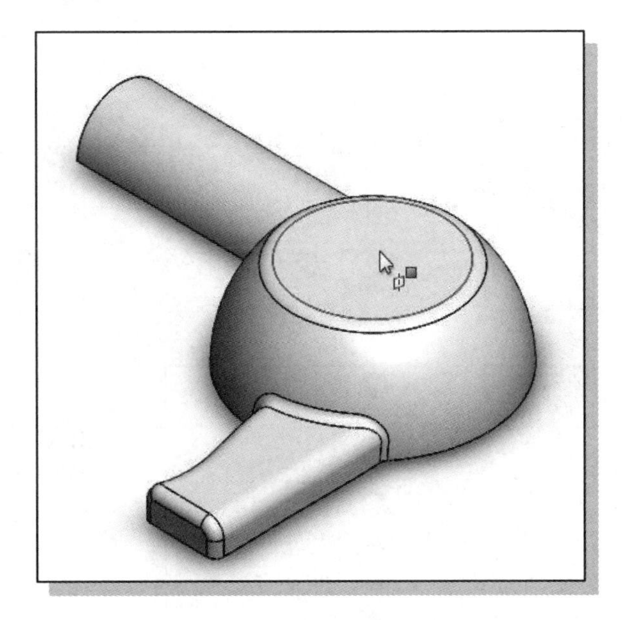

3. Pick the top face of the base feature as shown.

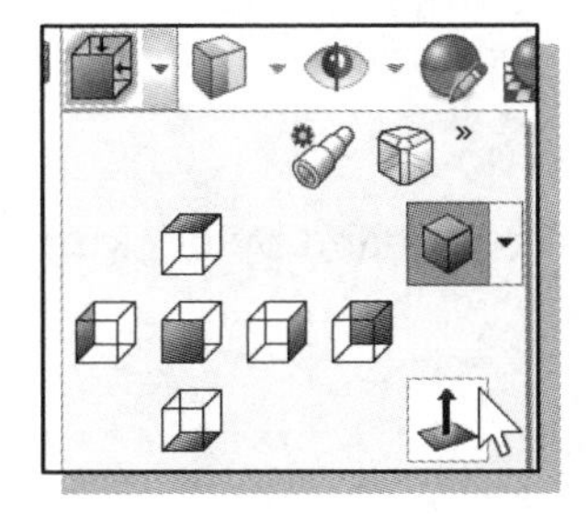

4. In the *View Orientation* pull-down menu on the *Heads-up View* toolbar, select the **Normal To** option.

5. On your own, create the **Rectangle** and add **Smart Dimensions** as shown in the figure below.

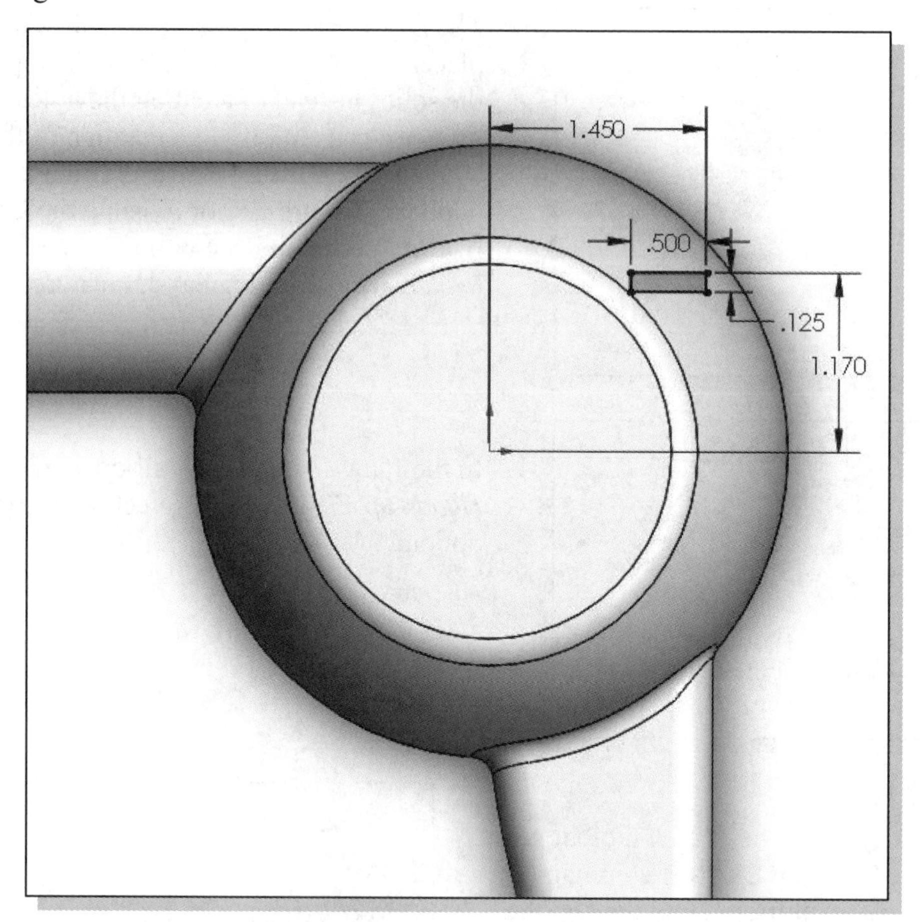

6. Select the **Exit Sketch** icon on the *Sketch* toolbar to exit the Sketch option.

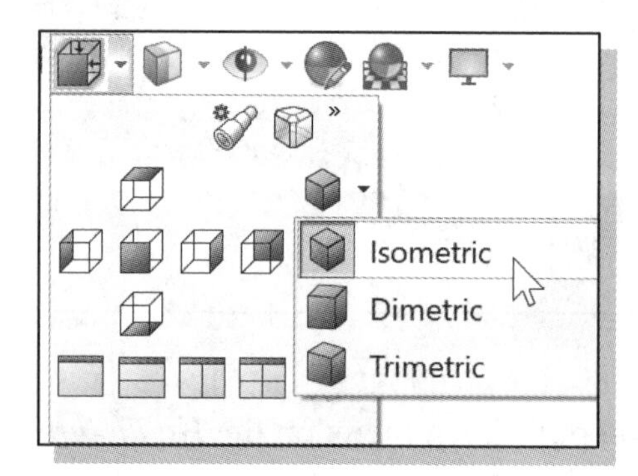

7. In the *View Orientation* pull-down menu on the *Heads-up View* toolbar, select the **Isometric** option.

8. In the *Features* toolbar, select the **Extruded Cut** command by clicking once with the left-mouse-button on the icon.

9. Create a Blind cut extrusion with a distance of **0.13 in**.

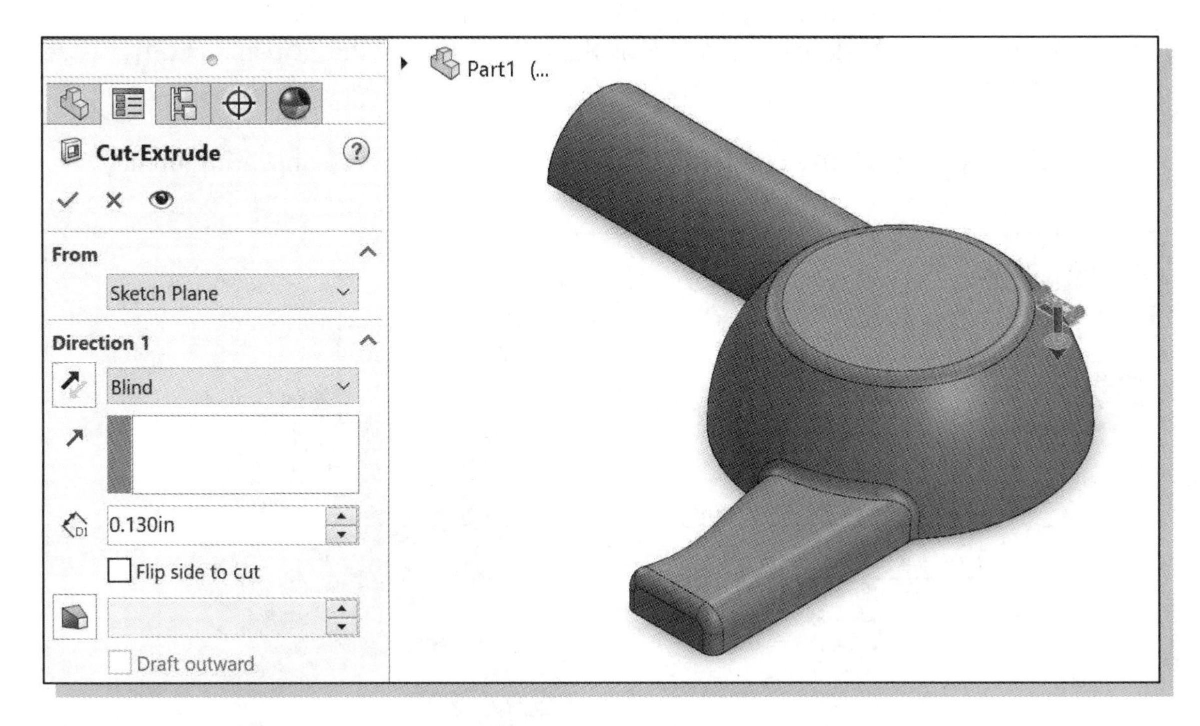

10. Select **OK** in the *PropertyManager* to create the extruded cut.

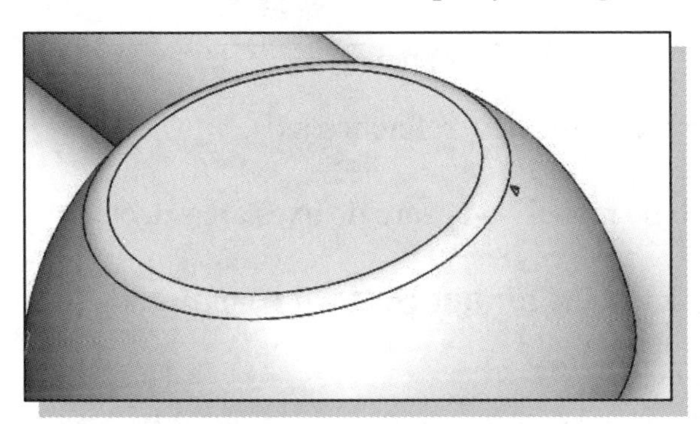

- Note that the pattern leader creates a fairly small cut on the solid model.

Creating a Linear Pattern

In SOLIDWORKS, existing features can be easily duplicated. The **Linear Pattern** command allows us to create rectangular arrays of features. The patterned features are parametrically linked to the original feature; any modifications to the original feature are also reflected on the arrayed features. An edge or an axis can be selected to define a pattern direction. We will create a reference axis and create a rectangular pattern using an edge to define one direction and the reference axis to define the other.

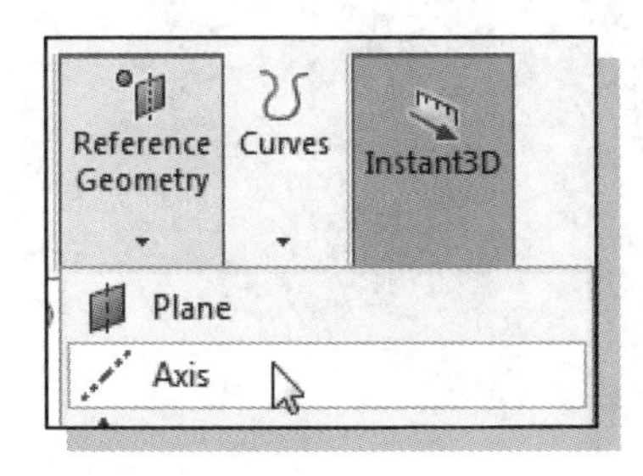

1. Select the **Reference Geometry** command from the *Features* toolbar, and select the **Axis** option from the pull-down menu.

2. In the *Selections* panel of the *Axis PropertyManager*, select the **Two Planes** option. (We will define the Z-axis as the intersection of the Top and Right Planes.)

3. Select the **Top Plane** and **Right Plane** in the *Design Tree* appearing in the graphics area.

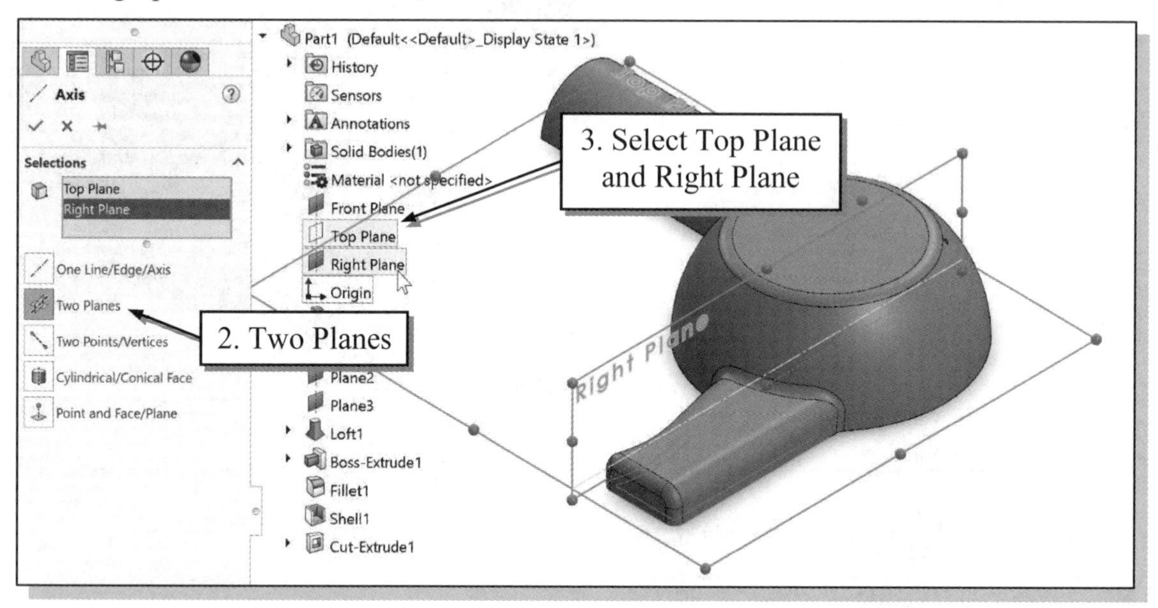

4. Click **OK** in the *PropertyManager* to create the new reference axis.

5. Click in the graphics area, away from the model, to ensure no axes are selected.

6. In the *Features* toolbar, select the **Linear Pattern** command by left-clicking once on the icon.

7. In the *Linear Pattern PropertyManager*, notice the *Pattern Direction* box in the *Direction 1* panel is activated. Click on an **edge** along the swept feature as shown in the figure.

➢ Notice Edge<1> appears in the *Pattern Direction* box.

8. In the *Direction 1* panel of the *PropertyManager*, click the **Reverse Direction** button (if necessary) to align the direction as shown in the figure.

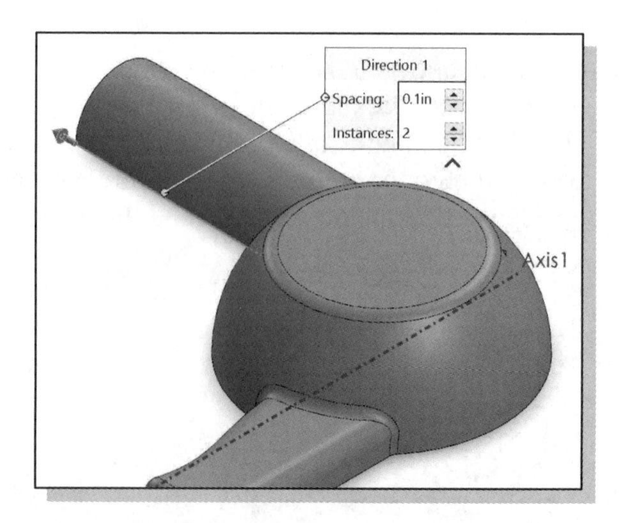

9. In the *Direction 1* panel, select the **Spacings and Instances** option and enter **0.6 in** for the *Spacing* and **5** for the *Number of Instances*.

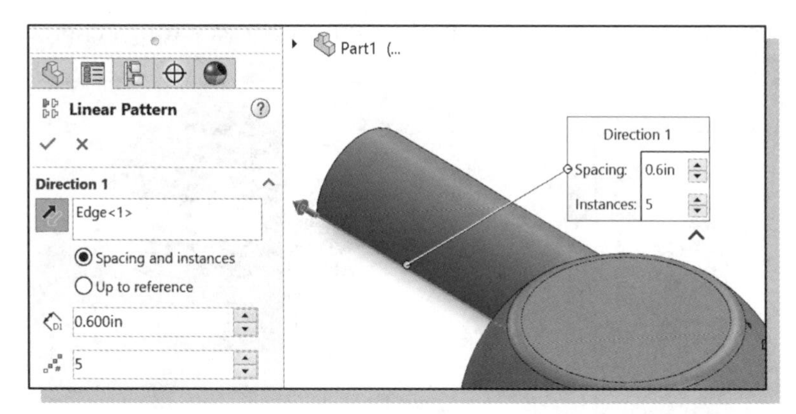

10. In the *Linear Pattern PropertyManager*, notice the *Pattern Direction* box in the *Direction 2* panel is now activated. Select the **reference axis** we just created by expanding the *design tree* in the graphics area and clicking on the **Axis1** icon.

➢ Notice Axis1 appears in the *Pattern Direction* box.

11. Check that the direction is aligned as shown in the figure below. **If necessary**, click the **Reverse Direction** button in the *Direction 2* panel of the *PropertyManager* to align the direction.

12. In the *Direction 2* panel, select the **Spacings and Instances** option and enter **0.25 in** for the *Spacing* and **10** for the *Number of Instances*.

13. In the *Linear Pattern PropertyManager*, notice the *Features to Pattern* box is now activated. Select **Cut-Extrude1**, the cut feature created in the last section, in the *Design Tree* appearing in the graphics area.

➢ Notice **Cut-Extrude1** appears in the *Features to Pattern* panel in the *Linear Pattern PropertyManager*.

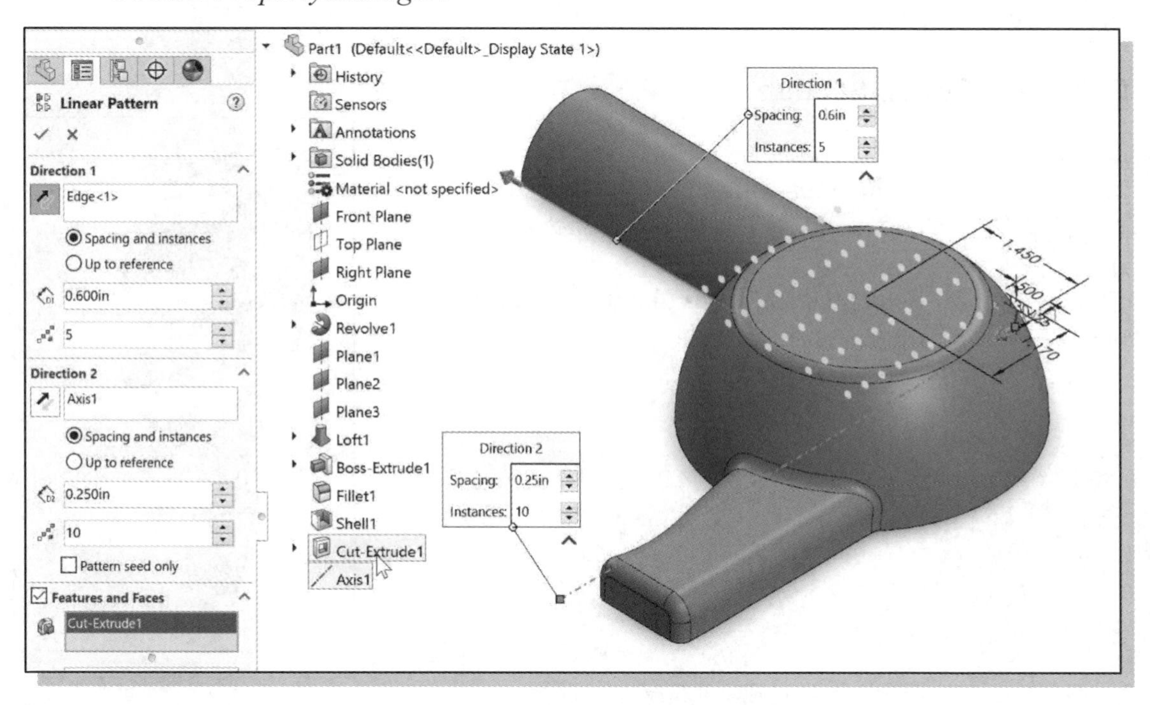

14. Click on the **OK** button to accept the settings and create the *linear pattern.*

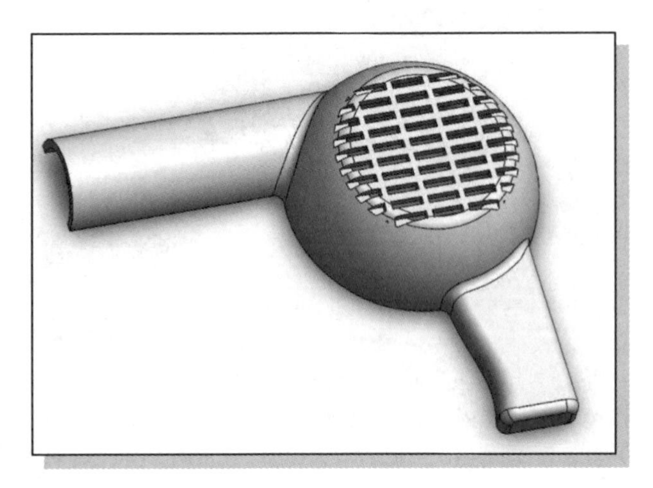

Creating a Swept Feature

The **Sweep** operation is defined as moving a planar section through a planar (2D) or 3D path in space to form a three-dimensional solid object. The path can be an open curve or a closed loop but must be on an intersecting plane with the profile. The **Extrusion** operation, which we have used in the previous lessons, is a specific type of sweep. The **Extrusion** operation is also known as a ***linear sweep*** operation, in which the sweep control path is always a line perpendicular to the two-dimensional section. Linear sweeps of unchanging shape result in what are generally called ***prismatic solids***, which means solids with a constant cross-section from end to end. In SOLIDWORKS, we create a ***swept feature*** by defining a 2D sketch of a cross section and a planar path. To create the path, we can use a sketch, existing model edges, or curves. The sketched profile is then swept along the planar path. The Sweep operation is used for objects that have uniform shapes along a trajectory. We will define a **Cut Sweep** feature using existing model edges to define the path.

Define the Sweep Section

1. In the *Menu Bar* select the **Sketch** command by left-clicking once on the icon.

2. Select the small circular surface of the model as shown.

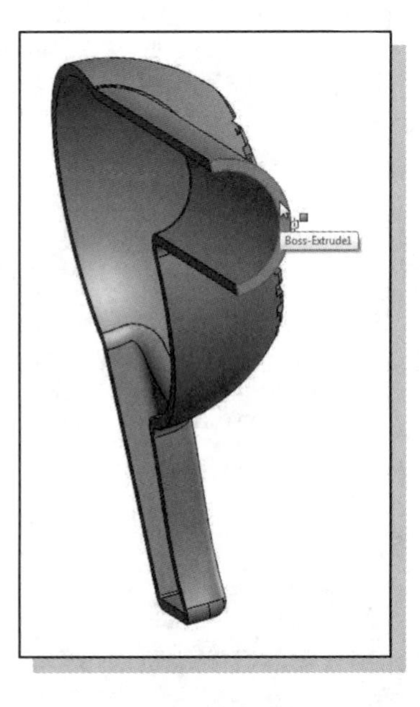

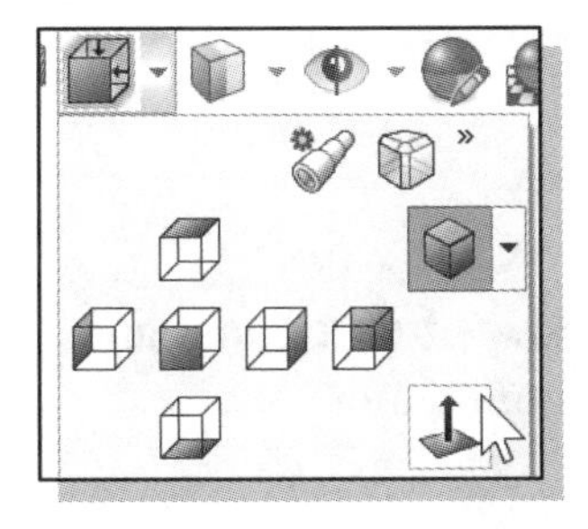

3. In the *View Orientation* pull-down menu on the *Heads-up View* toolbar, select the **Normal to** option.

4. On your own, create the **Rectangle** and add **Smart Dimensions** as shown in the figure below.

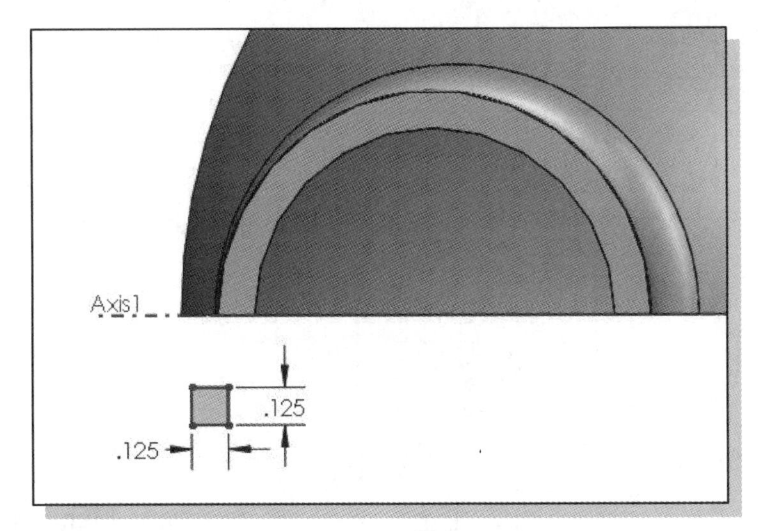

5. On your own, use the **Add Relation** command to add a **Midpoint** relation between the bottom edge of the rectangle and the corner of the half-cylinder, as shown below.

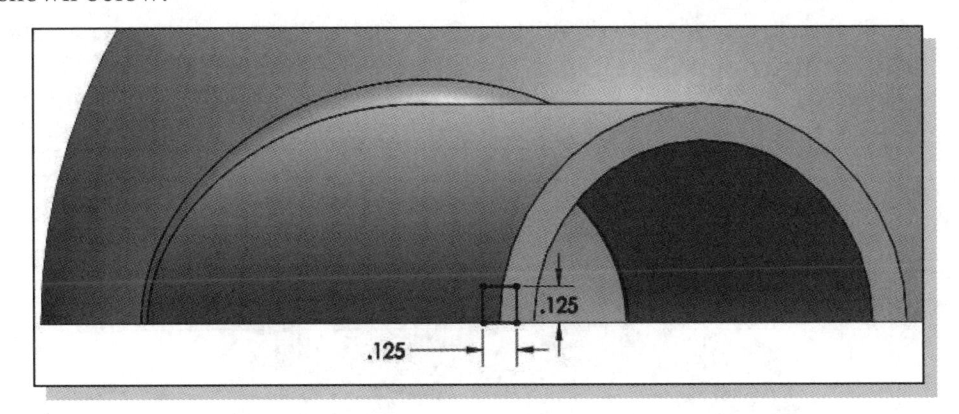

❖ **NOTE:** The sketched profile to be swept has been located in a manner to overlap the edge which will be selected for the sweep path. In theory this is not necessary. However, since our sweep path includes small radius curves (the fillets), rounding of calculated values sometimes leads to a thin shell being left on the curves of the swept path. Our sketch will cut a 0.0625 x 0.125 inch rectangular extrusion along the path. The 0.0625 inch overlap will ensure no shell is left.

6. The sketch should be Fully Defined as indicated in the *Status Bar* area. Select the **Exit Sketch** icon on the *Sketch* toolbar to exit the Sketch option.

Create the Swept Feature

1. Press the **[Esc]** key to ensure that no objects are selected.

2. In the *Features* toolbar, select the **Swept Cut** command. This will execute the **Cut-Sweep** command.

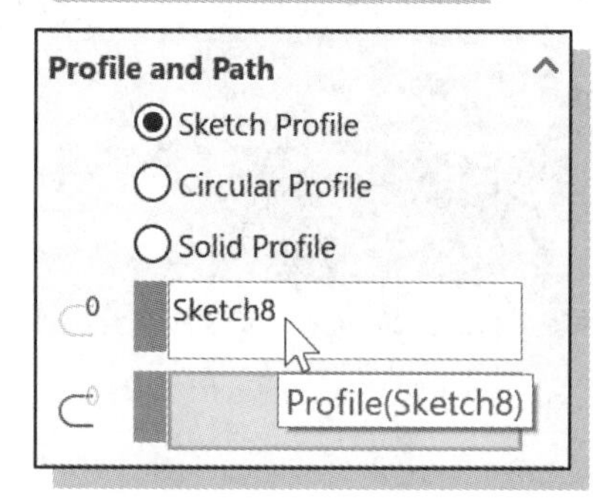

3. Select the new sketch from the *design tree* as the profile for the swept feature.

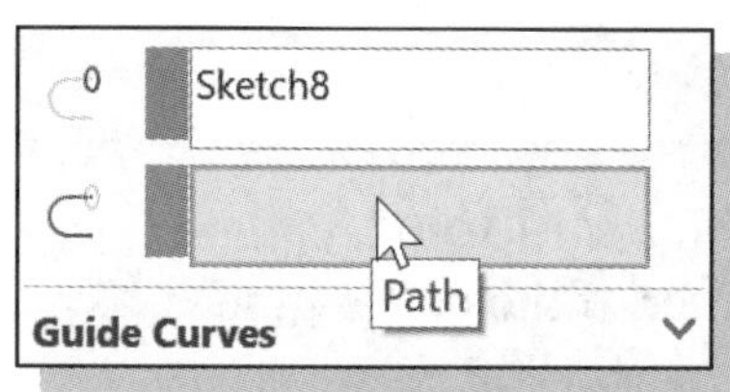

4. If necessary, highlight the *Path* selection box in the *Cut-Sweep PropertyManager* by clicking once with the **left-mouse-button**.

5. Move the cursor over the outer edge of the half-cylinder as shown and click once with the **right-mouse-button** to open the pop-up option menu. (NOTE: Do not select the edge – i.e., do not click the left mouse button.)

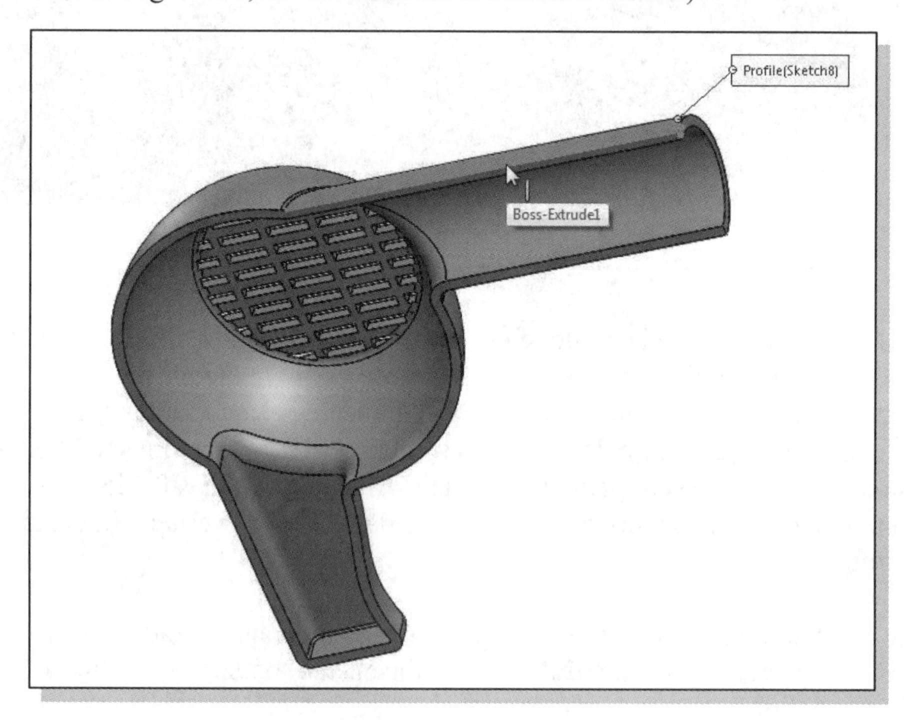

6. Select the **Selection-Manager** option from the pop-up menu.

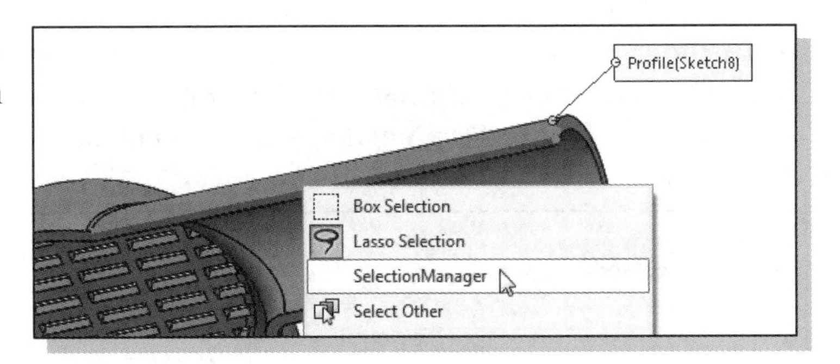

➢ The *SelectionManager* toolbar appears in the graphics area. The SOLIDWORKS **SelectionManager** enhances the selection of edges or guide curves for the creation of lofted or swept features. On your own, use the SOLIDWORKS Help option to view a description of the *SelectionManager* tools. We will use the **Select Group** tool.

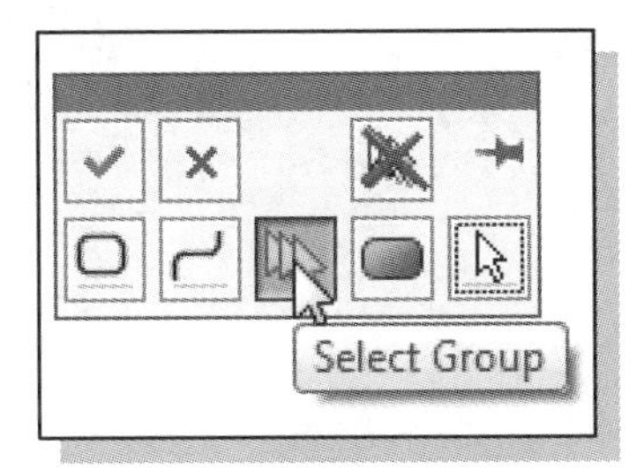

7. Select the **Select Group** icon on the *SelectionManager* as shown.

➢ Notice a preview of the cut-sweep path appears along the selected edge. A ***Tangent*** callout appears, indicating the next edges in the potential group are tangent.

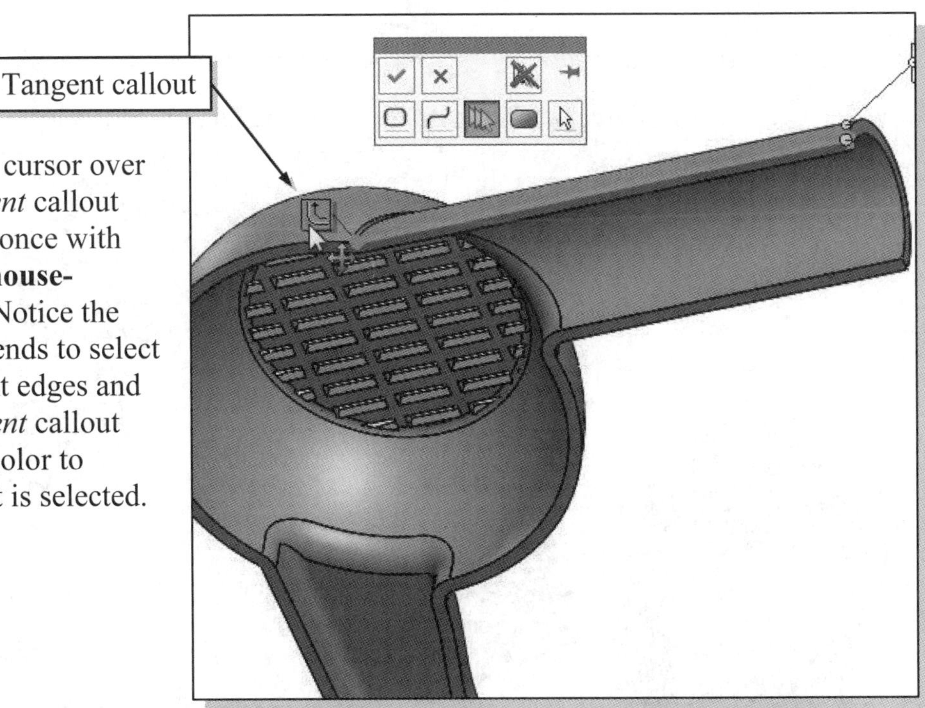

8. Move the cursor over the *Tangent* callout and click once with the **left-mouse-button**. Notice the chain extends to select all tangent edges and the *Tangent* callout changes color to indicate it is selected.

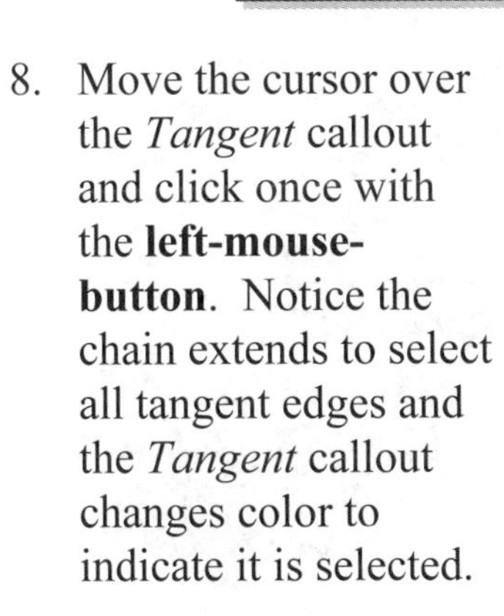

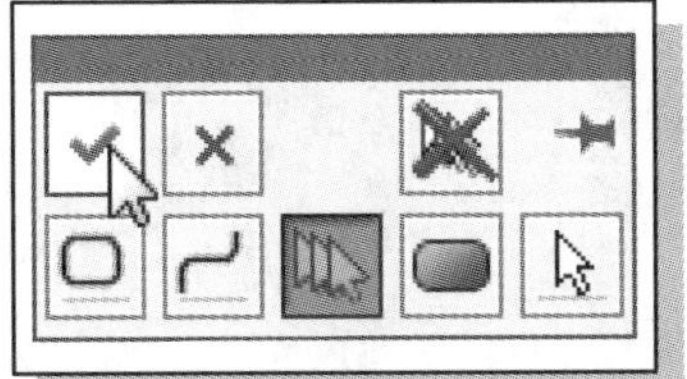

9. Select the **OK** icon on the *SelectionManager*.

10. **Open Group** appears in the *Path* selection window in the *Cut-Sweep PropertyManager*. **Uncheck Align with end faces** in the *Options* panel, then click **OK** in the *PropertyManager* to accept the settings and create the swept feature.

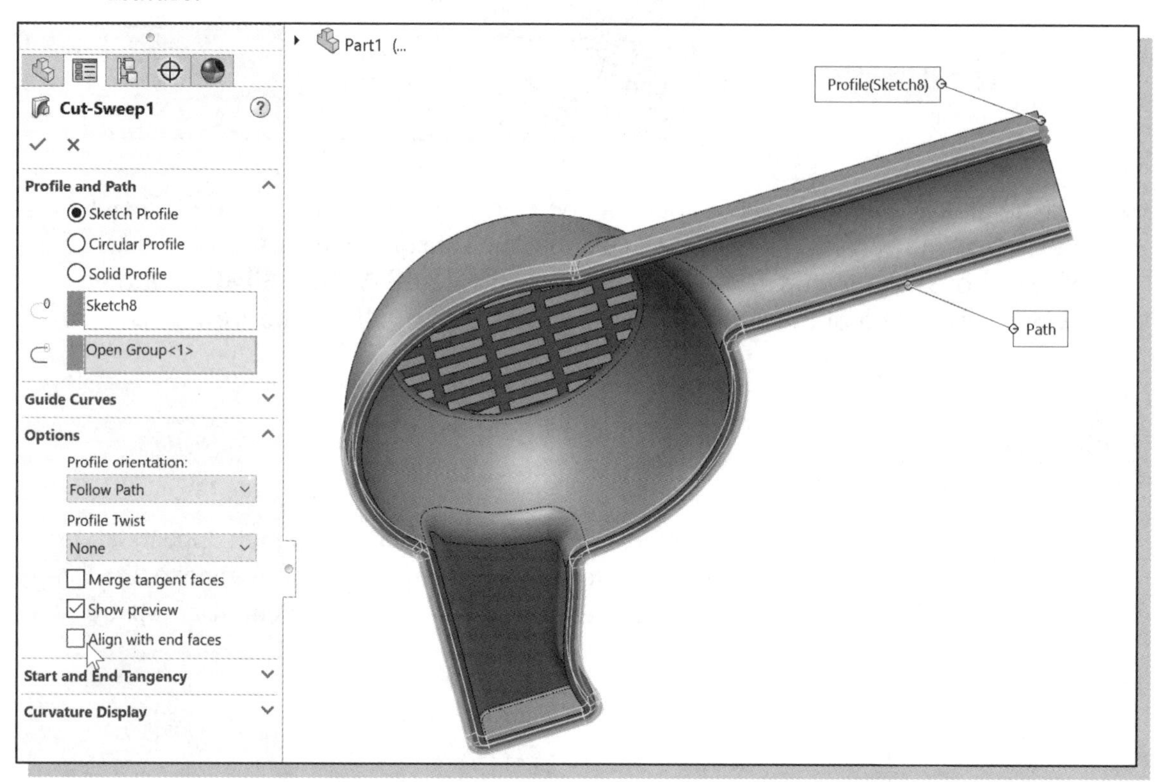

11. On your own, save the part with the filename **Dryer_Housing**.

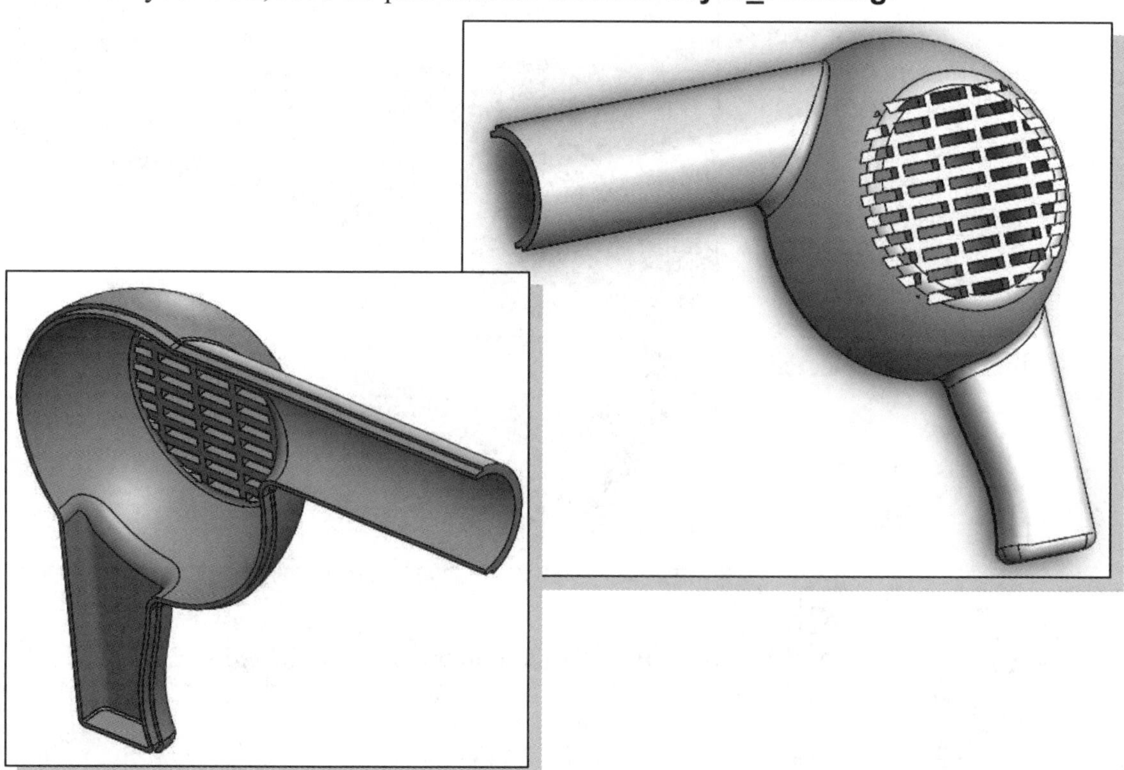

Using PhotoView 360, Scenes, and Appearances

Photorealistic renderings can be made of your model in SOLIDWORKS using **PhotoView 360**. The main steps for creating a rendering include defining a view, applying Appearances and defining Scenes. The view for the rendering can be one of the standard orthographic or perspective views or can be established using a camera view. **Appearances** can be used to make your model appear more realistic through the application of features including colors, material properties, transparency, illumination, and surface finish. **Scenes** and lighting can be applied and edited using the SOLIDWORKS library of default scenes and tools in the *DisplayManager*. In this lesson we will activate the PhotoView 360 add-in and create a rendering by applying an Appearance and a Scene and from the SOLIDWORKS Task Pane.

Activating PhotoView 360

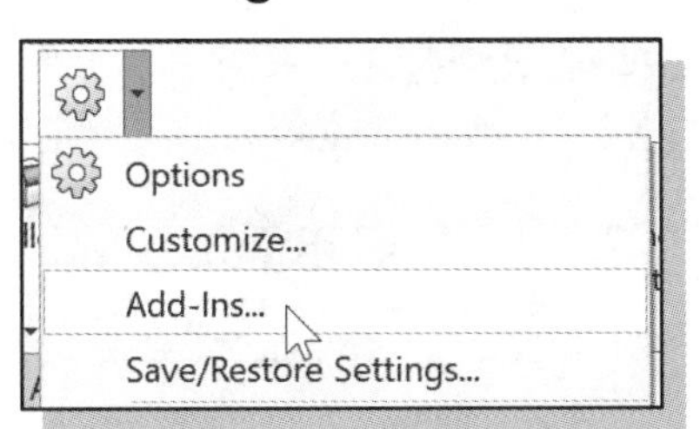

1. Select the **Add-Ins** option from the pull-down menu next to the *Options* icon.

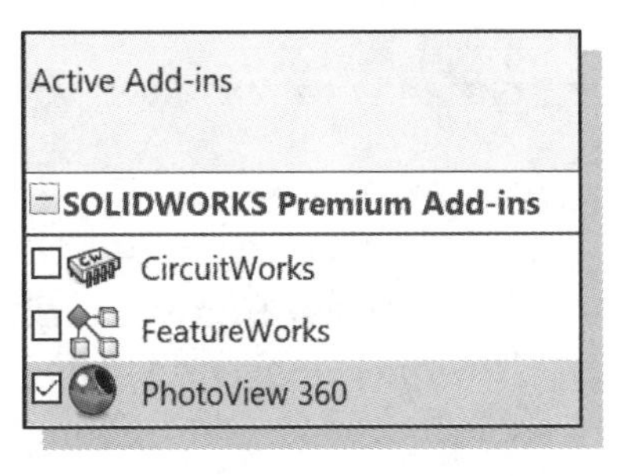

2. Select the **Photoview 360** option in the *Add-Ins* window and click **OK**.

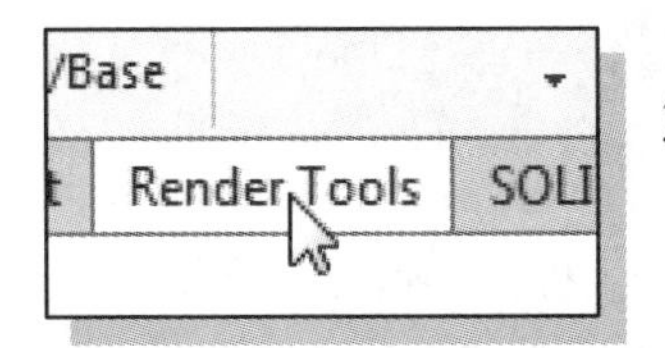

3. Select the **Render Tools** tab in the *CommandManager*, to reveal the *Render Tools* toolbar.

Adding an Appearance

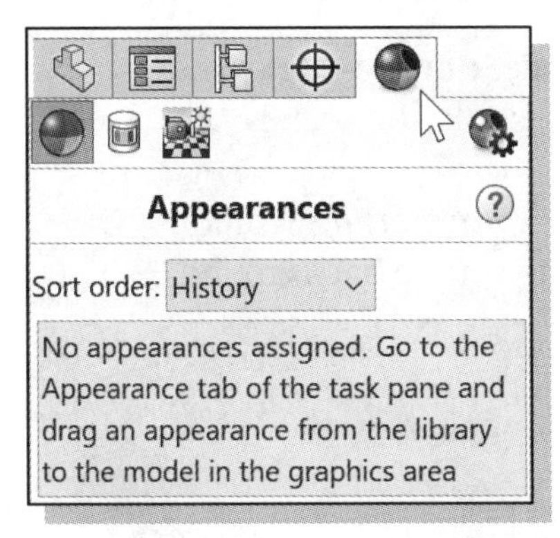

4. In the *Manager* pane, select the **DisplayManager** tab. Notice there are no appearances.

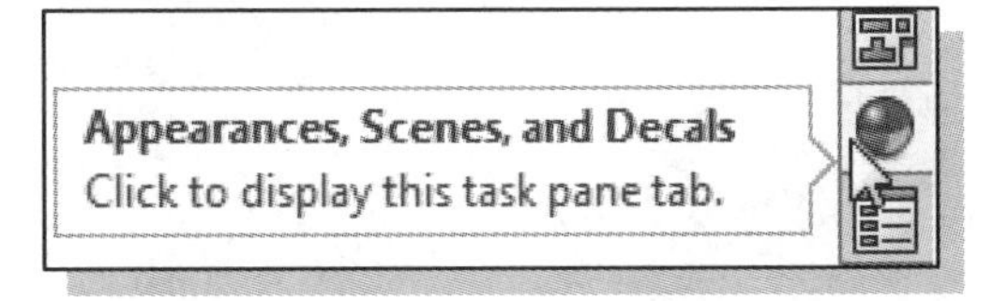

5. Left-click on the **Appearances, Scenes, and Decals** icon at the right of the graphics area to display the *Design Library* task pane.

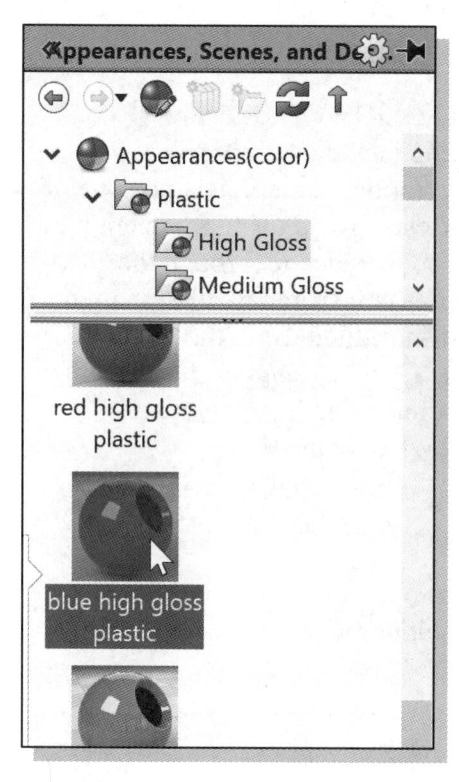

6. In *the Appearances, Scenes, and Decals* Task Pane, select **Appearances** > **Plastic** > **High Gloss**.

7. In the lower Task Pane, **double-click** on the **blue high gloss plastic** icon to apply the appearance to the entire part.

8. Notice the blue high gloss plastic appearance now appears in the *DisplayManager*.

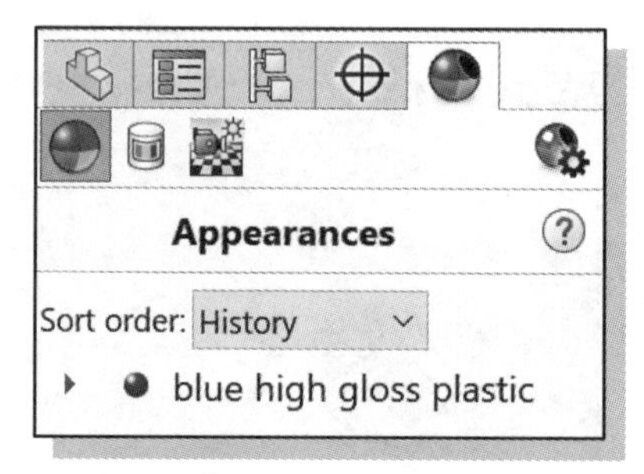

Adding a Scene

9. Select **View Scene, Lights, and Cameras** in the *DisplayManager*. The DisplayManager can be used to select scenes, edit lighting, and to add and adjust camera views. (We will use a default scene and our orthographic view rather than adjusting these settings.)

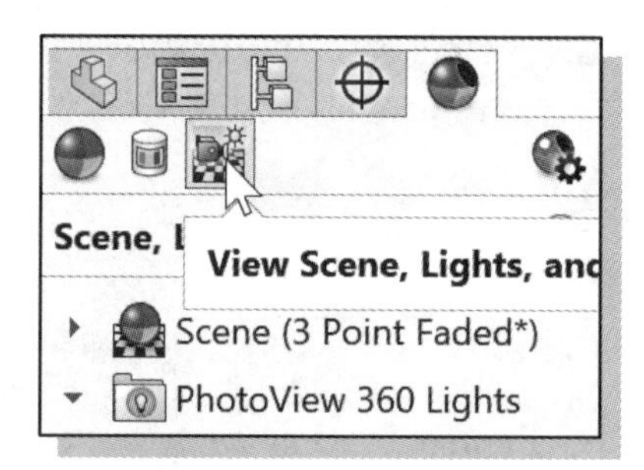

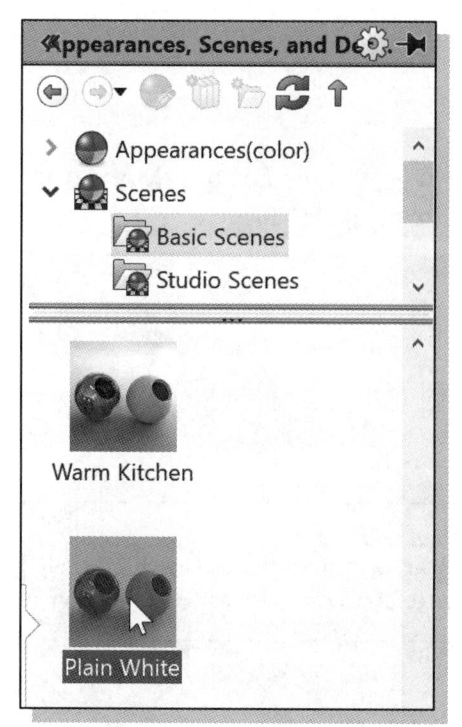

10. In *the Appearances, Scenes, and Decals* Task Pane (at the right of the graphics area), select **Scenes** > **Basic Scenes**.

11. In the lower Task Pane, **double-click** on the **Plain White** icon to apply the scene.

12. If the *Background Display Setting* window appears, select **Yes** to set the background appearance option.

13. Notice the scene now appears in the *DisplayManager*.

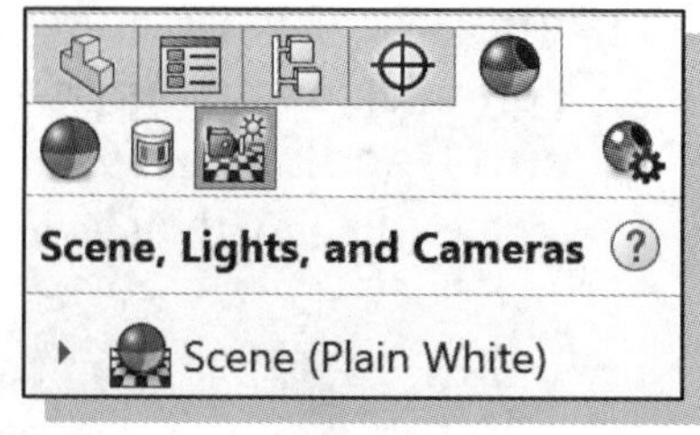

Performing a Final Render

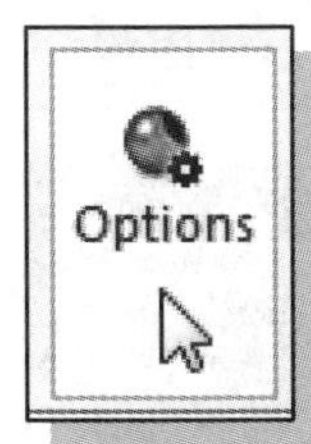

14. On the *Render Tools* toolbar, select the **Options** icon.

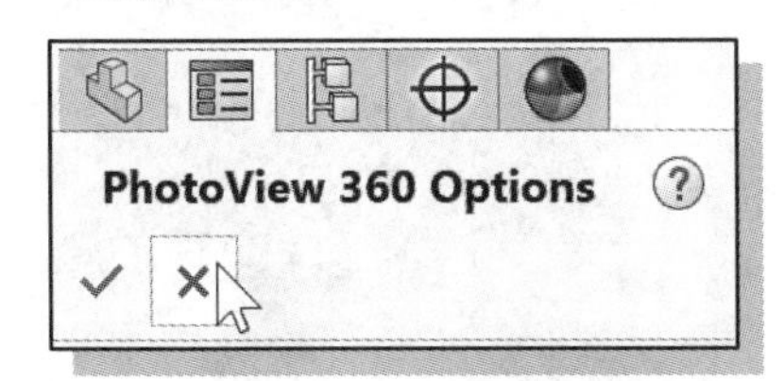

15. The PhotoView 360 Options *PropertyManager* appears. It can be used to adjust output image settings including image size, file format, and render quality. We will use the default values. Click on the **Cancel** option.

16. On your own, rotate the part to appear as shown.

17. On the *Render Tools* toolbar, select the **Final Render** icon.

18. Since no camera view was created, the Use Perspective Views in Renderings window may appear. We will create our rendering with the orthographic view rather than a perspective view or camera view. Select **Continue without Camera or Perspective**.

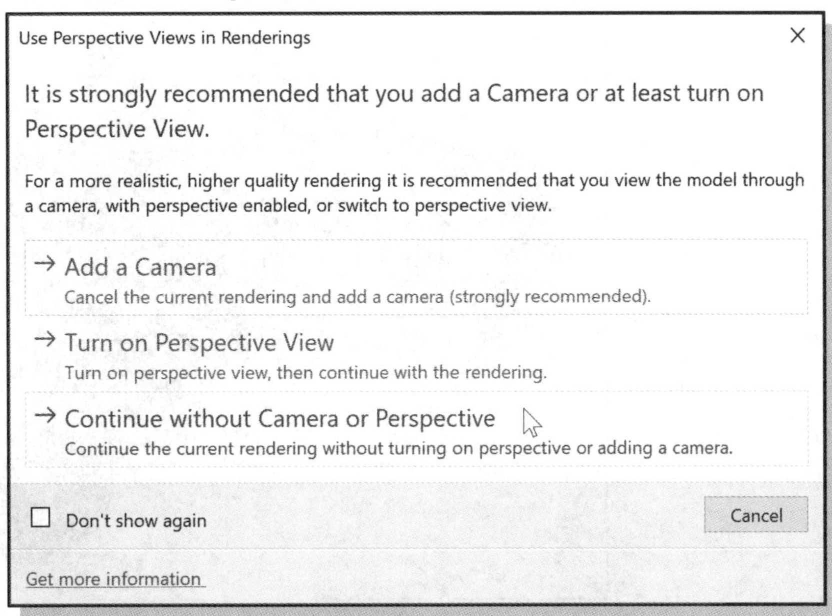

19. The *Final Render* window appears. When the render is complete, click **Save Image** and save the *.jpeg image with the filename **Dryer_Housing_Render**.

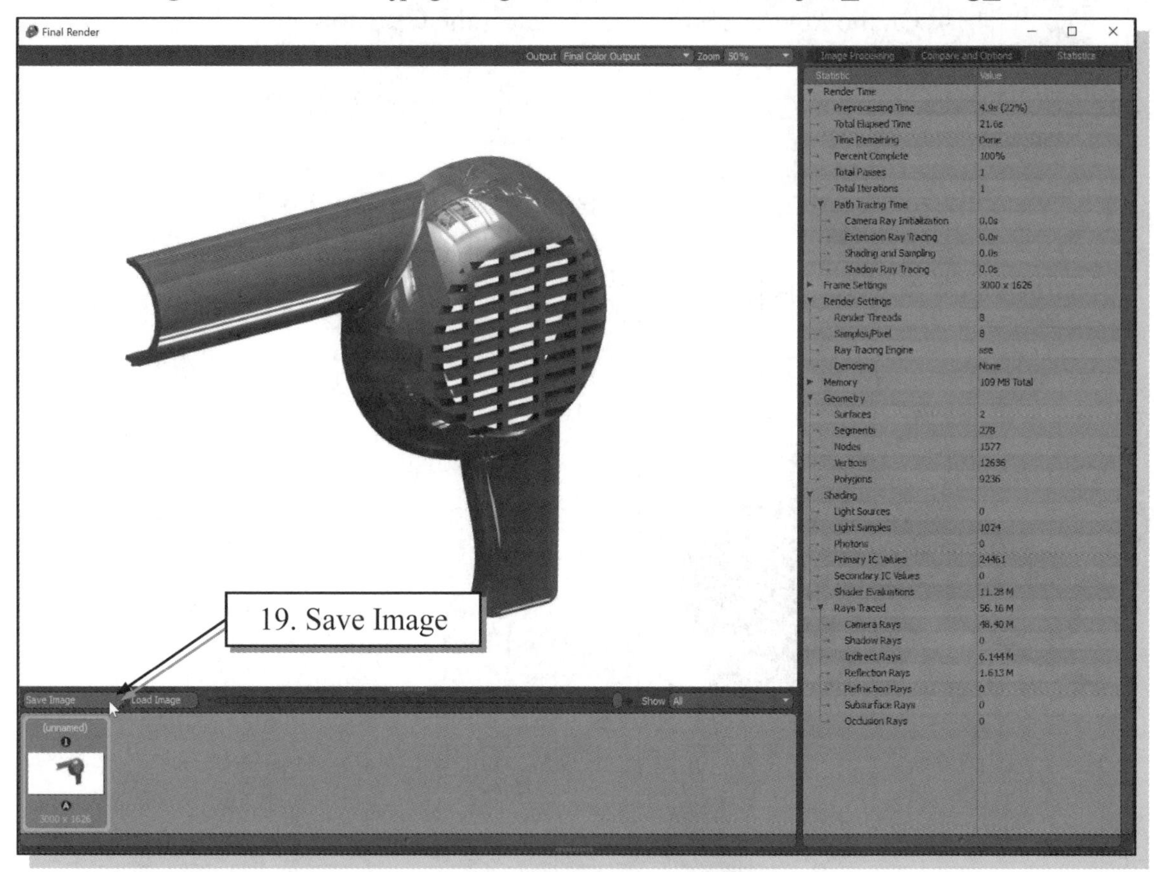

20. Close the *Final Render* window.

21. On your own, save and close the **Dryer_Housing** part file.

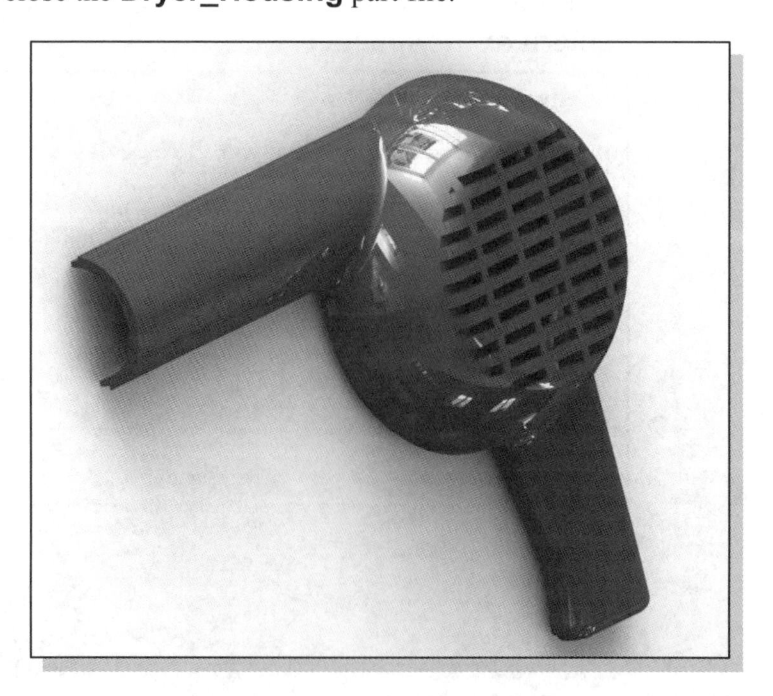

Questions:

1. Keeping the *Design Tree* in mind, what is the difference between *cut with a pattern* and *cut each one individually*?

2. What is the difference between Sweep and Extrude?

3. What are the advantages and disadvantages of creating fillets using the 3D Fillets command and creating fillets in the 2D profiles?

4. Describe the steps used to create the *Shell* feature in the lesson.

5. How do we modify the *Rectangular Pattern* parameters after the model is built?

6. Describe the elements required in creating a *Swept* feature.

7. Create sketches showing the steps you plan to use to create the model shown on the next page:

Exercises:

1. **Motor Housing** (Dimensions are in inches.)

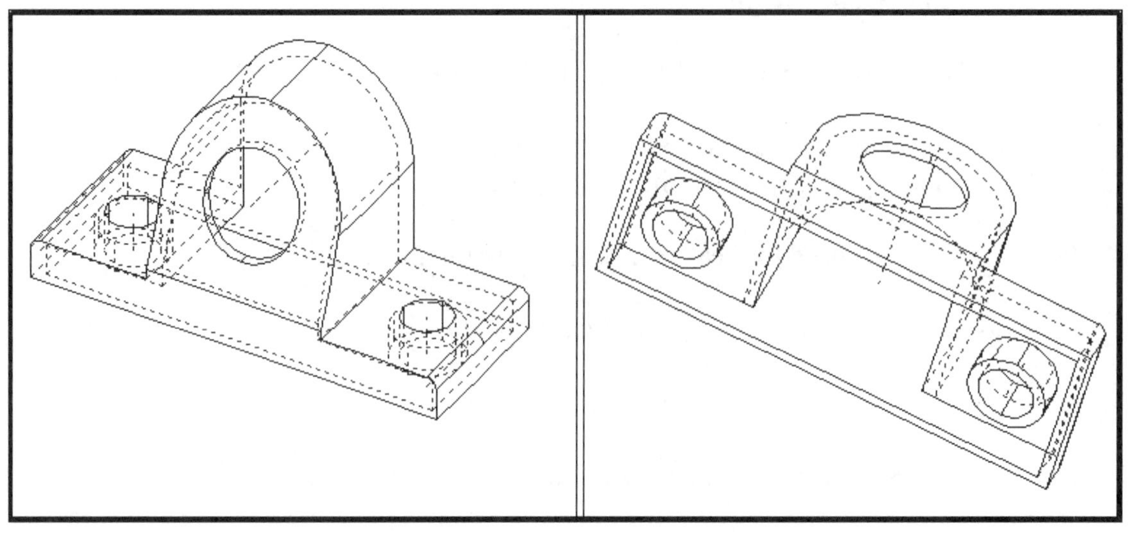

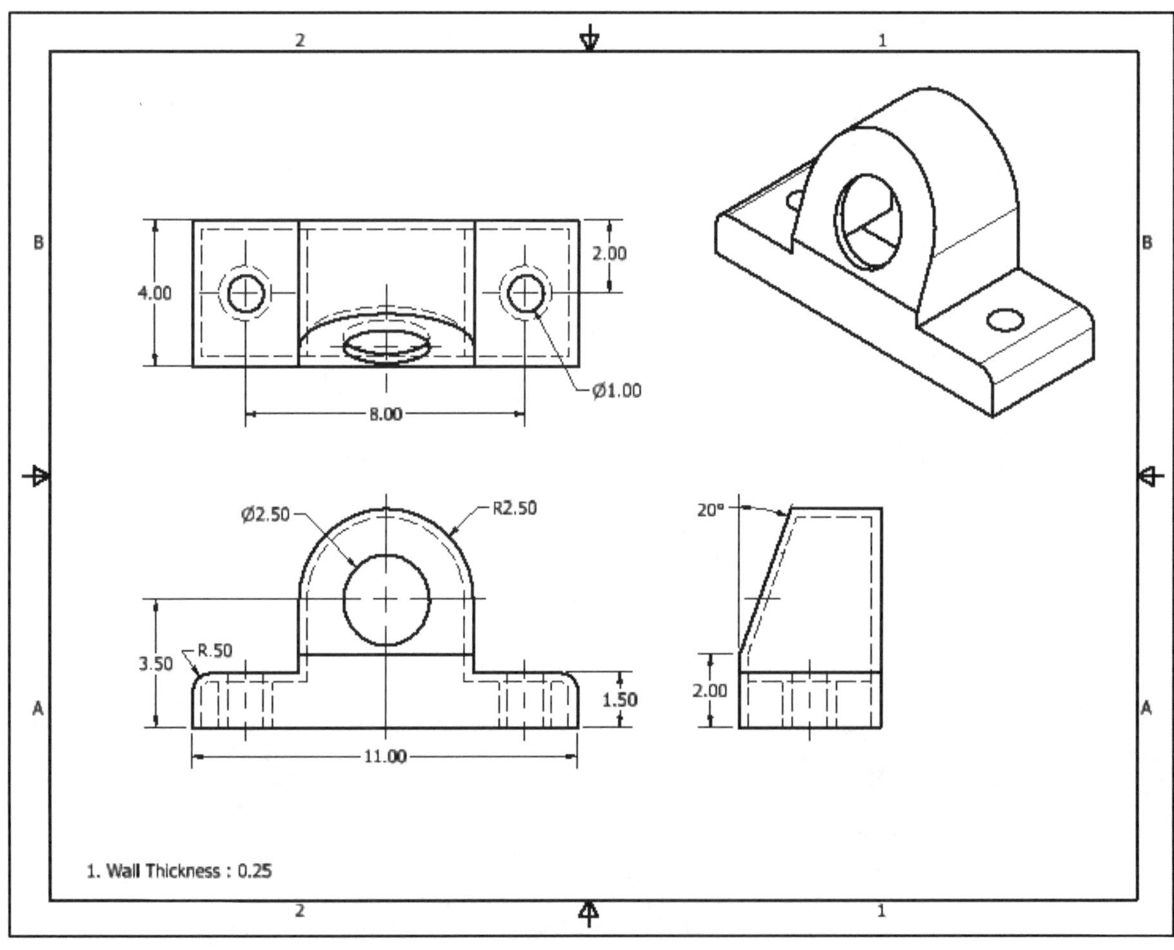

2. **Guide Base** (Dimensions are in mm.)

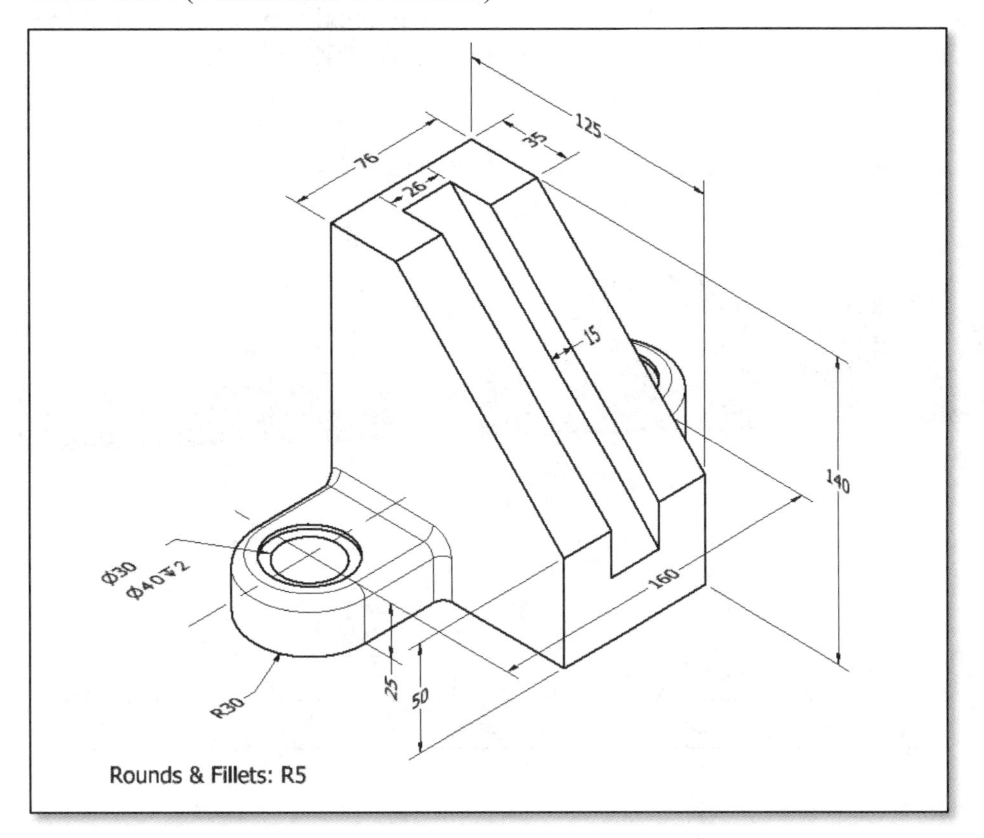

3. **Piston Cap** (Dimensions are in inches.)

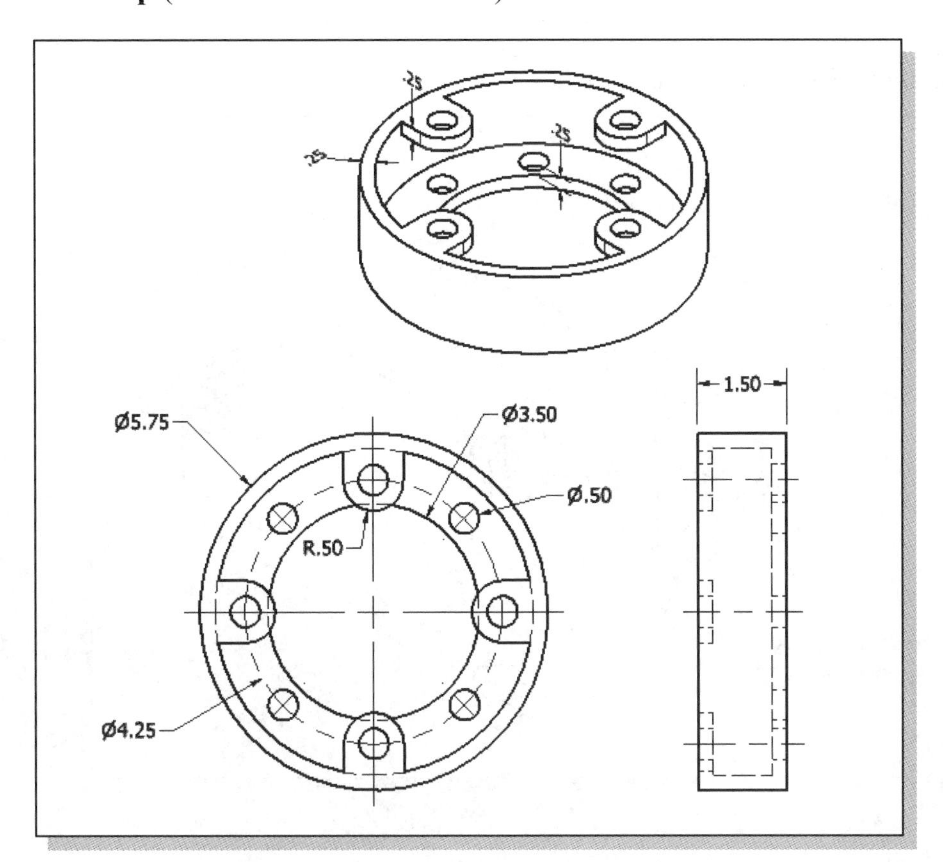

4. Using the same dimensions given in the tutorial, construct the other half of the dryer housing. Plan ahead, and consider how you would create the matching half of the design. (Save both parts so that you can create an assembly model once you have completed the assembly Chapter 13.)

5. **Intake Flange** (Dimensions are in inches. Thickness: .25 inches.)

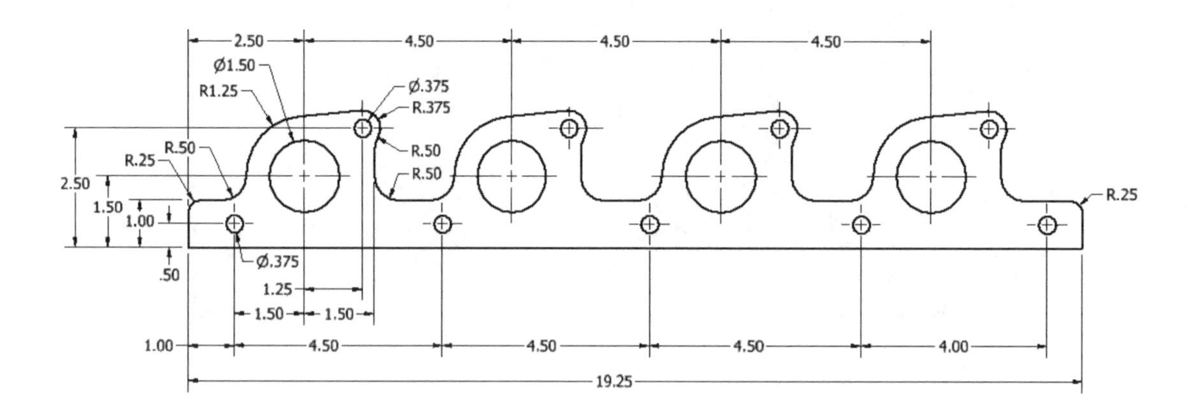

6. **Anchor Base** (Dimensions are in inches.)

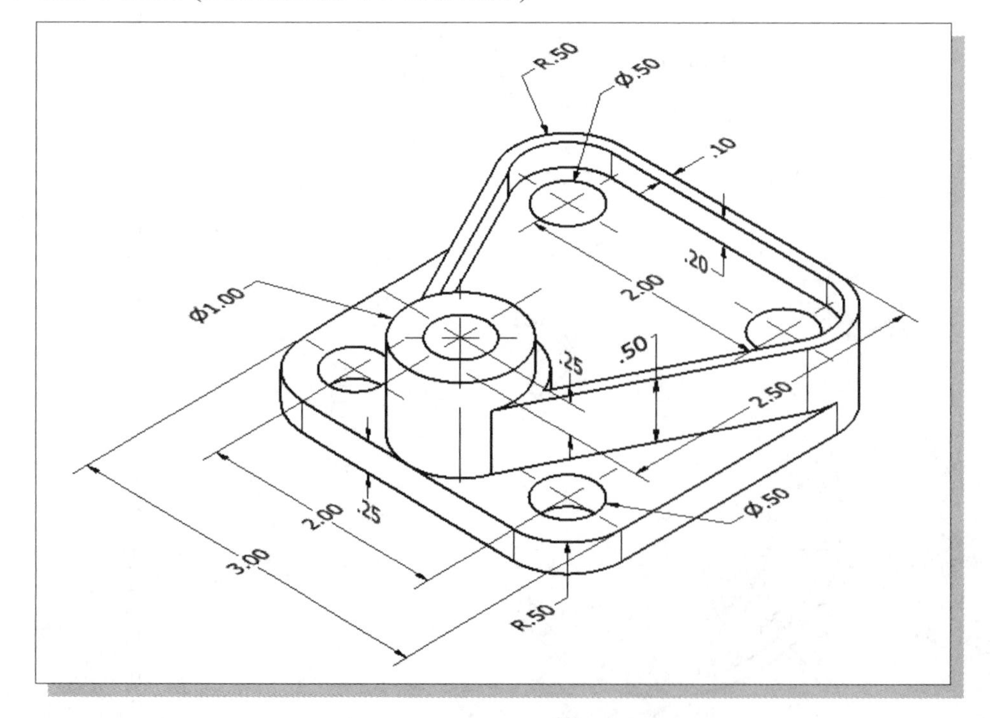

Chapter 13
Sheet Metal Designs

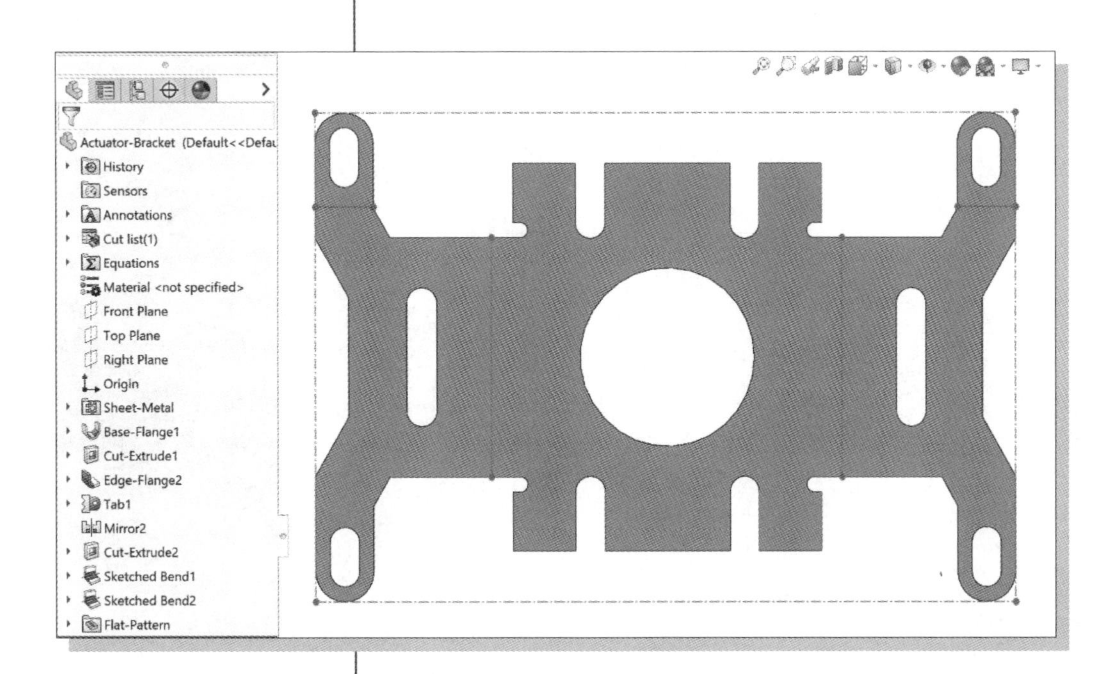

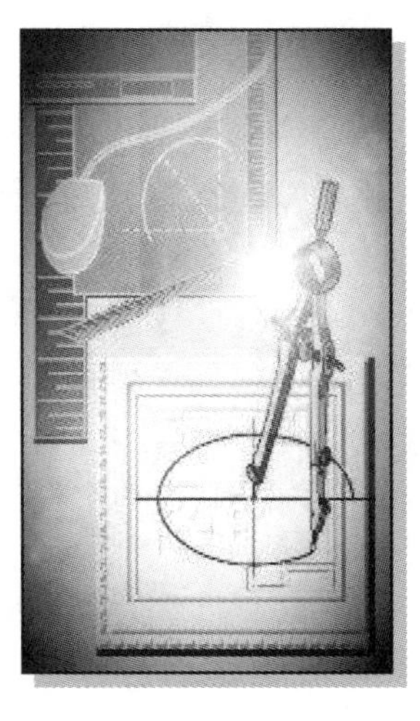

Learning Objectives

- **Understand the Sheet Metal Manufacturing Processes**
- **Understand the SOLIDWORKS Sheet Metal Modeling Methodology**
- **Create Sheet Metal Parts**
- **Utilize the SOLIDWORKS Sheet Metal Tools to Create Bends and Flanges**
- **Create Flat Pattern Layouts**

Certified SolidWorks Associate Exam Objectives Coverage

Drawing Sheets and Views

Objectives: Creating and Setting Properties for Drawing Sheets; Inserting and Editing Standard Views.

Sheet Metal Processes

Sheet metal is one of the most commonly used materials in our everyday life. Sheet metal is simply a thin and flat piece of metal, which can be cut and bent into a variety of different shapes. The thicknesses of sheet metal can vary significantly, but the thickness is generally between 0.006" and 0.250".

Sheet metal is generally produced by reducing the thickness of a work piece by compressive forces applied through a set of rolls. This process is known as rolling and has been around since 1500 AD. Sheet metal is identified by the thickness, or gauge, of the metal and is generally available as flat pieces or in coils. The gauge of sheet metal (see Appendix A) ranges from 30 gauge to about 6 gauge. The higher the gauge number, the thinner the metal is. Aluminum, brass, copper, cold rolled steel, tin, nickel and titanium are some of the more commonly available sheet metal materials. Typical sheet metal applications are seen in cars, boats, airplanes, casing for electronic devices and many other things.

The main feature of sheet metal is its ability to be formed and shaped by a variety of processes, such as **bending** and **cutting**. Different processes can be used to achieve the desired shape and form. Some of the more commonly used sheet metal processes include:

Drawing
Drawing forms sheet metal into parts by using a punch, where the punch presses a sheet metal blank into a die cavity. This process is generally used to create shallow or deep parts with relatively simple shapes. **Soft punches** can also be utilized to create more arbitrary shapes. **Deep drawing** is generally done by making multiple steps; this process is known as *draw reductions*.

Stretch forming
Stretch forming is a process where the sheet metal is clamped around its edges and stretched over a die. This process is mainly used for the manufacturing of large parts with shallow contours, such as aircraft wings, or automotive door and window panels.

Spinning
Spinning is the process used to make axis-symmetric parts by applying a work piece to a rotating mandrel with the help of rollers. *Spinning* is commonly used to make cylindrical shapes, such as missile nose cones and satellite dishes.

Stamping
Stamping is the general term used to describe a variety of operations, such as bending, flanging, punching, embossing, and coining. The main advantage of stamping is its speed; designs containing simple or complex shapes can be formed at relatively high production rates.

Flanging
Flanging is a process used to strengthen different sections of a sheet metal part and also to form various shapes. This process is commonly used for a variety of parts, for example, aluminum cans for soft drinks.

Bending

Bending is a process by which sheet metal can be deformed by plastically deforming the material and changing its shape. The material is stressed beyond the yield strength but below the ultimate tensile strength. With this process, the surface area of the material does not change much. *Bending* usually refers to deformation about one axis.

Bending is a flexible process by which many different shapes can be produced. Standard die sets are used to produce a wide variety of shapes. The material is placed on the die and positioned in place with *stops* and *gauges*. The material is held in place with *hold-downs*. The upper part of the press, the ram, with the appropriately shaped punch descends and forms the v-shaped bend.

Bending is usually done using ***press brakes***. The lower die of the press contains a V-shaped groove. The upper part of the press contains a punch that will press the sheet metal down into the v-shaped die, causing it to bend.

The most commonly used modern *Bending* method is the ***air bending*** method, where a sharper die angle is used; for example, an 85 degree angle is used for a 90 degree bend. *Air Bending* is done with the punch touching the work piece, and the work piece not bottoming in the lower die. By controlling the push stroke of the upper punch, the metal is pushed down to the required bend angle. The groove width of the lower die is typically 8 to 10 times the thickness of the metal to be bent. The *press brake* can also be computer controlled to allow the making of a series of bends to assure a high degree of accuracy in manufactured parts.

Cutting

Cutting sheet metal can be done in various ways, from using a variety of hand tools to very large powered shears. Today, computer-controlled cutting is also available for very precise cutting. Most modern computer-controlled sheet metal cutting operations are ***CNC laser cutting*** and ***CNC punch press***.

CNC laser cutting is done by moving the laser beam over the surface of the sheet metal. The sheet metal is heated and then burnt by the laser beam. The quality of the edge can be extremely smooth. *CNC punching* is performed by moving the sheet metal between the computer-controlled punch. The top punch mates with the bottom die, cutting a simple shape, such as a square, circle, or hexagon from the sheet. An area can be cut out by making several hundred small square cuts around the perimeter. A *CNC punch* is less flexible than a laser for cutting compound shapes, but it is faster for repetitive shapes. A typical *CNC punch* has a choice of up to 60 tools in a ***turret***. A modern *CNC punch* can run as fast as 600 blows per minute. A *CNC punch* or a *CNC laser* machine can typically cut a blank sheet into the desired shapes in less than 15 seconds, with very high precision.

Sheet Metal Modeling

In reality, a sheet metal part is made from a piece of flat metal sheet of uniform thickness by cutting out a flat pattern and then folding it into the desired shape. To construct a computer sheet metal part, we can (1) simulate the actual production methods and start with a flat pattern layout to make the model; (2) use the building block approach which concentrates on the different sections of the formed 3D design; or (3) construct a solid model first, then convert it into a sheet metal model. All three methods are applicable in modern parametric modeling software such as SOLIDWORKS.

Since the actual sheet metal manufacturing process requires a flat pattern layout, the accurate generation of the flat pattern layout in the computer modeling software is critical. The conversion between the 3D formed designs and 2D flat pattern layouts requires the use of the correct ***K-Factor***, which can be used to determine the required ***Bend Allowance***.

Bend allowance is the term used to describe how much material is needed between two panels to accommodate a given bend. Determining bend allowance is commonly referred to as ***Bend Development***.

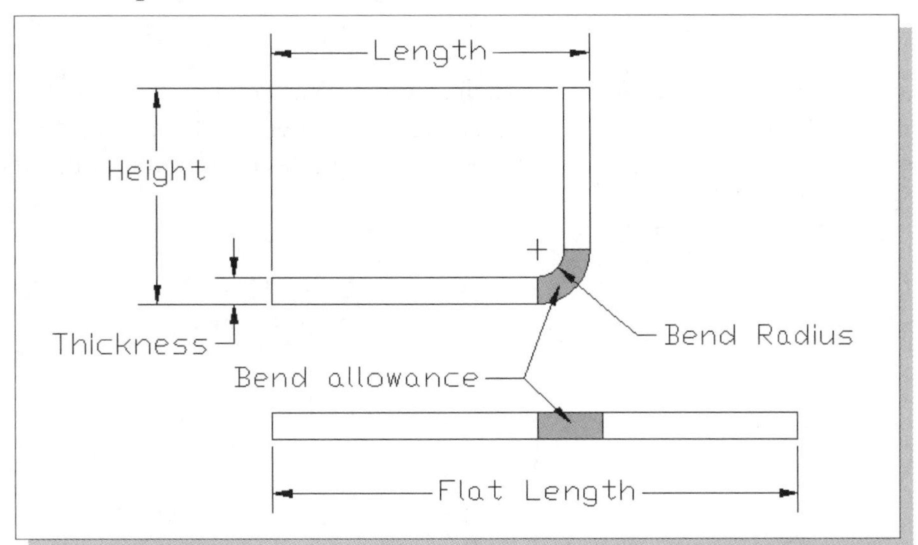

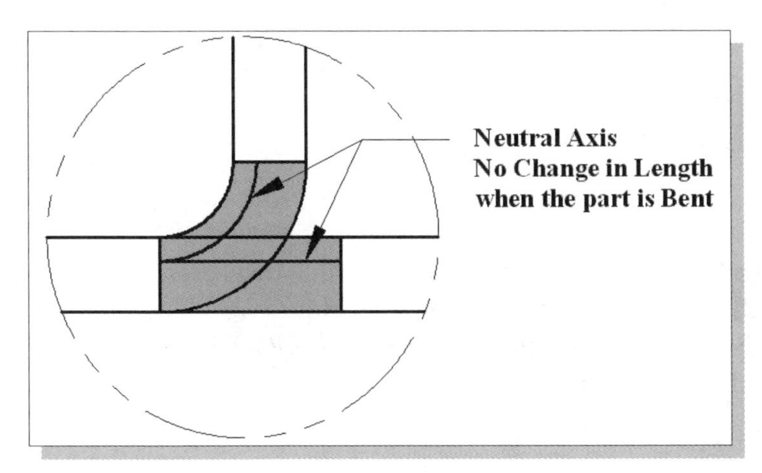

In sheet metal, the ***Neutral Axis*** is defined as the location where there is no change in length when the part is bent.

On the inside of the bend, above the neutral axis in the figure, the material is in compression, where the area below the neutral axis is in tension.

K-Factor

The location of the neutral axis in a bend is called the ***K-Factor***. Since the amount of inside compression is always less than the outside tension, the K-Factor can never exceed **0.50** in practical use. To the other extreme, a reasonable assumption is that the K-Factor cannot be less than **0.25**.

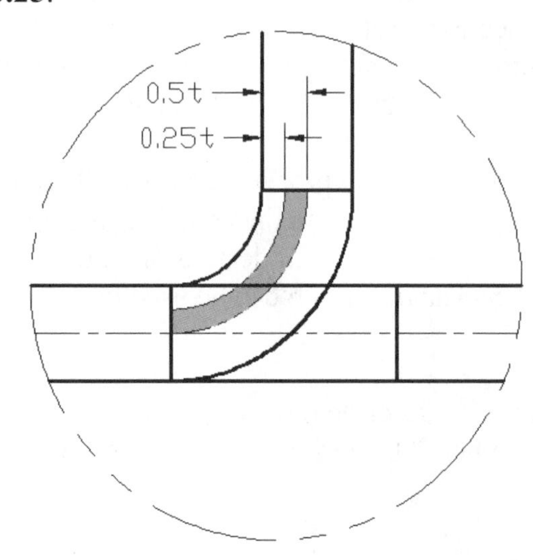

Several factors can change the K-Factor, such as the type of bending (free vs. constrained), tool geometry, rate of bend, material (Mild Steel, Cold Rolled Steel, Aluminum, etc.), and even grain direction. With some grades of aluminum, the age of the material can also be a factor.

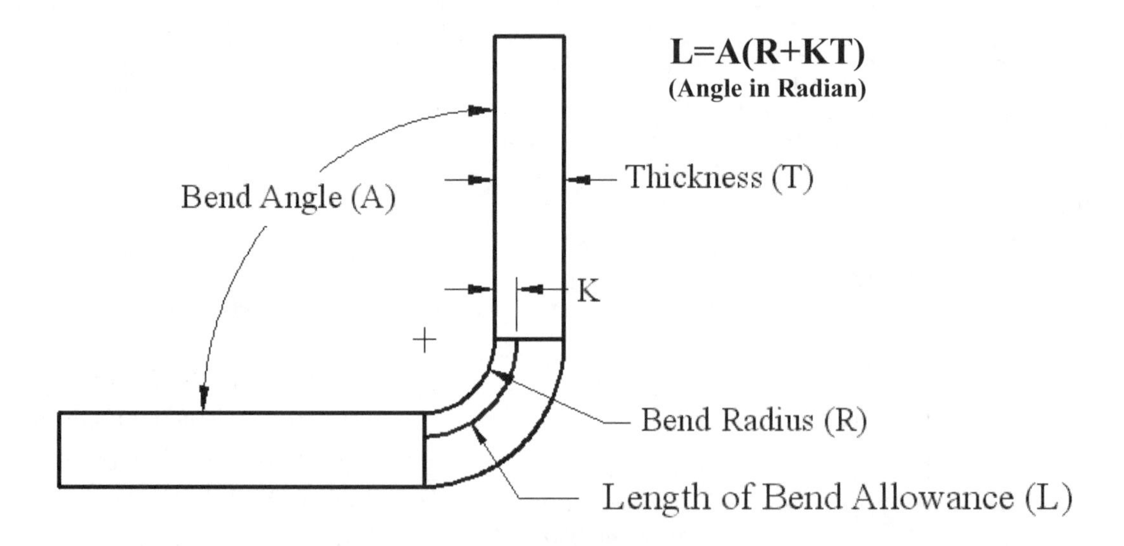

Sheet metal fabricators will typically have developed a K-Factor table (usually through trial and error) to use. SOLIDWORKS is set up to allow the K-Factor to be added to create material specific profiles. By using the correct data, SOLIDWORKS can be used to create fairly accurate and reliable flat patterns.

The *Actuator Bracket* Design

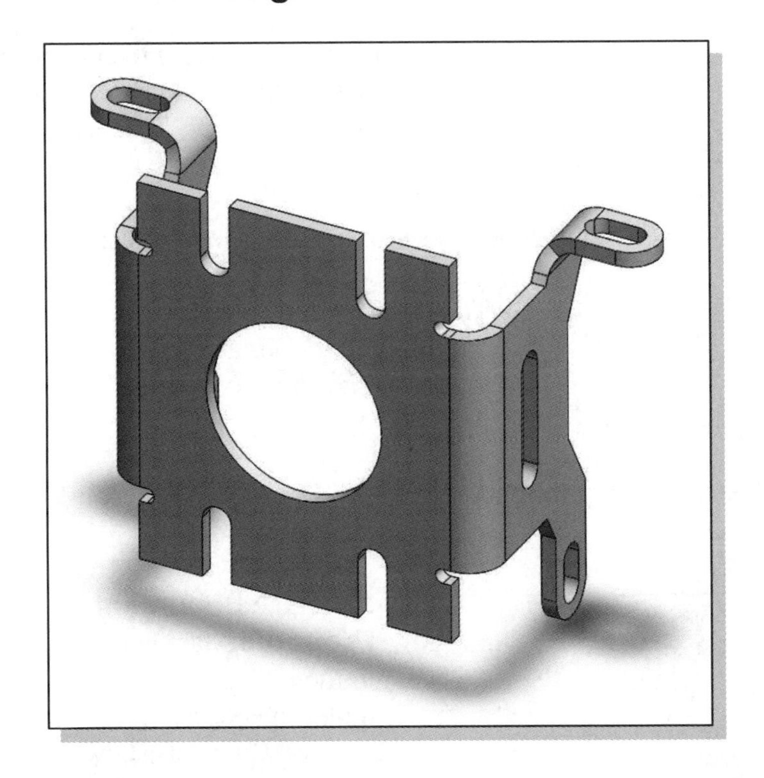

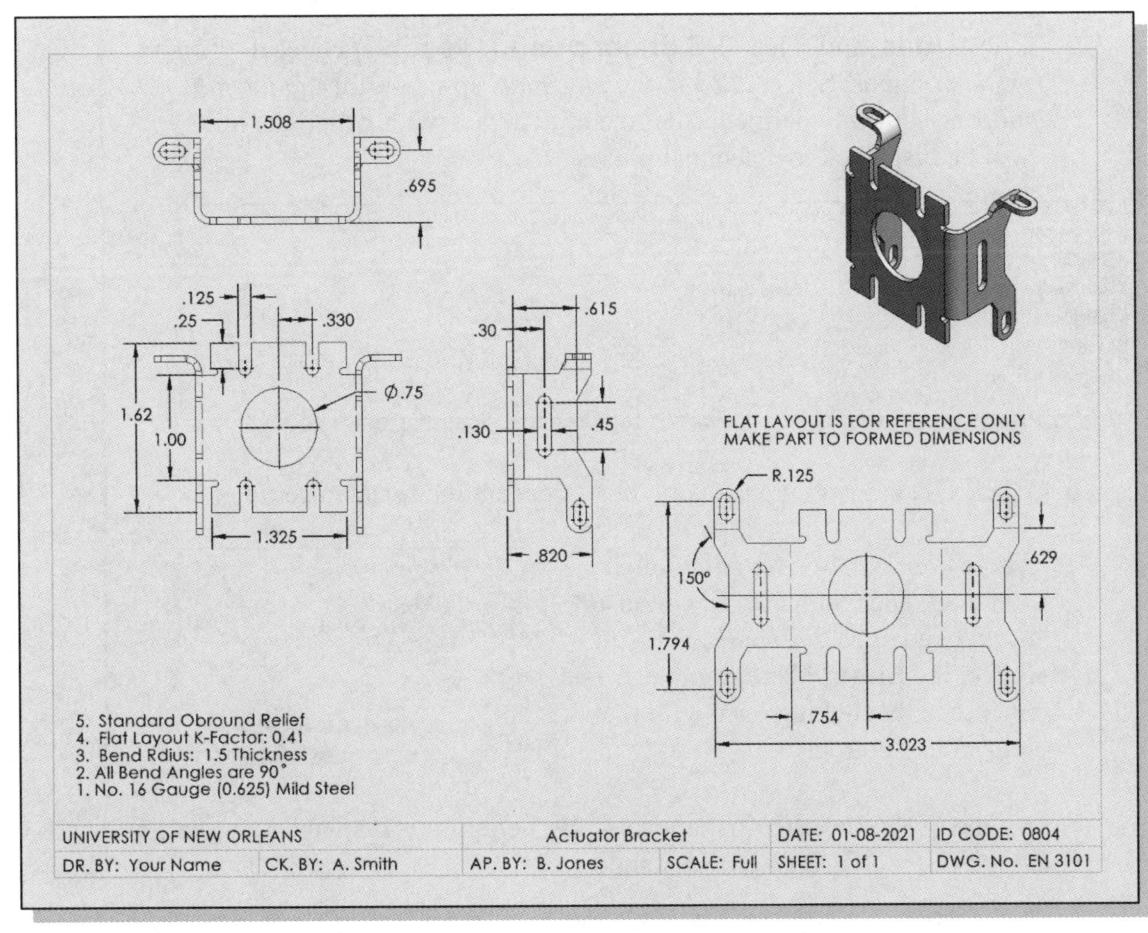

Starting SOLIDWORKS and Opening the Sheet Metal Toolbar

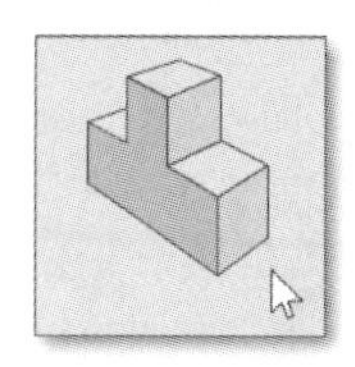

1. On your own, start a new part file using the default **Part** template in the SOLIDWORKS Templates folder.

IMPORTANT NOTE: In this lesson we will use the standard display of toolbars.

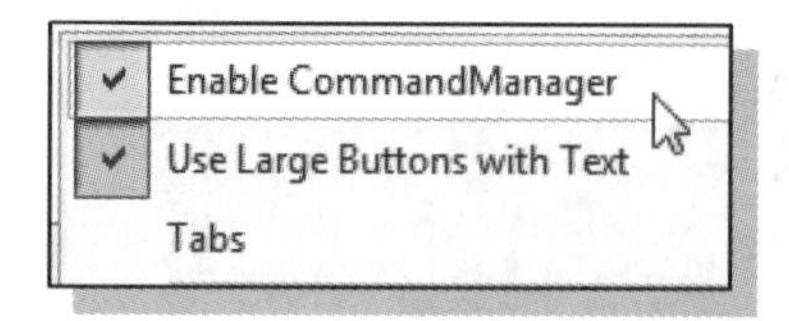

2. To turn ***OFF*** the *CommandManager*, right click on the *CommandManager* (or any other toolbar) and toggle the *Enable CommandManager* option *OFF*.

3. Select the **Options** icon from the *Menu Bar* to open the *Options* dialog box.

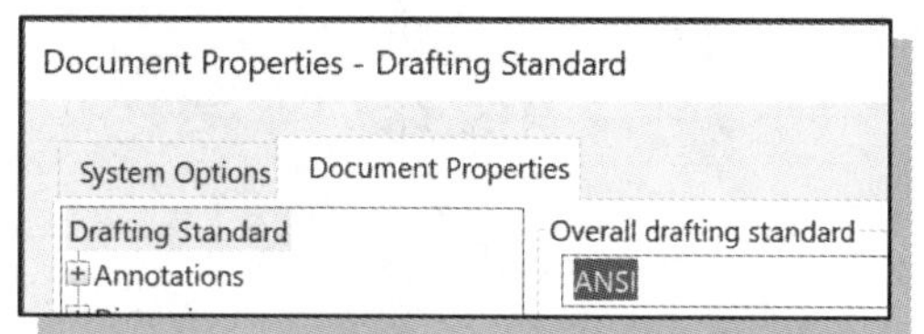

4. Select the **Document Properties** tab, and select **ANSI** in the pull-down selection window under the *Overall drafting standard* panel as shown.

5. Click **Units**, and select **IPS (inch, pound, second)** under the *Unit system* options. Select **.123** in the *Decimals* spin box for the *Length units* as shown to define the degree of accuracy with which the units will be displayed to 3 decimal places.

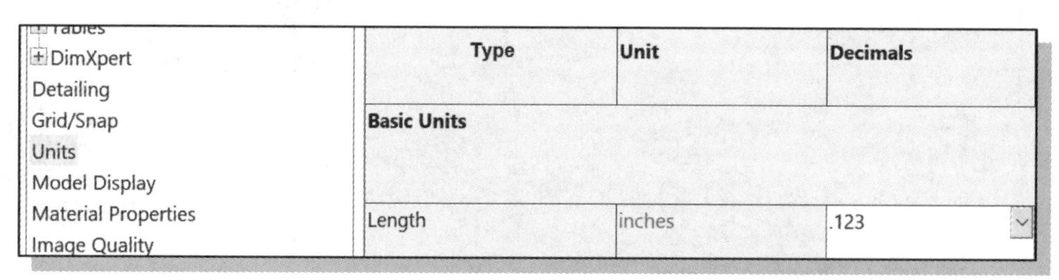

6. Click **OK** in the *Options* dialog box to accept the selected settings.

7. **Move** your cursor over any toolbar and click once with the right-mouse-button and open the *Toolbars* menu. Select the **Sheet Metal** toolbar from the menu by clicking once with the left-mouse-button.

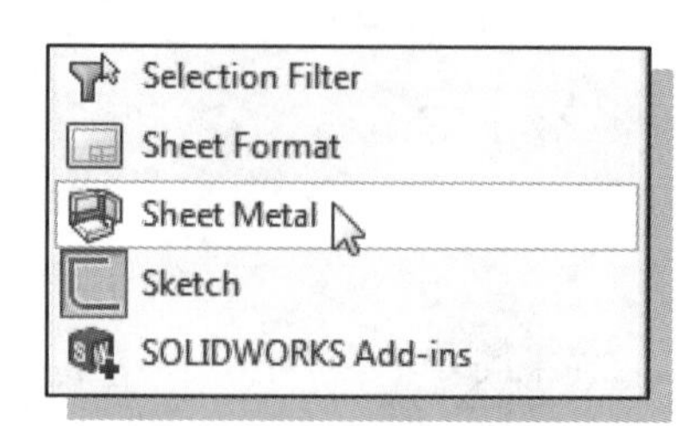

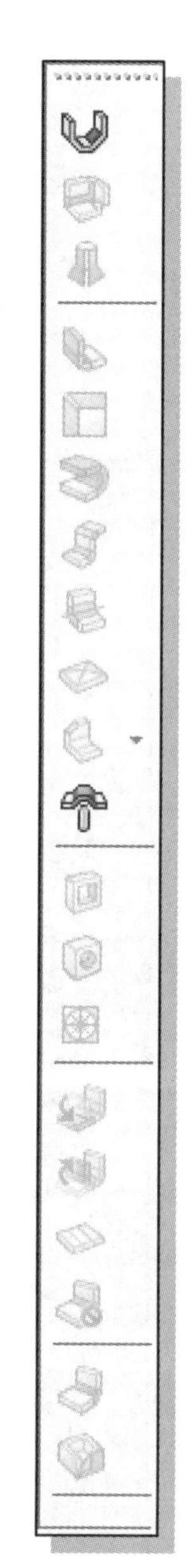

➢ Notice the *Sheet Metal* toolbar appears. By default it is displayed vertically at the left of the SOLIDWORKS window.

Creating the Base Feature of the Design

The main section of a sheet metal design is generally treated as the stationary portion of the design, to which all the other sections are added to form the final design. The main section is also typically the starting point of sheet metal modeling, and thus the base feature in parametric modeling.

1. Select the **Sketch** button on the *Sketch* toolbar to create a new sketch.

2. Select the **Front Plane** as the sketch plane for the new sketch.

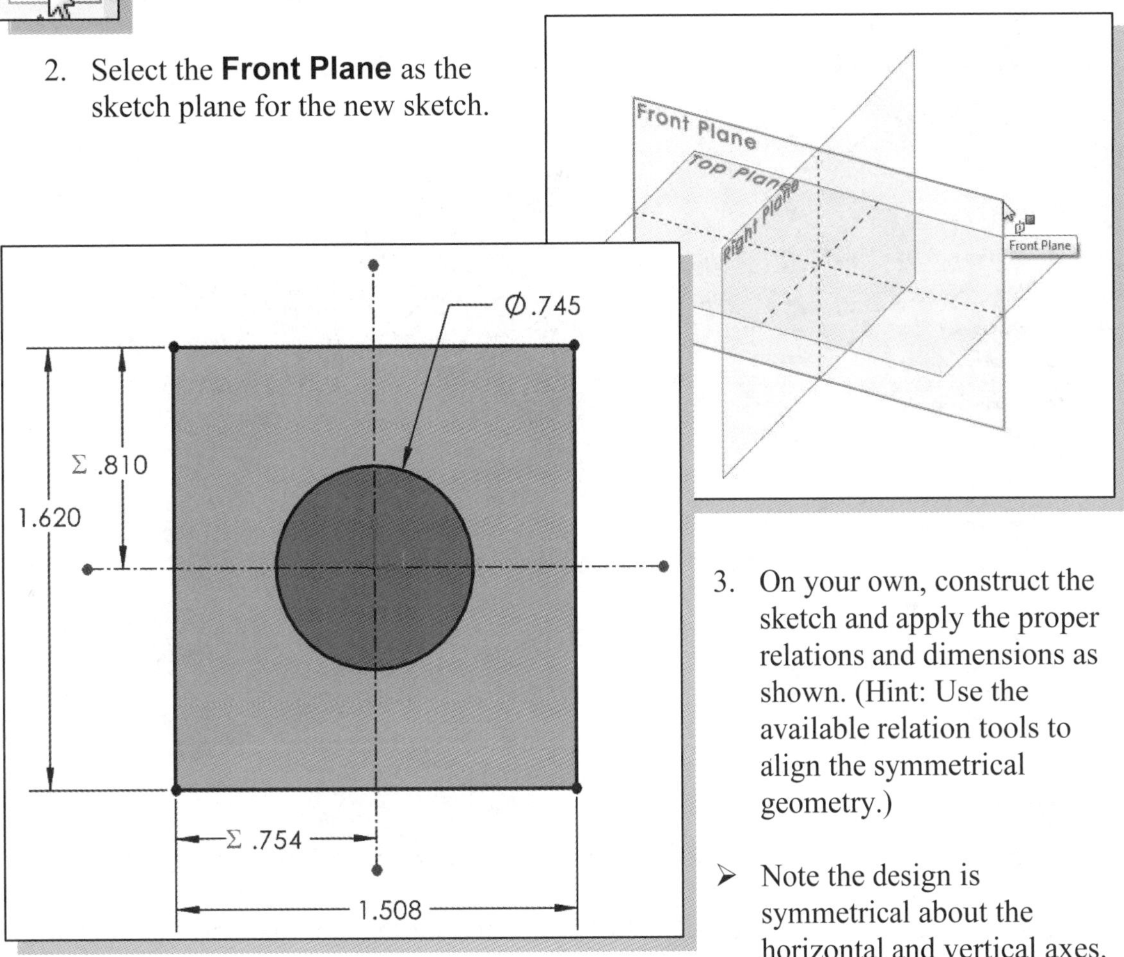

3. On your own, construct the sketch and apply the proper relations and dimensions as shown. (Hint: Use the available relation tools to align the symmetrical geometry.)

➢ Note the design is symmetrical about the horizontal and vertical axes.

4. Click once with the **left-mouse-button** on the **Sketch** icon on the *Sketch* toolbar to exit the Sketch option.

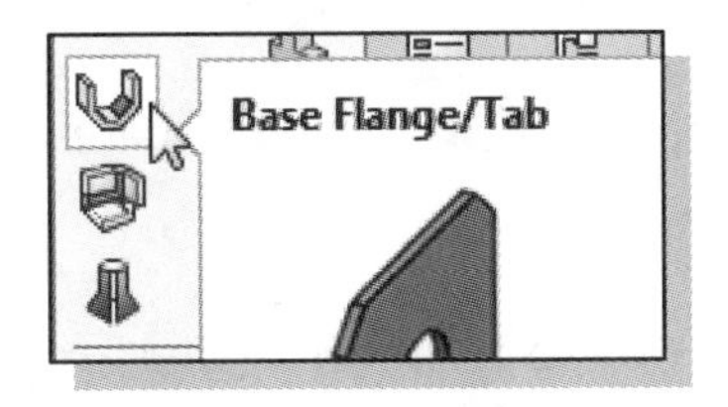

5. Verify that the sketch is pre-selected in the *FeatureManager Design Tree.*

6. In the *Sheet Metal* toolbar, select the **Base Flange/Tab** command by left-mouse-clicking the icon.

➤ Notice the *Base Flange PropertyManager* appears. The first panel in the *PropertyManager* is the *Sheet Metal Gauges* panel. Sheet metal gauge tables can be used to store properties for a designated material. Tables included in the SOLIDWORKS application, or tables created by the user, can be chosen to assign sheet metal parameters such as gauge thickness, allowable bend radii, and K-Factor. In this lesson, we will assign these values manually, rather than use a gauge table.

7. In the *Sheet Metal Parameters* panel on the *Base Flange PropertyManager*, enter **0.0625in** for the thickness. (This is the thickness of a 16 gauge mild steel component.)

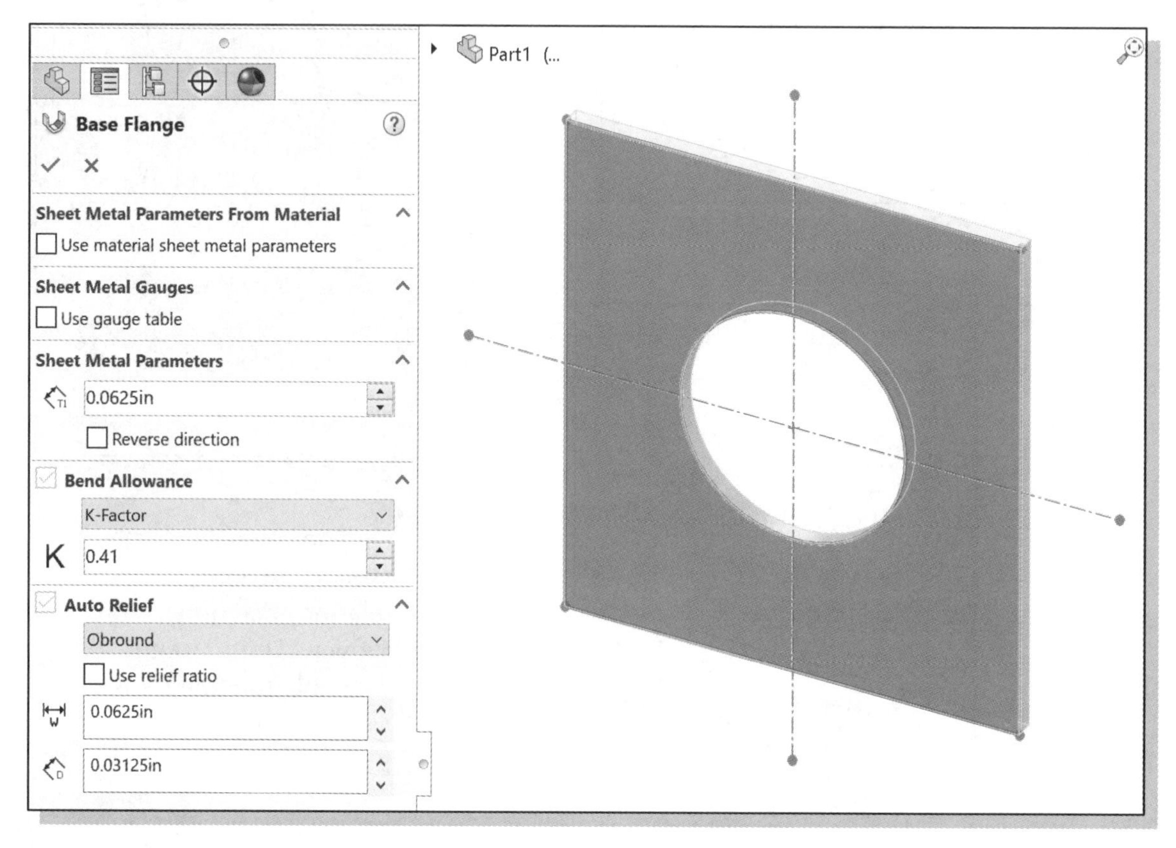

8. In the *Bend Allowance* panel, select **K-Factor** for the type and enter a value of **0.41**.

9. In the *Auto Relief* panel, select **Obround** for the type, **uncheck** the *Use relief ratio* box, and enter **0.0625in** (the gauge thickness) for the *Width* and **0.03125in** (0.5 × the gauge thickness) for the *Depth*.

10. Click **OK** (green check) in the *PropertyManager* to accept the settings and create the Base Flange.

➤ The *Bend Allowance* and *Auto Relief* settings have no effect on the creation of this base flange. However, these settings will become the default values for the sheet metal part.

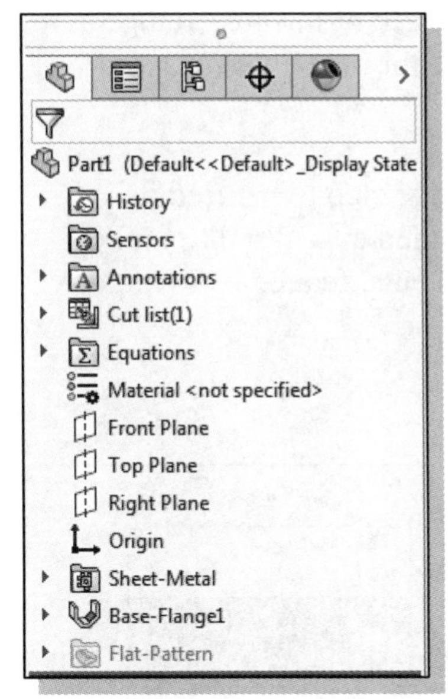

➢ The Base Flange/Tab command creates three features in the *FeatureManager Design Tree*:

- The **Sheet-Metal** feature contains the default bend parameters.
- The **Base-Flange1** feature represents the first solid feature of the sheet metal part.
- The **Flat-Pattern** feature flattens the sheet metal part. It is suppressed upon creation leaving the part in the bent condition. To flatten the part, unsuppress the Flat-Pattern feature. NOTE: When the Flat-Pattern feature is suppressed, new features are inserted above the Flat-Pattern feature.

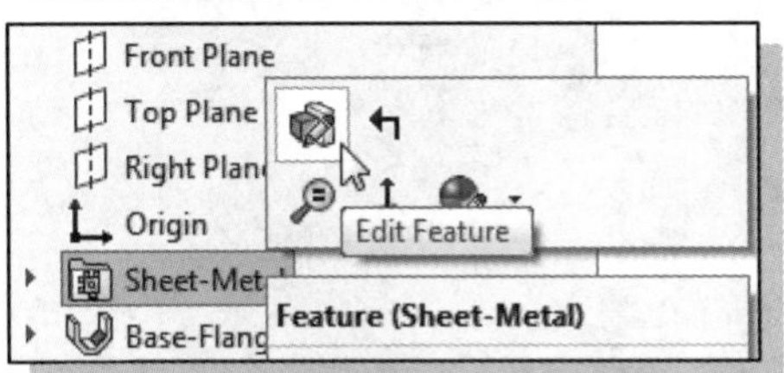

11. Right click on the **Sheet-Metal** icon in the *FeatureManager Design Tree* and select **Edit Feature** on the context toolbar.

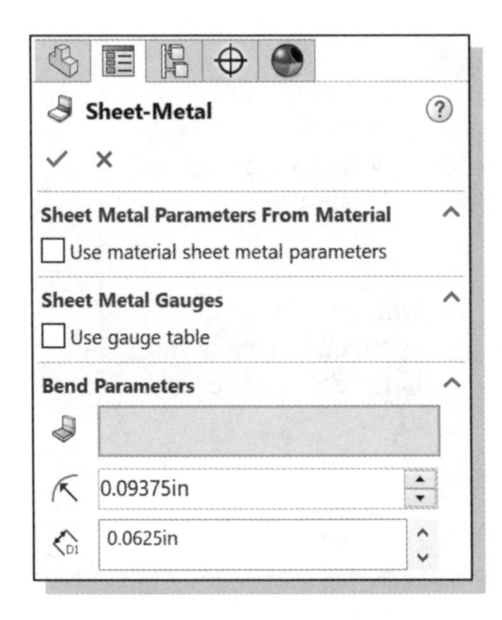

➢ The *Sheet-Metal PropertyManager* appears. This feature allows us to enter/modify the default bend parameters. We will use a bend radius equal to 1.5 × the gauge thickness.

12. In the *Bend Parameters* panel of the *Sheet-Metal PropertyManager*, enter **0.09375in** for the *Bend Radius* (1.5 × Gauge Thickness = 1.5×0.0625in = 0.09375in).

➢ Although these settings define the default parameters, many of these settings can be adjusted as the design is being constructed.

13. Click **OK** to accept the settings and exit the *Sheet-Metal PropertyManager*.

14. Select the **Sketch** button on the *Sketch* toolbar to create a new sketch.

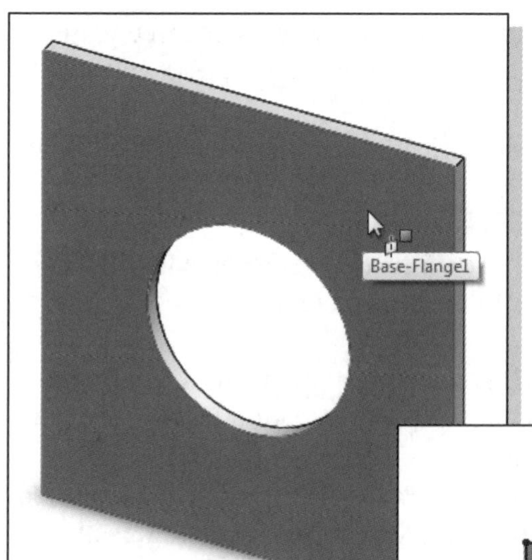

15. Select the front face of the base flange as the sketch plane for the new sketch as shown.

➢ We will create a sketch to be used in making an extruded cut. It will include eight individual cuts, related by symmetry.

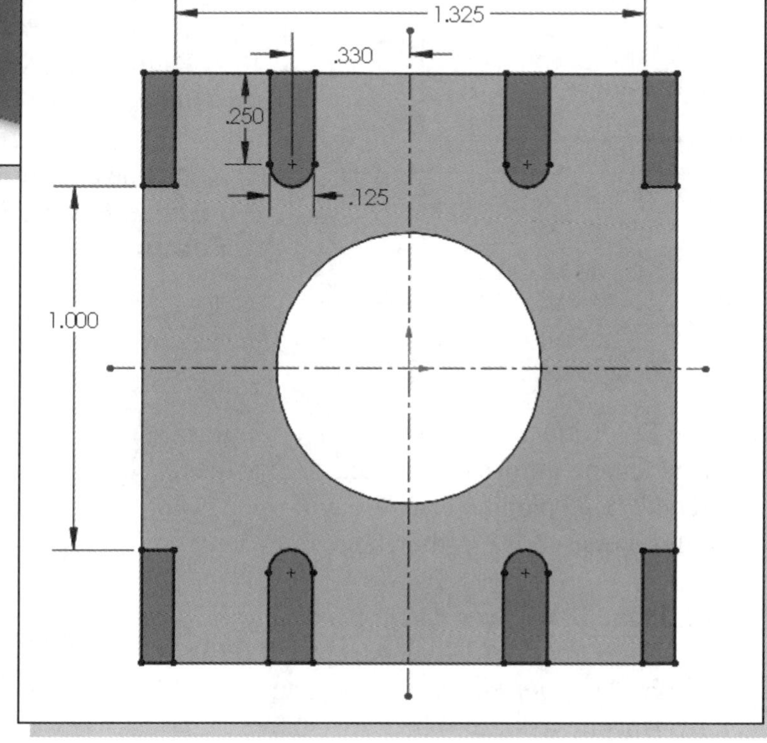

16. On your own, construct the sketch and apply the proper relations and dimensions as shown. (Hint: Make sure the sketch creates eight closed contours.)

➢ Note the design is symmetrical about the horizontal and vertical axes.

17. Click once with the **left-mouse-button** on the **Sketch** icon on the *Sketch* toolbar to exit the Sketch option.

18. Make sure the sketch is pre-selected and click once with the **left-mouse-button** on the **Extruded-Cut** icon on the *Sheet Metal* toolbar (or the same icon on the *Features* toolbar).

19. The *Cut-Extrude PropertyManager* appears. Verify that the eight individual regions are selected to create the extruded cut feature. If they are not, use the *Selected Contours* panel to define the regions to be extruded. Select the **eight individual contours** in the sketch to define regions for the extruded cut.

20. Select **Through All** as the end condition in the *Cut-Extrude PropertyManager*.

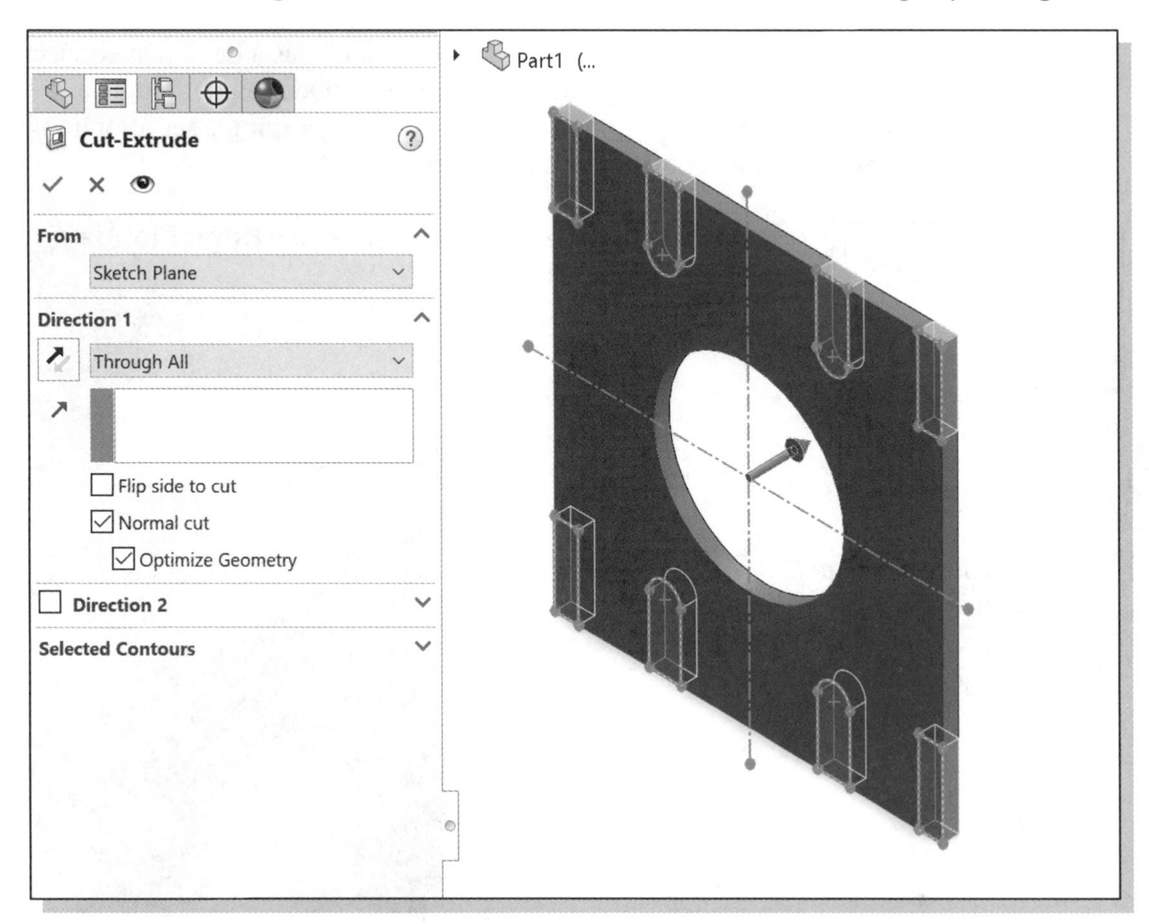

21. Click **OK** to accept the settings and create the *Extruded Cut* feature.

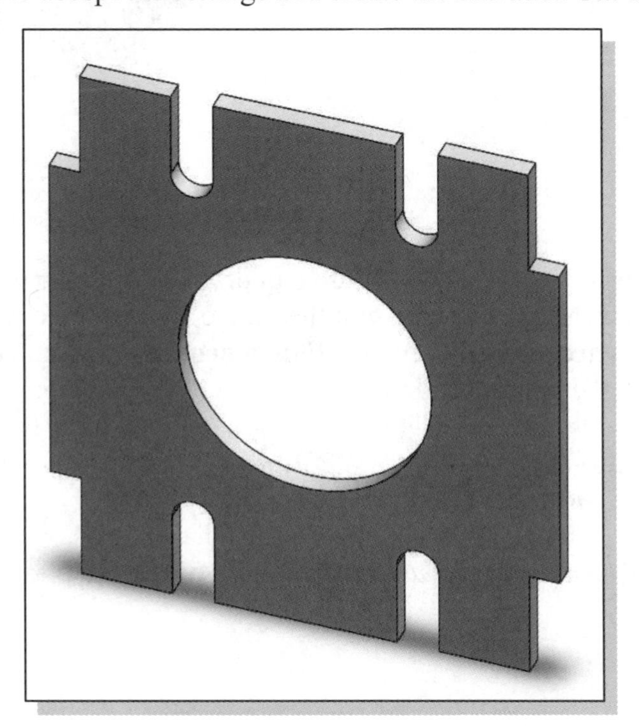

Creating an Edge Flange

➢ Sheet metal edge flange features consist of a flat face, which has a bend that connects the flat face to an existing straight edge of the sheet metal model. Edge flange features are added by selecting one or more edges and by specifying a set of options which determines the size and position of the material added.

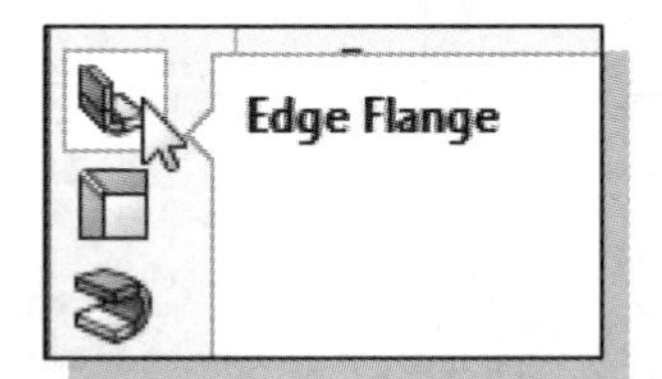

1. On the *Sheet Metal* toolbar, select the **Edge Flange** command as shown.

2. The *Edge Flange PropertyManager* appears with the edge selection window active. Select the **outer edge** of the base flange as shown.

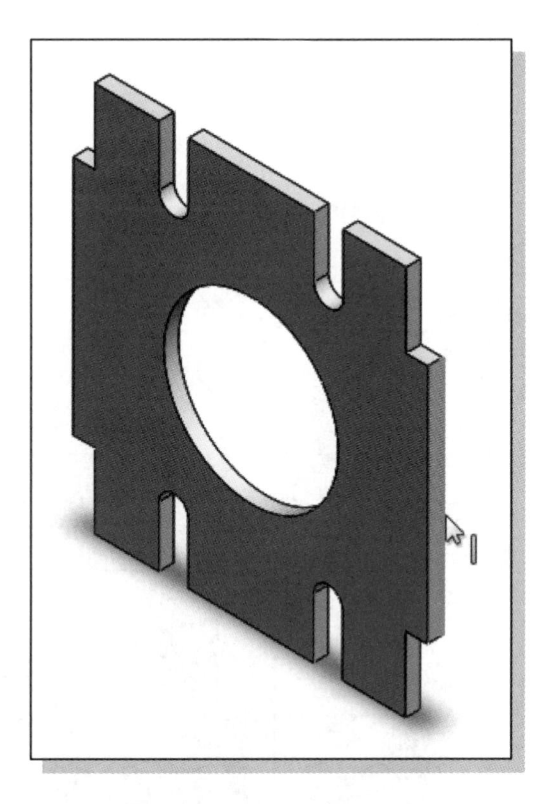

3. A preview appears, with the cursor active to drag the edge flange, setting the direction of the bend and the length of the flange. **Move the cursor to a location in the graphics area which creates a flange bending back** as shown below, and click once with the left-mouse-button.

4. In the *Flange Length* panel of the *Edge-Flange PropertyManager*, enter **0.500in** for the *Flange Length.*

5. In the *Flange Length* panel select the **Outer Virtual Sharp** length definition option.

6. In the *Flange Position* panel, select the **Material Outside** option. This option is used to control the position of the bend that connects to the new flange section.

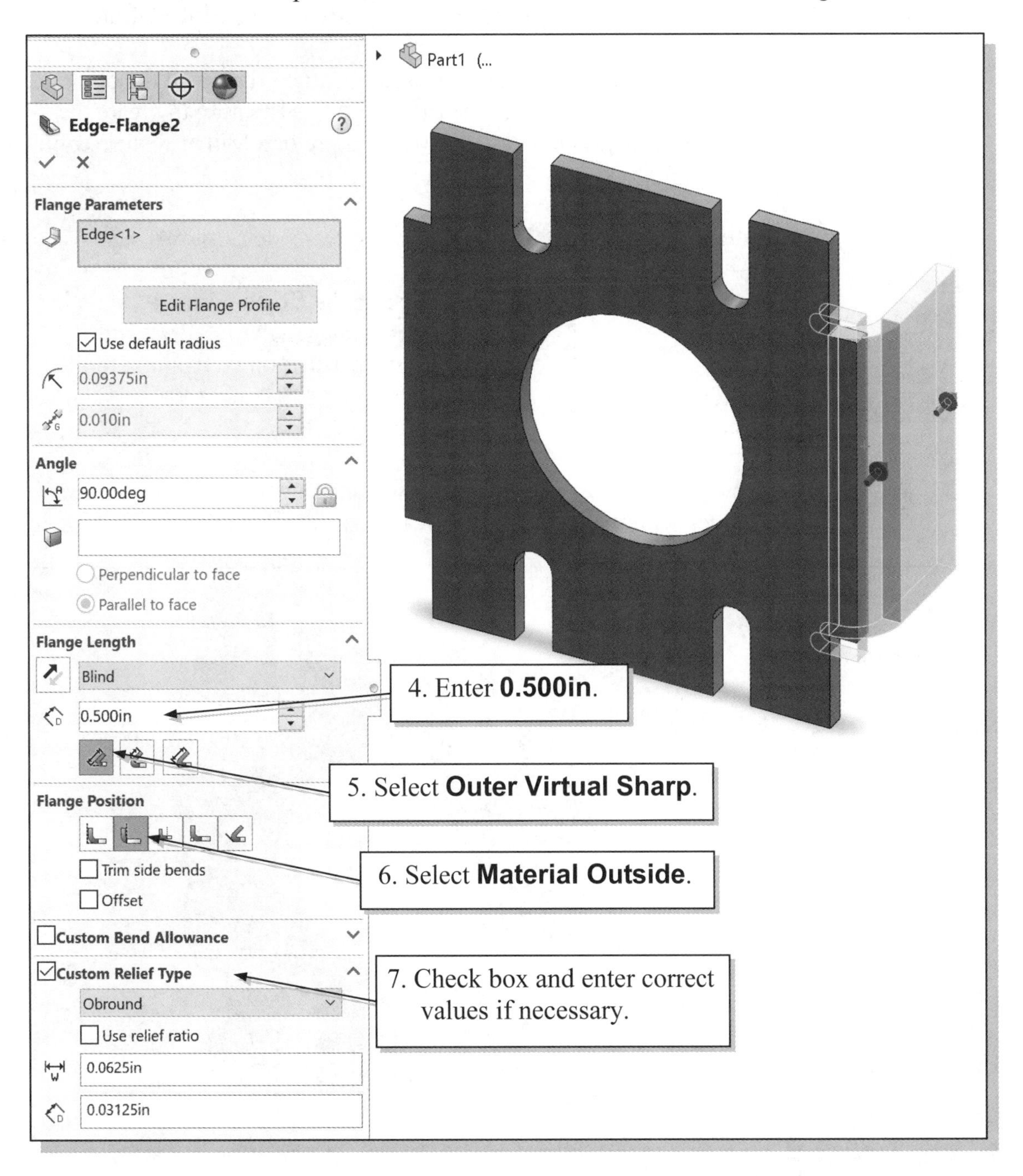

7. Notice the default values for *Bend Radius*, *Bend Allowance*, and *Relief Type* appear. These parameters can be modified by selecting the appropriate check box and entering custom parameters. (The values should be *Radius* = 0.09375in; *Bend Allowance:* K-factor = 0.41; *Relief Type:* Obound, Width = 0.0625in, Depth = 0.03125in.) If any default values are incorrect, check the appropriate box and enter values.

8. Click **OK** to accept the settings and create the *Edge Flange* feature. Notice the edge flange is created and Edge-Flange1 appears on the *FeatureManager Design Tree*. Also notice that Edge-Flange1 is inserted above Flat-Pattern1.

➢ The sheet metal design we are creating includes an identical flange on the other end of the base flange. We will modify the Edge-Flange1 feature to create the two flanges. Alternately, a second Edge-Flange feature or a Mirror feature could be added.

9. On your own, **rotate** the sheet metal part to view the rear side as shown.

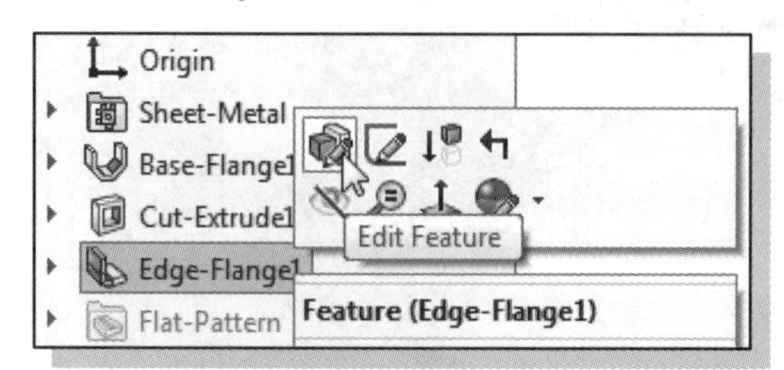

10. Right click on the **Edge-Flange1** icon in the *FeatureManager Design Tree* and select **Edit Feature** on the context toolbar.

11. With the *Edge* selection window active, move the cursor over the vertical edge as shown and click once with the **left-mouse-button** to select this edge.

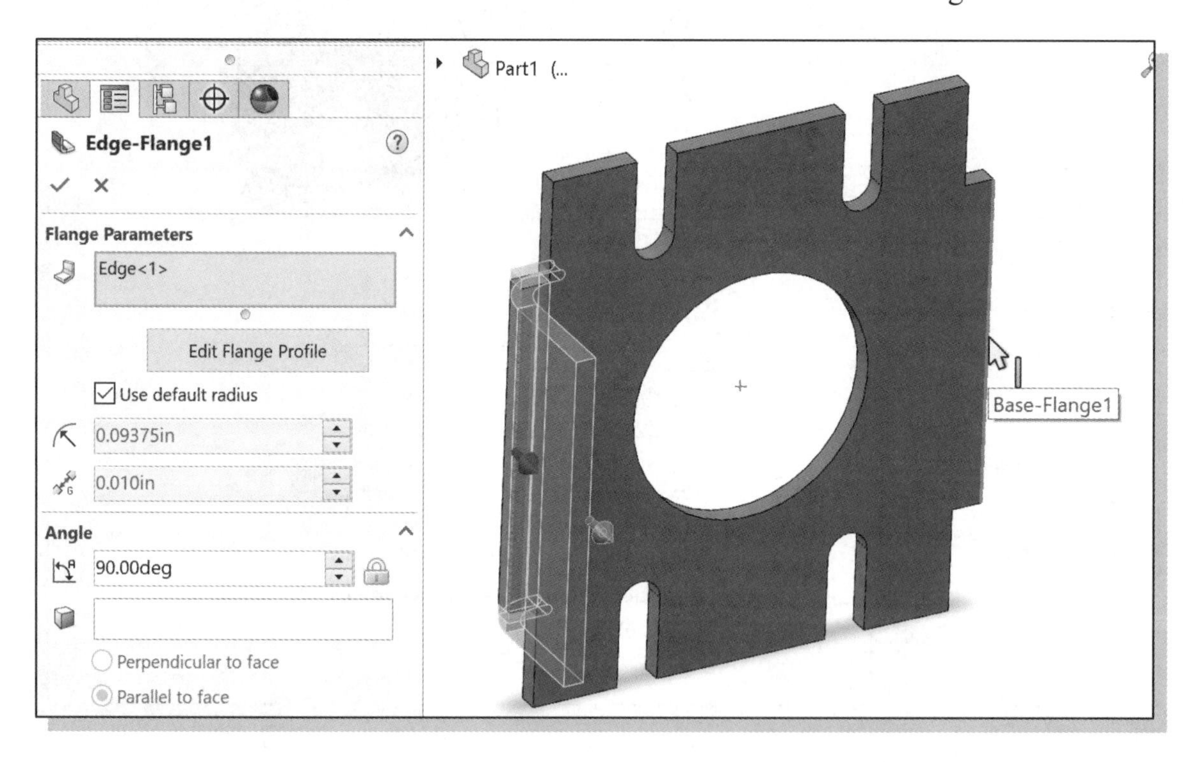

12. Notice Edge<2> is added to the *Edge* selection window and the preview now includes the two flanges. (Both flanges are created using the same parameters.) In the *Edge-Flange1 PropertyManager*, click **OK** to accept the selections and settings and create the two flanges.

➢ We will now perform measurements to verify the geometry of the sheet metal part.

13. Select the **Tools** pull-down menu, then select **Evaluate** and **Measure**.

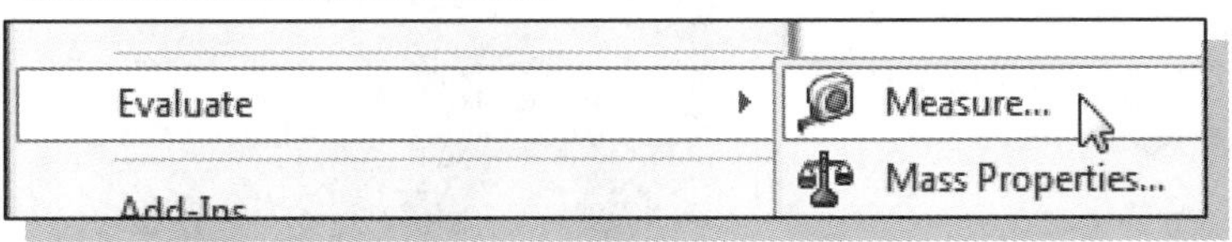

14. Select the two inside surfaces. Confirm the distance between the two inside surfaces of the two flanges is maintained as **1.508 in**. (Note: You may have to adjust the Units/Precision using the button shown.) This is the same as the dimension specified in the base feature. This is a result of selecting the Material Outside option for the *Flange Position*.

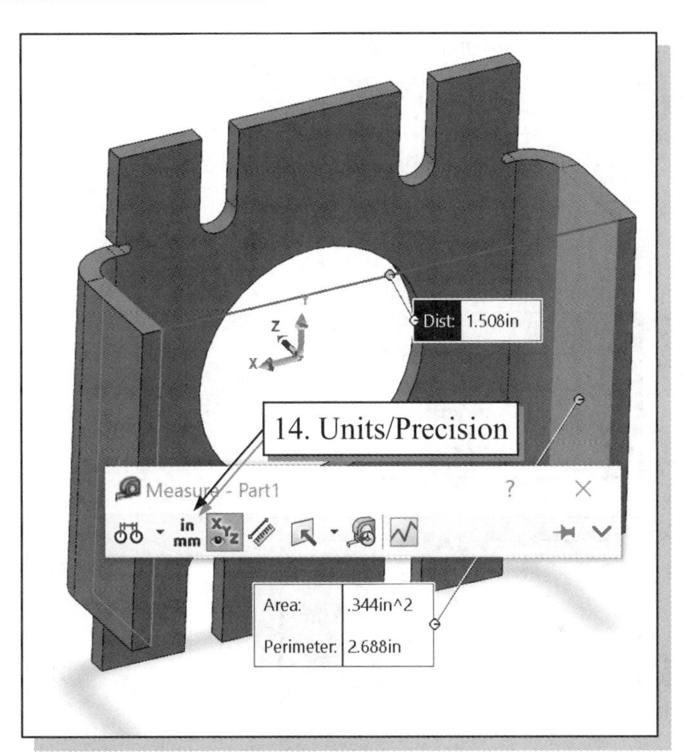

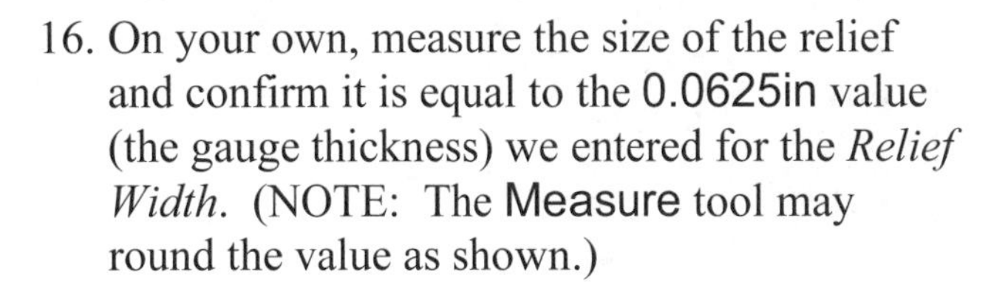

15. Click in the **graphics area**, away from the model, to clear the selections for the Measure tool.

➢ Note the relief cut is also added when the two flanges are created.

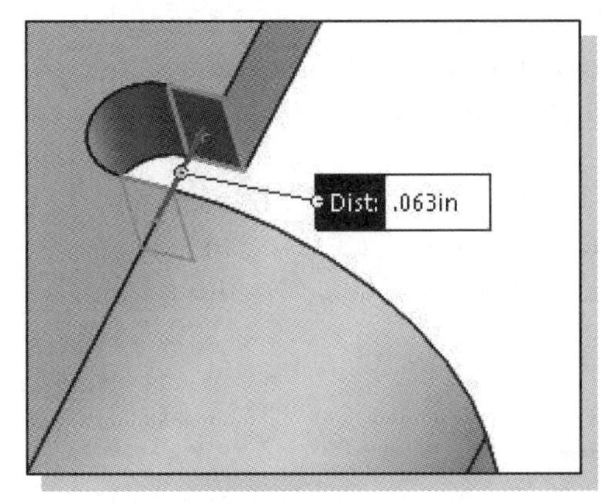

16. On your own, measure the size of the relief and confirm it is equal to the 0.0625in value (the gauge thickness) we entered for the *Relief Width*. (NOTE: The Measure tool may round the value as shown.)

17. Close the *Measure* pop-up toolbar or press the **[Esc]** key to exit the Measure command.

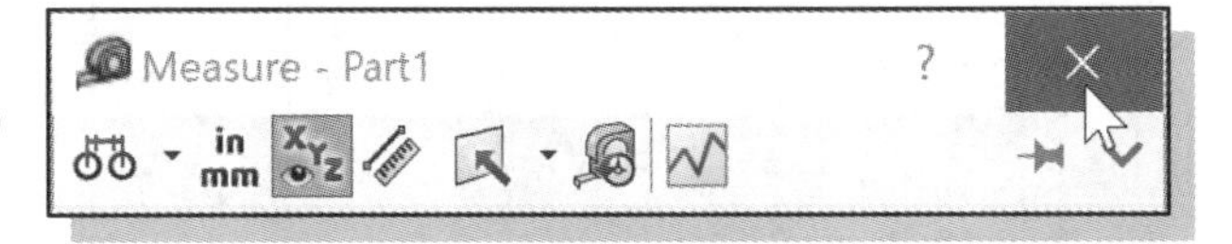

Adding a Tab

We will next add a tab with a more complex sketched section on the existing flange surface. Note that this new section could alternately have been created directly as part of the edge flange by using the **Edit Flange Profile** option in the *Edge-Flange PropertyManager*.

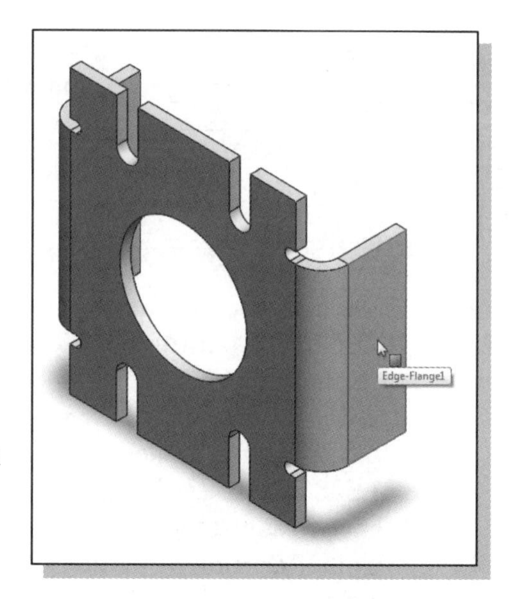

1. On your own, return to the standard isometric view.

2. Pre-select the outside surface of the flange by clicking once with the left-mouse-button as shown.

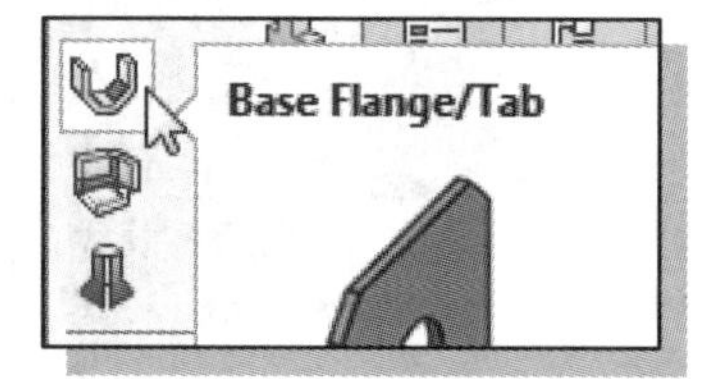

3. In the *Sheet Metal* toolbar, select the **Base Flange/Tab** command by left-mouse-clicking the icon.

4. A sketch opens on the selected face. On your own, construct the 2D sketch and apply the proper constraints and dimensions as shown. (Hint: Use the available constraint tools to align the symmetrical geometry.)

➤ Note the design is symmetrical about the horizontal axis and there are two sets of parallel inclined lines.

➤ Note: Be sure to sketch the line coincident with the existing flange edge; this is necessary to create a closed contour.

Be sure to include this line in your sketch.

Edges are parallel.

5. Click once with the **left-mouse-button** on the **Sketch** icon on the *Sketch* toolbar to exit the sketch.

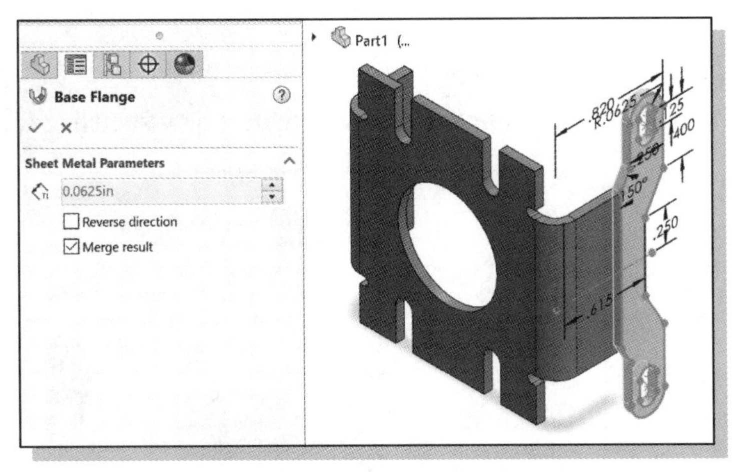

6. Upon exiting the sketch, the *Base Flange PropertyManager* appears. Click **OK** to accept the settings and create the Tab.

➢ Notice the tab is created and Tab1 appears on the *FeatureManager Design Tree*. We will now use the Mirror feature to create an identical tab on the other flange.

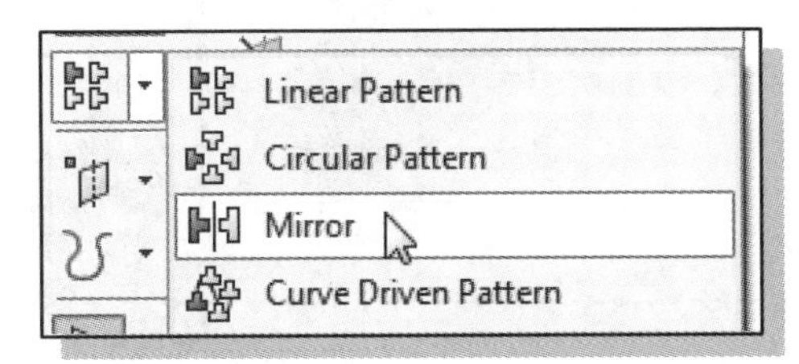

7. Select the **Mirror** option from the *Features* toolbar as shown.

8. Select the **Right Plane** as the *Mirror Face/Plane* and **Tab1** as the *Features to Mirror*, and click **OK** to create the Mirror feature.

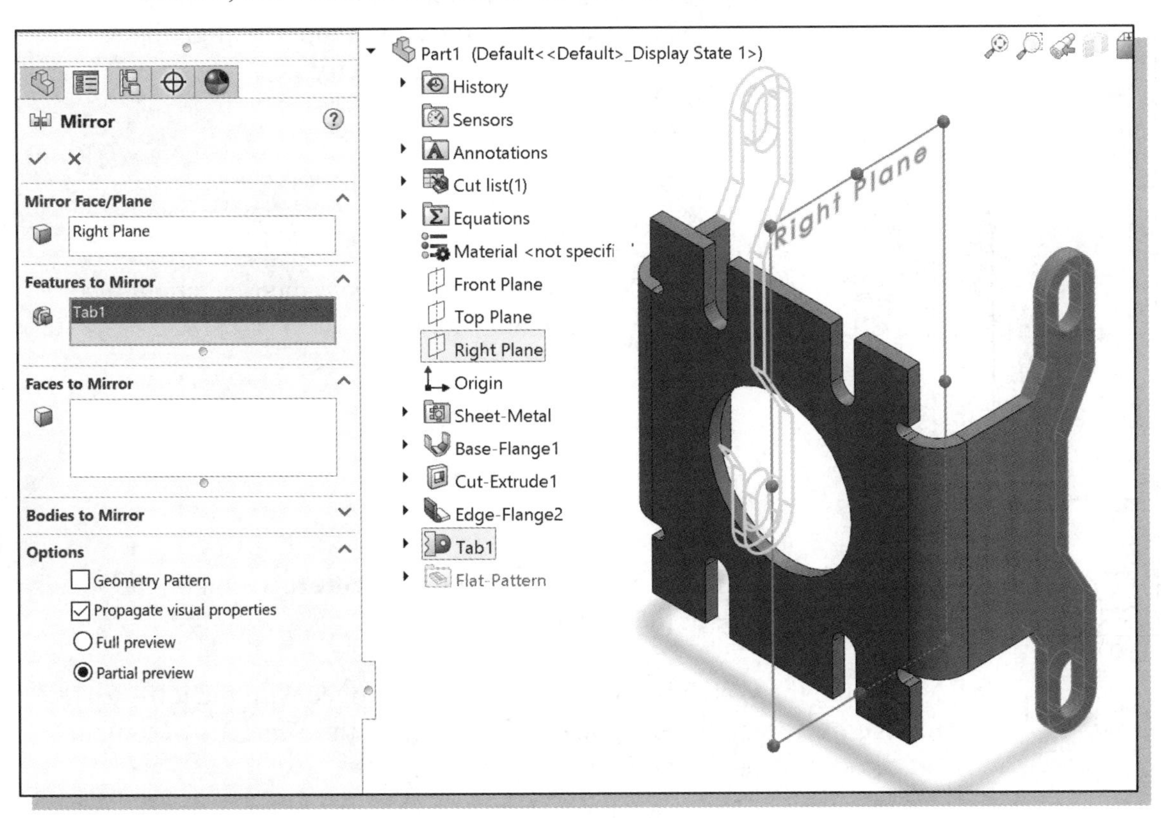

Creating a Cut Feature

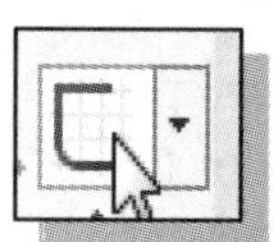

1. Select the **Sketch** button on the *Sketch* toolbar to create a new sketch.

2. Select the **Right Plane** as the sketch plane for the new sketch.

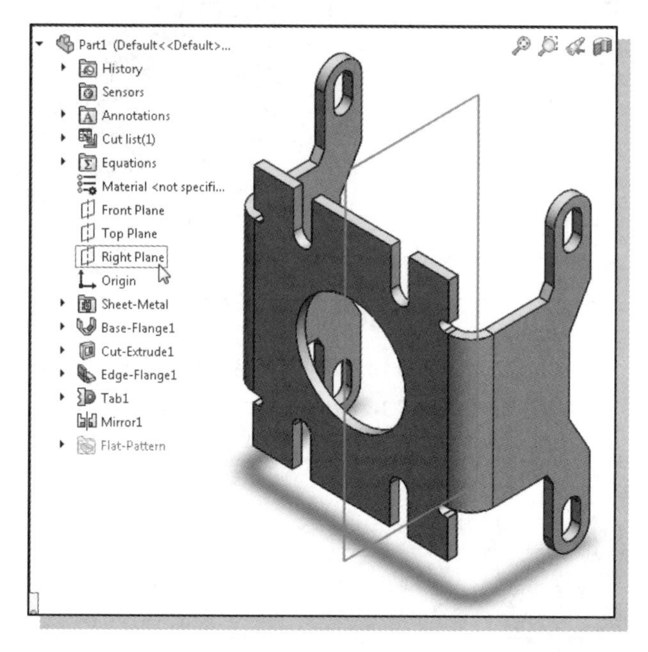

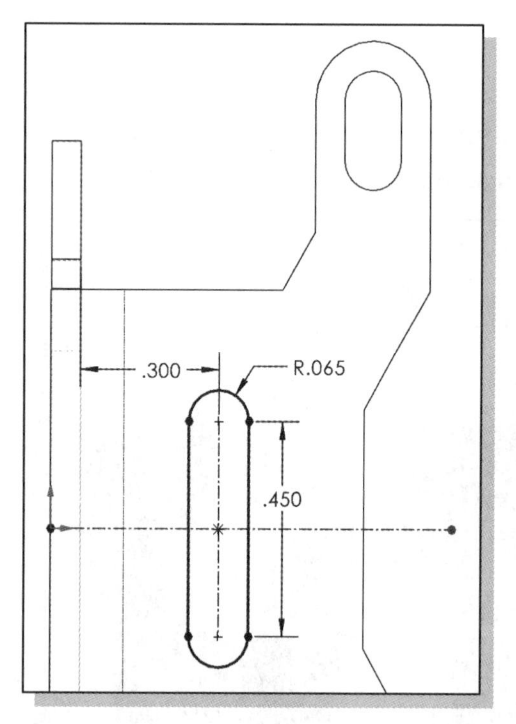

3. On your own, construct the 2D sketch and apply the proper constraints and dimensions as shown. (Hint: Use the available constraint tools to align the symmetrical geometry.)

➢ Note the design is symmetrical about the horizontal axis.

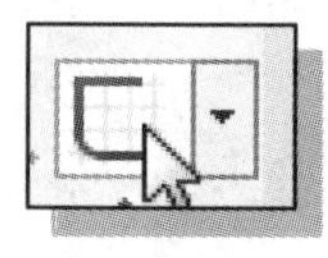

4. Click once with the **left-mouse-button** on the **Sketch** icon on the *Sketch* toolbar to exit the sketch.

5. Make sure the sketch is pre-selected and click once with the **left-mouse-button** on the **Extruded-Cut** icon on the *Sheet Metal* toolbar (or the same icon on the *Features* toolbar).

6. In the *Direction 1* panel of the *Cut-Extrude PropertyManager*, select **Through All** as the *End Condition*.

7. **Check** the check box next to the *Direction 2* panel heading to activate a second direction for the Cut-Extrude feature and select **Through All** as the *End Condition*.

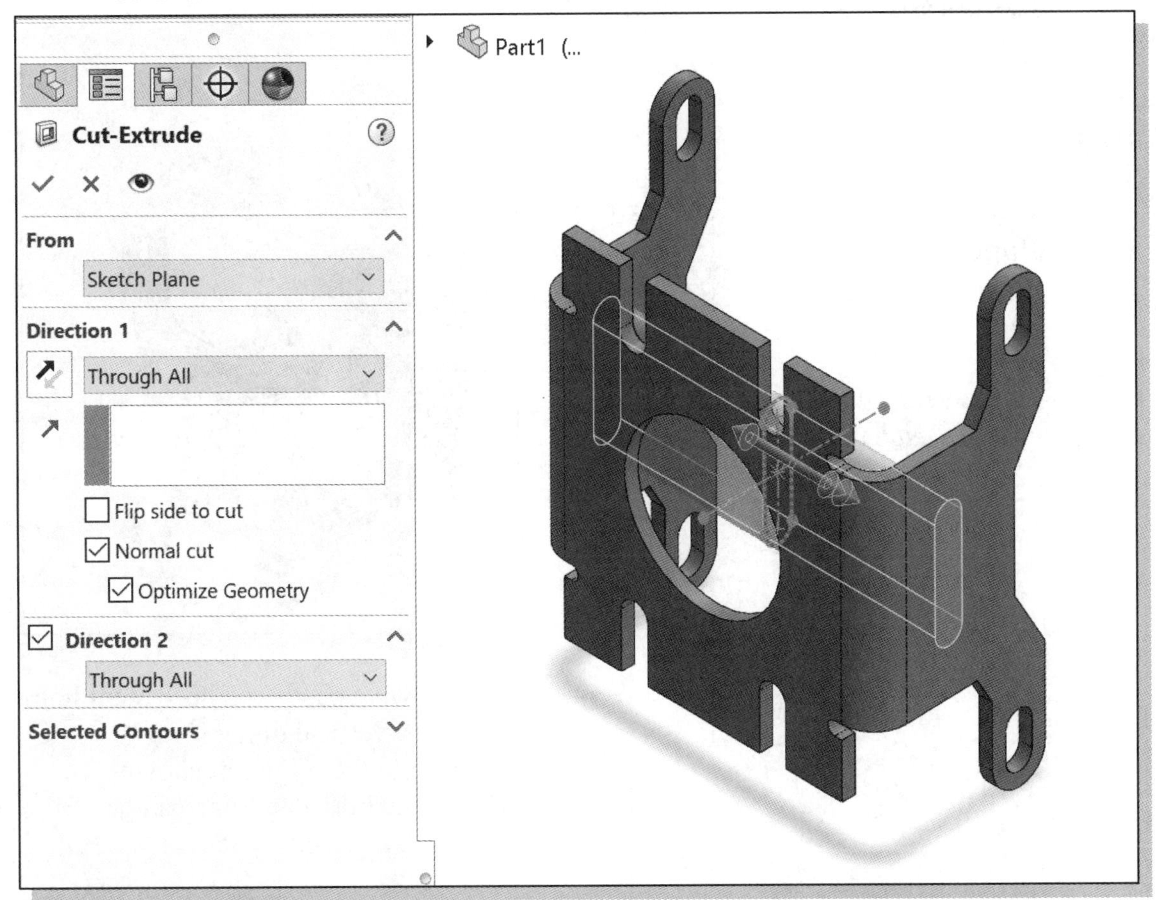

8. Click **OK** in the *PropertyManager* to accept the settings and create the Cut-Extrude feature.

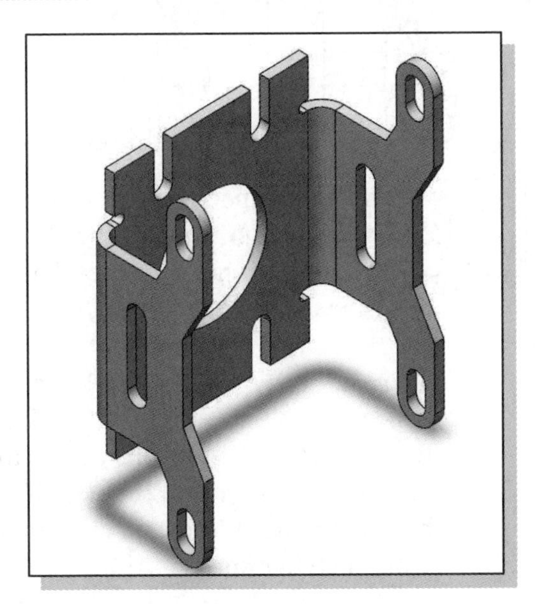

Creating a Bend

We will now bend the tab using the **Sketched Bend** command.

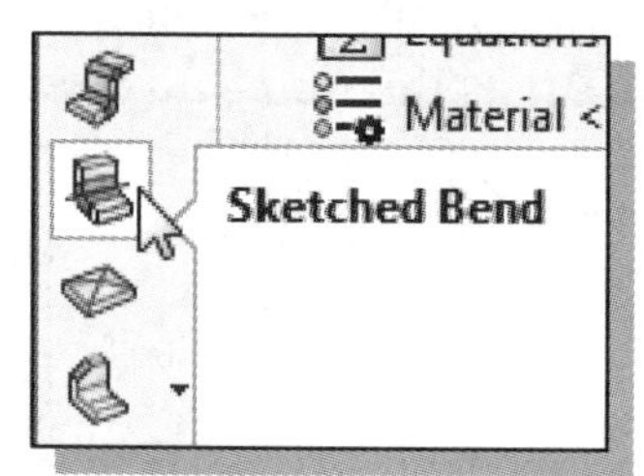

1. Select the **Sketched Bend** button on the *Sheet Metal* toolbar.

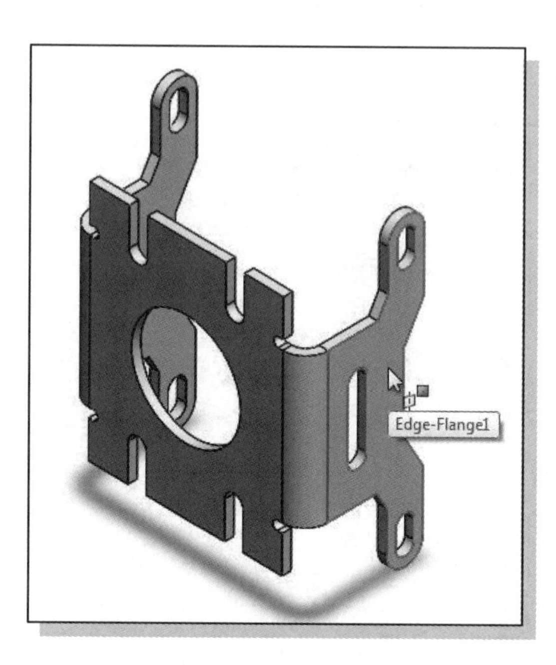

2. Select the outside surface of the **Edge-Flange** feature as shown.

3. Select the **Line** command on the *Sketch* toolbar.

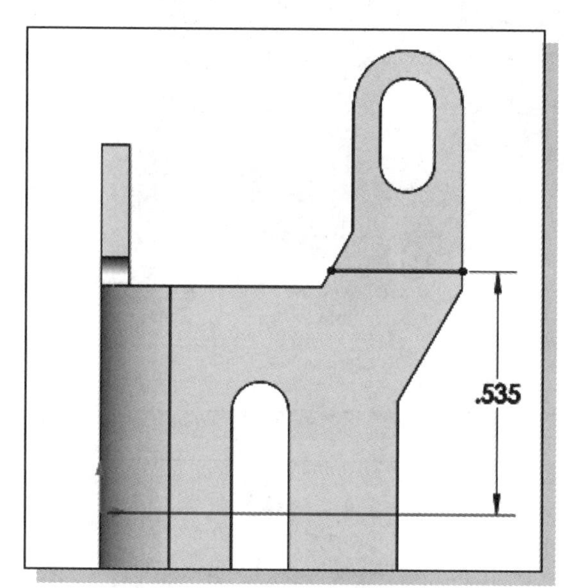

4. On your own, construct a horizontal line and apply a vertical dimension, measuring to the origin, as shown. (NOTE: The horizontal line can be drawn with an arbitrary length.)

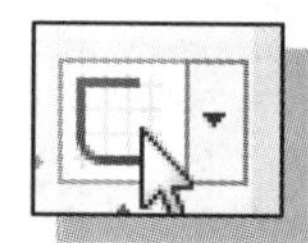

5. Click once with the **left-mouse-button** on the **Sketch** icon on the *Sketch* toolbar to exit the sketch.

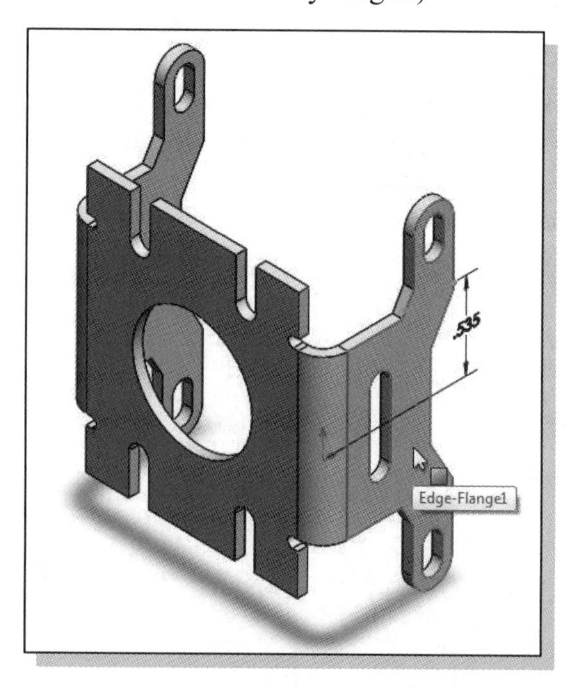

6. The *Sketched Bend PropertyManager* appears with the Fixed Face selection window active. Move the cursor to a location on the face of the sketch, **below the bend line**, and click once with the **left-mouse-button**.

➢ Notice Face<1> appears in the *Fixed Face* selection window and a circle appears in the graphics area marking the Fixed Face location.

7. On the *Bend Parameters* panel, under *Bend position:* select the **Bend Outside** option button.

8. Verify that the bend direction as indicated by the arrow in the graphics area is pointed in the correct direction (outward). If necessary, click the Reverse Direction button.

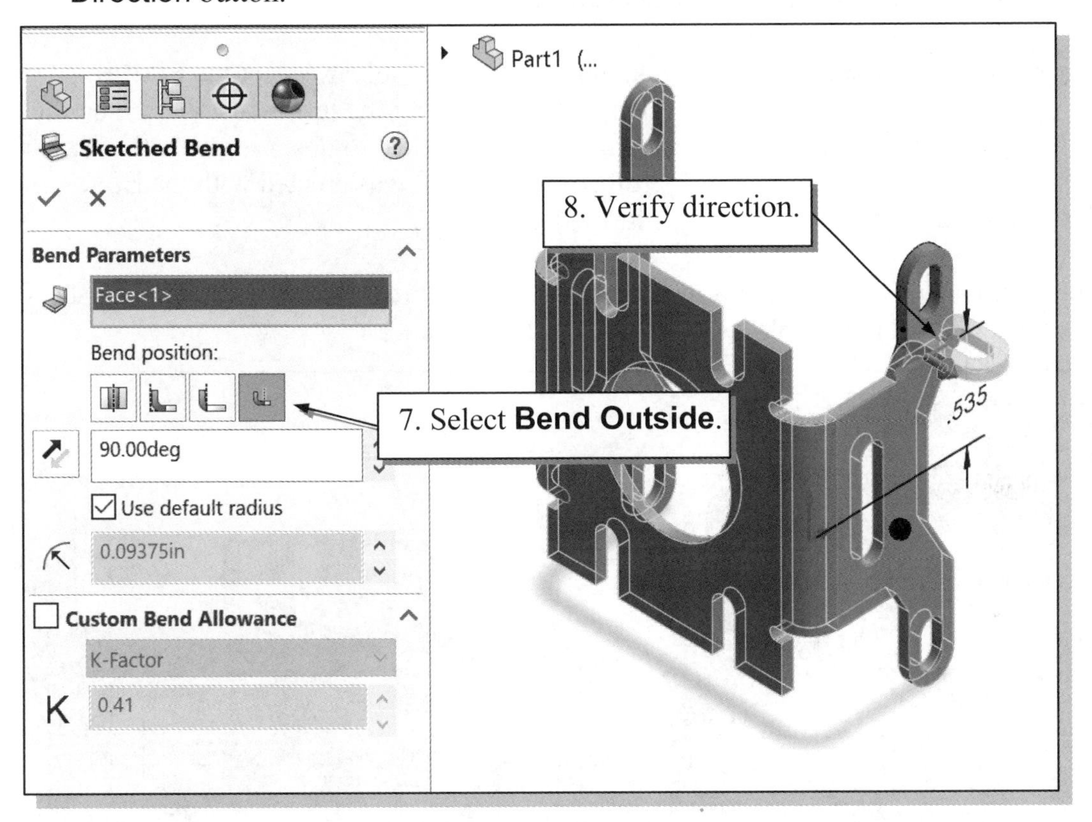

9. Click **OK** to accept the settings and create the Sketched Bend feature. (NOTE: We will use the default radius and K-Factor value we set earlier.)

10. On your own, repeat the steps above to create a second **Sketched Bend** feature as shown.

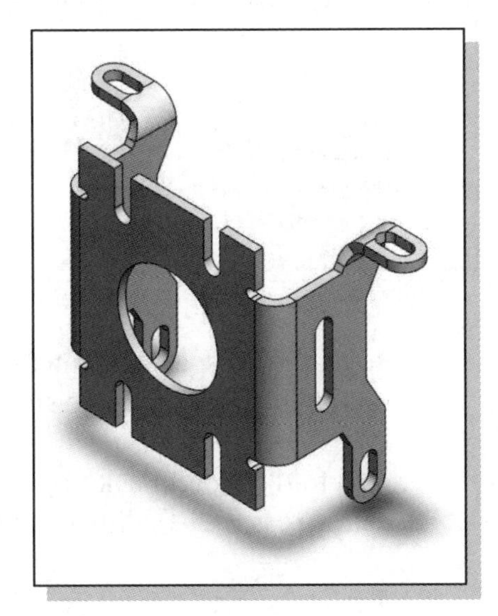

Flattening the Sheet Metal Part

A sheet metal flat pattern is the shape of the sheet metal part before it is formed. A flat pattern is required to create drawings for manufacturing. The flat pattern shows the shape of the sheet metal part before it is formed showing all the bend lines, bend zones, punch locations, and the shape of the entire part with all bends flattened and bend factors considered. A bounding box may also be shown indicating the smallest rectangle in which the flat pattern can fit. The SOLIDWORKS **Flatten** command calculates the material and layout required to flatten a 3D sheet metal model.

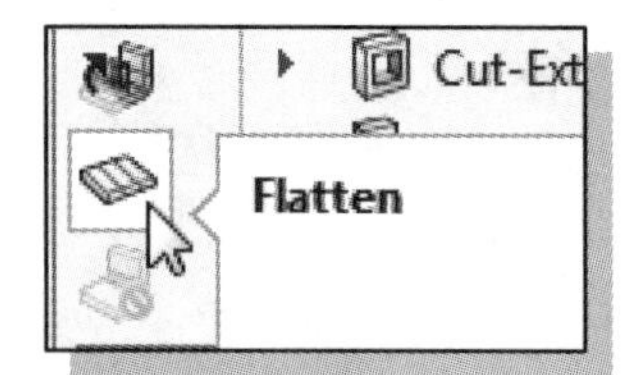

1. On the *Sheet Metal* toolbar, select the **Flatten** command by clicking the left-mouse-button on the icon. (This is identical to unsuppressing the Flat-Pattern feature that was created with the Base-Flange1 feature.)

❖ SOLIDWORKS calculates the material and layout required to flatten the 3D sheet metal model and displays the flat pattern as shown.

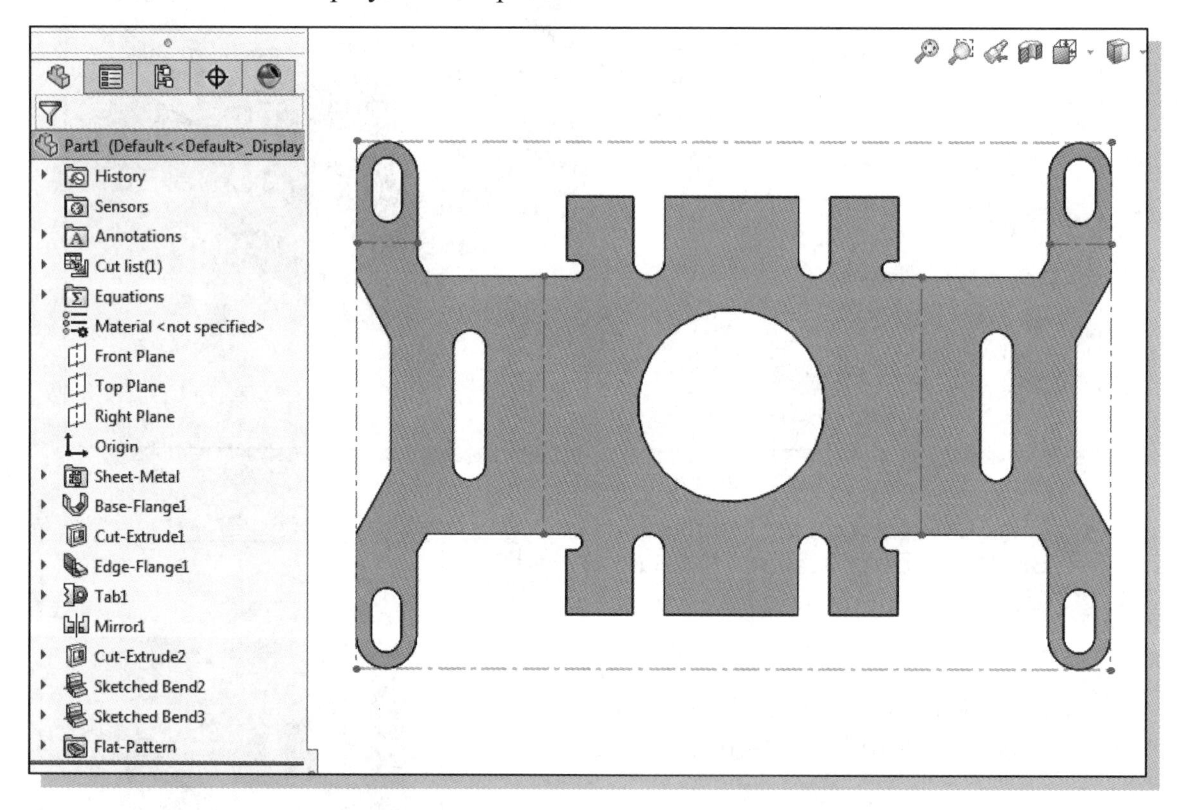

❖ Note that the displayed center lines identify the locations of the bends and the bounding box indicates the smallest rectangle in which the flat pattern can fit.

2. To fold the part back up, click **Flatten** on the *Sheet Metal* toolbar again, or **suppress** Flat-Pattern on the *FeatureManager Design Tree*.

Confirm the Flattened Length

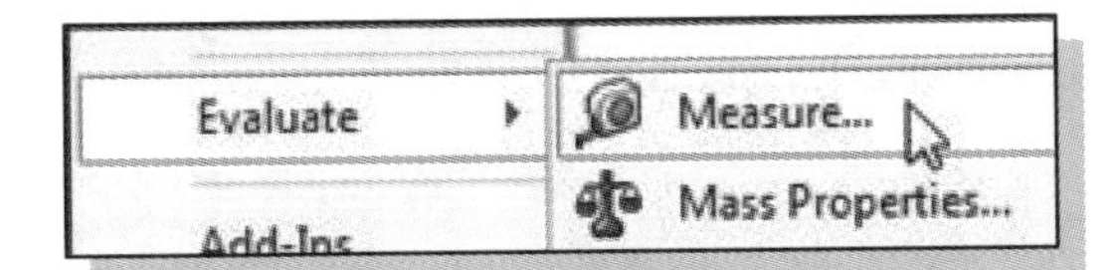

1. With the sheet metal part **flattened**, select the **Measure** tool from the *Tools* pull-down menu.

2. Select the **right edge**, as shown in the figure below, and the **large circle** to measure. (Select the **Center Dist** option as shown.)

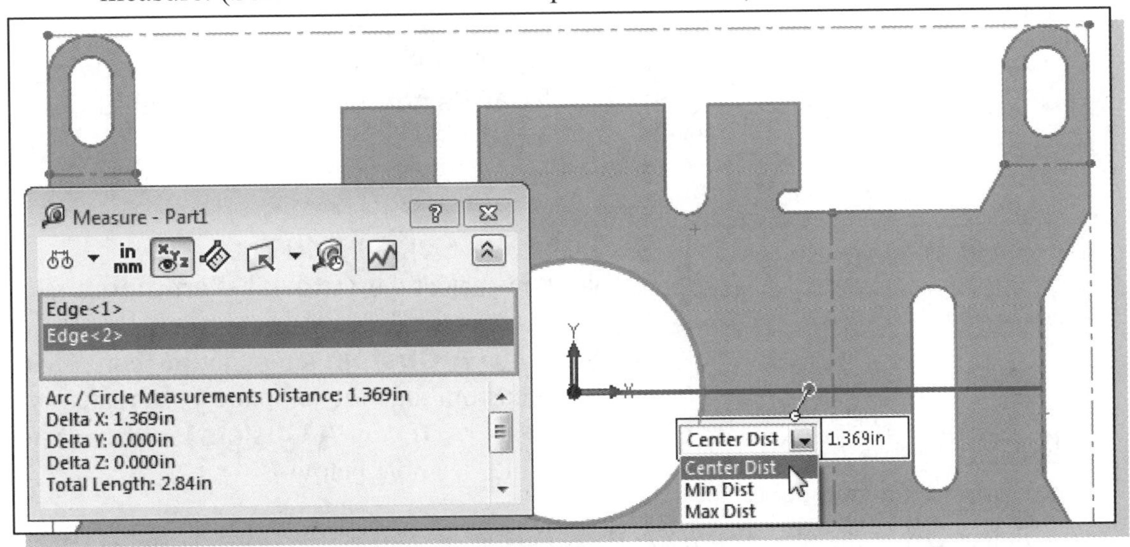

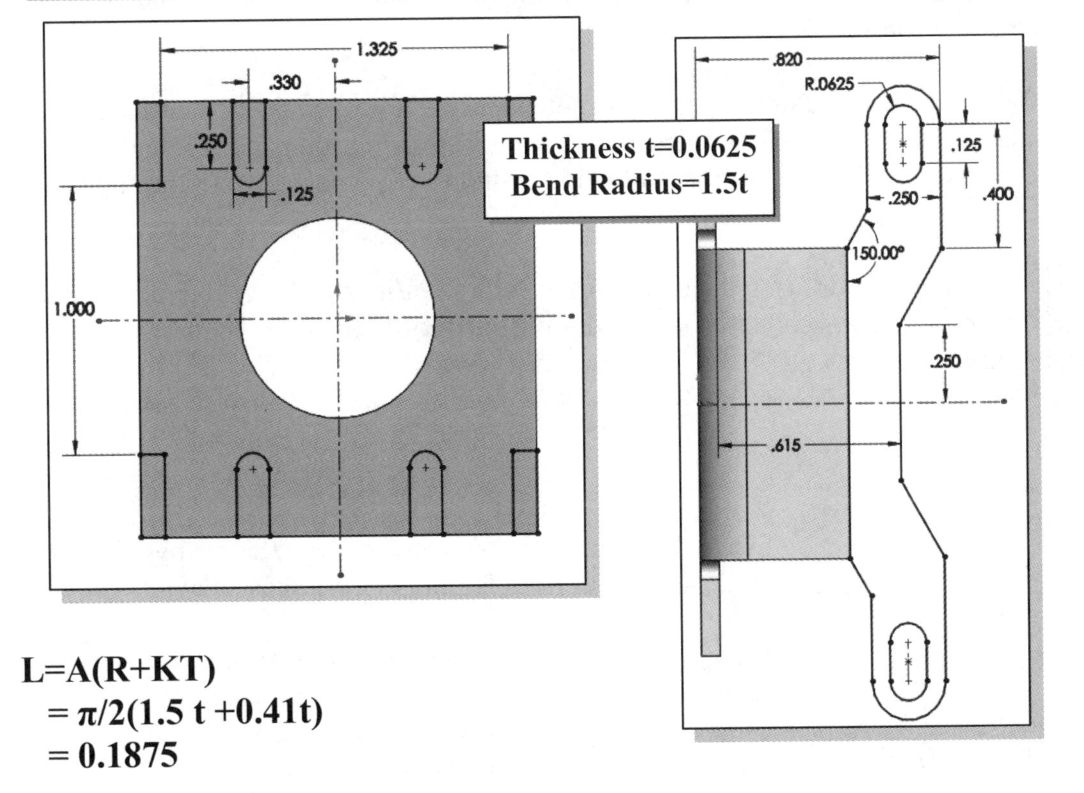

L=A(R+KT)

= π/2(1.5 t +0.41t)

= 0.1875

Flattened Length = W+H+L

=(1.508/2-1.5t)+(0.615-1.5t)+0.1875=1.3690

Creating a Sheet Metal Drawing

1. To fold the part back up, click **Flatten** on the *Sheet Metal* toolbar again, or **suppress** Flat-Pattern1 on the *FeatureManager Design Tree.*

2. On your own, save the design. Use the following filename: ***Actuator-Bracket.sldprt***, and enter ***Actuator Bracket*** as the description.

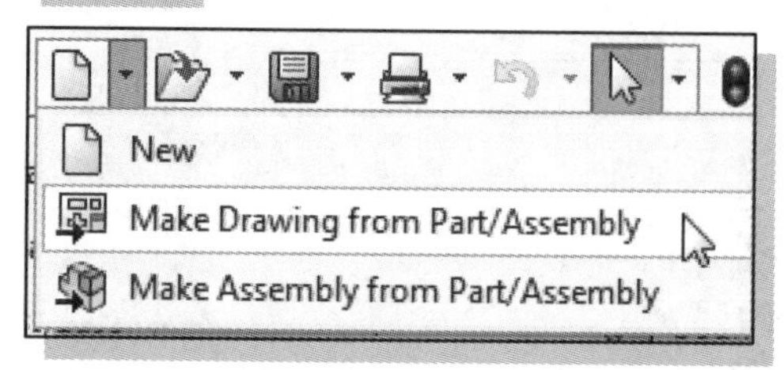

3. Click on the arrow next to the **New** icon on the *Menu Bar* and select **Make Drawing from Part/Assembly** in the pull-down menu.

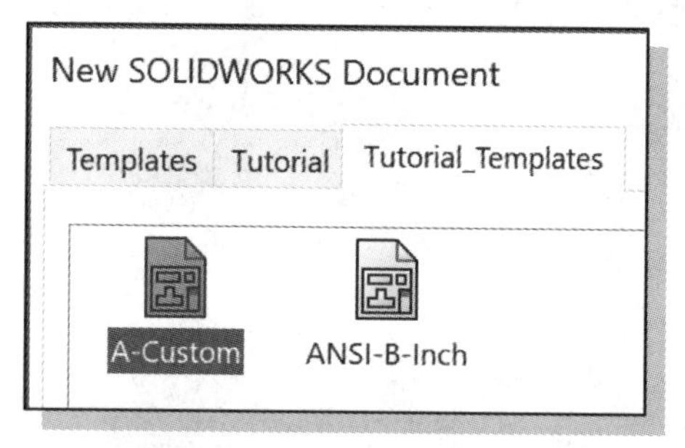

4. In the *New* SOLIDWORKS *Document* window, select the **Tutorial_Templates** tab.

5. Select the **A-Custom** template file icon. This is the custom drawing template you created in Chapter 11. (If your A-Custom template is not available, see note below.)

6. Select **OK** in the *New SOLIDWORKS Document* window to open the new drawing file.

❖ **NOTE:** If your A-Custom template is not available: (1) open the default Drawing template; (2) set document properties following instructions on page 8-7; and (3) set sheet properties and sheet format following instructions on pages 11-18 through 11-20.

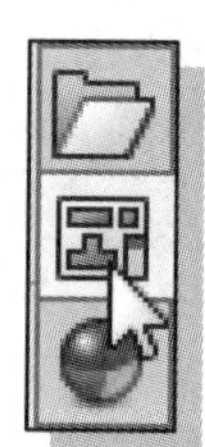

➢ The new drawing document opens with the **View Palette** active. (If the *View Palette* is collapsed, select the **View Palette** tab at the right of the graphics area.)

7. Select the **Isometric** view in the *View Palette* by clicking on the icon with the **left-mouse-button** as shown.

8. Click and hold down the **left-mouse-button** and **drag** the isometric view from the *View Palette* into the graphics window and place the **isometric view** near the upper right side of the border as shown.

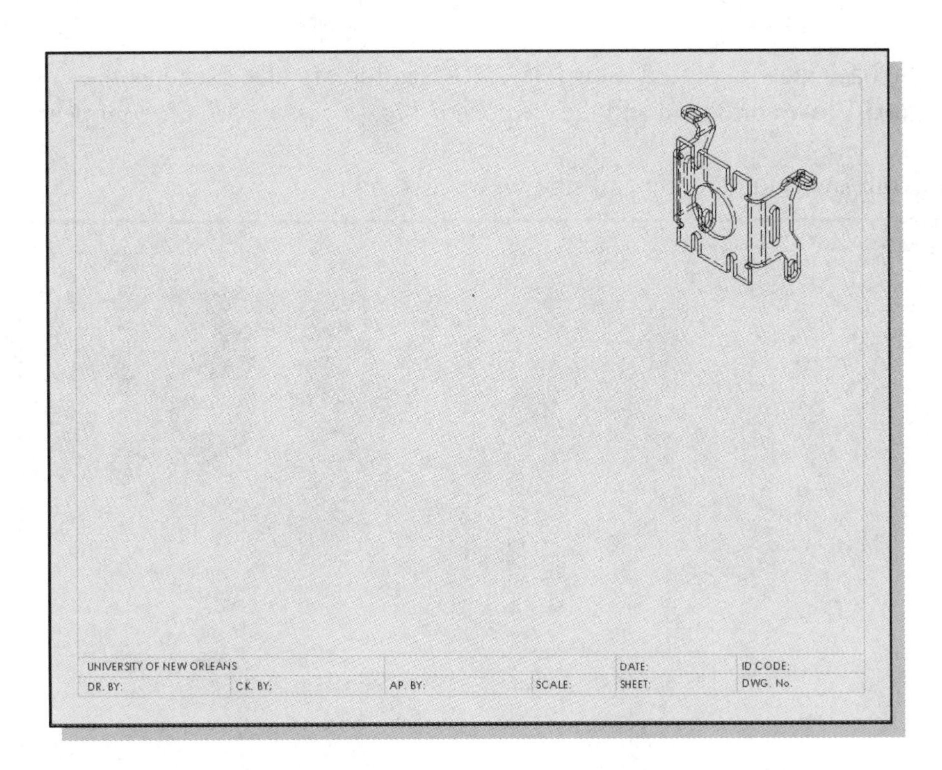

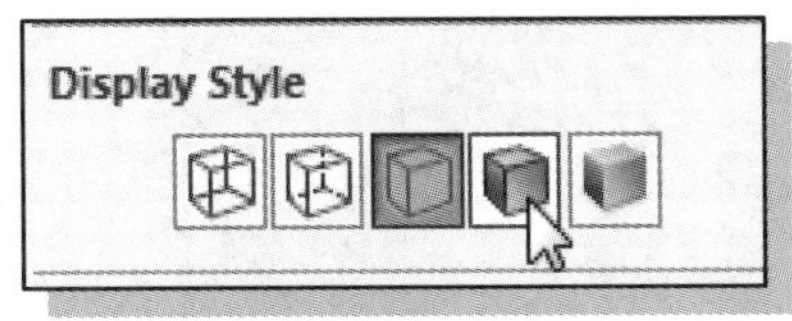

9. In the *Display Style* panel of the *Drawing View PropertyManager*, select **Shaded with Edges**.

10. Click **OK** in the *PropertyManager* to accept the setting.

11. Click on the **Model View** icon in the *Drawing* toolbar.

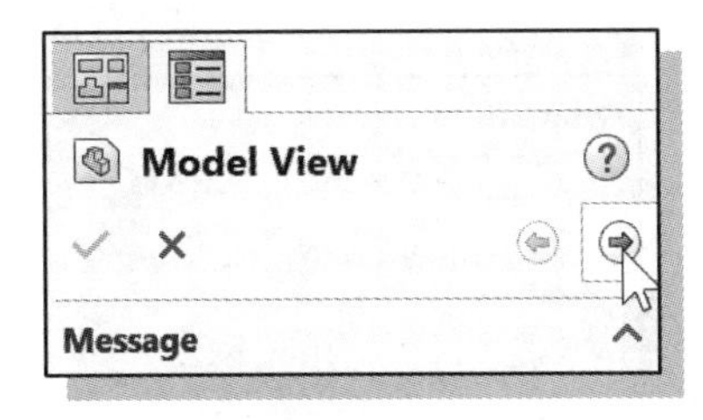

12. The *Model View PropertyManager* appears with the *Actuator-Bracket* part file selected as the part from which to create the base view. Click the **Next** arrow as shown to proceed with defining the base view.

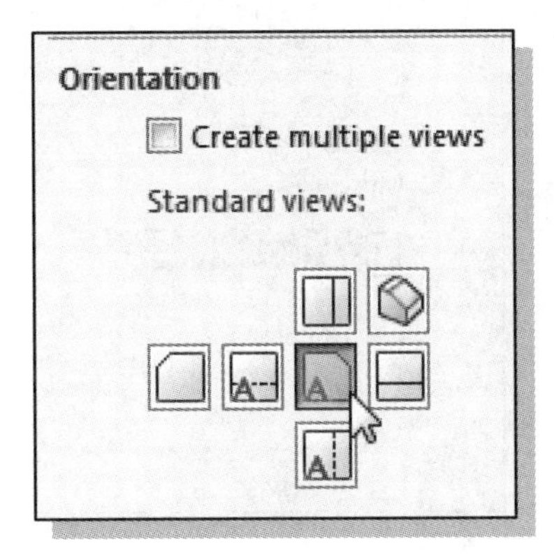

13. In the *Model View PropertyManager*, select the **Front View** for the *Orientation*, as shown. (Make sure the *Create multiple views* box is unchecked.)

14. Move the cursor inside the graphics area and place the **front** view, by clicking the left-mouse-button, near the left side.

➢ When the base view is placed, SOLIDWORKS automatically executes the **Projected View** command and the *Projected View PropertyManager* appears.

15. On your own, add the top and side views as shown.

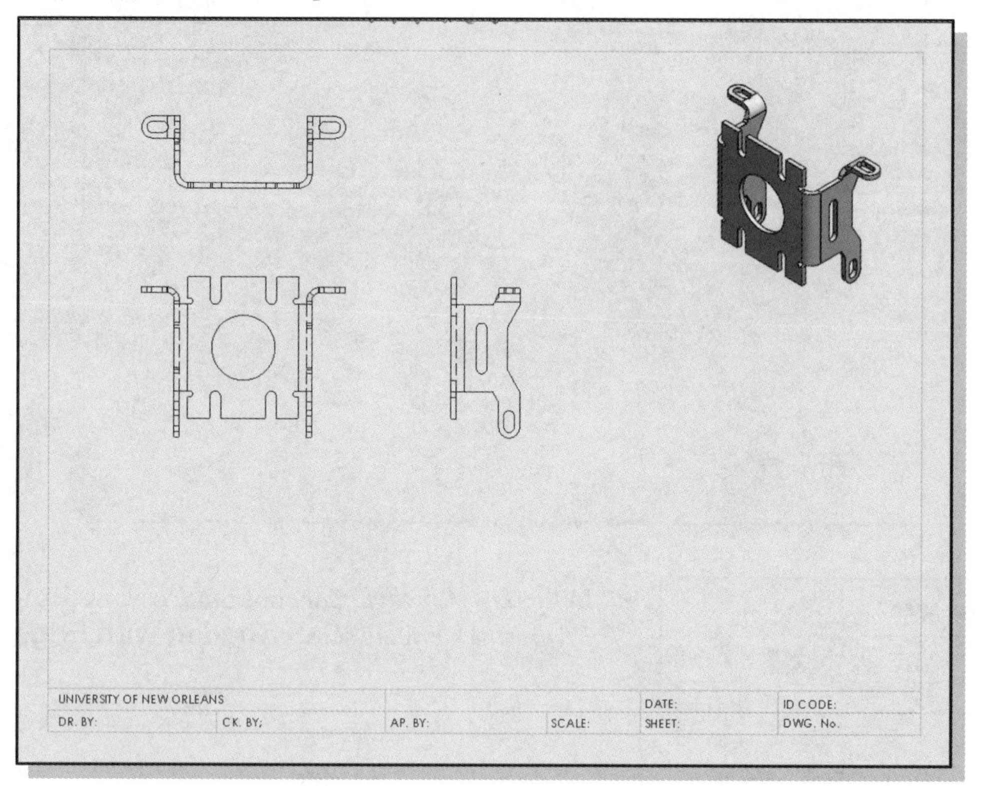

16. Press the **[Esc]** key to exit the **Projected View** command.

17. Click on the **Model View** icon in the *Drawing* toolbar.

18. In the *Model View PropertyManager*, click the **Next** arrow to proceed with inserting an additional view.

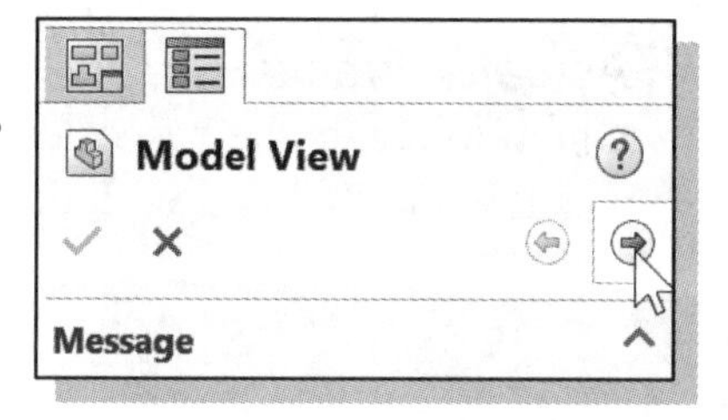

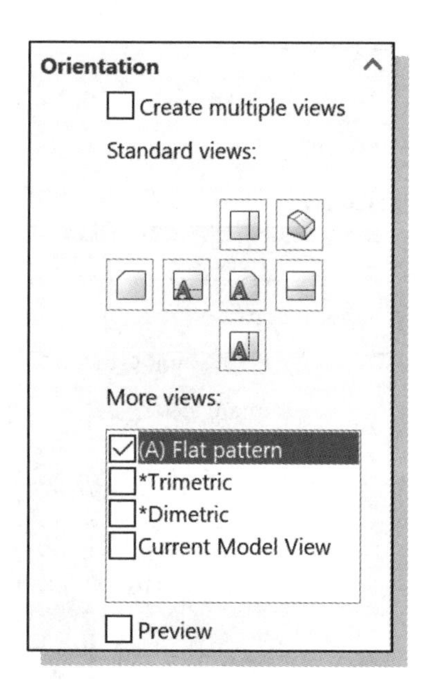

19. In the *Orientation* panel of the *Model View PropertyManager*, select the **(A) Flat pattern** view under *More views:* as shown.

20. Move the cursor inside the graphics window and place the **flat pattern** view, by clicking the left-mouse-button, below the *isometric* view as shown.

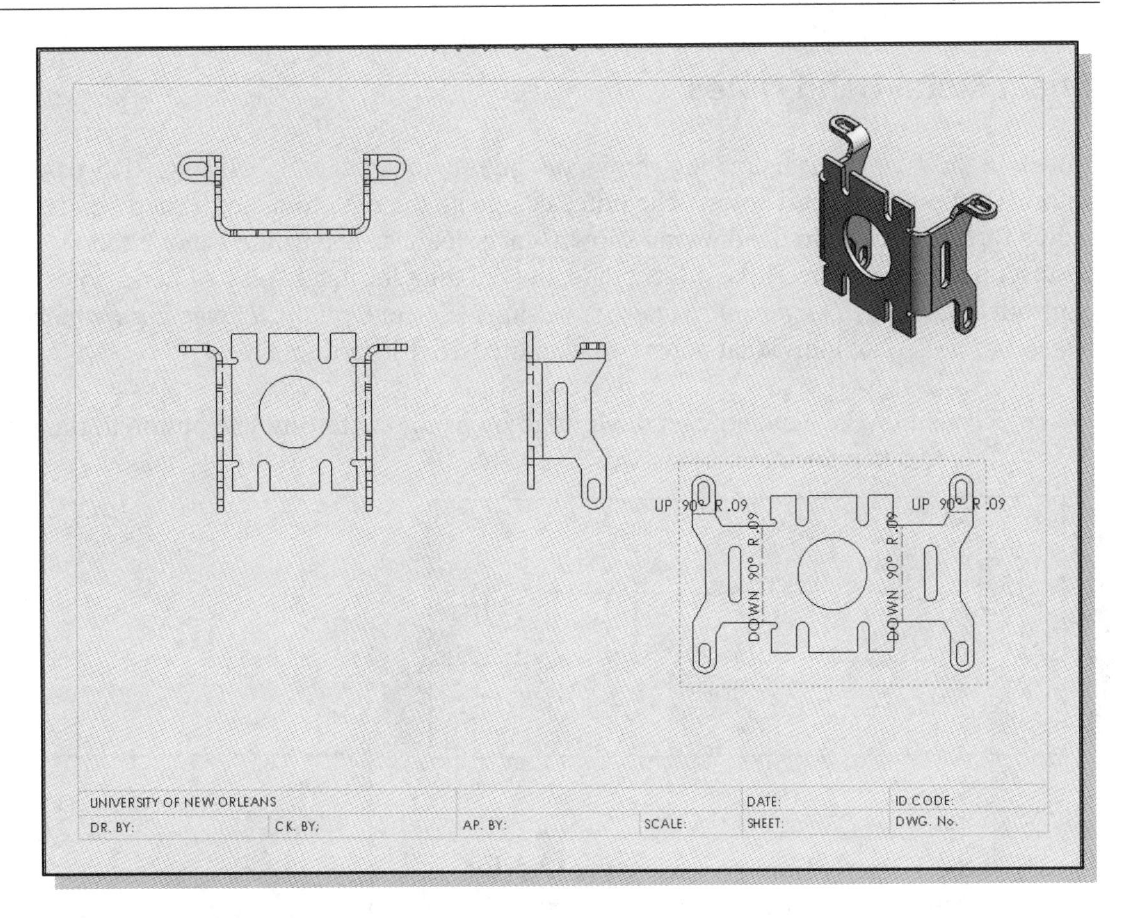

21. Press the **[Esc]** key to exit the **Projected View** command.

➢ The view orientation of the flat pattern created in the drawing view depends on such factors as the way the base flange was extruded. The view orientation can be adjusted as desired using the **Flat Pattern Display** option panel on the *Drawing View PropertyManager* or **Rotate View** command on the *Heads-up View* toolbar.

22. Select the **flat pattern view** by clicking on the view once with the left-mouse-button.

23. On your own, **experiment** with the Rotation Angle and Flip View option settings on the *Flat Pattern Display* panel on the *PropertyManager*.

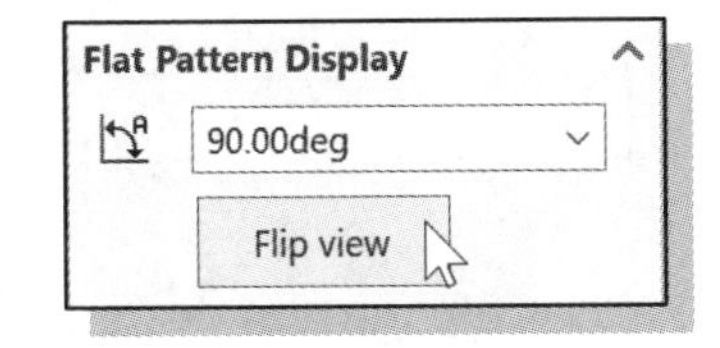

24. Return to the **original** *Flat Pattern Display* settings.

25. Click the **Rotate View** icon on the *Heads-up View* toolbar.

26. On your own, **experiment** with the effect of changing the angle in the dialog box.

27. Enter **0deg**, click **Apply**, and click **Close** to return to the **original** view orientation.

Sheet Metal Bend Notes

Notice in the *Flat Pattern* drawing shown on the previous page, SOLIDWORKS has automatically added bend notes. The notes designate the direction, angle, and bend radius for each bend. In the drawing shown, each note has been added above the corresponding bend line. Like dimensions, the defaults for the display of bend notes are controlled using the *Document Properties* settings (Select *Options*, *Document Properties*, *Sheet Metal*). Also, individual notes can be edited after insertion.

1. Reposition the bend note on the left tab by using the left-mouse-button to drag the note to a new location, as shown.

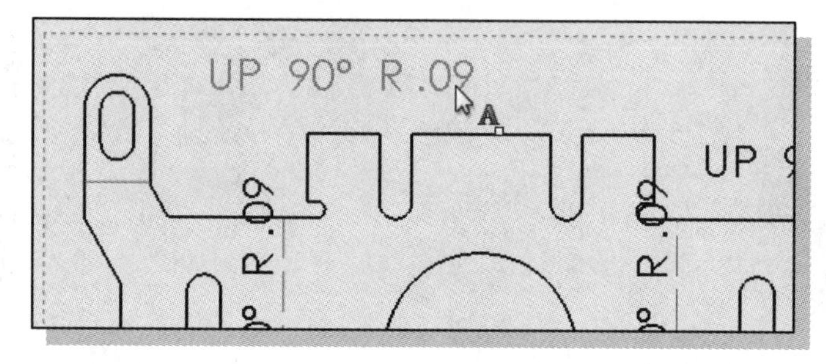

2. In the *PropertyManager*, select the **Leader** button on the *Leader* panel.

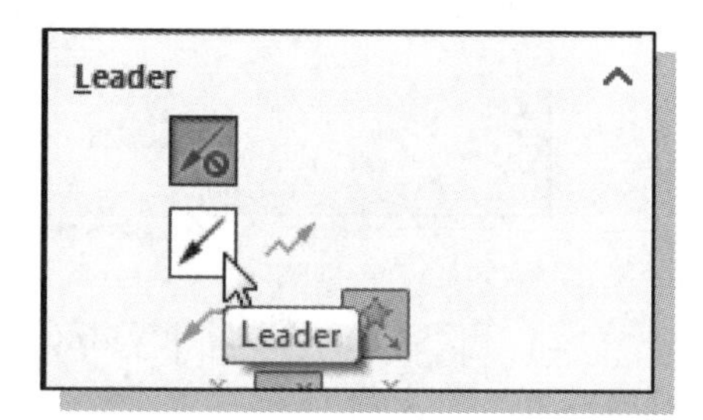

3. The leader appears. Reposition the note and leader as shown.

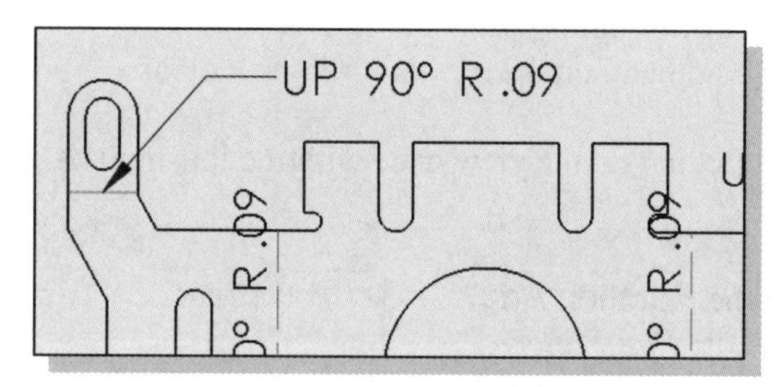

➢ For the current design, all bends use the same bend parameters. We will therefore opt to utilize one bend note for the drawing. The individual bend notes will be hidden.

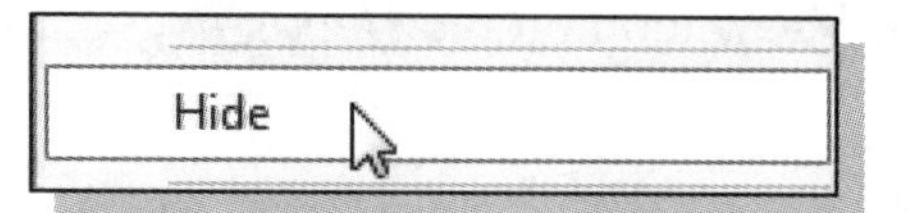

4. Move the cursor on top of the bend note and **right-mouse-click** once to bring up the option menu, then select **Hide** on the pop-up menu.

5. Repeat the previous step to **Hide** the other three bend notes.

Completing the Drawing

1. Click on the **Center Mark** button in the *Annotation* toolbar.

2. Click on the circle to place the associated center lines.

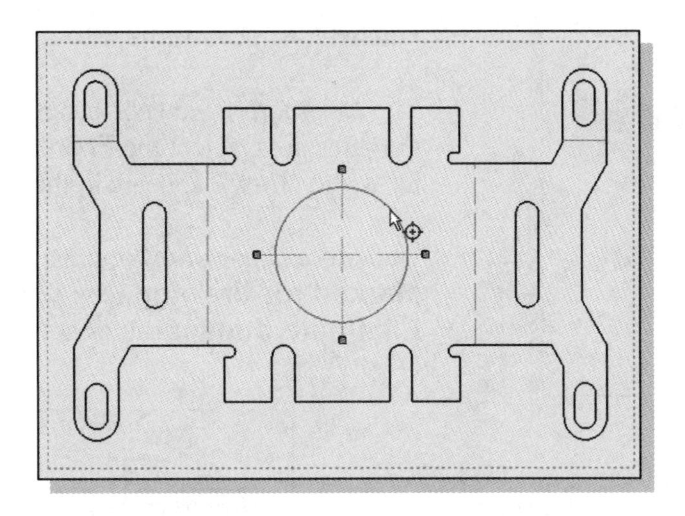

3. On your own, create and adjust the center marks for all of the circular features as shown.

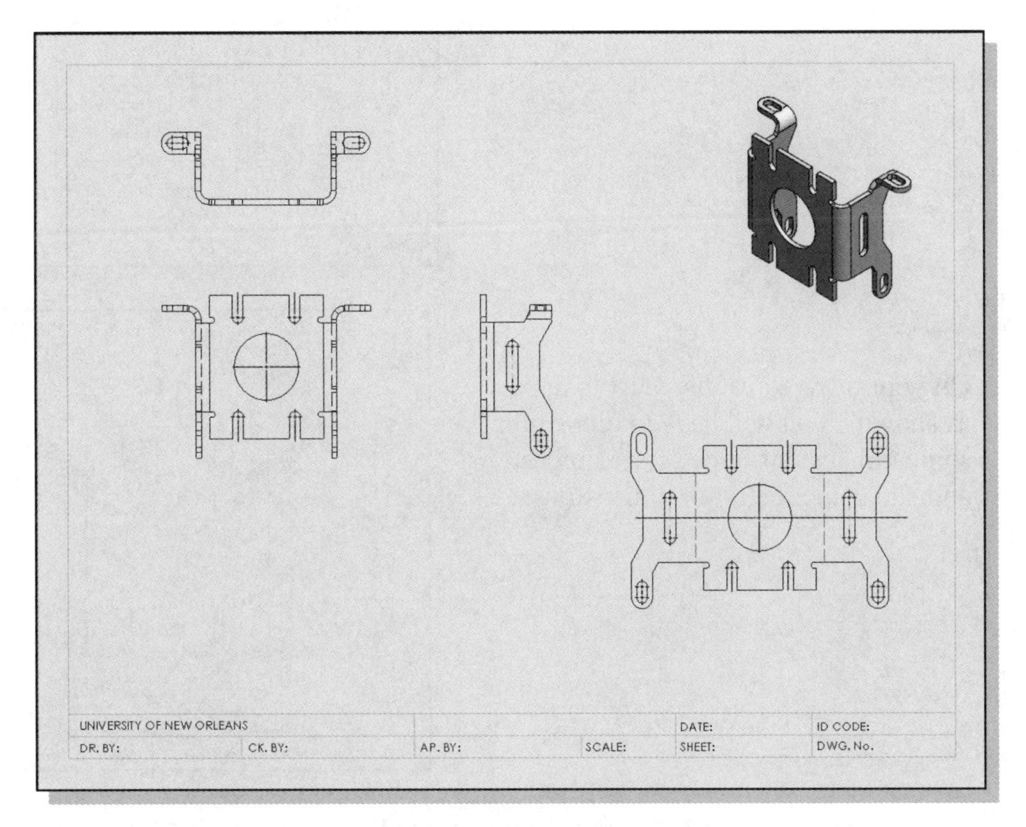

4. Left-mouse-click on the **Model Items** icon on the *Annotation* toolbar.

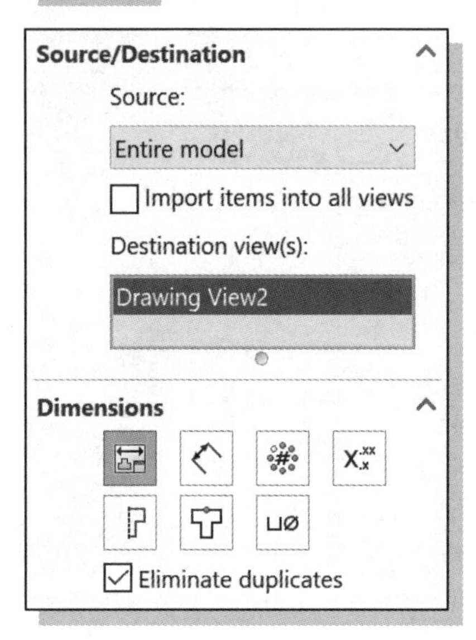

5. In the *Source/Destination* options panel of the *Model Items PropertyManager*, select the **Entire model** option under *Source*.

6. **Uncheck** Import items into all views as shown.

7. The *Destination view(s):* panel opens. In the graphics area, select the **Front View**. Notice Drawing View2 appears in the panel.

8. Under the *Dimensions* options panel, select the **Marked for drawing** icon and check **Eliminate duplicates** as shown.

9. Click the **OK** icon in the *Model Items PropertyManager*. Notice dimensions are automatically placed on the front drawing view.

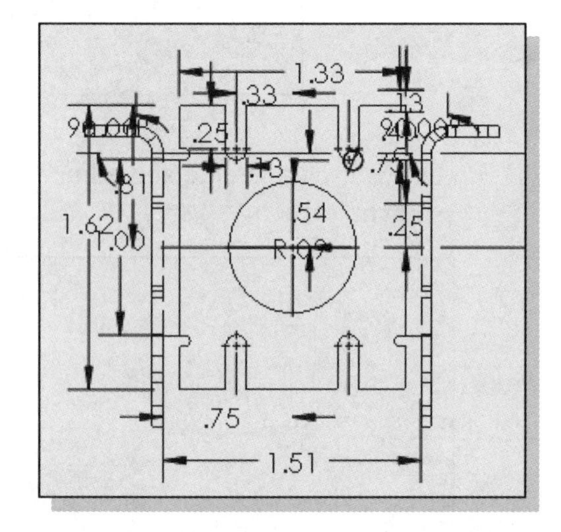

10. On your own, adjust the view to appear as shown. You will have to reposition some dimensions, hide some dimensions, and add some reference dimensions.

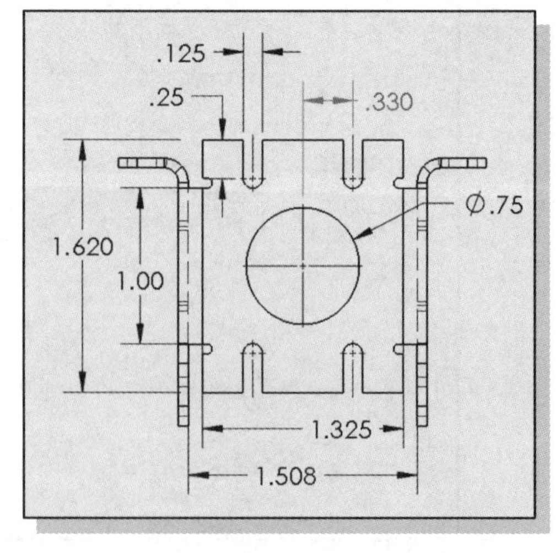

11. On your own, complete the drawing by adding the necessary dimensions and text.

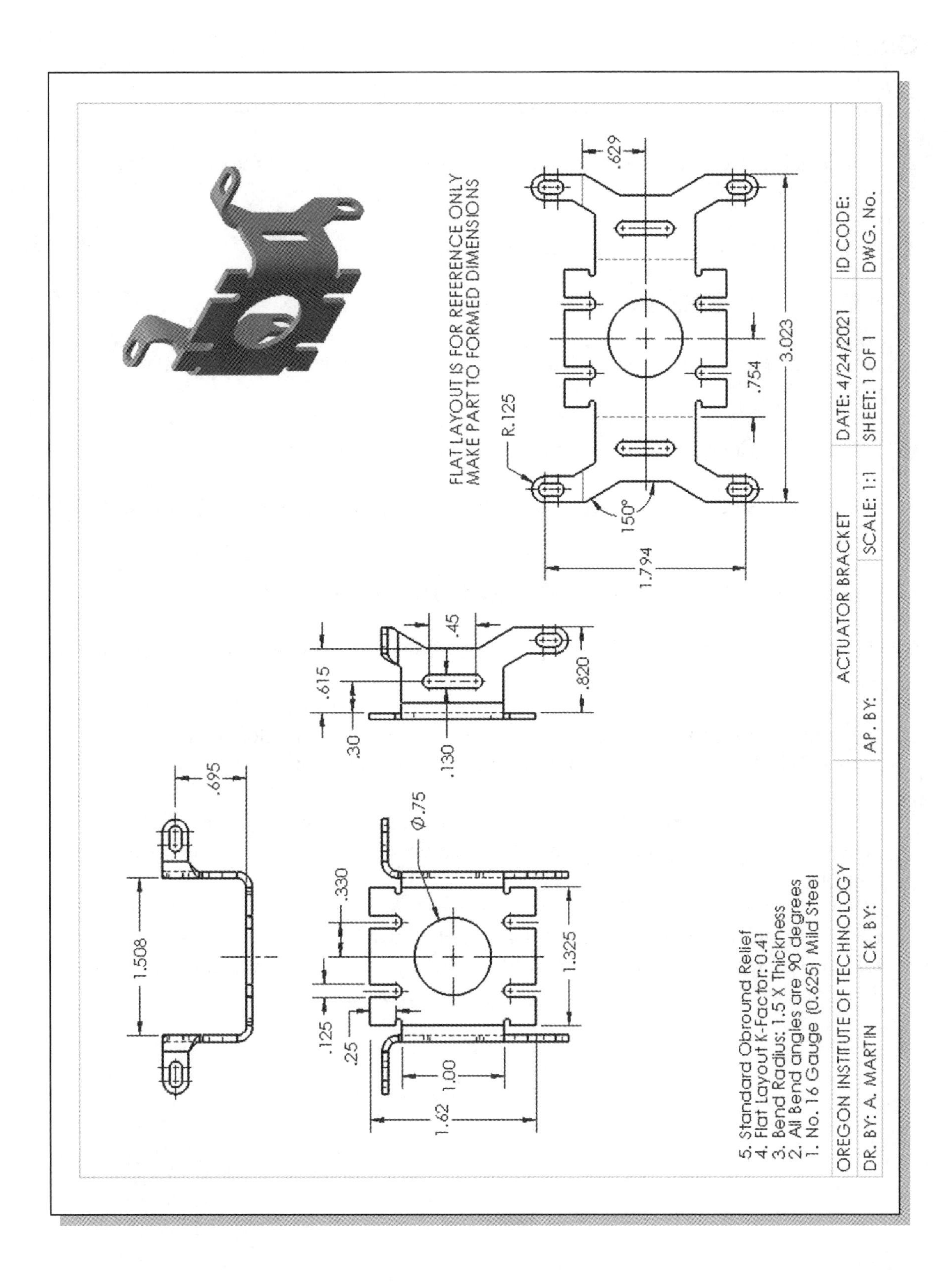
FLAT LAYOUT IS FOR REFERENCE ONLY
MAKE PART TO FORMED DIMENSIONS
R.125
.629
.754
3.023
150°
1.794
.45
.615
.820
.30
.130
.695
1.508
Ø.75
.330
.125
.25
1.325
1.00
1.62
5. Standard Obround Relief
4. Flat Layout K-Factor: 0.41
3. Bend Radius: 1.5 X Thickness
2. All Bend angles are 90 degrees
1. No. 16 Gauge (0.625) Mild Steel
OREGON INSTITUTE OF TECHNOLOGY
ACTUATOR BRACKET
DATE: 4/24/2021
ID CODE:
DR. BY: A. MARTIN
CK. BY:
AP. BY:
SCALE: 1:1
SHEET: 1 OF 1
DWG. No.

Questions:

1. List and describe two of the more commonly used sheet metal processes.

2. Is it possible to construct a solid model first, and then convert it into a sheet metal model in SOLIDWORKS?

3. How do we display the flat pattern of a 3D sheet metal design?

4. What is the **K-Factor** used in sheet metal processes?

5. How is the **K-Factor** used to calculate the flattened length in sheet metal flat patterns?

6. List and describe two of the factors that can change the K-Factor value.

7. List and describe two of the settings available in the Base Flange/Tab command in SOLIDWORKS.

8. Can the Model Items command be used on a ***flat pattern*** view?

9. What do the **Flat Pattern Display** options allow us to do when creating a *drawing view*?

10. In the SOLIDWORKS sheet metal module, can the feature-duplicating commands, such as Mirror and Pattern, be used on sheet metal features?

11. When does the Flat-Pattern item appear in the *FeatureManager Design Tree*? Where is it placed? Where are subsequent features placed relative to the Flat-Pattern?

12. Can we create a sheet metal feature that is at a 30 degree angle to the base face feature? Which command would you use if the new feature contains fairly complex 2D geometry?

Exercises:

(Create the 3D model and the associated 2D drawings. All dimensions are in inches.)

1. **Cooling Fan Cover**

1. No. 16 Gauge (0.0625) Mild Steel
2. Standard Straight Relief
3. Flat Layout K-Factor: 0.44
4. Bend Radius: Thickness
5. All Bend Angles are 90 degrees

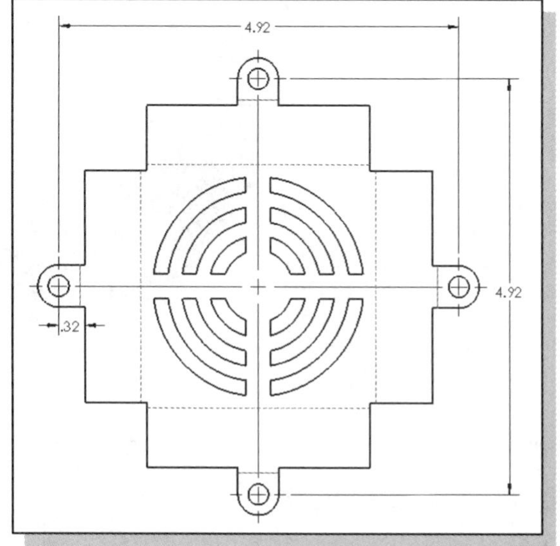

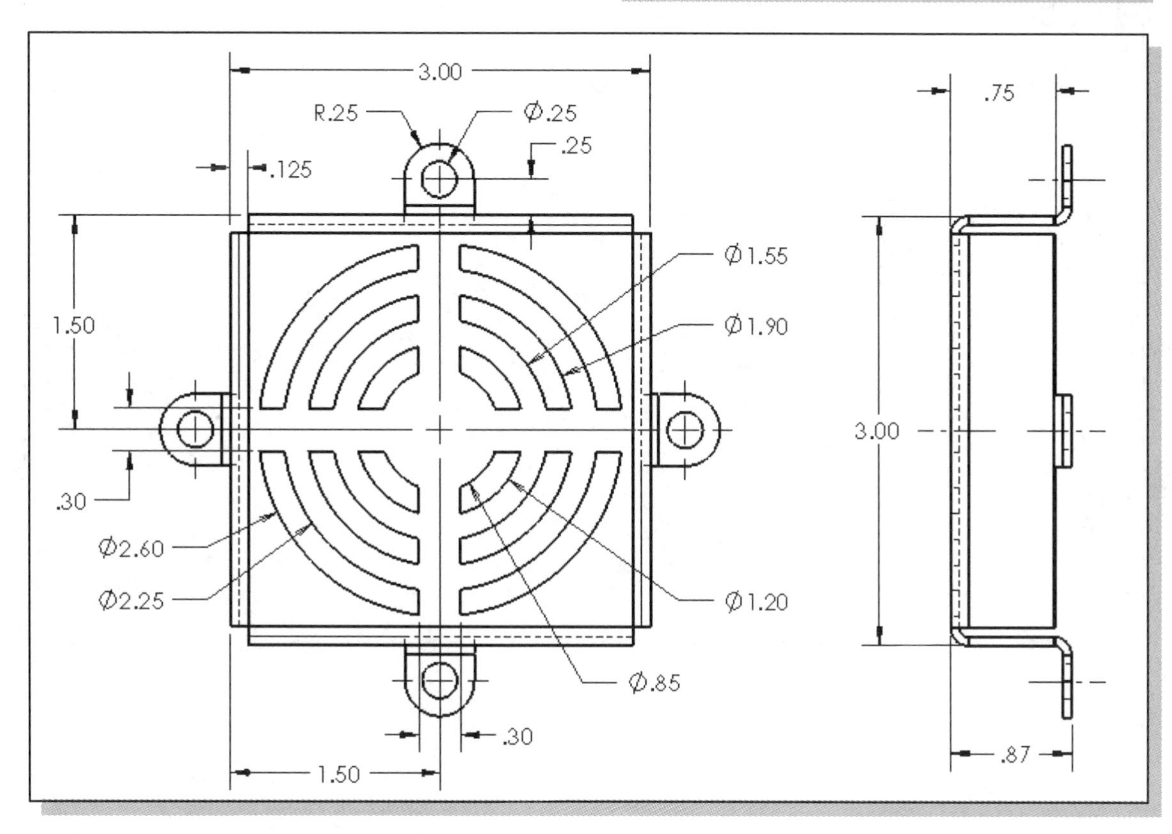

2. **Sheet Metal Rectangle to Square Transition** (Create and convert a Solid model to a Sheet Metal model.)

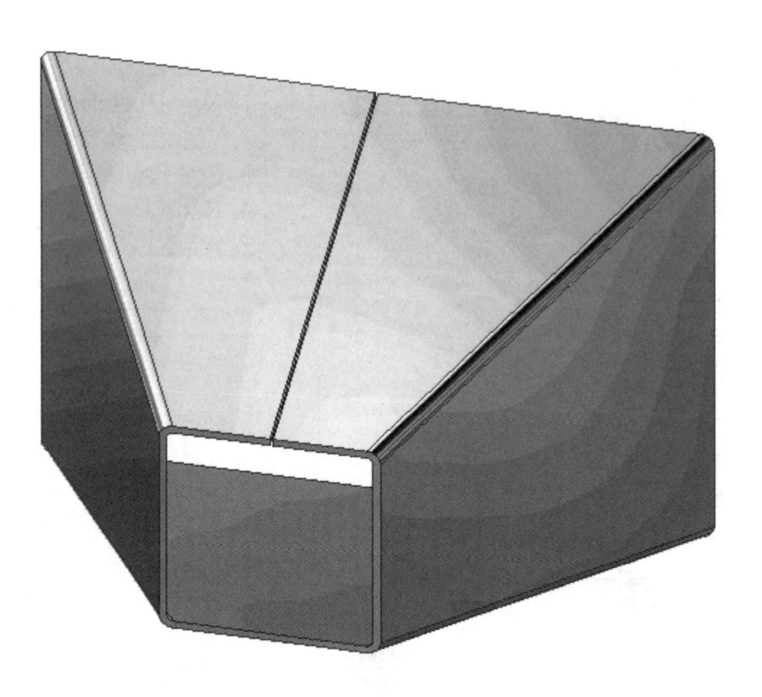

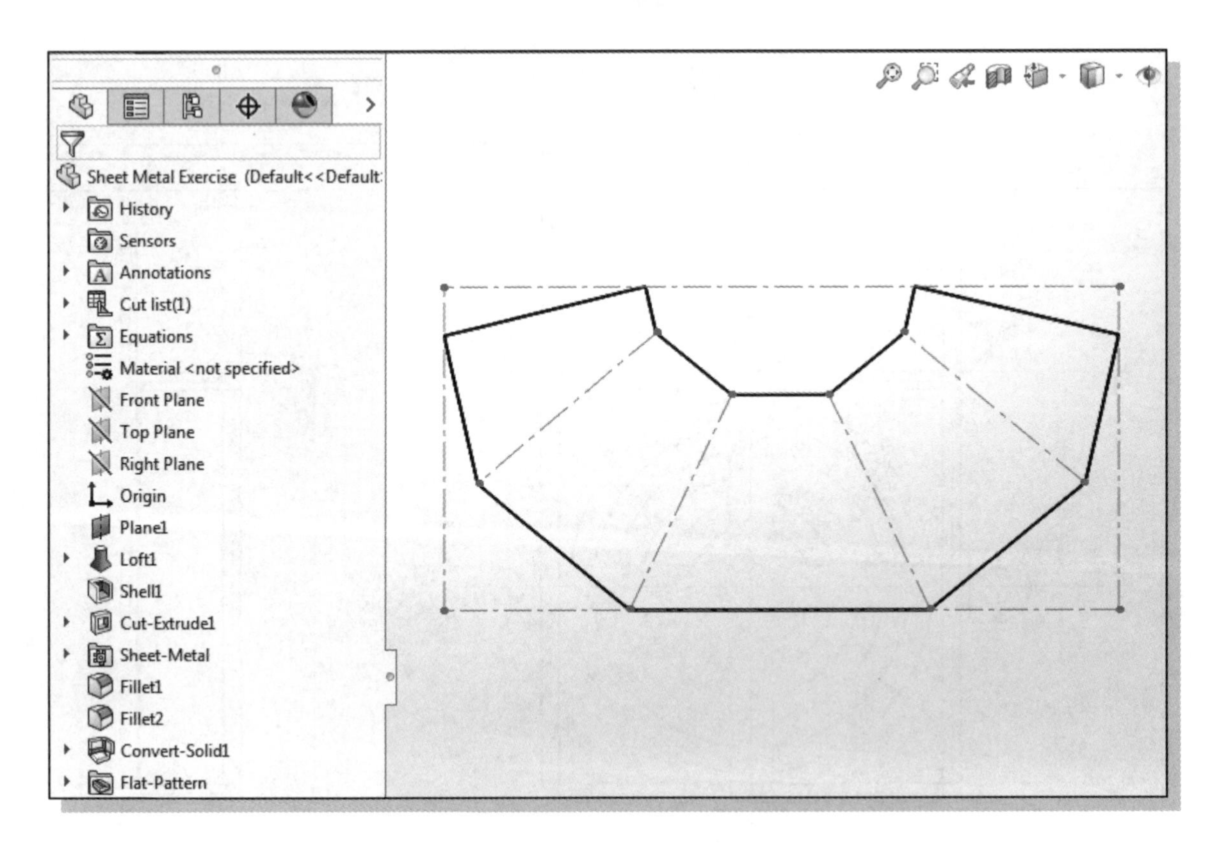

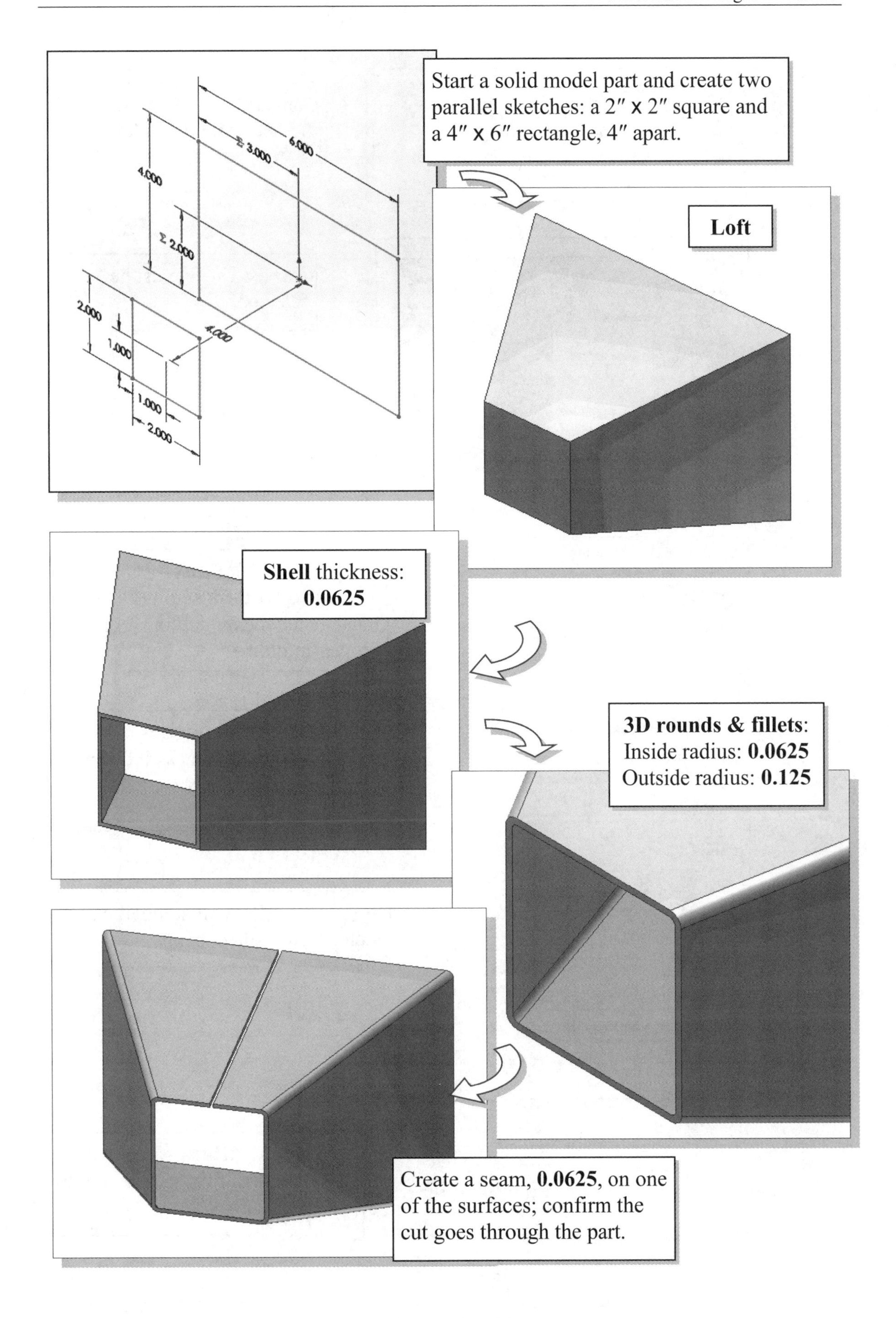
Start a solid model part and create two parallel sketches: a 2″ x 2″ square and a 4″ x 6″ rectangle, 4″ apart.
6.000
3.000
4.000
2.000
2.000
1.000
4.000
1.000
2.000
Loft
Shell thickness: 0.0625
3D rounds & fillets:
Inside radius: 0.0625
Outside radius: 0.125
Create a seam, 0.0625, on one of the surfaces; confirm the cut goes through the part.

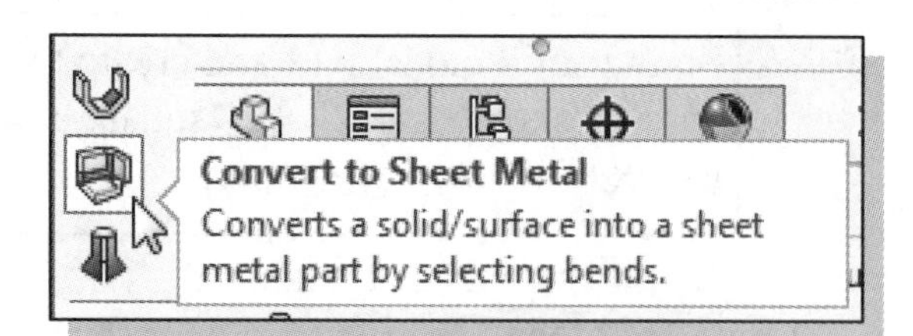

Convert the **Solid Model** to **Sheet Metal Model**.

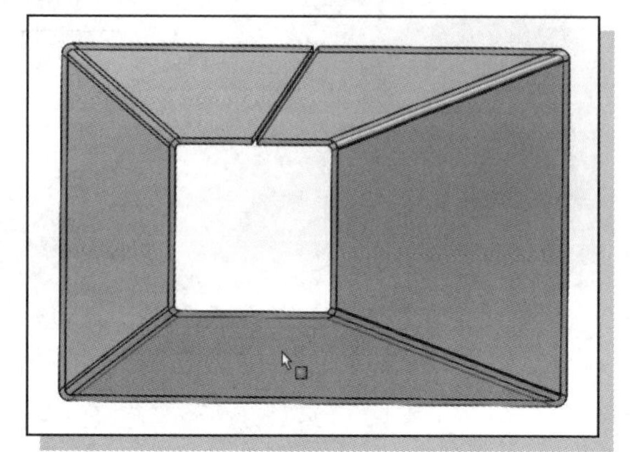

A fixed entity must be selected.

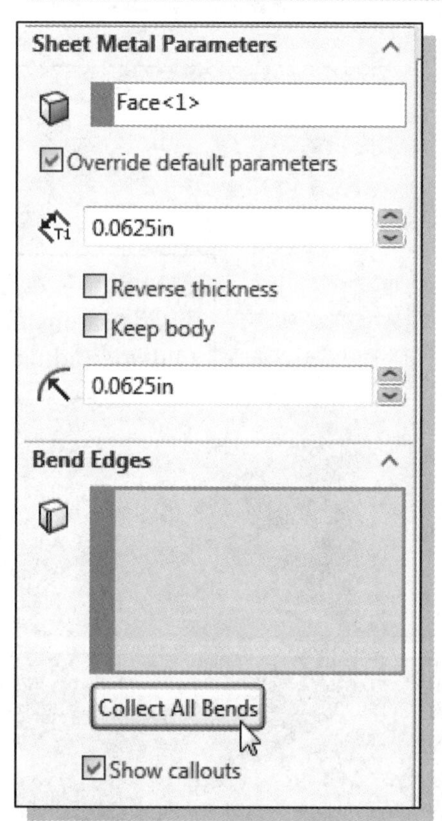

Sheet Metal **Thickness** and **Bend Radius** settings must match the model for proper unfolding.

Sheet Metal **Thickness** is set to match the shell thickness **0.0625**. The **Bend Radius** is set to the *Thickness* value.

SOLIDWORKS will automatically identify and collect bends.

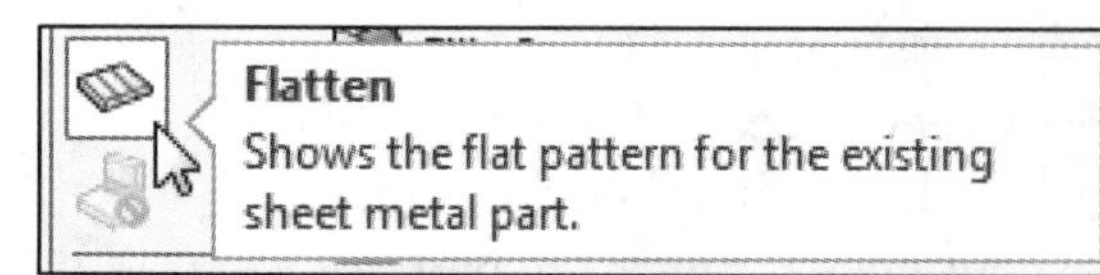

Flat Pattern can only be created if the parameters are set correctly.

Chapter 14

Assembly Modeling – Putting It All Together

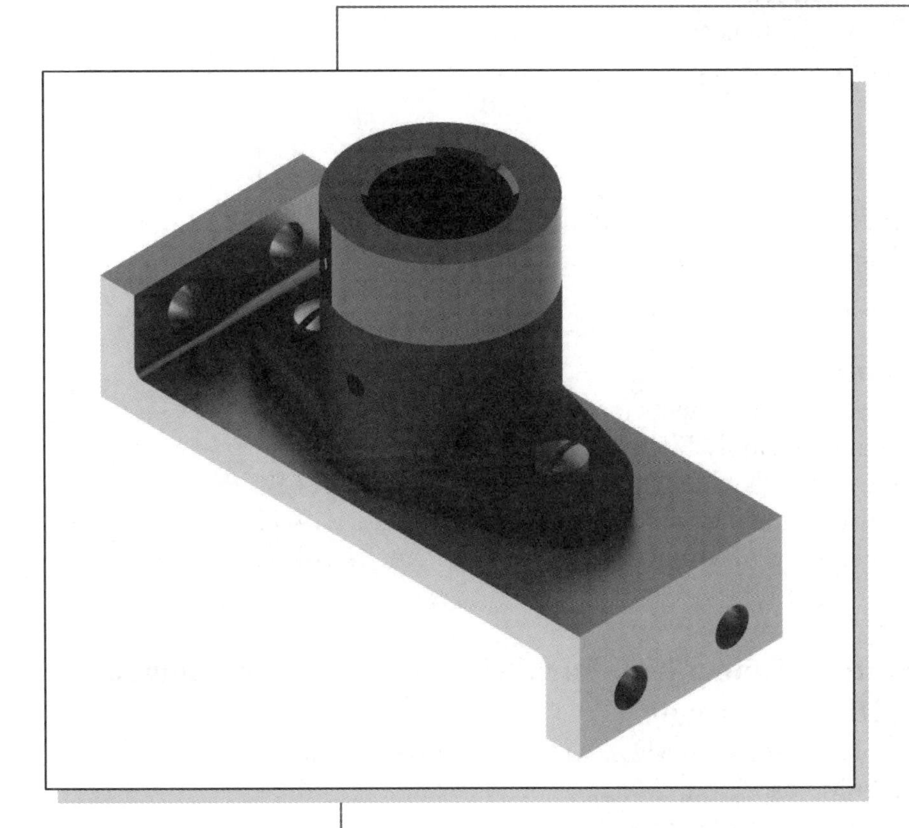

Learning Objectives

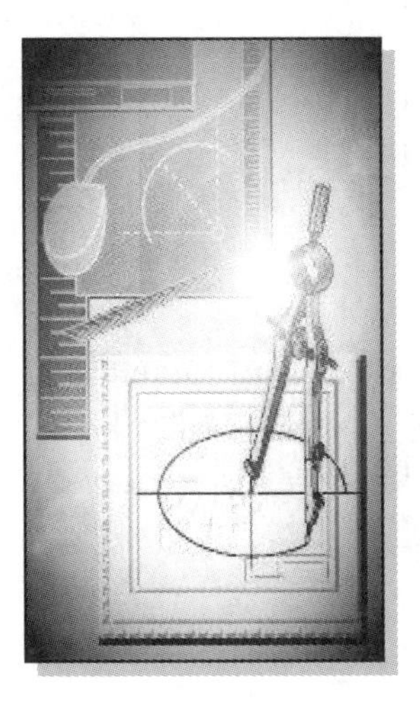

- ♦ **Understand the Assembly Modeling Methodology**
- ♦ **Understand and Control Degrees of Freedom for Assembly Components**
- ♦ **Understand and Utilize SOLIDWORKS Assembly Mates**
- ♦ **Place parts using SOLIDWORKS SmartMates**
- ♦ **Create Exploded Assemblies**
- ♦ **Create Assembly Drawings**
- ♦ **Create and Edit a Bill of Materials**

Certified SOLIDWORKS Associate Exam Objectives Coverage

Fillets and Chamfers

Objectives: Creating Fillets and Chamfer Features.

Materials

Objectives: Applying Material Selection to Parts.

Inserting Components

Objectives: Inserting Components into an Assembly.

Standard Mates – Coincident, Parallel, Perpendicular, Tangent, Concentric, Distance, Angle

Objectives: Applying Standard Mates to Constrain Assemblies.

Drawing Sheets and Views

Objectives: Creating and Setting Properties for Drawing Sheets; Inserting and Editing Standard Views.

Annotations

Objectives: Creating Annotations.

Introduction

In the previous lessons, we have gone over the fundamentals of creating basic parts and drawings. In this lesson, we will examine the assembly modeling functionality of SOLIDWORKS. We will start with a demonstration on how to create and modify assembly models. The main task in creating an assembly is establishing the assembly relationships between parts. It is a good practice to assemble parts based on the way they would be assembled in the actual manufacturing process. We should also consider breaking down the assembly into smaller subassemblies, which helps the management of parts. In SOLIDWORKS, a subassembly is treated the same way as a single part during assembling. Many parallels exist between assembly modeling and part modeling in parametric modeling software such as SOLIDWORKS.

SOLIDWORKS provides full associative functionality in all design modules, including assemblies. When we change a part model, SOLIDWORKS will automatically reflect the changes in all assemblies that use the part. We can also modify a part in an assembly. Bi-directional full associative functionality is the main feature of parametric solid modeling software that allows us to increase productivity by reducing design cycle time.

Assembly Modeling Methodology

The SOLIDWORKS assembly modeler provides tools and functions that allow us to create 3D parametric assembly models. An assembly model is a 3D model with any combination of multiple part models. SOLIDWORKS Mate Relations can be used to control relationships between parts in an assembly model.

SOLIDWORKS can work with any of the assembly modeling methodologies:

The Bottom Up approach

The first step in the *bottom up* assembly modeling approach is to create the individual parts. The parts are then pulled together into an assembly. This approach is typically used for smaller projects with very few team members.

The Top Down approach

The first step in the *top down* assembly modeling approach is to create the assembly model of the project. Initially, individual parts are represented by names or symbolically. The details of the individual parts are added as the project gets further along. This approach is typically used for larger projects or during the conceptual design stage. Members of the project team can then concentrate on the particular section of the project to which they are assigned.

The Middle Out approach

The *middle out* assembly modeling approach is a mixture of the bottom-up and top-down methods. This type of assembly model is usually constructed with most of the parts already created and additional parts are designed and created using the assembly for construction information. Some requirements are known and some standard components are used, but new designs must also be produced to meet specific objectives. This combined strategy is a very flexible approach for creating assembly models.

The different assembly modeling approaches described above can be used as guidelines to manage design projects. Keep in mind that we can start modeling our assembly using one approach and then switch to a different approach without any problems.

The Shaft Support Assembly

In this lesson, the *bottom up* assembly modeling approach is illustrated through the creation of the shaft support assembly shown in the figure. All of the parts (components) required to form the assembly are created first. SOLIDWORKS' assembly modeling tools allow us to create complex assemblies by using components that are created in part files or are created in assembly files. A component can be a subassembly or a single part, where features and parts can be modified at any time.

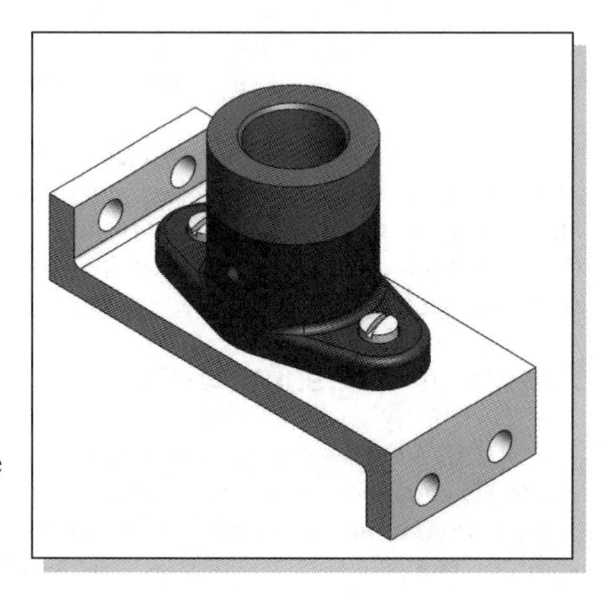

Parts

Four parts are required for the assembly: (1) ***Collar***, (2) ***Bearing***, (3) ***Base-Plate*** and (4) ***Cap-Screw***. Create the four parts as shown below, then save the models as separate part files: *Collar*, *Bearing*, *Base-Plate*, and *Cap-Screw*. (Close all part files or exit SOLIDWORKS after you have created the parts.)

Creating the Collar Using the Chamfer Command

(1) ***Collar***

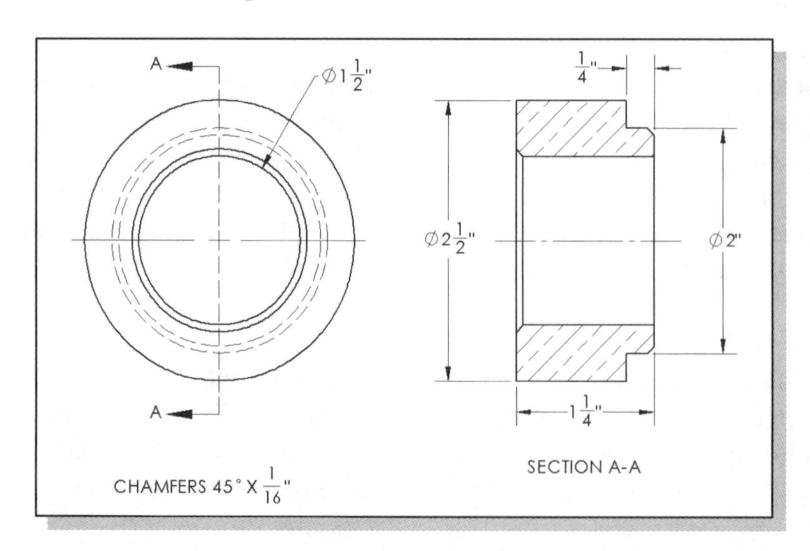

1. On your own, start a new part file using your **Part_IPS_ANSI** template and create the basic *Collar* without the chamfers, as shown below.

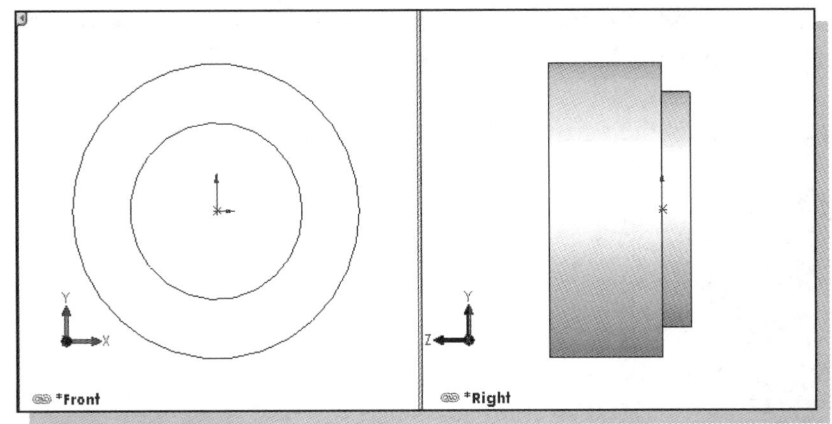

2. In the *Features* toolbar, left-click on the **arrow** next to the **Fillet** icon and select the **Chamfer** command from the pop-up menu.

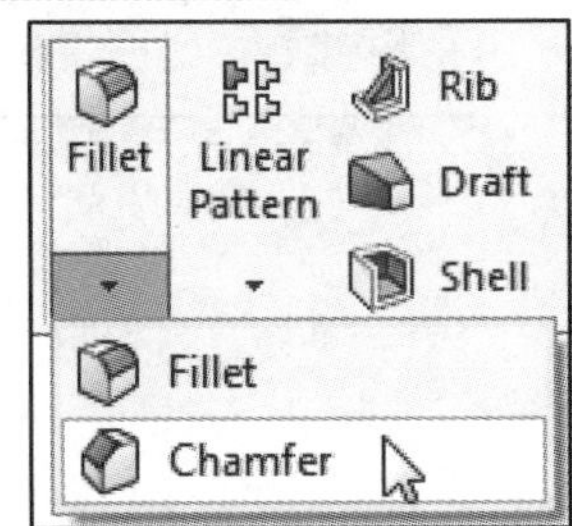

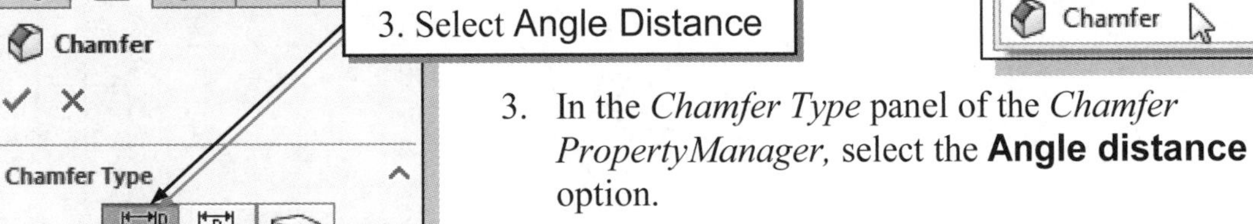

3. In the *Chamfer Type* panel of the *Chamfer PropertyManager,* select the **Angle distance** option.

4. Set the ***Distance*** to **1/16 in** (or 0.0625in).

5. Set the ***Angle*** to **45°**.

6. Select the two circular edges shown in the figures below.

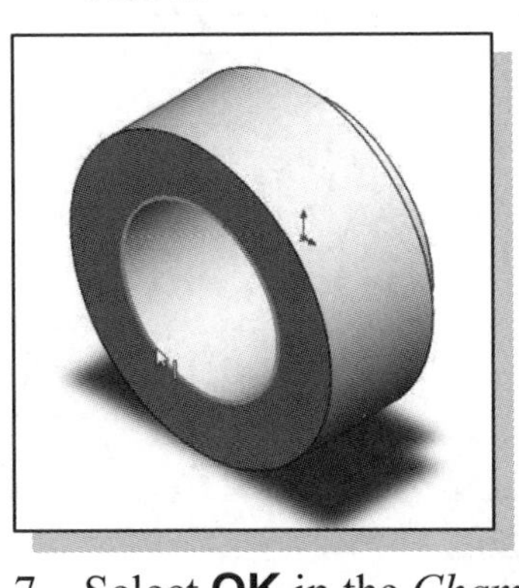

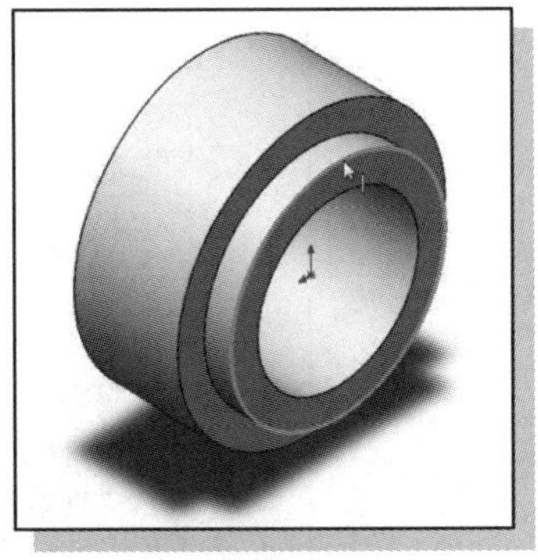

7. Select **OK** in the *Chamfer PropertyManager* to create the chamfers.

8. Save the part with the filename ***Collar***.

Creating the Bearing and Base-Plate

On your own, create and save the *Bearing* and *Base-Plate* part files based on the drawings below. Align the base of each part with the **Top Plane**.

(2) ***Bearing*** (Construct the part with the datum origin aligned to the bottom center.)

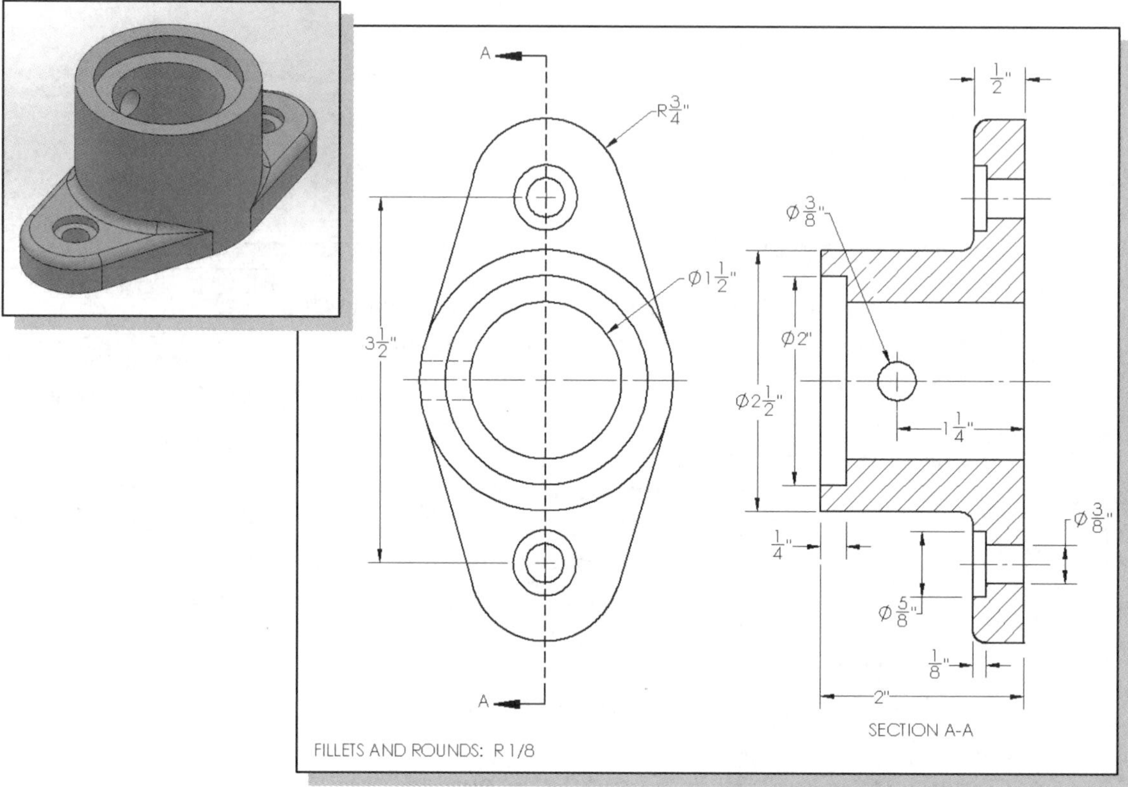

(3) ***Base-Plate*** (Construct the part with the datum origin aligned to the bottom center.)

NOTE: You can ignore the threads and simply create 3/8″ diameter holes.

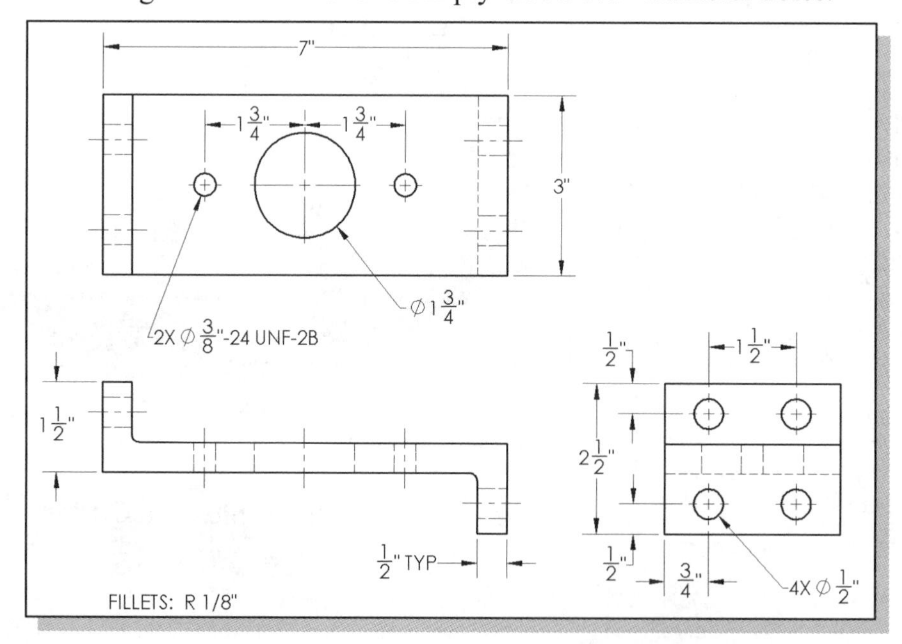

Creating the Cap-Screw

On your own, create and save the *Cap-Screw* part file based on the drawings below.
NOTE: Threads are not necessary.

(4) ***Cap-Screw***

- SOLIDWORKS provides two options for creating threads: **CosmeticThread** and **Helix**. The **Cosmetic Thread** command does not create true 3D threads; a pre-defined thread image is applied on the selected surface, as shown in the figure. The **Helix** command can be used to create true threads, which contain complex three-dimensional curves and surfaces. You are encouraged to experiment with the **Helix** command and/or the **Cosmetic Thread** command to create threads.
- To represent threads on a part using the **Cosmetic Thread** option, first make sure the shading option for Cosmetic Thread is turned on. Click Tools, Options. On the Document Properties tab, select Detailing. Make sure the Shaded cosmetic threads option is checked. Also right click on *Annotations* in the *FeatureManager Design Tree* and make sure Display Annotations is selected. To apply a cosmetic thread, click Insert, then Annotations, then Cosmetic Thread. Set the properties in the *PropertyManager* and select OK.

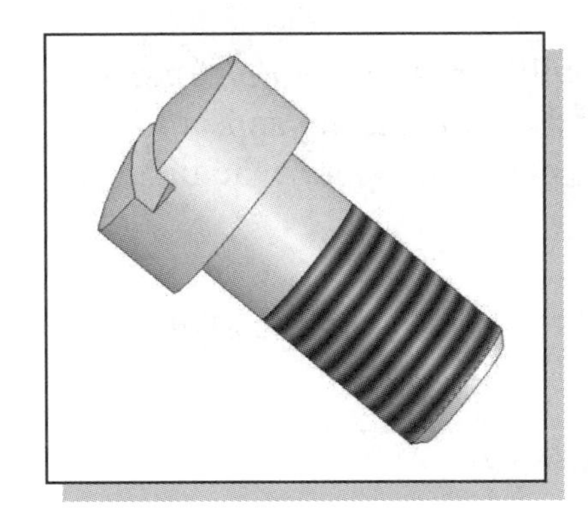

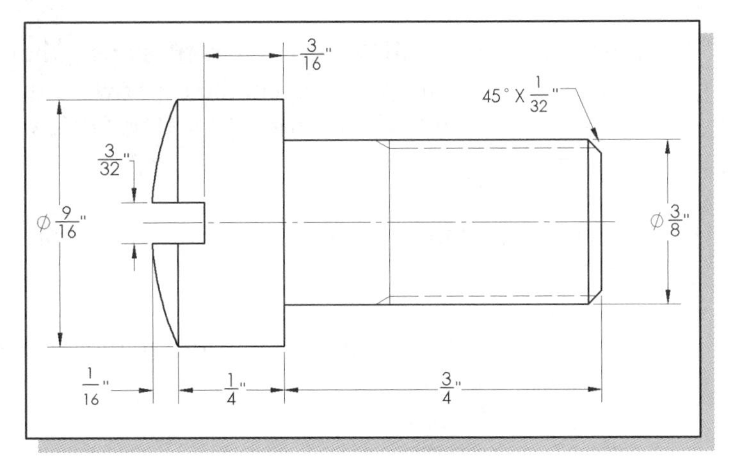

- **HINT:** Create the main shape with two cylindrical features; add a 1/16″ **Dome** feature (use Insert → Features → Dome); add a **Chamfer** feature; add the slot with an **Extruded Cut** feature.

OR

Create the profile shown (without the slot). Then create a revolved feature. Use the *Selected Contours* panel to select the upper half of the sketch to revolve.

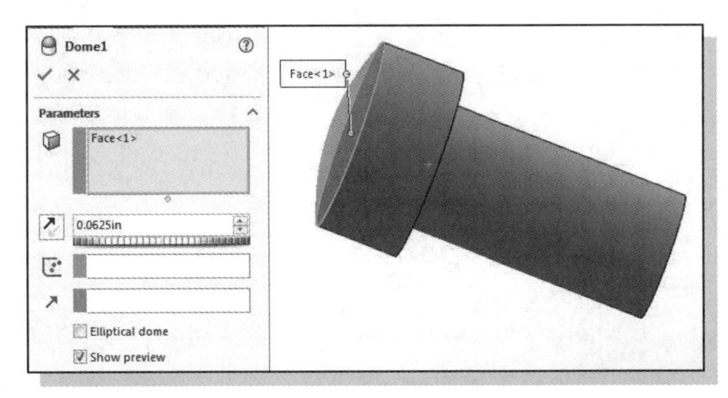

Starting SOLIDWORKS

1. Close all open SOLIDWORKS part files.

2. Select the **SOLIDWORKS** option on the *Start* menu or select the **SOLIDWORKS** icon on the desktop to start SOLIDWORKS.

3. Select the **New** icon with a single click of the left-mouse-button on the *Menu Bar*. The *New* SOLIDWORKS *Document* dialog box appears.

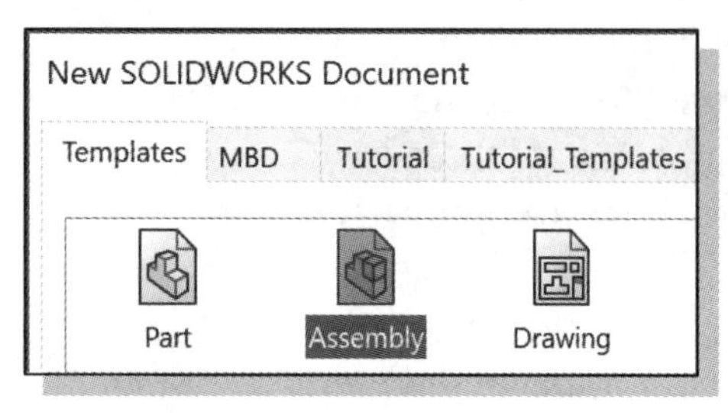

4. Select the **Templates** tab and select the **Assembly** icon as shown.

5. Click **OK** to open the new Assembly file.

❖ SOLIDWORKS opens an assembly file, and automatically opens the *Begin Assembly PropertyManager*. The message "*Select a part or assembly to insert* ..." appears in the *PropertyManager*.

❖ If there are any SOLIDWORKS part files open, they will appear in the *Part/Assembly to Insert* panel. If no part is open, the *Open* window will automatically appear to browse for the part to be inserted. SOLIDWORKS expects you to insert the first component.

❖ We will first set the document properties to match those of the previously created parts.

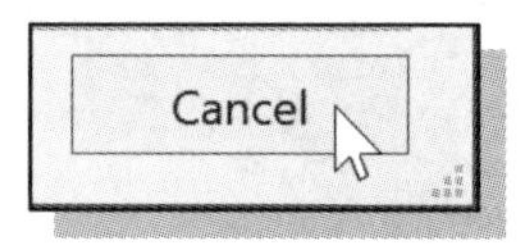

6. Click on **Cancel** in the *Open* window.

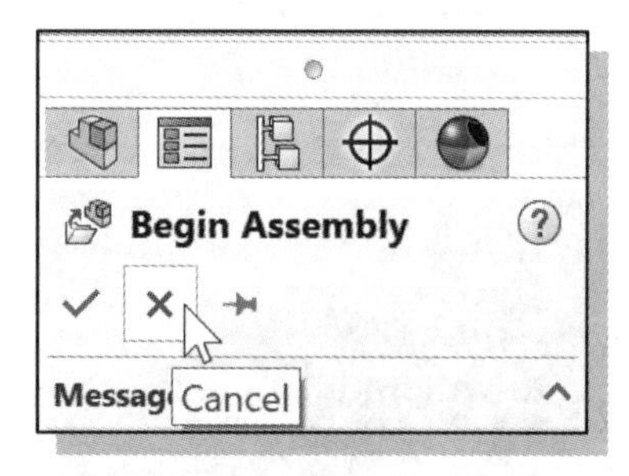

7. Cancel the Begin Assembly command by clicking once with the left-mouse-button on the **Cancel** icon in the *PropertyManager* as shown.

Document Properties

1. Select the **Options** icon from the *Menu Bar* to open the *Options* dialog box.

2. Select the **Document Properties** tab.

3. Select **ANSI** in the pull-down selection window under the *Overall drafting standard* panel as shown.

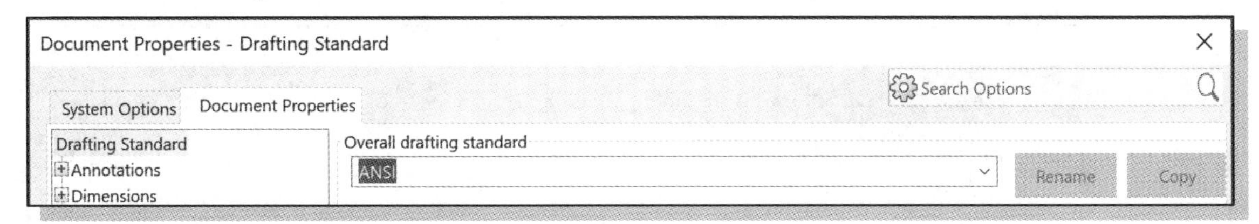

4. Click **Units**, select the **IPS (inch, pound, second)** unit system, and select **.123** in the *Decimals* spin box for the *Length units* as shown.

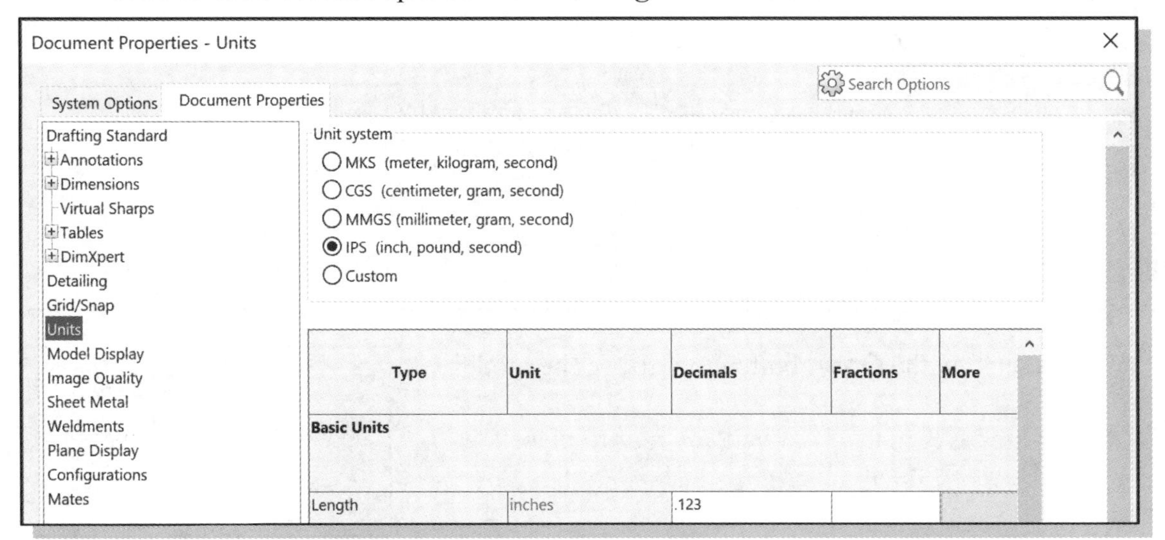

5. Click **OK** in the *Document Properties* window to accept the settings.

Inserting the First Component

The first component inserted in an assembly should be a fundamental part or subassembly. The first component in an assembly file sets the orientation of all subsequent parts and subassemblies. The origin of the first component is aligned to the origin of the assembly coordinates and the part is grounded (all degrees of freedom are removed). The rest of the assembly is built on the first component, the ***base component***. In most cases, this *base component* should be one that is **not likely to be removed** and **preferably a non-moving part** in the design. Note that there is no distinction in an assembly between components; the first component we place is usually considered as the *base component* because it is usually a fundamental component to which others are constrained. We can change the base component if desired. For our project, we will use the ***Base-Plate*** as the base component in the assembly.

1. In the *Assembly* toolbar, select the **Insert Components** command by left-mouse-clicking the icon.

2. Use the browser to locate and select the ***Base-Plate*** part file. HINT: Use the *Quick Filter* to browse SOLIDWORKS part files (*.sldprt) only.

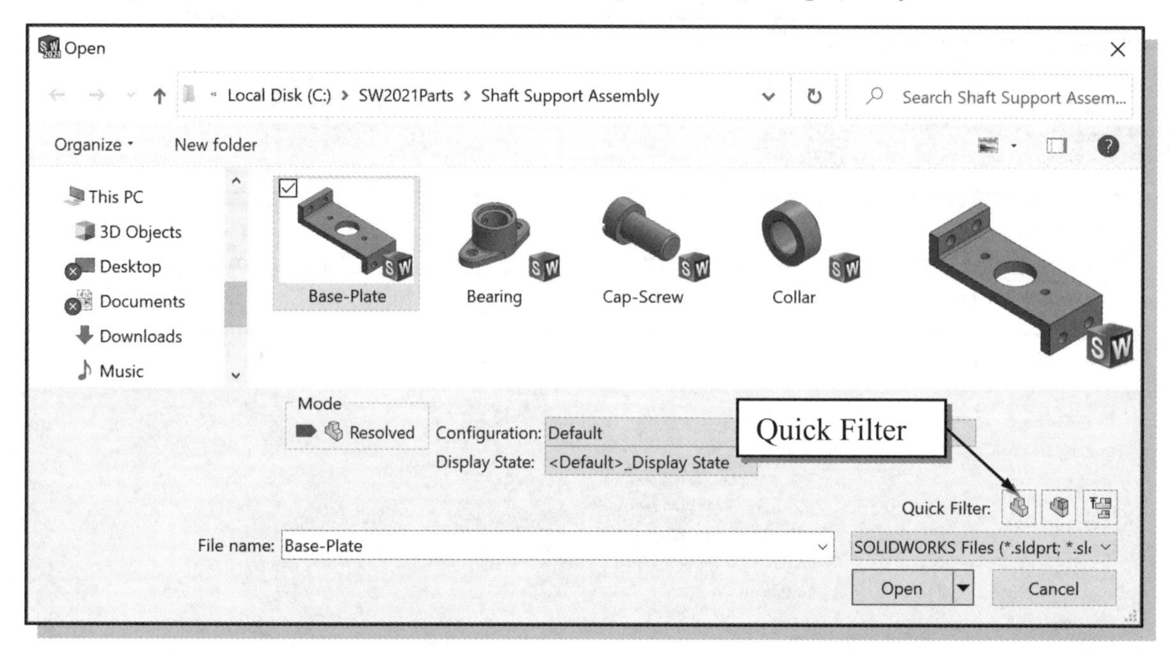

3. Click on the **Open** button to retrieve the model.

4. Click **OK** in the *PropertyManager* to place the *Base-Plate* at the origin.

Inserting the Second Component

We will retrieve the *Bearing* part as the second component of the assembly model.

1. In the *Assembly* toolbar, select the **Insert Components** command by left-mouse-clicking the icon.

2. Select the ***Bearing*** design (part file: **Bearing.sldprt**) in the browser. And click on the **Open** button to retrieve the model.

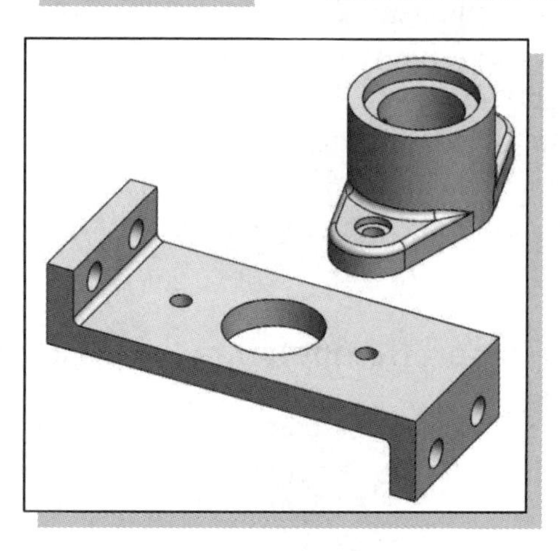

3. Place the *Bearing* toward the upper right corner of the graphics window, as shown in the figure. Click once with the **left-mouse-button** to place the component. (**NOTE:** Your *Bearing* may initially be oriented differently. We will apply assembly mates to orient the *Bearing*. Proceed with the instructions and the alignment of your *Bearing* will match those in the illustrations after applying the second assembly mate.)

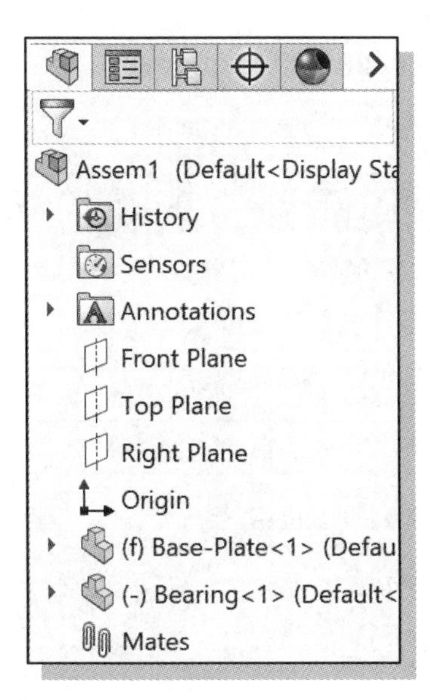

- Inside the *FeatureManager Design Tree*, the retrieved parts are listed in their corresponding order. The **(f)** in front of the ***Base-Plate*** filename signifies the part is fixed and all ***six degrees of freedom*** are restricted. The number behind the filename is used to identify the number of copies of the same component in the assembly model.

Degrees of Freedom

Each component in an assembly has six **degrees of freedom** (**DOF**), or ways in which rigid 3D bodies can move: movement along the X, Y, and Z axes (translational freedom), plus rotation around the X, Y, and Z axes (rotational freedom). *Translational DOF*s allow the part to move in the direction of the specified vector. *Rotational DOF*s allow the part to turn about the specified axis.

It is usually a good idea to fully constrain components so that their behavior is predictable as changes are made to the assembly. Leaving some degrees of freedom open can sometimes help retain design flexibility. As a general rule, we should use only enough assembly mate relations to ensure predictable assembly behavior and avoid unnecessary complexity.

Assembly Mates

We are now ready to assemble the components together. We will start by placing assembly constraints on the ***Bearing*** and the ***Base-Plate*** using SOLIDWORKS assembly **Mates**. Mates create geometric relationships between assembly components. Mates define the allowable directions of linear or rotational motion of the components, limiting the degrees of freedom. A component can be moved within its degrees of freedom, visualizing the assembly's behavior.

To assemble components into an assembly, we need to establish the assembly relationships between components. It is a good practice to assemble components the way they would be assembled in the actual manufacturing process. **Assembly Mates** create a parent/child relationship that allows us to capture the design intent of the assembly. Because the component that we are placing actually becomes a child to the already assembled components, we must use caution when choosing mate types and references to make sure they reflect the intent.

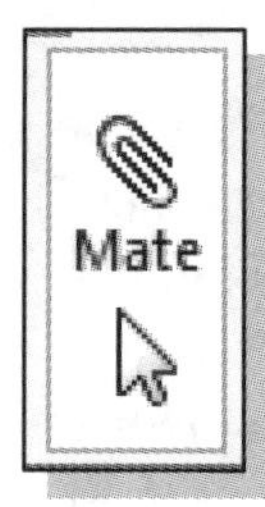

1. In the *Assembly* toolbar, select the **Mate** command by left-mouse-clicking once on the icon.

- The *Mate PropertyManager* appears on the screen.

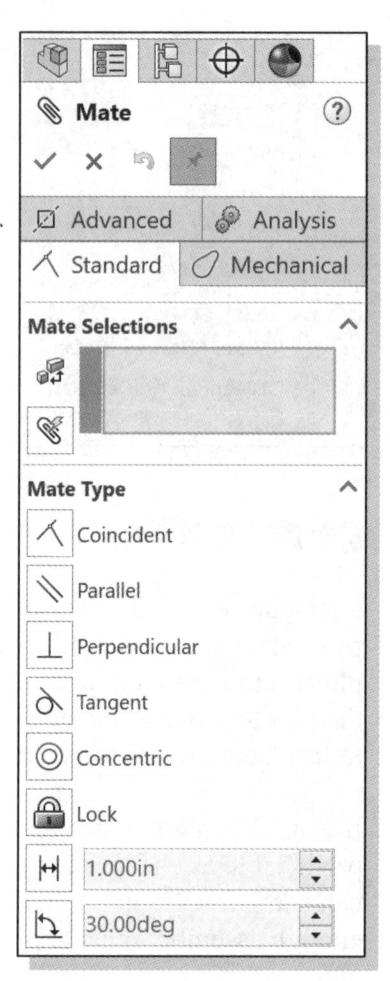

- The Mate relations are used to restrict the movement between parts. Mate relations eliminate rigid body degrees of freedom (**DOF**). A 3D part has *six degrees of freedom* since the part can rotate and translate relative to the three coordinate axes. Each time we add a Mate relation between two parts, one or more DOF is eliminated. The movement of a fully constrained part is restricted in all directions.

- Eight basic types of assembly mates, called Standard Mates, are available in SOLIDWORKS: Coincident, Parallel, Perpendicular, Tangent, Concentric, Lock, Distance, and Angle. Each type of mate removes different combinations of rigid body degrees of freedom. Note that it is possible to apply different constraints and achieve the same results.

- Advanced Mates including Gear and Cam Mates are also available.

The Standard Mates are described below:

- **Coincident** – positions selected faces, edges, and planes (in combination with each other or combined with a single vertex) so they share the same infinite line. Positions two vertices so they touch

- **Parallel** – places the selected items so they lie in the same direction and remain a constant distance apart from each other

- **Perpendicular** – places the selected items at a 90 degree angle to each other

- **Tangent** – places the selected items in a tangent mate (at least one selection must be a cylindrical, conical, or spherical face)

- **Concentric** – places the selections so that they share the same center point

- **Distance** – places the selected items with the specified distance between them

- **Angle** – places the selected items at the specified angle to each other

Apply the First Assembly Mate

1. Note that the *Mate Selections* panel in the *Mate PropertyManager* is highlighted. Select the **top horizontal surface** of the *Base-Plate* part as the first part for the Mate command.

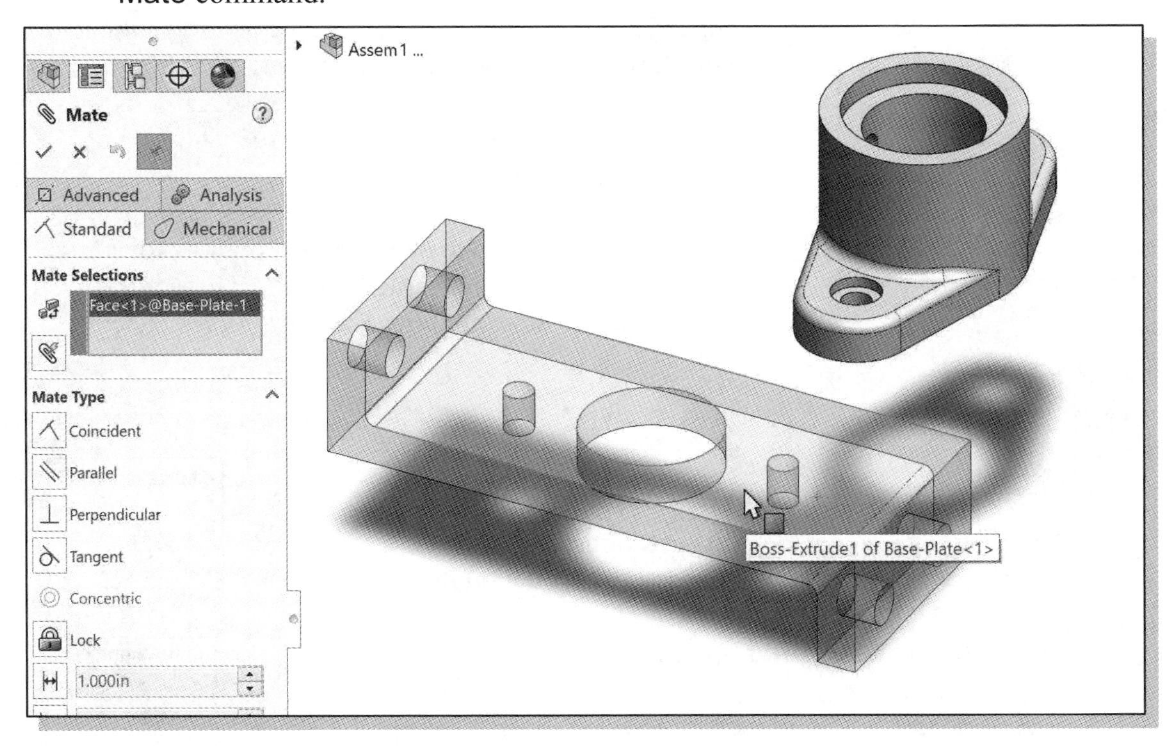

2. On your own, rotate the displayed model to view the bottom of the *Bearing* part, as shown in the figure below. (**HINT:** Use the up arrow [↑] on the keyboard.)

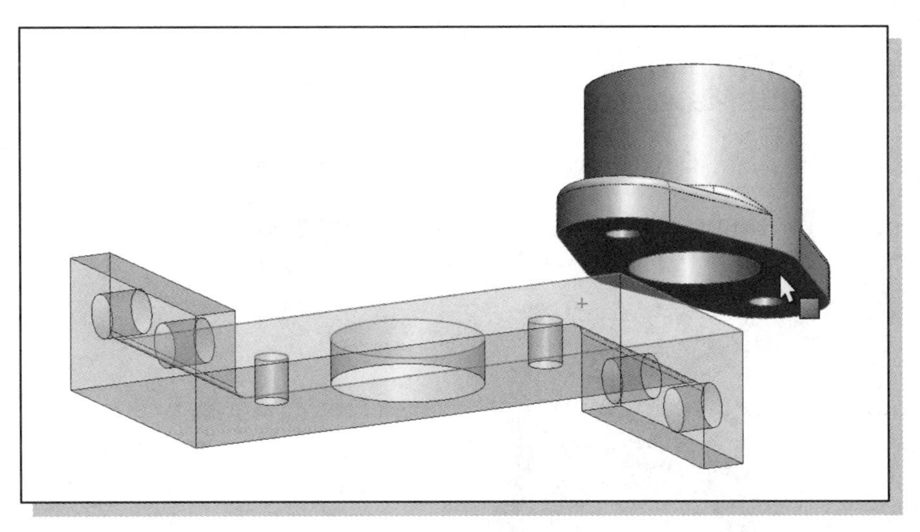

3. Click on the bottom face of the *Bearing* part as the second part selection to apply the relation. The *Bearing* moves to automatically apply the default Coincident mate.

➢ Notice the two faces selected appear in the *Mate Selections* panel of the *PropertyManager*.

➢ Notice the Tangent and Concentric options are not available. All the mate types are always shown in the *PropertyManager*, but only the mates that are applicable to the current selections are available.

➢ Notice the **Coincident** mate is selected. This is the relation we want to apply.

➢ For the Coincident mate there are two options for *Mate alignment* – Aligned and Anti-Aligned. These are displayed at the bottom of the *Standard Mates* panel.

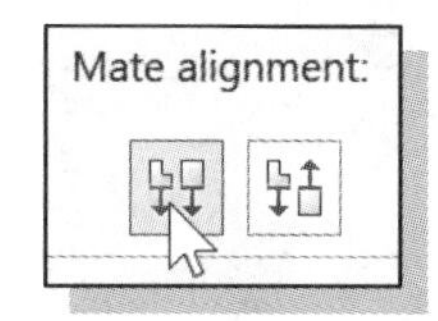

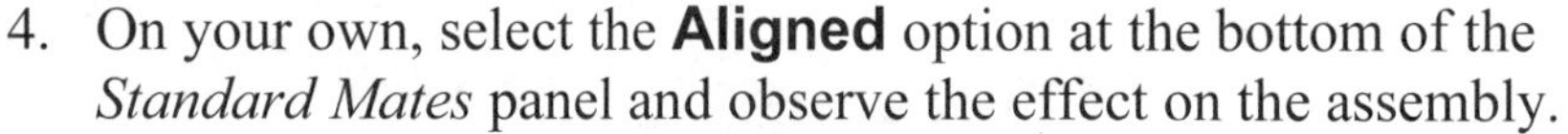

4. On your own, select the **Aligned** option at the bottom of the *Standard Mates* panel and observe the effect on the assembly.

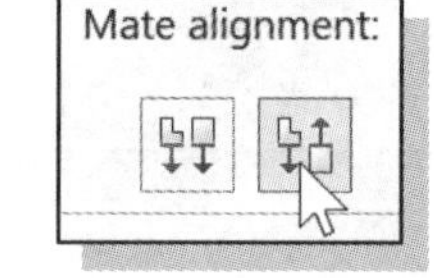

5. Select the **Anti-Aligned** option.

6. Click the **OK** button in the *PropertyManager* to accept the settings and create the **Anti-Aligned Coincident** mate.

7. Select the **Isometric View** to adjust the display of the assembly model.

Apply a Second Mate

The Mate command can also be used to align axes of cylindrical features.

1. On your own, turn *ON* the visibility of the **Temporary Axes**. (**HINT:** Use the *Hide/Show* pull-down menu on the *Heads-up View* toolbar.)

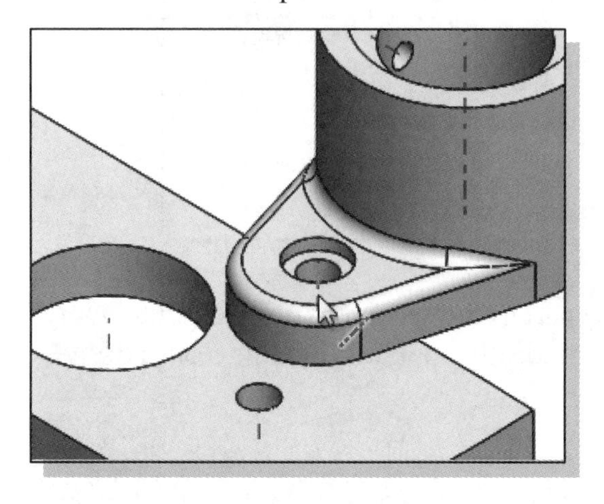

2. Note that the *Mate Selections* panel in the *Mate PropertyManager* is highlighted. (Select the Mate command again if you closed this.) Select the **axis** of the counter bore hole of the *Bearing* part.

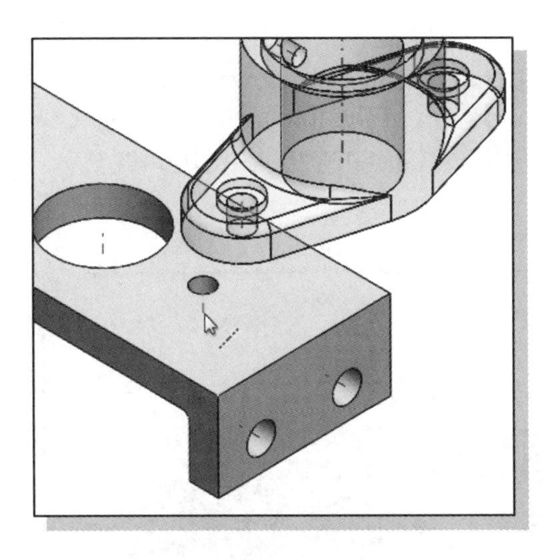

3. Select the **axis** of the small hole on the *Base-Plate* part, as shown. The *Bearing* moves to automatically apply the default **Coincident** mate.

4. Click **OK** in the *PropertyManager* to create the **Coincident** Mate.

5. In the *PropertyManager*, click on the **Close** button (the red **X** icon), or press the **[Esc]** key, to exit the **Mate** command.

❖ The *Bearing* part appears to be placed in the correct position. But the DOF for the *Bearing* are not fully restricted; the *Bearing* part can still rotate about the displayed vertical axis.

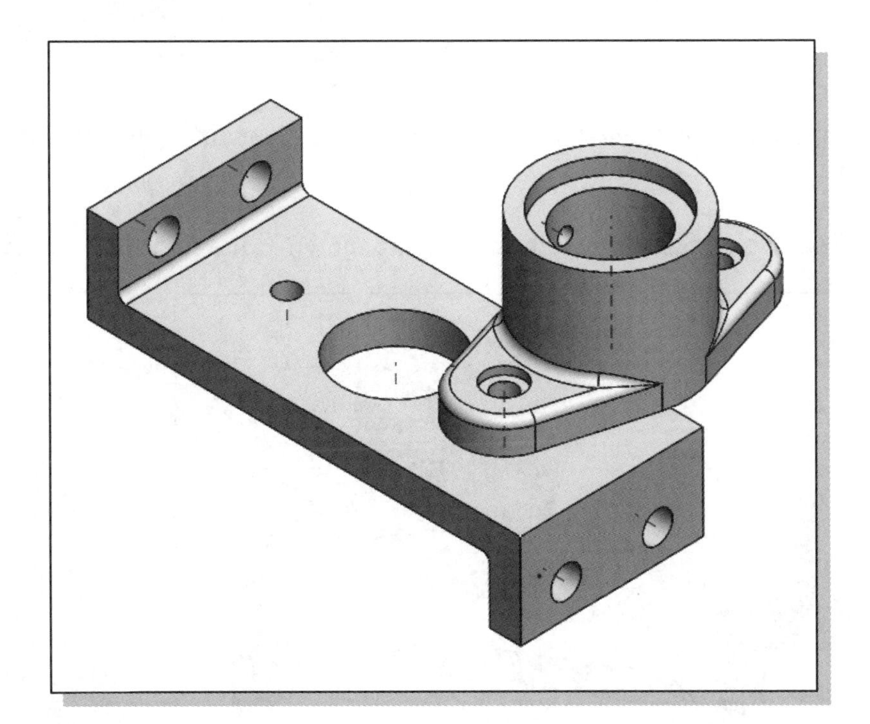

Constrained Move

To see how well a component is constrained, we can perform a constrained move. A constrained move is done by dragging the component in the graphics window with the left-mouse-button. A constrained move will honor previously applied assembly mates. That is, the selected component and parts mated to the component move together in their constrained positions. A fixed component remains fixed during the move.

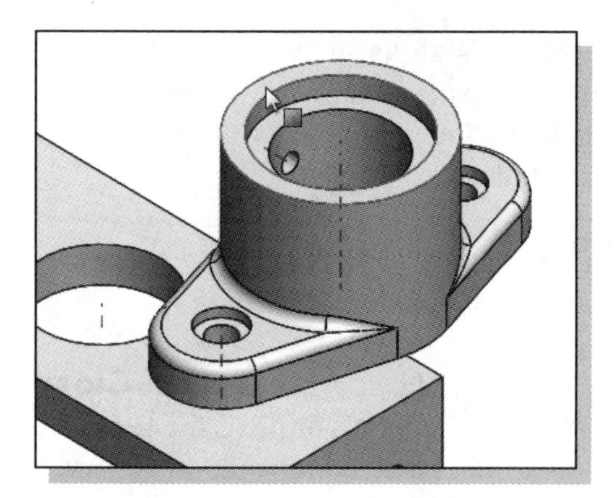

1. Inside the graphics window, move the cursor on top of the top surface of the *Bearing* part as shown in the figure.

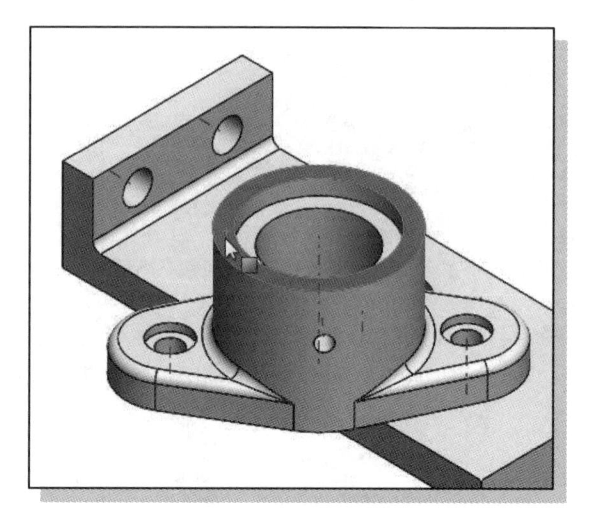

2. Press and hold down the left-mouse-button and drag the *Bearing* part downward.

❖ The *Bearing* part can freely rotate about the mated axis.

3. On your own, use the rotation commands to view the alignment of the *Bearing* part.

4. Rotate the *Bearing* part and adjust the display as shown in the figure.

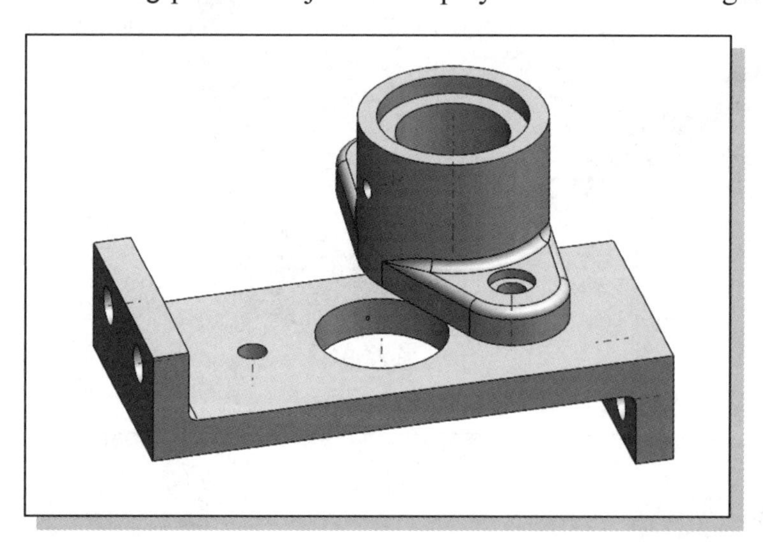

Apply a Third Mate

Besides selecting the surfaces of solid models to apply constraints, we can also select the established planes to apply the assembly mates. This is an additional advantage of using the *BORN technique* in creating part models. For the *Bearing* part, we will apply a Mate to two of the planes and eliminate the last rotational DOF.

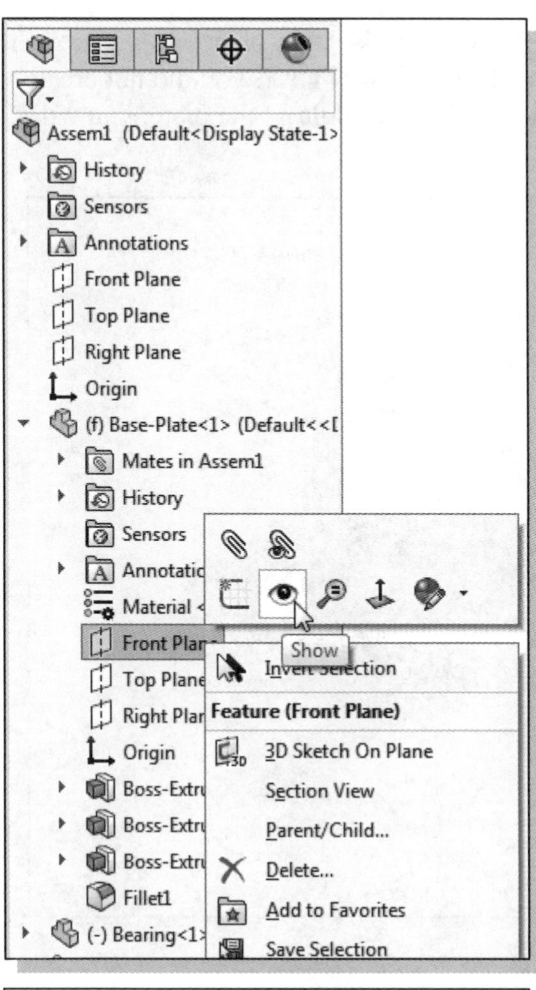

1. On your own, inside the *FeatureManager Design Tree* use the **Show** option to toggle *ON* the visibility for the planes that can be used for alignment of the *Base-Plate* and the *Bearing* parts. (**HINT:** Make sure the visibility of *Planes* is toggled *ON* in the *Hide/Show* pull-down menu.)

2. Press the **[Esc]** key to ensure that no object is selected.

3. In the *Assembly* toolbar, select the **Mate** command by left-mouse-clicking once on the icon.

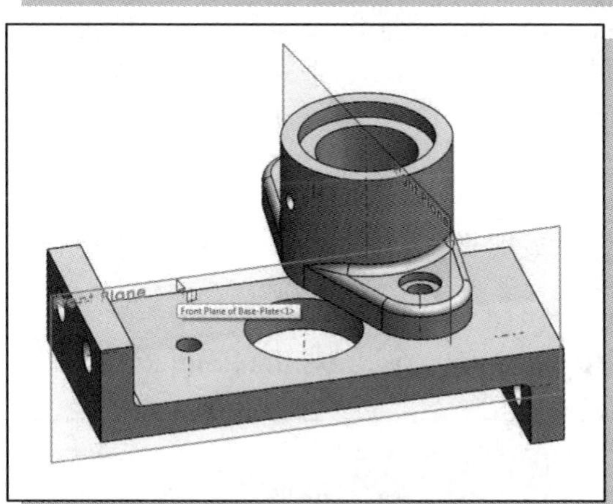

4. Select the *plane* of the *Base-Plate* part as the first part for the Mate command.

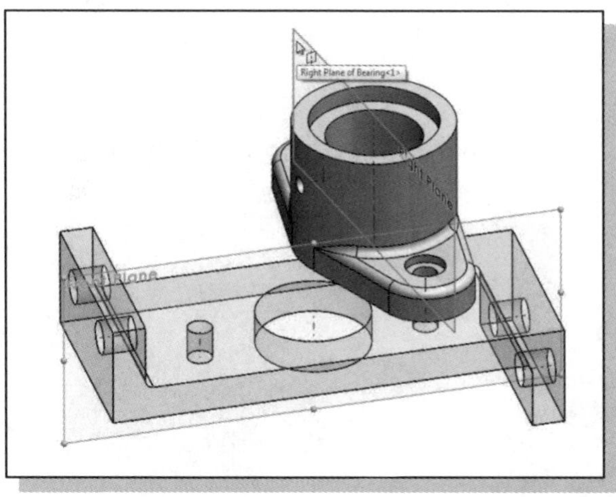

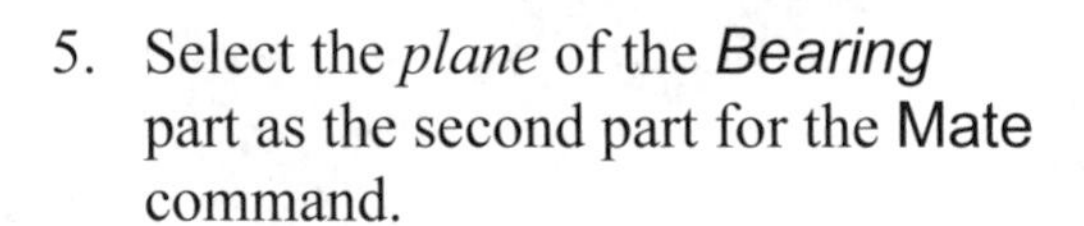

5. Select the *plane* of the *Bearing* part as the second part for the Mate command.

❖ The Coincident mate is executed by default and makes two planes coplanar with the **Anti-aligned** option automatically selected. (**NOTE:** If you did not create the base plate with the front plane aligned with the middle of the plate, you will use the Parallel mate.)

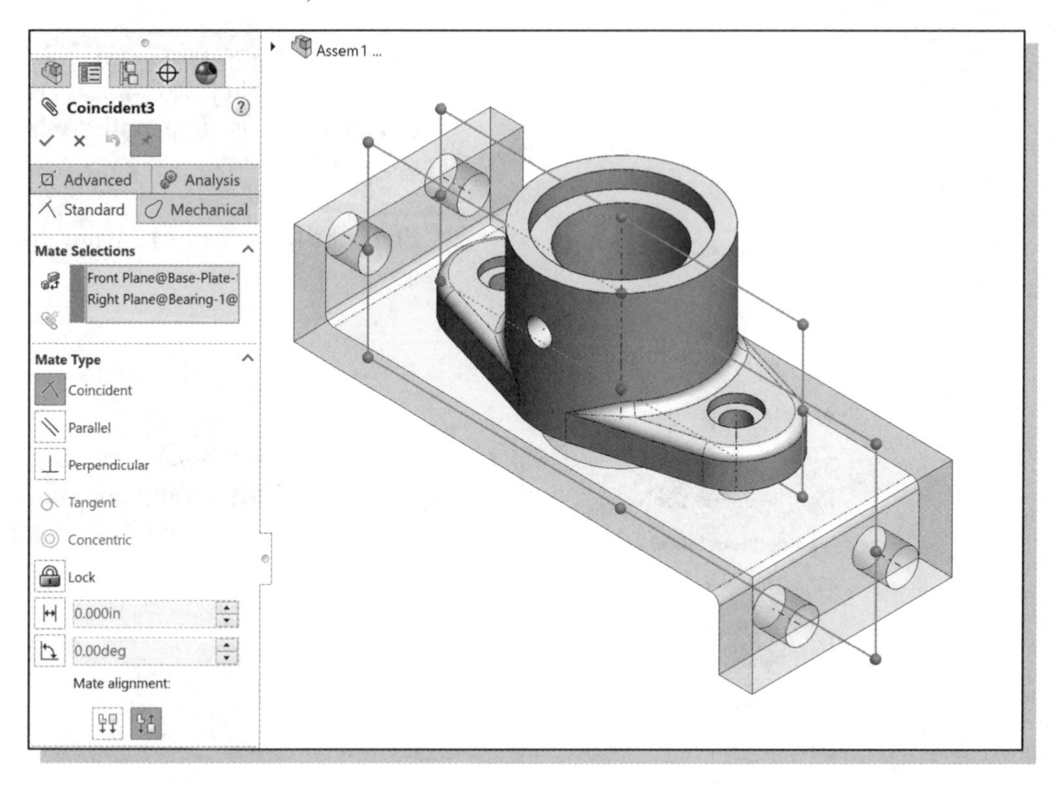

6. On your own, select the **Aligned** option at the bottom of the *Standard Mates* panel and observe the effect on the assembly.

7. Select the **Anti-aligned** option.

8. Click the **OK** button in the *PropertyManager* to accept the settings and create the **Anti-aligned Coincident** mate.

9. In the *PropertyManager*, click on the **Close** button (or press the **[Esc]** key) to exit the Mate command.

❖ Note the assembly is fully defined.

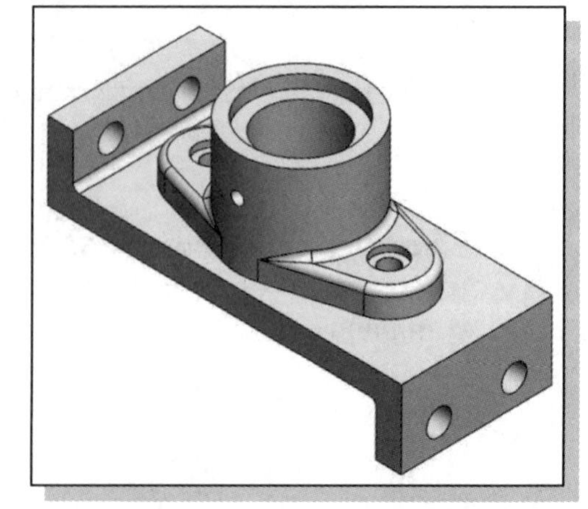

Inserting the Third Component

We will retrieve the *Collar* part as the third component of the assembly model.

1. In the *Assembly* toolbar, select the **Insert Components** command by left-mouse-clicking the icon.

2. Select the ***Collar*** design (part file: **Collar.sldprt**) in the browser and click on the **Open** button to retrieve the model.

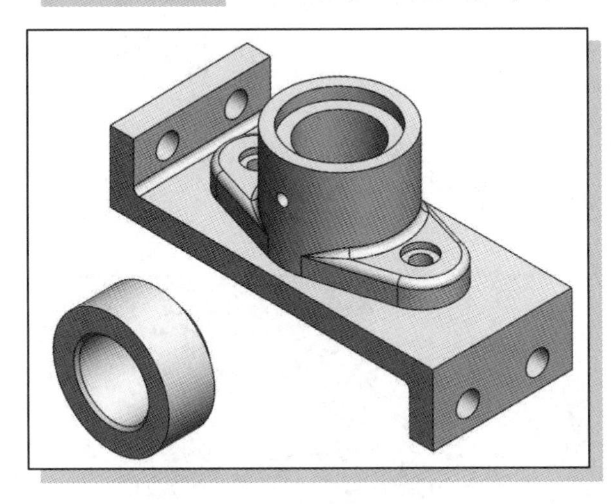

3. Insert the *Collar* part toward the lower left corner of the graphics window, as shown in the figure.

Applying Concentric and Coincident Mates

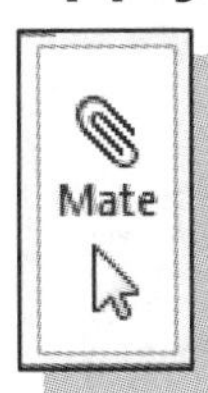

1. In the *Assembly* toolbar, select the **Mate** command by left-mouse-clicking once on the icon.

2. Select the external cylindrical surface of the *Collar* part as shown in the figure as the first part for the Mate command. (**HINT:** Use the arrow keys on the keyboard to rotate the view, if necessary.)

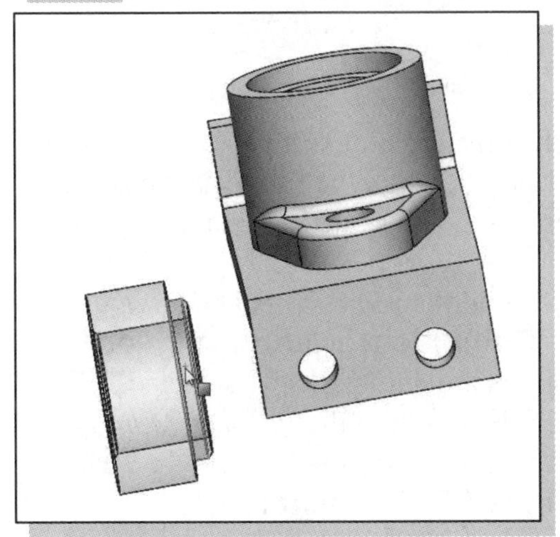

3. Select the internal cylindrical surface of the *Bearing* part as shown in the figure as the second part for the Mate command.

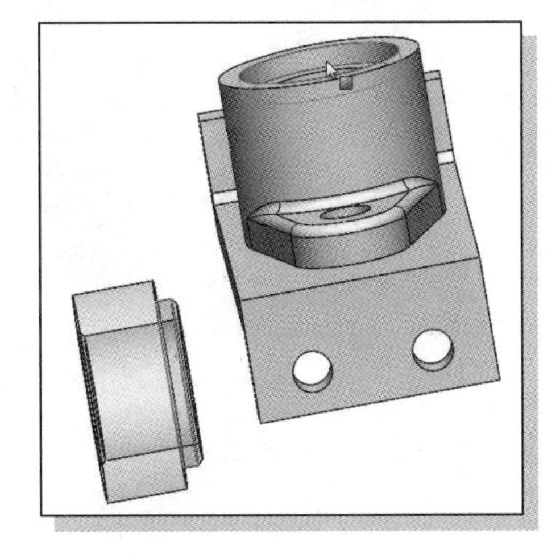

4. Notice in the *Mate PropertyManager* that the Concentric and Anti-Aligned options are automatically selected. On your own, **select the correct alignment** (Aligned or Anti-Aligned) for your assembly. Click the **OK** button in the *PropertyManager* to accept the settings and create the **Concentric** mate.

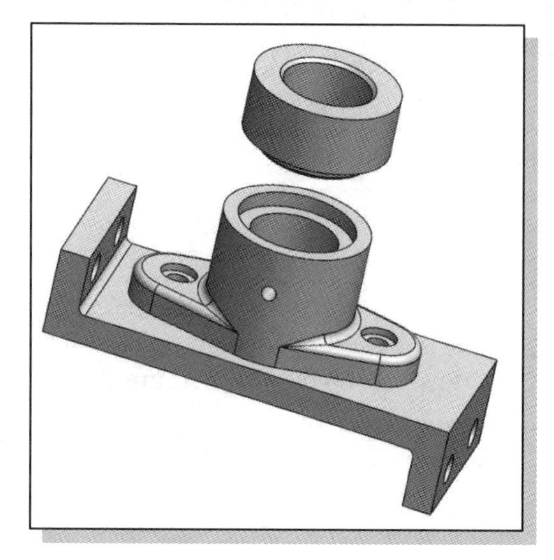

5. On your own, move the ***Collar*** part to a position like that in the figure. Notice that the move is constrained to maintain the concentric relationship.

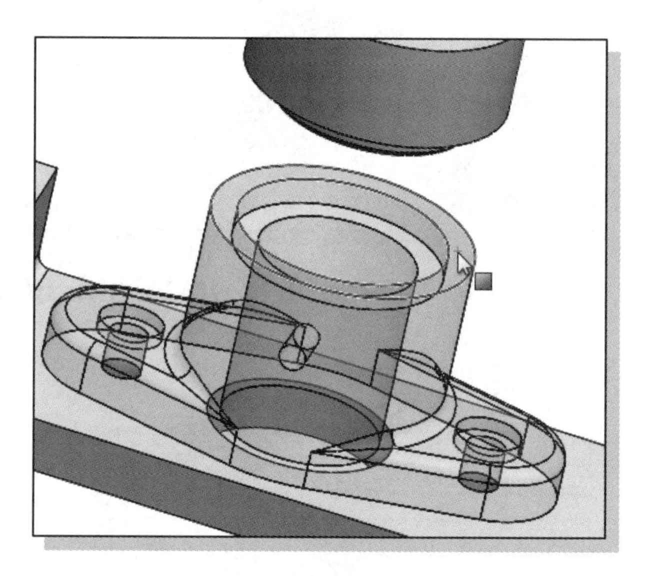

6. Select the top circular surface of the *Bearing* part as shown in the figure as the first part for a new Mate command.

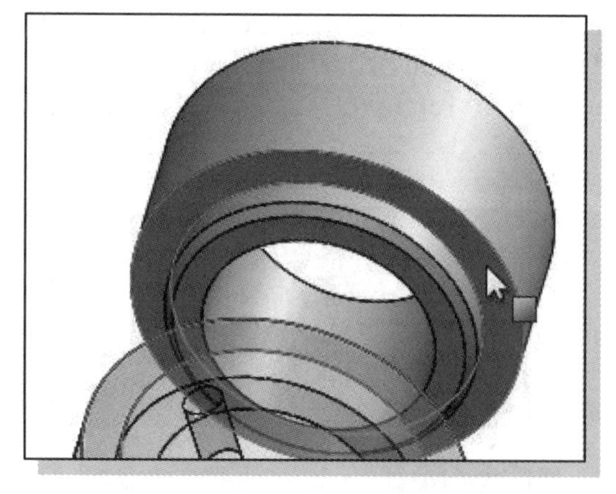

7. Select the circular surface of the *Collar* part as shown in the figure as the second part for a new Mate command.

8. Notice in the *Mate PropertyManager* that the **Coincident** and **Anti-Aligned** options are automatically selected. Click the **OK** button in the *PropertyManager* to accept the settings and create the **Anti-Aligned Coincident** mate.

9. In the *PropertyManager*, click on the **Close** button (or press the **[Esc]** key) to exit the Mate command.

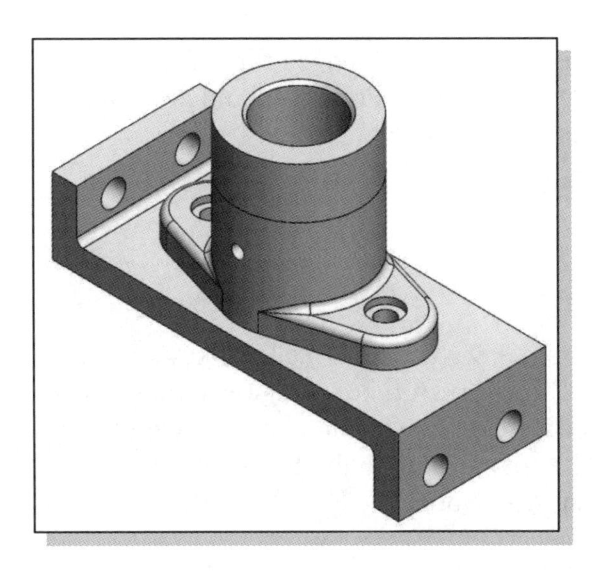

10. On your own, change the view to **Isometric**.

11. Note that the *Collar* part is not fully constrained. One rotational degree of freedom remains open; the *Collar* part can still freely rotate about the vertical axis. Click on the *Collar* and drag with the **left-mouse-button** to observe this constrained motion.

Assemble the Cap-Screws using SmartMates

We will place two *Cap-Screw* parts to complete the assembly model. We will use SOLIDWORKS **SmartMates** to apply the geometric relations. The SmartMates function allows us to create commonly-used mates without using the *Mate PropertyManager*. Application of a SmartMate involves dragging one part to a mating position with another. In most cases, one mate is created. The type of SmartMate created depends on the geometry used to drag the component, and the type of geometry onto which you drop the component.

The SmartMates functionality creates multiple mates under certain conditions. If the application finds circular edges to mate, a **Peg-In-Hole** SmartMate will be applied. Two mates are applied: a Concentric mate between cylindrical or conical faces, and a Coincident mate between the planar faces that are adjacent to the conical faces. This is the type of SmartMate we will apply when inserting the *Cap-Screw* parts.

- **Applying a SmartMate when two components are already in the assembly**

1. On your own, insert a ***Cap-Screw*** part into the assembly as shown.

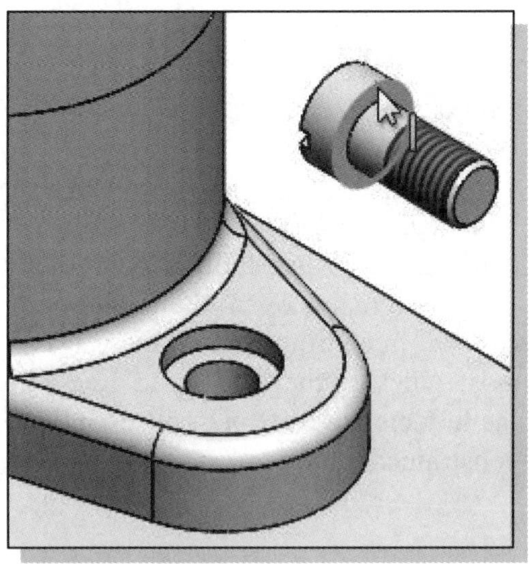

2. Select the circular edge of the *Cap-Screw* head as shown. Press and hold down the left-mouse-button.

3. Press down and hold the **[Alt]** key. While holding the **[Alt]** key, drag the circular edge of the *Cap-Screw* onto the circular face of the counter sink in the *Bearing* part as shown. Notice the **Peg-In-Hole** mate icon appears.

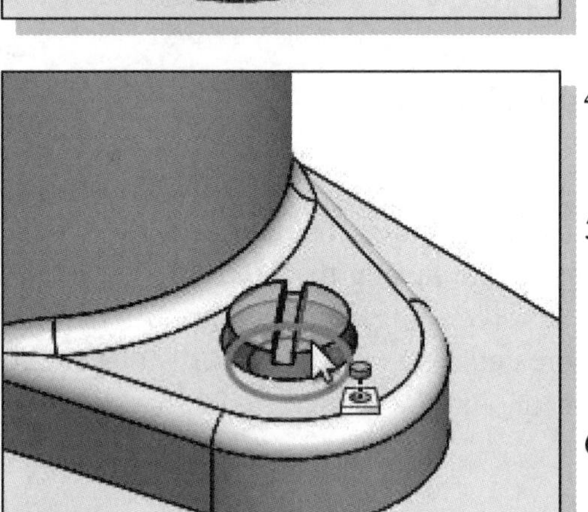

4. Release **[Alt]** while still holding the mouse key down.

5. If the alignment condition is incorrect (screw is upside down), press the **[Tab]** key while still holding the left mouse button down.

6. When the screw is aligned correctly, release the left mouse button to apply the SmartMate.

- **Applying a SmartMate to a new component while adding it to the assembly**

7. Click once with the **left-mouse-button** on the **Open** icon on the *Menu Bar*.

8. Select the ***Cap-Screw*** design (part file: **Cap-Screw.sldprt**) in the browser. And click on the **Open** button to retrieve the model.

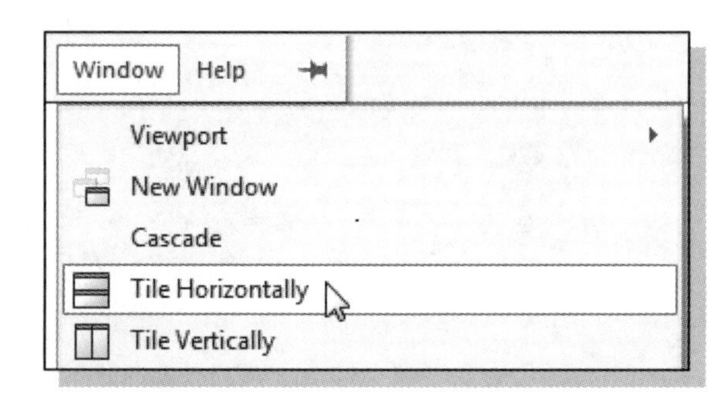

9. Select the **Tile Horizontally** option from the *Window* pull-down menu.

10. In the *Cap-Screw* window, rotate and zoom so that the circular edge of the screw head is visible as shown in the figure.

11. In the *Assembly* window, **Pan**, **Rotate**, and **Zoom** so that the counter bore of the *Bearing* appears as shown in the figure.

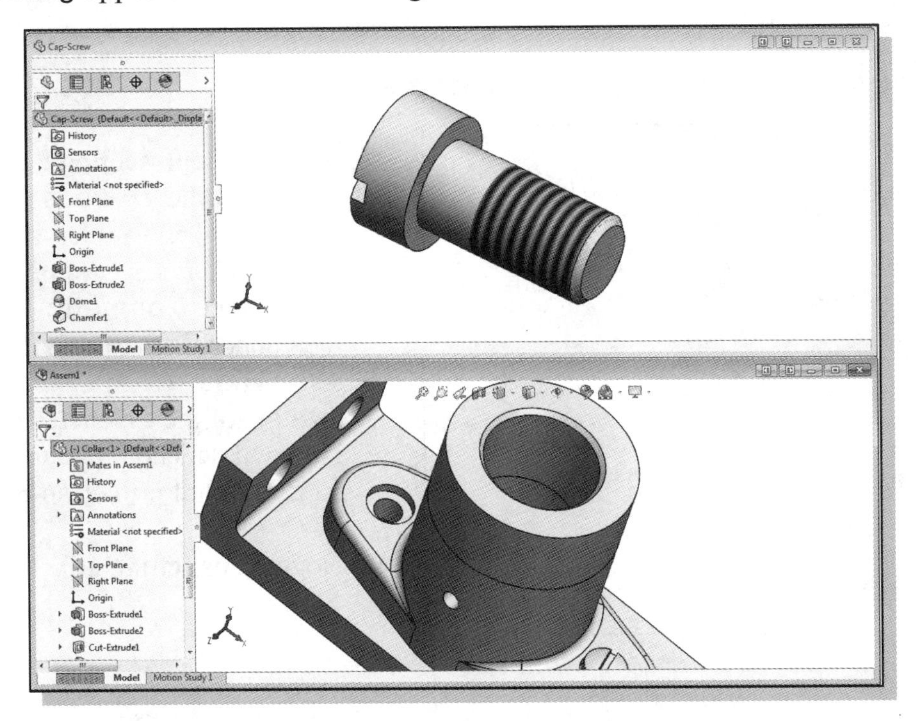

12. Using the left-mouse-button, click and drag the circular edge of the head of the *Cap-Screw* part into the assembly window. Release the mouse-button when the cursor is over the circular surface of the counter bore, applying the **Peg-In-Hole SmartMate**.

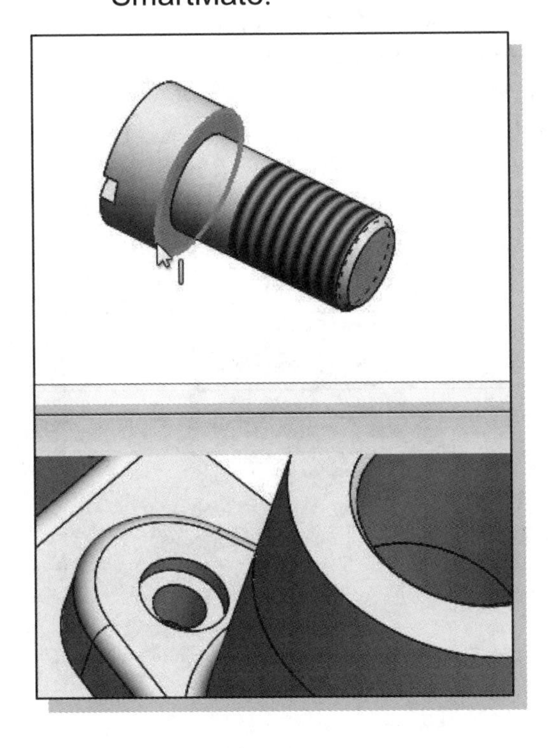

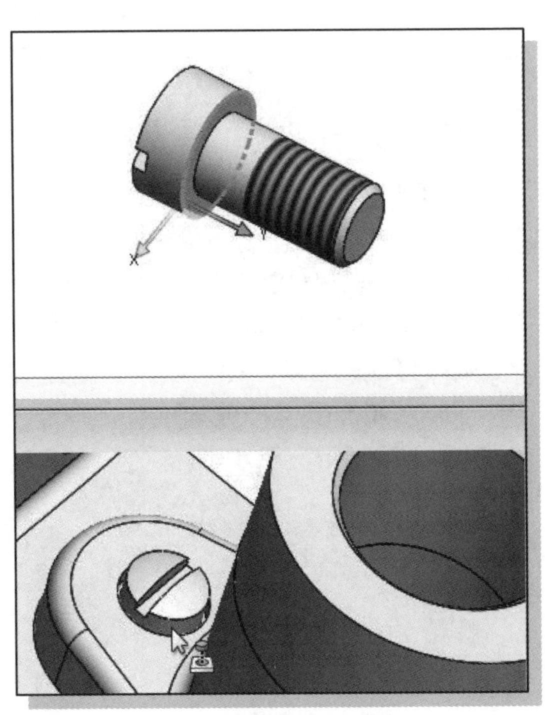

13. Close the *Cap-Screw* window.

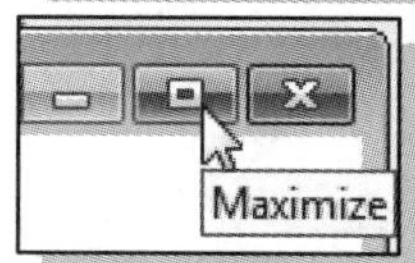

14. Maximize the *Assem1* window.

15. On your own, change the view to **Isometric**.

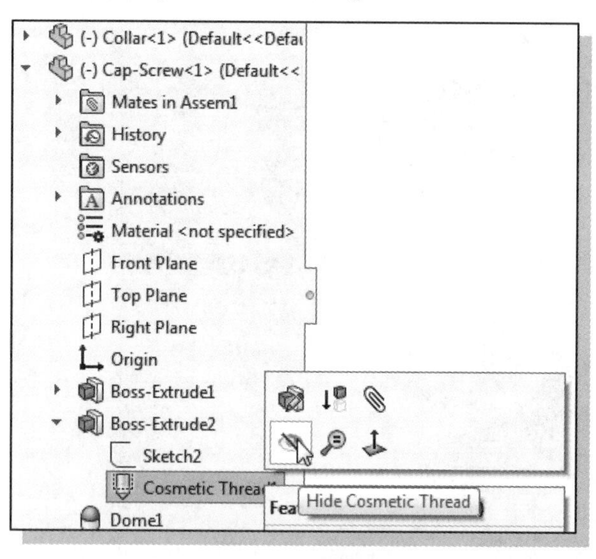

16. Isocircles representing the bottom of the **Cosmetic Threads** on the screws may be visible. In the *FeatureManager Design Tree*, **right-click** on the **Cosmetic Thread1** icon under **Cap-Screw(1)** and select **Hide**. (NOTE: The exact location of the Cosmetic Thread1 icon will depend on your construction for the Cap-Screw.)

❖ Notice, this setting affects both *Cap-Screws.*

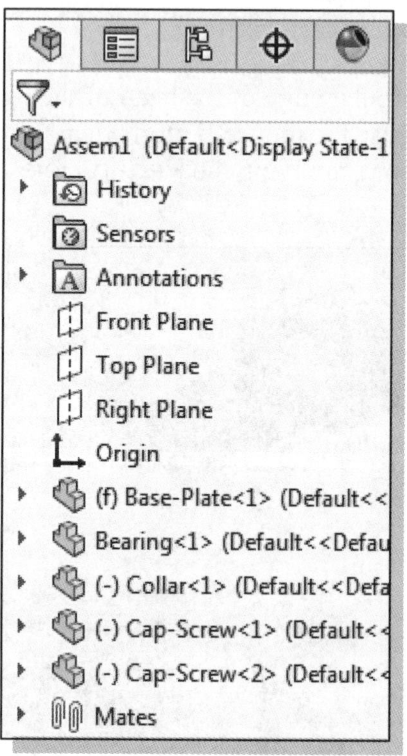

- Inside the *FeatureManager Design Tree*, the retrieved parts are listed in the order they are placed. The number behind the part name is used to identify the number of copies of the same part in the assembly model. Move the cursor to the last part name and notice the corresponding part is highlighted in the graphics window.

- The **(f)** symbol indicates the part is fixed. The **(-)** symbol indicates the part has at least one degree of freedom. The *Collar* and the two *Cap Screws* each have one rotational degree of freedom. Click and drag on each of them to observe the rotational freedom.

Exploded View of the Assembly

Exploded assemblies are often used in design presentations, catalogs, sales literature, and in the shop to show all of the parts of an assembly and how they fit together. In SOLIDWORKS, exploded views are created by selecting and dragging parts in the graphics area.

1. In the *Assembly* toolbar, select the **Exploded View** command by left-mouse-clicking once on the icon.

❖ Notice that the Explode *FeatureManager* opens and the **Explode Step1** is active.

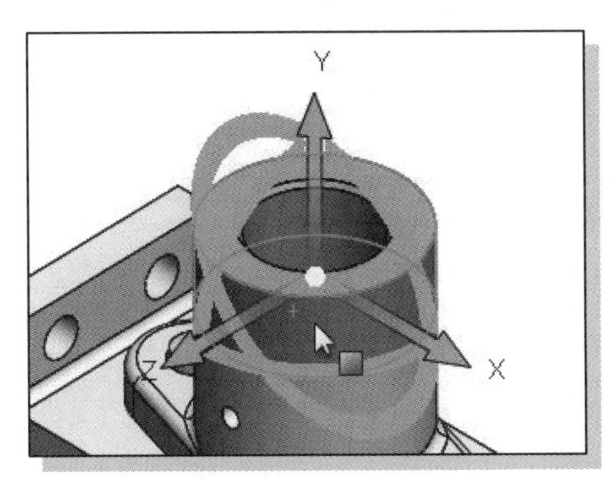

2. Select the *Collar* part by clicking on it with the left-mouse-button.

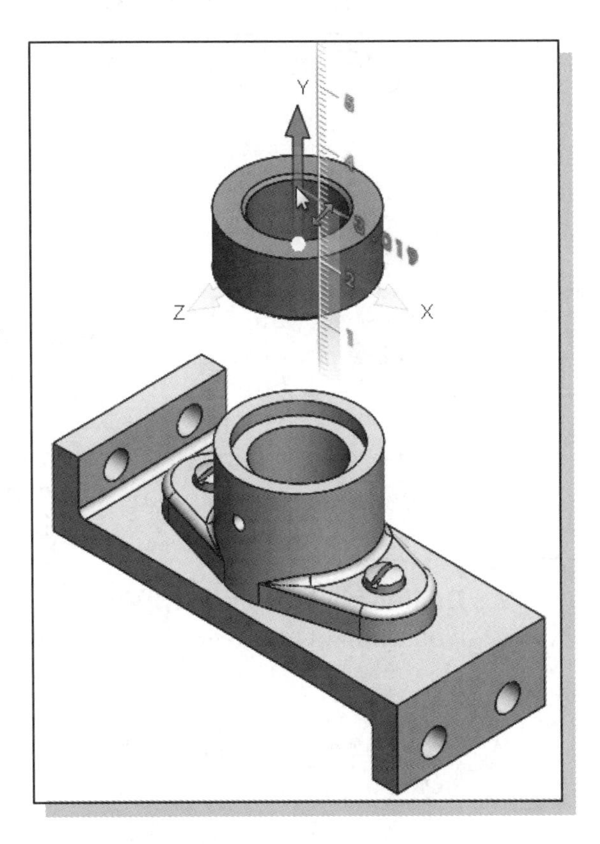

3. Notice a manipulator appears in the graphics area. Move the pointer over the manipulator handle that points in the upward direction and **drag** the manipulator handle to the location shown in the figure.

4. Click the **Done** button at the bottom of the *Editing Explode Step1* panel to complete the first step.

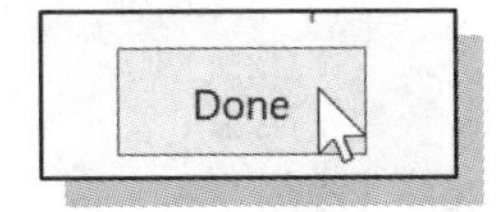

❖ The exploded view component movements can be made along the major axis directions by dragging the manipulator handle that points in the corresponding direction. Dragging the yellow sphere in the center of the manipulator will move the manipulator to a different location. The manipulator can be dropped on an edge or face and an axis of the manipulator will align with the edge or face.

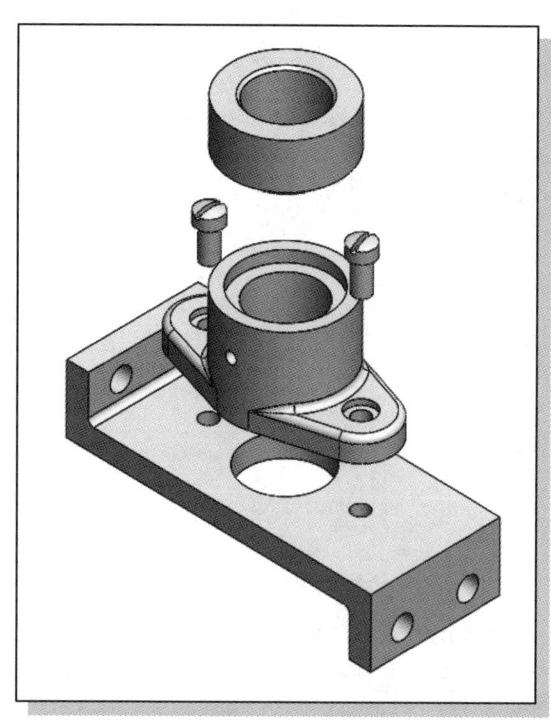

- When *Explode Step1* is completed, *Explode Step2* is automatically activated and the window for selection of component(s) is highlighted.

5. On your own, repeat the above steps and create an exploded assembly by repositioning the components as shown in the figure.

6. Click **OK** in the *PropertyManager* to close the Exploded View command.

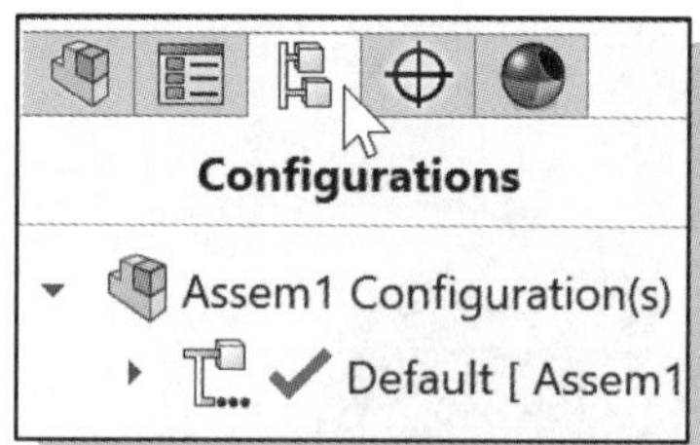

7. Select the **ConfigurationManager** tab.

8. In the *ConfigurationManager*, expand the tree to reveal the Exploded View icon labeled ExpView1. **Double-click** on the **ExplView1** icon to collapse the exploded view.

- Note that the components are reset back to their assembled positions, based on the applied assembly mates.

9. **Double-click** on the **ExplView1** icon again to return to the exploded view.

10. On your own, return to the collapsed view.

11. Select the **FeatureManager Design Tree** tab to return to the *Design Tree*.

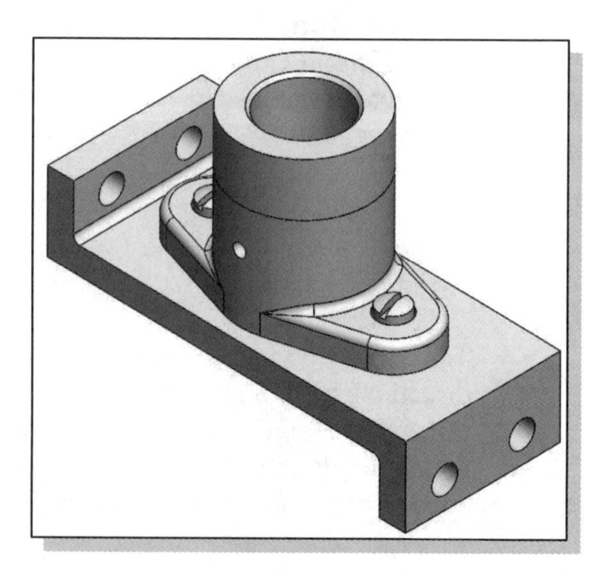

Save the Assembly Model

1. Select the **Save As** option from the *File* pull-down menu.

2. In the *Save As* pop-up window, enter **Support_Assembly** for the *File name* and **Shaft-Support Assembly** for the *Description*.

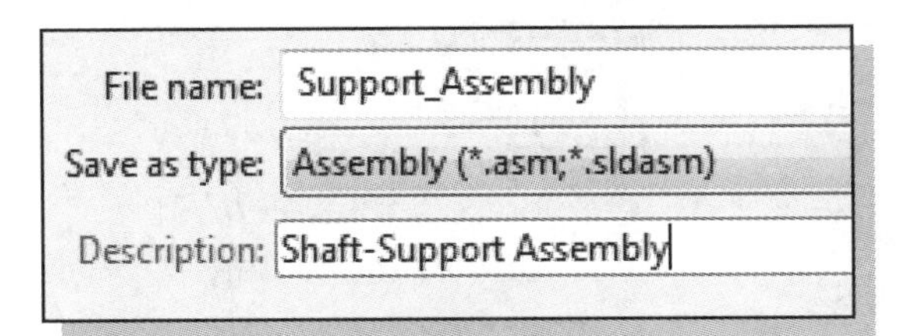

3. Click **Save** to save the file.

4. If the SOLIDWORKS window appears, click **Save All**.

Editing the Components

The associative functionality of SOLIDWORKS allows us to change the design at any level, and the system reflects the changes at all levels automatically.

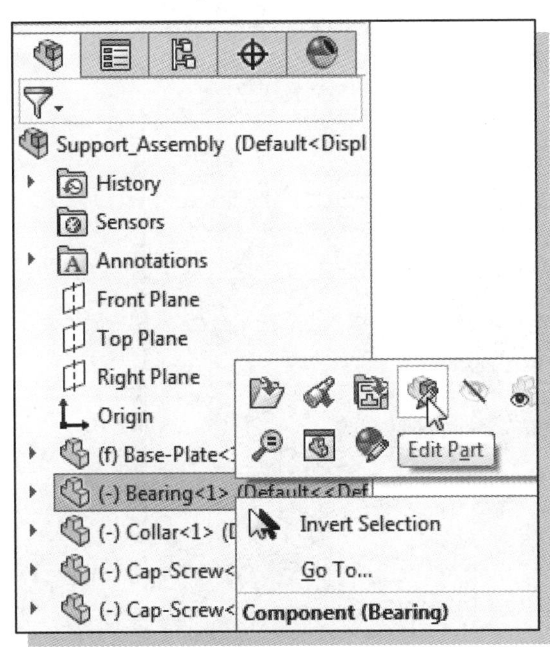

1. Inside the *FeatureManager Design Tree*, move the cursor on top of the **Bearing** part. Right-mouse-click once to bring up the option menu and select **Edit Part** in the option list.

❖ Note that we are automatically switched back to *Part Editing* mode. The Bearing component portion of the *FeatureManager Design Tree* appears blue to denote that the part model is active.

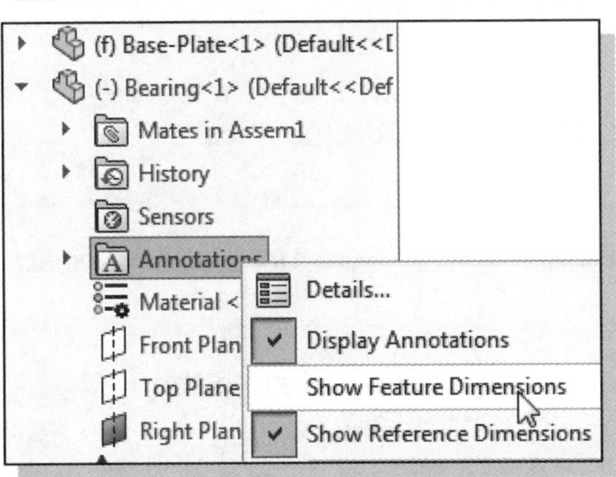

2. Expand the Bearing portion of the *Design Tree*. Move the cursor over the **Annotations** icon **under the Bearing**. Click once with the right-mouse-button to open the pop-up option menu. Select the **Show Feature Dimensions** option.

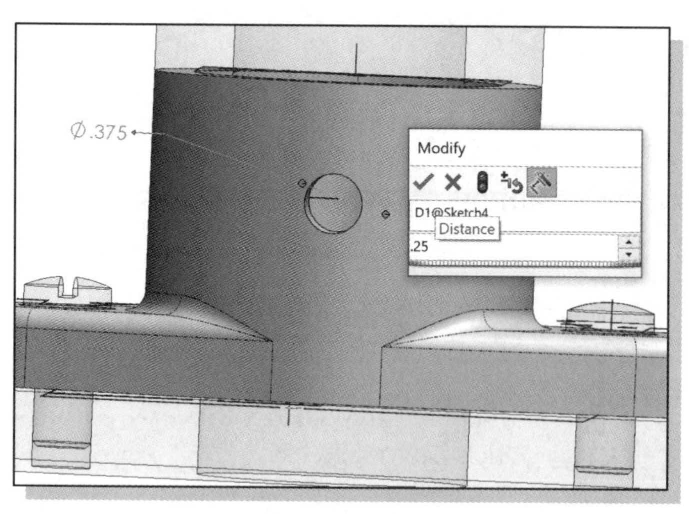

3. On your own, adjust the diameter of the small *Drill Hole* to **0.25** as shown.

4. Click on the **Rebuild** button in the *Menu Bar* to proceed with updating the model.

5. On your own, turn *OFF* the visibility of the feature dimensions.

6. Inside the **graphics window**, click once with the **right-mouse-button** to display the option menu.

7. Select **Edit Assembly:Support_Assembly** in the pop-up menu to exit *Part Editing* mode and return to *Assembly* mode.

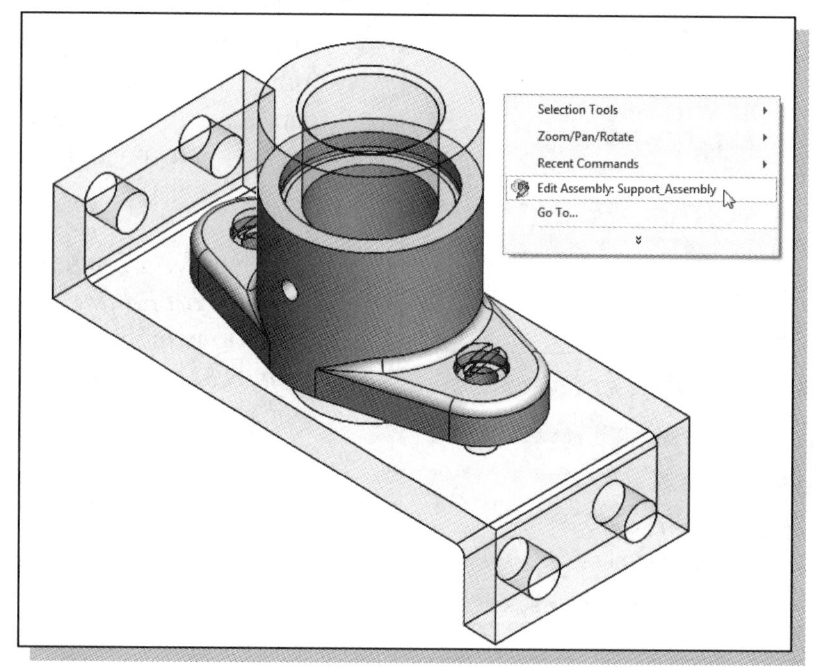

➢ SOLIDWORKS has updated the part in all levels and in the current *Assembly* mode. On your own, open the *Bearing* part to confirm the modification is performed.

8. On your own, save the Support_Assembly file. If the *Save Modified Documents* window appears, select **Save All**.

9. If the SOLIDWORKS window appears, select **Rebuild and save the document**.

Set up a Drawing of the Assembly Model

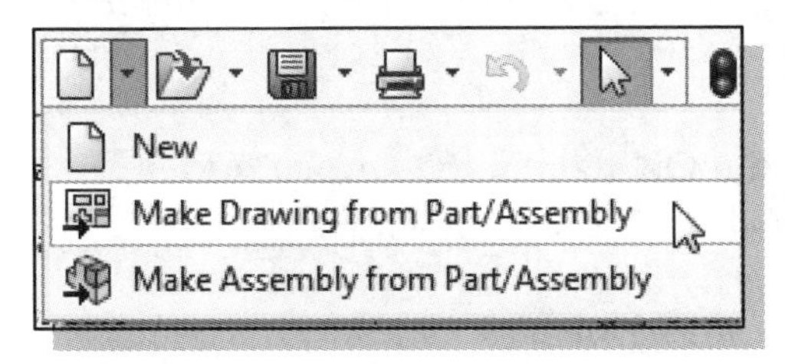

1. Click on the **arrow** next to the **New** icon on the *Menu Bar* and select **Make Drawing from Part/Assembly** in the pull-down menu.

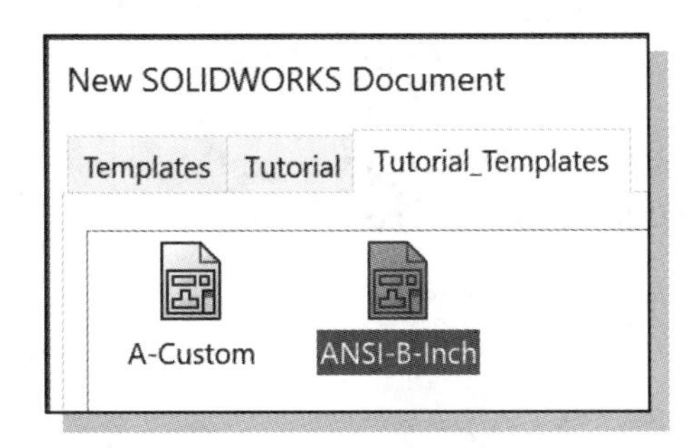

2. In the *New* SOLIDWORKS *Document* window, select the **Tutorial_Templates** tab.

3. Select the **ANSI-B-Inch** template file icon. This is the custom drawing template you created in Chapter 8. (If your ANSI-B-Inch template is not available, see note below.)

4. Select **OK** in the *New* SOLIDWORKS *Document* window to open the new drawing file.

➢ The new drawing file opens using the **ANSI-B-Inch.SLDDRW** template. The B-size drawing sheet appears.

❖ **NOTE:** If your ANSI-B-Inch template is not available: (1) open the default Drawing template; (2) set document properties following instructions on page 8-7; and (3) set sheet properties following instructions on page 8-10.

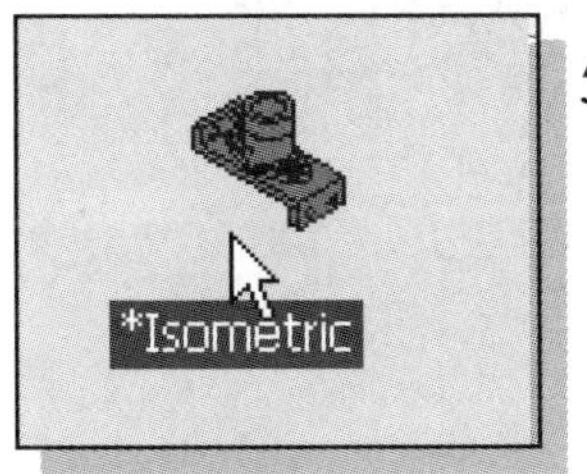

5. The *View Palette* appears in the *task pane*. Views can be dragged from the *View Palette* onto the drawing sheet to create a base drawing view. Click on the **Isometric** view in the *View Palette* as shown.

6. Hold the **left-mouse-button** down and **drag** to place the **isometric view** as shown.

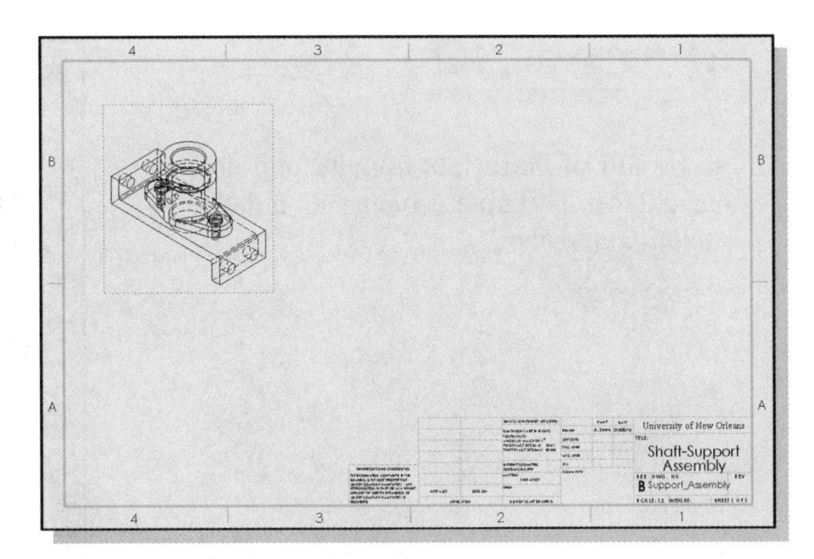

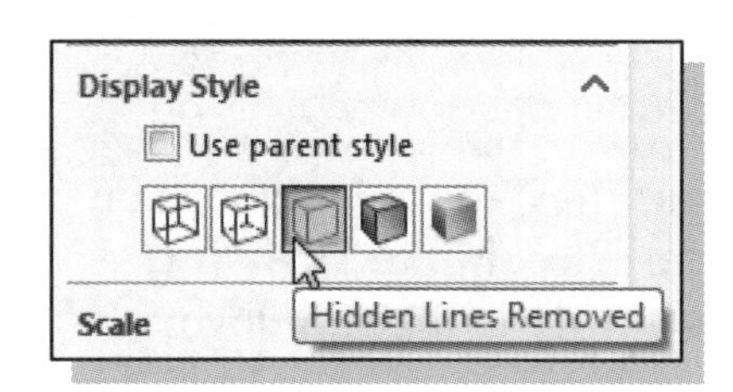

7. In the *Projected View PropertyManager*, select the **Hidden Lines Removed** option for the *Display Style*.

8. Left-click on the **OK** icon in the *Drawing View PropertyManager* to accept the view.

9. In the *FeatureManager Design Tree*, **right-click** on the **Sheet** icon to open the pop-up option menu and select **Properties**.

10. In the Sheet Properties window, set the scale to **1:1** and click **Apply Changes**.

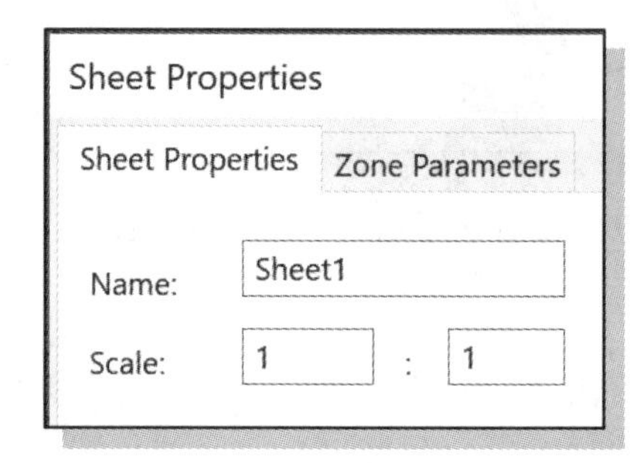

Creating a Bill of Materials

A *Bill of Materials (BOM)*, or parts list, can be automatically inserted into a drawing using SOLIDWORKS. A Bill of Materials is an example of a SOLIDWORKS *Annotation Table*.

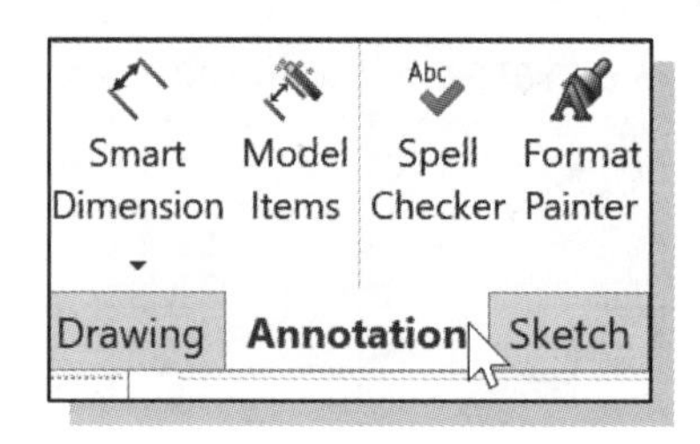

1. Select the **Annotation** tab on the *CommandManager* to display the *Annotation* toolbar.

2. Select **Bill of Materials** from the pull-down menu under the Tables command on the *Annotation* toolbar.

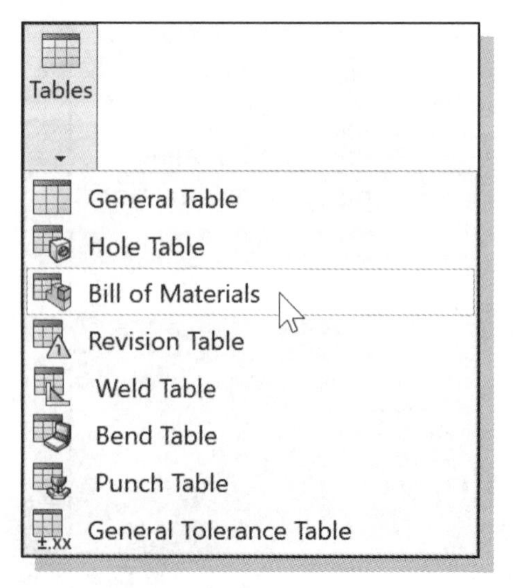

3. The message "*Select a drawing view to specify the model for creating the Bill of Materials*" appears in the *Bill of Materials PropertyManager*. Select on the shaft-support assembly view in the graphics area by clicking on it once with the **left-mouse-button**.

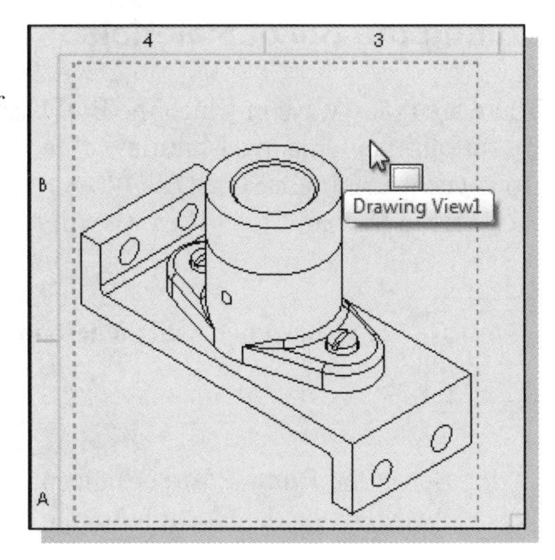

4. Click on **OK** in the *Bill of Materials PropertyManager* to insert the BOM. Locate the BOM so that it is aligned with the right border and the top of the title block as shown. Click the left-mouse-button to insert the BOM.

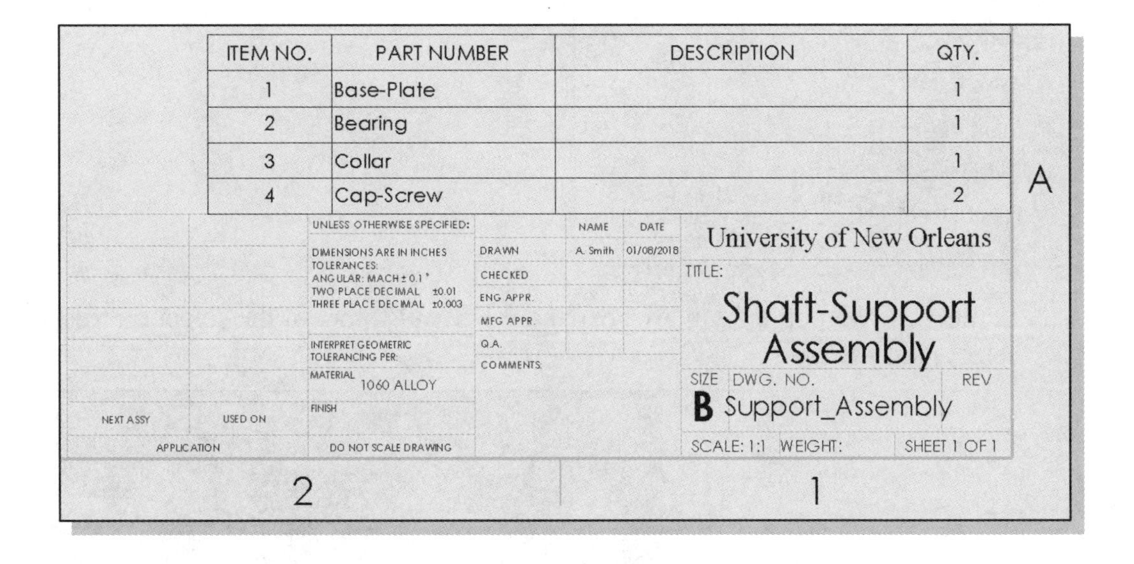

ITEM NO.	PART NUMBER	DESCRIPTION	QTY.
1	Base-Plate		1
2	Bearing		1
3	Collar		1
4	Cap-Screw		2

❖ The default SOLIDWORKS BOM appears. It includes four columns. The ITEM NO. column lists the individual parts inserted into the assembly. The PART NUMBER column includes Property Links to the part *filenames*. The DESCRIPTION column contains Property Links to the *description* entered in each part file (currently blank because no description was entered.) The QTY. column automatically lists the quantity of each part included in the assembly.

Editing the Bill of Materials

There are many ways in which the BOM table can be modified. We will change the Description column to a Material column. This could be done simply by entering text. However, we will use SOLIDWORKS Property Links. We will first set the material for each part in the part files. Then we will place the appropriate Property Links in the BOM.

1. Click once with the **left-mouse-button** on the **Open** icon on the *Menu Bar.*

2. Select the ***Base-Plate*** design (part file: **Base-Plate.sldprt**) in the browser. And click on the **Open** button to retrieve the model.

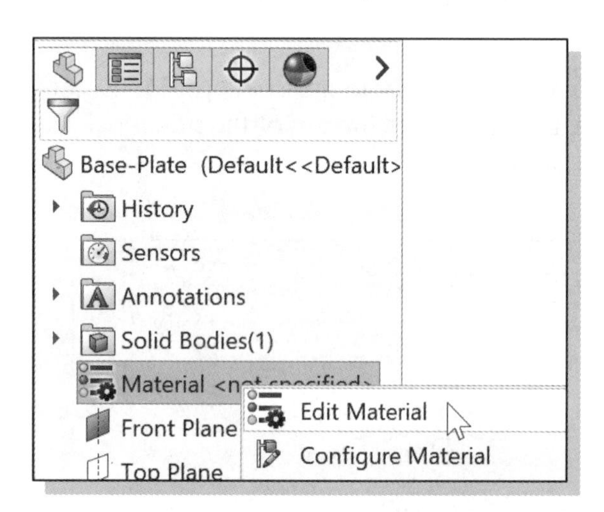

3. In the *FeatureManager Design Tree* for the part, right click on the **Material** icon, then select **Edit Material**.

4. In the *Material* pop-up window, expand the **Steel** folder of the selection tree, and select **Alloy Steel**.

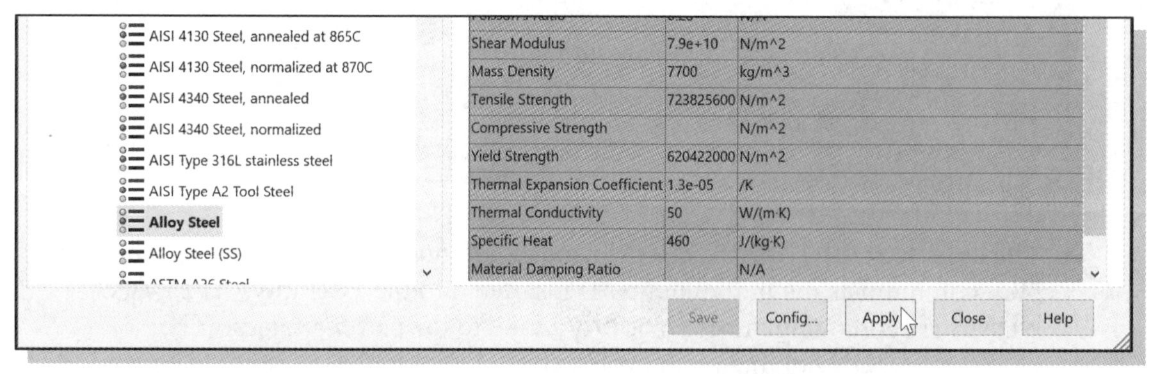

5. Click **Apply** in the *Material* pop-up window to accept the setting and select **Close** to close the window.

❖ The material has been selected for the *Base-Plate* and is defined as a SOLIDWORKS user-defined custom property.

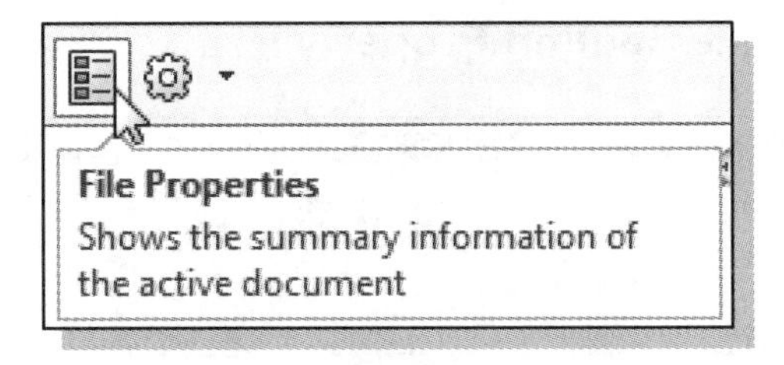

6. Select the **File Properties** option from the *Menu Bar* to open the *Summary Information* dialog box.

❖ The *Summary Information* window appears with the *Custom* tab selected.

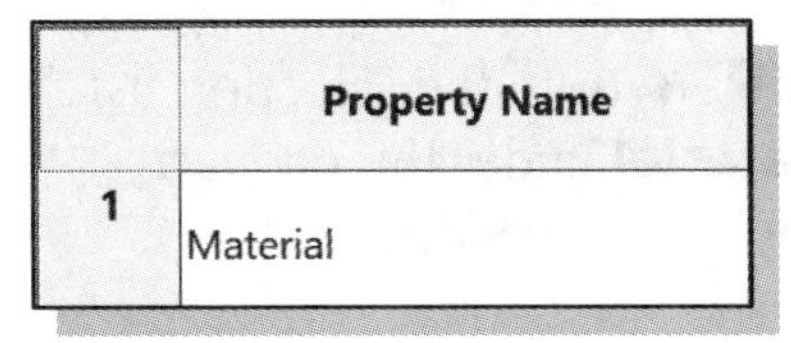

7. If Material does not already appear in the table, enter **Material** in the box under *Property Name*.

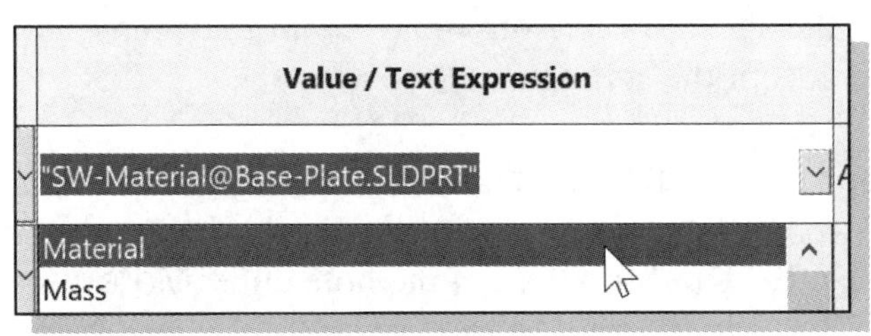

8. If necessary, click in the box under *Value/Text Expression* and use the pull-down option list to select **Material**.

❖ Notice the Material custom property appears and the Evaluated Value in the table automatically becomes Alloy Steel, the material selected for the part.

	Property Name	Type	Value / Text Expression	Evaluated Value
1	Material	Text	"SW-Material@Base-Plate.SLDPRT"	Alloy Steel

9. Click **OK** in the *Summary Information* window to accept the settings.

10. On your own, save and close the *Base-Plate* part file. If the SOLIDWORKS window appears, select **Yes**.

11. On your own, repeat the steps above to set the Material and the Material Custom Property for the *Bearing*, *Collar*, and *Cap-Screw* parts. Use the following: *Bearing* – Gray Cast Iron; *Collar* – Tin Bearing Bronze; *Cap-Screw* – Alloy Steel.

❖ The SOLIDWORKS custom property Material for each part is now available for use in Property Links. We will not use the Property Links in the BOM.

12. Move the cursor over the top of the third column in the BOM. When the **Column** icon appears, left-click to select the column.

	A	B	C	D
1	ITEM NO.	PART NUMBER	DESCRIPTION	QTY.
2	1	Base-Plate		1
3	2	Bearing		1
4	3	Collar		1
5	4	Cap-Screw		2

Column Icon

13. The pop-up *Column* toolbar appears. Select the **Column Property** icon.

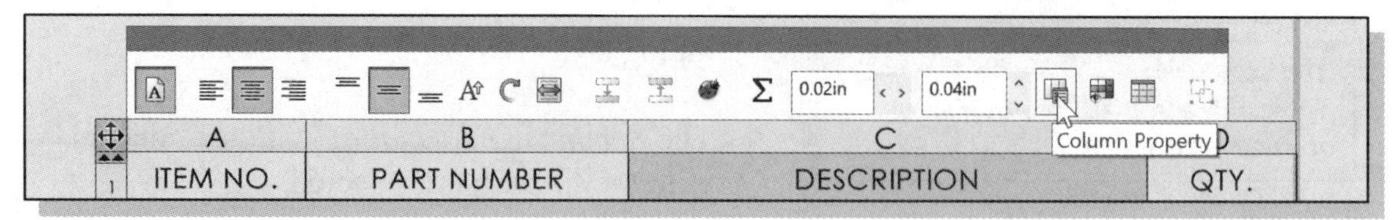

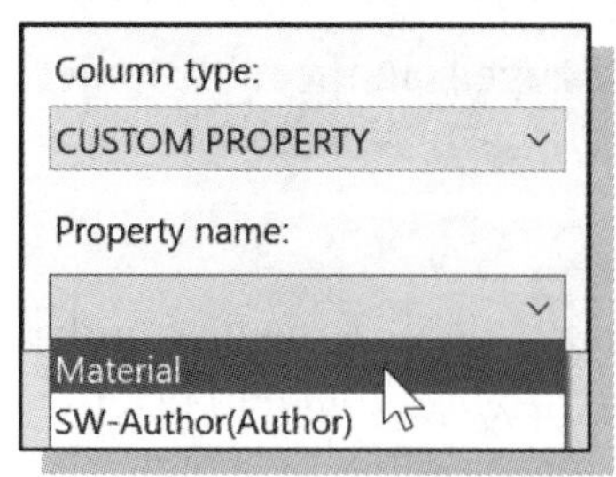

14. Select **CUSTOM PROPERTY** from the selection list under *Column type* and **Material** under the *Property name* option in the *Column Property* panel.

15. Click on the sheet, away from the *Column Property* panel, to accept the settings.

➤ Notice the corresponding material appears for each part due to the property link.

16. **Double-click** on the note entry **Material** in the new BOM.

17. Type over the word in all caps.

18. Press [**Enter**] to accept the change.

➤ Next, we will move the Parts List Header from top to bottom and invert the list.

19. Select the BOM table by clicking on the upper left corner of the table.

	A	B	C	D
1	ITEM NO.	PART NUMBER	MATERIAL	QTY.
2	1	Base-Plate	Alloy Steel	1
3	2	Bearing	Gray Cast Iron	1
4	3	Collar	Tin Bearing Bronze	1
5	4	Cap-Screw	Alloy Steel	2

20. In the pop-up toolbar, click on the **Table Header Top** icon to toggle the location of the Table Header.

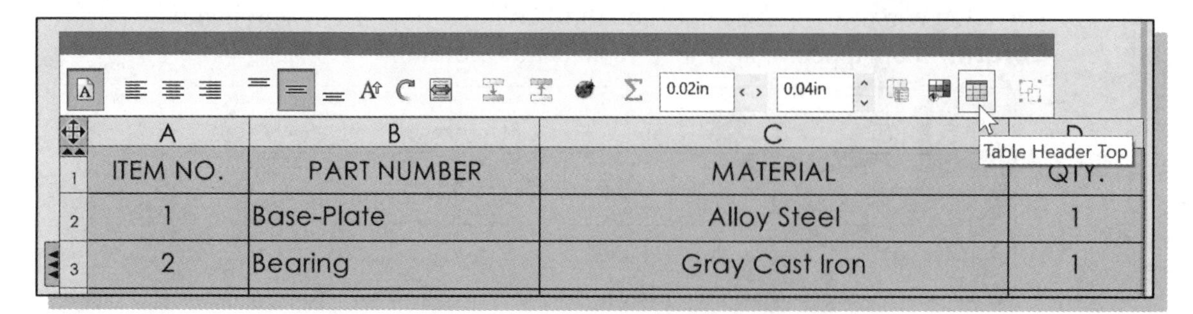

➢ The Table Header has been adjusted to the bottom and the Item Numbers are also adjusted accordingly as shown below.

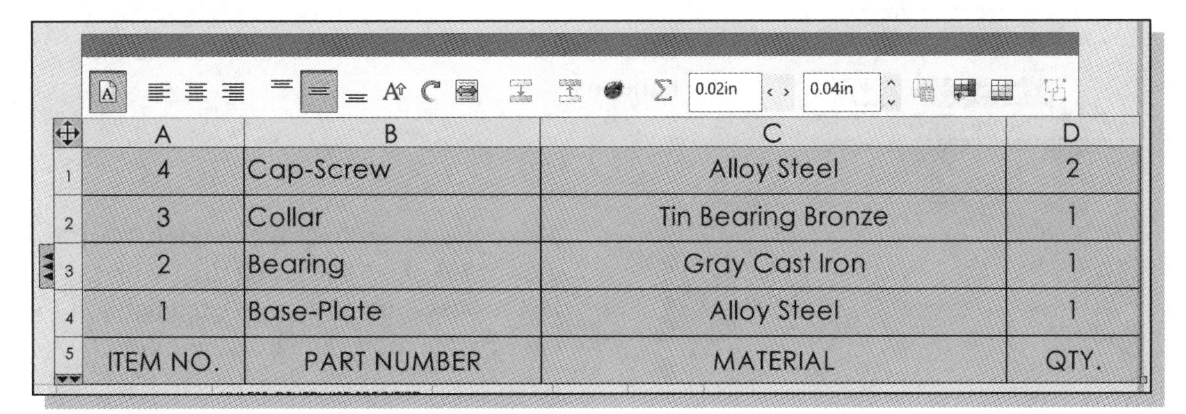

	A	B	C	D
1	4	Cap-Screw	Alloy Steel	2
2	3	Collar	Tin Bearing Bronze	1
3	2	Bearing	Gray Cast Iron	1
4	1	Base-Plate	Alloy Steel	1
5	ITEM NO.	PART NUMBER	MATERIAL	QTY.

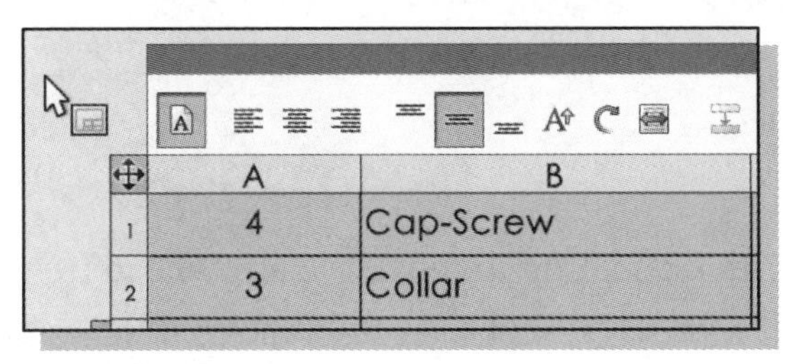

21. Click on the drawing sheet to exit the edit BOM table mode.

Completing the Assembly Drawing

1. Press the **[F]** key to fit the drawing to the graphics area.

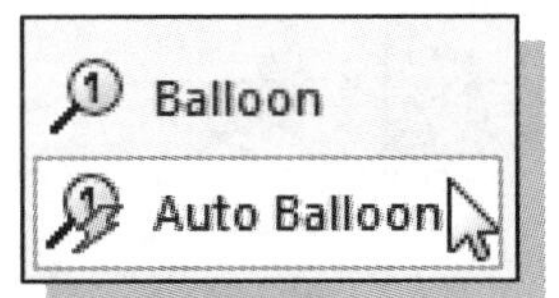

2. In the *Annotation* toolbar, click on the **AutoBalloon** button.

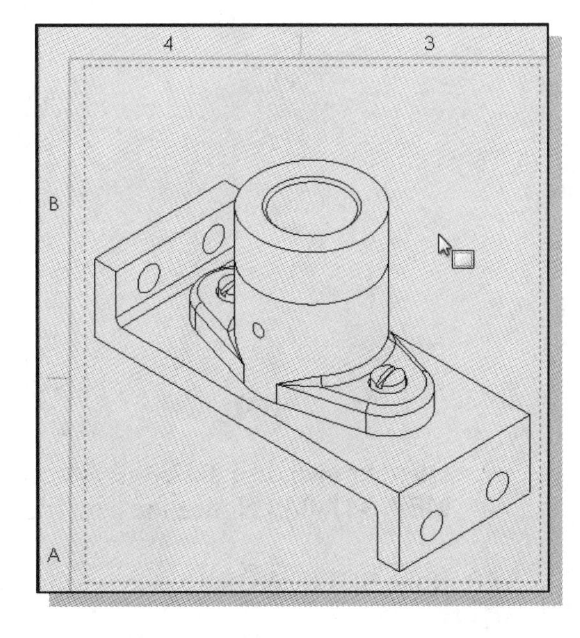

3. In the *AutoBalloon Property-Manager*, the message "*Select drawing sheet/view(s) to insert Auto Balloon*" is displayed. Click on the assembly view in the graphics area.

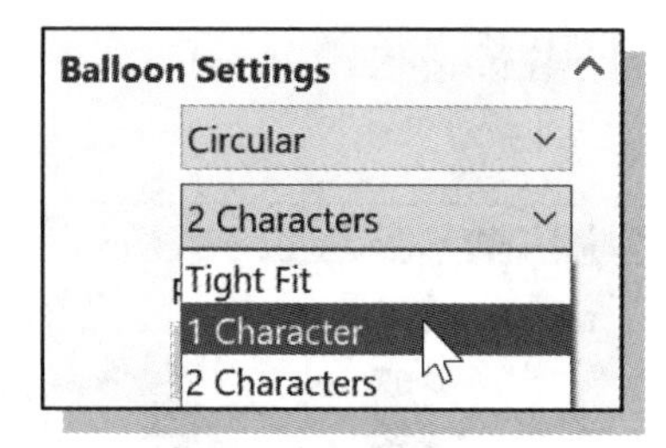

4. Select **1 Character** for the *Size* option on the *Balloon Setting* panel in the *PropertyManager*.

5. Click **OK** in the *PropertyManager* to create the balloons.

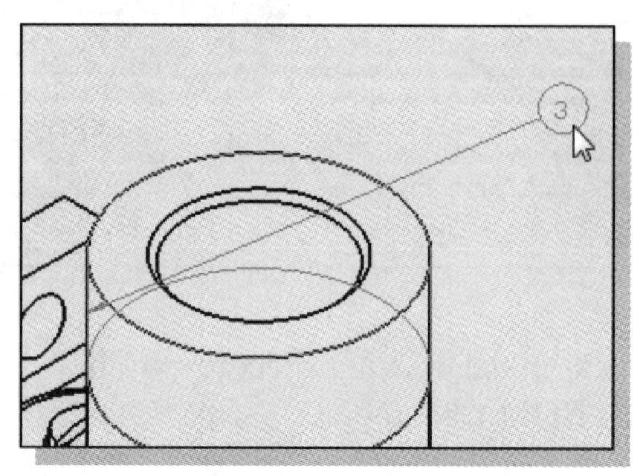

6. Move the cursor over the balloon labeling the *Collar* part. Click and drag with the **left-mouse-button** to move the balloon to the new location shown in the figure.

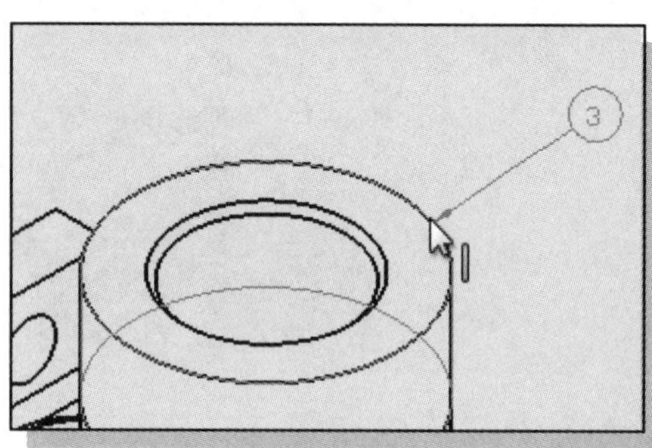

7. Click and drag to move the arrowhead to the new location shown in the figure.

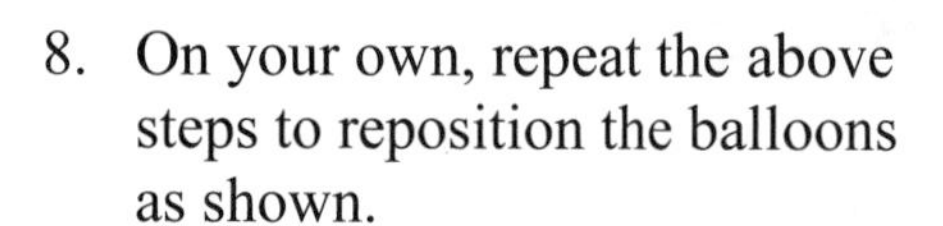

8. On your own, repeat the above steps to reposition the balloons as shown.

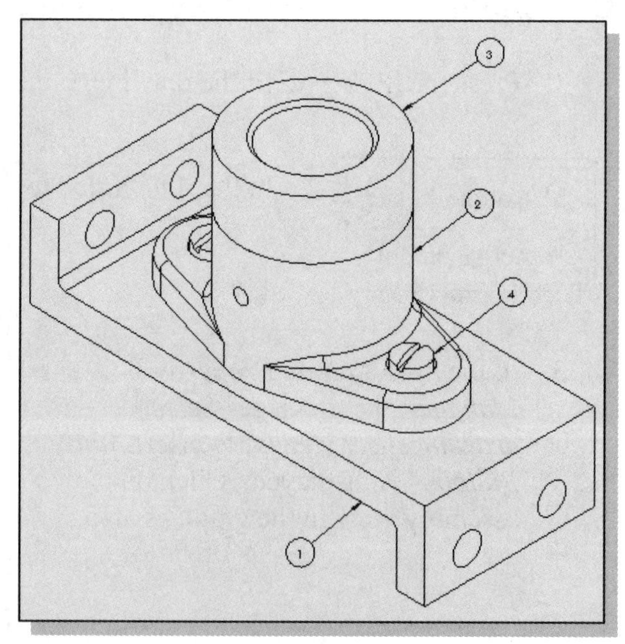

9. On your own, use the **Save As** command and save the drawing with the filename **MET 317-04**. Notice the new filename appears in the title block.

10. If the SOLIDWORKS window appears, select **Save All**.

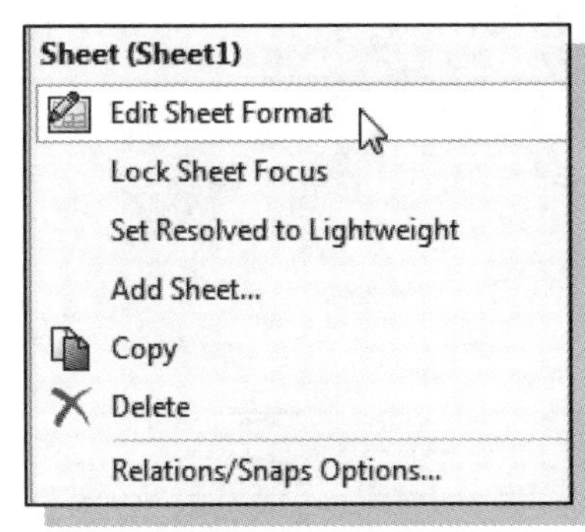

11. Move the cursor into the graphics area and **right-click** to open the pop-up option menu.

12. Select **Edit Sheet Format** from the pop-up option menu to change to *Edit Sheet Format* mode.

13. On your own, edit the title block to appear as shown in the figure.

UNLESS OTHERWISE SPECIFIED:		NAME	DATE	University of New Orleans
DIMENSIONS ARE IN INCHES TOLERANCES: ANGULAR: MACH± 0.1° TWO PLACE DECIMAL ±0.01 THREE PLACE DECIMAL ±0.003	DRAWN	A. Smith	01/08/2021	TITLE:
	CHECKED			Shaft-Support Assembly
	ENG APPR.			
	MFG APPR.			
INTERPRET GEOMETRIC TOLERANCING PER:	Q.A.			
MATERIAL	COMMENTS:			SIZE B · DWG. NO. MET 317-04 · REV
FINISH				
DO NOT SCALE DRAWING				SCALE: 1:1 · WEIGHT: · SHEET 1 OF 1

14. On your own, return to *Edit Sheet* mode.

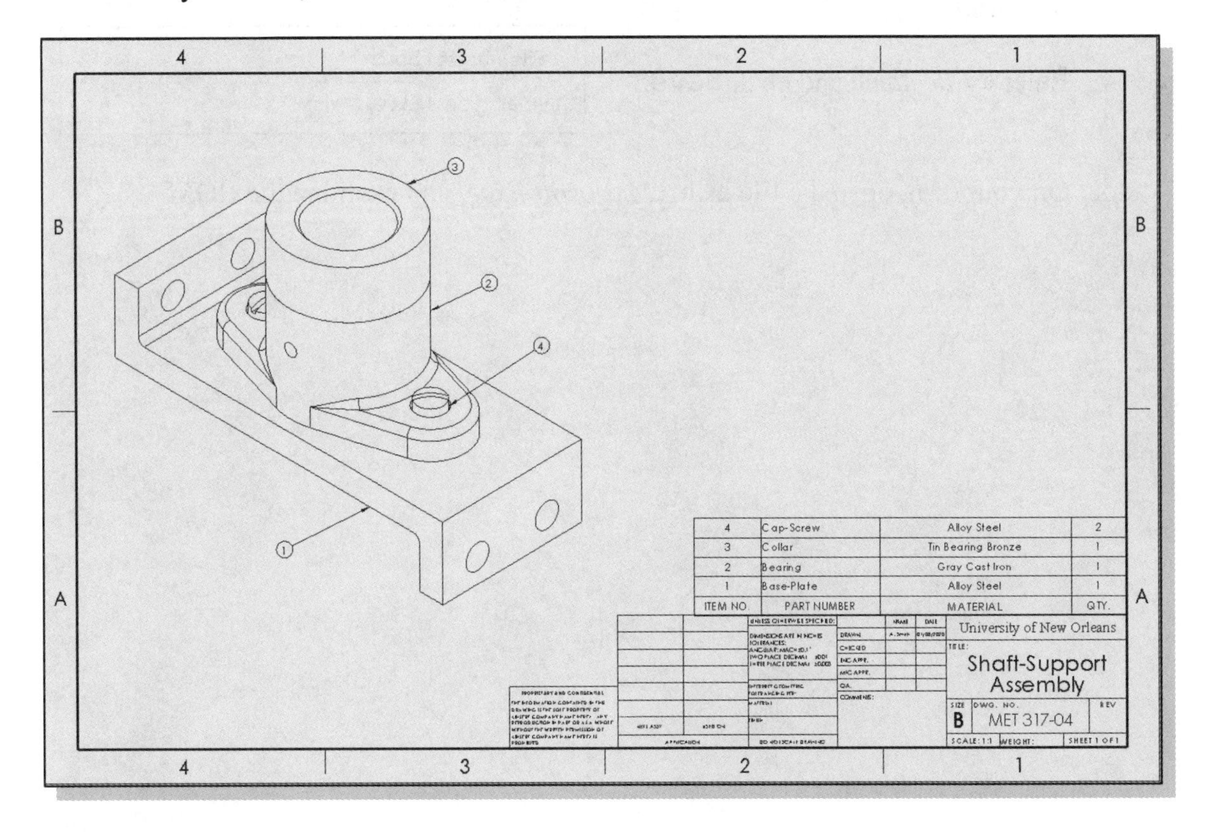

Exporting the Bill of Materials

❖ The bill of materials can be exported as an *Excel* file.

1. Click the bill of materials on the drawing sheet.

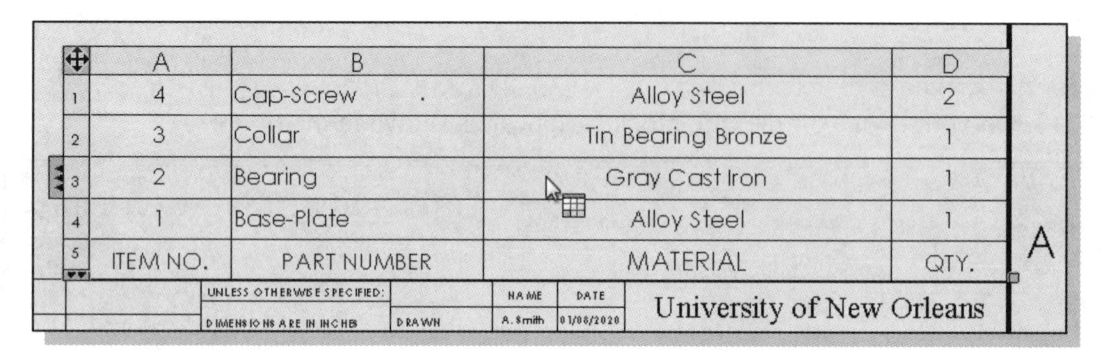

2. Select **Save As** from the pull-down menu.

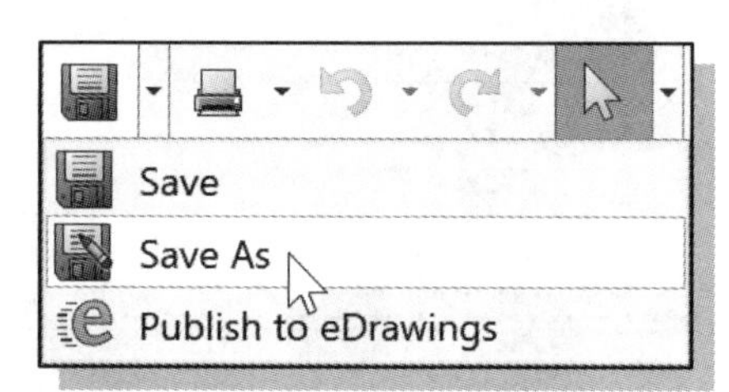

3. Select **Excel (*.xls)** or another appropriate Excel file type for the ***Save as type:*** option.

4. Enter a *File name* and click **Save**.

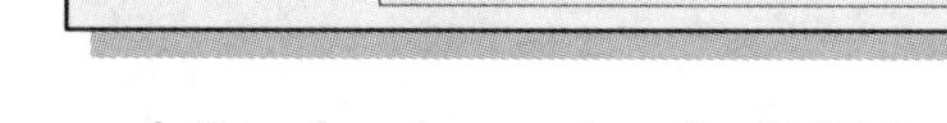

5. On your own, open the file using *Microsoft Excel* and examine the BOM.

Questions:

1. What is the purpose of using ***assembly mates***?

2. List three of the commonly used ***assembly mates***.

3. Describe the difference between the **Aligned** option and the **Anti-aligned** option.

4. In an assembly, can we place more than one copy of a part? How is it done?

5. How should we determine the assembly order of different parts in an assembly model?

6. How do we adjust the information listed in the **Bill of Materials** of an assembly drawing?

7. Create sketches showing the steps you plan to use to create the four parts required for the assembly shown on the next page:

Exercises:

Wheel Assembly (Create a set of detail and assembly drawings. All dimensions are in mm.)

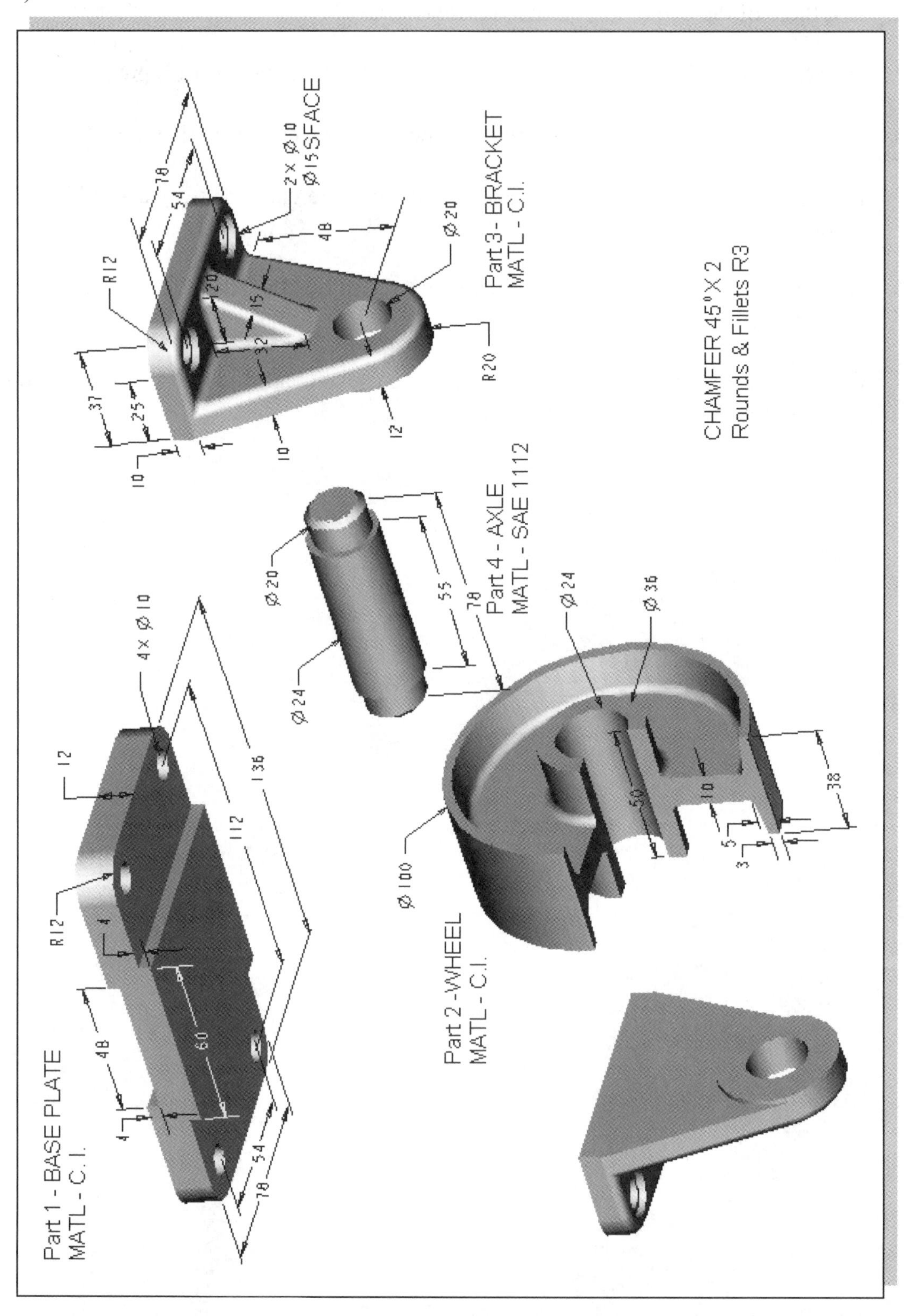

Chapter 15

Design Library and Basic Motion Study

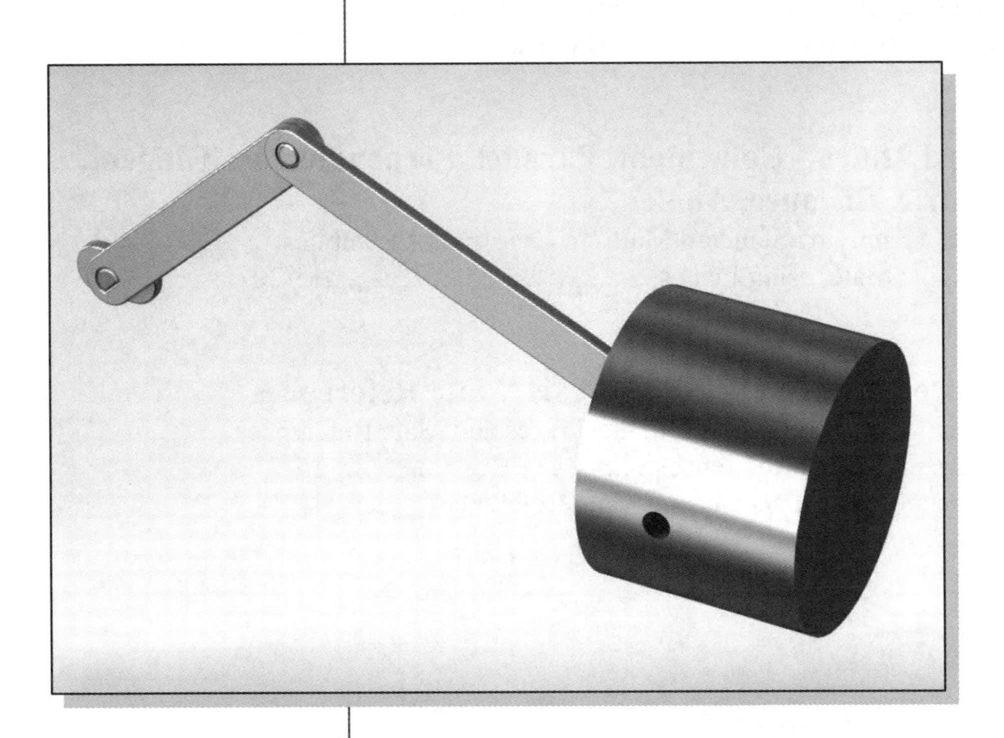

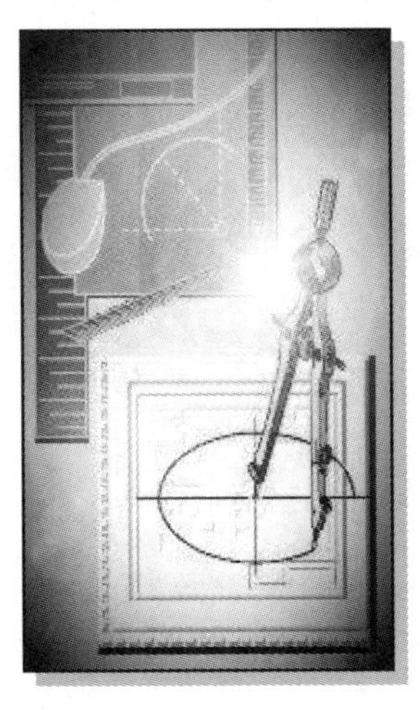

Learning Objectives

- ♦ Apply and Use Mate References
- ♦ Use the SOLIDWORKS Toolbox Libraries
- ♦ Use the Collision Detection Tool
- ♦ Use the Basic Motion Tool to Create a Basic Motion Study
- ♦ Output the Associated Simulation Video File

Certified SOLIDWORKS Associate Exam Objectives Coverage

Inserting Components

Objectives: Inserting Components into an Assembly.

Standard Mates – Coincident, Parallel, Perpendicular, Tangent, Concentric, Distance, Angle

Objectives: Applying Standard Mates to Constrain Assemblies.

Reference Geometry – Planes, Axis, Mate References

Objectives: Creating Reference Planes, Axes and Mate References.

Introduction

In this lesson, we will examine some of the procedures that are available in SOLIDWORKS to reuse existing 2D data and 3D parts. We will also build an assembly model and resolve any interference problems by performing a basic motion analysis.

In SOLIDWORKS, we also have the option of using the standard parts library through what is known as the ***Design Library***. The *Design Library* consists of multiple libraries of reusable elements such as parts, assemblies and sketches. It is accessed through the *Task Pane* at the right of the SOLIDWORKS work area. Significant amounts of time can be saved by using these parts. Note that we can also create and publish our libraries so that others can reuse our parts. The Design Library Task Pane divides the library into four major sections:

- **Design Library** – Contains subfolders populated by SOLIDWORKS with reusable items such as parts, blocks, and annotations. Additional folders and content can be added by the user.
- **Toolbox** – Contains multiple libraries of standard parts that have been created based on industry standards. To access the content, install and add in the SOLIDWORKS *Toolbox* browser.
- **3D ContentCentral** – Accesses 3D models from component suppliers and individuals via the internet. It includes *Supplier Content* (links to supplier web sites with certified 3D models) and *User Library* (links to models from individuals using 3D PartStream.NET).
- **SOLIDWORKS Content** – Contains additional SOLIDWORKS content for blocks, routing, and weldments in downloadable .zip files.

In the lesson, we will insert a part from the SOLIDWORKS *Toolbox*. The reader is encouraged to explore the other options in the *Design Library*.

One main advantage of using parametric parts and assemblies is the ability to check for potential problems without actually creating a physical prototype. In SOLIDWORKS, ***interference analysis*** can be performed in several different ways. The **Collision Detection** tool is a quick and easy option to examine interference between assembled components.

A ***Motion Study*** can be performed to visually confirm the proper assembly of the designs, and to check for any interference between mating parts and any other potential problems. In SOLIDWORKS, several tools are available to perform motion analysis, including the **Animation**, **Basic Motion**, and **Motion Analysis** tools. The Motion Analysis tool can be used to perform a very in-depth motion analysis, while the Basic Motion tool provides a relatively simple motion analysis that can be done in a matter of seconds.

In this chapter, the concepts and procedures of creating assemblies using the *Design Library* for standard parts are illustrated. The procedure for basic motion analysis of an assembly is also illustrated using the Basic Motion tool.

The *Crank-Slider* Assembly

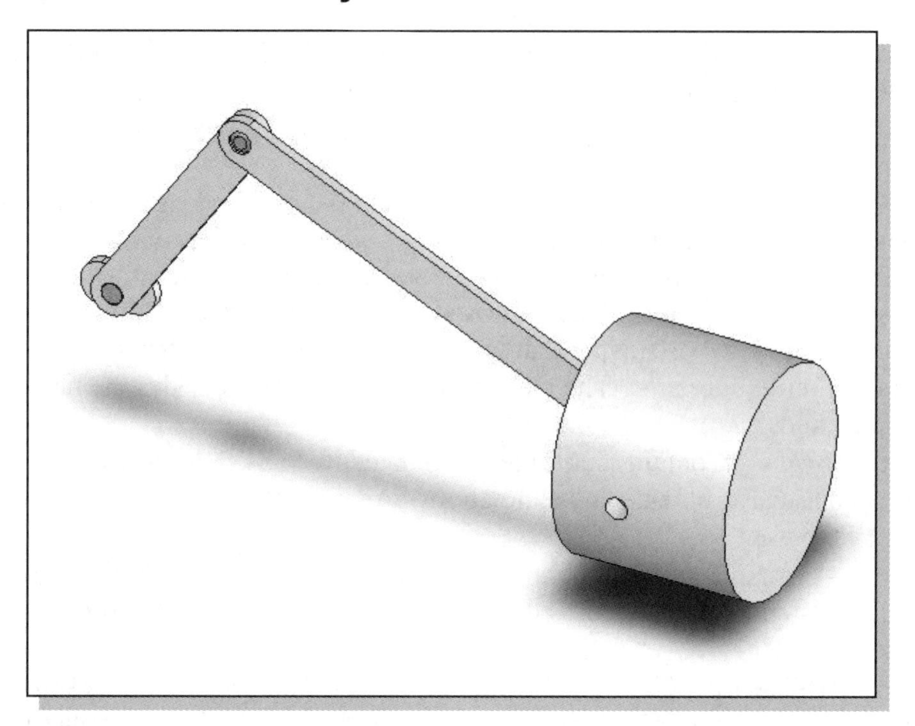

Creating the Required Parts

Six parts are required for this assembly: (1) ***Base***, (2) ***Crank***, (3) ***Connecting Rod***, (4) ***Slider***, (5) ***Main Pin***, and (6) ***Pin***. On your own, create the five parts shown below. Save the models as separate part files: *CS-Base*, *CS-Crank*, *CS-Connecting Rod*, *CS-Slider*, and *CS-Main Pin*. Units are in inches. (Note the location of the parts relative to the datum planes and close all part files or exit SOLIDWORKS after you have created the parts.) We will use a standard *Pin* that is available in the *Toolbox*.

(1) ***CS-Base*** (**NOTE:** Thickness = 0.250 in.)

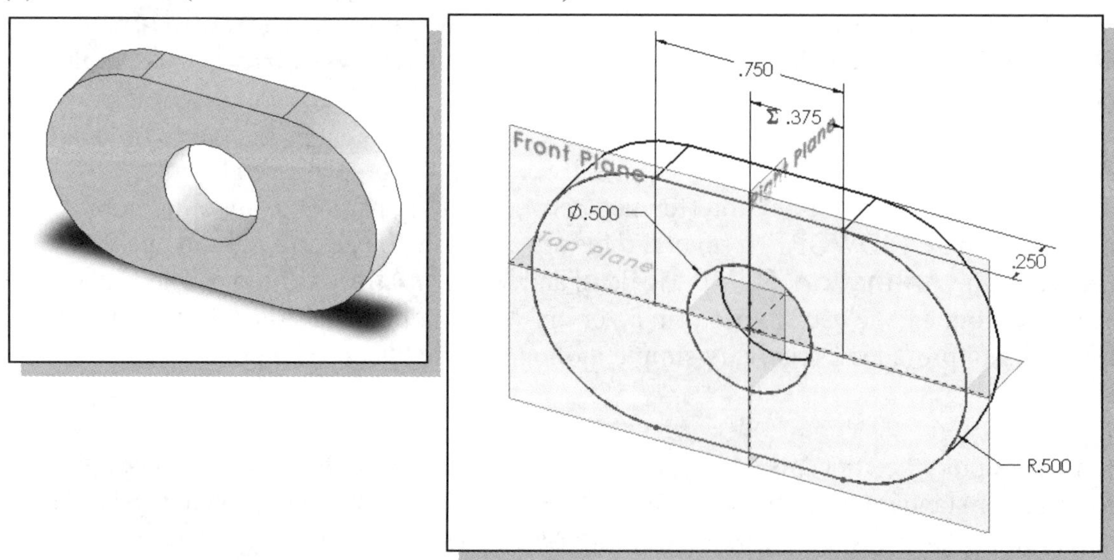

(2) ***CS-Crank*** (Construct the part with the datum origin aligned to the center of the part. **NOTE:** Thickness is 0.250 in.)

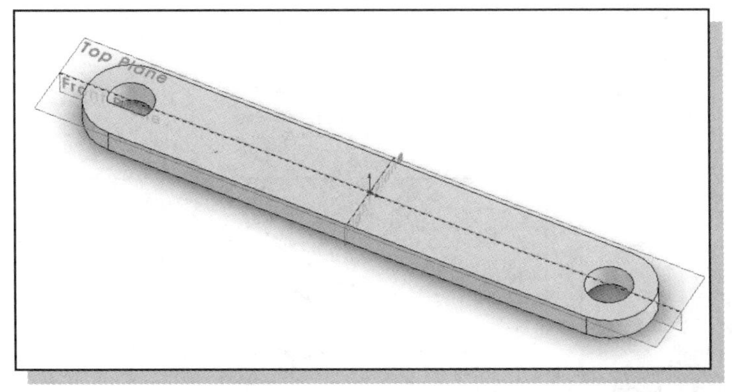

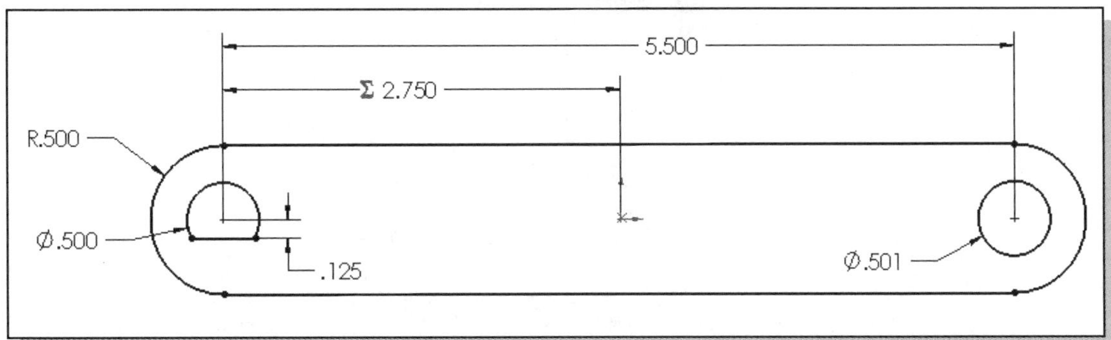

(3) ***CS-Connecting Rod*** (Construct the part with the datum origin aligned to the center of the part. **NOTE:** Thickness is 0.250 in.)

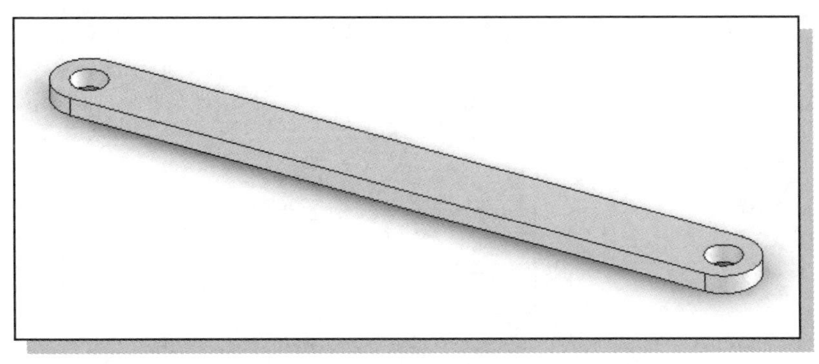

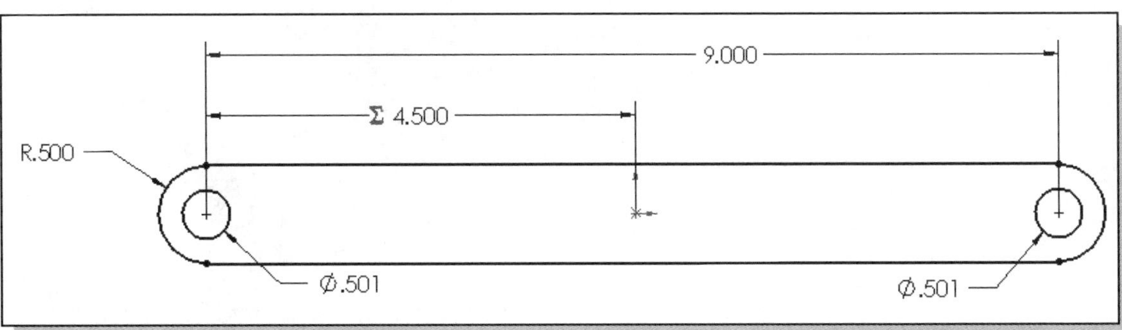

(4) ***CS-Slider*** (Construct the part with the datum planes aligned as shown. **NOTE:** Overall size is ∅ 5.000 in. x 4.000 in.)

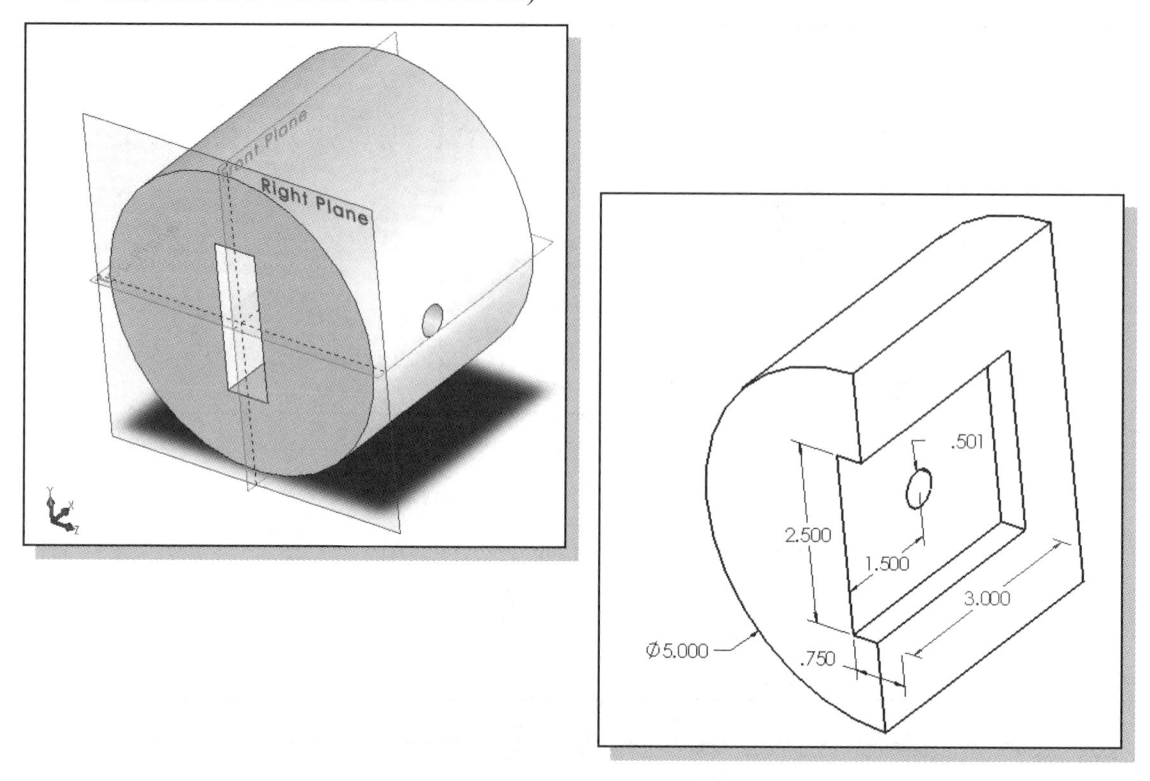

(5) ***CS-Main Pin*** (Construct the part with the datum planes aligned as shown. **NOTE:** Overall size is ∅ 0.500 in. x 1.000 in.)

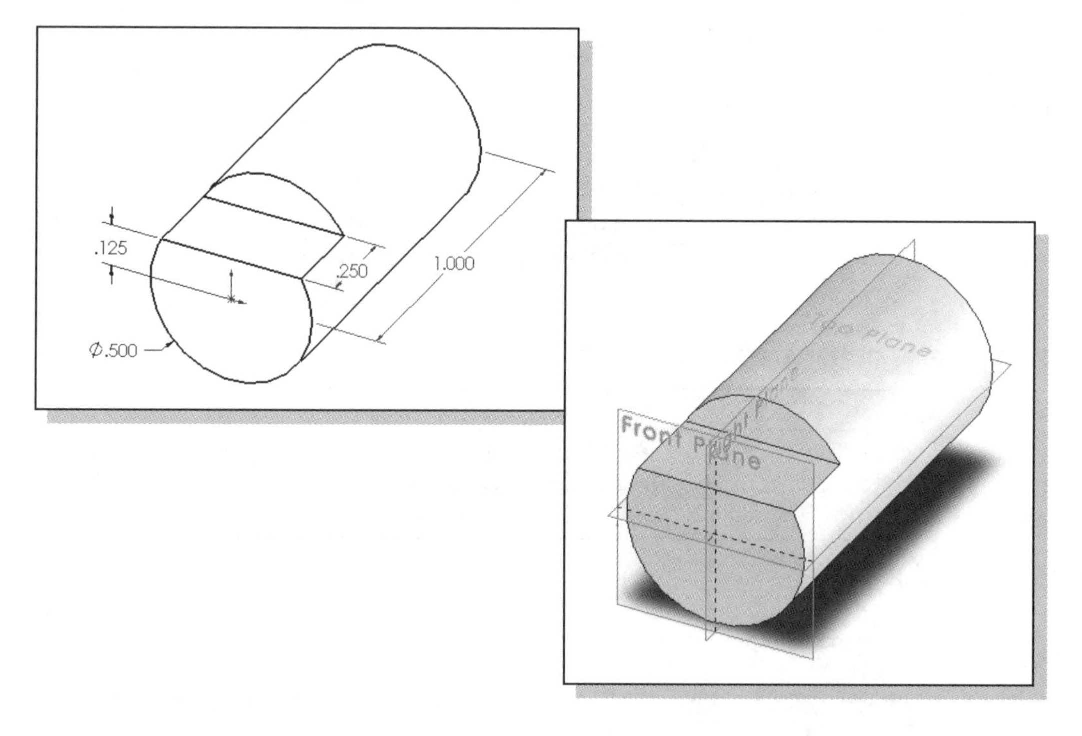

Mate References

SOLIDWORKS *Mate References* specify one or more entities of a component to use for automatic mating. When a component with a mate reference is dragged into an assembly, SOLIDWORKS looks for other combinations of the same mate reference name and mate type. If the mate name and type match, the mate can be added automatically.

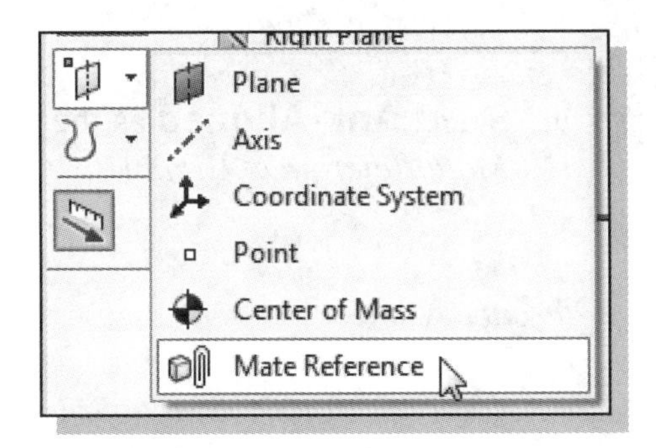

1. Open the ***CS-Crank*** part file you created and saved in the last section.

2. Select the **Reference Geometry** command from the *Features* toolbar, and select the **Mate Reference** option from the pull-down menu.

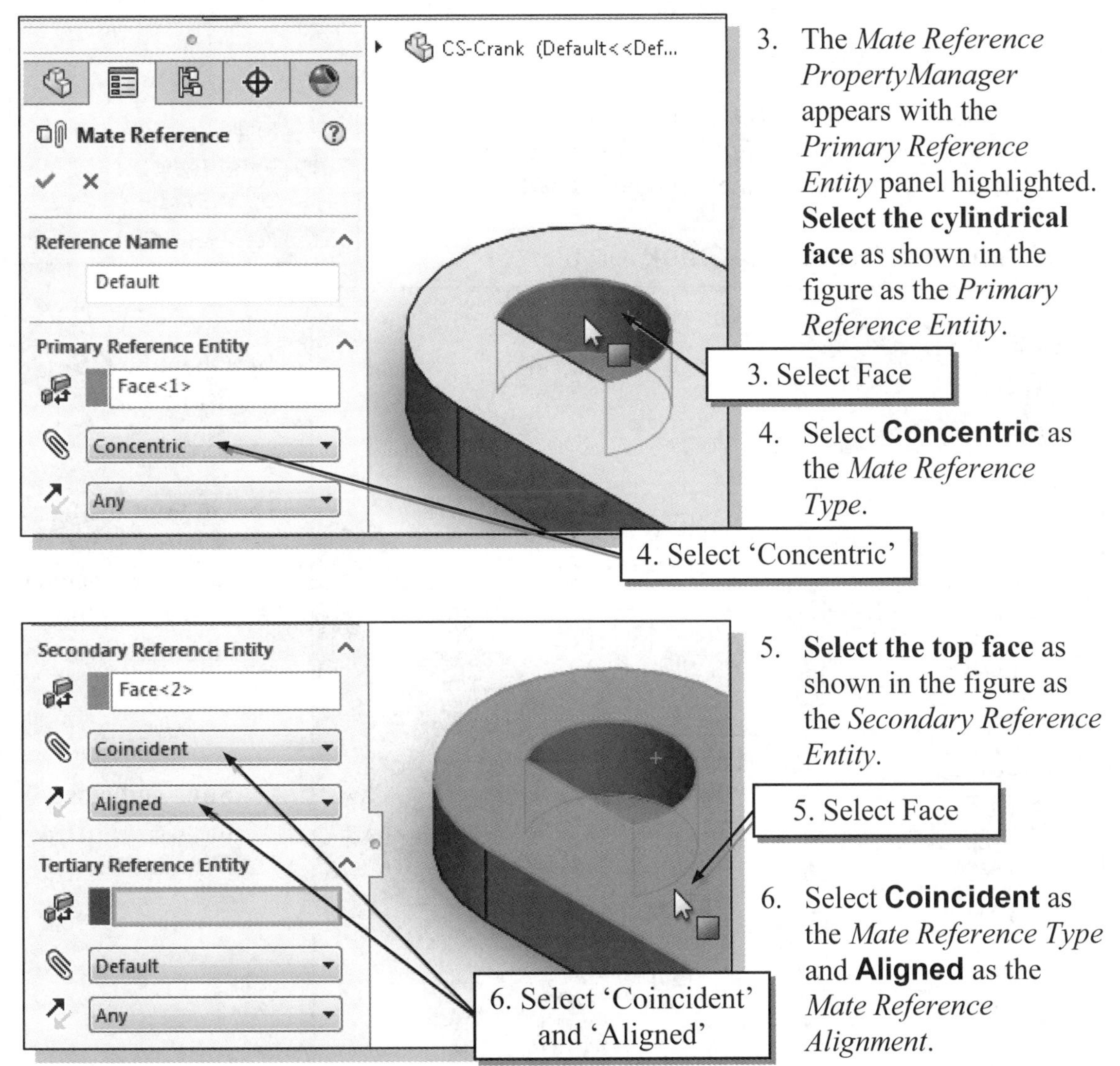

3. The *Mate Reference PropertyManager* appears with the *Primary Reference Entity* panel highlighted. **Select the cylindrical face** as shown in the figure as the *Primary Reference Entity.*

4. Select **Concentric** as the *Mate Reference Type.*

5. **Select the top face** as shown in the figure as the *Secondary Reference Entity.*

6. Select **Coincident** as the *Mate Reference Type* and **Aligned** as the *Mate Reference Alignment.*

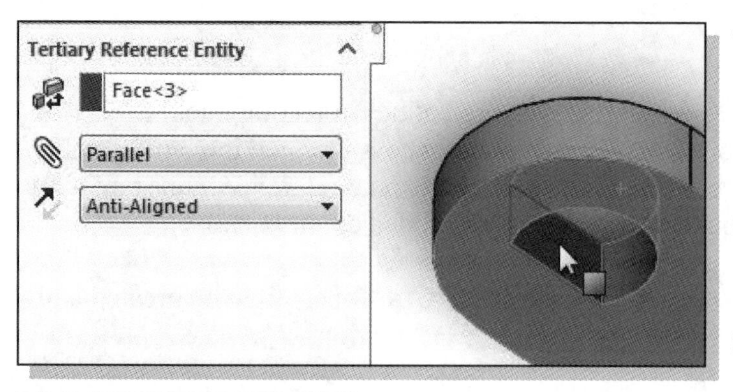

7. Rotate the view and **select** the flat face as shown in the figure as the *Tertiary Reference Entity*.

8. Select **Parallel** as the *Mate Reference Type*.

9. Select **Anti-Aligned** as the *Mate Reference Alignment*.

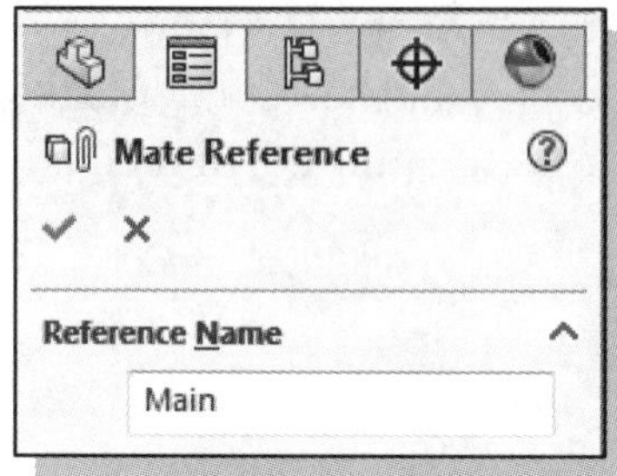

10. Enter ***Main*** as the *Reference Name*.

11. Select **OK** in the *Mate Reference PropertyManager* to accept the settings.

12. Notice the *MateReferences* folder appears in the *FeatureManager Design Tree*. **Expand** the folder and notice the only mate reference is *Main*.

Top Plane
Right Plane
Origin
MateReferences
Main-<1>
Boss-Extrude1

13. **Save** and **close** the ***CS-Crank*** part file.

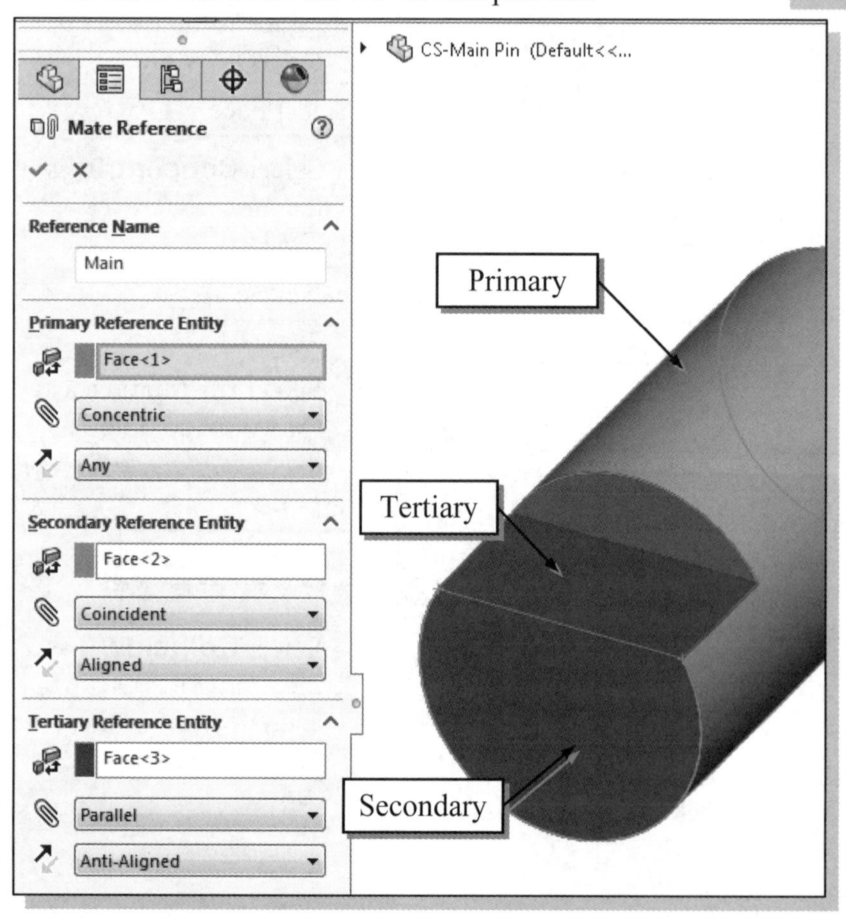

14. Open the ***CS-Main Pin*** part file.

15. On your own, create a *Mate Reference* named ***Main*** with the settings shown in the figure.

16. **Save** and **close** the ***CS-Main Pin*** part file.

Starting SOLIDWORKS

1. Select the **SOLIDWORKS** option on the *Start* menu or select the **SOLIDWORKS** icon on the desktop to start SOLIDWORKS. The SOLIDWORKS main window will appear on the screen.

2. Select the **New** icon with a single click of the left-mouse-button on the *Menu Bar*. The *New* SOLIDWORKS *Document* dialog box appears.

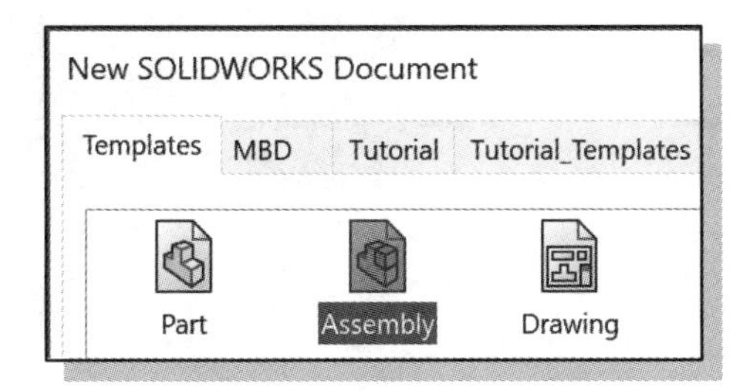

3. Select the **Templates** tab and select the **Assembly** icon as shown.

4. Click **OK** to open the new Assembly file.

❖ SOLIDWORKS expects you to insert the first component. We will first set the document properties to match those of the previously created parts.

5. Click on **Cancel** in the *Open* window.

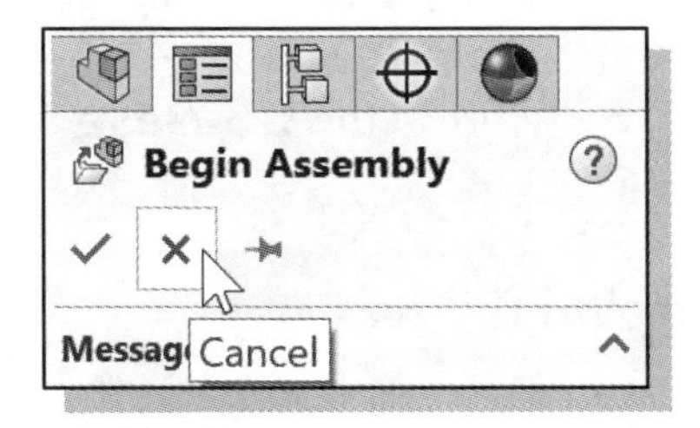

6. Cancel the Begin Assembly command by clicking once with the left-mouse-button on the **Cancel** icon in the *PropertyManager* as shown.

Document Properties

1. Select the **Options** icon from the *Menu Bar* to open the *Options* dialog box.

2. Select the **Document Properties** tab.

3. Select **ANSI** in the pull-down selection window under the *Overall drafting standard* panel as shown.

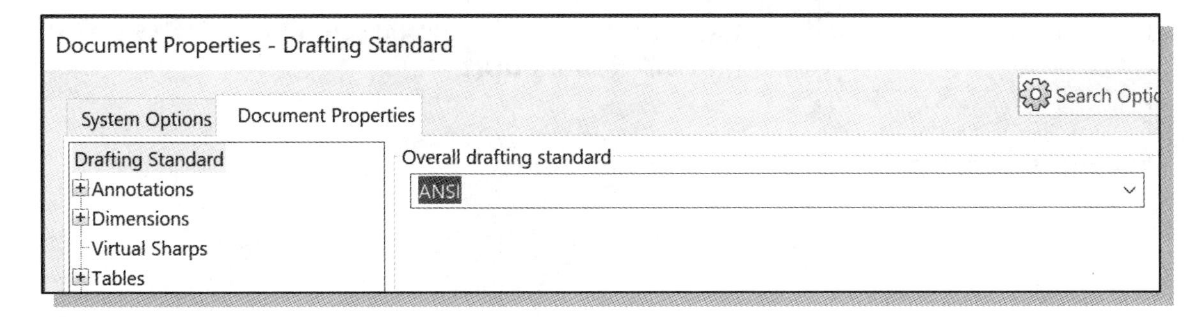

4. Click **Units**, select the **IPS (inch, pound, second)** unit system, and select **.123** in the *Decimals* spin box for the *Length units* as shown.

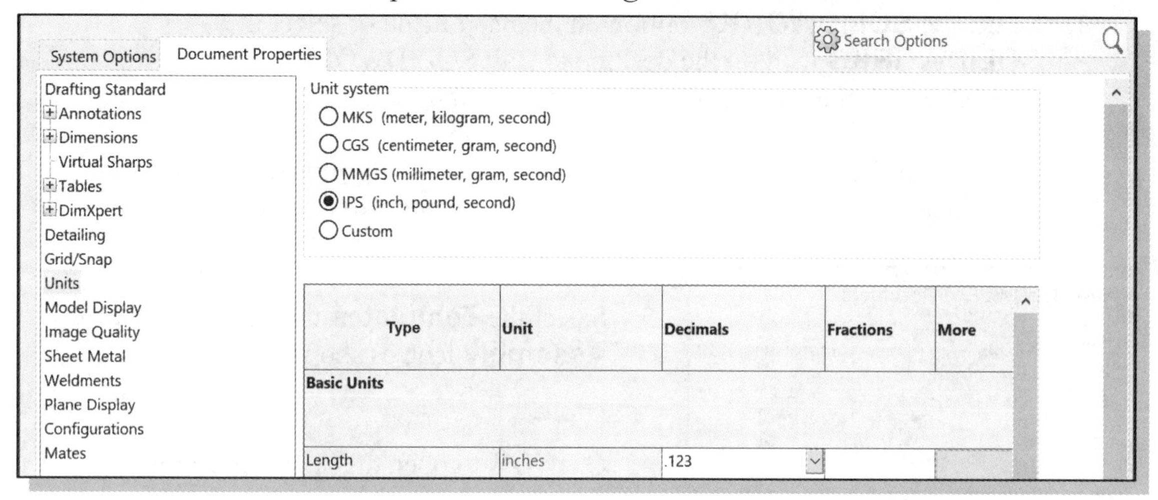

5. Click **OK** in the *Document Properties* window to accept the settings.

Inserting the First Component

The first component in an assembly file sets the orientation of all subsequent parts and subassemblies and therefore should be one that is **not likely to be removed** and **preferably a non-moving part** in the design. For our project, we will use the ***CS-Base*** as the base component in the assembly.

1. Right-click on the *Menu Bar* and open the *Assembly* toolbar (if necessary).

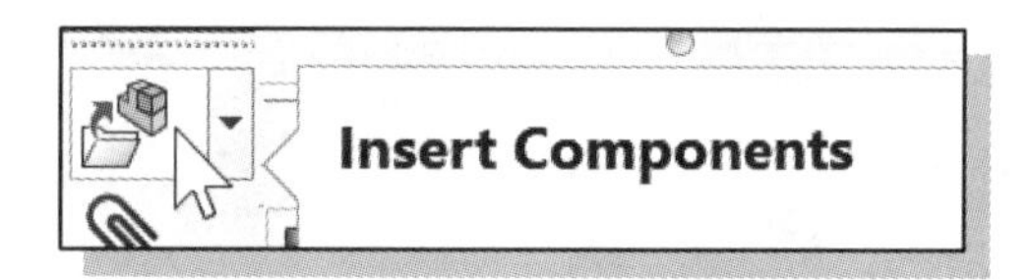

2. In the *Assembly* toolbar, select the **Insert Component** command by left-mouse-clicking the icon.

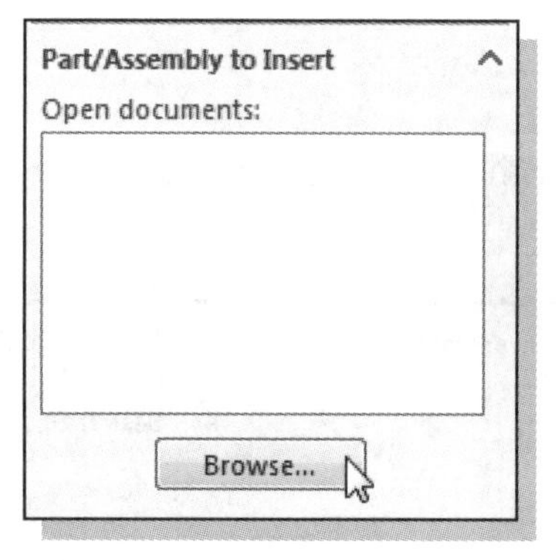

3. If necessary, click on the **Browse** button in the *Part/Assembly to Insert* panel in the *PropertyManager*.

4. Use the browser to locate and select the ***CS-Base*** part file: **CS-Base.sldprt**.

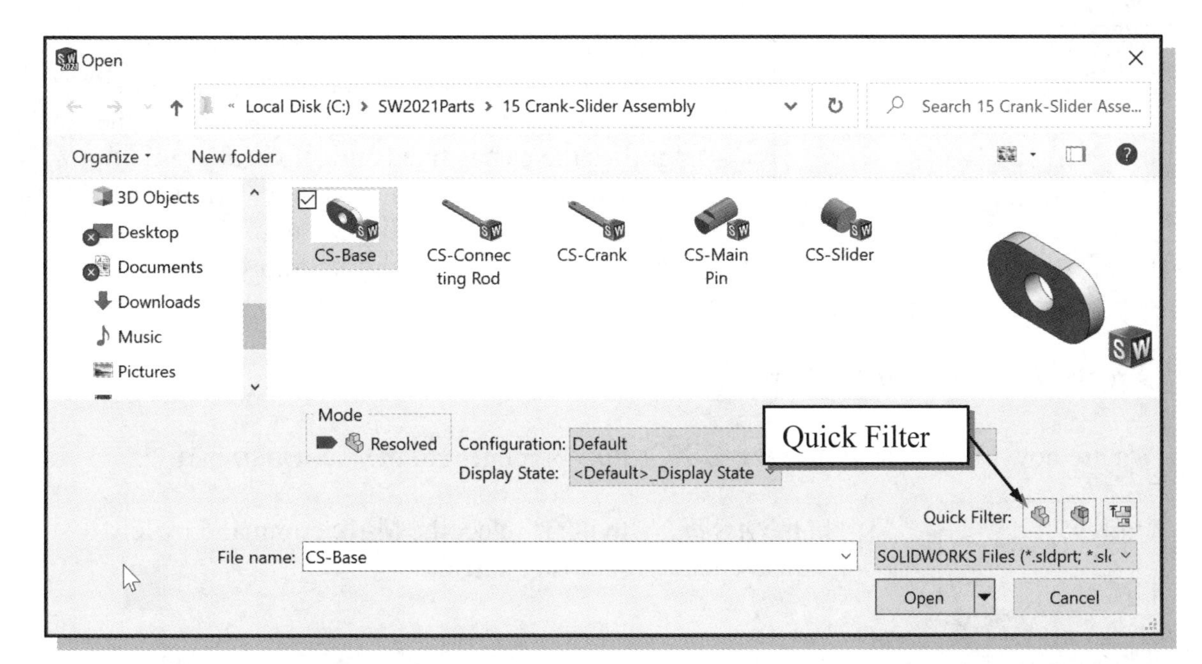

5. Click on the **Open** button to retrieve the model.
   HINT: Use the *Quick Filter* if necessary.

6. By default, the component is automatically aligned to the origin of the assembly coordinates. Click **OK** in the *PropertyManager* to place the *CS-Base* at the origin.

Inserting the Second Component

We will retrieve the *CS-Crank* part as the second component of the assembly model.

1. In the *Assembly* toolbar, select the **Insert Component** command by left-mouse-clicking the icon.

2. Select the ***CS-Crank*** design in the browser. Click on the **Open** button to retrieve the model.

3. Place the *CS-Crank* toward the front side of the CS-Base part, as shown in the figure. Click once with the **left-mouse-button** to place the component.

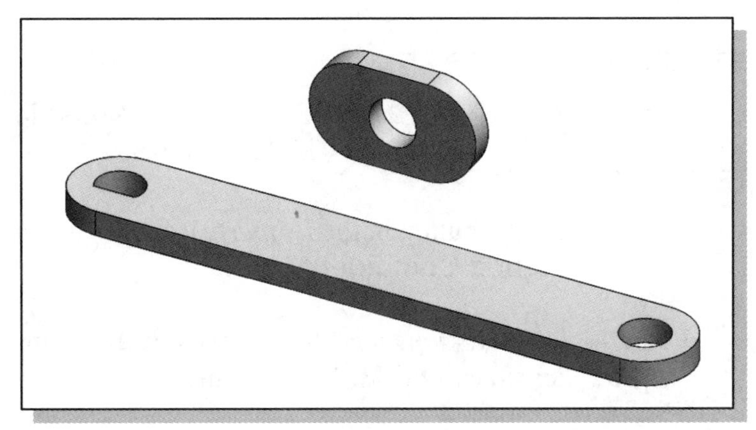

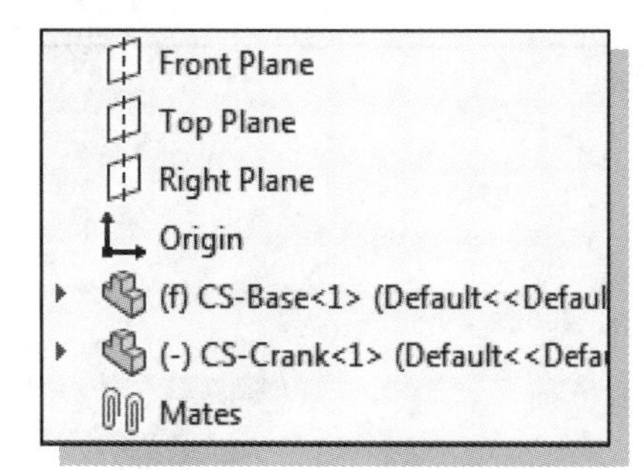

- Inside the *FeatureManager Design Tree*, the retrieved parts are listed in their corresponding order. The **(f)** in front of the *CS-Base* filename signifies the part is fixed and all ***six degrees of freedom*** are restricted. The number behind the filename is used to identify the number of copies of the same component in the assembly model.

Apply Assembly Mates

We are now ready to assemble the *CS-Crank* component to the *CS-Base* part.

1. In the *Assembly* toolbar, select the **Mate** command by left-mouse-clicking once on the icon.

2. Note that the *Mate Selections* panel in the *Mate PropertyManager* is highlighted. Select the **bottom** horizontal surface of the *CS-Crank* part as the first part for the Mate command, as shown.

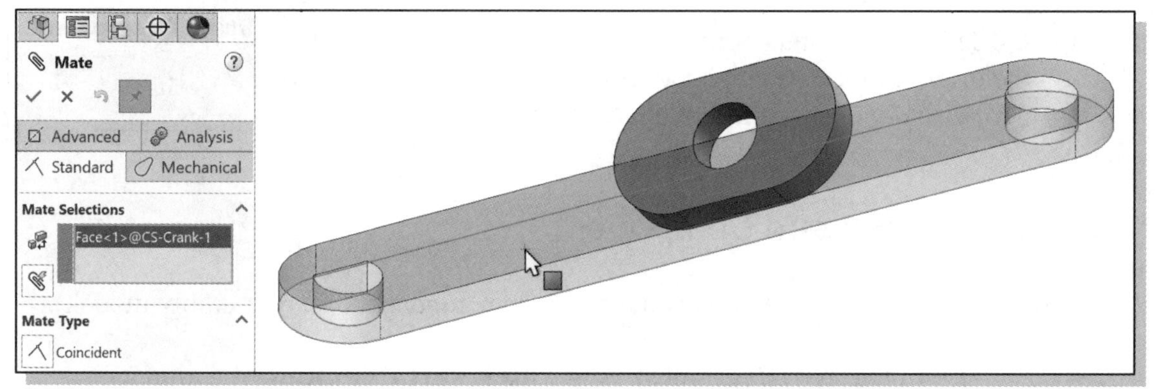

3. Click on the **front face** of the *CS-Base* part as the second part selection to apply the Mate. The *CS-Crank* moves to automatically apply the default Coincident mate.

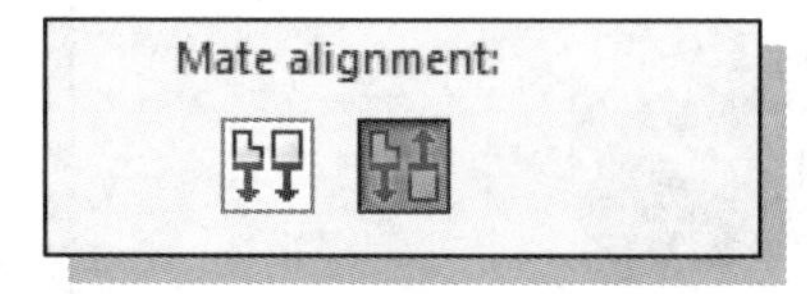

4. Check that the Anti-Aligned option is selected as the Mate alignment.

5. Click the **OK** button in the *PropertyManager* to accept the settings and create the **Anti-Aligned Coincident** mate.

6. In the *PropertyManager*, click on the **Close** button (the red **X** icon), or press the [**Esc**] key, to exit the Mate command.

Apply a Mate Using a Context Toolbar

Mates can be applied from a context toolbar. The toolbar appears when you press **[Ctrl]** and select multiple items in the graphics area. Mates that are appropriate for the selected items appear.

The **Mate** command can also be used to align axes of cylindrical features.

1. On your own, turn *ON* the visibility of the **Temporary Axes**. (**HINT:** Use the *View* pull-down menu.)

2. Move the cursor over the temporary axis of the left side hole of the *CS-Crank* part. Select the **axis** when it is highlighted as shown.

3. While holding down the **[Ctrl]** key, move the cursor over the temporary axis of the small hole on the *CS-Base* part. Select the **axis** when it is highlighted as shown.

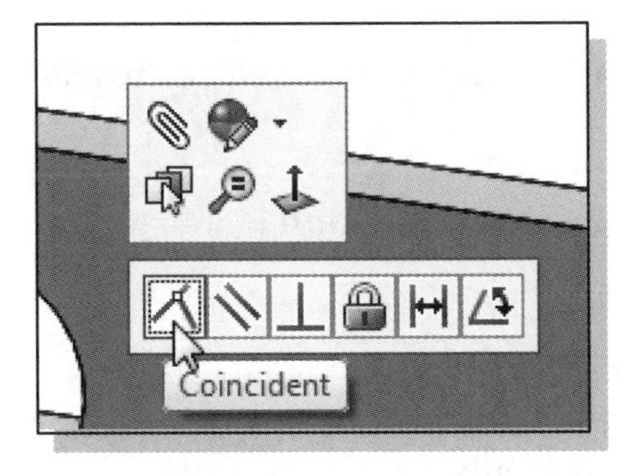

4. Release the **[Ctrl]** key. Notice the context toolbar appears.

5. Select the **Coincident** mate on the context toolbar as shown.

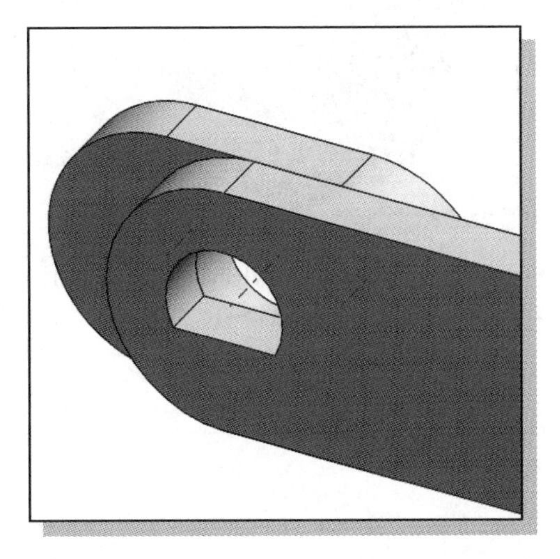

❖ The *CS-Crank* part is placed in the correct position, but the DOF for the *CS-Crank* are not fully restricted; the *CS-Crank* part can still rotate about the mated axis.

Constrained Move

A constrained move is done by dragging the component in the graphics window with the left-mouse-button. A constrained move will honor previously applied assembly mates. That is, the selected component and parts mated to the component move together in their constrained positions. A fixed component remains fixed during the move.

1. Press and hold down the left-mouse-button and drag the *CS-Crank* part and notice the *CS-Crank* part can freely rotate about the mated axis.

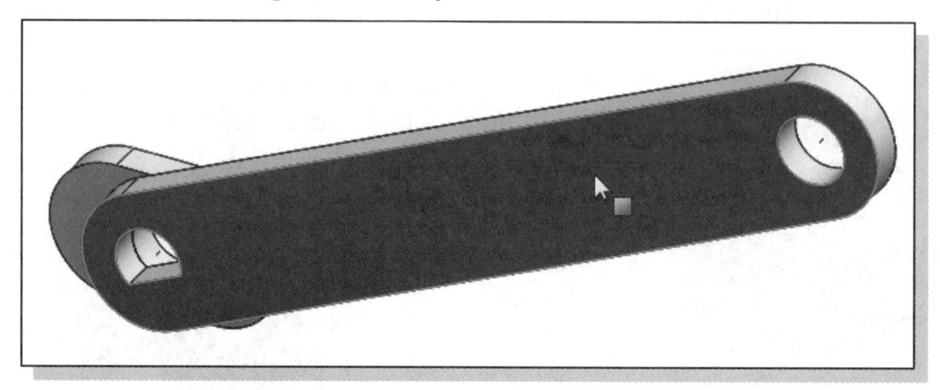

Placing the Third Component Using a Mate Reference

We will next use the **Mate Reference** we named ***Main*** to automatically apply mates upon inserting the *CS-Main Pin* as the third component of the assembly model.

1. In the *Assembly* toolbar, select the **Insert Component** command by left-mouse-clicking the icon.

2. Select the ***CS-Main Pin*** design in the browser. Click on the **Open** button to retrieve the model.

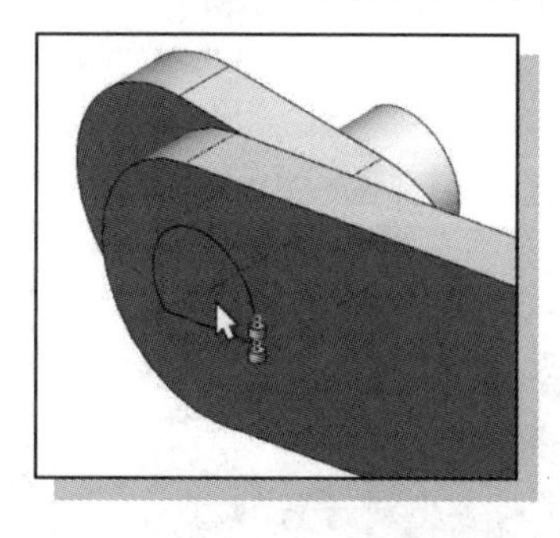

3. Move the cursor over the *CS-Crank* part in the graphics window. Notice the **Mate Reference** icon appears and the *CS-Main Pin* is automatically located using the reference mates you created and named *Main*. Click once with the **left-mouse-button** to insert the *CS-Main Pin* and apply the mates as shown.

4. Press and hold down the left-mouse-button and drag the *CS-Crank* part and notice the *CS-Crank* part still freely rotates about the axis, and that the *CS-Main Pin* rotates with it.

Assemble the CS-Rod Part

❖ Next we will place the *CS-Connecting Rod* part in the assembly model.

1. In the *Assembly* toolbar, select the **Insert Component** command by left-mouse-clicking the icon.

2. Select the ***CS-Connecting Rod*** design in the browser. Click on the **Open** button to retrieve the model.

3. Place the part toward the right side of the assembly as shown in the figure.

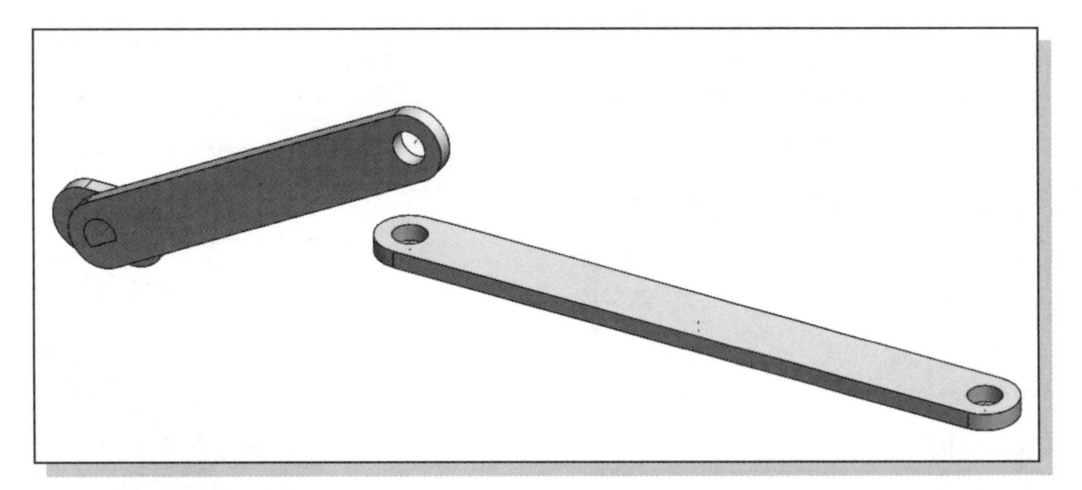

4. On your own apply mates to constrain the *CS-Connecting Rod* as shown in the figure below. Place (1) an anti-aligned Coincident mate between faces of the *CS-Connecting Rod* and the *CS-Crank* parts and (2) a Coincident mate between the axes of the holes on the same two parts.

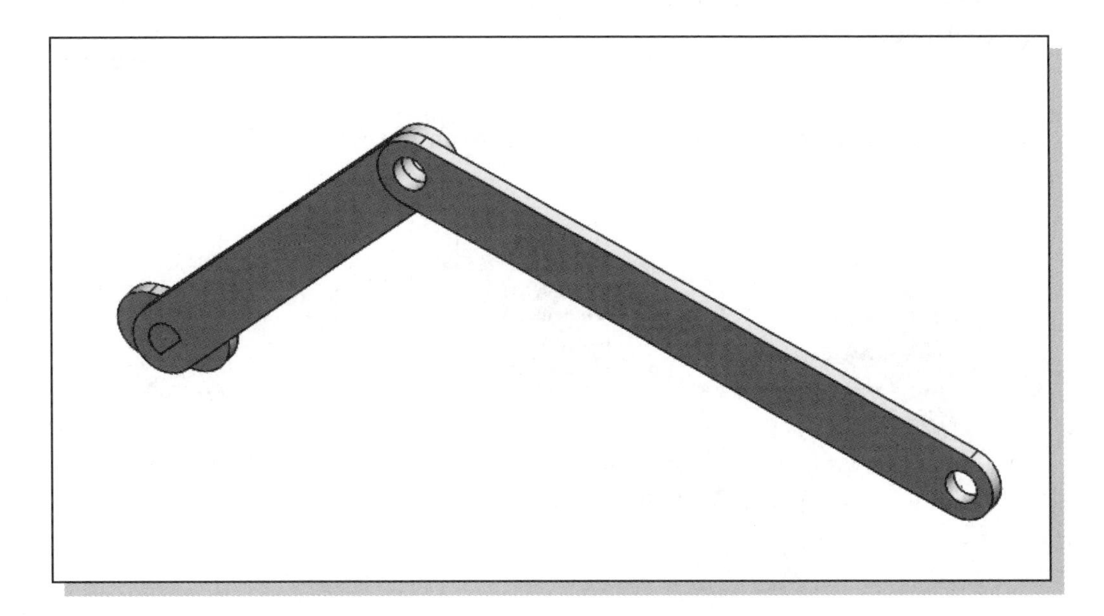

Inserting a Pin from the SOLIDWORKS Toolbox

The SOLIDWORKS *Toolbox* includes a library of standard parts. To access the content, the *Toolbox* library must be installed and the SOLIDWORKS *Toolbox Browser* must be activated as an **add-in** application.

We will add a *Pin* to our assembly from the SOLIDWORKS *Toolbox*. **NOTE:** If you do not have the SOLIDWORKS *Toolbox* library installed, simply create the *Pin* (∅0.5 in. x 0.5 in.) as a new part and insert it instead. Then proceed to Step 11 on the next page.

1. Left-click on the **Design Library** icon at the right of the graphics area to display the *Design Library* task pane.

2. Select **Toolbox**.

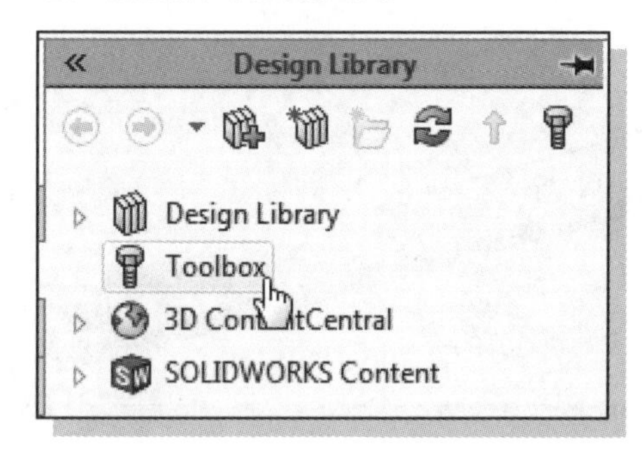

3. If *'Toolbox is not added in'* appears, select **Add in now** as shown. (**NOTE:** An alternate method to *Add in* the *Toolbox* is to select the Add-Ins option from the *Tools* pull-down menu and select the SOLIDWORKS Toolbox Browser.)

4. Double-click on the **ANSI Inch** folder from the *Toolbox* directory displayed in the task pane.

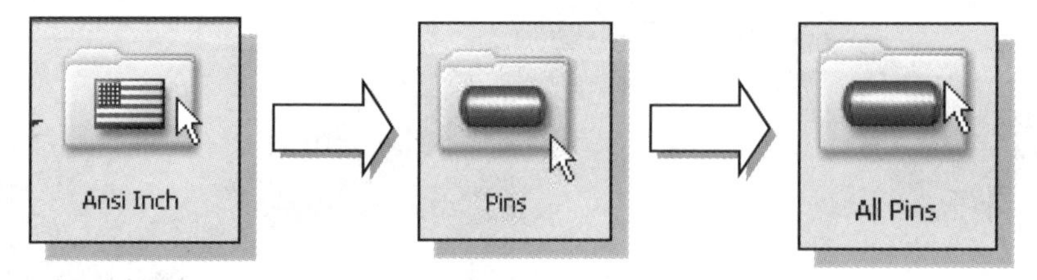

5. Scroll down to find the Pins folder. Double-click on the **Pins** folder. Then double-click on the **All Pins** folder.

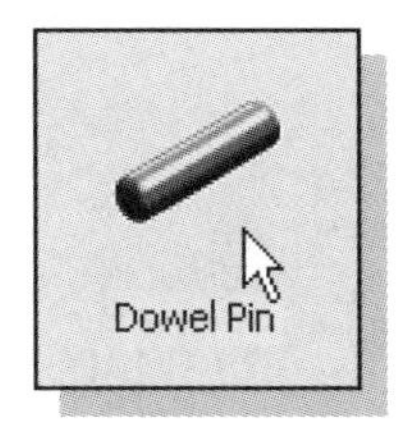

6. Move the cursor over the **Dowel Pin** icon. Press and hold down the left mouse button. Hold the left mouse button down and **drag** the *Dowel Pin* into the graphics area.

7. **Release** the mouse button to insert the *Dowel Pin.* Notice the *Dowel Pin* is inserted and the *Dowel Pin PropertyManager* appears.

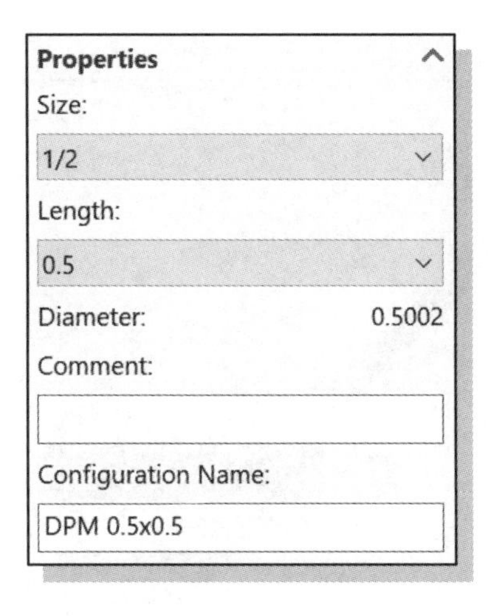

8. Select **1/2** for the *Size* and **0.5** for the *Length* in the *Properties* panel of the *Dowel Pin PropertyManager*.

➢ Notice the actual diameter of the 1/2″ nominal size *Dowel Pin* is 0.5002 inches (displayed in the *Properties* panel). The diameters of the holes in the *CS-Crank* and *CS-Connecting Rod* parts were defined at 0.501 in. to avoid interference.

9. Click **OK** in the *Dowel Pin PropertyManager* to accept the settings.

10. Notice the *Insert Components PropertyManager* appears, giving the opportunity to insert additional copies of this part. Click **Cancel** in the *PropertyManager* or hit the [**Esc**] key to exit the Insert Component command.

11. On your own, apply mates to constrain the *Dowel Pin* as shown in the figure below.

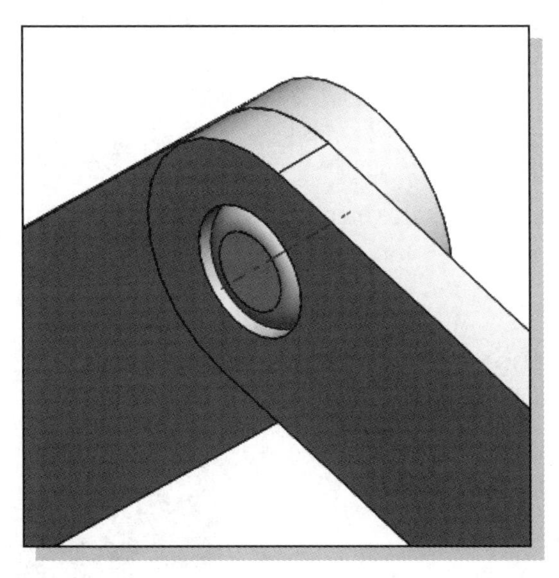

12. On your own, drag the different components in the graphics window with the left-mouse-button and confirm the parts are properly constrained.

Assemble the CS-Slider Part

We will insert the *CS-Slider* part to complete the assembly model.

1. In the *Assembly* toolbar, select the **Insert Component** command by left-mouse-clicking the icon.

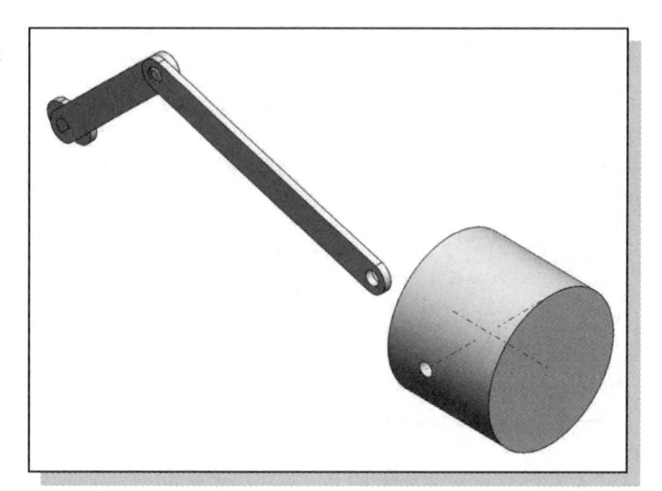

2. Select the ***CS-Slider*** design in the browser. Click on the **Open** button to retrieve the model.

3. Place the *CS-Slider* part on the right side of the assembly as shown in the figure.

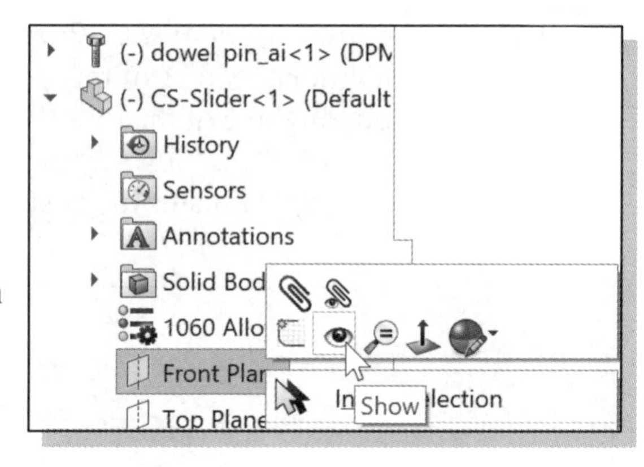

4. Expand the *CS-Slider* part in the *FeatureManager Design Tree*, right click on the Front Plane, and select **Show**.
(**NOTE:** It may be necessary to also turn *ON* the visibility by selecting the **Planes** option in the *View* pull-down menu.)

5. In the *Assembly* toolbar, select the **Mate** command by left-mouse-clicking once on the icon.

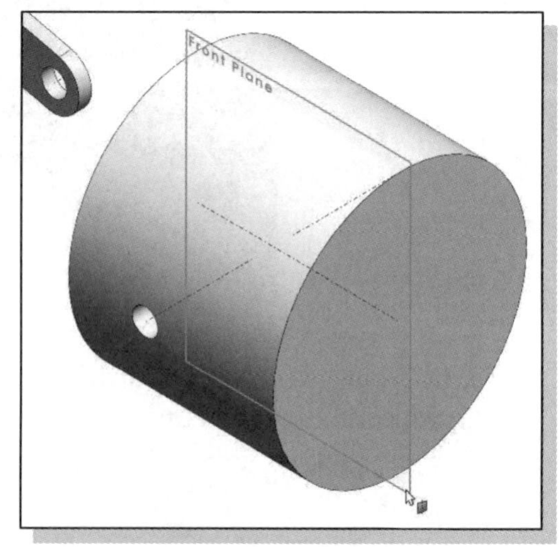

6. Select the Front Plane of the *CS-Slider* as the first surface for the Mate as shown.

7. Select the **back side** of the *CS-Connecting Rod* part as the second surface of the Mate as shown.

8. Check that the Anti-Aligned option is selected as the mate alignment.

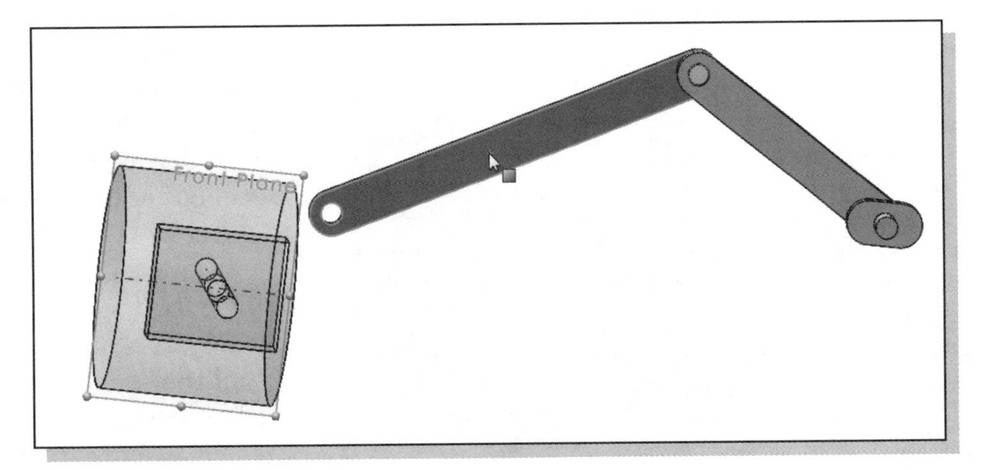

9. Click the **OK** button in the *PropertyManager* to accept the settings and create the **Anti-Aligned Coincident** mate.

10. On your own, apply another **Mate** constraint to align the center axes of the *CS-Rod* and the *CS-Slider* as shown.

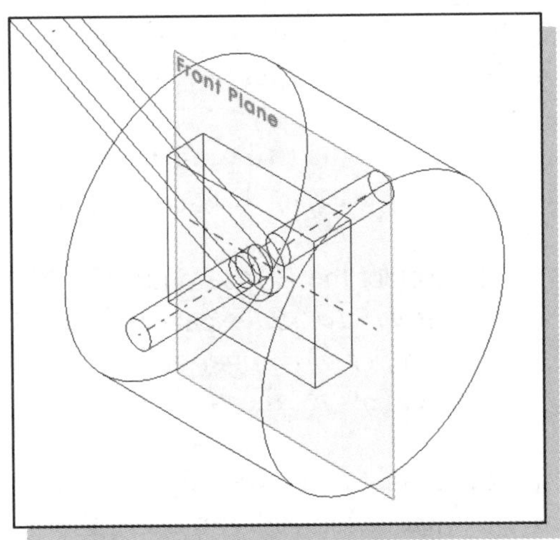

❖ The *CS-Slider* is not fully constrained; we will next restrict the motion of the slider to the horizontal direction only.

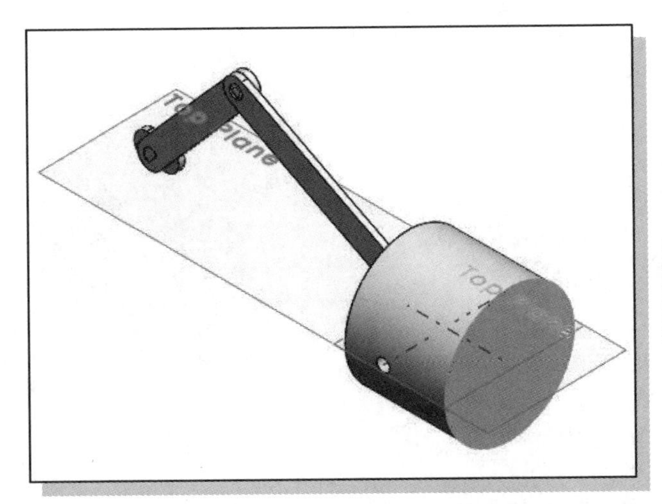

11. On your own, turn *ON* the visibility for the Top Plane of the *Assembly* ("Assem1" at the top of the *Design Tree*) and the Top Plane of the *CS-Slider* part.

12. **On your own**, create a **Coincident** mate between these two planes.

13. On your own, move the *CS-Slider* part and confirm the slider is properly constrained.

Adding an Angle Mate

The *Crank* is the driver of the assembly, and its position can be controlled by adding an Angle mate.

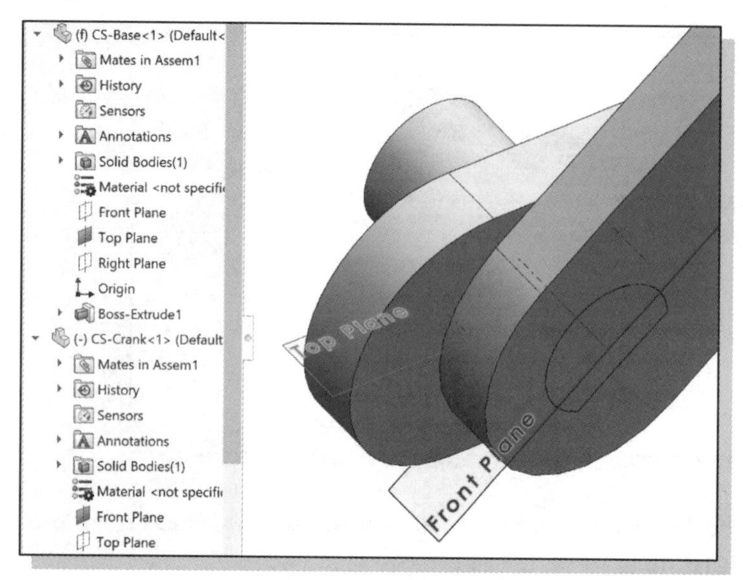

1. In the *FeatureManager Design Tree*, turn *ON* the visibility of the two datum planes that are in the length directions of the *CS-Base* and *CS-Crank* parts as shown in the figure. (These will be the Top Plane of the *CS-Base* part and the Front Plane of the *CS-Crank*; the parts were created as shown in the figures on pages 15-4 and 15-5.)

2. In the *Assembly* toolbar, select the **Mate** command by left-mouse-clicking once on the icon.

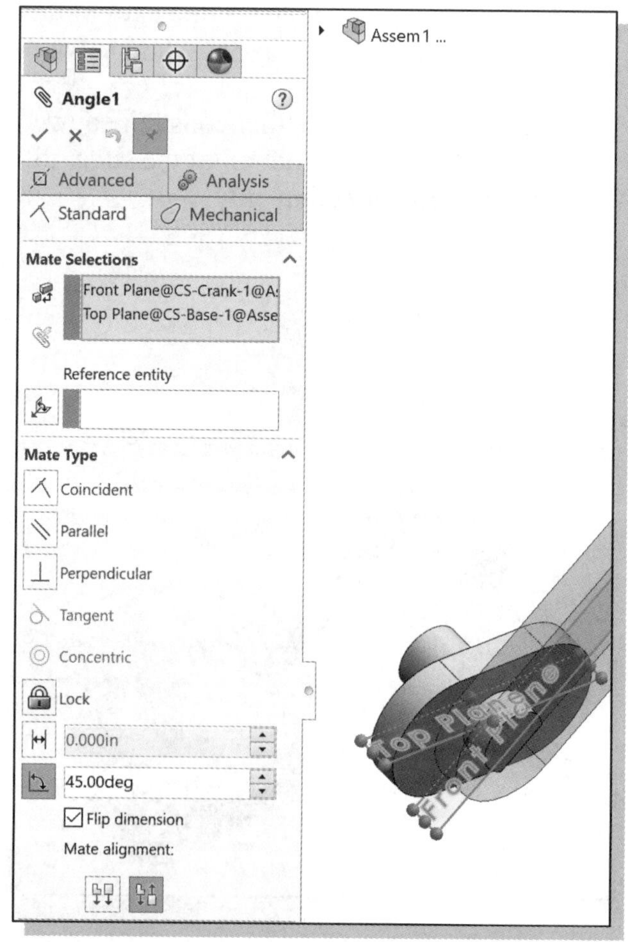

3. Select the **two planes** shown in Step 1 as the *Mate Selections*.

4. Select the **Angle** option in the *Standard Mates* panel of the *PropertyManager* and enter **45** degrees for the angle value.

5. Select the **Anti-Aligned** option under Mate alignment in the *Standard Mates* panel.

6. **If necessary** to produce the orientation in the figure, check the **Flip dimension** box in the *Standard Mates* panel.

7. Click **OK** to create the Angle mate.

8. Click on the **Close** button (the red **X** icon), or press the [**Esc**] key, to exit the Mate command.

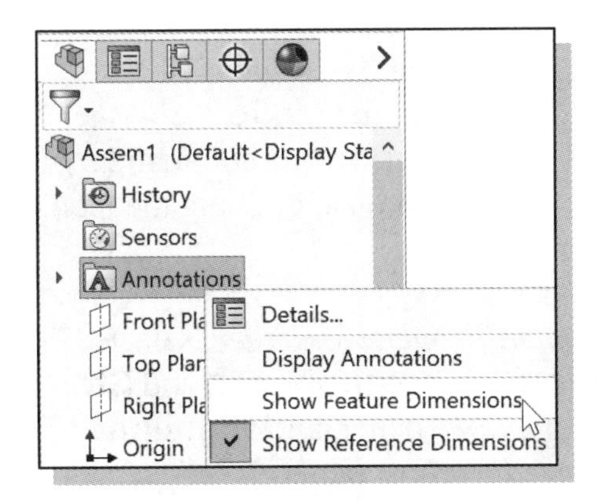

9. In the *FeatureManager Design Tree* window, right click on the **Annotations** folder directly below the Assem1 icon and select the **Show Feature Dimensions** option.

➢ Be sure to select the *Annotations* folder for the assembly and not one of the individual part *Annotations* folders. The Angle mate dimension should be the only feature dimension shown.

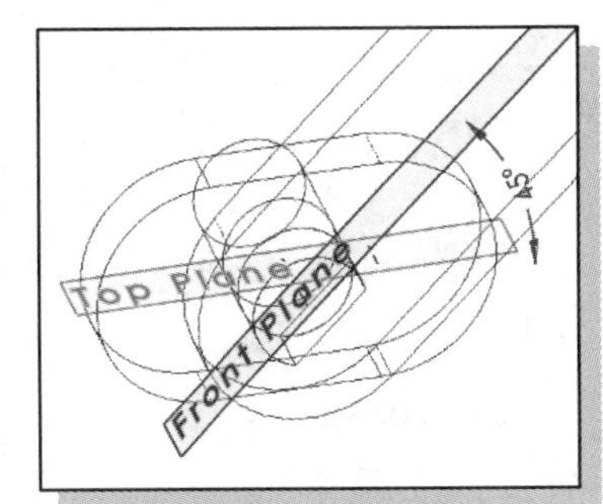

10. Notice the **45 deg** angle dimension from the Angle mate appears in the graphics area. (It may be necessary to move the cursor over the dimension location, or to change to wireframe display, to see the dimension.) **Double-click** on the dimension.

11. Change the dimension to **70.00 deg** and select **OK**.

12. Click on the **Rebuild** icon in the *Menu Bar* and the *Crank* position is adjusted accordingly.

13. On your own, adjust the angle to other values by repeating the above steps. Also confirm the *Crank*, *Connecting Rod* and *Slider* parts are properly constrained.

14. Change the angle to **15.00 deg** before continuing to the next section.

Collision Detection

SOLIDWORKS allows us to perform *Collision Detection* analysis based on the current assembled positions of the assembled parts. Collisions with other components can be detected when moving or rotating a component. The software can detect collisions with the entire assembly or a selected group of components. You can set the tool to detect collisions for all of the components that move as a result of mates or for a selected group of components.

We will first suppress the Angle Mate and allow the *CS-Crank* to rotate.

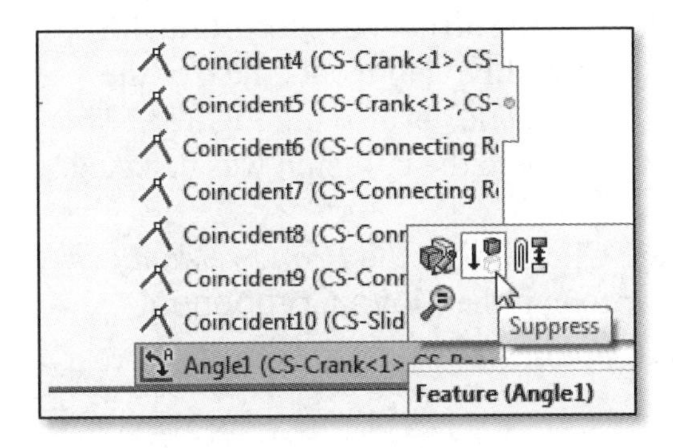

1. In the *FeatureManager Design Tree*, expand the Mates folder under the *CS-Crank* part.

2. **Right click** on the Angle mate.

3. Select the **Suppress** option from the pop-up toolbar.

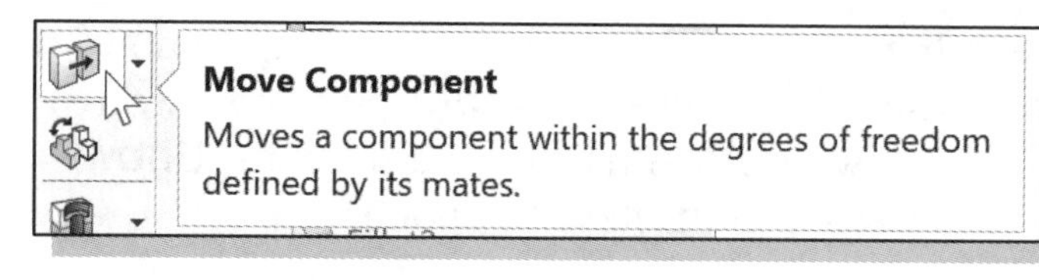

4. In the *Assembly* toolbar, select the **Move Component** command by left-mouse-clicking the icon.

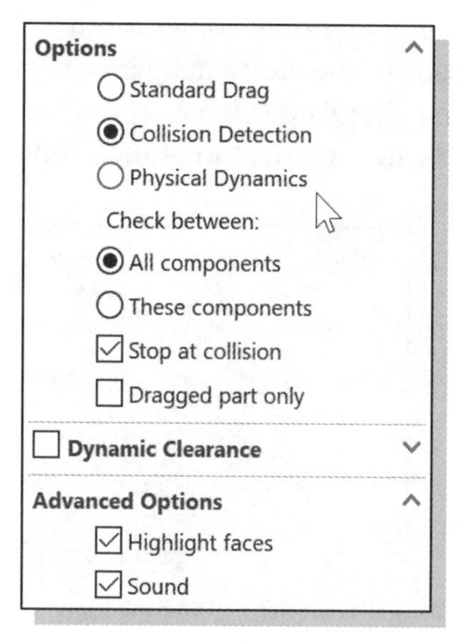

5. In the *Move Component PropertyManager*, select **Collision Detection**, **All components**, and **Stop at collision** on the *Options* panel, and **Highlight faces** on the *Advanced Options* panel.

6. Move the cursor over the *CS-Crank* part in the graphics area. Slowly **click-and-drag** the *CS-Crank* in the counter-clockwise direction.

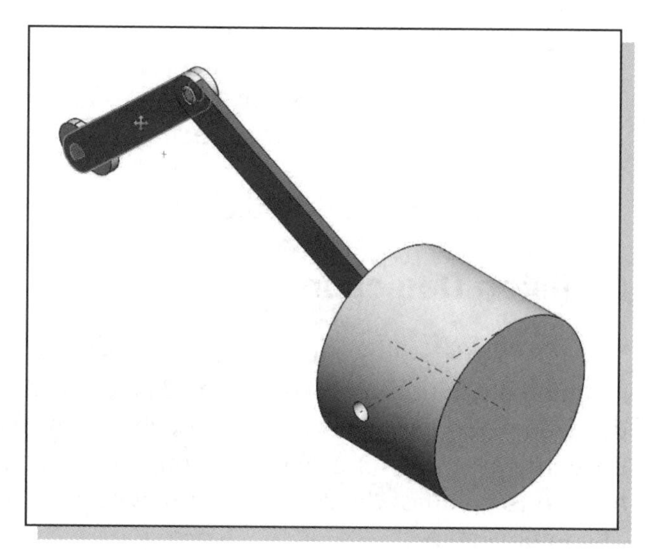

➢ Notice the motion stops at the position shown and the *CS-Connecting Rod* is highlighted.

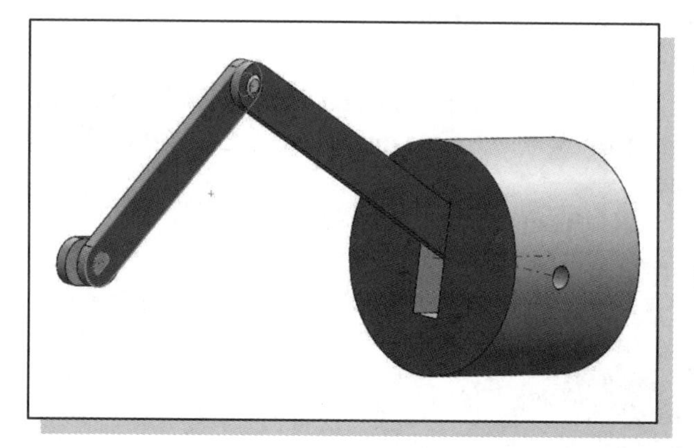

7. Use the arrow keys to **rotate** the assembly. Notice the face of the *CS-Slider* is also highlighted. This is where the collision was detected.

8. Click **OK** in the *PropertyManager* to exit the Move Component command.

Editing the CS-Slider Part in the Assembly

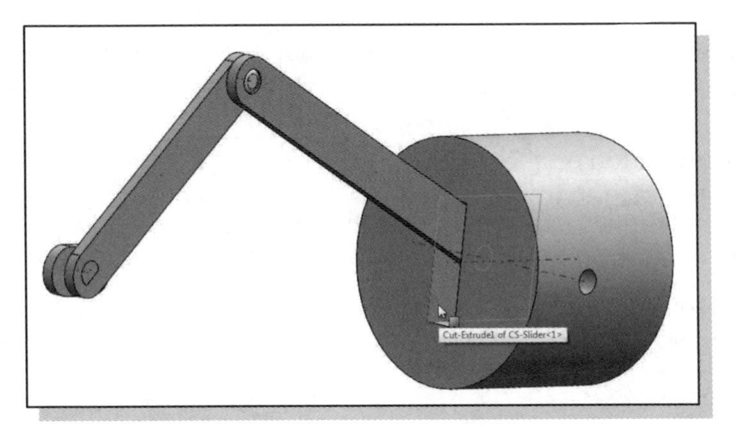

1. Move the cursor over the face on the interior of the slot in the *CS-Slider* part as shown.

2. Enter the *Edit* mode by **double-clicking** on the face.

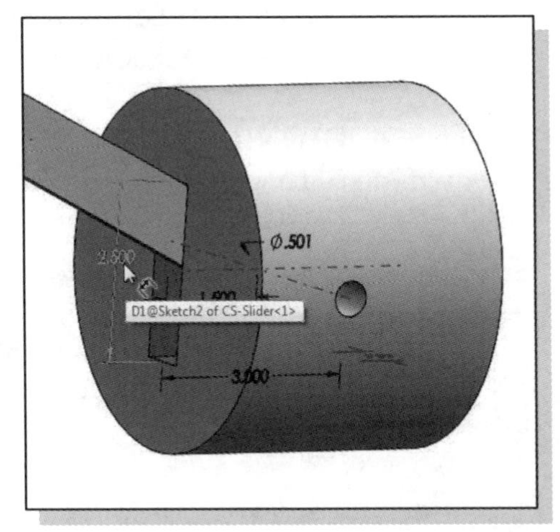

3. Notice the face is highlighted and the feature dimensions appear. **Double-click** on the 2.500 dimension of the rectangle.

4. Change the dimension to **3.75 in**.

5. Click **OK** to accept the new dimension.

6. Click on the **Rebuild** icon in the *Menu Bar* and the part and assembly are adjusted accordingly.

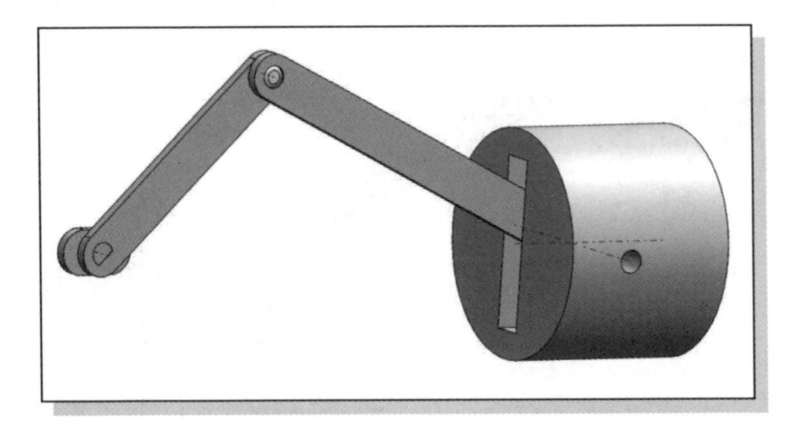

7. On your own, perform a **Collision Detection** analysis on the new assembly. (**HINT:** Repeat the steps in the Collision Detection section beginning on page 15-23.)

➢ The problem has been resolved and no collision occurs.

8. On your own, turn *OFF* the visibility of the Temporary Axes (and of the Planes if necessary) using the *View* pull-down menu.

Basic Motion Analysis

SOLIDWORKS' **Motion Study** tool allows us to perform basic motion analysis by creating simulations of assemblies with moving parts.

1. Select the **View Orientation** button on the *Heads-up View* toolbar and select the **Trimetric** view icon.

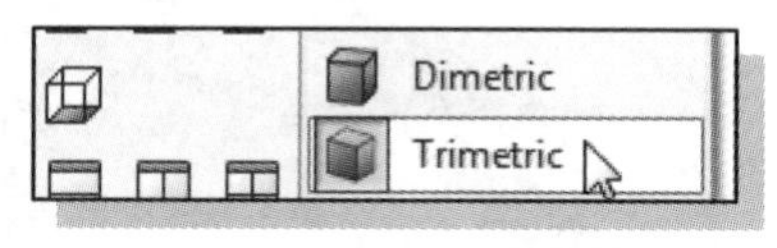

2. If necessary, move your cursor over any toolbar, right click, and toggle the **MotionManager** toolbar on.

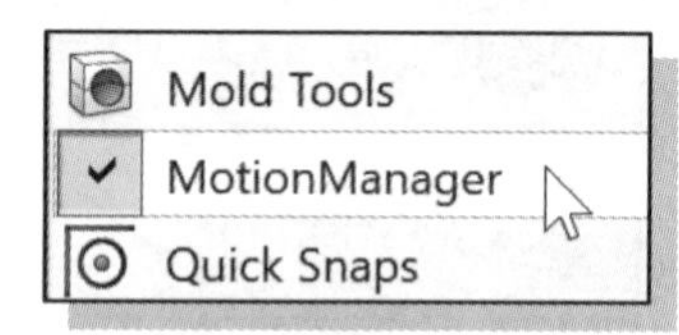

3. Select the **Motion Study** tab at the bottom of the SOLIDWORKS window.

❖ The SOLIDWORKS *MotionManager* appears. The *MotionManager* is a timeline-based interface with motion study tools located on the toolbar shown below.

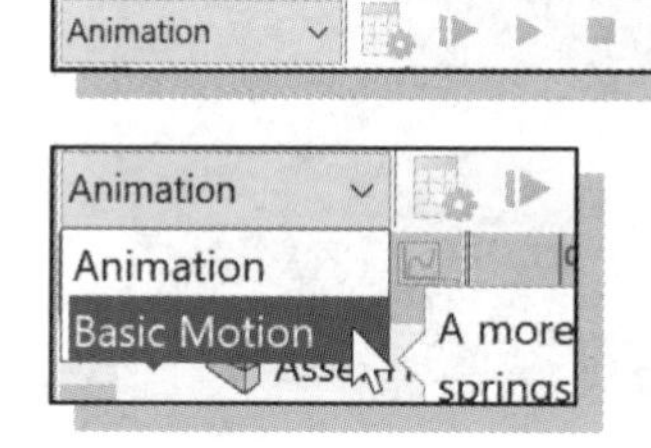

4. Select **Basic Motion** in the *Type of Study* selection window on the left end of the *MotionManager* toolbar.

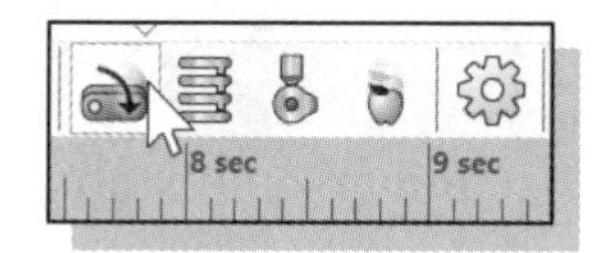

5. Move the cursor over the **Motor** command on the *MotionManager* toolbar and click once with the left mouse button to create a motor to drive the motion for the study.

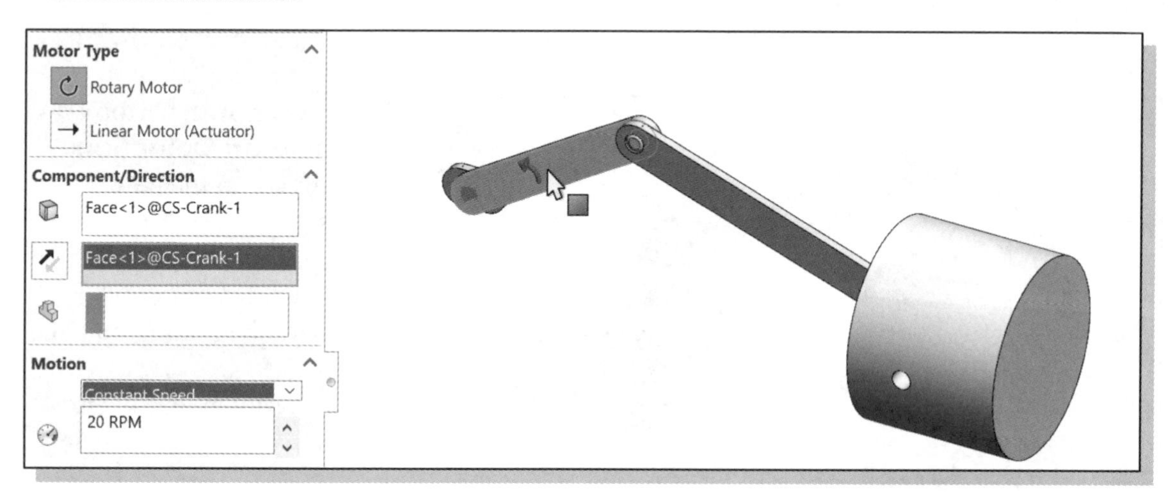

6. In the *Motor PropertyManager*, select **Rotary Motor** as the *Motor Type*, select the flat face of the ***CS-Crank*** as the *Motor Location* in the *Component/ Direction* panel. Select the **Constant Speed** option on the *Motion* panel and enter **20 RPM** for the speed.

7. Click **OK** in the *PropertyManager* to accept the setting and create the motor.

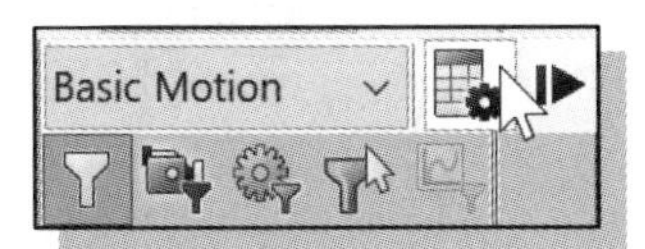

8. Move the cursor over the **Calculate** command on the *MotionManager* toolbar and click once with the left-mouse-button to calculate the *Motion Study*.

➢ Once it has been calculated, the *Motion Study* can be viewed using the **Play from Start**, **Play**, and **Stop** buttons on the *MotionManager* toolbar.

9. On your own, experiment with the **Play from Start**, **Play**, and **Stop** tools.

10. On your own, start the simulation and dynamically rotate the display (e.g., using the arrow keys) while the simulation is running.

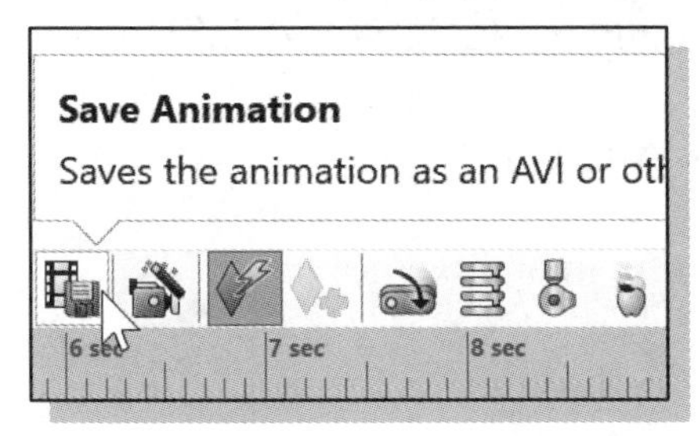

11. The simulation can also be saved as an AVI or other format file. Click the **Save Animation** button on the *MotionManager* toolbar as shown.

12. Enter ***CS-Assembly*** as the *File name* and click **Save** to accept the filename.

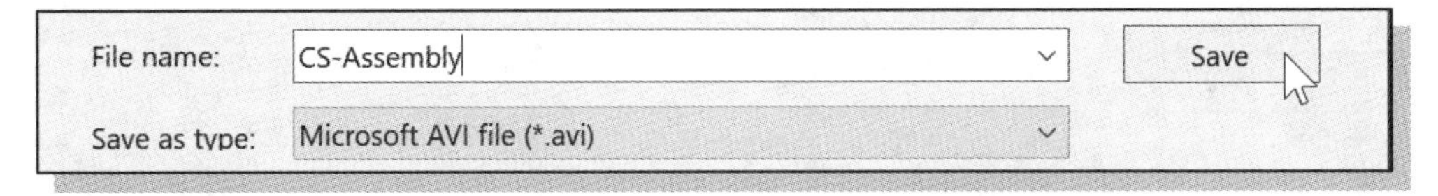

13. Select the *Video Compression* setting to set the quality and speed of the file and click **OK** to proceed.

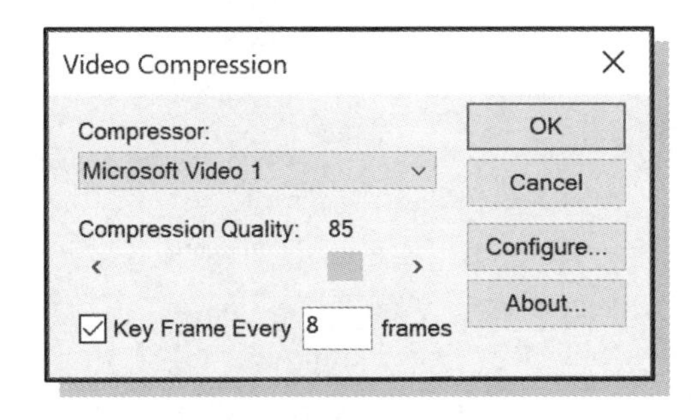

14. On your own, view the recorded video file.

15. Select the **Model** tab at the bottom of the SOLIDWORKS window to return to the *Assembly Editing* mode.

16. On your own, save the assembly with the filename ***CS-Assembly.SLDASM***.

17. If the SOLIDWORKS window appears, select the **Save All** option.

Questions:

1. What is included in the **SOLIDWORKS *Design Library***?

2. What is the usage of a **Mate Reference**?

3. How do you save the SOLIDWORKS motion study as an AVI movie?

4. How do you access the **Motor** command?

5. How do we enter the *Edit* mode in an assembly model?

6. Can we access the 2D sketch of a feature of a part in an assembly?

7. How do we end the *Edit* mode and return to the assembly model in SOLIDWORKS?

8. List and describe two methods to edit the angle of an **Angle** mate.

9. Describe the procedure to perform a *Collision Detection* in an assembly.

10. How do you control the speed of the simulation in SOLIDWORKS?

Exercises: (Create a set of detail and assembly drawings. All dimensions are in mm.)

1. **Leveling Assembly**

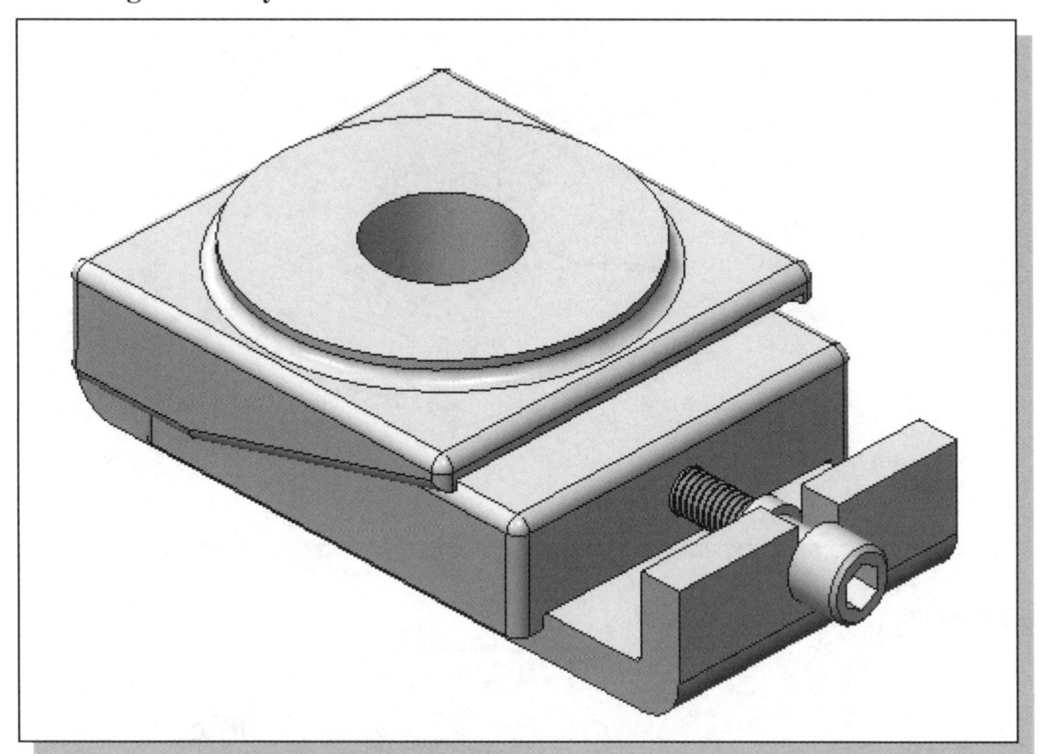

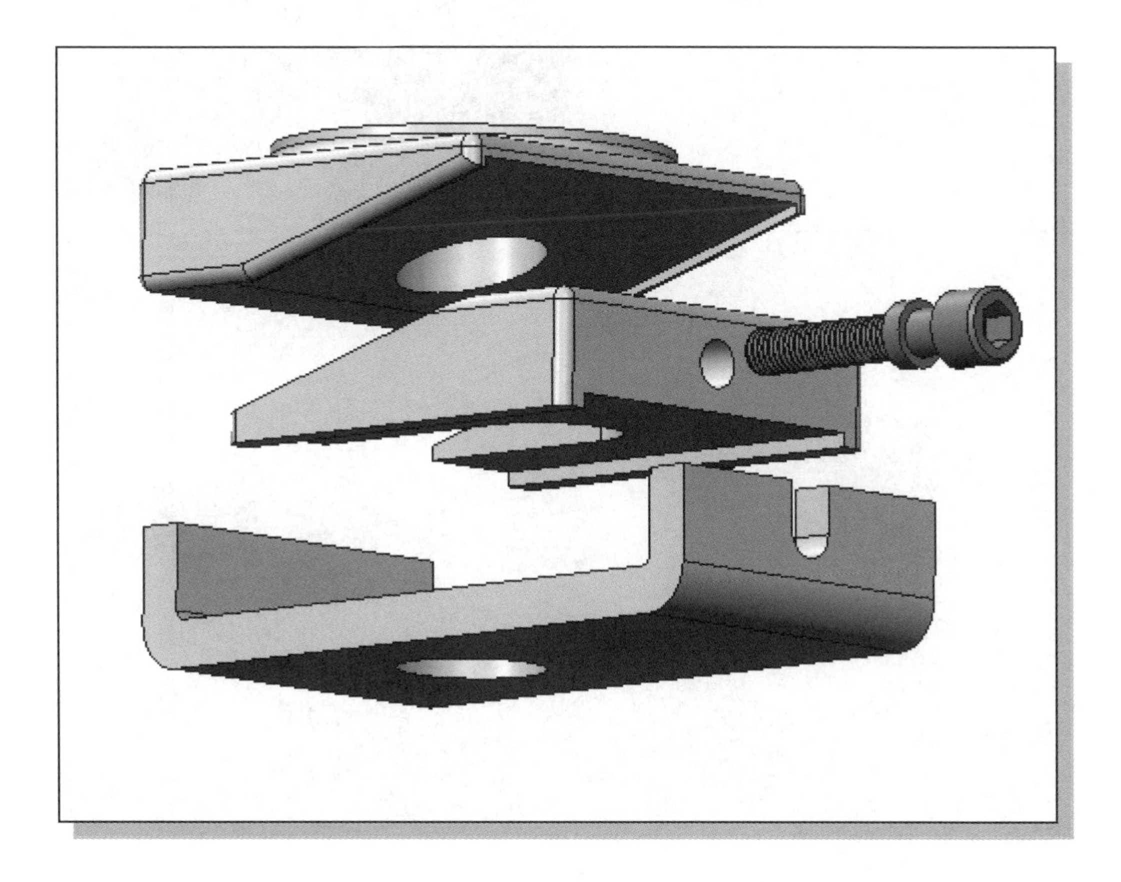

(a) **Base Plate**

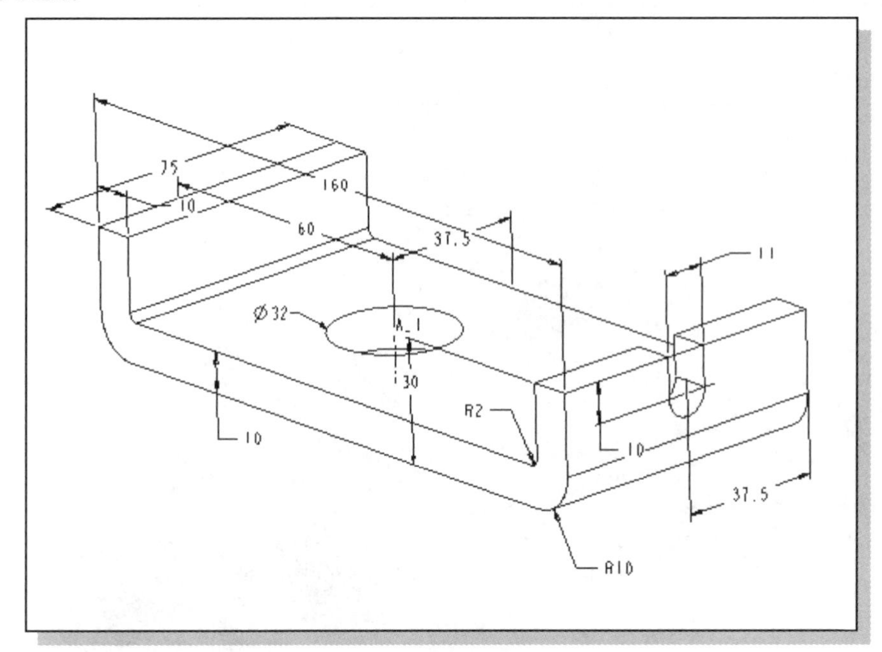

(b) **Sliding Block** (Rounds & Fillets: R3)

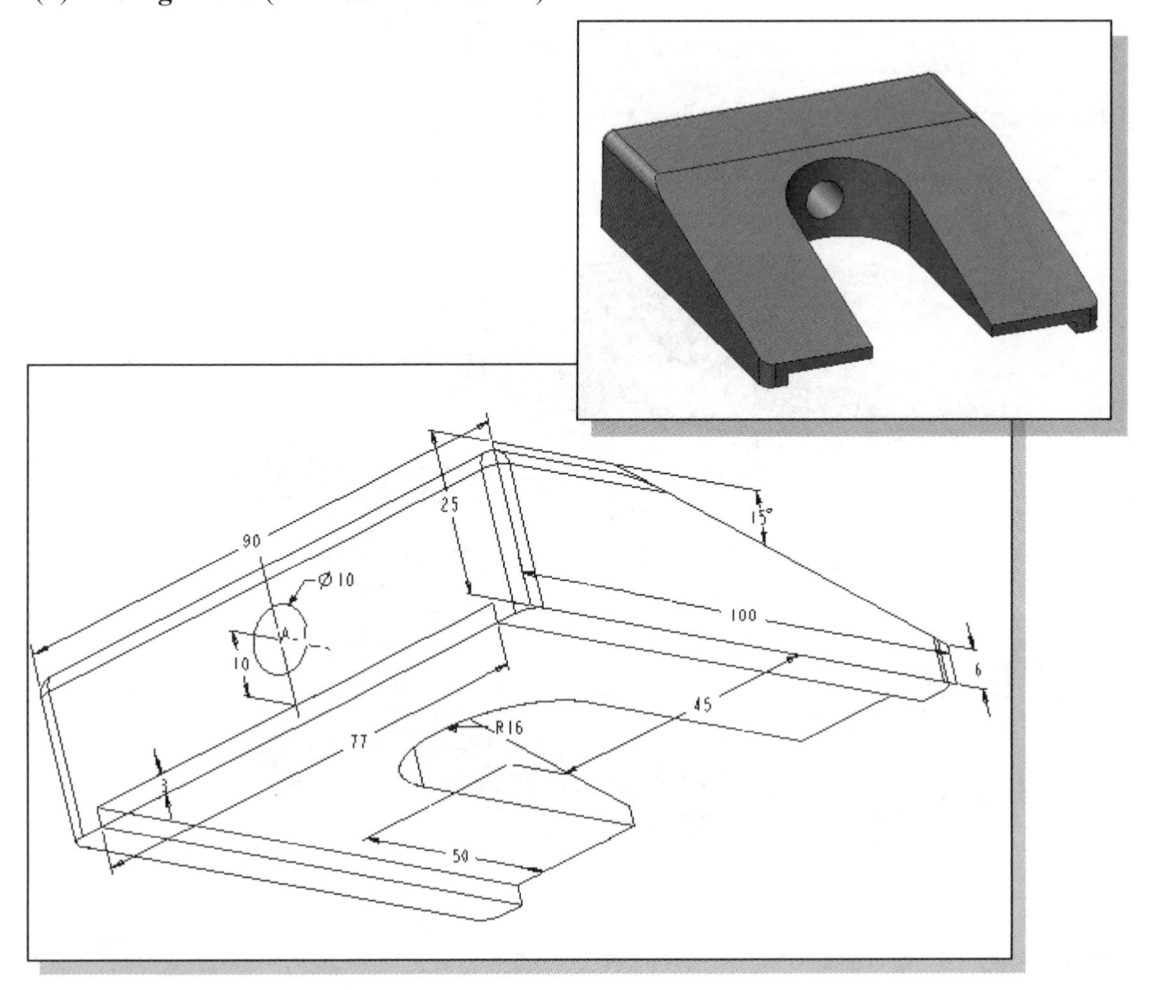

(c) **Lifting Block** (Rounds & Fillets: R3)

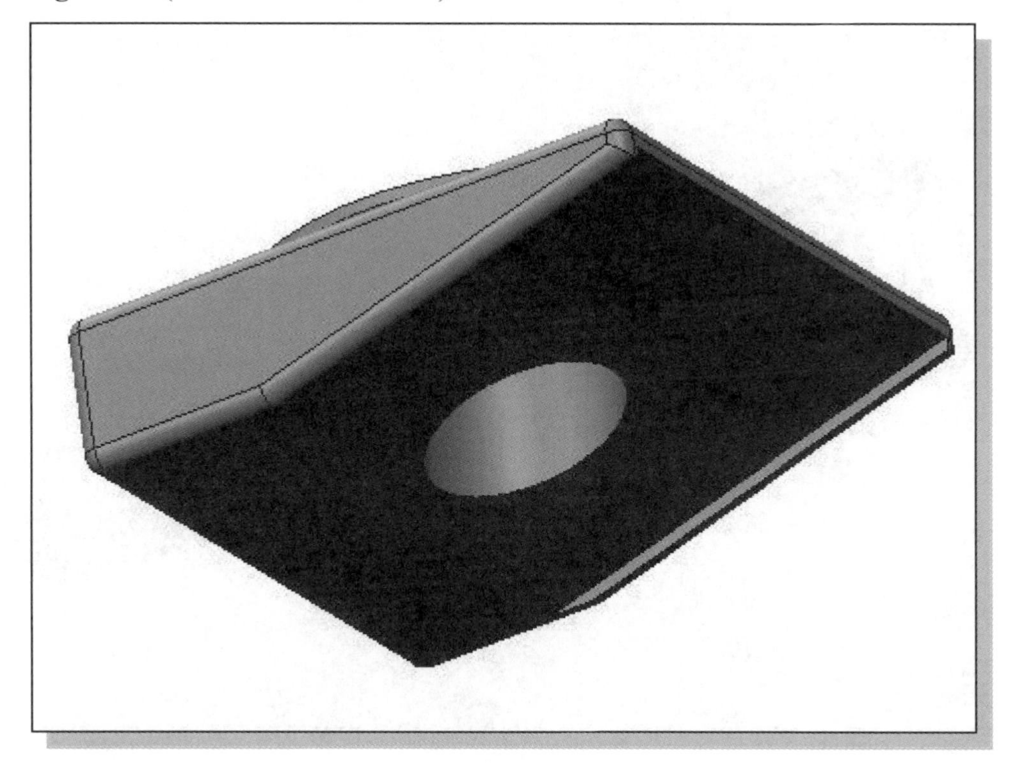

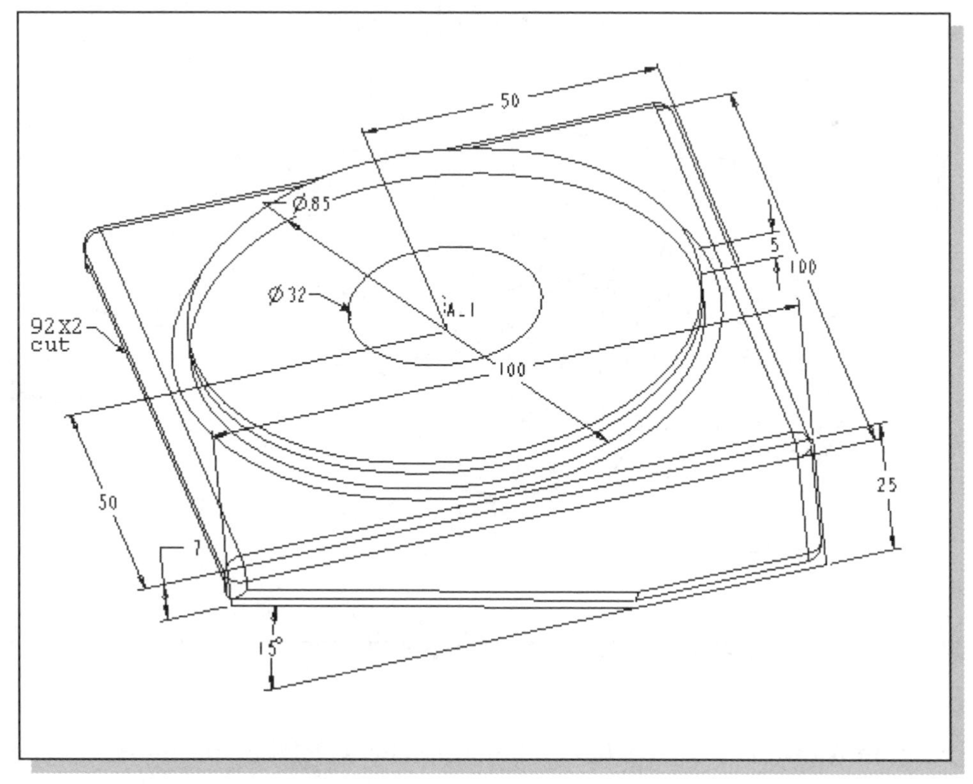

(d) **Adjusting Screw** (M10 × 1.5) (Use the **Threads** or **Coil** command to create the threads.)

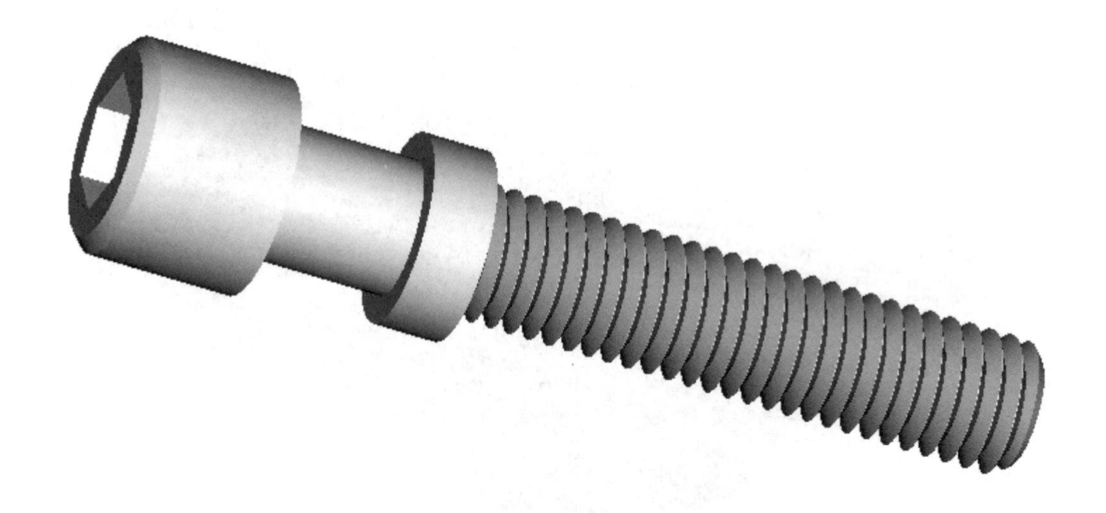

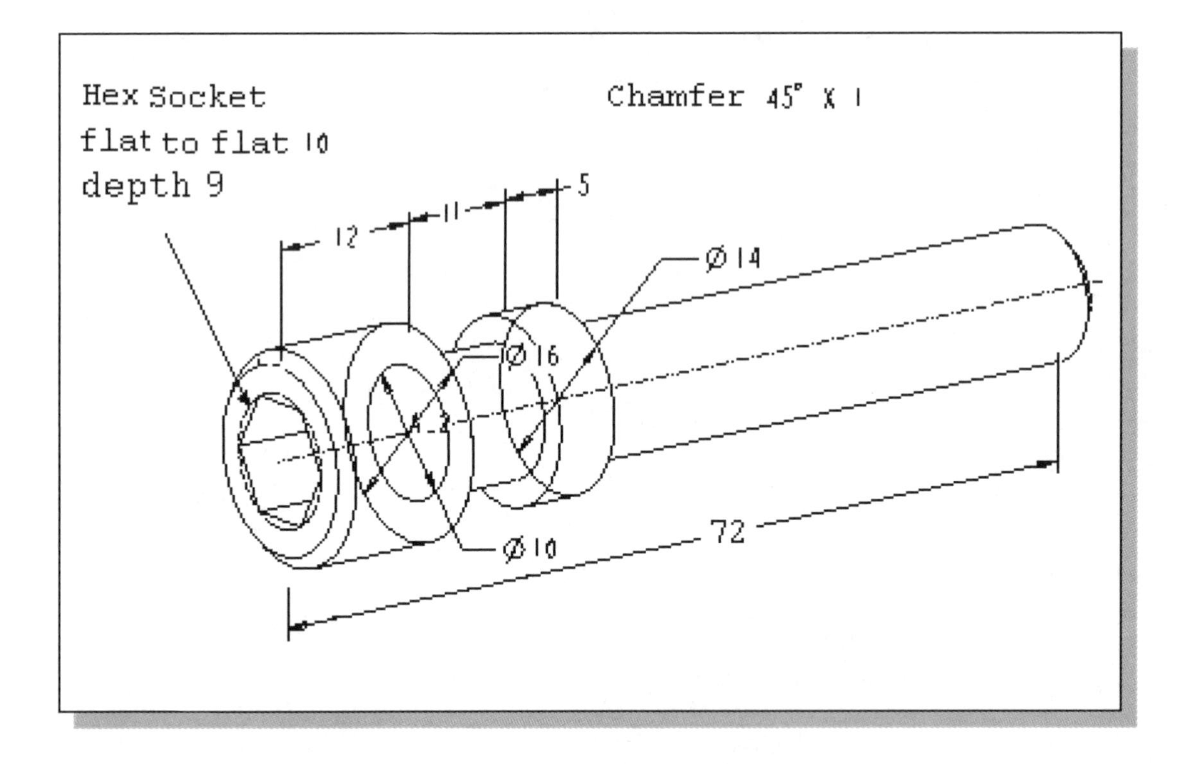

Basic Motion Analysis: Create a linear dimension between the **Sliding Block** and the **Base Plate** parts. Use the *Drive constraint* command to start the animation and perform the interference analysis.

2. **Toggle-Clamp Assembly**
(Create a set of detail and assembly drawings. All dimensions are in inches.)

(a) **Sheet Metal Base**

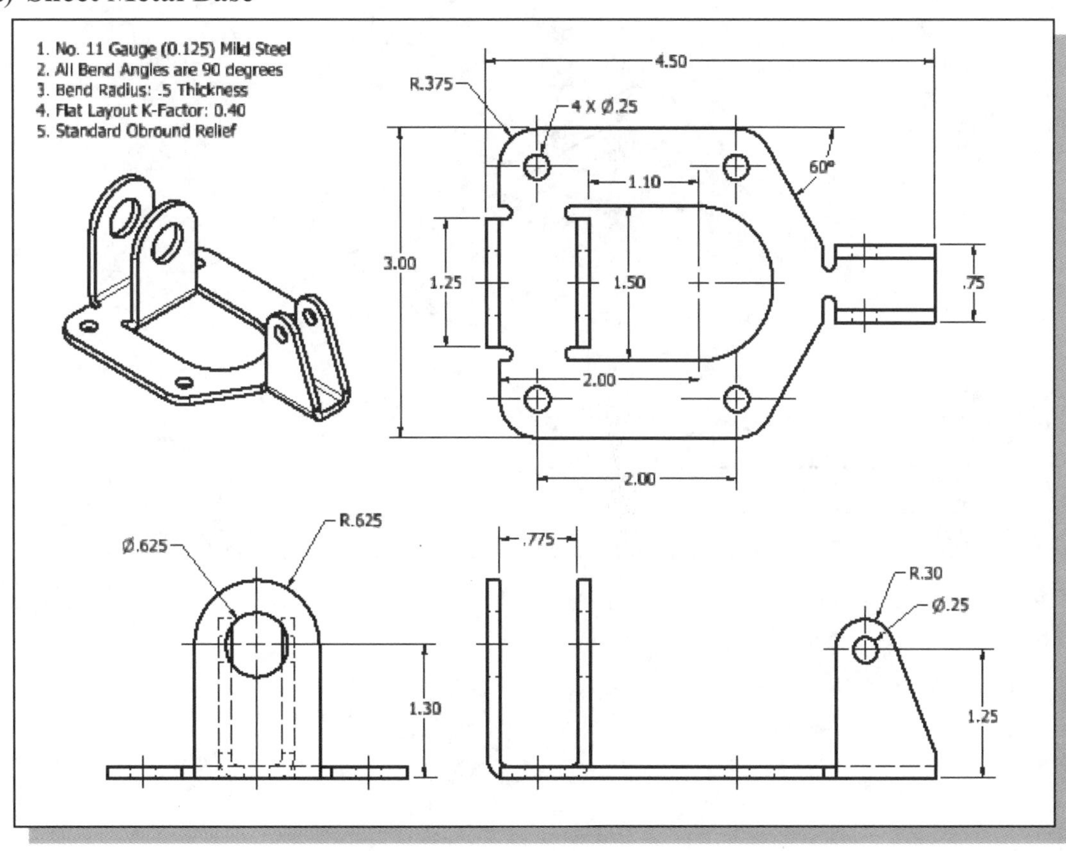

(b) **Connector**

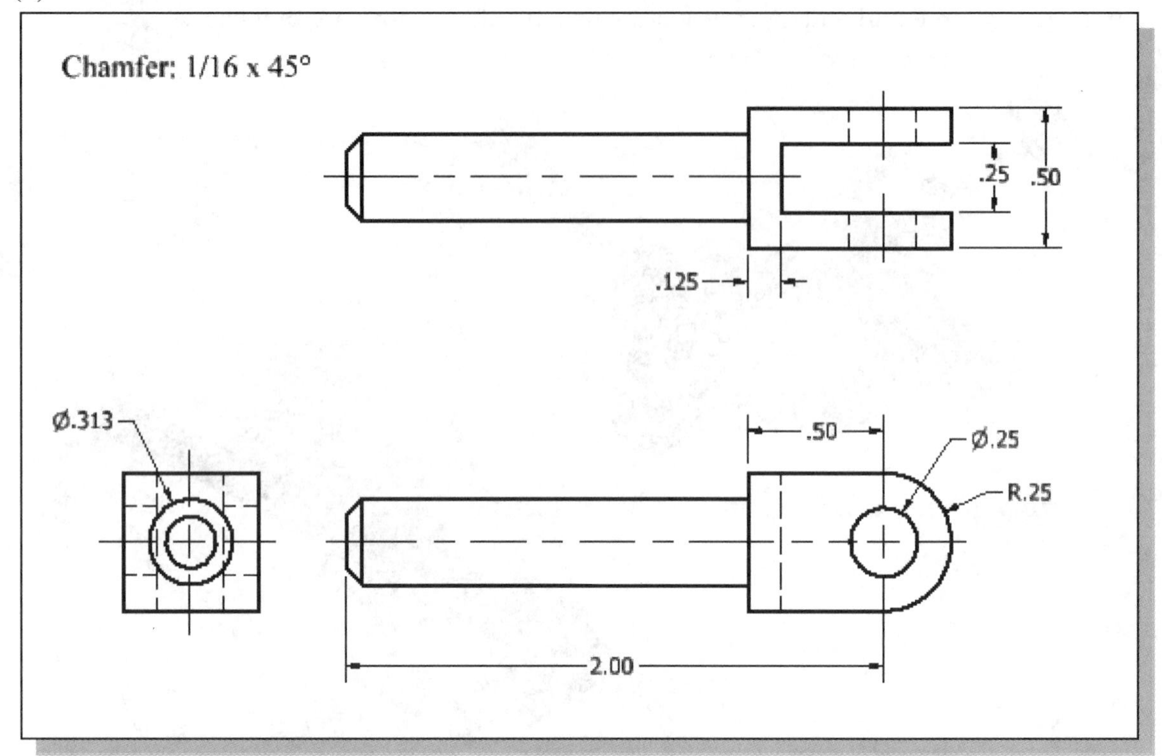

(c) **Handle**

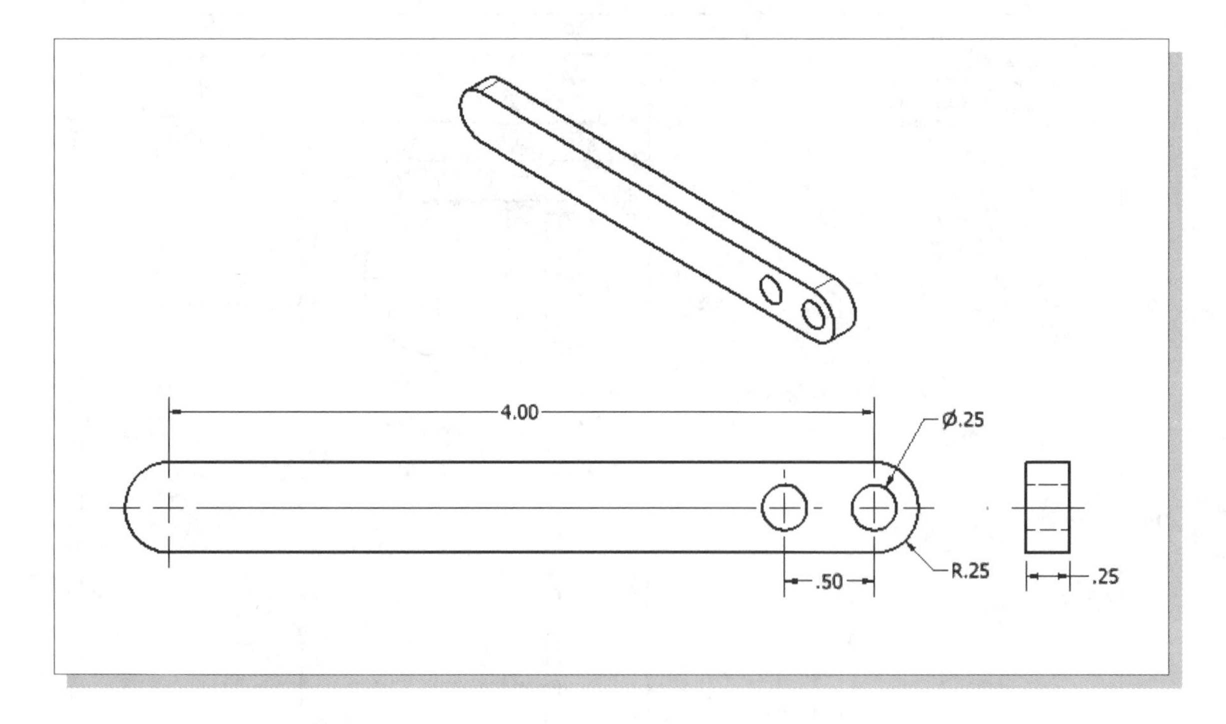

(d) **Joint Plate**

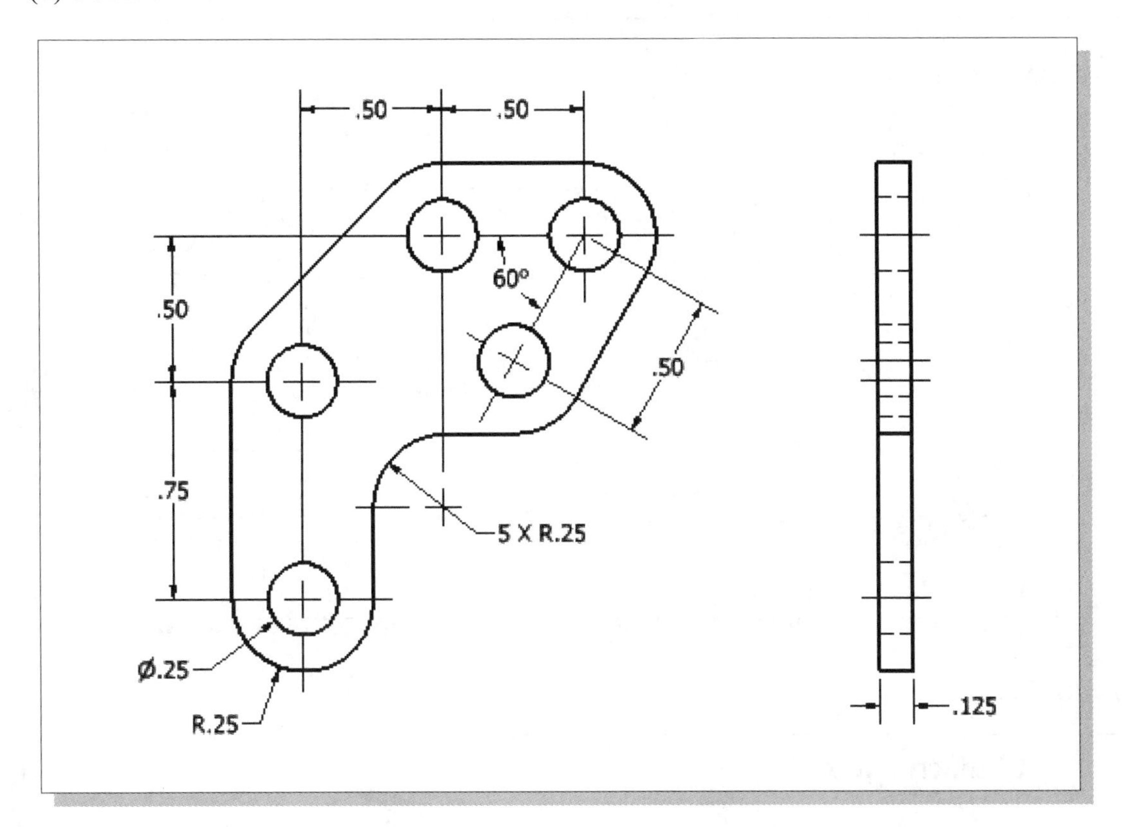

(e) **V-Link**

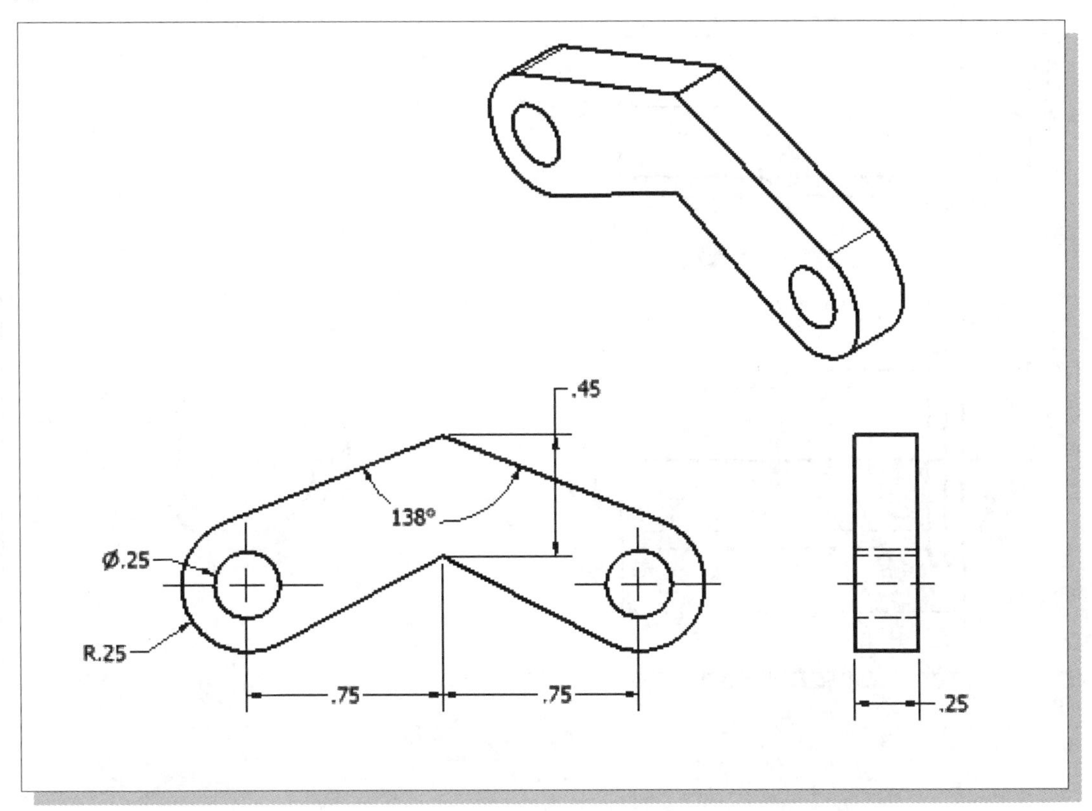

(f) **Rod**

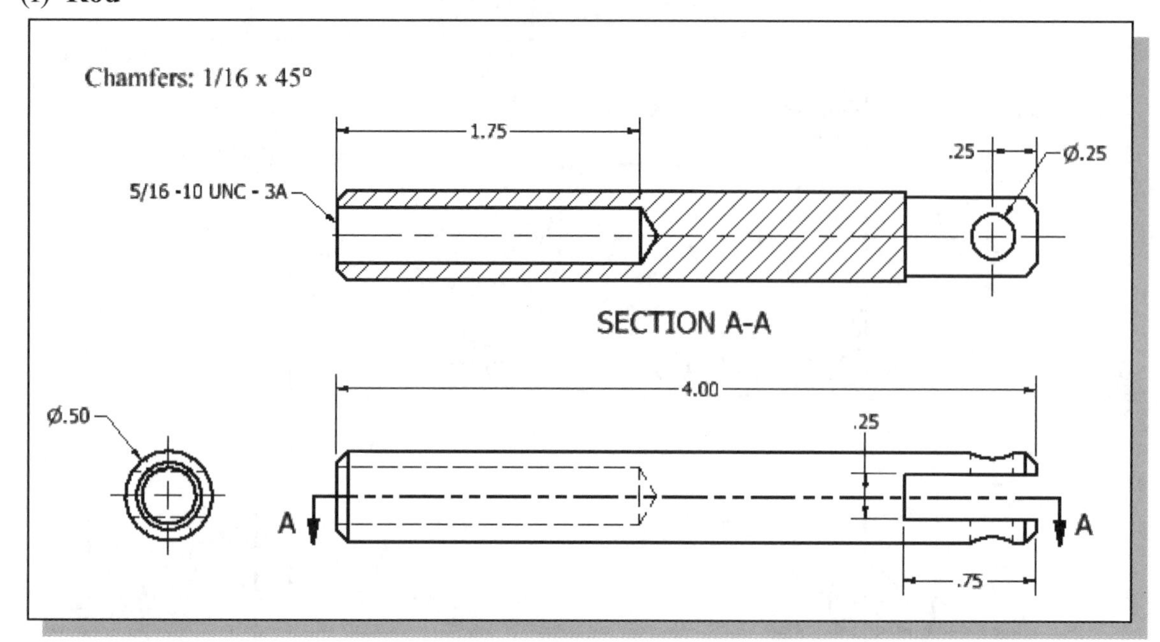

(g) **Bushing**

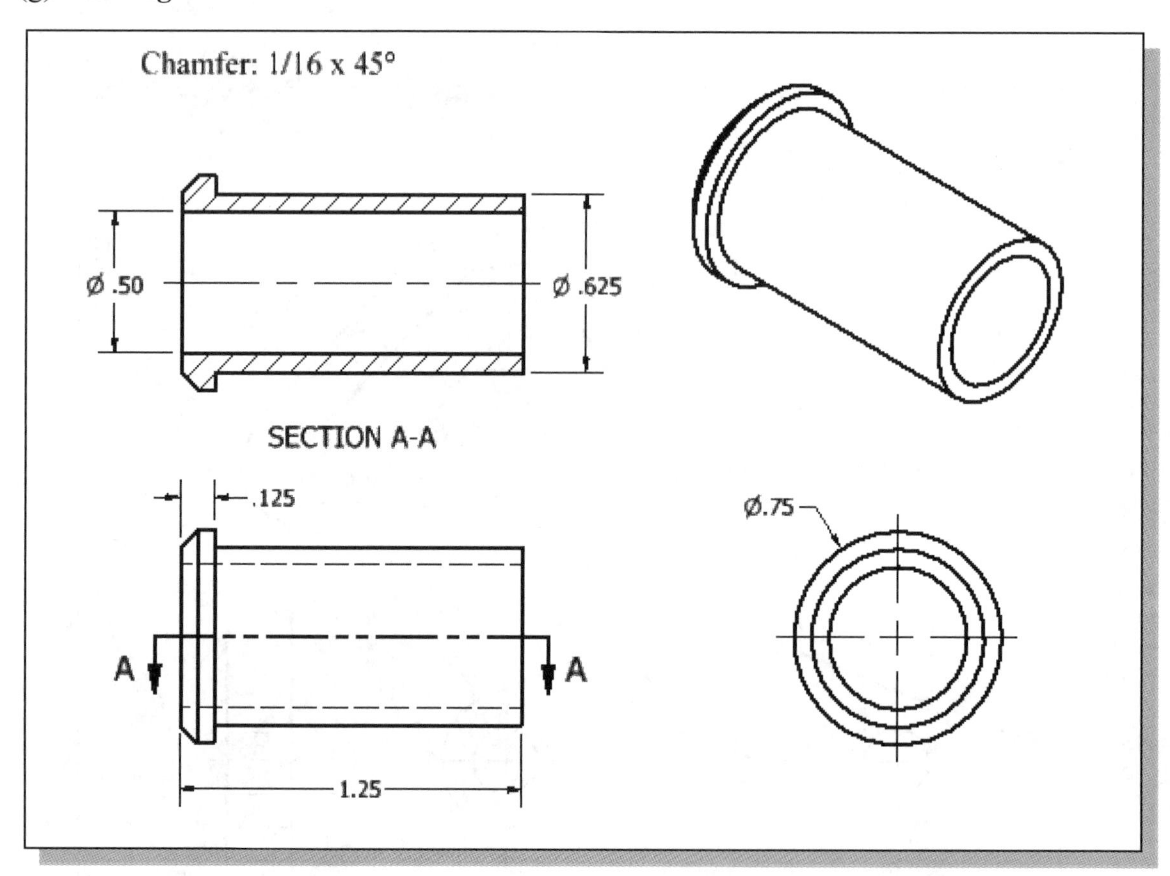

Chapter 16
Design Analysis with SimulationXpress

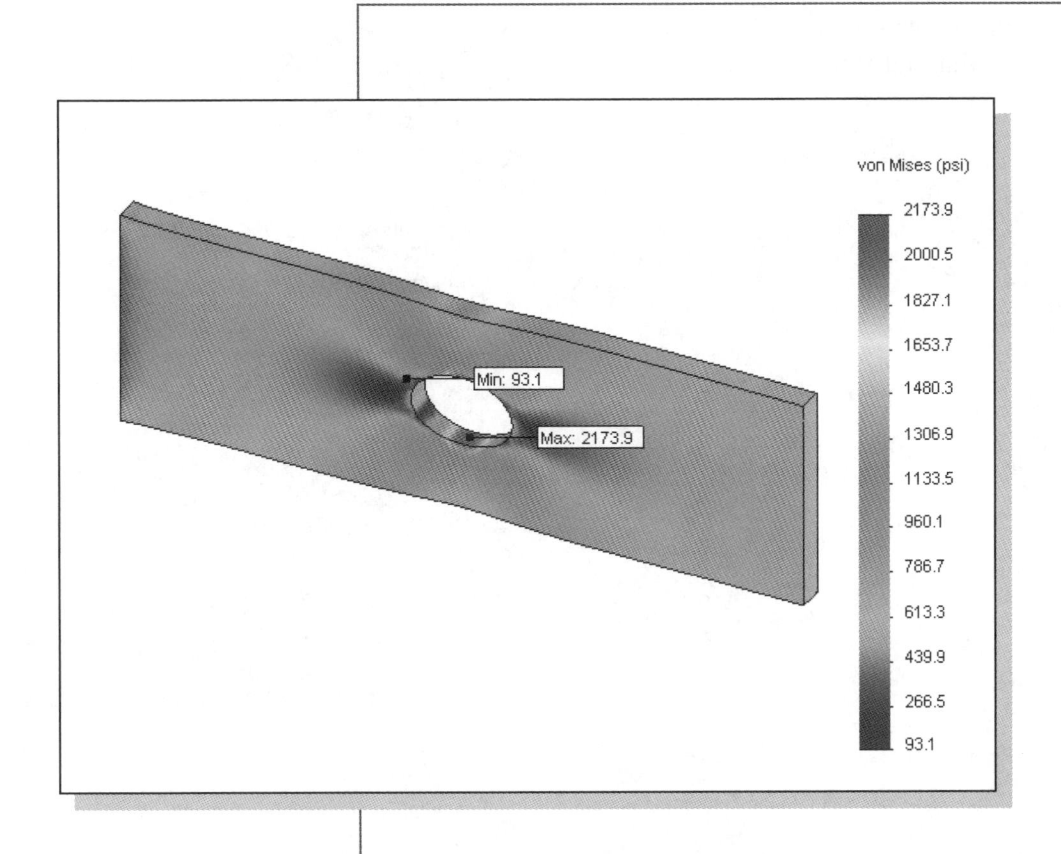

Learning Objectives

- Create SimulationXpress Study
- Apply Fixtures and Loads
- Perform Basic Stress Analysis
- View Results
- Assess Accuracy of Results
- Output the Associated Simulation Video File

Certified SOLIDWORKS Associate Exam Objectives Coverage

Materials

Objectives: Applying Material Selection to Parts.

Introduction

In this chapter we will explore basic design analysis using SOLIDWORKS *SimulationXpress*. *SimulationXpress* provides a tool for basic stress analysis, allowing the user to examine the effects of applied forces on a design. Displacements, strains, and stresses in a part are calculated based on material properties, fixtures, and applied loads. Stress results can be compared to material properties (e.g., yield strength) to perform failure analysis. The results can also be used to identify critical areas, calculate safety factors at various regions, and simulate deformation. *SimulationXpress* provides an easy-to-use method within the SOLIDWORKS environment to perform an initial stress analysis. The results can be used to improve the design.

In *SimulationXpress*, stresses are calculated using linear static analysis based on the *finite element method.* In linear static analysis, it is assumed that: 1) a linear relationship exists between the applied loads and the response, 2) the stress-strain relationship is linear, 3) no plastic deformation occurs, and 4) no suddenly applied loads are involved. The *finite element method (FEM)* is a numerical method for finding approximate solutions to partial differential equations. The technique is widely used for the solution of complex problems in engineering mechanics. Analysis using the method is called *finite element analysis (FEA).*

In the finite element method, a complex body (e.g., a part created in SOLIDWORKS) is modeled as an equivalent system of smaller bodies of simple shape, or *elements*, which are interconnected at common points called *nodes*. This process is called *discretization.* (An example is shown in the figures below.) Equations (e.g., for structural analysis) are formulated for each finite element and the resulting system of equations is solved simultaneously to obtain an approximate solution for the entire body. In general, a better approximation is obtained by increasing the number of elements (i.e., creating a finer mesh), but more computing time and resources are required.

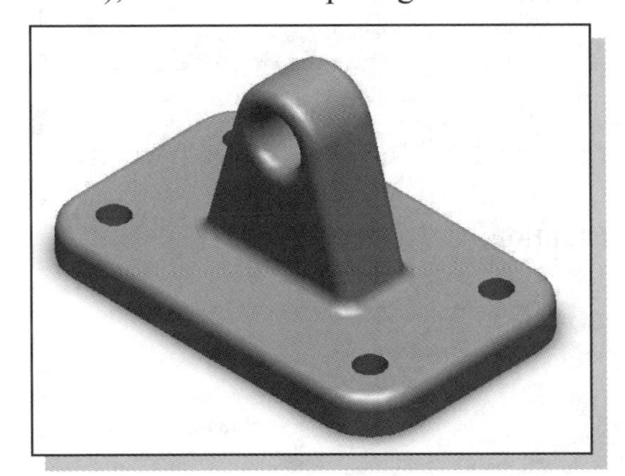

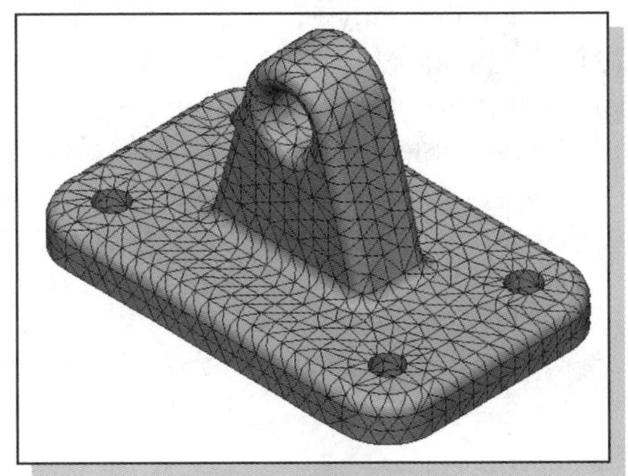

SimulationXpress utilizes ten-node tetrahedral elements for which the edges and faces can be curvilinear, facilitating the modeling of curved surfaces (as seen in the figure above). The behavior of these elements is analyzed using linear static analysis and the appropriate material properties to relate local coordinate nodal displacements to local forces. The motion of each node is described by displacements in the X, Y, and Z directions, called *degrees of freedom* (DOFs). The equations describing the behavior of each element are assembled into a global system of equations, incorporating compatibility requirements based on connectivity among elements. Using the known material properties, fixtures, and loads, *SimulationXpress* solves the system of equations for the unknown displacements at each node. These displacements are used in the results stage to calculate strains and stresses.

While *SimulationXpress* is a powerful and easy-to-use tool, it is important to appreciate that it is the designer's responsibility to properly assess the accuracy of the results. As stated above, a better FEA approximation is generally obtained by increasing the number of elements. An assessment must be made regarding whether the mesh used to discretize the model is adequate. There are other important factors affecting the accuracy of the results. The material properties used in the analysis must accurately characterize the behavior of the material. The fixtures and loads must be applied in a manner which accurately reflects the actual conditions. Proper meshing and application of boundary conditions often requires significant experience in FEA and may require tools and capabilities not available in *SimulationXpress*. *SimulationXpress* is an easy-to-use tool for a first-pass stress analysis. More advanced capabilities are available in the SOLIDWORKS *Simulation* design analysis application.

The SimulationXpress Wizard Interface

The wizard interface of *SimulationXpress* appears to the right of the graphics area when *SimulationXpress* is activated from the *Tools* pull-down menu. The wizard guides the user through a six-step process to perform an analysis:

(1) Fixtures – identify entities to be fixed in place during the analysis;
(2) Loads – apply loads;
(3) Material – specify material;
(4) Run – assemble equations and solve for the unknown degrees of freedom;
(5) Results – to view results, e.g., stress distribution; and
(6) Optimize – an optional step to find the optimal value for a parameter, e.g., a fillet radius, based on a selected criterion.

Problem Statement

Determine the maximum normal stress that loading produces in the Aluminum 1060 Alloy plate.

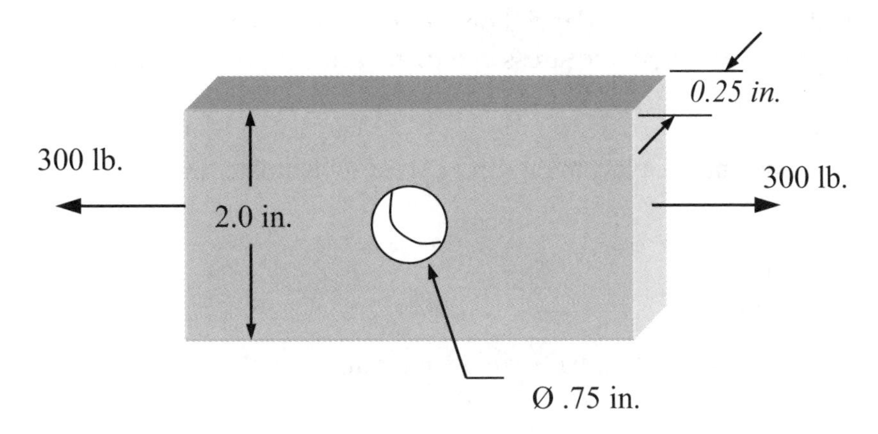

Preliminary Analysis

- **Maximum Normal Stress**

The nominal normal stress developed at the smallest cross section (through the center of the hole) in the plate is

$$\sigma_{nominal} = \frac{P}{A} = \frac{300}{(2 - 0.75) \times .25} = 960 \text{ psi.}$$

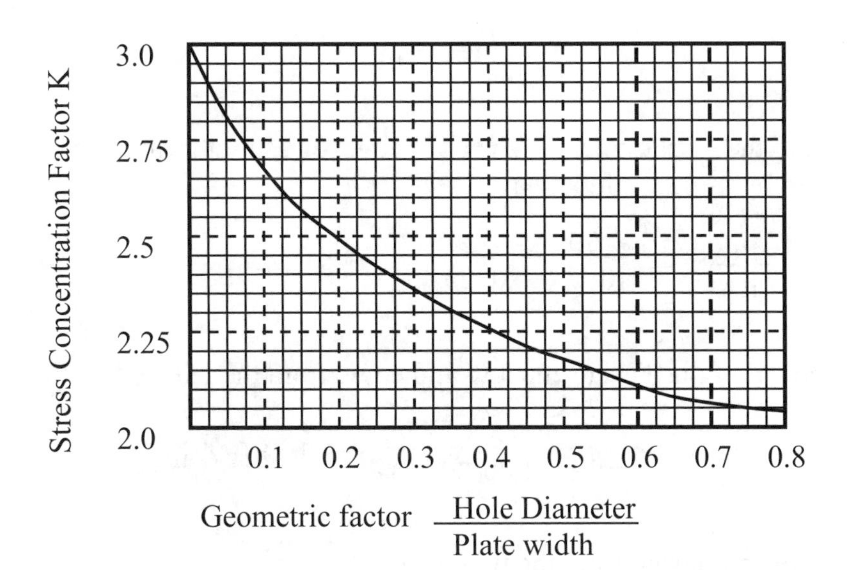

Geometric factor = .75/2 = 0.375

Stress concentration factor K is obtained from the graph, **K = 2.27**

$$\sigma_{MAX} = K\,\sigma_{nominal} = 2.27 \times 960 = 2180 \text{ psi.}$$

- **Maximum Displacement**

We will also estimate the displacement under the loading condition. For a statically determinant system, the stress results depend mainly on the geometry. The material properties can be in error and still the FEA analysis comes up with the same stresses. However, the displacements always depend on the material properties. Thus, it is necessary to always estimate both the stress and displacement prior to a computer FEA analysis.

The classic one-dimensional displacement can be used to estimate the displacement of the problem:

$$\delta = \frac{PL}{EA}$$

Where P=force, L=length, A=area, E= elastic modulus, and δ = deflection.

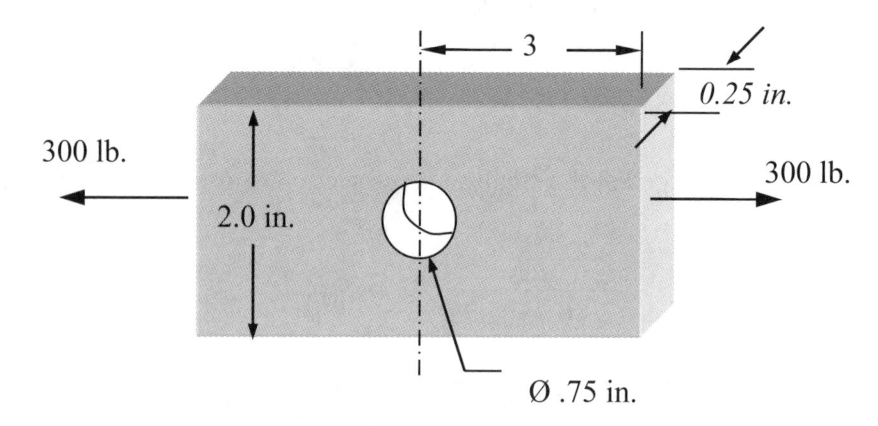

A lower bound of the displacement of the right edge, measured from the center of the plate, is obtained by using the full area:

$$\delta_{lower} = \frac{PL}{EA} = \frac{300 \text{ X } 3}{10E6X(2 \text{ X}0.25)} = 1.8E\text{-}4 \text{ in.}$$

and an upper bound of the displacement would come from the reduced section:

$$\delta_{upper} = \frac{PL}{EA} = \frac{300 \text{ X } 3}{10E6X(1.25 \text{ X}0.25)} = 2.88E\text{-}4 \text{ in.}$$

but the best estimate is a sum from the two regions:

$$\delta_{average} = \frac{PL}{EA} = \frac{300 \text{ X } 0.375}{10E6X(1.25X0.25)} + \frac{300 \text{ X } 2.625}{10E6X(2.0X0.25)}$$

$$= 3.6E\text{-}5 + 1.58E\text{-}4 = 1.94E\text{-}4 \text{ in.}$$

SOLIDWORKS SimulationXpress Study of the Flat Plate

In this lesson we will perform a stress analysis on the flat plate with a circular hole under axial tension using *SimulationXpress*. The *SimulationXpress* results will include values for maximum stress and maximum displacement calculated using finite element analysis. These results will be compared to the values for maximum stress and displacement calculated in the previous section.

This tutorial is performed using SOLIDWORKS *SimulationXpress*. *SimulationXpress* uses the same analysis methodology as SOLIDWORKS *Simulation* to perform stress analysis but has abbreviated capabilities and uses a different interface – the wizard described in the previous section. It is therefore necessary to ensure that the SOLIDWORKS *Simulation* add-in (if available) is de-activated in order to follow the instructions. If SOLIDWORKS *Simulation* is activated, the user will be presented with a different interface.

Getting Started – Create the SOLIDWORKS Part

1. On your own, start a new part file using the **Part_IPS_ANSI** template in the Tutorial_Templates folder. (You created this template in Chapter 4.)

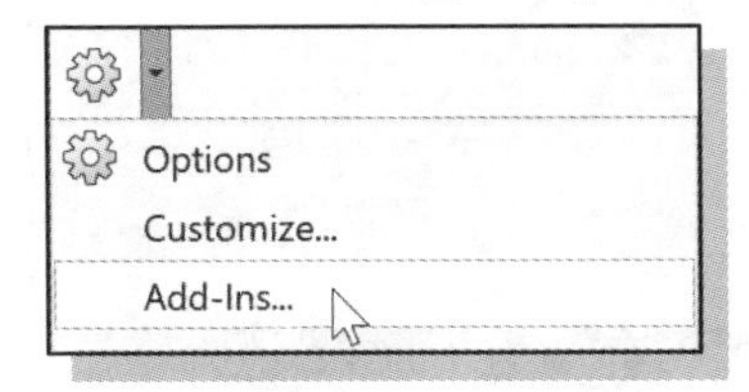

2. Select the **Add-Ins** option from the pull-down menu next to the *Options* icon.

3. In the *Add-Ins* pop-up window, make sure that the SOLIDWORKS *Simulation* application is **not** selected and click **OK**.

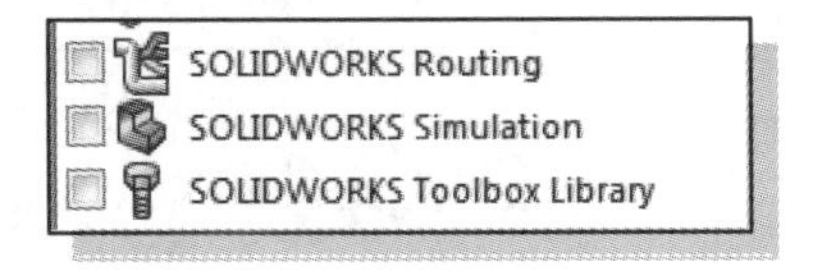

4. On your own, **create the sketch** shown below using the **Front Plane** as the sketch plane.

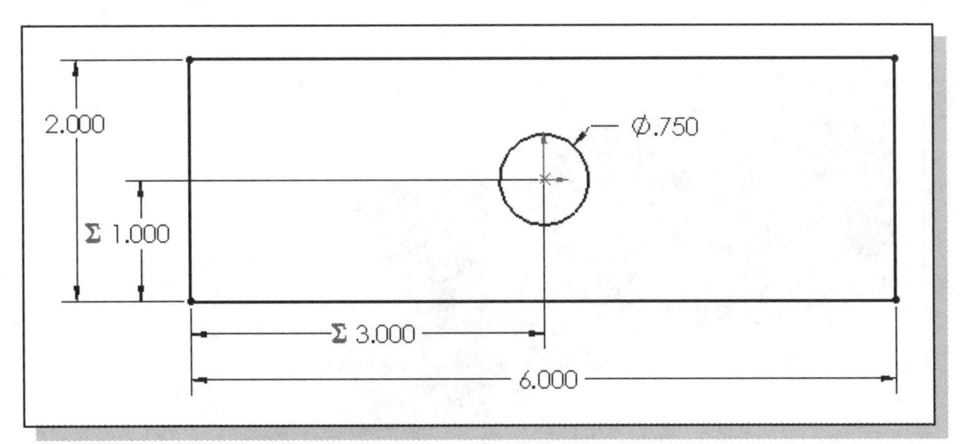

5. On your own, create an **Extruded Boss** feature with a thickness of **0.25 in**.

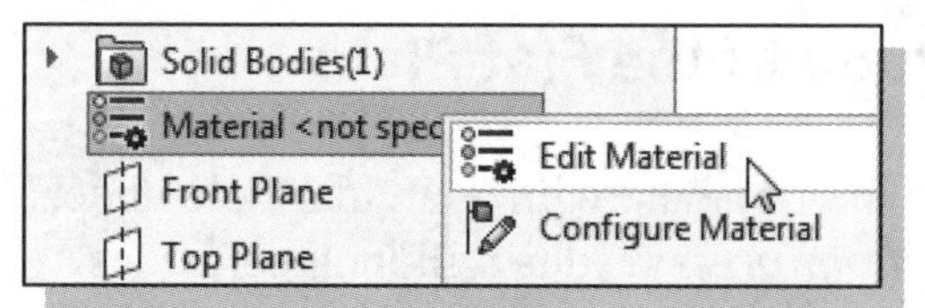

6. Right-click on the **Material** icon in the *Feature Manager Design Tree* and select **Edit Material** from the pop-up menu.

7. In the *Material* pop-up window, select **1060 Alloy** (in the *Aluminum Alloys* folder) as the material type. Notice the material properties (Young's modulus, etc.) listed. These properties will be used in the *SimulationXpress* analysis.

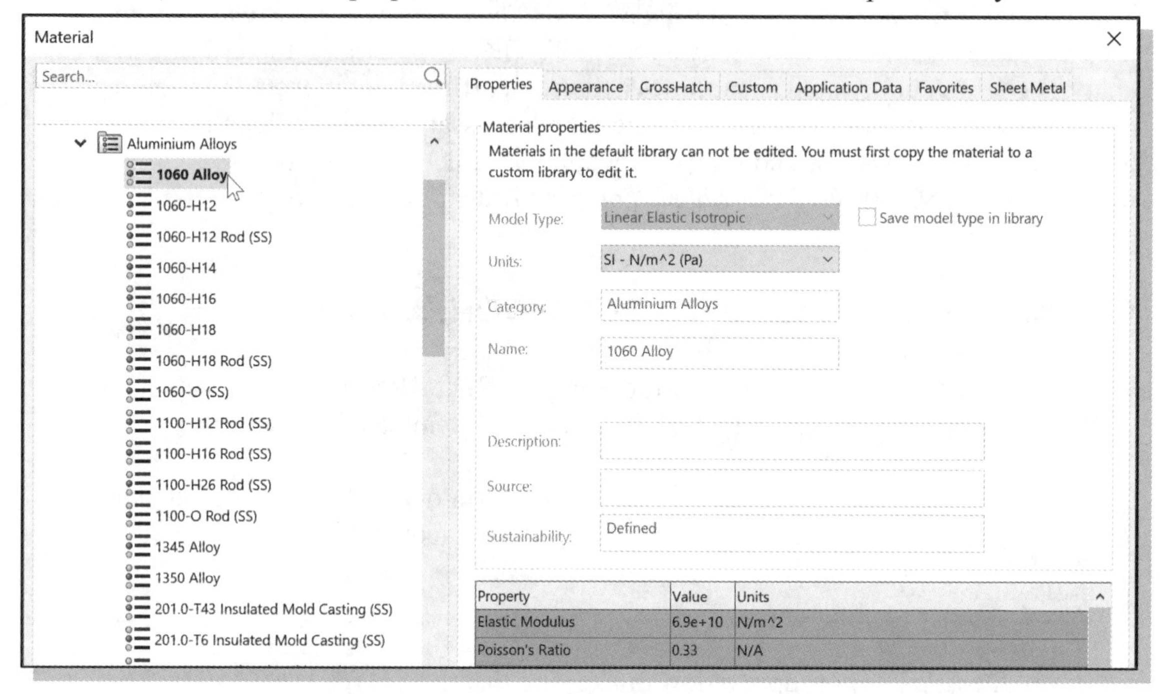

8. Click **Apply** to select the material and **Close** the *Material* window.

9. Save the part with the filename ***Aluminum_Plate***.

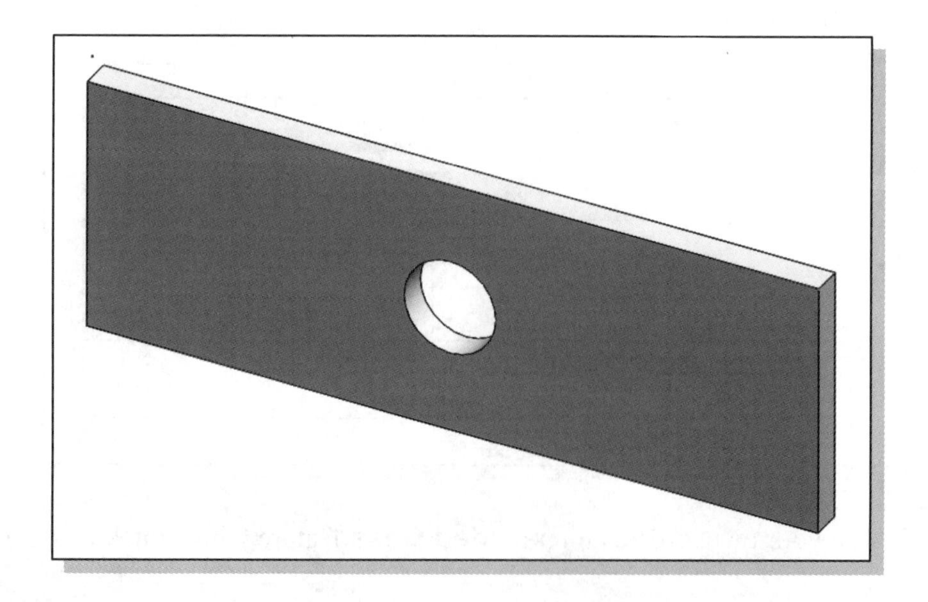

Create a *SimulationXpress* Study

1. Select the **SimulationXpress** command from the *Tools* pull-down menu.

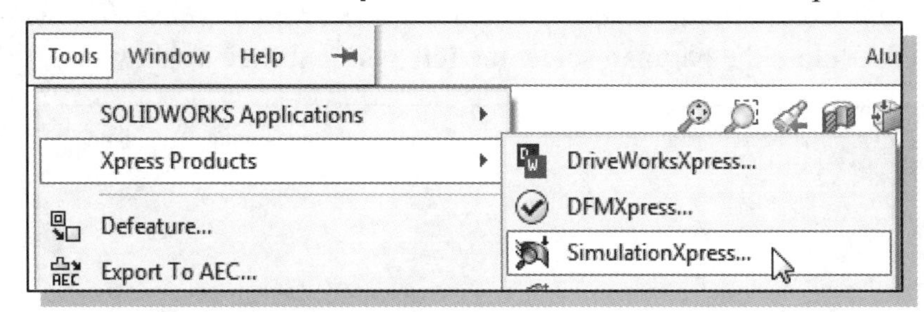

➢ NOTE: If the Enable SimulationXpress window appears, follow the instructions to obtain your Xpress product code and enable the tool.

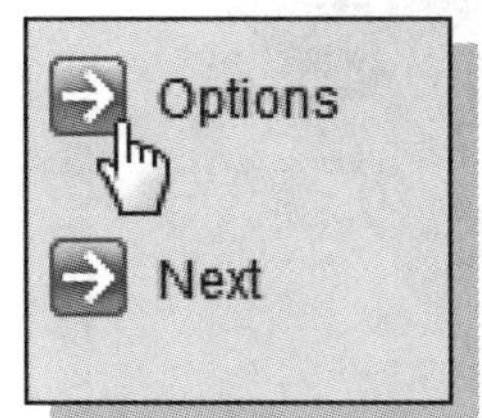

➢ The *SimulationXpress* wizard appears at the right of the screen.

2. Select the **Options** button on the *Welcome* screen of the *SimulationXpress* wizard.

3. Select **English (IPS)** in the *System of units:* option box.

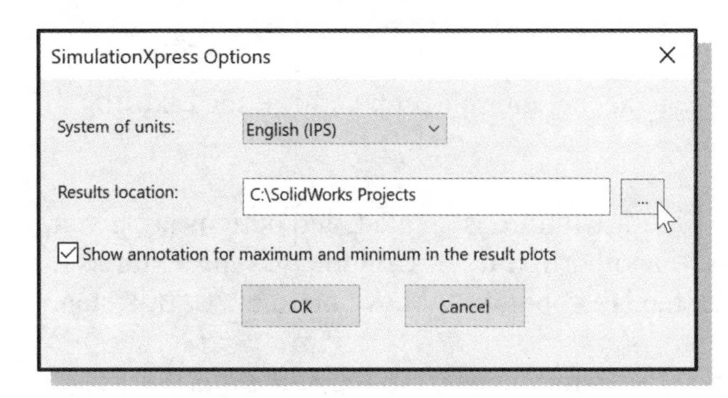

4. **Check** the **Show annotation for maximum and minimum in the results plots** box.

5. On the *Results location* option line, click the **Browse** button and select the folder in which you want to save the simulation results.

6. Select **OK** to accept the options and close the *SimulationXpress Options* window.

7. Click the **Next** button in the *SimulationXpress* wizard to move on to the Fixtures step.

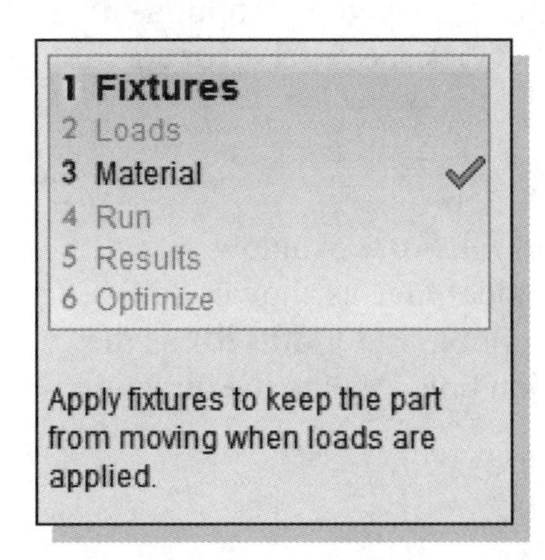

➢ The major steps in the *SimulationXpress* process appear, with the Fixtures step active. The check next to the Material step indicates that the material we applied to the part is recognized by the *SimulationXpress* wizard. An introduction to the Fixtures step is displayed.

8. Click the **Add a fixture** command in the task pane.

9. The *Fixture PropertyManager* appears with the *Faces for Fixture* window highlighted. Rotate the part and select the **left vertical face** as shown.

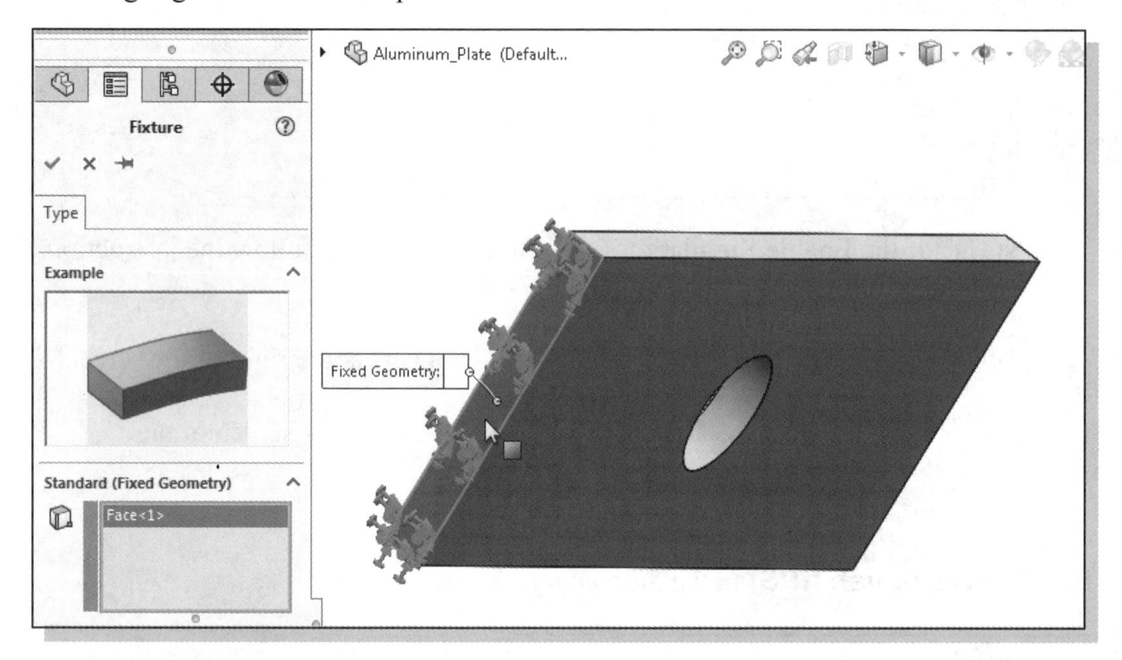

10. Click **OK** in the *PropertyManager* to accept the selection and exit the Fixture command.

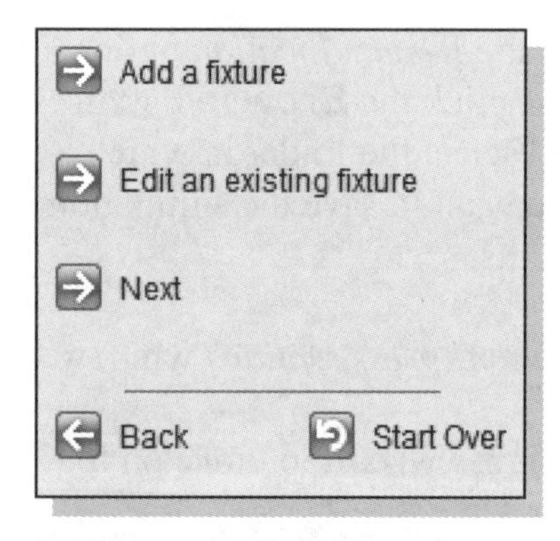

11. The Fixed-1 fixture is created, and options are given to Add another fixture or Edit the existing fixture. Click the **Next** button to move on to the Loads step.

➢ Notice the fixture appears in the *SimulationXpress Study Tree* to the left of the graphics area with the default name Fixed-1, and a checkmark appears next to Fixtures in the top pane of the *SimulationXpress* wizard, showing that the step is complete.

12. An introduction to the Loads step is displayed with options to apply a Force or a Pressure load. Click the **Add a force** command.

13. The *Force PropertyManager* appears with the *Faces for Force* window highlighted. Rotate the part and select the **right vertical face** as shown. Select the **Normal** direction option and the **English** unit option. Enter **300 lbf** as the force value and check the **Reverse direction** option box. Notice the direction of the force load is now outward.

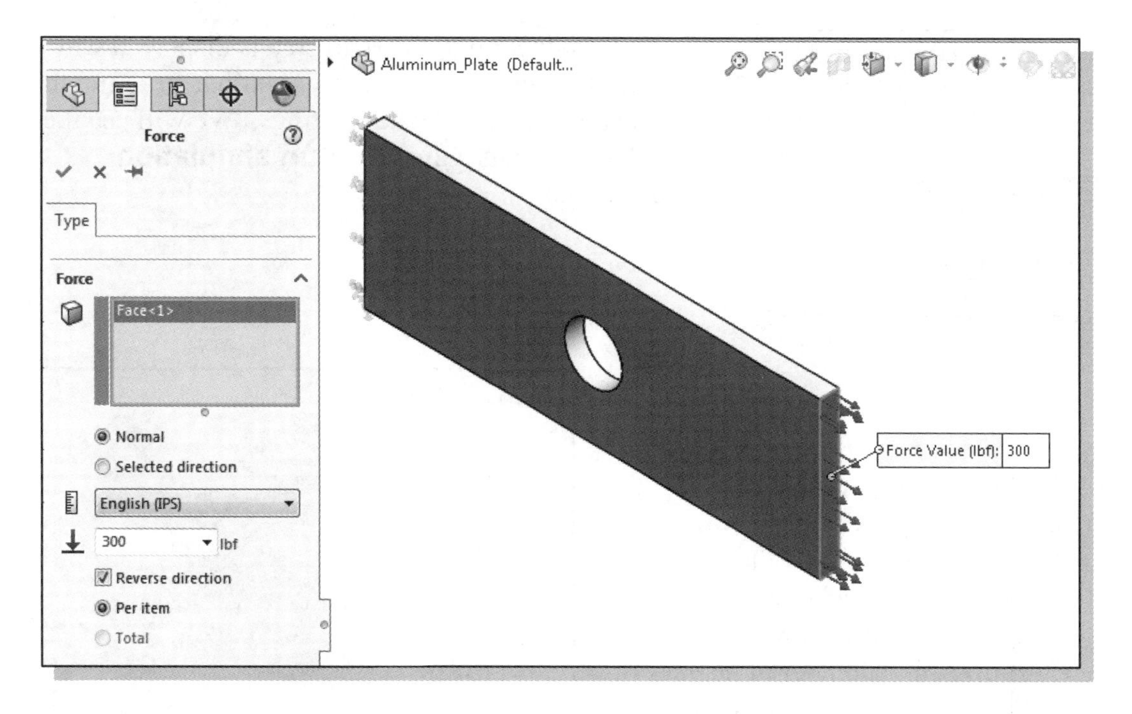

14. Click **OK** in the *PropertyManager* to accept the selection and exit the Force command.

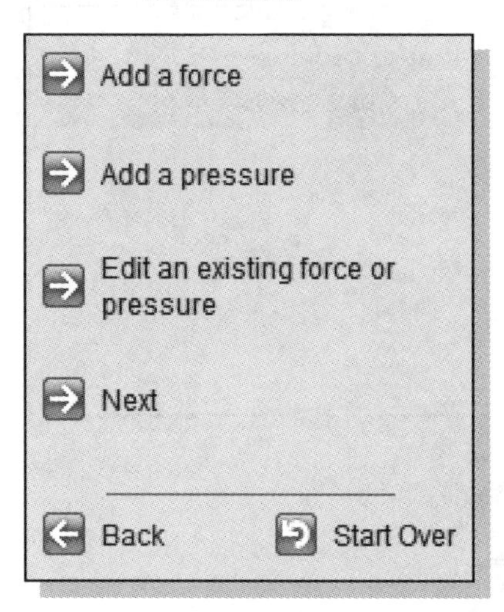

15. The Force-1 force load is created, and options are given to Add another load, or Edit the existing load. Click the **Next** button to move on to the Material step.

➢ Notice the force load appears in the *SimulationXpress Study Design Tree* to the left of the graphics area with the default name Force-1. A checkmark appears next to Loads in the top pane of the *SimulationXpress* wizard, showing that the step is complete.

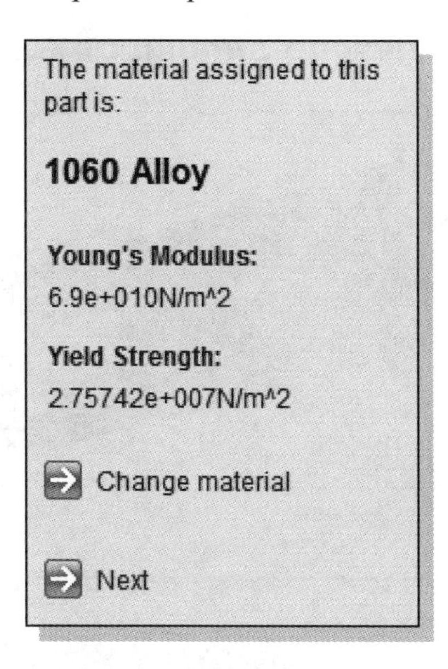

16. An introduction to the Material step is displayed. Notice the material we applied to the part is recognized by the *SimulationXpress* wizard. An option to change the material is given, but we will use the 1060 Alloy material properties. Click the **Next** button to move on to the Run step.

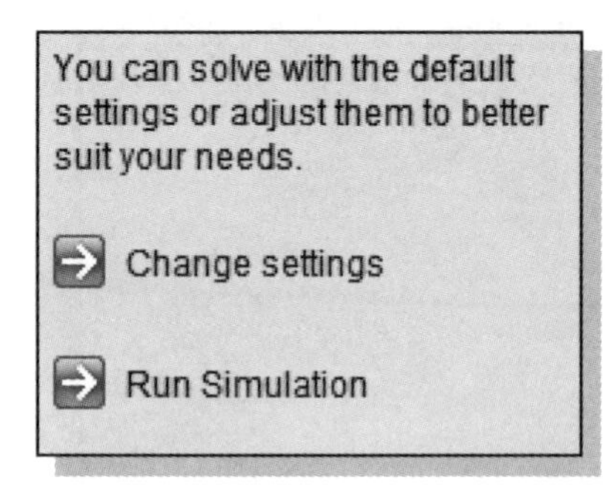

17. An introduction to the Run step is displayed with options to change the analysis settings or to run the analysis with the default settings. We will use the default settings. Click the **Run simulation** command.

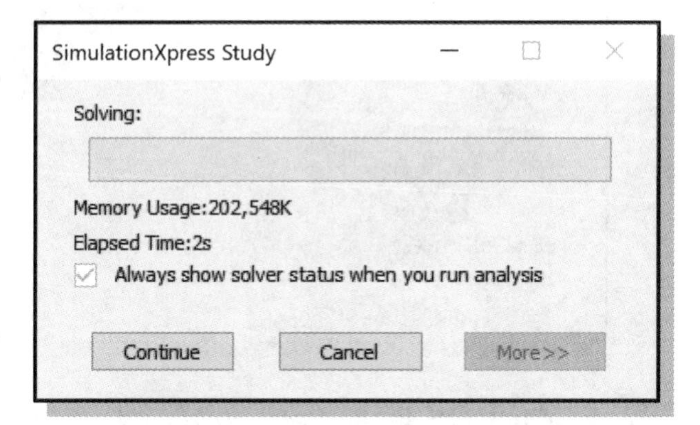

➢ While the solver is running, the *SimulationXpress Study* pop-up window appears briefly. This window can be used to display the number of Nodes, number of Elements, and number of degrees of freedom – D.O.F. – in the current discretization. For this analysis, there are 36,885 D.O.F., 12,440 nodes, and 7,140 elements. Your model may be slightly different. (**NOTE:** To see this view, click **Pause** and **More** in the pop-up window. It is not necessary to do this; simply continue to the Results step.)

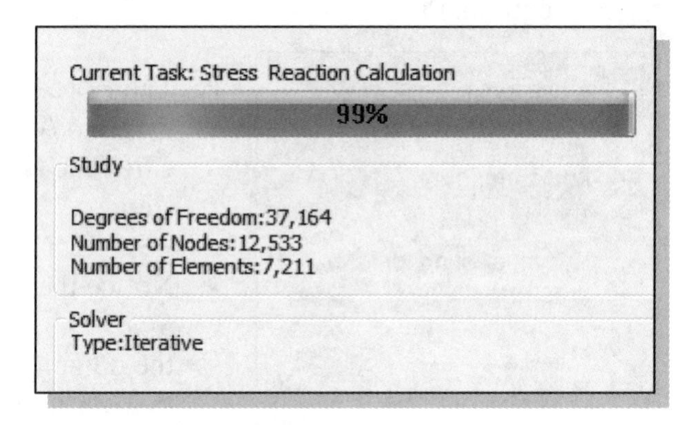

Viewing *SimulationXpress* Results

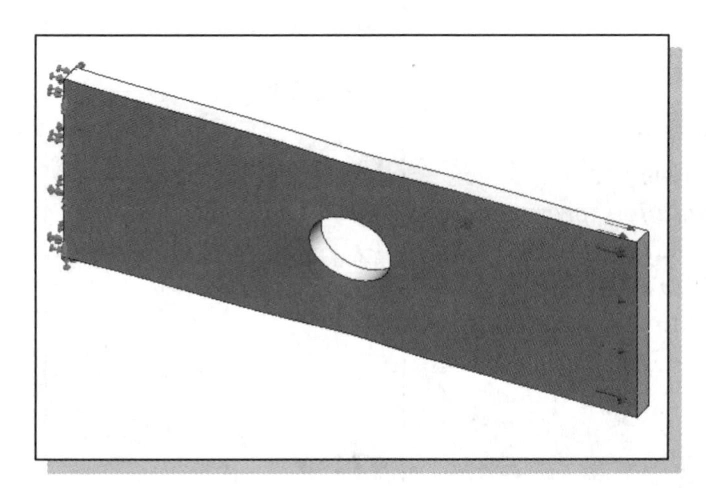

When the analysis is complete, the wizard moves on to the Results step and displays an animation of the part's response to the Fixtures and Loads entered in the previous steps. We will move on to view the stress results.

1. Verify that the deformation in the animation matches that in the figure shown above.

2. Click the **Stop animation** command in the *SimulationXpress* wizard.

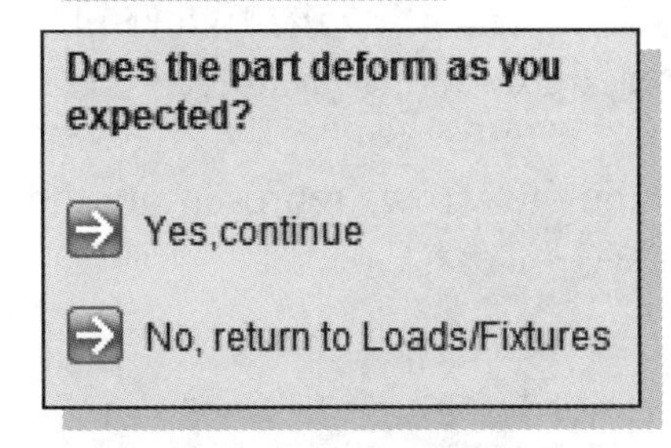

3. Below the question "*Does the part deform as you expected?*" click **Yes, continue**.

➢ We will now view the stress results.

Show von Mises stress

4. The wizard moves to the next Results screen. Click the **Show von Mises stress** command to display the stress distribution in the model.

➢ A plot of equivalent stress (von Mises stress) is generated and displayed in the graphics area.
➢ The stresses are plotted on the deformed shape of the part. **NOTE:** The actual deformation is very small. The deformation is exaggerated in the plot to better illustrate how the part deforms.
➢ A color-coded scale is used with the associated scale bar displayed at the right. Red represents the regions of highest stress.
➢ The maximum stress is 2113.2 psi (your result may be slightly different), well below the yield strength of the material (3999.3 psi), and is located at the stress concentration points at the upper and lower quadrants of the hole.
(**NOTE:** The trimetric view is used for the display shown here.)

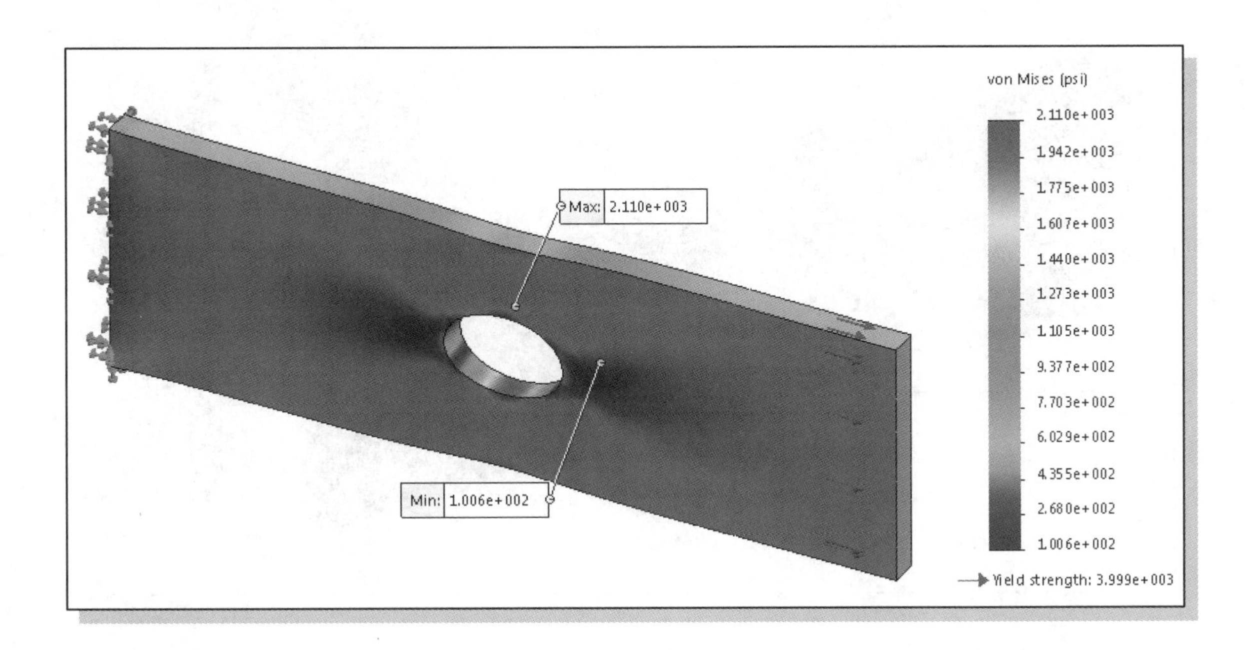

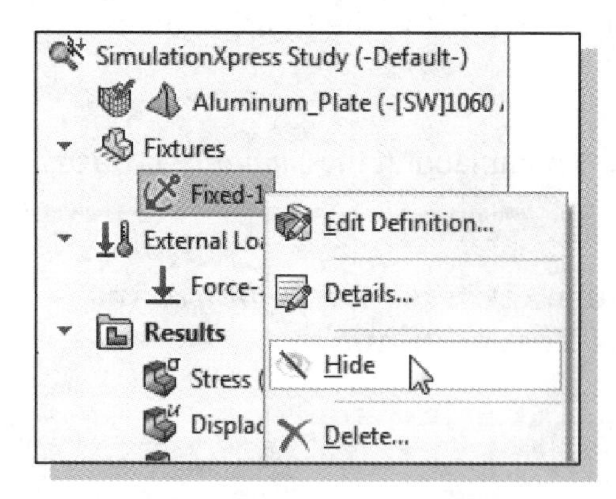

5. In the *SimulationXpress Study Tree*, **right-click** on the Fixed-1 fixture and select **Hide** from the pop-up menu.

➢ Notice the arrows representing the fixture in the graphics area are no longer visible.

6. On your own, hide the arrows representing the force load Force-1.

➢ We will now view the displacement results.

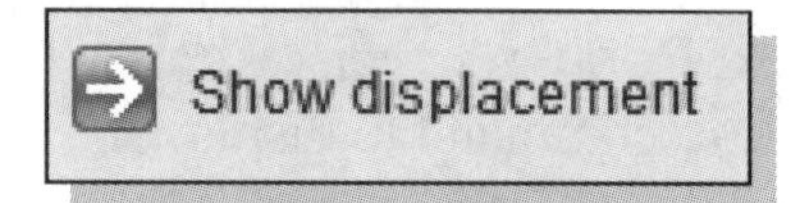

7. In the *SimulationXpress* wizard, click the **Show displacement** command to display the displacement distribution in the model.

➢ A plot of the displacement in the model is displayed. **NOTE:** The actual deformation is very small. The deformation is exaggerated in the plot to better illustrate how the part deforms.

➢ A color-coded scale is used with the associated scale bar displayed at the right. Red represents the regions of highest displacement.

➢ The maximum displacement is $4.040(10^{-4})$ in.
(**NOTE:** The trimetric view is used for the display shown here.)

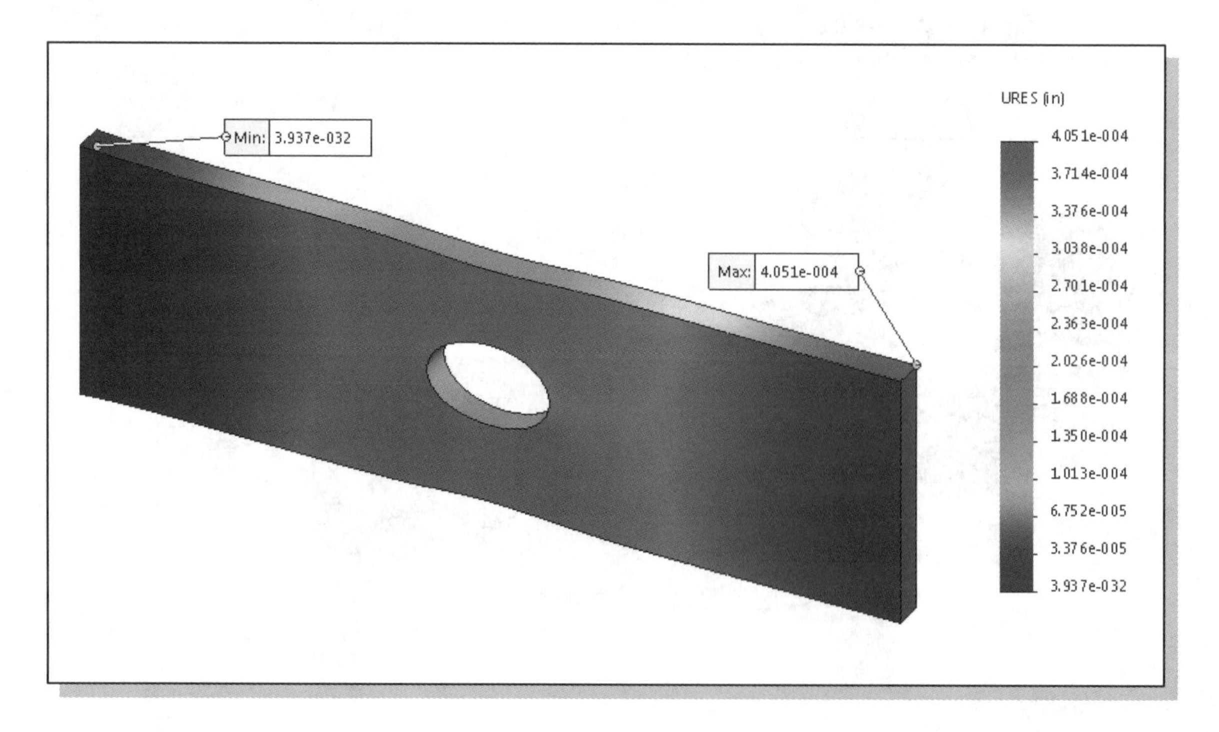

➤ We will now use the factor of safety result display option.

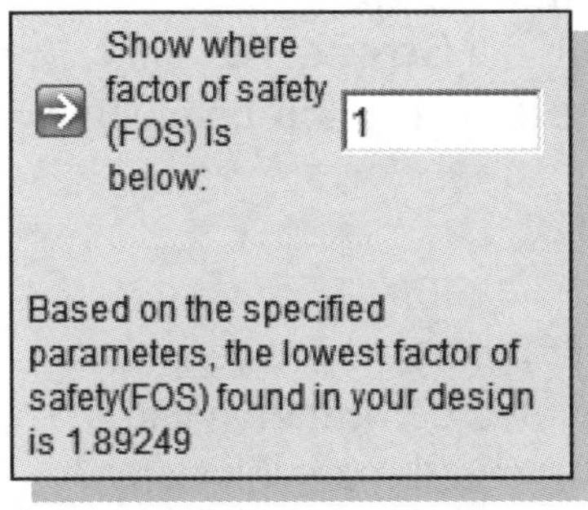

➤ The wizard indicates the lowest factor of safety (FOS) in the model, i.e., the FOS associated with the maximum stress.

➤ An option is given to select a FOS value and have *SimulationXpress* display regions which do not meet this criterion.

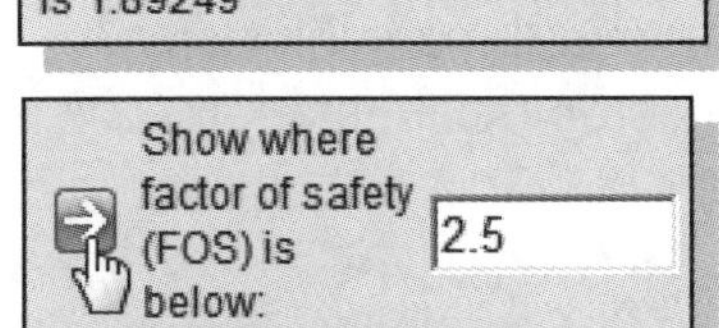

8. Enter **2.5** for the factor of safety and click the **Show FOS** button.

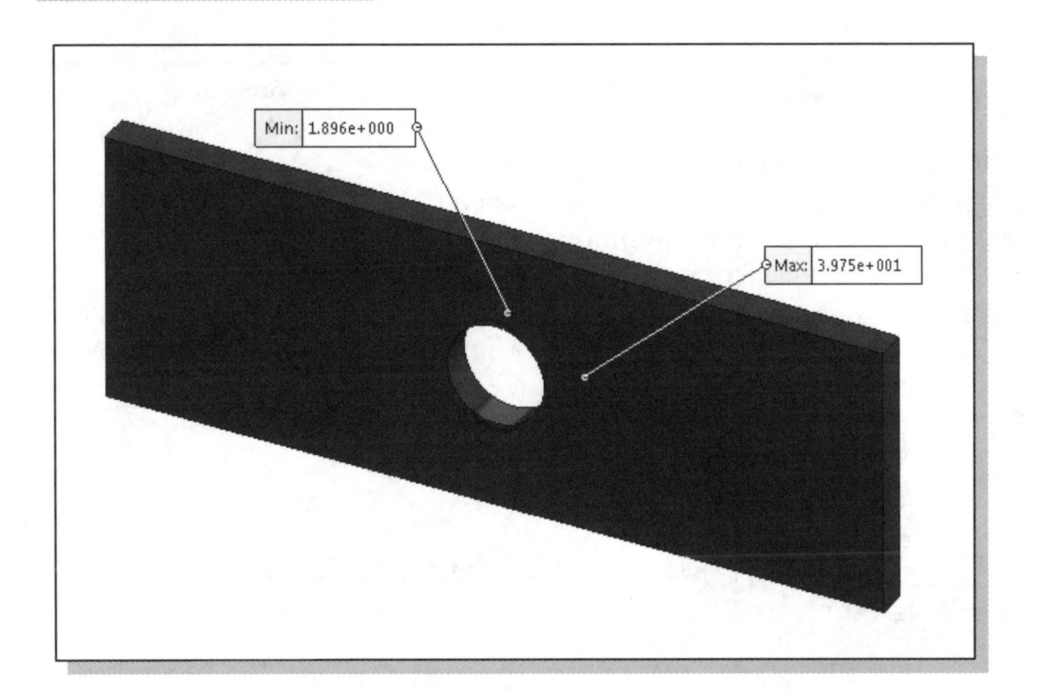

➤ For the factor of safety of 2.5, the plot displays the safe and unsafe regions in blue and red, respectively.

➤ We have used the *SimulationXpress* wizard to view the stress, displacement, and factor of safety results. An alternate method is to right-click on the desired view in the *SimulationXpress Study Design Tree* and select **Show**.

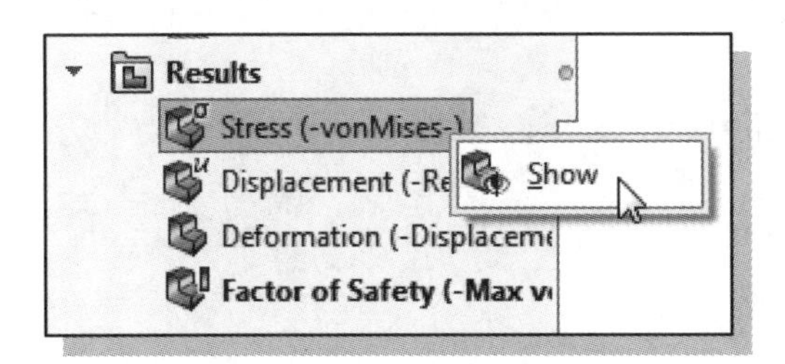

9. In the *SimulationXpress Study Tree*, expand the **Results** folder (if necessary), **right-click** on the **Stress** view, and select **Show** on the pop-up menu.

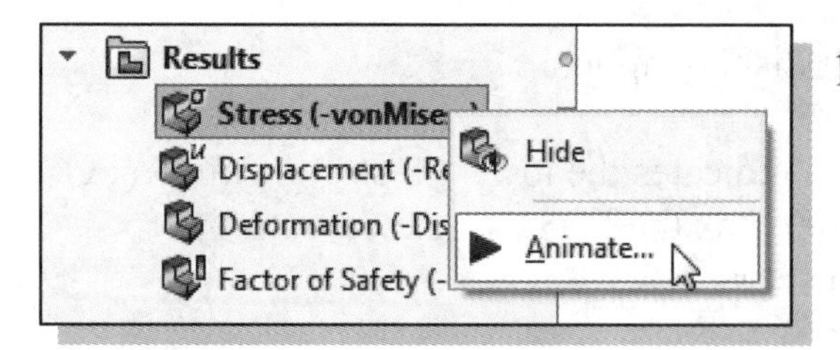

10. Notice the stress distribution is displayed. In the *SimulationXpress Study Tree*, **right-click** on the **Stress** view and select **Animate** on the pop-up menu.

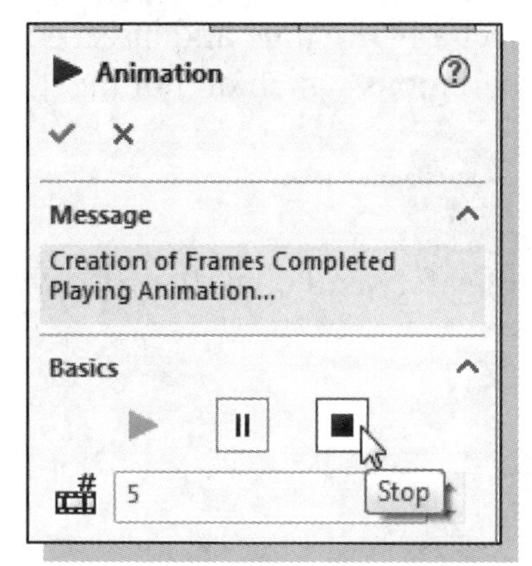

11. The *Animation PropertyManager* appears and an animation of the exaggerated deformation and associated stresses is active in the graphics area. Select the **Stop** button to stop the animation.

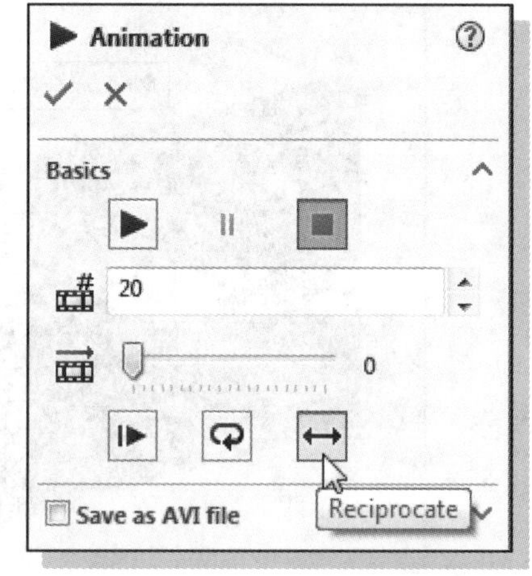

12. In the *Basics* panel of the *Animation PropertyManager* enter **20** for the *frames* option, move the slider to select the **minimum** *speed* option, and select the **Reciprocate** option.

13. Select the **Play** button to view the animation with the new settings.

14. Select the **Stop** button to stop the animation.

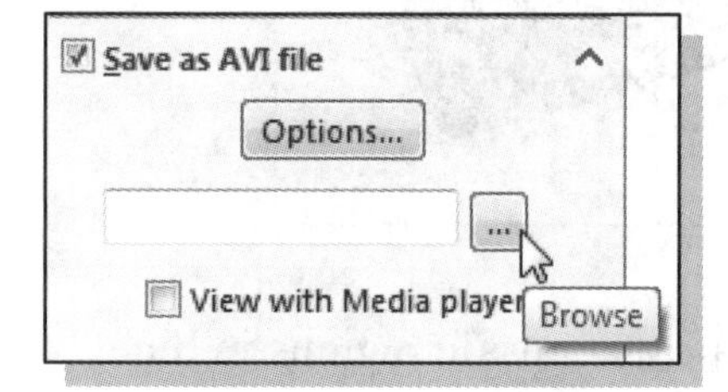

15. We will save this animation as an .AVI file. Check the **Save as AVI file** option box.

16. Select the **Browse** button to select the folder location.

17. In the *Save as* pop-up dialog box, select the folder location, enter the *File name* ***Plate_Animation_1***, and select the **Save** button.

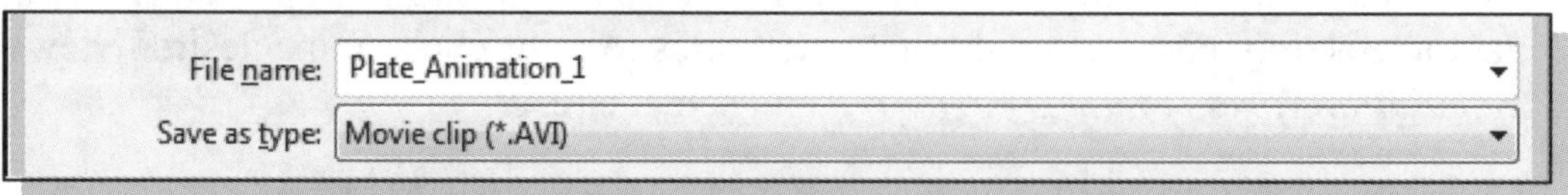

18. Select the **Play** button to view the animation and save the .AVI file.

19. On your own, view the animation by playing the .AVI file using an appropriate media playing application.

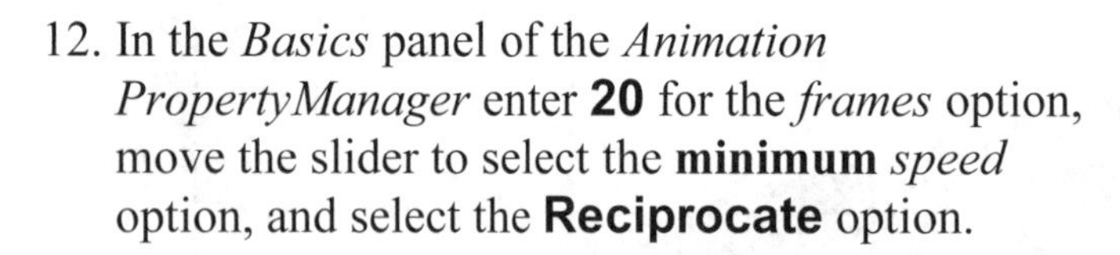

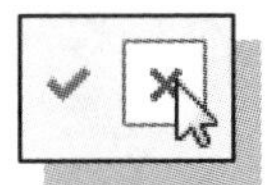

20. Select the **Cancel** button on the *PropertyManager* to exit the Animation command.

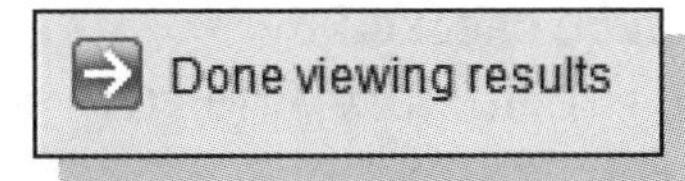

21. Click on the **Done viewing results** command.

Creating a Report and an eDrawings File

SimulationXpress includes options to create a report in Microsoft Word format which contains data related to the simulation and an *eDrawings* file for viewing the results. These options appear in the *SimulationXpress* wizard when the Done viewing results command is executed.

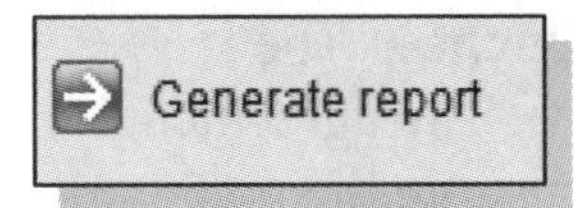

1. Select the **Generate report** option.

2. In the Report Settings window, enter text in the fields as desired (description of problem, your name, etc.) for inclusion in the report.

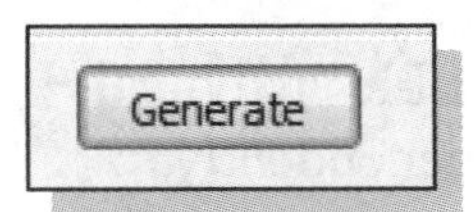

3. Click the **Generate** button to create the report file.

➢ *SimulationXpress* will create a document file and automatically open it in Microsoft Word. The file is saved in the folder selected earlier as the *Results location* for the *SimulationXpress* study.

4. On your own, read through the report. Notice that all the relevant data are recorded. From the report, note that for this simulation the number of elements is **~7,200** (see the *Mesh information – Details* table) and the maximum stress is **~2110 psi** (see the *Study Results* table). We will use this data in the next section.

5. On your own, close the report document.

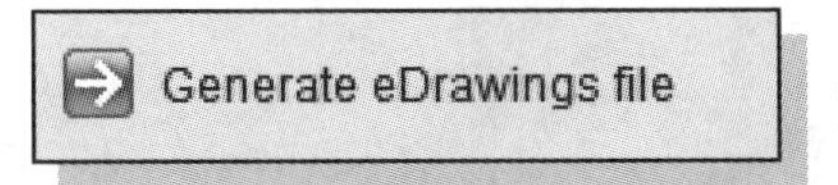

6. Select the **Generate eDrawings file** option.

7. In the *Save as* pop-up dialog box, select the **Save** button to save the file with the default name and location. (If the *Graphics Performance Check* window appears, select **OK**.)

➢ *SimulationXpress* will create the *eDrawings* file and open the *eDrawings* viewer. The equivalent stress plot is displayed by default, but the displacement, deformation, and factor of safety plots can also be viewed.

8. On your own, experiment with the *eDrawings* viewing options.

9. On your own, close the *eDrawings* window.

Accuracy of Results

The accuracy of the *SimulationXpress* results for this problem can be checked by comparing them to the analytical results presented earlier. In the Preliminary Analysis section (pages 16-5 – 16-6), we calculated the maximum stress using a stress concentration factor and obtained a value of **2180 psi**. The maximum stress obtained by finite element analysis using *SimulationXpress* (page 16-13) was **2110 psi**. The *SimulationXpress* result thus differs from the analytical result by 3.2%. (**NOTE:** The von Mises stress reported in the *SimulationXpress* result is the same as the axial stress for the uniaxial loading condition in this example.) In the Preliminary Analysis section, we estimated the maximum displacement to be 1.94E-4 inches, measured from the center of the hole to one end of the plate. The total displacement from one end of the plate to the other would therefore be **3.88E-4 inches**. The maximum displacement obtained by finite element analysis using *SimulationXpress* (page 16-14) was **4.051E-4 inches**. The *SimulationXpress* result thus differs from the analytical result by 4.4%. The agreement between the analytical results and those from *SimulationXpress* demonstrate the potential of *SimulationXpress* as a tool to perform an initial stress analysis within the SOLIDWORKS environment.

In general, FEA is applied to problems for which there is no readily available analytical solution. It is therefore necessary to assess the accuracy of the results independently. A standard method is to perform a ***convergence test***.

1. Click the **Back** button until the initial *Run* panel is displayed as shown below. (Alternately, click on Run in the top pane of the wizard.)

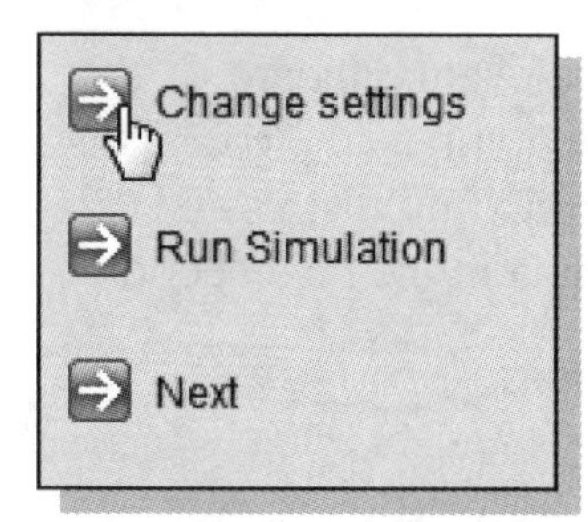

2. The message "*You can solve with the default settings or adjust them to better suit your needs*" appears. Select the **Change settings** command.

➢ An introduction to the mesh density is displayed with the option to change the mesh density settings or to continue with the default settings.

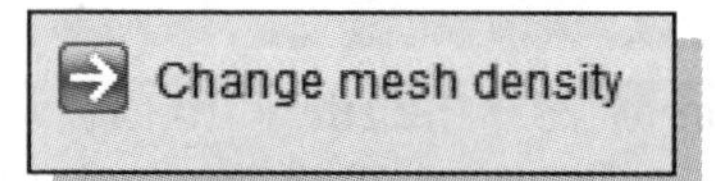

3. Click the **Change mesh density** command.

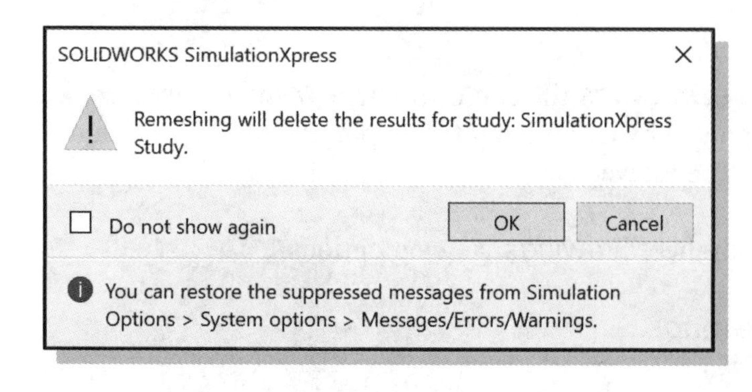

4. A pop-up window appears with the message "*Remeshing will delete the results for study*" appears. Click **OK**.

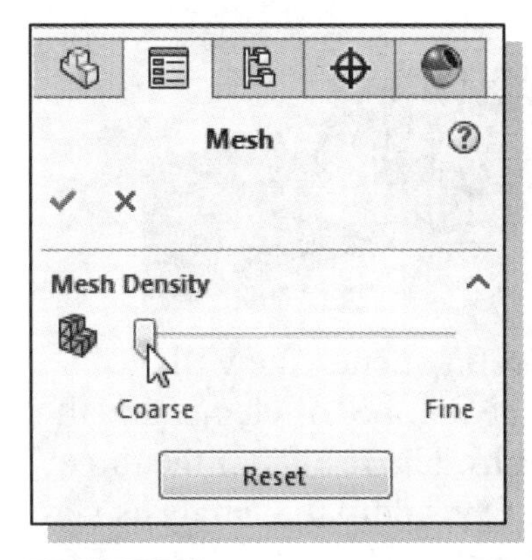

5. The *Mesh PropertyManager* appears and presents the option to adjust the fineness/coarseness of the mesh (i.e., adjust the size/number of elements used). **Move the slider** to the left-most position (this selects the coarsest mesh, i.e., the largest element size).

6. Click the **OK** button to remesh the model and exit the Mesh command. Notice a coarse mesh is displayed on the model in the graphics area.

7. Click **Next** in the *SimulationXpress* wizard to continue.

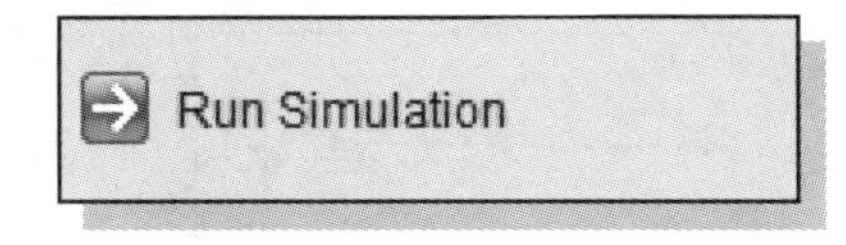

8. Click **Run Simulation** to run the analysis with the new mesh.

➤ When the analysis is complete, the wizard moves on to the Results step.

9. On your own, display the **stress distribution** results. (**NOTE:** This can be done either using the *SimulationXpress* wizard or the *SimulationXpress Study Tree*.)

➤ Notice the maximum stress is lower (~1,940 psi). The coarser mesh produced a less accurate approximation, as would be expected. To perform a convergence test, we refine the mesh in steps until the parameter of interest converges to an accurate value.

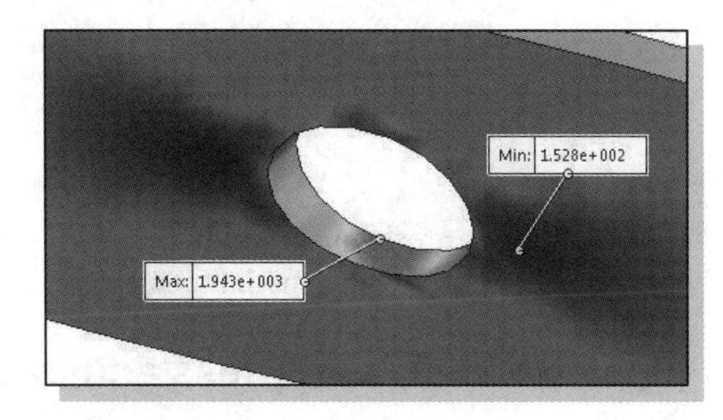

10. On your own, select **Done viewing results** and **Generate report**, and record the number of elements and maximum stress values in the report (e.g., Number of elements = 1,344; Maximum stress = 1,937.79).

11. On your own, go back to the Run panel, choose Change settings, and set the mesh to the **finest** setting as shown.

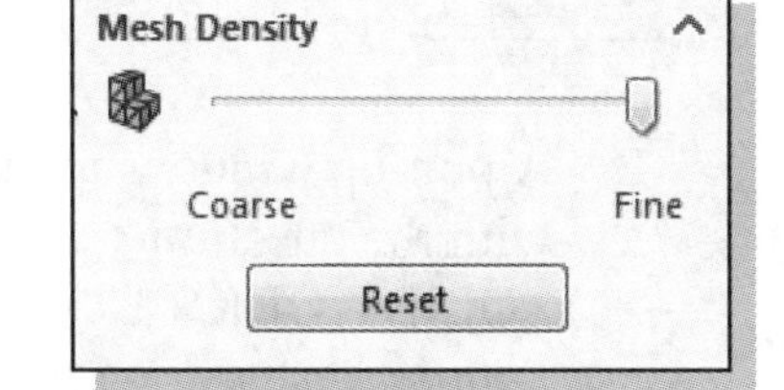

12. On your own, **Run** the analysis.

13. On your own, view the **stress distribution** and note the maximum stress.

14. On your own, generate the report and record the number of elements and maximum stress values in the report (e.g., Number of elements = 55,072; Maximum stress = 2,180.26).

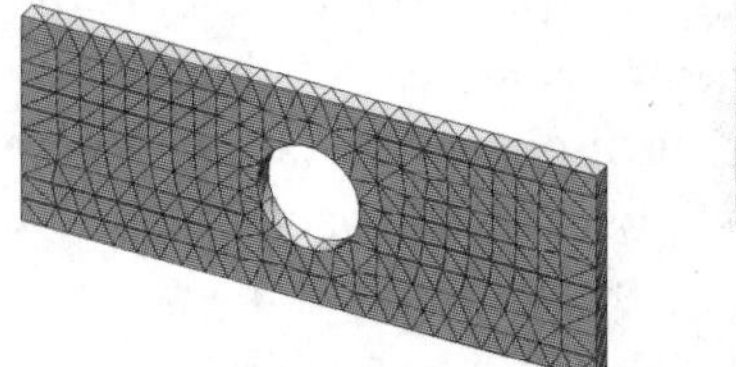

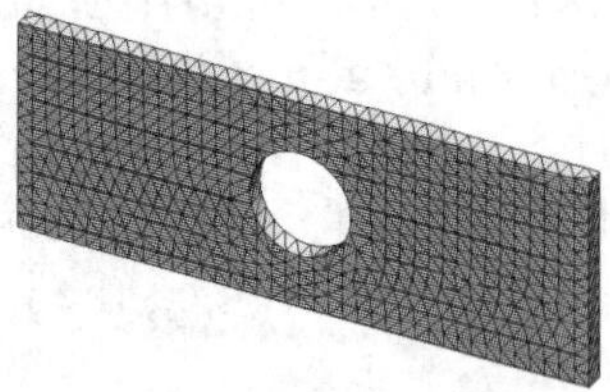

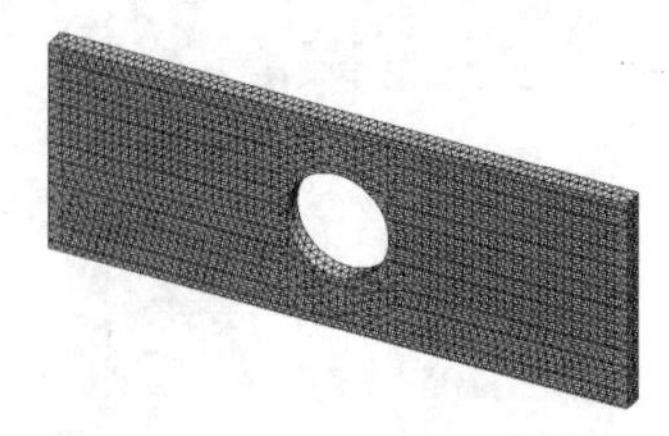

- We now have run three analyses with different levels of refinement. We will plot these to observe the effect of refinement on the result. The number of elements will be used as a parameter to describe the fineness of the mesh. The results of the three analyses are shown in tabular and graphical form below. One additional analysis was performed and included. (The numbers you obtain may vary.) The value for maximum stress can be seen to converge to the analytical result.

Elements	σ_{max} (psi)
1332	1940
7211	2110
30086	2177
55072	2180

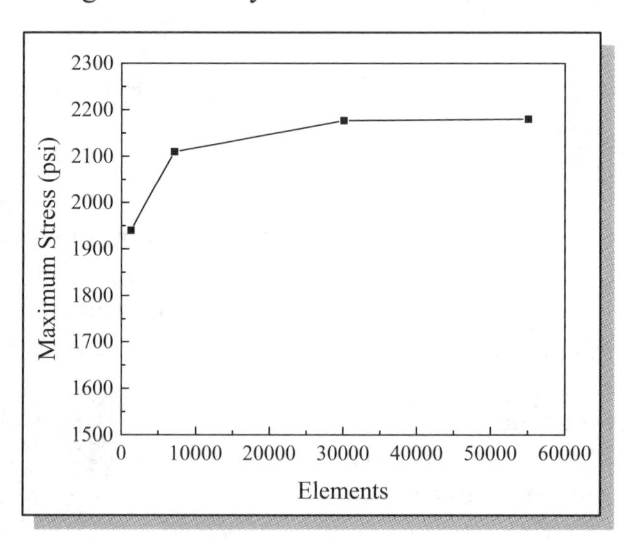

- When further refinement produces no significant change in the result, the solution is considered to have reached a 'converged' solution.

Closing SimulationXpress and Saving Results

1. Click on the **Close** button on the *SimulationXpress* wizard to exit the application.

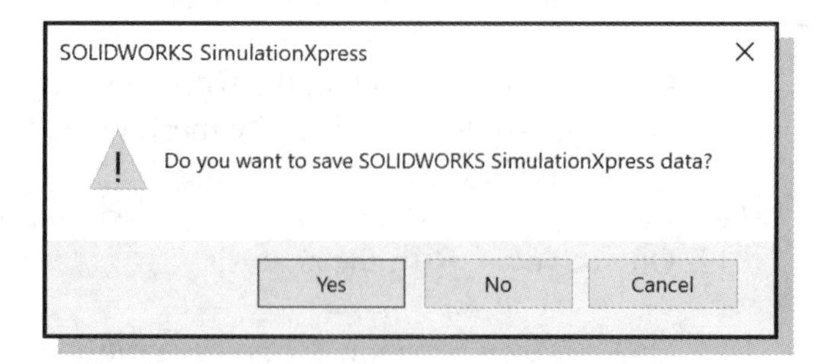

2. A pop-up window appears with the question "... *save* SOLIDWORKS *SimulationXpress data?*" Select **Yes** to save the data.

- The data is saved and, when you return to the *SimulationXpress* application, the restraints, loads, results, etc. will be loaded automatically.

3. On your own, **Save** and **Close** the *Aluminum_Plate* part file.

Questions:

1. What analysis method is used by SOLIDWORKS *SimulationXpress*?
2. Describe the steps in performing a stress analysis using *SimulationXpress*.
3. Describe two ways the material properties can be defined or edited.
4. How do we open the *SimulationXpress* wizard interface and create a *SimulationXpress* study?
5. What is meant by the term *Fixture* in *SimulationXpress*?
6. How do we control whether a load applied to a face is in the *outward* or *inward* direction?
7. Can we return to a previous step in the *SimulationXpress* study and make changes? If so, how?
8. Define degrees of freedom (DOF). How can you obtain the number of DOFs for a *SimulationXpress* analysis? How can you change the number of DOFs?
9. How is the 'lowest factor of safety' found?
10. How do we end the *SimulationXpress* study and return to the model in SOLIDWORKS?

Exercises:

1. The shaft shown below is fixed at the large end and a 25 kN force is applied to the small end. Find the maximum stress and maximum deflection in the shaft. The material is AISI 1020. (Dimensions in mm; Answer: max stress = 190MPa.)

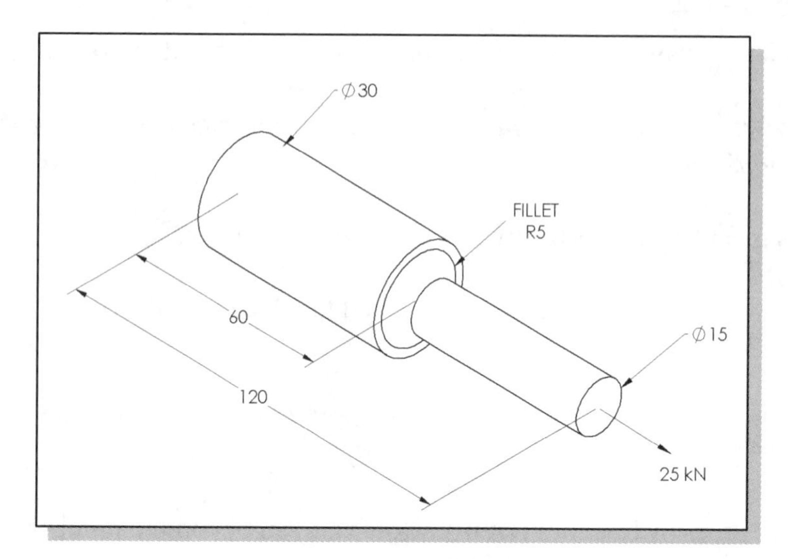

2. For the hanging bracket, the top face is fixed and a 100 psi pressure load is applied to the horizontal surface as shown. Find the max. stress and max. deflection in the bracket. The material is Alloy Steel. (Dim in inches; Answer: max stress = 39ksi.)

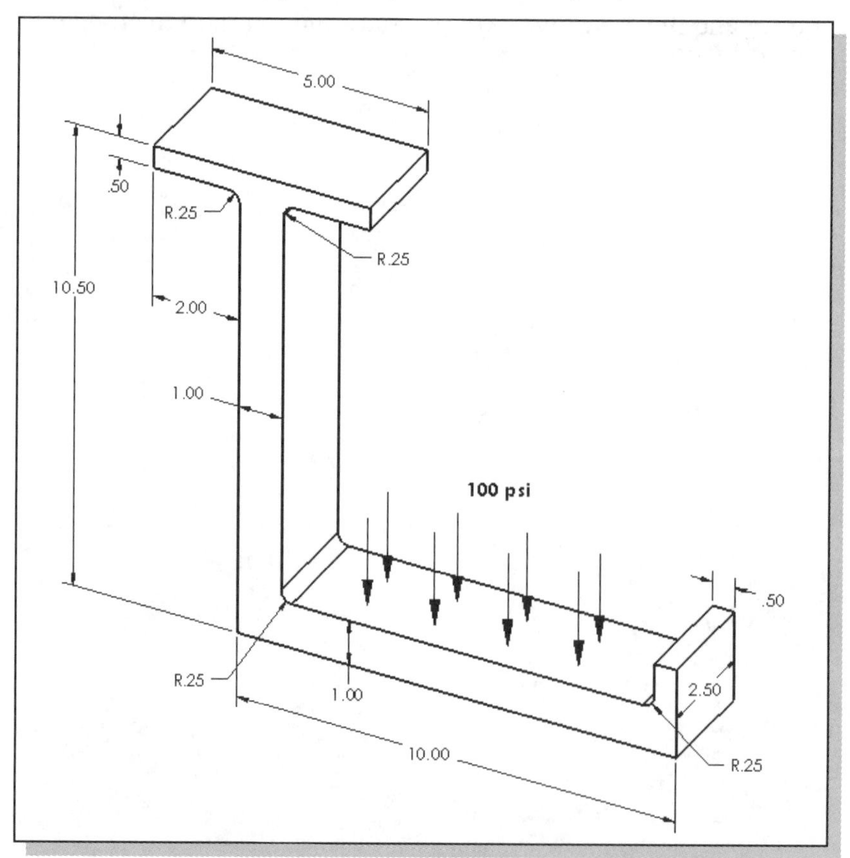

Chapter 17
CSWA Exam Preparation

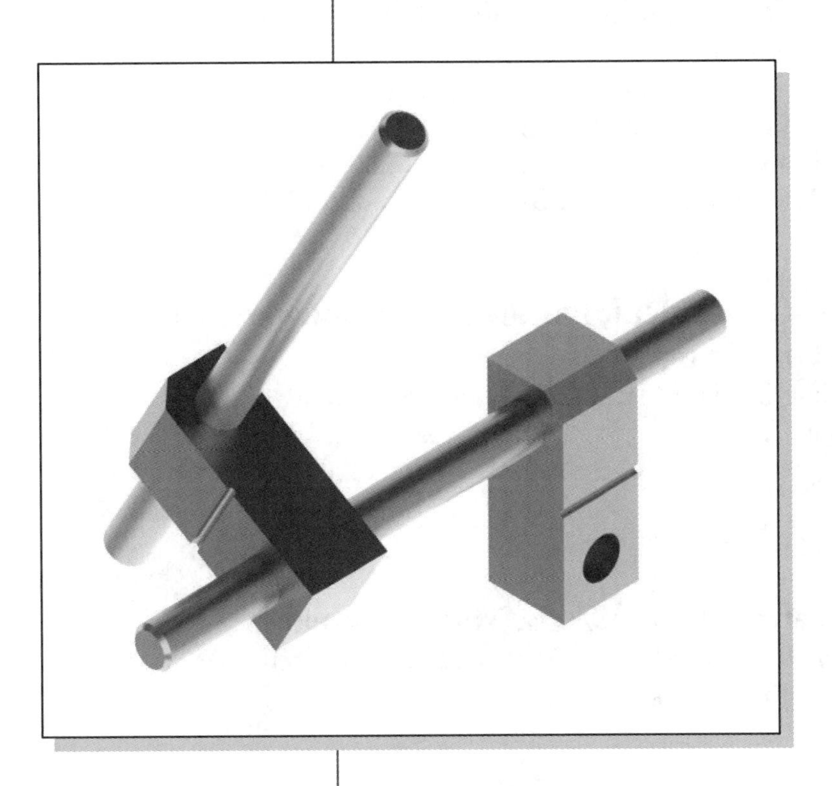

Learning Objectives

- Control Part Orientation
- Create a New View Orientation
- Use the SOLIDWORKS Mass Properties Tool
- Obtain the Mass and Center of Mass of a Part
- Obtain the Mass and Center of Mass of an Assembly
- Use Distance and Angle Mates
- Create and Use Reference Axes

Certified SOLIDWORKS Associate Exam Objectives Coverage

Mass Properties

Objectives: Obtaining Mass Properties for Parts and Assemblies.

Standard Mates – Coincident, Parallel, Perpendicular, Tangent, Concentric, Distance, Angle

Objectives: Applying Standard Mates to Constrain Assemblies.

Reference Geometry – Planes, Axis, Mate References

Objectives: Creating Reference Planes, Axes, and Mate References.

Tips about Taking the Certified SOLIDWORKS Associate (CSWA) Examination

1. **Study:** The first step to maximize your potential on an exam is to sufficiently prepare for it. You need to be familiar with the SOLIDWORKS package, and this can only be achieved by doing designs and exploring the different commands available. The Certified SOLIDWORKS Associate (CSWA) exam is designed to measure your familiarity with the SOLIDWORKS software. You must be able to perform the given task and answer the exam questions correctly and quickly.
2. **Make Notes**: Take notes of what you learn either while attending classroom sessions or going through study material. Use these notes as a review guide before taking the actual test.
3. **Time Management**: The examination has a time limit. Manage the time you spend on each question. Always remember you do not need to score 100% to pass the exam. Also keep in mind that some questions are weighed more heavily and may take more time to answer. You can flip back and forth to view different problems during the test time by using the arrow buttons. If you encounter a question you cannot answer in a reasonable amount of time, use the *Save As* feature in SOLIDWORKS to save a copy of the file, and move on to the next question. You can return to any question and enter or change the answer as long as you do not hit the [**I am done**] button.
4. **Use the SOLIDWORKS *Help System***: If you get confused and can't think of the answer, remember the SOLIDWORKS *Help System* is a great tool to confirm your considerations. In preparing for the exam, familiarize yourself with the help utility organization (e.g., Contents, Index, Search options).
5. **Use Internet Search**: Use of an internet search utility is allowed during the test. If a test question requires general knowledge, for example definitions of engineering or drafting concepts (stress, yield strength, auxiliary view, etc.), remember the internet is available as a tool to assist in your considerations.
6. **Use Common Sense**: If you are unable to get the correct answer and unable to eliminate all distracters, then you need to select the best answer from the remaining selections. This may be a task of selecting the best answer from amongst several correct answers, or it may be selecting the least incorrect answer from amongst several poor answers.
7. **Be Cautious and Don't Act in Haste:** Devote some time to ponder and think of the correct answer. Ensure that you interpret all the options correctly before selecting from available choices. Don't go into panic mode while taking a test. Use the ***Arrow Buttons*** to review each question. When you are confident that you have answered all questions, end the examination using the [**I am done**] button to submit your answers for scoring. You will receive a score report once you have submitted your answers.
8. **Relax before exam:** In order to avoid last minute stress, make sure that you arrive 10 to 15 minutes early and relax before taking the exam.

Introduction

In this lesson, we will create and examine part and assembly files in a manner consistent with the format of the CSWA exam. This is a multiple choice exam. You are asked to build complex models and assemblies and your work is assessed in a multiple choice format by asking the value of parameters such as the mass or center of mass. These can be retrieved using the SOLIDWORKS **Mass Properties** tool. In answering questions related to the location of the center of mass, it is important to ensure that the origin and coordinate system axes are consistent between your model and the exam problem as posed.

We will first create a part based on a dimensioned drawing with the origin and coordinate axes annotated. We will create the part, apply the correct material properties, and retrieve the mass and center of mass data. In building the model, we will ensure that it is oriented relative to the default coordinate system in the manner indicated in the problem. While the part is relatively simple, the procedure – taking care to locate the part, applying material properties, retrieving mass properties – is the same as should be followed for many CSWA exam questions. The part requires the creation of an inclined **Reference Plane** to use in creating a sketch.

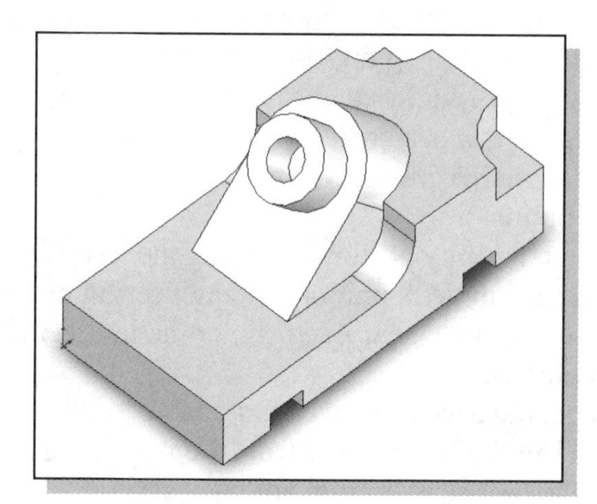

We will then create an assembly using multiple copies of two mating parts. The assembly will require the use of **Distance** and **Angle** mates, based on the dimensions given in the problem. In this case, we will not orient the parts or assembly based on the problem definition. Instead, after the assembly has been completed, we will create **Reference Axes** to match those in the problem statement and use these reference axes when retrieving the center of mass.

The Part Problem

We will model the part shown below and use the **Mass Properties** tool to determine the mass and center of mass of the part. The problem is presented below in a manner similar to that encountered on the CSWA exam.

Build the part shown below.
Unit system: MMGS (millimeter, gram, second)
Part origin: As shown.
Part material: Ductile iron Density: 0.0071 g/mm^3

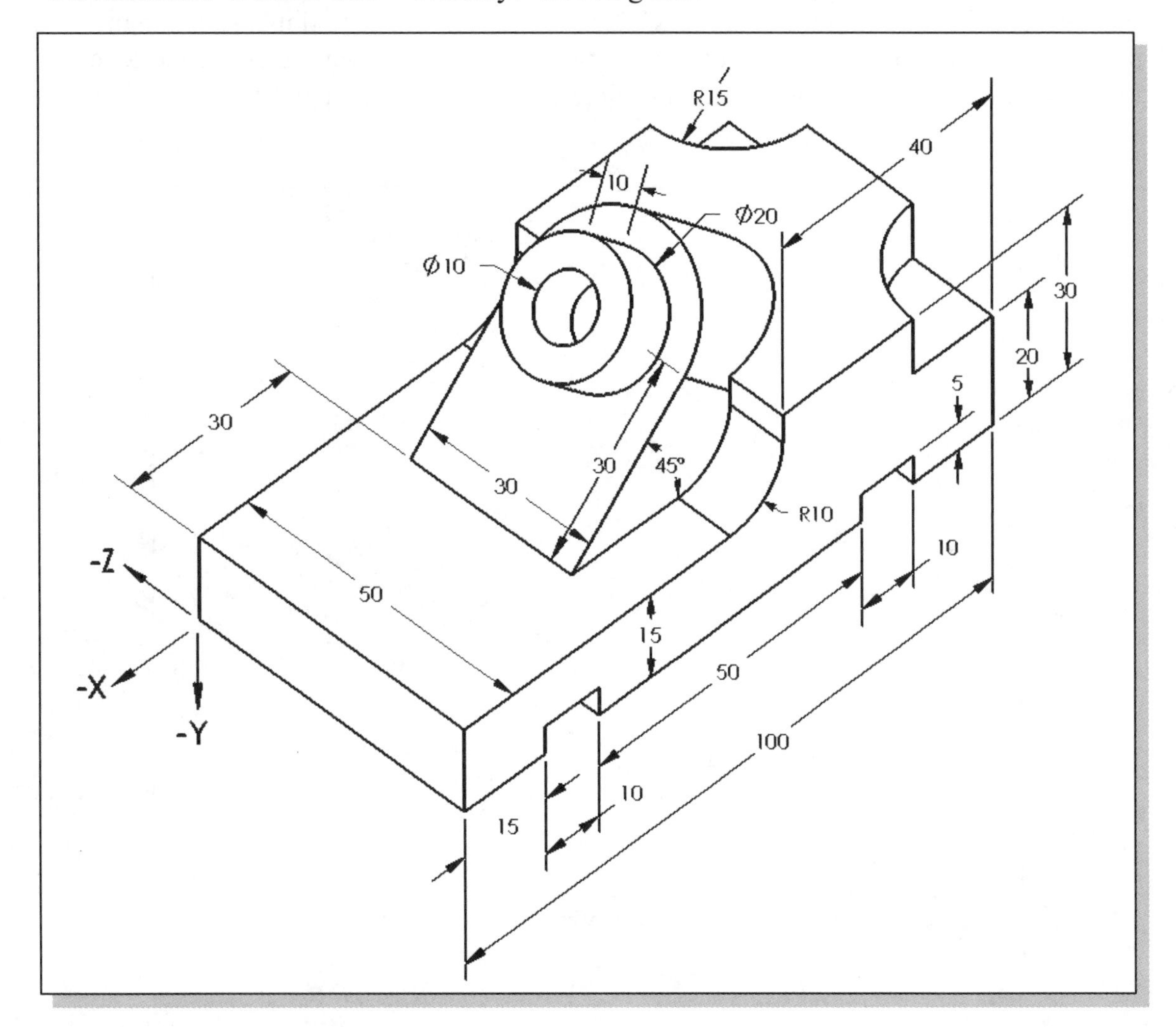

What is the overall mass of the part in grams?
What is the center of mass of the part?

Before starting to build the model, it is important to select a strategy for controlling the origin and orientation of the X, Y, and Z axes since the center of mass must be defined relative to a known reference system. In this case the axis directions are defined in the drawing. We can either (1) build the part aligned to the default or global coordinate system as specified in the figure, or (2) build the part independent of this constraint and, after the part is complete, create a reference coordinate system with axes aligned with the part as shown in the figure. We will use the first strategy for the part problem. The second strategy will be used in the assembly problem to be done later in the chapter.

Strategy for Aligning the Part to the Default Axis System

The first feature to be created will be the base feature of the part. We must determine which plane to use to define the first sketch in order to align the part correctly. In the dimensioned drawing defining the part, we can see that the base rests on the X-Z plane (i.e., the Top Plane) with the positive X-axis aligned with the long dimension and the positive Z-axis aligned with the short dimension. We therefore want to create the base feature as shown below, with the origin located at the * and the X and Z axes aligned as shown.

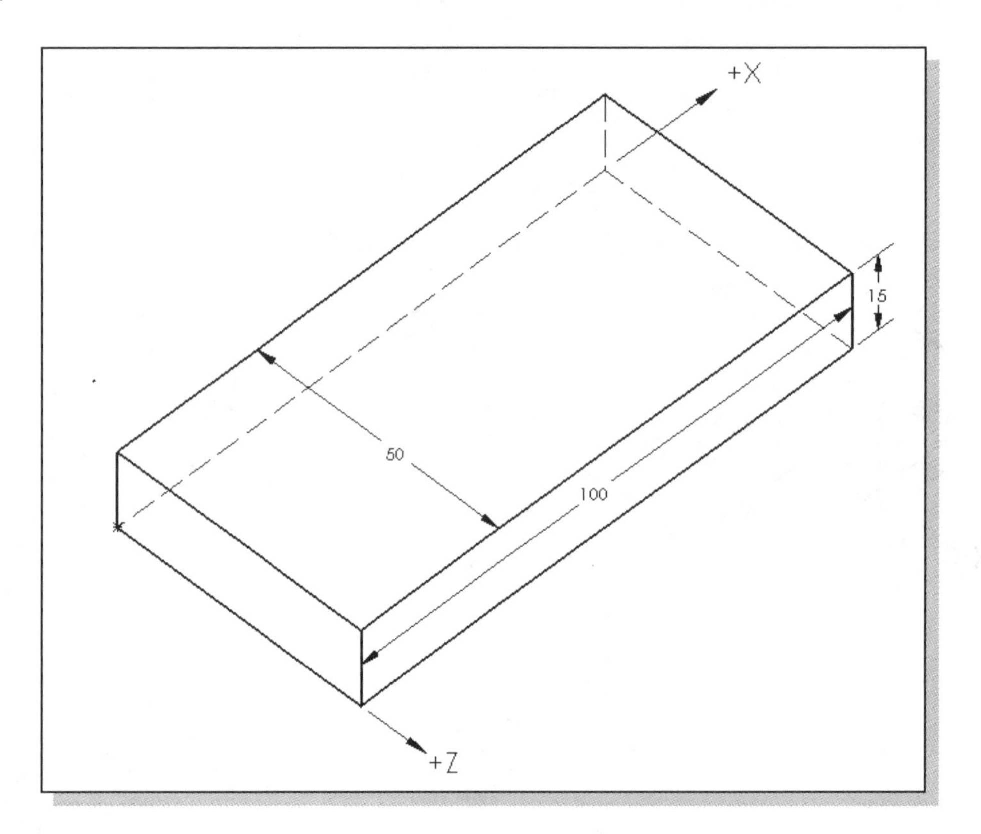

Creating the Base Feature

1. On your own, start a new part file using the default **Part** template in the SOLIDWORKS Templates folder.

2. Select the **Options** icon from the *Menu Bar* to open the *Options* dialog box. Select the **Document Properties** tab.

3. Select **ANSI** in the pull-down selection window under the *Overall drafting standard* panel as shown.

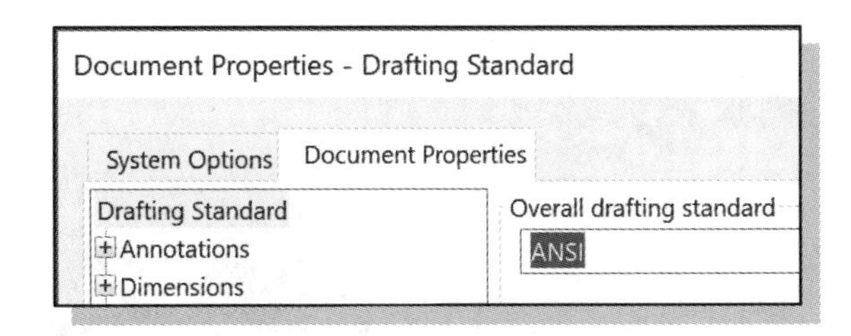

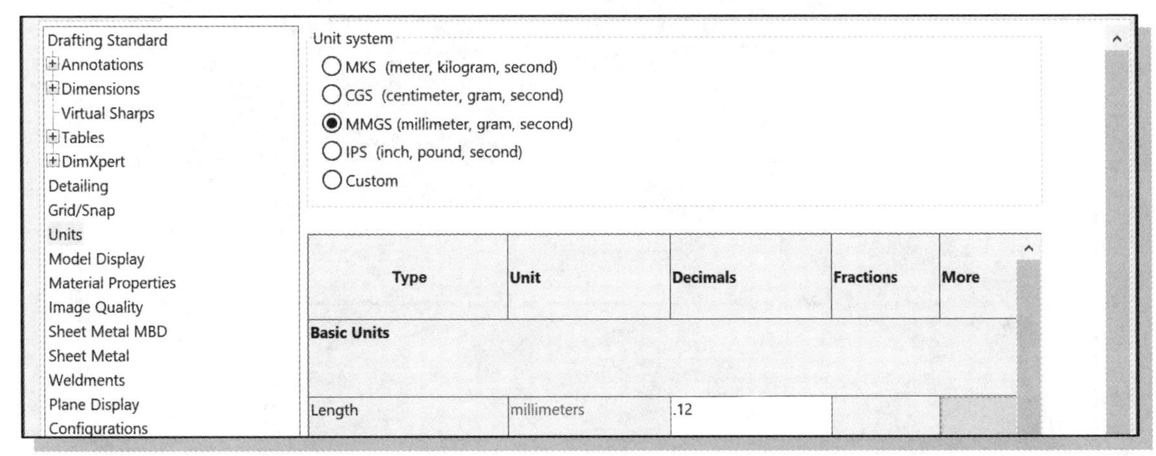

4. Click on the **Units** option. We will use the default **MMGS (millimeter, gram, second)** unit system as instructed in the problem definition.

5. Select **.12** in the *Decimals* spin box for the *Length units* as shown to define the degree of accuracy with which the units will be displayed to 2 decimal places.

6. Select the **Sketch** button on the *Sketch* toolbar to create a new sketch.

7. Select the **Top** (XZ) **Plane** as the sketch plane for the new sketch.

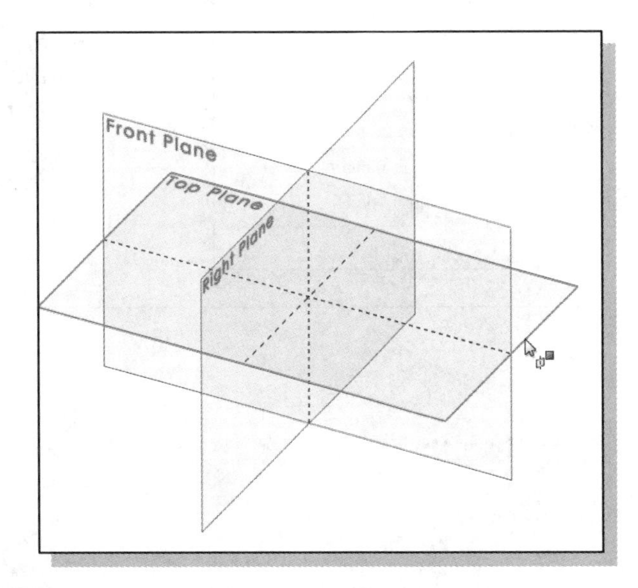

8. On your own, draw a **rectangle** with one corner coincident with the origin, the long edge aligned with the positive X-axis, and the short edge aligned with the positive Z-axis as shown below.

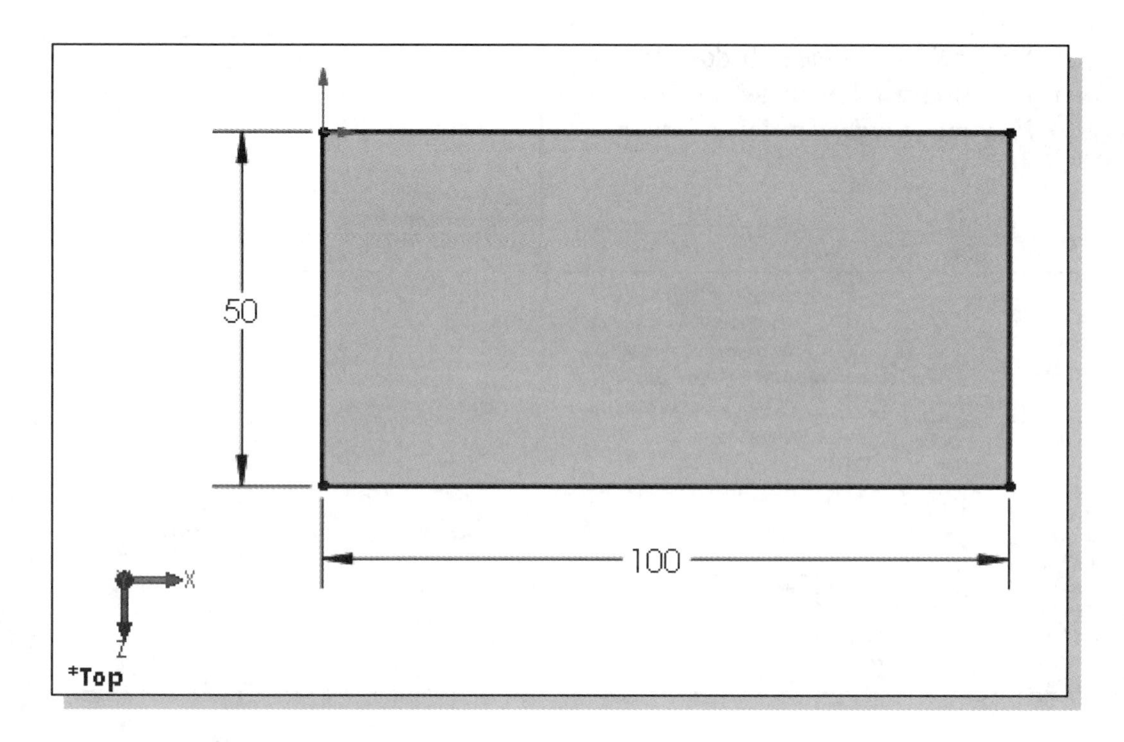

9. Apply the **Smart Dimensions** as shown above.

10. On your own, exit the sketch and create a **Boss Extrude** feature with a height of **15 mm**.

Creating a New View Orientation

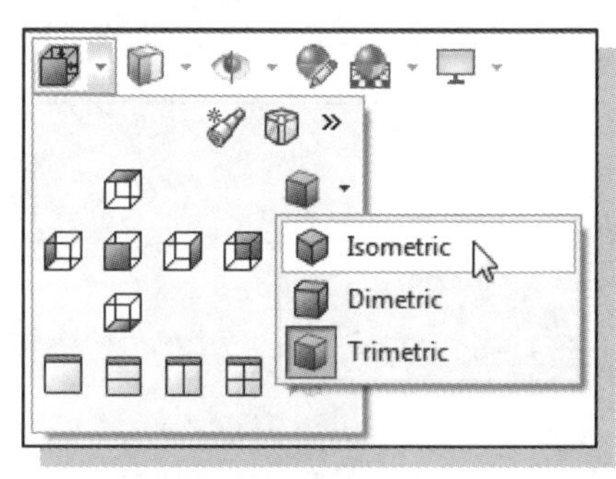

1. Left-click on the **View Orientation** icon on the *Heads-up View* toolbar to reveal the *View Orientation* pull-down menu. Select the **Isometric** command.

2. If necessary, use the *Hide/Show* pull-down menu on the *Heads-up View* toolbar to turn *ON* the origin's visibility.

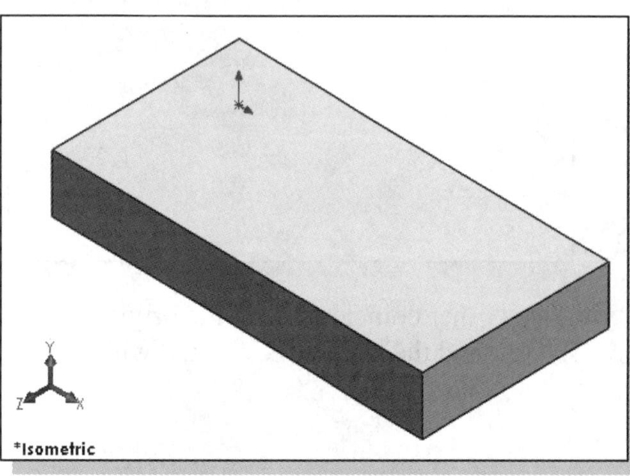

➤ Notice the axes are aligned correctly relative to the part, but the default isometric view does not correspond to the view orientation used in defining the problem. We will create a new view orientation which better matches the view used in defining the problem (page 17-5).

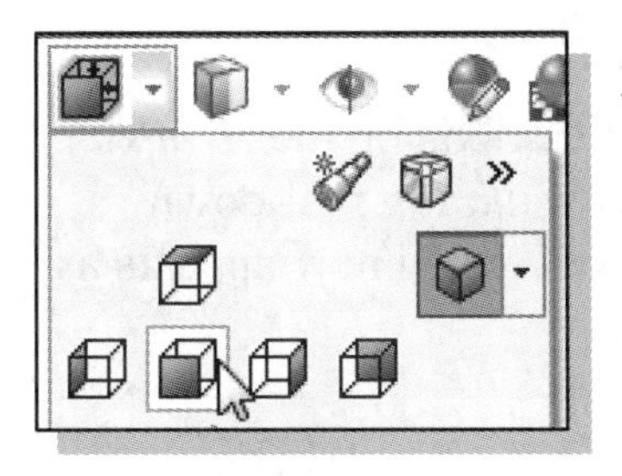

3. Left-click on the **View Orientation** icon on the *Heads-up View* toolbar to reveal the *View Orientation* pull-down menu. Select the **Front** command.

4. Click the right arrow button [→] three times to rotate the view 45°.

5. Click the down arrow button [↓] three times to rotate the view 45°.

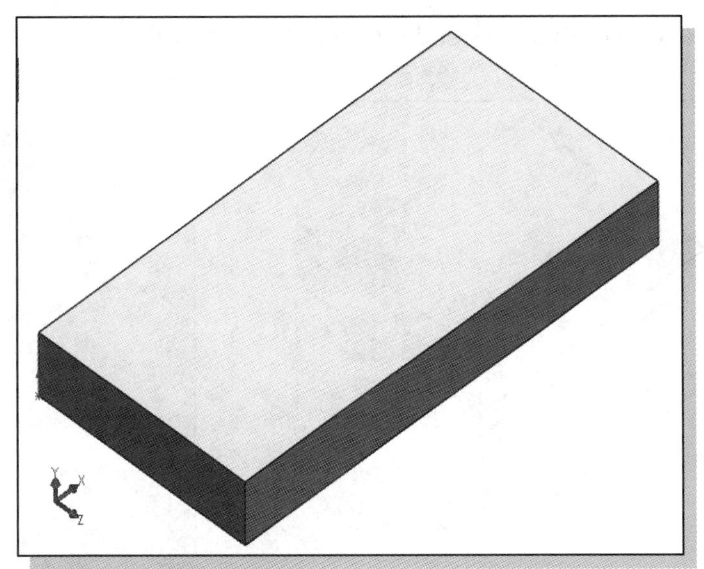

➢ The view should appear as shown here. This view orientation corresponds to the one in the problem definition. We will want to return to this view as we continue working on the part. SOLIDWORKS allows the user to define a new view orientation for this purpose.

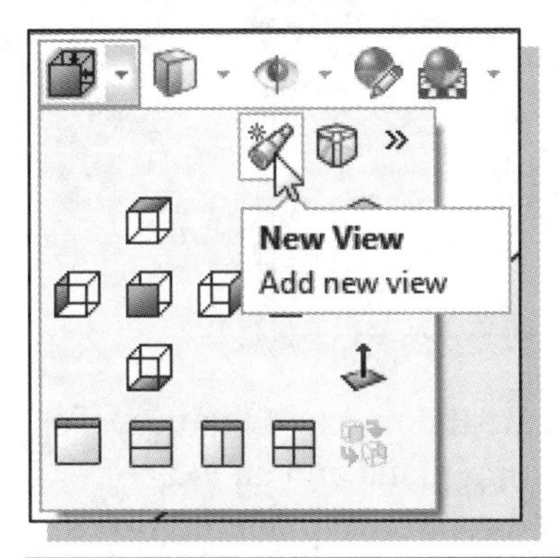

6. Left-click on the **View Orientation** icon on the *Heads-up View* toolbar to reveal the *View Orientation* pull-down menu. Select the **New View** command.

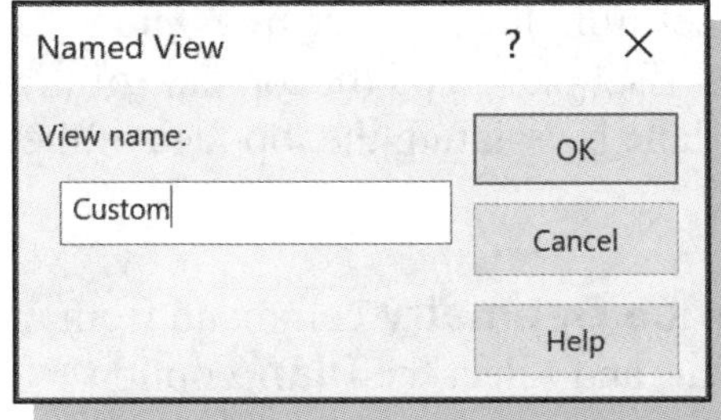

7. Enter **Custom** as the name for the new view and click **OK**.

8. Click anywhere in the graphics area to close the *Orientation Dialog Box*.

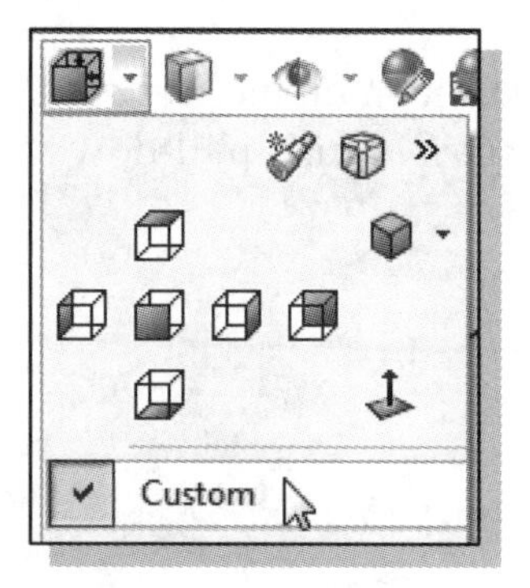

9. Left-click on the **View Orientation** icon on the *Heads-up View* toolbar to reveal the *View Orientation* pull-down menu. Notice the *Custom* view we created now appears as an option.

Completing the Part

1. On your own, create the second extruded boss and the fillet feature as shown below.

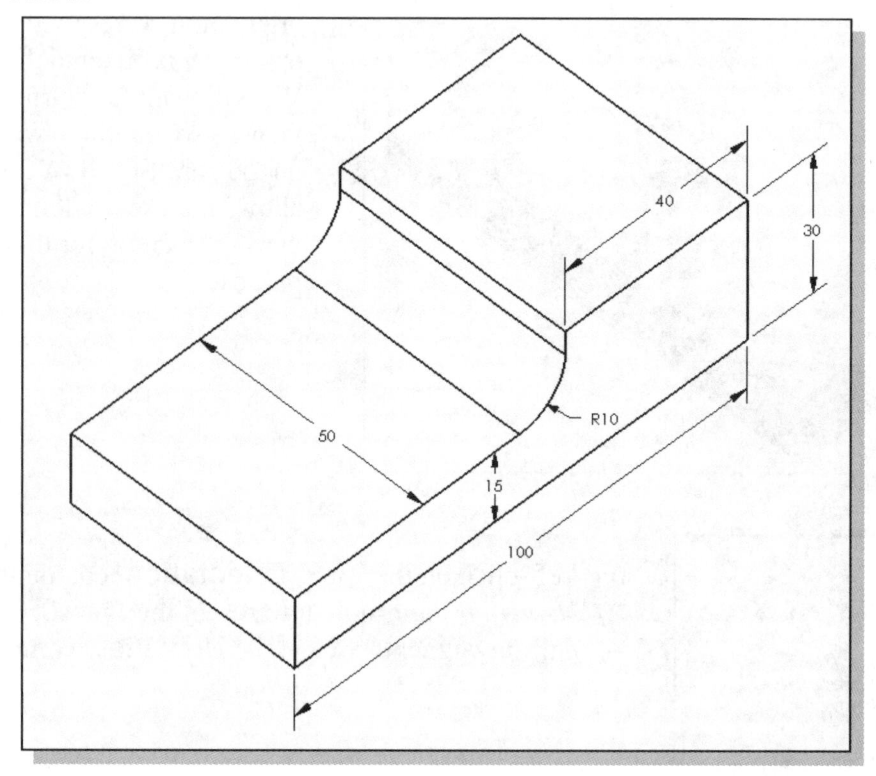

➢ We will now create the reference geometry needed to create the inclined feature. To create an inclined reference plane, we must select an existing plane and an axis of rotation. The strategy we will use for creating the inclined reference plane involves three steps: (1) create a vertical reference plane offset from the left face by 30 mm; (2) create a reference axis at the intersection of the offset plane and the top face of the base feature; and (3) create the inclined reference plane by rotating the top face of the base feature about the reference axis.

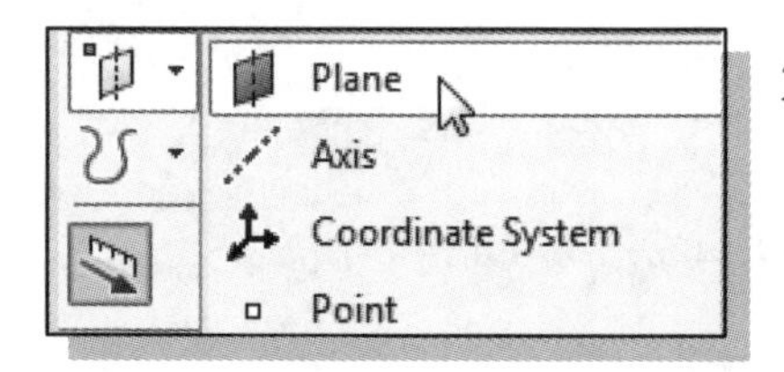

2. Select the **Reference Geometry** command from the *Features* toolbar, and select the **Plane** option from the pull-down menu.

3. Move the cursor over the left face of the base feature as shown below. Click once with the **left-mouse-button** to select the face.

4. Notice the face, called **Face<1>**, appears in the *First Reference* panel of the *Plane PropertyManager*, and the **Offset Distance** option is automatically selected.

5. Enter **30 mm** as the offset distance.

6. Check the **Flip** option box.

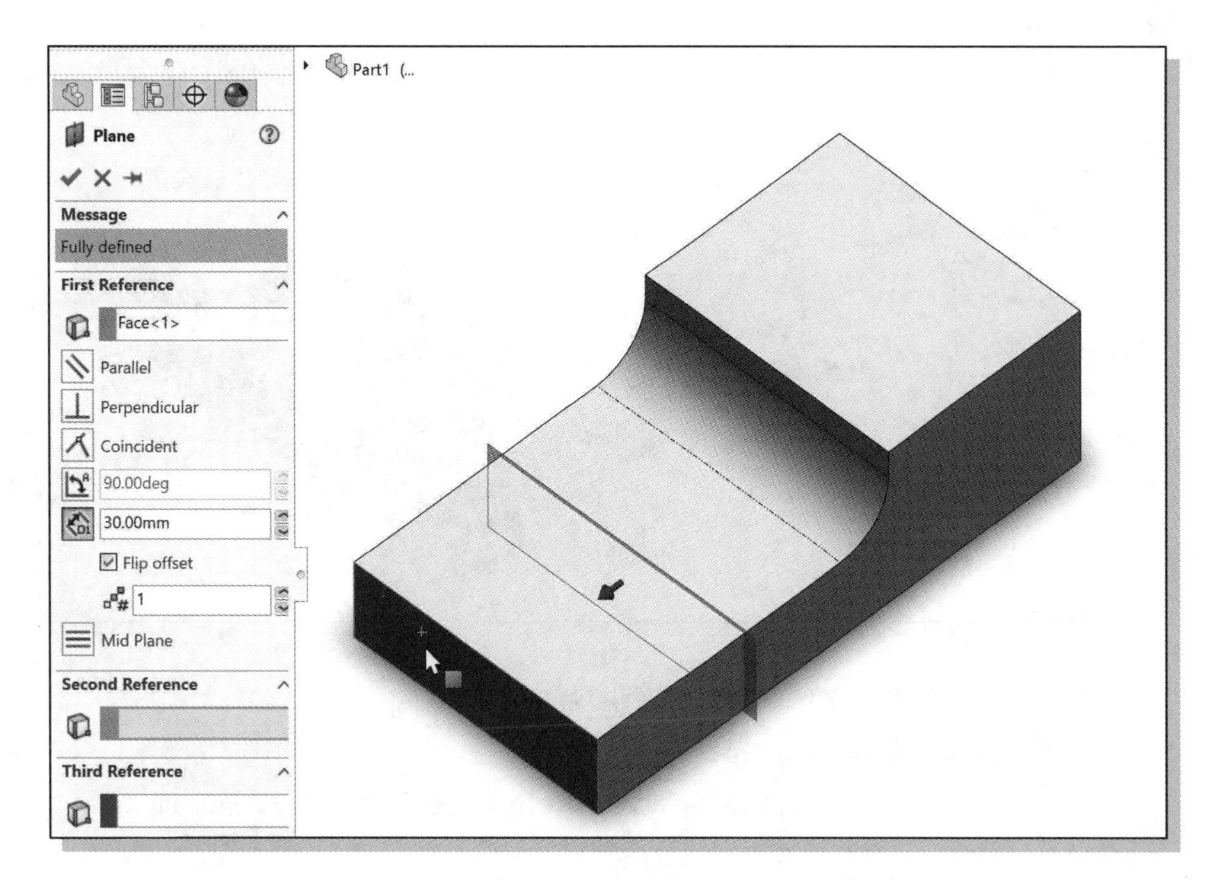

7. Click **OK** in the *PropertyManager* to accept the settings and create the new reference plane.

8. Press the **[Esc]** key to unselect the new plane.

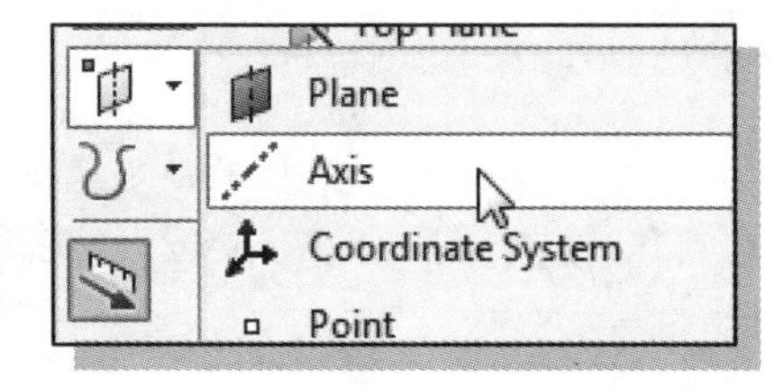

9. In the *Features* toolbar, select the **Reference Geometry** command by left-clicking the icon.

10. In the pull-down option menu of the Reference Geometry command, select the **Axis** option.

11. In the *Selections* panel of the *Axis PropertyManager*, select the **Two Planes** option. (We will define the reference axis as the intersection of the offset reference plane and top face of the base feature.)

12. Select the offset reference plane **Plane1** in the *Design Tree* (expand the *Design Tree* in the graphics area if necessary) and the **top face of the base feature** in the graphics area.

13. Click **OK** in the *PropertyManager* to create the new reference axis.

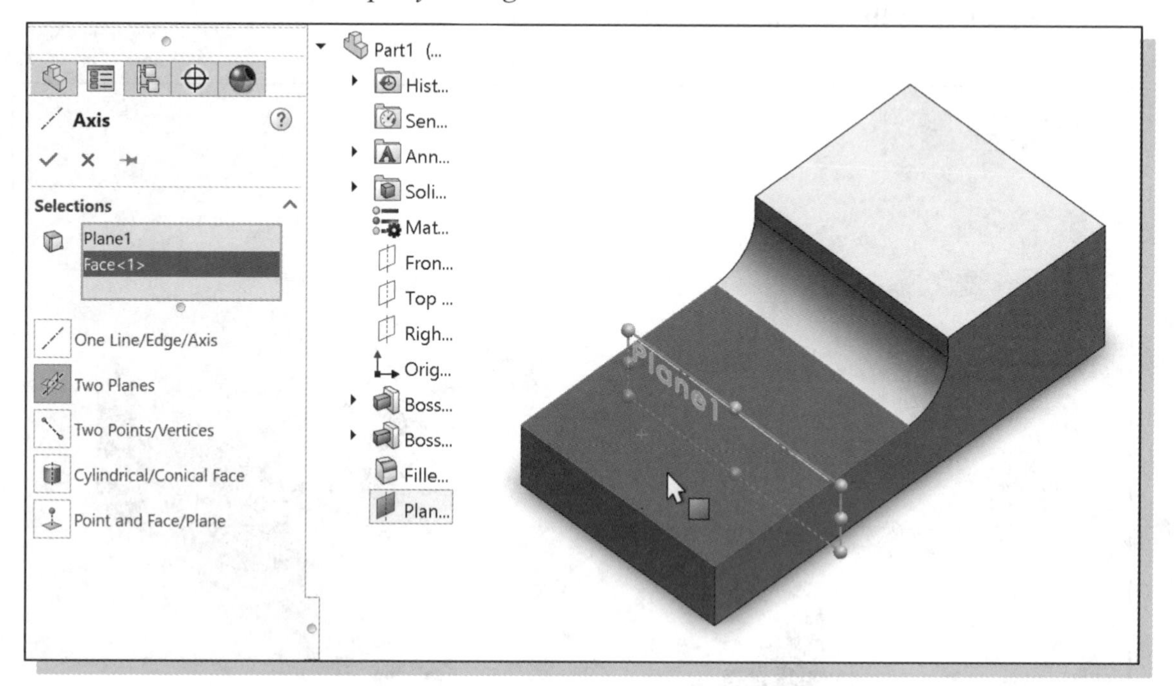

14. Press the [**Esc**] key to ensure no features are selected.

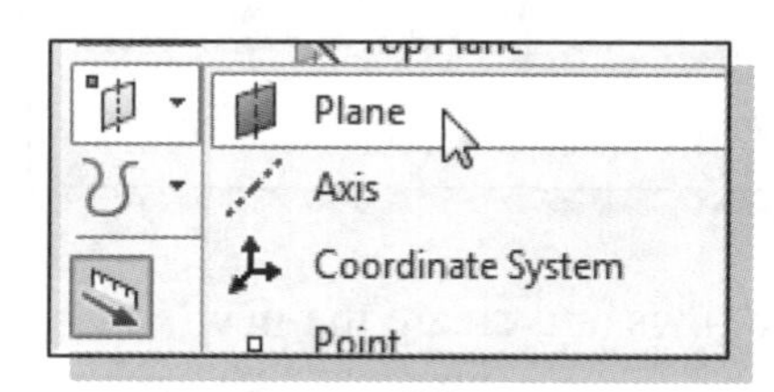

15. Select the **Reference Geometry** command from the *Features* toolbar, and select the **Plane** option from the pull-down menu.

16. Select **Axis1** (the reference axis created above) as the *First Reference* and the **top face of the base feature** as the *Second Reference*.

17. In the *Second Reference* panel of the *Plane PropertyManager*, select the **At Angle** option button.

18. Enter **45 deg** as the angle in the *PropertyManager*. The preview of the new plane should appear as in the figure below.

19. Click **OK** in the *PropertyManager* to create the new reference plane.

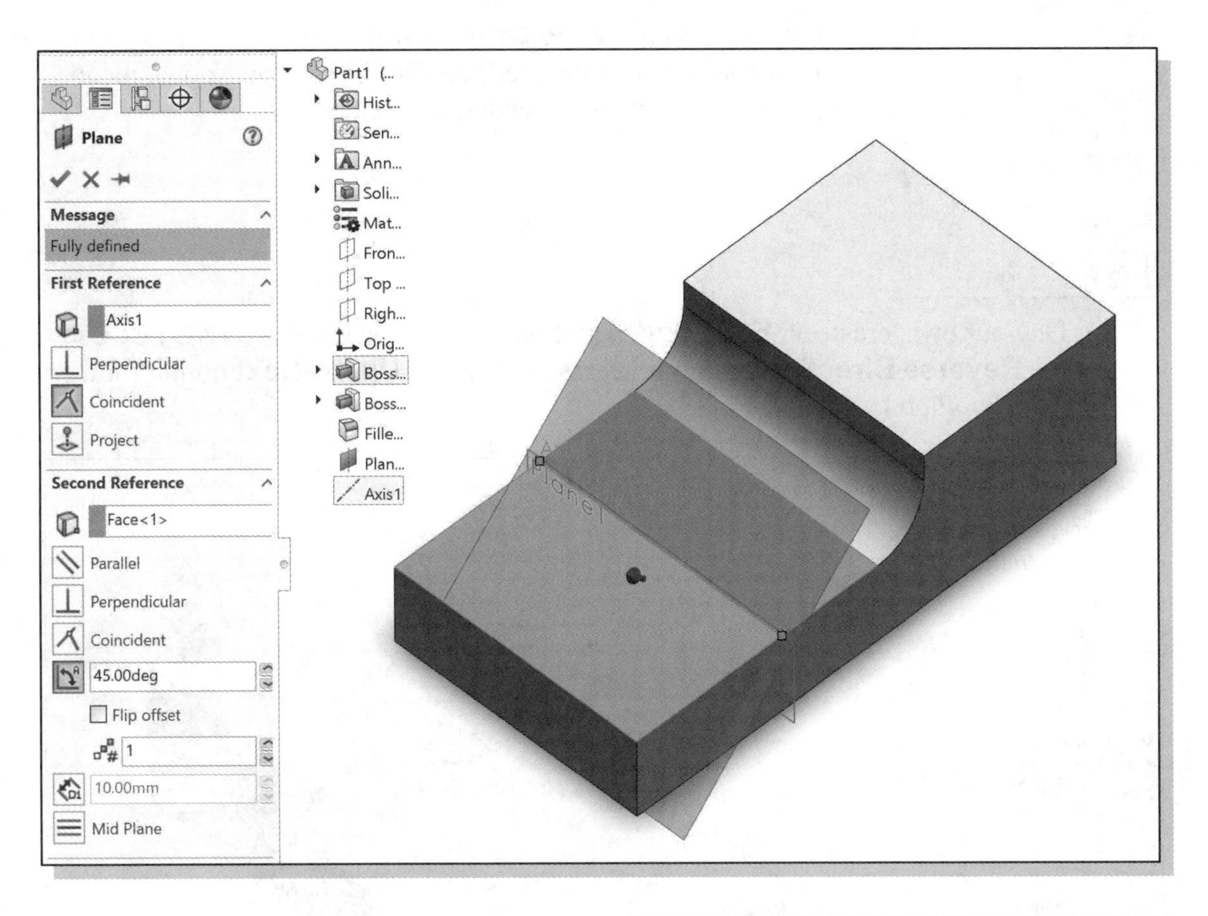

- We now have the correct inclined reference plane on which to sketch the inclined feature.

20. On your own, create the sketch shown using the inclined reference plane **Plane2** as the sketch plane. (**NOTE:** Use the **Normal to** view orientation to view the sketch plane as shown.)

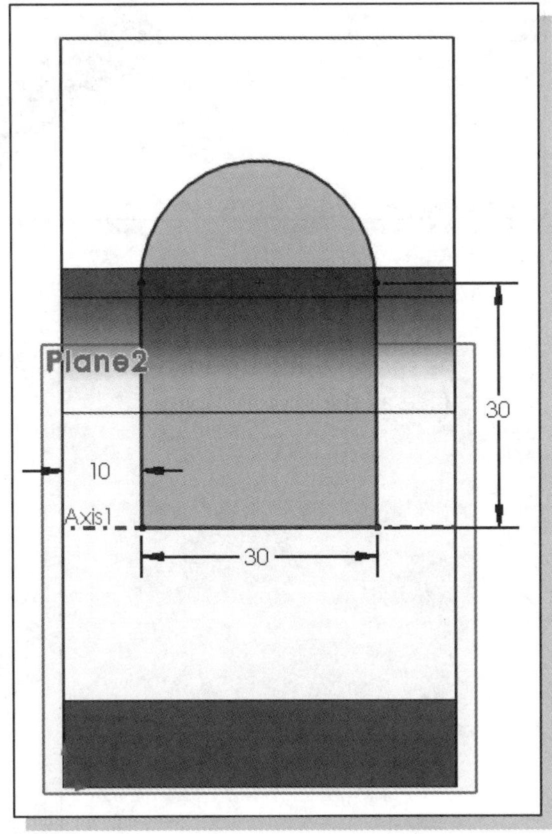

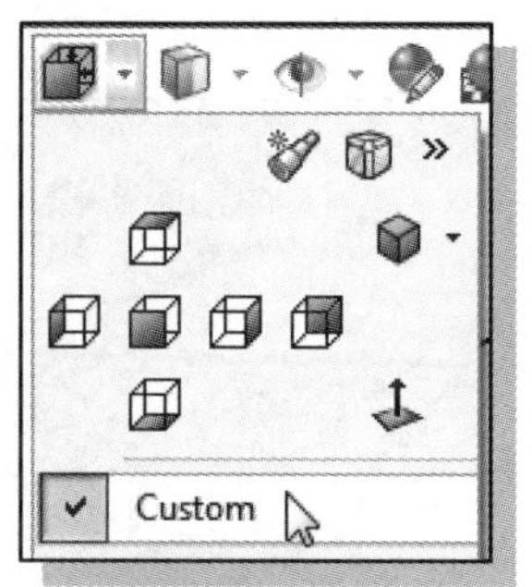

21. Left-click on the **View Orientation** icon on the *Heads-up View* toolbar to reveal the *View Orientation* pull-down menu. Select the **Custom** view that we created earlier.

22. On your own, create an **Extruded Boss** feature as shown below. (**NOTE:** Use the **Reverse Direction** option if necessary. Use the **Up To Next** option for the *end condition.*)

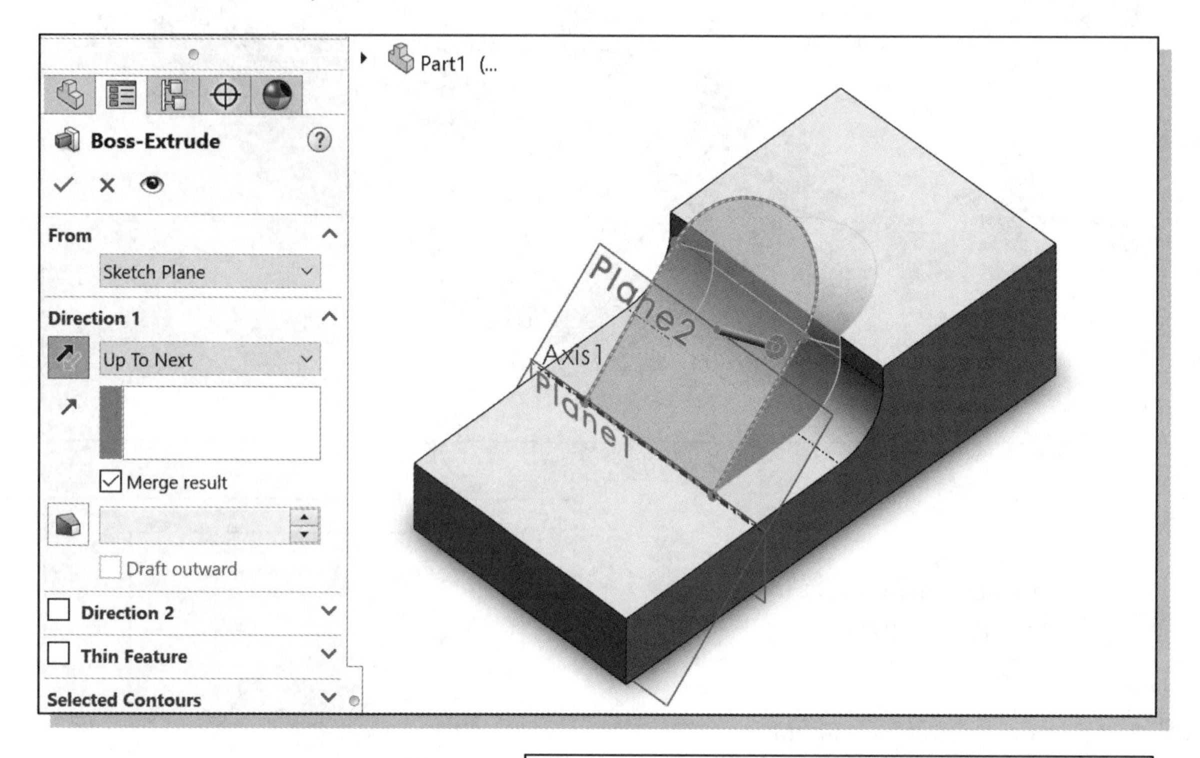

23. On your own, create a sketch as shown using the inclined face as the sketch plane.

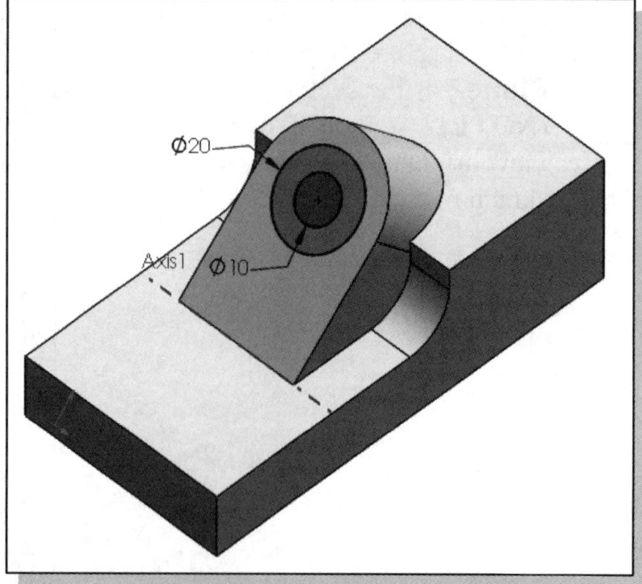

24. Create the 10 mm **Extruded Boss** feature as shown below.

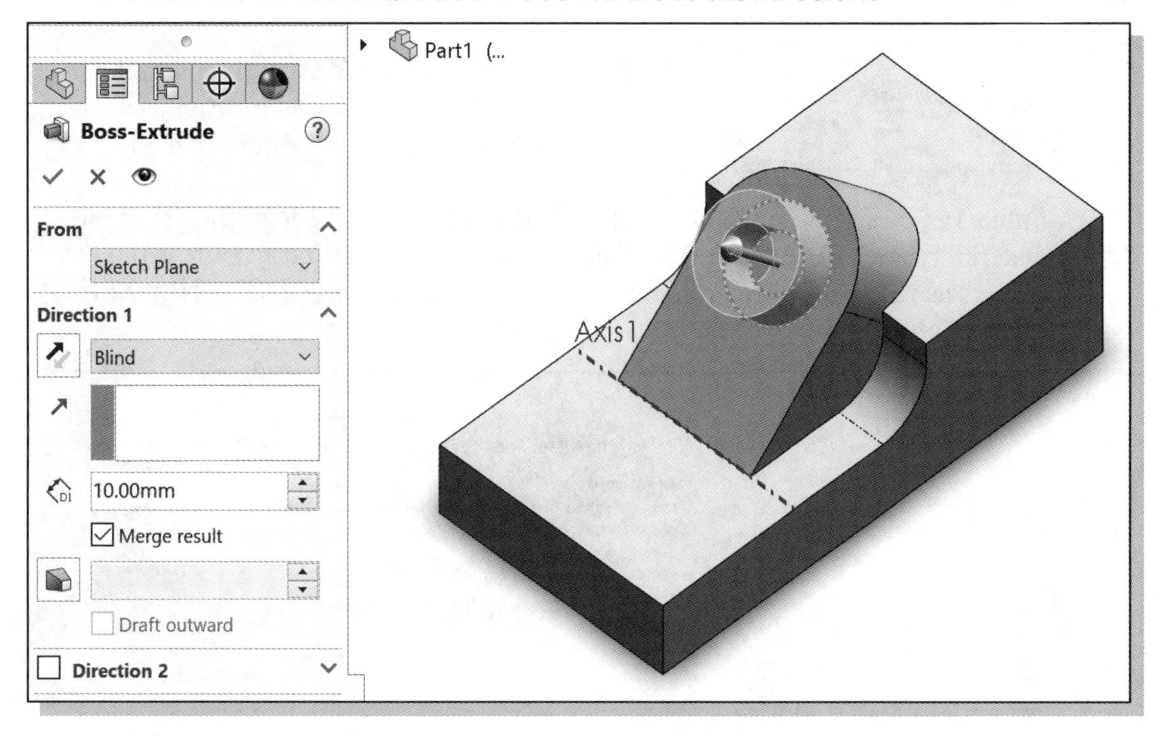

25. On your own, create the extruded cut features necessary to complete the part. (See page 17-5 for dimensions.)

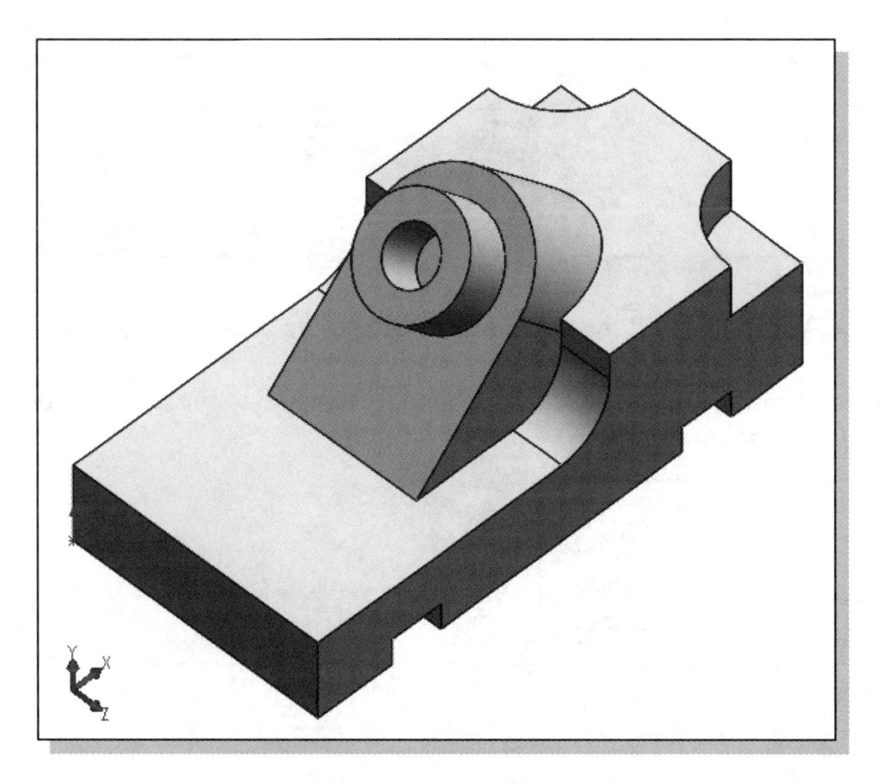

26. On your own, save the part with the filename ***Problem1.SLDPRT***.

Selecting the Material and Viewing the Mass Properties

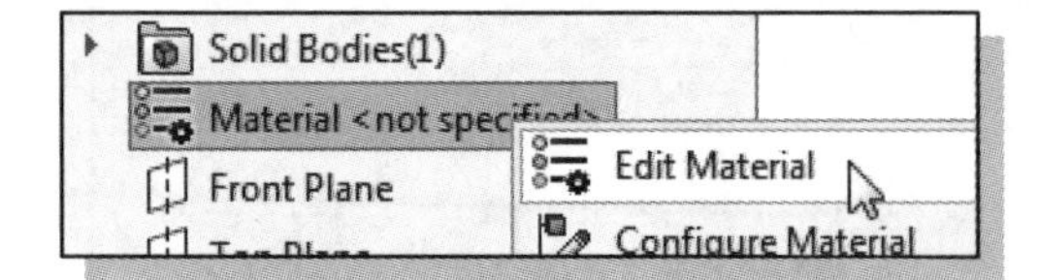

1. Right-click on the **Material** icon in the *FeatureManager Design Tree* and select **Edit Material** from the pop-up menu.

2. In the *Material* pop-up window, select **Ductile Iron** (in the Iron folder) as the material type. Notice the density listed in the *Material Property* table is 0.0071 g/mm^3, as required in the problem statement. (Set the *Units* to **SI – N/m^2 (Pa)**, if necessary).

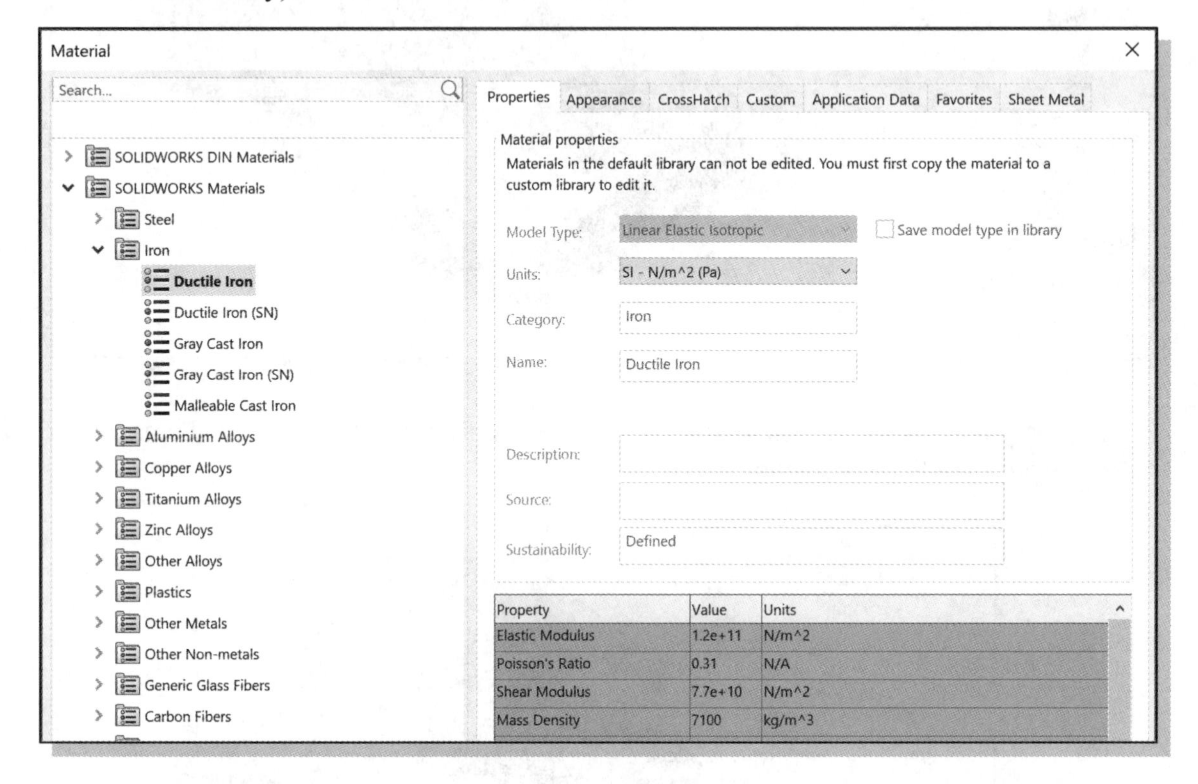

3. Click **Apply** to select the material and **Close** the *Material* window.

4. Select the **Mass Properties** command by expanding the *Evaluate* option on the *Tools* pull-down menu.

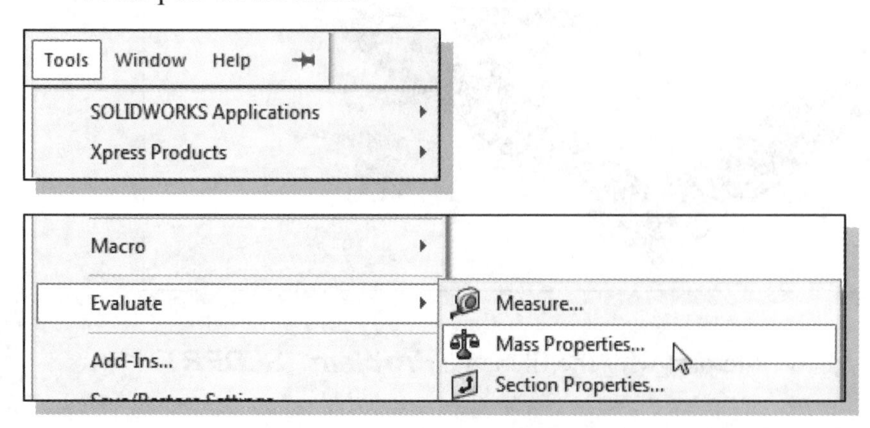

➤ The *Mass Properties* window is displayed. Notice the *Report coordinate values relative to:* window is set to the **default** system.

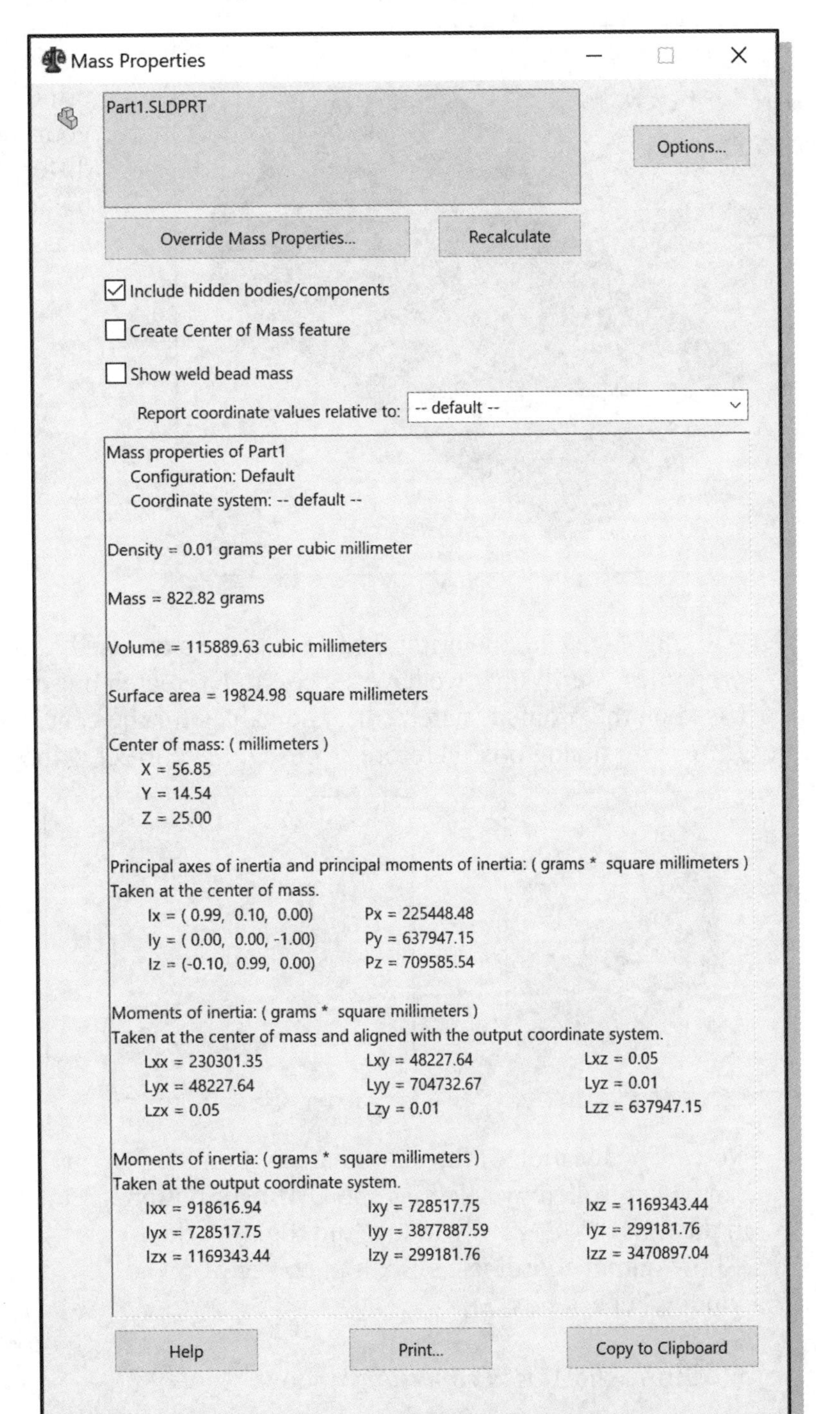

➤ Notice the *Mass* is **822.82 grams**.

➤ Notice the *Center of mass* is:
X = 56.85 mm
Y = 14.54 mm
Z = 25.00 mm

➤ Look again at the problem on page 17-5. Note that the *Center of mass* coordinates above are based on the correct origin and along the correct X, Y, Z directions.

➤ Notice that *Moments of Inertia* are also reported.

➢ Look at the part in the graphics area. (Minimize the *Mass Properties* window if necessary.)

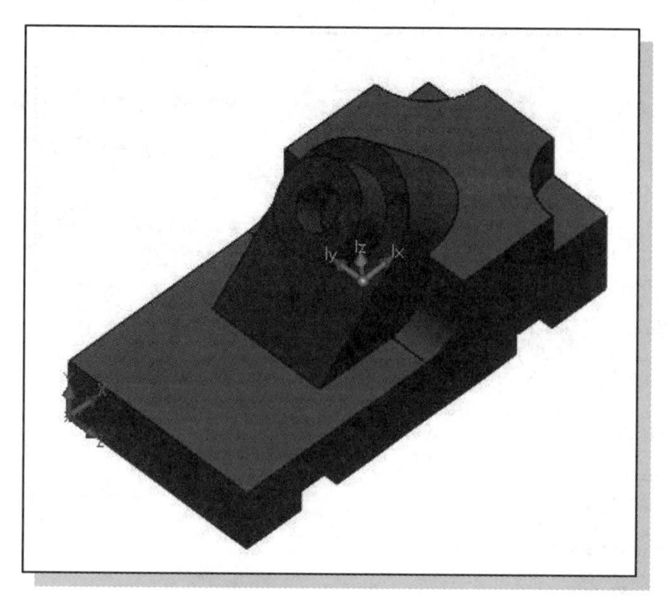

➢ Notice the center of mass and the principal axes are displayed. On your own, view the part from different directions and observe the location of the center of mass.

➢ Also displayed are the origin and coordinate axes used as the basis for the calculated properties. Verify that we have successfully created our part in the correct orientation based on the problem statement. The values for the center of mass in the *Mass Properties* dialog box therefore represent the correct values for the problem as stated.

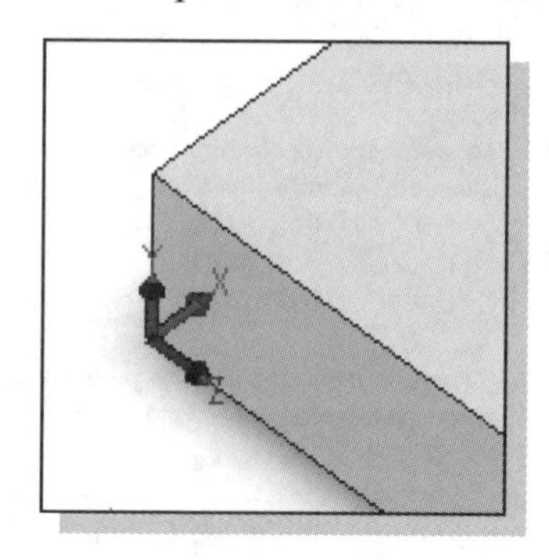

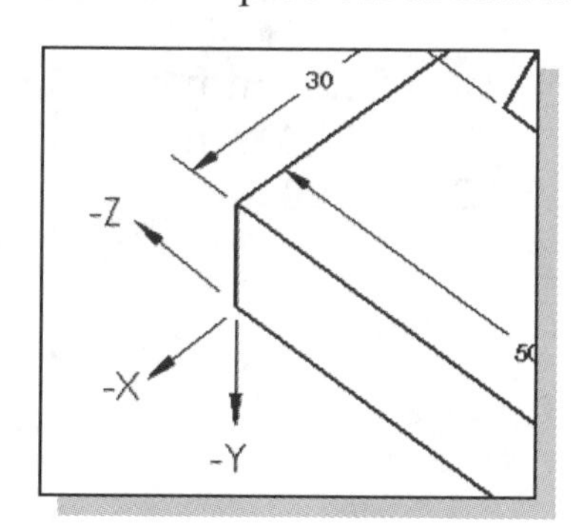

➢ Note: The location of the coordinate system icon can be controlled by selecting the **Options** button in the *Mass Properties* window and toggling the 'Show output coordinate system in corner of window' check box.

Show output coordinate system in corner of window

5. **Close** the *Mass Properties* window.

6. On your own, save the part with the filename ***Problem1.SLDPRT***.

➢ We have successfully completed the problem. The part was created using the coordinate axes defined in the problem statement and the mass and center of mass were obtained. The SOLIDWORKS Mass Properties tool provides an efficient method to obtain properties of parts and/or assemblies with complex shapes.

The Assembly Problem

We will model the assembly shown below and use the **Mass Properties** tool to determine the mass and center of mass of the assembly. The problem is presented below in a manner similar to that encountered on the CSWA exam.

Build the assembly shown.

It contains two ***Brackets*** and two ***Rods***.

Unit system: MMGS (millimeter, gram, second)

Assembly origin: as shown.

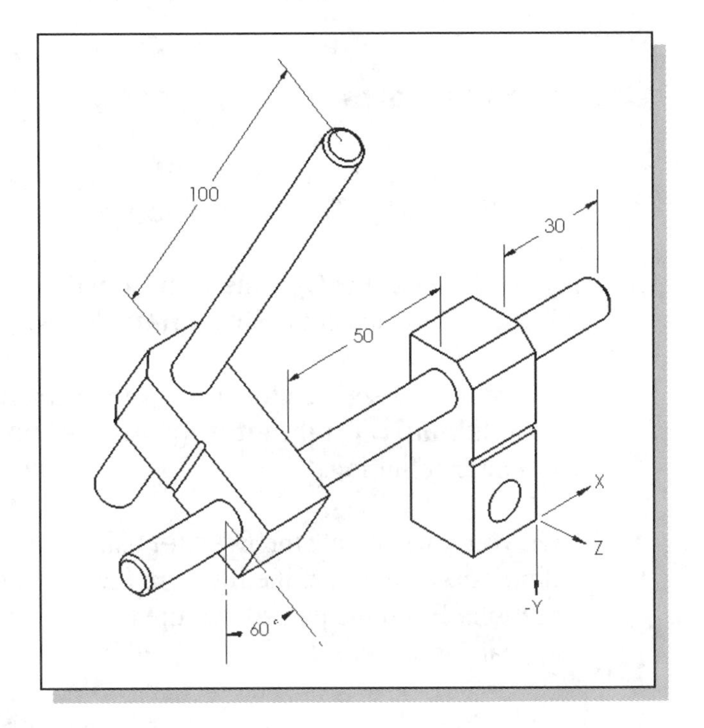

***Brackets**:* A dimensioned figure appears to the right (holes through all). The 1 mm radius notch is semicircular and located with its center 25 mm from the bottom edge of the bracket. Material: 1060 Alloy; Density = 0.0027 g/mm^3.

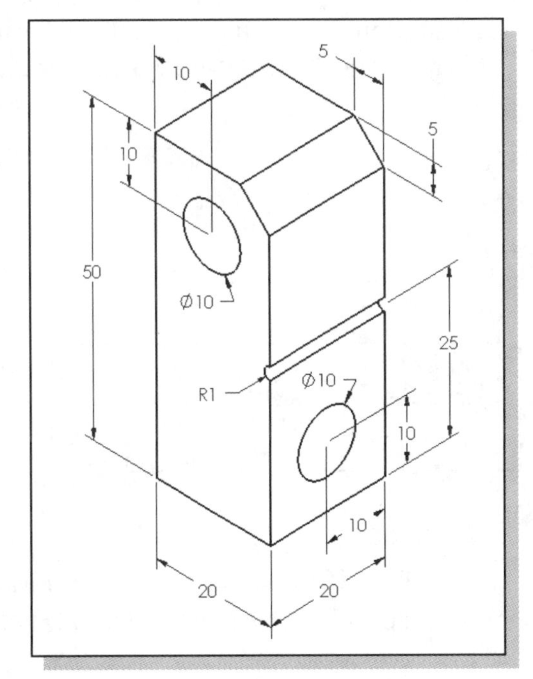

***Rods**:* 10 mm in diameter and 150 mm long, with 1mm x 45° chamfers at each end. Material: 201 Annealed Stainless Steel (SS); Density = 0.00786 g/mm^3.

What is the overall mass of the assembly in grams?

What is the center of mass of the assembly?

➢ Before starting to build the model, it is important to select a strategy for controlling the origin and orientation of the X, Y, and Z axes since the center of mass must be defined relative to a known reference system. The axis directions are defined in the drawing. We will build the part independent of this constraint and, after the part is complete, create a **reference coordinate system** aligned with the part as shown in the figure.

Creating the Parts

1. On your own, start a new part file using the default **Part** template in the SOLIDWORKS Templates folder.

2. Select the **Options** icon from the *Menu Bar* to open the *Options* dialog box. Select the **Document Properties** tab.

3. On your own, set the *Overall drafting standard* to **ANSI** and the *Units* to the default **MMGS (millimeter, gram, second)** unit system as instructed in the problem definition.

4. On your own, create the *Bracket* using the dimensions shown in the figure in the previous section. Build the part so that the isometric view appears as shown.

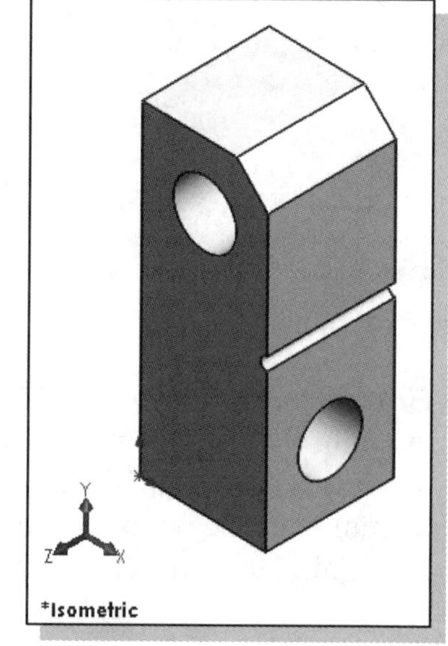

➢ The axis alignment is not the same as in the problem definition. We will address this later using the Reference Coordinate System tool.

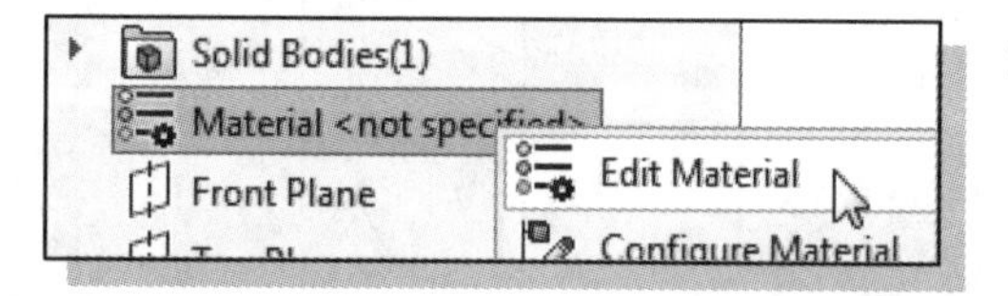

5. Right-click on the **Material** icon in the *FeatureManager Design Tree* and select **Edit Material** from the pop-up menu.

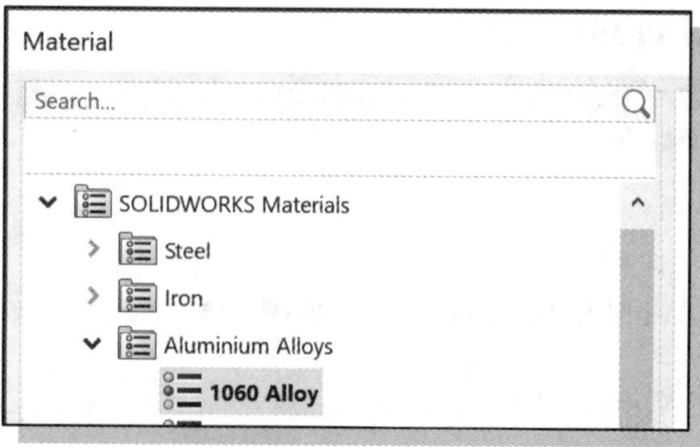

6. In the *Material* pop-up window, select **1060 Alloy** (in the Aluminum Alloys folder) as the material type.

7. Save the part with the filename ***Bracket.SLDPRT***.

8. On your own, start a new part file using the default **Part** template in the SOLIDWORKS Templates folder, set the *Overall drafting standard* to **ANSI** and the *Units* to the default **MMGS (millimeter, gram, second)** unit system.

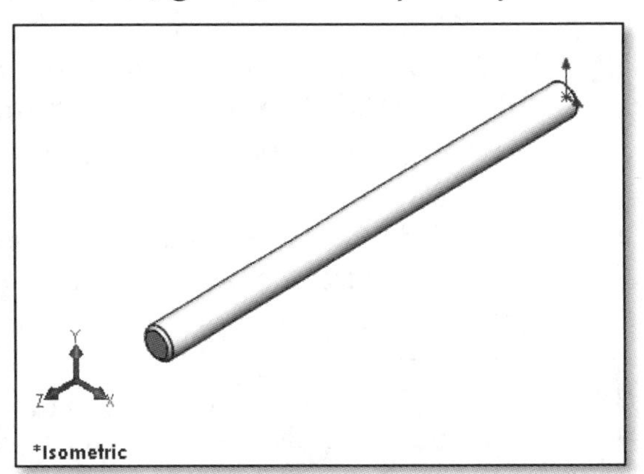

9. On your own, create the *Rod* using the description in the previous section. Build the part so that the isometric view appears as shown. (**HINT:** Draw a circle on the Front Plane and extrude.)

10. On your own, set the material to **201 Annealed Stainless Steel (SS)** (in the Steel folder).

11. Save the part with the filename ***Rod.SLDPRT***.

Creating the Assembly

1. Select the **New** icon with a single click of the left-mouse-button on the *Menu Bar*. The *New* SOLIDWORKS *Document* dialog box appears.

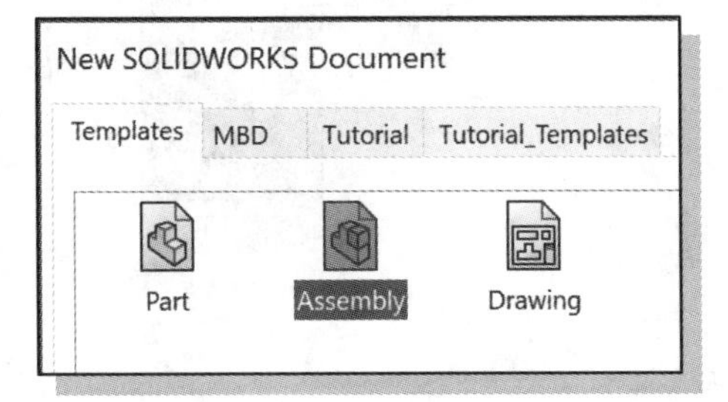

2. Select the **Templates** tab and select the **Assembly** icon as shown.

3. Click **OK** to open the new Assembly file.

➤ SOLIDWORKS opens an assembly file, and automatically opens the *Begin Assembly PropertyManager*. The message "*Select a part or assembly to insert ...*" appears in the *PropertyManager*. SOLIDWORKS expects you to insert the first component.

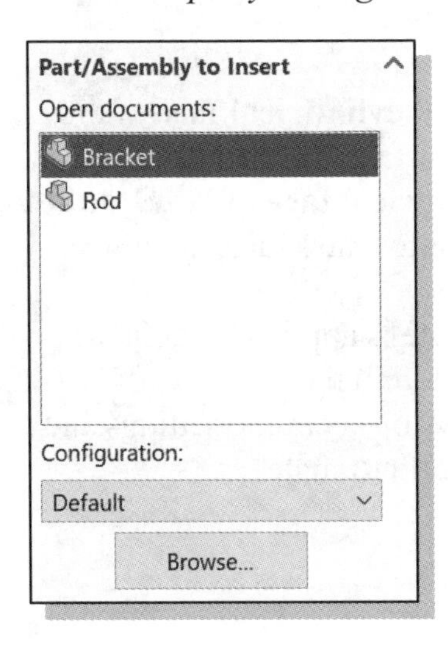

➤ The *Part/Assembly to Insert* panel displays *Open documents*, including the *Bracket* and *Rod* parts.

4. Select the *Bracket* part as the base part for the assembly by clicking on **Bracket** in the *Part/Assembly to Insert* panel. (**NOTE:** If you have closed the *Bracket* part, it will not appear. In that case, click the Browse button and use the browser to open the *Bracket* part file.)

5. By default, the component is automatically aligned to the origin of the assembly coordinates. Click **OK** in the *PropertyManager* to place the ***Bracket*** at the origin.

6. On your own, set the *Overall drafting standard* to **ANSI** and the *Units* to the default **MMGS (millimeter, gram, second)** unit system. (**HINT**: Select Tools → Option → Document Properties.)

7. In the *Assembly* toolbar, select the **Insert Component** command by left-mouse-clicking the icon.

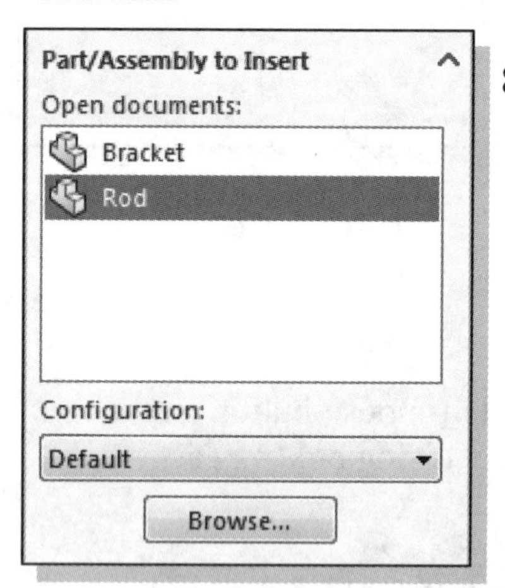

8. Select the ***Rod*** part in the *Part/Assembly to Insert* panel. (**NOTE:** If you have closed the *Rod* part, it will not appear. In that case, click the Browse button and use the browser to open the *Rod* part file.)

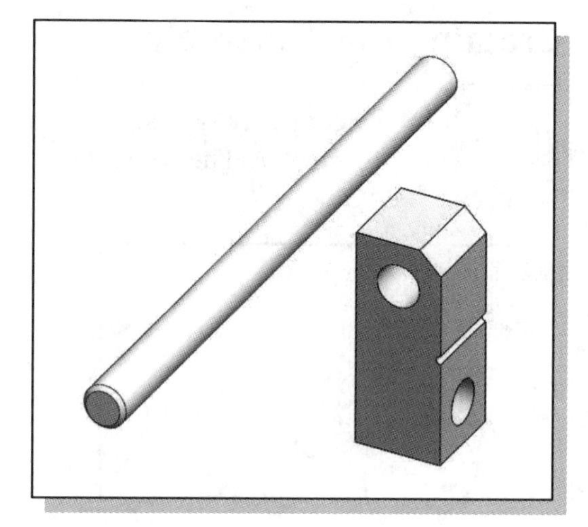

9. Place the *Rod* in the graphics area as shown in the figure. Click once with the **left-mouse-button** to place the component.

10. In the *Assembly* toolbar, select the **Mate** command by left-mouse-clicking once on the icon.

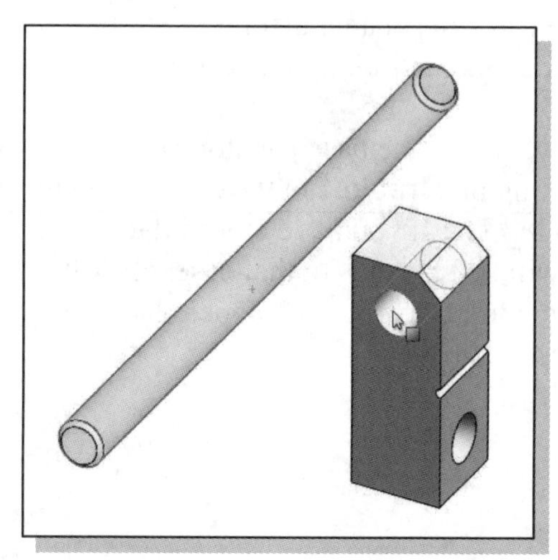

11. Select the **external cylindrical face** of the ***Rod*** as the first surface for the Mate, and the **interior cylindrical face** of the ***Bracket*** as the second face for the Mate, as shown.

12. A Concentric mate is applied by default. Click the **OK** button in the *PropertyManager* to accept the settings and create the **Concentric** mate.

❖ We will now add a **Distance** mate to define the position of the *Rod* relative to the *Bracket* as defined in the problem statement.

13. Rotate the view (e.g., using the arrow keys) to display the reverse side of the *Rod* and *Bracket* as shown below.

14. Select the **edge face** of the ***Rod*** and the **side face** of the ***Bracket*** as shown below.

➢ SOLIDWORKS will apply a Coincident mate by default. We want to apply a mate which is offset by a specific distance (30 mm as defined by the problem statement).

15. Select the *Distance* option by clicking the **Distance** button in the *Standard Mates* panel of the *PropertyManager*.

16. Enter **30 mm** as the distance.

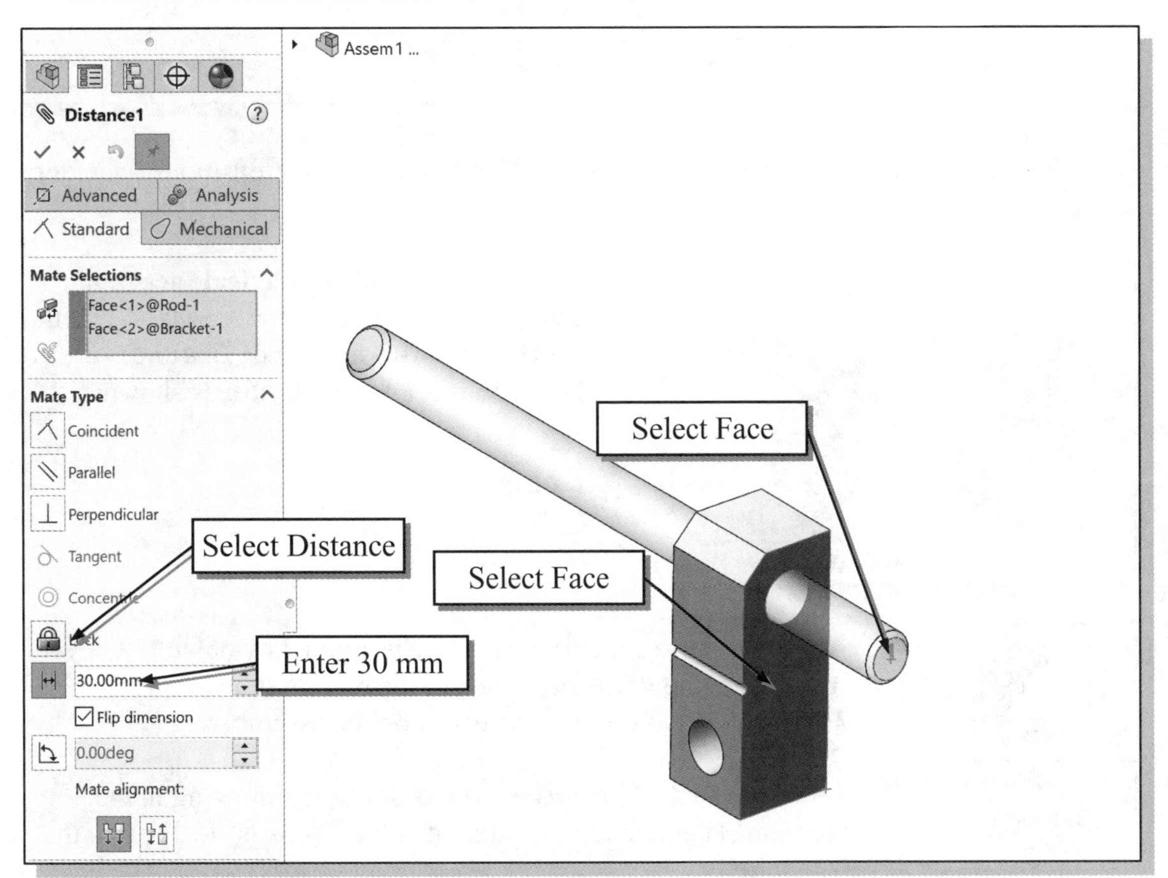

17. Click the **OK** button in the *PropertyManager* to accept the settings and create the **Distance** mate.

18. Click **OK** again (or hit the [**Esc**] key) to exit the Mate command.

19. On your own, return to the **Isometric** view.

20. In the *Assembly* toolbar, select the **Insert Component** command by left-mouse-clicking the icon.

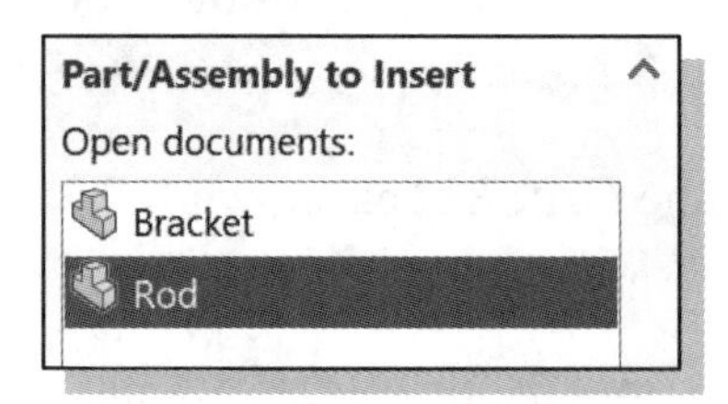

21. Select the *Bracket* part as the base part for the assembly by clicking on **Bracket** in the *Part/Assembly to Insert* panel.

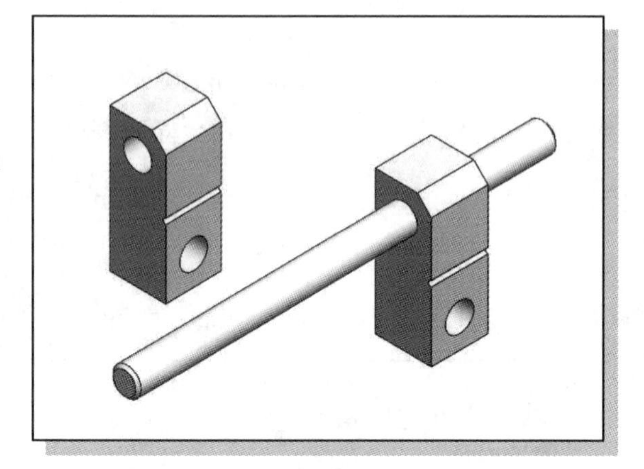

22. Place the *Bracket* in the graphics area as shown in the figure. Click once with the **left-mouse-button** to place the component.

23. In the *Assembly* toolbar, select the **Mate** command by left-mouse-clicking once on the icon.

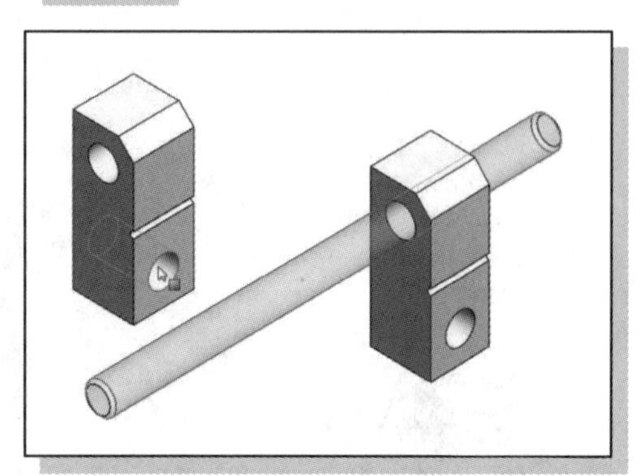

24. Select the **external cylindrical face** of the ***Rod*** as the first surface for the Mate, and the **interior cylindrical race** of ***Bracket<2>*** as the second face for the Mate, as shown.

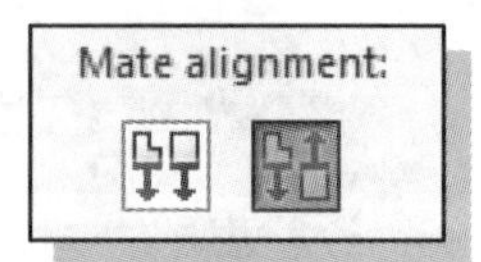

25. An Anti-Aligned Concentric mate is applied by default. On your own, select the **Aligned** option at the bottom of the *Standard Mates* panel and observe the effect on the assembly.

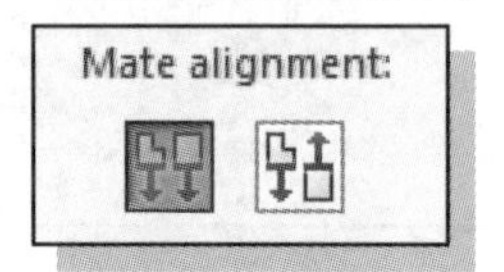

26. Select the **Anti-Aligned** option. Look at the drawing in the problem statement to verify that it is the Anti-Aligned option that is required.

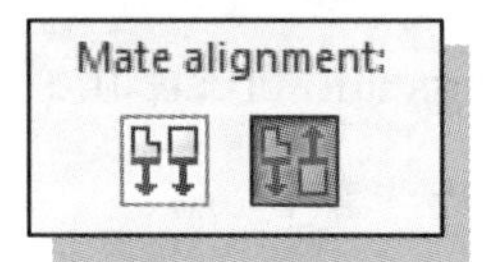

27. Click the **OK** button in the *PropertyManager* to accept the settings and create the **Anti-Aligned Concentric** mate.

❖ We will now add a Distance mate to define the position of *Bracket<2>* relative to *Bracket<1>* as defined in the problem statement.

28. Select the **face** of ***Bracket<1>*** as shown.

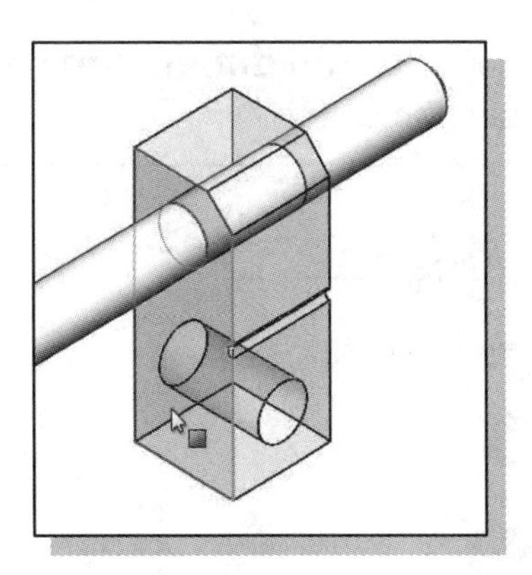

29. Rotate the view and select the **face** of ***Bracket<2>*** as shown.
30. Select the *Distance* option by clicking the **Distance** button in the *Standard Mates* panel of the *PropertyManager*.
31. Enter **50 mm** as the distance.

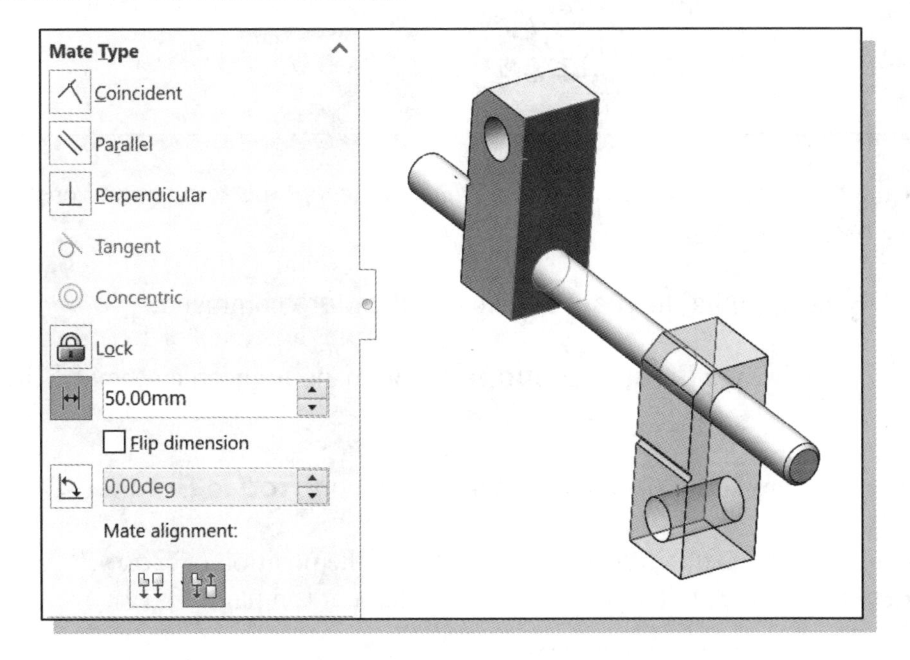

32. Click the **OK** button in the *PropertyManager* to accept the settings and create the **Distance** mate.

❖ We will now add an Angle mate to define the angular orientation of *Bracket<2>* relative to *Bracket<1>* as defined in the problem statement.

33. Select the **face** of ***Bracket<1>*** and the **face** of ***Bracket<2>*** as shown below.
34. Select the *Angle* option by clicking the **Angle** button in the *Standard Mates* panel of the *PropertyManager*.
35. Enter **60°** as the angle.

36. Check the **Flip dimension** box **if necessary** to set the angle as shown in the figure below.

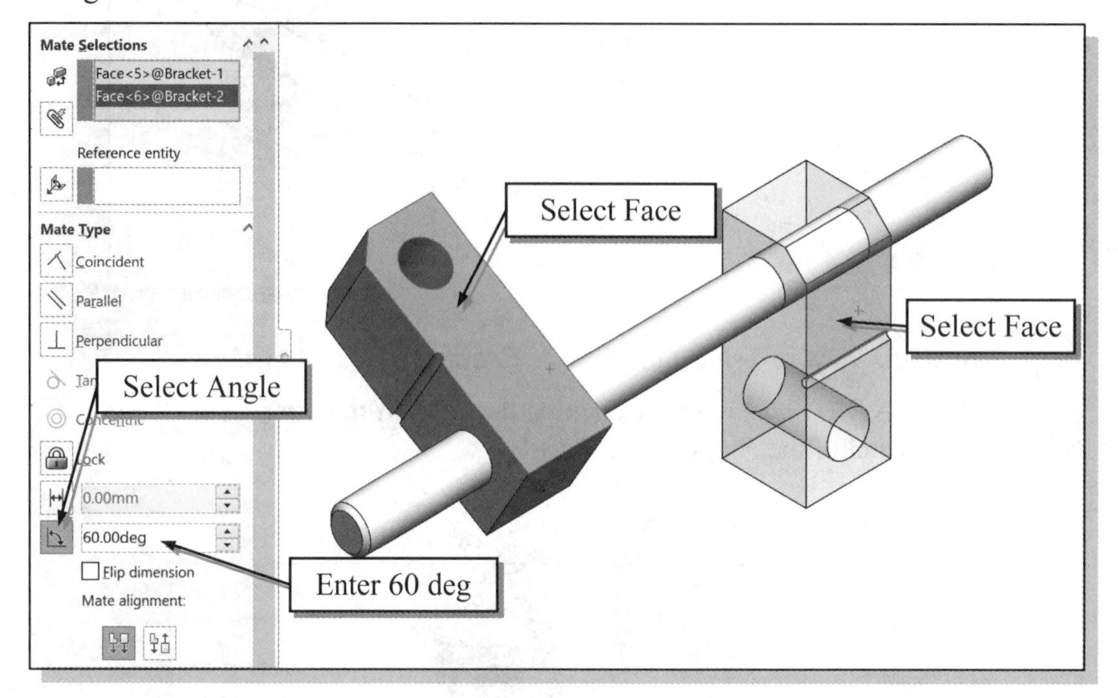

37. Click the **OK** button in the *PropertyManager* to accept the settings and create the **Angle** mate as shown.

38. Click **OK** again (or hit the [**Esc**] key) to exit the Mate command.

39. On your own, use the **Insert Component** command to insert a second ***Rod*** part.

40. On your own, create a **Concentric** mate to align the ***Rod*** to ***Bracket<2>***.

41. On your own, create a **Distance** mate to define the position of ***Rod<2>*** relative to ***Bracket<2>*** as defined in the problem statement (on page 17-19).

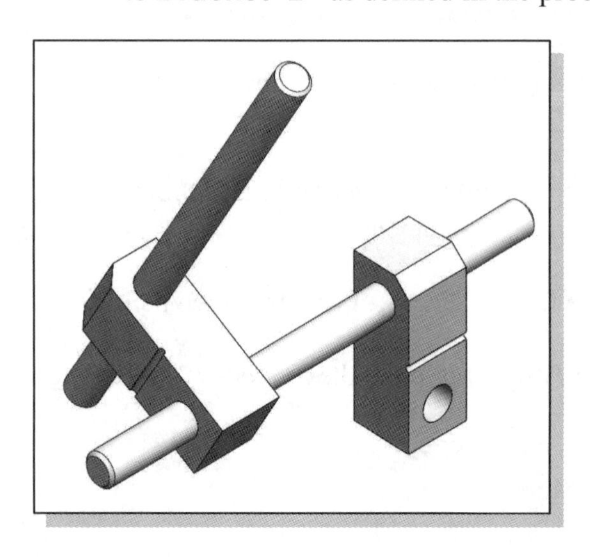

42. Compare the assembly to the drawing in the problem statement. Note that the locations and orientations of the brackets (particularly the locations of the chamfers and notches) and rods are correct.

43. On your own, save the assembly with the filename ***Problem2.SLDASM***.

Creating a Reference Coordinate System

We have created the assembly without aligning it with the origin and coordinate axes defined in the problem statement. In order to calculate the center of mass relative to the correct origin and coordinate axes we will create a Reference Coordinate System.

➤ The Reference Coordinate System will be defined to be aligned with the edges of the base bracket (*Bracket<1>* in our assembly) as shown in the figure defining the problem.

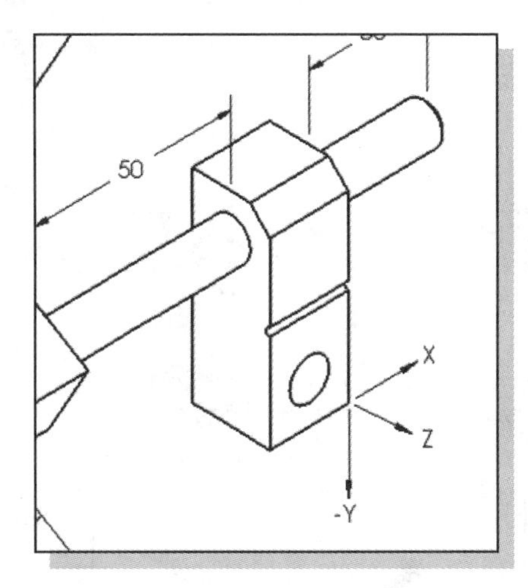

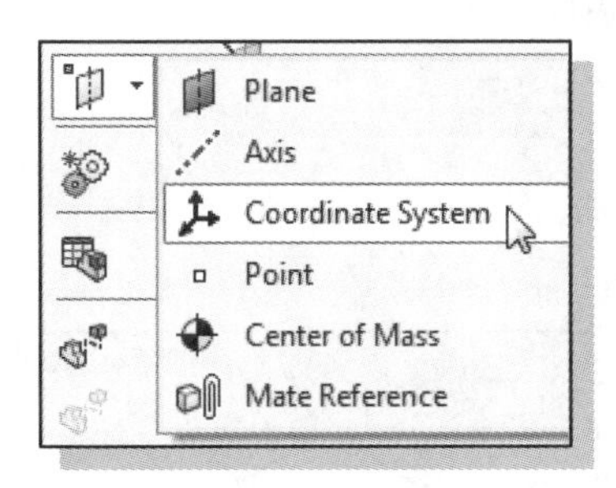

1. In the *Assembly* toolbar (or the *Features* toolbar), select the **Reference Geometry** command by left-clicking the icon.

2. In the pull-down option menu of the Reference Geometry command, select the **Coordinate System** option.

3. The *Coordinate System PropertyManager* opens with the *Origin* selection window active. Select the **lower right corner** of ***Bracket<1>*** as the origin, as shown.

➤ The origin selection appears in the *Property-Manager* as **Vertex<1>@Bracket-1**.

4. The *X-axis* selection window is now active. We want to align the X-axis with the **bottom edge on the right face** of ***Bracket<1>***. Move the cursor over the edge and click once with the left-mouse-button to **select** it.

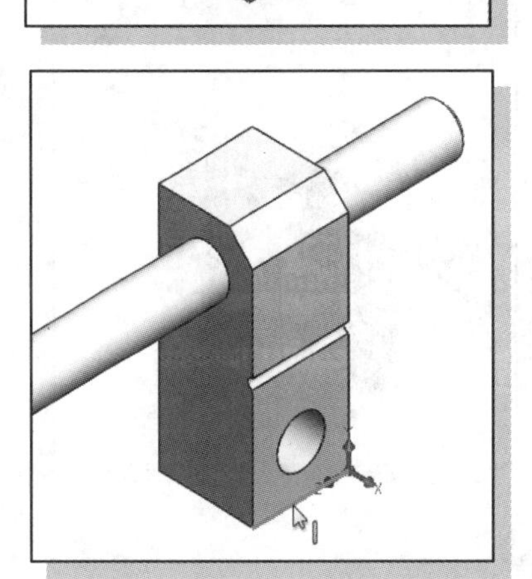

➢ The X-axis selection appears in the *PropertyManager* as Edge<1>@Bracket-1. Look at the coordinate icon which is now attached to the newly defined reference origin. Notice the X-axis in the figure is aligned with the correct edge, but the positive X-axis points in the wrong direction.

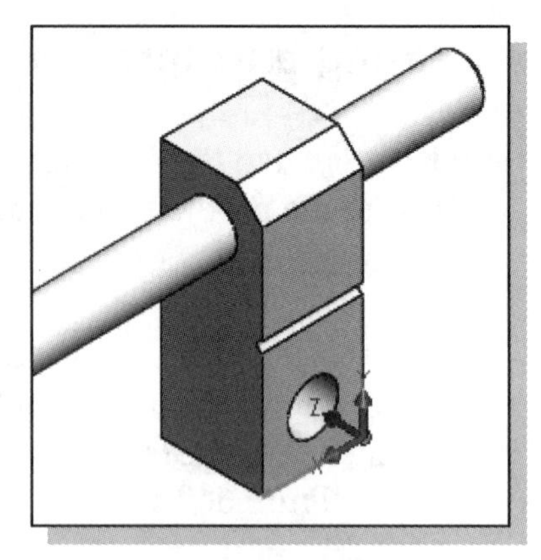

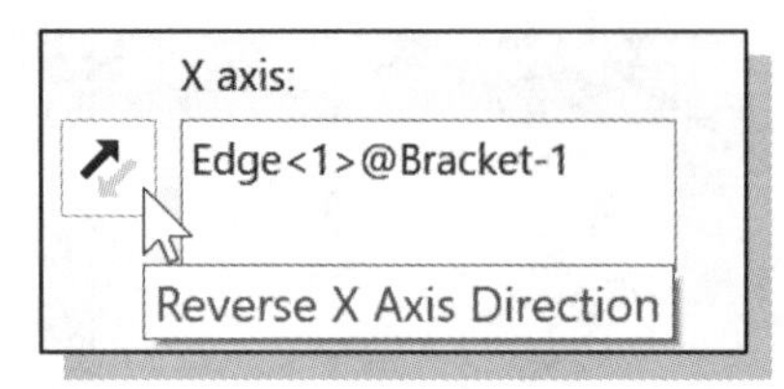

5. In the *PropertyManager*, select the **Reverse X Axis Direction** option by clicking once with the left-mouse-button on the toggle button located to the left of the *X axis* selection window (if necessary).

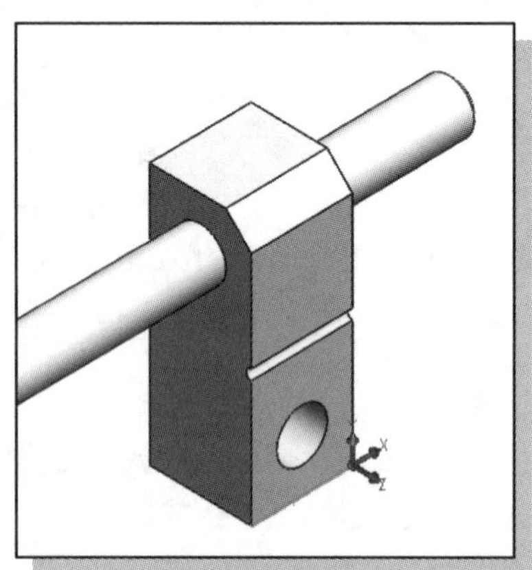

6. Verify that the X-axis is now aligned with the correct edge and points in the correct direction.

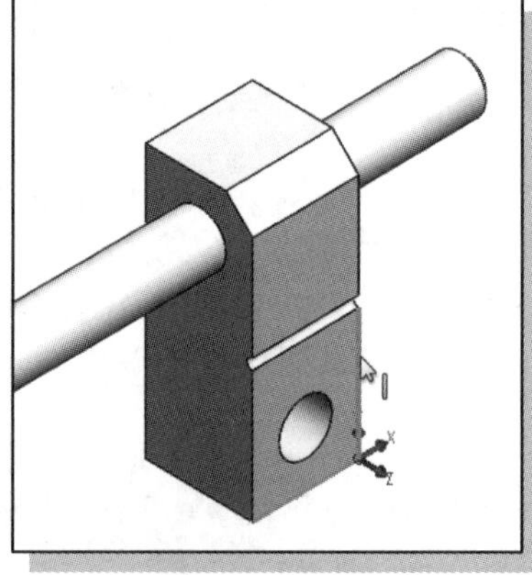

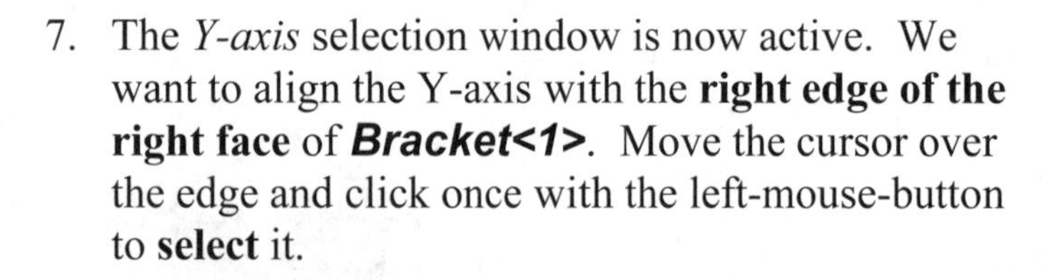

7. The *Y-axis* selection window is now active. We want to align the Y-axis with the **right edge of the right face** of ***Bracket<1>***. Move the cursor over the edge and click once with the left-mouse-button to **select** it.

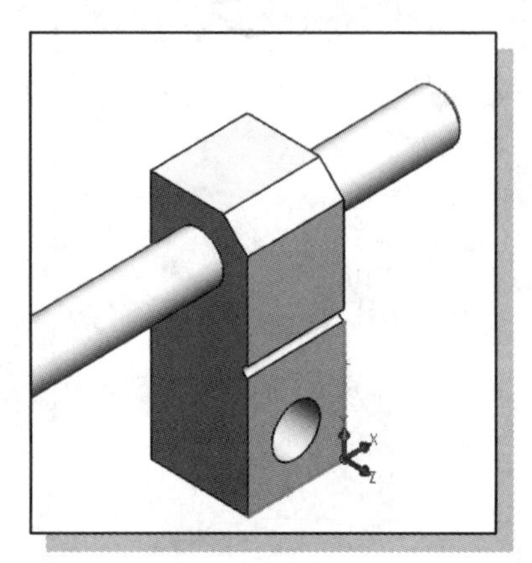

8. The Y-axis selection appears in the *PropertyManager* as Edge<2>@Bracket-1. Verify that the Y-axis is aligned with the correct edge, and the positive Y-axis points in the correct direction. (If the positive Y-axis points in the wrong direction, use the Reverse Y Axis Direction option to correct it.)

9. The origin and coordinate axes are now aligned as defined in the problem statement. Click **OK** in the *PropertyManager* to accept the setting and create the **Reference Coordinate System**.

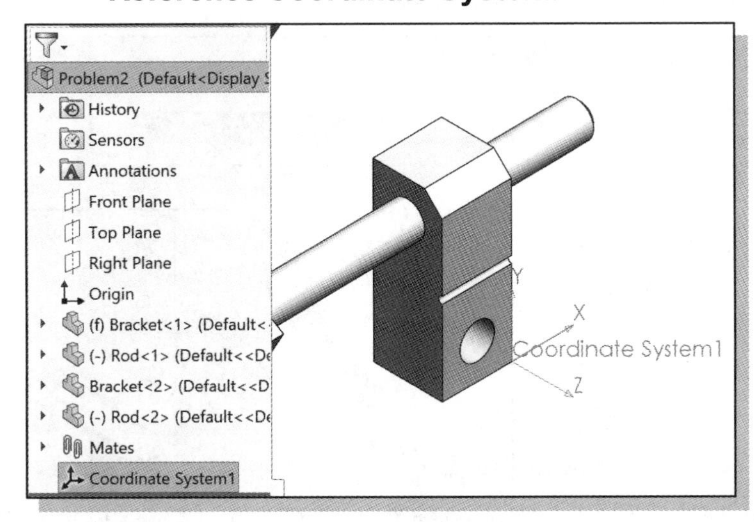

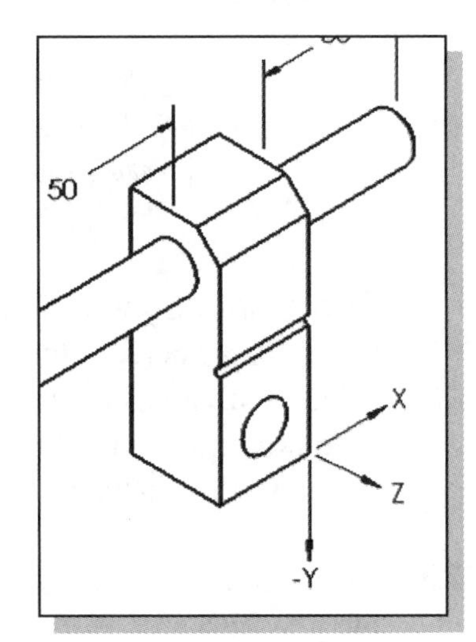

➢ Notice the Reference Coordinate System appears in the *FeatureManager Design Tree* as Coordinate System1, and that the reference axes correspond to the axes defined in the problem statement. We are now ready to calculate the *mass properties*.

View the Mass Properties

1. Select the **Mass Properties** command under the *Evaluate* option on the *Tools* pull-down menu.

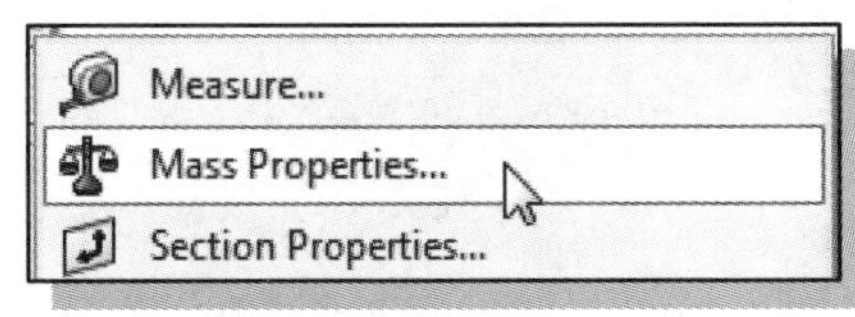

2. The *Mass Properties* window is displayed. Notice the *Report coordinate values relative to:* window is set to the default system.

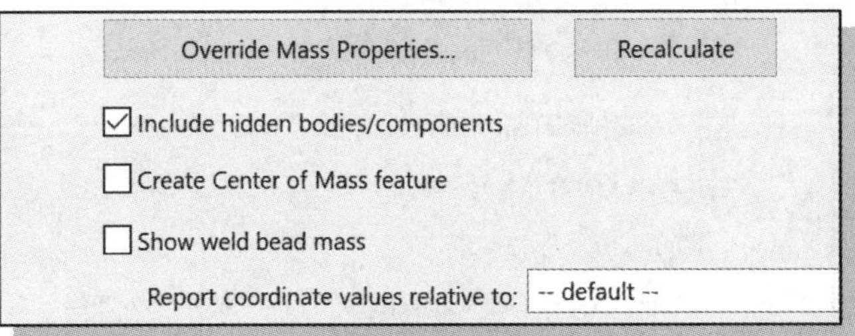

➢ In the graphics area, the default coordinate system is shown. The values appearing in the *Mass Properties* window were calculated relative to this coordinate system. These are **not** the results we need for the solution to the problem.

➢ We must change the *Report coordinate values relative to:* window to the Reference Coordinate System we created.

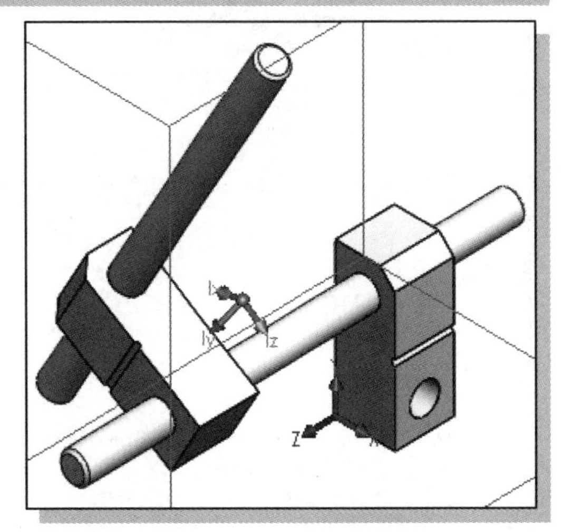

3. Inside the *Mass Property* window, **left-click** on the arrow at the right end of the *Report coordinate values relative to:* selection window, and select **Coordinate System1** from the selection list.

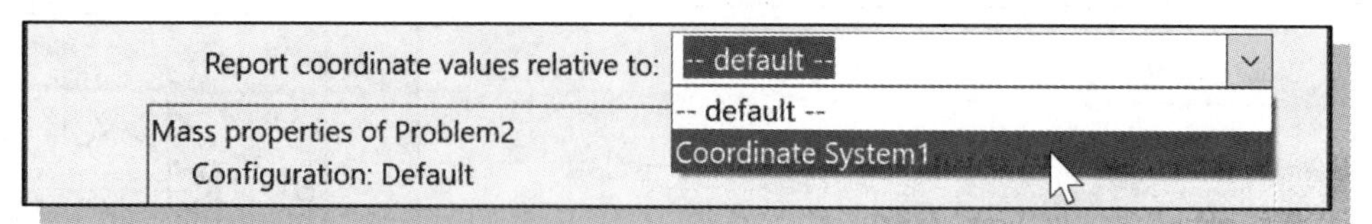

➢ Coordinate System1 is now the basis for the mass property calculations. Notice the calculated values for the center of mass (and other properties) change to those for the new coordinate system.

➢ Notice the coordinate system displayed in the graphics area is the correct system for the problem.

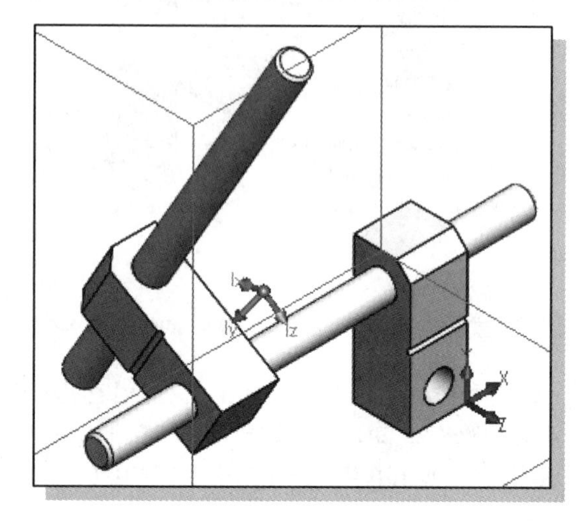

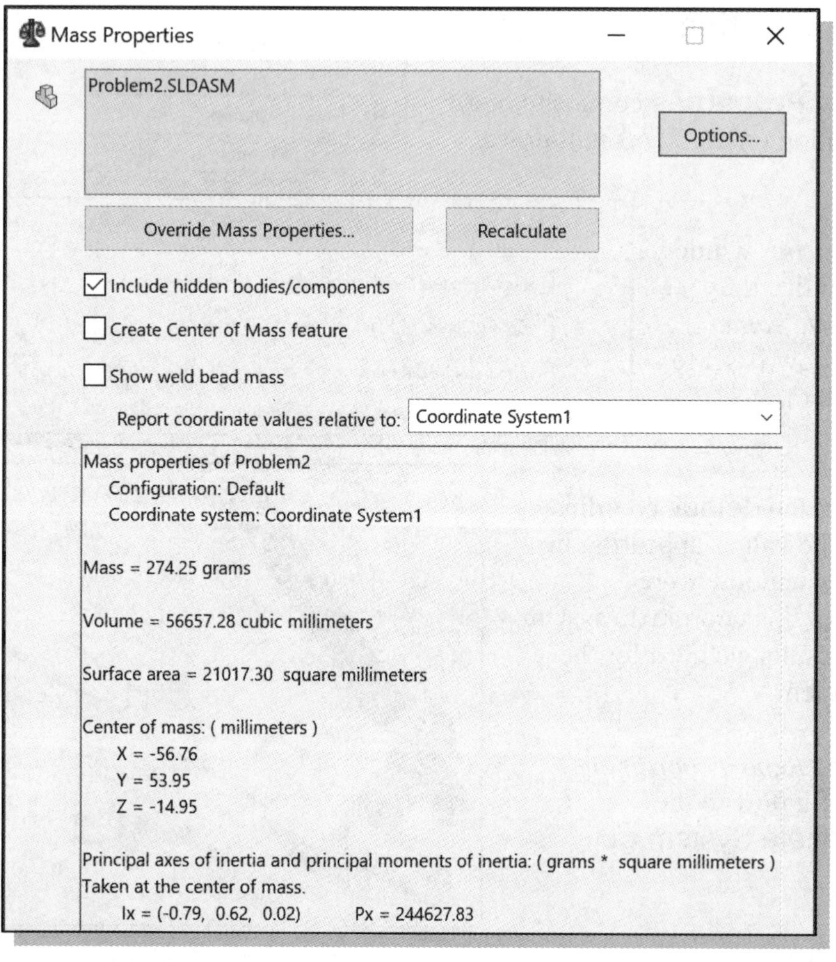

➢ Notice the *Mass* is

274.25 grams

➢ Notice the *Center of mass* is:

X = -56.76 mm
Y = 53.95 mm
Z = -14.95 mm

4. **Close** the *Mass Properties* window.

5. On your own, save the part with the filename ***Problem2.SLDASM***.

- We have successfully completed the problem. The correct alignment of the coordinate system was accomplished through the creation of a Reference Coordinate System.

- The SOLIDWORKS **Reference Coordinate System** and **Mass Properties** tools can be used in combination to obtain mass properties of complex parts and assemblies relative to a specific coordinate system.

Questions:

1. How do we save a new *View Orientation* for future use?

2. Describe the effect of using the *Up to Next* end condition for an extruded boss feature.

3. How can we view the mass of a part or assembly?

4. Describe the steps in creating a **Reference Coordinate System**.

5. What mass properties are dependent on the selection of the output coordinate system?

6. What mass properties are independent of the selection of the output coordinate system?

7. How do we control the output coordinate system used in calculating mass properties?

8. How can we apply an assembly mate establishing an offset distance between two planes or faces?

Exercises:

1. Build the part shown below.
   Unit system: IPS (inch, pound, second)
   Part origin: As shown.
   Part material: AISI 1020
   Density: 0.285406 lb/in^3

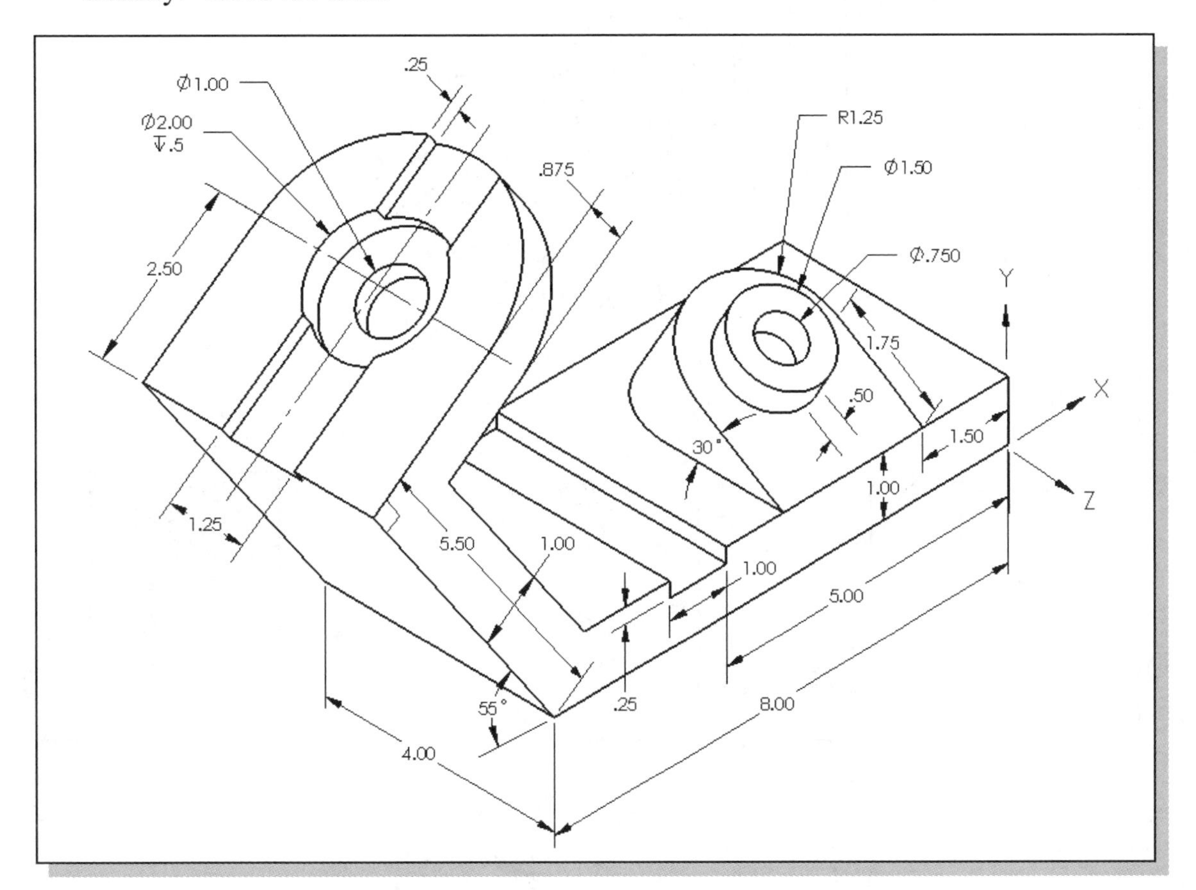

What is the overall mass of the part in pounds?

What is the center of mass of the part?

Answer:

Mass:
18.58 pounds

Center of mass:
X = -6.07 inches
Y = 1.91 inches
Z = -1.98 inches

2. Build the assembly in SOLIDWORKS. The assembly includes 8 components – 1 base, 1 jaw, 2 keys, 1 screw, 1 handle rod, 2 handle knobs.
   Use the IPS (inch, pound, second) unit system.

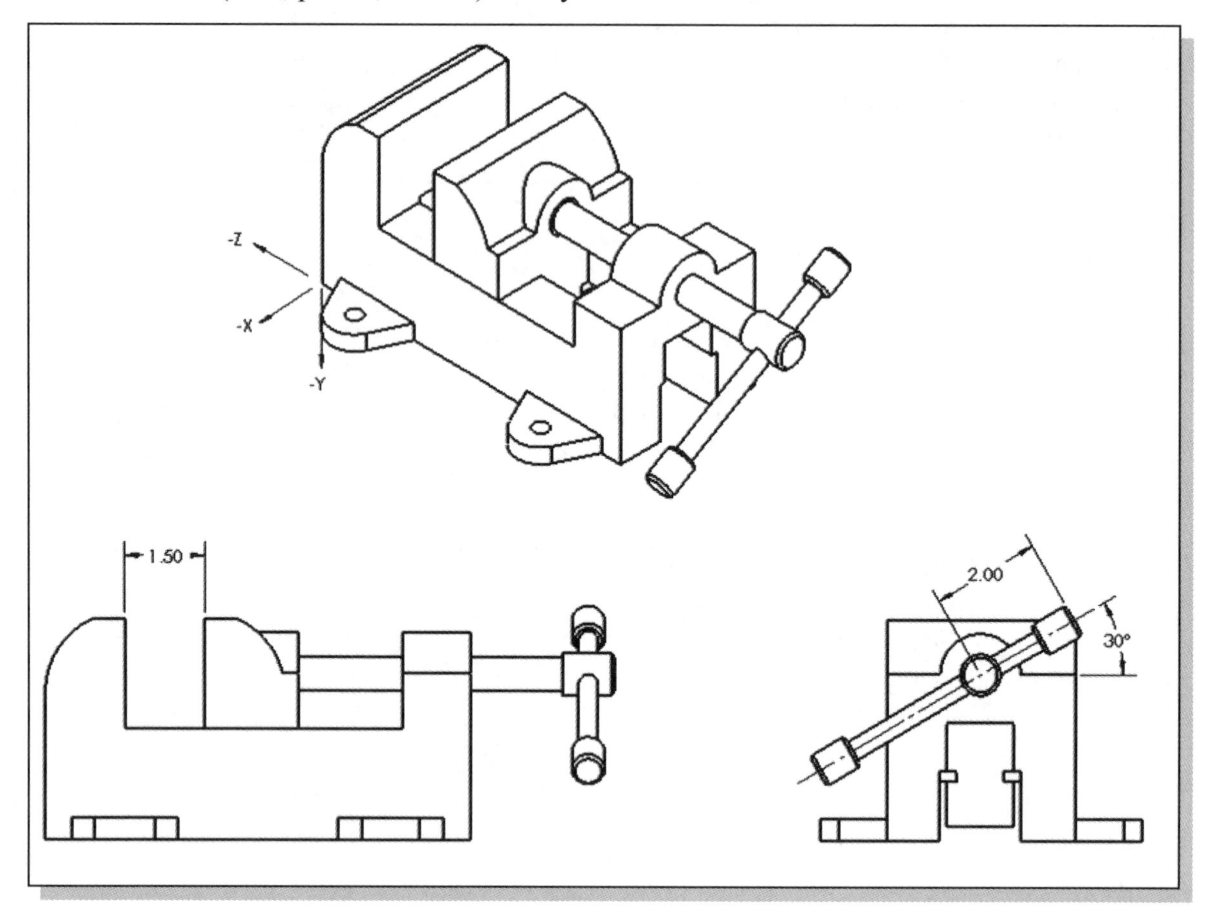

Base: A dimensioned figure appears to the right. The 1.5 inch wide and 1.25 inch wide slots are cut through the entire base.
Material: Gray Cast Iron
Density = 0.260116 lb/in^3

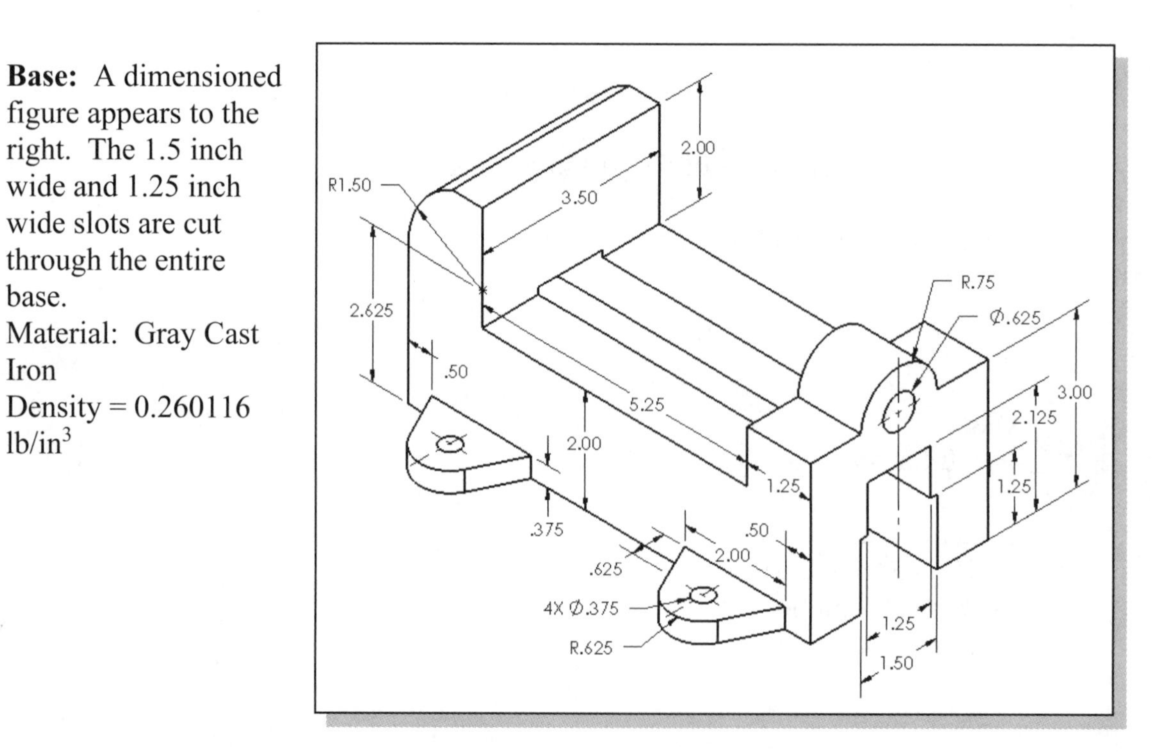

Jaw: A dimensioned figure appears to the right. The shoulder of the jaw rests on the flat surface of the base and the jaw opening is set to 1.5 inches.
Material: Gray Cast Iron
Density = 0.260116 lb/in^3

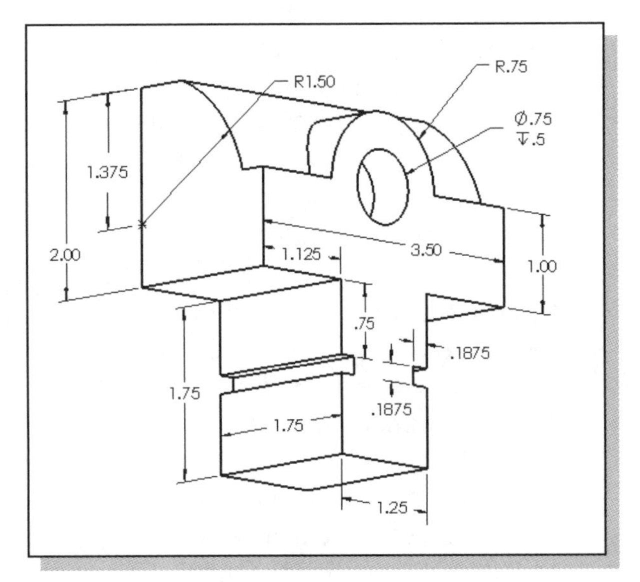

Key: 0.1875 inch H x 0.3125 inch W x 1.75 inch L. The keys fit into the slots on the jaw with the edge faces flush as shown in the sub-assembly to the right.
Material: Alloy Steel
Density = 0.27818 lb/in^3.

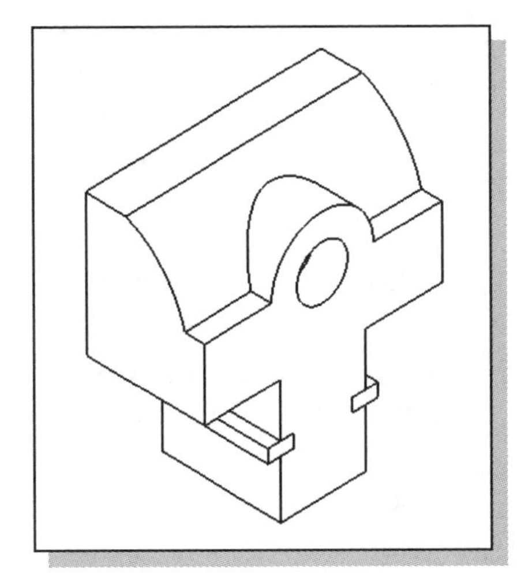

Screw: A dimensioned figure appears to the right. There is one chamfered edge (0.0625 inch x 45°). The flat ∅ 0.75″ edge of the screw is flush with the corresponding recessed ∅ 0.75 face on the jaw.
Material: Alloy Steel
Density = 0.27818 lb/in^3

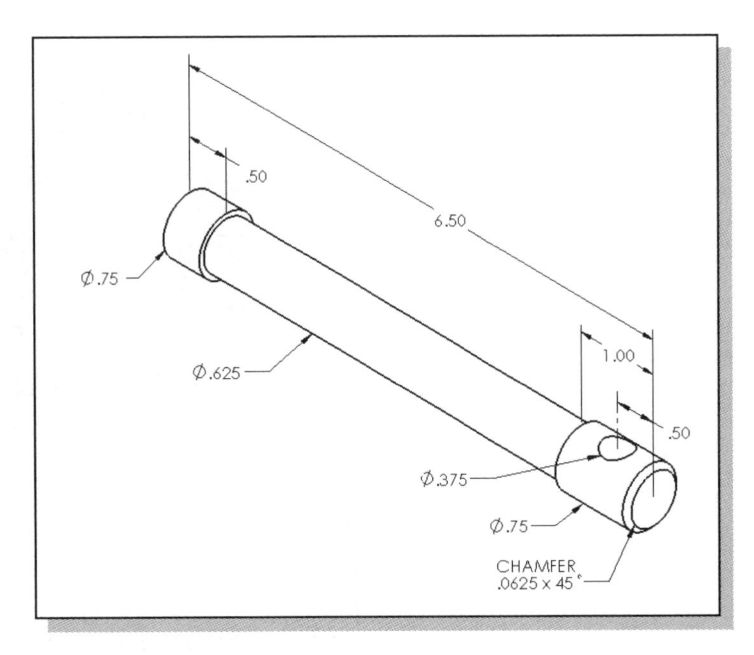

Handle Rod: ∅ 0.375″ x 5.0″ L. The handle rod passes through the hole in the screw and is rotated to an angle of 30° with the horizontal as shown in the assembly view. The flat ∅ 0.375″ edges of the handle rod are flush with the corresponding recessed ∅ 0.735 faces on the handle knobs.
Material: Alloy Steel
Density = 0.27818 lb/in^3.

Handle Knob: A dimensioned figure appears to the right. There are two chamfered edges (0.0625 inch x 45°). The handle knobs are attached to each end of the handle rod. The resulting overall length of the handle with knobs is 5.50″. The handle is aligned with the screw so that the outer edge of the upper knob is 2.0″ from the central axis of the screw. Material: Alloy Steel
Density = 0.27818 lb/in^3

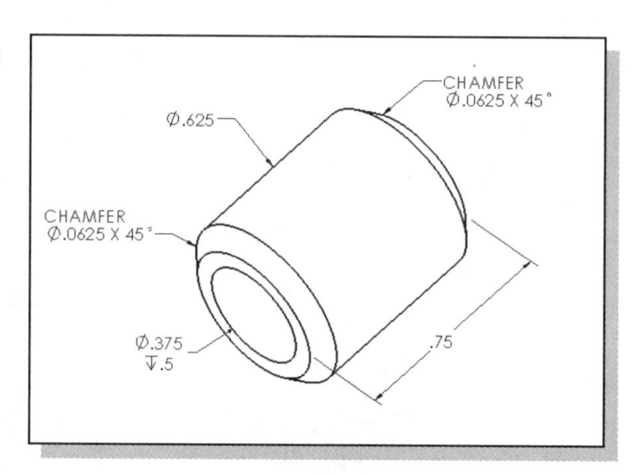

What is the mass of the assembly?

What is the center of mass of the assembly with respect to the coordinate system illustrated in the assembly view?

Answer:

Mass:
17.51 pounds

Center of mass:
X = 1.74 inches
Y = 1.75 inches
Z = 3.98 inches

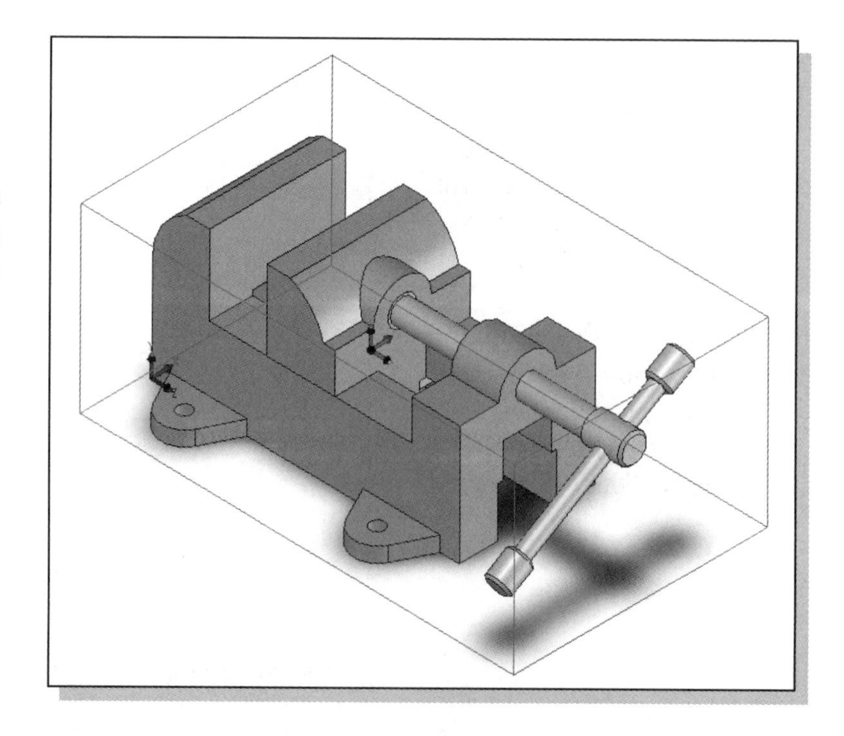

Appendix A

Sec. 206. Standard gauge for sheet and plate iron and steel
For the purpose of securing uniformity, the following is established as the only standard gauge for sheet and plate iron and steel in the United States of America, namely:

		Thickness				Weight		
Gauge	Frac. Inch	Dec. Inch	mm	oz/ft^2	lb/ft^2	kg/ft^2	kg/m^2	lb/m^2
0000000	1/2	.5	12.7	320	20.00	9.072	97.65	215.28
000000	15/32	.46875	11.90625	300	18.75	8.505	91.55	201.82
00000	7/16	.4375	11.1125	280	17.50	7.983	85.44	188.37
0000	13/32	.40625	10.31875	260	16.25	7.371	79.33	174.91
000	3/8	.375	9.525	240	15	6.804	73.24	161.46
00	11/32	.34375	8.73125	220	13.75	6.237	67.13	148.00
0	5/16	.3125	7.9375	200	12.50	5.67	61.03	134.55
1	9/32	.28125	7.14375	180	11.25	5.103	54.93	121.09
2	17/64	.265625	6.746875	170	10.625	4.819	51.88	114.37
3	1/4	.25	6.35	160	10	4.536	48.82	107.64
4	15/64	.234375	5.953125	150	9.375	4.252	45.77	100.91
5	7/32	.21875	5.55625	140	8.75	3.969	42.72	94.18
6	13/64	.203125	5.159375	130	8.125	3.685	39.67	87.45
7	3/16	.1875	4.7625	120	7.5	3.402	36.62	80.72
8	11/64	.171875	4.365625	110	6.875	3.118	33.57	74.00
9	5/32	.15625	3.96875	100	6.25	2.835	30.52	67.27
10	9/64	.140625	3.571875	90	5.625	2.552	27.46	60.55
11	1/8	.125	3.175	80	5	2.268	24.41	53.82
12	7/64	.109375	2.778125	70	4.375	1.984	21.36	47.09
13	3/32	.09375	2.38125	60	3.75	1.701	18.31	40.36
14	5/64	.078125	1.984375	50	3.125	1.417	15.26	33.64
15	9/128	.0703125	1.7859375	45	2.8125	1.276	13.73	30.27
16	1/16	.0625	1.5875	40	2.5	1.134	12.21	26.91
17	9/160	.05625	1.42875	36	2.25	1.021	10.99	24.22
18	1/20	.05	1.27	32	2	.9072	9.765	21.53
19	7/160	.04375	1.11125	28	1.75	.7938	8.544	18.84
20	3/80	.0375	.9525	24	1.50	.6804	7.324	16.15
21	11/320	.034375	.873125	22	1.375	.6237	6.713	14.80
22	1/32	.03125	.793750	20	1.25	.567	6.103	13.46
23	9/320	.028125	.714375	18	1.125	.5103	5.493	12.11
24	1/40	.025	.635	16	1	.4536	4.882	10.76

25	7/320	.021875	.555625	14	.875	.3969	4.272	9.42
26	3/160	.01875	.47625	12	.75	.3402	3.662	8.07
27	11/640	.0171875	.4365625	11	.6875	.3119	3.357	7.40
28	1/64	.015625	.396875	10	.625	.2835	3.052	6.73
29	9/640	.0140625	.3571875	9	.5625	.2551	2.746	6.05
30	1/80	.0125	.3175	8	.5	.2268	2.441	5.38
31	7/640	.0109375	.2778125	7	.4375	.1984	2.136	4.71
32	13/1280	.01015625	.25796875	6 1/2	.40625	.1843	1.983	4.37
33	3/320	.009375	.238125	6	.375	.1701	1.831	4.04
34	11/1280	.00859375	.21828125	5 1/2	.34375	.1559	1.678	3.70
35	5/640	.0078125	.1984375	5	.3125	.1417	1.526	3.36
36	9/1280	.00703125	.17859375	4 1/2	.28125	.1276	1.373	3.03
37	17/2560	.006640625	.168671875	4 1/4	.265625	.1205	1.297	2.87
38	1/160	.00625	.15875	4	.25	.1134	1.221	2.69

Reformatted from: **http://www4.law.cornell.edu/uscode/15/206.html**

INDEX

A

B

C

D

E

F

G

H

I

K

L

M

N

O

P

R

S

T

U

V

W

X

Y

Z